T0141064

COSMIC MAGNETIC FIELDS: FROM PLANETS, TO STARS AND GALAXIES

IAU SYMPOSIUM No. 259

COVER ILLUSTRATION: Magnetic fields in the solar interior.

A boundary-layer mean-field dynamo produces a dipole-like magnetic field structure. The simulation shows magnetic field lines of a dipole-like field structure. Blue means pointing inward, red means pointing outward. Individual flux-tube bundles may reach the stellar surface and then emerge as bipolar spot groups. The rotation axis is indicated. Simulation courtesy of Rainer Arlt, AIP.

INTERNATIONAL ASTRONOMICAL UNION
UNION ASTRONOMIQUE INTERNATIONALE

COSMIC MAGNETIC FIELDS: FROM PLANETS, TO STARS AND GALAXIES

PROCEEDINGS OF THE 259th SYMPOSIUM OF THE INTERNATIONAL ASTRONOMICAL UNION HELD IN PUERTO SANTIAGO, TENERIFE, SPAIN NOVEMBER 3–7, 2008

Edited by

KLAUS G. STRASSMEIER
Astrophysical Institute Potsdam, Potsdam, Germany

ALEXANDER G. KOSOVICHEV
W.W. Hansen Laboratory, Stanford University, Stanford, USA

and

JOHN E. BECKMAN
Instituto de Astrofisica de Canarias, La Laguna, Tenerife, Spain

CAMBRIDGE UNIVERSITY PRESS
The Edinburgh Building, Cambridge CB2 8RU, United Kingdom
32 Avenue of the Americas, New York, NY 10013-2473, USA
477 Williamstown Road, Port Melbourne, VIC 3207, Australia
Ruiz de Alarcón 13, 28014 Madrid, Spain
Dock house, The Waterfront, Cape Town 8001, South Africa

First published 2009

Printed in the United Kingdom at the University Press, Cambridge

Typeset in System LaTeX 2ε

A catalogue record for this book is available from the British Library

Library of Congress Cataloguing in Publication data

ISBN 9780521889902 hardback
ISSN 1743–9213

Table of Contents

Session 2: Multi-scale magnetic fields of the Sun; their generation in the interior, and magnetic energy release
Chair: Nigel O. Weiss

Posters

Session 3: Planetary magnetic fields and the formation and evolution of planetary systems and planets; exoplanets

Chair: Karl-Heinz Glassmeier

Posters

Session 4: Stellar magnetic fields: cool and hot stars
Chair: Swetlana Hubrig

Posters

Session 5: From stars to galaxies and the intergalactic space

Chairs: Dimitry Sokoloff and Bryan Gaensler

Posters

Session 6: Advances in methods and instrumentation for measuring magnetic fields across all wavelengths and targets

Chairs: Tom Landecker & Klaus G. Strassmeier

Preface

Understanding of the Universe is impossible without understanding cosmic magnetic fields, which span the enormous range of 24 magnitudes in strength and play a key role in the formation, structure and evolution of planets, stars and galaxies, and possibly the entire universe. Numerous active phenomena are associated with magnetic energy release. Magnetic fields of celestial bodies have now been studied for hundred years since the discovery of the first extraterrestrial magnetic field by George Ellery Hale in 1908, but the origin and evolution of cosmic magnetic fields is still an open question for fundamental physics and astrophysics alike. It becomes more and more clear that, despite the enormous differences in scales, the basic mechanisms of generation of magnetic fields, their evolution and dynamics may be quite similar.

The topic of this Symposium touched 10 out of 12 IAU Divisions, and is certainly of great interest to a broad astronomical community. Thus, the goal of this Symposium was to hold interdisciplinary sessions of the fundamental properties of cosmic magnetism, from planets and stars to galaxies and the early universe. This provided an interdisciplinary forum for exchange of new results, ideas, and future plans, which will help to better understand the magnetic effects in various objects. For example, the origin of a star's magnetic field is always related to the pre-existence of a seed field already in existence during star formation. The seed field is most likely the galactic magnetic field and its full understanding requires concerted stellar and galactic structural investigations.

The scientific topics of this Symposium covered today's most critical aspects of cosmic magnetism and included talks on magnetic fields in star-forming regions, the multi-scale field of the Sun and its interior, heliospheric and interplanetary fields, the Earth's magnetic field, surface fields of cool and hot stars and of degenerate objects, planetary-nebulae and Supernovae shaping by magnetic fields, jet and accretion-disk fields of very young stars, fields around black holes and magnetars, the magnetic field and dynamo of spiral galaxies, the primordial field of the early universe and, finally, instrumentation and techniques for measuring magnetic fields across all wavelengths, from the ground and space, with particular emphasis on two future facilities; the E-ELT and the SKA.

The Sun is certainly our Rosetta stone when it comes to magnetic-field studies in the entire universe. Solar magnetism is being studied in great details, from global fields of the interior by helioseismology, to the smallest resolved and even unresolved scales by new large solar telescopes from ground (GREGOR, ATST, SST and others) and from space (SOHO, *Hinode*, RHESSI, SDO, STEREO). However, in recent years the connections between the solar-physics community and night-time astronomy were pushed in the background partly due to increased emphasis on so-called "grant challenges" in cosmic vision programmes. Our Symposium not only restored and strengthened these links but established a new connection; the star-exoplanet relation. Much as a stellar wind can influence a planet, a magnetized planet may also have an impact on the stellar atmosphere. Enhanced stellar activity may result as a consequence of such a feedback interaction. However, much theoretical and observational work is required to detail this hypothetic interaction but it has the potential to impact on how we believe life has formed on Earth and other planets. In any case, we are keen to predict that the 21st century will become the century of cosmic magnetic-field research.

Klaus Strassmeier, Alexander Kosovichev and John Beckman, co-chairs SOC,
Potsdam, Stanford, La Laguna, January 1, 2009

THE ORGANIZING COMMITTEE

Scientific

Eduardo Battaner (Spain)
John E. Beckman (co-chair, Spain)
Andrew Collier-Cameron (U.K.)
Karl-Heinz Glassmeier (Germany)
A. G. Kosovichev (co-chair, USA)
Gauthier Mathys (ESO)
Hiromoto Shibahashi (Japan)
Lev Zeleny (Russia)
Rainer Beck (Germany)
Claude Catala (France)
Richard M. Crutcher (USA)
Karel van der Hucht (Netherlands)
Cristina Mandrini (Argentina)
Don Melrose (Australia)
Klaus G. Strassmeier (chair, Germany)
Shuang Nan Zhang (China)

Local

John E. Beckman (chair, Spain)
Rainer Arlt (Germany)
Valeria Buenrostro Leiter (Spain)
Katrin Götz (Germany)
Valentín Martínez Pillet (Spain)
Jorge A. Pérez Prieto (Spain)
Judith de Araoz (Spain)
Eva Bejarano (Spain)
Thorsten Carroll (Germany)
Christoph Kückein (Spain)
Klaus Puschmann (Spain)

Acknowledgements

The symposium is sponsored and supported by the IAU Divisions III (Planetary Systems Science), Division IV (Stars), Division V (Variable Stars), Division VI (Interstellar Matter), Division VII (Galactic System), Division VIII (Galaxies and the Universe), Division IX (Optical and IR Techniques), Division X (Radio Astronomy), and Division XI (Space & High-Energy Astrophysics).

The Local Organizing Committee operated under the auspices of the Instituto de Astrofísica de Canarias (IAC) and the Astrophysical Institute Potsdam (AIP).

Funding by the
International Astronomical Union,
Ministerio de Educación y Ciencia,
Gobierno de Canarias,
Bundesministerium für Bildung und Forschung,
Land Brandenburg,
European Space Agency ESA,
National Aeronautics and Space Administration NASA,
are gratefully acknowledged.

Participants

Alecian, Evelyne, Royal Military College of Canada, Canada evelyne.alecian@rmc.ca
Alves, Felipe, Institut de Cincies de l'Espai, Spain oliveira@ieec.uab.es
Amiri, Nikta, Joint Institute for VLBI in Europe (JIVE) /Leiden Observatory, Netherlands amiri@jive.nl
Arlt, Rainer, MHD, AIP, Germany rarlt@aip.de
Arshakian, Tigran, MPI für Radiastronomie, Germany tarshakian@mpifr-bonn.mpg.de
Auriere, Michel, Observatoire Midi Pyrenees, France michel.auriere@ast.obs-mip.fr
Balthasar, Horst, Astrophysikalisches Institut Potsdam, Germany hbalthasar@aip.de
Barker, Adrian, DAMTP, University of Cambridge, UK ajb268@cam.ac.uk
Battaner, Eduardo, University of Granada, Spain battaner@ugr.es
Beck, Rainer, MPI für Radioastronomie, Germany rbeck@mpifr-bonn.mpg.de
Beckman, John, Instituto de Astrofsica de Canarias, Spain jeb@iac.es
Berdyugina, Svetlana, Kiepenheuer Institute for Solar Physics, Germany sveta@astro.phys.ethz.ch
Bernal, Giovanny, IA UNAM, Mxico bernalcg@gmail.com
Bigot, Lionel, Observatoire de la Cote d'Azur, France lbigot@oca.eu
Blackman, Eric, University of Rochester, USA blackman@pas.rochester.edu
Bochkarev, Nikolai, Sternberg State Astronomical Institue, Russia boch@sai.msu.ru
Brandenburg, Axel, NORDITA, Sweden brandenb@nordita.org
Brown, Shea, University of Minnesota, USA brown@physics.umn.edu
Buenrostro-Leiter, Valeria, Instituto de Astrofsica de Canarias, Spain valeria@iac.es
Carroll, Thorsten A., Astrophysikalisches Institut Potsdam, Germany tcarroll@aip.de
Cenacchi, Elena, MPI für Radioastronomie, Germany cenacchi@mpifr.de
Della-Giustina, Daniella, University of Arizona, USA dellagiu@email.arizona.edu
Demidov, Mikhail, Institute of Solar-Terrestrial Physics, Russia demid@iszf.irk.ru
Denker, Carsten, Astrophysikalisches Institut Potsdam, Germany cdenker@aip.de
Dobbie, Peter, University of Sydney, Australia p.dobbie@physics.usyd.edu.au
Dolag, Klaus, Max-Planck-Institut fuer Astrophysik, Germany kdolag@mpa-garching.mpg.de
ud-Doula, Asif, Morrisville State College, USA uddoula@morrisville.edu
van Driel-Gesztelyi, Lidia, Mullard Space Science Lab. Univ. College London, U.K. Lidia.vanDriel@obspm.fr
Duez, Vincent, CEA Saclay - AIM, France vincent.duez@cea.fr
Durrer, Ruth, Univesite de Geneve, Switzerland Ruth.Durrer@unige.ch
Dzyurkevich, Natalia, Max-Planck Institute for Astronomy, Germany natalia@mpia-hd.mpg.de
Eislöffel, Jochen, Thüringer Landessternwarte, Germany jochen@tls-tautenburg.de
Elstner, Detlef, AIP, Germany elstner@aip.de
Fabas, Nicolas, GRAAL, Universite Montpellier II, France nicolas.fabas@gmail.com
Ferland, Gary, University of Kentucky, USA gjferland@googlemail.com
Fletcher, Andrew, Newcastle University, UK andrew.fletcher@ncl.ac.uk
Franco, Gabriel A. P., Universidade Federal de Minas Gerais, Brazil franco@fisica.ufmg.br
Gaensler, Bryan, The University of Sydney, Australia bgaensler@usyd.edu.au
Geisbuesch, Joern, University of Cambridge, UK joern@mrao.cam.ac.uk
Gil, Janusz, Kepler Institute of Astronomy, Poland jag@astro.ia.uz.zgora.pl
Glassmeier, Karl-Heinz, Technical University of Braunschweig, Germany kh.glassmeier@tu-bs.de
Gledhill, Tim, University of Hertfordshire, UK t.gledhill@herts.ac.uk
Gourgouliatos, Konstantinos, Institute of Astronomy, University of Cambridge, UK kng22@cam.ac.uk
de Gouveia Dal Pino, Elisabete, Universidade de Sao Paulo, IAG-USP, Brazil dalpino@astro.iag.usp.br
Grant, Julie, Centre for Radio Astronomy, Univ. Calgary, Canada jkgrant@ucalgary.ca
Gressel, Oliver, AIP, Germany ogressel@aip.de
Grunhut, Jason, Queen's University/Royal Military College of Canada, Canada Jason.Grunhut@rmc.ca
Hamilton-Drager, Catrina, Dickinson College, USA hamiltoc@dickinson.edu
Han, JinLin, National Astronomical Observatories of China, China hjl@bao.ac.cn
Hanasz, Michal, Centre for Astronomy, Nicolaus Copernicus University, Poland mhanasz@astri.uni.torun.pl
Harvey-Smith, Lisa, University of Sydney, lhs@usyd.edu.au
Heald, George, ASTRON, The Netherlands heald@astron.nl
Heesen, Volker, Astronomical Institute of the University of Bochum, Germany heesen@astro.rub.de
Heiles, Carl, University of California at Berkeley, USA heiles@astro.berkeley.edu
Henrichs, Huib, Astronomical Institute, Univ. Amsterdam, The Netherlands h.f.henrichs@uva.nl
Hernndez-Cervantes, Liliana, UNAM, Mexico liliana@astroscu.unam.mx
Herpin, Fabrice, LAB/OASU, France herpin@obs.u-bordeaux1.fr
Hubrig, Swetlana, ESO/Chile, Chile shubrig@eso.org
Hussain, Gaitee, ESO, Germany ghussain@eso.org
Ilyin, Ilya, AIP Potsdam, Germany, Germany ilyin@aip.de
Janson, Markus, University of Toronto, Canada janson@astro.utoronto.ca
Jardine, Moira, University of St Andrews, UK mmj@st-andrews.ac.uk
Johansen, Anders, Sterrewacht Leiden, The Netherlands ajohan@strw.leidenuniv.nl
Johns-Krull, Christopher, Rice University, USA cmj@rice.edu
Jordan, Stefan, ARI, University of Heidelberg, Germany jordan@ari.uni-heidelberg.de
Kandori, Ryo, National Astronomical Observatory of Japan, Japan r.kandori@nao.ac.jp
Karas, Vladimir, Astronomical Institute, Czech Republic vladimir.karas@cuni.cz
Karitskaya, Eugenia, Institute of Astronomy RAS, Russia karitsk@yandex.ru
Kawagoe, Shio, The University of Tokyo, Japan kawagoe@astron.s.u-tokyo.ac.jp
Kepley, Amanda, University of Virginia/NRAO, USA kepley@astro.wisc.edu
Khodachenko, Maxim, Space Research Institute, Acdy. of Sciences, Austria maxim.khodachenko@oeaw.ac.at
Kitiashvili, Irina, Stanford University, USA irinasun@stanford.edu
Kochukhov, Oleg, Uppsala University, Sweden Oleg.Kochukhov@fysast.uu.se
Kopacek, Ondrej, Astronomical Institute, Czech Republic kopacek@ig.cas.cz
Korhonen, Heidi, ESO, Germany hkorhone@eso.org
Kosovichev, Alexander, Stanford University, USA sasha@sun.stanford.edu
Kotarba, Hanna, University Observatory Munich, Germany kotarba@usm.lmu.de
Kothes, Roland, Dominion Radio Astrophysical Obs., Canada roland.kothes@nrc-cnrc.gc.ca

Koutchmy, Serge, Institut d'Astrophysique de Paris CNRS & UPMC, France koutchmy@iap.fr
Kovar, Jiri, Silesian University in Opava, Czech Republic jiri.kovar@fpf.slu.cz
Kowalik, Kacper, Nicolaus Copernicus University, Poland Kacper.Kowalik@astri.uni.torun.pl
Kramer, Michael, University of Manchester, UK Michael.Kramer@manchester.ac.uk
Krause, Marita, MPI für Radioastronomie, Germany mkrause@mpifr-bonn.mpg.de
Kronberg, Philipp, University of Toronto/LANL, USA kronberg@physics.utoronto.ca
Kuckein, Christoph, Instituto de Astrofisica de Canarias, Spain ckuckein@iac.es
Kudoh, Takahiro, National Astronomical Observatory of Japan, Japan kudoh@th.nao.ac.jp
Kudryavtsev, Dmitry, Special Astrophysical Obs. RAS, Russia dkudr@sao.ru
Kuiper, Rolf, MPI für Astronomie, Germany kuiper@mpia.de
Kulebi, Baybars, Astronomisches Rechen-Institut, Germany bkulebi@ari.uni-heidelberg.de
Kulpa-Dybel, Katarzyna, Astronomical Observatory of the Jagiellonian University, Poland kulpa@oa.uj.edu.pl
Kusakabe, Nobuhiko, National Astronomical Observatory of Japan, Japan nb.kusakabe@nao.ac.jp
Lai, Shih-Ping, Tsing-Hua University, Taiwan slai@phys.nthu.edu.tw
Landecker, Tom, Dominion Radio Astrophysical Observatory, Canada tom.landecker @nrc.gc.ca
Lèbre, Agnès, GRAAL, Universit Montpellier II, France lebre@graal.univ-montp2.fr
Liverts, Eduard, Ben-Gurion University of the Negev, Israel eliverts@bgu.ac.il
Loukitcheva, Maria, Astronomical Institute of Saint-Petersburg University, Russia marija@peterlink.ru
Maeder, Andr, Geneva Observatory, Switzerland andre.maeder@unige.ch
Martinez Pillet, Valentin, Instituto de Astrofisica de Canarias, Spain vmp@iac.es
Masada, Youhei, Institute of Astronomy and Astrophysics, Academia Sinica, Taiwan, R.O.C masada@asiaa.sinica.edu.tw
Mathis, Stephane, CEA/DSM/IRFU/SAp, France stephane.mathis@cea.fr
Matsumoto, Jin, Kyoto University, Japan jin@kusastro.kyoto-u.ac.jp
Meinel, Barbara, Meinel Group, USA b_meinel_j@yahoo.com
Melikidze, Giorgi, J. Kepler IoA, Univ. Zielona Gora, Poland gogi@astro.ia.uz.zgora.pl
Mendoza Torres, Jose Eduardo, INAOE, Mxico mend@inaoep.mx
Moreno-Insertis, Fernando, Instituto de Astrofsica de Canarias, Spain fmi@iac.es
Motschmann, Uwe, Technical University of Braunschweig, Germany u.motschmann@tu-bs.de
Neiner, Coralie, GEPI, Observatoire de Paris-Meudon, France coralie.neiner@obspm.fr
Nelson, Alistair, Cardiff University, UK nelsona@cf.ac.uk
Noutsos, Aristeidis, University of Manchester, UK aristeidis.noutsos@manchester.ac.uk
Otmianowska-Mazur, Katarzyna, Astronomical Observatory Jagiellonian University, Poland otmian@oa.uj.edu.pl
Petit, Vronique, Universit Laval, Canada veronique.petit.1@ulaval.ca
Petit, Pascal, OMP, France petit@ast.obs-mip.fr
Petr-Gotzens, Monika, ESO, Germany mpetr@eso.org
Popescu, Adrian Sabin, Astronomical Institute of the Romanian Academy, Romania sabinp@aira.astro.ro
Puschmann, Klaus G., Instituto de Astrofisica de Canarias, Spain kgp@iac.es
Raison, Frederic, ESAC, SPAIN frederic.raison@sciops.esa.int
Reich, Wolfgang, MPI für Radioastronomie, Germany wreich@mpifr-bonn.mpg.de
Reiners, Ansgar, Georg-August-Universitt Gttingen, Germany Ansgar.Reiners@phys.uni-goettingen.de
Roberts, Paul, UCLA, USA roberts@math.ucla.edu
Robishaw, Tim, UC Berkeley, USA robishaw@physics.usyd.edu.au
Romanyuk, Iosif, Special Astrophysical Observatory of RAS, Russia roman@sao.ru
Ruiz-Granados, Beatriz, Universidad de Granada, Spain bearg@ugr.es
Rüdiger, Guenther, AIP, Germany gruediger@aip.de
Salmin, Vladimir, Siberian Federal University, Russia vsalmin@gmail.com
Santillan, Alfredo J., UNAM, Mexico alfredo@astroscu.unam.mx
Santos de Lima, Reinaldo, Universidade de So Paulo - IAG, Brazil rlima@astro.iag.usp.br
Sarma, Anuj, DePaul University, USA asarma@depaul.edu
Schlickeiser, Reinhard, Ruhr-University Bochum, Germany rsch@tp4.rub.de
Shulyak, Denis, IoA, Vienna University, Austria denis@jan.astro.univie.ac.at
Siejkowski, Hubert, Astronomical Observatory of the Jagiellonian University, Poland siejkowski@gmail.com
Silvester, James, Queen's University / RMC, Canada james.silvester@rmc.ca
Sokoloff, Dmitry, Moscow State University, Russia sokoloff@dds.srcc.msu.su
Spruit, Hendrik, Max Planck Institute for Astrophysics, Germany henk@mpa-garching.mpg.de
Stasyszyn, Federico, MPI für Astrophysic, Germany fstasys@mpa-garching.mpg.de
Steffen, Matthias, Astrophysikalisches Institut Potsdam, Germany msteffen@aip.de
Stenflo, Jan, ETH Zurich, Switzerland stenflo@astro.phys.ethz.ch
Stepanov, Rodion, Institute of Continuous Media Mechanics UB RAS, Russia rodion@icmm.ru
Stil, Jeroen, University of Calgary, Canada stil@ras.ucalgary.ca
Strassmeier, Klaus G., AIP, Germany kstrassmeier@aip.de
Su, Yi-Jiun, University of Texas at Arlington, USA yijiun@uta.edu
Trujillo-Bueno, Javier, Instituto de Astrofsica de Canarias, Spain jtb@iac.es
Tuominen, Ilkka, Observatory, University of Helsinki, Finland Ilkka.Tuominen@Helsinki.fi
Vidotto, Aline, University of Sao Paulo, Brazil aline@astro.iag.usp.br
Wade, Gregg, RMC, Canada wade-g@rmc.ca
Weber, Michael, AIP, Germany mweber@aip.de
Weiss, Nigel, University of Cambridge, UK now@damtp.cam.ac.uk
Wielebinski, Richard, MPI für Radioastronomie, Germany rwielebinski@mpifr-bonn.mpg.de
Wolleben, Maik, Dominion Radio Astrophysical Obs., Canada maik.wolleben@nrc.gc.ca
Woltanski, Dominik, Nicolaus Copernicus University, Poland minikwolt@astri.uni.torun.pl
Zinnecker, Hans, Astrophysikalisches Institut Potsdam, Germany hzinnecker@aip.de
Zolotova, Nadezhda, Institute of Physics of St. Petersburg State Univ., Russia ned@geo.phys.spbu.ru, libra2001@inbox.ru

Session I

Interstellar magnetic fields, star-forming regions, and the death valley

Cosmic Magnetic Fields:
From Planets, to Stars and Galaxies
Proceedings IAU Symposium No. 259, 2008
K.G. Strassmeier, A.G. Kosovichev & J.E. Beckman, eds.

doi:10.1017/S1743921309030014

Measuring interstellar magnetic fields by radio synchrotron emission

Rainer Beck

Max-Planck-Institut für Radioastronomie, Auf dem Hügel 69, 53121 Bonn, Germany
email: rbeck@mpifr-bonn.mpg.de

Abstract. Radio synchrotron emission, its polarization and its Faraday rotation are powerful tools to study the strength and structure of interstellar magnetic fields. The total intensity traces the strength and distribution of total magnetic fields. Total fields in gas-rich spiral arms and bars of nearby galaxies have strengths of 20–30 μGauss, due to the amplification of turbulent fields, and are dynamically important. In the Milky Way, the total field strength is about 6 μG near the Sun and several 100 μG in filaments near the Galactic Center. – The polarized intensity measures ordered fields with a preferred orientation, which can be regular or anisotropic fields. Ordered fields with spiral structure exist in grand-design, barred, flocculent and even in irregular galaxies. The strongest ordered fields are found in interarm regions, sometimes forming "magnetic spiral arms" between the optical arms. Halo fields are X-shaped, probably due to outflows. – The Faraday rotation of the polarization vectors traces coherent regular fields which have a preferred direction. In some galaxies Faraday rotation reveals large-scale patterns which are signatures of dynamo fields. However, in most galaxies the field has a complicated structure and interacts with local gas flows. In the Milky Way, diffuse polarized radio emission and Faraday rotation of the polarized emission from pulsars and background sources show many small-scale and large-scale magnetic features, but the overall field structure in our Galaxy is still under debate.

Keywords. Techniques: polarimetric – ISM: magnetic fields – galaxies: magnetic fields – galaxies: spiral – radio continuum: galaxies

1. Introduction

Interstellar magnetic fields were discovered already in 1932 by Karl Guthe Jansky who first detected diffuse low-frequency radio emission from the Milky Way, but the explanation as synchrotron emission was given only in 1950 by Karl Otto Kiepenheuer. The sensitivity of radio observations has improved by several orders of magnitude in the past decades, and synchrotron emission was detected from the interstellar medium (ISM) in almost all star-forming galaxies, in galaxy halos and the intracluster medium, proving that a large fraction of the Universe is permeated by magnetic fields. However, in spite of our increasing knowledge on interstellar magnetic fields, many important questions are unanswered, especially their first occurrence in young galaxies, their amplification when galaxies evolved, and their effect on galaxy dynamics.

As magnetic fields need illumination by cosmic-ray electrons to become observable by synchrotron emission, which are generated in star-forming regions or intracluster shocks, we do not know yet whether magnetic fields also exist in radio-quiet elliptical or dwarf galaxies or in the general intergalactic medium (IGM). Progress can be expected from using Faraday rotation which does not need cosmic rays, only magnetic fields and thin ionized gas. One of the research areas of the forthcoming radio telescopes (LOFAR, ASKAP, SKA) will be the search for Faraday rotation in these objects against polarized background sources (Gaensler, this volume). The SKA will also be needed to detect magnetic fields in young galaxies (Beck & Gaensler 2004, Arshakian *et al.* 2008).

Figure 1. Total radio emission (contours) and B–vectors of M 51, combined from observations at 6 cm wavelength with the VLA and Effelsberg telescopes and smoothed to 15" resolution (Fletcher et al., in prep.), overlaid onto an optical image from the HST (Copyright: MPIfR Bonn and *Hubble Heritage Team*. Graphics: *Sterne und Weltraum*).

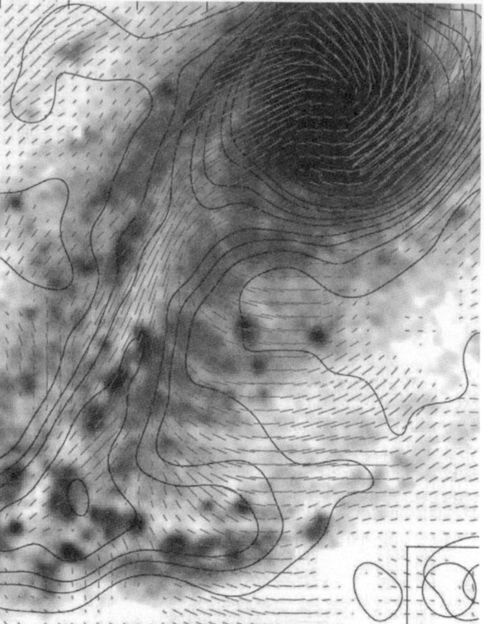

Figure 2. Total radio emission (contours) and B–vectors of the barred galaxy NGC 1097, observed at 6 cm wavelength with the VLA and smoothed to 10" resolution (Beck *et al.* 2005). The background optical image is from Halton Arp (Copyright: MPIfR Bonn and Cerro Tololo Observatory).

2. Tools to measure interstellar magnetic fields

Most of what we know about galactic and intergalactic magnetic fields comes through the detection of radio waves. *Zeeman splitting* of radio spectral lines is the best method to directly measure the field strength (Heiles, this volume).

The intensity of *synchrotron emission* is a measure of the number density of cosmic-ray electrons in the relevant energy range and of the strength of the total magnetic field component in the sky plane. Polarized emission emerges from ordered fields. As polarization "vectors" are ambiguous by 180°, they cannot distinguish *regular fields* with a constant direction within the telescope beam from *anisotropic fields* which are generated from turbulent magnetic fields by compressing or shearing gas flows and frequently reverse their direction on small scales. Unpolarized synchrotron emission indicates *turbulent fields* with random directions which have been tangled or generated by turbulent gas flows.

The intrinsic degree of linear polarization of synchrotron emission is about 75%. The observed degree of polarization is smaller due to the contribution of unpolarized thermal emission, which may dominate in star-forming regions, by *Faraday depolarization* along the line of sight and across the beam (Sokoloff *et al.* 1998), and by depolarization due to variations of the field orientation within the beam and along the line of sight.

At short (e.g. centimeter) radio wavelengths the orientation of the observed B–vector is generally parallel to the field orientation, so that the magnetic patterns of many galaxies could be mapped directly (Beck 2005). The orientation of the polarization vectors is changed in a magnetized thermal plasma by *Faraday rotation*. The rotation angle increases with the plasma density, the strength of the component of the field along the line of sight and the square of the observation wavelength. As the rotation angle is sensitive to the sign of the field direction, only regular fields can give rise to Faraday rotation, while anisotropic and random fields do not. For typical plasma densities and regular field strengths in the interstellar medium of galaxies, Faraday rotation becomes significant at wavelengths larger than a few centimeters. Measurements of the Faraday rotation from multi-wavelength observations allow to determine the strength and direction of the regular field component along the line of sight. Its combination with the total intensity and the polarization vectors can yield the three-dimensional picture of the magnetic field and allows to distinguish the three field components: *regular, anisotropic and random.*

3. Total synchrotron emission: Tracer of star formation

The integrated flux densities of total radio continuum emission at centimeter wavelengths (frequencies of a few GHz), which is mostly of non-thermal synchrotron origin, and far-infrared (FIR) emission of star-forming galaxies are tightly correlated, first reported by de Jong *et al.* (1985). This correlation achieved high popularity in galactic research as it allows to use radio continuum emission as a extinction-free tracer of star formation. The correlation holds for starburst galaxies (Lisenfeld *et al.* 1996b) as well as for blue compact and low-surface brightness galaxies (Chyży *et al.* 2007). It extends over five orders of magnitude (Bell 2003) and is valid to redshifts of at least 3 (Seymour *et al.* 2008), so that radio emission serves as a star formation tracer in the early Universe. Only galaxies with very recent starbursts reveal significantly smaller radio-to-FIR ratios because the timescale for the acceleration of cosmic rays and/or for the amplification of magnetic fields is longer than that of dust heating (Roussel *et al.* 2003).

Strongest total synchrotron emission (tracing the total, mostly turbulent field) generally coincides with highest emission from dust and gas in the spiral arms: The correlation also holds for the local radio and FIR or mid-IR (MIR) intensities *within galaxies* (e.g. Beck & Golla 1988, Hoernes *et al.* 1998, Walsh *et al.* 2002, Tabatabaei *et al.* 2007a). The highest correlation of all spectral ranges is found between the total intensity at $\lambda 6$ cm and the mid-infrared dust emission, while the correlation with the cold gas (traced by the CO(1-0) transition) is less tight (Frick *et al.* 2001, Walsh *et al.* 2002, Nieten *et al.* 2006). A wavelet cross-correlation analysis for M 33 showed that the radio–FIR correlation holds for all scales down to 1 kpc (Tabatabaei *et al.* 2007a). The correlation breaks down below scales of about 50 pc (Hughes *et al.* 2006) and in radio halos, probably due to the smoothing effect on synchrotron intensity by cosmic-ray propagation.

If the thermal and non-thermal radio components are separated with help of Hα and FIR data (e.g. within M 33, Tabatabaei *et al.* 2007b), an almost perfect correlation is found between thermal radio and infrared intensities at all scales. The nonthermal–FIR correlation is less pronounced but highly significant. The polarized synchrotron intensity, tracing the ordered field, is anticorrelated or not correlated with all tracers of star formation (Frick *et al.* 2001).

It is not obvious why the nonthermal synchrotron and the thermal FIR intensities are so closely related. The intensity of synchrotron emission depends not only on the density of cosmic-ray electrons (CRE), but also on about the square of the strength of the total magnetic field $B_{\rm t}$ (its component in the sky plane, to be precise). The radio–FIR correlation requires that magnetic fields and star-formation processes are connected. If $B_{\rm t}$ is strong, most of the cosmic-ray energy is released via synchrotron emission within the galaxy, the CRE density decreases with $B_{\rm t}^2$ and the integrated radio synchrotron luminosity depends on the CRE injection rate, not on B_t. If most thermal energy from star formation is also emitted within a galaxy via far-infrared emission by warm dust, this galaxy can be treated as a "calorimeter" for thermal and nonthermal emission. Prime candidates for "calorimeter" galaxies are those with a high star-formation rate (SFR). If B_t increases with SFR according to $B_t \propto SFR^{0.5}$, a linear radio–FIR correlation for the integrated luminosities is obtained (Lisenfeld *et al.* 1996a,b). However, the calorimeter model cannot explain the local correlation within galaxies.

In galaxies with low or moderate SFR and B_t, synchrotron lifetime of CRE is sufficiently large to leave the galaxy. To obtain a global or local radio–FIR correlation, coupling of magnetic fields to the gas clouds is needed. A scaling $B_t \propto \rho^{1/2}$ was proposed (Helou & Bicay 1993, Niklas & Beck 1997, Hoernes *et al.* 1998) where ρ is the average density of the neutral gas. A nonlinear correlation (with a slope of about 1.3) between the nonthermal radio luminosity and the FIR luminosity from warm dust is achieved by further assuming energy equipartition between magnetic fields and cosmic rays and a Schmidt law of star formation ($SFR \propto \rho^{1.5}$) (Niklas & Beck 1997). In this model the total magnetic field strength and the star-formation rate SFR are related via $B_t \propto SFR^{0.3}$.

The radio–FIR correlation indicates that *equipartition* between the energy densities of the total magnetic field and the total cosmic rays is valid, at least on spatial scales larger than about 100 pc (Stepanov *et al.*, this volume) and on timescales of larger than the CRE acceleration time (a few 10^6 years). Then the strength of the total magnetic field can be determined from the intensity of the total synchrotron emission, assuming a ratio K between the numbers of cosmic-ray protons and electrons in the relevant energy range (usually $K \simeq 100$). In regions where electrons lost already a significant fraction of their energy, e.g. in strong magnetic fields or radiation fields or far away from their places of origin, K is > 100 and the standard value of 100 yields an underestimate (Beck & Krause 2005). On the other hand, in case of field fluctuations along the line of sight or across the telescope beam, the equipartition value is an overestimate (Beck *et al.* 2003).

The typical average equipartition strength of the total magnetic field in spiral galaxies is about $10\,\mu$G. Radio-faint galaxies like M 31 and M 33, our Milky Way's neighbors, have weaker total magnetic fields (about $5\,\mu$G), while gas-rich galaxies with high star-formation rates, like M 51 (Fig. 1), M 83 and NGC 6946, have total field strengths of 20–30 μG in their spiral arms. The degree of radio polarization within the spiral arms is only a few %; hence the field in the spiral arms must be mostly tangled or randomly oriented within the telescope beam, which typically corresponds to a few 100 pc. Turbulent fields in spiral arms are probably generated by turbulent gas motions related to supernovae (de Avillez & Breitschwerdt 2005), stellar winds, spiral shocks (Dobbs & Price 2008) or a small-scale turbulent dynamo (Beck *et al.* 1996).

The mean energy densities of the magnetic field and of the cosmic rays in NGC 6946 and M 33 are $\simeq 10^{-11}$ erg cm^{-3} and $\simeq 10^{-12}$ erg cm^{-3}, respectively (Beck 2007, Tabatabaei *et al.* 2008 and this volume), about 10 times larger than that of the ionized gas, but similar to that of the turbulent gas motions across the whole star-forming disk. The magnetic energy possibly dominates in the outer disk of NGC 6946.

The strongest total fields of 50–100 μG are found in starburst galaxies, like M 82 (Klein *et al.* 1988) and the "Antennae" NGC 4038/9 (Chyży & Beck 2004), and in nuclear starburst regions, like in the centers of NGC 1097 and other barred galaxies (Beck *et al.* 2005). In starburst galaxies, however, the equipartition field strength per average gas surface density is much lower than in normal spirals. This indicates strong energy losses of the cosmic-ray electrons, so that the equipartition field strength is probably underestimated by a factor of a few (Thompson *et al.* 2006). This was recently confirmed by Zeeman measurements of OH maser lines (Robishaw *et al.* 2008).

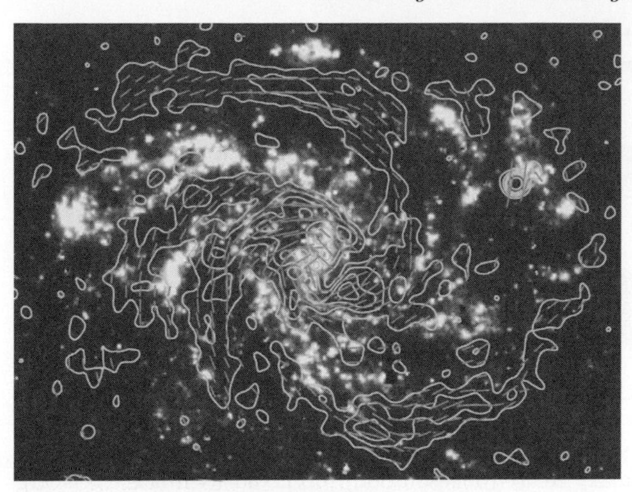

Figure 3. Polarized radio emission (contours) and B-vectors of NGC 6946 (15" resolution), combined from observations at 6 cm wavelength with the VLA and Effelsberg 100m telescopes (Beck & Hoernes 1996). The background image shows the Hα emission (Ferguson *et al.* 1998) (Copyright: MPIfR Bonn. Graphics: *Sterne und Weltraum*).

In case of energy equipartition, the scale length of the total field in the disk of galaxies is at least $(3-\alpha)$ times larger than the synchrotron scale length of typically 4 kpc (where $\alpha \simeq -1$ is the synchrotron spectral index). The resulting value of $\simeq$16 kpc is a lower limit because the CRE lose their energy with distance from the star-forming disk and the equipartition assumption yields too small values for the field strength. The galactic fields extend far out into intergalactic space. The same argument holds for the vertical extent of radio halos around galaxies which is also limited by CRE energy losses. The dumbbell-shaped halo around edge-on galaxies like NGC 253 (Heesen *et al.* 2008 and this volume) is the result of enhanced synchrotron and Inverse Compton losses in the inner region.

As proposed by Battaner & Florido (2000), the magnetic field energy density may generally reach the level of global rotational gas motion and affect galaxy rotation in the outermost parts of spiral galaxies. At GHz frequencies the measured extent of the radio disks of galaxies is limited by energy loss of cosmic-ray electrons, so that measurements at low frequencies (where energy losses are smaller) are needed, e.g. with LOFAR (Beck 2008). Faraday rotation towards polarized background sources may allow to measure weak fields to even larger distances from the star-forming disks.

4. Polarized synchrotron emission: Tracer of ordered fields

The ordered (regular and/or anisotropic) fields traced by the polarized synchrotron emission are generally strongest (10–15 μG) in the regions *between* the optical spiral arms, oriented parallel to the adjacent optical spiral arms. In some galaxies the field forms *magnetic arms* between the optical arms, like in NGC 6946 (Fig. 3). These are probably generated by a large-scale dynamo. In galaxies with strong density waves some of the ordered field is concentrated at the inner edge of the spiral arms, e.g. in M 51 (Patrikeev *et al.* 2006), but the arm–interarm contrast of the ordered field is small, much smaller than that of the random field.

The ordered magnetic field forms spiral patterns in almost every galaxy (Beck 2005), even in ring galaxies (Chyży & Buta 2008) and in flocculent galaxies without an optical spiral structure (Soida *et al.* 2002). Hence, the field lines generally do *not* follow the (almost circular) gas flow and need dynamo action to obtain the required radial field components. Spiral fields with large pitch angles are also observed in the central regions of galaxies and in circum-nuclear gas rings (Beck *et al.* 2005).

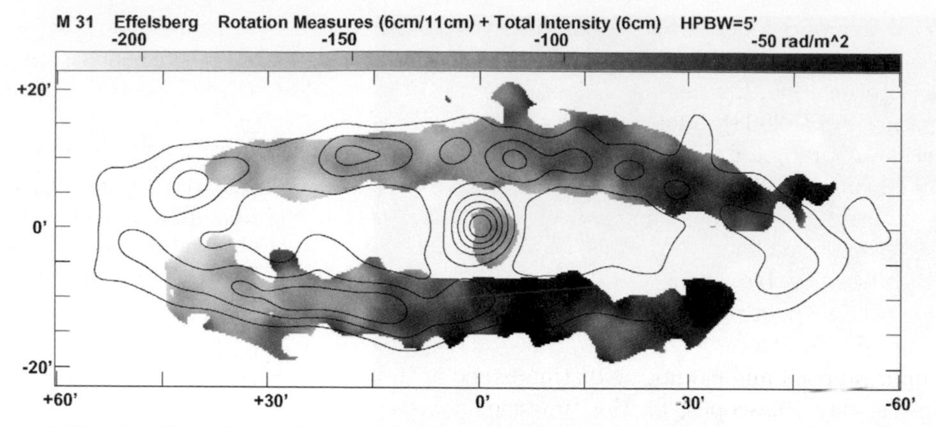

Figure 4. Total radio emission (contours) at 6 cm and Faraday rotation measures (RM) between 6 cm and 11 cm of the Andromeda galaxy M 31 (5' resolution), observed with the Effelsberg telescope (Berkhuijsen *et al.* 2003) (Copyright: MPIfR Bonn).

In galaxies with massive bars the field lines follow the gas flow (Fig. 2). As the gas rotates faster than the bar pattern of a galaxy, a shock occurs in the cold gas which has a small sound speed, while the flow of warm, diffuse gas is only slightly compressed but sheared. The ordered field is also hardly compressed. It is probably coupled to the diffuse gas and strong enough to affect its flow (Beck *et al.* 2005, Fletcher *et al.*, this volume). The polarization pattern in spiral arms and bars can be used as a tracer of shearing gas flows in the sky plane and hence complements spectroscopic measurements.

Nearby galaxies seen edge-on generally show a disk-parallel field near the disk plane (Dumke *et al.* 1995), so that polarized emission can also be detected from distant, unresolved galaxies (Stil *et al.* 2008). High-sensitivity observations of edge-on galaxies like NGC 891 (Krause 2008) and NGC 253 (Heesen *et al.*, this volume) revealed vertical field components in the halo forming an X-shaped pattern (Krause, this volume). This is inconsistent with the predictions from standard dynamo models. The field is probably transported from the disk into the halo by an outflow emerging from the disk. The similarity of scale heights of radio halos in different galaxies indicates that the outflow velocity increases with the star-formation rate (Krause, this volume). Interestingly, a recent model for global outflows from galaxy disks (neglecting magnetic fields) shows an X-shaped velocity field (Dalla Vecchia & Schaye 2008). Improved outflow models including magnetic fields and dynamo action are needed.

5. Faraday rotation: Tracer of regular dynamo fields

Faraday rotation is a signature of regular (coherent) fields which could be generated by the mean-field (or large-scale) *dynamo* (Elstner *et al.* 1992, Beck *et al.* 1996; Elstner, this volume). Dynamo fields are described by modes of different azimuthal symmetry in the disk plane and vertical symmetry (even or odd parity) perpendicular to the disk plane. Several modes can be excited in the same object. In flat, rotating objects like galaxy disks, the strongest mode S0 consists of a toroidal field of *axisymmetric spiral* shape within the disk, without sign reversals across the equatorial plane, and a weaker poloidal field of quadrupolar structure with a reversal of the vertical field component across the plane. Antisymmetric (A–type) fields are generated preferably in spherical objects like halos (Moss & Sokoloff 2008). The magneto-rotational instability (MRI) supports symmetric fields while primordial seed fields support antisymmetric fields (Krause & Beck 1998).

Spiral dynamo modes can be identified from the pattern of polarization angles and Faraday rotation measures (RM) from multi-wavelength radio observations of galaxy disks (Krause 1990, Elstner *et al.* 1992) or from RM data of polarized background sources (Stepanov *et al.* 2008). The disks of a few spiral galaxies indeed reveal large-scale RM patterns, as predicted. The Andromeda galaxy M 31 hosts a dominating axisymmetric disk field (mode S0) (Fig. 4 and Fletcher *et al.* 2004). Other candidates for a dominating axisymmetric disk field are the nearby spiral IC 342 (Krause *et al.* 1989) and the irregular Large Magellanic Cloud (LMC) (Gaensler *et al.* 2005). The magnetic arms in NGC 6946 can be described by a superposition of two azimuthal dynamo modes which are phase shifted with respect to the optical arms (Beck 2007). However, in many observed galaxy disks no clear patterns of Faraday rotation were found. Either several dynamo modes are superimposed and cannot be distinguished with the limited sensitivity and resolution of present-day telescopes, or the timescale for the generation of large-scale modes is longer than the galaxy's lifetime. Dynamo models predict a rapid amplification of small-scale fields already in protogalaxies, while the generation of fully coherent large-scale fields takes several Gyrs, depending on the galaxy size (Arshakian *et al.* 2008 and this volume). Furthermore, anisotropic fields dominate over dynamo modes if shearing gas flows are strong (see Sect. 4).

Vertical fields as predicted by dynamo models should generate large-scale RM patterns around edge-on galaxies, but so far indication for an antisymmetric poloidal field was found only in the halo of NGC 253 (Heesen *et al.*, this volume). Previous indirect evidence for symmetric fields from the dominance of inward-directed fields (Krause & Beck 1998) is no longer supported by the increased sample of galaxies.

Faraday rotation in the direction of QSOs allows to determine the field pattern in an *intervening galaxy* (Stepanov *et al.* 2008). This method can be applied to much larger distances than the analysis of RM of the polarized emission from the foreground galaxy itself. Faraday rotation of QSO emission in distant, intervening galaxies revealed significant regular fields of several μG strength (Bernet *et al.* 2008, Kronberg *et al.* 2008) which is a challenge for classical dynamo models (Arshakian *et al.* 2008).

The recently developed method of *RM Synthesis*, based on multi-channel spectro-polarimetry, transforms the spectral data cube into a data cube of maps in Faraday depth space (Brentjens & de Bruyn 2005, Heald, this volume). RM components from distinct regions along the line of sight can be distinguished by their positions and widths in Faraday depth. Faraday screens appear as sharp lines, while emitting and rotating layers form broad structures in Faraday depth. The transformation of Faraday depth into geometrical depth (*Faraday tomography*) needs modeling.

6. Magnetic fields in the Milky Way

Surveys of the total synchrotron emission from the Milky Way yield equipartition strengths of the total field of 6 μG near the Sun and about 10 μG in the inner Galaxy (Berkhuijsen, in Wielebinski 2005), consistent with Zeeman splitting data of low-density gas clouds (Heiles, this volume). In the nonthermal filaments near the Galactic center the field strength may reach several 100 μG (Reich 1994). Milligauss fields were found in pulsar wind nebulae from the break in the synchrotron spectrum (e.g., Kothes *et al.* 2008).

The observed degree of radio and optical polarization in the local Galaxy implies a ratio of ordered to total field strengths of $\simeq 0.6$ (Heiles 1996). For a total field of 6 μG this gives 4 μG for the local ordered field component (including anisotropic fields). Faraday RM and dispersion measure data of pulsars give an average strength of the local coherent

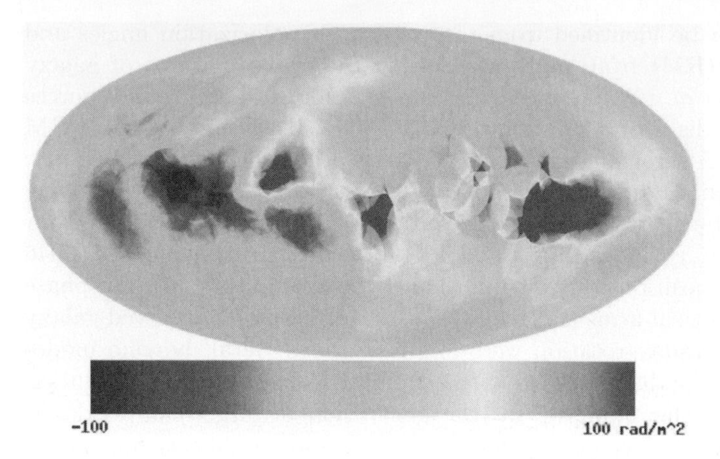

Figure 5. All-sky map of rotation measures in the Milky Way, generated from data towards about 800 polarized extragalactic sources compiled by Johnston-Hollitt *et al.* (2004) and from about 1800 data of the new Effelsberg RM survey (from Sun *et al.* 2008).

regular field of $1.4 \pm 0.2\,\mu$G (Rand & Lyne 1994). In the inner Norma arm, the average strength of the coherent regular field is $4.4 \pm 0.9\,\mu$G (Han *et al.* 2002).

The all-sky maps of polarized synchrotron emission at 1.4 GHz from the Milky Way from DRAO and Villa Elisa and at 22.8 GHz from WMAP, and the new Effelsberg RM survey of polarized extragalactic sources were used to model the large-scale Galactic field (Sun *et al.* 2008). One large-scale reversal is required about 1–2 kpc inside the solar radius, which also agrees with the detailed study of RMs from extragalactic sources near the Galactic plane (Brown *et al.* 2007, Kothes & Brown, this volume). Recent pulsar RM values indicate reversals at several Galactic radii (Han, this volume). None of the models for a simple large-scale field structure of the Milky Way survived a statistical test (Men *et al.* 2008). Similar to external galaxies, the Milky Way's regular field probably has a complex structure which can be revealed only based on a larger sample of pulsar and extragalactic RM data.

RMs of extragalactic sources and of pulsars reveal no large-scale reversal across the plane for Galactic longitudes l=90°–270° (Fig. 5): the local disk field is part of a large-scale symmetric (quadrupolar) field structure. However, towards the inner Galaxy (l=270°–90°) the RM signs are *opposite* above and below the plane (Fig. 5). This may indicate a global antisymmetric mode in the halo (Han *et al.* 1997), similar to the results for the spiral galaxy NGC 253 (Heesen *et al.*, this volume). Note that a superposition of a disk field with even parity and a halo field with odd parity cannot be explained by classical dynamo theory (Moss & Sokoloff 2008).

While the large-scale field is much more difficult to measure in the Milky Way than in external galaxies, Galactic observations can trace magnetic structures to much smaller scales (Wielebinski 2005, Reich & Reich, this volume). Small-scale and turbulent field structures are enhanced in spiral arms compared to interarm regions, as indicated by the different slopes of their *RM structure functions* (Haverkorn *et al.* 2006). The all-sky and the Galactic plane polarization surveys at 1.4 GHz (Reich & Reich and Landecker et al., this volume) reveal a wealth of structures on pc and sub-pc scales: filaments, canals, lenses and rings. Their common property is to appear only in the maps of polarized intensity, but not in total intensity. Some of these are artifacts due to strong depolarization of background emission in a foreground Faraday screen, called *Faraday ghosts*, but carry valuable information about the turbulent ISM (Fletcher & Shukurov 2006). Faraday rotation in foreground objects (e.g. supernova remnants, planetary nebulae, pulsar wind nebulae or photo-dissociation regions of molecular clouds) embedded in diffuse polarized Galactic emission may generate a *Faraday shadow* or *Faraday screen* which enable to estimate the regular field strength by modeling the radiation transfer (Wolleben & Reich 2004, Ransom *et al.* 2008, Reich & Reich, this volume).

7. Outlook

Future radio telescopes will widen the range of observable magnetic phenomena. High-resolution, future deep observations at high frequencies with the Extended Very Large Array (EVLA) and the Square Kilometre Array (SKA) (Gaensler, this volume) will directly show the detailed field structure and the interaction with the gas. The SKA will also measure the Zeeman effect in much weaker magnetic fields in the Milky Way and nearby galaxies. Forthcoming low-frequency radio telescopes like the Low Frequency Array (LOFAR) and the Murchison Widefield Array (MWA) will be suitable instruments to search for extended synchrotron radiation at the lowest possible levels in outer galaxy disks, halos and clusters, and the transition to intergalactic space. At low frequencies we will get access to the so far unexplored domain of weak magnetic fields in galaxy halos (Beck 2008). The detection of radio emission from the intergalactic medium will allow to probe the existence of magnetic fields in such rarified regions, measure their intensity, and investigate their origin and their relation to the structure formation in the early Universe. Low frequencies are also ideal to search for small Faraday rotation from weak interstellar and intergalactic fields. "Cosmic magnetism" is the title of Key Science Projects for LOFAR and SKA.

References

Arshakian, T. G., Beck, R., Krause, M., & Sokoloff, D. 2008, *A&A*, in press, arXiv:0810.3114

Battaner, E. & Florido, E. 2000, *Fund. Cosmic Phys.* 21, 1

Beck, R. 2005, in: R. Wielebinski, & R. Beck (eds.), *Cosmic Magnetic Fields*, Lecture Notes in Physics (Berlin: Springer) vol. 664, p. 41

Beck, R. 2007, *A&A* 470, 539

Beck, R. 2008, *Rev. Mexicana AyA*, in press, arXiv:0804.4594

Beck, R. & Gaensler, B. M. 2004, *New Astron. Revs* 48, 1289

Beck, R. & Golla, G. 1988, *A&A* 191, L9

Beck, R. & Hoernes, P. 1996, *Nature* 379, 47

Beck, R. & Krause, M. 2005, *AN* 326, 414

Beck, R., Brandenburg, A., Moss, D., Shukurov, A., & Sokoloff, D. 1996, *ARA&A* 34, 155

Beck, R., Shukurov, A., Sokoloff, D., & Wielebinski, R. 2003, *A&A* 411, 99

Beck, R., Fletcher, A., Shukurov, A., *et al.* 2005, *A&A* 444, 739

Bell, E. F. 2003, *ApJ* 586, 794

Berkhuijsen, E. M., Beck, R., & Hoernes, P. 2003, *A&A* 398, 937

Bernet, M. L., Miniati, F., Lilly, S. J., Kronberg, P. P., & Dessauges-Zavadsky, M. 2008, *Nature* 454, 302

Brentjens, M. A. & de Bruyn, A. G. 2005, *A&A* 441, 1217

Brown, J. C., Haverkorn, M., Gaensler, B. M., *et al.* 2007, *ApJ* 663, 258

Chyży, K. T. & Beck, R. 2004, *A&A* 417, 541

Chyży, K. T. & Buta, R. J. 2008, *ApJ* 677, L17

Chyży, K. T., Bomans, D. J., Krause, M., *et al.* 2007, *A&A* 462, 933

Dalla Vecchia, C. & Schaye, J. 2008, *MNRAS* 387, 1431

de Avillez, M. A. & Breitschwerdt, D. 2005, *A&A* 436, 585

de Jong, T., Klein, U., Wielebinski, R., & Wunderlich, E. 1985, *A&A* 147, L6

Dobbs, C. L. & Price, D. J. 2008, *MNRAS* 383, 497

Dumke, M., Krause, M., Wielebinski, R. & Klein, U. 1995, *A&A* 302, 691

Elstner, D., Meinel, R., & Beck, R. 1992, *A&AS* 94, 587

Ferguson, A. M. N., Wyse, R. F. G., Gallagher, J. S., & Hunter, D. A. 1998, *ApJ* 506, L19
Fletcher, A. & Shukurov, A. 2006, *MNRAS* 371, L21
Fletcher, A., Berkhuijsen, E. M., Beck, R., & Shukurov, A. 2004, *A&A* 414, 53
Frick, P., Beck, R., Berkhuijsen, E. M., & Patrikeev, I. 2001, *MNRAS* 327, 1145
Gaensler, B. M., Haverkorn, M., Staveley-Smith, L., *et al.* 2005, *Science* 307, 1610
Han, J. L., Manchester, R. N., Berkhuijsen, E. M., & Beck, R. 1997, *A&A* 322, 98
Han, J.L., Manchester, R. N., Lyne, A. G., & Qiao, G. J. 2002, *ApJ* 570, L17
Haverkorn, M., Gaensler, B. M., Brown, J. C., *et al.* 2006, *ApJ* 637, L33
Heesen, V., Beck, R., Krause, M., & Dettmar, R.-J. 2008, *A&A*, in press, arXiv:0812.0346
Heiles, C. 1996, in: W. G. Roberge, & D. C. B. Whittet (eds.), *Polarimetry of the Interstellar Medium*, ASP Conf. Ser. (San Francisco: Astron. Soc. Pac.), vol. 97, p. 457
Helou, G. & Bicay, M. D. 1993, *ApJ* 415, 93
Hoernes, P., Berkhuijsen, E. M., & Xu, C. 1998, *A&A* 334, 57
Hughes, A., Wong, T., Ekers, R., *et al.* 2006, *MNRAS* 370, 363
Johnston-Hollitt, M., Hollitt, C. P., & Ekers, R.D. 2004, in: B. Uyaniker, W. Reich & R. Wielebinski (eds.), *The Magnetized Interstellar Medium* (Katlenburg: Copernicus), p. 13
Kiepenheuer, K. O. 1950, *Phys. Rev.* 79, 738
Klein, U., Wielebinski, R. & Morsi, H. W. 1988, *A&A* 190, 41
Kothes, R., Landecker, T. L., Reich, W., Safi-Harb, S., & Arzoumanian, Z. 2008, *ApJ* 687, 516
Krause, F. & Beck, R. 1998, *A&A* 335, 789
Krause, M. 1990, in: R. Beck, R. Wielebinski, & P. P. Kronberg (eds.), *Galactic and Intergalactic Magnetic Fields* (Dordrecht: Kluwer), p. 187
Krause, M. 2008, *Rev. Mexicana AyA*, in press, arXiv:0806.2060
Krause, M., Hummel, E., & Beck, R. 1989, *A&A* 217, 4
Kronberg, P. P., Bernet, M. L., Miniati, F., *et al.* 2008, *ApJ* 676, 70
Lisenfeld, U., Völk, H. J., & Xu, C. 1996a, *A&A* 306, 677
Lisenfeld, U., Völk, H. J., & Xu, C. 1996b, *A&A* 314, 745
Men, H., Ferrière, K., & Han, J. L. 2008, *A&A* 486, 819
Moss, D. & Sokoloff, D. 2008, *A&A* 487, 197
Nieten, C., Neininger, N., Guélin, M., *et al.* 2006, *A&A* 453, 459
Niklas, S. & Beck, R. 1997, *A&A* 320, 54
Patrikeev, I., Fletcher, A., Stepanov, R., *et al.* 2006, *A&A* 458, 441
Rand, R. J. & Lyne, A. G. 1994, *MNRAS* 268, 497
Ransom, R. R., Uyaniker, B., Kothes, R., & Landecker, T. L. 2008, *ApJ* 684, 1009
Reich, W. 1994, in: R. Genzel, & A. I. Harris (eds.), *The Nuclei of Normal Galaxies* (Dordrecht: Kluwer), p. 55
Robishaw, T., Quataert, E. & Heiles, C. 2008, *ApJ* 680, 981
Roussel, H., Helou, G., Beck, R., *et al.* 2003, *ApJ* 593, 733
Seymour, N., Dwelly, T., Moss, D., *et al.* 2008, *MNRAS* 386, 1695
Soida, M., Beck, R., Urbanik, M. & Braine, J. 2002, *A&A* 394, 47
Sokoloff, D. D., Bykov, A. A., Shukurov, A., Berkhuijsen, E. M., Beck, R., & Poezd, A. D. 1998, *MNRAS* 299, 189, and Erratum in *MNRAS* 303, 207
Stepanov, R., Arshakian, T. G., Beck, R., Frick, P., & Krause, M. 2008, *A&A* 480, 45
Stil, J. M., Krause, M., Beck, R., & Taylor, A. R. 2008, *ApJ*, in press, arXiv:0810.2303
Sun, X. H., Reich, W., Waelkens, A., & Enßlin, T. A. 2008, *A&A* 477, 573
Tabatabaei, F., Beck, R., Krause, M., *et al.* 2007a, *A&A* 466, 509
Tabatabaei, F., Beck, R., Krügel, E., *et al.* 2007b, *A&A* 475, 133
Tabatabaei, F., Krause, M., Fletcher, A., & Beck, R. 2008, *A&A* 490, 1005
Thompson, T. A., Quataert, E., Waxman, E., Murray, N., & Martin, C.L. 2006, *ApJ* 645, 186
Walsh, W., Beck, R., Thuma, G., *et al.* 2002, *A&A* 388, 7
Wielebinski, R. 2005, in: R. Wielebinski, & R. Beck (eds.), *Cosmic Magnetic Fields*, Lecture Notes in Physics (Berlin: Springer) vol. 664, p. 89
Wolleben, M. & Reich, W. 2004, *A&A* 427, 537

Discussion

MAEDER: To what extend can the magnetic field you are observing contribute to the dynamics of the external regions of galaxies? In particular, is the field contribution to the flat rotation curve negligible or not?

BECK: Magnetic fields in external regions will certainly affect outflows and tide gas flows, as evidenced from existing data. However, the field strengths seem to be insufficient to affect outer rotation curves. Low-frequency observations with e.g. LOFAR should allow to measure magnetic fields to much larger radii.

HEILES: How good is the equipartition assumption? In particular, the derived field strength depends on the cosmic ray proton/electron ratio. How serious is this?

BECK: The equipartition field strength depends on the cosmic-ray proton/electron ration K as about $K^{1/4}$, so that variations from the standard value of K=100 by a factor of several are not serious. However, in case of strong energy losses of electrons, K may become very large, and the equipartition estimate underestimates the field strength. We need independent data on cosmic-ray protons and electrons, e.g. from γ-ray data.

HEILES: Why are synchrotron-derived field strengths so much larger than RM-derived fields? How about assumptions about equipartition?

BECK: Equipartition field strengths derived from total synchrotron intensity include turbulent fields, anisotropic (sheared) fields and regular (coherent) fields, whereas Faraday rotation traces regular fields only. In case of significant field fluctuations ($\delta B^2 \geqslant B^2$), the equipartition field strength is an overestimate. In case of fluctuations in regular field and ionized gas density, the Faraday-derived field strength is an underestimate if $\delta \mid B_{\rm reg} \mid$ and $\delta n_{\rm e}$ are anticorrelated (see Beck *et al.* 2003 for details).

RÜDIGER: Rainer, would you add, please, some comments about the equatorial parity of the ordered magnetic field? Is the quadrupolar parity now better established than in the past?

BECK: There is still no clear indications for quadrupolar or dipolar field patterns in galaxies because Faraday rotation measures in the halos of edge-on galaxies do not show large-scale patterns (but errors are large). We need better RM data, e.g. from LOFAR. In the Milky Way the local field is vertically symmetric (quadrupole), but the halo field is possibly antisymmetric (see Sun *et al.* 2008).

DE GOUVEIA DAL PINO: Could you comment on the present status of the observations towards the potential bi-symmetric pattern in the Milky Way?

BECK: Please see the talks by A. Noutsos, J.L. Han and W. Reich for the present status of the interpretation of the existing data. In external galaxies a bisymmetric field was found only in M81 some time ago (Krause *et al.* 1989), but this is presently re-investigated with new data by M. Krause & A. Fletcher.

Rainer Beck

Aristeidis Noutsos

Cosmic Magnetic Fields:
From Planets, to Stars and Galaxies
Proceedings IAU Symposium No. 259, 2008
K.G. Strassmeier, A.G. Kosovichev & J.E. Beckman, eds.

doi:10.1017/S1743921309030026

Measuring ISM fields using Pulsars

Aristeidis Noutsos†
University of Manchester, Jodrell Bank Centre for Astrophysics, Alan Turing Building, Manchester M13 9PL.
email: aristeidis.noutsos@manchester.ac.uk

Abstract. The sample of available Galactic pulsar rotation measures has proven an invaluable tool for measuring the direction and magnitude of the interstellar magnetic fields of our Galaxy. In this review, I present highlights of recent efforts to measure and map the Galactic magnetic field using pulsars. I give an overview of the analysis methods that were used by previous authors and underline the key results that have given us a clear picture of the magnetic field in certain regions of the Galaxy. This review also lays out the limitations of the present analysis methods and the observational difficulties that have so far hindered the study of the Galactic magnetic field with pulsars. Despite these difficulties, the continuous discovery of new pulsars in more and more sensitive surveys offer a continuous improvement on the existing knowledge of the Galactic magnetic field.

Keywords. Galaxy: structure – ISM: magnetic fields – pulsars: general – techniques: polarimetric

1. Introduction

The shape and strength of the Galactic magnetic field (GMF) has been the subject of research since the birth of radio astronomy, in the 1930s. There are many interesting astrophysical processes that are connected with the GMF: e.g star formation, the deflection of ultra-high energy cosmic rays, etc. However, despite the many efforts to produce a clear picture of the GMF – both in the field of radio astronomy, but also by studying the optical and infrared radiation through the interstellar medium (ISM) – the subject is still under discussion and there is little consensus over most of the field's properties.

1.1. *Sizing up the Galactic Magnetic Field*

Many of the previous studies of the GMF have tried to distinguish between a large-scale, regular component of the field, with typical scales of ~ 1 kpc, and a smaller-scale, turbulent component, with scales $\sim 10 - 100$ pc. This is, of course, an artificial classification, aimed at simplifying the view of the ISM field, whereas in reality the ISM field everywhere is the inseparable combination of all of its components.

In general, the regular field of the Galaxy is thought to have originated from a primordial field that has been strengthened via dynamo action and shaped by the Galactic gas motions or perhaps by the interaction with a companion galaxy. The turbulent component is usually attributed to the fields of localised structures, like supernova bubbles and ionised HII regions or even to frozen fields in molecular HI clouds. Well known examples of this type are the "North Polar Spur" and the "Gum Nebula", which are parts of supernova features.

† Present address: University of Manchester, Jodrell Bank Centre for Astrophysics, Alan Turing Building, Manchester M13 9PL.

1.2. *The shape and direction of the regular magnetic field*

To date, there is no conclusive evidence for the current shape of the Galactic disc field, and the subject of the field's origin remains open. The spiral shape of the luminous matter of the Galaxy has made the study of spiral forms for the regular component of the GMF an attractive possibility. In addition, observational data — mainly from pulsars — have suggested the existence of large-scale discontinuities in the field direction, known as "field reversals". Beyond the disc field, the Galaxy maintains a thick-disc or halo component with a measured typical scale-height of ~ 1.5 kpc: comparable to that of the free electron distribution (Beuermann *et al.* 1985; Han & Qiao 1994). The Galactic rotation has almost certainly stretched any primordial field across the GP. As a result, the vertical component of the GMF is about an order of magnitude weaker than the planar components.

1.3. *Models of the regular magnetic field*

There are three main classes of model for the regular field of the GMF: the concentric-ring field, the axisymmetric spiral (ASS) and the bisymmetric spiral (BSS).

The concentric-ring models have no radial component, i.e. $B_r = 0$, but their azimuthal component, B_θ, can vary with Galactocentric radius, r (see figure 1a). These models find support in the density-wave theory, which predicts that gas should follow circular orbits around the Galactic centre (GC).

The ASS models are compatible with theories of dynamo action on a primordial field. These models have both radial and azimuthal components, which vary independently only with Galactocentric distance, r, but not with azimuth, θ (see figure 1b). Although pure ASS models do not naturally predict magnetic-field reversals — and those that do usually restrict their location between the Perseus and Crux–Scutum arm — under certain conditions, i.e. that the primordial field exhibits strong reversals, reversals are compatible with the dynamo theory.

BSS models are described by radial and azimuthal components that are sinusoidal functions of θ (figure 1c). They can easily incorporate field reversals in both the radial and azimuthal components.

A property of all the above models is the symmetry of the field with respect to the GP: antisymmetric (odd) field models are oppositely directed either side of the GP, whereas symmetric (even) field models maintain the same direction.

Finally, most of the above models adopt a field-strength variation with r, with B increasing towards the GC. However, the actual function, $B(r)$, has not been conclusively estimated yet.

1.4. *Observational tracers of ISM fields*

The methods that have been employed so far for the detection and measurement of ISM fields are only able to explicitly produce the value of the field component perpendicular or parallel to our line-of-sight (LOS), but not the full, three-dimensional $\boldsymbol{B}$ vector.

Zeeman splitting of the spectral lines measures the parallel component to the LOS. Although this method is successful in measuring the local fields in molecular clouds, etc., it is difficult to relate these measurements with the large-scale magnetic field.

Polarization of dust emission at infrared, mm and sub-mm wavelengths can reveal the perpendicular-to-the-LOS (transverse) component of the magnetic field. Currently, this method only works well with bright molecular clouds. Hence, the only relevant region to the large-scale GMF that can be studied this way is the central molecular ring of the Galaxy.

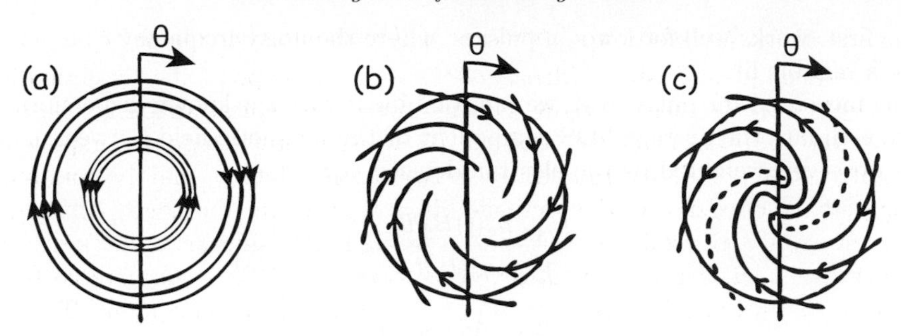

Figure 1. Three classes of regular Galactic magnetic-field model: (a) concentric rings (b) axisymmetric spiral and (c) bisymmetric spiral.

Polarization of starlight is another tracer of the magnetic field that can measure the sky-projected component of the GMF. Unfortunately, since the stars that can be measured this way are at most ~2 kpc from the Sun, the field further afar cannot be explored with this method.

Maps of the transverse component of the GMF can also be constructed from polarization observations of the synchrotron emission of the Galactic relativistic-electron population. However, the anisotropic, random fields of large-angular-scale features, like the North Polar Spur, also appear in such maps, so that it can be difficult to separate them from the regular GMF.

Finally, Faraday rotation of the polarised emission of pulsars and extragalactic sources (EGRS) can be used to measure the LOS component of the GMF. Since the magnitude of this effect represents the integrated (along the LOS) interaction between the magnetised ISM and the polarised emission, pulsars at different distances can reveal the field's strength and direction across different depths through the Galaxy.

2. Pulsars as probes of the Galactic Magnetic Field

Soon after the discovery of pulsars, in 1967, it was realized that their polarized emission can be used to retrieve spatial information about the strength and direction of the GMF. The linearly polarized pulsar emission on its way to the observer interacts with the magnetized ISM, which causes the rotation of the plane of linear polarization ($\Delta\Psi$): the well-known Faraday effect. The magnitude of this effect is a quadratic function of the emission wavelength (λ) and proportional to the Rotation Measure (RM): the latter is a function of the pulsar distance and the radial profiles of the free-electron density and the interstellar magnetic field along the LOS to the pulsar. I.e.

$$\Delta\Psi = \lambda^2 \cdot 0.812 \int_{\mathrm{PSR}}^{\oplus} n_e(l) B_{\parallel}(l) dl = \lambda^2 \cdot \mathrm{RM} \qquad (2.1)$$

Using a receiver of a known, finite bandwidth, $c/\lambda_1 - c/\lambda_2$, one could measure the difference in the amount of rotation between the edges of the band, $\Delta\Psi(\lambda_2) - \Delta\Psi(\lambda_1)$, and calculate RM. Given a strongly polarized pulsar, this is usually done in any of the two following ways: (a) the recorded Stokes parameters of the polarized signal, Q and U, are combined to calculate $\Psi = 0.5 \arctan(U/Q)$ for each frequency channel across the band, and the rotation of Ψ with frequency is fitted with the quadratic function of Eq. 2.1 to obtain the best value of RM. (b) Alternatively, one can combine Q and U to calculate the linear polarization, $L = (Q^2 + U^2)^{1/2}$, in each frequency channel and then sum the Ls of all channels together by first de-rotating Q and U with a candidate RM. This process can be repeated for a range of candidate RMs and should produce the maximum L at the correct RM. The second method, although evidently more computationally expensive than the first, works well for low-s/n pulsars, where the noisy frequency channels cannot produce a reliable fit.

Having measured the pulsar RM, we can combine it with the known dispersion measure (DM) to estimate the average LOS component of the magnetic field between the pulsar and the observer, weighted by the electron density, n_e:

$$\langle B_{\parallel} \rangle = \frac{\int_{\mathrm{PSR}}^{\oplus} n_e(l) B_{\parallel} dl}{\int_{\mathrm{PSR}}^{\oplus} n_e(l) dl} = 1.232 \, \frac{\mathrm{RM}}{\mathrm{DM}} \tag{2.2}$$

2.1. *Advantages*

The efforts to determine the three-dimensional structure of the GMF are hindered by our location being inside the volume we are trying to probe. The use of pulsars in this attempt is however advantageous for a number of reasons: (a) pulsar emission carries, in general, a high degree of linear polarization, which allows us to measure RMs with great accuracy; (b) since pulsar magnetospheres contain ultra-relativistic plasma that has a minor, if any at all, contribution to the measured RMs, the Faraday effect on pulsar emission is a direct consequence of solely the properties of the ISM; (c) the pulsar population is distributed across the entire galactic volume, which allows us to sample the ISM properties over a wide range of Galactic longitudes and distances (figure 2a); (d) last but not least, the pulse dispersion caused by the ISM (given by the pulsar DM), not only allows us to estimate their distances but also, when combined with their RMs, to directly measure the average value of the interstellar magnetic field between the pulsar and the observer.

Most pulsars are found close to the Galactic plane and there is an appreciable concentration of pulsars along the spiral arms. The large sample of pulsars at low latitudes has proven beneficial for the studies of the thin-disc component of the GMF, whereas those that are found at high latitudes allow us to study the magnetic field of the Galactic halo (Han & Qiao 1994).

2.2. *Using Pulsar RMs to reveal the structure of the Galactic Magnetic Field*

The aforementioned advantages of pulsars in studies of the GMF have encouraged many workers to use them to map the large-scale component of the field. The early work by Manchester (1972) and Manchester (1974) used Eq. 2.2 to estimate the field direction, averaged over the entire LOS to the pulsar, as a function of Galactic longitude. Later, Lyne & Smith (1989) were the first to advance this method by exploiting the RM–DM gradients along given LOS. They used pairs of pulsars, close to each other in the sky but at different distances, to estimate the local variations of $B_{\parallel}$ as a function of distance: i.e. $\langle B_{\parallel} \rangle_{d_1 - d_2} \propto (RM_2 - RM_1)/(DM_2 - DM_1)$. Using this method for different LOS revealed the field variations in different directions, across nearly 200° of Galactic longitude.

As the number of pulsars with RMs reached nearly 200, the reverse logic could be applied to uncover the structure of the regular field: it became possible to dream up various models of the field's configuration and test their viability by fitting the models' predicted RM values to those observed (e.g Rand & Kulkarni 1989; Rand & Lyne 1994); the model's goodness-of-fit would then be the decisive factor. Such multivariate approaches have been followed by many authors in the recent years (see e.g. Han *et al.* 2006; Brown *et al.* 2007; Noutsos *et al.* 2008; Vallée 2008; Men *et al.* 2008).

The latest methodology in pulsar studies of the GMF is the use of wavelets to analyse the currently hundreds of available pulsar RMs (Frick *et al.* 2001; Stepanov *et al.* 2002). Wavelets are self-similar functions that can be used to describe the spatial and frequency distribution of the RM sample. Using wavelets, one can fit not the RM data themselves but the wavelet transform of the RM data to the magnetic-field model. A significant advantage of wavelet analysis over alternative methods of studying the regular GMF is its ability to filter out the smaller scales (due to the turbulent component), thus potentially improving the fits to the large-scale GMF models.

2.3. *Fundamental problems and limitations of analyses using Pulsars*

General: All of the above methods have been successfully used with the sample of RMs that were available to the different workers. However, despite their usefulness, there are shortcomings associated with each one, which limits their reliability.

The simple approach of dividing the pulsar RMs by their DMs to characterize the magnetic field in a certain direction in the sky, for example, assumes implicitly that the average values of n_e and $B_{\|}$ are a good representation of their actual profiles along the LOS: that is to say, the electron density and magnetic field are uniform between the observer and the pulsar. We know, of course, that this is far from the truth, as the magnitude of both can change along the LOS; and there is good evidence that the the magnetic field's direction changes as well.

Those methods that used pairs of pulsars, located close to each other in the sky, suffer from the limitation that dominates all pulsar studies of the magnetic field: the number of available RMs is not sufficiently high for those methods to be able to track the fluctuations of the magnetic field along different LOS. Not only does the RM sample not have adequate density for a reliable mapping of the ISM fields to small scales, but the sparseness and irregularity of the pulsar positions results in very noisy maps of the magnetic field, when the latter is calculated from the gradients $\Delta\mathrm{RM}/\Delta\mathrm{DM}$ (Ruzmaikin *et al.* 1988; Stepanov *et al.* 2002).

Methods that rely on RM–DM gradients to represent magnetic-field variations are sensitive to the presence of localized structures: e.g. if a magnetized HI cloud lies between a pair of pulsars along our LOS, it will contribute little in DM across its volume but may have a substantial contribution in RM (Manchester 1974). There are certainly observations of neighboring pulsars with very different, and even sometimes opposite RMs (e.g. Han *et al.* 2006).

An added complication comes from the pulsar distances that are derived from their DMs: this requires a model of the free-electron density along the LOS. Earlier work, based on the limited information available at the time, assumed a uniform electron density (e.g. Lyne & Smith 1989), which almost certainly introduced significant errors into their distance measurements. But even nowadays, with the use of the more accurate free-electron density model of Cordes & Lazio (2002), there is still a 10–20% error in the estimation of pulsar distances in the Galactic disc, whereas a much higher error is introduced at high latitudes (Gaensler *et al.* 2008).

The multivariate approach of fitting an ad hoc model to the data may also be influenced by the effects mentioned above, which makes it sometimes impossible to be certain of which model fits the data best. In addition, the a priori restrictions that these models impose on the GMF are not always represented in actuality.

Wavelet transform is certainly an improvement on previous analyses. However, it still requires a sufficiently dense sample of pulsars: the current RM-sample density fails to satisfy this criterion beyond 3 kpc. Application of wavelets on the currently available RM database has produced rather confusing and difficult-to-interpret maps of the local GMF (Stepanov *et al.* 2002).

Finally, caution is required when combining RM data sets from different epochs in order to perform a global fit to magnetic-field models. There are many examples of pulsars whose derived RM is seen to change significantly over a period of years (van

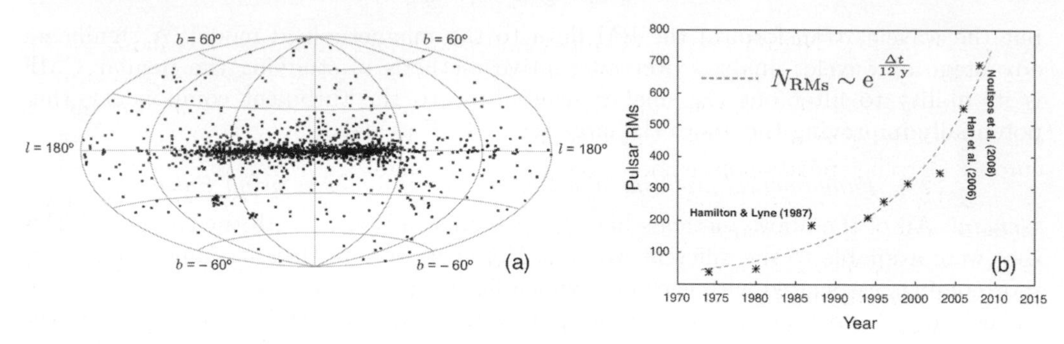

Figure 2. (a) Hammer–Aitoff projection of all discovered pulsars. (b) Measured number of pulsar RMs as a function of time.

Ommen *et al.* 1997; Han *et al.* 1999; Han *et al.* 2006 and Noutsos *et al.* 2008). These changes can occur as a result of magnetic-field or electron-density variations along the LOS — either due to changes in the ISM or due to different LOS as a consequence of pulsar proper motions. Nevertheless, it has been suggested that the changes of several rad m^{-2} observed are most likely because of ISM-field variations (Weisberg *et al.* 2004).

RM variations are also observed at much shorter time scales. Recent work by Ramachandran *et al.* (2004) clearly showed RM variations of tens of rad m^{-2} across the pulse phase, for a number of strong pulsars. It is yet unclear if there is a single reason for these variations, although ISM scattering is thought to play an important role (Karastergiou 2008, *submitted*).

The turbulent field: Any method that tries to map the regular component of the GMF should be capable of identifying the large-scale trends of the field in a forest of small-scale, turbulent fields. The turbulent fields are highly organized in a range of scales, as is observed e.g. for the fields of the North Polar Spur and "Region A": parts of extended supernova remnants (Rand & Kulkarni 1989). The typical strength of such fields has been estimated to be as much as twice that of the regular field: i.e. 3–7 μG (Beck *et al.* 2003). The impact of the turbulent component of the GMF on our RM measurements is significant: for example, its effect is evident in RM–DM plots, where it dominates the RM scatter – by an order of magnitude – over measurement errors.

As was shown by Beck *et al.* (2003), the fluctuations of the magnetic field in a turbulent medium are correlated with those of the electron density. Since the measured values of $\langle B_{\|} \rangle$ are based on the weighted average of Eq. 2.2, which requires that σ_{n_e} and $\sigma_{B_{\|}}$ vary independently, there is an additional systematic error on the estimates of the regular field: if these quantities are anticorrelated, the regular component of the field is underestimated, whereas positive correlation leads to overestimation of $\langle B_{\mathrm{reg}} \rangle$.

Finally, the work of Mitra *et al.* (2003) showed that LOS that translate HII regions exhibit anomalous RM and DM fluctuations, which calls for the need to exclude such LOS from any analysis that aims at recovering the regular component of the GMF. However, they point out the value of using pulsars behind such HII regions to study the local value of the turbulent magnetic field. In the recent years, many workers have noticed the effects on RM data from such regions, like e.g. the anomalous RMs obtained in the direction of the Carina arm (e.g. Han *et al.* 2006; Noutsos *et al.* 2008).

3. The current picture of the Galactic Magnetic Field from Pulsars

3.1. *The Pulsar sample*

Pulsar studies of the GMF in the 1970s had to rely on a very poor sample of only a few tens of RMs, most of which were within 2 kpc from the Sun (Manchester 1972; Manchester

1974; Thomson & Nelson 1980). An important step in this direction was made in the late 1980s with the acquisition of 163 pulsar RMs by Hamilton & Lyne (1987). Up until the end of the last decade, some 320 RMs had been measured (Costa *et al.* 1991; Rand & Lyne 1994; Qiao *et al.* 1995; van Ommen *et al.* 1997; Han *et al.* 1999). To date, a number of recent pulsar-polarization surveys have brought the RM count to 690 (Mitra *et al.* 2003; Weisberg *et al.* 2004; Han *et al.* 2006; Noutsos *et al.* 2008). Figure 2b shows the accumulation of measured pulsar RMs as a function of time.

As was mentioned before, since most pulsars are found near the Galactic disc, naturally most RM measurements correspond to low Galactic latitudes. In addition, the higher pulsar density in the inner Galactic quadrants compared to that beyond the solar radius means that the first (Q1) and fourth quadrant (Q4) of the disc have a richer RM sample than the other two. An observational bias that is caused by enhanced scattering and generally lower detectability at larger distances has resulted in a drop of the density of the RM sample with increasing distance; an exception is the high-latitude pulsars, like e.g. those in the LMC and SMC, which are not obscured by the dense environment of the Galactic disc and can be detected even at large distances.

An additional nuisance factor in RM measurements is that not all pulsars have enough intrinsic polarization to be measurable with the instruments' current sensitivity. It is reasonable to assume that there will always be a fraction of the discovered pulsar sample, whose RMs will not be measurable: at the moment, the fraction of pulsars with a measured RM stands a little over 40%; but, with more sensitive instruments planned for the future (e.g. LOFAR, FAST, SKA), it is bound to increase.

3.2. *What have we learnt so far?*

So far, the majority of pulsar work on the GMF has concluded that the solar neighborhood is permeated by a nearly azimuthal, $\sim$2 μG field, with a CW direction (as seen from the Galactic north). Initial publications concluded that the local field is directed towards $\ell \sim 90°$ (Manchester 1972; 1974; Thomson & Nelson 1980), advocating for a ring-like configuration of the regular field. Following work, however, noted that those early studies may have been biased by selection effects, mainly due to poor sampling (Han & Qiao 1994), and that the most likely direction of the field lies somewhere between $\ell = 75° - 85°$, i.e. close but perhaps at a significant angle to the Orion spur. Similar values for the direction of the local field are nowadays commonly adopted when spiral models of the field are assumed (Brown *et al.* 2007; Noutsos *et al.* 2008; Sun *et al.* 2008).

Moreover, other methods of measuring the local field (e.g. Zeeman splitting, diffuse synchrotron emission) have resulted in fields of $\sim$ 4 μG for the regular component. According to Beck (2008), depending on the properties of the turbulent medium, the pulsar-based result may be an underestimate of the actual field strength, or the other methods may be overestimating that value.

Also, many investigators have provided convincing evidence for a reversal of the local field in Q1, near the Carina–Sagittarius arm, within 1 kpc from the Sun (Lyne & Smith 1989; Rand & Kulkarni 1989; Han & Qiao 1994; Rand & Lyne 1994; Frick *et al.* 2001; Weisberg *et al.* 2004; Han *et al.* 2006; Noutsos *et al.* 2008). The reversal changes the CW direction of the local field to CCW in the Carina–Sagittarius arm.

The current estimate of the the vertical component of the local magnetic field is $B_z \approx$ 0.4 μG, with a direction pointing from the South to the North Galactic Pole (Han & Qiao 1994; Han *et al.* 1999). The result was drawn from RMs of high-latitude EGRs ($|\, b \,| > 60°$) and was checked for consistency with 8 pulsars in the north polar region. From modelling, Indrani & Deshpande (1999) found that B_z remains more or less constant in the solar vicinity.

In addition to the local reversal towards the Carina–Sagittarius arm, Brown *et al.* (2007) combined 120 pulsar RMs with 148 EGRS RMs and provided convincing evidence for a second reversal between the Sagittarius–Carina and the Scutum–Crux arm, in Q4 (towards $\ell \sim 312°$). This find was confirmed by Noutsos *et al.* (2008), based on 150 southern-pulsar RMs. The reversal changes the CW field of the Carina–Sagittarius arm, in Q4, to CCW in the Scutum–Crux arm. The total of only two reversals confirmed, contrasts previous pulsar studies that reported reversals in every arm or interarm region, including reversals exterior to the solar circle (Weisberg *et al.* 2004; Han *et al.* 2006).

3.3. *Compelling evidence*

A number of studies of RMs of both EGRS and high-latitude pulsars ($|\,b\,| > 8°$) have revealed an antisymmetry in the signs of pulsar RMs, with respect to the Galactic equator, towards the inner Galaxy ($\ell = 270° - 90°$) (Han 2007; Sun *et al.* 2008). Based on such antisymmetry, it has been claimed that the Galactic halo maintains an azimuthal field that is oppositely directed below and above the plane, with an estimated strength of ~ 1 μG (Han *et al.* 1997; Han *et al.* 1999). This is the signature of an A0 dynamo operating in the halo of our Galaxy. If these finds are confirmed by more data, it will be the first time that such a signature is clearly identified.

Previous studies of the regular GMF strength have suggested that it decreases with Galactocentric radius (Rand & Lyne 1994; Han *et al.* 2002; Han *et al.* 2006): e.g. based on pulsar RMs in the Norma arm, Han *et al.* (2002) estimated a field of ~ 4 μG, which is twice as strong as the local field. Despite those clues, the functional form of the field's strength with r remains uncertain: earlier models assumed a $1/r$ and $1/r^2$ dependence on the Galactocentric radius; the more recent work of Han *et al.* (2006) used a larger sample of RMs and tried to parameterize the magnetic-field strength as an exponential function of r.

Many models of the regular GMF assume an exponential suppression of the planar field, $B(x, y)$, with Galactic height, z (see e.g. Thomson & Nelson 1980; Kachelrieß *et al.* 2007). From pulsar work, we have very little, if any at all, information about the field's dependence on z. Under the assumption of a frozen-in GMF in the magneto-ionic slab of the Galaxy, Han & Qiao (1994) concluded that the field should exponentially decrease as a function of Galactic height, with an exponential scale-height similar to that of the free electrons.

3.4. *Controversial results*

In the recent years, the efforts to describe the regular GMF with one of the three aforementioned classes of model has led to controversy. The BSS models have been favored in earlier work, e.g. by Simard-Normandin & Kronberg (1980) and Indrani & Deshpande (1999), and more recently by Han *et al.* (2006). The concentric-ring models, despite having received heavy criticism (e.g. Han *et al.* 2006), have survived the test of time (Rand & Kulkarni 1989; Rand & Lyne 1994; Vallée 2005; Vallée 2008). Even though ASS models follow naturally from the dynamo theory, they have received limited support until now (Vallée 1996). This type of model was challenged by the recent results of Han *et al.* (2006) and Weisberg *et al.* (2004), showing reversals near the GC and exterior to the Perseus arm, since it only allows for reversals between the Crux-Scutum and Perseus arms.

Nevertheless, the most recent work on the large-scale field of the Galactic disc, which utilized a larger and more reliably measured sample of RMs, has rejected any of the three model classes as being a good description of the observed pulsar RMs. Men *et al.* (2008) tried a fit of all three types of model to 482 pulsar RMs and concluded that none of them was acceptable, noting however that a BSS model was the least unacceptable; the authors conceded that the true form of the GMF is much more complex than the

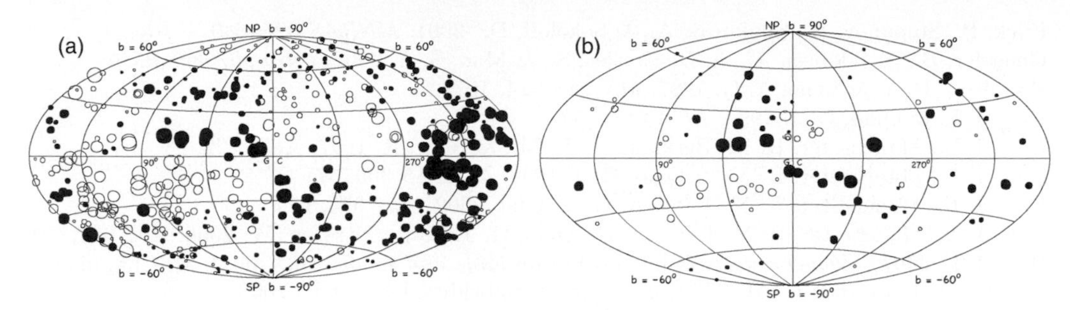

Figure 3. (From Han *et al.* 1997) The antisymmetric patterns with respect to the Galactic plane of the RMs of (a) extragalactic sources and (b) pulsars, seen at high latitudes towards the Galactic centre.

existing models lead us to believe. Moreover, a similar analysis by Noutsos *et al.* (2008), using 150 well-measured southern-pulsar RMs, also rejected any of the three types of BSS model as being significantly better than the rest. On the other hand, Vallée (2008) tried a concentric ring and the ASS model of Brown *et al.* (2007) on 554 pulsar RMs and found them reasonable. On those grounds, he urged future investigators to consider the concentric ring model in their analyses.

So far, current modelling has fallen short of a good description of the GMF, and perhaps a combination of all the above models is required — and with possibly only specific regions of the Galaxy being compatible with any one model.

4. Future studies of ISM fields with Pulsars

In the last decade, most published work on the study of the GMF with pulsars has used data either from the Parkes 64m radio-telescope, in the southern hemisphere, or from northern-hemisphere telescopes, like Effelsberg (100m) and Arecibo (305m). However, experiments planned for the near future, like the LOw-Frequency ARray (LOFAR) in the Netherlands, and those that are expected to be completed in the next decade, like the Square Kilometre Array (SKA), will almost certainly discover a large number of pulsars and subsequently measure their RMs. One of the scientific goals of LOFAR is to perform an all-sky survey of the northern sky at 150 MHz, in search for nearby and high-latitude pulsars. LOFAR will then perform follow-up polarimetric observations on the new pulsar sample and, having the advantage of operating at frequencies where pulsar spectra peak and the ISM Faraday effect is strong, accurately measure new pulsar RMs: some 300–500 new pulsar RMs is currently estimated to emerge from such a survey. The new RM sample, together with the use of novel analysis methods, like the RM synthesis (Brentjens & de Bruyn 2005), will help us probe the local fields with unprecedented resolution, contributing to our knowledge of the small-scale, turbulent component of the GMF; it will also help us map the vertical profile of the field and trace its extent.

References

Beck, R., Shukurov, A., Sokoloff, D., & Wielebinski, R. 2003, *A&A* 411, 99

Beuermann, K., Kanbach, G., & Berkhuijsen, E. M. 1985, *A&A* 153, 17

Beck, R. 2008, in: F. A. Aharonian, W. Hofmann, & F. M. Rieger (eds.), *High Energy Gamma-Ray Astronomy*, AIP Conf. Proc., in press

Brentjens, M. A. & de Bruyn, A. G. 2005, *A&A* 441, 1217

Brown, J. C., Haverkorn, M., Gaensler, B. M., Taylor, A. R., Bizunok, N. S., McClure-Griffiths, N. M., Dickey, J. M., & Green, A. J. 2007, *ApJ* 663, 258

Cordes, J. M. & Lazio, T. J. W. 2002, *astro-ph/0207156*

Costa, M. E., McCulloch, P. M., & Hamilton, P. A. 1991, *MNRAS* 252, 13

Frick, P., Stepanov, R., Shukurov, A., & Sokoloff, D. 2001, *MNRAS* 325, 649
Gaensler, B. M., Madsen, G. J., Chatterjee, S., & Mao, S. A. 2002, *astro-ph/0808.2550*
Hamilton, P. A. & Lyne, A. G. 1987, *MNRAS* 224, 1073
Han, J. L. & Qiao, G. J. 1994, *A&A* 288, 759
Han, J. L., Manchester, R. N., Berkhuijsen, E. M., & Beck, R. 1997, *A&A* 322, 98
Han, J. L., Manchester, R. N., & Qiao, G. J. 1999, *MNRAS* 306, 371
Han, J. L., Manchester, R. N., Lyne, A. G., & Qiao 2002, *ApJ* 570, L17
Han, J. L., Manchester, R. N., Lyne, A. G., Qiao, G. J., & van Straten, W. 2006, *ApJ* 642, 868
Han, J. L. 2007, *Magnetic fields in our Galaxy on large and small scales*, in: J. M. Chapman & W. A. Baan (eds.), IAU 242 Conf. Proc. (Cambridge, UK: CUP) 2008
Indrani, C. & Deshpande, A. A. 1999, *New Astronomy* 4, 33
Lyne, A. G. & Smith, F. G. 1989, *MNRAS* 237, 533
Manchester, R. N. 1972, *ApJ* 172, 43
Manchester, R. N. 1974, *ApJ* 188, 637
Men, H., Ferriére, K., & Han, J. L. 2008, *A&A* 486, 819
Mitra, D., Wielebinski, R., Kramer, M., & Jessner, A. 2003, *A&A* 398, 993
Noutsos, A., Johnston, S., Kramer, M., & Karastergiou, A. 2008, *MNRAS* 386, 1881
Qiao, G., Manchester, R. N., Lyne, A. G., & Gould, D. M. 1995, *MNRAS* 274, 572
Ramachandran, R., Backer, D. C., Rankin, J. M., Weisberg, J. M., & Devine, K. E. 2004, *ApJ* 606, 1167
Rand, R. J. & Kulkarni, S. R. 1989, *ApJ* 343, 760
Rand, R. J. & Lyne, A. G. 1994, *MNRAS* 268, 497
Ruzmaikin, A., Sokolov, D., & Shukurov, A. 1988, in: Ruzmaikin et al. (eds.), *Magnetic fields of galaxies*, ApSS Library (Moscow: Izdatel'stvo Nauka), vol. 133, p. 280
Simard-Normandin, M. & Kronberg, P. P. 1980, *MNRAS* 242, 74
Stepanov, R., Frick, P., Shukurov, A., & Sokoloff, D. 2002, *A&A* 391, 361
Sun, X. H., Reich, W., Waelkens, A., & Enßlin, T. A. 2008, *A&A* 477, 573
Thomson, R. C. & Nelson, A. H. 1980, *MNRAS* 191, 863
Vallée, J. P. 1996, *A&A* 308, 433
Vallée, J. P. 2005, *ApJ* 619, 297
Vallée, J. P. 2008, *ApJ* 681, 303
van Ommen, T. D., D'Alessandro, F., Hamilton, P., & McCulloch, P. 1997, *MNRAS* 287, 307
Weisberg, J. M., Cordes, J. M., Kuan, B., Devine, K. E., Green, J. T., & Backer, D. C 2004, *ApJS* 150, 317

Discussion

BECKMAN: Have the RM measurements been used to quantify local fields, and if so what were the volumes derived?

NOUTSOS: Rand & Kulkarni (1989) used the residuals to a pulsar-RM fit to the concentric ring model to estimate the field of local structures (¡ 3kpc). They found $B_r \approx 5\ \mu$G. But they used a *single* coherence scale, which they conclude is not representative of the actual multi-scale turbulent field. Mitra *et al.* (2003) estimated the random field towards $\ell \approx 149°$ from the scatter of pulsar RMs. They give an estimate of 5.7 μG (in agreement with RK89). Both analyses attribute the random fields to HII regions in the directions investigated. The general magnetic spectrum from pulsar data was calculated by Han *et al.* (2001). They investigated scales between 0.5 and 15 kpc. The pulsar RM sample is not dense enough yet to calculate the field of individual structures (HI, HII regions).

Cosmic Magnetic Fields:
From Planets, to Stars and Galaxies
Proceedings IAU Symposium No. 259, 2008
K.G. Strassmeier, A.G. Kosovichev & J.E. Beckman, eds.

doi:10.1017/S1743921309030038

Magnetic fields, stellar feedback, and the geometry of H II regions

Gary J. Ferland

Department of Physics & Astronomy, University of Kentucky, Lexington, KY 40506, USA
email: gjferland@gmail.com

Abstract. Magnetic pressure has long been known to dominate over gas pressure in atomic and molecular regions of the interstellar medium. Here I review several recent observational studies of the relationships between the H^+, H^0 and H_2 regions in M42 (the Orion complex) and M17. A simple picture results. When stars form they push back surrounding material, mainly through the outward momentum of starlight acting on grains, and field lines are dragged with the gas due to flux freezing. The magnetic field is compressed and the magnetic pressure increases until it is able to resist further expansion and the system comes into approximate magnetostatic equilibrium. Magnetic field lines can be preferentially aligned perpendicular to the long axis of quiescent cloud before stars form. After star formation and pushback occurs ionized gas will be constrained to flow along field lines and escape from the system along directions perpendicular to the long axis. The magnetic field may play other roles in the physics of the H II region and associated PDR. Cosmic rays may be enhanced along with the field and provide additional heating of atomic and molecular material. Wave motions may be associated with the field and contribute a component of turbulence to observed line profiles.

Keywords. ISM: magnetic fields – HII regions – ISM: individual (M17,M42) – cosmic rays – ISM: molecules

---

1. Introduction – The magnetic field of a quiescent cloud

Magnetic fields play pivotal roles in star-forming environments. Many aspects of this rich topic are covered in other papers in this book, and the review by Heiles & Crutcher (2005) is essential reading.

The first of the many influences of the field is in the formation of the molecular cloud itself. The crucial physics is the coupling between the magnetic field and even weakly ionized gas. This so-called flux freezing means that there is a relationship, set by the geometry of any expansion or contraction that occurs, between the gas and field density. This means that while gas is free to move along field lines, gas motions perpendicular to the field will magnify or weaken the field.

Figure 1, taken from Heiles (1988), shows the dark cloud L204. The orientation of the magnetic field, as deduced from starlight linear polarization, is indicated by the black lines. Field lines tend to lie perpendicular to the long axis of the filament. The Pipe Nebula (Alves *et al.* 2008) is another example. This geometry is not uncommon (Heiles & Troland 2005). One interpretation is that the field is strong enough to guide contraction along field lines so that clouds tend to form as sheets or filaments (Heitsch, Stone & Hartmann 2009). Gravitational contraction along the filament may further strengthen the field.

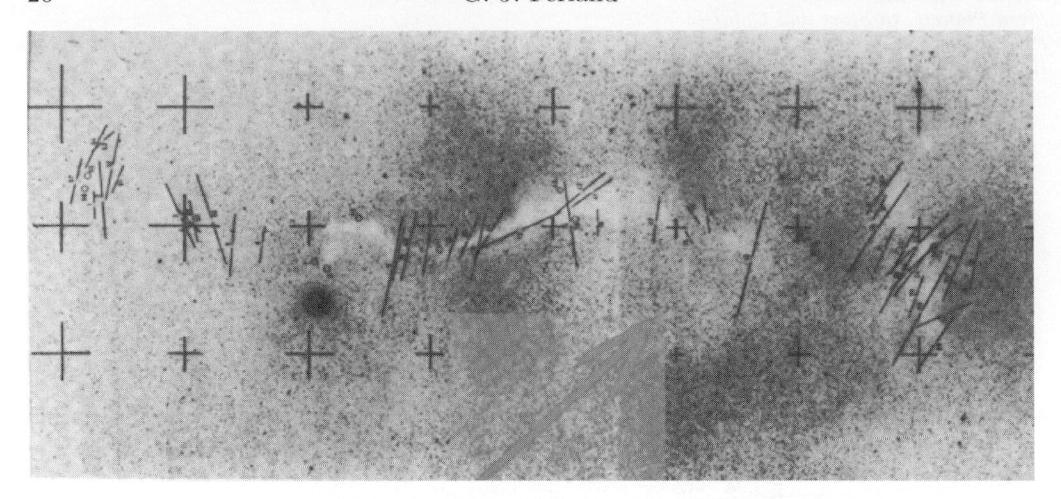

Figure 1. Figure 1, from Heiles (1988), showing the dark cloud Lynds 204. Magnetic field lines, shown as black lines, line roughly orthogonal to the long axis of the filament.

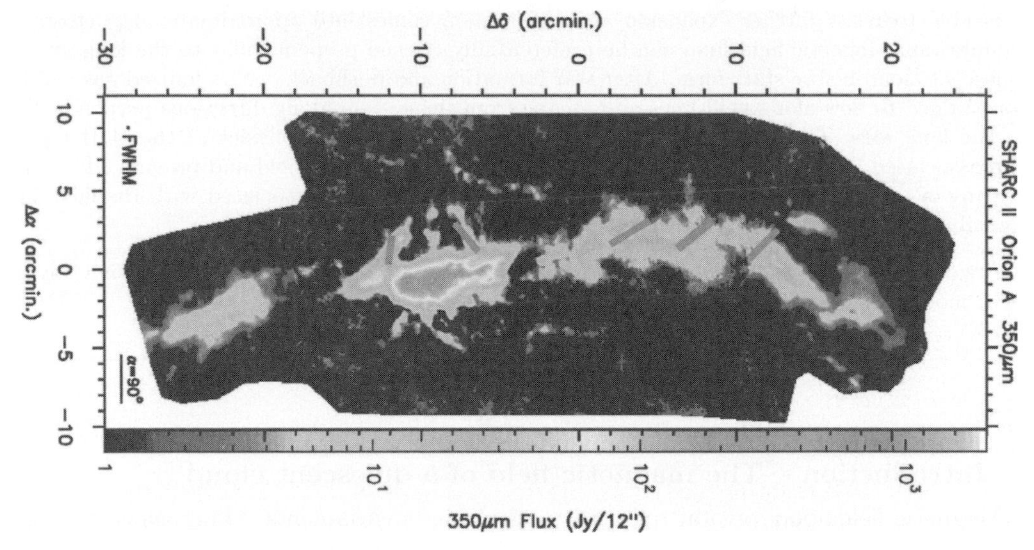

Figure 2. The Orion Molecular Cloud as imaged at 350μm by Houde *et al.* (2004). The figure has been rotated so that north is to the right. The red lines show the direction of the magnetic field deduced from the linearly polarized dust emission. The Trapezium cluster is centered on the bright region to the left.

2. Orion - an active star-forming region

The Orion complex is more complicated because of the feedback associated with active star formation. Figure 2, taken from Houde *et al.* (2004), shows the geometry of the field as revealed by linearly polarized dust emission. The surface brightness of the $\lambda 350\mu$m thermal emission is shown by the colored scale. The red lines indicate the deduced field direction. The bright region to left of center is warm dust in molecular material surrounding the Trapezium cluster. Active star formation and associated starlight cause the dust to be radiatively heated and glow brightly near the young stars.

The field lines tend to lie perpendicular to the long axis of the molecular cloud in the relatively quiescent northern regions. This is reminiscent of the geometry of Lynds 204 shown in Figure 1.

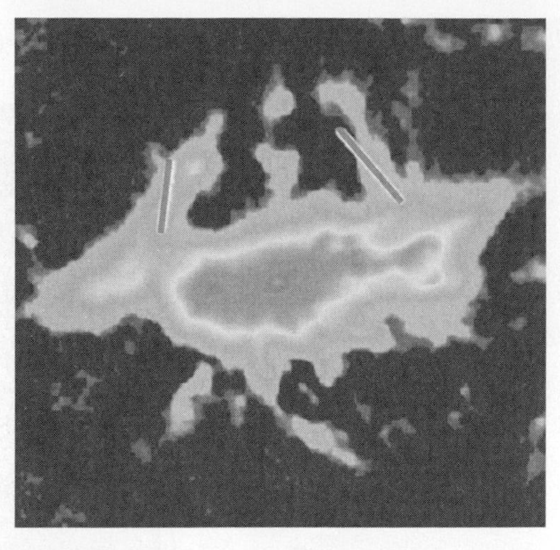

Figure 3. A zoom into Figure 2 from Houde *et al* (2004) showing emission from the parent molecular cloud in the region around the Trapezium cluster. The orientation of the image is the same as in Figure 2. The orientation of the magnetic field is indicated by the red lines.

The region surrounding the Trapezium is more complicated due to the presence of massive stars and their associated radiation pressure. Figure 3 shows a zoom of the region near the Trapezium. Here the field lines tend to lie perpendicular to the line between the point and the Trapezium.

This review centers on the relationships between stars and the surrounding H^+, H^0, and H_2 layers, the so-called H II region, PDR, and molecular cloud. Figure 4 shows the geometry of the regions near the young star cluster and bright H II region in the Orion complex. This is adopted from Osterbrock & Ferland (2006).

Feedback from Trapezium cluster has strongly affected the cloud geometry. Stellar winds and associated shocks produce a bubble of hot gas that has recently been detected in the X-rays (Güdel *et al.* 2007). A combination of thermal gas pressure and starlight momentum pushes cooler gas away from the star cluster and results in the blister geometry shown in Figure 4.

Most of the extinction seen in the HST image of the Orion H II Region arises in the Veil, the layer of predominantly atomic gas that lies on this side of the hot bubble (O'Dell 2001a). The Veil has been extensively studied at H I 21 cm. The thermal continuum emission produced by the H^+ region is used to probe the Veil, where atomic gas produces a 21 cm absorption line. Zeeman measurements of the line of sight magnetic field show it to be surprisingly strong, approaching 50 μG, roughly 1 dex stronger than the field in the diffuse ISM (Troland *et al.* 1989). Why is the field so strong?

Abel *et al.* (2004; 2006) combined optical and UV measurements of absorption lines formed in the Veil to derive its density, kinetic temperature (and so its gas pressure) and its distance from the Trapezium. They found that magnetic pressure greatly exceeded the gas pressure, as is typical of the ISM, and that the magnetic pressure exceeded even the turbulent pressure in one of the two Veil components. The Veil is a thin sheet which we view roughly face on.

3. M17 and magnetostatic equilibrium

The M17 star-forming region is much larger and more luminous than Orion but also much further away. Similar Zeeman polarization measurements of the magnetic field in

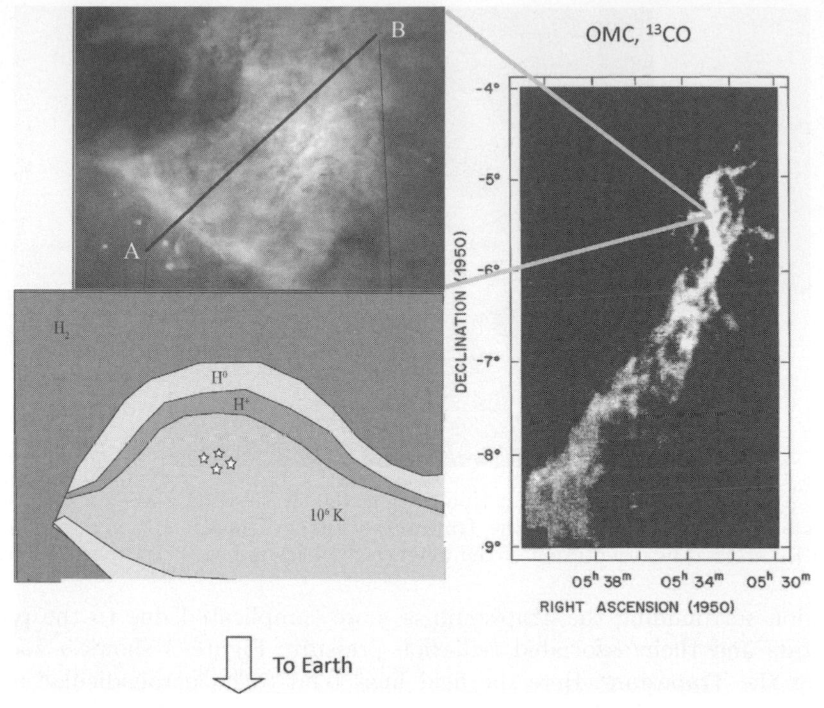

Figure 4. The geometry of the Orion H II region, PDR, and molecular cloud. This Figure, adopted from Osterbrock & Ferland (2006), shows a CO image of the molecular cloud to the right and a zoom into the HST image of the H II region at upper left. The geometry of the cut shown as the black line from A to B in the HST image is shown in the lower left panel. The star cluster is surrounded by hot gas produced by stellar winds. The bright H II region is mainly an ionized layer on the surface of the background molecular cloud. Much of the extinction visible in the HST image arises in Orion's "Veil", the layer of predominantly atomic gas that lies on the near side of the star cluster. The Veil is the region where H I 21 cm circular polarization measures the line-of-sight magnetic field.

the atomic hydrogen region have been performed (Brogan & Troland 2001) and an even stronger field, approaching 700 μG, was found.

Pellegrini *et al* (2007 hereafter P07) combined a broad range of spectral observations to make a coherent picture of the geometry of M17. M17 has an overall geometry that is similar to Orion, but viewed from a a different angle, as shown in Figure 5. Some of these ideas are further outlined in Ferland (2008). The ionizing star cluster and the intrinsically brightest part of the H II region are hidden behind a layer of atomic and molecular gas. The Zeeman magnetic field is stronger in the regions to the right (West) of the star cluster then along directions more towards the cluster.

P07 showed that the H^+ H^0 and H_2 layers were in a state of quasi-magnetostatic equilibrium. The outward force of starlight, mainly ionizing radiation acting on gas and dust, is resisted by the magnetic pressure in deeper regions of the cloud. The picture they proposed is that the combination of thermal gas pressure from the hot bubble and radiation pressure due to starlight has pushed back surrounding gas, strengthening the magnetic field, until the magnetic pressure could resist further compression.

The gas pressure in the hot wind-blown bubble is close to the pressure near the illuminated face of the H II Region. The absorption of the outward-flowing starlight pushes the H II region away from the cluster increasing the gas density and magnetic pressure.

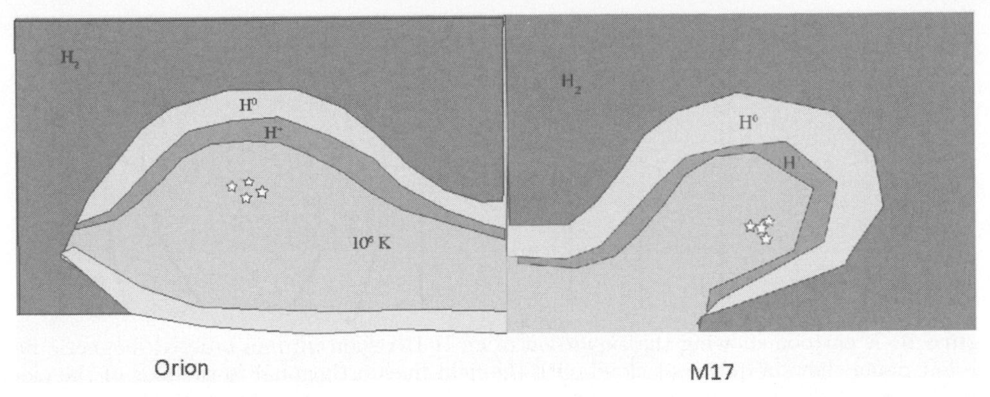

Figure 5. A comparison of the geometry of Orion (left) and M17 (right). This is not drawn to scale - M17 is roughly 1 dex larger than Orion. We view both geometries from the bottom looking up. The atomic layer in front of the star cluster and H II region is translucent in Orion. In M17 the layer has a large extinction and has atomic, ionized, and molecular constituents. The Zeeman magnetic field measurements use the emission of the H^+ layer as the continuum source. The radio absorption measurements probe the line of sight component of the magnetic field in the H^0 layer between the observer and the H^+ region.

The magnetic pressure increases as the square of the field, so for many geometries the magnetic pressure will increase faster than the gas pressure. Most starlight is absorbed by the outermost parts of the H_2 layer due to the large dust extinction. Magnetic pressure dominates at this point so the total luminosity of the star cluster and the magnetic field at this point are related.

P07 gave a simple relationship between the total luminosity of the star cluster and the magnetic pressure in deeper regions of the cloud. In this picture the strong magnetic fields associated with active regions of star formation are directly related to the luminosity of the central stars. This provides a natural explanation for why strong fields are found near star-forming regions.

4. The magnetic field and the geometry of the M17 H II region

The linear polarization measurements (Figures 1 & 2) suggest that the magnetic field may be roughly perpendicular to the axis of a filamentary molecular cloud. What happens when star formation occurs in such a geometry? Two forces act to guide the resulting expansion of the ionized gas.

The first is the effects of an ordered magnetic field as shown in Figure 6. The left panel shows a segment of a filamentary molecular cloud with an ordered field. The right panel shows the expanded hot bubble and compressed field lines. The magnetic field increases perpendicular to the "equator" of the cloud. Expansion can be halted by the field in this direction. There is no increase along the "poles" and gas may be free to move in this direction.

Figure 7 shows a composite image of the Orion molecular cloud, star cluster, and H II Region. An image with higher resolution data is given in Henney (2008). Figure 7 shows that the visible H II region opens in the direction below the molecular cloud in the image. This is the direction where the X-ray emission discovered by Güdel *et al.* (2008) occurs and is the likely direction of outflow of the hot gas. The geometry of the magnetic field shown in Figure 6 may account for the direction of the extended optical emission since gas will only be free to move in directions perpendicular to the filament.

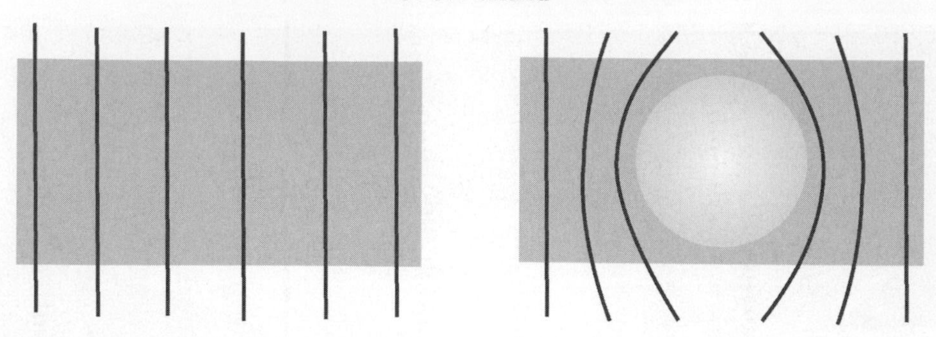

Figure 6. A cartoon showing the expansion of an H II region into an ordered magnetic field. The left panel shows a quiescent cloud with the field lines orthogonal to the axis of the cloud. The right panel shows the expanded ionized gas with associated changes in the field geometry.

Linear polarization shows the direction of the magnetic field. Figure 3, which is a zoom into a portion of the larger map given in Figure 2, suggests that the field lines near the Trapezium are roughly orthogonal to the direction towards the star cluster. This field geometry is consistent with the picture shown in Figure 6.

Zeeman circular polarization measurements detect the line of sight component of the magnetic field. For the geometry shown in Figure 6 observations in the direction of the star cluster along the equator of the slab will detect the smallest field. The line-of-sight field should increase as the telescope beam moves away from the cluster and the impact parameter of the passing ray increases. This is the general sense of the measured magnetic field of Orion and M17. The magnetic field tends to be stronger in directions that do not center on the star cluster. But the field does not go to zero in the direction of the cluster in Orion. (Measurements do not exist in this direction in M17.) This may indicate that the field has a disorganized component in addition to the ordered component detected by linear polarization.

The mass shaping due to the large mass of the molecular cloud is a second effect influencing the expansion of the H II region. In both Orion and M17 star formation appears to have occurred near a surface of the molecular cloud and expanding x-ray emitting gas is freely expanding out into the general ISM along along an open path. There is relatively little material along our line of sight to the Trapezium while the main outflow appears to be towards the Southwest (towards the bottom in Figure 7). In M17 there is a large column density of material along our line of sight towards the cluster, which is optically obscured, while the outflowing hot gas is directed towards the East (to the left in the right panel of Figure 5). In both cases a large mass of molecular gas is present in some directions but not in others. This mass shaping must also affect the geometry of the ionized bubble.

5. Are cosmic rays amplified as well?

Cosmic rays are in approximate energy equipartition with the magnetic field in the diffuse ISM (Webber 1998). Energy equipartition is usually assumed when observations of radio synchrotron emission are interpreted in terms of a cosmic ray density or magnetic field, as reviewed in other papers in this volume. Such equipartition does not occur because of any direct microphysical coupling between the energy reservoirs in the environment. Rather it is a minimum energy configuration that can be established if the system exists for a long enough time to have become relaxed. The equipartition value of the cosmic ray density does provide a reference point to which I will come back below.

Figure 7. William J. Henney's composite of an optical image of the Orion H II region in the center and a radio image showing the distribution of carbon monoxide molecules in red. Optical image obtained as part of the Two Micron All Sky Survey (2MASS), a joint project of the University of Massachusetts and the Infrared Processing and Analysis Center/California Institute of Technology, funded by the National Aeronautics and Space Administration and the National Science Foundation. The CO image is derived from Plume *et al.*(2000). The x-ray emission, which is not shown, occurs below the "fan" of optical emission which points below the molecular cloud.

Low-energy cosmic ray electrons are strongly trapped by the magnetic field because their gyro radius is much smaller than the physical scales of the molecular cloud or H II region. The compression of field lines shown in Figure 6 will also increase the cosmic ray density. Cosmic rays will be free to drift along field lines unless their pitch angle is perfectly aligned with the field. They should drift out of the environment relatively quickly. This drift would be inhibited if the field has a disorganized component. Low-energy cosmic rays can be locally enhanced by MHD waves (Padoan & Scalo 2005) so spatial variations are expected. This is clearly a complicated topic.

The galactic cosmic ray density is usually derived from observations of the ion chemistry (Spitzer 1978). Recent observations of the fraction of gas in H_3^+ along a number of sight lines suggest that the galactic background cosmic ray ionization rate has been underestimated by about 1 dex, and that this rate varies from one sight line to another (Indriolo *et al.* 2007). This shows that the cosmic ray density of even quiescent regions is open to question.

Cosmic rays both heat and ionize gas. The proportion of energy that goes into heating or ionization is determined by the degree of ionization of the gas (Osterbrock & Ferland 2006). Heating by cosmic rays is not insignificant in atomic regions even with the old lower background rate (Tielens & Hollenbach 1985). If the cosmic ray rates are significantly higher than previously thought then they will be even more important in heating atomic and molecular regions.

Giammanco & Beckman (2005) proposed that a high flux of cosmic rays may explain the "temperature fluctuations" problem in H II regions. The basic conundrum is that different measures of the kinetic temperature in the ionized gas do not agree with one another. The disagreement is systematic and suggests that a range of kinetic temperatures is present. It is hard to understand how this can occur due to the rapid increase of the gas cooling function with increasing temperature (Ferland 2003) and the fact that a Boltzmann distribution is established so quickly in an ionized gas (Spitzer 1978). Giammanco & Beckman argue that the supplemental heating provided by enhanced cosmic rays would produce the observed effects.

Orion's "Bar", a linear feature obvious on most optical images of the H II region (O'Dell 2001a), is thought to be an escarpment on the surface of the background molecular cloud. Ionizing radiation from the Trapezium strikes the Bar in such a way that we can observe the transition between H^+ H^0 and H_2 regions nearly edge on. Because of its special geometry, the Bar has become a decisive test for the physics of the interface between molecular and ionized regions.

Pellegrini *et al.* (2008) and Shaw *et al.* (2009) studied spectral variations across the Bar with the goal of reproducing the observed emission profiles of various tracers of ionized, atomic, and molecular gas. They found that cosmic rays in equipartition with the observed strong magnetic field reproduced the observations. P08 also required an enhanced flux of cosmic rays to account for the atomic and molecular emission in Orion.

6. Magnetic fields and non-thermal line widths

ISM spectral lines, both emission and absorption, are usually found to have widths too large to be due to purely thermal motions. These non-thermal line widths were often interpreted as a form of non-dissipative wave motion with the kinetic and magnetic energies in rough energy equipartition. Numerical MHD simulations suggest that such waves should be damped, which then requires that some energy source drive them, but in any case the turbulently broadened lines can be taken as empirically motivated. Heiles & Crutcher (2005) review the situation.

It is successively more difficult to measure non-thermal line widths in PDRs and H II regions than in molecular clouds due to the increasing temperature and decreasing mass per particle. A given turbulent motion, which might be associated with a particular magnetic field, is a smaller fraction of the thermal line width in hot ionized gas than it would be in a cold molecular medium. Non-thermal line widths are seen in the spectra of the H II regions and PDRs. PDR lines have long been known to be turbulently broadened (Tielens & Hollenbach 1985) and Roshi (2007) has recently suggested that non-thermal line widths of carbon radio recombination lines, which should form in the PDR, may be associated with MHD waves.

Optical emission lines from the H II region also have larger-than-expected line widths. These widths are observed to scale with the luminosity of the star cluster and have been studied extensively because of their possible use as a standard candle (Melnick *et al.* 1988). For low-luminosity objects like Orion the observation is difficult because the expansion of the ionized gas away from the molecular cloud, which occurs at roughly the speed of sound in the H II region ($\sim$10 km s^{-1}), is similar to the turbulent speed. However, very careful studies have been done (O'Dell *et al* 2005; García-Díaz et a. 2008) and find a significant component of turbulence. There is no agreed-upon model for the origin of the turbulence seen in the optical lines but MHD waves are a possibility (Ferland 2001; O'Dell 2001a, 2001b). This is important because the turbulent energy is a significant part of the energy budget in Orion. If this turbulence is dissipative, as suggested by numerical MHD simulations, then it would act to heat the gas (Pan & Padoan 2008).

Line widths in excess of 100 km s^{-1} can be found in luminous extragalactic H II regions. Beckman & Relaño (2004) argue that these line widths may be due to MHD effects associated with the magnetic field. This would imply that relationships between magnetic field, turbulent, and gravitational energies extend over many orders of magnitude of mass. Again, if the waves are dissipative then, depending on the damping timescale, they may constitute a significant heating source.

7. Conclusions

(1) A number of studies of the H^+, H^0 / H_2 regions of active star-forming regions, chosen to have high-quality Zeeman 21 cm detections of the magnetic field in the H^0 region, have been conducted. In both Orion's Veil and the H^0 region along the line of sight to the H^+ region in M17 the magnetic pressure is much greater than the gas pressure, as is commonly found in the ISM. The magnetic pressure exceeds the turbulent pressure in one component of the Veil.

(2) The strongest magnetic fields in diffuse gas are found near regions of active star formation. The field in the atomic hydrogen region of M17, $\sim 700\mu$G, is among the strongest observed. This corresponds to a magnetic pressure roughly 3 dex larger than is found in the diffuse ISM.

(3) Simulations of the emission-line spectra of the H^+, H^0 / H_2 regions in M17 which reproduce the observed magnetic field show that the geometry is approximately in magnetostatic equilibrium. The outward radiation pressure due to starlight is balanced by the magnetic pressure in well-shielded regions of the cloud. This accounts for the large field that is observed.

(4) Linear polarization studies suggest that the field is often oriented perpendicular to the long axis of filamentary clouds. This may be the result of the magnetic field guiding the formation of the cloud or of the subsequent gravitational contraction along the filament (Heitsch *et al.* 2009).

(5) Cosmic rays are found to be in energy equipartition with the magnetic field in the diffuse ISM, and equipartition is often assumed to hold for synchrotron emitting regions. Cosmic rays both heat and ionize atomic and molecular gas. If cosmic rays are enhanced along with the field then they would constitute an important heating source for atomic and molecular regions.

(6) Non-thermal line widths are often found in emission line spectra of star-forming regions. Several recent studies have suggested that these may be due to MHD waves associated with the magnetic field. If the waves are dissipative then this would constitute a gas heating mechanism.

Acknowledgements

Support by the NSF (AST 0607028) and NASA (ATFP07-0124) is gratefully acknowledged. I thank Carl Heiles and Will Henney for providing figures and Bob O'Dell and Tom Troland for comments.

References

Abel, N. P., Brogan, C. L., Ferland, G. J., ODell, C. R., Shaw, G., & Troland, T. H. 2004, *ApJ* 609, 247

Abel, N. P., Ferland, G. J., ODell, C. R., Shaw, G., & Troland, T. H. 2006, *ApJ* 644, 344

Alves, F. O., Franco, G. A. P., & Girart, J. M. 2008, *A&A* 486, L13

Beckman, J. E. & Relaño, M. 2004, *Ap&SS* 292, 111

Brogan, C. L., & Troland, T. H. 2001, *ApJ* 560, 821
Ferland, G. J. 2001, *PASP* 113, 41
Ferland, G. J. 2001, *PASP* 113, 41
Ferland, G. J. 2003, *ARAA* 41, 517
Ferland, G. J. 2008, EAS Publications Series, Volume 31, 2008, pp.53-56
García-Díaz, M. T., Henney, W. J., López, J. A., & Doi, T., 2008, *RMxAA* 44, 181
Giammanco, C. & Beckman, J. E., 2005, *A&A* 437, L11
Güdel, M., Briggs, K. R., Montmerle, T., Audard, M., Rebull, L., & Skinner, S. L., 2008, *Science 319* 309
Heiles, C. 1988, *ApJ* 324, 321
Heiles, C. & Crutcher, R. 2005, chapter in Cosmic Magnetic Fields. Edited by Richard Wielebinski and Rainer Beck. Lecture notes in Physics Volume 664
Heiles, C. & Troland, T. H. 2005, *ApJ* 624, 773
Heitsch, F., Stone, J. M., & Hartmann, L. W., 2009, *ApJ*, in press arXiv:0812.3339v1)
Henney, W., 2008, La Nebulosa de Orin en cuatro dimensiones Dr. William Henney Boletn de la UNAM Campus Morelia, No. 16, Julio/Agosto 2008, p. 1 http://www.csam.unam.mx/vinculacion/Julio-Agosto%202008.pdf
Houde, M., Dowell, C. D., Hildenbrand, R. H., Dotson, J. L. Vaillancourt, J. E., Phillips, T. G., Peng, R., & Bastien, P. 2004, *ApJ* 604, 717
Indriolo, N., Gaballe, T., Oka, T., & McCall, B. 2007, *ApJ* 671, 1736
Melnick, J., Terlevich, R., & Moles, M. 1988, *MNRAS* 235, 297
O'Dell, C. R. 2001a, *ARAA* 39, 99
O'Dell, C. R. 2001b, *PASP* 113, 29
O'Dell, C. R., Peimbert, M., & Peimbert, A. 2003, *AJ* 125, 2590
Osterbrock, D. E., & Ferland, G. J., 2006, *Astrophysics of Gaseous Nebulae & Active Galactic Nuclei*, 2nd edition, Mill Valley; University Science Press
Padoan, P. & Scalo, J., 2005, *ApJ* 624, L97
Pan, L. & Padoan, P. 2008, *ApJ*, in press, arXiv:0806.4970
Pellegrini, E., Baldwin, J., Brogan, C., Hanson, M., Abel, N., Ferland, G., Nemala, H., Shaw, G., & Troland, T. 2007, *ApJ* 668, 1119
Pellegrini, E., Baldwin, J., Ferland, G., Shaw, G., & Heathcote, S. 2008, *ApJ*, in press (arXiv:0811.1176)
Plume, R. *et al* 2000, *ApJ* 539L, 133
Roshi, D. A. 2007, *ApJ* 658, L41
Shaw, G., Ferland, G. J., Henney, W. J., Stancil, P. C., Abel, N. P., Pellegrini, E. W., Baldwin, J. A., & van Hoof, P. A. M. 2009, *ApJ*, submitted
Spitzer, L. 1978, *Physical Processes in the Interstellar Medium*, New York: Wiley
Tielens, A. G. G. M., & Hollenbach, D. 1985, *ApJ* 291, 722
Troland, T. H., Heiles, C., & Goss, W. M. 1989, *ApJ* 337, 342
Webber, W. R. 1998, *ApJ* 506, 329

Cosmic Magnetic Fields:
From Planets, to Stars and Galaxies
Proceedings IAU Symposium No. 259, 2008
K.G. Strassmeier, A.G. Kosovichev & J.E. Beckman, eds.

doi:10.1017/S174392130903004X

Magnetic fields in Planetary Nebulae: paradigms and related MHD frontiers

Eric G. Blackman

Department of Physics and Astronomy, University of Rochester, Rochester, NY 14627, USA
email: blackman@pas.rochester.edu

Abstract. Many, if not all, post AGB stellar systems swiftly transition from a spherical to a powerful aspherical pre-planetary nebula (pPNE) outflow phase before waning into a PNe. The pPNe outflows require engine rotational energy and a mechanism to extract this energy into collimated outflows. Just radiation and rotation are insufficient but a symbiosis between rotation, differential rotation and large scale magnetic fields remains promising. Present observational evidence for magnetic fields in evolved stars is suggestive of dynamically important magnetic fields, but both theory and observation are rife with research opportunity. I discuss how magnetohydrodynamic outflows might arise in pPNe and PNe and distinguish different between approaches that address shaping vs. those that address both launch and shaping. Scenarios involving dynamos in single stars, binary driven dynamos, or accretion engines cannot be ruled out. One appealing paradigm involves accretion onto the primary post-AGB white dwarf core from a low mass companion whose decaying accretion supply rate owers first the pPNe and then the lower luminosity PNe. Determining observational signatures of different MHD engines is a work in progress. Accretion disk theory and large scale dynamos pose many of their own fundamental challenges, some of which I discuss in a broader context.

Keywords. Planetary Nebulae – magnetic fields – accretion – MHD – stars: AGB – stars: binaries

---

1. Introduction

Asymmetries were observed in planetary nebulae PNe before the Hubble Space Telescope but their ubiquity, the rapidity with which they develop, and collimated outflow power have led to a cultural change in the field over the past decade (e.g. Balick & Frank 2002). Asymmetric p/PNe have become the standard and understanding how these asymmetries arise is now fundamental rather than anecdotal. Perhaps most dramatic is that collimated momenta of the pre-PNe (pPNe) (the reflection nebulae precursor to the ionized nebulae of PNe) exceed that which can be supplied by radiation pressure alone (Bujarrabal *et al.* 2001). This is reminiscent of the young stellar object (YSOs) subject decades ago (Reipurth & Bally(2001)). The recent cultural change toward the view that the end states of stars (including supernovae e.g. Wang & Wheeler 2008) are asymmetric, offers a plethora of research opportunity.

The beginning and ending of stars share the ingredients of in-fall, collapse, turbulence, and angular momentum transport. The associated increased free energy in differential rotation can be tapped to amplify fields and produce outflows, so common underlying MHD principles likely at work. Focusing on pPNe/PNe, I review the role of magnetic fields in these systems, and current evidence for their influence. I also address broader fundamental questions in magnetic field generation and angular momentum transport for which pPNe are yet another laboratory. The role of large scale fields and the need to connect different MHD engines with observations are central to the theme.

2. Basic properties of p/PNe

Generally, pPNe exhibit a fast bipolar outflow embedded within a slow spherically symmetric wind from the AGB star Bujarrabal *et al.* 2001. Presently, data do not rule out all pPNe having gone through a strongly asymmetric outflow and all PNe having gone through an asymmetric pNE phase. Though AGB stars produce spherically symmetric outflows, the initial pPNe asymmetry emerges within $\leqslant 100$ yr (e.g. Imai *et al.* 2005). PNe could reflect the late stages of the mechanism that produces pPNe with less asymmetry at very late times as supersonic motions damp.

For pPNe (Bujarrabal *et al.* 2001), each fast wind has a typical age $\Delta t \sim 10^2 - 10^3$yr, speed $\sim$ 50km/s, mass $M_f \sim 0.5 M_\odot$, outflow rate, $\dot{M}_f \sim 5 \times 10^{-4} M_\odot$/yr, momentum $\Pi \sim 5 \times 10^{39}$g.cm/s, and mechanical luminosity $L_{m,f} \geqslant 8 \times 10^{35}$erg/s (can be as high as 10^{37}erg/s). The slow pPNe wind has an age $\Delta t \sim 6 \times 10^3$yr, a speed $v_w \sim$ 20km/s a mass $M_s \sim 0.5 M_\odot$, outflow rate, $\dot{M}_s \sim 10^{-4} M_\odot$/yr, momentum $\Pi_s \sim 2 \times 10^{39}$g cm/s, and mechanical luminosity $L_{m,s} \sim 10^{34}$erg/s. For PNe, observations suggest (Balick & Frank 2002) an age $\Delta t \sim 10^4$yr a slow wind of speed $v_s \sim$ 30km/s of mass $M_s \sim 0.1 M_\odot$, outflow rate, $\dot{M}_s \sim 10^{-5} M_\odot$/yr, momentum $\Pi_s \sim 6 \times 10^{38}$g cm/s, and mechanical luminosity $L_{m,s} \sim 3 \times 10^{33}$erg/s.v PNe have fast winds of speed as high as $v_f \sim$ 2000km/s, mass $M_f \sim 10^{-4} M_\odot$, outflow rate, $\dot{M}_f \sim 10^{-8} M_\odot$/yr, momentum $\Pi_f \sim 4 \times 10^{37}$g cm/s, and mechanical luminosity $L_{m,f} \sim 1.3 \times 10^{34}$erg/s.

The pPNe phase demands the most power and the linear momenta of fast bipolar pPNe outflows seems too large for radiation driving Bujarrabal *et al.* 2001. This motivates the need to tap rotational energy, either from redistribution of angular momentum within a single star or via deposition of angular rotational energy from binaries.

3. Observations of magnetic fields

Detection of magnetic fields in p/PNe is not by itself a proof of their dynamical importance as the relative strength of the field and the local kinetic energies must be considered. Complementarily, a weak magnetic field at large distances from the outflow engine does not rule out a magnetically dominated launch at the engine.

Magnetic field detection is technically challenging but careful analysis has led to estimate of magnetic field strengths and geometries in several important classes of late AGB and post-AGB systems by different techniques. In water maser sources such as W43 (an evolved AGB star with a precessing jet signaturing the beginning of the pPNe stage) magnetic fields of $\sim$ 85mG have been inferred at radii of 500AU from the engine via interpretation of circular polarization of H_2O masers (Vlemmings *et al.* 2006). The 85mG fields at 500AU and their inferred geometry seem consistent with what is needed to dynamically influence the flow. (See Sabin *et al.* 2007 for field geometry in other sources). The precession suggests a binary interaction, but whether the field is dynamo produced and/or connected to a companion within the envelope or the AGB core is uncertain. Vlemmings (2007) reviews maser field measurements of giant stars highlighting that the H_2O masers probe field on $\sim 10^2$AU scales, OH masers probe 10^3 AU scales and SiO masers probe $\leqslant$ 10AU. The complied measurements show that, statistically, fields fall off faster than r^{-1} but slower than r^{-3}. This is a weak constraint on models.

From VLT spectropolarimeteric analysis of several central stars of PNe (CSPN), Jordan *et al.* (2005) detected magnetic field strengths of order kG. The scale of the photosphere for these system is $\sim 2 \times 10^{10}$cm. These field strengths are are consistent with the larger scale scalings presented in Vlemmings *et al.* (2007) and the strengths of fields needed to power the pPNe outflows by Poynting flux, and fields produced by dynamos in the AGB engines (Nordhaus *et al.* 2007).

Soker and Zoabi (2002) and Soker (2006b) have argued for weaker fields and rather than a primary driver of outflows, an indirect shaper of outflows possibly via influence on the dust distribution and geometry of radiative driven winds. But this would leave unsolved the large collimated momenta of pPNe engines Bujarrabal *et al.* 2001) for which radiation driving is insufficient.

4. Dynamical magnetic shaping and/or launching

The "launch" region of MHD outflows is the region where the magnetic force dominates the flow. This extends to a height typically no greater than $z_c \sim 50R_i$, where R_i is the inner-most radial scale of the engine. (e.g. the inner radius of an accretion disk). In the launch region the bulk flow is sub-Alfénic below z_c. The "propagation" region describes $z > z_c$ where the poloidal flow speed exceeds the Alfén speed, eventually approaching its asymptotic speed. Presently, only the propagation regions are observationally spatially resolved. Also, because there are ~ 2 or more orders of magnitude between the engine where dynamos and jets formation operates and the asymptotic propagation region the computational demands prohibit simulatating the combined physics of the launch region, jet formation and asymptotic propagation. As emphasized in sec 7., even the nonlinear physics of just an accretion disk has been prohibitively computationally expensive. We must patch together different pieces of physics from different approaches and scales to extract a complete picture.

4.1. *Shaping with imposed fields*

In a semi-analytic approach, Chevalier and Luo (1994), imposed a magnetic field geometry such that the toroidal field falls off more slowly than the poloidal field and at large distances and a spherical wind shocks the ambient gas increasing the toroidal field via compression. For their parameters, the magnetic shaping occurs only after the shock.

García-Segura (1997) and García-Segura *et al.* (1999) were the first to simulate the direct effect of imposed toroidal fields on PNe shaping by driving a hydrodynamic wind into a pre-magnetized medium. The field strengths were consistent with those of sec. 3 above. and their influence (along with stellar rotation, and photoionization) on the shaping was summarized in simple schematic (García-Segura 1999) . Rotation + magnetic fields leads to strong collimation compared to the weaker collimation of just the rotational influence on radiatively driven winds. Gardiner & Frank (2001) discussed that the shaping observed in such simulations were outside the restrictive parameter regime of Chevalier & Luo (1994) and that the influence of reasonable strength fields should generically affect the outflow before the shock compression. .

4.2. *Launching and shaping with imposed initial fields*

Because the observations require both a mechanism of launch and shaping, it is necessary to consider the role of magnetic fields in both. García-Segura *et al.* (2005) imposed a toroidal field the surface of the AGB star to launch an outflow via the field pressure gradient and studied the propagation and asymptotic collimation over 2 orders of magnitude in scale. Given field strengths of $\sim 40G$ (consistent with observations) at the surface of the AGB star, outflows consistent with the required power and shaping were observed.

Matt *et al.* (2006) added additional physics in demonstrating how an initially weak field can grow linearly in time by extracting rotational energy of its anchoring base, and produce a rapid bipolar outflow from the gradient in toroidal field pressure. They took a gravitating spherical core surrounded by an initially hydrostatic envelope of ionized gas. A dipolar magnetic field was anchored on the core, threading the envelope. The core was set to rotate initially at 10% of the escape speed. The toroidal field amplified from the differential rotation between core and envelope and when the toroidal field pressure gradient overcame the envelope binding energy, envelope material was rapidly expelled in a quadrupolar outflow: The wound up diploe field had maximum toroidal field at intermediate poloidal angles in each hemisphere, so material was both along the poles also squeezed out from the equator.

4.3. *Launching via dynamo engines*

The models of the previous two subsections do not fully address where the dynamically important large scale fields come from. Even the Matt *et al.* (2006) paper invoked linear, laminar field growth and treating the rotator base as a boundary condition.

Papers such as those by Pascoli (1993,1997) Blackman *et al.* (2001a,b), Nordhaus *et al.* (2007) use flavors of a mean field dynamo theory for stellar or disk engines to estimate field strengths and Poynting fluxes that could arise. These papers are not simulations and do not track global field geometry, or follow the production and evolution of outflows from the field production region. Furthermore, the nonlinear physics of dynamos (see sec. 7) require approximations when emedding them in astrophysical scenarios. Despite their shortcomings, such toy models do produce promising results with respect to field strengths and Poynting fluxes

5. Single star vs. binary models

5.1. *Single star models not yet ruled out*

Binaries can easily supply the needed free energy to amplify fields via differential rotation (ordhaus & Blackman 2006). Isolated star MHD models cannot be ruled out, but there are caveats (Soker 2006b; Nordhaus *et al.* 2007) as I now discuss.

Blackman *et al.* (2001) investigated an isolated interface dynamo model operating at the base of the AGB convective zone in which angular momentum is conserved on spherical shells as the star evolves off the main sequence and the resulting rotation profile provides the available differential rotational energy from which the field is amplified. To drive bipolar pPN/PN, the corresponding dynamo must operate through the lifetime of the AGB phase (10^5yr) until radiation pressure has bled most of the envelope material away. Only then can the Poynting flux unbind the remaining material. But the dynamo would drain differential rotation on time-scales short compared to the AGB lifetime. Only if differential rotation is extracted from deep within the core or re-seeded by convection can the drain be overcome (Nordhaus *et al.* 2007). Viability of the single star dynamo outflow model, depends on whether or not the differential rotation can be re-supplied over the AGB lifetime via convection as in the Sun.

5.2. *Common envelope binary scenarios*

In a common envelope (CE) Iben & Livio (1993) the companion drags on the envelope of the primary, transferring angular momentum and energy. If the envelope cooling time is long enough, a fraction $\alpha \geqslant 0.1$ of the loss in gravitational energy can spin up and unbind the envelope. Reyes-Ruiz & Lopez (1999) discuss the secondaries for which accretion disks will form around the primary core. Brown dwarf (BD) ($0.003M_\odot < M_{crit} \sim 0.07M_\odot$) radii increase with decreasing mass while their Roche radii decrease with decreasing mass. Such objects unstably lose mass. Since the circularization radius lies outside of the primary's core, a disk can form within a few orbits. This contrasts the $M > M_{crit}$ case for which the stellar radius decreases with decreasing mass more strongly than the Roche radius. Supercritical companions have a circularization radius *within* the primary's core. Material leaving the secondary would then initially spiral swiftly into the primary rather than orbit quasi-stably. Nordhaus & Blackman (2006) show that accretion from planets and low mass stars may also be important (and perhaps more common because of the brown dwarf desert Grether & Lineweaver(2006)) and can supply the needed accretion power for pPNe. Planets are of sub-critical mass and a disk could form as per a BD, before unbinding the envelope.

A companion star is supercritical so if stellar incurs Roche lobe overflow after CE, sustained accretion requires the binary to lose angular momentum. This could happen as the overflowing secondary drags on residual inner envelope material. Even though the circularization radius for $M_2 > M_{crit}$ is inside the core, the in-spiraling material still incurs differential rotation and could amplify magnetic fields. A significant energy release via accretion in the $M_2 > M_{crit}$ case could occur on impact to the inner core, producing dwarf novae bursts masked by the stellar envelope. If this initial accretion phases can drop M_2 below M_{crit}, keep M_2 filling its Roche lobe, and leave enough angular momentum to form a Keplerian disk, then accretion could proceed as for the initial $M_2 < M_{crit}$ case.

Huggins (2007) suggests a $\sim O(100)$ year delay between ejection of cicumbinary dust tori and jets in pPNe. If a fraction of ejected envelope material becomes the dust torus, the delay could be the time for the companion to lose enough mass to move the circularization radius outside the core for the supercritical case, or a viscous time.

6. Accretion disk outflows in pPNe and PNe

Accretion disk outflows have a mechanical power Blackman *et al.* (2001b)

$$L_m \sim \frac{GM_*\dot{M}_a\epsilon}{2R_i} = 4.5 \times 10^{36}\epsilon_{-1}\frac{M_*\dot{M}_{-4}}{R_{i,10}}, \tag{6.1}$$

where ϵ is the efficiency of conversion from accretion to outflow, R_i is the inner disk radius, $\dot{M}_a$ is the accretion rate, G is Newton's constant, M_* is the central stellar mass, $R_{i,10} \equiv R_i/10^{10}$cm, $\epsilon_{-1} \equiv \epsilon/0.1$ and $\dot{M}_{a,-4} = \dot{M}_a/10^{-4}M_\odot$/yr. For an MHD outflow, Eq. (6.1) equals the Poynting flux at the launch surface. Propagation into a low density medium produces an asymptotic outflow speed $\sim \Omega r_A$ where Ω is the angular speed of field anchor point and r_A is the radius where the poloidal outflow speed equals the Alfvén speed. This product is typically a few times the escape speed of the inner most radius of a disk and is thus at least $v_{out} \sim v_{esc} = 1600\left(\frac{M_*}{R_{*,10}}\right)^{1/2}$ km/s

6.1. *Accretion onto primary*

The v_{out} above depends only weakly on $\dot{M}_a$ via R_i, but strongly on the inertia of material blocking the outflow: Momentum conservation gives

$$v_{obs} = \frac{M_f v_{out}}{f_\Omega M_{env} + M_f} \sim 80\text{km/s}, \tag{6.2}$$

where $M_f/M_\odot = 3.3 \times 10^{-4}\epsilon_{-1}\dot{M}_{a0,-3}\int_1^{1000}\tau^{-5/4}d\tau$ is the mass in one of the fast collimated outflows, M_{env} is the envelope mass, $f_\Omega \sim 0.2$ is the solid angle fraction intercepted by the collimated outflow, and $\tau \equiv t/1yr$ is used to incorporate $\dot{M}_a \propto t^{-5/4}$ of Reyes-Ruiz & Lopez (1999). The numbers are scaled to pPNe so $M_{env} >> M_f$ and for an envelope of mass $2M_\odot$, the intercepted mass is 0.2 $M_\odot$ for $f_\Omega = 0.2$.

Eq. (6.2) is the observed speed of the fast when blocked and loaded by the envelope. By the end of the pPNe phase, the envelope is quite extended, reducing the optical depth and revealing material moving at the "free streaming" fast wind speed. Assuming a dust-to-gas mass ratio of 1/100 and micron sized grains of density of 2g/cm^3, the optical depth from dust is $\tau_d \sim 2.5 \times 10^{-3} \left(\frac{n_d}{2.5\times10^{-13}\text{cm}^{-3}}\right)\left(\frac{\sigma_d}{10^{-8}\text{cm}^2}\right)\left(\frac{R}{10^{18}\text{cm}}\right)$, scaled for PNe. For pPNe, the density increases by $\geqslant 10^4$ and R is down by a factor of 10, so $\tau_d \geqslant 2.5$. The different optical depths of pPNe and PNe can thus explain why observed PNe fast winds can have $v_f = v_{out} > 1600\text{km/s}$, whilst those of pPNe have $v_f = v_{out} < 100\text{km/s}$.

Time dependent accretion outflows described with Eqs. (6.1) and (6.2) are consistent with the high pPNe outflow mechanical luminosity and the fast PNe wind speed of Sec. 1 when M_*/R_* corresponds to a WD. Reyes-Ruiz & Lopez (1999) considers a companion of mass $M_2 \sim 0.03M_\odot < M_{crit}$ and a Shakura-Sunyaev (Shakura & Syunyaev(1973)) viscosity parameter $\alpha_{ss} \sim 0.01$, for which the accretion rate then decays as $\dot{M}_a \sim 1.6 \times 10^{-3} t^{-5/4} M_\odot/\text{yr}$.. Using this in (6.1) with $\epsilon = 0.1$ for $t = 100$ yr with $R_i = 2 \times 10^9$cm and $M_1 = 0.6M_\odot$ gives $L_{m,f} \sim 4.3 \times 10^{39}(t/1\text{yr})^{-5/4}$. This provides the needed power demands of Sec. 1 for pPNe after 1000 yr and for PNe after 10^4 yr.

A more careful analysis of the predicted jet speed evolution is needed for specific outflow models as the envelope evolves in order to test the idea contained within the rough estimates just discussed. See also Garicia-Diaz *et al.* (2008) for a non-accretion based time dependent comparison of jet outflow speed from simulations with observations.

6.2. *Accretion onto secondary*

It is also possible for accretion disks to form around the secondary (Soker & Livio 1994; Mastrodemos & Morris 1998; Soker 2005). The Bondi wind accretion rate is $\frac{\dot{M}}{\dot{M}_s} = \left(\frac{M_2}{M_1}\right)^2 \frac{(v/v_s)^4}{[1+(v/v_s)^2]^{3/2}}$, where v is the orbital speed of the secondary and v_w is the slow wind speed from the primary. In general, for $M_2 < M_1$, reasonable parameters provide an accretion rate compatible PNe luminosities if the companion is either main sequence or compact star, but the ubiquity of high collimated fast wind pPNe powers (Bujarrabal *et al.* 2001) would require an overabundance of accreting WD companions.

7. Key issues in large scale MHD dynamo and accretion theory

7.1. *Large scale dynamo theory: recent developments, open questions*

Large scale dynamo (LSD) theory describes the sustenance of magnetic fields on time or spatial scales larger than turbulent scales. Whether large scale fields for jets are advected or LSD produced has been debated but but ultimately, the equations that include a the competition between turbulent transport, flux accretion, as well as LSD action need to be solved. Field reversals in the sun prove that an LSD can and must operate therein.

For ~ 50 years, a problem with textbook LSD theory (e.g. Moffatt (1978)) has been the absence of a proper saturation theory that predicts how strong the large scale fields get before non-linearly quenching via the backreaction of the field on the driving flow kicks in. But substantial progress toward a nonlinear mean field theory has emerged in the last decade via a symbiosis between analytical and numerical work. Coupling the dynamical evolution of magnetic helicity into the dynamo equations turns out to be fundamental for predicting the saturation seen in simulations. For recent extensive reviews see (Brandenburg & Subramanian 2005; Blackman 2007).

Much of the work in LSD theory, has focused on systems that are initially globally reflection asymmetric (GRA). This means a global pseudoscalar is imposed by the boundary conditions–for example, rotation and stratification lead to the kinetic helicity pseudoscalar, common to the standard textbook (Moffatt (1978)) "α_{dyn} effect" of mean field dynamos. However magnetic helicity is actually the unifying quantity for LSD. There are two classes of GRA LSD for which an electromotive force aligned with the mean magnetic field is essential (see e.g. Blackman 2007). The first is flow driven helical dynamos (FDHD) which occur inside of astrophysical rotators. Here the initial energy is dominated by flows and the field responds. These are linked to a corona by magnetic buoyancy. In coronae, the second type of LSD, the magnetically driven helical dynamo (MDHD) can operate. This characterizes relaxation of magnetic structures to larger (jet mediating) scales in a magnetically dominated environment subject to the injection of smaller scale magnetic helicity. The MDHD is directly analogous to laboratory plasma dynamos that occur in reverse field pinches (RFPs) and Spheromaks.

LSDs always involve some helical growth of the large scale field which is coupled to a helical scale fields of opposite sign. When small scale magnetic or currently helicity evolution is coupled to the large scale field growth, the modern mean field 'dynamo α_{dyn} effect that predicts the correct saturation becomes the difference between kinetic helicity and current helicity: For a FDHD simulated in a closed box (Brandenburg 2001), the current helicity builds up as the large scale field grows and quenches the FDHD (Blackman & Field 2002). Complementarily, for an MDHD, the system is first dominated by the current helicity and a growing kinetic helicity then acts as the backreaction (Blackman & Field 2004). Both FDHD and MDHD are accessible with in the same framework, all unified by tracking magnetic helicity evolution, and aided by thinking of the field as ribbons rather than lines (Blackman & Brandenburg 2003). More work on how the fields evolve from within the rotator to produce the global scale fields in coronae which in turn produce jets are needed as most work on dynamo theory has focused on the FDHD.

Because the buildup of small scale magnetic (or current) helicity quenches the LSD, preferential ejection of small scale helicity vs. large scale helicity through a boundary can in principle alleviate this quenching (Blackman & Field 2000; Vishniac Cho 2001; Sur *et al.* 2007). Numerical simulations support this general notion, particularly when shear is present and when surfaces of shear align toward open boundaries (Brandenburg & Sandin 2004; Kapyla *et al.* 2008), thereby allowing needed helicity fluxes.

LSD action has also been observed in non GRA simulations (e.g. Yousef *et al.* 2008; Lesur & Ogilvie 2008) implying that the minimum global ingredients for this class of analytic LSD is shear, plus turbulence that feeds azimuthal field back to toroidal field. There is work in progress to understand the simulations guided by analytic models (e.g. Vishniac & Brandenburg 1997; Blackman 1998; Kleeorin & Rogachevskii 2003; Schekochihin *et al.* 2008). The non GRA LSDs grow large scale fields on scales larger than the turbulence but smaller than the global scale. In coherence regions, there is a field aligned electromotive force (EMF), and thus an intermediate scale source of magnetic helicity that may switch signs between coherence regions and globally average to zero. It may be that the non-GRA LSD action always involves a local helicity flux between coherence regions.

7.2. *Accretion disks: more questions and the need for large scale fields*

The magneto-rotational instability (MRI) has emerged as a leading candidate for angular momentum transport in accretion disks (e.g. Balbus & Hawley 1998) Although the MRI exists without the LSD, they are likely coupled in nature. I explain this below.

There is a disconnect between what shearing box simulations have told us about the MRI vs. how the instability might operate in nature. To date, simulations have primarily told us that the MRI is plausible but do not produce a robust theory of saturation or robust values of transport coefficients for modelers. This may frustrate, but patience (for several more decades) is required, as the computational and conceptual demands are substantial. For example, to achieve better resolution, most first generation MRI simulations (except Brandenburg *et al.*1995) did not use explicit viscosity or magnetic diffusivity (see discussion in e.g. Fromang *et al.* 2007). The magnetic Prandtl number affects the magnetic spectrum and the transport coefficients. In addition, the value of the angular momentum transport coefficient α_{ss} depends strongly on the box size and the strength of the initially imposed weak mean field strength (Pessah *et al.* 2007). Interestingly, α_{ss} varies ~ 4 orders between simulations, but $\alpha_{ss}\beta$, where β is the ratio of thermal to magnetic pressure, is nearly a constant (Blackman *et al.* 2007).

In addition to the need for explicit diffusivities, perhaps the most important frontier is actually role of large scale magnetic fields in angular momentum transport and thus, non-local contributions to magnetic stresses that transport angular momentum. Its importance is motivated from three different paths:

(1) Large scale fields are evidenced from theory, simulation, and observations of coronae and jets. Large field structures more easily survive the buoyant rise to coronae without being shredded by turbulence within the disk. Plausibly, the integrated stress for structures on scales above that which survives the buoyant rise would contribute to the large scale magnetic stress. But shearing boxes that study only a local region non-local large scale magnetic stress is excluded. In a real system this could be the most important part.

(2) Shearing box simulations artificially impose a steady-state because differential rotation is imposed as a steady forcing not subject to the backreaction of the amplified field. Hubbard and Blackman (2008b), argue that this may be more restrictive that previously recognized: In a real disk, energy in differential rotation is susatined only by accretion itself. If a steady-state is to be maintained via turbulent transport alone then there must be 100% power throughput from differential rotation to the turbulent cascade. This is not guaranteed if large scale fields drain power. An accessible steady-state solution must then incorporate stresses from large-scale fields in addition to turbulent transport.

(3) The role of large scale magnetic fields is consistent with the implications of (Pessah *et al.* 2007) which shows that MRI stresses in simulation boxes where the radial extent is of order the vertical scale height scale with the ratio of box size to scale height. The contribution of large scale fields would increase as the radial and vertical scales are increased, highlighting a strongly non-local contribution to stresses that transport angular momentum. Evidence for non-local MRI behavior is also seen in Bodo *et al.* (2008). Box sizes for thin disks must be extended in the radial direction and accordingly in the vertical direction as buoyant loops tend to have radial scales comparable to vertical scales.

Ultimately, mean field accretion disk theory should be coupled to an LSD theory in a real disk as they are actually artificially separated components of what should be a single mean field theory. Note that for any turbulent disk, any assumption of axisymmetry for bulk quantities automatically requires that theory to be a mean field theory. This is often veiled by a mere replacement of the actual viscosity with a turbulent viscosity as in the Shakura-Sunyaev approach (Shakura & Syunyaev(1973)). Hubbard & Blackman (2008a) for example, suggest that in a formal mean field theory, a term involving fluctuations of density and velocity might be interpreted as an additional transport coefficient in the mean surface density equation, restricting available steady states. Balbus *et al.* (1994)incorporated this term into a redefinition of accretion rate.

8. Much work needed to link theory and observation

It remains a major challenge to rigorously couple the engine physics of field generation and accretion to jet launch and jet propagation in a unified theory or simulation that make distinct observational predictions for specific engines. This enterprise spans several subfields of theoretical astrophysics, *let alone* the specific application to p/PNe.

Here however, is a list of possible lower hanging fruit for linking theory with observations: (1) Evaluate the kinematic constraints/predictions of outflows from the scenario of accretion onto the primary and compare the distribution of inferred fast outflow speeds to what would be expected from known binary statistics of low mass stars. This would help constrain the commonality of accretion onto the primary vs. secondary as the latter has a broader range of masses. (2) The more massive the companion, the more CE models would predict mostly Oxygen rich rather than Carbon rich post AGB systems because the binding energy for the early AGB phases is higher. Low mass companions like planets may terminate the AGB only in the thermal pulse phase. (3) Crystalline dust in post-AGB systems can be produced if a binary induced spiral shock anneals silicates (Edgar *et al.*(2008)). Is this universal? (4) CE evolution would predict equatorial outflows that precedes any accretion driven poloidal jet. Is this consistent with the delay of Huggins (2007) and the geometry of equatorial outflows? (6) Are fast outflows contaminated by material that could represent accretion disk residue of shredded low mass companions? (7) Are time scales of observed outflow precession consistent with the gravitational influence of a binary? (8) Can double peaked line profiles be detected to identify accretion disks within the launch region? (9) Can shrouded novae outbursts from a $M_2 > M_{crit}$ companion feeding the primary be detected in X-rays? (10) Improved statistics on the fraction of bipolar pPNe, the fraction of suitable precursor binaries for CE, and the fraction of stars which evolve to be pPNe will improve evaluation as to whether all PNe incur asymmetric phase. (12) Can the approach of Ferreira *et al.* (2006) be generalized to pPNe outflows?

References

Balbus, S. A. & Hawley, J. F. 1998, *Reviews of Modern Physics* 70, 1
Balbus, S. A., Gammie, C. F. & Hawley, J. F. 1994, *MNRAS* 271, 197
Balick, B. & Frank, A. 2002, *ARAA* 40, 439
Blackman, E. G. 1998, *ApJ* 496, L17
Blackman, E. G., & Field, G. B. 2000, *MNRAS* 318, 724
Blackman, E. G., Frank, A., Markiel, J. A., Thomas, J. H., & Van Horn, H. M. 2001a, *Nature* 409, 485
Blackman, E. G., Frank, A., & Welch, C. 2001b, *ApJ* 546, 288
Blackman, E. G. & Field, G. B. 2002, *Phys. Rev. Letters* 89, 265007
Blackman, E. G. & Brandenburg, A. 2003, *ApJ* 584, L99
Blackman, E. G. 2007, *New Journal of Physics* 9, 309
Blackman, E. G., Penna, R. F. & Varnière, P. 2008, *New Astronomy* 13, 244
Bodo, G., Mignone, A., Cattaneo, F., Rossi, P., & Ferrari, A. 2008, *A&A* 487, 1
Brandenburg, A., Nordlund, A., Stein, R. F., & Torkelsson, U. 1995, *ApJ* 446, 741
Brandenburg, A. 2001, *ApJ* 550, 824
Brandenburg, A. & Sandin, C. 2004, *A&A* 427, 13
Brandenburg, A. & Subramanian, K. 2005, *Phys. Rep.* 417, 1
Bujarrabal, V., Castro-Carrizo, A., Alcolea, J., & Sánchez Contreras, C. 2001, *A&A* 377, 868
Chevalier, R. A. & Luo, D. 1994, *ApJ* 421, 225
Edgar, R. G., Nordhaus, J., Blackman, E. G., & Frank, A. 2008, *ApJ* 675, L101
Ferreira, J., Dougados, C., & Cabrit, S. 2006, *A&A* 453, 785

Fromang, S., Papaloizou, J., Lesur, G., & Heinemann, T. 2007, *A&A* 476, 1123
García-Segura, ,G., 1997, *ApJ* 489, L189
García-Díaz, M. T., López, J. A., García-Segura, G., Richer, M. G., & Steffen, W. 2008, *ApJ* 676, 402
García-Segura, G., Langer, N., Różyczka, M., & Franco, J. 1999, *ApJ* 517, 767
García-Segura, G., López, J. A., & Franco, J. 2005, *ApJ* 618, 919
Gardiner, T. A. & Frank, A. 2001, *ApJ* 557, 250
Grether, D. & Lineweaver, C. H. 2006, *ApJ* 640, 1051
Herpin, F., Baudry, A., Thum, C., Morris, D., & Wiesemeyer, H. 2006, *A&A* 450, 667
Hubbard, A., & Blackman, E. G. 2008a, *MNRAS* 390, 331
Hubbard, A. & Blackman, E. G. 2008b, *MNRAS* , submitted
Huggins, P. J. 2007, *ApJ* 663, 342
Iben, I. J. & Livio, M. 1993, *PASP* 105, 1373
Imai, H., Nakashima, J.-I., Diamond, P. J., Miyazaki, A., & Deguchi, S. 2005, *ApJ* 622, L125
Jordan, S., Werner, K., & O'Toole, S. J. 2005, A&A 432, 273
Kapyla, P. J., Korpi, M. J., & Brandenburg, A. 2008, *A&A* 491, 353
Lesur, G. & Ogilvie, G. I. 2008, *A&A* 488, 451
Matt,S., Frank A. & Blackman, E.G., 2006, *ApJ* 647, L45
Mastrodemos, N. & Morris M., 1998, *ApJ* 497, 303
Moffatt, H. K. 1978, *Magnetic field generation in electrically conducting fluids*, Cambridge University Press, p.353
Nordhaus, J. & Blackman, E. G., 2006, *MNRAS* 370, 2004
Nordhaus,J., Blackman, E. G. & Frank, A., 2007, *MNRAS* 376, 599
Pascoli, G. 1993, *Journal of Astrophys. and Astr.* 14, 65
Pascoli, G. 1997, *ApJ* 489, 946
Pessah, M. E., Chan, C.-K., & Psaltis, D. 2007, *ApJ* 668, L51
Pudritz, R. E., 2004, Les Houches Summer School, 78, 187
Reipurth, B. & Bally, J. 2001, *ARAA* 39, 403
Reyes-Ruiz, M. & Lopez, J. A. 1999 *ApJ* 524, 952
Rogachevskii, I. & Kleeorin, N. 2003, *Phys. Rev. E* 68, 036301
Sabin, L., Zijlstra, A. A., & Greaves, J. S. 2007, *MNRAS* 376, 378
Schekochihin, A. A., *et al.*, J. C., Rogachevskii, I., & Yousef, T. A. 2008, arXiv:0810.2225
Shakura, N. I. & Syunyaev, R. A. 1973, *A&A* 24, 337
Steffen, W., García-Segura, G., & Koning, N. 2008, arXiv:0809.5263
Soker,N., 2005, *AJ* 129, 947
Soker,N, 2006a, *ApJ* 645, L57
Soker, N. 2006b, *PASP* 118, 260
Soker, N., & Livio, M. 1994, *ApJ* 421, 219
Soker, N. & Zoabi, E. 2002, *MNRAS* 329, 204
Sur, S., Shukurov, A., & Subramanian, K. 2007, *MNRAS* 377, 874
Tout, C. A. & Pringle, J. E. 1992, *MNRAS* 256, 269
Vishniac, E. T. & Brandenburg, A. 1997, *ApJ* 475, 263
Vishniac, E. T. & Cho, J. 2001, *ApJ* 550, 752
Vlemmings, W. H. T. 2007, in Astrophysical Masers and their Environments, IAU Symposium 242, p.37
Vlemmings, W. H. T., Diamond, P. J., & Imai, H. 2006, *Nature* 440, 58
Vlemmings, W. H. T., & van Langevelde, H. J. 2008, *A&A* 488, 619
Wang, L. & Wheeler, J. C. 2008, *ARAA* 46, 433
Yousef, T. A. *et al.*, 2008, *Phys Rev. Lett.* 100, 184501

Discussion

De Gouveia dal Pino: You have mentioned a sort of magneto-centrifugal process out of a magnetized accretion disk as a possible driving mechanism for PNe winds, or in other words, the same mechanism proposed for jets. But this mechanism predicts a very collimated bipolar outflow and in general PNe winds do not have a very collimated morphology. So how do you explain this in terms of the process above? Is it due to a combination with the outer envelope?

BLACKMAN: I should emphasize that the main reason for considering MHD bipolar outflow models is because the observations of all pre-planetary nebulae and many planetary nebulae do in fact have a strongly collimated or bipolar component as revealed by observations in the past decade. That being said, in all of these objects one also always has the radiation driven stellar wind in addition to whatever MHD outflow might be present. The radiation driven wind will always provide a quasi-spherical or less strongly collimated component to the morphology.

RÜDIGER: You suggested that differential rotation may be reseeded by convection. However, in accretion disks the rotation vector $\vec{\Omega}$ is parallel to the density gradient associated with the convectively instable direction whereas I think one would need the rotation vector perpendicular to this for such a mechanism. Can you address this?

BLACKMAN: Yes. The convective reseeding of differential rotation that I was speaking of was in the context of an isolated star not for a thin disk. The point was that an AGB star would need a mechanism to reseed the differential rotation to maintain a magnetic field long enough to power a strong bipolar pPNe outflow in the absence of a binary companion. For a star there is a significant $\vec{\Omega} \perp \nabla \rho$. The presence of an accretion disk from a binary would circumvent this need for reseeding if the jet came from the disk.

Eric Blackman

Gary Ferland

Fabrice Herpin

Cosmic Magnetic Fields:
From Planets, to Stars and Galaxies
Proceedings IAU Symposium No. 259, 2008
K.G. Strassmeier, A.G. Kosovichev & J.E. Beckman, eds.

doi:10.1017/S1743921309030051

Magnetic fields in AGB stars and (proto-) Planetary Nebulae

Fabrice Herpin[1,2], A. Baudy[1,2], E. Josselin[3], C. Thum[4] and H. Wiesemeyer[4]

[1]Université de Bordeaux, Lab. d'Astrophysique de Bordeaux, F-33000 Bordeaux, France
email: herpin@obs.u-bordeaux1.fr

[2]CNRS/INSU, UMR 5804, BP 89, 33271 FLOIRAC cedex, France

[3]GRAAL, Université Montpellier II - ISTEEM, CNRS, Place Eugène Bataillon, F-34095 Montpellier Cedex, France

[4]Institut de Radio Astronomie Millimétrique, 300 rue de la Piscine, F-38406 Saint Martin d'Hères, France

Abstract. During its quick transition to the Planetary Nebula stage, the Asymptotic Giant Branch star will completely change its geometry. This AGB stellar evolution stage is characterized by a high mass loss driven by the radiation pressure. Strong magnetic field may rule the mass loss geometry and the global shaping of these objects. Following our previous work on the polarization of the SiO maser emission in a representative sample of O-rich evolved stars, we present here a study towards C-rich objects and PPN/PN objects to obtain unbiased conclusions. Using Xpol at the IRAM-30 m telescope, we have conducted CN N=1-0 observations to investigate the Zeeman effect in this molecule and draw conclusion on the evolution of the magnetic field and its influence during the transition of an AGB star to the PN stage. Following the analysis described by Crutcher *et al.* (1996) we derive an estimate of the magnetic field.

Keywords. Magnetic field – stars: evolution – radio lines: stars – ISM: molecules

1. Introduction

The prodigious mass loss observed in the numerous and widespread evolved stars make these objects the main recycling agents of the interstellar medium, and thus one of the most important objects in the Universe. During its quick transition to the Planetary Nebula (hereafter PN) stage, the Asymptotic Giant Branch (hereafter AGB) star will completely change its geometry: the quasi-spherical object becomes axisymmetrical, point symmetrical or even shows more high-order symmetries (e.g. Sahai & Trauger 1998). The classical or generalized *Interacting Stellar Winds* models (cf. Kwok 2000) try to explain this shaping, but have serious difficulties in producing complicated structures with peculiar jets or ansae and do not fully address the origin of the wind.

Strong magnetic field may rule the mass loss geometry and could thus determine the global shaping of these objects. Some recent studies tend to demonstrate the importance of magnetic field in evolved objects. Bujarrabal *et al.* (2001) show that in 80% of the PPNe from their sample the fast molecular flows have too high momenta to be powered by radiation pressure (1000 times larger in some cases) what may be explained by magnetic field. Recently, magnetic field was discovered for the first time in central stars of PN (Jordan *et al.* 2005) and estimated to be at the kiloGauss level, but polarimetric observations toward AGB stars are needed to constrain the magnetic field strength. Moreover, new models involving the magnetic field B were developed where B plays the role of a catalyst and of a collimating agent.

2. Magnetic field in AGB stars

The circumstellar envelope of evolved stars can be probed at different depths through the study of the maser emission of three different molecules, OH, H_2O and SiO, located at different distances from the central star, respectively at 1000-10000, a few 100 and 5-10 AU (one stellar radius $R_\star \sim 1$ AU).

Measurement of the SiO maser radiation polarization can lead to an estimation of the value of magnetic field, $B_{//}$ on the line of sight (for a single dish antenna), or can reveal the structure of the magnetic field (interferometric observations). Until now numerous polarimetric observations of OH masers have been done, several of H_2O masers, but few of SiO maser emission. It must be stressed that SiO is a non-paramagnetic species. The Zeeman splitting exists but the sublevels overlap; the effect is thus undetectable and hence only net polarization can be used to trace magnetic fields. The current state of the knowledge of B is:

- between 1000-10000 AU, $B_{//} \sim 5 - 20$ mG (OH masers, e.g. Kemball & Diamond 1997, Szymczak & Cohen 1997),
- at a few 100 AU from the star, $B_{//} \sim$ a few 100 mG (H_2O masers, e.g. Vlemmings, Diamond & van Langevelde 2001),
- at 5-10 AU, $B_{//} \sim 5 - 10$ G (SiO masers; Kemball & Diamond 1997, VLBI observations in TX Cam).

3. The SiO maser polarization results

Simultaneous spectroscopic measurement of the 4 Stokes parameters (cf. Fig. 1) were carried out towards 57 O-rich evolved stars by Herpin *et al.* (2006) via observations of the SiO (v=1, J=2-1) line at 86.243 GHz. These observations were performed with the IF polarimeter installed at the IRAM 30m telescope at Pico Veleta, Spain (Thum *et al.* 2003). From the Stokes parameters measurements one deduces for each velocity channel:

- the circular polarization rate $p_C = V/I$
- the linear polarization rate $p_L = \sqrt{Q^2 + U^2}/I$
- the polarization angle $\chi = \frac{\arctan(U/Q)}{2}$

Assuming Elitzur *et al.* (1996) maser theory, we calculate the mean value of the magnetic field $B_{//}$ for each SiO maser component (cf. Herpin *et al.* 2006). $B_{//}$ is between 0 and 18 Gauss, with a mean value of 3.5 G. This value combined with the strength of the field in more outer layers of the envelope (given by OH and H_2O masers) agrees with a variation law for B in $1/r$.

The main bias in this study was the source sample, because SiO maser emission is only present in O-rich evolved objects and disappears soon after the star has reached the end of the AGB (Nyman *et al.* 1998), this method cannot be used in C-rich objects or PPN/PN. Moreover, no OH/H_2O maser lines are detected toward C-rich stars (e.g., Szczerba *et al.* 2002). As a consequence, none of the maser molecules can be used to estimate the magnetic field in these objects. Nevertheless, the same type of study we did in O-rich stars should also be conducted within C-rich objects to obtain unbiased conclusions on AGB stars and to investigate the evolution of the magnetic field and its influence during the transition of an AGB star to the PN stage.

4. CN as a tracer of the field in carbon stars ?

CN seems to be a good molecular tracer to perform such studies in carbon stars. First of all, the N=1 $\rightarrow$ 0 and N=2 $\rightarrow$1 lines have already been observed and easily detected

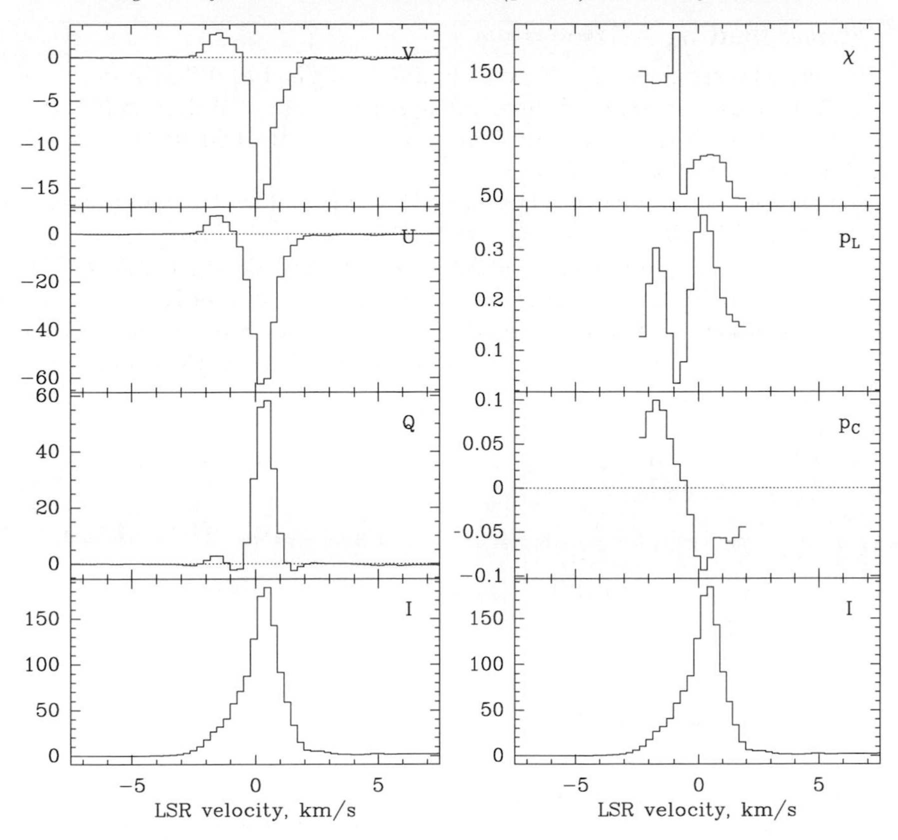

Figure 1. *Left*: V, U, Q et I Stokes parameters for R Leo (given in Kelvins T_{mb}; SiO maser observations). *Right*: derived position angle of polarization (χ) in degrees, linear (p_L) and circular (p_C) polarization levels and intensity for R Leo.

Table 1. Zeeman Splittings for CN N=1→ 0 (Crutcher *et al.* 1996). R.I. stands for *Relative Intensity* in LTE conditions.

(N', J', F')→(N, J, F)	ν_0 (GHz)	Z (Hz μG^{-1})	R.I.	Z × R.I.
1. (1, 1/2, 1/2) → (0, 1/2, 3/2)	113.14434	2.18	8	17.4
2. (1, 1/2, 3/2) → (0, 1/2, 1/2)	113.17087	-0.31	8	2.5
3. (1, 1/2, 3/2) → (0, 1/2, 3/2)	113.19133	0.62	10	6.2
4. (1, 3/2, 3/2) → (0, 1/2, 1/2)	113.48839	2.18	10	21.8
5. (1, 3/2, 5/2) → (0, 1/2, 3/2)	113.49115	0.56	27	15.1
6. (1, 3/2, 1/2) → (0, 1/2, 1/2)	113.49972	0.62	8	5.0
7. (1, 3/2, 3/2) → (0, 1/2, 3/2)	113.50906	1.62	8	13.0

at the 30m by Bachiller *et al.* (1997a,b) and Josselin & Bachiller (2003) towards these objects. Moreover, CN is a paramagnetic species, thus exhibiting Zeeman splitting in its 3mm N=1-0 line emission when the spectral line-forming region is permeated by a field B. The CN Zeeman signal is thermal (as opposed to SiO) and therefore free of the idiosyncrases of maser theory.

CN N=1-0 line has a total of 9 hyperfine components (splitted in two groups, one around 113.17 GHz, the other around 113.49 GHz), with 7 main lines. Of those 7, 4 exhibit strong Zeeman effect (see Table 1, line 1,4, 5 and 7). Crutcher *et al.* (1996) developed an analysis procedure consisting of a least-squares fit in frequency to all seven

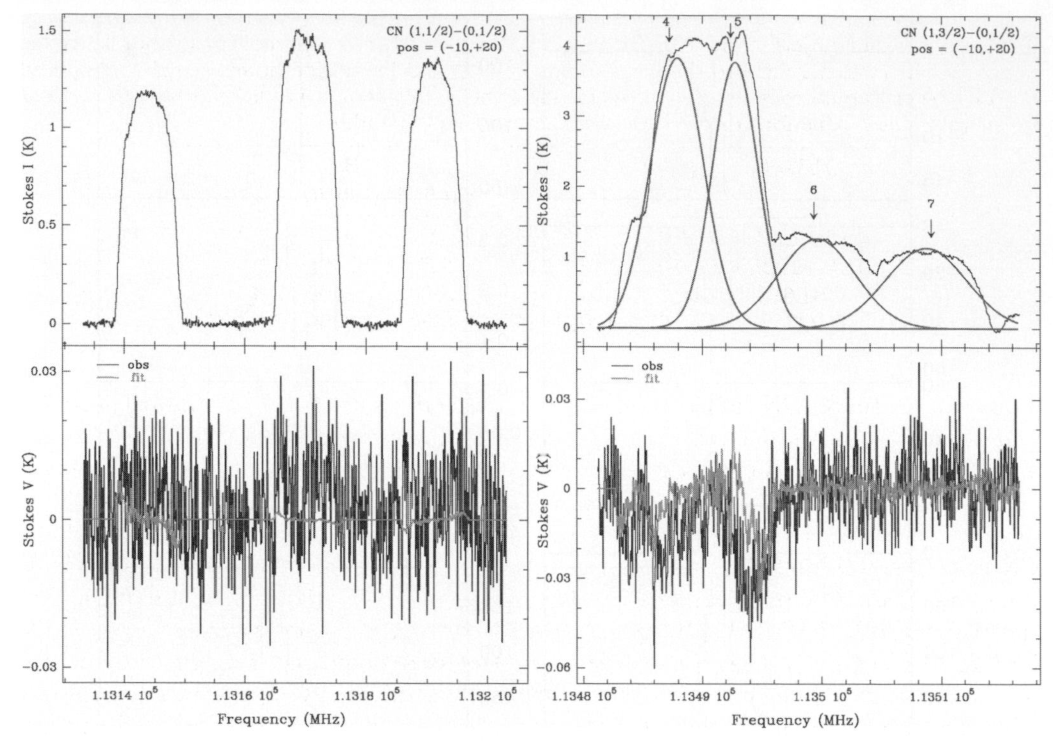

Figure 2. *Left*: CN $(1,1/2) \rightarrow (0,1/2)$ Stokes I (*Top*) and V (*Bottom*) spectra for IRC+10216 (position -10",+20"); *Right*: same for the CN $(1,3/2) \rightarrow (0,1/2)$. Observations, gaussian fits for I and least-squares fits for V are respectively in black, blue and red. No Stokes V signal is detected toward the non(low)-paramagnetic hyperfines, demonstrating that the detected features are indeed due to Zeeman.

hyperfine transitions of the line V spectra, making possible to separate the spurious and Zeeman signals.

The main purpose of this work was to measure and analyze the polarization of the CN N=1-0 line emission in a sample of evolved stars: two C-rich AGB stars (RW LMi and IRC+10216 at a distance of respectively 440 and 120 pc), one PPN (AFGL618, at 900 pc),and one PN object (NGC7027, at 880 pc). As the diameter of the CN ring around IRC+10216 is larger than the 30m beam, our observations were focused on two selected positions for that source.

Simultaneous spectroscopic measurement of the 4 Stokes parameters for the seven hyperfine transitions given in Table 1 were made in november 2006 with the XPol polarimeter (Thum *et al.* 2008) at the IRAM-30m. Following the analysis procedure described by Crutcher *et al.* (1996), we do a least-squares fit in frequency simultaneously to all 7 hyperfine line V spectra (i=1,7):

$$V_i(\nu) = C_1 I_i(\nu) + C_2 (dI_i(\nu)/d\nu) + C_3 Z_i (dI_i(\nu)/d\nu) \tag{4.1}$$

This method enables us to distinguish between the Zeeman effect ($C_3 = B_{//}/2$) and instrumental effects ($C_{1,2}$).

5. CN results and consequences

From the fit of the observed Stokes V spectra (see Fig. 2 for IRC+10216), we then estimate the magnetic field strength along the light-of-sight within the CN layer for the

Table 2. For each object in our sample are given for the CN layer the molecular abundance (χ, relative to H_2), its distance to the central star in AU (and in its size in arcseconds). $B_{8r_\star}$ and $B_{r_\star}$ are the extrapolated strengths of the magnetic field (following a $1/r$ law) at 8 $r_\star$ (equivalent to the SiO maser zone for O-rich objects) and at one stellar radius.

Object	χ(CN)	d_{CN} [AU]	$B_{8r_\star}$ [Gauss]	$B_{r_\star}$ [Gauss]	$r_\star$ [AU]
RW LMi	8.3 10^{-6}	2200 (5")	2	16	1
IRC+10216	6.2 10^{-7}	2500 (21")	2.7	21.6	1
CRL618	2.1 10^{-6}	5000 (5")	<300	<2400	2.3 10^{-3}
NGC7027	2.3 10^{-7}	10000 (11")	<250	<2000	2.3 10^{-3}

4 objects: 7.2 (poor S/N ratio) and 8.6 Gauss, respectively for RW LMI and IRC+10216. Only upper limits are derived for CRL618 ($\leqslant$ 0.9 Gauss) and NGC7027 ($\leqslant$ 0.3 Gauss) as no Stokes V signal is detected.

The objects from our sample have been already observed in CN lines by some authors, making hence possible to estimate the size (and distance to the central object) of the CN layer (see Table 2) that we are investigating through our polarization observations: Lindqvist *et al.* (2000), Lucas *et al.* (1995), Josselin *et al.* (private communication) respectively for RW LMi, IRC+10216 and NGC7027. Only HCN observations are available for CRL618, but HCN being a molecule of photospheric origin that gets photodissociated by the ambient UV-field into CN (e.g. Huggins & Glassgold 1982), we can assume CN molecules are surrounding the HCN envelope, leading to a rough estimate of the CN envelope size.

First of all, the detection or non-detection of the Stoked V signal cannot be explained by the CN abundances as shown in Table 2. Therefore, we checked any instrumental effect or bias that could have lead to a non-detection, but the instrumental polarization is very low and the same for all sources. Furthermore, the size of the CN layer is smaller than the 30m beam ($21''$ at that frequency), and the side-lobe contamination has been well constrained.

We also tried to verify that our magnetic field estimates are consistent with previous works. Hence, following Herpin *et al.* (2006) conclusion, we have applied a $1/r$ variation law for the magnetic field strength across the envelope. We have therefore extrapolated the B value at a distance of 8 stellar radius, roughly the same distance than the SiO layer in O-rich objects and at one stellar radius. For AGB objects, the derived strength of the magnetic field is between 2 and 2.6 Gauss, entirely compatible with the estimate of B by Herpin *et al.* (2006) in O-rich objects from SiO maser observations. In PPN and PN objects, we have estimated upper limits for B of 2000 and 2400 kGauss at 1 $r_\star$ respectively in CRL618 and NGC7027, in agreement with Jordan *et al.* (2005), who measured magnetic field of a few kGauss at that distance to the central star.

We then conclude that our CN results are completely reliable and consistent. Moreover, even for AGB stars, the strength of the magnetic field at the stellar surface is strong enough to allow B to play a major role in the object shaping.

6. Conclusion

We have extended the magnetic field study in evolved stars made by Herpin *et al.* (2006) in order to study its evolution and its influence during the transition of an AGB star to the PN stage. We have observed the Zeeman effect in CN line emission in C-rich objects and more evolved stars (one PPN and one PN). Our B estimates are consistent with previous studies and confirm that the magnetic field varies as $1/r$ accross the circumstellar envelope.

The magnetic field is sufficiently strong to be determinant during the evolution of the AGB star: magnetic field seems to be able to help in the process of collimating, and more generally, of shaping in AGB objects, but is not the main agent at this stage. The increasing strength of the magnetic field in the following stages of evolution, as shown by our results and Jordan *et al.* (2005), will make B a determinant shaping agent.

References

Bachiller, R. *et al.* 1997a, *A&A* 319, 235
Bachiller, R. *et al.* 1997b, *A&A* 324, 1123
Bujarrabal, V., Castro-Carrizo, A., Alcolea, J., & Sánchez Contreras, C. 2001, *A&A* 377, 868
Crutcher, R. M. *et al.* 1996, *ApJ* 456, 217
Elitzur, M. 1996, *ApJ* 457, 415
Herpin, F., Baudry, A., Thum, C., Morris, D., & Wiesemeyer, H. 2006, *A&A* 450, 667
Huggins, P. J. & Glassgold, A. E. 1982, *AJ* 87, 1828
Jordan, S., Werner K., & O'Toole, S. J. 2005, *A&A* 432, 273
Josselin, E. & Bachiller, R. 2003, *A&A* 397, 659
Kemball, A. J. & Diamond, P. J. 1997 *ApJ* 481, L111
Kwok, S. 2000, in *The origin and Evolution of Planetary Nebulae*, Cambridge Astrophysics Series 31, Cambridge University Press edition
Lindqvist, M. *et al.* 2000, *A&A* 361, 1036
Lucas, R. *et al.* 1995, *Ap&SS* 224, 293
Nyman, L.-A., Hall, P. J., & Olofsson, H. 1998, *A&ASS* 127, 185-200
Sahai, R. & Trauger, J. T. 1998, *AJ* 116, 1357
Soker, N. 2002, *MNRAS* 337, 1038
Szczerba, R. *et al.* 2002, *A&A* 381, 491
Szymczak, M. & Cohen, R. J. 1997, *MNRAS* 288, 945
Thum, C., Wiesemeyer, H., Morris, D., Navarro, S., & Torres, M. 2003 *SPIE* 4843, 272
Thum, C. *et al.* 2008, *PASP* 120, 777
Vlemmings, W., Diamond, P. J., & van Langevelde, H. J. 2001, *A&A* 375, L1

Discussion

JORDAN: Did you see any correlation between magnetic field strength and object's asymmetry? If not, this will indicate that the magnetic field does not play any role in the object's shaping.

HERPIN: Actually, the shape of the AGB stars is only known for few objects. It is therefore difficult to conclude on any correlation. Moreover, the shaping occurs – or more exactly is seen – after the star has left the AGB phase.

BECKMAN: How accurate/reliable are the magnetic field strength estimates you make?

HERPIN: For SiO masers, the magnetic field estimate is indeed model-dependent, but the values we derived, assuming Elitzur's theory, are consistent with former water and OH observations, i.e. **B** varying in $1/r$ across the envelope. The location of SiO maser cells is also very well known. The Zeeman effect observed for the CN line emission is well understood too.

Cosmic Magnetic Fields: From Planets, to Stars and Galaxies
Proceedings IAU Symposium No. 259, 2008
K.G. Strassmeier, A.G. Kosovichev & J.E. Beckman, eds.

doi:10.1017/S1743921309030063

The magnetic field structure in the multi-source magnetized core NGC 2024 FIR 5

Felipe de O. Alves[1], J. M. Girart[1], S.-P. Lai[2], R. Rao[3] and Q. Zhang[4]

[1]Institut de Ciències de l'Espai (IEEC–CSIC), Bellaterra, Catalunya 08193, Spain
email: oliveira@ieec.uab.es

[2]Institute of Astronomy and Department of Physics, National Tsing Hua University, Hsinchu 30043, Taiwan
email: slai@phys.nthu.edu.tw

[3]Institute of Astronomy and Astrophysics, Academia Sinica, P.O. Box 23–141, Taipei 10617, Taiwan
email: rrao@sma.hawaii.edu

[4]Harvard-Smithsonian Center for Astrophysics, 60 Garden Street, Cambridge, MA 02138, USA
email: qzhang@cfa.harvard.edu

Abstract. This work reports high resolution SMA polarimetric observations toward NGC 2024 FIR 5, a magnetized core previously found to harbour protostars. Our 345 GHz data indicates the presence of an extended dust emission associated with the dense core where the protostars are embedded. The 3σ polarized intensity shows depolarization toward the peak of Stokes I emission. This diminishing polarized flux implies that the alignment efficiency of the core dust grains is low within higher column densities where grain properties are likely different. The derived magnetic field geometry exhibits pinched field lines which are typical in evolved supercritical clouds where the magnetic field no longer support the core from collapsing. As a consequence for protostars, the gravitational pulling along the disk's long axis makes an equatorial bend to the field lines that, in turn, results in a hourglass shape. The SMA field structure agrees perfectly with the BIMA map. However, models are still necessary to provide a complete description of the evolutionary scenario of FIR 5.

Keywords. ISM: individual (NGC 2024) – ISM: magnetic fields – polarization – stars: formation – techniques: interferometric

1. Introduction

Understanding the evolution of molecular clouds and protostellar cores is one of the outstanding concerns of modern astrophysics. Particularly, efforts are concentrated in determine which physical agents are mainly responsible to control the dynamical properties of dense cores. It is already accepted by the astronomical community that magnetic fields must be taken into account in evolutionary models of collapsing protostellar cores (Shu, Allen, Shang, *et al.* (1999)). Although some theories claim that turbulent supersonic flows drives star formation in the interstellar medium (Elmegreen & Scalo (2004), Mac Low & Klessen (2004)), some new results demonstrate that the ambipolar diffusion collapse theory reproduces properly observed molecular cloud lifetimes and star formation timescales (Tassis & Mouschovias (2004), Mouschovias, Tassis & Kunz (2006)).

Recent advances in instrumentation for astronomical polarimetry improved significantly the techniques to measure the Galactic magnetic field. Several telescopes and interferometers are now capable of doing polarimetric observations at distinct wavebands. At mm/submm wavelengths, the Sub-millimeter Array (Hawaii) is the only instrument that allows to observe the polarization flux at high sensitivity and at the same time high angular and spectral resolution.

2. The Sub-millimeter Array

2.1. *SMA capabilities*

The Sub-millimeter Array is installed at the Mauna Kea summit, in Hawaii. Its superior site results in a very good atmospheric transmissivity for sub-millimeter waves. SMA has a bandwidth correlator composed by 2 sidebands with a width of 2 GHz each one. As a consequence, the mm/submm observations increase sensitivity in detection of thermal emission (from approximately 10 to a few hundred K) from dust and cool gas in molecular clouds, dense cores and stellar envelopes.

The instrument was designed to achieve sub-arcsecond spatial resolutions, reaching values as high as $0.15''$ when operating at the highest frequencies in the very extended configuration. The high resolution of the SMA makes possible to trace the thermal emission of dust grains at physical scales of 10^2 astronomical units (for objects closer than 1 kpc) and, therefore, is able to resolve compact dust cores into multiple components. For a detailed description of SMA, please refer to Ho, Moran & Lo (2004).

2.2. *Polarization with SMA*

Assuming that dust grains are aligned perpendicularly with respect to the interstellar magnetic field, the thermal emission detected by the telescopes must have a degree of polarization. Cold dust emits mainly at far-IR, sub-mm and mm, optically thin wavelengths not affected by scattering or absorption. As described in the previous section, SMA has the suitable instrumentation for observations at sub-mm and mm bands. In general, the polarized flux of a source is only a small fraction of the total emission, but the high sensitivity of SMA represents an improvement in polarimetric observations compared to other interferometers (e.g. OVRO, BIMA).

SMA receivers at 345 GHz are single and linearly polarized. By using a quarter wave plate attached to each receiver, incoming radiation is converted to circular polarization (L,R). The SMA correlator combines the circular polarization vectors (RR, LL, RL, LR) into the Stokes parameters needed to calculate the polarization fraction and the position angle of the polarization vectors. A complete description of the polarimetric system of SMA can be found in Marrone & Rao (2008).

2.3. *Science with SMA: first results in polarization*

Distinct classes of protostellar cores were already observed with SMA in mode of polarization. Recently, the polarimetric properties of G5.89–0.39, a massive ultra compact HII region, were obtained with SMA at high angular resolution ($3''$) (Tang, Ho, Girart, *et al.* (2008)). The extended polarized flux detected around the ionized gas is related to the kinematics of the region, where the magnetic field morphology seems to be disturbed by the expansion of the HII region and outflows produced by young stellar objects in G5.89.

The textbook case of a pinched magnetic field of a low mass young stellar system is NGC 1333 IRAS 4A (Girart, Rao & Marrone (2006)). With a resolution of $1.6'' \times 1.0''$, the dust emission traces a physical scale of 300-1000 AU's, associated to a circumbinary envelope. The magnetic field strength for this source was estimated in 5 mG and the derived mass-to-flux ratio was 2, indicating that the core is collapsing. Such supercritical stage is reflected in the SMA polarization maps (Fig. 1), which indicates a clear hourglass morphology for the plane-of-sky magnetic field, in agreement with theories of collapse of magnetized clouds (Fiedler & Mouschovias (1993), Galli & Shu (1993)). Recently, these data was used to test models of collapsing magnetized cores from Galli & Shu (1993) and from Shu, Galli, Lizano, *et al.* (2006). The agreement between the SMA data and the models leads to the assumption that field dissipation must have occurred in the central parts of IRAS 4A (Gonçalves, Galli & Girart (2008)). This could explain the observed fragmentation (Reipurth, Rodríguez, Anglada, *et al.* (2002), Looney, Mundy & Welch (2000)) that may be occurring in this core, since magnetic braking would be alleviated.

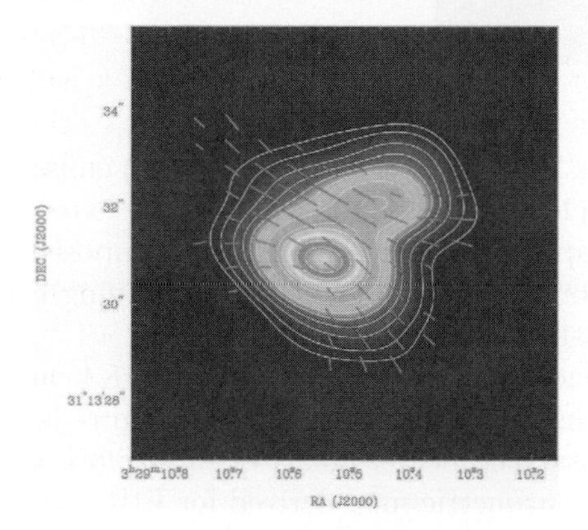

Figure 1. SMA dust emission polarization maps of low mass protostars: NGC 1333 IRAS 4A. Contours and color scale refers to the 870 μm dust continuum emission. The emission is resolved into a binary system (Girart, Rao & Marrone (2006)).

3. NGC 2024: polarization observations

NGC 2024 is a massive star forming region at 415 pc from the Sun, in the Orion B giant molecular cloud. The region contains a luminous HII region where several reflection nebulae are found. Also a north-south molecular ridge is seen toward its center and corresponds to a dust lane in the optical image. Along this ridge, several dust cores were detected and catalogued at 1300 and 350 μm (Mezger, Chini, Kreysa, *et al.* (1988), Mezger *et al.* (1992)). These cores were assigned as FIR (from Far–InfraRed) cores and, in this work, we study the brightest and most evolved of them, FIR 5. A highly collimated outflow extended $\sim 5'$ south of the core (Richer, Hills & Padman (1992)) is associated to this core which has an intermediate mass and multiple components (Wiesemeyer, Güsten, Wink, *et al.* (1997)).

Near-infrared imaging polarimetry was conducted by Kandori, Tamura, Kusakabe, *et al.* (2007) toward NGC 2024. They found a prominent and extended polarized nebula over NGC 2024 and constrained the location of the illuminating source through an analysis of polarization vectors (Fig. 2). A massive star, IRS 2b, with spectral type of O8–B2, is located at the center of the centrosymmetric vector pattern observed in J, H and K.

3.1. *BIMA polarization maps of FIR 5*

The first results on the polarized thermal emission from FIR 5 was done by Lai, Crutcher, Girart, *et al.* (2002). These authors obtained BIMA dust continuum polarization maps combining three different configurations. With the highest resolution ever obtained

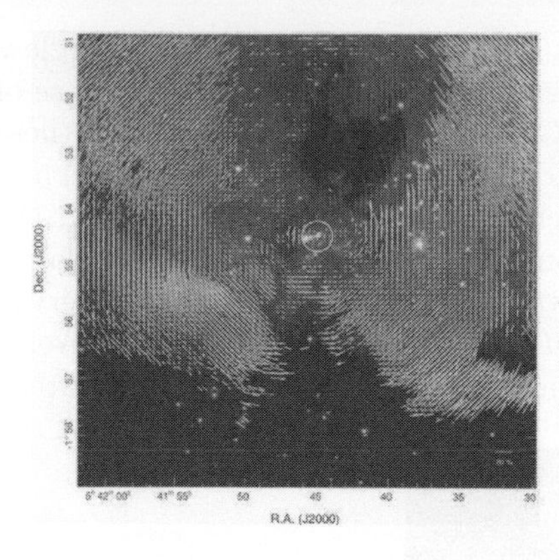

Figure 2. Polarization generated by scattering toward NGC 2024. Vectors exhibits a centrosymmetric symmetry. Normal lines of each vector intersect at a position near the center of the cloud ($\alpha = 54^{\rm h}\ 41^{\rm m}\ 45.01^{\rm s}$, $\delta = -1^{\rm o}\ 54'\ 27''.7$). The massive star IRS 2b is probably the illuminating source of the region since its position coincides with the center of symmetry of the polarization vectors (Kandori, Tamura, Kusakabe, *et al.* (2007)).

($1.6'' \times 1.2''$, P.A. = 11°), they could resolve the 1.3 mm emission into 7 clumps with flux peak stronger than 7 σ (1 $\sigma \sim 3.3$ mJy beam^{-1}). The extended dust emission harbors four of the detected clumps, and for this reason it is supposed to represent a circumstellar disk with multiple young objects. The three remaining clumps are supposed to be collapsing toward this disk.

The polarization emission extends over an area of $\sim$ 8 beam sizes along the direction perpendicular to the elongated continuum emission. Figure 3, left panel, shows polarization vectors where the observed linearly polarized intensity is greater than 3 σ_{I_P} (1 $\sigma_{I_P} \sim$ 2.1 mJy beam^{-1}). The polarimetric map derived for FIR 5 was fitted to a set of parabolas with the same focus point. It means that the magnetic fields lines in the core are systematically curved, consistent with the hourglass morphology predicted by theoretical works. Although some missing polarized flux makes the hourglass shape incomplete (probably due to a lower column density to the east of the region), these authors used the Chandrasekhar-Fermi formula to estimate the magnetic field strength at the core from the fit residuals between the observed and synthetic position angles. As a result, the plane-of-sky component of the magnetic field was calculated as $\sim$ 2 mG. The turbulent-to-magnetic energy was estimated as less than 0.14, meaning that the magnetic field likely dominates the turbulent motions in the core.

3.2. *SMA polarization maps of FIR 5*

The SMA observations of FIR 5 were carried out in the compact configuration at an LO frequency of 345 GHz. The resulting map have a spatial resolution of 2.9" × 1.7" with a beam position angle of -37°. The peak of continuum intensity is 1.3 Jy/beam and the rms $\sim$ 15 mJy/beam. The polarized intensity have a peak intensity of 66 mJy/beam. At this position, the polarization degree reaches $\approx$ 19 %.

The resulting polarization pattern is consistent with the previous one obtained by Lai, Crutcher, Girart, *et al.* (2002) with BIMA, however it is more compact. In both cases, the polarized flux has a peak to the north of the continuum peak and decreases in magnitude toward it. The observed depolarization could be explained by the lower dust alignment efficiency in high density regions, although the same effect could be produced also by geometrical purposes, due simply by the projection of field lines on the line-of-sight (Gonçalves, Galli & Walmsley (2005)). Figure 3, right panel, exhibits SMA continuum emission contours at 10 σ threshold and polarized vectors at a 3-σ level.

SMA continuum emission is elongated at the east-west direction, as found with BIMA, and it is possibly tracing a protostellar core. The main contribution of the polarized flux has a position angle perpendicular to this putative disk but, in contrast to BIMA maps, there is a larger contribution at the eastern part of the core.

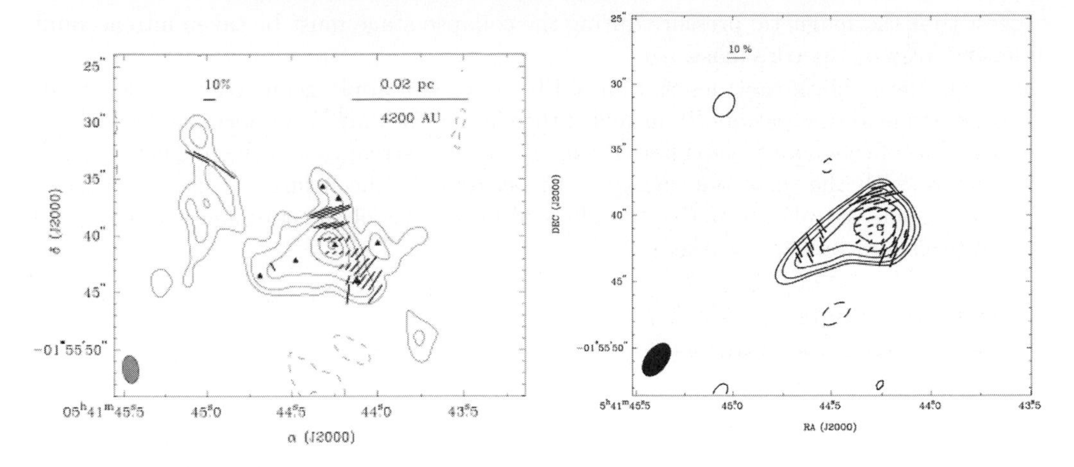

Figure 3. Dust polarization maps of NGC 2024 FIR 5. For both maps, contours are the continuum emission and the vectors refer to the polarized flux. *Left panel*: BIMA map obtained by Lai, Crutcher, Girart, *et al.* (2002). *Right panel*: SMA map of FIR 5. Both maps have the same global patterns, but SMA is more sensitive to the polarized flux at the east part of the core.

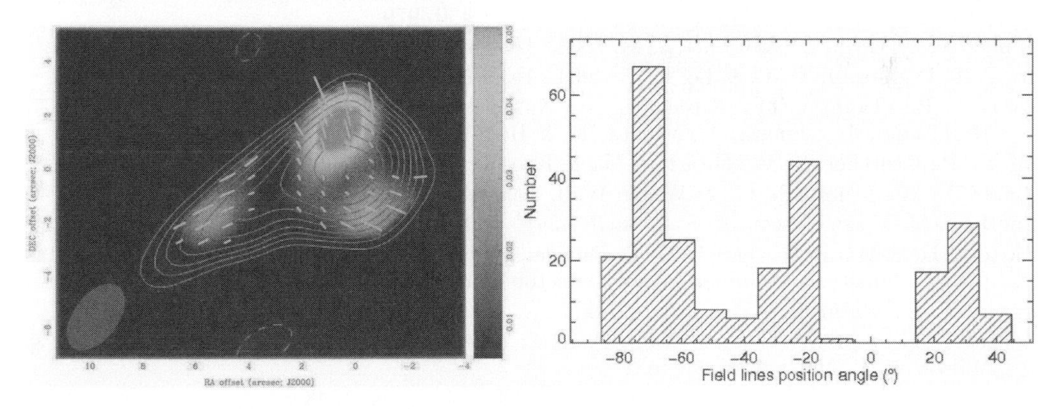

Figure 4. *Left panel*: magnetic field lines (yellow bars) for a 2σ polarization intensity level. Color scale indicates de polarized flux and red contours the continuum emission. *Right panel*: histogram of field lines position angle.

After a 90° rotation of the polarization vectors, we obtain the plane-of-sky component of the magnetic field lines toward the core. As expected, a partial hourglass field morphology is obtained at scales of 10^3 AU's. An histogram of the field lines position angle (Fig. 4, right panel) reproduces partially the preferential orientations for the field lines seen in Fig. 4, left panel. There is an abrupt change (by almost 90°) in the direction of the magnetic field east of the hourglass shape.

4. Conclusions and future work

NGC 2024 FIR 5 has all signatures of a magnetized core under gravitational collapse driven by ambipolar diffusion. The magnetic field geometry is characterized by bended field lines likely produced by the gravitational pull along the protostellar disk, typical of collapsing cores which just achieve a supercritical regime. The magnetic field morphology observed for FIR 5 seems to be reproduced in different physical scales, what may suggest that the magnetic pressure during the collapse stage must be taken into account independently of the core mass.

For a better evolutionary description of FIR 5, we will apply geometrical models to fit the observed field orientation. By means of the Chandrasekhar-Fermi formula, we will use the residuals of the model to estimate the magnetic field strength and the magnetic energy to compare with the turbulent energy of the system. In the future, by use of radiative transfer models, we plan to derive the physical properties of the core and compare with observations and previous work.

Acknowledgements

We would like to thank the SMA staff for the support during observations.

References

Elmegreen, B. G. & Scalo, J 2004, *ARAA* 42, 211
Fiedler, R. A. & Mouschovias, T. Ch. 1993, *ApJ* 415, 680
Galli, D. & Shu, F. H. 1993, *ApJ* 417, 243
Girart, J. M., Rao, R., & Marrone, D. P. 2006, *Science* 313, 812
Gonçalves, J., Galli, D., & Walmsley, M. 2005, *A&A* 430, 979
Gonçalves, J., Galli, D., & Girart, J. M. 2008, *A&A* 490, L39
Ho, P. T. P., Moran, J. M., & Lo, K. Y. 2004, *ApJ* 616, L1
Kandori, R., Tamura, M., Kusakabe, N., Nakajima, Y., Nagayama, T., Nagashima, C., Hashimoto, J., Ishihara, A., Nagata, T., & Hough, J. H. 2007, *PASJ* 59, 487
Lai, S. -P., Crutcher, R. M., Girart, J. M., & Rao, R. 2002, *ApJ* 566, 925
Looney, L. W., Mundy, L. G., & Welch, W. J. 2000, *ApJ* 529, 477
Mac Low, M.-M. & Klessen, R. S. 2004, *Reviews of Modern Physics* 76, 125
Marrone, D. P. & Rao, R. 2008, in W. D. Duncan, W. S. Holland, S. Withngton & J. Zmuidzinas (eds),Millimeter and Submillimeter Detectors and Instrumentation, *SPIE* 7020
Mezger, P. G., Chini, R., Kreysa, E., Wink, J. E., & Salter, C. J. 1988, *A&A* 191, 44
Mezger, P. G., Sievers, A. W., Haslam, C. G. T., Kreysa, E., Lemke, R., Mauersberger, R., & Wilson, T. L. 1992, *A&A* 256, 631
Mouschovias, T. C., Tassis, K., & Kunz, M. W. 2006, *ApJ* 646, 1043
Reipurth, B., Rodríguez, L. F., Anglada, G., & Bally, J. 2002, *AJ* 124, 1045
Richer, J. S., Hills, R. E., & Padman, R. 1992, *MNRAS* 254, 525
Shu, F. H., Allen, A., Shang, H., Ostriker, E. C., & Li, Z. 1999, in: C. J. Lada & N. D. Kylafis (eds.), *in The Origin of Stars and Planetary Systems*, p.193
Shu, F. H., Galli, D., Lizano, S., & Cai, M. 2006, *ApJ* 647, 382
Tang, Y. -W., Ho, P. T. P., Girart, J. M., Rao, R., Koch, P. M., & Lai, S. -P. 2008, *ApJ*, submitted
Tassis, K. & Mouschovias, T. C. 2004, *ApJ* 616, 238
Wiesemeyer, H., Güsten, R., Wink, J. E., & Yorke, H. W. 1997, *A&A* 320, 287

Discussion

De Gouveia dal Pino: In one of the binary targets you showed there was a very organized **B**-field structure in each of the sources. Is there any evidence for an hour-glass shape in each of them?

Alves: For the brightest source, one can see clearly the magnetic field with an hour-glass morphology. The other component, that is unresolved in previous observations with JCMT, doesn't exhibit bended field lines. At higher resolutions, maybe we can detect an hour-glass morphology if this is the case of a supercritical collapsing core.

ZINNECKER: The obvious question would be: is the NGC2024 FIR S core supercritical or subcritical? Is the core collapsing?

ALVES: Yes. Previous BIMA observations resolved FIR S into 7 components by Lai *et al.* (2002) (These authors estimated a **B** strength of 2 mG and a turbulent to magnetic energy ratio of 0.14). In addition, a highly collimated outflow is observed toward it. These star-forming signatures, combined with our SMA hourglass magnetic field for this core, indicate that it is collapsing and maybe stare to experience fragmentation, in agreement with modern theoretical models of core collapse including magnetic support.

Felipe Alves

Eric Blackman talking to Amanda Kepley. In the background: M. Khodachenko, U. Motschmann and A. Reiners (from left to right)

Elisabete de Gouveia dal Pino asking, Christoph Küker (in the background) listening.

Cosmic Magnetic Fields:
From Planets, to Stars and Galaxies
Proceedings IAU Symposium No. 259, 2008
K.G. Strassmeier, A.G. Kosovichev & J.E. Beckman, eds.

doi:10.1017/S1743921309030075

The source of magnetic fields in (neutron-) stars

Hendrik C. Spruit

Max-Planck-Institut für Astrophysik, Postfach 1317, D-85741 Garching, Germany
email: henk@mpa-garching.mpg.de

Abstract. Some arguments, none entirely conclusive, are reviewed about the origin of magnetic fields in neutron stars, with emphasis of processes during and following core collapse in supernovae. Possible origins of the magnetic fields of neutron stars include inheritance from the main sequence progenitor and dynamo action at some stage of evolution of progenitor. Inheritance is not sufficient to explain the fields of magnetars. Energetic considerations point to differential rotation in the final stages of core collapse process as the most likely source of field generation, at least for magnetars. A runaway phase of exponential growth is needed to achieve sufficient field amplification during relevant phase of core collapse; it can probably be provided by a some form of magnetorotational instability. Once formed in core collapse, the field is in danger of decaying again by magnetic instabilities. The evolution of a magnetic field in a newly formed neutron star is discussed, with emphasis on the existence of stable equilibrium configurations as end products of this evolution, and the role of magnetic helicity in their existence. A particularly puzzling problem is the large range of field strengths observed in neutron stars (as well as in A stars and white dwarfs). It implies that a single, deterministic process is insufficient to explain the origin of the magnetic fields in these stars.

Keywords. Magnetic fields – (magnetohydrodynamics:) MHD

---

1. Sources of magnetic field

At all stages of stellar evolution some percentage of stars appears to be strongly magnetic: the magnetic A- and B -stars, magnetic white dwarfs, and neutron stars. Do these fields have a common origin or are different mechanisms at work? What kinds of mechanisms do we know of, and which would be the most plausible one for neutron stars? Three possible origins have been discussed for pulsar magnetic fields:

- the fossil field hypothesis, also called 'flux conservation',
- fields generated by a dynamo process in the progenitor star,
- the thermomagnetic effect in the neutron star crust.

The last one will not be discussed here, since the effect is probably not enough to produce pulsar and magnetar fields of the required strength. See, however, Ho *et al.* (2004).

1.1. *Fossil fields*

The simplest and most popular hypothesis, flux conservation, is that neutron star magnetic fields are simple remnants of the field of their main sequence (MS) progenitors. How does this work out quantitatively? The strongest fields known in MS stars are around 10^4 G. For a MS progenitor of radius $4\,10^{11}$ cm, ($10\mathrm{M}_\odot$) the star would contain a magnetic flux of $\sim 5\,10^{27}\,\mathrm{G\,cm}^2$. A neutron star with the same flux would have surface field strength of $5\,10^{15}$ G, sufficient for a magnetar. But this is an unrealistically optimistic estimate. The neutron star contains only some 15% of the progenitor's mass, and the inner $1.4\mathrm{M}_\odot$ of the progenitor occupy only some 2% of the star's cross section. If the field is roughly uniform, the flux contained in the core would correspond to 10^{14} G on a neutron star, insufficient for a magnetar.

The second problem is statistics: only a very small fraction of the progenitors has magnetic fields as large as 10^4 G. Magnetars, on the other hand, are born frequently. As shown in Woods (2008), their birth rate is probably comparable to that of normal neutron stars.

The fossil hypothesis is therefore not very attractive, at least not for magnetars. Magnetar fields might have a different origin than normal pulsars, of course, but that raises the obvious question why a process operating in magnetar progenitors could not work, at lower strength, in those of pulsars as well.

In any case, the 'flux conservation' of a fossil field would have interesting consequences for the rotation of pulsars, since this scenario implies that a strong magnetic field is present in the star during its entire evolution. This includes the giant/supergiant phases in which the envelope rotates very slowly. Since this envelope also contains almost all the moment of inertia of the star, any coupling between core and envelope will quickly spin the core down to the same low rotation rate. The internal magnetic field needed to explain the magnetic flux of pulsars is more than enough to provide such coupling. Except in the very latest stages of evolution, when the Alfvén travel time through the star becomes longer than the evolution time scale, the core will corotate with the envelope. The angular momentum left in the core is then much too low to explain to rotation rates of pulsars.

This would disqualify the fossil field hypothesis unless the rotation of pulsars is due to something else, for example 'birth kicks', as proposed by Spruit and Phinney (1998). Either the angular momentum, or the magnetic field of pulsars can be fossil, but not both.

Birth kicks would generically lead to correlation between spin axis and direction of proper motion (independent of whether the magnetic field is a fossil or not). Interestingly, such correlations are now being found (Rankin 2006). As in the case of the Crab and Vela pulsars, there appears to be a general preference for alignment of proper motion and spin axis on the plane of the sky. In the analysis of Spruit and Phinney, this means that the kicks are of relatively long duration, long compared with the spin period of the (proto-)neutron star at the time of the kicks (Romani 2005, Wang *et al.* 2006).

1.2. *Dynamos*

A magnetic field might be generated in the core of a SN progenitor at some earlier stage of its evolution, and subsequently amplified during core collapse (for example simply by 'flux conservation'). Possible processes are:

1.2.1. *Convection*

Large scale magnetic fields generated by dynamo processes are traditionally associated with convective zones. Field strengths of such dynamos are often estimated by assuming equipartition of magnetic energy density with the kinetic energy density of convection, $B^2 = 4\pi\rho v^2$ where v is the convective velocity. The energy flux in convection is $F \approx \rho v^3$, and if this carries the star's luminosity L, the equipartition field strength is found as

$$B_{\rm e} \approx M^{1/6} L^{1/3} r^{-7/6} \approx 10^9\, G\, (\frac{M}{M_\odot})^{1/6} L_{38}^{1/3} r_8^{7/6}, \tag{1.1}$$

where the mean density $3M/4\pi r^3$ has been assumed for ρ. Compressing such a field from $r = 10^8$ to 10^6 cm would yield $B = 10^{13}$ G. This would be sufficient for a normal pulsar.

It is not at all clear that this is a reasonable estimate, however. In the Sun, the *large scale* (dipole) field is some 300 times smaller than equipartition with convection would suggest. If this carries over to a stellar interior, convection-generated fields in the progenitor would be insufficient for producing pulsars fields.

It is also quite unclear if a field generated by a convective zone, even if it were strong enough, would leave a permanent dipole moment in the core when it becomes stably stratified again. Does the magnetic field generated in a convective region simply retreat, together with the convection, or can it leave a net magnetic field behind? Since expansion and retreat of a convective zone is slow compared with the characteristic dynamo reversal time, it is likely to leave behind, if anything a series of magnetic zones of different orientation instead of a net large scale field (much like the famous magnetic stripes along mid-ocean ridges in the earth's crust).

1.2.2. *Field generation in stable zones*

Convection is not an essential ingredient in magnetic field generation in stars. The energy of differential rotation is sufficient. 'Closing of the dynamo loop' can be achieved by instabilities in the magnetic field itself, in much the same way as 'magnetorotational' fields are generated in accretion disks (Hawley *et al.* 1995).

Contrary to conventional wisdom, field generation can thus take place also in stably stratified zones of stars (Spruit 2002, Braithwaite and Spruit 2006). The stable stratification, however, strongly limits the radial length scale on which the process operates (just as it does with hydrodynamic processes, cf. Zahn 1983, 1992). For this reason it is unlikely to produce a large scale field.

As in the case of convection it is unlikely that such a process would produce a net dipole moment of much significance for the neutron star descendant. It would, however, be of critical importance for the evolution of the angular momentum distribution in the progenitor, since even weak magnetic fields with small radial length scales can exert torques that are significant on evolutionary time scales (Heger *et al.* 2005).

1.2.3. *Neutrino-driven convection*

The convective velocities in the above estimates are driven by the (radiative) luminosity of the star. Much more powerful convection takes place during core collapse, driven by the much higher neutrino flux. The luminosity is then of order 10^{52} erg/s, the size of the core some $3\,10^6$, which yields (from Eq. 1.1) $B_\mathrm{e} \sim 10^{15}$ G, i.e. in the magnetar range (Thompson & Duncan 1993). However, if the actual dipole field generated is as small, compared with this equipartition number, as it is in the Sun, it would be 300 times smaller. Even after contraction from $3\,10^6$ to 10^6 cm, this would produce a field of only 10^{14} G, marginal for a magnetar, though sufficient for ordinary pulsars.

1.3. *Magnetic fields during collapse: energy estimates*

The arguments above suggest that field generation in the progenitor followed by compression during core collapse is not the most promising scenario for the production of pulsar fields. The field generation processes in the progenitor are important for providing a 'seed field', but it still has to be amplified by a large factor.

Suppose a useful initial magnetic field has somehow been generated in the pre-SN core. If the collapse proceeds with flux conservation, $B \sim 1/R^2$, the magnetic energy $E_B \approx B^2R^3$ increases as $1/R$, i.e. as a constant fraction of the gravitational binding energy E_G of the core, $E_\mathrm{G} \sim GM^2/R$. Simple flux conservation therefore cannot increases a dynamically insignificant initial field into a dynamically significant field. A field of of 10^{15} G in the newly formed neutron star ($R = 10^6$) would require a field of 10^{11} G in a pre-collapse core of $R = 10^8$ cm. As I have argued above, neither a fossil field nor an internally generated (pre-collapse) field is likely to be this high.

It is possible to do better, at least in simple energetic terms, by exploiting differential rotation of the star. Under angular momentum conservation the rotational energy varies as

$$E_\Omega = \frac{1}{2}I\Omega^2 = \frac{1}{2}J^2/I \sim 1/R^2. \tag{1.2}$$

This increases more rapidly than the gravitational energy. For the generation of a magnetic field, only the energy fraction in *differential* rotation is relevant:

$$E_{\Delta\Omega} = (\frac{\Delta\Omega}{\Omega})^2 E_\Omega \sim 0.1E_\Omega, \sim 1/R^2 \tag{1.3}$$

where I have assumed $\Delta\Omega \approx 0.3\Omega$ as a plausible rate of differential rotation.

Assuming there is a process that can convert all energy of differential rotation into magnetic energy $E_B = B^2 R_{\rm NS}^3 \sim E_{\Delta\Omega}$, this would produce a maximum field strength inside the star

$$B_{\rm max} \approx 10^{17}\frac{\Delta\Omega}{\Omega}\frac{\Omega_{\rm NS}}{\Omega_{\rm K}}, \tag{1.4}$$

where $\Omega_{\rm NS}$ is the spin of the neutron star end product and $\Omega_{\rm K}$ the Kepler frequency at the neutron star surface. The observable surface field strength would be lower than this internal field, and the conversion of differential rotation into magnetic energy less than 100% efficient. These efficiency factors are hard to guess, but it suggests that field strengths of the order of a magnetar field are plausible, provided the collapsing core contains enough angular momentum.

How large can the angular momentum of the collapsing core be? An upper limit is set by the average energy of observed core collapse supernovae, of the order 10^{51} erg. If the rotational energy of the neutron star formed is larger than this, and the the star strongly magnetic, spindown by pulsar emission will be so fast that the rotational energy is dumped already into the supernova itself. This sets an upper limit of 10^{51} erg to the rotational energy, corresponding to a rotation period of 4ms for a neutron star of $1.4\mathrm{M}_\odot$, or $\Omega_{\rm NS}/\Omega_{\rm K} \approx 0.1$. The maximum field strength from (1.4) is then $3\,10^{15}$ G, assuming again $\Delta\Omega/\Omega = 0.3$. This shows that magnetar field strengths, up to 10^{15}, are possible, but require a fairly efficient conversion of differential rotation into magnetic energy.

A way of estimating the angular momentum available during core collapse is to follow the evolution of the internal rotation in the progenitor up to core collapse. This requires faith in one's quantitative understanding of the angular momentum transport processes in the star. Internal torques due to magnetic fields are probably more important than purely hydrodynamic processes alone. Heger *et al.* (2005) have computed the initial spins of neutron stars with the prescription for magnetic torques from Spruit (2002). Rotation periods in the range 8-10ms were obtained, somewhat independent of the initial (main sequence) spin. This is comfortably below the upper limit given above, but still rapid enough for magnetar fields of 10^{15} G.

1.3.1. *Consequences and extrapolations*

If differential rotation is an essential ingredient for the production of a strong magnetic field, as suggested above, at least magnetars must be formed spinning rapidly. If a magnetar with $B = 3\,10^{14}$ G is formed in this way, and becomes visible through the supernova debris after a couple of years, say, it would spin at about $P \sim 0.1$s, and would be more luminous than the Crab pulsar. It is not clear if this is compatible with observations of supernovae in nearby galaxies.

On the other hand, it is clear from pulsars with independent age estimates (independent from the spindown time scale) that normal pulsars are formed with a range of initial spins, from $\sim$ 15ms (Crab) to 400 ms (Camilo *et al.* 2007, Gotthelf & Halpern 2007, see also Gotthelf, this volume). This shows that there is either a significant spread in angular momentum of the collapsing core, or that an efficient post-collapse spindown process operates, such as friction against a fall-back disk (cf. Li & Jiang 2007, Wang *et al.* 2007).

Some of the slow-born pulsars have rather weak fields (Camilo et al. 2007, Gotthelf & Halpern 2007). This would agree with angular momentum of the pre-SN core playing an important role in determining the field strength of a neutron star, as suggested above.

The rapid initial spin required for magnetar fields can be extrapolated to higher rotation rates. At sufficiently rapid rotation, the spindown of the nascent, highly magnetized neutron star could power a Gamma-ray burst. This magnetar-GRB scenario has recently become popular, cf. Kommissarov & Barkov (2007), Bucciantini *et al.* (2007), Yu & Dai (2007) and references therein.

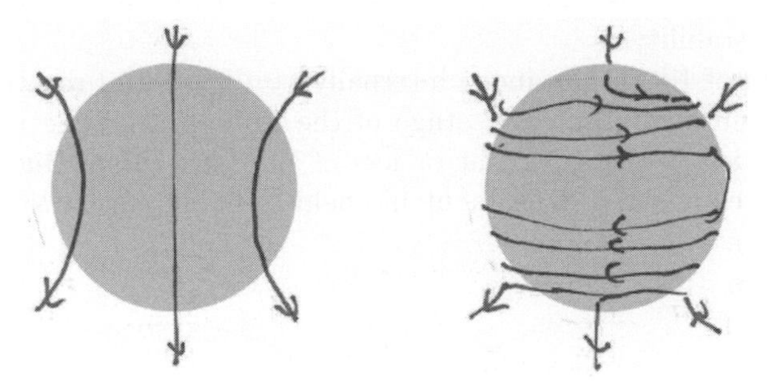

Figure 1. 'Winding up' of a poloidal field by differential rotation. The field lines are stretched into a stronger azimuthal field inside the star. The surface field remains unchanged.

2. Requirements on the field-production process

To explain the strength of the dipole component of a pulsar magnetic field, not just any field generated inside in a SN core will do. An illustrative example is the 'winding-up' of an initial magnetic field (or 'seed field') by differential rotation.

Suppose the field is initially a purely poloidal field $\mathbf{B}_\mathrm{p}$ (field lines in meridional planes, no azimuthal field component), as sketched in fig 1. Differential rotation stretches the field lines in azimuthal direction, creating an azimuthal field component B_ϕ that increases linearly with time, $B_\phi \sim B_\mathrm{p}\Delta\Omega t$. The surface field, and its dipole component remain *unchanged* in this process.

2.1. *Stratification*

To produce an interesting pulsar dipole, something else has to happen: the strong internal field B_p somehow has to be brought to the surface. The obvious process is magnetic buoyancy: a strong field reduces the gas pressure and density in it, so that loops of the azimuthal field tend to float upward against the direction of gravity (see fig 2). The stretch of the loop that finds itself outside the star forms a vacuum field with a dipole moment. In this way an initially weak poloidal field can be amplified by differential rotation into a field with a strong dipole.

In a stable stratification, a magnetic field can float to the surface only if it is strong enough to overcome the stratification. This is the case, approximately, when the Alfvén speed V_A exceeds HN, where H the pressure scale height and N the buoyancy frequency

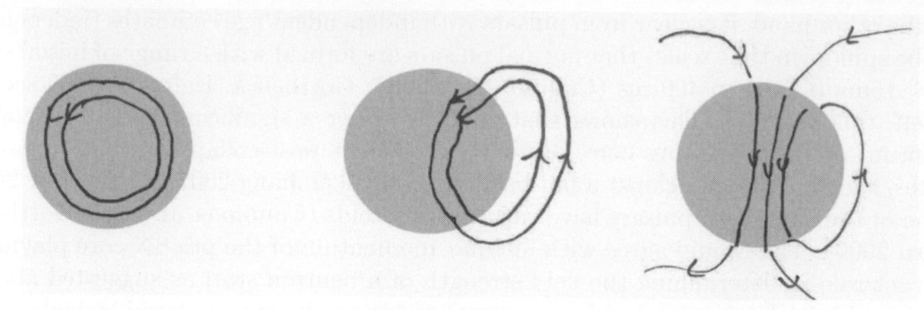

Figure 2. Magnetic buoyancy of an azimuthal field loop causes part of it to rise, provided the field is strong enough to overcome the stratification. This creates a dipole component at the surface.

of the stratification. Once the neutron star has formed, the stratification of the neutron to proton density ratio is able to prevent fields of up to 10^{17} G from emerging by such a direct buoyancy instability.

A magnetic field of 10^{15} G produced internally by differential rotation must therefore have reached the surface in an early stage of the formation of the neutron star, when the buoyancy frequency was a percent or less of the final value. Buoyant rise is aided, however, by the presence of a dense neutrino field (see Thompson & Murray 2001 for a detailed discussion).

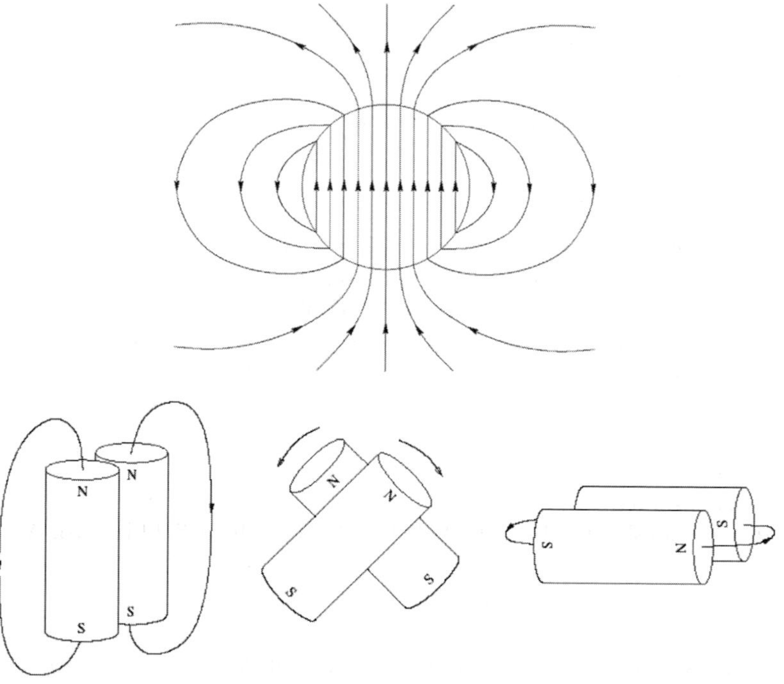

Figure 3. Instability of a poloidal field in a star. (From Flowers and Ruderman, 1977)

2.2. *Instability of poloidal fields*

A poloidal field, such as that created by the buoyant process of Fig. 2, cannot be the dipole field seen in pulsars, since it is highly unstable. All purely poloidal fields in a star are unstable (Wright, 1973, Markey and Tayler 1974, Flowers & Ruderman 1977).

A simple example is that of Flowers and Ruderman, shown in Fig. 3. In this example a uniform field inside the star connects to a vacuum field outside. This external field is that of a point dipole centered on the star (cf. Jackson E&M). Cutting the star in two halves without cutting field lines is possible along the axis of the field. Rotating one half in place by 180° does not change any of the energy components in the interior of the star (thermal, gravitational, magnetic). But the external magnetic energy changes, in the same way the energy of two parallel bar magnets changes when one is rotated by 180°. The instability proceeds on the time it takes an Alfvén wave to cross the star.

How would this bar magnet process work in practice, in a gravitating fluid sphere like a star? This can be answered in some detail with numerical simulations. Fig. 4 shows the result of calculations by Jonathan Braithwaite. The star does not flip hemispheres as Fig. 3 suggests. Instead, the instability sets in at a smaller azimuthal length scale. The evolution of the surface field distribution resembles the mushroom-like mixing patterns observed in Rayleigh-Taylor instability. Both properties are characteristic of interchange type instabilities. [The radial field component at the surface shows the instability nicely, but of course the entire interior of the star takes part in it.]

In these simulations, the field decayed completely, within the numerical uncertainties. This can be understood from the properties of *magnetic helicity*, discussed below. The conclusion is that buoyant rise of a loop of strong field is not enough, at least not in the simple form of fig 2. At least one more ingredient is needed to end up with a magnetic field configuration that is useful for explaining magnetar fields.

2.3. *The role of magnetic helicity*

The magnetic helicity $\mathcal{H}$ of a field configuration is a global quantity,

$$\mathcal{H} = \int \mathbf{B} \cdot \mathbf{A} \, \mathrm{d}V, \tag{2.1}$$

where $\mathbf{A}$ is a vector potential of $\mathbf{B}$, and the integral is over the volume of the magnetic field. [The gauge dependence of $\mathbf{A}$ is physically significant. It causes helicity to be global: $\mathbf{B} \cdot \mathbf{A}$ has no meaning as a local 'helicity density']. In a perfectly conducting fluid with fixed boundary conditions magnetic helicity is a conserved quantity. In practice perfect conduction is not a realistic situation, since rapid reconnection can take place even at very high conductivity, especially in a dynamically evolving field configuration. Nevertheless, in laboratory experiments helicity is often observed to be at least approximately conserved.

The consequence is the existence of stable equilibrium configurations. To the extent that helicity is conserved, a non-equilibrium or unstable magnetic field with a finite helicity cannot decay completely, since the helicity of a vanishing field is zero. An experiment like that of Fig. 4, a field starting with a finite amount of helicity therefore decays initially, but eventually settles into a stable equilibrium. [In laboratory plasma physics this process of settling is called Taylor relaxation].

Two factors are thus necessary for a stable magnetic equilibrium in a fluid star: a stable stratification (otherwise the field would just escape through to the surface by its buoyancy) and a finite magnetic helicity.

Stellar models with such equilibria have been found by Braithwaite and Nordlund (2006). They usually consist of an approximately axisymmetric torus of twisted field lines inside the star, and a bundle of field lines feeding through the 'donut hole' and through the surface of the star, see Fig. 6. The existence of stable configurations of this general form in stars has been surmised several times in the past (Prendergast 1956,

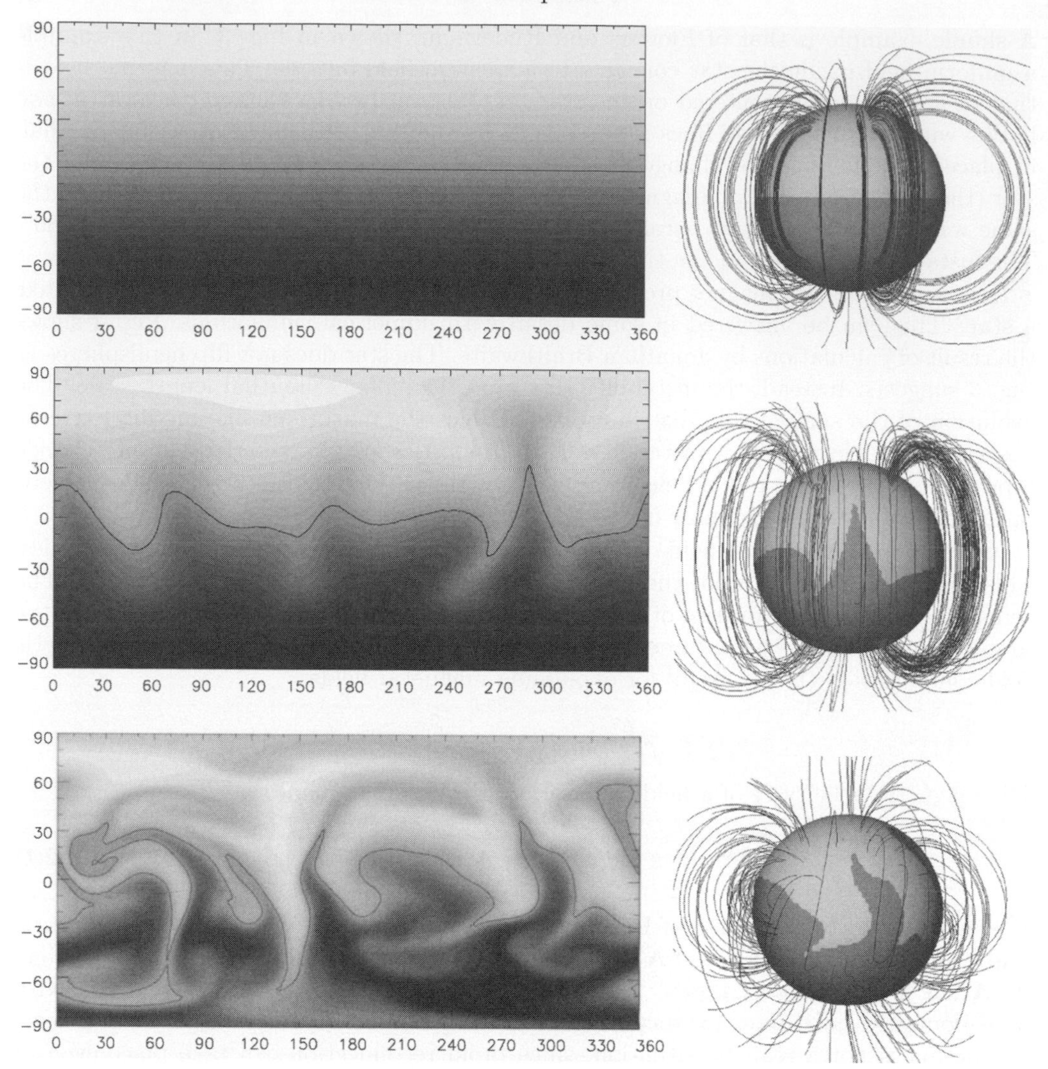

Figure 4. Numerical simulation of Flowers-Ruderman instability in a star. Left panels: radial component of the field strength on the star's surface, in the initial dipolar state and at two times during the nonlinear evolution of the instability. The magnetic equator (radial component zero) is shown as a dark line (from Braithwaite and Spruit 2006)

Kamchatnov 1982, Mestel 1984). It is quite possible that there also exist stable fields with more complex configurations.

2.4. *Field amplification in core collapse*

To make use of the energy of differential rotation (Eq. 1.3), a process has to be found that can convert it to magnetic energy, and can do so fast enough. Most of the rotational energy becomes available only towards the end of the collapse, when the stratification begins to develop its stable buoyancy and starts to suppress emergence of the field to the surface. This leaves a window of perhaps a few seconds for an observable surface magnetic field to grow from rotational energy.

To see what this implies, approximate the collapse as a two stage process: an instantaneous collapse to neutron star size $R = 10^6$ cm with angular momentum conservation

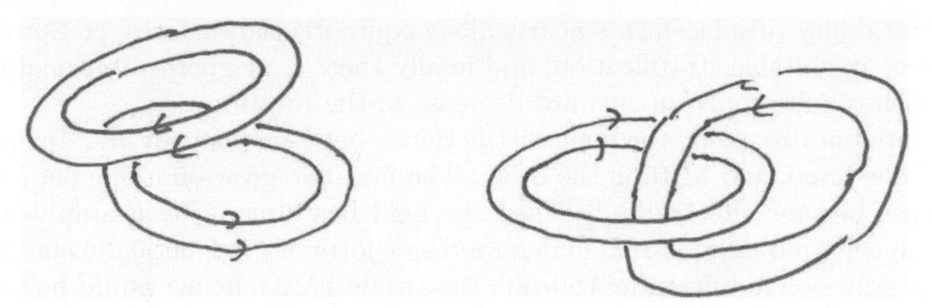

Figure 5. Magnetic helicity. A configuration consisting of separated loops of untwisted magnetic field (left sketch) has zero helicity. A 'linked poloidal toroidal' field made of the same loops (right) is helical.

Figure 6. Equilibrium found as end state of the decay of a random field with finite helicity in a stably stratified star (idealized sketch). The surface field is determined by the poloidal field lines feeding through the hole of the torus. On their own they would be unstable as in fig 3; they are stabilized by the twisted torus surrounding them. (From Braithwaite and Nordlund 2006)

and flux freezing, followed by a phase of field amplification by differential rotation. Consider first the simple winding-up process of Fig. 1. With a nominal rotation period of 10ms, $\Delta\Omega/\Omega = 0.3$, and a time available of 3s, the field can be amplified by a factor 100. To reach a field strength of 10^{15} G, the initial field would have to be 10^{13}, corresponding to 10^9 in a pre-collapse core of 10^8 cm. This is marginally achievable with differential rotation-generated fields (Heger *et al.* 2005), but does not leave much room for inefficiencies. In addition, as shown in Fig. 1, something more has to happen to convert a wound-up internal field into something observable (i.e. a field with a dipole moment at the surface).

This something is likely to be an instability affecting the magnetic field. If the instability operates on a magnetic time scale $R/v_{\rm A}$, it will have noticeable effects only when this time scale becomes significantly shorter than our time window of a few seconds. If such an instability increases the poloidal field strength, the azimuthal field strength produced by winding-up in the continuing differential rotation will then increases more rapidly than before, and the instability will grow faster. The net effect is a runaway amplification.

There are three instabilities that can in principle be involved in such a scenario. Buoyant instability (Fig. 2) will be most effective when the stratification is convective;

a Tayler instability (displacements nearly along equipotential surfaces, cf. Spruit 1999) is also effective in a stable stratification, and finally there is magneto-rotational instability (MRI) which requires only an outward decrease of the rotation rate.

For the present discussion the main distinction is between buoyant and Tayler instability on the one hand, and MRI on the other. The first two grow on a magnetic time scale $R/v_{\rm A}$, hence become effective only when the field has already been amplified to some extent by linear winding-up. MRI, understood as a form of ('magnetically enabled') shear instability, grows on a differential rotation time scale $1/\Delta\Omega$, hence would be expected to set in immediately. This gives MRI modes a clear advantage when time is of the essence.

2.4.1. *Numerical simulations*

An example of such a process is given by the simulations of Akiyama *et al.* (2003), Ardeljan *et al.* (2005), Moiseenko *et al.* (2006). For an instability with a fixed growth rate (independent of the field strength) the time t needed to reach a given amplitude B from an initial field strength B_0 scales as $t \sim \ln(B/B_0)$. This is roughly consistent with the evidence given by Moiseenko *et al.* (2006) and suggests that some form of MRI is responsible.

The standard case presented by Moiseenko *et al.* (a ratio of magnetic to gravitational energy of 10^{-6}), corresponds to a wildly optimistic initial field strength ($\sim 10^{15}$ G at $R = 10^6$, or 10^{11} at $R = 10^8$ cm). But because of the logarithmic dependence on B_0 the process still works with a more realistic initial field 10^3 times smaller. It is possible that it will work even better in 3 dimensions, because of the additional modes of instability available in 3D.

2.5. *The first three minutes*

Suppose a field of pulsar or magnetar strength has been formed during core collapse, and that the stratification of the proto-neutron star has become stable enough to prevent the field from escaping by simple buoyant rise. After this dynamical phase the field is still out of equilibrium or at least far from stable. At the same time the star is still fluid: a crust forms only after some 100s, and initially is probably more ductile and deformable than a cooled-down neutron star crust. The evolution of the field over the first few minutes is therefore critical for its survival as a pulsar or magnetar field.

The field will decay by whatever magnetic instabilities are available to it. Without rotation, these instabilities will operate on time scales of the order of the Alfvén crossing time through the star, $t_{\rm A} \approx 3{\rm s}\ B_{14}^{-1}$. Such instabilities continue to exist in a rapidly rotating star because the Coriolis force is perpendicular to the velocity, hence does not contribute an additional energy barrier to be overcome by instability (like a stable stratification does).

The Coriolis force restricts the range of modes that are unstable when the rotation rate is larger than the Alfvén frequency $1/t_{\rm A}$. But usually there are some modes that remain unaffected by the Coriolis force (a trivial example being modes with displacements nearly parallel to the rotation axis). Most modes suffer only a reduction of their growth rate, characteristically by a factor $\Omega t_{\rm A}$, while some can actually become stable (Pitts and Tayler 1986).

A simple survival estimate can be made by assuming that the growth of the relevant unstable modes is reduced by the characteristic factor $\Omega t_{\rm A}$ (Flowers and Ruderman 1977). Without the helicity constraint, the field would then decay with time t such that the growth time of the instability is of the order t: $t_{\rm A}^2\Omega \approx t$. After 200s, this yields as maximum surviveable field strength in the absence of a helicity constraint:

$$B(200) \approx 5\,10^{12}\ {\rm G}\ P^{-1/2}, \tag{2.2}$$

where P is the spin period in s. For an initial spin of 10ms, fields of the order of observed pulsar strength could thus survive until the crust forms and saves them from further decay.

Fields higher than $\sim 5\,10^{13}$, i.e. magnetar fields, can survive only if their helicity is sufficient to reach a stable equilibrium. In fact, this holds independent of crust formation, since 10^{13} G is about the maximum field strength believed to be supportable by a crustal lattice. For the survivability of magnetar-strength fields, the neutron star can be treated as fluid, with or without crust. [But the crust of course affects the actual decay of magnetar fields, modulating the energy releases into a series of starquakes.]

The effect of rotation on field decay can again be tested by numerical simulation. Braithwaite (2007) finds that rotation does indeed slow down the decay, but not quite in the simple way assumed above. Starting with the uniform internal field of the Flowers-Ruderman argument, oriented at some angle to the rotation axis, there is first a stage of decay which proceeds at the Alfvén crossing time, independent of the rotation rate. The implication is that this phase is due to the modes that are not affected by the Coriolis force. The further decay is slowed down by rotation, but within numerical uncertainty no stable equilibria were found, consistent with the vanishing helicity of the initial state.

3. Summary

Three popular scenarios for the origin of magnetic fields in pulsars and magnetars are i) the fossil field hypothesis, ii) fields generated internally in the progenitor, and iii) field generation during core collapse. Fossil fields are ruled out for magnetars (required field strength of the MS progenitor too large, number of magnetic MS progenitors too small). If a fossil field plays a role for normal pulsars, magnetic coupling between core and envelope of the progenitor will remove the angular momentum from the core, producing neutron stars with extremely long rotation periods. The spin of pulsars must then be due to the 'birth kicks' that also give rise to their proper motions.

Internally generated fields, followed by core collapse under flux conservation, is unattractive because of the rather low field strengths that can plausibly be produced. The most plausible source of energy for field generation is the differential rotation during core collapse; it increases more rapidly than even the gravitational energy. This predicts that the most magnetic stars are also formed rotating intrinsically rapidly (but suffering rapid spindown by a pulsar wind already during the supernova, cf. Metzger *et al.* 2007).

The most critical factor for the production of a strong field is the finite time available in the final collapse phase, on the order of seconds. At earlier times in the collapse the energy in differential rotation is not yet high enough, at later times (when the neutrinos have left) there may still be enough energy in differential rotation, but the stratification has then become so stable that even fields of magnetar strength cannot reach the surface any more.

Combined with the lowish initial field strengths in the pre-collapse core, this implies the existence of a very efficient field amplification process during collapse. Linear winding-up in differential rotation is too slow except possibly for normal pulsars. Runaway growth of a magnetic field is possible by the combination of differential rotation with magnetically driven or shear-driven instabilities. Magnetorotational instability works fastest because it is essentially a shear-driven instability operating on the short time scale of differential rotation. The simulations of Akiyama *et al.* (2003) and Ardeljan *et al.* (2005) illustrate the feasibility of such a process.

Most simulations of magnetic core collapse (e.g. Shibata *et al.* 2006, Obergaulinger *et al.* 2006, Sawai *et al.* 2007, Burrows *et al.* 2007, see also the more analytical considerations in Uzdensky and MacFadyen 2007) start with rather optimistic assumptions on the strength and configuration of the pre-collapse field (an aligned dipole of 10^{11} G at $R = 10^8$ cm for example). Such initial states produce interesting results in themselves. But to explain magnetar fields, or magnetically powered supernovae, *let alone* the magnetic field strengths needed to produce magnetically powered GRB jets, much more has to happen than the flux freezing collapse of a dipole field. It is likely that pre-collapse fields with more realistic configurations and strengths, and processes like those seen in Moiseenko *et al.* (2006), will produce magnetic fields that behave rather differently. In particular it is still an open question how the ordered, rotation-aligned magnetic field configurations are formed that work so well in producing jets (or if they form at all).

Since the field of the neutron star is formed in a highly dynamic event, it is likely to be far from equilibrium, or at least far from stable. For the same reason it is also likely to be dominated by higher multipoles rather than the dipole component that determines pulsar phenomenology. For field strengths of 10^{13} G and less magnetic instabilities are slow enough for the field to survive until it can be anchored by the newly formed crust (rapid rotation helps here, since the Coriolis force reduces the growth rate of instabilities).

Fields of magnetar strength decay rapidly, and are too strong to be anchored by the crust. Their apparent survival implies that they have found a stable equilibrium configuration within seconds after their formation. The formation of such stable equilibria from complex, strongly unstable initial fields can be demonstrated with numerical simulations. The key ingredients for the existence of such equilibria are i) the stable stratification of a neutron star, and ii) a non-vanishing magnetic helicity.

3.1. *The 'range in field strengths' problem*

An intrinsic difficulty affects any mechanism proposed for the origin of the magnetic field in neutron stars. A convincing scenario, in the usual astrophysical sense, would advocate a well-defined, deterministic physical mechanism, yielding a characteristic field strength for given initial conditions. Since observed field strengths vary by several orders of magnitude, with comparable numbers per decade in field strength any such scenario will necessarily fail for the majority of observed objects. Since a very similar spread is observed in main sequence A stars (and in white dwarfs), it is tempting to make the 'frozen field' or 'flux conservation' assumption: that the fields of neutron stars are inherited from their main sequence progenitors. As I have argued above, this is unlikely at least for fields of magnetar strength, and would create difficulties for the interpretation of rotation rates of the end products of stellar evolution. It would also be a temporary solution, just moving the question to the star formation stage. The unattractive possibility would be that a range of different physical mechanisms is involved, their combined effect somehow conspiring to produce an approximately uniform-in-the-log distribution. This would imply that there is no simple answer. An elegant mechanism would be one that somehow has an intrinsic stochastic indeterminacy. There do not seem to be any good examples of such mechanisms in astrophysics.

References

Akiyama, S., Wheeler, J. C., Meier, D. L., & Lichtenstadt, I. 2003, *ApJ* 584, 954

Ardeljan, N. V., Bisnovatyi-Kogan, G. S., & Moiseenko, S. G. 2005, *MNRAS* 359, 333

Braithwaite, J. 2007, *A&A* 469, 275

Braithwaite, J. & Nordlund, Å. 2006, *A&A* 450, 1077

Braithwaite, J. & Spruit, H. C. 2006, *A&A* 453, 1097

Bucciantini, N., Quataert, E., Arons, J., Metzger, B. D., & Thompson, T. A. 2007, *MNRAS* 380, 1541
Burrows, A., Dessart, L., Livne, E., Ott, C. D., & Murphy, J. 2007, *ApJ* 664, 416
Camilo, F., Ransom, S. M., Halpern, J. P., & Reynolds, J. 2007, *ApJ* 666, L93
Flowers, E. & Ruderman, M. A., 1977, *ApJ* 215, 302
Gotthelf, E. V. & Halpern, J. P. 2007, *ApJ* 664, L35
Halpern, J. P., Gotthelf, E. V., Camilo, F., & Seward, F. D. 2007, *ApJ* 665, 1304
Hawley, J. F., Gammie, C. F., & Balbus, S. A. 1995, *ApJ* 440, 742
Heger, A., Woosley, S. E., & Spruit, H. C. 2005, *ApJ* 626, 350
Ho, W. C. G., Blandford, R. D., & Hernquist, L. 2004, Bulletin of the American Astronomical Society, 36, 917
Kamchatnov, A. M., 1982, *Zh. Eksp. Teor. Fiz.* 82, 117.
Li, X.-D. & Jiang, Z.-B. 2007, *ApSS* 308, 525
Markey, P. & Tayler, R.J., 1974, *MNRAS* 168, 505.
Mestel, L. 1984, *AN* 305, 301
Metzger, B. D., Thompson, T. A., & Quataert, E. 2007, *ApJ* 659, 561
Moiseenko, S. G., Bisnovatyi-Kogan, G. S. & Ardeljan, N. V. 2005, *Mem. Soc. Astr. It.* 76, 575
Obergaulinger, M., Aloy, M. A., & Müller, E. 2006, *A&A* 450, 1107
Pitts, E. & Tayler, R. J., 1985, *MNRAS* 216, 139
Prendergast, K. H. 1956, *ApJ* 123, 498
Rankin, J. M. 2007, *ApJ* 664, 443
Romani, R. W., 2005, Binary Radio Pulsars, ASP Conference Series, Vol. 328, p.337
Sawai, H., Kotake, K., & Yamada, S. 2007, ArXiv e-prints, 709, arXiv:0709.1795
Shibata, M., Liu, Y. T., Shapiro, S. L., & Stephens, B. C. 2006, *Phys. Rev. D* 74, 104026
Spruit, H. C., 1999, *A&A* 349, 189
Spruit, H. C., 2002, *A&A* 381, 923
Spruit, H. C. & Phinney, E. S. 1998, *Nature* 393, 139
Thompson, C. & Duncan, R. C. 1993, *ApJ* 408, 194
Thompson, C. A. & Murray, N. 2001, *ApJ* 560, 339
Uzdensky, D. A. & MacFadyen, A. I. 2007, *ApJ* 669, 546
Zahn, J.-P. 1974, in Stellar Instability and Evolution, IAU Symposium 59, p185
Zahn, J.-P. 1992, *A&A* 265, 115
Wang, Z., Kaplan, D. L., & Chakrabarty, D. 2007, *ApJ* 655, 261
Wang, C., Lai, D., & Han, J. L. 2006, *ApJ* 639, 1007
Woods, P. 2008, in *40 years of pulsars*, C.G. Bassa *et al.* (eds.), AIP conference proceedings Volume 983, p.227
Wright, G. A. E., 1973, *MNRAS* 162, 339
Yu, Y. W. & Dai, Z. G. 2007, *A&A* 470, 119

Discussion

WADE: We don't know the fraction of massive stars with fields of kG or tens of kG. What fraction would be required to explain magnetars?

SPRUIT: It has been estimated from the number of magnetars. This is small bit they also live much shorter (10^3 – 10^4 yrs) than ordinary pulsars. Pete Woods in the Montreal conference ("40 yrs of pulsars") finds that a significant fraction of neutron stars (>10%) must be born as magnetars.

Hendrik Spruit

Roland Kothes

Cosmic Magnetic Fields:
From Planets, to Stars and Galaxies
Proceedings IAU Symposium No. 259, 2008
K.G. Strassmeier, A.G. Kosovichev & J.E. Beckman, eds.

doi:10.1017/S1743921309030087

Probing interstellar magnetic fields with Supernova remnants

Roland Kothes[1] **and Jo-Anne Brown**[2]

[1]National Research Council of Canada, Herzberg Institute of Astrophysics, Dominion Radio Astrophysical Observatory, P.O. Box 248, Penticton, British Columbia, V2A 6J9, Canada
email: Roland.Kothes@nrc-cnrc.gc.ca

[2]Department of Physics and Astronomy, University of Calgary, 2500 University Drive N.W., Calgary, AB, Canada
email: jocat@ras.ucalgary.ca

Abstract. As Supernova remnants expand, their shock waves are freezing in and compressing the magnetic field lines they encounter; consequently we can use Supernova remnants as magnifying glasses for their ambient magnetic fields. We will describe a simple model to determine emission, polarization, and rotation measure characteristics of adiabatically expanding Supernova remnants and how we can exploit this model to gain information about the large scale magnetic field in our Galaxy. We will give two examples: The SNR DA530, which is located high above the Galactic plane, reveals information about the magnetic field in the halo of our Galaxy. The SNR G182.4+4.3 is located close to the anti-centre of our Galaxy and reveals the most probable direction where the large-scale magnetic field is perpendicular to the line of sight. This may help to decide on the large-scale magnetic field configuration of our Galaxy. But more observations of SNRs are needed.

Keywords. Magnetic fields – polarization – ISM: individual (DA 530, G182.4+4.3) – ISM: magnetic fields – Supernova remnants

---

1. Introduction

Recently, there have been several studies of the Milky Way's large-scale magnetic field utilizing observations of the rotation measure of compact polarized objects like extra-galactic point sources (e.g. Brown *et al.* 2007) or pulsars (see A. Noutsos and J.-L. Han, this proceedings). However, we still do not know the large-scale magnetic field configuration or even the number of field reversals within our Galaxy. One difficulty is that through Faraday rotation studies of compact extra-galactic sources we only derive the average magnetic field parallel to the line of sight $B_{||}$ through our Galaxy weighted by the electron density n_e. Here, the rotation measure RM is given by

$$RM = 0.81 \int_l B_{||} n_e dl. \tag{1.1}$$

Here, RM is given in rad/m^2, B in μG, n_e in cm^{-3}, and the pathlength l in pc. In addition extragalactic sources may suffer from intrinsic Faraday rotation of unknown magnitude. Faraday rotation studies of pulsars average $B_{||}$ between us and the pulsar weighted by n_e. One significant problem of the averaging procedure is that there could be numerous field reversals along the line of sight, which would be averaged out. This ambiguity could be solved if we had anchor points for the magnetic field structure within our Galaxy. We propose to determine these anchor points with polarization and Faraday rotation studies of Supernova remnants (SNRs), since these can be used as magnifying glasses of their ambient magnetic field.

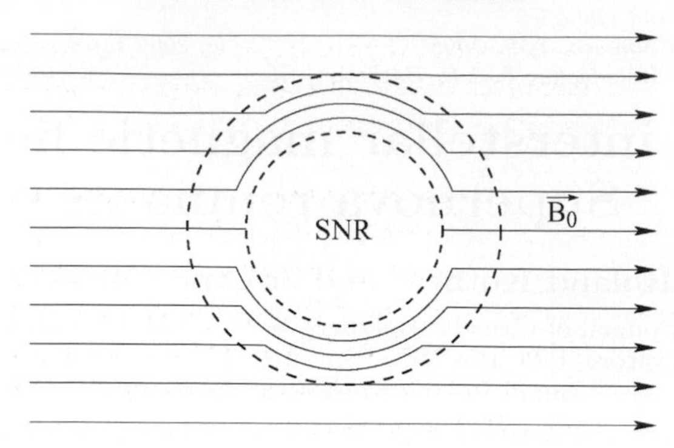

Figure 1. Simple model of the magnetic field inside an adiabatically expanding SNR.

2. A simple model of Supernova remnants

We assume a Supernova remnant, which is adiabatically expanding into an ambient medium of constant density n_0 and constant homogeneous magnetic field $\vec{B}_0$. In our model the SNR is spherical and has a shell with a width of about 10 % of the SNR's radius in which the ambient medium and its magnetic field are compressed by a factor of 4. While the electron density in the shell is assumed to be constant the magnetic field changes with the angle between the SNR's expansion direction and the ambient magnetic field (see Fig. 1). It is proportional to the "number" of magnetic field lines, which are swept up by the expanding shock wave, hence the magnetic field inside the shell is strongest where the SNR is expanding perpendicular to the ambient magnetic field, and zero where it is parallel. The synchrotron emission is integrated along the line of sight from the back of the SNR to the front and appropriately Faraday rotated (see Equation 1.1). The synchrotron flux density S produced by a spectrum of relativistic electrons $N(E)$ at frequency ν is given by

$$S_\nu \propto K B_\perp^{\frac{1}{2}(\delta+1)} \nu^{-\frac{1}{2}(\delta+1)}, \quad N(E)dE = KE^{-\delta}dE. \tag{2.1}$$

Here, E is the energy and the values for K and δ are defined by $N(E)$.

In Fig. 2 we display the simulated emission structure in total intensity and polarized intensity and the internal rotation measure that would be observed from a typical Supernova remnant at different viewing angles Θ. In this simulation the ambient magnetic field is pointing away from us from the front left to the back right and Θ is the angle between the plane of the sky and the magnetic field lines the SNR is expanding into. For negative Θ, the magnetic field would point towards us from the back left to the front right. In the simulation shown in Fig. 2 the ambient magnetic field is $4\,\mu$G, the ambient density is $1\,\mathrm{cm}^{-3}$, the distance to the SNR is 2 kpc, and its diameter is 20 pc. The simulation was convolved with a beam of 2.5'.

In emission we find the typical bilateral structure of SNRs in total power and polarized intensity with a tangential magnetic field up to a Θ of about 60°, beyond which the SNR becomes circular and thick-shelled with a radial magnetic field. The surface brightness of the SNR is decreasing from $\Theta = 0°$ to $\Theta = 90°$ by a factor of almost 10. The RM structure reveals a few very interesting characteristics. The internal rotation measure in the centre of the SNR is always zero, which indicates that from observations we can determine the foreground RM there. The entire SNR except for $\Theta = 0°$ is dominated by RM of one sign; in Fig. 2 it is mostly negative. This indicates whether the ambient magnetic field is pointing away from us or towards us. There is also a smaller area of opposite sign. Its location indicates the orientation of the ambient magnetic field and its

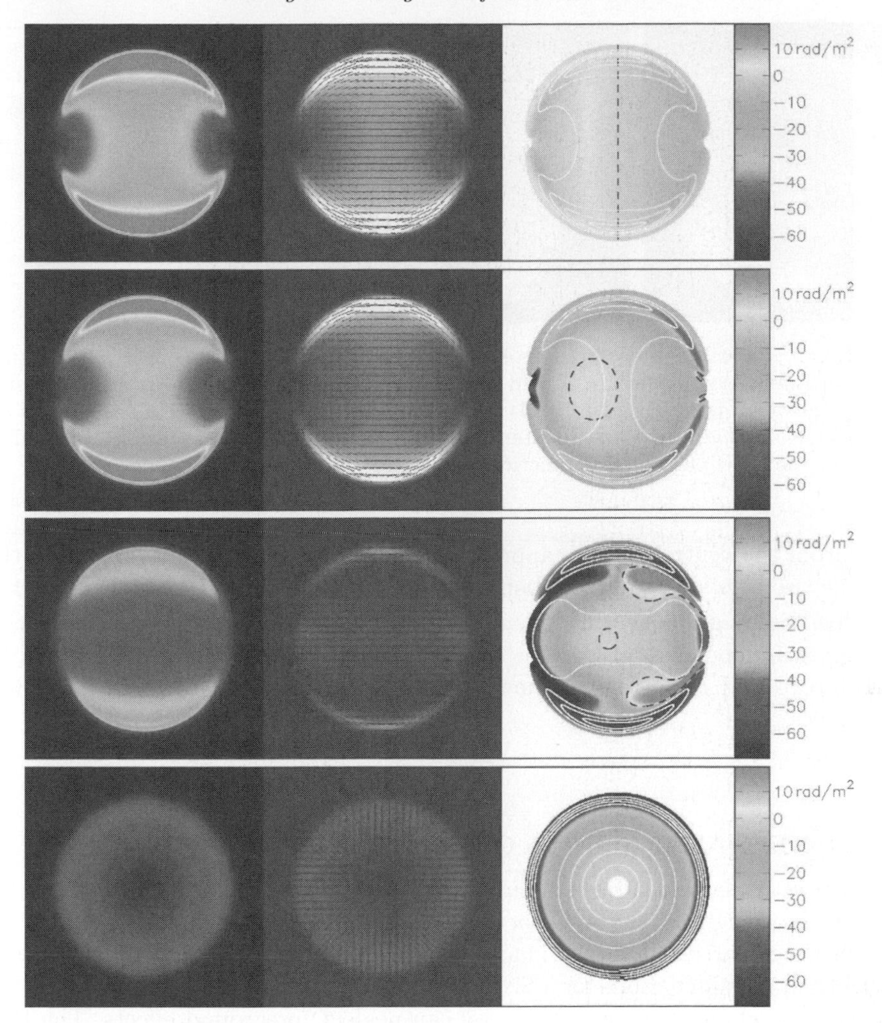

Figure 2. Simulated emission structure in Stokes I (left column), polarized intensity (centre column) with overlaid B-vectors, and RM (right column) with overlaid white contours indicating the polarized emission for different angles Θ (from top to bottom: $\Theta = 0°$, $30°$, $60°$, and $90°$).

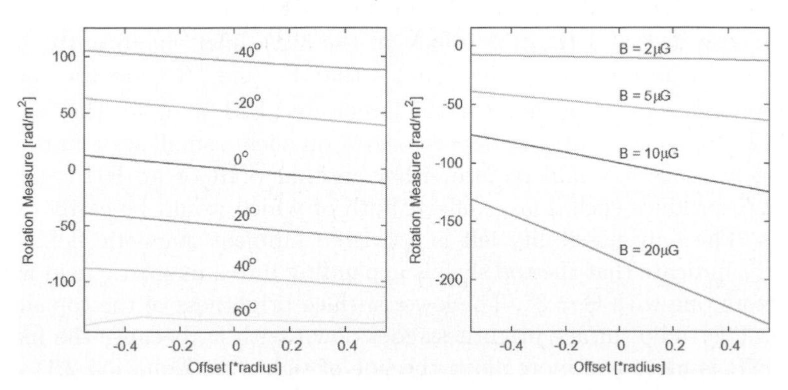

Figure 3. The rotation measure gradient on the arcs of the SNR as a function of the distance from their centres in the simulation described in section 2.2. Left: Θ varies from $-40°$ (top) to $+60°$ (bottom). Right: $\Theta = 20°$ and $|\vec{B}_0|$ varies from 2μG (top) to 20μG (bottom).

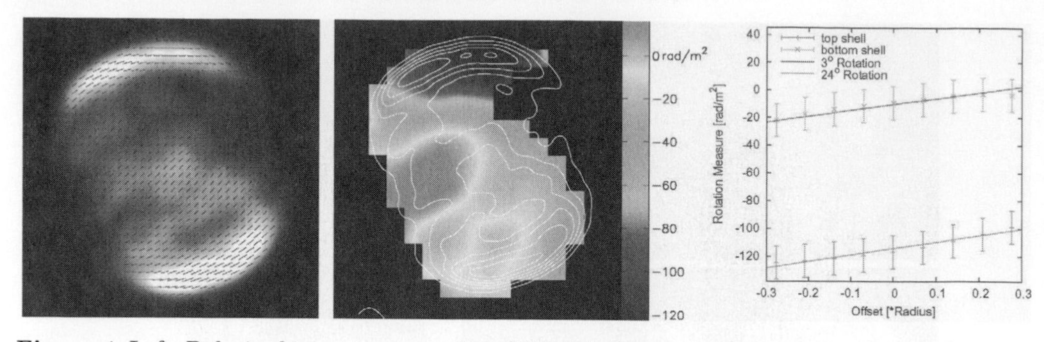

Figure 4. Left: Polarized intensity map of the SNR DA 530 at 4.85 GHz, observed with the 100 m telescope in Effelsberg. Vectors in B-field direction are overlaid. Centre: Rotation measure map calculated between 4.85 GHz and 10.45 GHz (100 m Effelsberg) of SNR DA 530 with overlaid white contours indicating polarized intensity. Right: Observed rotation measure gradients on the shells of the SNR DA 530 as a function of distance from the centre of the shell.

distance from the centre gives an approximate value for Θ. In addition we find that the RM on the two arcs shows a linear behaviour up to about $\Theta = 60°$ (see Fig. 3). This can be described by its gradient, which depends entirely on $|\vec{B_0}| \times n_0$ and the size of the SNR and is almost independent of Θ and the RM in the middle, which is determined from Θ and the foreground RM. To show how we can exploit these characteristics we will give two examples.

3. The SNR DA 530 and the Galactic halo

In Fig. 4 we display a polarized intensity and rotation measure map of the SNR DA 530, which is located high above the Galactic plane at a latitude of about $+7°$. The overlaid vectors in the polarized intensity image indicate a tangential magnetic field, which we expected from the simulations for a SNR with a Θ of less than 60°. The rotation measure in the centre is about $0\,\mathrm{rad/m^2}$, hence we can neglect foreground effects. The RM map is dominated by negative rotation measures, but there is a small area of positive RM to the left of the centre. This indicates that the SNR is expanding into a magnetic field, which is pointing away from the front left to the back right.

As can be seen in Fig. 4 the RM values on the arcs differ significantly, however, the gradient seems to be the same. This implies that n_0 and $|\vec{B_0}|$ are the same for both shells. The difference could be either in the foreground RM or $\vec{B_0}$ for the two arcs have a different Θ. To find such a large difference in RM on such a small scale in the foreground is very unlikely, since it would require either another SNR or an HII region along the line of sight to produce such a large effect, both of which would be easily detectable by other means. The only possibility left is a twisted ambient magnetic field. Simulations for both shells indicate that the top shell is expanding into a magnetic field with $\Theta = 24°$ and the bottom one with $\Theta = 3°$. The lower surface brightness of the top shell supports this finding. The radio surface brightness goes down with Θ, because the magnetic field inside the SNR is more and more along the line of sight (see Equation 2.1).

Radio observations of other galaxies show twisted magnetic spurs emerging from star forming regions (e.g. Review by Beck 2008). DA 530 is located above an area of the Milky Way, which is rich in star forming regions, HII regions, and SNRs. Is DA 530 expanding inside these twisted magnetic spurs?

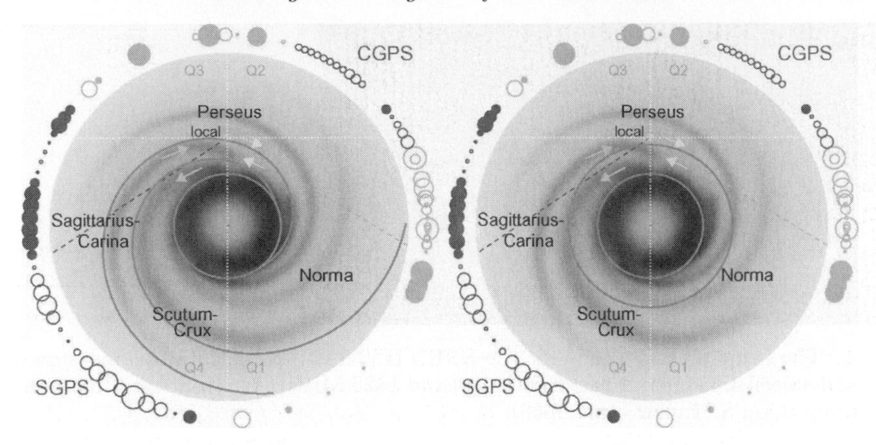

Figure 5. Extragalactic rotation measures and the large-scale magnetic field of the Milky Way - a birds-eye-view. Filled symbols indicate positive RM, open circles indicate negative RM, and the size of the circles are proportional to the RM. The gray-scale is the electron density model of Cordes & Lazio (2002). The arrows in the first quadrant (Q1) indicate the magnetic field directions commonly accepted, while the arrows in the fourth quadrant (Q4) indicate the magnetic field directions as determined from Brown *et al.* (2007). The solid lines are suggested spiral field lines (left) and circular field lines (right). The common names of the spiral arms are also labeled.

4. The SNR G182.4+4.3 and the Galactic anti-centre

In terms of the large-scale Galactic magnetic field, it is generally accepted that the field is directed clockwise in our local arm, as viewed from the North Galactic pole, and that there is at least one large-scale magnetic field reversal (a region of magnetic shear where the field is seen to reverse directions by roughly 180°) in the inner Galaxy between the local and Sagittarius-Carina arms (Simard-Normandin & Kronberg 1979; Thomson & Nelson 1980), as indicated by the filled arrows in quadrant 1 (Q1) in Fig. 5. Further to the existence and location of additional magnetic field reversals is the question of field alignment with the optical spiral arms. It is often assumed that the field is closely aligned with the spiral arms. Brown *et al.* (2007) demonstrated conclusively that the field in the fourth quadrant (Q4) of the Sagittarius-Carina arm is directed clockwise, the same as the local field. This presents a continuity problem if the field does indeed follow the optical arms, since the field is known to be counter-clockwise in Q1 of the Sagittarius-Carina arm, as shown in Fig. 5 (left). However, if the field is much less inclined than the optical spiral arms, perhaps even purely azimuthal as suggested by Vallée (2005), the continuity problem is resolved, as shown in Fig. 5 (right). Additional information regarding the field alignment may be found towards the anti-centre region of the Milky Way. If the large-scale magnetic field of our Galaxy is azimuthal the magnetic field component parallel to the line of sight $B_{\|}$ would be 0 towards a Galactic longitude of 180°; if the magnetic field follows the spiral arms we would expect $B_{\|} = 0$ towards Galactic longitude between 165° and 170°.

The SNR G182.4+4.3 (Fig. 6) is conveniently located close to the anti-centre at a Galactic longitude of 182.4°. Unfortunately only one shell of this SNR is visible, likely because the bottom shell is expanding towards the Galactic plane into higher density and the top shell away from the plane into very low density. Without emission from the centre of the SNR we cannot easily determine the foreground *RM*. On the bottom shell we find a RM gradient which goes from about 0 in the right to about 50 rad/m^2 on the left (Fig. 6). This implies that the ambient magnetic field is pointing towards us from back left to front right. This is not surprising, since the magnetic field in the outer

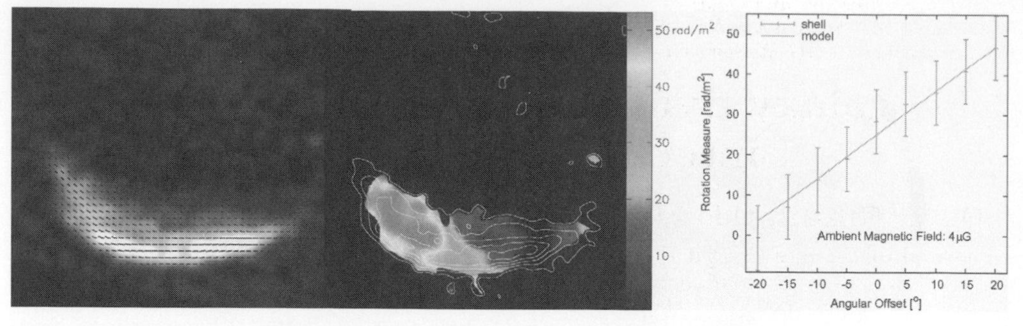

Figure 6. The same as Fig. 4 but for the SNR G182.4+4.3. The rotation measure map was calculated between 4.85 GHz (100 m Effelsberg) and 1420 MHz taken from the Canadian Galactic Plane Survey (CGPS, Taylor *et al.* 2003).

Galaxy should be directed from higher to lower longitudes and we are here at a longitude which is higher than the $B_{\|} = 0$ point for both possible magnetic field configurations. Neglecting the foreground contribution for the RM we can determine an upper limit for Θ by simulating a SNR for the RM map of G182.4+4.3. The result is independent of the distance to G182.4+4.3. We derive $\Theta \geqslant -4°$, hence the Galactic longitude at which $B_{\|} = 0$ must be larger than 178°. If we now assume that the foreground magnetic field has the same Θ as the SNR's ambient field and the foreground field is constant at $4\,\mu$G, which would be a reasonable assumption, we would find $\Theta = -2°$. This supports the assumption that the Milky Way has an azimuthal magnetic field structure.

Acknowledgements

The Dominion Radio Astrophysical Observatory is a National Facility operated by the National Research Council Canada. The Canadian Galactic Plane Survey is a Canadian project with international partners, and is supported by the Natural Sciences and Engineering Research Council (NSERC). This research is based on observations with the 100-m telescope of the MPIfR at Effelsberg.

References

Beck, R. 2008, *astro-ph* 0711.4700

Brown, J. C., Haverkorn, M., Gaensler, B. M., Taylor, A. R., Bizunok, N. S., McClure-Griffiths, N. M., Dickey, J. M., & Green, A. J. 2007, *ApJ* 663, 258

Cordes, J. M. & Lazio, T. J. W. 2002, *astro-ph* 0207156

Simard-Normandin, M. & Kronberg, P. P. 1979, *Nature* 279, 115

Taylor, A. R., Gibson, S. J., Peracaula, M., *et al.* 2003, *AJ* 124, 3145

Thomson, R. C. & Nelson, A. H. 1980, *MNRAS* 191, 863

Vallée, J. P. 2005, *ApJ* 619, 297

Discussion

GAENSLER: Does your technique only work for old SNRs in which synchrotron emission and polarizations comes from swept-up gas? Most observable SNRs are younger and have emissions coming from turbulent reverse shock.

KOTHES: It should work for all Supernova remnants, which show the typical shell structure with a tangential magnetic field. It will not work for very young SNRs expanding in complex media.

Cosmic Magnetic Fields:
From Planets,to Stars and Galaxies
Proceedings IAU Symposium No. 259, 2008
K.G. Strassmeier, A.G. Kosovichev & J.E. Beckman, eds.

doi:10.1017/S1743921309030099

A galaxy dynamo by Supernova-driven interstellar turbulence

Oliver Gressel, Udo Ziegler, Detlef Elstner and Günther Rüdiger

Astrophysikalisches Institut Potsdam, An der Sternwarte 16, 14482 Potsdam, Germany
email: ogressel,uziegler,elstner,gruediger@aip.de

Abstract. Supernovae are the dominant energy source for driving turbulence within the interstellar plasma. Until recently, their effects on magnetic field amplification in disk galaxies remained a matter of speculation. By means of self-consistent simulations of supernova-driven turbulence, we find an exponential amplification of the mean magnetic field on timescales of a few hundred million years. The robustness of the observed fast dynamo is checked at different magnetic Reynolds numbers, and we find sustained dynamo action at moderate Rm. This indicates that the mechanism might indeed be of relevance for the real ISM.

Sensing the flow via passive tracer fields, we infer that SNe produce a turbulent α effect which is consistent with the predictions of quasilinear theory. To lay a foundation for global mean-field models, we aim to explore the scaling of the dynamo tensors with respect to the key parameters of our simulations. Here we give a first account on the variation with the supernova rate.

Keywords. Turbulence – ISM: supernova remnants – turbulence – magnetic fields

---

We here present new results on our local box simulations of a differentially rotating, vertically stratified, turbulent interstellar medium threaded by weak magnetic fields (Gressel *et al.* 2008a). In our model, we apply optically thin radiative cooling and heating to account for the heterogeneous, multi-phase nature of the ISM. Improving over existing models, we compute a radiatively stable initial solution to avoid the transient collapse seen in models applying an isothermal stratification. The central feature of our simulations is the driving of turbulence via several thousand localized injections of thermal energy, which well resemble the kinetics of the supernova feedback. Unlike for artificial forcing, the energy and distribution of the SNe are determined by observable parameters.

In a preceding paper, we have shown that the turbulence created by SNe does in fact exponentially amplify the mean magnetic field (Gressel *et al.* 2008b). Here we extend this work towards a broader parameter base. The organisation of this article is as follows: in Section 1, we discuss the possible relevance of the discovered effect for realistic Reynolds numbers and compare the kinetically driven SN dynamo with the cosmic ray dynamo found by Hanasz *et al.* (2004). In Section 2, we then report on the influence of the supernova rate on the measured dynamo parameters. For a short review on mean-field modelling of the galactic dynamo, we refer the reader to Elstner *et al.* (this volume).

1. Slow versus fast dynamo

In laminar dynamos, diffusion sets the relevant timescale for magnetic reconnection, thus defining an upper limit for the allowable growth rate of the mean magnetic field. Because the microscopic diffusivity is usually low, these dynamos are commonly referred to as "slow dynamos". Within the ISM, the diffusion time $\tau_{\rm d} = L^2/\eta$ (related to the microscopic value $\eta \simeq 10^8\,{\rm cm^2 s^{-1}}$) by far exceeds the Hubble time. This means that the field amplification mechanism in galaxies needs to be a "fast dynamo" in the sense

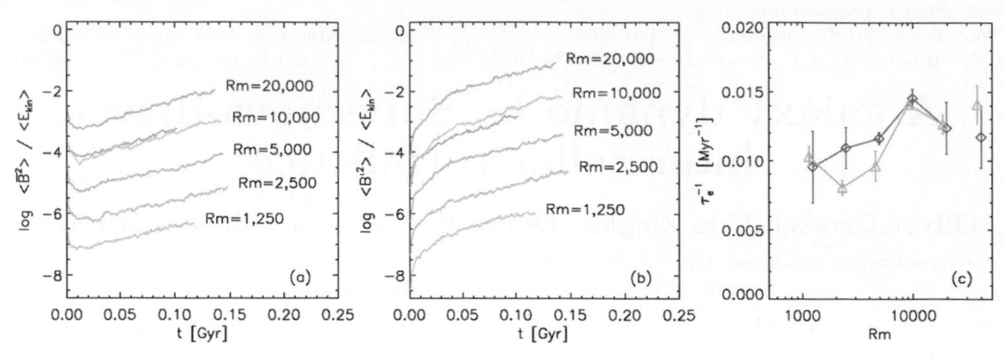

Figure 1. Evolution of the regular (a) and fluctuating (b) magnetic field strength for different Rm. For clarity, the ordinates of the different models have been offset by an order of magnitude each. In panel (c), we compare the growth rates for the turbulent (diamonds) and regular (triangles) magnetic field. The unconnected data points to the right correspond to a run with $\eta = 0$ and provide an indication for the level of numerical diffusivity.

that it works on a timescale different than $\tau_{\rm d}$ (Lazarian & Vishniac 1999). Accordingly, one assumes that the galactic dynamo is determined by some sort of effective turbulent diffusivity $\eta_{\rm t}$. Although the dynamo may still be limited by topological changes via reconnection, this process can be considerably faster due to the much higher value of $\eta_{\rm t}$ compared to η. Turbulent reconnection has recently been studied numerically by Otmianowska-Mazur, Kowal & Lazarian (this volume).

According to the definition of $\tau_{\rm d}$, in the laminar case one expects the dynamo growth rate to increase with η. The picture, however, changes significantly as soon as the Reynolds number is high enough to allow for developed turbulence. Structures of integral length scale L are now efficiently broken down to the Kolmogorov microscale where the atomic diffusion takes over. Varying the microscopic value of η does only change the extent of the inertial range towards higher wavenumbers but is no longer relevant to the large-scale flow. This implies that a turbulent dynamo should be insensitive to variations in η as soon as a critical value $\rm Rm_c$ is exceeded (cf. Lazarian & Vishniac 1999).

In Fig. 1, we plot the evolution of the mean and turbulent magnetic field (normalized to the kinetic energy) for different magnetic Reynolds numbers $\rm Rm = L^2\, q\Omega/\eta$. The values for Rm are obtained by varying η while keeping the rotation rate Ω fixed. The kinematic viscosity ν is furthermore adopted to keep the magnetic Prandtl number $\rm Pm = \nu/\eta = 2.5$ constant. Irrespective of the value of Rm, we observe a nice and steady exponential amplification of both the regular and turbulent fields. To estimate the influence of the finite grid resolution, we have performed a fiducial run (dark grey lines in panels (a) and (b) of Fig. 1) at double the grid spacing for $\rm Rm = 10,000$. The obtained values are consistent with the higher resolved run and provide a first indication that the simulation results are reasonably converged at this level of dissipation and below.

In panel (c) of Fig. 1, we compare the growth rates for the turbulent (diamonds) and regular (triangles) magnetic field as a function of the magnetic Reynolds number. The unconnected data points to the right correspond to a run with $\eta = 0$, i.e., the case of (formally) infinite Rm. As can be seen from a comparison with these points, above $\rm Rm \simeq 10,000$ we are limited by the finite value of the inherent numerical diffusivity of our code, i.e., better resolved runs become mandatory to study the regime of higher Rm. With respect to the reliably converged runs, we observe growth rates that increase with Rm – suggesting that the effect is of genuinely turbulent nature.

In their models of the cosmic-ray-driven buoyant instability, Otmianowska-Mazur *et al.* (2007) observe that their dynamo crucially relies on the presence of a "microscopic"

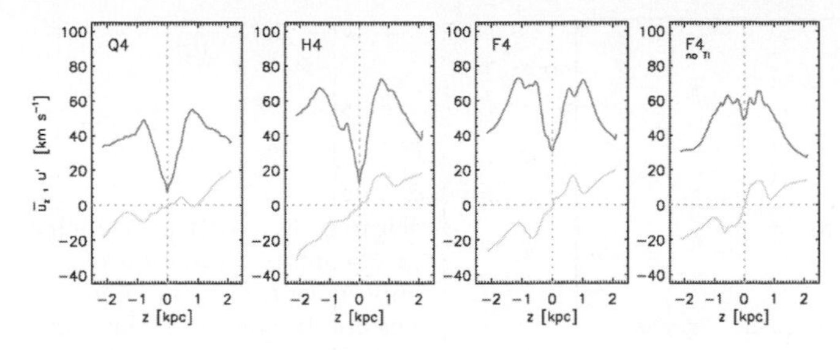

Figure 2. Vertical profiles of the mean flow $\bar{u}_z$ (light) and the turbulent velocity u' (dark colour) at quarter, half, and full supernova rate σ/σ_0.

diffusivity η. The authors actually find the efficiency of the field amplification to scale with this parameter. Moreover, they state that no scale separation is manifest in the spectra of their simulations. In this respect, it remains disputable whether the high value for η might rather be interpreted as an effective turbulent dissipation η_t. If so, the simulations would have to be regarded as large eddy simulations, i.e., simulations assuming turbulent effects rather than having them emerge from first principles.

2. The dependence on supernova rate

Quasilinear theory (Krause & Rädler 1980) is a powerful tool in predicting the dynamo α effect from the underlying kinetic structure of the turbulent flow. Unfortunately, as far as galaxies are concerned, little has been known with certainty about the vertical profiles of the turbulent velocity $u'(z)$ and the mean flow $\bar{u}_z(z)$. In particular, $\bar{u}_z$ had to be ignored in analytical derivations (see e.g. Fröhlich & Schultz 1996) assuming hydrostatic equilibrium. In that regard, direct simulations can give important new insights, especially if one is interested in the dependence on factors like the supernova frequency.

2.1. *Vertical structure & wind*

In the following, we measure the supernova rate σ in units of the galactic value $\sigma_0 = 30\,\mathrm{Myr}^{-1}\,\mathrm{kpc}^{-2}$, representative of type-II SNe. All models include sheared galactic rotation with $|q\Omega| = 100\,\mathrm{km\,s}^{-1}\,\mathrm{kpc}^{-1}$. The time averaged turbulent and mean velocity profiles of three models with $\sigma/\sigma_0 = 0.25$, 0.5, and 1.0 are depicted in Fig. 2: The vertical structure of the turbulent velocity u' shows a distinct M-shape, which peaks at ± 1 kpc. The positive gradient of u' in the central disk strongly suggests an inward transport of the mean magnetic field. The inner part of the profiles is similarly shaped as the ones obtained from MRI turbulence (Dziourkevitch *et al.* 2004; Piontek & Ostriker 2007) but considerably steeper. Crudely extrapolating the fall-off in $u'(z)$, we estimate that the MRI might become important in maintaining the observed velocity dispersions above galactic heights of $|z| \simeq 3$ kpc.

While the overall amplitude of the turbulence increases with the SN intensity, its gradient is only weakly affected. This can be seen in Fig. 3, where we plot the scaling of the fitted velocity gradients. Unlike the turbulent velocity gradient $\mathrm{d}u'/\mathrm{d}z$, which goes into saturation for $\sigma \simeq 0.1\,\sigma_0$, the wind profile shows a distinct scaling with the supernova frequency. We thus estimate the wind profile from kinetic feedback as

$$\bar{u}_z(z) \simeq 15.\,\mathrm{km\,s}^{-1} \left(\frac{\sigma}{\sigma_0}\right)^{0.4} \frac{z}{1.\,\mathrm{kpc}}, \tag{2.1}$$

which, of course, neglects the characteristic modulation of the mean flow within the V-shaped region of $u'(z)$, where the kinetic pressure counteracts the thermal pressure.

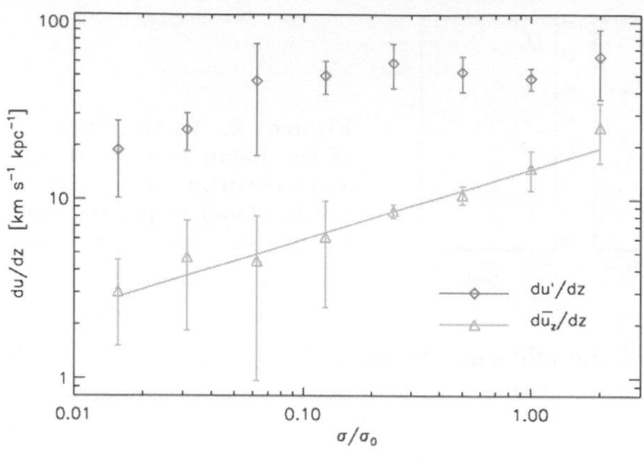

Figure 3. Vertical velocity gradients as a function of the supernova rate σ/σ_0. The values are derived from the inner disk for u', and from the full domain for $\bar{u}_z$ (cf. Fig. 2); the overplotted regression shows a logarithmic slope of 0.4.

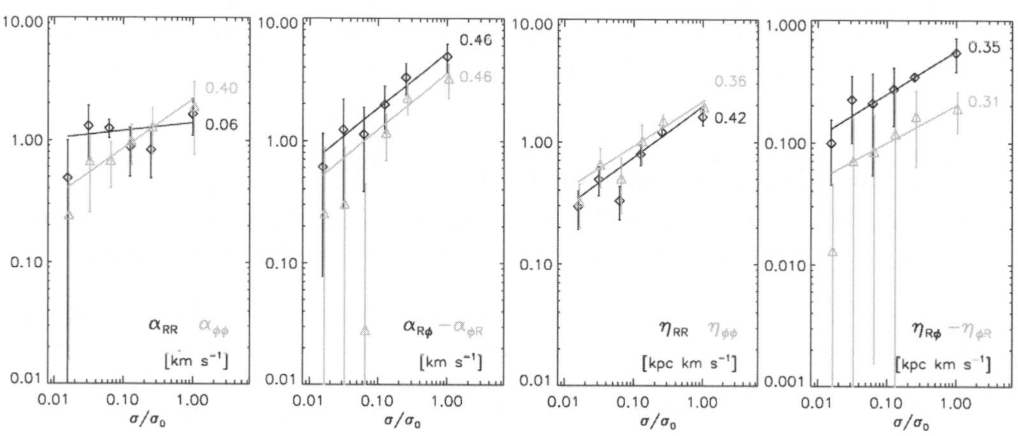

Figure 4. Coefficients of the dynamo α and $\tilde{\eta}$ tensor as obtained with the test-field method.

2.2. *Dynamo tensors & growth rates*

One main focus of our work is the derivation of mean-field closure parameters from direct simulations. We here assume a standard parameterisation of the turbulent EMF

$$\mathcal{E}_i = \alpha_{ij}\bar{B}_j - \tilde{\eta}_{ij}\varepsilon_{jkl}\partial_k\bar{B}_l\,, \quad i,j \in \{R,\phi\}\,, k=z\,, \tag{2.2}$$

which enters the mean-field induction equation via $\nabla\times\mathcal{E}$. To measure these coefficients, we apply the test-field method of Schrinner *et al.* (2005, 2007). From the thus derived profiles $\alpha(z)$ and $\tilde{\eta}(z)$, we compute vertically averaged amplitudes. In Fig. 4, we plot these integral mean values as a function of the supernova rate σ. Notably, for each of the data points we had to perform a separate 3D simulation, covering a few hundred million years to obtain reasonable statistics. Despite the strong scatter in the plotted values, we observe a quite robust scaling with respect to σ. An exception to this are the coefficients α_{RR}, which display irregular scaling in some of the runs and will be subject to further investigations.

The main result from Fig. 4 is that all coefficients (except α_{RR}) scale in a similar way. In particular, this implies that the dynamo number

$$C_\alpha = \alpha_{\phi\phi}H/\eta_t\,, \tag{2.3}$$

remains approximately constant with σ. Furthermore, as we have already demonstrated in an earlier work, the effects of the off-diagonal elements $\alpha_{R\phi}$ and $\alpha_{\phi R}$ (which are responsible for the diamagnetic pumping) are approximately balanced by the mean flow

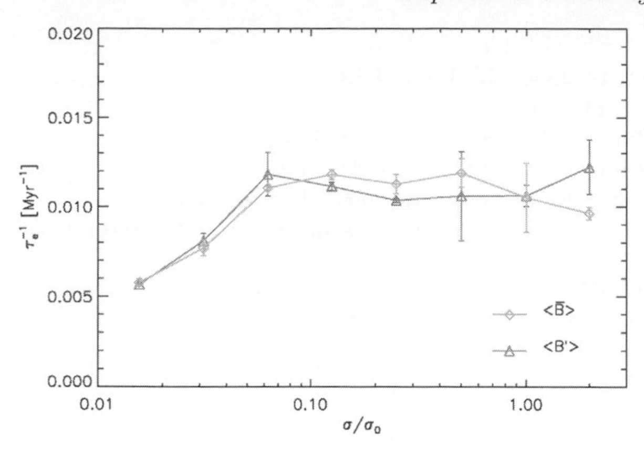

Figure 5. Growth rate τ_e^{-1} of the (vertically integrated) regular field $\langle \bar{B} \rangle$ (diamonds) and the turbulent field $\langle B' \rangle$ (triangles) as a function of σ/σ_0.

$\bar{u}_z$. Comparing the slope shown in Fig. 3 with the ones in the second panel of Fig 4, we see that this balance of forces is roughly independent of σ and therefore seems to be of rather general nature.

With the constant ratio of the α effect versus the turbulent diffusion η_t on one hand, and the mutual balance of the turbulent pumping and mean flow on the other hand, we expect the overall growth rates to be largely insensitive to the applied supernova rate. For sufficiently high values of σ this is indeed the case, as can be seen from Fig. 5 above – we want to point out that these growth rates are obtained from the direct simulations and hence do not depend on the α prescription. With constant growth rates over more than one order of magnitude in star formation activity, the kinetically driven SN dynamo proves to be rather universal. Accordingly, it may well be applicable to a wide range of scenarios – covering dwarf galaxies as well as starbursts.

3. Conclusions

We have presented recent simulation results on the variation of the key dynamo parameters with the supernova frequency. Yet more work is needed to understand how the vertical structure of the galactic disk is linked to the emergence of the dynamo effect and the associated transport processes. As a primary question, one is of course interested in finding a simple explanation for the observed scaling relations – which are somewhat flatter than the square-root of the supernova rate σ.

From quasilinear theory, it is understood that the dynamo effect depends on vertical gradients both in the density ρ and the turbulence intensity u'. The disk structure, on the other hand, is determined by the balance of the kinetic pressure from the forcing and the gravitation of the stellar component – here equipartition arguments can yield a first approximation. To connect kinematic properties with dynamo parameters, we however need to specify the correlation time τ_c of the turbulent flow. By comparing direct simulations with SOCA predictions, it will eventually become possible to determine how this quantity depends on the supernova frequency.

To connect the measured dynamo parameters α and $\tilde{\eta}$ with the growth rates observed in the simulations – and thereby check the applicability of the approach – more mean-field modelling is necessary. We are, however, confident that the classical framework of MF-MHD remains a valuable tool in the quest of understanding galactic magnetism.

Acknowledgements

This project was kindly supported by the Deutsche Forschungsgemeinschaft (DFG) under grant Zi-717/2-2. All computations were performed at the AIP `babel` cluster.

References

Dziourkevitch, N., Elstner, D., & Rüdiger, G. 2004, *A&A* 423, L29
Fröhlich, H.-E. & Schultz, M. 1996, *A&A* 311, 451
Gressel, O., Ziegler, U., Elstner, D., & Rüdiger, G. 2008a, *AN* 329, 619
Gressel, O., Elstner, D., Ziegler, U., & Rüdiger, G. 2008b, *A&A* 486, L35
Hanasz, M., Kowal, G., Otmianowska-Mazur, K., & Lesch, H. 2004, *ApJ* 605, L33
Krause, F. & Rädler, K. H. 1980, *Mean-field magnetohydrodynamics and dynamo theory* (Oxford: Pergamon Press)
Lazarian, A. & Vishniac, E. T. 1999, *ApJ* 517, 700
Otmianowska-Mazur, K., Kowal, G., & Hanasz, M. 2007, *ApJ* 668, 110
Piontek, R. A. & Ostriker, E. C. 2007, *ApJ* 663, 183
Schrinner, M., Rädler, K.-H., Schmitt, D., Rheinhardt, M., & Christensen, U. 2005, *AN* 326, 245
Schrinner, M., Rädler, K.-H., Schmitt, D., Rheinhardt, M., & Christensen, U. R. 2007, *GApFD* 101, 81

Discussion

JOHANSEN: Do you see any signs of Parker instability in the azimuthal fields?

GRESSEL: Typically, our simulations are seeded with dynamically weak magnetic fields ($\beta_{\rm P} = 2 \times 10^7$). If we, however, start with stronger fields and approach equipartition strengths, we clearly see buoyancy effects.

DE GOUVEIA DAL PINO: 1) You have obtained the appropriate **B** fields through a SN-driven dynamo but what were the coherence scales? 2) How could you compare your 3D MHD galaxy SN-wind model with those by Avillez and colleagues?

GRESSEL: 1) From the inferred velocity structure functions, we derive a coherence-length of 80–100 pc, which coincides with the outer scale of SNRs but is considerably smaller than the typical sizes of super-bubbles. The scale of the observed mean field is the scale-height of the inner thick disk, which is of the same order. 2) The models are very similar. However, we include galactic rotation and shear, which are not present in their models and are a mandatory prerequisite for the excitation of a mean-field dynamo.

BRANDENBURG: I agree with you that shear also produced an α effect, but it has the opposite effect than the effect caused by rotation; although it is not as strong. What seems to matter is $2\Omega + \overline{\mathbf{W}}$, with $\overline{\mathbf{W}} = \nabla \times \overline{\mathbf{U}}$ is the vorticity of the shear flow. As you say, the Parker instability becomes important for strong magnetic fields, and so does the MRI. In that case α really reverses sign and should be proportional to $\overline{b_x b_y}$. Would you agree?

GRESSEL: The α effect caused by shear is only *somewhat* weaker than the effect caused by the rotation. Otherwise I agree with your statement.

Cosmic Magnetic Fields:
From Planets, to Stars and Galaxies
Proceedings IAU Symposium No. 259, 2008
K.G. Strassmeier, A.G. Kosovichev & J.E. Beckman, eds.

doi:10.1017/S1743921309030105

Dynamically dominant magnetic fields in the diffuse interstellar medium

Andrew Fletcher[1], M. Korpi[2] and A. Shukurov[1]

[1]School of Mathematics and Statistics, Newcastle University, NE1 7RU, UK
email: andrew.fletcher@ncl.ac.uk, anvar.shukurov@ncl.ac.uk

[2]Observatory, Tähtitorninmäki (PO Box 14), FI-00014 University of Helsinki, Finland
email: maarit.korpi@helsinki.fi

Abstract. Observations show that magnetic fields in the interstellar medium (ISM) often do not respond to increases in gas density as would be naively expected for a frozen-in field. This may suggest that the magnetic field in the diffuse gas becomes detached from dense clouds as they form. We have investigated this possibility using theoretical estimates, a simple magneto-hydrodynamic model of a flow without mass conservation and numerical simulations of a thermally unstable flow. Our results show that significant magnetic flux can be shed from dense clouds as they form in the diffuse ISM, leaving behind a magnetically dominated diffuse gas.

Keywords. ISM: clouds – ISM: magnetic fields – MHD

---

1. The problem: regular magnetic field resists strong shocks

Magnetic fields in the interstellar medium of the Milky Way and other disc galaxies are, to a first approximation, frozen-in to the interstellar gas: put another way, the magnetic Reynolds number of the medium is sufficiently high that advection dominates diffusion in the magnetic induction equation. One would therefore expect that strong velocity shear, such as that which occurs in barred galaxies, and large-scale shocks, like the density wave induced shocks along spiral arms, would lead to a strengthening of the magnetic field when the large-scale magnetic field is approximately perpendicular to the shear or parallel to the shock, as is commonly observed in barred (Beck, Shoutenkov, Ehle *et al.* 2002) and normal spiral galaxies (Beck, Brandenberg, Moss *et al.* 1996).

Surprisingly, observations of the barred galaxies NGC 1097 and NGC 1365 (Beck, Fletcher, Shukurov *et al.* 2005) show that the polarized radio emission, tracing the *regular* magnetic field, is hardly affected by the velocity shear of about $200\,\mathrm{km\,s^{-1}\,kpc^{-1}}$, increasing by a factor of 1–7 whereas theoretically one would expect an increase by a factor of about 60. It is important to note that similar calculations are able to accurately predict the observed increase in the *total* radio emission, as a result of compression and shear of the turbulent magnetic field in the bar. In normal spiral galaxies a similar, but less pronounced, discrepancy is observed. For example in M51 the neutral gas density increases on average by a factor of 4 in the spiral arms, which should produce an increase by a factor of at least 16 in polarized emission if the magnetic field were frozen-in and depolarization remains constant, whereas on average there is no observed increase in polarized emission in the spiral arms (Fletcher *et al.* in prep.).

2. The solution: dense clouds detach from the regular magnetic field

We suggest that the weaker than expected increase in regular magnetic field strength in regions of strong shocks and shear is due to the multi-phase nature of the ISM. The

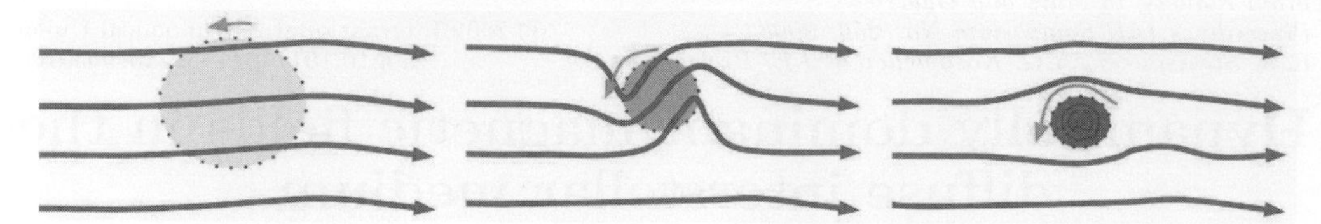

Figure 1. Cartoon showing three stages in the condensation of a dense cloud in a diffuse medium threaded by a large-scale magnetic field. As the cloud collapses its rotation velocity increases.

regions in bars and spiral arms where the ISM density increases are also the regions where dense clouds of molecular gas ($n \gtrsim 100\,\mathrm{cm}^{-3}$) are rapidly formed from diffuse interstellar gas ($n \lesssim 1\,\mathrm{cm}^{-3}$). As clouds collapse in the turbulent flow, they will rotate faster, winding up the magnetic field lines that thread the cloud. The clouds can then become detached from the background magnetic field in the diffuse ISM by flux expulsion (Weiss 1966) via ambipolar diffusion (Mestel & Paris 1984) or magnetic reconnection. The cartoon in Figure 1 shows the basic principle.

The number of rotations a cloud makes can be bracketed by the following limits. Assuming angular momentum conservation and turbulent flow, an initial density of $1\,\mathrm{cm}^{-3}$ in a proto-cloud of radius $r = 10\,\mathrm{pc}$ rotating at $v = v_0(r/l_0)^{1/3} \approx 5\,\mathrm{km\,s}^{-1}$, where $v_0 = 10\,\mathrm{km\,s}^{-1}$ and $l_0 = 100\,\mathrm{pc}$ are typical values for the largest turbulent eddies, a cloud will rotate about 40 times in a typical lifetime of 3-5 Myr or 100 times in a formation time of 12-20 Myr. On the other hand, observations of *magnetically braked* mature clouds suggest about 2 rotations in the formation time (Bodenheimer 1995). These estimates suggest that sufficient rotations for magnetic reconnection to occur are possible and the model is worth examining in more detail.

We have carried out numerical simulations, in 2D and 3D, of clouds forming via the thermal instability in the diffuse, magnetized ISM. We measure the fraction of magnetic flux threading gas at different densities once the instability has fully developed (after approximately 60 Myr). The simulations show that magnetic flux is detached from the forming clouds, with the rate of detachment dependent on the magnetic diffusivity. Furthermore, when the perturbations that trigger cloud formation have non-zero rotation the critical density at which flux detachment becomes important scales inversely with the rotation rate.

If the dense clouds become detached from the regular magnetic field in the diffuse ISM, then the mass-to-flux ratio will decrease for the diffuse gas. Thus the regular magnetic field becomes more important in the dynamics of the diffuse gas and eventually will be strong enough to resist shearing and shocks in the diffuse ISM. This dynamical importance will continue until the dense clouds are dissipated (for example by star formation) and their remaining gas reloads the regular magnetic field in the diffuse ISM.

References

Beck, R., Brandenberg, A., Moss, D., Shukurov, A., & Sokoloff, D. 1996, *ARA&A* 34, 155

Beck, R., Shoutenkov, V., Ehle, M., Harnett, J. I., Haynes, R. F., Shukurov, A., Sokoloff, D., & Thierbach, M. 2005, *A&A* 444, 739

Beck, R., Fletcher, A., Shukurov, A., Snodin, A., Sokoloff, D., Ehle, M., Moss, D., & Shoutenkov, V. 2005, *A&A* 444, 739

Bodenheimer, P. 1995 *ARA&A* 33, 199

Mestel, L. & Paris, R. B. 1984 *A&A* 136, 98

Weiss, N. 1966, *Proc. Roy. Soc. A.* 293, 310

Cosmic Magnetic Fields:
From Planets, to Stars and Galaxies
Proceedings IAU Symposium No. 259, 2008
K.G. Strassmeier, A.G. Kosovichev & J.E. Beckman, eds.

doi:10.1017/S1743921309030117

GMIMS: the Global Magneto-Ionic Medium Survey

Maik Wolleben[1,12], T. L. Landecker[1], E. Carretti[2], J. M. Dickey[3], A. Fletcher[4], B. M. Gaensler[5], J. L. Han[6], M. Haverkorn[7], J. P. Leahy[8], N. M. McClure-Griffiths[9], D. McConnell[9], W. Reich[10] and A. R. Taylor[11]

[1]NRC Herzberg Institute of Astrophysics, DRAO, Penticton, BC, V2A6J9, Canada
[2]INAF Istituto di Radioastronomia, Via Gobetti 101, 40129 Bologna, Italy
[3]Physics Department, University of Tasmania, Hobart, TAS 7001, Australia
[4]School of Mathematics and Statistics, Newcastle University, NE1 7RU, U.K.
[5]Institute of Astronomy, School of Physics, The University of Sydney, NSW 2006, Australia
[6]National Astronomical Observatories, Chinese Academy of Sciences, Beijing 100012, China
[7]Jansky Fellow, National Radio Astronomy Observatory, Charlottesville, VA 22903, USA
[8]School of Physics & Astronomy, The University of Manchester, Cheshire SK11 9DL, UK
[9]Australia Telescope National Facility, CSIRO, PO Box 76, Epping, NSW 2121, Australia
[10]Max-Planck-Institut für Radioastronomie, Auf dem Hügel 69, 53121 Bonn, Germany
[11]Centre for Radio Astronomy, University of Calgary, Calgary, AB T2N 1N4, Canada
[12]Covington Fellow

Abstract. The Global Magneto-Ionic Medium Survey (GMIMS) is a project to map the diffuse polarized emission over the entire sky, Northern and Southern hemispheres, from 300 MHz to 1.8 GHz. With an angular resolution of 30–60 arcmin and a frequency resolution of 1 MHz or better, GMIMS will provide the first spectro-polarimetric data set of the large-scale polarized emission over the entire sky, observed with single-dish telescopes. GMIMS will provide an invaluable resource for studies of the magneto-ionic medium of the Galaxy in the local disk, halo, and its transition.

Keywords. ISM: magnetic fields – ISM: structure – polarization

1. Motivation

Polarization of the diffuse synchrotron emission from the Galaxy provides a sensitive probe of the magnetic fields and warm and hot plasmas making up the Galactic magneto-ionic medium (MIM). Polarization maps at decimeter wavelengths thus reveal intriguing rotation measure structures that are otherwise undetectable. These structures contain information about the Galactic magnetic field and the distribution of the ionized gas. In short, the physics of the MIM is imprinted on the diffuse polarized emission. So far, single-frequency (or narrow-band) polarization surveys have only allowed rather qualitative studies due to ambiguities in the data. Wide-band polarization data, as they are now underway through various survey projects, can resolve these ambiguities and will allow quantitative analysis of the diffuse polarized emission of the Galaxy.

Figure 1 summarizes our current knowledge of the northern polarized sky at 408 MHz (left) and 1.4 GHz (right). The polarization vectors at 408 MHz were measured more than 30 years ago. Similar data at 1.4 GHz were only recently superseded by the new DRAO Low-Resolution Polarization Survey. Data with higher angular resolution have been observed by, for example, the EMLS (Reich *et al.* 2004) and CGPS (Landecker *et al.*, in prep.) in the northern sky, as well as SGPS (Haverkorn *et al.* 2006) and S-PASS

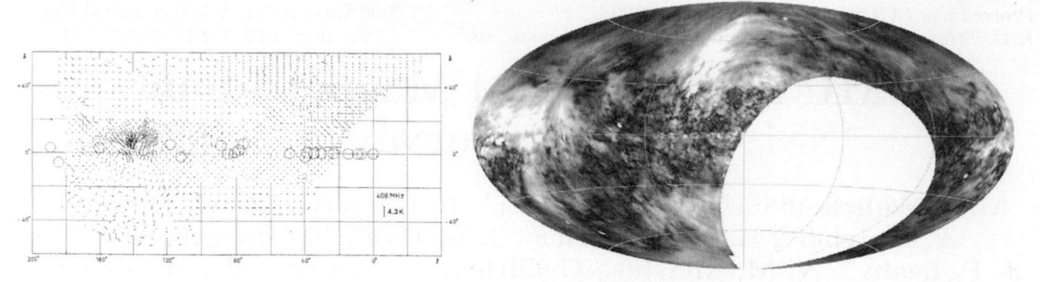

Figure 1. Polarization maps of the northern sky in Galactic coordinates at 408 MHz (left, from Brouw & Spoelstra 1976) and 1.4 GHz (right, from Wolleben *et al.* 2006). See Reich & Reich (this issue) for an all-sky version of the 1.4 GHz survey.

(Carretti *et al.* 2008) on the southern sky. GMIMS aims at superseding and complementing these data through its frequency coverage (or, more importantly for rotation measure studies, through its λ^2-coverage).

2. Strategy

The goal of GMIMS is to provide absolutely calibrated Stokes I, U, and Q data cubes of the entire sky. The intended frequency band (300 MHz to 1.8 GHz) has been split into Low-, Mid-, and High-Band. GMIMS thus comprises six *component surveys*, three in the North and three in the South. The first of these, the High-Band North (1.3 - 1.8 GHz), uses the DRAO 26-m Telescope and is now 25% complete. Its southern counterpart is the STAPS project (PI M. Haverkorn), currently being carried out with the Parkes telescope. A very successful pilot study for the Low-Band South (300 - 900 MH) survey has recently been completed with the Parkes telescope. The other component surveys are in the planning stages.

3. From 2-D to 3-D Polarimetry

GMIMS is made possible by new developments and technologies in radio astronomy: new wide-band feeds, wide-band digital polarimeters, and rotation measure synthesis (RMS); all of which only recently became available to polarization science. The design of wide-band antennas benefits from modern EM-simulation software and is motivated by the SKA; field programmable gate arrays (FPGA) are now powerful enough to provide the basis for digital wide-band polarimeters; and Brentjens & de Bruyn (2005) demonstrated the potential of RMS in analyzing wide-band polarization data.

References

Brentjens, M. A. & de Bruyn, A. G. 2005, *A&A* 441, 1217
Brouw, W. N. & Spoelstra, T. A. T. 1976, *A&AS* 26, 129
Carretti, E., Haverkorn, M., McConnell, D., Bernardi, G., Cortiglioni, S., McClure-Griffiths, N. M., & Poppi, S. 2008, arXiv:0806.0572
Haverkorn, M., Gaensler, B. M., McClure-Griffiths, N. M., Dickey, J. M., & Green, A. J. 2006, *ApJS* 167, 230
Reich, P., Reich, W., & Testori, J. C. 2004 in B. Uyanker, W. Reich & R. Wielebinski (eds.), The Magnetized Interstellar Medium, (Katlenburg-Lindau: Copernicus GmbH), p. 63
Wolleben, M., Landecker, T. L., Reich, W., & Wielebinski, R. 2006, *A&A* 448, 411

Cosmic Magnetic Fields:
from Planets, to Stars and Galaxies
Proceedings IAU Symposium No. 259, 2008
K.G. Strassmeier, A.G. Kosovichev & J.E. Beckman, eds.

doi:10.1017/S1743921309030129

The role of the random magnetic fields in the ISM: HVC numerical simulations

Alfredo Santillán[1]†, J. Kim[2], F. J. Sánchez–Salcedo[3], J. Franco[3] and L. Hernández–Cervantes[3]

[1]Dirección General de Servicios de Cómputo Académico, UNAM, 04510, Mexico City, Mexico
email: alfredo@astrosu.unam.mx

[2]Korea Astronomy and Space Science Institute, 61–1, Hwaam–dong, Yuseong–gu, Daejeon, Republic of Korea 305-348

[3]Instituto de Astronomía , UNAM, 04510, Mexico City, Mexico

Abstract. We know that the galactic magnetic field possesses a random component in addition to the mean uniform component, with comparable strength of the two components. This random component is considered to play important roles in the evolution of the interstellar medium (ISM). In this work we present numerical simulations associated with the interaction of the supersonic flows located at high latitude in our Galaxy (High Velocity Clouds, HVC) with the magnetized galactic ISM in order to study the effect that produces a random magnetic field in the evolution of this objects.

Keywords. Magnetic fields – ISM: clouds – ISM: magnetic fields – ISM: structure

1. Introduction

Numerical simulations of the evolution of HVC collisions with the Milky Way have been performed for more than two decades by different authors. The details of resulting supersonic flows depend of on the model assumptions, and the intensity and initial configuration of the magnetic field. Santillán *et al.* (1999) made models that illustrate the effects of magnetic pressure, and differentiate them from those due to magnetic tension. The evolution of the interaction of a HVC with a magnetized interstellar medium, is studied by setting a random magnetic field that satisfies the divergence–free constraint ($\nabla.\mathbf{B}=0$) at all times. To mimic the average galactic magnetic field, that is oriented parallel to the disk, the horizontal component dominates over the vertical component.

2. Results

We perform simulations of HVCs interacting with the interstellar medium without and with magnetic field, ISM–1 and ISM–2, respectively, using the MHD code ZEUS–3D (Stone & Norman 1992a,b). The ISM–models are plane parallel and have constant density and temperature, $n = 1$ cm^{-3} and $T = 1000$ K. The galactic gas is initially at rest. For ISM–2 case the total intensity of the magnetic field is 2 μG. The enter position of the HVC is located 3 kpc above from the midplane and has a velocity of 100 km/s. In the figure 1, the density is shown in logarithmic gray–scale plots and the magnetic field is indicated by arrows.

The early times evolution for both models display the same characteristics, e.g. the interaction between the HVC and the halo gas creates a strong galactic shock directed

† Present address: Unidad de Investigación en Cómputo Aplicado–DGSCA, UNAM, 04510, Mexico City, Mexico

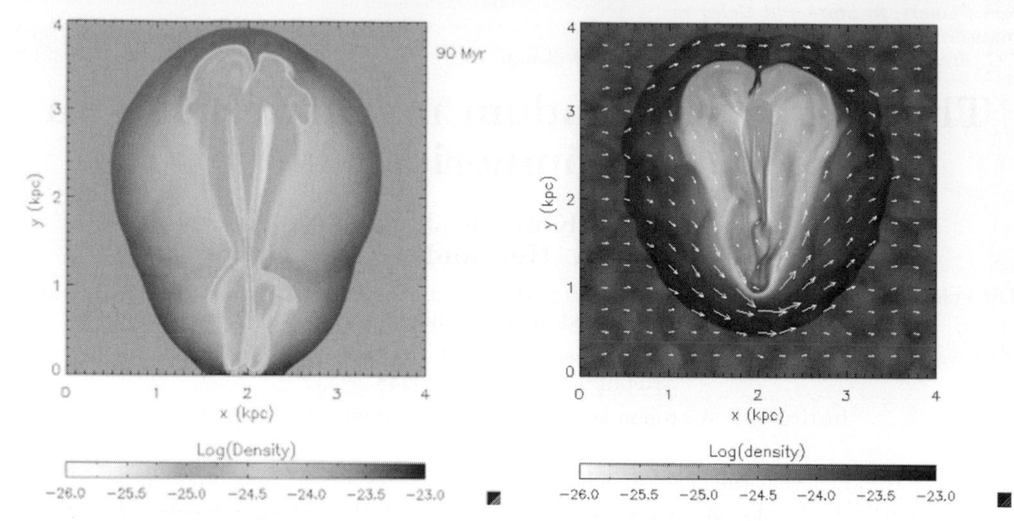

Figure 1. Evolution of HVC in two medium: without and with magnetic field. The figure shows the density (*gray logarithmic scale*) and the magnetic field indicated by *arrows* at 90 Myr.

downwards, along with reverse shock that penetrates into the cloud. The galactic shock tends to move radially away from the location of impact (Santillán *et al.* 1999). As time proceeds, the structure of the HVC changed slightly; in both models a large fraction of the original cloud mass remains locked up in the shocked layer, and a small amount of it re–expanded back into the rear wake and tail. However, in the HD–case, the clouds moves a greater distance than the MHD–case, due to the effect of the magnetic tension of the horizontal–component of random magnetic field. In the magnetic case, the HVC distorts and compresses the B–field lines during the evolution, increasing both the field pressure and the tension, and forming a magnetic barrier for the moving gas. Finally, as seen in figure 1, the late times structure produced by hydrodynamic simulations is completely different from that of its magnetic counterpart. In the case non–magnetic case, ISM–1, the interaction of HVC with ambient medium creates a thin structure and the size of perturbation region at 90 Myr has grown to nearly 3 kpc. On the other hand, for the magnetic case, the evolution of the cloud at the same evolutionary time creates a thick structure of ∼2 kpc and a great amount of gas collects in a magnetic valley formed by the interaction with the horizontal–component of the random magnetic field.

The numerical calculations were performed using UNAM's supercomputers.

Acknowledgements

This work has been partially supported from DGAPA-UNAM grant IN104306 and CONACyT proyect CB2006–60526.

References

Santillán A., Franco, J., Martos, M., & Kim, J. 1999, *ApJ* 515, 657
Stone, J. M. & Norman, M. L. 1992a, *ApJS* 80, 753
Stone, J. M. & Norman, M. L. 1992b, *ApJS* 80, 791

Cosmic Magnetic Fields:
From Planets, to Stars and Galaxies
Proceedings IAU Symposium No. 259, 2008
K.G. Strassmeier, A.G. Kosovichev & J.E. Beckman, eds.

doi:10.1017/S1743921309030130

Relative distributions of cosmic ray electrons and magnetic fields in the ISM

Rodion Stepanov[1], A. Fletcher[2], A. Shukurov[2], R. Beck[3], L. La Porta[3], and F. S. Tabatabaei[3]

[1]Institute of Continuous Media Mechanics, 1 Korolyov St., Perm 614013, Russia
[2]School of Mathematics and Statistics, Newcastle University, NE1 7RU, UK
[3]Max-Planck-Institut für Radioastronomie, Auf dem Hügel 69, 53121 Bonn, Germany

Abstract. We calculate the relative magnitudes of the fluctuations in total synchrotron intensity in the interstellar medium, both from observations and from theory under various assumptions about the correlation or anticorrelation between cosmic rays and interstellar magnetic fields. The results are inconsistent with local energy equipartition between cosmic rays and magnetic fields. The distribution of cosmic rays must be rather uniform at scales of order 1 kpc, whereas interstellar magnetic fields vary at much smaller scales.

Keywords. Cosmic rays – magnetic fields – radio continuum

The concept of energy equipartition between cosmic rays and magnetic fields is often used in the analysis of radio astronomical observations (Beck & Krause 2005). There is no sound basis for this assumption, but as cosmic rays are confined by magnetic fields, it seems natural to expect some relation between them, though this does not emerge from cosmic ray propagation models (e.g., Padoan & Scalo 2005). Here we test the equipartition hypothesis using models of the non-thermal interstellar medium (ISM) and radio observations. We show that equipartition is inconsistent with observations at scales of order 0.1 kpc (assuming that cosmic ray protons and electron are similarly distributed).

We used radio maps of the Milky Way and the nearby galaxy M 33 to calculate the standard deviation and mean value of the total synchrotron intensity in a range of scales. The data used are: (i) the 408 MHz all-sky survey of Haslam *et al.* (1982) at the resolution of $50'$ (upper left panel in Fig. 1), (ii) the 22 MHz survey of Roger *et al.* (1999) at a resolution of approximately $1^\circ \times 2^\circ$, and (iii) the synchrotron 1.4 GHz map of M 33 (Tabatabaei *et al.* 2007) at a resolution of $90''$ shown in Fig. 1 (right-hand panel).

At a distance of 1 kpc, the angular diameter of a turbulent cell of 0.1 kpc in size is 6°. Thus, the nearest turbulent cells are resolved in both Milky Way maps used. At both 408 MHz and 22 MHz, the emission is predominantly synchrotron radiation. At the distance of M 33, the resolution of our data is 0.4 kpc. We considered eight rectangular regions which are free of bright sources (Fig. 1). Since the regions are several kpc across, we removed large-scale linear and quadratic trends in synchrotron intensity in each region.

The standard deviation σ_I and the mean value $\overline{I}$ of the synchrotron intensity were calculated for each map using sliding averaging with a Gaussian kernel of angular radius a. For random fluctuations, the ratio $\sigma_I/\overline{I}$ is independent of a for $a \gg a_0$, where a_0, the angular radius of a turbulent cell, corresponds to 0.05 kpc. This is indeed the case and, remarkably, the relative fluctuations of the synchrotron intensity have rather similar magnitudes in all the data sets considered, $\sigma_I/\overline{I} \lesssim 0.2$. We believe that this value is not significantly affected by either regular trends in the radio intensity, nor by discrete radio

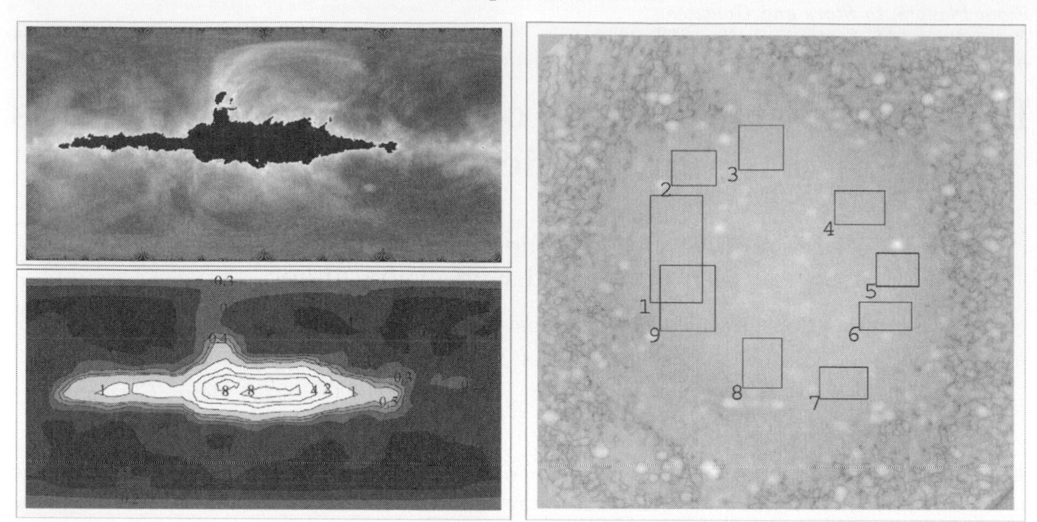

Figure 1. Upper left: the 408 MHz all-sky map with the strongest discrete sources removed by La Porta *et al.* (2008). Lower left: $\sigma_I/\overline{I}$ from the data smoothed at the scale (radius) $a = 7°$. Right panel: the synchrotron 1.4 GHz map of M 33 with the regions used in our analysis. Note the smoothness of the images out of the Milky Way plane and across M 33: here $\sigma_I/\overline{I} \lesssim 0.2$.

sources or thermal emission. Anyway, the above estimate is a firm *upper* limit of the level of fluctuations in synchrotron intensity arising in the ISM of the Milky Way and M 33.

We model the non-thermal ISM by specifying a divergence-free magnetic field $\mathbf{B}$, with a given power spectrum of isotropic fluctuations and a mean value B_0, in terms of a Fourier series with randomly chosen parameters. Cosmic ray electron density n (also represented by a mean value and fluctuations) is then similarly introduced, with a given cross-correlation coefficient C between n and B^2, with $|C| \leqslant 1$. Synthetic radio maps are then calculated with a range of C and a range of relative fluctuation amplitudes in $\mathbf{B}$ and n. The overall conclusion is that $\sigma_I/\overline{I}$ is as observed only if $-1 \lesssim C < 0$.

Thus, observations appear to rule out any positive correlation between magnetic fields and cosmic rays at scales of order 0.1 kpc. Interpretations that rely on this or similar assumptions seriously underestimate magnetic field fluctuations and overestimate those in cosmic rays. However, some form of equipartition may still be maintained at scales of 1 kpc or larger. We suggest an alternative approach to the interpretation of radio maps, where they are first smoothed to a linear resolution of about 1 kpc, and the large-scale distributions of magnetic fields and cosmic rays may be obtainable from the equipartition argument. The cosmic ray distribution thus obtained can be used together with the original synchrotron data to recover a full-resolution picture of magnetic fields.

This work was supported by the STFC grant F003080/1, the NSF Grant NSF PHY05-51164 and by grant RFBR-08-02-92881.

References

Beck, R. & Krause, M. 2005, *AN* 326, 414

Haslam, C. G. T., Salter, C. J., Stoffel, H., & Wilson, W. E. 1982, *A&AS* 47, 1

La Porta, L., Burigana, C., Reich, W., & Reich, P. 2008, *A&A* 479, 641

Padoan, P. & Scalo, J. 2005, *ApJ* 624, L97

Roger, R. S., Costain, C. H., Landecker, T. L., & Swerdlyk, C. M. 1999, *A&AS* 137, 7

Tabatabaei, F. S., Beck, R., Krause, M., *et al.* 2007, *A&A* 475, 133

Cosmic Magnetic Fields:
From Planets, to Stars and Galaxies
Proceedings IAU Symposium No. 259, 2008
K.G. Strassmeier, A.G. Kosovichev & J.E. Beckman, eds.

doi:10.1017/S1743921309030142

Scattered OH Maser sources in the direction of W49N

Jose E. Mendoza-Torres[1], W. M. Goss[2], S. Streb[3], A. A. Deshpande[4], and R. Ramachandran[5]

[1]Instituto Nacional de Astrofisica, Optica y Electronica, Mexico
[2]NRAO, Socorro, NM, USA
[3]Salpointe Catholic High School, Tucson, AZ, USA
[4]Raman Research Institute, Bangalore, India
[5]San Jose, CA, USA

Abstract. OH masers spots are observed with the VLBA at 1612, 1665 and 1667 MHz. The orientation of the ellipses resulting from scattering are all nearly aligned perpendicular to the galactic plane. These ordered orientation could be due to the galactic magnetic field.

Keywords. OH masers – W49N – interstellar scattering

1. Observations and results

Using VLBA observations of 2005 October 6 towards W49N at the 1612 MHz, 1665 MHz and 1667 MHz OH frequencies, 215 maser spots were detected (Fig.1). LCP and RCP are observed simultaneously in 240 spectral channels with a resolution of 0.1 km/sec in a range of about 22 km/sec. The beam size is about 20 mas by 15 mas at a position angle of 84 degrees. The galactic plane is at a position angle of 30 degrees. The magnetic field at the maser emitting region obtained based on Zeeman pairs gives values from -4 to 6 mGauss. All for a region of about one arcsecond square near the center of the field. The deconvolved major axis (mjax) of the spots are between 30 to 70 mas and the ratio of mjax to minor axis (mnax) range from about 1.5:1 to 3:1.

The center of expansion (CE) of Gwinn *et al.* (1992) for H_2O maser spots at W49N is denoted by + in Fig. 1. The velocities of the nearest spots are slightly higher or smaller than 9 km/s, a value near the middle of the entire velocity range that we use as reference in Fig. 1a). Red shifted spots predominantly appear at the right (as also observed by Kent & Mutel (1982)) and blue shifted spots are located at the left, slightly to the north respect the red shifted spots, as occurs for H_2O spots (Gwinn *et al.* 1992). However, the left hand side H_2O spots are red shifted while the OH of the same side are blue shifted and vice versa.

The principal axis (PA) takes values from about 80 to 130 degrees, i.e., even though the spots are spread more than $7{\times}10^4$ AU, their PAs are all nearly aligned. Most of the ellipses of the right group are oriented as if they would be traced from the CE but the ellipses of the left hand side group do not follow any given orientation respect the CE.

The larger PA values take place at the larger mnax (Fig. 2) and a similar correlation is observed between PA and mjax. However, there is no correlation between PA and the majax/minax ratio. This suggests that it is the size of the spots and not the ratio of their axes that correlates with PA.

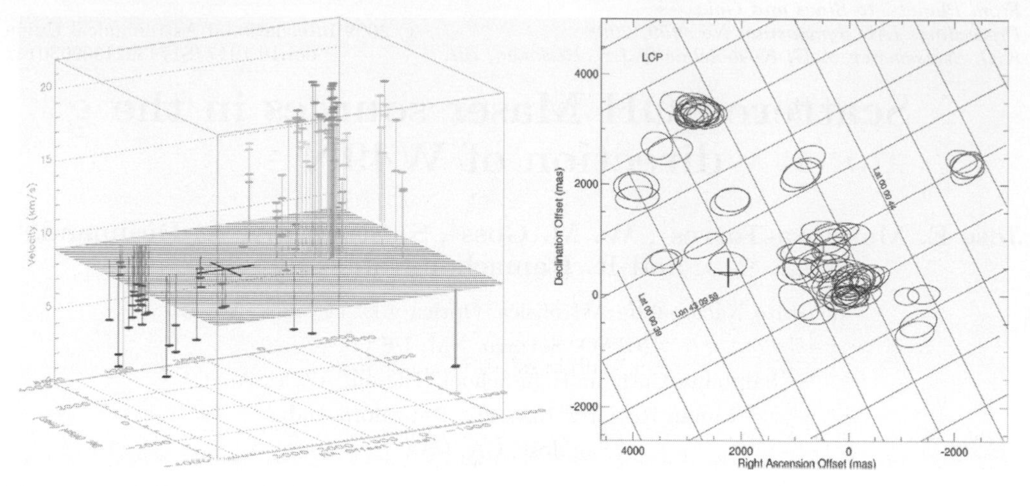

Figure 1. a) Scatter 3D plot for the spots at 1667 MHz, a plane around the central velocity is plotted as reference. b) Distribution of LCP 1667 MHz spots. The locations are denoted by ellipses multiplied by a factor of 15. A grid of lines of constant Galactic Longitude and constant Galactic Latitude are shown. The CE of Gwinn *et al.* (1992) is denoted by +.

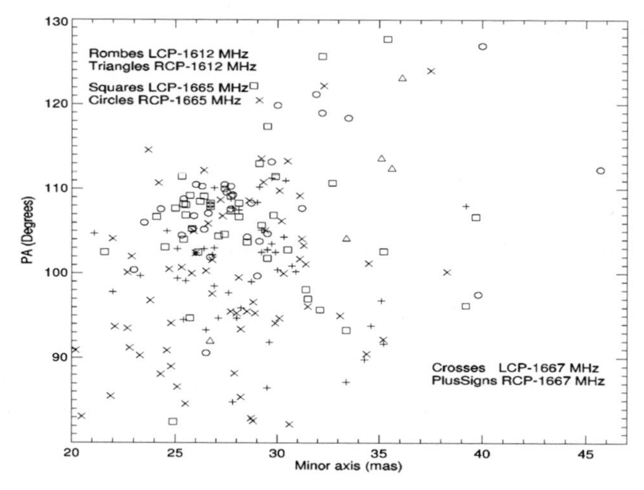

Figure 2. The scatter plot of the PA and the minor axis of all the spots.

2. Discussion

The observed size and orientation of the spots are not related to CE. Density inhomogeneities produced by turbulence are expected to be elongated along the galactic magnetic field (Desai *et al.* 1994) and produce ellipses whose apparent sizes are larger in the direction of the short axes (Narayan *et al.* 1988). The magnetic field could order the density inhomogeneities of the interstellar medium at such scales. In this scheme, the scattering of the inhomogeneities could determine the largest observed sizes.

References

Desai, K. M., Gwinn, C. R., & Diamond, P. J. 1994, *Nature* 372, 754
Gwinn, C. R., Moran, J. M., & Reid, M. 1992, *ApJ* 393, 149
Kent, S. R. & Mutel R. L. 1982, *ApJ* 263, 145
Narayan, R. & Hubbard, W. B. 1988, *ApJ* 325, 503

Cosmic Magnetic Fields:
From Planets, to Stars and Galaxies
Proceedings IAU Symposium No. 259, 2008
K.G. Strassmeier, A.G. Kosovichev & J.E. Beckman, eds.

doi:10.1017/S1743921309030154

Magnetic fields in massive star forming regions: wide-field NIR polarimetry of M 42 and Mon R2

Nobuhiko Kusakabe, Motohide Tamura, Ryo Kandori, and the IRSF/SIRPOL group

National Astronomical Observatory, 2-21-1 Osawa, Mitaka, Tokyo 181-8588
email: nb.kusakabe@nao.ac.jp

Abstract. Magnetic fields are believed to play an important role in star formation. We observed M42 and Mon R2 massive star forming regions using the wide-field ($8' \times 8'$) near-infrared imaging polarimeter SIRPOL in South Africa. Magnetic fields are mapped on the basis of dichroic polarized light from hundreds of young stars embedded in the regions. We found "hourglass shaped" magnetic field structure toward OMC-1 region, which is very consistent with magnetic fields traced by using dust emission polarimetry at sub-mm to FIR wavelengths. In the Mon R2 region, we found "S-shaped" magnetic field structure across the massive protostar IRS 1 and IRS 2. We will present the results of comparison of magnetic fields at NIR with those at other wavelengths.

Keywords. Stars – ISM: individual (M42, Mon R2) – polarization – stars: formation

1. Introduction

Wide-field polarimetry of star forming regions, particularly at near-infrared wavelengths, is a powerful tool to reveal their magnetic field structure through the measurements of dichroic extinction due to magnetically aligned dust grains. However, NIR imaging polarimetry has been conducted in relatively small regions (e.g., near IRc2 or BN - Simpson *et al.* 2006). We conducted wide-field NIR imaging polarimetry of M42 and Monoceros R2 core region using a near-infrared imaging polarimeter SIRPOL (Kandori *et al.* 2006) on the IRSF 1.4-m telescope in South Africa.

2. Magnetic Field in M42

We carried out software aperture polarimetry of 314 point sources on the combined intensity images for each wave plate angle (I_{0°, $I_{22.5^\circ}$, I_{45°, $I_{67.5^\circ}$) in M42. Polarization degrees P and angles θ were obtained by the Stokes parameters.

We found good consistency between H and 350μm vectors (see. fig 1 left). The average position angles show almost same angle ($\sim$ 125 degree). Both of these are, mainly, due to dichroic absorption in this field and suggest the magnetic field direction projected on the sky. Consequently, wide-field NIR aperture polarimetry can trace the magnetic field.

3. Magnetic Field in Mon R2

As we discussed M42, we carried out software aperture polarimetry of 321 point sources in Mon R2. Based on aperture photometry data of each angle image, polarization degree P and angle θ for each source were derived by the Stokes parameters. Then we found "S-character" structure. Figure 1 (Right) clearly shows the magnetic field structure to

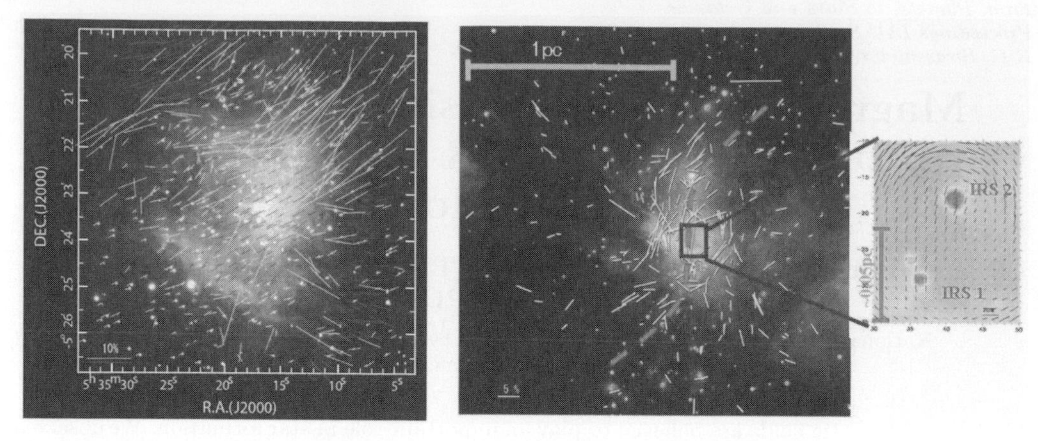

Figure 1. Left — H band polarization vectors (yellow) and 350μm polarization vectors (red) are superposed on the $JHKs$ color composite image of M42 (Kusakabe *et al.* 2008). Right — H band aperture polarization vectors (yellow) on the JHKs color composite image. The dashed red line indicate "S-character" geometry. The close-up view of the central region is also shown in the right side.

be twisted around the center where there are IRS 1 and IRS 2 which has been alleged to result from gravitational distortion caused by the IRS 2 and/or IRS 1. Although some vectors are aligned in north-east or south-west direction, the vectors seem to associate with the foreground molecular cloud with elongated shape located north-east to the Mon R2 core. To check the "S-character" geometry, we calculate vectors which are located within 135 arcsec around IRS 1 using simple third function, $y = g + Cx^3$, and the parameter C is -10^{-5}. Then we found a histogram of $\theta_{\rm obs} - \theta_{\rm cal} = \Delta\theta$ clearly shows a peak around 0 degree. This explains the "S-character" structure well.

We connected outer magnetic fields (observed by optical polarization) to the deep inside of Mon R2. Furthermore, the close up view (see. Fig.1, right) is the close up polarization map of the central region containing IRS 1 and IRS 2. The small degrees of polarizations (<10%) are seen at the central region. Although the overall vector pattern is more or less center-symmetric around IRS 2, the vectors are aligned north-south at the center of the close up image. Such a structure of the central region could not be due to circumstellar structures of IRS 1 and IRS 2. We suggest that the aligned polarization patterns show dichroic polarization of the emission from the HII region, thus representing the magnetic field structure of the central region.

References

Genzel, R. & Stutzki, J. 1989, *ARAA* 27, 41

Houde, M., Dowell, C. D., Hildebrand, R. H., Dotson, J. L., Vaillancourt, J. E., Phillips, T. G., Peng, R., & Bastien, P. 2004, *ApJ* 604, 717

Jarrett, T. H., Novak, G., Xie, T., & Goldsmith, P. F. 1994, *ApJ* 430, 743

Kandori, R., *et al.* 2006, *SPIE* 6269, 159

Kusakabe, N., Tamura, M., Kandori, R., Hashimoto, J., Nakajima, Y., Nagata, T., Nagayama, T., Hough, J., & Lucas, P., 2008, *AJ* 136, 621K

Simpson, J. P., Colgan, S. W. J., Erickson, E. F., Burton, M. G., & Schultz, A. S. B., 2006, *ApJ* 642, 339

Cosmic Magnetic Fields:
From Planets, to Stars and Galaxies
Proceedings IAU Symposium No. 259, 2008
K.G. Strassmeier, A.G. Kosovichev & J.E. Beckman, eds.

doi:10.1017/S1743921309030166

Observations of magnetic fields toward the star forming region S88B

Anuj P. Sarma[1], C. L. Brogan[2], T. Bourke[3], and M. Eftimova[1,4]

[1]Physics Department, DePaul University, Chicago IL, USA
email: asarma@depaul.edu

[2]National Radio Astronomy Observatory, Charlottesville, VA

[3]Harvard-Smithsonian Center for Astrophysics, Cambridge, MA

[4]Current Address: Dept. of Physics & Astronomy, Univ. of North Carolina, Chapel Hill, NC

Abstract. We present observations of the high mass star forming region S88B taken with the VLA with the aim of measuring magnetic fields via the Zeeman effect. By observing thermal absorption lines of OH at 1665 and 1667 MHz, we obtain magnetic fields between 90 and 210 μG. We find these magnetic fields to be dynamically significant in this region.

Keywords. H II regions – ISM: clouds – ISM: individual (S88B) – ISM: kinematics and dynamics – ISM: magnetic fields – ISM: molecules

1. Introduction

The importance of magnetic fields in the formation of stars has long been acknowledged. Despite remarkable advances in theory and instrumentation, however, observational data on magnetic fields is still scarce. Observations of the Zeeman effect in absorption lines with interferometers like the Very Large Array (VLA) provide an excellent method of *mapping* the magnetic field in regions that are along the line of sight toward strong background continuum sources (e.g., Brogan & Troland 2001; Sarma *et al.* 2000). With these considerations in view, we have observed the star forming region S88B with the VLA for the Zeeman effect in thermal absorption lines of OH at 1665 and 1667 MHz.

2. Observations & Data Reduction

The observations were carried out with the Very Large Array (VLA) of the NRAO in 2003 in the B-configuration, and combined with C-configuration data observed in 1997. Both right (RCP) and left (LCP) circular polarizations and both main lines (1665 and 1667 MHz) were observed simultaneously. The data were calibrated and imaged using standard procedures in the AIPS package (NRAO) while the magnetic fields were determined using routines in the MIRIAD software package (BIMA/CARMA).

3. Results & Discussion

Magnetic field strengths were determined by fitting a numerical frequency derivative of the Stokes I = (RCP + LCP)/2 spectrum to the Stokes V = (RCP − LCP)/2 spectrum. The technique is described in detail in Roberts *et al.* (1993). The results of the fits give the line-of-sight component of the magnetic field, $B_{\rm los}$. Figure 1 shows the resulting $B_{\rm los}$ map (from the 1665 MHz data for which the derived value of $B_{\rm los}$ is greater than the 3σ level). In the area enclosed by the ellipse in Fig. 1, the detected field is above the 3σ

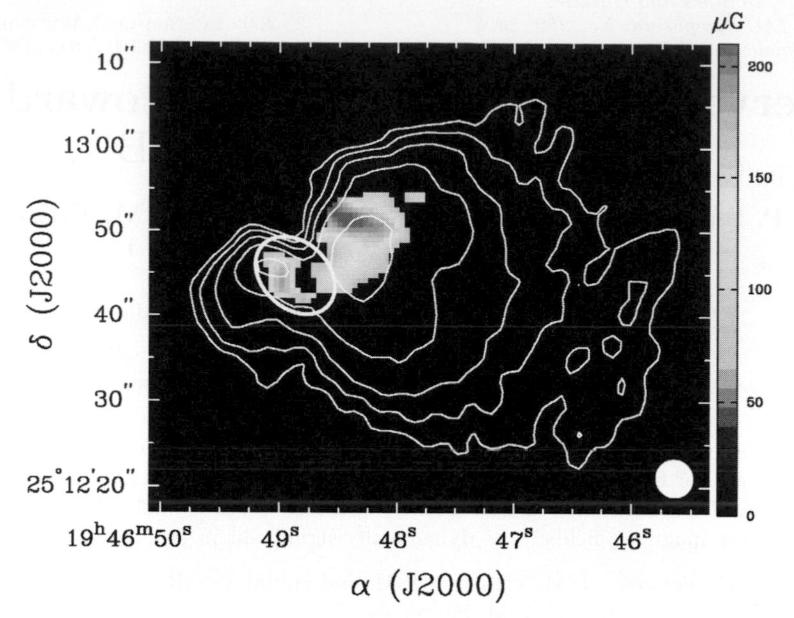

Figure 1. Gray-scale image of the detected $B_{\rm los}$ toward S88B. The filled ellipse in the bottom right represents the beam for the OH absorption line observations ($4.5'' \times 4.3''$, PA = 14°). The contours depict the 18 cm continuum, and are at 6, 12, 24, 48, 96 mJy beam^{-1}.

level only in the 1665 MHz line, whereas in the rest of the displayed $B_{\rm los}$ map, the field is above the 3σ level in both 1665 and 1667 MHz lines.

In order to estimate of the importance of the magnetic field in a star forming cloud, we use the relation

$$B_{S,\ \rm crit} = 5 \times 10^{-21}\ N_p \quad \mu\mathrm{G},$$

where N_p is the average proton column density in the cloud. The relation has been obtained by equating the static magnetic energy of the cloud to its gravitational energy, and $B_{S,\ \rm crit}$ is then the average static magnetic field in the cloud that would completely support it against self-gravity. The field can be judged to be dynamically important to the region even if it is less than $B_{S,\ \rm crit}$, but is comparable to it. Using relevant physical parameters from our observations and the literature, we find $N_p = 1.4 \times 10^{23}$ cm^{-2}. The equation above then gives $B_{S,\ \rm crit} = 700$ μG. Following Crutcher (1999), we use total (static) magnetic field strength equal to 2 times $B_{\rm los}$. For our average adopted value for $B_{\rm los} = 150$ μG from Fig. 1, we find that the observed magnetic field is less than, but comparable to, the critical field. Therefore, the magnetic field should be dynamically significant, providing an important source of support against self gravity.

Acknowledgements

APS thanks the Chair of the Physics Department (Dr. Jesús Pando) and DePaul University for travel funds.

References

Brogan, C. L. & Troland, T. H. 2001, *ApJ* 560, 821
Crutcher, R. M. 1999, *ApJ* 520, 706
Roberts, D. A., Crutcher, R. M., Troland, T. H., & Goss, W. M. 1993, *ApJ* 412, 675
Sarma, A. P., Troland, T. H., Roberts, D. A., & Crutcher, R. M. 2000, *ApJ* 533, 271

Cosmic Magnetic Fields:
From Planets, to Stars and Galaxies
Proceedings IAU Symposium No. 259, 2008
K.G. Strassmeier, A.G. Kosovichev & J.E. Beckman, eds.

doi:10.1017/S1743921309030178

Magnetic fields in the Pipe nebula region†

Gabriel A. P. Franco[1], **Felipe O. Alves**[2] **and J. M. Girart**[2]

[1]Departamento de Física – ICEx – UFMG, Brazil
email: franco@fisica.ufmg.br

[2]Institut de Ciències de l'Espai (CSCI–IEEC), Campus UAB–Facultat de Ciències, Torre C5–Parell 2[a], 08193 Bellaterra, Catalunya, Spain
email: [oliveira; girart]@ieec.uab.es

Abstract. Magnetic fields are supposed to play an important role in the formation and support of self-gravitating clouds and the formation and evolution of protostars in such clouds. We used R-band linear polarimetry collected for about 12 000 stars in 46 fields with lines of sight toward the Pipe nebula to investigate the properties of the magnetic fields acting on this dark cloud complex.

Keywords. ISM: clouds – ISM: Pipe nebula – ISM: magnetic fields – techniques: polarimetric

1. Introduction

The Pipe nebula is a massive cloud (10^4 M⊙, Lombardi, Alves & Lada 2006) located at the solar vicinity (145 pc; Alves & Franco 2007) which presents an apparently quiescent nature. The only known active star forming site, Barnard 59 (B 59), corresponds only to a small fraction of the entire cloud mass. The lack of star forming regions motivated us to investigate the magnetic properties that may be involved in the cloud evolution.

R band linear polarimetry collected for about 12 000 stars was used to investigate the properties of the polarization across the Pipe nebula dark cloud complex. Mean polarization vectors show that the magnetic field is locally perpendicular to the large filamentary structure of the Pipe nebula (the 'stem'), indicating that the global collapse may have been driven by ambipolar diffusion. The northwestern end of the nebula (B 59 region) is found to have a low degree of polarization and high dispersion in polarization position angle, while at the other extreme of the cloud (the 'bowl') we found mean degrees of polarization as high as 15% and a low dispersion in polarization position angle.

2. Magnetic Fields

The obtained dispersion in polarization angles were used to estimate the magnetic field strength for the observed areas by applying the modified Chandrasekhar-Fermi formula (Chandrasekhar & Fermi 1953, Ostriker, Stone & Gammie 2001). The volume density and line width of the molecular line emission associated with the dust that produces the observed optical polarization and extinction were estimated from the molecular data available in the literature. We found that the magnetic field strength in the B 59 region, stem, and bowl, in the plane of the sky, are about 17, 30, and 65 μG, respectively (Alves, Franco & Girart 2008). The almost perpendicular alignment between the magnetic field and the main axis of the Pipe nebula's stem indicates clearly that this part of the cloud contracted in the direction of the field lines. Indeed, the magnetic pressure ($P_{mag} = \frac{B^2}{8\pi}$)

† Based on observations collected at Observatório do Pico dos Dias, operated by Laboratório Nacional de Astrofísica (LNA/MCT, Brazil).

of the diffuse part of the cloud (where is most of the mass) is the dominant source of pressure in the direction perpendicular to the field lines (12×10^5 and 2.6×10^5 K cm^{-3} for the bowl and stem, respectively), being higher than the pressure due to the weight of the cloud ($P_{cloud}/k = 10^5$ K cm^{-3}, according to Lada, Muench, Rathborne *et al.* 2008).

3. Conclusions

• The global polarimetric properties of the Pipe nebula can be described as an increasing polarization degree along the filamentary structure from B 59 towards the bowl, while the dispersion in polarization angles decreases along this way.

• The results appear to indicate that there exist three regions in the Pipe nebula of distinct evolutionary stages:

(*a*) B 59, the only active star-forming site in the cloud. For the observed fields, we measure a large dispersion in polarization angle, low polarization degree, and a estimated magnetic field strength of 17 μG.

(*b*) The 'stem', which collapsed by means of ambipolar diffusion but has not yet given birth to stars. It appears to represent a transient evolutionary state between B 59 and the bowl. We estimated an average magnetic field strength of 30 μG for this part of the cloud.

(*c*) The 'bowl', which shows the highest values of mean polarization and lowest values of dispersion in polarization angle. The dust grains in the bowl seems to be highly aligned by the rather strong magnetic fields that permeates this part of the cloud ($B \approx 65$ μG).

Further details will be published elsewhere (Franco, Alves & Girart 2009).

Acknowledgements

We thank the staff of the Observatório do Pico dos Dias (LNA/MCT, Brazil) for their hospitality and invaluable help during our observing runs. Drs. A. M. Magalhães and A. Pereyra are acknowledged for providing the polarimetric unit and the software used for data reductions. This research has been partially supported by CEX APQ-1130-5.01/07 (FAPEMIG, Brazil) and AYA2005-08523-C03 (Ministerio de Ciencia e Innovación, Spain).

References

Alves, F. O. & Franco, G. A. P. 2007, *A&A* 470, 597
Alves, F. O., Franco, G. A. P., & Girart, J. M. 2008, *A&A* 486, L13
Chandrasekhar, S. & Fermi, E. 1953, *ApJ* 118, 113
Franco, G. A. P., Alves, F. O., & Girart, J. M. 2009, in preparation
Lada, C. J., Muench, A. A., & Rathborne, J. *et al.* 2008, *ApJ* 672, 410
Lombardi, M., Alves, J., & Lada, C. J. 2006, *A&A*, 454, 781
Ostriker, E. C., Stone, J. M., & Gammie, C. F. 2001, *ApJ* 546, 980

Cosmic Magnetic Fields:
From Planets, to Stars and Galaxies
Proceedings IAU Symposium No. 259, 2008
K.G. Strassmeier, A.G. Kosovichev & J.E. Beckman, eds.

doi:10.1017/S174392130903018X

Radiative magneto-hydrodynamics in massive star formation and accretion disks

Rolf Kuiper, Mario Flock and Hubert Klahr

Max-Planck Institut für Astronomie,
Königstuhl 17, D-69117 Heidelberg, Germany
email: kuiper@mpia.de

Abstract. We briefly overview our newly developed radiation transport module for MHD simulations and two actual applications. The method combines the advantage of the speed of the Flux-Limited Diffusion approximation and the high accuracy obtained in ray-tracing methods.

Keywords. Radiative transfer – MHD – stars: formation – accretion disks – methods: numerical

---

1. Radiative MHD

Aim of the development of the radiation transport module described here is to achieve a fast method for approximative frequency-dependent radiation transport for (magneto-) hydrodynamical simulations in studies of the environment around a single object source.

The radiation transport module provides a Flux-Limited Diffusion (hereafter FLD) solver in cartesian, cylindrical and spherical coordinates with a potentially non equidistant grid spacing in the first dimension, e.g. logarithmic in the radial coordinate. Additionally to the FLD approximation the module consists of a first order ray-tracing technique along the first dimension to account for single source irradiation (e.g. stellar heating of a proto-planetary disk or radiative feedback in massive star formation).

The local radiation field is split into two parts, originated from an externally irradiated (frequency-dependent) flux $\vec{F}(\nu)$ and a diffuse radiation energy density E_R, which are in equlibrium with the radiation from the dust grains:

$$aT^4 = E_R + \frac{1}{c\,\kappa_P(T)} \int \kappa(\nu) \left|\vec{F}(\nu)\right| d\nu, \tag{1.1}$$

where a is the radiation constant, c is the speed of light, $\kappa(\nu)$ and $\kappa_P(T)$ represent the frequency-dependent and the Planck mean opacity for a given temperature T respectively. Gas and dust temperatures are assumed to be the same. The irradiation is treated as an instantaneous source of additional energy, calculated via a first order ray-tracing as a function of the optical depth $\tau_\nu(r)$ and distance r from the central star:

$$\vec{F}(\nu, r) = \vec{F}_*(\nu) \left(\frac{R_*}{r}\right)^2 e^{-\tau_\nu(r)} \tag{1.2}$$

The evolution of the radiation energy is described by a Flux-Limited Diffusion equation

$$\partial_t E_R = -f_c \left(\vec{\nabla} \cdot \left(D\vec{\nabla} E_R + \int \vec{F}(\nu, r) d\nu\right) - Q^+\right) \tag{1.3}$$

with $f_c = (c_v \rho / 4aT^3 + 1)^{-1}$, $D = \lambda c / \kappa_R \rho$ and $Q^+ = -p\vec{\nabla} \cdot \vec{v}$+additional source terms from hydrodynamics. The flux-limiter λ is chosen according to Levermore & Pomraning (1981), c_v is the specic heat capacity, ρ, p and $\vec{v}$ the gas density, thermal pressure and dynamical velocity and κ_R specifies the Rosseland mean opacity. Scattering is neglected.

2. Application I: Massive Star Formation

Due to the short Kelvin-Helmholtz contraction timescale of a massive star (Shu *et al.* 1987) the accretion process onto such a star is described by the interaction of the gravitationally forced inflow of matter with the radiative force escaping from the newly born star. The conservation of angular momentum leads to the formation of an accretion disk as well as polar cavities. At present we're studying this collapse scenario with respect to the effect of dimensionality (from 1 to 3D) and different applied physics (isothermal and adiabatic test runs, realistic cooling and frequency-averaged as well as frequency-dependent radiative feedback).

3. Application II: Accretion Disks

We perform accretion disk simulations with a hydrodynamical stable stratified disk model $\left(\frac{H}{R} = 0.05\right)$ and an initial toroidal magnetic field. The initial plasma beta β is constant everywhere in the disk at a value of 25, the runs by Fromang & Nelson (2006) (hereafter FN). The convergence tests have a radial extension from 4 to 6 AU, $\pm$ 1 scale height as vertical and $\pi/3$ as azimuthal extension. We choose perodic boundaries in vertical and azimuthal direction. Small velocity perturbations in radial and vertical direction about $10^{-4}c_0$ initiate the non-linear MRI evolution. With the used high-order MHD Riemann solver (Mignone, *et al.* 2007) we get a converged alpha value of about 10^{-2} presented in figure 1. In opposite, FN got an increasing alpha value from low $\alpha = 10^{-3}$ to high $\alpha = 5 * 10^{-3}$ (yet still not converged) with increasing resolution for a non-Riemann solver. In the future we will use the radiative module to calculate proper

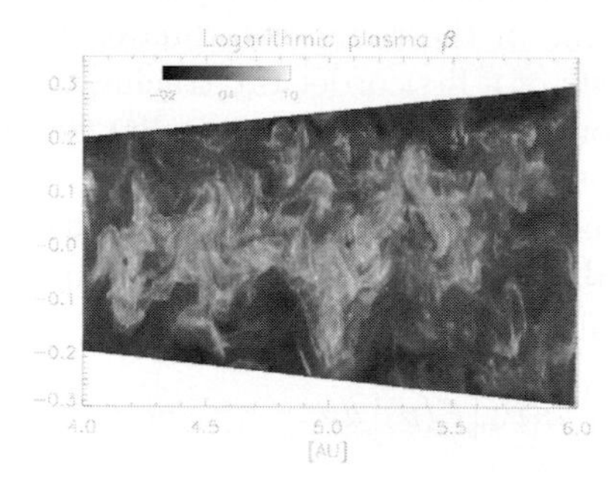

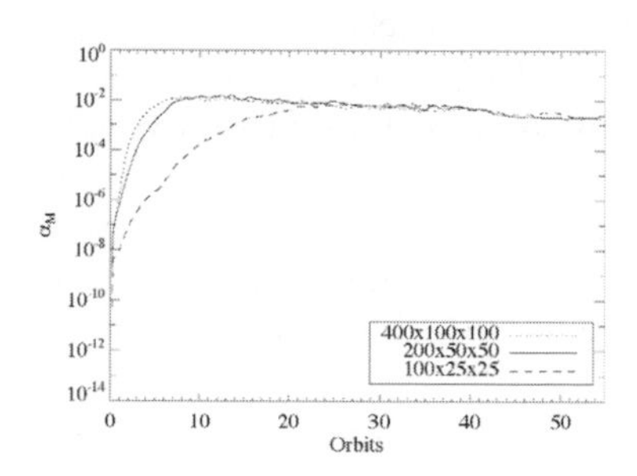

Figure 1. Left: Azimuthally averaged logarithmic plasma beta in the turbulent phase. Right: Evolution of Maxwell alpha at different resolution, converging against each other.

resistive terms $\eta(T, \rho)$ and to handle correctly the resulting heating and cooling processes of MHD turbulence and the behavior for the alpha stresses.

References

Levermore, C. D. & Pomraning, G. C. 1981, *ApJ* 248, 321

Fromang, S. & Nelson, A. P. 2006, *A&A* 457, 343

Mignone, A., Bodo, G., Massaglia, S. Matsakos, T., Tesileanu, O., Zanni, C., & Ferrari, A. 2007, *ApJ* 170, 228

Pascucci, I., Wolf, S., Steinacker, J., *et al.* 2004, *A&A* 417, 793

Shu, F. H., Adams, F. C., & Lizano, S. 1987, *ARAA* 25, 23

Cosmic Magnetic Fields:
From Planets, to Stars and Galaxies
Proceedings IAU Symposium No. 259, 2008
K.G. Strassmeier, A.G. Kosovichev & J.E. Beckman, eds.

doi:10.1017/S1743921309030191

Local star formation triggered by SN shocks in magnetized diffuse neutral clouds

M. R. M. Leão[1], E. M. de Gouveia Dal Pino[1], D. Falceta-Gonçalves[2,3], C. Melioli[1,4] and F. G. Geraissate[1]

[1]Instituto de Astronomia e Geofísica - Universidade de São Paulo, Brazil
email: mrmleao@astro.iag.usp.br, dalpino@astro.iag.usp.br

[2]Núcleo de Astrofísica Teórica, Universidade Cruzeiro do Sul, Brazil

[3]Astronomy Department, University of Wisconsin, Madison, USA

[4]Dipartimento di Astronomia, Universitá di Bologna, Bologna, Italy

Abstract. Considering that the main source of turbulence in the ISM medium are the explosions of supernovae (SNe), we explore here the role of SN shock front interactions with clouds on the star formation triggering in presence and in absence of magnetic field.

Keywords. Stars: star formation — ISM: clouds – Supernova remnants.

We consider a supernova remnant (SNR) either in its adiabatic or in its radiative phase impacting with an initially homogeneous diffuse neutral cloud and show that it is possible to derive analytically a set of conditions that constrain a domain in the relevant parameter space where these interactions may lead to the formation of gravitationally unstable, collapsing structures.

One of these conditions determines the Jeans mass limit for the shocked cloud material due to the SNR impact. A second condition establishes that the shock front that propagates inside the cloud must have energy enough to sweep the entire cloud before stalling inside it due to radiative losses. In this way, the shock will be able to inject a maximum possible energy into the cloud material and compress it efficiently. A third condition establishes that the same shock front should not be too strong, otherwise it could destroy the cloud completely making the gas to disperse in the interstellar medium before becoming gravitationally unstable.

Using these conditions, we have built diagrams of the SNR radius, R_{SNR}, versus the initial cloud density, n_c. This work is an extension to previous study performed without considering magnetic fields (Melioli *et al.* 2006). The diagrams are also tested with fully 3-D MHD simulations involving a SNR and a self-gravitating cloud and we find that the numerical analysis is consistent with the results predicted by the diagrams, see Fig. (1).

The inclusion of a homogeneous magnetic field approximately perpendicular to the impact velocity of the SNR with an intensity $\sim 1\ \mu$G within the cloud results only a small shrinking of the star formation triggering zone in the diagram relative to that without magnetic field, a larger magnetic field ($\sim 10\ \mu$G) causes a significant shrinking, as expected. Though derived from simple analytical considerations these diagrams provide a useful tool for identifying sites where star formation could be triggered by the impact of a SN blast wave.

Applications of them to a few regions of our own galaxy have revealed that star formation in those sites could have been triggered by shock waves from SNRs for specific values of the initial neutral cloud density and the SNR radius. Finally, we have also evaluated the effective star formation efficiency for this sort of interaction.

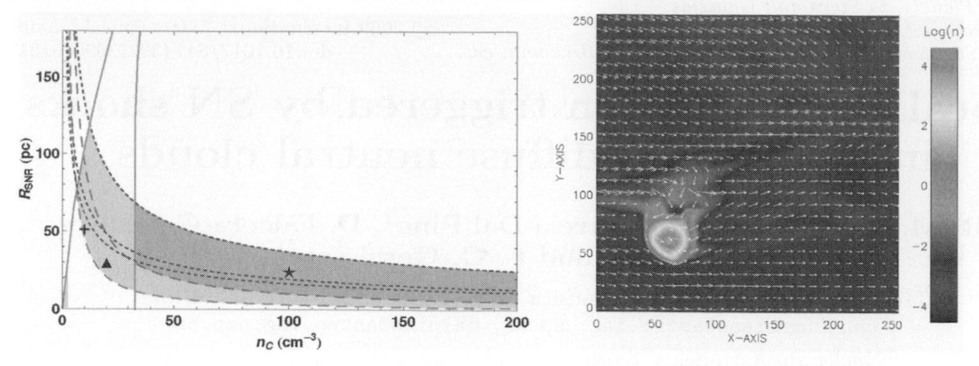

Figure 1. Left panel: Constraints on the SNR radius versus cloud density for a cloud with $r_c =$ 10 pc. Dashed (green) line: upper limit for complete cloud destruction; solid (red) line: upper limit for the shocked cloud to reach the Jeans mass; dotted (blue) lines: upper limits for the shock front to travel into the cloud for different values of the cooling function $\Lambda(T'_{sh})$. The shaded zones define the region where star formation can be induced by a SNR-cloud interaction. The light shaded zone defines the SF region domain for a cloud with $B_c = 0$ while the dark one defines the SF region domain for $B = 1$ μG. The symbols in the panels indicate the initial conditions assumed for the clouds in numerical simulations. Right panel: Example of a colour-scale map of the mid-plane density distribution and magnetic field vectors for a simulated model with $n_c = 100$ cm^{-3}, $R_{SNR} = 25$ pc, and $B = 1\mu$G (the star in the left diagram) at time 2.5×10^6 yr. (Leão *et al.* 2008)

Conclusions

A uniform magnetic field (1 μG) plays some role over the Jeans constraint only, causing a drift of the allowed SF zone to higher cloud densities in the diagram when compared to the case with no magnetic field. When larger intensities of magnetic fields are considered ($5 - 10$ μG), the shrinking of the allowed SF zone in the diagrams is much more significant.

The sfe for these SNR-cloud interactions is generally much smaller than the observed values in our own Galaxy (sfe ~ 0.01–0.3). This result is consistent with previous work in the literature and also suggests that the mechanism presently investigated, though very powerful to drive structure formation, supersonic turbulence and eventually, local star formation, does not seem to be sufficient to drive *global* star formation in normal star forming galaxies, not even when the magnetic field in the neutral clouds is neglected.

Acknowledgements

E.M.G.D.P., M.R.M.L., D.F.G. and F.G.G. acknowledge financial support from grants from the Brazilian Agencies FAPESP, CNPq and CAPES.

References

Falceta-Goncalves D., Lazarian A., & Kowal G. 2008, *ApJ* 679, 537
Kowal G. & Lazarian A. 2007, *ApJ* 666, L69
Leão, M. R. M., de Gouveia Dal Pino, E. M., Falceta-Gonalves, D., Melioli, C., & Geraissate, F. G., 2008, *MNRAS* accepted.
Melioli, C., de Gouveia Dal Pino, E. M., & Raga, A. 2005, *A&A* 443, 495
Melioli, C., de Gouveia Dal Pino, E. M., de la Reza, R., & Raga, A. 2006, *MNRAS* 373, 811

Cosmic Magnetic Fields: From Planets, to Stars and Galaxies
Proceedings IAU Symposium No. 259, 2008
K.G. Strassmeier, A.G. Kosovichev & J.E. Beckman, eds.

doi:10.1017/S1743921309030208

Distortion of magnetic fields in the pre-stellar core Barnard 68

Ryo Kandori, Motohide Tamura, Ken-ichi Tatematsu, Nobuhiko Kusakabe, Yasushi Nakajima, and the IRSF/SIRPOL group

National Astronomical Observatory, 2-21-1 Osawa, Mitaka, Tokyo 181-8588
email: r.kandori@nao.ac.jp

Abstract. Magnetic fields are believed to play an important role in controlling the stability and contraction of molecular cloud cores. In the present study, magnetic fields of a cold pre-stellar core, Barnard 68, have been mapped based on wide-field near-infrared polarimetric observations of background stars. A distinct "hourglass-shaped" magnetic field is identified toward the core, as the observational evidence of magnetic field structure distorted by mass accumulation in a pre-stellar core. Our findings on the geometry of magnetic fields as well as the mass-to-magnetic flux ratio are presented.

Keywords. Stars: formation – ISM: globules – ISM: magnetic fields – polarization

1. Introduction

The characteristics of newborn stars are thought to be determined primarily by the properties of the molecular cloud core prior to the onset of gravitational collapse, and the magnetic field pervading the core is believed to play an important role in controlling the stability and contraction of the pre-stellar core (e.g., Shu, Adams, & Lizano 1987). It is therefore of importance to investigate the magnetic fields affecting the cores in the pre-stellar phase to clarify the initial conditions of star formation. The direction of magnetic fields projected onto the sky can be inferred from polarimetric observations of dust emission (B $\perp$ E) and/or dichroic extinction (B $\parallel$ E, e.g., Lazarian *et al.* 2007). Dust emission polarimetry, particularly at sub-millimeter to far-infrared wavelengths, has proven to be a powerful technique for tracing magnetic field structures in dense regions, such as in protostellar envelopes (e.g., Girart *et al.* 2006). However, polarized dust emission from cold pre-stellar cores is too weak to be detected by present instruments except for dense central regions (e.g., Ward-Thompson, 2000). *There are thus no observations of pre-stellar cores tracing magnetic fields from the center to the outermost regions.*

2. Magnetic Properties of Barnard 68

In the present study, a magnetic field map of the pre-stellar core Barnard 68 (hereafter B68) is constructed on the basis of near-infrared polarimetry toward background stars suffering dichroic extinction by magnetically aligned dust in the core. Observations were made in 2007 June using the near-infrared imaging polarimeter SIRPOL (Kandori *et al.* 2006) on the IRSF 1.4-m telescope in South Africa. IRSF/SIRPOL provides deep (18.6 mag in the H band, 5 σ in one hour exposure) and wide-field ($7.7' \times 7.7'$) polarization images. In Figure 1, the H band polarization vectors of 125 background stars (yellow lines) toward B68 are shown. The white lines show the magnetic field configuration inferred from the fitting using a parabolic function. The fitting appears reasonable, since the rms of residual angles of polarization vectors in the fitting is smaller for the parabolic

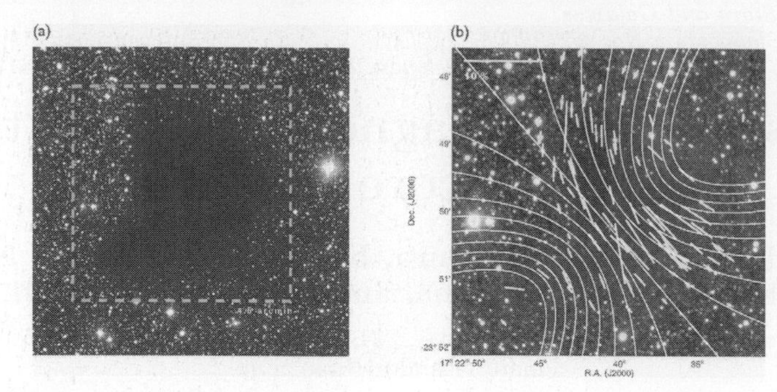

Figure 1. (a) — Optical image of B68 (Alves *et al.* 2001). (b) — H band polarization vectors of 125 background stars (yellow lines) on the intensity image. The white lines show the most probable magnetic field configuration inferred from the fitting using a parabolic function, $y = g + gCx^2$, with the magnetic axis position angle of $\sim 40^\circ$ and $C \sim 2 \times 10^{-4}$ arcsec^{-2}.

field case than for the linear case. The magnetic field seems to follow a distinct not linear axi-symmetric shape reminiscent of an hourglass. The orientation of the magnetic axis ($\sim$40°) of B68 is nearly perpendicular to the major axis of the core elongation ($\sim$130°) in the dense regions (A_V >10 mag). These geometrical magnetic properties are consistent with theoretical suggestions that (1) initially straight frozen-in magnetic fields pervading a core can be dragged toward the center by the contracting medium and distorted into an hourglass-shaped structure, and (2) a disk-like structure can eventually be formed in the center by mass accretion along the magnetic field lines, with the major axis of the disk aligned perpendicular to the magnetic axis (e.g., Galli & Shu 1993).

Given the magnetic field structure and other properties known for B68 (e.g., Alves *et al.* 2001), the strength of the plane-of-sky magnetic field was obtained as 20 ± 5 μG based on the Chandrasekhar-Fermi method (Chandrasekhar & Fermi, 1953) with a theoretically suggested correction factor of 0.5 (Ostriker *et al.* 2001). The comparison of the observed mass-to-magnetic flux to a critical value suggested by theory (Nakano & Nakamura 1978) resulted in $(M/\Phi)_{\rm obs}/(M/\Phi)_{\rm critical} \sim 2.3$ (magnetically supercritical). The result does not necessarily mean the gravitational collapse, because B68 could be marginally stable from the support by thermal and (small) turbulent pressures (e.g., Lada *et al.* 2003). The magnetic fields can provide an extra support to retain the core to be stable. Finally we speculate the origin of hourglass field structure. Since B68 seems stable in the present phase, the structure is interpreted as an imprint of mass condensation during the development of the core under the influence of the gravity and turbulence.

References

Alves, J. F., Lada, C. J., & Lada, E. A. 2001, *Nature* 409, 159
Chandrasekhar, S. & Fermi, E. 1953, *ApJ* 118, 113
Galli, D. & Shu, F. 1993, *ApJ* 417, 220
Girart, J. M., Rao, R., & Marrone, D. P. 2006, *Science* 313, 812
Kandori, R., *et al.* 2006, *Proc. SPIE* 6269, 159
Lada, C. J., Bergin, E. A., Alves, J. F., & Huard, T. L. 2003, *ApJ* 586, 286
Lazarian, A. 2007, *J. Quant. Spectrosc. Rad. Trans.* 106, 225
Nakano, T. & Nakamura, T. 1978, *PASJ* 30, 671
Ostriker, E. C., Stone, J. M., & Gammie, C. F. 2001, *ApJ* 546, 980
Shu, F., Adams, F. C., & Lizano, S. 1987, *ARA&A* 25, 23
Ward-Thompson, D., *et al.* 2000, *ApJ* 537, 135

Cosmic Magnetic Fields: From Planets, to Stars and Galaxies
Proceedings IAU Symposium No. 259, 2008
K.G. Strassmeier, A.G. Kosovichev & J.E. Beckman, eds.

doi:10.1017/S174392130903021X

The magnetic field of the evolved star W43A

Nikta Amiri[1,3], Wouter Vlemmings[2] and Huib Jan van Langevelde[1,3]

[1]Joint Institute for VLBI in Europe (JIVE), The Netherlands
email: amiri@jive.nl, langevelde@jive.nl

[2]Argelander Institute for Astronomy, University of Bonn, Germany
email: wouter@astro.uni-bonn.de

[3]Leiden Observatory, Leiden University, The Netherlands

Abstract. Planetary nebulae (PNe) often show large departures from spherical symmetry. The origin and development of these asymmetries is not clearly understood. The most striking structures are the highly collimated jets that are already observed in a number of evolved stars before they enter the PN phase. The aim of this project is to observe the Zeeman splitting of the OH maser of the W43A star and determine the magnetic field strength in the low density region. The 1612 MHz OH masers of W43A were observed with MERLIN to measure the circular polarization due to the Zeeman splitting of 1612 OH masers in the envelope of the evolved star W43A. We measured the circular polarization of the strongest 1612 OH masers of W43A and found a magnetic field strength of $\sim 100\mu G$. The magnetic field measured at the location of W43A OH masers confirms that a large scale magnetic field is present in W43A, which likely plays a role in collimating the jet.

Keywords. Planetary nebula – ISM: magnetic fields – stars: magnetic fields

1. Introduction

Planetary nebulae (PNe), supposedly formed out of the ejected outer envelopes of AGB stars, often show large departures from spherical symmetry. The origin and development of these asymmetries is not clearly understood. The most striking structures are the highly collimated jets that are already observed in a number of evolved stars before they enter the PN phase. However, the origin of the collimation of the jet is still puzzling. Theoretical models have shown that collimated jets can be caused by magnetic field in evolved stars (Blackman *et al.* 2001). Studying the large scale magnetic field through polarization observations of different maser species in the circumstellar envelope (CSE) of these stars provides a unique tool to understand the role of the magnetic field in the process of jet collimation. The aim of this project is to observe Zeeman splitting of the OH maser of W43A and determine the magnetic field strength in the low density region. H_2O maser polarization observations have revealed that the jet of W43A is likely magnetically collimated (Vlemmings *et al.* 2006). OH masers arising from the material surrounding the jet are predicted to also have a detectable magnetic field and our observation will thus be able to confirm the role of the magnetic field in collimating the jet in proto-PNe.

The 1612 MHz OH masers of W43A were observed with MERLIN in June 2007. The observations were done with maximum possible 256 spectral resolution since all 4 polarization channels were required. The source 3C84 was used for bandpass and polarization calibration. The resulting noise in the emission free channel was 65 mJy/Beam.

2. Results

From the image plane we have obtained the typical double peak spectrum of the OH maser region of W43A. Fig. 1 shows the velocity profile of the integrated flux of each channel in the I, V and P (polarization intensity) data cubes. The velocity spectrum of the total intensity data cube is a typical double peak profile with velocity in the range 27-43 km/s; despite the fact that W43A is a water fountain source and has much higher velocities in the H_2O maser region (–53 to 126 km/s). The brightest peak is redshifted and the blueshifted peak has a much lower brightness; only 3 % of the redshifted brightness. The peaks in the polarization intensity and circular polarization spectra are 10% linearly and 12% circularly polarized. The relatively low level of linear polarization excludes most non-Zeeman interpretations of the circular polarization (Fish & Reid 2006).

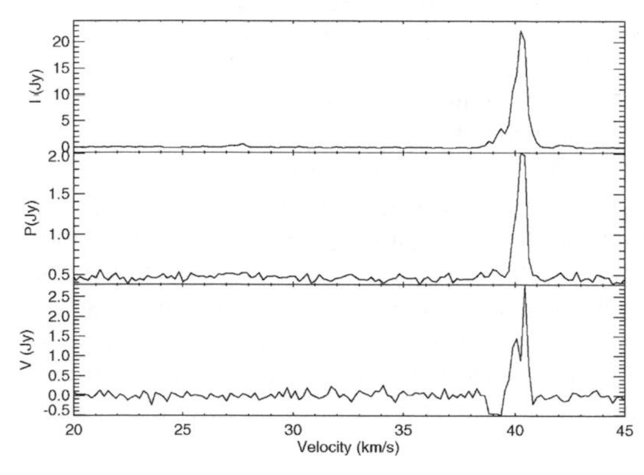

Figure 1. The 1612 MHz spectra of W43A.

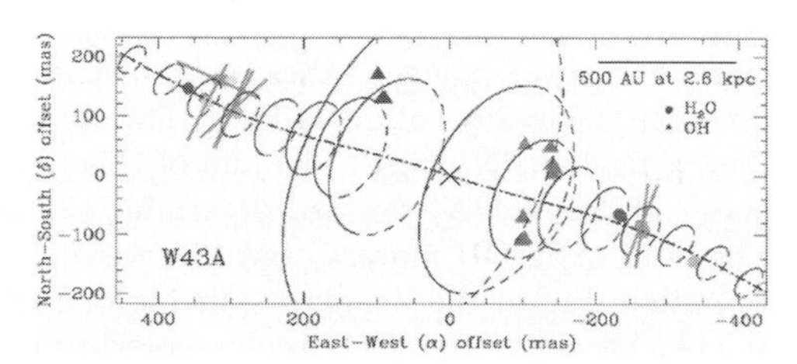

Figure 2. The spatial distribution of the OH and H_2O maser features of W43A.

Figure 2 shows the spatial distribution of the OH maser features of W43A together with H_2O maser positions (Vlemmings *et al.* 2006). The offset positions are with respect to the reference feature. H_2O maser features are indicated by filled circles. OH maser features are shown as triangles. Red and blue colors show the redshifted and blueshifted features. This map clearly shows that H_2O maser occurs at the tips of a jet and the OH maser has a projected distance much closer to the star. The actual measurement of the magnetic field strength may depend on a number of conditions, but is estimated to be 100 μG. This value is consistent with the expected magnetic field strength of 70 μG extrapolated from H_2O maser observations (Vlemmings *et al.* 2006). A lower OH magnetic field is expected compared to H_2O due to the higher density and shocked nature of the H_2O masers at the tips of the jet.

3. Conclusions

The magnetic field and jet characteristics of W43A have previously been reported from H_2O maser polarization observations in relation to the formation of aspeherical planetary nebulae. Here we present the detection of the magnetic field of $\sim$ 100 μG in the OH maser region surrounding the collimated jet. This confirms that a large scale magnetic field is present in W43A, which likely plays a role in collimating the jet.

References

Blackman, E. G., Frank, A., Markiel, J., Thomas, J., & Van Horn, H. M. 2001, *Nature* 409, 485
Fish, V. L. & Reid, M. J. 2006, *ApJ* 164, 99
Vlemmings, W. H. T., Diamond, P. J., & Imai, H. 2006, *Nature* 440, 58

Cosmic Magnetic Fields: From Planets, to Stars and Galaxies
Proceedings IAU Symposium No. 259, 2008
K.G. Strassmeier, A.G. Kosovichev & J.E. Beckman, eds.

doi:10.1017/S1743921309030221

On global stability of thin ionized disks immersed in an external magnetic field

Edward Liverts and Michael Mond

Department of Mechanical Engineering, Ben-Gurion University of the Negev, P.O. Box 653, Beer-Sheva 84105, Israel
email: eliverts@bgu.ac.il, mond@bgu.ac.il

Abstract. The problem of the global stability of rotating magnetized thin disks is considered. The appropriate boundary value problem (BVP) of the linearized MHD equations is solved by employing the WKB approximation to describe the dynamical development of an initial perturbation. The eigenfrequencies as well as eigenfunctions are explicitly obtained and are verified numerically. The importance of considering the initial value problem (IVP) as well as the question of global stability for finite systems is emphasized and discussed in detail. It is further shown that thin enough disks are stable (global stability) but as their thickness grows increasing number of unstable modes participate in the solution of the IVP. However it is demonstrated that due to the localization of the initial perturbation the growth time of the instability may be significantly longer than the calculated inverse growth rate of the individual unstable eigenfunctions.

Keywords. Accretion disks – magnetic fields – MHD – instabilities

---

We consider geometrically thin disks and weak magnetic fields $V_A < c_s$ where $V_A = B_z/\sqrt{4\pi\rho_0}$ is the Alfvén velocity and c_s is the isothermal sound velocity. In order to model the finite thickness of the disk the following profile $\rho(r,z) = \rho_0(r)\exp(-z^2)$ is assumed for the density (the disk is assumed to be isothermal). It is assumed that the typical length scale for the variation of the physical quantities in the radial direction is much longer than that in the vertical direction. For that reason the disk model incorporates the vertical stratification while the radial structure of the disk ignored excepting of the velocity shear. Thus, due to the independence of the perturbations on the radial direction r is merely a parameter. It should be noted however that in the classical works of Velikhov (1959), Chandrasekhar (1960), and Balbus & Hawley (1991), an infinite cylinder has been considered and the effects of the boundary conditions as well as vertical structure of the disk were neglected. Therefore, in those works, naturally, the thickness of the disk does not play any role and consequently cannot influence the extent of the domains of instability.

To perform the stability analysis of such stratified disks the common practice of taking the Laplace transform for arbitrary initial conditions (IC), consequently reducing the set of linear equations describing dynamical system to the single inhomogeneous ordinary differential equation (ODE), and solving the latter subject to certain boundary conditions (BC), is employed [see for example the well-known work of Landau (1946)].

The BC of the considered problem are obtained due to the requirement that at $z \to \pm\infty$, where $\rho \to 0$ and $B_z = const$, the energy flux of the perturbation is finite.

The solution of such problems is facilitated by obtaining the natural frequencies of the bounded system ω_n for which the state vector of the linearized dynamical system may be written as $\mathbf{u_n}(z,t) = \mathbf{A}_n(z)e^{-i\omega_n t}$, where $\mathbf{A}_n(z)$ are the eigenfunctions of the BVP subjected to specific BC. See also Sano & Miyama (1999), Coppi & Coppi (2001), Coppi & Keyes (2003).

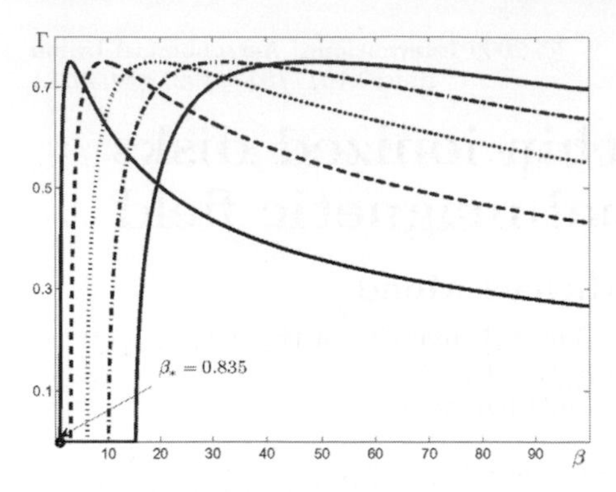

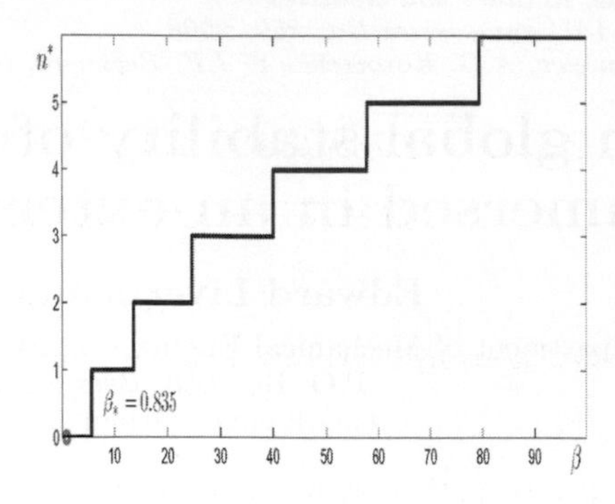

Figure 1. *Left panel.* Growth rate Γ vs plasma beta parameter for the first five eigenfunctions ($n = 0, .., 4$). *Right panel.* Axial standing wave number of the most unstable mode as a function of the plasma beta.

The full solution of the dynamical evolution in time of a localized Gaussian wave packet using such techniques has been examined recently by Liverts & Mond 2008.

We summarize our results as follows.

• It has been shown analytically and numerically that in the case of sufficiently thin disk in the sense of the disk thickness c_s/Ω and the Alfvén length V_A/Ω ratio which are defined also by the plasma beta $2c_s^2/V_A^2$ ($\beta < \beta_*$ Fig. 1), there is no instability at all and the disk is globally stable. It is instructive to note that the importance of studying the influence of the finite size of the system on its stability has been demonstrated by Budker (1956), Sturrock (1958) who reported that the size of system plays a stabilizing role for the two stream instability. If the disk is thin enough, then the localized initial perturbations do not grow in time, they just move back and forth with the local Alfvén velocity (Liverts & Mond 2008).

• If the thickness of the disk increases ($\beta > \beta_*$ Fig. 1,left panel) so does the number of unstable modes. Asymptotic estimation reveals that n^* (Fig. 1,right panel) is proportional to $\sqrt{\beta}$. This implies that the wavelength of the most unstable mode is of the order of $h/n^* \sim V_A/\Omega$ (where h is the thickness of the disk).

• Considering the boundary effects may have a large impact on the design of laboratory experiments (Couette flows) to model MRI. See Noguchi *et al.* (2002), Rüdiger *et al.* (2006), Ji *et al.* (2006).

References

Balbus, S. A. & Hawley, J. F. 1991, *ApJ* 376, 214
Budker, G. I. 1956, *Atomic Energy* 1, 673
Chandrasekhar, S. 1960, *Proc. Nat. Acad. Sci.* A46, 223
Coppi, B. & Coppi, P. S. 2001, *Phys. Rev. Lett.* 87, 051101
Coppi, B. & Keyes, E. A. 2003, *ApJ* 595, 1000
Ji, H. T., Burin, M., Schartman, E., & Goodman, J. 2006, *Nature* 444, 343
Landau, L. D., 1946, *Zh. Eksp. Teor. Fiz.* 16, 574 [1946, *J. Phys. USSR*, 10,25]
Liverts, E. & Mond, M. 2008, *MNRAS* (in press), astro-ph 0810.1864v1
Noguchi, K., Pariev, V. I., Colgate, S. A., Beckley, H. F., & Nordhaus, J. 2002, *ApJ* 575, 1151
Rüdiger, G., Hollerbach, R., Stefani, F., Gundrum, T., Gerbeth, G., & Rosner, R. 2006, *ApJ* 649, L145
Sano, T. & Miyama, S. M. 1999, *ApJ* 515, 776
Sturrock, P. A. 1958, *Phys. Rev. A* 112, 1488
Velikhov, E. P. 1959, *Zh. Eksp. Teor. Fiz.* 36, 1398 [1959, *Sov. Phys. JETP* 36, 995].

Cosmic Magnetic Fields:
From Planets, to Stars and Galaxies
Proceedings IAU Symposium No. 259, 2008
K.G. Strassmeier, A.G. Kosovichev & J.E. Beckman, eds.

doi:10.1017/S1743921309030233

Magnetic braking and ambipolar diffusion in metal-poor protostars

Hans Zinnecker

Astrophysikalisches Institut Potsdam, An der Sternwarte 16, D-14482 Potsdam, Germany
email: hzinnecker@aip.de

Abstract. In this short note, we raise the issue how magnetic braking of prestellar and protostellar condensations depends on the metallicity of the molecular gas cloud. We suggest that the degree of ionization, which determines the timescale for redistribution of magnetic flux (ambipolar diffusion) in the dense cloud core, depends linearly on the heavy element and dust abundance of the gas. This implies that magnetic braking is less efficient in metal-poor condensations, and hence more of the angular momentum problem of star formation must be solved by other means, such as fragmentation into (wider) binary systems. Observations of orbital periods (separations) of binary systems among the metal-poor Galactic halo stars could test this prediction.

Keywords. Magnetic fields – diffusion – protostars – ISM: molecules

---

1. Angular momentum problem in star formation

Magnetic braking of rotating, slowly contracting, self-gravitating protostellar cores via torsional Alfven-waves to the outside cloud medium can help to solve the angular momentum problem in low-mass star formation (Ebert *et al.* 1960, Mestel 1965, Mouschovias and Paleologou 1979). In the case of an aligned magnetic rotator up to two orders of magnitude in specific angular momentum (J/M) can be lost by magnetic braking, before dynamical collapse sets in due magnetic flux redistribution (*ambipolar diffusion*) in the core, at central gas densities $> 10^4$ to $10^6 cm^{-3}$). We refer to the key papers of Basu and Mouschovias (1994, 1995ab) and the review by Mouschovias (1996).

2. Ambipolar diffusion

The timescale for ambipolar diffusion to drive the core unstable is (Mouschovias 1996):

$$t\,(AD) = 2\left(x_i/10^{-7}\right) Myr$$

where x_i is the fractional degree of ionisation ($x_i = n_i/n_{tot}$) which scales as $n_{tot}^{-1/2}$ for total molecular gas density n_{tot} (Elmegreen 1979, Nakano 1979). The scaling constant is proportional to the metal abundance and the dust content of the molecular gas (with Fe, Mg, Na being the dominant donors of free electrons and most grains negatively charged). Cosmic rays (100 MeV protons) provide a steady source of ionizing flux (Elmegreen 1979).

3. Observational prediction for metal-poor stars

The theoretical prediction is that the ambipolar diffusion timescale is shorter for lower-metallicity molecular gas (owing to the lower fractional degree of ionisation x_i). Thus the collapse of protostellar cores in metal-poor conditions (such as in the LMC/SMC and in the early Galactic halo) would result in higher angular momentum systems, i.e. wider binaries and/or faster rotating stars. This prediction can be tested observationally (cf. Mouschovias 1977; Zinnecker *et al.* 2004).

4. Magnetic collapse: from subcritical to supercritical cores

Strong magnetic fields inhibit molecular cloud collapse and star formation. There is a critical mass which must be exceeded for a given magnetic field strength B and cloud gas density n

$$M\,(crit) = 5\,(B/10\mu G)^3 \,/\, \left(n/10^4 cm^{-3}\right)^2 M_\odot$$

for gravitational instability to begin. We call a molecular cloud core "supercritical" when the magnetic field is not strong enough to prevent collapse, "subcritical" when it can prevent collapse. Even in the latter case, a molecular cloud core evolves through quasi-static contraction through the motion or slow drift of neutral gas past the ionized component (ambipolar diffusion). It is only the ionized component (electrons, charged grains) which is held up by magnetic forces while the neutral gas is coupled to the ionized component by frequent ion-neutral collisions. If this coupling is perfect, the magnetic flux is frozen into the cloud. At high densities (typically in excess of 10^6 cm^{-3}) the degree of ionization in cloud cores starts to drop sufficiently (from 10^{-6} to 10^{-8}, say) so that the assumption of flux freezing breaks down. Then ambipolar diffusion starts to redistribute the magnetic flux $\Phi = BR^2$ in the center of the cloud core, thereby increasing the central mass-to-flux ratio until it exceeds the critical value for collapse: the core goes "supercritical". An inside-out dynamical collapse (somewhat retarded compared to free-fall without magnetic forces) sets in; the prestellar core turns into a real protostar, including a magnetized rotating disk and bipolar jet!

3D MHD simulations of collapsing rotating magnetic 1 $M_\odot$ cloud cores, using adaptive mesh refinement techniques, have been published (Ziegler 2005) which show that these cores can fragment into wide binary systems, when the core mass-to-flux ratio exceeds the value for supercritical collapse by a factor of about 2, even if the magnetic flux is frozen (i.e. without ambipolar diffusion).

Acknowledgements

The author thanks John Beckman and Klaus Strassmeier for making my attendance possible. It is a pleasure to thank T. Mouschovias and U. Ziegler for extensive discussions.

References

Basu, Sh. & Mouschovias T.Ch. 1994, *ApJ* 432, 720
Basu, Sh. & Mouschovias T.Ch. 1995a, *ApJ* 452, 386
Basu, Sh. & Mouschovias T.Ch. 1995b, *ApJ* 453, 271
Ebert, R., v. Hoerner S., & Temesvary, St. 1960, *in: Die Entstehung von Sternen durch Kondensation diffuser Materie*, Springer Verlag, pp. 290
Elmegreen, B. G. 1979, *ApJ* 232, 729
Mestel, L. 1965, *Quarterly Journal of the Royal Astronomical Society* 6, 265
Mouschovias, T.Ch. 1977, *ApJ* 211, 147
Mouschovias, T.Ch. & Paleologou E. V. 1979, *ApJ* 230, 204
Mouschovias, T.Ch. 1985, *A&A* 142, 41
Mouschovias, T.Ch. 1996, *in: Solar and Astrophysical Magnetohydrodynamic Flows*, eds. K. C. Tsinganos, Kluwer Academic Publishers, p. 505
Nakano, T. 1979, *PASJ* 31, 697
Ziegler, U. 2005, *A&A* 435, 385
Zinnecker, H., Köhler R., & Jahreiß, H. 2004, *RevMexAA* 21, 33

Cosmic Magnetic Fields:
From Planets, to Stars and Galaxies
Proceedings IAU Symposium No. 259, 2008
K.G. Strassmeier, A.G. Kosovichev & J.E. Beckman, eds.

doi:10.1017/S1743921309030245

Three-dimensional MHD simulations of molecular cloud fragmentation regulated by gravity, ambipolar diffusion, and turbulence

Takahiro Kudoh[1] **and Shantanu Basu**[2]

[1]Division of Theoretical Astronomy, National Astronomical Observatory of Japan,
Mitaka, Tokyo 181-8588, Japan
email: kudoh@th.nao.ac.jp

[2]Department of Physics and Astronomy, University of Western Ontario,
London, Ontario N6A 3K7, Canada
email: basu@astro.uwo.ca

Abstract. We find that the star formation is accelerated by the supersonic turbulence in the magnetically dominated (subcritical) clouds. We employ a fully three-dimensional simulation to study the role of magnetic fields and ion-neutral friction in regulating gravitationally driven fragmentation of molecular clouds. The time-scale of collapsing core formation in subcritical clouds is a few $\times 10^7$ years when starting with small subsonic perturbations. However, it is shortened to approximately several $\times 10^6$ years by the supersonic flows in the clouds. We confirm that higher-spacial resolution simulations also show the same result.

Keywords. MHD – stars: formation – ISM: magnetic fields

1. Introduction

When the gravitational energy is larger than the magnetic energy (supercritical), the cloud is fragmented by gravitational instability in a free-fall time of the cloud ($\sim 10^6$ years). On the other hand, when the gravitational energy is smaller than the magnetic energy (subcritical), the cloud is gravitationally stable because magnetic field prevents the contraction of the cloud. However, because the molecular cloud contains a lot of neutrals as well as some ions, magnetic diffusion induced by ion-neutral friction (amibipolar diffusion) occurs in the cloud. Due to this effect, gravitational instability develops gradually over the diffusion time even when the cloud is subcritical.

It has been suggested that subcritical clouds have a time-scale problem: the age spreads of young stars in nearby molecular clouds is often $\sim 10^6$ years, while the ambipolar diffusion time is typically $\sim 10^7$ years. Recently, however, Li & Nakamura (2004) have shown that the time-scale of mildly subcritical cloud fragmentation is reduced by supersonic turbulence to $\sim 10^6$ years by performing 2D simulations in the thin-disk approximation. In this paper, we study the 3D extension of the model by including the self-consistent calculation of vertical structure of the cloud.

2. Results

The initial cloud is a gas layer that is in a self-gravitational equilibrium along uniform magnetic field lines. The initial cloud is assumed to be slightly subcritical (the magnetic field strength is 2 times larger than the critical value). An initial random velocity perturbation (v_a) is input perpendicular to the magnetic field. 3D-MHD simulation with

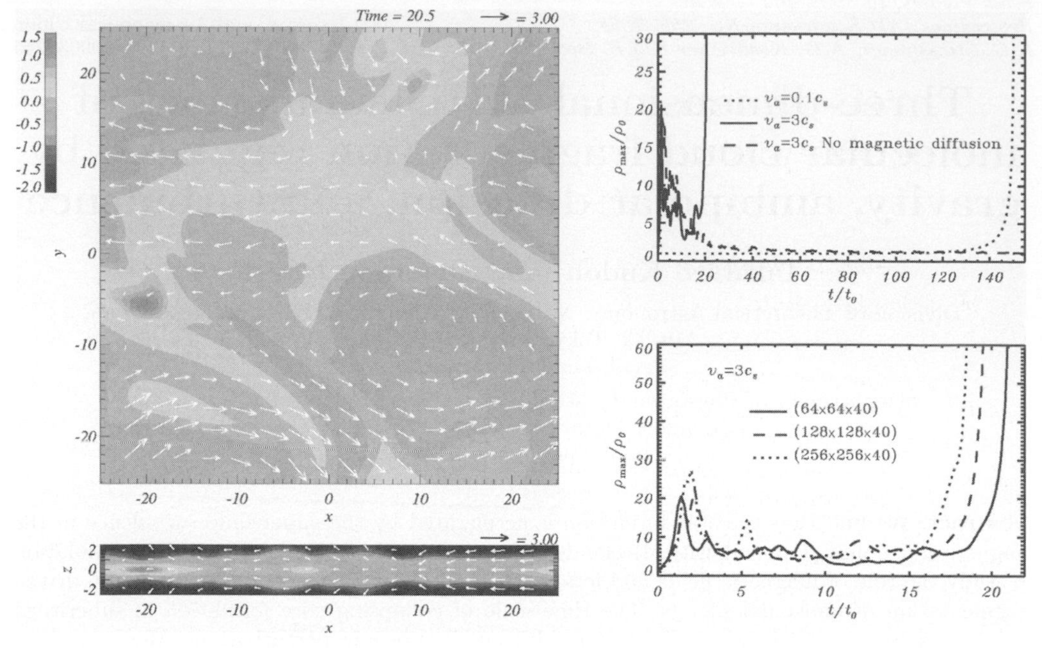

Figure 1. (Left panel) Logarithmic density image. It shows at $t = 20.5t_0$ ($\sim 4\times10^6$ years) for the case of initially supersonic perturbation ($v_a = 3c_s$), where c_s is initial sound speed. A collapsing core is located in the vicinity of $x = -20H_0$ and $y = -6H_0$, where H_0 is the scale height of the gas layer ($H_0 \sim 0.05$pc). ; (Right upper panel) Time evolution of the maximum density for the different strength of the perturbation. The unit time (t_0) is $t_0 = H_0/c_s \sim 2 \times 10^5$ year. The timescale of collapsing core formation for the supersonic perturbations case ($v_a = 3c_s$) is much shorter than that for the subsonic perturbation case ($v_a = 0.1c_s$). The dashed line shows that there is no collapsing core formation in the subcritical cloud without ambipolar diffusion; (Right lower panel) Time evolution of the maximum density for the different spatial resolutions of numerical simulations. (The left and right upper panels corresponds to the results of $(N_x, N_y, N_z) = (64 \times 64 \times 40)$ grid sizes.) Each line shows the evolution for an initially supersonic perturbation ($v_a = 3c_s$). Higher resolution simulations confirm the same result.

ambipolar diffusion is performed (see details in Kudoh *et al.* 2007 and Kudoh & Basu 2008).

Results are summarized in figure 1. Our three-dimensional MHD simulations have shown that the supersonic nonlinear flows significantly reduce the timescale of collapsing core formation in subcritical clouds. It is of order several $\times10^6$ years for typical parameters, or ~ 10 times less than found in the linear initial perturbation studies. The short timescale of the core formation is caused by the small ambipolar diffusion time when the turbulent compression creates structure with small scale length (Kudoh & Basu 2008).

Acknowledgements

The numerical computations were done mainly on the VPP5000 and SX-9 at the National Astronomical Observatory of Japan.

References

Kudoh, T., Basu, S., Ogata, Y., & Yabe, T. 2007, *MNRAS* 380, 499

Kudoh, T. & Basu, S. 2008, *ApJ* 679, L97

Li, Z.-Y. & Nakamura, F. 2004, *ApJ* 609, L83

Cosmic Magnetic Fields:
From Planets, to Stars and Galaxies
Proceedings IAU Symposium No. 259, 2008
K.G. Strassmeier, A.G. Kosovichev & J.E. Beckman, eds.

doi:10.1017/S1743921309030257

3D global MHD simulations of a proto-planetary disk: dead zone and large-scale magnetic fields

Natalia Dzyurkevich, Mario Flock and Hubert Klahr

Max-Planck Institute for Astronomy, Königstuhl 17, D-69117 Heidelberg, Germany
email: natalia@mpia.de

Abstract. We present 3D global MHD simulations of proto-planetary disks calculated with the ZeusMP code. We focus on gas dynamics; the magnetic diffusivity and temperature are fixed during the simulation. A zone with low gas ionization at the midplane is included within ± 2 scale heights. We mimic the 'snow'-line radius with one order of magnitude jump in magnetic diffusivity η. Resulting turbulent Maxwell and Reynolds stresses are present at the midplane despite of low ionization. We find no radial inhomogeneities in turbulent α stress for a mild ionization contrast at the 'snow'-line. A smooth azimuthal magnetic field is produced in the dead zone which may be a driving force for a weak accretion flow.

Keywords. MHD – turbulence – accretion disks – protoplanetary disks

---

Magnetic fields play a key role in extracting excess angular momentum in proto-planetary disks, making accretion of the material onto star possible. Radial inhomogeneities in accretion rates, thus building up pressure ridges, have been suggested to be a solution for one of the major problems in the theory of planet formation - building of planetesimals (Kretke & Lin 2007). In ideally ionized disk, the magneto-rotational instability (MRI) provides velocities high enough to prevent the dust grains from growing. Special events like pressure bumps and absence of radial drift may help to overcome the growth barrier, yet the most severe reason for rapid fragmentation is the turbulent motion itself (Brauer *et al.* 2008). Therefor we focus on MRI in poorly ionized disk. The properties of global magnetic fields are worth to be investigated in global 3D simulations. The launching of outflows and jets is very sensitive to field topology (Beckwith *et al.* 2008).

We consider the MRI turbulence in partly-ionized circumstellar disk with a 'dead' zone, including the water sublimation front called 'snow'-line. We solve the set of MHD equations using 3D global simulations on a spherical grid (r, Θ, ϕ),

$$\frac{\partial \vec{B}}{\partial t} = \nabla \times [\vec{u} \times \vec{B} - \eta(r,\Theta)\nabla \times \vec{B}], \tag{0.1}$$

$$\frac{\partial \vec{u}}{\partial t} = -\nabla P + \rho\nabla\Psi + \frac{1}{4\pi}[\nabla \times \vec{B}] \times \vec{B}, \tag{0.2}$$

$$\frac{\partial \rho}{\partial t} + \nabla \cdot (\rho\vec{u}) = 0. \tag{0.3}$$

The notation is the usual one. Ψ is a point-mass gravitational potential. We use the locally isothermal approach to describe vertically stratified disk, adopting $P = c_s^2(r)\rho(r,\Theta)$ with a pressure scale height $H/R = 0.07$. We choose a purely azimuthal magnetic 'seed' field, which is force-free and $P_{\rm gas}/P_{\rm mag} = 25$. The models are listed in Table 1. Ionization profiles used in model **m1** are based on Turner *et al.* (2007). The mid-plane values of η depend on the presence of ice dust grains, so that $\eta_a = 0.016$ in the presence of 1μ

Table 1. Model properties and midplane α-stresses before (α_b) and behind (α_a) 'snow'-line.

Model	Domain Size	Resolution	η_b	η_a	α_b	α_a
m1	[8 AU : 8.4H: $\pi/4$]	[256:128:64]	7.10^{-4}	0.016	5.10^{-5}	5.10^{-5}
m2	[8 AU : 8.4H: $\pi/4$]	[256:128:64]	0	0.016	10^{-3}	10^{-6}

Model	Domain Size	Resolution	η_b	η_a	α_b	α_a
m1	[8 AU : 8▷4H: π◁4]	[256:128:64]	7▷10^{-4}	0▷016	5▷10^{-5}	5▷10^{-5}
m2	[8 AU : 8▷4H: π◁4]	[256:128:64]	0	0▷016	10^{-3}	10^{-6}

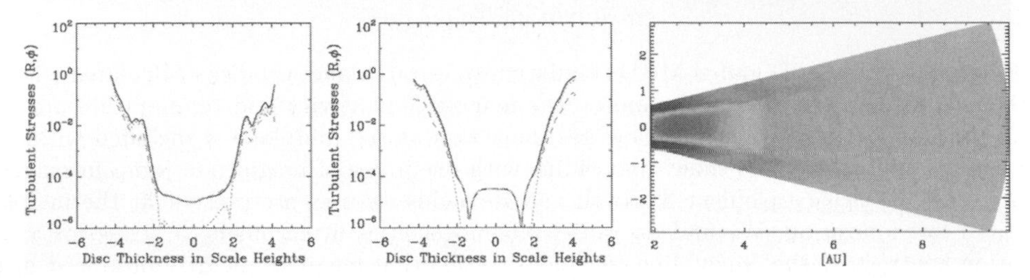

Figure 1. Left: Turbulent stresses between 2AU and 4 AU. Middle: Turbulent stresses between 6AU and 8AU. Red dotted line shows Maxwell stress, red dashed line is Reynolds stress, black line is total stress. Right: azimuthal magnetic field after $t = 240$ orbits, red-green-blue colors the change of sign from positive to negative.

ice particles, and $\eta_b = 7.10^{-4}$ if the temperature is high enough for water sublimation. 'Snow'-line is chosen to be at 4.5 AU. Model **m2** is made for less realistic situation when gas is totally ionized before 'snow'-line (Table 1).

In case of one order of magnitude jump in ionization (model **m1**), the total alpha-stress is around $5 \cdot 10^{-5}$ at the mid-plane, when calculated separately for locations before and after the 'snow'-line (Fig. 1). Model **m2** with extreme high ionization jump (from ideally ionized matter into resistive medium with 1μ-size dust) shows the expected steep jump in stress α at the 'snow'-line (Table 1). This leads to the question which ionization thresholds exist in the real accretion disks. Taking into account the recombination on grains larger then 1μ may lead to more realistic radial α profile, filling the gap between the models **m1** and **m2**. It is note-worthy that the accretion in partly ionized midplane is determined by Reynolds stresses, whereas the magnetic pressure is indeed significantly reduced. The turbulent velocity dispersion is about 0.01 (in local sound speed units) in mid-plane layer, and 0.2 in upper layers. At 4 AU, this corresponds to the velocities of about 10 m/s and 200 m/s, making the dust growth through coagulation possible.

In the 'dead'-zone ($\pm 2H$) a smooth azimuthal magnetic field B_ϕ is produced (Fig. 1), confirming the result of Turner & Sano (2008). For perfectly ionized disk we observe the periodical change of sign in B_ϕ with respect to the midplane. When a 'dead'-zone is included, only weak periodical change of sign in radial velocity remains. The field symmetry and effects of the boundary conditions, as well as extension of models **m1** and **m2**, will be discussed in the following publication.

References

Kretke, K. A. & Lin, D. N. C. 2007, *ApJ* 664, L55

Brauer, F., Henning, Th., & Dullemond, C. P. 2008, *A&A* 487, L1

Beckwith, K., Hawley, J. F., & Krolik, J. H. 2008, *ApJ* 678, 1180

Turner, N. J., Sano, T., & Dziourkevitch, N. 2007, *ApJ* 659, 729

Turner, N. J. & Sano, T. 2008, *ApJ* 679, 131

Cosmic Magnetic Field:
From Planets, to Stars and Galaxies.
Proceedings IAU Symposium No. 259, 2008
K.G. Strassmeier, A.G. Kosovichev & J.E. Beckman, eds.

doi:10.1017/S1743921309030269

Dead zone formation and non-steady hyperaccretion in collapsar disks

Youhei Masada

Institute of Astronomy and Astrophysics, Academia Sinica
Roosevelt Rd. Sec. 4 Taipei 10617, Taiwan R.O.C.
email: masada@asiaa.sinica.edu.tw

Abstract. In ultra dense and hot region realized in stellar core-collapse, neutrino takes major role in energy and momentum transports. We investigate the growth of magnetorotational instability (MRI) in neutrino viscous matter by using linear theory. It is found from the local linear analysis that the neutrino viscosity can suppress the MRI in the regime of weak magnetic field ($B \ll 10^{14}$G). This suggest that MHD turbulence sustained by the MRI might not be driven efficiently in the neutrino viscous media. Applying this result to collapsar disk, which is known as the central engine of gamma-ray burst (GRB), we find that the MRI can be suppressed only in its inner region. Based on this finding, a new evolutionary scenario of collapsar disk, "Episodic Disk Accretion Model" are proposed.

Keywords. Gamma-rays: bursts – accretion disks – MHD

1. Introduction

Gamma-ray bursts (GRBs) are the most energetic event in the universe. GRBs are generally considered to be powered by hyperaccretion onto a stellar-mass black hole (Woosley 1993). The hyperaccretion rate is an order of $0.1M_{\rm sun}s^{-1}$ and the release of gravitational energies powers the burst.

A key process for releasing the gravitational energy of disk systems is angular momentum transport. As in the case of the other disk systems, the MRI-driven turbulence is believed to play an essential role in the angular momentum transport of the hyperaccretion disk. Physical conditions of the disk are quite different from the others. Because it is ultra-dense and hot like supernova core, the energy and momentum are mainly transported by the neutrino in neutrino-opaque regions. Masada *et al.* (2007a) investigate the effect of neutrino transport on the MRI in the supernova core and show that the neutrino viscosity can suppress the MRI in weak magnetized conditions. We can thus apply these results to the hyperaccretion disk here.

The growth time of the MRI in the absence of viscosity is given by λ/v_A, where λ is the wavelength of a perturbation and $v_A = B/(4\pi\rho)^{1/2}$ is the Alfvén speed. The MRI is suppressed dramatically if the growth time is longer than the viscous damping time $\sim \lambda^2/\nu$, where ν is the kinematic viscosity. Then a large enough viscosity can reduce the linear growth rate of the MRI. Since the typical wavelength of the MRI is $\lambda \sim v_A/\Omega$ with angular velocity Ω, the condition for the efficient growth of the MRI can be given as

$$R_{\rm MRI} \equiv LU/\nu = v_A^2/\nu\Omega \gtrsim 1 , \qquad (1.1)$$

where $R_{\rm MRI}$ is the Reynolds number for MRI. We choose v_A/Ω as the typical length scale L and v_A as the velocity scale U. In the neutrino-opaque matter, the neutrino viscosity becomes quite large, so that the condition (1.1) would not be satisfied.

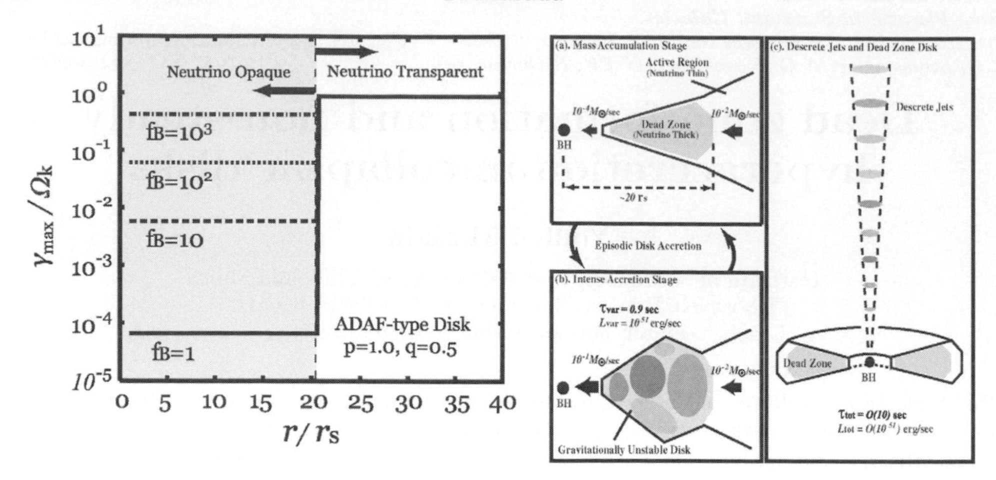

Figure 1. (a). The maximum growth rate of the MRI in ADAF-type collapsar disks as a function of disk radius. (b). Episodic Disk Accretion Model proposed in Masada *et al.* 2007b.

2. Episodic Disk Accretion Model

We investigate where the MRI operates in hyperaccreting collapsar disks focusing on the neutrino viscous effect. As the radial structure of the collapsar disk, we adopt the ADAF-type simple power law model with the surface density $\Sigma(r) = \Sigma_0 \hat{r}^{-0.5}$, the disk temperature $T(r) = T_0 \hat{r}^{-1}$ and magnetic field $B(r) = f_B B_0 \hat{r}^{-1}$, where $\hat{r}$ is the disk radius normalized by Schwarzschild one r_s, $\Sigma_0 = 10^{18}\mathrm{gcm}^{-2}$, $T_0 = 4 \times 10^{11}\mathrm{K}$, and $B_0 = 10^{11}\mathrm{G}$ (see Masada *et al.* 2007b for more details.).

Figure 1a shows the maximum growth rate of the MRI for the cases with different field parameter $f_B = 1$–10^3. The units of vertical and horizontal axes are the Keplerian velocity and Schwarzschild radius. The critical radius dividing the neutrino-thick and thin regions locates at $r_{\mathrm{crit}} \simeq 20 r_s$. Gray shaded area represents the neutrino opaque region. The strong field more than 10^{14}G is found to be necessary for the efficient growth of the MRI in collapsar disks. When the magnetic field is weaker than the critical value, the inner disk can be "dead zone" where the MRI is suppressed by the neutrino viscosity.

Finally, we consider a dead zone formed around the inner part of a collapsar disk. Then the angular momentum in the dead zone can be transported by the neutrino viscosity itself with the amplitude $\alpha_v \simeq 10^{-4}$. On the other hand, in the active region, the turbulent viscosity sustained by the MRI should take the angular momentum transport. Nonlinear studies tell us that the α-parameter of the MRI-driven turbulence would be $\alpha_t \simeq 10^{-2}$. The baryonic matter is thus expected to be accumulated into the dead zone from outer active region. If the mass accumulation continues, the inner dead zone becomes gravitationally unstable at some stage. Then the gravitational torque causes intermittent mass accretion and drives discrete jets, which are thought to be the origin of short-term variability in the prompt emission of GRBs. The prospective evolutionary scenario of collapsar disks "Episodic Disk Accretion Model" is depicted schematically in Figure 1b, and which can explain various observational features of GRBs qualitatively (see Masada *et al.* 2007b for more details).

References

Masada, Y., Sano, T., & Shibata, K. 2007a, *ApJ* 655, 447
Masada, Y., Kawanaka, N., Sano, T., & Shibata, K. 2007b, *ApJ* 663, 437
Woosley, S. E. 1993, *ApJ* 405, 273

Cosmic Magnetic Field:
From Planets, to Stars and Galaxies.
Proceedings IAU Symposium No. 259, 2008
K.G. Strassmeier, A.G. Kosovichev & J.E. Beckman, eds.

doi:10.1017/S1743921309030270

Axisymmetric magneto-rotational instability in viscous accretion disks

Youhei Masada[1] and Takayoshi Sano[2]

[1]Institute of Astronomy and Astrophysics, Academia Sinica
email: masada@asiaa.sinica.edu.tw

[2]Institute of Laser Engineering, Osaka University; email: sano@ile.osaka-u.ac.jp

Abstract. Axisymmetric MRI in viscous accretion disks is investigated. The linear growth of the viscous MRI is characterized by the Reynolds number $R_{\rm MRI} \equiv v_A^2/\nu\Omega$, where v_A is the Alfvén velocity, ν is the kinematic viscosity, and Ω is the angular velocity of the disk. Although the linear growth of the MRI is suppressed as the Reynolds number decreases, its nonlinear behavior is found to be almost independent of $R_{\rm MRI}$. At the nonlinear stage, the channel flow grows and the Maxwell stress increases even though $R_{\rm MRI}$ is much smaller than unity. Nonlinear behavior of the MRI in the viscous regime can be explained by the characteristics of the linear dispersion relation. Applying our results to the case with both viscosity and resistivity, it is anticipated that the critical value of the Lundquist number $S_{\rm MRI} \equiv v_A^2/\eta\Omega$ for active turbulence would depend on the magnetic Prandtl number $Pm \equiv \nu/\eta$, where η is the magnetic diffusivity.

Keywords. Turbulence – accretion disks – methods: numerical

---

1. Introduction

The magnetic Prandtl number Pm takes a wide range of values in astrophysical disk systems. In protoplanetary disks, the magnetic Prandtl number is much smaller than unity because of their low ionization degree (Nakano 1984). In the disks of compact X-ray sources and AGNs, it ranges from $\simeq 10^{-3}$ to 10^3 depending on the disk radius (Balbus & Henri 2008). More systematic study on the MRI in the presence of both viscosity and resistivity is thus quite important for understanding the disk accretion.

One important unsettled matter is the role of the viscosity at the nonlinear stage of the MRI. In general, the viscosity as well as the resistivity can suppress the linear growth of the MRI. However the dependence of nonlinear outcome on the Prandtl number indicates that the role of the viscosity in MRI turbulence could be different from that of the resistivity (Lesur & Longaretti 2007; Fromang et al. 2007). Focusing on two non-dimensional parameters, Reynolds number $R_{\rm MRI}$ and Lundquist number $S_{\rm MRI}$, we clarify the difference in nonlinear features of the MRI between the viscous and resistive systems.

2. Results

The time- and volume-averaged $\alpha_{\rm tot}$ $[\equiv (\langle v_x \delta v_y \rangle - \langle B_x B_y \rangle/4\pi)/\langle P \rangle]$ at the nonlinear stage is depicted as a function of the initial $R_{\rm MRI}$ and $S_{\rm MRI}$ in Fig. 1. The diamonds show the results in the viscous fluid, and the crosses are those in the resistive one. The upward arrow over-plotted on the symbols denotes that the value is the lower limit, and the downward arrow stands for decaying models and thus the stress is the upper limit.

The stresses at the nonlinear stage are almost the same when $R_{\rm MRI} \gtrsim 1$ or $S_{\rm MRI} \gtrsim 1$. However, a huge difference can be seen in the highly diffusive regime. In the presence of the ohmic dissipation, the stress rapidly decreases with decreasing $S_{\rm MRI}$. For the models

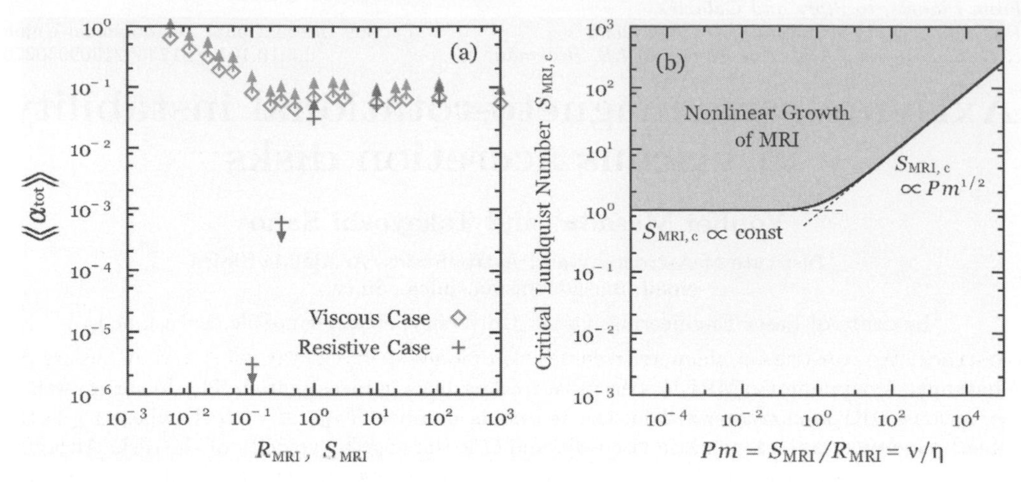

Figure 1. (a) Time- and volume-averaged α_{tot} at the nonlinear stage as a function of the initial Reynolds number R_{MRI} and Lundquist number S_{MRI}. (b) The critical Lundquist number for active MRI-driven turbulence as a function of the magnetic Prandtl number Pm.

with the kinematic viscosity, on the other hand, it increases with the decrease of R_{MRI}. The inverse correlation between $\langle\langle\alpha_{tot}\rangle\rangle$ and R_{MRI} for the cases with large viscosity could be originated from stable growth of a channel flow [see Masada & Sano 2008 (MS08)].

3. Discussion

We focus on the critical wavelength obtained from the linear theory to briefly explain the nonlinear behavior of the MRI. In the viscous fluid, the critical wavelength is given by $\lambda_{crit} \simeq v_A/\Omega$ despite the size of R_{MRI}. Even if R_{MRI} is much smaller than unity, the critical wavelength thus shifts to larger scale as the instability grows. Then the system always evolves toward a less dissipative state and is not saturated. On the other hand, in the resistive case, there is a critical point at which the critical wavelength switches from the decreasing function of the field strength to the increasing one. The MHD turbulence can thus decay in resistive fluid if $S_{MRI} \lesssim 1$ (see MS08 for more details).

Our results suggest that the nonlinear stage of the MRI can be anticipated by the shape of the critical wavelength. We can thus predict the nonlinear behavior of the MRI-driven turbulence in the system with both viscosity and resistivity from the linear dispersion relation (Pessah & Chan. 2008). In Fig. 1b, the critical Lundquist number $S_{MRI,c}$ is plotted as a function of Pm. This diagram implies that the MRI turbulence would be suppressed if S_{MRI} is less than a critical value. In the regime of $Pm \gg 1$, it is proportional to the square root of Pm and remains to be constant in the range $Pm \ll 1$. It is interesting that this can reproduce the result of the three-dimensional study for the doubly diffusive MRI performed by Fromang *et al.* (2007) qualitatively.

References

Balbus, S. A. & Henri, P. 2008, *ApJ* 674, 408
Fromang, S., Papaloizou, J., Lesur, G., & Heinemann, T. 2007, *A&A* 476, 1123
Lesur, G. & Longaretti, P.-Y. 2007, *MNRAS* 378, 1471
Nakano, T. 1984, *Fundamentals of Cosmic Physics* 9, 139
Masada,Y. & Sano, T. 2008, *ApJ* 689, 1234
Pessah, M. E. & Chan, C.-K. 2008, *ApJ* 684, 498

Cosmic Magnetic Field:
From Planets, to Stars and Galaxies.
Proceedings IAU Symposium No. 259, 2008
K.G. Strassmeier, A.G. Kosovichev & J.E. Beckman, eds.

doi:10.1017/S1743921309030282

Large solar-type magnetic reconnection model for magnetar giant flare

Youhei Masada

Institute of Astronomy and Astrophysics, Academia Sinica
email: masada@asiaa.sinica.edu.tw

Abstract. We construct a magnetic reconnection model for magnetar giant flare in the framework of solar flare/coronal mass ejection theory. As is the case with the solar flare, the explosive magnetic reconnection plays a crucial role in the energetics of the magnetar flare. A key physics controlling the energy transport in the system, on the other hand, is the radiative process unlike that in the solar flare. After the release of the magnetic energy via the magnetic reconnection, the radiative heat flux drives the baryonic evaporation. Our model can predict that the baryonic matter evaporated in the preflare stage would be the origin of the radio emitting ejecta observed in association with the giant flare on 2004 December 27 from SGR1806-20.

Keywords. Gamma rays: bursts – stars: individual (SGR 1806-20) – MHD – stars: neutron

1. Introduction

There has recently been growing evidence that soft gamma-ray repeaters (SGRs) is the population of strongly magnetized neutron star, so called Magnetar (Thompson & Duncan 1995). Activities in this object are believed to be powered by the dissipation of strong magnetic fields. Here we focus on the giant flare with enormous energy and long bursting duration which is observed from SGRs.

The giant flare from SGR1806-20 is the most recent and energetic event. It is characterized by an ultra-luminous hard spike with the energy 10^{46} erg, which decays rapidly into a soft pulsating tail. A preflare activity is additionally detected before the main burst. The exceptional phenomenon observed in association with the flare is the radio emitting ejecta loaded by the baryonic matter of 10^{24}g. The origin of the baryon-load is, however, not understood in the context of existing model (Thompson & Duncan 1995).

Analogous mass ejection events are observed accompanying with solar flare, that is called coronal mass ejection (CME). Observational features of the solar flare/CME are similar to those of the magnetar flare (Lyutikov 2006). The solar flare/CME theory is applied for constructing an alternative model for the magnetar giant flare.

2. Magnetic Reconnection Model

In the magnetar flare, the balance between the reconnection heating and the radiative cooling gives the flare temperature unlike the solar flare controlled by the conductive cooling (Shibata & Yokoyama 1999). Considering that the blackbody cooling retains the thermal balance of the system, the balance equation is, with the flare duration Δt; $E_{\rm GF} = (4\pi R^2 \sigma_B T_{\rm f}^4)\Delta t$, where $E_{\rm GF} = (B^2/4\pi)RL^2$ is the reconnection-induced energy.

After the flare begins, the released energy is transported to the magnetar crust by the photon flux. Then the incident radiative heat flux should counterbalance with the outgoing enthalpy flux of the evaporation flow. The baryon number density of the evaporation flow $n_{\rm ev}$ is thus obtained from the balance equation; $F_{\rm GF} \simeq (n_{\rm ev} k_B T_{\rm f}) v_{\rm ev}$, where

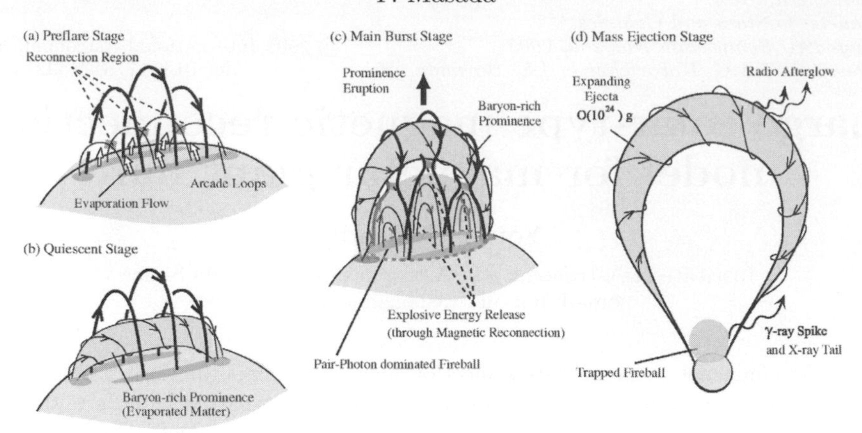

Figure 1. Solar-type Magnetic Reconnection Model for Magnetar Giant Flare

$F_{\rm GF} = E_{\rm GF}/(4\pi R^2 \Delta t)$ is the radiative heat flux, $v_{\rm ev}$ is the velocity of the evaporation flow. For given parameters $E_{\rm GF}$, Δt and R, we can obtain $T_{\rm f}$, $v_{\rm ev}$, and $n_{\rm ev}$ by solving these equations. The evaporated mass can be thus given by a simple scaling relation,

$$M_{\rm ev} = 3.4 \times 10^{24} (R/10^6 {\rm cm})^2 (T_{\rm f}/10^8 {\rm K})^3 (\Delta t/1.0 {\rm sec}) \ .$$

Our model suggest that the preflare stage with $T_{\rm f} \simeq 10^8$ K and $\Delta t \simeq 1.0$ sec is suitable for suppling the baryonic matter loaded in the radio emitting ejecta observed with the giant flare from SGR1806-20 (see Masada *et al.* 2008 for more details).

3. A Possible Scenario for the Giant Flare from SGR 1806-20

In the preflare stage, the magnetic energy stored in the active region is released via the magnetic reconnection. The radiative heat flux transports the released energy and heats the magnetar crust up. Then the pressure of the crustal matter increases drastically and drives the upward evaporation flow. A hot and dense prominence trapped by the post-flare loop is created by the evaporated matter with the mass $M_{\rm ev} \simeq 10^{24}$g.

After the preflare stage, the prominence eruption causes global field reconfiguration and triggers main burst. The released energy in the main burst stage is converted into the kinetic energy of the erupting matter and the radiative energy of the post-flare loop. The ejected matter would be observed as the radio emitter. The remained post-flare loop can be the origin of the trapped fireball, and which produces the luminous γ-ray spike and the subsequent hard X-ray pulsating tail.

Our proposed scenario for the Magnetar giant flare is summarized in Figure 1 schematically. An important prediction from our model is that the preflare activity would be the origin of the baryonic matter loaded in the radio emitting ejecta. This suggest that the radio afterglow is expected to be observed only in a giant flare with preflare activity.

References

Lyutikov, M. 2006, *MNRAS* 367, 1594
Masada, Y., Nagataki, S., Shibata, K., & Terasawa, T. 2008, arXiv:0803.3818
Shibata, K. & Yokoyama, T. 1999, *ApJL* 526, L49
Thompson, C. & Duncan, R. C. 1995, *MNRAS* 275, 255

Cosmic Magnetic Fields:
From Planets, to Stars and Galaxies
Proceedings IAU Symposium No. 259, 2008
K.G. Strassmeier, A.G. Kosovichev & J.E. Beckman, eds.

doi:10.1017/S1743921309030294

Off-equatorial circular orbits in magnetic fields of compact objects

Zdeněk Stuchlík[1], **Jiří Kovář**[1] **and Vladimír Karas**[2]

[1]Institute of Physics, Faculty of Philosophy and Science, Silesian University in Opava, Bezručovo nám. 13, CZ-746 01, Opava, Czech Republic
email: Jiri.Kovar@fpf.slu.cz

[2]Astronomical Institute, Academy of Sciences, Boční II, CZ-141 31, Prague, Czech Republic
email: Vladimir.Karas@cuni.cz

Abstract. We present results of investigation of the off-equatorial circular orbits existence in the vicinity of neutron stars, Schwarzschild black holes with plasma ring, and near Kerr-Newman black holes and naked singularities.

Keywords. Black hole physics – stars: neutron – magnetic fields – stellar dynamics

---

The motivation of our work was the analysis of charged particle motion in a pure dipole magnetic field done by Störmer (1955), the so-called 'classical Störmer problem'. It provides a description leading to the understanding of radiation belts composed of individual ions surrounding Earth and other magnetized planets. This problem was generalized by Dullin *et al.* (2002) to the dynamics of charged dust grains in planetary magnetospheres, when there are smaller charge to mass ratios, and the planetary gravity and co-rotating electric field play a role. Such studies point out the existence of dust grains stable off-equatorial circular orbits of constant radius and latitude, the so-called stable halo orbits. As was shown by Howard *et al.* (1999), the orbits take place near, e.g., Saturn.

We focused on the halo orbits existence in strong gravitational fields, i.e., in the vicinity of three qualitatively different kinds of compact objects, endowed with a dipole-type magnetic field. These are a neutron star, a Schwarzschild black hole with a current loop, and a Kerr-Newman black hole or naked singularity. The first case corresponds to the original Störmer's problem, but with the strong gravity described within the pseudo-Newtonian approach by the Paczyński-Wiita potential. The second case, investigated within the general relativity, may serve as a toy model of the black-hole spacetime with a magnetized accretion disc. In the last case, the solution was found by de Felice (1979) and Calvani *et al.* (1982). But the authors only simply conclude that the halo orbits do exist in Kerr-Newman spacetimes, regardless of the stability of the orbits and Kerr-Newman black-hole and naked-singularity space-time classes.

Our study (see the paper Kovář *et al.* (2008)) points out the stable halo orbits existence near all of the three studied kinds of compact objects. However, in the case of Kerr-Newman black hole, the stable halo orbits are of marginal astrophysical importance, being hidden under the inner event horizon. Above the outer event horizon, only unstable halo orbits can appear, while the stable circular orbits are allowed in the equatorial plane only. In principle, as follows from our analysis, stable halo orbits become accessible in the Kerr-Newman naked-singularity spacetimes. On the other hand, in the field of neutron stars or near black holes with current loops, the stable halo orbits can be located outside the body and can be astrophysically relevant. Discussion and searching for the halo orbits were performed in the way of searching for minima of 2-dimensional effective potential

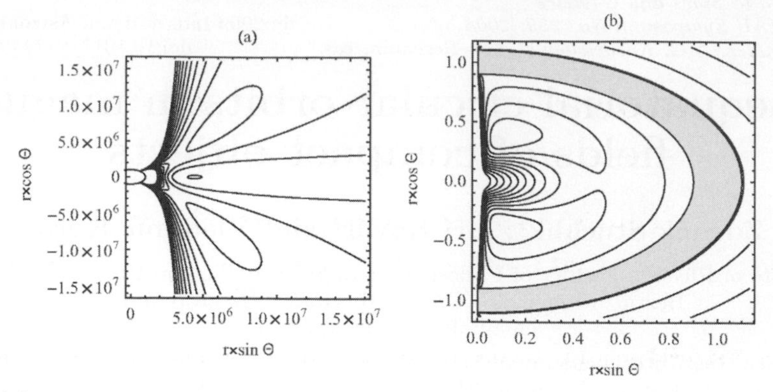

Figure 1. (a) Contours of pseudo-Newtonian effective potential for co-rotating charged dust grains with $q = -10^{-14}$C, $m = 10^{-17}$kg moving near a neutron star with $\Omega = 10^3$rad/s, $B = 10^8$G and $R = 1.2 \times 10^6$cm. (b) Countours of general relativistic effective potential for motion of charged particles with $\tilde{L} = 1$ and $\tilde{q} = -5$ in the Kerr-Newman black-hole spacetime with $a^2 = 0.04$ and $e^2 = 0.95$ (in geometric units and units of M). The gray region match the black-hole spacetime region between the outer and inner event black-hole horizons; the potential minima correspond to the stable halo orbits positions.

for the charged particles motion and related potential wells, necessary for the charged particles collections (see figure 1).

Halo orbits could have outstanding astrophysical consequences for radiation of accreting compact objects. This is because the halo orbits form two lobes of stable motion - one above and the other one below equatorial plane - through which the accretion disc radiation has to pass on its way towards a distant observer. Therefore, the halo lobes could act as a part of radiation reprocessing corona. The existence of the lobes can have profound consequences even for plasma oscillations. It allows us to imagine a situation where the bulk motion is along the halo orbit in the center of the lobe, around which a fraction of trapped particles oscillate. The oscillations modulate the electromagnetic signal from the plasma cloud, observed as a mode of oscillations in the detected radiation. The halo orbits in the magnetic field of compact stars and black holes could be related to the oscillatory motion with 'halo radial' and 'halo vertical' epicyclic frequencies, which could be considered as a complementary model of the QPOs observed in the binary system of compact stars and microquasars.

Further details can be found in the paper of Kovář *et al.* (2008).

Acknowledgements

The work was supported by the Czech GrantsMSM4781305903 and AV0Z10030501.

References

Calvani, M. *et al.* 1982, *Nuovo Cimento B Serie* 67, 1
Dullin, H. R., Horányi, M., & Howard, J. E. 2002, *Physica D* 171, 178
de Felice, F. 1979, *Physics Letters A* 69, 307
Howard, J. E., Horányi M., & Stewart, G. R. 1999, *Physical Review Letters* 83, 3993
Kovář, J., Stuchlík, Z., & Karas, V. 2008, *Class. Quantum Grav.* 25, 095011
Störmer, C. 1955, *The Polar Aurora*, Clarendon Press, Oxford

Cosmic Magnetic Fields:
From Planets, to Stars and Galaxies
Proceedings IAU Symposium No. 259, 2008
K.G. Strassmeier, A.G. Kosovichev & J.E. Beckman, eds.

doi:10.1017/S1743921309030300

Electro-magnetic fields around a drifting Kerr black hole

Ondřej Kopáček and Vladimír Karas

Astronomical Institute of the Academy of Sciences of the Czech Republic
email: kopacek@ig.cas.cz, vladimir.karas@cuni.cz

Abstract. A test field solution is constructed, describing the electromagnetic field around a Kerr black hole which is drifting in an arbitrary direction. An asymptotically uniform magnetic field is assumed with a general orientation (with respect to the rotation axis). Lines of force of electric and magnetic intensities are explored, as measured by an observer orbiting around the black hole (above the innermost stable circular orbit, ISCO) or freely falling (below the ISCO) down to the horizon. Magnetic null point is found.

Keywords. Black hole physics – magnetic fields – relativity

---

1. Introduction

Electromagnetic (EM) test field solutions of Maxwell equations in a curved spacetime play an important role in astrophysics because they show the effect of strong gravitational fields. On the other hand we can usually suppose that astrophysically relevant EM fields are weak enough, so that their influence upon background geometry may be neglected.

We are interested in the solutions describing an originally uniform magnetic field under the influence of the Kerr black hole. Since the Kerr metric is asymptotically flat, this EM field reduces to the original homogenous magnetic field in the asymptotic region. First, such a test field solution was given by Wald (1974) for the special case of perfect alignment of the asymptotically uniform magnetic field with the symmetry axis. More general solution for arbitrary orientation of the asymptotic field was given by Bičák *et al.* (1985). We use this solution to construct the EM field around the Kerr black hole which is drifting through the asymptotically uniform magnetic field.

2. Lines of force

Natural definition of the lines of force (of magnetic and electric fields), as measured by a given observer equipped with the orthonormal tetrad $e^{\mu}_{(\alpha)}$, is their identification with the lines along which magnetic/electric charge connected to this observer would accelerate due to the presence of the EM field $F^{\mu}_{\ \nu}$. Coordinate components of the magnetic (electric) field intensities are determined by (coordinate components of) the Lorentz force felt by the unit magnetic (electric) charge (Hanni, 1973):

$$B^{\mu} = {}^{*}F^{\mu}_{\ \nu}u^{\nu}, \quad E^{\mu} = F^{\mu}_{\ \nu}u^{\nu},$$

where u^{ν} represents a 4-velocity of the charge.

Tetrad components of the vector field determining desired lines of force are given as the spatial part of the projection onto $e^{\mu}_{(\alpha)}$:

$$B^{(i)} = B_{(i)} = e^{(i)}_{\ \mu}\ {}^{*}F^{\mu}_{\ \nu}u^{\nu}, \quad E^{(i)} = E_{(i)} = e^{(i)}_{\ \mu}F^{\mu}_{\ \nu}u^{\nu},$$

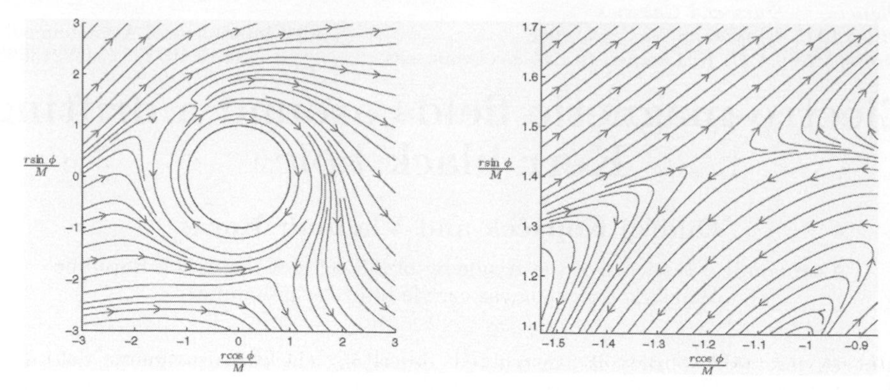

Figure 1. Counter-rotating observer around a drifting ($v_x = 0.5\,c$ and $v_y = -0.7\,c$) extreme Kerr black hole measures magnetic field (in the first panel). We notice the null point of the magnetic field (zoomed in the second panel). Original homogenous magnetic field is chosen to be parallel to the horizontal axis of these figures which show portion of the equatorial plane.

where $e^{(\alpha)}_{\ \mu}$ are 1-forms dual to the tetrad vectors $e^{\mu}_{(\alpha)}$. Lowering/rising spatial tetrad indices doesn't matter since the tetrad is supposed to be orthonormal: $g_{(\mu)(\nu)} = \eta_{(\mu)(\nu)}$.

Components of electromagnetic tensor $F^{\mu}_{\ \nu}$ describing the field around a drifting Kerr black hole were derived from the non-drifting solution given by Bičák *et al.* (1985). General Lorentz transformation (e.g. Jackson, 1998) was used to correctly balance the asymptotic values of the components to acquire the field of desired properties.

As we are primarily interested in astrophysically relevant situations we choose the orthonormal tetrad $e^{(\alpha)}_{\ \mu}$ carried by the observer on the circular Keplerian orbit around the black hole. Such an orbit is specified by the values of constants of motion – by specific angular momentum $\tilde{L} \equiv u_\phi$ and specific energy $\tilde{E} \equiv -u_t$. However we don't find stable circular orbits all the way down to the event horizon. It can be shown (Bardeen *et al.* 1972) that there exists a radial boundary $r_{\rm ms}$ – marginally stable orbit.

For $r < r_{\rm ms}$ there are no stable circular orbits. Thus we suppose that the orbiting observer who reaches this limit performs a free fall to the black hole keeping the values of the constants of motion corresponding to the marginally stable orbit at $r_{\rm ms}$. He is falling with $\tilde{E}_{\rm ms} \equiv \tilde{E}(r_{\rm ms})$ and $\tilde{L}_{\rm ms} \equiv \tilde{L}(r_{\rm ms})$.

3. Conclusions

This work represents initial steps of our investigation of the asymptotically uniform electromagnetic test fields around a drifting Kerr black hole. Components of the electromagnetic tensor $F_{\mu\nu}$ describing such a field were derived in the terms of the nondrifting solution. The existence of the null point in Fig. 1 suggests that the magnetic reconnection may be initiated by the interplay of electromagnetic and gravitational fields. Further details can be found in Karas & Kopáček (2008).

References

Wald, R. M. 1974, *Phys. Rev. D* 10, 1680
Bičák, J. & Janiš, V. 1985, *MNRAS* 212, 899
Hanni, R. S. & Ruffini, R. 1973, *Phys. Rev. D* 8, 3259
Jackson, J. D. 1998, *Classical electrodynamics*, 3rd ed, John Wiley & Sons
Bardeen, J. M., Press, W. H., & Teukolsky, S. A. 1972, *ApJ* 178, 347
Karas, V. & Kopáček, O. 2008, *CQG*, in press (arXiv:0811.1772)

Cosmic Magnetic Fields:
From Planets, to Stars and Galaxies
Proceedings IAU Symposium No. 259, 2008
K.G. Strassmeier, A.G. Kosovichev & J.E. Beckman, eds.

doi:10.1017/S1743921309030312

New global 3D MHD simulations of black hole disk accretion and outflows

Peter B. Dobbie[1], **Zdenka Kuncic**[1], **Geoffrey V. Bicknell**[2] **and Raquel Salmeron**[2]

[1]School of Physics, University of Sydney, NSW 2006, Australia
email: p.dobbie@physics.usyd.edu.au

[2]Research Centre for Astronomy and Astrophysics, Australian National University, ACT 2611, Australia

Abstract. It is widely accepted that quasars and other active galactic nuclei (AGN) are powered by accretion of matter onto a central supermassive black hole. While numerical simulations have demonstrated the importance of magnetic fields in generating the turbulence believed necessary for accretion, so far they have not produced the high mass accretion rates required to explain the most powerful sources. We describe new global 3D simulations we are developing to assess the importance of radiation and non-ideal MHD in generating magnetized outflows that can enhance the overall rates of angular momentum transport and mass accretion.

Keywords. Accretion – accretion disks – magnetic fields – MHD – radiative transfer – turbulence – methods: numerical – galaxies: active, jets, magnetic fields, nuclei – (galaxies:) quasars: general – x-rays: binaries

1. Introduction

Accretion onto a black hole is widely believed to power the central engine of high-energy astrophysical objects such as active galactic nuclei (AGN) and some X-ray binaries. Since the standard theory of astrophysical disk accretion was formulated over 30 years ago (Shakura & Sunyaev 1973, Novikov & Thorne 1973), arguably the most important advances in understanding these systems have come from computational modelling. Numerical simulations demonstrate unequivocally that the magnetorotational instability (MRI) can produce magnetohydrodynamic (MHD) turbulence and the outward transport of angular momentum required for accretion to proceed (see Balbus 2003 for a review).

Notwithstanding the important advances made using MHD simulations over the last two decades, they have so far been unable to resolve two key outstanding issues: 1. How are the high mass accretion rates inferred in the most powerful sources achieved? and 2. How are the outflows and jets observed across the mass spectrum of accreting sources produced? Our new work will test the hypothesis that these issues may be connected and mutually resolved by a generalized model for MHD disk accretion.

2. MHD Disk Accretion Theory

The analytical foundation for our new work is based on the generalized accretion disk model of Kuncic & Bicknell (2004) which is schematically illustrated in Fig. 1. The model includes contributions from both radial and vertical transport of angular momentum to the overall mass accretion rate: angular momentum is transported radially outwards by internal MHD stresses and vertically outwards by both mass outflows in a wind and MHD stresses acting over the disk surface. This vertical transport of energy and momentum

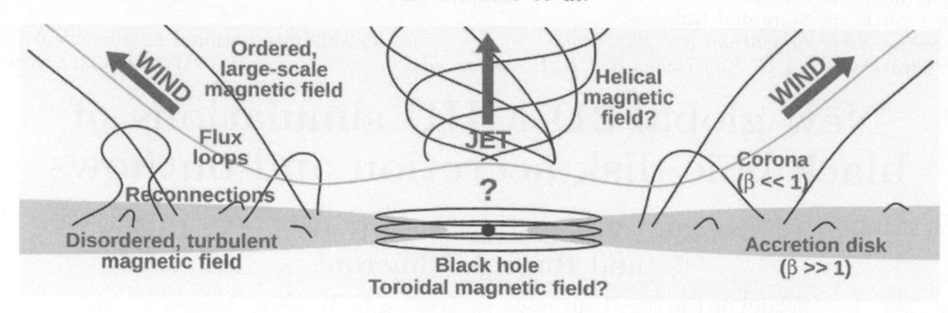

Figure 1. A schematic illustration of the inner regions of an MHD accretion flow around a black hole. β is the usual plasma beta (the ratio of gas to magnetic pressures). The wind can drive mass loss from the disk, while the jet may be dominated by Poynting flux.

is not modelled in standard accretion disk theory. Our new simulations are designed to compare its effects with the radial turbulent transport of angular momentum and to test the prediction of a modified emission spectrum (see e.g. Kuncic & Bicknell 2007a,b).

3. New MHD Accretion Simulations

To implement our model, we are using and extending FLASH† (Fryxell *et al.* 2000), the public MHD code developed at the University of Chicago. FLASH implements adaptive mesh refinement and solves the equations of MHD in conservative form thus explicitly conserving energy.

To date, most simulations of the MRI in accretion disks have been conducted in a shearing box approximation but this approach introduced complications and pitfalls (see e.g. Regev & Umurhan 2008). We will instead conduct global 3D simulations.

By explicitly modelling a finite resistivity, we will explore its effect on the evolution of the magnetic field topology, particularly the self-consistent emergence of a significant z-component necessary to produce outflows and high mass accretion rates.

We will include radiation in our simulations to directly compare our results to the observational data. This will allow us to test whether the blackbody emission from the disk is modified by outflows as predicted. Radiative transfer is also be required to transport the internal energy dissipated in the disk plasma at the end of a turbulent cascade, i.e. to cool the disk. In addition, it may be that regions of the disk where radiation dominates may be thermally unstable (Shakura & Sunyaev 1976), thus affecting the dynamics.

References

Balbus, S. A. 2003, *ARAA* 41, 555

Fryxell, B., Olson, K., Ricker, P., Timmes, F. X., Zingale, M., Lamb, D. Q., MacNeice, P., Rosner, R., Truran, J. W., & Tufo, H. 2000, *ApJS* 131, 273

Kuncic, Z. & Bicknell, G. V. 2004, *ApJ* 616, 669

Kuncic, Z. & Bicknell, G. V. 2007a, *Ap&SS* 311, 127

Kuncic, Z. & Bicknell, G. V. 2007b, *Mod. Phys. Lett. A* 22, 1685

Novikov, I. D., Thorne, K. S. 1973, *Black Holes*, C. DeWitt, & B. DeWitt (New York: Gordon and Breach), 343

Regev, O. & Umurhan, O. M. 2008, *A&A* 481, 21

Shakura, N. I. & Sunyaev, R. A. 1973, *A&A* 24, 337

Shakura, N. I. & Sunyaev, R. A. 1976 *MNRAS* 175, 613

† FLASH is freely available at http://flash.uchicago.edu.

Cosmic Magnetic Fields:
From Planets, to Stars and Galaxies
Proceedings IAU Symposium No. 259, 2008
K.G. Strassmeier, A.G. Kosovichev & J.E. Beckman, eds.

doi:10.1017/S1743921309030324

Surface magnetic fields in pulsars

George I. Melikidze and Janusz Gil

Institute of Astronomy, University of Zielona Gora, Lubuska 2, 65-265, Zielona Gora, Poland
email: gogi@astro.ia.uz.zgora.pl, jag@astro.ia.uz.zgora.pl

Abstract. Observations of hot-spot thermal X-ray emission from radio pulsars implicate that surface magnetic field (SMF) at the polar cap is much stronger than the conventional dipolar component estimated from the pulsar spin-down. This strongly suggests that SMF is dominated by the crust anchored small scale magnetic field. We present the observed values of black body temperature and bolometric luminosity of X-ray emission from hot polar caps of a number of pulsars. In all cases the inferred value of SMF is close to 10^{14} G.

Keywords. Stars: neutron – stars: magnetic field – pulsars: PSR J0108-1431

Thermal X-ray emission seems to be a quite common feature of radio pulsars. XMM - Newton and Chandra observations demonstrated that not only relatively young and hot pulsars could be seen in thermal X-rays. In older and cooler pulsars this radiation can originate on the surface of hot polar caps Pavlov *et al.* (2008). Therefore, temperature and bolometric luminosity of black body radiation from hot spots allow us to shed some light on the main features of the polar cap region in radio pulsars. The most unexpected result was estimation of the hot spot surface area (derived from the spectral fit with a hot-spot black body model), which appears to be much smaller than the conventional polar cap surface area $A_0(P) \simeq 6.6 \times 10^8 P^{-1}$ [cm^2]. Here P is the pulsar period. The only natural explanation of this observational feature is an assumption that the magnetic field at the stellar surface differs essentially from the pure dipole field. The standard model for pulsars assumes that the source of the pulsar activity is associated with the inner acceleration region (called gap) above the polar cap, where the electric field has a component along the magnetic field lines. Particles (electrons and positrons) are accelerated in both directions: outwards and towards the stellar surface. Consequently, outstreaming particles generate the magnetospheric X-ray emission while the backstreaming particles heat the surface and provide necessary energy for the thermal emission from hot polar cap.

The black body fit allow us to obtain directly the temperature $T_{\rm BB}$ and the surface area of the hot spot. As mentioned above, in most cases $A_{\rm BB}$ is much less than A_0 (Zhang *et al.* (2005); Kargaltsev*et al.* (2006)) and it can be easily explained by assuming that the surface magnetic field of pulsars differs significantly from the pure dipole one. One can estimate an actual surface magnetic field by the magnetic flux conservation law as $b = A_0/A_{\rm BB} = B_s/B_d$. Here $B_d \simeq 2 \times 10^{12}(P\dot{P}_{-15})^{0.5}$ [G] is a dipolar magnetic field and $P\dot{P}_{-15}$ is a period derivative. Then $B_s \simeq 1.3 \times 10^{21} A_{\rm BB}^{-1}(P\dot{P}_{-15})^{0.5}$ G. In most cases $b \gg 1$ implying $B_s \gg B_d$, and $T_{\rm BB} \simeq (2 \div 4) \times 10^6$ K (see figure 1).

For such a scenario Gil, Melikidze & Geppert (2007) have proposed the model for the inner acceleration region, which assumes that the gap operates in such a way that surface temperature is always very close the certain, unambiguously defined by the surface magnetic field critical temperature T_i above which iron ions are thermally ejected from the surface up to the maximum co-rotational limited amount. This model is known as the Partially Screened Gap (PSG) model, as it assumes that the potential drop near the stellar surface is partially screened by the positive ions. The screening factor η can be defined as $\eta = 1 - \rho_i/\rho_{\rm GJ} = 1 - \exp(C_i - \varepsilon_c/kT_s)$. Here ρ_i and $\rho_{\rm GJ}$ are the actual and

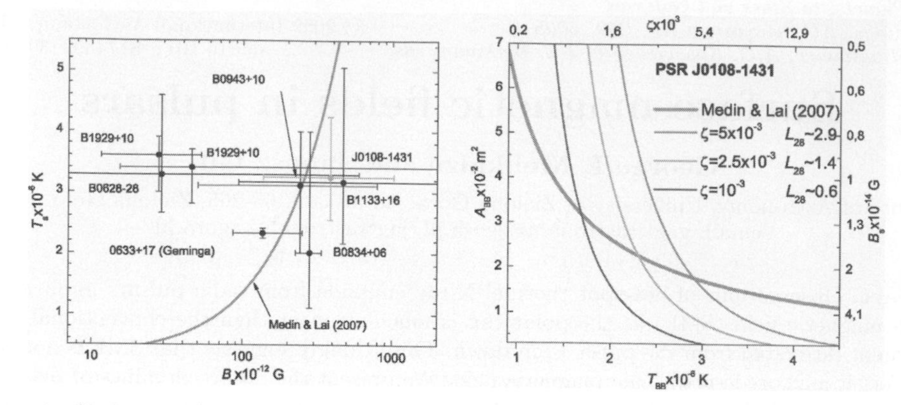

Figure 1. (Left panel) Positions of a number pulsars (with the known BB spectra fitting parameters) on the $T_{\rm BB} - B_s$ diagram. The thick red line represents the Medin & Lai dependence $T_i = Ti(B_s)$ for $C_i = 30$. Recently observed very old PSR J0108-1431 is marked by a green color. (Right panel) Lines of the constant bolometric luminosity for PSR J0108-1431.

co-rotational charge densities, ε_c is the cohesive energy of ions, T_s is the surface temperature and the parameter $C_i \approx 30$ depends on magnetic field strength and temperature. Then we can define the critical temperature as a temperature when $\rho_i = \rho_{\rm GJ}$. Thus $T_i = \varepsilon_c / C_i k$.

In order to estimate T_i we need to know the cohesive properties of the surface matter. This problem has been recently examined by Medin & Lai (2007). They have estimated the critical temperature of the neutron star's iron crust and concluded that if the surface magnetic field is about 10^{14} Gauss, then the gap can be formed even if the surface temperature reaches few million K, i.e., the condition $\rho_i < \rho_{\rm GJ}$ is satisfied even for such a high temperature. The dependence of $T_i = T_i(B_s)$ obtained by Medin & Lai (2007) is shown as red lines in both panels of figure 1.

Let us emphasize once more that within PSG model the hot spot temperature should be very near to the critical temperature $T_{\rm BB} \simeq T_i$. Therefore, while fitting the hot spot BB spectra $A_{\rm BB}$ and $T_{\rm BB}$ should not be treated as an independent parameters. The pulsars, which X-ray spectra can be fitted to the black body emission from the hot spot should obey the $T_i = T_i(B_s)$ dependence. Assuming $T_i = T_{\rm BB}$, we can rewrite the function $T_i = T_i(B_s)$ and obtain for a given pulsar

$$A_{\rm BB} = F(P, \dot{P}_{-15}, T_{\rm BB}).$$

Using PSG model we can also estimate the X-ray luminosity $L_X \approx 4 A_{\rm BB} \sigma T_{\rm BB}^4$, and its efficiency as $\zeta = L_X / L_{\rm SD}$, where $L_{\rm SD} = 3.9 \times 10^{31} \dot{P}_{-15} P^{-3}$ [erg s^{-1}] is the pulsar spin-down energy loss rate. The lines of the constant efficiency for PSR J1702-19 are shown in figure 1 (right panel).

Acknowledgements

The work was supported by Grants NN203 2738 33 (JG) and GNSF ST06/4-096 (GM).

References

Gil, J., Melikidze, G. I., & Geppert, U. 2003, *A&A* 407, 315
Kargaltsev, O., Pavlov, G. G., & Garmire, G. P. 2006, *ApJ* 636, 406
Medin, Z., Lai, D. 2007, *MNRAS* 382, 1833
Pavlov, G. G., Kargaltsev, O., Wong, J. A., & Garmire, G. P. 2008, astro-ph/0803.0761
Zhang, B., Sanwal, D., & Pavlov, G. G. 2005, *ApJ* 624, L109

Cosmic Magnetic Fields:
From Planets, to Stars and Galaxies
Proceedings IAU Symposium No. 259, 2008
K.G. Strassmeier, A.G. Kosovichev & J.E. Beckman, eds.

doi:10.1017/S1743921309030336

Evidence of a magnetic sheath around a jet from NGC 6543

Aden B. Meinel and Barbara Meinel

2548 Eclipsing Stars Drive Henderson, NV 89044, USA
email: ammeinel@cox.net; ameinel@optics.arizona.edu

Abstract. We present observational evidence that NGC 6543 produced a jet of cosmic rays that irradiated the Earth recorded as cosmogenic10Be found ice cores. This identification shows that the jet was accompanied by a magnetic field of sufficient strength to travel 220pc and retain evidence of the celestial coordinates of the source object.

Keywords. Stars: AGB and post-AGB – accretion – planetary nebulae

---

The basic data of the flux of cosmogenic isotope 10Be is from the Greenland GRIP data archive and is corrected for the dilution effect of variable annual ice accumulation. Our attention was drawn to a set of sinusoidal oscillations indicated by the five black dots.

Fig. 1(A). We recognized that the change of angle of the Earth with regard to the sky as the result of precession would produce no signature of precession on an the isotropic flux of Galactic cosmic rays. The 21 ky period of lunar-solar precession therefore indicated the presence of a significant flux of cosmic rays from a point source.

Fig. 1(B). We calculated the effect of the combined 21 ky lunar-solar plus 120 ky apsidal precession as a function of Right Ascension of the point source. Fig. 1(B) shows the significant changes in the attenuation.

Fig. 1(C). The data for 10Be has been corrected in Fig. 1(C) for changing attenuation as the geomagnetic field. Note that the prominent surge at 40 kyBP in Fig. 1(A) has merged imperceptibly into the general pattern of wide variations of the flux of 10Be, but that the several oscillations are unchanged. The best fit of the calculated attenuation patterns is for a source close to 18h Right Ascension and +65 Declination This is very close to the position of NGC 6543. Our finding is that this planetary nebula was the source of the cosmic rays that irradiated the Earth. Our calculation of the absolute fractional variation of 10Be best-fitted in Fig. 1(C) also defines the invariant flux of Galactic cosmic rays of 0.5 10Be particles/square centimeter per year.

The data pertaining to NGC 6543 of importance in our paper are:
A parallax of 0.045 ± 0.021 arcseconds was determined from three years of trigonometric observations by Gatewood and his associates at Allegheny Observatory (2008), thus a distance of 220 parsecs. The radius of curvature of a typical cosmic ray particle in the Galactic magnetic field however is less than 100th parsec. Thus no cosmic rays could reach the solar vicinity from NGC 6543 unless the cosmic rays were protected by a magnetic field accompanying the jet of cosmic rays.

The HST image in Fig. 1(C)a shows the central active region of NGC 6543 in negative format to show better the pairs of emission bubbles ejected in opposite directions as shown in Fig. 1(C)b. The schematic diagram (top) shows the orbital geometry consistent with our identification of NGC 6543 as the source of continual jetting of cosmic rays.

We find that the central object of NGC 6543 must consist of an Aging Giant Branch (AGB) star (S1) emitting plasma feeding a compact companion (S2), a neutron or white

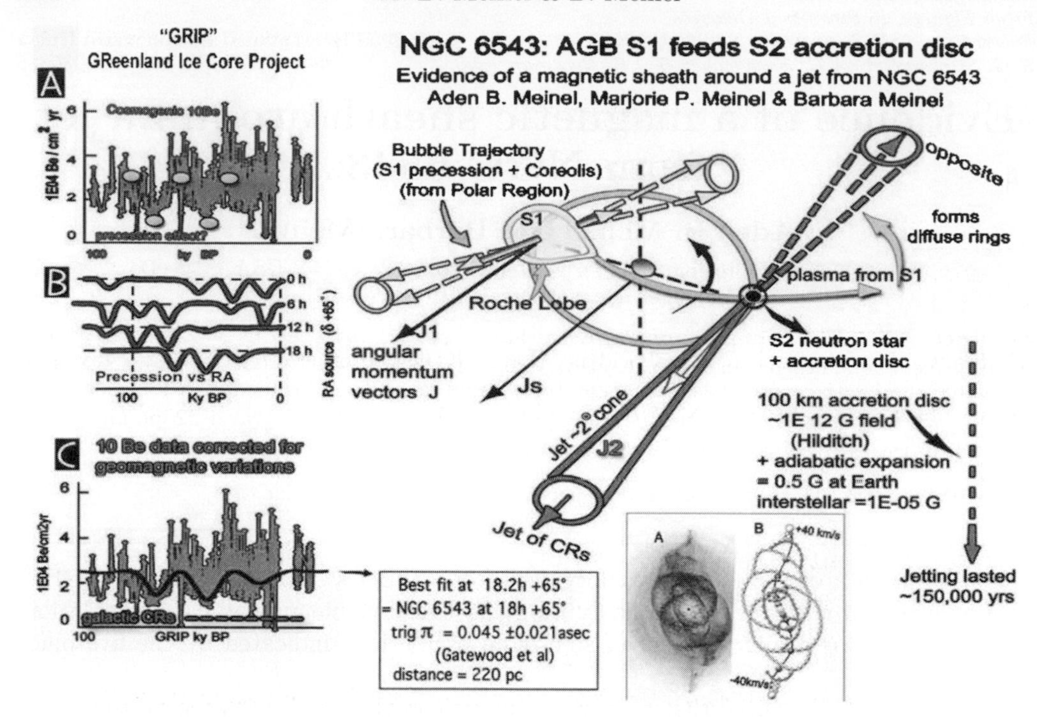

Figure 1. NGC 6543. AGB star S1 feeds S2 accretion disk.

dwarf star, plus an accretion disc. Part of this stream of plasma is acquired by the accretion disc, the balance escaping (Boyarchuk *et al.*) to be spread by orbital rotation into the diffuse rings shown in the HST image (Fig. 1(C)a).

The rotation axes of both stars and the system are parallel and indicated by their invariant angular momentum vectors J1, Js, and J2. The jet is aimed throughout its orbit toward the observer. The outflow of pairs of bubbles is from the polar regions of S1. The direction of outflow is affected by both precession of S1 and the Coriolis force. The direction of successive pairs of bubbles relative to the observer are unaffected by motion of S1 around its binary orbit, thus forming the vertical chain of bubbles shown in Fig. 1(C)b.

The survival of the jet in traversing the interstellar magnetic field depends on the very strong magnetic field of either a white dwarf or neutron star of 10^{12} Gauss. If the divergence angle of the jet is 2°, adiabatic expansion of the magnetic energy in the jet after a travel of 220 parsecs will still be 0.5 Gauss, much stronger than the 10^{-5} Gauss of the interstellar field.

Thus our conclusion is that a jet of cosmic rays from NGC 6543 can survive to reach the Earth and still indicate from whence they came. The astrophysical surprise is that the jetting of NGC 6543 lasted longer than 150,000 years.

Cosmic Magnetic Fields:
From Planets, to Stars and Galaxies
Proceedings IAU Symposium No. 259, 2008
K.G. Strassmeier, A.G. Kosovichev & J.E. Beckman, eds.

doi:10.1017/S1743921309030348

Accretion and magnetic field submergence in neutron star surface

C. Giovanny Bernal, Dany Page and William H. Lee

Instituto de Astronomía, Universidad Nacional Autónoma de Mèxico, Mexico DF 04510

Abstract. We study the effects of hypercritical accretion onto a neutron star surface. The magnetic field submergence in the neutron star crust and the possible rediffusion is investigated.

Keywords. Stars: neutron – SN 1987A – methods: numerical – accretion – magnetic fields

1. Introduction

Supernova 1987A in the Large Magellanic Cloud was the first nearby supernova in over 400 years. Fortunately, the LMC has been well studied, and the precursor star has been identified: SK–69 202 was a 19-$M_{\odot}$ B3I supergiant star with a helium core mass of 6 $M_{\odot}$, an iron core mass of 1.45 $M_{\odot}$ that resulted in a 1.40 $M_{\odot}$ neutron star. Nevertheless, the compact remnant is not detected yet.

We present HD and MHD simulations of hypercritical accretion onto the compact object in SN 1987A. The submergence of the field by any amount of accreted matter can easily explain the absence of PSR activity in recent SNe, but the very existence of PSRs leads to the conclusion that in some fraction of SNe the fall-back accretion is weak enough that it allows a re-diffusion of the field in $10^3 - 10^4$ years. Stellar ejecta at the center of the SNR might still be optically thick in X-rays. Compare the observed 3-8 keV band images before and after adding simulated point sources (with various count rates) at the center of the SNR in order to determine an upper limit (90 %) to point source contribution. Point source spectrum: $\Gamma = 1.7 - 3.0$, $N_H = 2 \times 10^{21} - 10^{24}$ cm^{-2} are assumed. Based on the image taken on July 22, 2004, a point source upper limit at 3-10 keV is $L_x \approx 5 \times 10^{33} - 3 \times 10^{35}$ erg s^{-1} (Park *et al.* 2005). A numerical approximation of MHD hypercritical accretion in SN1987A can help us to resolve this mystery.

2. Hypercritical accretion onto SN1987A core

Analytical. In the hypercritical accretion onto compact objects scenarios, Chevalier (1989) argued that if the accretion mass rate ($\dot{M}$) exceeds the Eddington $\dot{M}_{\rm Eddington}$, then some of the accretion energy must be removed by means other than photons. In this case, the accretion is called hypercritical. Chevalier (1989) took in account that neutrinos can carry away accretion energy and developed self-consistent solutions for hypercritical accretion.

When $\dot{M} > 2 \times 10^{22}$ g s^{-1}, the ram pressure at the neutron star surface is larger than the pressure of a 10^{12} G magnetic field, which means that the accretion flow is purely hydrodynamical. This critical $\dot{M}$ is also the accretion rate at which hypercritical accretion is stopped by radiation pressure and becomes unstable, at which time about 0.1 $M_{\odot}$ has been accreted in the case of SN1987A. The analytical solution of envelope structure is: $P = P_{sh}(\frac{r}{R_{sh}})^{-4}$, $\rho = \rho_{sh}(\frac{r}{R_{sh}})^{-3}$, and $v = v_{sh}(\frac{r}{R_{sh}})$, where $R_{sh}, P_{sh}, \rho_{sh}, v_{sh}$ are

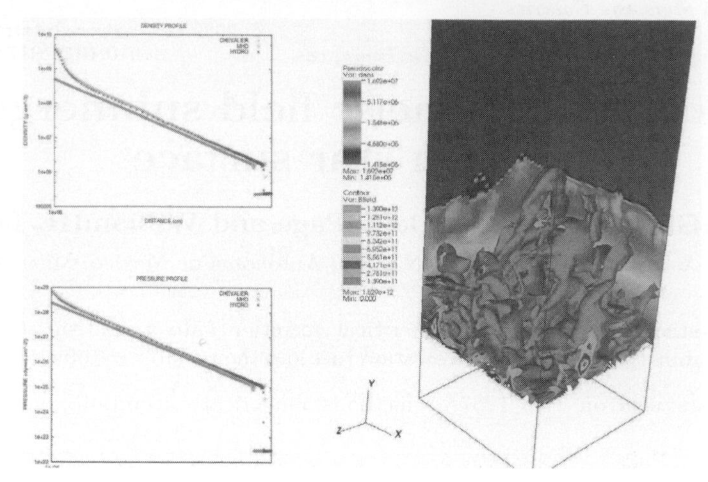

Figure 1. Left: The radial density (top) and pressure profiles (bottom). Right: Submergence and re-diffusion of the magnetic field in the SN-1987A scenario.

the values in the accretion shock. For the SN1987A scenario the values are: $v_0 = \sqrt{\frac{2GM}{R_{sh}}} \simeq 2.1 \times 10^9$ cm s^{-1}, $\rho_0 = \frac{\dot{M}}{4\pi R_{sh}^2 v_0} \simeq 108.75$ g cm^{-3}, $\rho_{sh} = 7\rho_0 \simeq 761.75$ g cm^{-3}, $v_{sh} = -\frac{1}{7}v_0 \simeq -3 \times 10^8$ cm s^{-1}, $P_{sh} = \frac{7}{8}\rho_0 v_0^2 \simeq 4.17 \times 10^{20}$ dyn cm^{-2}.

Numerical: Above analytical approximation does not take into account the magnetic field nor the piling up from material in the neutron star surface. We used the AMR FLASH2.5 code. The absence of PSR activity in recent SNe unfortunately depends on the poorly understood late evolution of the envelope of massive stars, but the nature of the progenitor of SN1987A shows that this fraction is not vanishingly small (Muslimov & Page 1995).

The radial profiles match very good with the Chevalier (1989) profiles, and the magnetic field onto the neutron star surface is submerged in the crust. This does not depend on the magnetic field configuration but on the intensity and the accretion rate. Good agreement between the analytical solution and numerical approach is achieved but we have gained the crust of the neutron star. The 3D simulation show the submergence and the possible re-diffusion of the magnetic field.

3. Conclusions

The accretion onto object compact in SN1987A is hypercritical ($\dot{M} \gg \dot{M}_{\rm Eddington}$). The diffusion of radiation from central region causes the later accretion to be ballistic, if the luminosity of the central source can be neglected. Steady state accretion envelopes are plausible for the mass accretion rates of SN1987A. The submergence of the magnetic field can be hiding the pulsar in SN1987A. The numerical simulations with FLASH2.5 code in KANBALAM cluster (128 processors), support this hypothesis.

The software used in this work was in part developed by the DOE-supported ASC Alliance Center for Astrophysical Thermonuclear Flashes at the University of Chicago.

References

Chevalier, R. A. 1989, *ApJ* 346, 847

Muslimov, A. & Page, D. 1995, *ApJ* 440, L77

Park, S., Zhekov, S. A., Burrows, D. N., Garmire, G., & McCray, D. 2005, *Adv. Space Res.* 35, 991

Cosmic Magnetic Fields:
From Planets, to Stars and Galaxies
Proceedings IAU Symposium No. 259, 2008
K.G. Strassmeier, A.G. Kosovichev & J.E. Beckman, eds.

doi:10.1017/S174392130903035X

The magnetic field in the X-ray binary Cyg X-1

Eugenia A. Karitskaya[1], N. G. Bochkarev[2], S. Hubrig[3], Yu. N. Gnedin[4], M. A. Pogodin[4], R. V. Yudin[4], M. I. Agafonov[5] and O. I. Sharova[5]

[1]Astronomical Institute of RAS, 48 Pyatnitskaya str., Moscow 119017, Russia, email: karitsk@sai.msu.ru

[2]Sternberg Astronomical Institute 13 Universitetskij pr., Moscow 119991, Russia

[3]ESO, Santiago, Chile

[4]Pulkovo Observatory RAS, St. Petersburg 196140, Russia

[5]Radiophysical Research Institute, Nizhny Novgorod 603950, Russia

Abstract. VLT FORS 1 observations indicate the presence of a variable significant magnetic field in the X-ray binary Cyg X-1. The importance of this investigation comes from the fact that it rules one of the most significant BH manifestations: the X-ray millisecond flickering, usually related to reconnection of magnetic lines in the innermost part of the accretion disc.

Keywords. X-ray binaries – black hole physics – magnetic fields – spectropolarimetry – accretion

1. Introduction

The first prediction of the presence of a magnetic field in Cyg X-1 was done by V. F. Shvartsman (1971). Many previous attempts to measure the magnetic field have been unsuccessful providing only the upper limit for the strength of the magnetic field (e.g. Gnedin *et al.* 2003).

2. Observations and analysis

Between June 18 and July 9, 2007 we obtained in service mode six FORS 1 spectropolarimetric spectra. We used the GRISM 1200B and a 0.4 arcsec slit to obtain a spectral resolution of 4000. This grism covers the spectral range 3680 - 5129 Å which includes Balmer lines from Hβ to Hϵ. At that time Cyg X-1 was in X-ray hard state. For each observation we usually took a number of continuous series of two exposures. More details on the observing technique with FORS 1 and data reduction can be found elsewhere (Hubrig *et al.* 2004 and references therein). Using the method described in the same place we obtained the mean longitudinal magnetic field $\langle B_z \rangle$ averaged over the stellar hemisphere visible at the time of observation of the component of the magnetic field parallel to the line of sight, weighted by the local emergent spectral line intensity.

As an important step, before the assessment of $\langle B_z \rangle$, we removed all the spectral features not belonging to the photosphere of the Cyg X-1 optical component (O9.7 Iab): telluric and interstellar lines, CCD defects, the He II λ4686 Å emission line, lines with strong P Cyg components. The measurements of the longitudinal magnetic field at different orbital phases are presented in Fig. 1. It is quite possible that near the orbital

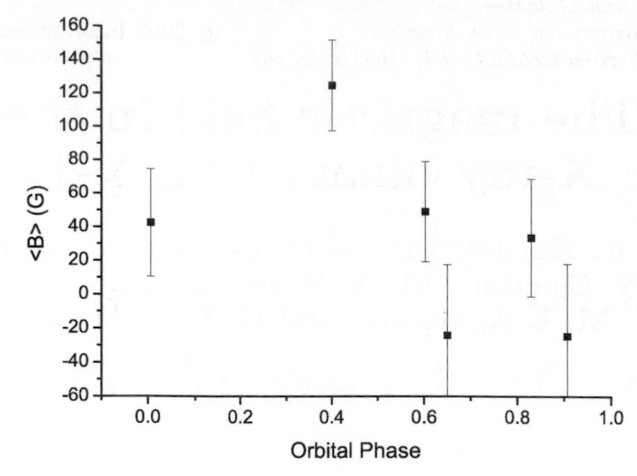

Figure 1. The longitudinal magnetic field plotted over the orbital phase (P = 5.6 days) of Cyg X-1.

phase 0.5 (having the X-ray source in front) we are observing the optical component in the vicinity of the magnetic pole with the $\langle B_z \rangle$ value close to 130 G on 5 σ level.

On the second step of our research we investigated $HeII\lambda 4686\mathring{A}$ spectral line separately. The accuracy of magnetic field measuring over one line lower than from all lines. Estimation on 2 sigma level was found only for orbital phase 0.65 $\langle Bz \rangle = -607 \pm 310$ G. $\langle Bz \rangle$ reality is confirmed: by Zeeman S-wave in V-spectrum and by its correspondence to the $dI(\lambda)/d\lambda$ wave.

3. Discussion

The Doppler tomogram of the system based on the HeII $\lambda 4686$ Åline using FORS 1 Stokes I spectra shows that the line emission component originates in the outer regions of the accretion structure. We used the Doppler tomogram building technique of Agafonov, (2004), so-called radio astronomical approach. The magnetic field strength of 600 G is an expected value since the gas stream from the star carrying the magnetic field onto the accretion structure is compressed by interaction with its outer rim by a factor of 6-10 at a distance from the black hole $D = 6 * 10^{11}$ cm. According to the standard model of the disc accretion, such a result would correspond to the near-Black-Hole magnetic field value of $\sim 10^8 - 10^9$ G and thus the value of its energy would be sufficient to explain Cyg X-1 X-ray millisecond flickering.

Acknowledgements

We would like to acknowledge ESO for grants for VLT observations in service mode. The work was supported by RFBR by grants 06-02-16234, 06-02-16843, 07-02-00535 and Sci.Schole 6110.2008.2.

References

Agafonov, M. I. 2004, *AN* 325, 259
Gnedin, Yu. N. *et al.* 2003, astro-ph/4158G
Hubrig, S., Kurtz, D. W., Bagnulo, S. *et al.* 2004, *A&A* 415, 661
Shvartsman, V. F. 1971, *AZh* 48, 479

Cosmic Magnetic Fields:
From Planets, to Stars and Galaxies
Proceedings IAU Symposium No. 259, 2008
K.G. Strassmeier, A.G. Kosovichev & J.E. Beckman, eds.

doi:10.1017/S1743921309030361

Two-dimensional numerical study for relativistic outflow from strongly magnetized neutron stars

Jin Matsumoto[1,2], Youhei Masada[3], Eiji Asano[1] and Kazunari Shibata[1]

[1]Kwasan and Hida Observatories, Kyoto University, Kyoto, Japan
email: jin@kusastro.kyoto-u.ac.jp

[2]Department of Astronomy, Kyoto University, Kyoto, Japan.

[3]Institute of Astronomy and Astrophysics, Academia Sinica, Taiwan, R.O.C.

Abstract. Using special relativistic magnetohydrodynamic simulation, the nonlinear dynamics of the magnetized outflow triggered on the magnetar surface is investigated. It is found that the strong shock propagates in the circumstellar medium in association with the expanding outflow. The shock velocity $v_{\rm sh}$ depends on the strength of the dipole field anchored to the stellar surface $B_{\rm dipole}$ and is described by a simple scaling relation $v_{\rm sh} \propto B_{\rm dipole}{}^{0.5}$. In addition, the outflow-driven shock can be accelerated self-similarly to the relativistic velocity when the density profile of the circumstellar medium is steeper than the critical density profile, that is $\alpha \equiv {\rm d}\log\rho({\rm r})/{\rm d}\log{\rm r} \lesssim \alpha_{\rm crit} = -5.0$, where the density is set as a power law distribution with an index α and r is the cylindrical radius. Our results suggest that the relativistic outflow would be driven by the flaring activity in a circumstellar medium with a steep density profile.

Keywords. Relativity – MHD – stars: neutron – methods: numerical

1. Introduction

Explosive magnetic energy release is widely accepted to play a crucial role in the flaring activity observed on ultra-strongly magnetized neutron stars, so called "magnetars" ($B \gtrsim 10^{14}$ G; Thompson & Duncan 1995). The physical mechanism of the energy release on the magnetar surface and the characteristics of the magnetically-driven outflow are not fully understood. Here we investigate, using axisymmetric special relativistic magnetohydrodynamic simulation, the nonlinear dynamics of the magnetized outflow triggered by the magnetic explosion on the magnetar surface.

As an initial setting, we set the hydrostatic circumstellar medium to have a density $\rho(r) \propto r^{\alpha}$ and a dipole magnetic field, B_r, $B_\theta \propto B_0 r^{-3}$, where α is the power-law index, B_0 is the surface field strength, and r is the cylindrical radius. To initiate the expanding outflow, an azimuthal shearing motion is added around the equatorial surface, and which produces the azimuthal component of the magnetic field (Mikic & Linker 1994).

2. Results

Figure 1 shows the density contour and the magnetic field lines in the meridional plane for the case where $\alpha = 3.44$ and $B_0 = 10^{14} G$. The twisted magnetic field created around the equatorial surface, by the shearing motion, drives the supersonically expanding outflow. It produces a strong shock wave propagating through the circumstellar medium. The relation between the strength of the initial dipole field and the shock velocity is

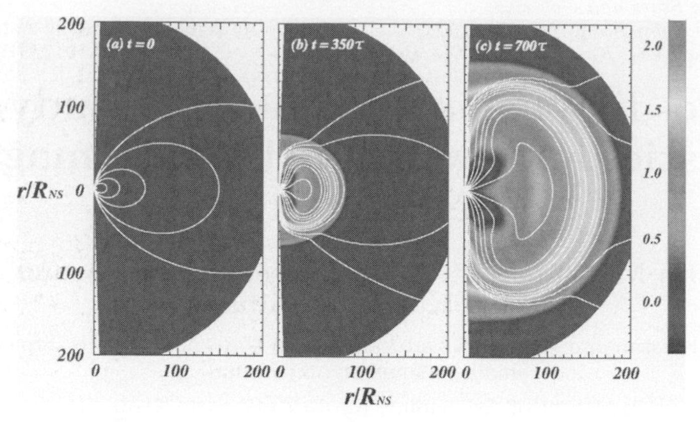

Figure 1. The time evolution of the density distribution (log scale) and magnetic field lines. The left, middle and right panels are corresponding to those in $t/\tau = 0$, 350, and 700 respectively, where $\tau = R_{\rm NS}/c \sim 4.5 \times 10^{-5}$sec is the dynamical timescale.

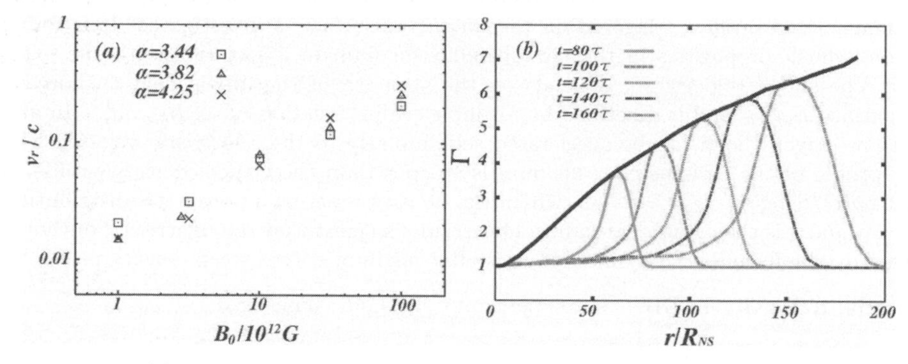

Figure 2. Panel (a): The relation between the strength of the dipole magnetic field and the shock velocity. Panel (b): The time evolution of the Lorentz factor in the case with $\alpha = -7.29$. A black solid line represents the Lorentz factor of the plasma measured in the comoving frame of the shock surface.

examined by varying B_0 from 10^{12} to 10^{14}G, and which is shown in Figure 2a. We can find from this figure, the scaling relation $v_{\rm sh} \propto B_{\rm init}{}^{0.5}$ when $\alpha = 3.44, 3.82$, and 4.25.

The physical properties of the shock wave can be changed due to the initial density profile of the circumstellar medium (i.e., index α). It is accelerated rapidly to the relativistic velocity when α is smaller than ≈ -5. Note that a smaller index α corresponds to a steeper density profile. The relativistic shock wave can thus be accelerated through the circumstellar medium only with a steep density profile. Figure 2b illustrates the time evolution of the shock velocity in the relativistic expansion case with $\alpha = -7.29$. The vertical axis represents the Lorentz factor of the plasma. The horizontal axis is the cylindrical radius. The curves indicate the cases where $t/\tau = 80$, 100, 120, 140, and 160 respectively. It can be seen from this figure that the outflow must be accelerated self-similarly to the relativistic velocity. Our nonlinear study suggests that the expanding outflow is accelerated in association with the flaring activity on the magnetar surface.

References

Mikic, Z. & Linker, J. A. 1994, *ApJ* 430, 898
Thompson, C. & Duncan, R. C. 1995, *MNRAS* 275, 255

Cosmic Magnetic Fields:
From Planets, to Stars and Galaxies
Proceedings IAU Symposium No. 259, 2008
K.G. Strassmeier, A.G. Kosovichev & J.E. Beckman, eds.

doi:10.1017/S1743921309030373

Spectropolarimetry of supernova remnant G296.5 + 10.0

Lisa Harvey-Smith, Bryan M. Gaensler, C.-Y. Ng and Anne J. Green

Sydney Institute for Astronomy (SIFA), School of Physics,
The University of Sydney, NSW 2006, Australia.
email: lhs@usyd.edu.au

Abstract. Radio continuum emission from the supernova remnant G296.5 + 10.0 was observed using the Australia Telescope Compact Array. Using a 104 MHz bandwidth split into 13 × 8 MHz spectral channels, it was possible to produce a pixel-by-pixel image of Rotation Measure (RM) across the entire remnant. A lack of correlation between RM and X-ray surface brightness reveals that the RMs originate from outside the remnant. Using this information, we will characterise the smooth component of the magnetic field within the supernova remnant and attempt to probe the magneto-ionic structure and turbulent scale sizes in the ISM and galactic halo along the line-of-sight.

Keywords. Polarization – magnetic fields – ISM: supernova remnants – ISM: radio continuum – X-rays: ISM

1. Observations

The supernova remnant G296.5 + 10.0 was observed using the Australia Telescope Compact Array at 1.4 GHz with an angular resolution of 30 arcseconds. Signals from the dual linear polarized feeds were combined to produce images of the remnant in linear polarisation. The Rotation Measure synthesis method (Brentjens & de Bruyn, 2005) was employed to obtain an RM spectrum at each pixel in the image (Figure 1). The 104 MHz bandwidth was split into 13 × 8 MHz spectral channels, giving sufficient coverage in λ^2 to remove any nπ ambiguities in the RM values. X-ray data in the energy range 0.1–2.4 keV were taken from the ROSAT archive, via NASA's SkyView. These complimentary data allow us to estimate the electron density in the remnant.

2. Determination of n_e and $\mathbf{B}_{||}$

In order to disentangle the electron density (n_e) and line-of-sight magnetic field ($\mathbf{B}_{||}$) we measured the X-ray intensity at every point in the ROSAT image, after regridding and smoothing them to match the ATCA image. RM is defined as the integral of n_e and $\mathbf{B}_{||}$ along the line of sight. The thermal X-ray surface brightness, S_x, is proportional to ${\mathrm{n}_e}^2$. Therefore if RM and ${\mathrm{S}_x}^{1/2}$ are correlated, the Faraday rotation is occurring within the thermal plasma (Matsui et al. 1984). A plot of RM vs. ${\mathrm{S}_x}^{1/2}$ (Figure 2) shows that there is no clear correlation between RM and ${\mathrm{S}_x}^{1/2}$ in the remnant, which means that the RMs originate from the ISM/halo along the line-of-sight. Therefore the RM variations across the remnant may reflect the structure of the magnetised interstellar medium (Haverkorn *et al.* 2004) and the smooth component may tell us about the overall magnetic field morphology within the remnant (Ransom *et al.* 2008). This will form the basis of our further analysis of these data.

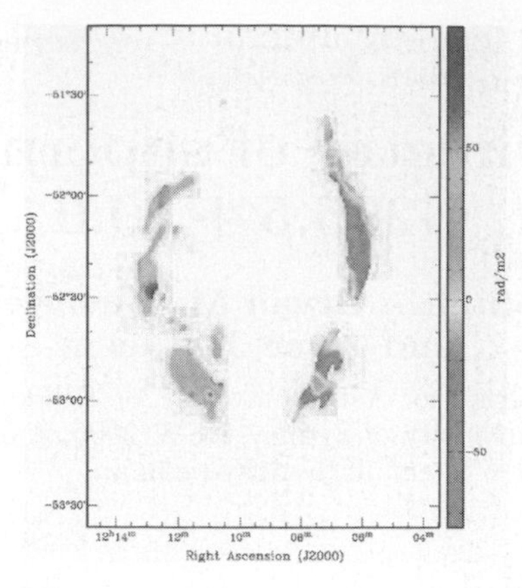

Figure 1. Rotation-Measure image of G296.5 + 10.0

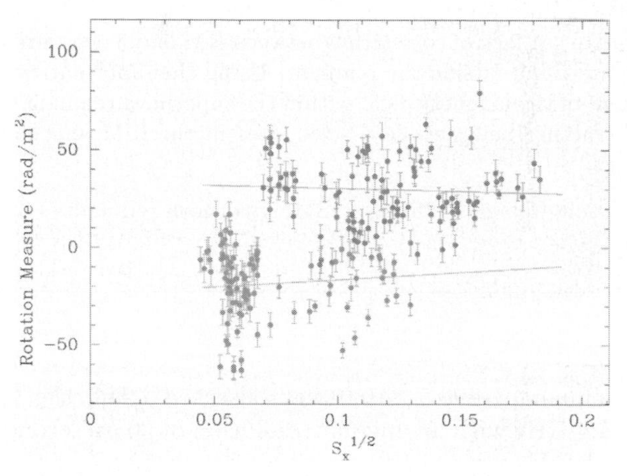

Figure 2. Plots of (X-ray surface brightness)$^{1/2}$ against Rotation Measure for the east (positive) and west (negative) sides of the remnant. A least-squares best fit is plotted onto each data set.

3. On-going work

The following analytical steps will be taken: (1) Modelling of the RM structure across the remnant to determine the large-scale structure and morphology of the magnetic field (i.e. linear, spiral, toroidal) in the remnant (2) Analysis of smaller-scale RM variations across the remnant, to probe the turbulent scale-sizes in the ISM and galactic halo.

Acknowledgements

Lisa Harvey-Smith acknowledges the IAU for a travel grant for this symposium.

References

Brentjens, M. A. & de Bruyn, A. G. 2005, *ApJ* 441, 1217
Haverkorn, M., Katgert, P., & de Bruyn, A. G. 2004, *A&A* 427, 169
Matsui, Y., Long, K. S., Dickel, J. R., & Greisen, E. W. 1984, *ApJ* 287, 295
Ransom, R. R., Uyaniker, B., Kother, R., & Landecker, T. L. 2008, *ApJ* 684, 1009

Welcome speech: John Beckman

Welcome speech: Klaus Strassmeier

Rainer Beck thinking, Detlef Elstner (left) working

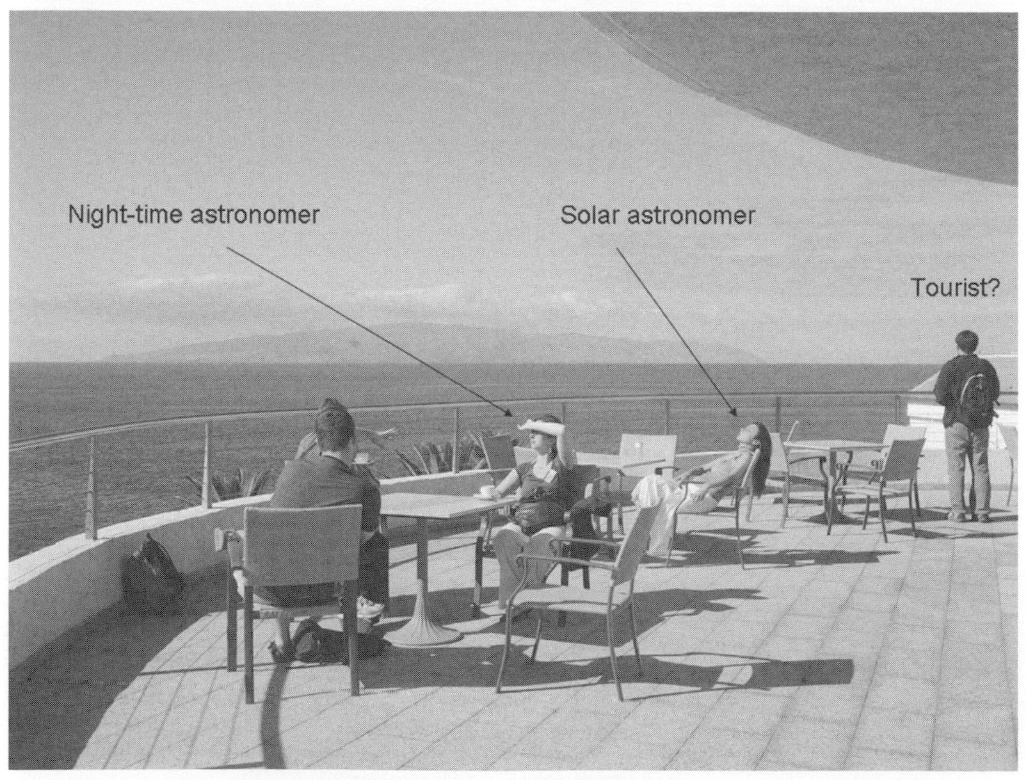

Outside the lecture room

Session II

Multi-scale magnetic fields of the Sun; their generation in the interior, and magnetic energy release

Cosmic Magnetic Fields:
From Planets, to Stars and Galaxies
Proceedings IAU Symposium No. 259, 2008
K.G. Strassmeier, A.G. Kosovichev & J.E. Beckman, eds.

doi:10.1017/S1743921309030397

Magnetic fields and dynamics of the Sun's interior

Alexander G. Kosovichev

W. W. Hansen Experimental Physics Lab, Stanford University, Stanford, CA 94305, USA
email: sasha@sun.stanford.edu

Abstract. Advances in helioseismology provide new knowledge about the origin of solar magnetic activity. The key questions addressed by helioseismology are: what is the physical mechanism of the solar dynamo, how deep inside the Sun are the magnetic fields generated, how are they transported to the surface and form sunspots? Direct helioseismic signatures of the internal magnetic fields are weak and difficult to detect. Therefore, most of the information comes from observations of dynamical effects caused by the magnetic fields. I review results of recent helioseismic observations of the magnetohydrodynamics of the solar interior on various scales, including global dynamics associated with the dynamo processes, and formation of sunspots and active regions.

Keywords. Sun: activity – Sun: helioseismology – Sun: magnetic fields – Sun: interior – sunspots

1. Introduction

Solar and stellar magnetic fields are produced by a dynamo process in the stellar interiors, in which the magnetic field is maintained by turbulent motions against Ohmic dissipation. In astrophysical objects, dynamo can exist when plasma consists of seed magnetic field and flow fields. However, sufficient conditions for dynamo are not well-determined. Mean-field MHD theories of solar and stellar dynamos predict the cyclic behavior, resembling observed properties of solar cycles such as the butterfly diagram for sunspot formation zone and polar field polarity reversals (Fig. 1). It is well-established from observations that during a solar cycle the zone of bipolar magnetic region emergence migrates from the mid latitudes towards the equator (forming the famous "butterfly" diagram) and that the magnetic flux of the following polarity of the bipolar regions migrates toward the poles causing the polar field reversals. However, our understanding of the underlying physical processes is still schematic.

It is quite clear that despite the long history of observations of solar and stellar activity more systematic and detailed studies, both observational and theoretical, are need to advance our understanding of the dynamo processes. From the observational point of view, it is important to use advances of helioseismology to determine the links between the interior dynamics and surface and coronal phenomena: emergence and evolution of active regions, magnetic flux transport, polar field reversals and magnetic flux dissipation and escape. From the theoretical point, it is necessary to link the dynamo models with observational data, directly related to the magnetic field evolution, such as variations of the differential rotation rate in the form of 'torsional oscillations' and variations of the meridional flow.

One of the puzzling features of solar magnetism is its multi-scale spatial and temporal behavior. High-resolution observations reveal that magnetic field on the Sun's surface is very structured and consists of small, rapidly evolving magnetic elements, the ultimate scale of which is still unresolved. They form active regions and sunspots, which seem to

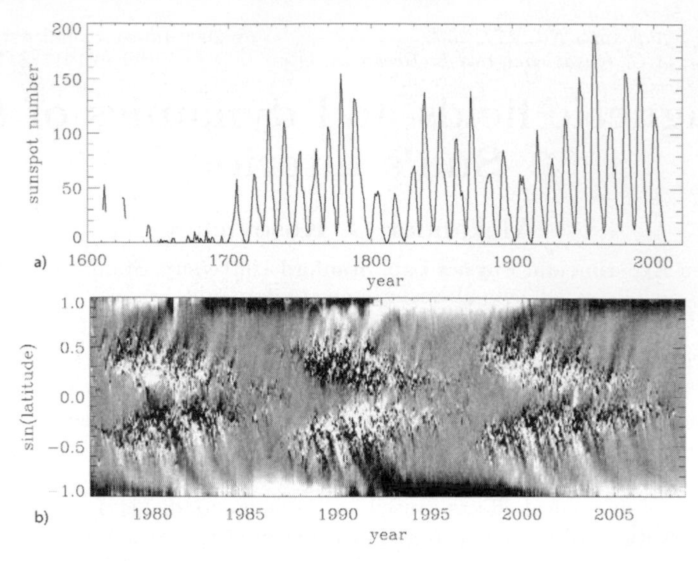

Figure 1. a) The sunspot number record from 1610 to 2008. b) Azimuthally averaged magnetic field as a function of latitude and time for three solar cycles of 1976-2008, observed at the Kitt Peak Observatory. Magnetic field of the positive and negative polarities is shown in white and black colors. Toroidal magnetic field produces bipolar active regions in mid- and low-latitude zones forming the characteristic "butterfly" diagrams. The magnetic field of the following polarity of the active regions is transported to the polar regions causing periodic polarity reversals and forming the poloidal field, which is believed to be a source of toroidal field of future cycles.

emerge randomly. At the same time, solar magnetic fields show a remarkable degree of organization on the global scale, displaying the 'butterfly' diagram and polarity reversals quite regularly with 11-year sunspot cycle (Fig. 1).

Many key questions of solar magnetism remain unanswered. How deep is the dynamo? What is the role of the two radial shear layers: tachocline and subsurface shear layer? What is the relationship between the internal dynamics and magnetic fields? What is the origin of the torsional oscillations? Are there any solar-cycle (11-year) variations in the tachocline? What is the structure of the meridional flows? How do they change with the solar cycle? Why is the solar magnetic field is concentrated in sunspots? How deep are the roots of sunspots? What is the structure and dynamics of the solar plasma beneath sunspots?

2. Probing solar magnetism and dynamics by helioseismology

The basic mechanisms of solar and stellar magnetism operate below the visible surface, and thus unaccessible for direct astronomical observations. However, helio- and asteroseismology provides tools for probing the physical conditions inside the Sun and stars. In particular, a great deal of information about the structure and dynamics of the solar interior has been obtained. This unique knowledge about the dynamo process and formation of magnetic structures in the conditions typical for astrophysical plasmas is extremely important for understanding not only the solar magnetic activity but the origin of cosmic magnetism in general.

Helioseismology is based on observations and analysis of solar acoustic waves. These waves are stochastically excited by turbulent convection in a shallow subsurface layer and travel through the solar interior. Methods of helioseismology are generally divided into global, based on analysis on oscillation frequencies of resonant normal modes, and

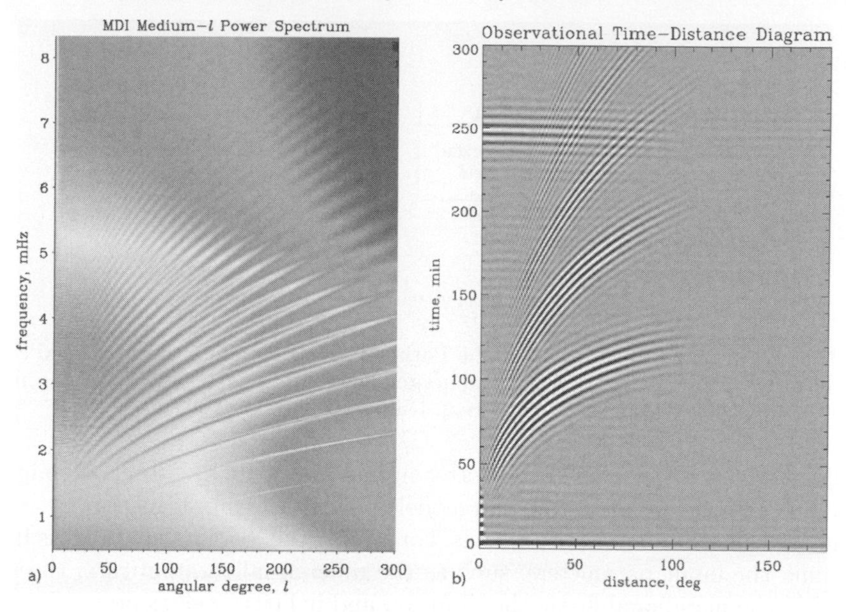

Figure 2. a) Power spectrum of solar oscillations, obtained from SOHO/MDI data. The ridges (yellow and red) correspond to global oscillation modes. The lowest faint ridge is the signal of the surface gravity wave (f-mode). The higher ridges correspond to acoustic waves (p-modes). b) The cross-covariance function of solar oscillations (time-distance diagram). The lowest ridge corresponds to acoustic wave packets traveling directly between two surface points through the interior. The higher ridges correspond to waves arriving after intermediate reflections (bounces) from the surface.

local, based on measurements of properties of acoustic waves (such as travel times and dispersion) in local areas. Figure 2a shows the oscillation power spectrum used for global helioseismology diagnostics. Figure 2b shows the time-distance diagram, which is calculated as cross-covariance of the oscillation signals observed at various locations as a function of distance between the measurement points and lag time. This diagram is used to measure travel times of acoustic wave packets traveling through the interior. In magnetic regions these measurements correspond to travel times of fast magneto-acoustic waves. The travel times are used to infer perturbations of the wave speed and flow velocities using a 3D tomographic inversion procedure (Kosovichev 1999). The methods of helioseismology are being extensively developed and tested by the use of numerical simulations.

3. Solar dynamics and dynamo

3.1. *Dynamo models*

Most of our current understanding of solar magnetism comes from turbulent dynamo models developed on the principles of mean-field electrodynamics. The dynamo process in these so-called $\alpha - \Omega$ models consists of cyclic transformations between predominantly poloidal and toroidal states of solar magnetic fields (Parker 1955) as illustrated in Fig. 3. The models predict that the toroidal magnetic field, which is the primary source of magnetic active regions and sunspots, is generated from the poloidal component by the Sun's differential rotation (Ω-effect), and that the poloidal field is produced by a helical turbulence (α-effect). In the $\alpha - \Omega$ dynamo model the equator-ward migration of sunspots is explained in terms of dynamo waves, and the polar-ward flux transport is a result

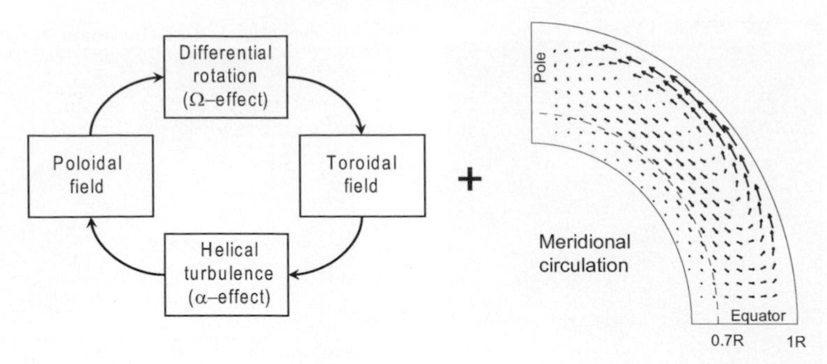

Figure 3. a) A schematic illustration of the Parker's $\alpha - \Omega$ dynamo. b) meridional circulation model. The combination of the $\alpha - \Omega$ dynamo and magnetic flux transport by the meridional circulation gives a flux-transport dynamo (e.g. Dikpati & Gilman 2008).

of turbulent flux diffusion. To get a better agreement with the observed migration of magnetic flux into the polar regions the models include the flux transport by meridional flows, and are called flux-transport models. For testing the dynamo models it is important to determine the input parameters, such as the differential rotation rate, the structure and speed of the meridional flows, the diffusion and helicity coefficients.

3.2. *Internal differential rotation and zonal flows*

The internal differential rotation is inferred from splitting of frequency multiplets of normal acoustic modes of the Sun. The results of these measurements (Fig. 4a) reveal two radial shear layers at the bottom of the convection zone (so-called tachocline) and in the upper convective boundary layer. A common assumption is that the solar dynamo operates in the tachocline area (interface dynamo) where it is easier to explain storage of magnetic flux than in the upper convection zone because of the flux buoyancy. However, there are theoretical and observational difficulties with this concept. First, the magnetic field in the tachocline must be quite strong, $\sim 60 - 160$ kG, to sustain the action of the Coriolis force transporting the emerging flux tubes into high-latitude regions (D'Silva & Choudhuri 1993). The magnetic energy of such field is above the equipartition level of the turbulent energy. Second, the back-reaction such strong field should suppress turbulent motions affecting the Reynolds stresses. Since these turbulent stresses support the differential rotation one should expect significant changes in the rotation rate in the tachocline. However, no significant variations with the 11-year solar cycle are detected. Third, magnetic fields often tend to emerge in compact regions on the solar surface during long periods lasting several solar rotations. This effect is known as "complexes of activity" or "active longitudes". However, the helioseismology observations show that the rotation rate of the solar tachocline is significantly lower than the surface rotation rate. Thus, magnetic flux emerging from the tachocline should be spread over longitudes (with new flux lagging the previously emerged flux) whether it remains connected to the dynamo region or disconnected. It is well-known that sunspots rotate faster than surrounding plasma. This means that the magnetic field of sunspots is anchored in subsurface layers. Observations show that the rotation rate of magnetic flux matches the internal plasma rotation in the upper shear layer (Fig. 4b) indicating that this layer is playing an important role in the solar dynamo, and causing a shift in the dynamo paradigm (Brandenburg 2005).

Variations in solar rotation clearly related to the 11-year sunspot cycle are observed in the upper convection zone. These are so-called 'torsional oscillations' which represent bands of slower and faster rotation, migrating towards the equator as the solar cycle

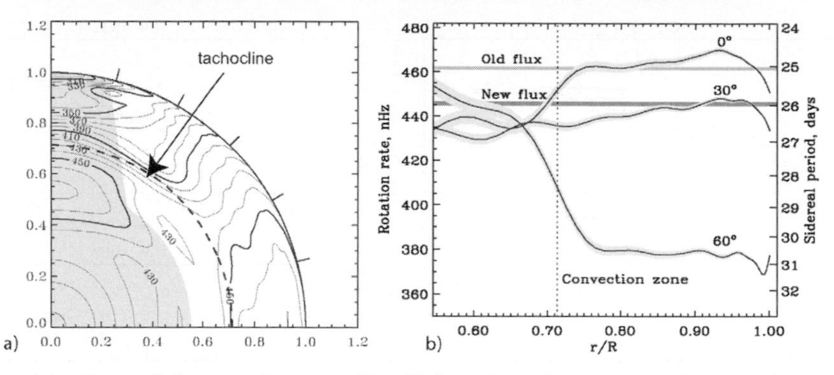

Figure 4. a) Isolines of the rotation rate (in nHz) in the solar interior obtained from helioseismology inversion results (Schou *et al.* 1998). The gray area is where the results are uncertain. Dashed curve shows the location of the bottom of the convection zone and tachocline. b) The rotation rate as a function of radius at three latitudes. The horizontal lines indicate the rotation rate of the surface magnetic flux at the end of solar cycle 22 ("old flux") and at the beginning of cycle 23 ("new flux") (Benevolenskaya, Hoeksema, Kosovichev, & Scherrer 1999)

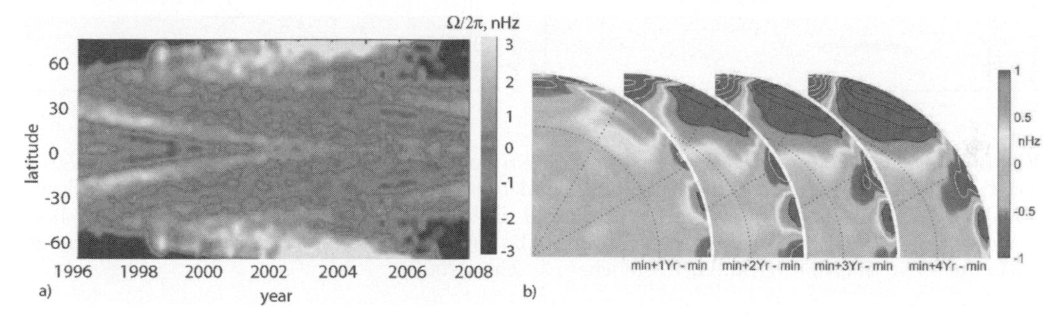

Figure 5. a) Migration of the subsurface zonal flows with latitude during solar cycle 23 from SOHO/MDI data (Howe 2008). Red shows zones of faster rotation, green and blue show slower rotation. b) Variations of the zonal flows with depth and latitude during the first 4 years after the solar minimum. (Vorontsov *et al.* 2002)

progresses (Fig. 5). The torsional oscillations were first discovered on the Sun's surface (Howard and Labonte 1980), and then were found in the upper convection zone by helioseismology (Kosovichev and Schou 1997; Howe *et al.* 2000). The depth of these evolving zonal flows is not yet established. However, there are indications that they may be persistent through most of the convection zone, at least, at high latitudes (Vorontsov *et al.* 2002). The physical mechanism is not understood. Nevertheless, it is clear that these zonal flows are closely related to the internal dynamo mechanism that produces toroidal magnetic field. On the solar surface, this field forms sunspots and active regions which tend to appear in the areas of shear flows at the outer (relative to the equator) part of the faster bands. Thus, the torsional flows are an important key to understanding the solar dynamo, and one of the challenges is to establish their precise depth and detect corresponding variations in the thermodynamic structure of the convection zone. Recent modeling of the torsional oscillations by the Lorentz force feedback on differential rotation showed that the poleward-propagating high-latitude branch of the torsional oscillations can be explained as a response of the coupled differential rotation/meridional flow system to periodic forcing in midlatitudes of either mechanical (Lorentz force) or thermal nature (Rempel 2007). However, the main equatorward-propagating branches cannot be explained by the Lorenz force, but maybe driven by thermal perturbations

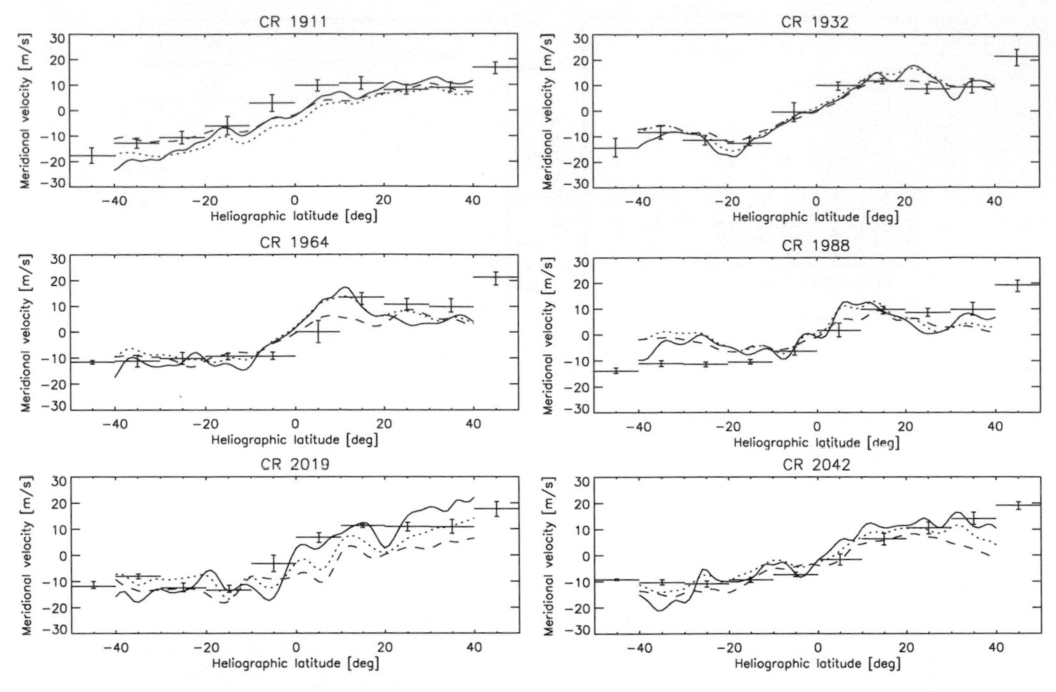

Figure 6. The longitudinally averaged meridional flow speed measured for a set of Carrington rotations by time-distance helioseismology. Solid line plots the time-distance mean meridional flow at 3–4.5 Mm depth, dotted at 6–9 Mm, and dashed at 9–12 Mm. The dots with error-bars represent the 10-degree-bin-averaged values of the flux transport speed derived from the magnetic butterfly diagram (Švanda, Kosovichev, & Zhao 2007).

caused by magnetic field (Spruit 2003). It is intriguing that starting from 2002, during the solar maximum, the helioseismology observations show new branches of "torsional oscillations" migrating from about 45° latitude towards the equator (Fig. 5a). They indicate the start of the next solar cycle, number 24, in the interior, and are obviously related to magnetic processes inside the Sun. However, magnetic field of the new cycle appeared on the surface only in 2008.

3.3. *Meridional flows and flux transport*

The meridional flows of the solar plasma have been reliably measured by local helioseismology only in the upper convection zone. These flows are directed from the equator to the polar regions. The return meridional flow expected from mass conservation has not been detected. The helioseismology measurements from SOHO/MDI and GONG show that the speed of the flows significantly varies during the solar cycle. In particular, during the activity maximum the speed decreases. This seems to be explained by the effect of large-scale inflows developed around magnetic active regions (Haber, Hindman, Toomre, & Thompson 2004; Zhao & Kosovichev 2004). This decrease may affect the magnetic flux transport in to the polar regions and the polar field reversals. However, the relationship between the meridional flows and the magnetic flux transport is not straightforward (Švanda, Kosovichev, & Zhao 2007). While, despite a general correspondence there are significant differences between the speeds of the meridional flow flux transport (Fig. 6). There is also an indication that the inflows are predominantly developed around the leading polarity areas of active regions, where magnetic field is stronger and more stable. Helioseismic measurements of the meridional flows are difficult (Giles, Duvall,

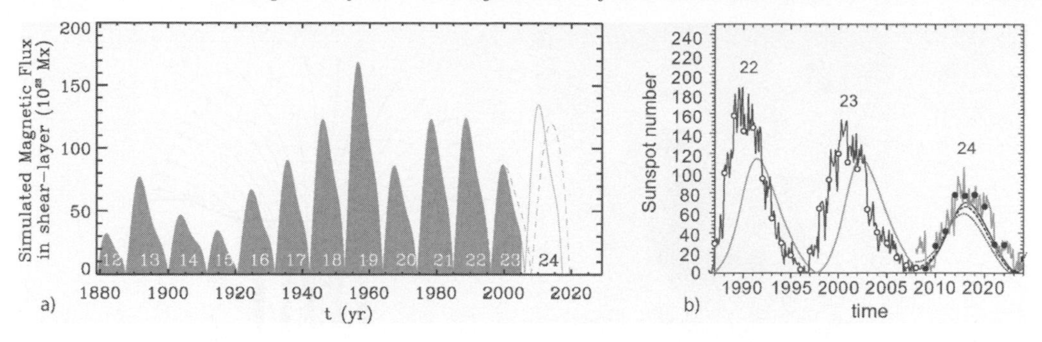

Figure 7. Predictions of solar cycle 24 by using data assimilations and dynamo models: a) flux-transport model (Dikpati, de Toma, & Gilman 2006), b) model with magnetic helicity balance (Kitiashvili & Kosovichev 2008).

Scherrer, & Bogart 1997; Braun & Fan 1999; Haber *et al.* 2002; Zhao & Kosovichev 2004; Krieger, Roth, & von der Lühe 2007; Mitra-Kraev & Thompson 2007) because of their low speed, $\sim 10-20$ m/s, but definitely most critical for understanding the solar dynamo mechanism.

3.4. *Predictions of solar cycles*

Because of the insufficient knowledge of the solar dynamics and without a reliable dynamo model physics-based predictions of solar cycles are difficult. Using simple arguments that the new toroidal field (which determines the sunspot magnetic flux) is generated from the poloidal field by differential rotation, and thus the poloidal field strength determines the strength of the sunspot cycle, Schatten (2005) and Svalgaard *et al.* (2005) predicted that the next solar cycle will be lower than the current one. However, the calculations of Dikpati, de Toma, & Gilman (2006), based on the calibrated flux-transport dynamo model, suggest that such direct relationship does not hold, and the next cycle will be significantly stronger (Fig. 7a). The recent efforts are to apply data assimilation methods for incorporating observational data in the dynamo models. This approach commonly based on Kalman-filter methods is used, for instance, for weather forecasts. It is proved to be useful for estimating the current state of a system from a set of observations and for predicting future states even when the precise nature of the system is unknown. The data assimilation methods allow to account for the uncertainties in both observations and theory. Kitiashvili & Kosovichev (2008, 2009a) applied the Ensemble Kalman Filter (EnKF) method to assimilate the sunspot number data (Fig. 1a) into a simple dynamical model of the solar dynamo, which in addition to the standard $\alpha-\Omega$ mechanism considers the evolution of large- and small-scale magnetic helicity (Kitiashvili & Kosovichev 2009b). The results predict a weak cycle 24 (Fig. 7b), but also indicate that the sunspot number data alone do not provide sufficient constraints, and suggest that synoptic magnetic field data used for further development of this approach along with more detailed dynamo modeling.

4. Subsurface structure and dynamics of sunspots

One of the great puzzles of solar magnetism is the concentration of the dynamo-generated magnetic field in compact strong-field structures, sunspots. The sunspots have a fibril structure (Fig. 8a), and detailed observations showed that sunspots represent bundles of magnetic tubes and that the plasma can flow inside sunspots between these bundles (Severny 1965). Following these observations Parker (1979) suggested a cluster

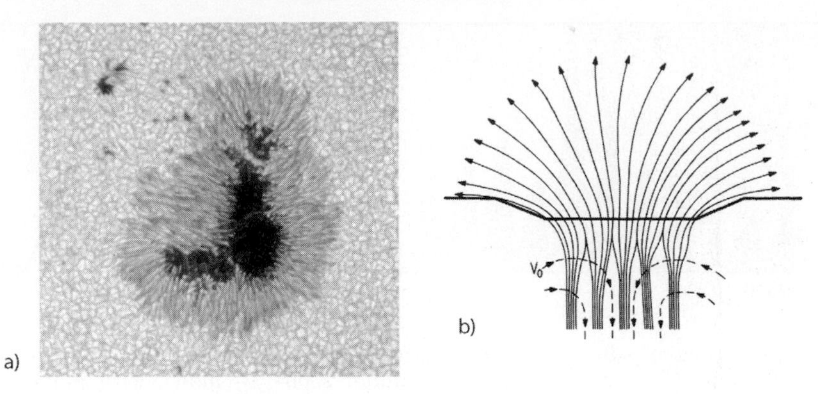

Figure 8. a) High-resolution image of a sunspot from Hinode spacecraft. b) Cluster model of sunspots (Parker 1979).

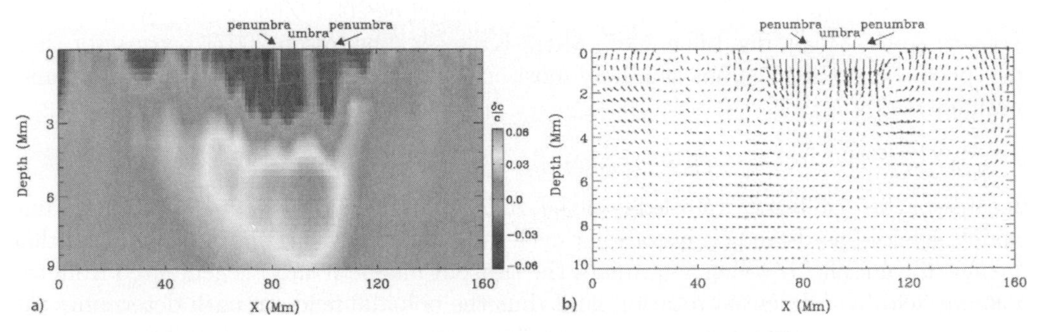

Figure 9. Variations of the acoustic wave speed (a) and the flow field (b) beneath the sunspot shown in Fig. 8a, obtained by time-distance helioseismology from Hinode data (Zhao, Kosovichev, & Sekii 2009). Red color shows regions where the wave speed is higher than in surrounding plasma, and blue color shows where the wave speed in lower. The longest arrow corresponds to the plasma velocity of ~ 0.5 km/s.

model (Fig. 8b), in which the magnetic structure of sunspots is maintained by inflows of plasma. Such inflows can be a part of the large-scale circulation around sunspots originating from the downdraft motions of the cool plasma beneath sunspots. Time-distance helioseismology measurements have shown the existence of the downdrafts and inflows (Duvall *et al.* 1996; Zhao, Kosovichev, & Duvall 2001). They also showed that beneath sunspots there is a 4-5 Mm deep region of relatively low wave speed (presumably due to the low plasma temperature), and that in the deeper interior the wave speed becomes higher (Kosovichev, Duvall, & Scherrer 2000). Recent high-resolution observations from Hinode spacecraft (Kosugi *et al.* 2007; Tsuneta *et al.* 2008) have provided a remarkable confirmation of these results and clearly revealed the sunspot cluster structure in the subphotosphere (Zhao, Kosovichev, & Sekii 2009). Figure 9 shows a vertical structure of the acoustic wave-speed perturbations and flows of the sunspot shown in Fig. 8a. The high-resolution helioseismology observations open new perspectives of studying the sunspot dynamics in much great detail.

5. Formation and evolution of active regions

Recent observations and modeling reveal some interesting features of the properties of emerging magnetic flux and associated dynamics on the solar surface and in the upper convection zone. In particular, the local helioseismology results obtained by both, the

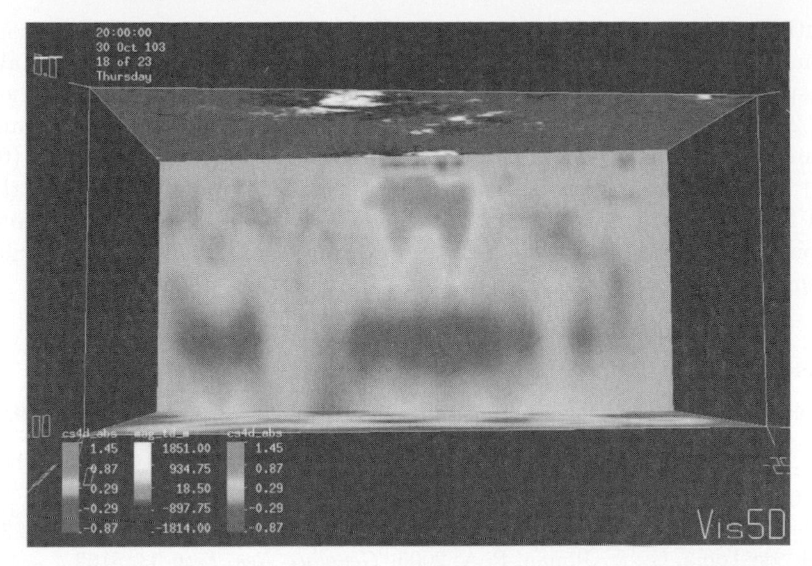

Figure 10. Acoustic wave speed image of active region NOAA 10488 in the solar interior. (Kosovichev 2009)

time-distance and ring-diagram techniques, show large-scale outflows beneath the surface during most of the emergence phase, and also formation of converging flows around the magnetic structure of sunspots. However, the structure of the vertical flows remains unclear. There is an indication of upflows mixed with downflows at the beginning of emergence, but then the downflows dominate. In the case of AR 10488, there were two or three major flux emergence events. The photospheric magnetic flux rate and subsurface flow divergence show two or three peaks, which are not in phase, but it is unclear if the flux rate precedes the variation of the flow divergence or follows it (Kosovichev 2009).

From the observations it is obvious that the multiple flux emergence events over several days plays important role in the formations and maintaining the magnetic structure of the large active region. This reminds the idea of a common 'nest' in the deep interior (Castenmiller, Zwaan, & van der Zalm 1986). However, such nests have not been found in the helioseismic images of the subphotospheric magnetosonic wave speed variations, which are currently obtained up to the depth of 40-50 Mm (Fig 10). The wave speed images reveal that the emerging magnetic flux structures travel very fast in the upper convection zone, with a speed of at least 1 km/s. This makes very difficult the detection of these structures before the magnetic field becomes visible on the surface. However, it should be possible to use the measurements of both, the wave speed variations and flow velocities, for predicting the growth and and decay of active regions and, perhaps, the complexity of their magnetic structure. This task will require a substantial statistical analysis of emerging active regions by methods of local helioseismology.

6. Conclusion and outlook

Helioseismology has uncovered intriguing dynamics of plasma in the Sun's interior. Many of the initial results are unexpected and counter-intuitive and are not explained by the current theoretical models. For understanding the basic mechanisms of solar activity, magnetic energy generation, storage and release, it is important to investigate in detail the processes of the dynamics of the tachocline, meridional flows, torsional oscillations, emergence and evolution of active regions, and their relationship to formation and evolution of magnetic field structures in the corona, causing magnetic reconnections and plasma eruptions. Helioseismic and Magnetic Imager (HMI) instrument along with the Atmospheric Imaging Assembly (AIA) on Solar Dynamics Observatory scheduled for launch in 2009 will measure Doppler velocity and vector magnetic field, also image high-temperature solar corona, almost uninterruptedly, providing high-resolution (0.5 arcsec per pixel) data for these investigations. It is equally important to develop theoretical models of the solar dynamo and sunspot formation, as well as, investigate the magnetic activity on other stars, using long-term monitoring, Doppler imaging techniques and asteroseismology, and understand the place of the Sun in cosmic magnetism.

References

Benevolenskaya, E. E., Hoeksema, J. T., Kosovichev, A. G., & Scherrer, P. H. 1999, *ApJ* 517, L163

Brandenburg, A. 2005, *ApJ* 625, 539

Braun, D. C. & Fan, Y. 1999, *ApJ* 510, L81

Castenmiller, M. J. M., Zwaan, C., & van der Zalm, E. B. J. 1986, *Sol. Phys.* 105, 237

Dikpati, M., de Toma, G., & Gilman, P. A. 2006, *Geophys. Res. Lett.* 33, 5102

Dikpati, M. & Gilman, P. A. 2008, *J. Astrophys. Aastron.* 29, 29

D'Silva, S. & Choudhuri, A. R. 1993, *A&A* 272, 621

Duvall, T. L. J., D'Silva, S., Jefferies, S. M., Harvey, J. W., & Schou, J. 1996, *Nature* 379, 235

Giles, P. M., Duvall, T. L., Jr., Scherrer, P. H., & Bogart, R. S. 1997, *Nature* 390, 52

González Hernández, I., Kholikov, S., Hill, F., Howe, R., & Komm, R. 2008, *Sol. Phys.* 252, 235

Haber, D. A., Hindman, B. W., Toomre, J., & Thompson, M. J. 2004, *Sol. Phys.* 220, 371

Haber, D. A., *et al.* 2002, *ApJ* 570, 855

Howard, R., Labonte, B. J. 1980, *ApJ* 239, L33

Howe, R. *et al.* 2000, *Science* 287, 2456

Howe, R. 2008, *Adv. Sp. Res.* 41, 846

Kitiashvili, I., & Kosovichev, A. G. 2008, *ApJ* 688, L49

Kitiashvili, I. N., & Kosovichev, A. G. 2009a, this proceedings

Kitiashvili, I. N., & Kosovichev, A. G. 2009b, *Geophys. Astrophys. Fluid Dyn.* 103, 53

Komm, R., Howe, R., Hill, F., & Morita, S. 2008, *ASPC Conf. Ser.* 383, 83

Kosovichev, A. G. & Schou, J., 1997, *ApJ* 482, L207

Kosovichev, A. G. 1999, *J. Comp. Appl. Math.* 109, 1

Kosovichev, A. G., Duvall, T. L. Jr., & Scherrer, P. H. 2000, *Sol. Phys.* 192, 159

Kosovichev, A. G., 2009, *Sp. Sci. Rev.*, in press

Kosugi, T. *et al.* 2007, *Sol. Phys.* 243

Krieger, L., Roth, M., & von der Lühe, O. 2007, *AN* 328, 252

Mitra-Kraev, U. & Thompson, M. J. 2007, *AN* 328, 1009

Parker, E. N. 1955, *ApJ* 122, 293

Parker, E. N. 1979, *ApJ* 230, 905

Rempel, M., 2007, *ApJ* 655, 651

Schou, J., *et al.* 1998, *ApJ* 505, 390

Severny, A. B. 1965, *Soviet Astronomy* 9, 171

Schatten, K., 2005, *Geophys. Res. Lett.* 32, 21106

Spruit, H. C. 2003, *Sol. Phys.* 213, 1

Svalgaard, L., Cliver, E. W., Kamide, Y. 2005, *Geophys. Res. Lett.* 32, 1104

Švanda, M., Kosovichev, A. G., & Zhao, J. 2007, *ApJ* 670, L69

Tsuneta, S. *et al.* 2008, *Sol. Phys.* 249, 167

Vorontsov, S. V. *et al.* 2002, *Science* 296, 101

Zhao, J., Kosovichev, A. G., & Duvall, T. L., Jr. 2001, *ApJ* 557, 384

Zhao, J. & Kosovichev, A. G. 2004, *ApJ* 603, 776

Zhao, J., Kosovichev, A. G., & Sekii, T., 2009, *Proc. Hinode-2 conf., ASP Ser.*, in press

Discussion

MAEDER: The circulation you are finding at the convective solar surface is indeed different from the usual meridional circulation, which results from a breakdown of radiative equilibrium in radiative regions. Your circulation velocities are indeed orders of magnitude larger than the "usual" meridional circulation.

KOSOVICHEV: The meridional circulation in the convection zone is produced by the interaction of turbulent convection with rotation. This mechanism is different from the mechanism of the meridional circulation in the radiative zone. Therefore, the mean velocity of the meridional flows in the convection zone is higher than in the radiative zone. Current theoretical models of the meridional flows in the convection zone show that these flows are highly dynamic, and often consist of multiple cells. The multiple cell structure has not been detected by helioseismology but if it exists it will have important implications for the flux-transport dynamo models. Observational and theoretical studies of meridional flows on the Sun are of primary importance for our understanding of solar dynamo.

DE GOUVEIA DAL PINO: Comment: There is a poster (P.26) with my PhD student Gustavo Guerrero where we have performed mean field dynamo simulations where we found that the turbulent pumping is dominant over the overall deep meridional circulation to provide the right flux transport to the right latitudes and right depths. Also, if combined with near surface shear, it also provides the appropriate butterfly diagram without requiring a deep meridional flow (Guerrero & de Gouveia dal Pino, A & A, 2008).

The chairman, left, Nigel Weiss, and the speaker, right, Alexander Kosovichev, eagerly expecting questions

More detailed answering. A. Kosovichev and Irina Kitiashvili.

Axel Brandenburg

Cosmic Magnetic Fields:
From Planets, to Stars and Galaxies
Proceedings IAU Symposium No. 259, 2008
K.G. Strassmeier, A.G. Kosovichev & J.E. Beckman, eds.

doi:10.1017/S1743921309030403

Paradigm shifts in solar dynamo modeling

Axel Brandenburg

Nordita, Roslagstullsbacken 23, 10691 Stockholm, Sweden

Abstract. Selected topics in solar dynamo theory are being highlighted. The possible relevance of the near-surface shear layer is discussed. The role of turbulent downward pumping is mentioned in connection with earlier concerns that a dynamo-generated magnetic field would be rapidly lost from the convection zone by magnetic buoyancy. It is argued that shear-mediated small-scale magnetic helicity fluxes are responsible for the success of some of the recent large-scale dynamo simulations. These fluxes help in disposing of excess small-scale magnetic helicity. This small-scale magnetic helicity, in turn, is generated in response to the production of an overall tilt in each Parker loop. Some preliminary calculations of this helicity flux are presented for a system with uniform shear. In the Sun the effects of magnetic helicity fluxes may be seen in coronal mass ejections shedding large amounts of magnetic helicity.

Keywords. MHD – turbulence – Sun: coronal mass ejections (CMEs) – Sun: magnetic fields

1. Introduction

Unlike the geodynamo, which is widely regarded a solved problem (e.g. Glatzmaier & Roberts 1995), the solar dynamo problem remains unsolved in that there is no single model that is free of theoretical shortcomings and that actually reproduces the Sun. The flux-transport dynamo models (e.g. Dikpati & Charbonneau 1999; Küker *et al.* 2001; Chatterjee *et al.* 2004) are currently the only models that display some degree of realism, but there are aspects in their design that are arguably not sufficiently plausible. One problem is the assumption of a rather low turbulent magnetic diffusivity that needs to be assumed in a *ad hoc* fashion. On theoretical grounds the turbulent magnetic diffusivity should be comparable with the turbulent kinematic viscosity (Yousef *et al.* 2003), which is not the case on these flux-transport dynamo models.

The purpose of this review is to collect recent findings that are relevant in solving the solar dynamo problem. On the one hand, there are direct simulations of convective dynamo action in spherical shells by Brun & Toomre (2002), Brun *et al.* (2004, 2006), Browning *et al.* (2006), and Brown *et al.* (2007). These models are already quite realistic and should ultimately be able to reproduce the solar dynamo. Their main shortcomings range from still insufficient resolution (even though it is already exhausting current capabilities) to the negligence of certain physical features of the model. Examples of shortcomings include the still insufficient degree of stratification as well as the simplified treatment or even the neglect of surface and tachocline boundary layers. On the other extreme, there are phenomenologically motivated mean-field models that are constructed based on their success in reproducing the Sun. These models tend to utilize a subset of known turbulent transport effects with *ad hoc* amplitudes and prescriptions of their radial profiles. Such models are made nonlinear by simple α quenching terms. However, this type of prescription that is still commonly used in mean field models does not correctly describe the saturation behavior of large-scale magnetic fields as known from three-dimensional turbulence simulations. This is true even for much simpler models, for example those where the turbulence is driven at a specific length scale via a forcing function (Brandenburg 2001).

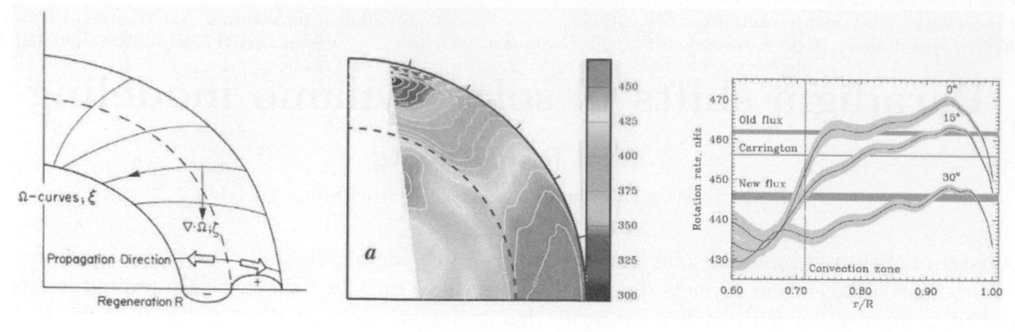

Figure 1. Comparison of the differential rotation contours that were originally expected by Yoshimura (1975) based on solar dynamo model considerations (left) with those by Thompson *et al.* (2003) using helioseismology (middle). Note the similarities between the contours on the left (over the bulk of the convection zone) and those on the right (over the outer 5% of the solar radius). In the right hand panel we show radial profiles of angular velocity as given by Benevolenskaya *et al.* (1999). Note the sharp negative radial gradient near the surface.

2. Negative radial shear?

In the 1970s a number of solar dynamo models were discussed that were based on an α effect that is positive in the northern hemisphere (and negative in the southern hemisphere) and a differential rotation that had a negative radial gradient, i.e. $\partial\Omega/\partial r < 0$ (see the left panel of Fig. 1). Such models produce an equatorward migration of magnetic activity toward the equator. Examples of such models in full spherical geometry include the papers Steenbeck & Krause (1969), Roberts & Stix (1972), Köhler (1973), and Yoshimura (1975). During the 1980s these models faced two difficulties, a conceptual one and one based on a conflict with observations. The conceptual difficulty has to do with the idea that the magnetic field may be fibral (Parker 1984), i.e. it is distributed in the form of many flux tubes with a small filling factor and relatively large field strength. However, such fields would easily be unstable to magnetic buoyancy instabilities. This and other considerations led Spiegel & Weiss (1980) to propose that the solar dynamo might instead work at the bottom of the solar convection zone and not, as assumed until them, in a distributed fashion in the bulk of the convection zone. In fact, in an earlier paper by Parker (1975) this possibility was already discussed in some detail. The conflict with observations had to do with the fact that helioseismology began to put tight constraints on the form of the Sun's internal angular velocity profile. This seemed to rule out a negative radial Ω gradient.

These two difficulties led to a new type of dynamo model that works in the lower overshoot layer, where α has the opposite sign (DeLuca & Gilman 1986, 1988). Not only would magnetic flux tubes presumably be stabilized against magnetic buoyancy instabilities, but at that depth the sign of the α effect would be reversed and thus dynamo waves would again propagate equatorward even though $\partial\Omega/\partial r > 0$. However, several problems with this approach have been discussed (phase relation, number of toroidal flux belts, etc.). Yet another model is the flux transport model, where a meridional circulation (equatorward at the bottom of the convection zone) is chiefly responsible for driving the dynamo wave equatorward. However, it is now quite clear that there is a strong shear layer near the surface where $\partial\Omega/\partial r < 0$, just as originally anticipated. The difference is that now this shear layer only extends over the outer 5% of the solar radius. In the past the relevance of this near-surface shear layer was discarded mainly on the grounds that magnetic buoyancy effects would lead to a rapid loss of magnetic field. Another reason is that near the surface the local turnover time is still short compared

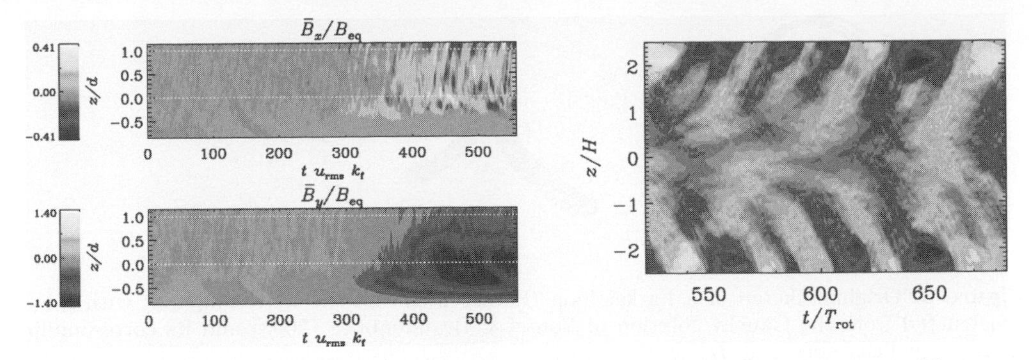

Figure 2. Horizontally averaged magnetic fields $\overline{B}_x$ (upper panel on the left) and $\overline{B}_y$ (lower panel on the left), as functions of time and z for a run with shear and no rotation (Käpylä *et al.* 2008), as well as $\overline{B}_y$ from a local model of an accretion disc (right hand panel), where $z = 0$ corresponds to the midplane. On the left-hand panel the dotted white lines show top ($z = d$) and bottom ($z = 0$) of the convection zone.

with the rotation period. However, simulations of compressible convection did show some time ago that magnetic buoyancy is effectively overpowered by the effects of turbulent downward pumping. Regarding the timescales, rotational effects are likely to play a role at the bottom of the near-surface shear layer, which is at a depth of about 40 Mm. The relative importance of rotation and convection is determined by the Coriolis number,

$$\mathrm{Co} = 2\Omega\tau, \tag{2.1}$$

where Ω is the solar rotation rate and $\tau = H_p/u_{\mathrm{rms}}$ is the turnover time with H_p being the pressure scale height. Using $\Omega = 3 \times 10^{-6}\,\mathrm{s}^{-1}$ for the solar rotation rate, $u_{\mathrm{rms}} = 50\,\mathrm{m/s}$ for the rms velocity at a depth of 40 Mm, and $H_p = 13\,\mathrm{Mm}$ for the pressure scale height at that depth we find $\tau = 1.3\,\mathrm{d}$, and hence $\mathrm{Co} = 1.4$, suggesting that rotation should become important at the bottom of the near-surface shear layer.

3. Quenching and examples of large-scale dynamos

Since the early 1990s there has been the notion that catastrophic α quenching has been a major concern in dynamo theory (Vainshtein & Cattaneo 1992). This led Parker (1993) to suggest that the solar dynamo might work in a segregated fashion where shear acts only near the lower overshoot layer where the turbulent magnetic diffusivity is low. In the bulk of the convection zone, on the other hand, the magnetic field is supposed to be weak, and so there was hope that the α effect would not be catastrophically quenched. However, only later the origin of catastrophic quenching was understood to be due to magnetic helicity conservation. As a consequence, catastrophic quenching could not be alleviated by spatial rearrangements, but only by magnetic helicity fluxes out of the domain. Indeed, there are now a number of simulations that show successful generation of large-scale magnetic fields; see Fig. 2. All these simulations have in common that there are contours of constant shear crossing the outer surface. This is believed to be important because the magnetic helicity flux is expected to be directed along the contours of constant shear (Brandenburg & Subramanian 2005a).

4. α effect and production of small-scale magnetic helicity

The α effect is central to many mean-field approaches of the solar dynamo. Its presence is usually associated with the systematic action of the Coriolis force on convecting

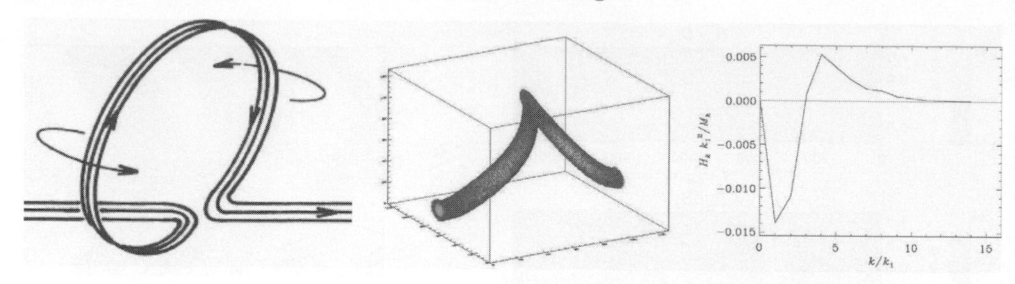

Figure 3. Original sketch of a Parker loop (Parker 1957) on the left, compared with a loop constructed from the Cauchy solution of Yousef & Brandenburg (2003) and its corresponding magnetic helicity spectrum H_k on the right.

fluid elements. As such an element rises vertically, it expands and counter rotates due to the Coriolis force, thereby attaining a downward oriented vorticity in the northern hemisphere, and an upward oriented vorticity in the southern hemisphere. As the magnetic field is dragged along with the gas, a systematic poloidal field component is being produced from a toroidal field. An early sketch of this process was given by Parker (1957), where a rising loop attains a systematic tilt; see the left hand panel of Fig. 3.

In recent years the nonlinear saturation of this process has been understood to be due to the internal twist that is being induced as the loop begins to tilt. This is already quite evident from the early sketch by Parker (1957) suggesting the presence of a spatially extended structure of the loop with oppositely oriented twisting motions on both ends of the loop; see the round arrows on the tube in the left hand panel of Fig. 3.

In the middle panel of Fig. 3 we show a visualization of an analytically generated representation of a Parker loop using the Cauchy solution for ideal magnetohydrodynamics together with the corresponding numerically constructed magnetic helicity spectrum (Yousef & Brandenburg 2003); see the right hand panel of Fig. 3. This spectrum shows quite clearly a negative magnetic helicity at low wavenumbers $k/k_1 \approx 1$ (where $k_1 = 2\pi/L$ is the smallest wavenumber in a box of scale L) and positive magnetic helicity at larger wavenumbers around $k/k_1 \approx 4$. This confirms an early finding of Seehafer (1996) that the α effect corresponds to a segregation of magnetic helicity in scale and that magnetic helicity of smaller scale and opposite sign is inevitably being produced by the α effect. It is this what quenches the α effect in a potentially catastrophic manner. Alleviating this type of quenching requires that we have to get rid of this smaller-scale magnetic helicity through helicity fluxes. This is something that the Sun seems to accomplish through coronal mass ejections (Démoulin *et al.* 2002, Blackman & Brandenburg 2003). In the following we describe in more detail the magnetic helicity flux and its relation to the amount of shear.

5. Magnetic helicity flux

There are a number of analytic calculations of magnetic helicity fluxes relevant to astrophysical dynamos (Kleeorin *et al.* 2000, 2002, 2003a,b; Vishniac & Cho 2001; Subramanian & Brandenburg 2004, 2006; Brandenburg & Subramanian 2005a,b). Here we present preliminary numerical calculations in a simple system. We calculate the small-scale current helicity flux, $\overline{\boldsymbol{\mathcal{F}}}_C^{\rm SS}$, in homogeneous turbulence in the presence of shear, S, and a uniform magnetic field $\boldsymbol{B}_0$. Since the system is completely uniform, it makes sense to consider full volume averages, denoted here by an overbar. The wavenumber of the averaged quantities is zero, so we have infinite scale separation and can then force the turbulence at the scale of the system, i.e. at wavenumbers between 1 and 2 times the

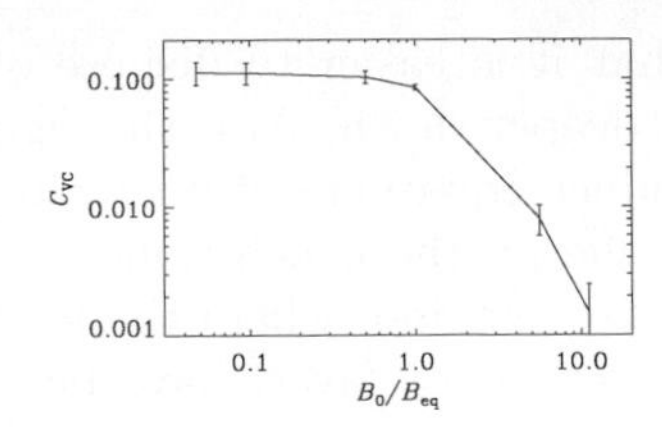

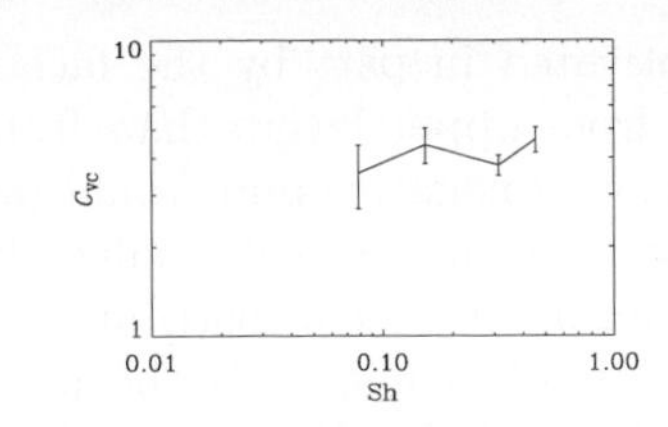

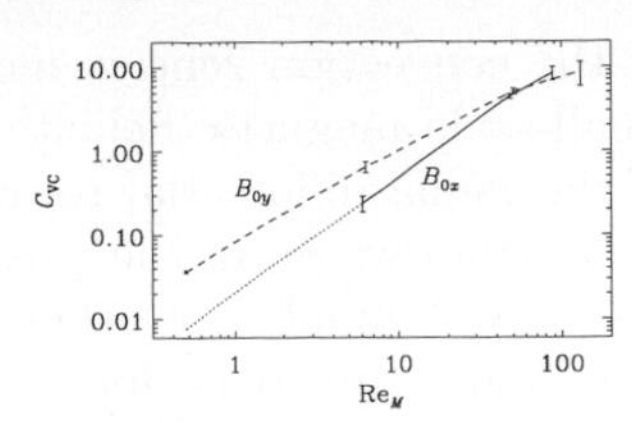

Figure 4. B dependence of $-C_{\rm VC}$ for $\boldsymbol{B}_0 = B_{0y}\hat{\boldsymbol{y}}$ with $R_{\rm m} = 1.4$, $\mathrm{Sh} \approx -2$ (left), S dependence of $C_{\rm VC}$ for $R_{\rm m} = 30$, $B_0/B_{\rm eq} \approx 0.2$ (middle), and $R_{\rm m}$ dependence of $C_{\rm VC}$ for $\mathrm{Sh} \approx 0.5$, $B_0/B_{\rm eq} \approx 0.2$.

smallest finite wavenumber, k_1. Our average forcing wavenumber is therefore $k_{\rm f}/k_1 = 1.5$. We consider the dependence on the parameters $B_0/B_{\rm eq}$, $R_{\rm m}$ and Sh, where

$$B_{\rm eq}^2 = \mu_0 \langle \rho \boldsymbol{u}^2 \rangle, \quad R_{\rm m} = u_{\rm rms}/\eta k_{\rm f}, \quad \mathrm{Sh} = S/\eta k_{\rm f}^2. \tag{5.1}$$

The small-scale current helicity flux is calculated from the expression (Brandenburg & Subramanian 2005b)

$$\overline{\boldsymbol{\mathcal{F}}}_C^{\rm SS} = 2\overline{\boldsymbol{e} \times \boldsymbol{j}} + \overline{(\boldsymbol{\nabla} \times \boldsymbol{e}) \times \boldsymbol{b}}, \tag{5.2}$$

where

$$\boldsymbol{e} = -\boldsymbol{u} \times (\boldsymbol{b} + \boldsymbol{B}_0) + \eta \mu_0 \boldsymbol{j} \tag{5.3}$$

is the small-scale electric field and $\boldsymbol{j} = \boldsymbol{\nabla} \times \boldsymbol{b}/\mu_0$ is the small-scale current density with μ_0 being the vacuum permeability. In order to avoid taking more than two derivatives we integrate the second expression in equation (5.2) by parts, i.e.

$$\overline{(\boldsymbol{\nabla} \times \boldsymbol{e}) \times \boldsymbol{b}} = \overline{(\boldsymbol{\nabla} \boldsymbol{b})^T \boldsymbol{e}}, \tag{5.4}$$

where we have used the fact that $\boldsymbol{\nabla} \cdot \boldsymbol{b} = 0$. The i component of this term can also be written as $\overline{b_{j,i} e_j}$, where a comma denotes partial differentiation.

Based on calculations using the minimal τ approximation the small-scale current helicity flux is expected to be given by (Brandenburg & Subramanian 2005a)

$$\overline{\boldsymbol{\mathcal{F}}}_C^{\rm SS} = C_{\rm VC} (\overline{\mathsf{S}} \overline{\boldsymbol{B}}) \times \overline{\boldsymbol{B}}, \tag{5.5}$$

where $C_{\rm VC}$ is a non-dimensional coefficient that is of the order of St^2, where $\mathrm{St} = \tau u_{\rm rms} k_{\rm f}$ is the Strouhal number. Throughout this work we use a uniform shear flow, i.e. $\overline{\boldsymbol{U}} = (0, Sx, 0)$, so the z component of $\overline{\boldsymbol{\mathcal{F}}}_C^{\rm SS}$ is

$$\overline{\mathcal{F}}_{Cz}^{\rm SS} = \tfrac{1}{2} C_{\rm VC} S \left(\overline{B}_y^2 - \overline{B}_x^2 \right). \tag{5.6}$$

In the following we quote values of $C_{\rm VC}$ that are computed by imposing a uniform field either in the x or in the y direction. The numerical resolution is only 32^3. In the left hand panel of Fig. 4 we show the dependence on B_0 for a small value of $R_{\rm m}$ ($R_{\rm m} = 1.4$) and fixed shear parameter $\mathrm{Sh} \approx -2$ (for $S < 0$). So far there is no indication that the flux depends on Sh; see the middle panel of Fig. 4. However, the flux shows a strong increase with $R_{\rm m}$; see the right hand panel of Fig. 4. It should be noted that, since we keep the forcing unchanged while changing the viscosity, the resulting rms velocity also changes, and so the resulting values of Sh and $B_0/B_{\rm eq}$ also change somewhat.

6. Discussion

In this paper we have presented selected aspects of the solar dynamo problem where there has been recent progress. The idea that the solar dynamo may operate in the bulk of the convection zone is motivated in part by the fact that it is easier to dispose of small-scale magnetic helicity from upper layers than from deeper down. Also the sign of the radial differential rotation is negative and would produce equatorward migration of dynamo waves in the presence of a positive α effect in the northern hemisphere of the Sun. A number of other arguments for a distributed solar dynamo, where the field that makes sunspots does not solely come from the lower overshoot layer, have been discussed elsewhere (Brandenburg 2005). However, what is not yet well addressed is the production of sunspots within active regions, and perhaps even the active regions themselves. It is plausible, and it has indeed been argued, that sunspots can be the result of negative turbulent magnetic pressure effects (Kleeorin & Rogachevskii 1994) or a turbulent magnetic collapse phenomenon (Kitchatinov & Mazur 2000). Both processes rely on turbulent transport processes that could be verified numerically and whose effects could also be demonstrated directly in suitably arranged simulations.

Acknowledgements

The computations have been carried out at the National Supercomputer Centre in Linköping and at the Center for Parallel Computers at the Royal Institute of Technology in Sweden. This work was supported in part by the Swedish Research Council.

References

Benevolenskaya, E. E., Hoeksema, J. T., Kosovichev, A. G., & Scherrer, P. H. 1999, *ApJ* 517, L163
Blackman, E. G. & Brandenburg, A. 2003, *ApJ* 584, L99
Brandenburg, A. 2001, *ApJ* 550, 824
Brandenburg, A. 2005, *ApJ* 625, 539
Brandenburg, A. & Subramanian, K. 2005a, *AN* 326, 400
Brandenburg, A. & Subramanian, K. 2005b, Phys. Rep., 417, 1
Brown, B. P., Browning, M. K., Brun, A. S., *et al.* 2007, AIPC, 948, 271
Browning, M. K., Miesch, M. S., Brun, A. S., & Toomre, J. 2006, *ApJ* 648, L157
Brun, A. S. & Toomre, J. 2002, *ApJ* 570, 865
Brun, A. S., Miesch, M. S. & Toomre, J. 2004, *ApJ* 614, 1073
Brun, A., Miesch, M., & Toomre, J. 2006, ApJ, 614, 1073
Chatterjee, P., Nandy, D., & Choudhuri, A. R. 2004, *A&A* 427, 1019
DeLuca, E. E., Gilman, P. A. 1986, *Geophys. Astrophys. Fluid Dyn.* 37, 85
DeLuca, E. E., Gilman, P. A. 1988, *Geophys. Astrophys. Fluid Dyn.* 43, 119
Démoulin, P., Mandrini, C. H., van Driel-Gesztelyi, L., *et al.* 2002, *ApJ* 382, 650
Dikpati, M. & Charbonneau, P. 1999, *ApJ* 518, 508
Glatzmaier, G. A. & Roberts, P. H. 1995, *Nature* 377, 203
Käpylä, P. J., Korpi, M. J., & Brandenburg, A. 2008, *A&A* 491, 353
Kitchatinov, L. L. & Mazur, M. V. 2000, *Solar Phys.* 191, 325
Kleeorin, N. & Rogachevskii, I. 1994, *Phys. Rev. E* 50, 2716
Kleeorin, N., Moss, D., Rogachevskii, I., Sokoloff, D. 2000, *A&A* 361, L5
Kleeorin, N., Moss, D., Rogachevskii, I., Sokoloff, D. 2002, *A&A* 387, 453
Kleeorin, N., Moss, D., Rogachevskii, I., Sokoloff, D. 2003a, *A&A* 400, 9
Kleeorin, N., Kuzanyan, K., Moss, D., *et al.* 2003b, *A&A* 409, 1097
Köhler, H. 1973, *A&A* 25, 467
Küker, M., Rüdiger, G., & Schultz, M. 2001, *A&A* 374, 301
Parker, E. N. 1957, *Proc. Nat. Acad. Sci.* 43, 8
Parker, E. N. 1975, *ApJ* 198, 205
Parker, E. N. 1984, *ApJ* 283, 343
Parker, E. N. 1993, *ApJ* 408, 707

Seehafer, N. 1996, *Phys. Rev. E* 53, 1283
Steenbeck, M. & Krause, F. 1969, *AN* 291, 49
Roberts, P. H. & Stix, M. 1972, *A&A* 18, 453
Spiegel, E. A. & Weiss, N. O. 1980, *Nature* 287, 616
Subramanian, K. & Brandenburg, A. 2004, *Phys. Rev. Lett.* 93, 205001
Subramanian, K. & Brandenburg, A. 2006, *ApJ* 648, L71
Thompson, M. J., Christensen-Dalsgaard, J., Miesch, M. S., & Toomre, J. 2003, *ARA&A* 41, 599
Vainshtein, S. I. & Cattaneo, F. 1992, *ApJ* 393, 165
Vishniac, E. T. & Cho, J. 2001, *ApJ* 550, 752
Yoshimura, H. 1975, *ApJS* 29, 467
Yousef, T. A. & Brandenburg, A. 2003, *A&A* 407, 7
Yousef, T. A., Brandenburg, A., & Rüdiger, G. 2003, *A&A* 411, 321

It was a long way up; during the excursion to Teide National Park

Spectacular Tenerife. Find the Izana observatory on the south-eastern ridge of the Teide!

Günther Rüdiger

Cosmic Magnetic Fields:
From Planets, to Stars and Galaxies
Proceedings IAU Symposium No. 259, 2008
K.G. Strassmeier, A.G. Kosovichev & J.E. Beckman, eds.

doi:10.1017/S1743921309030415

Magnetic pinch-type instability in stellar radiative zones

Günther Rüdiger[1], **Leonid L. Kitchatinov**[1,2] **and Marcus Gellert**[1]

[1]Astrophysikalisches Institut Potsdam, An der Sternwarte 16, D-14482 Potsdam, Germany
email: gruediger@aip.de

[2]Institute for Solar-Terrestrial Physics, P.O. Box 291, Irkutsk, 664033, Russia

Abstract. The solar tachocline is shown as hydrodynamically stable against nonaxisymmetric disturbances if it is true that no $\cos^4\theta$ term exists in its rotation law. We also show that the toroidal field of 200 Gauss amplitude which produces the tachocline in the magnetic theory of Rüdiger & Kitchatinov (1997) is stable against nonaxisymmetric MHD disturbances – but it becomes unstable for rotation periods slightly slower than 25 days. The instability of such weak fields lives from the high thermal diffusivity of stellar radiation zones compared with the magnetic diffusivity. The growth times, however, result as very long (of order of 10^5 rotation times). With estimations of the chemical mixing we find the maximal possible field amplitude to be ~500 Gauss in order to explain the observed lithium abundance of the Sun. Dynamos with such low field amplitudes should not be relevant for the solar activity cycle.

With nonlinear simulations of MHD Taylor-Couette flows it is shown that for the rotation-dominated magnetic instability the resulting eddy viscosity is only of the order of the molecular viscosity. The Schmidt number as the ratio of viscosity and chemical diffusion grows to values of ~ 20. For the majority of the stellar physics applications, the magnetic-dominated Tayler instability will be quenched by the stellar rotation.

Keywords. Sun: rotation – stars: interiors – instabilities – turbulence

1. Introduction

We ask for the stability of differential rotation in radiative stellar zones under the presence of magnetic fields. If the magnetic field is aligned with the rotation axis then the answer is simply 'magnetorotational instability' (MRI). If the field is mainly toroidal (the rule rather than the exception) the answer is more complicated. Then the Rayleigh criterion for stability against axisymmetric perturbations of Taylor-Couette (TC) flow with the rotation profile $\Omega = \Omega(R)$ reads

$$\frac{1}{R^3}\frac{\mathrm{d}}{\mathrm{d}R}(R^2\Omega)^2 - \frac{R}{\mu_0\rho}\frac{\mathrm{d}}{\mathrm{d}R}\left(\frac{B_\phi}{R}\right)^2 > 0. \tag{1.1}$$

Hence, almost uniform fields or fields with $B_\phi \propto 1/R$ (a current-free field) are *stabilizing* the TC flow and no new instability appears.

More interesting is the question after the stability against nonaxisymmetric perturbations. Tayler (1973) found the necessary and sufficient condition

$$\frac{\mathrm{d}}{\mathrm{d}R}(RB_\phi^2) < 0 \tag{1.2}$$

for stability of an ideal fluid against nonaxisymmetric perturbations. Now almost homogenous fields are unstable while the fields with $B_\phi \propto 1/R$ are stable.

We have probed the interaction of such stable toroidal fields with stable flat rotation laws and found, surprisingly, the Azimuthal Magnetorotational Instability (AMRI) which for small magnetic Prandtl number scales with the magnetic Reynolds number Rm of the global rotation similar to the standard MRI (Rüdiger *et al.* 2007).

In the following, as an astrophysical application of these nonaxisymmetric instabilities the magnetic theory of Rüdiger & Kitchatinov (1997, 2007) of the solar tachocline is presented. In the last Section we return to a TC flow under the presence of strong enough toroidal fields presenting first results of the eddy viscosity and the turbulent diffusion of chemicals for the Tayler instability (TI).

2. Solar tachocline

The tachocline is the thin shell between the solar convection zone and the radiative interior of the Sun where the rotation pattern dramatically changes.

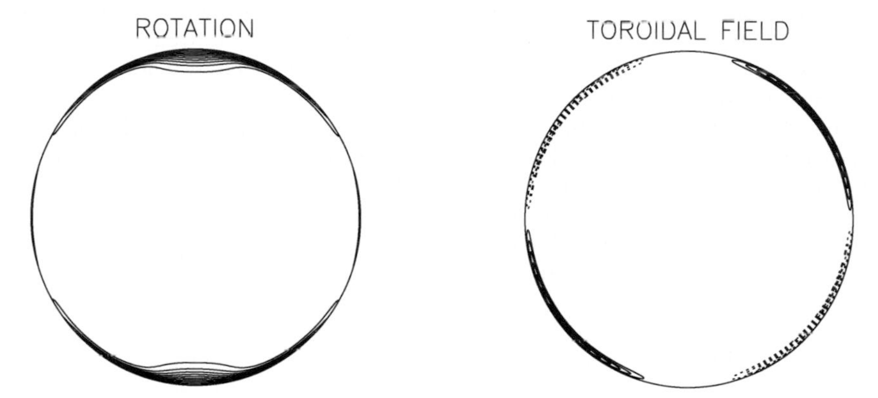

Figure 1. The tachocline formation on the basis of a fossil poloidal field of 10^{-4} Gauss confined in the radiative solar interior (Rüdiger & Kitchatinov 1997). Left: the stationary rotation profile; right: the isolines of the resulting toroidal field with its amplitude of 200 Gauss.

The nonuniform rotation of the solar convection zone – which is due to the interaction of the convection with the global rotation – has no counterpart in the solar core but the convection zone rotates in the average with the same angular velocity as the interior does. The radial coupling is thus large. This phenomenon cannot be explained by viscous coupling (the viscosity below the convection zone is by more than 10 orders of magnitude smaller) but it can be explained with a weak fossil poloidal field which is confined in the solar radiative interior. For the amplitude of this field only values of order mGauss are necessary resulting in a tachocline thickness of about 5% of the solar radius (Fig. 1). The resulting toroidal field amplitude inside the tachocline of about 200 Gauss mainly depends on the magnetic Prandtl number (and the rotation velocity) for which $\mathrm{Pm} = 5 \cdot 10^{-3}$ has been used in the model (Rüdiger & Kitchatinov 1997, 2007). One can estimate the resulting toroidal field in terms of the Alfvén velocity $V_{\mathrm{A}} = B_\phi / \sqrt{\mu_0 \rho}$ simply as

$$V_{\mathrm{A}} = \sqrt{\mathrm{Pm}} U_0 \tag{2.1}$$

with U_0 the linear velocity of rotation, $U_0 = R\Omega$. For $\mathrm{Pm} \simeq 1$ the resulting toroidal field amplitude would be of order 100 kGauss which is certainly unstable. For $\mathrm{Pm} \simeq 10^{-4}$ the field strength is reduced to only 1 kGauss so that we have carefully to check its stability. The magnetic theory holds for two main conditions: i) the field must completely be confined in the radiation zone and ii) the magnetic Prandtl number must be small enough.

(h)

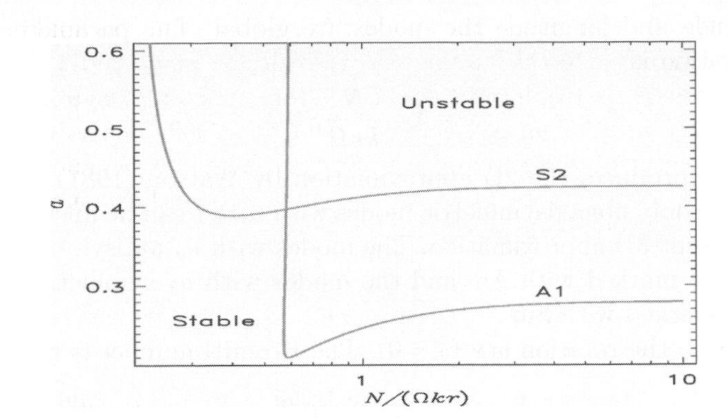

Figure 2. Neutral hydrodynamic instability lines for the rotation law without $\cos^4\theta$-term in 3D. Only A1 and S2 modes are unstable. The A1 mode for wavenumber $k \to 0$ reproduces the Watson (1981) result $a = \delta\Omega/\Omega_0 \simeq 0.286$.

There are several possibilities to fulfill the first condition (cf. Garaud 2007; Rüdiger & Kitchatinov 2007) which, however, shall not be discussed in the present paper.

2.1. *Hydrodynamic stability*

To fulfill the second condition the radiative tachocline must *hydrodynamically* be stable. On the first view this should not be a problem. As the differential rotation in latitude forms a shear flow its amplitude, $\delta\Omega = \Omega_{\rm eq} - \Omega_{\rm pole}$, decides the stability properties. By use of a 2D approximation which ignores the radial coordinate Watson (1981) derived for ideal fluids the condition $\delta\Omega/\Omega < 0.286$ for stability. This rather large value would lead to a stable tachocline. The radial velocity components, however, are small but not zero. Also the latitudinal profile of the angular velocity is more complicate than the simple $\cos^2\theta$-law used by Watson. We have thus to rediscuss the stability of shear flows with

$$\Omega(\theta) = \Omega_0 \left(1 - a((1-f)\cos^2\theta + f\cos^4\theta)\right) \tag{2.2}$$

where $\delta\Omega/\Omega_0 = a$; f is the contribution of the $\cos^4\theta$ term which describes the shape of the rotation law in midlatitudes. At the solar surface we have $a \simeq 0.286$ and $f = 0.55$.

The radial density gradient forms a 'negative' buoyancy leading to damped oscillations with the frequency

$$N = \left(\frac{g}{C_p}\frac{\partial S}{\partial r}\right)^{\frac{1}{2}}. \tag{2.3}$$

S is the entropy. The frequency N in the upper solar core is by more than a factor of 100 larger than the rotation frequency Ω. So the equation system

$$\begin{aligned} \frac{\partial \boldsymbol{u}}{\partial t} + (\boldsymbol{U}\nabla)\boldsymbol{u} + (\boldsymbol{u}\nabla)\boldsymbol{U} &= -\left(\frac{1}{\rho}\nabla p\right)' + \nu\Delta\boldsymbol{u} \\ T\left(\frac{\partial s}{\partial t} + (\boldsymbol{U}\nabla)s + (\boldsymbol{u}\nabla)S\right) &= C_p\chi\Delta T' \end{aligned} \tag{2.4}$$

must be solved for the flow perturbation $\boldsymbol{u}$ and the entropy fluctuation s. It is $\mathrm{div}\boldsymbol{u} = 0$; $s = -C_p\rho'/\rho$. T' and ρ' are the fluctuations of the temperature and the density, resp. The mean flow $\boldsymbol{U}$ is given by (2.2). The equations are solved with a Fourier expansion $\exp(\mathrm{i}(kr + m\phi - \omega t))$ in the short-wave approximation $kr \gg 1$ with m as the azimuthal wave number. As usual, in latitude a series expansion after Legendre polynomials is used.

In both latitude and longitude the modes are global. The parameter including the density stratification is

$$\hat{\lambda} = \frac{N}{kr\Omega}, \tag{2.5}$$

so that $\hat{\lambda} \to \infty$ reproduces the 2D approximation by Watson (1981) and Cally (2001). They showed that only nonaxisymmetric modes with $m=1$ can be unstable and the same is true in the present 3D approximation. The modes with u_θ antisymmetric with respect to the equator are marked with Am and the modes with u_θ symmetric with respect to the equator are marked with Sm .

Let us start with the rotation law ($f=0$). The Prandtl number is fixed as

$$\mathrm{Pr} = \frac{\nu}{\chi} = 2 \cdot 10^{-6}, \tag{2.6}$$

where ν and χ are viscosity and thermal conductivity.

The main result is given in Fig. 2 which shows the neutral-stability lines for various $a = \delta\Omega/\Omega$. Only A1 and S2 are obtained as unstable (S1 is stable!). For $k \to 0$ the Watson result $a = 0.286$ is reproduced. With radial stratification, however, this critical value is reduced to $a = 0.21$. For ideal fluids Cally (2003) found instability for $a = 0.24$ which also fits our result. Hence, for $a < 0.21$ the solar tachocline remains hydrodynamically stable. There is thus no shear-induced turbulence. Note also that for $N \to 0$ (mimicking the convection zone) no instability exists.

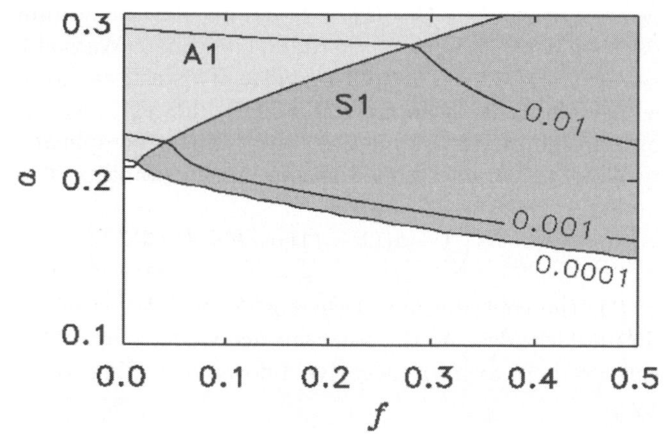

Figure 3. Hydrodynamic-stability map for various values f for the power of the $\cos^4\theta$-term in the rotation law (2.2). This term destabilizes the S1 mode. For high f already rather small shear values become unstable.

The calculations have been repeated with the $\cos^4\theta$-term included in the rotation law. Then the S1 mode becomes dominant and reduces the critical shear. For $f=0.5$ the maximum shear for stability results to $a=0.16$ (Fig. 3).

This is a rather small value. Both our modifications of the Watson approach are destabilizing the differential rotation. If the shape of the surface rotation with $f=0.5$ would be conserved through the convection zone *and* the tachocline then only 50% of the surface shear can indeed produce hydrodynamic turbulence in the tachocline. If the shear is not conserved ($f=0$) then the minimum shear for turbulence is 0.21, large enough to ensure stability.

A dramatic *stabilization*, however, of the latitudinal shear results from the inclusion of the tachocline rotation law as observed with its large radial gradients into the calculations. With a 3D code without stratification one can show that in this case all rotation laws with $a < 0.5$ should be stable (Arlt *et al.* 2005).

In order to decide the stability problem of the solar tachocline informations are needed about the exact shape of the rotation law there. Charbonneau *et al.* (1999) analyzed the helioseismic data and found $a \leqslant 0.15$ and $f \simeq 0$. After our analysis (and also after theirs) such a rotation law is hydrodynamically *stable*. Obviously, we have to know in detail the space dependence (and time dependence) of the internal solar rotation.

2.2. *MHD stability*

We consider the tachocline as hydrodynamically stable. The magnetic Prandtl number is thus the microscopic one for which we use $\mathrm{Pm} = 0.005$ as a characteristic value. For the instability of toroidal fields the ratio of Pm and Pr, the Roberts number

$$\mathrm{q} = \frac{\chi}{\eta} \tag{2.7}$$

plays a basic role. For the Sun the typical value 2500 is used. The interaction of rotation and toroidal magnetic fields (of a simplified structure) will be demonstrated with these parameters. The general result is that for small q the field must be strong to become unstable while for $\mathrm{q} \gg 1$ much weaker fields become unstable. However, the growth times (in units of the rotation period) for the weak-fields are much longer (by orders of magnitudes) than the growth times of strong fields (of order of the Alfvén period).

The left plot of Figs. 4 concerns a field with only one belt which peaks at the equator. The Alfvén frequency Ω_{A} is defined by

$$B_\phi = r \sin\theta \sqrt{\mu_0 \rho}\, \Omega_{\mathrm{A}}, \tag{2.8}$$

and is considered as constant (see Cally 2003). The global rotation is assumed as rigid. For this case Fig. 4 (left) reveals the fields with $\Omega_{\mathrm{A}} > \Omega$ as always unstable with growth rates $\gtrsim \Omega_{\mathrm{A}}$. Toroidal fields with amplitudes $V_{\mathrm{A}} \gtrsim U_{\mathrm{eq}}$, i.e. $\sim 10^5$ Gauss (for the Sun) cannot stably exist in the radiative solar core. Even weaker fields with $\Omega_{\mathrm{A}} \gtrsim 0.005\Omega$ can be unstable but only for very high heat-conductivity. For $\mathrm{q} = 0$ the weak fields are stable. For $\mathrm{q} \gg 1$ they are indeed unstable but with growth rates smaller than 10^{-4}. The growth time τ_{growth} in units of the rotation period is

$$\frac{\tau_{\mathrm{growth}}}{\tau_{\mathrm{rot}}} = \frac{1}{2\pi\gamma} \tag{2.9}$$

with the normalized growth rate $\gamma = \Im(\omega)/\Omega_0$. The maximum growth time is then > 1000 rotation periods (but much shorter than the diffusion time). Even for very large q there is a magnetic limit below the magnetic field is always stable. From Fig. 4 (one or two belts) we find the minimum field is $\Omega_{\mathrm{A}} \simeq 0.005\Omega$. For the Sun, therefore, the maximum stable field in the model is ~ 600 Gauss.

The model is insofar correct as the microscopic diffusivity values (viscosity, magnetic diffusivity, heat-conductivity) have their real amplitudes (for ideal fluids, see Cally 2003). The model, however, is insofar not correct as the radial profiles of the fields are assumed as nearly uniform. Arlt *et al.* (2007) work with a 3D code without buoyancy ($\mathrm{q} \to \infty, \mathrm{Pm} = 0.01$) for toroidal field belts with strong radial gradients and find instability for weak fields with amplitudes of order 10 Gauss.

Not surprisingly, for two belts with equatorial antisymmetry (the field vanishes at the equator) there are some differences to the one-belt model, but the maximal stable field amplitudes are always of the same order (Fig. 4, right).

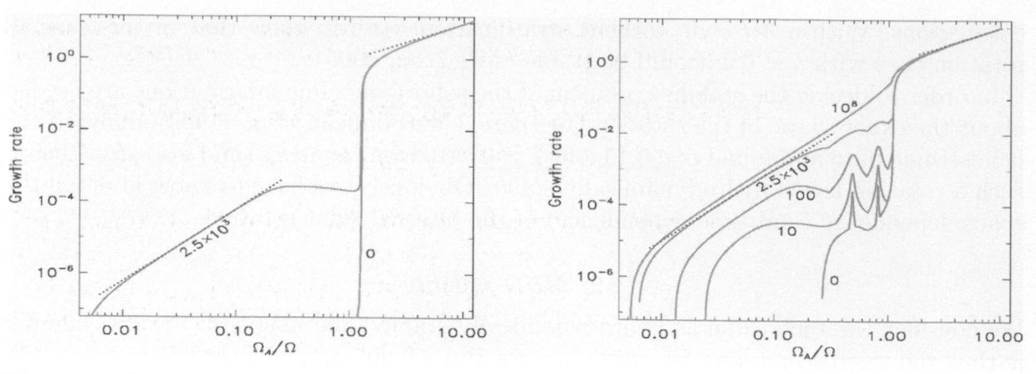

Figure 4. Normalized growth rates for the magnetic instability in rigidly rotating stars. The curves are marked with their values of the Roberts number (2.7). Left: one magnetic belt peaking at the equator; right: two magnetic belts with zero-field at the equator. There are two instability domains: for strong magnetic fields ($\Omega_A > \Omega$) the growth rates are high ($\simeq \Omega_A/\Omega$) and for weak magnetic fields ($\Omega_A < \Omega$) they are very small ($\simeq (\Omega_A/\Omega)^2$). The latter domain only exists for q $\gg$ 1 (perfect heat conduction).

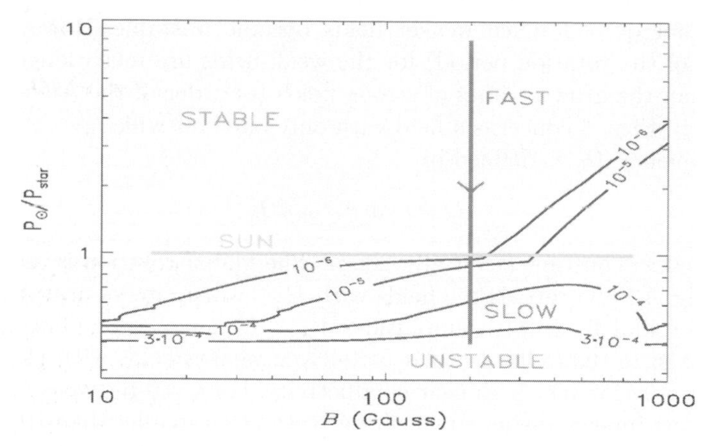

Figure 5. A star spinning down moves from the top (stable area) to the bottom (unstable area). It is shown that a toroidal magnetic field of 200 Gauss is stable for the solar rotation period but the stability is lost for slower rotation.

2.3. *Effect of stellar spin-down*

The question arises whether a solar-type star is always able to form a tachocline. The older the star the slower its rotation. Hence the rotational quenching of the Tayler instability becomes weaker and weaker for older stars so that the instability becomes more efficient. By its spin-down the star moves to the right along the abscissa of both the Figs. 4. The (slow) magnetic decay goes in the opposite direction; this effect is still neglected. We assume that the total amount of the latitudinal differential rotation remains constant during the star's spin-down. This is a well-established assumption (see Kitchatinov & Rüdiger 1999; Küker & Stix 2001).

Figure 5 shows the results. We have computed the normalized growth rates of the magnetic instability of rotating stars with a toroidal field of 200 Gauss and a differential rotation of $\delta\Omega \simeq 0.06$ day^{-1} (the solar value). The rotation period is normalized with 25 days in Fig. 5 so that the horizontal yellow line represents the Sun. In the upper part of the plot the 200 Gauss are stable while in the lower part of the plot they are unstable. Obviously, the Sun lies in the stable area but very close to the instability limit. We are thus tempted to predict that G2 stars older than the Sun (or better: of slower rotation) should not have a tachocline. When the toroidal field becomes unstable then the resulting turbulence is able to destroy the tachocline rather fast.

2.4. *Chemical mixing*

The flow pattern of the magnetic instability also mixes passive scalars like temperature and chemical concentrations. The instability, therefore, could be relevant for the so-called lithium problem. In order to explain the observed lithium concentration at the solar surface one needs a turbulent mixing beneath the convection zone which enhances the microscopic value of the diffusion coefficient of 30 cm^2/s by (say) two orders of magnitude. Note the smallness of this quantity; only a very mild turbulence can provide such a small value of the diffusion coefficient

$$D_{\mathrm{T}} \simeq \langle u'^2 \rangle \tau_{\mathrm{corr}}. \tag{2.10}$$

This relation is used here as a rough estimate, a quasilinear theory of turbulent mixing has been established by Rüdiger & Pipin (2001) also for rotating turbulences. For a correlation time of the order of the rotation period the desired mixing velocity is only 1 cm/s.

One can estimate the characteristic time by $\tau_{\mathrm{corr}} \simeq l^2/D_{\mathrm{T}}$ with l as the radial scale of the instability and $D_{\mathrm{T}} \simeq 10^4$ cm^2/s. For the radial scale the value 1000 km has been found by Kitchatinov & Rüdiger (2008). With this value it follows $\tau_{\mathrm{corr}} \simeq 10^{12}$ s which corresponds to a very small normalized growth rate of $\tau_{\mathrm{rot}}/\tau \simeq 10^{-6}$. The resulting toroidal field which fulfills this condition is smaller than 600 Gauss (Fig. 6). Stronger fields would produce a too strong mixing which would lead to much smaller values for the lithium abundance in the solar convection zone than observed.

Our result in connection with the observed lithium values also excludes the possibility that some dynamo works in the upper part of the solar radiative core. If such a ('Tayler-Spruit') dynamo exists then the resulting toroidal fields with less than 600 Gauss are much too weak to influence the magnetic activity of the Sun with magnetic fields exceeding 10 kGauss.

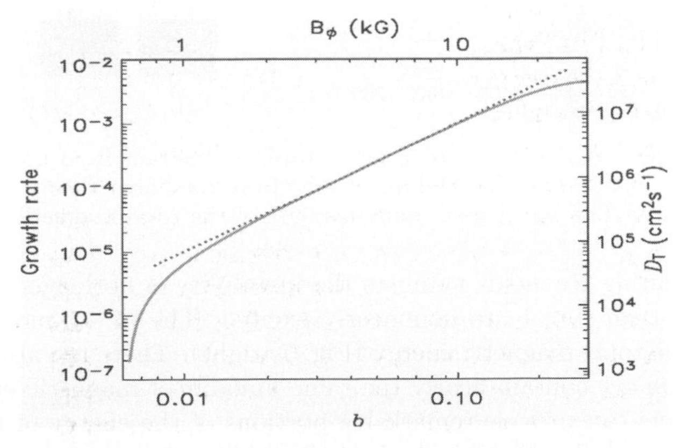

Figure 6. Growth rates in units of Ω for the two-belts model and for a fixed radial scale of 1000 km. The right-hand scale gives the estimated values for the diffusion coefficient for passive scalars and the uppermost scale gives the magnetic field amplitude in kGauss.

3. Tayler instability in Taylor-Couette systems

To simplify matters we consider the pinch-type instability in a Taylor-Couette system filled with a conducting fluid which is nonstratified in axial direction. The stationary rotation law between the cylinders is $\Omega = a + b/R^2$ where a and b are given by the fixed rotation rates of the cylinders. In a similar way the stationary toroidal field results as $B_\phi = AR + B/R$. In the following we have fixed the values at the cylinders to $\Omega_{\rm out} = 0.5\Omega_{\rm in}$ and mostly $\mu_B = B_{\rm out}/B_{\rm in} = 1$ is used. The outer cylinder radius is fixed to $2R_{\rm in}$. Reynolds number Rm and Hartmann number Ha are defined as

$$\mathrm{Rm} = \frac{\Omega_{\rm in} R_{\rm in}^2}{\eta}, \qquad \mathrm{Ha} = \frac{B_{\rm in} R_{\rm in}}{\sqrt{\mu_0 \rho \nu \eta}}, \tag{3.1}$$

the magnetic Prandtl number here is always put to unity.

A detailed description of the used nonlinear MHD code for incompressible fluids is given by Gellert *et al.* (2007). In the vertical direction periodic boundary conditions are used to avoid endplate problems. In this approximation the endplates rotate with the same rotation law as the fluid does. The height of the virtual container is assumed as $6D$ with D the gap width between the cylinders. The cylinders are considered as perfect conductors. The code was first tested for the nonaxisymmetric AMRI which appears if stable rotation laws and stable toroidal fields (current-free, $\mu_B = 0.5$) are combined (Rüdiger *et al.* 2007). Figure 7 (left) shows the instability domain (solid) which for given (supercritical) magnetic field always lies between a lower Reynolds number and an upper Reynolds number. For too slow rotation the nonaxisymmetric modes are not yet excited, but for too fast rotation the nonaxisymmetric instability modes are destroyed so that the field becomes stable again.

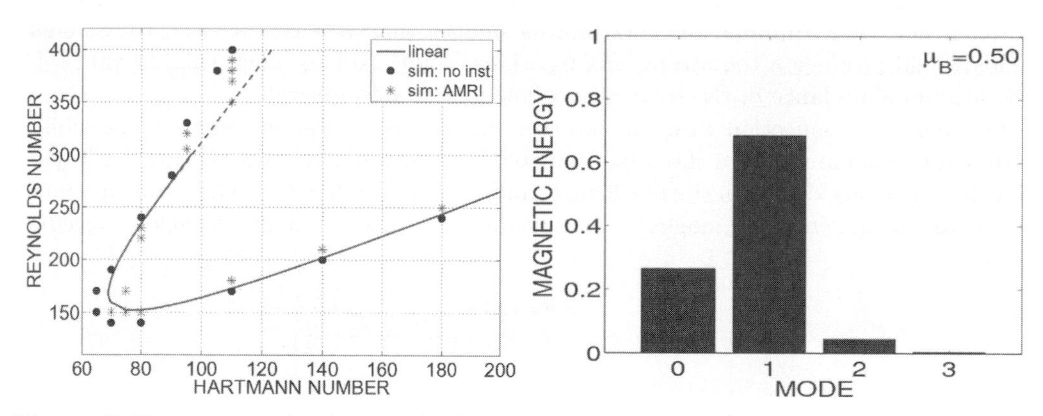

Figure 7. $\mathrm{Pm} = 1$, $\mu_B = 0.5$ (current-free). Left: stability diagram from nonlinear simulations. Dots: stability, stars: instability; the solid line results from the linear theory. Right: m-spectrum for $\mathrm{Re} = 250$, $\mathrm{Ha} = 110$. The $m = 1$ mode contains 69% of the total magnetic energy.

Between the limiting Reynolds numbers the instability is no longer monochrome but also other modes than $m = 1$ are nonlinearly excited. The $m = 1$ mode often contains the majority of the total magnetic energy (Fig. 7, right). There are also cases, however, where $m = 0$ and $m = 1$ contain nearly the same amount of magnetic energy. Note that the nonlinear effects can provide remarkable portions of the energy of the instability in form of axisymmetric rolls although the basic instability is a nonaxisymmetric one.

Now with $\mu_B = 1$ positive axial currents between the cylinders are allowed. Again there are two different instability domains for the nonaxisymmetric modes with $m = 1$. They are separated by a stable domain (Fig. 8, left). The upper one is for fast rotation and

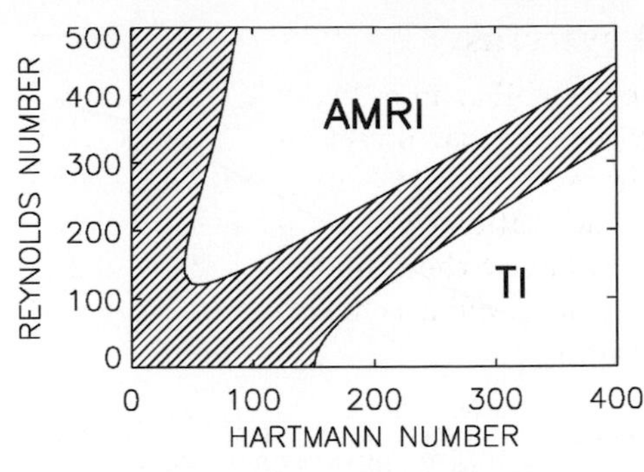

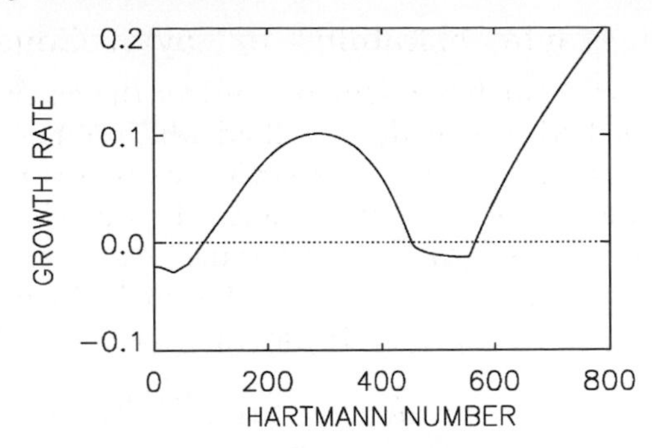

Figure 8. Left: The instability map for $\mu_\Omega = 0.5$, $\mu_B = 1$, $\mathrm{Pm} = 1$. The hatched area is stable. Right: The growth rate (normalized with the rotation rate of the inner cylinder) along the horizontal line at $\mathrm{Re} = 500$. The instability in the upper domain (AMRI) grows slowly while in the lower domain (TI) it grows much faster.

weak fields (Re > Ha) and the lower one is for strong fields and slow rotation (Ha > Re). The growth rates in both domains are very different: The weak-field instability is slow and the strong-field instability is very fast (Fig. 8, right). The rotation-dominated instability disappears for rigid rotation while the magnetic-dominated instability even exists without rotation (Re = 0). The latter one is the Tayler instability (TI) under the stabilizing influence of the basic stellar rotation (Pitts & Tayler 1985). The rotation-dominated ('upper') instability appears to be the AMRI which also exists under the modifying influence of weak axial currents in the fluid. Note again that i) too fast rotation finally stops both the instabilities, and ii) its growth rate is very small. One must also stress that the presented results only concern the most simple case of Pm = 1. For very strong fields the stable domain between AMRI and TI disappears.

The condition for the existence of TI as given in Fig. 8 (left) is $B_\phi > \sqrt{\mu_0 \rho} R\Omega$ while the condition for AMRI is $R\Omega > B_\phi/\sqrt{\mu_0 \rho}$. Both the instabilities, however, only exist if the rotation is not too fast. Nonaxisymmetric modes are *always stabilized* by sufficiently fast rotation.

Ap stars (with 10 kGauss and a rotation period of several days) and neutron stars (with 10^{12} Gauss and a rotation period of 10 ms) are rotation-dominated. Their instabilities are not of the Tayler-type. In the following we have thus considered the AMRI in more detail. For given Ha (= 500) the eddy viscosity, the diffusion coefficient D_T and the Schmidt number

$$\mathrm{Sc} = \frac{\nu_\mathrm{T}}{D_\mathrm{T}} \tag{3.2}$$

are computed. The eddy viscosity ν_T is the ratio of the angular momentum transport by Reynolds stress and Maxwell stress and the differential rotation. We find a ν_T of the order of the microscopic value (Fig. 9, left). The maximum exists as the instability – as mentioned – disappears for too fast rotation. The averaging procedure concerns the whole container so that the values in Fig. 9 are lower limits.

So far the diffusion coefficient for chemicals could only be estimated by $D_\mathrm{T} \simeq \langle u_R'^2 \rangle / \Omega$. It proves to be much smaller than the viscosity. Brott *et al.* (2008) have shown that for too strong mixing the stellar evolution is massively affected. A better theory must solve the diffusion equation.

Accepting this approximation the resulting Schmidt number (3.2) reaches values of 20 ... 30 (Fig. 9, right). Obviously, the angular momentum is mainly transported by the Maxwell stress while the diffusion of passive scalars is due to only the Reynolds stress which is much smaller. Quite similar results have been obtained by Carballido *et al.*

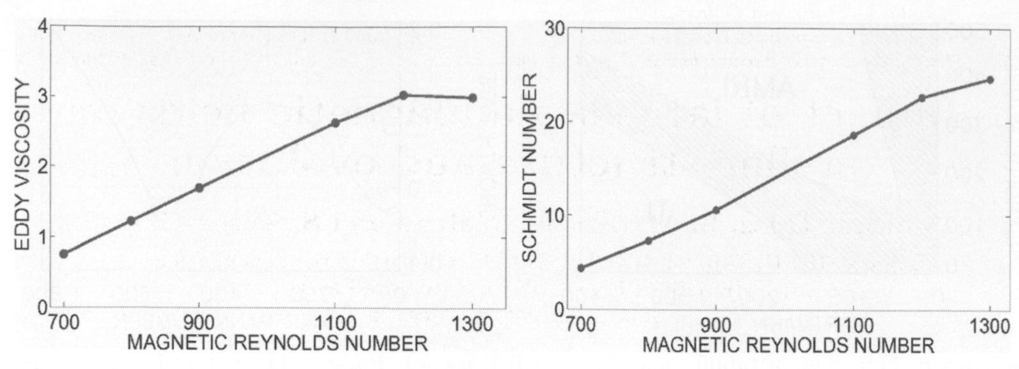

Figure 9. Left: the eddy viscosity in units of the microscopic viscosity for $\mu_\Omega = 0.5$, $\mu_B = 1$, $\mathrm{Pm} = 1$ and $\mathrm{Ha} = 500$. It grows for faster rotation. Right: the Schmidt number (3.2).

(2005) and Johansen *et al.* (2006) for the Schmidt number of the standard MRI. Maeder & Meynet (2005) for a hot star with 15 solar masses and for 20 kGauss find much higher values of the Schmidt number (10^6). Also Heger *et al.* (2005) work with magnetic amplitudes of 10 kGauss for which $\Omega_\mathrm{A} < \Omega$ hence the growth rates are small.

References

Arlt, R., Sule, A., & Rüdiger, G. 2005, *A&A* 441, 1171
Arlt, R., Sule, A., & Rüdiger, G. 2007, *A&A* 461, 295
Brott, I., Hunter, I., Anders, P., & Langer, N. 2008, in: B. W. O'Shea *et al.* (eds.), *FIRST STARS III*, AIPC 990, p. 273
Cally, P. S. 2001, *SP* 199, 231
Cally, P. S. 2003, *MNRAS* 339, 957
Carballido, A., Stone, J. M., & Pringle, J. E. 2005, *MNRAS* 358, 1055
Charbonneau, P., Dikpati, M., & Gilman, P. A. 1999, *ApJ* 526, 523
Garaud, P. 2007, in: D. W. Hughes *et al.* (eds.), *The Solar Tachocline*, p. 147
Gellert, M., Rüdiger, G., & Fournier, A. 2007, *AN* 328, 1162
Heger, A., Woosley, S. E., & Spruit, H. C. 2005, *ApJ* 626, 350
Johansen, A., Klahr, H., & Mee, A. J. 2006, *MNRAS* 370, 71
Kitchatinov, L. L. & Rüdiger, G. 1999, *A&A* 344, 911
Kitchatinov, L. L. & Rüdiger, G. 2008, *A&A* 478, 1
Küker, M. & Stix, M. 2001, *A&A* 366, 668
Maeder, A. & Meynet, G. 2005, *A&A* 440, 1041
Pitts, E. & Tayler, R. J. 1985, *MNRAS* 216, 139
Rüdiger, G. & Kitchatinov, L. L. 1997, *AN* 318, 273
Rüdiger, G. & Pipin, V. V. 2001, *A&A* 375, 149
Rüdiger, G. & Kitchatinov, L. L. 2007, *New J. Phys.* 9, 302
Rüdiger, G., Hollerbach, R., Schultz, M., & Elstner, D. 2007, *MNRAS* 377, 1481
Tayler, R. J. 1973, *MNRAS* 165, 39
Watson, M. 1981, *GAFD* 16, 285

Cosmic Magnetic Fields:
From Planets, to Stars and Galaxies
Proceedings IAU Symposium No. 259, 2008
K.G. Strassmeier, A.G. Kosovichev & J.E Beckman, eds.

doi:10.1017/S1743921309030427

Impact of large-scale magnetic fields on stellar structure and evolution

Vincent Duez, S. Mathis, A. S. Brun and S. Turck-Chièze

DSM/IRFU/SAp, CEA Saclay, F-91191 Gif-sur-Yvette Cedex, France;
AIM, UMR 7158, CEA - CNRS - Université Paris 7, France
email: vincent.duez@cea.fr

Abstract. We study the impact on the stellar structure of a large-scale magnetic field in stellar radiation zones. The field is in magneto-hydrostatic (MHS) equilibrium and has a non force-free character, which allows us to study its influence both on the mechanical and and on the energetic balances. This approach is illustrated in the case of an A_p star where the magnetic field matches at the surface with an external potential one. Perturbations of the stellar structure are semi-analytically computed. The relative importance of the magnetic physical quantities is discussed and a hierarchy, aiming at distinguishing various refinement degrees in the implementation of a large-scale magnetic field in a stellar evolution code, is established. This treatment also allows us to deduce the gravitational multipolar moments and the change in effective temperature associated with the presence of a magnetic field.

Keywords. Stars: interiors – stars: magnetic fields – stars: evolution

1. Introduction

Nowadays it is well known from spectropolarimetric measurements that a non negligible fraction of the A-type stars, the peculiar ones (representing about 5 % of the population) exhibit magnetic fields organised over large scales, and whose strengths can reach several kG at their surface. Moreover, it has been shown by Alecian, *et al.* (2008) (see also her contribution in these proceedings) that some so-called Herbig Ae/Be stars are magnetic. Hence it is likely that a fossil magnetic field, already present before the early stages of stellar evolution, could have survived during the pre-main sequence phase and influenced the evolutionary track of their hosts.

We propose here to look at the effects that such a large-scale fossil field could have on the stellar structure by considering a magneto-hydrostatic (MHS) equilibrium in an A_p type star, based on a Grad-Shafranov model. This magnetic field in non force-free, presents a mixed poloidal-toroidal (twisted) configuration and spreads across the whole volume of the star; at its surface it matches with a potential, dipolar field with a 8 kG strength.

Based on a simplified stability analysis, we provide some elements tending to prove that the configuration found is likely to be stable.

The physical quantities likely to modify the stellar structure are then semi-analytically derived and illustrated in the case of interest.

Then, perturbations of the gravitational potential, density, pressure and radius are computed throughout the whole radius up to the surface. In particular, the gravitational multipolar moments induced by the presence of a magnetic field are obtained.

Finally, we establish the change in temperature owing to the perturbation of pressure and density; we investigate the energetical quantities perturbations generated by ohmic heating, Poynting's flux, and by the change of nuclear reaction rates induced by modification of the hydrostatic balance.

This allows us to propose a hierarchy of the various effects associated with the magnetic field and likely to act over evolution timescales.

2. The Non Force-Free Magneto-Hydrostatic Equilibrium

We here look for a large-scale magnetic field geometry likely to exist in the stellar radiative zone of A_p-type stars, at the surface of which has been observed (see Wade *et al.*, 2000) dipolar, roughly axisymmetric configurations which are probably remnants of a fossil magnetic field.

Since we know from Tayler (1973) that purely toroidal fields are unstable, and from Markey & Tayler (1973, 1974) that purely poloidal fields are also unstable, a mixed poloidal-toroidal (twisted) configuration is needed for the field to survive over evolution timescales.

Furthermore, if force-free MHS equilibria are currently observed in plasma experiments, especially in spheromaks ones, the conditions of pressure in stellar interiors make the problem quite different: in the former case the plasma is in the low-β regime. In the latter, as the Lorentz force is a perturbation compared with the gravitational one and the gaseous pressure gradient (high-β regime), the magnetic field is constrained to be in non force-free equilibrium.

Owing to these facts, we focus on magnetic field configurations such that the field is dipolar, in magneto-hydrostatic equilibrium and non force-free.

2.1. *The Axisymmetric Magnetic Field*

Let us express the magnetic field $\boldsymbol{B}(r,\theta)$ in the axisymmetric case as a function of a poloidal flux $\Psi(r,\theta)$ and a toroidal potential $F(r,\theta)$ such that it remains automatically divergence-free :

$$\boldsymbol{B} = \frac{1}{r\sin\theta}\nabla\Psi\times\hat{\mathbf{e}}_\varphi + \frac{1}{r\sin\theta}\,F\,\hat{\mathbf{e}}_\varphi, \tag{2.1}$$

where in spherical coordinates the poloidal direction is in the meridional plane $(\hat{\mathbf{e}}_\mathbf{r}, \hat{\mathbf{e}}_\theta)$ and the toroidal direction is along the azimuthal one $\hat{\mathbf{e}}_\varphi$.

2.2. *Non Force-Free Condition*

Let us now write the magneto-hydrostatic (MHS) equilibrium as follows:

$$\rho\,\boldsymbol{g} - \nabla P_{gas} + \boldsymbol{F_{\mathcal{L}}} = \boldsymbol{0}, \tag{2.2}$$

where ρ is the density, $\boldsymbol{g}$ the local gravity field, $P_{\rm gas}$ the gas pressure, and $\boldsymbol{F_{\mathcal{L}}} = \boldsymbol{j}\times\boldsymbol{B}$ the Lorentz force, $\boldsymbol{j}$ being the current density.

In the toroidal direction, the Lorentz force $F_{\mathcal{L}_\varphi}$ vanishes everywhere, since in lack of rotation there is no other force in this direction to compensate for the equilibrium deviation. This condition writes as $\partial_r\Psi\partial_\theta F - \partial_\theta\Psi\partial_r F = 0$. The non trivial values for F are obtained by setting $F(r,\theta) = F(\Psi)$. Looking at the first order case such that the azimuthal magnetic field is regular, we have $F(\Psi) = \lambda_1\,\Psi$ where λ_1 is a real constant. According to (2.1) and to the Ampère's law $\nabla\times\boldsymbol{B} = \mu_0\,\boldsymbol{j}$ (in the classical MHD approximation; μ_0 being the vacuum permeability), the Lorentz force can finally be concisely stated as†

$$\boldsymbol{F_{\mathcal{L}}} = \mathcal{A}(r,\theta)\,\nabla\Psi \quad \text{where} \quad \mathcal{A}(r,\theta) = -\,\frac{1}{\mu_0\,r^2\,\sin^2\theta}\left(\lambda_1^2\Psi + \Delta^*\Psi\right) \tag{2.3}$$

† Notice that written in this way, we see immediatly that when $\Delta^*\Psi = -\,\lambda_1^2\Psi$, the field is force-free and corresponds to the solution described by Chandrasekhar (1956), and generalized later by Marsh (1992).

and where we introduce the so-called Grad-Shafranov operator in spherical coordinates

$$\Delta^* \Psi \equiv \frac{\partial^2 \Psi}{\partial r^2} + \frac{\sin\theta}{r^2}\frac{\partial}{\partial\theta}\left(\frac{1}{\sin\theta}\frac{\partial\Psi}{\partial\theta}\right). \tag{2.4}$$

Taking the curl of the MHS equation divided by the equilibrium density ρ_0 we have

$$\nabla \times \left(\frac{1}{\rho_0}\nabla P_{\rm gas} - \boldsymbol{g}\right) = \nabla \times \left(\frac{1}{\rho_0}\boldsymbol{F_{\mathcal{L}}}\right), \tag{2.5}$$

which, assuming that the Lorentz force is a weak perturbation to the density and assuming the barotropic equilibrium, vanishes. We can then write using eq. (2.3)

$$\nabla\left(\frac{\mathcal{A}}{\rho_0}\right) \times \nabla\Psi = \boldsymbol{0}. \tag{2.6}$$

This projects only along $\hat{\mathbf{e}}_\varphi$ as $\partial_r\left(\mathcal{A}/\rho_0\right)\partial_\theta\Psi - \partial_\theta\left(\mathcal{A}/\rho_0\right)\partial_r\Psi = 0$ so that there exists a function G of Ψ such that $\mathcal{A}/\rho_0 = G(\Psi)$, that reduces in the simplest, linear case to $G(\Psi) = \beta_0$. Then, Eq. (2.3) leads to the Grad-Shafranov-like linear partial differential equation

$$\Delta^*\Psi + \left(\frac{\lambda_1}{R}\right)^2 \Psi = -\mu_0 r^2 \sin^2\theta\, \rho_0\, \beta_0. \tag{2.7}$$

Using Green's function method (Morse & Feschbach, 1953; Payne & Melatos, 2004), the equation above can be solved analytically. The expression for Ψ in terms of the density profile is found to be:

$$\begin{aligned} \Psi\left(r,\theta\right) &= -\mu_0\,\beta_0\frac{\lambda_1}{R}\sin^2\theta\Big\{r j_1\left(\lambda_1\frac{r}{R}\right)\int_r^R\left[y_1\left(\lambda_1\frac{\xi}{R}\right)\rho_0\xi^3\right]\mathrm{d}\xi\; + \;\ldots \\ &\ldots\; + r\, y_1\left(\lambda_1\frac{r}{R}\right)\int_0^r\left[j_1\left(\lambda_1\frac{\xi}{R}\right)\rho_0\xi^3\right]\mathrm{d}\xi\Big\} \end{aligned} \tag{2.8}$$

where j_1 and y_1 are respectively the spherical Bessel and Neumann functions of latitudinal order $l = 1$; the eigenvalue λ_1 is given by the boundary conditions at $r = R$ and the constant parameter β_0 is constrained by the magnetic field strength. The iso-Ψ surfaces (normalized to its maximum), tangent to the poloidal magnetic field, and the corresponding radial component of the Lorentz force (normalized to $B_0^2/\mu_0 R_*$) are represented in Fig. 1 in the meridional plane, in the case of a dipolar surface field with a mean surface magnetic field of 8 kG presenting a potential behaviour ($\lambda_1 = \pi/2$). It shows that the magnetic force has a centrifugal behaviour below 0.3 R_* and a centripetal, but much weaker in the external part of the star.

3. Stability Analysis

Following Reisenegger (2008), we perform a first-order stability analysis. The variational principle of minimizing the magnetic energy is introduced (see Bernstein *et al.*, 1958). The variation of magnetic energy under an arbitrary lagrangian displacement ξ is given by:

$$\delta W_B = \frac{1}{2\mu_0}\delta\left[\int_V \boldsymbol{B}^2 \mathrm{d}V\right] = \frac{1}{\mu_0}\int_V \left[\boldsymbol{B}\cdot\nabla\times(\xi\times\mathbf{B})\right]\mathrm{d}V \tag{3.1}$$

In the case of stellar radiation zones, due to the strong stable stratification, the anelastic approximation can be adopted for ξ so that $\nabla\cdot(\rho_0\,\xi) = 0$. Then, it is possible to introduce

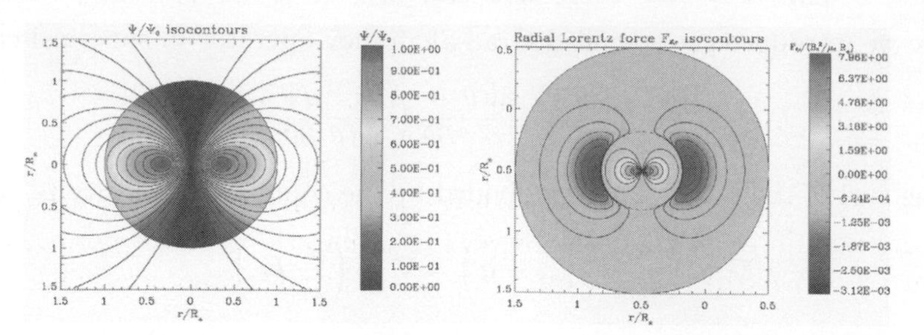

Figure 1. Left: Ψ isocontours (normalized). Right: radial Lorentz force isocontours (normalized).

an arbitrary vector field $\mathbf{a}$ such that: $\nabla\times\mathbf{a}=\rho_0\,\xi$. Eq. (3.1) then becomes

$$\delta W_B \quad = \quad \int_V \nabla\cdot\left[\frac{(\mathbf{j}\times\mathbf{B})\times\mathbf{a}}{\rho_0}\right]\mathrm{d}V - \int_V\left[\mathbf{a}\cdot\nabla\times\left(\frac{\mathbf{j}\times\mathbf{B}}{\rho_0}\right)\right]\mathrm{d}V. \tag{3.2}$$

Furthermore, since the anelastic approximation is used, we assume that $\hat{\mathbf{n}}\cdot\xi=0$ at the surface of the star, giving $\hat{\mathbf{n}}\times\mathbf{a}=\boldsymbol{0}$. The first integral thus cancels while the second one vanishes by construction. The considered equilibrium seems therefore to be stable since the total magnetic energy variation is nul.

4. Influence on the Stellar Structure

4.1. *Mechanical Balance*

4.1.1. *Magnetic Pressure Force vs. Magnetic Tension Force*

We can write the Lorentz force as the sum of the gradient of a magnetic pressure and of a magnetic tension force:

$$\boldsymbol{F}_{\mathcal{L}}=\boldsymbol{F}_{\mathcal{T}}-\nabla P_{\mathrm{mag}}. \tag{4.1}$$

From Fig. 2 (Left panel), it appears that the magnetic pressure gradient has a predominant role in the internal part of the star over the magnetic tension. However, the latter's strength is of the order of the former in particular on the symmetry axis and in the vicinity of the surface, where both ones counterbalance each other. This leads to a force-free state, that cannot be achieved by considering the magnetic pressure as the only effect.

4.1.2. *Lorentz Force Perturbations on the Stellar Structure*

Let us then project the Lorentz force components on the Legendre polynomials $P_l(\cos\theta)$ (of order $l=0$ and $l=2$ in the case of a dipolar field), assuming it is a perturbation around the stellar non-magnetic state:

$$F_{\mathcal{L},r}(r,\theta)=\sum_l \mathcal{X}_{\boldsymbol{F}_{\mathcal{L}};l}(r)\,P_l(\cos\theta)\,,\qquad F_{\mathcal{L},\theta}(r,\theta)=-\sum_l \mathcal{Y}_{\boldsymbol{F}_{\mathcal{L}};l}(r)\,\partial_\theta P_l(\cos\theta) \tag{4.2}$$

which gives us at the surface the gravitational potential $J_l=(R_*/GM_*)\,\widehat{\phi_l}\,(r=R_*)$. We can then deduce the gravitational potential perturbation $\widehat{\phi_l}$ to the non-magnetic state

ϕ_0, from Sweet's equation†

$$\frac{1}{r}\frac{\mathrm{d}^2}{\mathrm{d}r^2}\left(r\widehat{\phi}_l\right) - \frac{l(l+1)}{r^2}\widehat{\phi}_l - \frac{4\pi G}{g_0}\frac{\mathrm{d}\rho_0}{\mathrm{d}r}\widehat{\phi}_l = \frac{4\pi G}{g_0}\left[\mathcal{X}_{\boldsymbol{F}_{\mathcal{L}};l} + \frac{\mathrm{d}}{\mathrm{d}r}\left(r\mathcal{Y}_{\boldsymbol{F}_{\mathcal{L}};l}\right)\right]. \tag{4.3}$$

where g_0 is the equilibrium gravity and where we have $\phi\left(r,\theta\right) = \phi_0 + \sum_l \widehat{\phi}_l\left(r\right) P_l\left(\cos\theta\right)$.

After numerical integration of the Sweet's equation, the density perturbation ρ_l and the pressure one P_l for the mode l can respectively be computed according to

$$\widehat{\rho}_l = \frac{1}{g_0}\left[\frac{\mathrm{d}\rho_0}{\mathrm{d}r}\widehat{\phi}_l + \mathcal{X}_{\boldsymbol{F}_{\mathcal{L}};l} + \frac{\mathrm{d}}{\mathrm{d}r}\left(r\mathcal{Y}_{\boldsymbol{F}_{\mathcal{L}};l}\right)\right] \quad \text{and} \quad \widehat{P}_l = -\,\rho_0\widehat{\phi}_l - r\mathcal{Y}_{\boldsymbol{F}_{\mathcal{L}};l}. \tag{4.4}$$

Diagnosis from the stellar radius variation induced by the magnetic field can be established. The radius of an isobar is given by

$$r_P\left(r,\theta\right) = r\left[1 + \sum_{l\geqslant 0} c_l(r)P_l(\cos\theta)\right] \text{ with } c_l = -\frac{1}{r}\frac{\widehat{P}_l}{\mathrm{d}P_0/\mathrm{d}r} = \frac{\rho_0}{\mathrm{d}P_0/\mathrm{d}r}\left(\frac{1}{r}\widehat{\phi}_l + \frac{\mathcal{Y}_{\boldsymbol{F}_{\mathcal{L}};l}}{\rho_0}\right). \tag{4.5}$$

Finally, it can be interesting to look for temperature perturbations. Following Kippenhahn & Weigert (1990), we introduce the general equation of state for the stellar plasma $\mathrm{d}\rho/\rho = \alpha_s \mathrm{d}P/P - \delta_s \mathrm{d}T/T + \varphi_s \mathrm{d}\mu_s/\mu_s$ where $\alpha_s = \left(\partial\ln\rho/\partial\ln P\right)_{T,\mu_s}$, $\delta_s = -\left(\partial\ln\rho/\partial\ln T\right)_{P,\mu_s}$ and $\varphi_s = \left(\partial\ln\rho/\partial\ln\mu_s\right)_{P,T}$. For a perturbative Lorentz force, the stellar temperature (T) and the mean molecular weight (μ_s) can be expanded like P, ρ and ϕ according to $\mu_s\left(r,\theta,t\right) = \mu_{s;0}\left(r\right) + \sum_{l\geqslant 0}\widehat{\mu}_{s;l}\left(r,t\right)P_l\left(\cos\theta\right)$. Linearizing the equation of state, we finally obtain

$$\widehat{T}_l = \frac{T_0}{\delta_s}\left[\alpha_s\frac{\widehat{P}_l}{P_0} - \frac{\widehat{\rho}_l}{\rho_0} + \varphi_s\frac{\widehat{\mu}_{s;l}}{\mu_{s;0}}\right]. \tag{4.6}$$

Results for the normalized perturbations of gravitational potential $\tilde{\Phi}_l$, density $\tilde{\rho}_l$, pressure $\tilde{P}_l$, temperature $\tilde{T}_l$ and radius c_l are shown in Fig. 2 for the modes $l=0$ and $l=2$ (resp. middle and right panel). At the surface the effective temperature change is found from the $l=0$ temperature perturbation: it is $\widehat{T}_0 = +\,1.45\times10^{-4}T_{\mathrm{eff}}$, i.e. for the considered case $T_{\mathrm{eff}} = 8422$K instead of 8421K. The gravitational multipolar moments are $J_0 = -\,1.31\times10^{-7}$ and $J_2 = -\,2.54\times10^{-8}$.

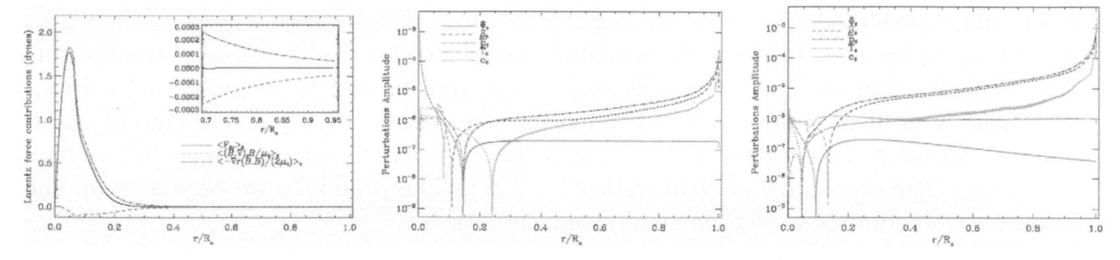

Figure 2. Left: radial Lorentz force (normalized), with the relative contributions of the magnetic pressure gradient and the magnetic tension. Middle and Right: perturbations of mode $l=0$ and $l=2$ (in log. scale) respectively. Bold lines represent positive values whereas thin lines represent negative ones. The spikes corresponds to the vanishing of source terms for the equations 4.3, 4.4, 4.5 and 4.6.

† Let us recall that Sweet (1950) was the first to derive this result for the most general perturbing force, Moss (1974) having introduced the special case of the Lorentz force in the case of a poloidal field, while later Mathis & Zahn (2005) treated the general axisymmetric case

4.2. Energetic Balance

4.2.1. Poynting's Flux and Ohmic Heating

The Ohmic heating is defined by

$$Q_{\mathrm{Ohm}}(r,\theta)=\mu_0\,\eta\,\boldsymbol{j}^2(r,\theta), \tag{4.7}$$

where η is the magnetic diffusivity that can be evaluated with the temperature-dependent law from Spitzer (1962)

$$\eta=5.2\times10^{11}\ \log\Lambda\,T^{-3/2}\ \mathrm{cm^2s^{-1}}. \tag{4.8}$$

The Poynting's flux is given by $F_{\mathrm{Poynt}}=\nabla\cdot(\boldsymbol{E}\times\boldsymbol{B}/\mu_0)$. In the static case, the simplified Ohm's law $\boldsymbol{j}=\sigma\boldsymbol{E}$ together with the identity $\eta=(\mu_0\sigma)^{-1}$, reduces the Poynting's flux expression to

$$F_{\mathrm{Poynt}}=\nabla\cdot(\eta\,\boldsymbol{F}_{\mathcal{L}})\,. \tag{4.9}$$

4.2.2. Perturbation of the Energetic Balance

Here again a perturbative approach is adopted. The luminosity is expanded as

$$L=L_0+\widehat{L}_{\mathrm{tot}}. \tag{4.10}$$

$\widehat{L}_{\mathrm{tot}}$ is the luminosity perturbation due to the magnetic terms:

$$\widehat{L}_{\mathrm{tot}}(r)=L_{\mathrm{Ohm}}(r)+L_{\mathrm{Poynt}}(r)+\widehat{L}_{\mathrm{nuc}}(r)\,, \tag{4.11}$$

which are respectively the Ohmic heating contribution, the Poynting's flux one, and the one related to the induced modification of the specific energy production rate.

First, we integrate the Ohmic heating and the Poynting's flux over the volume delimited by r

$$L_{\mathrm{Ohm}}(r)=\int_0^r\int_\Omega Q_{\mathrm{Ohm}}(r',\theta')\,\mathrm{d}\Omega\,r'^2\mathrm{d}r'; \tag{4.12}$$

$$L_{\mathrm{Poynt}}(r)=\int_0^r\int_\Omega F_{\mathrm{Poynt}}(r',\theta')\,\mathrm{d}\Omega\,r'^2\mathrm{d}r', \tag{4.13}$$

where $\mathrm{d}\Omega=\sin\theta'\mathrm{d}\theta'\mathrm{d}\phi'$, r' thus ranging from 0 to r, θ' from 0 to π and ϕ' from 0 to 2π. Then, to be able to conclude we finally consider the modification of the specific energy production rate (ε), which depends on ρ and T, due to magnetic field. First, the logarithmic derivative of ε is expanded like the one of ρ (cf. the equation of state, and see Mathis & Zahn, 2004 and references therein): $\mathrm{d}\ln\varepsilon=\lambda\,\mathrm{d}\ln\rho+\nu\,\mathrm{d}\ln T$, where $\lambda=(\partial\ln\varepsilon/\partial\ln\rho)_T$ and $\nu=(\partial\ln\varepsilon/\partial\ln T)_\rho$. Then, like ρ and T, we expand ε on the Legendre polynomials so that we finally end up with

$$\varepsilon(r,\theta)=\varepsilon_0(r)+\sum_{l\geqslant0}\widehat{\varepsilon_l}(r)\,P_l(\cos\theta)\quad\text{where}\quad\widehat{\varepsilon_l}=\varepsilon_0\left[\lambda\frac{\widehat{\rho_l}}{\rho_0}+\nu\frac{\widehat{T_l}}{T_0}\right]. \tag{4.14}$$

The luminosity perturbation induced by the MHS equilibrium over the nuclear reaction rates is

$$\widehat{L}_{\mathrm{nuc}}(r)=\int_0^r\int_\Omega\widehat{\varepsilon}_0\,\rho_0\,r'^2\mathrm{d}r'\mathrm{d}\Omega=4\pi\int_0^r\left\{\varepsilon_0\left[\lambda\frac{\widehat{\rho_0}}{\rho_0}+\nu\frac{\widehat{T_0}}{T_0}\right]\right\}\rho_0\,r'^2\mathrm{d}r'. \tag{4.15}$$

The values found at the stellar surface are $\widehat{L}_{\rm nuc} = -6.06 \times 10^{29}$erg.s, $L_{\rm Ohm} = 5.71 \times 10^{23}$erg.s and $L_{\rm Poynt} = -5.97 \times 10^{22}$erg.s, whereas the total luminosity is $L_0 = 1.59 \times 10^{35}$erg.s.

5. Conclusion

We have shown that at a first glance the non force-free, barotropic MHS equilibria are stable. This type of configuration is thus relevant to model initial conditions for evolutionary calculations involving large-scale, long-time evolving fossil fields in stellar radiation zones as well as in degenerate objects such as white dwarfs or neutron stars (see Payne & Melatos, 2004). More particulary it can be used to initiate MHD rotational transport in dynamical stellar evolution codes where it is implemented (cf. Mathis & Zahn, 2005; Duez *et al.*, 2008) since axisymmetric transport equations that have been derived are devoted to the stable axisymmetric component of the magnetic field, the magnetic instabilities being treated using phenomenological prescriptions (see Spruit, 1999; Maeder & Meynet, 2004) that have to be verified or improved by numerical experiments (see Braithwaite, 2006 and subsequent works; Zahn, Brun & Mathis, 2007).

In the context of implementing the magnetic field's effects in a stellar evolution code, the qualitative importance of the magnetic tension has been underlined.

In the case exposed here, the perturbative approach has shown that the direct contribution of the magnetic field to the change in the energetic balance through Ohmic heating or through Poynting's flux is weak compared with the indirect modification to the energetic balance induced by the change in pressure and density over the nuclear reaction rate. In the case studied here, this contribution amounts to -3.8×10^{-6} of the total luminosity. A first approach, consisting in limiting the impact of a large-scale magnetic field only to its impact upon the hydrostatic balance will therefore be justified.

Acknowledgements

We would like to acknowledge the IAU for the grant allocation delivered, that supported our participation to the conference.

References

Alecian, E., Wade, G. A., *et al.* 2008, in: C. Neiner & J. -P. Zahn (eds.), *Stellar Magnetism*
Bernstein, I. B., Friemann, E. A., Kruskal, M. D., & Kulsrud, R. M. 1958, in *RSL Proc. Series A* 244, 17
Braithwaite, J. & Nordlund, A. 2006, *A&A* 450, 1077
Chandrasekhar, S. 1956, *PNAS* 42, 1
Duez, V., Brun, A.-S., Mathis, S., Nghiem, P. A. P. & Turck-Chièze, S. 2008, in *Mem. S.A.It.* 79, 716
Kippenhahn, R. & Weigert, A. 1990, Stellar Structure and Evolution, Springer-Verlag, Berlin
Maeder, A. & Meynet, G. 2004, *A&A* 422, 225
Markey, P. & Tayler, R. J. 1973, *MNRAS* 163, 77
Markey, P. & Tayler, R. J. 1974, *MNRAS* 168, 505
Marsch, G. E. 1992, *Phys. Rev. A* 45, 7520
Mathis, S. & Zahn, J.-P. 2004, *A&A* 425, 229
Mathis, S. & Zahn, J.-P. 2005, *A&A* 440, 653
Morse, P. M. & Feshbach, H. 1953, Method of Theoretical Physics, McGraw Hill Book Company, New York
Moss, D. L. 1974, *MNRAS* 168, 61
Payne, D. J. B. & Melatos, A. 2004, *MNRAS* 351, 569
Reisenegger, A. 2008, *A&A*, submitted
Spitzer, L. 1962, Physics of Fully Ionized Gases (New York : Interscience)

Spruit, H. C. 1999, *A&A* 349, 189
Sweet, P. A. 1950, *MNRAS* 110, 548
Tayler, R. J. 1973, *MNRAS* 161, 365
Wade, G. A., Kudryavtsev, D., Romanuyk, I. I. , Landsreet, J. D., & Mathys, G. 2000, *A&A* 355, 1080
Zahn, J.-P., Brun, A.-S., & Mathis, S. 2007, *A&A* 474, 145

Discussion

De Gouveia dal Pino: Can you make any predictions and/or diagnostics based on the model for the COROT satellite?

Duez: It seems unlikely that COROT may provide any constructs on the geometry of the internal field; first because it will perform asteroseismology measurements in the range of frequencies usually devoted to probe the external layers; second, because the frequencies shifts due to a magnetic field, if ever they would be detected, should allow to determine eventually the amplitude of the magnetic field, regarding the few number of non-radial modes detected. Moreover splittings are an integrated information. More appropriate tools are spectropolarimeters which give quantitative dues about both the field's strength and field's geometry at the stellar surface.

Beckman: Could you simply tell us whether the presence of a magnetic field increases or decreases the surface temperature for a star of the Sun's mass?

Duez: I apologize: in fact the difference is positive instead of being negative as I said after may talk; the difference between the bold lines and the thin lines was not clear enough on the projection screen! For stars of same mass, the net effect of a large-scale magnetic field upon the stellar structure, especially on the effective temperature is an increase compared to the star without magnetic field. For example for a $2.75\text{-}M_{\odot}$ Ap-type star, the increase in temperature is approximately +13 K (for an effective temperature of 8926 K).

Vincent Duez

Cosmic Magnetic Fields:
From Planets, to Stars and Galaxies
Proceedings IAU Symposium No. 259, 2008
K.G. Strassmeier, A.G. Kosovichev & J.E. Beckman, eds.

doi:10.1017/S1743921309030439

On the relation between photospheric magnetic field and chromospheric emission in the quiet Sun

Maria A. Loukitcheva[1,2], Sami K. Solanki[2] and Stephen M. White[3]

[1]Astronomical Institute, St. Petersburg University, 198504 St. Petersburg, Russia
email: marija@astro.spbu.ru

[2]Max-Planck-Institut für Sonnensystemforschung, D-37191 Katlenburg-Lindau, Germany

[3]Astronomy Department, University of Maryland, College Park, MD 20742

Abstract. In this contribution we present an observational study of the interaction of the photosphere with different chromospheric layers. We study the correlations between emissions at varying temperature from the temperature minimum region (UV continuum at 1600 Å from TRACE) through the low chromosphere (CaII K-line from BBSO) to the middle chromosphere (continuum at 3.5 mm from BIMA) and photospheric magnetic field from MDI/SOHO. For the first time millimeter observational data are included in such analysis.

We report a high degree of correlation between considered emissions formed at different heights in the chromosphere. A power law is found to be a good representation for the relationship between photospheric magnetic field and chromospheric emissions at all considered wavelengths. Our analysis shows that the dependence of chromospheric intensities on magnetic field is different for the network and internetwork regions. In the network a power law provides the best fit with the exponent being close to $0.5 - 0.6$, while almost no dependence of chromospheric intensity on magnetic flux is found for the cell interiors. The obtained results support the idea of different heating mechanisms acting in the network (magnetic) and cell interiors (acoustic).

Keywords. Sun: magnetic fields – Sun: chromosphere – Sun: radio radiation – Sun: photosphere

1. Introduction

In the quiet-Sun chromosphere, even far from the activity complexes, magnetic field is believed to be responsible for chromospheric structuring, which displays a variety of inhomogeneities at different spatial scales. The primary quiet-Sun brightness patterns, revealed from chromospheric images, include chromospheric network, located at the boundaries of supergranular cells, and bright points in the interior of the cells – internetwork grains. For the primary chromospheric diagnostic, which is the emission in the cores of the calcium K and H resonance lines, a strong correlation between the presence and amount of the quiet-Sun magnetic flux and excess core emission is well established (e.g. Leighton 1959; Schrijver *et al.* 1989). And naturally these resonance lines commonly serve as indicators of changes in the chromospheric structure related to global magnetic activity and heating of the outer atmosphere of the Sun and cool stars (Rutten & Uitenbroek 1991; Schrijver *et al.* 1989).

Apart from the chromospheric emission related to the magnetic field, Schrijver (1987, 1992) has suggested that a part of the radiative flux from solar and stellar chromospheres is of nonmagnetic origin, it is referred to as basal flux, and is believed to be due to acoustic heating of the chromosphere. Consequently, these two components – basal and strong-field, are the basic components, which determine the properties of the solar brightness

Table 1. Overview of data taken on May 18, 2004

Instrument	λ	$\Delta\lambda$(Å)	Spatial resolution (arcsec)	Pixel size (arcsec)
MDI/SOHO	6768 Å	0.094	4	1.96
TRACE	1600 Å	275	1	0.5
BBSO	3933 Å	0.6	2-4	0.645
BIMA	3.5 mm		12	3.0

structures. Thus the chromospheric network is known to display a one-to-one spatial correspondence with regions of enhanced photospheric magnetic field (Skumanich *et al.* 1975; Sivaraman *et al.* 1982; Nindos & Zirin 1998), while the origin (magnetic or non-magnetic) of the internetwork grains is a controversial issue.

Quantitative studies of the relationship between the chromosphere in calcium, both active and quiet, and magnetic field were carried out by a number of authors and naturally fall into two distinct classes. In a number of studies (see, e.g., Schrijver *et al.* 1989; Harvey & White 1999) it was obtained that the relation between the K-line excess flux density and magnetic field flux density is best described by a power law with an exponent around 0.5 – 0.6. This conclusion refers to different types of brightness structures, including active regions, decaying active regions, the enhanced network and quiet (weak) network. However, for quiet-Sun chromosphere Skumanich *et al.* (1975) and Nindos & Zirin (1998) found no evidence of non-linear behaviour and obtained a linear correlation.

2. Observational Data

In this work to study the relation between the photosphere and different chromospheric layers we analyzed photospheric magnetograms from the Michelson Doppler Imager (MDI, Scherrer *et al.* 1995) on board of SOHO, UV images at 1600Å from the Transition Region and Coronal Explorer (TRACE, Handy *et al.* 1999), CaII K-line filtergrams from the Big Bear Solar Observatory (BBSO) and millimeter images at 3.5 mm from the Berkeley–Illinois–Maryland Array (BIMA), obtained on May 18, 2004. This is the first analysis that includes millimeter observational data. A quiet-sun region close to the disk center was observed for 3.5 hours. In Table 1 we give the overview of the data (including instruments, operating wavelengths, spatial resolution and image pixel size). We based our correlation analysis on the mean (averaged in time) images to avoid scatter introduced by chromospheric oscillations. Thus, by averaging the images we reduce significantly the scatter in the relationship and the noise in the magnetograms. In this work we studied images, averaged over the whole observational run of 3.5 hours and over half-hour time interval. The latter was used to check the stability of the derived structures. Prior to averaging careful spatial co-alignment of the observational data cubes was performed. Finally, the full disk MDI images were chosen as the coordinate reference against which all other mean images were co-aligned. The remaining image displacements do not exceed 1" and do not affect the results of the analysis. More details about the data and its reduction and alignment can be found in Loukitcheva *et al.* (2009).

3. Results

The resulted images of the solar chromosphere from the temperature minimum (TRACE 1600 Å image) through the low chromosphere (CaII K-line filtergram) to the

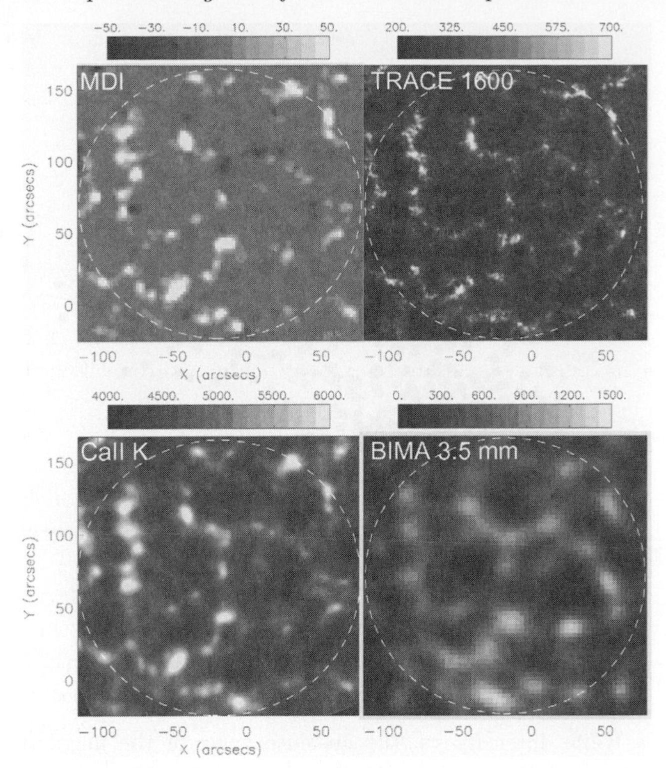

Figure 1. The solar chromosphere at 4 different wavelengths on May 18, 2004. From top left to bottom right: MDI longitudinal photospheric magnetogram, TRACE 1600 Å image, CaII K-line center image from BBSO and BIMA image at 3.5 mm. The images are created by averaging over 3.5h observational run. Dashed circles mark the 96 arcsec BIMA FOV containing flux. X and Y axes are in arcsec from the disk center.

middle chromosphere (BIMA image at 3.5 mm) at the original resolution of the instruments and the corresponding photospheric magnetogram can be seen in Fig. 1. All chromospheric maps and (unsigned) MDI magnetograms reveal clearly the same pattern, although at different spatial resolutions, most prominently on spatial scales of the enhanced magnetic flux and chromospheric network. The degree of spatial agreement between radio and other chromospheric radiation and photospheric magnetic field is more clearly visible in maps degraded to the BIMA resolution of 12 arcsec (not shown).

Firstly, we computed pixel-to-pixel scatter plots for the pairs of images globally in the field of view and searched for the best linear and power-law fits (Fig. 2). By comparison of the least-square deviations of the best fits, we were able to determine confidently the non-linearity of the behaviour for all 3 considered chromospheric magnetic-intensity relationships. The exponents of the best power-law fits are listed in the last column of Table 2. The tightest relation is displayed by the calcium K-line, which is shown plotted in Fig. 2. The power law exponent of the relation, being 0.31, is smaller than those found by Schrijver *et al.* (1989) and Harvey & White (1999) (see Sect. 1). This probably has to do with the smaller B range available in the quiet Sun, although it cannot be ruled out that the calcium intensity does not increase so rapidly in the quiet Sun as in active regions. It is seen from Fig. 2 that the non-linearity of the relation is influenced mostly by weak chromospheric emission, that reveals a strong dependence on magnetic field. Thus, if we set a B threshold of 20 G we cannot distinguish between the quality of the linear and power-law fits for the relation between calcium intensity and magnetic field strength, as

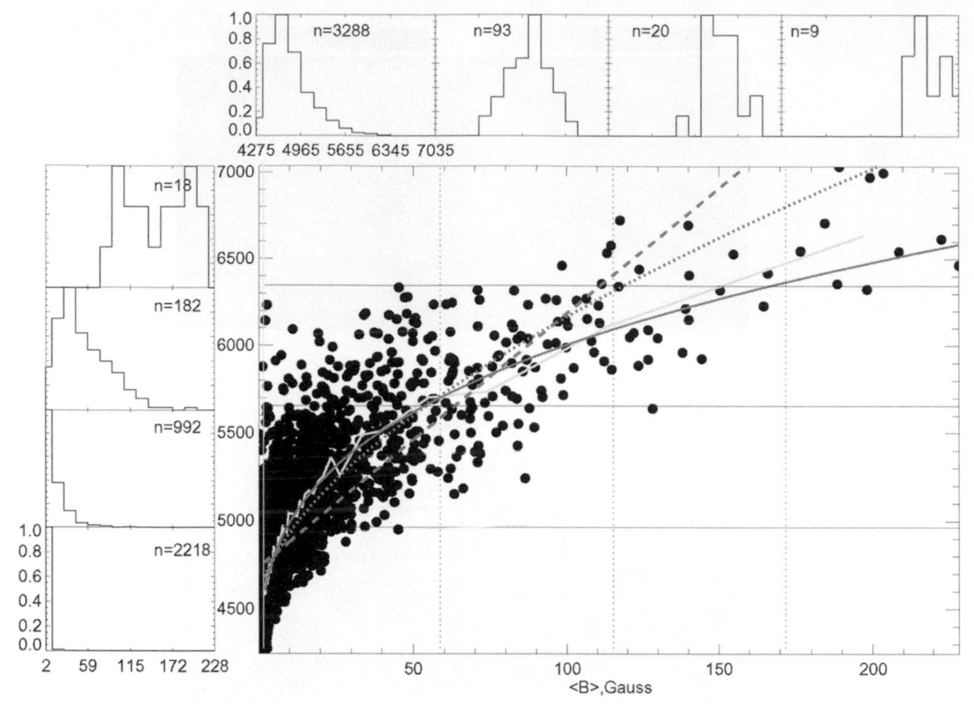

Figure 2. The Ca K-line intensity vs. the absolute value of the magnetic flux density for a 2 arcsec pixel size. The images are averaged in time over 3.5h. The vertical line at 1.5 G represents the 1σ noise limit of the magnetogram signal. The fit results are represented by a solid curve for a power-law fit with free exponent, dotted curve for a power-law fit with the exponent fixed at 0.6 and dashed for a linear fit. The light grey solid curve outlines the CaII K intensity binned in 70 intervals containing an equal number of points. Histograms on the left show the distribution of magnetic field strength in 4 subranges of the CaII K-line flux. Histograms on top of the central panel depict the distribution of calcium intensity in 4 subranges of the magnetic flux. For each histogram the corresponding range of considered values is marked in the main figure: by horizontal solid lines for magnetic flux distributions and by vertical dotted lines for the CaII K-line flux histograms. The number of counts, n, represented in the histograms are indicated.

well as for the relations between UV intensity at 1600 Å and magnetic flux, and between 3.5 mm brightness and magnetic flux. Regarding the dependence of millimeter brightness on magnetic field strength (not shown), the data also strongly suggests that it is non-linear and among the considered fitting functions is best described by a power law, but the sizeable scatter does not allow the exponents of the fit to be determined reliably.

Next we employed a brightness threshold technique for calcium intensity image and created a binary map discriminating between the areas representing cell interiors or internetwork (IN) and chromospheric network (NW). We studied the relations between the chromospheric emission and the magnetic flux separately for IN and NW locations. On the whole, chromospheric emission displays a rather different dependence on the network and internetwork magnetic field (see Fig. 3 for calcium intensity). For the network, again as in the case of the global FOV, the relations between all three chromospheric quantities and magnetic field are best described by a power-law. In Table 2 we list the exponents from the best-fit power laws for the images at the resolution of the MDI magnetograms (4 arcsec) and of the BIMA images (12 arcsec). The exponents of the relation for calcium intensity are found to be similar to $0.5-0,6$, cited by Schrijver *et al.* (1989) for the active region and by Harvey & White (1999) for the quiet-Sun network.

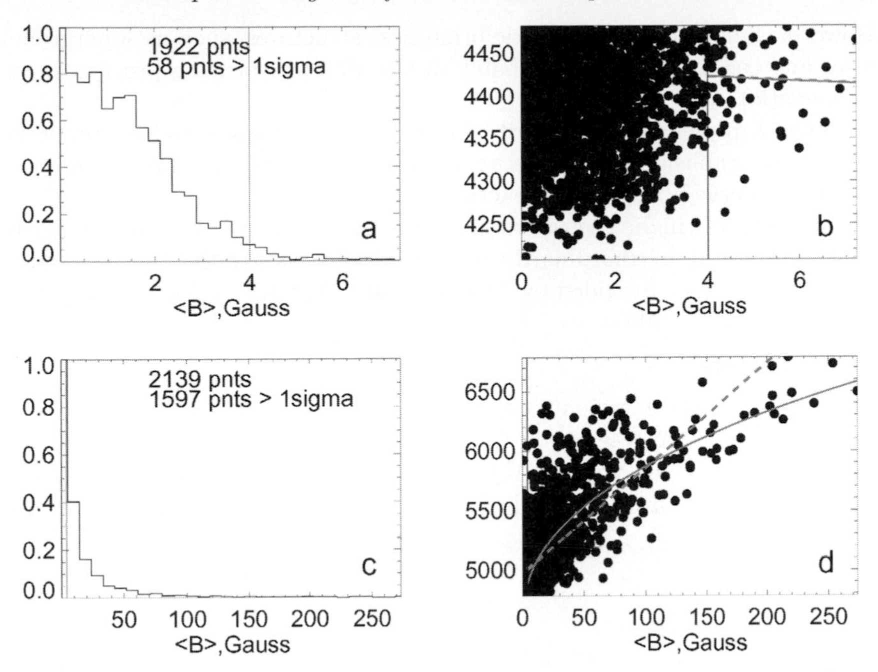

Figure 3. Comparison of IN and NW emission averaged over 0.5h: histograms of **a)** IN and **c)** NW magnetogram signal, **b, d)** CaII K-line flux vs. absolute value of the IN and NW magnetic field, respectively. The solid vertical lines represent 1σ-noise of the magnetogram signal.

Table 2. Dependence of the network emission on the magnetic flux: exponents of the power-law fits calculated for the images at 4 arcsec and 12 arcsec resolution, averaged over 3.5h and 0.5h. In the last column we list for comparison the exponents of the power-law fits calculated globally in the field of view.

NW flux	4 arcsec resolution		12 arcsec resolution		FOV flux
	3.5h	0.5h	3.5h	0.5h	
$\|B\| - I_{CaIIK}$	0.6	0.43	0.54	0.47	0.31
$\|B\| - I_{1600\mathring{A}}$	0.49	0.53	0.66	0.44	0.36

In the internetwork regions almost no dependence of the calcium, UV at 1600 Å and 3.5 mm intensity on the magnetogram signal was found. However, due to the very low B values in the internetwork, as seen in Fig. 3, top panel, only a few percent of IN structures are associated with magnetogram signal exceeding noise, and this makes a confident study difficult. Nonetheless, our result confirms the conclusions drawn in the recent paper by Rezaei *et al.* (2007), and supports the idea that bright locations in the internetwork are due to nonmagnetic heating (e.g., acoustic heating).

4. Summary and Conclusions

(*a*) Summarizing, clear spatial correlations found between photospheric magnetograms, calcium K-line, UV intensity at 1600 Å and 3.5 mm emission indicate that heating in the quiet-Sun lower and middle chromosphere maps out the underlying photospheric magnetic field rather well. This implies that on the chromospheric heights covered by the analyzed here wavelengths we deal with the same heating mechanism.

(*b*) However, different chromospheric brightness structures, such as supergranular cell boundaries and cell interiors, are probably due to different heating processes (magnetic and nonmagnetic).

(*c*) In general, present data on both chromospheric emission and magnetic field are not sufficient to establish physical mechanisms that are acting at chromospheric heights leading to the observed brightness structures.

(*d*) In this respect, higher sensitivity magnetograms are needed, and regarding the chromospheric data an extraordinarily powerful tool to study the thermal structure of the chromosphere will be provided by Atacama Large Millimeter Array (ALMA, Bastian 2002; Loukitcheva *et al.* 2008).

Acknowledgements

Millimeter–wavelength astronomy at the University of Maryland is supported by NSF grant AST–0540450. Solar research at the University of Maryland is supported by NASA grants NNX06–AC18G and NNX08–AQ48G. M. Loukitcheva acknowledges support from MK-1853.2007.2 of the Grant Council of the President of the Russian Federation.

References

Bastian, T. S. 2002, *AN* 323, 271
Handy, B. N. & others 1999, *Sol.Phys.* 187, 229
Harvey, K. & White, O. 1999, *ApJ* 515, 812
Leighton, R. B. 1959, *ApJ* 130, 366
Loukitcheva, M., Solanki, S. K., & White, S. M. 2008, *Ap&SS* 313, 197
Loukitcheva, M., Solanki, S. K., & White, S. M. 2009, *A&A*, *in press*
Nindos, A. & Zirin, H. 1998, *Solar Phys.* 179, 253
Rezaei, R., Schlichenmaier, R., Beck, C. A. R., Bruls, J., & Schmidt, W. 2007, *A&A* 466, 1131
Rutten, R. J. & Uitenbroek, H. 1991, *Solar Phys.* 134, 15
Scherrer, P. H. & others 1995, *Solar Phys.* 162, 129
Schrijver, C. J. 1987, *A&A* 172, 111
Schrijver, C. J. 1992, *A&A* 258, 507
Schrijver, C. J., Cotè, J., Zwaan, C., & Saar, S.H. 1989, *ApJ* 337, 964
Sivaraman, K. R. & Livingston, W. C., 1982, *Solar Phys.* 80, 227
Skumanich, A., Smythe, C., & Frazier, E. N. 1975, *ApJ* 200, 747

Discussion

KOUTCHMY: Two comments. 1) Intranetwork magnetic elements are small bi-poles (both polarities are present) which emerge and migrate towards the boundary of the cell. 2) Network elements are rather uni-polar magnetic elements collected at the vertex of converging flows due to super granulation. I guess this makes the difference.

LOUKITCHEVA: Yes, intranetwork & network elements are of different nature. We are discriminating between these two and study their emission and dependence on magnetic flux separately. And we obtain different relationships for magnetic elements and chromospheric emission. To take into account the dynamic behavior of intra-network elements, their proper motions, we studied them from very short time averages. And for network elements, which are more stable in time, longer time integration was applied.

Cosmic Magnetic Fields:
From Planets, to Stars and Galaxies
Proceedings IAU Symposium No. 259, 2008
K.G. Strassmeier, A.G. Kosovichev & J. Beckman, eds.

doi:10.1017/S1743921309030440

Magnetic reconnection and energy release on the Sun and solar-like stars

Lidia van Driel-Gesztelyi[1,2,3]

[1]University College London, Mullard Space Science Laboratory, Holmbury St. Mary, Dorking, Surrey, RH5 6NT, U.K.

[2]Konkoly Observatory of Hungarian Academy of Sciences, Budapest, Hungary

[3]Observatoire de Paris, LESIA, FRE 2461(CNRS), F-92195 Meudon Principal Cedex, France
email: Lidia.vanDriel@obspm.fr

Abstract. Magnetic reconnection is thought to play an important role in liberating free energy stored in stressed magnetic fields. The consequences vary from undetectable nanoflares to huge flares, which have signatures over a wide wavelength range, depending on e.g. magnetic topology, free energy content, total flux, and magnetic flux density of the structures involved. Events of small energy release, which are thought to be the most numerous, are one of the key factors in the existence of a hot corona in the Sun and solar-like stars. The majority of large flares are ejective, i.e. involve the expulsion of large quantities of mass and magnetic field from the star. Since magnetic reconnection requires small length-scales, which are well below the spatial resolution limits of even the solar observations, we cannot directly observe magnetic reconnection happening. However, there is a plethora of indirect evidences from X-rays to radio observations of magnetic reconnection. I discuss key observational signatures of flares on the Sun and solar-paradigm stellar flares and describe models emphasizing synergy between observations and theory.

Keywords. Sun: flares – Sun: magnetic fields – stars: activity – stars: flare – stars: magnetic fields

1. Introduction

Magnetic reconnection is a topological restructuring of a magnetic field, which *per definition*, leads to a change in the connectivity of its field lines. It allows the release of free magnetic energy, i.e. above the potential, zero-current, energy, which is stored in a force-free magnetic system in the form of field-aligned electric currents. During magnetic reconnection free magnetic energy is converted into kinetic energy of fast particles, mass motions and radiation across the entire electromagnetic spectrum manifested in sudden brightening i.e. a flare.

The Sun is the best-observed star in the Universe and solar physics is in a privileged position of having a fleet of spacecraft and ground-based telescopes which are able to resolve details in the solar photosphere and chromosphere as small as $0.2''$ ($\approx$140 km) and coronal structures of about $1''$ ($\approx$700 km). However, even with such fine spatial resolution nobody has been able to *directly observe* magnetic reconnection or the energy release site in the solar atmosphere. This is mainly due to the small size and low brightness of the magnetic reconnection region. There are very small length-scales needed for the break-down of ideal magneto-hydrodynamic (MHD) conditions in a low plasma-β environment, like the solar corona, which are satisfied in thin current sheets, whose scale is orders of magnitudes lower than that of our finest spatial resolution. Furthermore, the reconnection region has low plasma density and therefore low emission measure, making its observation very difficult in the optically thin corona.

However, there is no doubt that magnetic reconnection is taking place on the Sun at all scales: There are plenty of *indirect evidences* provided by multi-wavelength observations to prove it. In this paper I select some of the highlights of solar observational signatures of magnetic reconnection and discuss observations and models of the consequent energy release events. Furthermore, I use the solar paradigm to draw parallels between solar and stellar flares and coronal heating on the Sun and stars, pointing out the differences which may originate in the differences of magnetic flux, flux density, complexity and perhaps levels of non-potentiality between the Sun and solar-like stars.

2. Solar and stellar flares - observations and models

2.1. *Classification of solar flares*

Energy released up to 10^{25} J = 10^{32} erg in the largest solar flares. Many more much smaller flare-like events occur (micro-flares, nano-flares), down to energies of 10^{24} erg and even less. GOES (Geostationary Environmental Operational Satellites) soft X-ray (SXR) classification is most common these days due to the readily available, long and continuous dataset of solar X-ray emission. The flux in the 1- 8 Å = 0.1-0.8 nm range is classified with the letters of X, M, C, B indicating flux of $10^{-4}, 10^{-5}, 10^{-6}, 10^{-7}$ W m^2, respectively. According to this classification an X2 flare has a flux of 2×10^6 Wm^{-2}.

2.2. *Origin and storage of free magnetic energy*

There is plenty of observational evidence indicating that magnetic flux emerges twisted i.e. in a non-potential state from the solar interior, carrying free energy which is ready to be released (Leka *et al.* 1996; for a recent review see Démoulin 2007a). MHD simulations have shown that untwisted flux cannot even make it through the convection zone because it gets eroded by vortexes forming in its wake (Schüssler 1979; Longcope *et al.* 1996), however, sufficient twist can prevent significant fragmentation (Moreno-Insertis & Emonet 1996). These simulation results imply that all the large-scale flux that has crossed the convection zone must be twisted.

However, the observed twist is not strong, as it was deduced from photospheric current helicity measurements (Longcope *et al.* 1999). The coronal helicity content of ARs also appears to be modest being equivalent to that of a twisted flux tube having 0.2 turn ($\mathbf{H}_{max}$(AR) $\approx 0.2\Phi^2$, where Φ is the total magnetic flux of the AR; Démoulin 2007a). More recent simulations by Fan (2008) also suggest that in order to comply with Joy's law (a systematic deviation from the east-west alignment of bipolar ARs with the leading spots being closer to the equator on both solar hemispheres which results from the action of Coriolis force) the twist-induced tilt in rising flux ropes, which is in the opposite sense than that of the Coroilis force, cannot be as high as expected. In order for the emerging tube to show the tilt direction consistent with observations, the initial twist rate of the flux tube needs to be less than half of that needed for a cohesive rise. Under such conditions, severe flux loss was found during the rise, with less than 50% of the initial flux remaining in the Ω-tube by the time it reaches the surface.

Before the discovery that flux is emerging twisted from the solar interior, the generally accepted idea was that free magnetic energy in an active region is generated by shearing flows, which move opposite polarity footpoints in anti-parallel directions on both sides of a magnetic neutral line. There are indeed large-scale flows in the solar photosphere (e.g. differential rotation) and local deviations are clearly seen from the mean differential rotation rate in flow-maps of the solar surface (e.g. Sobotka 1999 and Meunier 2005) as well as in the solar interior, especially around active regions (Zhao & Kosovichev 2004; Švanda *et al.* 2008). Surface shearing flows can result from the emergence of a flux

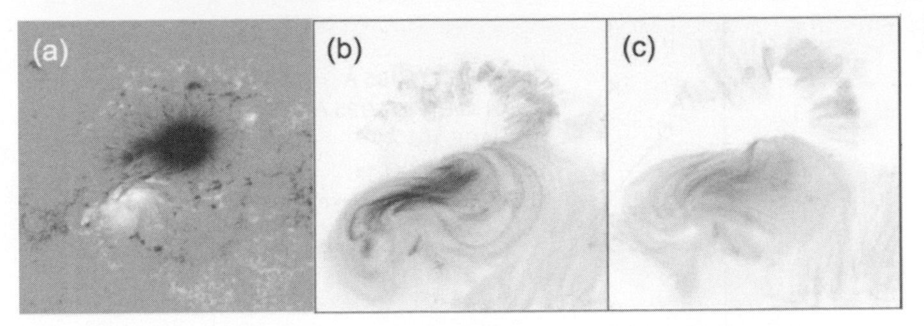

Figure 1. (*a*) *Hinode*/SOT magnetogram of AR 10930 on 13 December 2006. Integrated electric currents in the AR (*b*) before and (*c*) after an X3.4 flare. Note the current filaments' organization into an apparent flux-rope structure along the IL (adapted from Schrijver *et al.* 2008).

rope (Démoulin & Berger 2003) due to the Lorenz force arising from the nonuniform expansion of the magnetic field in a highly pressure-stratified atmosphere (Manchester 2007). Flux rope emergence has many caveats, *e.g.* in the concave-up parts under the flux rope axis plasma accumulates, leading to a fragmentation of the emerging flux rope (e.g., Magara 2004; Manchester *et al.* 2004) which can only emerge through many small-scale reconnections (Pariat *et al.* 2004). Nevertheless, characteristic magnetic patterns in emerging flux regions originating from the changing azimuthal component of a flux rope while crossing the photosphere, the so-called "magnetic tongues" (López-Fuentes *et al.* 2000; Démoulin & Pariat 2008) indicate that there is an overall organization in the emerging flux tube, which is compatible with a global twist. Besides organised motion patterns random magnetic footpoint motions (shuffling) are considered important to entangle field lines leading to the formation of small-scale current sheets.

Magnetic free energy is stored relatively low in an AR $\leqslant 20$ Mm above the photosphere and may mainly be concentrated along the magnetic inversion line in the filament channel in form of current filaments (Figure 1; Schrijver *et al.* 2008). This is supported by a strong connection found between high-gradient, strong-field magnetic inversion lines and flaring in active regions by Schrijver (2007), who introduced a new metric, R, the summed unsigned magnetic flux of the overlap of positive and negative magnetic field areas, where $B \leqslant 150$ Mx cm^{-2} with kernels of $6'' \times 6''$. R characterises newly emerged highly non-potential magnetic fields, and appears successful in forecasting major flares. If $R \geqslant 2 \times 10^{21}$ Mx ($logR \geqslant 4.8$), the probability of M or X-flare occurrence was found to be ≈ 1, while it was almost zero if $R \leqslant \times 10^{19}$ Mx ($logR \leqslant 2.8$).

2.3. *Confined flares - quadrupolar reconnection*

Reconnection takes place (i) at nullpoints (X-points) (ii) at separatrices and their intersection, the separator and (iii) at quasi-separatrix layers (QSLs, which are the non-zero-thickness generalisations of separatrixes) even in the absence of nullpoints. As a result, four flare kernels or ribbons appear at the footpoints of reconnected loops or in the vicinity of drastic field line connectivity changes, respectively. For detailed treatment of reconnection topologies see the book by Priest & Forbes (2000) and for a recent review see Démoulin (2007b).

Along QSLs field line mapping is continuous but steep gradients are present (Démoulin *et al.* 1996). Reconnection along QSLs also occurs in a continuous manner. Field lines may slip across each other, as shown in MHD simulations (Aulanier *et al.* 2006). A recent observational confirmation in Hinode/XRT data of slip-running reconnection was shown by Aulanier *et al.* (2007).

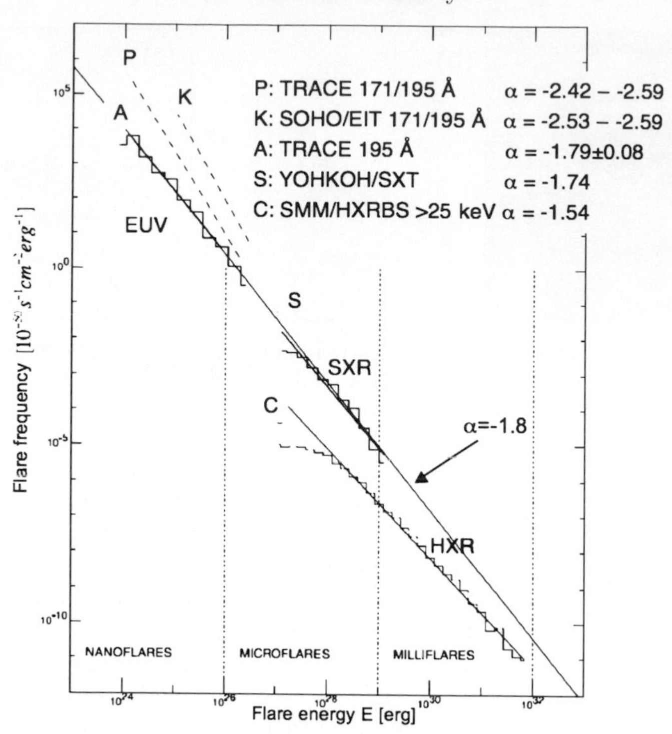

Figure 2. Powerlaws of solar flare number *versus* energy distribution adapted from Aschwanden *et al.* (2000).

2.4. *Flare frequency spectrum*

It feels natural that there are more smaller flare events than large ones. Quantifying this expectation showed that the number of flares N falls off with increasing energy E as a flat power law with a slope of $\alpha \approx -1.8$ (HXR, SXR, EUV, microwave bursts, optical flares; see Hudson 1991 and references therein). $dN/dE = A \times E^{\alpha}$ (erg s)$^{-1}$, where A normalisation factor varies with the level of activity. For the Sun, α values were found to be between -1.5 and -2.6 (Figure 2; Aschwanden *et al.* 2000 and references therein). Large flares do not supply sufficient energy to heat the solar corona (Hudson 1991). The significance of the α value is that if $\alpha \geqslant 2$, then smaller flares contribute enough to provide heating for the corona equal to its radiative losses.

Parenti *et al.* (2006) showed that there is an important effect of the coronal energy transport on event distribution. They studied statistical properties of solar coronal loops, subject to turbulent heating, to test whether the plasma response simply transmits the statistical distribution of events, with no modification. They found that for EUV lines the power-law index of the output distribution was strongly modified by becoming steeper (due to the dominance of radiative cooling), while for hotter lines ($T \approx 10^7$ K), where conductive cooling dominates the distribution is well preserved. This may explain the higher absolute values of α found for the very small events observed in EUV (cf. Figure 2) suggesting that for the Sun $\alpha \leqslant 2$.

On the other hand, similar flare statistics for solar-like stars give α values between -2.2 and -2.8 (Güdel 2007), suggesting that for stellar coronae flare heating is a viable option. Examples of stellar powerlaw distribution of a sample of 29 flares observed over a period of almost seven days on two solar-like stars 47 Cas and EK Dra are shown in

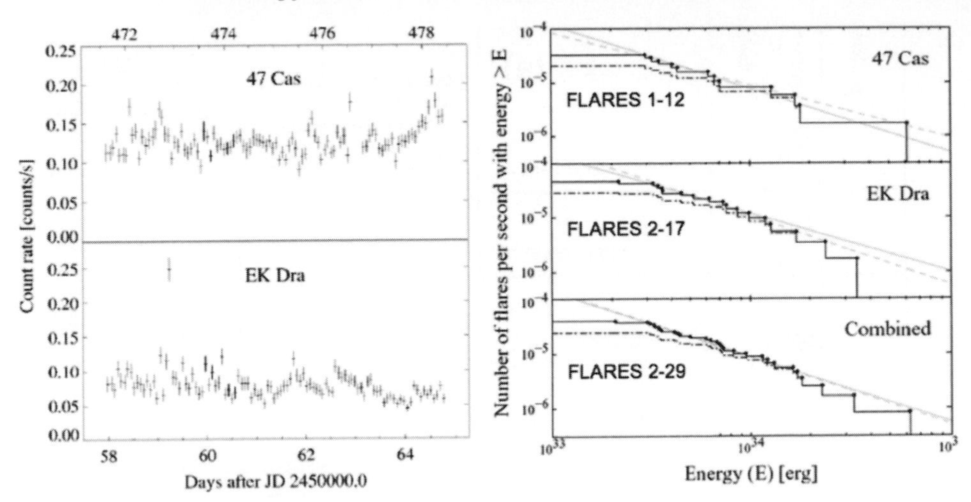

Figure 3. (*a*) Light-curves taken in the 40–190 Å range, using the *Extreme Ultraviolet Explorer* (EUVE) of two solar-like stars 47 Cas and EK Dra showing flaring. The bin-size corresponds to one EUVE orbit (96 min) (*b*) Powerlaws ($dN/dE \sim E^2$, with $\alpha \approx 2.2 \pm 0.2$) of stellar flare differential distribution *vs.* the total X-ray energy in the range between 3×10^{33} and 6×10^{34} ergs (adopted from Audard *et al.* 1999).

Figure 3, from Audard *et al.*(1999). Stellar flare observations are obviously sampling the highest-energy end of the flare distribution, less modified by transport effects.

2.5. *Standard model for eruptive solar flares*

The best-studied solar flares are the two-ribbon flares. Their unified model is often referred to as the CSHKP model (Carmichael 1964; Sturrock 1966; Hirayama 1974; Kopp & Pneuman 1976). However, the scenario has gone through important evolution since the publication of its main elements in the sixties and seventies (see *e.g.* Forbes 2001). A cartoon of the model and characteristic multi-wavelength light-curves are shown in Figure 4, while observations of a typical event is shown in Figure 5.

Two-ribbon flares occur in a dominantly bipolar magnetic configuration, with a filament along the magnetic inversion line (IL). As magnetic shear is increasing in the active region (with a concentration of shear in the vicinity of the IL) the filament is slowly rising (reaching a new equilibrium), stretching the enveloping bipolar magnetic arcade, leading to the formation of a current sheet under it (Figure 4b). There is, however, a point when new equilibrium is impossible and the filament starts accelerating and erupts. In the current sheet, which had formed under the erupting filament, magnetic reconnection takes place. Field lines break and change connectivity and magnetic field dissipates. The tension force of the reconnected field lines then accelerates the plasma out of the dissipation region. A subsequently in-flowing plasma carries the ambient magnetic field lines into the dissipation region. These field lines continue the reconnection cycle. Through this process, the magnetic energy stored near the current sheet or magnetic null-point is released to become the thermal and bulk-flow energy of the plasma. Magnetic energy is converted into heat and kinetic energy. Large electric fields are created in the dissipation region as well as shock waves, which help to accelerate particles. Since fast particles and heat conduction are mainly directed along the field lines, they will be channelled along the newly reconnected loops (Figure 4b). Electrons accelerated in the reconnection process gyrate along magnetic field lines emitting gyrosynchrotron radiation. Collisions in the dense chromosphere emit bremstrahlung observed in hard X-rays

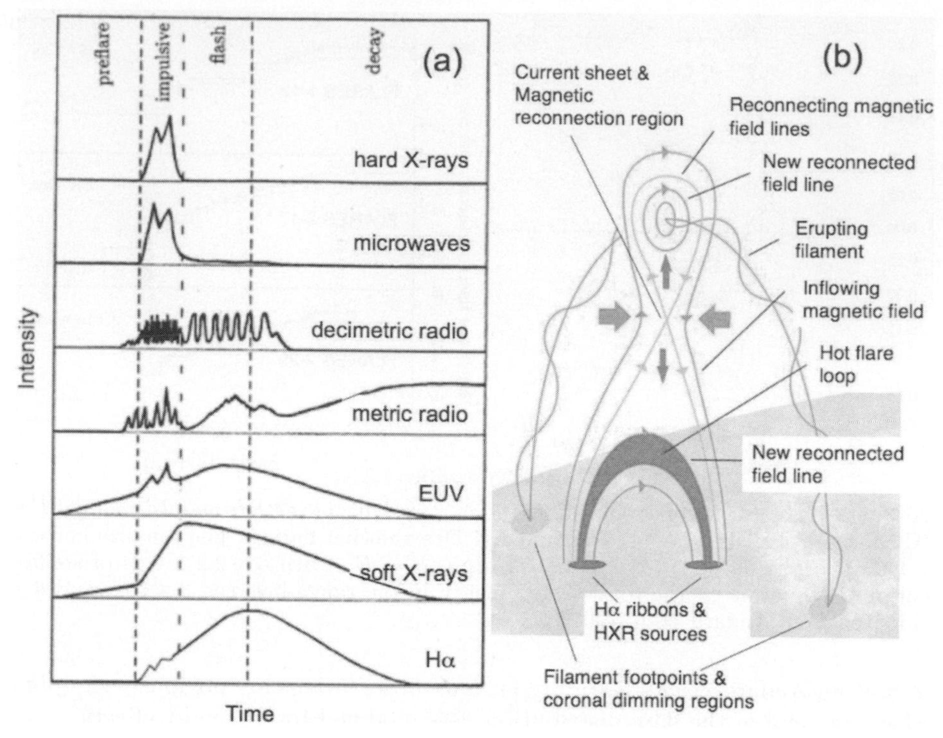

Figure 4. (*a*) Typical light-curves of a two-ribbon eruptive flare (LDE & CME) representing emissions from the chromosphere to the corona, adapted from Benz (2008). (*b*) Cartoon of a two-ribbon eruptive flare, adapted from Shibata (1998).

(HXRs; $\geqslant$ 20 keV). Accelerated electrons impulsively heat the chromosphere leading to optical and UV emission (e.g. flare ribbons seen in Hα). Heated chromospheric plasma expands upward, increasing the density and temperature in the reconnected coronal loops, leading to the formation of hot flare loops observed in EUV and SXRs. Continuing reconnection between field lines more and more distant from the IL leads to the formation of flare loops of increasing height while the flare ribbons at their footpoints are moving apart. Flares described by the CSHKP model are so-called long duration events (LDEs), which last for up to tens of hours (Figure 4a) and are invariably associated with coronal mass ejections (CMEs).

A few tenths of the total flare energy is released within a few minutes in the *impulsive phase* over a broad spectrum in HXRs, white light, UV, microwaves, etc. (Figure 4a). The radiations have an intermittent, bursty profile indicating patchy reconnection events along the long current sheet. During this phase important role is played by non-thermal electrons. In the decay phase reconnection is still taking place at increasing height along the lengthening current sheet at a gradually decreasing rate. The role of non-thermal particles is diminished, conduction fronts gaining importance instead.

In flares there is a very important relationship between HXRs and SXRs called the *Neupert effect* (Neupert, 1968). It expresses that SXRs mainly originate from plasma heated by the accumulated energy deposited by accelerated electrons from flare start(t_0).

$$F_{SXR}(t) \approx \int_{t_0}^{t} F_{HXR}(t)dt. \tag{2.1}$$

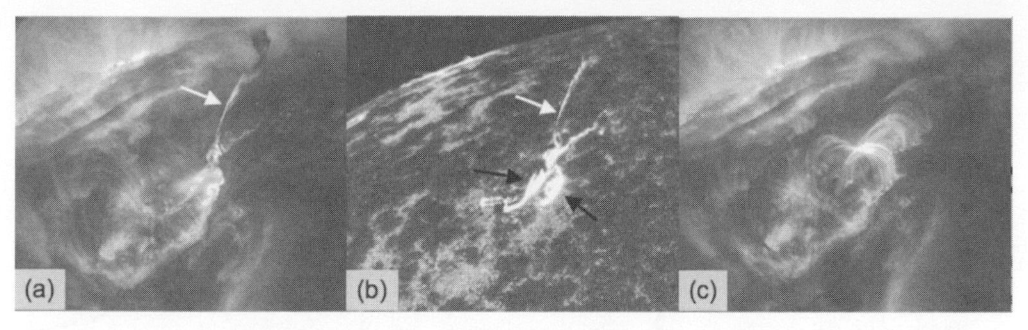

Figure 5. These TRACE spacecraft images were taken on 25 June 2000, (*a*) and (*b*) around 07:37UT and (*c*) 2h 10m later. The images were rotated, so that north is to the left. (*a*) White arrow points at a filament in the process of being ejected from the Sun, with cool (dark) and hot (bright; $\approx 1.5\,10^6$ degrees) material at opposite ends of the long, nearly vertical structure. (*b*) A 1600 Å image represents plasma of about 10^5 degrees, where the heated-up filament material (white arrow). The brightest features are flare ribbons (black arrows). (*c*) This 195 Å image shows a rapidly cooling arcade of flare loops in the late phase of the flare, when the flare ribbons at the footpoints of the flarc loops have decreased in brightness.

The expression can be reversed, i.e. the derivative of SXR radiation shows similar time profile of that of the HXR or microwave radiation.

There are hundreds of papers providing evidence for the flare model described above. Here I can only mention a few which I consider to be the most important.

Using SOHO/EIT and Yohkoh/SXT data, an exceptionally clear evidence for the existence of the reconnection inflow was discovered by Yokoyama *et al.* (2001). They observed an eruptive flare on the solar limb, which displayed a geometry and scenario (e.g. filament eruption, cusp, X-point) highly resembling to the 2D reconnection cartoon shown in Figure 4(b). Following the filament eruption, a clear plasma motion with v = 1.0-4.7 km s^{-1} was observed towards the reconnection region (X-point). The reconnection rate, which is defined as the ratio of the inflow speed to the estimated Alfvén speed, derived from this observation was M_A = 0.001-0.03, which is roughly consistent with Petchek's (1964) fast reconnection model.

Observations of outflow from the reconnection region provide another key evidence. Forbes and Acton (1996) showed how newly formed cusped flare loops shrink and relax into a roughly semi-circular shape due to magnetic tension. McKenzie and Hudson (1999) discovered supra-arcade down-flows in LDEs at speeds between 40 and 500 km s^{-1}, confirming the existence of such outflow, indicating a patchy and intermittent reconnection process. Asai *et al.* (2004) showed that the start of these downflows are associated with HXR emission and microwave bursts observed with RHESSI and the Nobeyama radioheliograph, respectively. In radio wavelengths double type III bursts (due to electron beams propagating with $v \approx 0.2 - 0.6c$) are frequently observed propagating upward and downward from a common source at $\approx 0.9 - 5 \times 10^5$ km height in the corona (Benz, 2008 and references therein).

Multi-wavelength observations of flare ribbons also support the picture arising from the CSHKP model. RHESSI observations provided evidence that in flares HXR and γ-ray footpoint sources are co-spacial with the bright Hα and UV flare ribbons (Benz 2008 and references therein) and co-spatial with the highest magnetic flux densities and highest magnetic reconnection rates along the ribbons (Asai *et al.* 2002). Czaykowska *et al.* (1999) and Harra *et al.* (2005) using SOHO/CDS EUV spectroheliograms found bright down-flowing plasma to coincide with the ends of flare loops. The dimmer plasma on the outer side of the ribbons showed strong blue-shifts (upflows). Since the outer edge

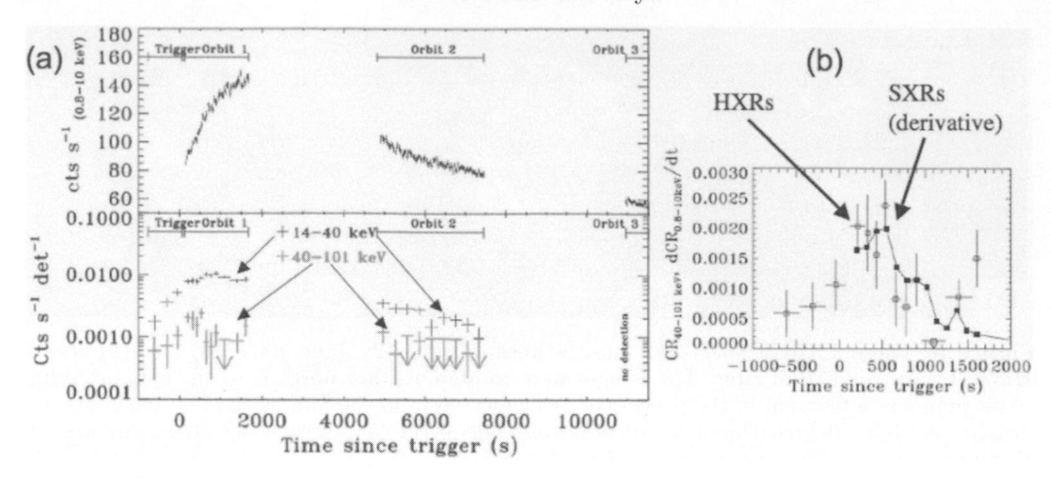

Figure 6. Swift observations of a stellar flare from II Peg on 16 December 2005 from the trigger at 11:21:52 UT from Osten *et al.* (2007). (*a*) Light curves from XRT, 0.8-10 keV (top) and two HXR energy bands from the BAT (14-40 and 40-101 keV; bottom). (*b*) Plot of 40-101 keV BAT light curve (open squares with error bars), along with scaled derivative of the XRT light curve for the first 2500 s of the flare. The correlation for roughly 1000 s after the trigger is consistent with the *Neupert effect.*

of flare ribbons are mapping to more recently reconnected loops than the inner edge, evaporation (blueshift) was observed at the outer edges, while downflows in the cooling loops mapped along the inner edge of flare ribbons.

LDEs are invariably associated with coronal mass ejections, of which the erupting flux rope containing the filament forms the core. Coronal mass ejections carry away typically 10^{15} g of solar material embedded in $10^{20} - 10^{22}$ Mx magnetic flux. Due to fast-decreasing pressure with height in the solar corona, CMEs expand. As suggested by Attrill *et al.* (2007), their magnetic fields meet with non-parallel fields of surrounding magnetic structures, current sheets form and dynamic reconnections take place. As a consequence, a large part of the Sun become CME constituent, supplying mass to the CME (van Driel-Gesztelyi *et al.* 2008 and references therein). Similar reconnection mechanism may lead to small flares in the wake of big stellar flares (Kővári *et al.* 2007).

2.6. *Solar-paradigm stellar flare*

Though stellar flares can release energy six orders of magnitude higher than that of a large solar flare, the energy release mechanism may not be very different of their solar equivalents. Osten *et al.* (2007) reported Swift observations of a large flare from the II Pegasi active binary system, which released a total energy of 10^{38} ergs (Figure 6). The SXR lightcurve of the flare (0.8-10 keV) closely resembles that of a solar LDE shown in Figure 4(a). The temperature of ≈ 80 MK was several factors higher than that of a big solar flare. The novelty of this observation is that the non-thermal component of a stellar flare was observed for the first time with Swift's high-energy channels (14-40 keV and 40-101 keV). The derivative of the SXR emission has a similar time-profile as the HXR emission, suggestive of the Neupert effect found for solar flares, implying that (coronal) SXRs mainly originate from plasma heated by accelerated particles impacting in denser (chromospheric) layers. For comparison, if II Pegasi was at a distance of 1 AU, the GOES classification of this flare would have been X4.4 $\times 10^6$, while the largest solar flare ever observed was about X30 (the exact number was impossible to determine due to saturation of the detector)!

3. Conclusions

Though all our observational proofs are indirect, reconnection seems to be at work at all scales in the solar atmosphere, releasing magnetic free energy mainly brought up by twisted flux emergence from the solar interior. To further understand details of the energy release process, we need high-resolution (vector) magnetic field measurements not only in the non-force-free photosphere, but in the chromosphere and the corona. Though simple 2D models are good starting points, we have to keep in mind that the events are in 3D. In order to understand observations we have to focus more on 3D modeling, embracing more complicated 3D reconnection models, e.g. the QSL concept.

For understanding stellar flares the solar paradigm is powerful, but it cannot cover all the physical conditions on active stars. On the Sun, flaring probability and importance increase with total magnetic flux, electric current and magnetic shear in the active region (e.g. Leka and Barnes, 2007) and with the flux in the vicinity of its high-gradient IL (Schrijver, 2007). On active stars, where large magnetic filling factors have been found (up to $\approx$80%; Berdyugina, 2005), active region flux may be two orders of magnitude higher than their solar counterparts. We have no information, however, on the non-potentiality of stellar magnetic fields and probably we underestimate their complexity. Flare energies exceed that of the solar flares by five-six orders of magnitude, therefore stellar flares in spite of sharing at least some of the essential physics with solar flares, cannot be regarded as their simple scaled-up versions. What makes the difference? The answer may differ for different stars.

Acknowledgements

The author thanks the TRACE Team for the data and acknowledges Hungarian government grant OTKA T048961.

References

Asai, A., Masuda, S., Yokoyama, T., Shimojo, M., Isobe, H., *et al.* 2002, *ApJ* 578, L91
Asai, A., Yokoyama, T., Shimojo, M., & Shibata, K. 2004, *ApJ* 605, L77
Aschwanden, M. J., Tarbell, T. D., Nightingale, R. W., *et al.* 2000, *ApJ* 535, 1047
Attrill, G. D. R., Harra, L. K., van Driel-Gesztelyi, L., & Démoulin, P. 2007, *ApJ* 656, L101
Audard, M., Güdel, M., & Guinan, E. F. 1999, *ApJ* 513, L53
Aulanier, G., Pariat, E., Démoulin, P., & DeVore, C. R. 2006, *Solar Phys.* 238, 347
Aulanier, G., Golub, L., DeLuca, E. E., Cirtain, J. W., Kano, R., *et al.* 2007, *Science* 318, 1588
Benz, A. O. 2008, *Living Rev. Solar Phys.* 5, 1
Berdyugina, S. V. 2005, *Living Rev. Solar Phys.* 2, 8
Carmichael, H. 1964, in W. N. Hess (ed.): *Physics of Solar flares*, NASA SP-50, 451
Czaykowska, A., De Pontieu, B., Alexander, D., & Rank, G. 1999, *ApJ* 521, L. 75.
Démoulin, P. 2007a, *Adv. Space Res.* 39, 1674
Démoulin, P. 2007b, *Adv. Space Res.* 39, 1367
Démoulin, P. & Berger, M. A. 2003, *Solar Phys.* 215, 203
Démoulin, P. & Pariat, E. 2008, *Adv. Space Res.*, submitted.
Démoulin, P., Priest, E. R., & Lonie, D. P. 1996, *JGR* 101, A4, 7631
Fan, Y. 2008, *ApJ* 676, 680
Forbes, T. G. 2000, *JGR* 105, A10, 23153
Forbes, T. G. & Acton, L. W. 1996, *ApJ* 459, 330
Güdel, M. 2007, *Living Rev. Solar Phys.* 4, 3
Harra, L. K., Démoulin, P., Mandrini, C. H., Matthews, S. A., van Driel-Gesztelyi, L., Culhane, J. L., & Fletcher, L. 2005, *A&A* 438, 1099
Hirayama, T. 1974, *Solar Phys.* 34, 323
Hudson, H. S. 1991, *Solar Phys.* 133, 357
Kopp, R. A. & Pneuman, G. W. 1976, *Solar Phys.* 50, 85

Kővári, Zs., Vilardell, F., Ribas, I., Vida, K., van Driel-Gesztelyi, L., Jordi, C., & Oláh, K. 2007, *AN* 328, 904
Leka, K. D. & Barnes, G. 2007, *ApJ* 656, 1173
Leka, K. D., Canfield, R. C., McClymont, A. N., & van Driel-Gesztelyi, L. 1996, *ApJ* 462, 547
Longcope, D. W., Linton, M. G., Pevtsov, A. A., Fisher, G. H., & Klapper, I. 1999, in M. R. Brown, R. C. Canfield, A. A. Pevtsov (eds.): *Magnetic helicity in Space and Laboratory Plasmas*, (Geophys. Monograph. 111, Washington D.C.: AGU), 93
López-Fuentes, M., Démoulin, P., Mandrini, C.H., & van Driel-Gesztelyi, L. 2000, *ApJ* 544, 540
Magara, T. 2004, *ApJ* 605, 480
Manchester, W., IV 2007, *ApJ* 666, 532
Manchester, W., IV, Gombosi, T., DeZeeuw, D., & Fan, Y. 2004, *ApJ* 610, 588
McKenzie, D. E. & Hudson, H. S. 1999, *ApJ* 519, 93
Meunier, N. 2005, *A&A* 443, 309.
Moreno-Insertis, F. & Emonet, T. *ApJ* 472, L53
Osten, R. A., Drake, S., Tueller, J., Cummings, J., Perri, M., Moretti, A., & Covino, S. 2007, *ApJ* 654, 1052
Pariat, E., Aulanier, G., Schmieder, B., *et al.* 2004, *ApJ* 614, 1099
Petchek, H. E. 1964, in *Physics of Solar flares*, ed. W. N. Hess, NASA SP-50, 425
Priest, E. R. & Forbes, T. 2000, *Magnetic Reconnection* Cambridge, UK: Cambridge University Press
Schrijver, C. J. 2007, *ApJ* 655, L117
Schrijver, C. J., DeRosa, M. L., Metcalf, T., Barnes, G., Lites, B. *et al.* 2008, *ApJ* 675, 1637
Schüssler, M. 1979, *A&A* 71, 79
Shibata, K. 1998, *Astrophys. Space Sci.* 264, 129
Sobotka, M., Vázquez, M., Bonet, J. A., Hanslmeier, A., & Hirzberger, J. 1999, *ApJ* 511, 436
Sturrock, P. A. 1966, *Nature* 211, 695
Švanda, M., Kosovichev, A. G., & Zhao, J. 2008, *ApJ* 680, L161
van Driel-Gesztelyi, L., Attrill, G. D. R., Démoulin, P., Mandrini, C. H., & Harra, L. K. 2008, *Ann. Geophys.* 26, 3077
Yokoyama, T., Akita, K., Morimoto, T., Inoue, K., & Newmark, J. 2001, *ApJ* 546, L69
Zhao, J. & Kosovichev, A. G. 2004, *ApJ* 603, 776.

Discussion

STRASSMEIER: You mentioned that no magnetic reconnection has ever be seen and that all evidence is indirect. Is this also the case for the most recent TRACE 171 Å movies?

VAN DRIEL-GESZTELYI: Only *consequences* of magnetic reconnection or changes due to reconnection are seen even in the highest-resolution observations. The reconnection region is small and it has low emission measure being difficult to observe. However, long current sheets where reconnection can happen have been observed in SOHO/UVCS and more recently in *Hinode*/XRT data.

ZINNECKER: My question focuses on young solar-mass stars before the Main Sequence, the T Tauri stars. We know that they have much larger X-ray luminosity than Main Sequence stars, due to gigantic loops filled with plasma. They also have enhanced flare activity. Now, what is the reason for such increased stellar activity, what is different in their magnetic field structure?

VAN DRIEL-GESZTELYI: Though the majority of models assume a simple bipolar field for the accreting T Tauri stars, in reality the field topology may well be much more complex. Complexity combined with high magnetic flux density would create long high-flux, high-gradient magnetic inversion lines, which have been shown to scale with flare activity on the Sun.

Cosmic Magnetic Fields:
From Planets, to Stars and Galaxies
Proceedings IAU Symposium No. 259, 2008
K.G. Strassmeier, A.G. Kosovichev & J.E. Beckman, eds.

doi:10.1017/S1743921309030452

X-ray jets and magnetic flux emergence in the Sun

Fernando Moreno-Insertis[1,2]

[1]Instituto de Astrofisica de Canarias, 38200 La Laguna (Tenerife), Spain
email: fmi@iac.es

[2]Dept of Astrophysics, Univ. La Laguna, Tenerife, Spain

Abstract. Magnetized plasma is emerging continually from the solar interior into the atmosphere. Magnetic flux emergence events and their consequences in the solar atmosphere are being observed with high space, time and spectral resolution by a large number of space missions in operation at present (e.g. SOHO, Hinode, Stereo, Rhessi). The collision of an emerging and a preexisting magnetic flux system in the solar atmosphere leads to the formation of current sheets and to field line reconnection. Reconnection under solar coronal conditions is an energetic event; for the field strengths, densities and speeds involved in the collision of emerging flux systems, the reconnection outflows lead to launching of high-speed (hundreds of km/s), high-temperature (10^7 K) plasma jets. Such jets are being observed with the X-Ray and EUV detectors of ongoing satellite missions. On the other hand, the spectacular increase in computational power in recent years permits to carry out three-dimensional numerical experiments of the time evolution of flux emerging systems and the launching of jets with a remarkable degree of detail.

In this review, observation and modeling of the solar X-Ray jets are discussed. A two-decade long computational effort to model the magnetic flux emergence events by different teams has led to numerical experiments which explain, even quantitatively, many of the observed features of the X-ray jets. The review points out that, although alternative mechanisms must be considered, flux emergence is a prime candidate to explain the launching of the solar jets.

Keywords. Sun: X-rays – Sun: flares – Sun: corona – Sun: atmosphere – magnetic fields – magnetohydrodynamics – methods: numerical

---

1. Introduction

A fundamental driver of the general dynamics of the solar atmosphere is the emergence of new magnetic flux from the solar interior. Episodes of magnetic flux emergence occur in the Sun on a bewildering variety of length- and timescales. Best known are those associated with large active regions or activity complexes which take place especially frequently toward the maximum of the solar activity cycle. When a large magnetic bipolar region or activity complex is formed on the surface, the simplest interpretation is that a bunch of magnetic flux ropes has emerged from the deep solar interior into the atmosphere (Zwaan 1978; Schrijver & Zwaan 2000). The global length scales of such magnetic complexes can be as large as 10^5 km, indicating that the associated flux ropes were formed in the main bulk of the convection zone (or at the bottom thereof) and have traversed it before appearing at the surface (Moreno-Insertis 1992, 1997). Going down from the largest scales, we find active regions in a continuum of sizes down to ephemeral active regions (Hagenaar *et al.* 2003, 2008). These appear at the center of meso- and supergranular cells; the two polarities of the region separate toward the boundaries of the cell, therefore reaching separations of order 10^4 km. At even smaller scales, recent high-resolution observations with the Hinode satellite have allowed to see flux emergence episodes in granules: Centeno *et al.* (2007) and Orozco Suárez *et al.* (2008) have clearly detected the emergence of magnetic tubes at the granular scale, with a length-scale of several hundreds of km. All of the foregoing indicates that turbulent convection in stars, in all its dominant scales, is at the root of the formation and fragmentation of the magnetic tubes that emerge at the surface. One should also expect flux emergence episodes on multiple scales to be a common occurrence in stars with magnetized convection in their envelope.

The emergence of magnetic flux from the interior causes important changes in the solar atmosphere, all the way from the photosphere to the corona. The corona is particularly affected since its structure and dynamics are dominated by the magnetic field. In the corona, two major types of change can occur following flux emergence. First, preexisting magnetic structures in equilibrium can become destabilized by the newly emerged magnetized plasma. Filament eruptions and coronal mass ejections may result from this kind of destabilization. Second, even when no large-scale destabilization takes place, *flaring* (possibly with ejection of collimated jets) can result from the flux emergence event (see Heyvaerts *et al.* 1977; Forbes & Priest 1984; Yokoyama & Shibata 1995; Shibata 1999): when the upcoming and preexisting magnetic systems get in contact, they press each other at a mutual interface. The magnetic field in general will be almost discontinuous across the interface: a concentrated current sheet is thus formed, which is a natural site for reconnection (Priest & Forbes 2000; Biskamp 2000). In the reconnection, field lines from either side of the interface are cut and new connections established, whereby the field lines from one side now become connected to those from the other side. This has profound implications for the structure of the corona. On the one hand, the general magnetic topology is modified. On the other, magnetic dissipation, which is proportional to the second derivatives of the field components, is important in the current sheet, given the large magnetic gradients. Magnetic energy is thus converted into heat at the reconnection site, and the plasma can become very hot, easily tens of 10^6 K. Further, according to the standard models of reconnection the plasma is ejected from the reconnection site with velocity of order the Alfven velocity, v_A. In the corona, v_A is high, say $v_A \gtrsim 100$ km s^{-1}. Hence, hot, X-ray emitting plasma is ejected with high speeds. Finally, the reconnection region is a natural site for the acceleration of microscopic particles (see references in Hannah & Fletcher 2006; Dalla & Browning 2008), which are then launched along the field lines both downward, toward the chromosphere, and also outward into more external regions of the corona or possibly into interplanetary space. Flux emergence is, therefore, a natural cause for the production of *flares* and their associated phenomena

The study of solar eruptions has received an important boost through the solar satellite missions with X-ray and EUV imaging and/or spectroscopic capabilities of the past ten years, like SOHO, YOHKOH, TRACE, RHESSI and Hinode. Of particular interest have been the observations of X-ray jets: these were discovered in the past decade using the soft X-ray telescope (SXT) onboard the Yohkoh satellite; a good summary and statistics of the properties of the jets deduced at the time was given by Shimojo *et al.* (1996). The Hinode mission, launched in September 2006, is monitoring these jets with a much higher spatial resolution in X-rays (close to 1 arcsec); it also has EUV spectroscopic capabilities through the EIS spectrometer. Particularly spectacular has been the observation of jets in coronal holes (see Cirtain *et al.* 2007), i.e., in the large coronal regions with field lines that extend out into the interplanetary medium. Fig. 1 shows an XRT image of a coronal hole in the polar regions of the Sun. In it, numerous sets of hot coronal loops are visible as bright features; also, a thin, elongated jet is visible toward the center of the figure. Detailed inspection reveals that many such jets have an inverted-Y shape and are flanked on one side by a compact set of hot loops (deep dark in the figure). A statistics of polar jets obtained in the first year of the mission contained interesting surprises: Savcheva

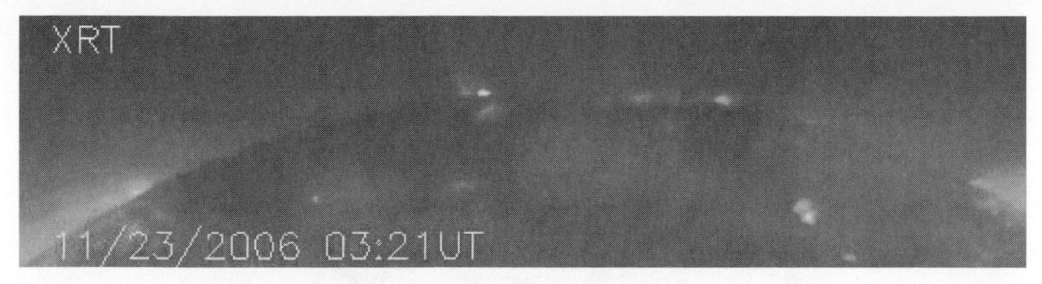

Figure 1. Jet observed in a polar coronal hole by the X-Ray Telescope (XRT) onboard Hinode. The figure shows an image taken with XRT only a few months after the launch of the mission; a jet features prominently at the center of the image. A number of hot coronal loop systems appear in the coronal hole. Courtesy: Hinode mission / Monica Bobra.

et al. (2007) showed that there was a much higher number of jet occurrences (60 jets per day) than expected on the basis of the earlier data. Those authors also showed that the most probable outward velocities of the jets were between about 100 and 300 km s^{-1}. The histogram for jet lifetimes peaked at some 10 min; that for their sizes peaked at 8 Mm (width) and 50 Mm (height).

Following the abundant observational prompts, there has been a large amount of theoretical work trying to understand the generation and evolution of X-ray jets in the Sun, especially based on computer modeling. In the following, a summary of the theoretical and modeling effort is provided, with special emphasis in the most recent developments.

2. The theoretical understanding of the launching of X-ray jets in the Sun: 2D models

In the past twenty years a large effort was devoted to the two-dimensional modeling of emerging magnetic flux regions in and out of the solar interior (as exemplified by, for instance, Forbes & Priest 1984; Shibata *et al.* 1989, 1992; Nozawa *et al.* 1992; Yokoyama & Shibata 1995, 1996; Moreno-Insertis & Emonet 1996; Emonet & Moreno-Insertis 1998; Fan *et al.* 1998; Krall *et al.* 1998; Magara 2001; Miyagoshi & Yokoyama 2004; Nishizuka *et al.* 2008). The papers by Forbes & Priest (1984) and Yokoyama & Shibata (1995, 1996) were particularly illuminating for understanding the details of the ejection of X ray jets following flux emergence. The latter authors started with a magnetized sheet below the surface which was unstable to the *Parker instability*: bending the field lines in the vertical direction with, say, a sinusoidal shape of sufficiently large wavelength leads to excess evacuation at the tops and to the development of a buoyancy instability. In their calculations, the corona had a preexisting field either in the horizontal direction or inclined with some arbitrary angle. The rising top of the sheet, upon entering the atmosphere, bumps against the ambient coronal field. A current sheet is formed and reconnection ensues (see Fig. 2); the reconnected field lines are ejected as part of alfvenic outflows. Those impinge upon the surrounding matter, the plasma in the outflows goes through a fast shock and is diverted and launched along the field lines both downward, toward the surface, as well as outward into other coronal regions. The latter constitutes a hot, high-speed jet, which is a promising candidate to explain the actual jets observed in the Sun.

The model of Yokoyama & Shibata had a number of limitations, in part forced by the computing hardware available at the time. Their model had a rather dense corona (this allows to ameliorate the problem of the small timestep imposed by the high Alfven speed);

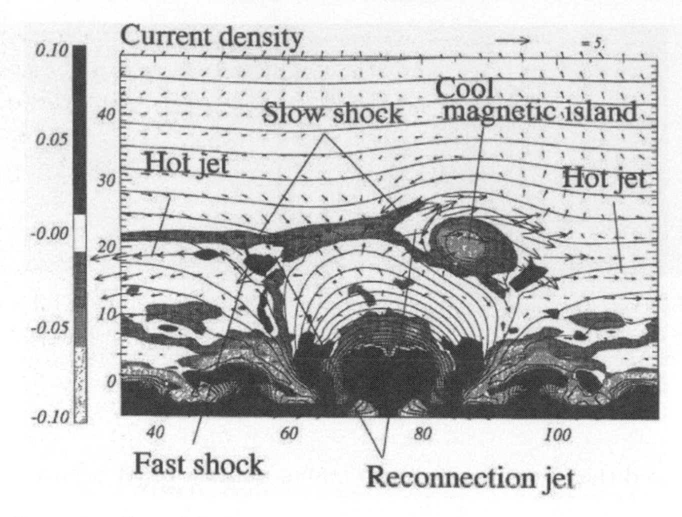

Figure 2. Two-dimensional model of reconnection and jet launching as a result of flux emergence from the solar interior. The emerged magnetized plasma collides with the overlying ambient coronal field and a current sheet is formed. Two hot jets are launched sideways along the coronal field lines. A thick plasmoid (cool island) is being formed in the current sheet. From Yokoyama & Shibata (1996)

also, the authors used a first-generation numerical scheme (Lax-Wendroff) which has the advantage of its simplicity, but can easily be fraught with high numerical diffusion. Yet, the results were an excellent first step toward a more complete modeling, possibly in three dimensions and with more sophisticated numerical tools.

3. Three-dimensional models

3.1. *General properties*

A new generation of more realistic, three-dimensional models of the launching of hot jets in the corona has been published in the past few years. Galsgaard *et al.* (2005, 2007), Archontis *et al.* (2005), Archontis & Török (2008) and Moreno-Insertis *et al.* (2008) have all provided detailed modeling of the reconnection phenomenon and the ensuing jet emission in 3D. Galsgaard *et al.* (2005, 2007) and Archontis *et al.* (2005) studied the results of reconnection between an emerging magnetic flux tube and a preexisting coronal field that points in the horizontal direction. These authors found the current sheet to have the shape of a concentrated, arch-like ribbon on top of the rising emerging plasma, as visible in Fig. 3. The jets are launched from the sides of the current sheet and the general geometry of sheet and jets is as shown in the right-hand panel. In the left panel, the color map corresponds to the plasma velocities; the jet is seen to have velocities of about 100 km s^{-1}. The temperature in the jet was typically several times 10^6 K. Galsgaard *et al.* (2007) studied the dependence of the results of the experiment on the mutual orientation of the magnetic field in the emerging and preexisting systems. The actual rate of rise of the current sheet is seen to be widely independent of the mutual orientation; reconnection is passive in these experiments and adapts to the dynamics imposed by the emergence process. Instead, the actual *reconnection rate* and the global change of connectivity between the ambient corona and the emerging system do depend sensitively on the mutual orientation. For initially almost antiparallel systems (the most favorable case for reconnection), some 30 min after emerging at the surface approximately 70% of

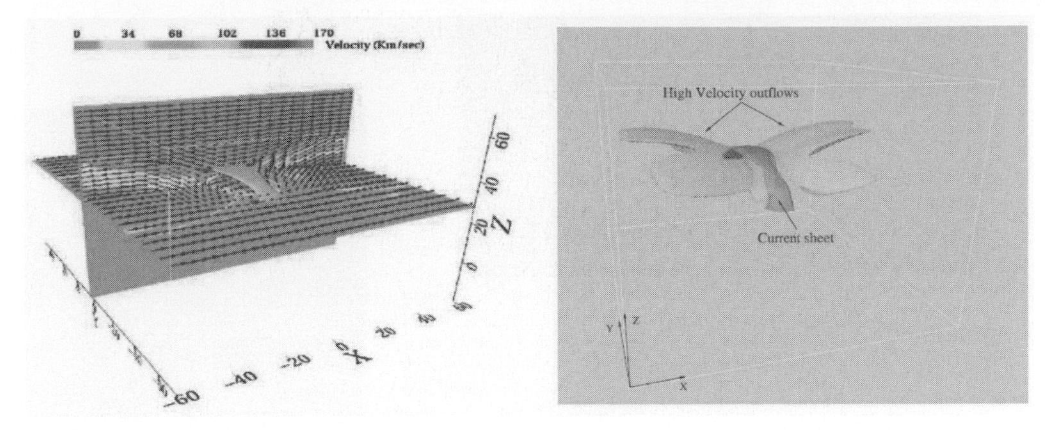

Figure 3. Structure of the current sheet and jet in a 3D experiment of flux emergence. From Archontis *et al.* (2005).

the initial flux in the upcoming system had changed connectivity and was now linked to the corona. In turn, in the least favorable case (magnetic field vectors close to parallel), very little reconnection occurs and the two systems remain virtually unconnected along the duration of the experiments.

Detailed conclusions concerning the actual reconnection process were provided by Archontis *et al.* (2005). The experiments revealed fully three-dimensional reconnection taking place across the current sheet. For instance, the reconnection occurred even though no prominent null points were apparent in the sheet. Also, in line with theoretical expectations (Pontin *et al.* 2005; Priest 2003; Priest *et al.* 2003) the reconnection in those experiments was shown to be a continuous process, instead of a one-off event, like in the classical 2D theory. In the 3D case, by pursuing a given field line in time one could see that it changed connectivity for as long as it was traversing the current sheet.

3.2. *Jets in coronal holes: theory and observation*

Of particular interest are the jets observed in coronal holes, for which there are now abundant, high-quality observational data, as explained in Sec. 1. The magnetic field in coronal holes links directly the Sun to the heliosphere, so the consequences of the violent reconnection and jet-launching in them can be felt directly in the interplanetary medium. A particular difficulty is the low density in the corona: 10^8 atoms cm^{-3} are expected there (Wilhelm 2006; Wilhelm *et al.* 2002), yielding typical Alfven velocities of order 1000 km s^{-1}, thus imposing very strict limits to the advance in time in the numerical experiments. In fact, the experiments described in the last section (3.1) were done with coronal density at least one or two orders of magnitude above that value (even higher densities were used for those of Sec. 2). The first 3D experiment of X-ray jets that included low coronal-hole densities was carried out by Moreno-Insertis *et al.* (2008) as part of a project aiming at comparing SOHO and Hinode data with 3D modeling. The results are summarized in the following.

Moreno-Insertis *et al.* (2008) start by using observational data of a coronal hole jet obtained with SOHO/MDI (magnetogram), Hinode/XRT (X-ray images) and Hinode/EIS (spectroscopic data). The jet appeared coinciding with a flux emergence episode at the surface. The 4-panel mosaic on the left in Fig. 4 shows the results of the analysis. The subpanel at the top left shows the XRT data in the form of a difference image (Open/Ti_poly). The jet appears as an intense (i.e., dark) feature around coordinates $(-25, -110)$. Next to it, slightly below and to the right, a roundish dark feature

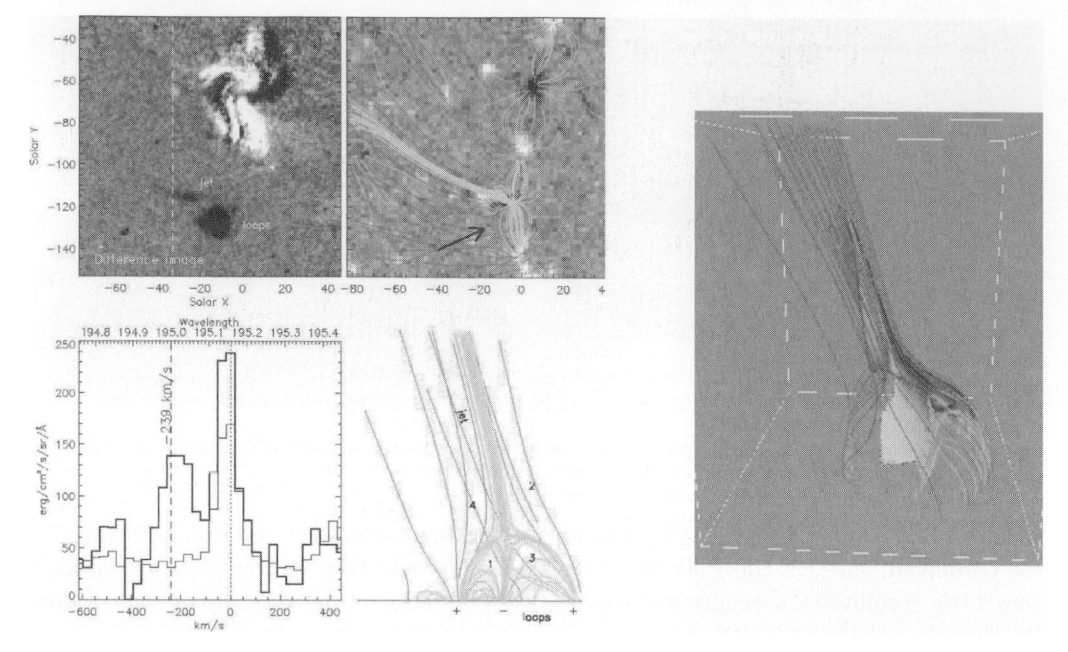

Figure 4. Left: Mosaic of observational results for a jet event occurred on 10th March 2007. The mosaic combines X-ray images from Hinode XRT, EUV spectroscopic data from Hinode EIS and magnetograms from SOHO MDI. Right: 3D perspective of a jet event obtained in the 3D numerical experiment of Moreno-Insertis *et al.* (2008)

corresponds to a set of hot coronal loops. In the subpanel immediately to the right, the result of carrying out a force-free magnetic field extrapolation on MDI magnetogram data of that same region is presented (the gray scale on the surface corresponds to the signed MDI magnetic flux measurement). A bunch of open fieldlines is seen stemming from the hot loop region and traversing the domain where the jet is located. The vertical perspective of the field line configuration, as seen from the direction indicated by the arrow, is given in the subpanel right below. Finally, the velocity values obtained on the basis of EIS spectra are given in the bottom-left subpanel. The prominent maximum at 240 km s^{-1} corresponds to the jet.

The numerical experiments in the paper by Moreno-Insertis *et al.* (2008) match the observational data remarkably well, both those just mentioned as well as the statistical analysis in the paper by Savcheva *et al.* (2007) discussed at the end of Sec. 1. In the experiments, the emergence of a twisted flux tube from below the photosphere leads to the formation of a current sheet, the start of reconnection, and the launching of a fast, hot jet. Fig. 5 contains a velocity (left) and temperature (right) map on a vertical cut around the time of peak jet activity. The jet clearly has the appearance of an *anemone* jet, in the notation of Yokoyama & Shibata (1996). Maximum temperatures and velocities are reached at the reconnection site itself (400 km $s^{-1}$and $3\,10^7$ K, respectively). In the jet, 200 km $s^{-1}$and 10^7 K are reached.

The field line configuration is worth noticing: below the reconnection site, two dome-shaped volumes are visible. The left dome contains the emerging material whose magnetic field is being brought up to reconnect with the overlying coronal system. The right dome contains hot, reconnected coronal loops. This configuration is strongly reminiscent of the field line extrapolation from observed surface data of Fig. 4 (bottom-right panel in the mosaic). To facilitate the comparison, Fig. 4 (right) contains a 3D view of the field lines

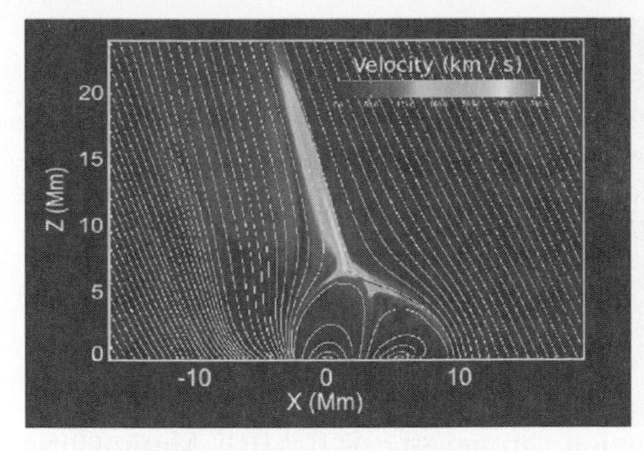

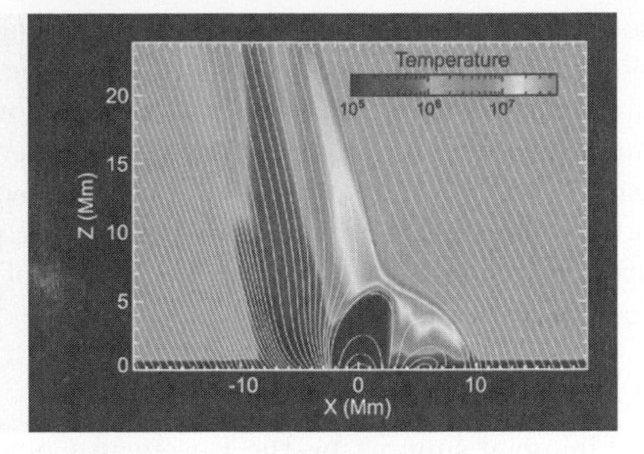

Figure 5. Velocity and temperature maps in a vertical cut of the three-dimensional experiment of Moreno-Insertis *et al.* (2008) around the time of peak jet activity

around the current-sheet. An isosurface of the current intensity is given in light blue as well as an isosurface of the temperature (pink) for $T = 6.5\ 10^6$ K. The similarity in the geometry and topology between the extrapolation and the simulation is apparent. A further result is the *drift* of the jet seen in the experiments: if Fig. 5 were a movie, we would see the position of the jet shift toward the right as time advances. This is associated with the gradual loss of plasma and magnetic flux from the emerged dome and the corresponding growth of the reconnected-loop dome. As a final note, the values of size, duration, ejection velocity and drift velocity of the jet in this experiment fall well within the ranges of variation of the jet parameters published by Savcheva *et al.* (2007).

4. Discussion and outlook

The study of X-ray jets, both observationally and theoretically, is undergoing rapid progress at present. The impressive series of solar space missions of the past 10 years have revealed that fast, hot jets are highly frequent in the Sun, especially in coronal holes, and have permitted to obtain a good idea of their main features. In this review an overview of results from jet models that occur as a consequence of emergence of magnetic regions from the solar interior has been provided. The most recent models include comparison with observational results obtained using X-ray, EUV and magnetogram data. An excellent match is obtained both in terms of field geometry and topology of the event as well as concerning values of physical quantities of the jet (duration, size, velocity, temperature, density and drift motion).

Additionally to the emergence of new magnetic flux, other mechanisms have been proposed that can cause the launching of jets like those observed by Hinode (Pariat *et al.* 2009; Patsourakos *et al.* 2008; Schmieder *et al.* 2008). Pariat *et al.* (2009) focus their attention on the helical structure and untwisting exhibited by a fraction of the observed jets. In their model, free energy is built up by twisting an initial dipole configuration that has an axisymmetric null-point + spine topology. If perfectly axisymmetric, reconnection would not be allowed in such a configuration. When finally the symmetry is broken, the energy is suddenly released, a burst is produced and a jet with obvious helical features is launched. The explanation of the twisting motions of some of the observed jets is a nice feature of this model. The match to the observed features in the jets perhaps needs to be improved, to reach the good level obtained by the flux emergence models.

Various important aspects of the jet phenomenon remain to be studied. A few of the 2D models mentioned in Sec. 2 took into account heat conduction. This is an important ingredient for the dynamics of any coronal phenomenon and, specifically, to study the so-called *evaporation* resulting from dumping thermal energy originating in the reconnection site down onto the dense chromospheric heights. To our knowledge, no 3D jet models up to date have included thermal conduction or optically thin cooling. A still larger step must be taken to properly include the thermodynamics of the low-atmospheric layers: models with proper treatment of the interaction between plasma and radiation field are needed for that. Fortunately, radiation-magnetohydrocodes dealing with all these aspects are already available, so a new generation of still more comprehensive jet models is likely to appear in the not too distant future.

Acknowledgements

Financial support by the European Commission through the SOLAIRE Marie Curie Network (MTRN-CT-2006-035484) and by the Spanish Ministry of Education through projects AYA2007-66502 and CSD2007-00050 is acknowledged. The computer resources, technical expertise and assistance provided by the MareNostrum (BSC/CNS, Spain) and LaPalma (IAC/RES, Spain) supercomputer installations are also acknowledged with thanks. Thanks are also due to the *HINODE* mission for their open policy regarding the use of figures. *HINODE* is a Japanese mission developed and launched by ISAS/JAXA, with NAOJ, NASA and STFC as international partners. It is operated by these agencies in co-operation with ESA and the NSC (Norway). The author is grateful to Drs Yokoyama and Shibata for permission to reproduce a figure from their 1996 article.

References

Arber, T. D., Haynes, M., & Leake, J. E. 2007, *ApJ* 666, 541
Archontis, V., Galsgaard, K., Moreno-Insertis, F., & Hood, A. 2006, *ApJ (Letters)* 645, L161
Archontis, V., Moreno-Insertis, F., Galsgaard, K., & Hood, A. 2005, *ApJ* 635, 1299
Archontis, V. & Török, T. 2008, *A&A* 492, L35
Biskamp, D. 2000, Magnetic Reconnection in Plasmas (Cambridge: Cambridge U. P.)
Centeno, R., Socas-Navarro, H., Lites, B., Kubo, M., Frank, Z., Shine, R., Tarbell, T., Title, A., Ichimoto, K., Tsuneta, S., Katsukawa, Y., Suematsu, Y., Shimizu, T., & Nagata, S. 2007, *ApJ (Letters)* 666, L137
Cirtain, J. W., Golub, L., Lundquist, L., van Ballegooijen, A., Savcheva, A., Shimojo, M., DeLuca, E., Tsuneta, S., Sakao, T., Reeves, K., Weber, M., Kano, R., Narukage, N., & Shibasaki, K. 2007, *Science* 318, 1580
Dalla, S. & Browning, P. K. 2008, *A&A* 491, 289
Emonet, T. & Moreno-Insertis, F. 1998, *ApJ* 492, 804
Fan, Y., Zweibel, E. G., & Lantz, S. R. 1998, *ApJ* 493, 480
Forbes, T. G. & Priest, E. R. 1984, *Solar Phys.* 94, 315
Galsgaard, K., Archontis, V., Moreno-Insertis, F., & Hood, A. 2007, *ApJ* 666, 516
Galsgaard, K., Moreno-Insertis, F., Archontis, V., & Hood, A. 2005, *ApJ* 618, 153
Hagenaar, H., DeRosa, M., & Schrijver, C. 2008, *ApJ* 678, 541
Hagenaar, H., Schrijver, C., & Title, A. 2003, *ApJ* 584, 1007
Hannah, I. G. & Fletcher, L. 2006, *Solar Phys.* 236, 59
Heyvaerts, J., Priest, E. R., & Rust, D. M. 1977, *ApJ* 216, 123
Krall, J., Chen, J., Santoro, R., Spicer, D. S., Zalesak, S. T., & Cargill, P. J. 1998, *ApJ* 500, 992
Magara, T. 2001, *ApJ* 549, 608
Martínez-Sykora, J., Hansteen, V., & Carlsson, M. 2008, *ApJ* 679, 871
Miyagoshi, T. & Yokoyama, T. 2004, *ApJ* 614, 1042
Moreno-Insertis, F. 1992, in Thomas, J. & Weiss, N. (eds.), *Sunspots, Theory and Observations*, Kluwer
Moreno-Insertis, F. 1997, *Mem Soc Astr It* 68, 429
Moreno-Insertis, F. & Emonet, T. 1996, *ApJ* 472, L53

Moreno-Insertis, F., Galsgaard, K., & Ugarte-Urra, I. 2008, *ApJ (Letters)* 673, L211
Nishizuka, N., Shimizu, M., Nakamura, T., Otsuji, K., Okamoto, T. J., Katsukawa, Y., & Shibata, K. 2008, *ApJ (Letters)* 683, L83
Nozawa, S., Shibata, K., Matsumoto, R., Sterling, A. C., Tajima, T., Uchida, Y., Ferrari, A., & Rosner, R. 1992, *ApJS* 78, 267
Orozco Suárez, D., Bellot Rubio, L. R., del Toro Iniesta, J. C., & Tsuneta, S. 2008, *A&A* 481, L33
Pariat, E., Antiochos, S. K., & DeVore, C. R. 2009, *ApJ* 691, 61
Patsourakos, S., Pariat, E., Vourlidas, A., Antiochos, S. K., & Wuelser, J. P. 2008, *ApJ (Letters)* 680, L73
Pontin, D. I., Galsgaard, K., Hornig, G., & Priest, E. R. 2005, *Phys Plasmas* 12, 052307
Priest, E. R. 2003, *Adv. Sp. Res.* 32, 1021
Priest, E. R. & Forbes, T. 2000, Magnetic Reconnection (Cambridge: Cambridge University Press)
Priest, E. R., Hornig, G., & Pontin, D. I. 2003, *Journal of Geophysical Research (Space Physics)* 108, 6
Savcheva, A., Cirtain, J., Deluca, E. E., Lundquist, L. L., Golub, L., Weber, M., Shimojo, M., Shibasaki, K., Sakao, T., Narukage, N., Tsuneta, S., & Kano, R. 2007, *PASJ* 59, 771
Schmieder, B., Török, T., & Aulanier, G. 2008, in *Exploring the solar system and the universe*, AIP Conference Proceedings, Vol. 1043, 260
Schrijver, C. J. & Zwaan, C. 2000, *Solar and Stellar Magnetic Activity*, Cambridge University Press
Shibata, K. 1999, *Ap&SS* 264, 129
Shibata, K., Nozawa, S., & Matsumoto, R. 1992, *PASJ* 44, 256
Shibata, K., Tajima, T., Matsumoto, R., Horiuchi, T., Hanawa, T., Rosner, R., & Uchida, Y. 1989, *ApJ* 338, 471
Shimojo, M., Hashimoto, S., Shibata, K., Hirayama, T., Hudson, H. S., & Acton, L. W. 1996, *PASJ* 48, 123
Tortosa-Andreu, A. & Moreno-Insertis, F. 2009, *A&A* , in preparation
Wilhelm, K. 2006, *A&A* 455, 697
Wilhelm, K., Dammasch, I. E., & Hassler, D. M. 2002, *Ap&SS* 282, 189
Yokoyama, T. & Shibata, K. 1995, *Nature* 375, 42
Yokoyama, T. & Shibata, K. 1996, *PASJ* 48, 353
Zwaan, C. 1978, *Solar Phys.* 60, 213

Discussion

JARDINE: Is the horizontal drift rate of the jets seen in coronal holes determined by the rate at which the reconnection site moves, rather than the velocity of the footpoints?

MORENO-INSERTIS: Yes, that is indeed the case. In some sense the footpoints play a passive role in the experiment and do not *move* much: as it changes connectivity, a given footpoint stops being part of the emerged *dome* and goes over either to be part of the set of hot reconnected loops or to be the root of one of the open field lines along which the jet hurries away. In doing so, the footpoint does not physically move. The reconnection site, on the other hand, is moving and changing shape as part of the process, and this causes the horizontal drift of the jet.

KHODACHENKO: As far as I can see, you use an assumption of a fully ionized hydrogen plasma in your numerical simulations. This is probably a good approximation in the corona, but in the chromosphere and especially in photosphere, solar plasma is known to be partially ionized. In photosphere $n_n/n_i \sim 10^4$, which completely changes the physics of magnetic field dynamics in the low solar atmosphere. Electric current dissipation rate is also thousands of times higher in the partially ionized plasmas, due to ion-neutral collisions. Could you somehow comment the validity of your model in the low solar atmosphere?

MORENO-INSERTIS: Complete ionization is not the only strong simplification of this type of models for the low atmosphere. All models discussed in this review (Sec. 2 and 3) disregard the interaction of the plasma with the radiation field. Like the partial ionization problem you correctly point out, the radiative cooling/heating may importantly affect the rising plasma especially in the photosphere. There are magnetic flux emergence models which already take into account either aspect like those by Arber *et al.* (2007), for the partial ionization, or Martínez-Sykora *et al.* (2008) and Tortosa-Andreu & Moreno-Insertis (2009) for the radiative aspects. However, to my knowledge no specific X-ray jet experiments have been carried out including those aspects so far.

DE GOUVEIA DAL PINO: You have detected in the reconnection simulations continuous jets. What would the changes be in the initial conditions in order to see plasmons (or CMEs), or more intermittent ejections, since the same sort of reconnection phenomenon is expected to produce them?

MORENO-INSERTIS: In our previous jet experiments (see Archontis *et al.* 2005, 2006), we already saw the production of plasmoids which were ejected out of the current sheet toward the first part of the jet production process. We are investigating the possibility of intermittent behavior in the jets, but I cannot provide any conclusive reply on this at present.

KOUTCHMY: Two short questions: (a) Is gravity taken into account in your numerical simulation which means buoyancy is a driving force in the scenario proposed? (b) What about the interpretation of the apparent motions in transverse direction of the jet part(s)? Do field lines also move like in the case of an Alfven wave or a kink wave?

MORENO-INSERTIS: Gravity is certainly taken into account in our model: the emergence of the magnetized plasma is driven by the buoyancy force. The transverse motion of the jet is unlikely to be a phase motion associated with a wave. It rather seems to be due to the displacement and deformation of the reconnection site as more and more emerged flux is converted into hot coronal loops

OTMIANOWSKA-MAZUR: The initial state. Do you apply the forced reconnection?

MORENO-INSERTIS: The reconnection appearing in the experiment is forced in the sense that the emerging plasma is pressing hard against the preexisting coronal field. That it does so is solely due to the magnetic forces that try to make the emerging material rise and expand in the corona and encounter some resistance in the ambient coronal medium. On a different score, the reconnection is facilitated by the hyperdiffusive resistivity we used in the model: it concentrates the diffusivity wherever there are large gradients and keeps the rest of the domain at a low-resistivity level, thus allowing for much larger Reynolds numbers than otherwise possible.

Cosmic Magnetic Fields:
From Planets, to Stars and Galaxies
Proceedings IAU Symposium No. 259, 2008
K.G. Strassmeier, A.G. Kosovichev & J.E. Beckman, eds.

doi:10.1017/S1743921309030464

The second solar spectrum and the hidden magnetism

Jan O. Stenflo

Institute of Astronomy, ETH Zurich, HIT J 23.6, CH-8093 Zurich
email: stenflo@astro.phys.ethz.ch

Abstract. Applications of the Hanle effect have revealed the existence of vast amounts of "hidden" magnetic flux in the solar photosphere, which remains invisible to the Zeeman effect due to cancellations inside each spatial resolution element of the opposite-polarity contributions from this small-scale, tangled field. The Hanle effect is a coherency phenomenon that represents the magnetic modification of the linearly polarized spectrum of the Sun that is formed by coherent scattering processes. This so-called "Second Solar Spectrum" is as richly structured as the ordinary intensity spectrum, but the spectral structures look completely different and have different physical origins. One of the new diagnostic uses of this novel spectrum is to explore the magnetic field in previously inaccessible parameter domains. The earlier view that most of the magnetic flux in the photosphere is in the form of intermittent kG flux tubes with tiny filling factors has thereby been shattered. The whole photospheric volume instead appears to be seething with intermediately strong fields, of order 100 G, of significance for the overall energy balance of the solar atmosphere. According to the new paradigm the field behaves like a fractal with a high degree of self-similarity between the different scales. The magnetic structuring is expected to continue down to the 10 m scale, 4 orders of magnitude below the current spatial resolution limit.

Keywords. Sun: magnetic fields – Sun: photosphere – scattering – turbulence – atomic processes – line: formation – techniques: polarimetric

---

1. Why magnetic flux on the Sun is hidden

The impressive advances in angular resolution, from space with Hinode, from ground with the help of adaptive optics, allow us to explore ever smaller structures on the Sun. This brings fundamental insights into the basic astrophysical processes that govern the structuring, dynamics, and heating of stellar atmospheres. As however the structuring continues on scales that are far smaller than can be resolved in any foreseeable future even in solar observations, there is always a need for methods that allow us to extract information about the physics on scales that are too small to be resolved. This need has of course always been familiar to stellar physics, since stellar disks are normally unresolved (except for Doppler-Zeeman imaging of rapid rotators).

1.1. *The standard model*

Due to their finite angular resolution, all solar telescopes smear the true solar image with a smoothing window, the spatial resolution element. Solar magnetograms represent smoothed maps of the circular polarization produced by the longitudinal Zeeman effect. The early magnetograph recordings back in the 1960s showed that the apparent field strengths increased with the angular resolution of the instrument, so the question arose what the strength would be if we would have infinite resolution. With the introduction of the line-ratio technique in the early 1970s (Stenflo 1973) this question could be answered and it became clear that most (more than 90 %) of the magnetic flux on the quiet Sun seen in the magnetograms of that time (with a resolution of a few arcsec) came from intrinsically strong, 1-2 kG fields. Since the apparent field strengths in these line-ratio observations were of order 1-10 G, this finding implied that the kG flux elements were far smaller than the available resolution, so the interpretation had to be based on a parameterized model for the magnetic field. The simple 2-component model that was introduced was found to give consistent results not only for the line-ratio data (with various line combinations), but also for the rich sets of constraints provided by the FTS (Fourier Transform Spectrometer) circular-polarization spectra (Stenflo *et al.* 1984) and for the resolved Zeeman splittings in the near infrared, (cf. Rüedi *et al.* 1992) (with occasional extensions of the 2-component to a 3-component model).

The 2-component model assumed that we have one magnetic component with a certain filling factor, while the other component was assumed to be non-magnetic. The magnetic, kG component soon found its theoretical counterpart in MHD models of magnetic flux tubes (e.g. Spruit 1976). Including MHD constraints like self-consistent field expansion with height in the 2-component model, it became possible to construct semi-empirical flux-tube models with increasing levels of sophistication (cf. Solanki 1993).

It was however obvious already when the 2-component model was introduced nearly four decades ago that the concept of a "non-magnetic" component is non-physical and only introduced for the sake of mathematical and interpretational convenience. Due to the enormously high electrical conductivity of the turbulent photosphere it is physically inconceivable that anything can be "non-magnetic". What the term "non-magnetic" means in the diagnostic context is that this component does not contribute significantly to the net longitudinal Zeeman-effect polarization integrated over the resolution element.

The next challenge therefore became to find out what the intrinsic magnetic properties of the "non-magnetic" component are. Since the longitudinal Zeeman effect is "blind" to this component, another spectral signature had to be found, which could provide a glimpse of its elusive properties. Magnetic line broadening had insufficient sensitivity and could only set an upper limit to the field strength (Stenflo & Lindegren 1977). The breakthrough came with the application of the Hanle effect (cf. Sect. 2.1) to an interpretational model based on a random, turbulent field. This led to the discovery of the existence of "hidden" (with respect to the Zeeman effect), turbulent fields with strengths in the range 10-100 G (Stenflo 1982).

The combination of line-ratio and Hanle data led to a "standard model" or "paradigm" for solar magnetism, as expressed in Fig. 1. The flux tube component expands rapidly with height, to satisfy the requirement of pressure balance, until the field fills the coronal volume. This component provides the flux that is visible in magnetograms (the circular-polarization maps of the longitudinal Zeeman effect). The component between the flux tubes contains a weaker, chaotically tangled "turbulent" field. Since the opposite magnetic polarities of this tangled field are mixed on subresolution scales, the positive and negative contributions to the circular polarization cancel out, so that this field becomes "hidden", i.e., invisible to the Zeeman effect.

2. Second Solar Spectrum and the Hanle effect

In contrast to the Zeeman effect, the Hanle effect is a coherence phenomenon that only occurs in coherent scattering processes. Such scattering produces polarization also in the absence of any magnetic fields, a familiar example being the polarization of the blue sky by Rayleigh scattering on molecules. The Sun's spectrum is polarized by coherent scattering, but the degree of polarization is small due to the small degree of anisotropy of the incident radiation field for scattering processes inside the Sun's atmosphere. It

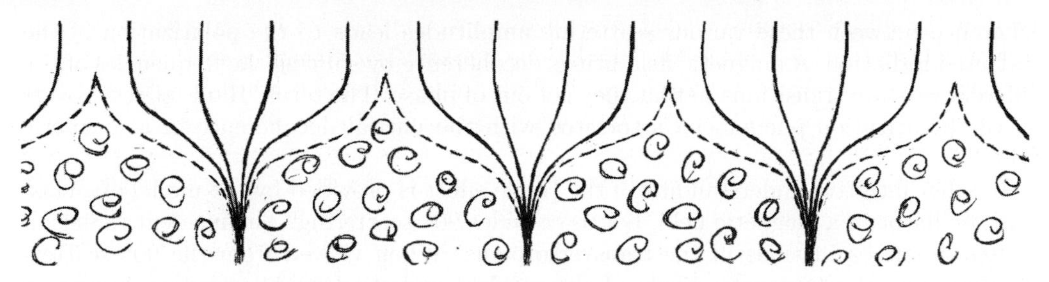

Figure 1. Standard model of quiet-sun magnetism (here illustrated for the case when the different flux tubes have the same polarity). This dualistic scenario describes the atmosphere in terms of two components, one representing the flux tubes, the other the tangled fields in between. With the Zeeman effect we see the flux tubes but not the turbulent field, with the Hanle effect the situation is the opposite. The dashed lines mark the canopy and interface between the two components.

was therefore only with the advent of highly sensitive polarimeters that this type of polarization could be fully revealed. The breakthrough came with the implementation of the ZIMPOL (Zurich Imaging Polarimeter) technology in 1994, which allowed imaging polarimetry with a precision of 10^{-5} in the degree of polarization (Povel 1995, 2001; Gandorfer *et al.* 2004). At this level of sensitivity everything is polarized, even in the absence of magnetic fields. It came as a big surprise, however, that the polarized spectrum was so richly structured, as richly as the ordinary intensity spectrum but without resembling it. It was as if a new spectral face of the Sun had been unveiled, and we had to start over to identify the various spectral structures and their physical origins. It was therefore natural to call this new and unfamiliar spectrum the "Second Solar Spectrum" (Ivanov 1991; Stenflo & Keller 1997). A spectral atlas of the Second Solar Spectrum from 3160 to 6995 Å has been published in three volumes (Gandorfer 2000, 2002, 2005).

Most of the fundamental physical processes that govern the polarized structures in the Second Solar Spectrum have by now been identified in terms of a multitude of previously unfamiliar phenomena, although we still have a long way to go for detailed quantitative modelling (cf. Stenflo 2004). Examples of identified physics are quantum-mechanical interference between atomic states of different total angular momentum quantum numbers (Stenflo 1980), hyperfine structure and isotope effects (Stenflo 1997), optical pumping that creates ground-state atomic polarization (Trujillo Bueno & Landi Degl' Innocenti 1997; Manso Sainz & Trujillo Bueno 2003, 2007), and molecular scattering (Stenflo & Keller 1996, 1997; Berdyugina, Stenflo, & Gandorfer 2002). Still there are polarization features that seem to be at odds with quantum mechanics as we know it, and which so far have eluded all attempts at a physical explanation, like the polarization peak observed at the cores of the D_1 lines of sodium and barium, lines that are supposed to be intrinsically unpolarizable (Stenflo 2008).

2.1. *Signatures of the Hanle effect*

The Second Solar Spectrum exists regardless of whether there are magnetic fields around or not. The shapes and amplitudes of the various structures are however modified by magnetic fields, and it is this modification that goes under the name "Hanle effect", discovered by Wilhelm Hanle in Göttingen in 1923 (Hanle 1924; Moruzzi & Strumia 1991). His discovery played a significant role in the early development of quantum mechanics, since it demonstrated the fundamental concept of linear superposition of quantum states. The Second Solar Spectrum is the astrophysical playground for the Hanle effect.

A scattering transition between different J states consists of a coherent superposition of the scattering transitions between all the possible m-state combinations. The phase coherence between these various scattering amplitudes leads to the polarization of the scattered radiation. A magnetic field brings decoherence by shifting the frequencies of the different m-state transitions so that they get out of phase. The term "Hanle effect" covers all the polarization phenomena associated with the partial decoherence in a magnetic field.

A rather intuitive understanding of the Hanle effect is provided by the model of a classical oscillator in a magnetic field. Let us consider 90° scattering. The incident radiation excites dipole oscillations in the transverse plane. When viewed from the 90° scattering direction, the transverse plane of the incident radiation projects out to become a line. Therefore, when projected on the transverse plane of the scattered radiation, the dipole oscillation is limited to the direction perpendicular to the scattering plane, implying that the emitted radiation becomes 100 % linearly polarized along that direction. If we introduce a magnetic field along the scattering direction, the dipole oscillation performs a Larmor precession around the magnetic field while being radiatively damped. The trajectory of the oscillation then forms a Rosette pattern as illustrated in Fig. 2. The spectral polarization properties of the scattered radiation are obtained by transforming the Rosette patterns to the Fourier domain, which is the domain used by quantum mechanics with its energy levels.

The form of the Rosette pattern depends on the relative magnitudes of the precession and damping rates. For weak magnetic fields the precession is slow, so the oscillation decays before much precession has happened, and the scattering polarization does not deviate much from the non-magnetic case. For strong magnetic fields the precession has time to randomize the pattern and make it more isotropic before the oscillator has decayed. The result is depolarization, reduction of the amount of scattering polarization. In the intermediate case when the precession and damping rates are similar in magnitude, we get partial depolarization combined with a net rotation of the plane of linear polarization. The Hanle effect manifests itself differently for other orientations of the magnetic field, but qualitatively the main two effects are depolarization and rotation of the plane of polarization.

Figure 2 also illustrates what Hanle polarization signatures really look like in the Sun's spectrum, and how they differ from the Zeeman effect signatures. Since the Hanle effect is a coherency phenomenon that only affects the aspect of the line formation process that is due to coherent scattering, the Hanle effect is prominent in lines for which coherent scattering plays a dominating role. In the illustrated example, recorded with ZIMPOL at NSO / Kitt Peak, the strong, resonant Ca I 4227 Å line represents such a case, while the surrounding blend lines do not show any significant scattering polarization but instead depolarize the Ca I line wings and the continuum polarization. While the 4227 Å line is subject to both the Hanle and Zeeman effects, the surrounding lines only show the magnetic signatures of the Zeeman effect.

The sensitivity range for the Hanle effect depends on the ratio between the Zeeman splitting and the damping width, while the polarization response of the Zeeman effect is related to the comparison of the Zeeman splitting with the spectral line width, which is mainly determined by the Doppler width. Since the damping width is smaller than the Doppler width by typically a factor of 30, the Hanle effect is sensitive to much weaker fields than the Zeeman effect.

2.2. *Resolved fields and unresolved mixed-polarity fields*

Dramatic variations of the scattering polarization along the spectrograph slit in both Stokes Q and U, as shown in Fig. 2, are only found in strong chromospheric lines. In contrast, the scattering polarization in photospheric lines like the often used Sr I 4607 Å

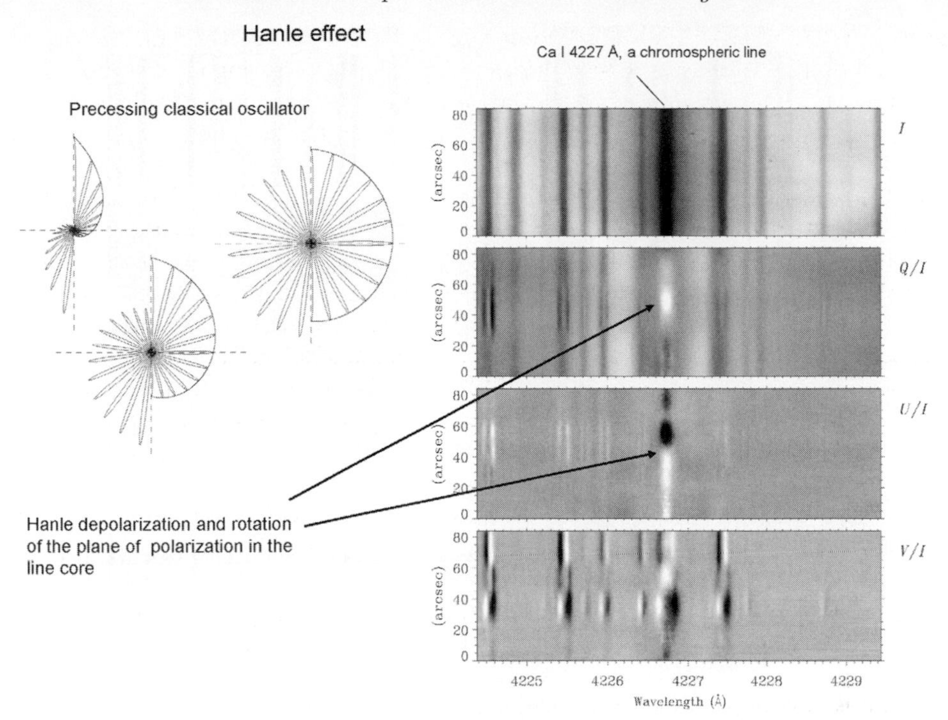

Figure 2. Left diagram: Rosette patterns formed by the trajectories of a classical dipole oscillator in a magnetic field along the line of sight, illustrating the Hanle depolarization and rotation effects. As the field strength increases, the pattern becomes more isotropic. Right diagram: Recording of the Stokes vector (the four parameters intensity I and the fractional polarizations Q/I, U/I, and V/I) with the spectrograph slit across a weakly magnetic region 20 arcsec inside and parallel to the west solar limb (at $\mu = 0.20$, where μ is the cosine of the heliocentric angle). The Hanle signatures are seen in Stokes Q and U in the core of the Ca I 4227 Å line, while the surrounding lines exhibit the characteristic signatures of the transverse Zeeman effect. In Stokes V all the lines show the anti-symmetric signatures of the longitudinal Zeeman effect.

line or molecular lines like the CN lines shown in Fig. 3, exhibits little if any spatial variations in Stokes Q, and rarely displays significant signatures of Hanle rotation in Stokes U. The reason is not that the chromosphere is more magnetically structured than the photosphere. On the contrary, detailed analysis reveals that the photosphere is full of magnetic fields in the sensitivity range for the Hanle effect, but that these fields are structured mainly on spatial scales much smaller than the angular resolution of current Hanle-effect observations (a few arcsec, due to the long integration times needed to reach the required polarimetric precision).

Let us consider the hypothetical case of a small-scale turbulent magnetic field, for which the many unresolved magnetic elements within each resolution element have a random, isotropic distribution of their field vectors. While such a field may contain a large amount of magnetic energy, it does not carry much net magnetic flux after averaging over the spatial resolution elements of our instrument. It would therefore be invisible in magnetograms, due to cancellations of the plus and minus contributions to the circular polarization from the opposite polarities. Such cancellation also occurs for the Hanle rotation signatures in Stokes U, since the Hanle rotation effect has the same symmetry between plus and minus as the longitudinal Zeeman effect. In contrast, the Hanle depolarization effect does not suffer from such cancellations, since the depolarization only

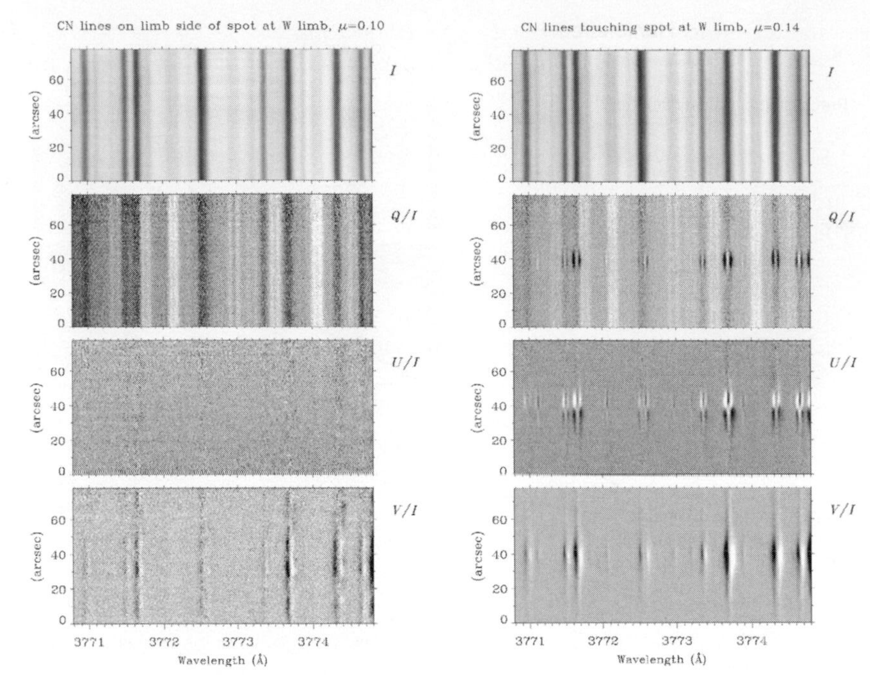

Figure 3. Example of the behavior of molecular lines in the Second Solar Spectrum. The bright, emission-like bands in Stokes Q/I are due to scattering polarization by the CN molecule. Note that there is no scattering polarization in U/I and no significant variation of Q/I along the slit, in contrast to the surrounding atomic lines, which exhibit the familiar signatures of the transverse and longitudinal Zeeman effect. The recording was made with ZIMPOL at NSO / Kitt Peak inside the west solar limb (Stenflo 2007).

has one sign (reduction of the scattering polarization amplitude), regardless of the sign of the field orientation. The "hidden" field can therefore be revealed if we can determine the amount of Hanle depolarization that it causes.

Detailed analysis of Hanle-effect observations in photospheric lines (cf. Sect. 3 below) show that we indeed have lots of Hanle depolarization indicating the presence of large amounts of photospheric flux that is not seen in magnetograms. The circumstance that the same observations show very little sign of any Hanle rotation signals in Stokes U, and little variation of Stokes Q along the slit, implies that the determined Hanle depolarization effect must be due to a spatially unresolved distribution of magnetic fields with magnetic elements that are much smaller than the spatial resolution used. If the tangled field were partially resolved, then the cancellation effects within each spatial resolution element would be incomplete, and we would expect to see net Hanle rotation effects and varying net depolarization effects along the slit. The observed absence of these effects is evidence for tangled field structuring on scales much smaller than a few arcsec.

3. Diagnostics of the hidden magnetic flux

To determine the amount of Hanle depolarization from a measured polarization amplitude we need to relate the observed polarization to the non-observed amplitude that we would have found in the absence of depolarizing magnetic fields. There are different ways to approach this problem: (1) Theoretically predict the expected non-magnetic scattering polarization amplitude by solving the polarized radiative transfer problem for a realistic model atmosphere. (2) Apply the *differential Hanle effect* (Stenflo, Keller & Gandorfer 1998; Manso Sainz, Landi Degl'Innocenti & Trujillo Bueno 2004; Berdyugina & Fluri 2004), namely to record the scattering polarization simultaneously in several spectral lines with different sensitivities to the Hanle effect. (3) From the observed statistical distribution of polarization amplitudes, let the upper envelope to the distribution represent the non-magnetic values, since magnetic fields will only reduce, not enhance the polarization (Bianda *et al.* 1998, 1999).

Once we have managed to determine the empirical value of the Hanle depolarization with one of the above-mentioned methods, we can use this depolarization to constrain the properties of the hidden magnetic field. This step needs an assumed interpretational model. Since we only have one observable (if a single spectral line is used), the Hanle depolarization, our model is not allowed to contain more than one free parameter. With combinations of simultaneously observed lines, more free parameters may be possible (Stenflo, Keller & Gandorfer 1998), which offers an avenue for future use of more sophisticated model constraints. With one free parameter the simplest and most natural choice has been the field strength of a randomly oriented field with an isotropic angular distribution, as we have mentioned above. It is however clear that the real turbulent field is far from being single-valued and should instead be governed by a continuous probability distribution function (PDF) that extends over a wide range of field strengths. The problem has been to characterize such an unknown PDF if we only have one free parameter at our disposal.

This problem is now being solved with guidance from empirical PDFs determined from magnetograms, representing the spatially resolved scales (Stenflo & Holzreuter 2002, 2003) and from theoretical PDFs determined from numerical simulations of magnetoconvection, representing scales that are not yet resolved (Nordlund & Stein 1990; Cattaneo 1999). The shapes of the empirical and theoretical PDFs are very similar to each other, although not identical. It is possible to represent the typical PDF by an analytical function that is characterized by one free scaling parameter that governs the stretching of the field-strength scale, and let this parameter be constrained by the requirement that the observed amount of Hanle depolarization should be reproduced.

The most elaborate effort with the highest degree of realism to interpret the available Hanle depolarization data has been based on 3-D radiative transfer with model atmospheres generated by numerical simulations of granular convection (Trujillo Bueno, Shchukina & Asensio Ramos 2004). The magnetic field was constrained through fits of available data for the Sr I 4607 Å line and C_2 molecular lines. A field model assuming a single-valued volume-filling turbulent distribution gave a field strength of 60 G, substantially higher than previous estimates, which implies a significant amount of magnetic energy contained in this hidden field. However, when a much more realistic, continuous functional shape of the PDF was used instead, similar to the PDF shapes obtained from numerical simulations of magnetoconvection, the Hanle depolarization constraint leads to still stronger fields (of order 100 G) with correspondingly higher average energy densities, so high that the tangled field might even dominate the energy balance of the solar atmosphere (Trujillo Bueno, Shchukina & Asensio Ramos 2004). Due to the model dependence, however, the question whether the turbulent field really dominates the energy balance remains controversial.

While the empirical values of the Hanle depolarization in the Sr I 4607 Å line indicate such large magnetic energy densities for the hidden flux, the molecular lines indicate much smaller values. A solution to this apparent contradiction has been found through 3-D modelling (based on numerical hydrodynamic simulations) of the spatial distribution of the molecular abundance, which shows that the molecular lines are preferentially

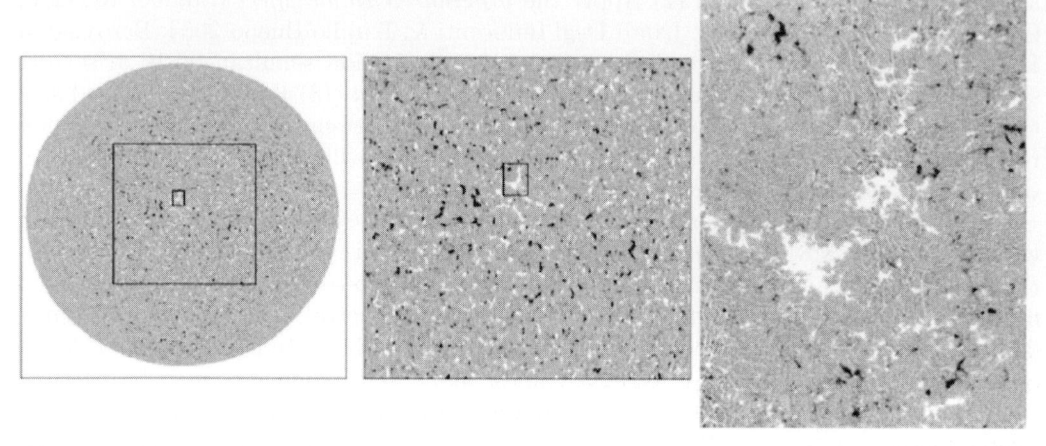

Figure 4. Zooming in on the magnetic pattern of the quiet Sun observed on 9 February 1996. The two left maps are from a Kitt Peak full-disk magnetogram, while the right, high-resolution magnetogram was recorded the same day at the Swedish La Palma Observatory (courtesy Göran Scharmer). The La Palma magnetogram covers an area that is only 0.35 % of the map next to it.

formed inside the hot granules and are largely absent in the intergranular lanes (Trujillo Bueno, Shchukina & Asensio Ramos 2004). A consistent interpretation of the observed difference in the Hanle depolarization response between the atomic and molecular lines then implies that the turbulent field is weak inside the hot granules, while the large Hanle depolarization seen in lines like Sr I 4607 Å predominantly comes from the intergranular lanes. If this is indeed the case, then the turbulent field in the intergranular lanes with a relatively small filling factor must be stronger than the field derived before for a filling factor of unity. The introduction of a filling factor in the Hanle interpretation thus leads to an increase in the average magnetic energy density and strengthens the argument for an important role of the hidden magnetic flux in the overall energy balance of the Sun's atmosphere.

4. Beyond the standard model

The Second Solar Spectrum became accessible to observations only through the development of sufficiently sensitive imaging Stokes polarimeters, in particular with the introduction of the ZIMPOL technology, which allows the two main noise sources, seeing noise and gain table noise, to be completely eliminated. Since the only remaining noise source is the fundamental photon noise of Poisson statistics, we need larger telescope apertures to improve the S/N ratio that we can reach with a certain temporal (integration time) and angular resolution. It is not possible to optimize spectral, spatial, temporal resolution, and polarimetric precision simultaneously, but major trade-offs between them are necessary, regardless of the size of the telescope (Stenflo 2001). The high polarimetric precision required for Hanle-effect explorations is not compatible with diffraction-limited resolution. Still the resolution of the solar granulation pattern in maps of the scattering-polarization and the Hanle effect is within reach, (cf. Trujillo Bueno & Shchukina 2007), but it still remains an unattained near-future challenge.

The previous dualistic magnetic-field paradigm or two-component "standard model" with a kG flux tube component with small filling factor, and a turbulent field component that is filling the remaining part of the volume, is now being replaced by a picture characterized by probability distribution functions (PDFs). While the strong-field tail of the distribution contains the "flux tubes" of the standard model, the bulk of the PDF corresponds to the "turbulent field" component. Instead of using two different interpretational models for the Zeeman and Hanle effects when diagnosing the spatially unresolved domain, it appears logical to apply a single, unified interpretational model based on PDFs in both cases. This has yet to be done in a consistent way.

This task is complicated by various factors. Thus the different Hanle behavior of atomic and molecular lines suggests that the PDFs are different between the inside of the granules and the intergranular lanes. To clarify this situation we need to resolve the solar granulation in Hanle effect observations. Another very important problem is that we know much less about the PDF for the angular distribution of field vectors than we know about the PDF for the vertical field strengths. Figure 4 illustrates the fractal appearance as we zoom in on the quiet-sun magnetic pattern at the center of the solar disk. The pattern looks similar on all scales, with a coexistence of weak and strong fields over a wide dynamic range. The PDF for the vertical field-strength component is nearly scale invariant and can be well represented by a Voigt function with a narrow Gaussian core and "damping wings" extending to kG values (Stenflo & Holzreuter 2002, 2003). A fractal dimension of 1.4 has been found from both observations and numerical simulations (Janssen, Vögler & Kneer 2003). The simulations indicate that this fractal behavior extends well into the spatially unresolved domain.

The richly structured Second Solar Spectrum with its many novel magnetic-field effects opens a new window to explorations of previously inaccessible aspects of the Sun. With the vast amounts of hidden magnetic energy in the spatially unresolved magneto-convective spectrum, the determination of the properties of the hidden field has become a central task for contemporary solar physics.

References

Berdyugina, S. V. & Fluri, D. M. 2004, *A&A* 417, 775

Berdyugina, S. V., Stenflo, J. O., & Gandorfer, A. 2002, *A&A* 388, 1062

Bianda, M., Solanki, S. K., & Stenflo, J. O. 1998, *A&A* 331, 760

Bianda, M., Stenflo, J. O., & Solanki, S. K. 1999, *A&A* 350, 1060

Cattaneo, F. 1999, *ApJ* 525, L39

Gandorfer, A. 2000, *The Second Solar Spectrum*, Vol. I: 4625 Å to 6995 Å, ISBN no. 3 7281 2764 7 (Zurich: VdF)

Gandorfer, A. 2002, *The Second Solar Spectrum*, Vol. II: 3910 Å to 4630 Å, ISBN no. 3 7281 2855 4 (Zurich: VdF)

Gandorfer, A. 2005, *The Second Solar Spectrum*, Vol. III: 3160 Å to 3915 Å, ISBN no. 3 7281 3018 4 (Zurich: VdF)

Gandorfer, A. M., Povel, H. P., Steiner, P., Aebersold, F., Egger, U., Feller, A., Gisler, D., Hagenbuch, S., & Stenflo, J. O. 2004, *A&A* 422, 703

Hanle, W. 1924, *Z. Phys.* 30, 93

Ivanov, V. V. 1991, in: L. Crivellari, I. Hubeny, & D. G. Hummer (eds.), *Stellar Atmospheres: Beyond Classical Models* (Dordrecht: Kluwer), Proc. NATO, pp. 81

Janssen, K., Vögler, A., Kneer, F. 2003, *A&A* 409, 1127

Manso Sainz, R. & Trujillo Bueno, J. 2003, *Phys. Rev. Letters* 91, 111102

Manso Sainz, R. & Trujillo Bueno, J. 2007, in: P. Heinzel, I. Dorotovic, & R. J. Rutten (eds.), *The Physics of Chromospheric Plasmas*, ASP Conf. Ser. (San Francisco: ASP), vol. 368, 155

Manso Sainz, R., Landi Degl'Innocenti, E., & Trujillo Bueno, J. 2004, *ApJ* 614, L89

Moruzzi, G. & Strumia, F. (eds.) 1991, *The Hanle Effect and Level-Crossing Spectroscopy* (New York: Plenum)

Nordlund, Å. & Stein, R. F. 1990, in: J. O. Stenflo (ed.), *Solar Photosphere: Structure, Convection, and Magnetic Fields*, IAU Symp. 138, 191–211
Povel, H. 1995, *Optical Engineering* 34, 1870
Povel, H. 2001, in: G. Mathys, S. K. Solanki & D. T. Wickramasinghe (eds.), *Magnetic Fields Across the Hertzsprung-Russel Diagram*, ASP Conf. Ser. (San Francisco: ASP), vol. 248, pp. 543–552
Rüedi, I., Solanki, S. K., Livingston, W., Stenflo, J. O. 1992, *A&A* 263, 323
Solanki, S. K. 1993, *Space Sci. Rev.* 63, 1
Spruit, H. 1976, *SP* 50, 269
Stenflo, J. O. 1973, *SP* 32, 41
Stenflo, J. O. 1980, *A&A* 84, 68
Stenflo, J. O. 1982, *SP* 80, 209
Stenflo, J. O. 1997, *A&A* 324, 344
Stenflo, J. O. 2001, in: G. Mathys, S. K. Solanki & D. T. Wickramasinghe (eds.), *Magnetic Fields Across the Hertzsprung-Russel Diagram*, ASP Conf. Ser. (San Francisco: ASP), vol. 248, pp. 639
Stenflo, J. O. 2004, *Rev. Mod. Astron.* 17, 269
Stenflo, J. O. 2007, *Memorie della Societa Astronomica Italiana* 78, 181
Stenflo, J. O. 2008, in: S. Berdyugina, K. N. Nagendra, & R. Ramelli (eds.), *Solar Polarization*, Proc. 4th SPW, ASP Conf. Ser. (San Francisco: ASP), in press
Stenflo, J. O. & Holzreuter, R. 2002, in: H. Sawaya-Lacoste (ed.), *Magnetic Coupling of the Solar Atmosphere*, ESA Publ. SP-505, 101
Stenflo, J. O. & Holzreuter, R. 2003, in: A. A. Pevtsov & H. Uitenbroek (eds.), *Current Theoretical Models and High Resolution Solar Observations*, Proc. 21st International NSO/SP Workshop, ASP Conf. Ser. (San Francisco: ASP), vol. 286, 169
Stenflo, J. O. & Keller, C. U. 1996, *Nature* 382, 588
Stenflo, J. O. & Keller, C. U. 1997, *A&A* 321, 927
Stenflo, J. O. & Lindegren, L. 1977, *A&A* 59, 367
Stenflo, J. O., Harvey, J. W., Brault, J. W., & Solanki, S. K. 1984, *A&A* 131, 333
Stenflo, J. O., Keller, C. U., & Gandorfer, A. 1998, *A&A* 329, 319
Trujillo Bueno, J. & Landi Degl' Innocenti, E. 1997, *ApJ* 482, L183
Trujillo Bueno, J. & Shchukina, N. 2007, *ApJ* 664, L135
Trujillo Bueno, J., Shchukina, N., & Asensio Ramos, A. 2004, *Nature* 430, 326

Discussion

STRASSMEIER: If you were to observe the "Sun as a Star", wouldn't the Stokes-Q signal due to the coherent scattering cancel out? And, if you would ignore it, would it cross talk into Stokes V?

STENFLO: For stars with axially symmetric stellar disks the effect would indeed cancel out. However, for objects with significant deviations from axial symmetry there may be net observable scattering polarization signatures. Apart from cross talk produced by the instrument used, there should not be any other kind of Q vs. V cross talk that one needs to be concerned about.

KOUTCHMY: Is it possible to observe the Second Solar Spectrum using FTS instruments? If not, why? If yes, why it is not done?

STENFLO: My first encounter with the Second Solar Spectrum was in 1978-79 when I made a survey at the scattering polarization from the deep UV (3160 Å) to the near infrared. For the wavelengths above 4200 Å I used the FTS at Kitt Peak. However, as the noise level was about 0.1% in the degree of linear polarization, I only saw the "tips of the icebergs" at the Second Solar Spectrum, the majority of the structures were down in the noise. It did not seem possible to push the noise much below 0.1% with the FTS when it was used, as we did, with 1000-Å wide prefilters. Limiting the simultaneous spectral coverage would help the noise, but this avenue was never pursued (which is unfortunate).

Jan Stenflo and ...

... a very interested auditorium

Maria Loukitcheva

Ilya llyin (left) and Lidia van Driel-Gesztelyi

Cosmic Magnetic Fields:
From Planets, to Stars and Galaxies
Proceedings IAU Symposium No. 259, 2008
K.G. Strassmeier, A.G. Kosovichev & J.E. Beckman, eds.

doi:10.1017/S1743921309030476

Mini-filaments – small-scale analogues of solar eruptive events?

Carsten Denker[1] and Alexandra Tritschler[2]

[1]Astrophysikalisches Institut Potsdam, An der Sternwarte 16, D-14482 Potsdam, Germany
email: cdenker@aip.de

[2]National Solar Observatory/Sacramento Peak, Sunspot, NM 88349, USA
email: ali@nso.edu

Abstract. Mini-filaments are a small-scale phenomenon of the solar chromosphere, which frequently occur across the entire disk (see e.g. Wang, Li, Denker, *et al.* 2000). They share a variety of characteristics with their larger-scale cousins and may serve as a proxy for more complex systems. They play an important role in the energy and mass supply to the corona. In the case of small-scale eruptive filaments, only a single, small-scale loop system is involved. Furthermore, they are supported by simple magnetic field configurations (see Livi, Wang & Martin 1985), either magnetic bipoles or well-defined multipoles, easing their theoretical description. Since mini-filaments are small (just a few tens of seconds of arc) but highly dynamic (eruptions can occur within just a few minutes), they are an ideal target for high-resolution two-dimensional spectroscopy. We present a preliminary analysis of two-dimensional Hα spectroscopic data accompanied by broad-band speckle-restored images to demonstrate that chromospheric small-scale phenomena can serve as building blocks for our understanding of solar eruptive events such as filament/prominence eruptions and even coronal mass ejections (CMEs).

Keywords. Sun: photosphere – Sun: chromosphere – Sun: filaments – Sun: magnetic fields

1. Observations

The observations were obtained with the *Göttingen Fabry-Pérot Interferometer* at the Vacuum Tower Telescope located at Observatorio del Teide, Tenerife. About 50 data sets were taken from 07:55 to 08:32 UT on 7 August 2008. The target was a small emerging magnetic dipole at heliographic coordinates 21.7° East and 1.1° South. The dipole was short-lived (less than a day) showing up as an inconspicuous brightening in Hα full-disk images. The data consist of narrow-band (3.5 pm) filtergrams equidistantly spaced (2.21 pm) spanning the chromospheric absorption line Hα and simultaneous broad-band (5 nm) images at 630 nm for post-facto image restoration. The broad-band images are restored using the speckle masking method and the narrow-band filtergrams are improved using speckle deconvolution. For this reason, multiple frames have to be taken per wavelength position resulting in a total of $8 \times 61 = 488$ frames per scan. The data acquisition rate was about 15 frames/s resulting in a cadence of about 34 s. The image scale was $0.11''$/pixel corresponding to a field-of-view (FOV) of $77'' \times 58''$.

2. Results

Speckle masking imaging is a powerful tool to recover spatial information down to the diffraction limit of solar telescopes. The speckle-restored broad-band image in Fig. 1d shows a *quiet Sun* area, which contains a small emerging magnetic bipole. Small-scale brightenings down to sizes of about $0.25''$ indicate the presence of individual flux tubes or small clusters thereof. Small-scale dark features, so called *magnetic knots*, point to more

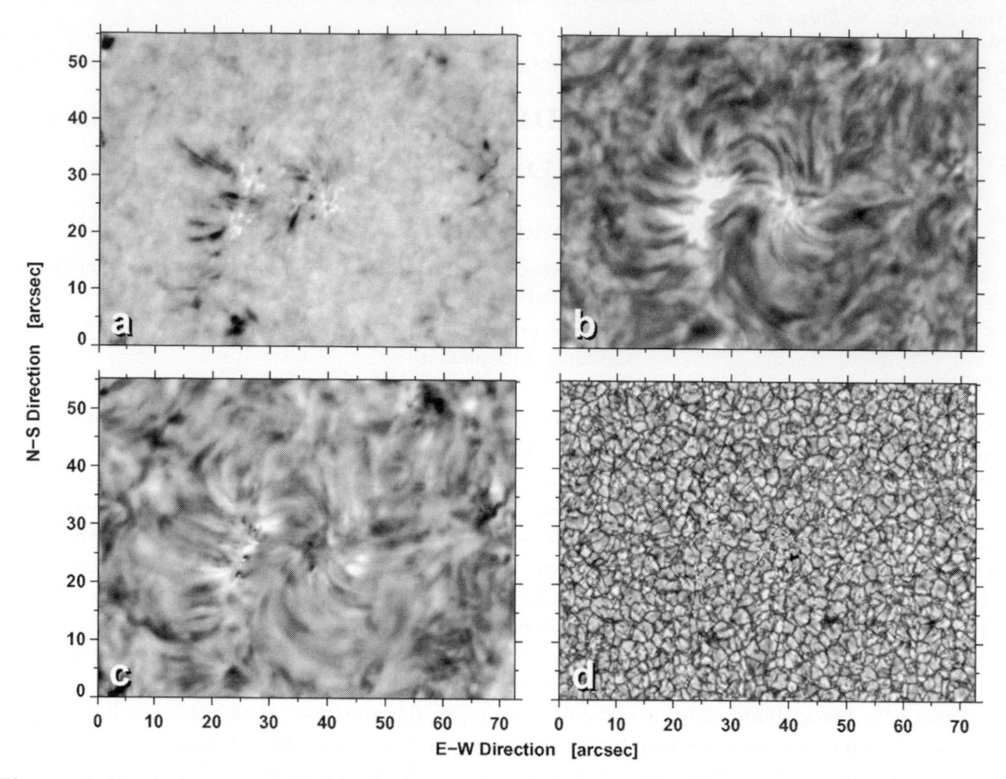

Figure 1. Emerging magnetic bipole observed on 7 August 2008. Narrow-band Hα line wing (a) and line core (b) filtergram. (c) Hα Doppler velocity map and (d) speckle-restored broad-band image showing small-scale magnetic features (filigree and magnetic knots).

concentrated magnetic fields. Hα line core filtergrams (Fig. 1b) show the mini-filament loop system connecting the emerging magnetic bipole. Small-scale brightenings, so called *Ellerman bombs*, are clearly visible in the far blue wing of the Hα line (Fig. 1a). These features are cospatial with small-scale brightenings in the speckle-restored broad-band image. The highest velocities exceeding 5 km/s (downward flows) are encountered in the vicinity of the mini-filament's footpoints (see the Hα line core Doppler velocity map in Fig.1c). Upward flows occur along the neutral line dividing the Hα brightenings.

We presented a glimpse at preliminary results concerning small-scale emerging flux elements, which induce the formation of mini-filaments. Even though small-scale magnetic bipoles and multipoles are involved in the formation of mini-filaments, not all configurations are conducive to their formation. In our example the mini-filament did not form along the magnetic neutral line of the emerging dipole, it much more resembled arch filament systems observed in emerging flux regions. The footpoints of the mini-filament are rooted in the regions of opposite polarity and show brightenings in the line core Hα. The observed mini-filaments in this magnetic configuration appeared to be stable, while the flux emergence lasted. We are currently analyzing further data sets to find magnetic configurations, which lead to more active mini-filaments and possibly to their eruption.

References

Livi, S. H. B., Wang, J., & Martin, S. F. 1985, *AuJPh* 38, 855

Wang, J., Li, W., Denker, C., Lee, C., Wang, H., Goode, P. R., McAllister, A., & Martin, S. F. 2000, *ApJ* 530, 1071

Cosmic Magnetic Fields:
From Planets, to Stars and Galaxies
Proceedings IAU Symposium No. 259, 2008
K.G. Strassmeier, A.G. Kosovichev & J.E. Beckman, eds.

doi:10.1017/S1743921309030488

The three-dimensional structure of the magnetic field of a sunspot

Horst Balthasar[1] and Peter Gömöry[2]

[1]Astrophysikalisches Institut Potsdam, An der Sternwarte 16, D–14482 Potsdam, Germany
email: hbalthasar@aip.de

[2]Astronomical Institute of the Slovak Academy of Science, 05960 Tatranská Lomnica, Slovakia
email: gomory@astro.sk

Abstract. Spectro-polarimetric observations in several spectral lines allow to determine the height variation of the magnetic field of a small sunspot throughout the solar photosphere. The full Stokes-vector is measured with high spatial resolution. From these data we derive the magnetic field vector. The magnetic field strength decreases with height everywhere in the spot, even in the outer penumbra where some other authors have reported the opposite. The precise value of this decrease depends on the exact position in the spot. Values vary between 0.5 and 2.2 $\mathrm{G\,km^{-1}}$ when they are determined from an iron and a silicon line in the near infrared. The magnetic field is less inclined in the higher layers where the silicon line is formed. Once the magnetic vector field is known, it is straight forward to determine current densities and helicities. Current densities exhibit a radial structure in the penumbra, although it is still difficult to correlate this with the structure seen in the intensity continuum. In spite of this, current densities have a potential to serve as diagnostic tools to understand the penumbra, at least with the spatial resolution of the upcoming telescopes. The mean infered helicity is negative, as expected for a spot in the northern hemisphere. Nevertheless, there are locations inside the spot with positive helicity.

Keywords. Sun: sunspots – Sun: magnetic fields – polarization

---

1. Introduction

The magnetic field of sunspots was discovered one hundred years ago by Hale (1908), but up to now, many details of the sunspots' magnetic structure are not well understood. There are controversies how fast the magnetic field decreases with height and by how much the penumbral field is inclined. Twist of the magnetic field might be important to balance the magnetic pressure in the upper layers of the photosphere and above. Electric currents have a potential to distinguish between different penumbra models. To shed more light on these issues, we measured the magnetic vector field of a small sunspot in two different spectral lines, Fe I 1078.3 nm and Si I 1078.6 nm. On May 27, 2006 the spot was very close to the central meridian, and we observed it with the Tenerife Infrared Polarimeter (TIP, see Collados, Lagg, Díaz García, *et al.*, 2007) at the German Vacuum Tower Telescope (VTT) on Tenerife. From the Stokes profiles we determine the magnetic vector field with the SIR-code (Stokes Inversion based on Response functions) of Ruiz Cobo & del Toro Iniesta (1992). More detailed results have already been published by Balthasar & Gömöry (2008).

2. Results

We find that the mangetic field inside the spot is always higher in the deeper layers where the iron line is formed. In the first step, we devide the difference of the magnetic field from the two lines by the height difference. The total magnetic field strength decreases by about 2.2 G km^{-1} in the umbra and by 0.5 G km^{-1} in the penumbra. Our umbral value is smaller than that of Mathew, Lagg, Solanki, *et al.* (1995). Looking at the vertical component B_z, we find only a slightly smaller value in the umbra, while B_z increases in the outer penumbra. The latter is explained by the high inclination to the vertical in the outer penumbra, and this inclination decreases with height, which is in contrast to Mathew, Lagg, Solanki, *et al.* (1995).

The height dependence of B_z can also be calculated from $\operatorname{div} B = 0$. We determine the required derivatives from differences of the neighboring pixels. The values of this height dependence of B_z are smaller than those from the quotients of the differences, but they are slightly larger than previous results by Hofmann & Rendtel (1989). Uncertainties of the formation heights of the spectral lines could be one reason, but the solar atmosphere is not extended enough to explain all discrepencies this way. Our main concern is the accuracy of the horizontal derivatives, because the neighboring pixels might represent different fine structures and different geometrical heights.

Despite these inaccuracies, we also determine the vertical component of the electric current density from the horizontal derivatives of the magnetic field strength using

$$(\nabla \times \boldsymbol{B})_z = \frac{\partial B_y}{\partial x} - \frac{\partial B_x}{\partial y} = \mu J_z. \tag{2.1}$$

We find radial structures of current densities in the penumbra, but they are not correlated with intensity structures. Nevertheless, our result demonstrates the diagnostic power of current densities.

In a similar way we also calculate current helicities

$$H_z = B_z \cdot (\nabla \times \boldsymbol{B})_z. \tag{2.2}$$

Helicity is an indicator of twist in magnetic field lines which might be very important to keep the magnetic field together in the upper photosphere and above. In agreement with the result of Seehafer (1990), we find a dominance of negative helicity as expected for a spot in the northern hemisphere. However, there are some locations inside the spot with positive helicity, as it was reported previously by Socas Navarro (2005) for another spot.

3. Prospects

With a new generation of solar telescopes such as the 1.5 m GREGOR, which will be commissioned in 2010, we will increase the spatial resolution significantly. This is very important to calculate more reliable derivatives of the magnetic field.

References

Balthasar, H. & Gömöry, P. 2008, *A&A* 488, 1085

Collados, M., Lagg, A., Díaz García, J. J., Hernández Suárez, E., López López, R., Páez Maña, E., & Solanki, S. K. 2007, in P. Heinzel, I. Dorotovič, R. J. Rutten (eds.) *The Physics of Chromospheric Plasmas*, Astron. Soc Pac. Conf. Ser. 368, p. 611.

Hale, G. E. 1908, *ApJ* 28, 315

Hofmann, A. & Rendtel, J. 1989, *AN* 310, 61

Mathew, S. K., Lagg, A., Solanki, S. K., Collados, M., Borrero, J. M., Berdyugina, S., Krupp, N., Woch, J., & Frutiger, C. 1995, *A&A* 410, 695

Ruiz Cobo, B. & del Toro Iniesta, J.C. 1992, *ApJ* 398, 375

Seehafer, N. 1990, *SP* 125, 219

Socas Navarro, H. 2005, *ApJ* 631, L67

Cosmic Magnetic Fields:
From Planets, to Stars and Galaxies
Proceedings IAU Symposium No. 259, 2008
K.G. Strassmeier, A.G. Kosovichev & J.E. Beckman, eds.

doi:10.1017/S174392130903049X

Isolated quasi-axisymmetric sunspots

Serge Koutchmy and Vincent Le Piouffle

Institut d'Astrophysique de Paris, CNRS, UPMC, 98 Bis Bd Arago, F-75014 Paris, France
email: koutchmy@iap.fr

Abstract. We briefly review the question of the origin, during a sunspot cycle, of well isolated sunspots. This includes big sunspots like the one observed in Nov. 2006. An overall axi-symmetric morphology is not perfectly observed when the morphological details of both the umbra and of the penumbra are considered. This is especially the case of umbral dots always present inside the core of a sunspot and also of penumbral filaments with non radial parts. However, the distribution of the surrounding fields, including deep layers, the occurrence of persistent coherent running penumbral waves, the magnetic moat behavior, the bright ring phenomena, etc. seem to justify a revival of the naive former but revised (converging motions are considered) Larmor model of a sunspot (as suggested by Lorrain *et al.* 2006). To discuss the "emergence" of single isolated sunspots from deep layers we performed a quasi-statistical analysis limited to cycle 23. It is based on MDI data taken in the continuum, using the accompanying magnetograms to check our assertion. Surprisingly, single sunspots are definitely and preferably found to occur at low latitude and during the descending branch of the cycle. To explain our observations we speculate about the behavior of the deeply seated magnetic loop, following the original idea of H. Alfven (with whirl rings which follow the global dipolar field when approaching the surface). It could lead to a closed loop approximately orthogonal to the local radius, similar to "smoke rings" arriving at the surface of the Sun and sometimes also called a plasmoid. The ring will only very weakly feel the destabilizing Coriolis force, when emerging at very low latitudes, which seems consistent with our observations.

Keywords. Sunspot – magnetic fields – Sun: photosphere

---

1. Introduction

Single sunspots were observed since the very beginning of visual observations of dark spots on the Sun. The father C. Scheiner and Galileo Galilee, starting in 1611, reported observations of single sunspots rotating together with the Sun, etc. This was done when obviously nobody was suspecting they are magnetic of nature and evidently, no polarity was assigned nor polarity rules were suggested to exist. Later, in his classical book printed in 1873, the father A. Secchi was still describing sunspots by selecting single sunspots (SSs) as a prototype of this complex solar feature. In the modern era, sunspots are considered as the main indicator of the solar magnetic activity and sunspot pairs of different polarities are rather preferred. In the famous Zurich classification of sunspot "groups", the SS is only at the very end of the list, making doubtful that it represents something consistent with the existing theories of the origin of sunspots. Indeed today solar physicists concentrate their attention to sunspot regions and active centers and the systematic observation of single "unipolar" and quasi-symmetric sunspots is widely neglected, although theoreticians like to consider a SS when they discuss the details of their physics, especially when the umbral core and the penumbral filaments are analyzed. In this short note there is no space to discuss the parameters of a SSs which includes an umbra, a penumbra and a surrounding moat with an opposite polarity. We just briefly mention one result of statistical significance to conclude on a possible scenario to explain this result.

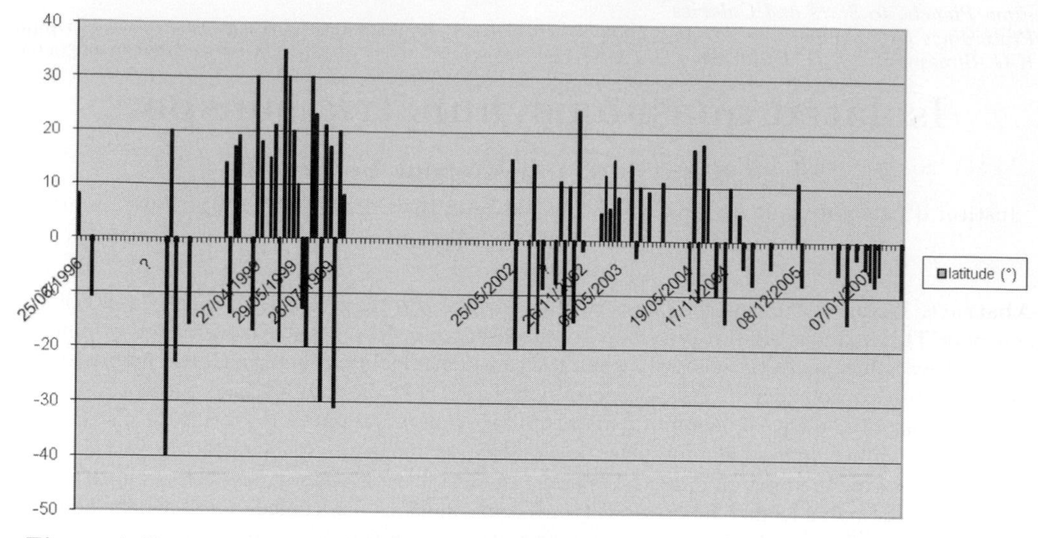

Figure 1. Dates and positions of unipolar isolated sunspots reported in this analysis of the solar cycle 23.

2. Analysis of Unipolar Isolated (single) Sunspots for Cycle 23

We used the whole available data basis from MDI (SoHO) observations to look at the distribution over the disk of SSs. W-L images were examined together with the corresponding magnetograms to confirm the selection. We were interested by the frequency of appearance and the value of the latitude of the SSs. It appears that during the Years of sunspot maxima it is impossible to select the SSs, presumably because the "overlapping" effect of sunspot regions. Figure 1 gives an overview of our results shown using the usual latitude to time diagram. We further performed a finer analysis which revealed the preference of SS occurring at low latitudes. It is difficult to claim if this is due to the predominance of SSs at the end of the preceding cycle 22 or if it is a definite property of the current cycle 23.

The occurrence of SSs at low latitude suggests that the Coriolis force could be negligible during their process of emergence. It is interesting to apply this idea to the former model suggested by Alfven (1951) when considering the path of a whirl along the global dipolar magnetic field of the Sun. Near the equator field lines are parallel to the surface and the whirl emerges almost parallel to the surface, like a smoke ring. Following the suggestion of Lorrain *et al.*, converging motions toward the axis of the whirl will enhance any radial magnetic field which further, would form an umbra by radiative cooling, and later a penumbra. The absence of any destabilizing force could explain the occurence of really big stable SSs.

References

Alfven, H. 1951, *Cosmical Electrodynamics*, Oxford at the Clarendon Press, p.116

Lorrain, P. *et al.* 2006, *Magneto-Fluid Dynamics*, A&A Library, Springer, p.205

Cosmic Magnetic Fields:
From Planets, to Stars and Galaxies
Proceedings IAU Symposium No. 259, 2008
K.G. Strassmeier, A.G. Kosovichev & J.E. Beckman, eds.

doi:10.1017/S1743921309030506

The Sun as a magnetic star: on the manifestation of different surface structures in disk-integrated observations

Mikhail L. Demidov

Institute of Solar-Terrestrial Physics, P.O. Box 291, 664033 Irkutsk, Russia
email: demid@iszf.irk.ru

Abstract. A reality of solar global magnetic fields Stokes-meter observations with polarimetric accuracy as high as $10^{-5} - 10^{-6}$, and the dependence of Stokes V profiles on distribution of surface magnetic fields in the central zone of the disk are demonstrated. A possibility to use solar disk-integrated and spatially resolved magnetic fields measurements for testing of magnetic fields mapping methods on Sun-like stars is suggested.

Keywords. Sun: magnetic fields – Sun: activity – polarization

---

The strength of solar magnetic fields strongly depends on the spatial resolution and place of observations. In the sunspot umbra we see thousands of Gauss, in the plages - hundreds, in the quite atmosphere - only a few Gauss usually. But if we'll have a look on the Sun in the integrated radiation (in parallel beam of light or using defocusing of the image), we'll measure almost nothing. This "almost nothing" is the Solar Mean Magnetic Field (SMMF) or the magnetic field of the Sun as-a-star. Tremendous weakening of the measured strength originates from the low magnetic filling factor and non-coherency of solar magnetic fields, when opposite polarities cancel each other.

Cancel, but not perfectly, and if the accuracy of measurements is good enough, we can detect reliable Zeeman signal and explore the properties of the Sun as a magnetic star. Big scientific significance of such data is the reason why several observatories provide the regular or episodic measurements of disk-integrated solar magnetic fields. Up to now the only instrument which measures the SMMF in Stokes-meter regime (detection of the Stokes I and V profiles in several spectral lines simultaneously) is the STOP telescope at the Sayan Solar Observatory (Demidov *et al.*, 2002). If the integration time is big enough, then a very weak signals could be detected (figure 1). Comparison of observations in different spectral lines shows that the SMMF strengths can differ by factor two or even more. Due to the limb darkening and some other reasons the main "contributor" to the SMMF signal is the central part of solar disk with radius ≈ 0.5 of the full radius, and effect from limb zones, including polar caps, is negligible.

As other global solar parameters, the SMMF could (and must!) be used for testing of stellar observations. Indeed, recently a significant progress in measurements of magnetic fields on the other stars, including solar-type ones, is achieved (Petit *et al.*, 2008). There are some promising attempts to map the surface distribution of stellar magnetic fields, using rotational modulation of the Stokes V profiles. Simultaneous observations of the SMMF and full-disk solar magnetograms provide a unique chance for testing of such mapping algorithms. Figure 2 illustrates a strong dependence of the SMMF Stokes V spectral lines profiles on the distribution across the solar disk (especially in the central zone) of the large-scale magnetic fields.

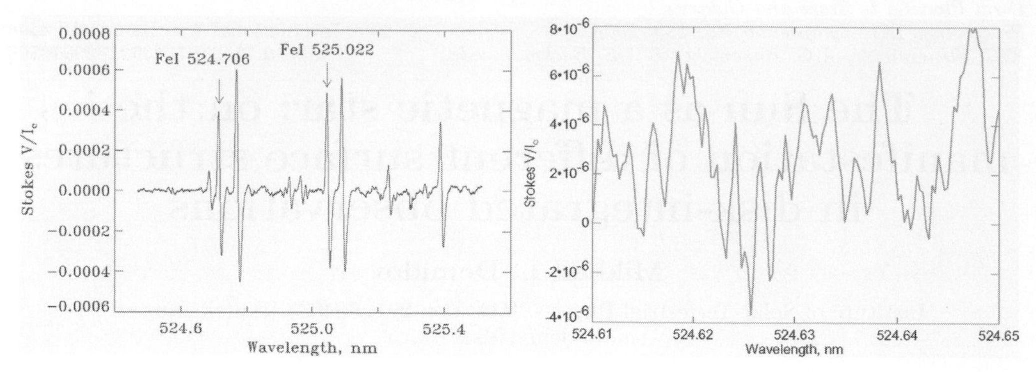

Figure 1. Left panel: the distribution of the Stokes V/I_c parameter in the vicinity of spectral line Fe I λ525.02 nm in spectral band with width ≈ 1 nm. Right panel: the distribution of the same parameter in the continuum. It is possibly to see, that the noise level is very small and such high degree of circular polarization as some units of 10^{-6} could be measured. Duration of SMMF observations is about 5 hours.

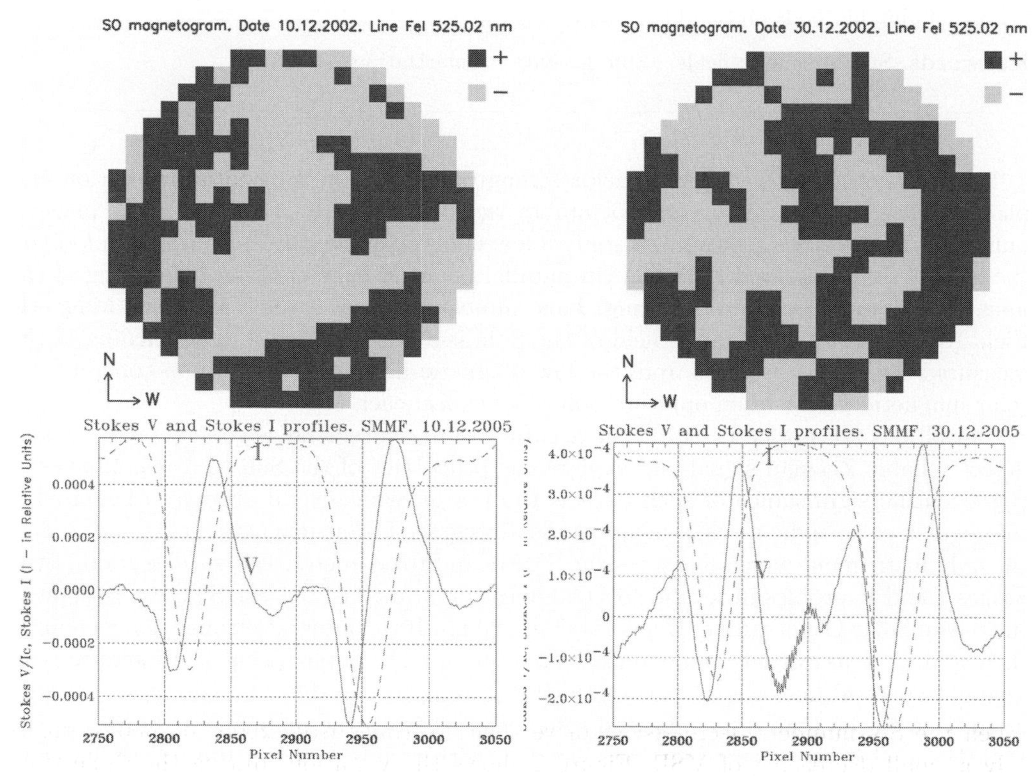

Figure 2. Illustration of the SMMF Stokes V profiles (two bottom pictures) dependence on the structure of surface large-scale magnetic fields (two top pictures).

Acknowledgements

The author is grateful to the IAU S259 SOC and to the RFBR for financial support.

References

Demidov, M. L., Zhigalov, V. V., Peshcherov, V. S., & Grigoryev, V. M. 2002, *SP* 209, 217

Petit, P., Ditrans, B., Solanki, S. K., *et al.* 2008, *MNRAS* 388, 80

Cosmic Magnetic Fields:
From Planets, to Stars and Galaxies
Proceedings IAU Symposium No. 259, 2008
K.G. Strassmeier, A.G. Kosovichev & J.E. Beckman, eds.

doi:10.1017/S1743921309030518

On the dependence of magnetic line ratios on time and spatial scales

Mikhail L. Demidov

Institute of Solar-Terrestrial Physics, P.O. Box 291, 664033 Irkutsk, Russia
email: demid@iszf.irk.ru

Abstract. Comparison of magnetic fields measurements made in different spectral lines and observatories is an important tool for diagnostics of magnetohydrodynamic conditions in the solar atmosphere. But there is a deficit of information about the dependence of results on detailed position on the solar disk, spatial resolution and time. In this study these issues are discussed in application to the solar large-scale and Sun-as-a-star magnetic fields observations.

Keywords. Sun: magnetic fields – sunspots – Sun: activity

Comparisons of solar magnetic field observations in different spectral lines (*e.g.* Howard & Stenflo, 1972; Stenflo, 1973; Ulrich, 1992) and in different observatories (Wenzler *et al.*, 2004; Tran *et al.*, 2005; Demidov *et al.*, 2008) are of crucial importance for understanding of the solar magnetism nature. According to different studies there are some differences in the values of magnetic strength ratios (MSR) even when the same combinations of spectral lines are used. There are some results concerning the dependence of MSR for several cases on filling factor, amount of flux, or center-to-limb angle. Complicated and sometimes rather unexpected character of spatial distribution of MSR across the solar disk when different combinations of spectral lines and observatories are used, is shown in the recent paper (Demidov *et al.*, 2008). The main objective of this study is to explore the possible dependence of MSR on time and spatial resolution in application to the large-scale (LSMF) and solar mean (SMMF) magnetic fields.

Comparison of SMMF measurements, made in different combinations of four spectral lines in the vicinity of Fe I λ525.02 nm is shown in Fig. 1 (left panel). To obtain this result, the Stokes-meter simultaneous observations made at the STOP telescope at the Sayan Solar Observatory (SSO) during 1999-2006 years were used. The sunspot numbers (SN) data (http://www.ngdc.noaa.gov) are shown in the right panel of figure 1. For all combinations of spectral lines there are only very small variations of MSR with time. When the SN numbers decreased in more then six times from 2000 to 2006, there is only a small decreasing of MSR. Figure 2 shows the variations during the same time interval of the mean (absolute values) SMMF strength in the case of the J.Wilcox Solar Observatory (WSO) measurements (left panel) and of the B_{WSO}/B_{SSO} SMMF strength ratio (right panel). Naturally, SSO data are in the Fe I λ525.02 nm spectral line.

There are some indications that MSR could depend on spatial resolution used in observations (Ulrich, 1992; Demidov *et al.*, 2008). According to Ulrich (1992), the reduction factor (by which the MWO magnetic fields observed in the Fe I λ525.02 nm spectral line should be multiplied in comparison to observations in the Fe I λ523.29 nm) is increasing with decreasing of spatial resolution: $R = 4.5$ for the disk center for the resolution of 20 arc seconds, and $R = 3.5$ for the case of 5 arc seconds. But this result contradicts to the Howard and Stenflo (1972) and recent author's results. According to Howard and

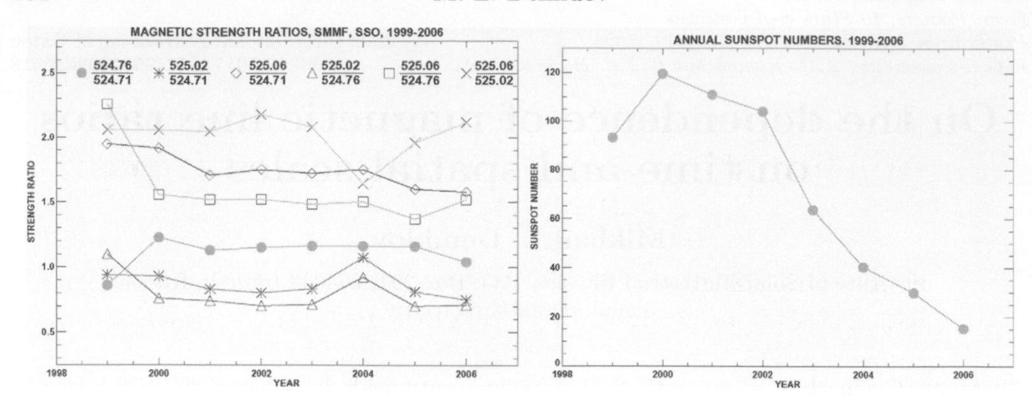

Figure 1. Left panel: strength ratios for the SMMF Stokes-meter observations (Sayan Solar Observatory STOP telescope) in four spectral lines near Fe I λ525.02 nm. Right panel: the sunspot numbers for 1999-2006 years.

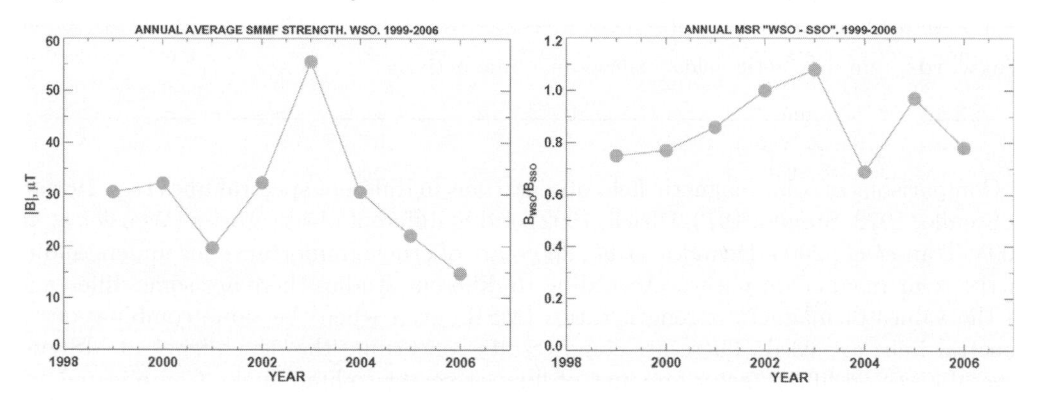

Figure 2. Variation with time of the mean absolute SMMF strength according to WSO data (left panel), and of the WSO/SSO SMMF strength ratio (right panel).

Stenflo (1972) $R = 1.8$ for the disk center, and $R = 3.0$ (spatial resolution about 100 arc seconds) according to the STOP SSO observations.

Numerous LSMF and SMMF observations on the STOP telescope give the following MSR numbers for the Fe I λ524.70 nm - Fe I λ525.02 nm combination: $R = 0.83$ for the LSMF, and $R = 0.93$ for the SMMF. In the paper (Demidov *et al.*, 2008) the possibility is suggested that the dependence of MSR on spatial resolution is the consequence of non-linearity of regression coefficients on magnetic field strength (filling factor).

Acknowledgements

The author is grateful to the IAUS 259 SOC and to the RFBR for partial financial support.

References

Demidov, M. L., Golubeva, E. M., Balthasar, H., Staude, J., & Grigoryev, V. M. 2008, *SP* 250, 279

Howard, R. & Stenflo, J. O. 1972, *SP* 22, 402

Stenflo, J. O. 1973, *SP* 32, 41

Tran T., Bertelle L., Ulrich, R. K., & Evans, S. 2005, *ApJS* 156, 295

Ulrich, R. K. 1992, *ASP Conf. Ser.* 26, 265

Wenzler, T., Solanki, S. K., Krivova, N. A., & Fluri, D. M. 2004, *A&A* 427, 1031

Cosmic Magnetic Fields: From Planets, to Stars and Galaxies
Proceedings IAU Symposium No. 259, 2008
K.G. Strassmeier, A.G. Kosovichev & J.E. Beckman, eds.

doi:10.1017/S174392130903052X

Near-surface stellar magneto-convection: simulations for the Sun and a metal-poor solar analog

Matthias Steffen[1], H.-G. Ludwig[2], and O. Steiner[3]

[1]Astrophysikalisches Institut Potsdam, An der Sternwarte 16, D-14482 Potsdam, Germany, email: msteffen@aip.de

[2]GEPI, Observatoire de Paris, CNRS, Université Paris Diderot, 92195 Meudon Cedex, France

[3]Kiepenheuer-Institut für Sonnenphysik, Schöneckstrasse 6, D-79104 Freiburg, Germany

Abstract. We present 2D local box simulations of near-surface radiative magneto-convection with prescribed magnetic flux, carried out with the MHD version of the CO^5BOLD code for the Sun and a solar-like star with a metal-poor chemical composition (metal abundances reduced by a factor 100, [M/H] = −2). The resulting magneto-hydrodynamical models can be used to study the influence of the metallicity on the properties of magnetized stellar atmospheres. A preliminary analysis indicates that the horizontal magnetic field component tends to be significantly stronger in the optically thin layers of metal-poor stellar atmospheres.

Keywords. Convection – Sun: atmosphere – stars: atmospheres – MHD

1. Introduction

Up to now, numerical simulations of near-surface magneto-convection are essentially restricted to the Sun (see e.g. Carlsson *et al.* 2004; Vögler *et al.* 2005; Stein & Nordlund 2006; Schüssler & Vögler 2008; Steiner *et al.* 2008). The idea of the present work is to study the influence of metallicity on the structure and distribution of magnetic flux concentrations in stellar atmospheres. For this purpose, realistic local box simulations of near surface magneto-convection with prescribed magnetic flux have been carried out for the Sun and a metal-poor solar analog, allowing a differential comparison of the thermal, dynamical, and magnetic properties of the photospheric layers.

2. 2D Numerical simulations

We compare two sets of simulations, both with $T_{\rm eff} = 5690$ K, and $\log g = 4.44$, but with different chemical composition ([M/H] = 0 and [M/H] = −2). All models are computed on the same 2D Cartesian grid (400×165 cells, $\Delta x = 28$ km, 12 km $< \Delta z <$ 28 km, periodic side boundaries and open/transmitting boundaries at bottom/top). Gray radiative transfer with realistic opacities for the respective chemical composition are used together with an appropriate equation of state that accounts for partial ionization of H and He. Note that the main effect of the reduced metallicity is a general decrease of the opacity; as a consequence, the gas pressure at the visible surface ($\tau = 1$) is about 3 times higher in the metal-poor model. Figure 1 shows the equipartition magnetic field strength $B_{\rm eq} \equiv \sqrt{4\pi\langle\rho\vec{u}^2\rangle}$ (where $\langle\rho\vec{u}^2\rangle/2$ is the horizontally averaged kinetic energy density) on the geometrical and optical depth scale, as computed from the two non-magnetic simulations.

A homogeneous unipolar vertical magnetic field of four different strengths ($B_{z,0} = 10$ (quiet) 20, 40, and 80 (network) Gauss, respectively, was superimposed on the non-magnetic models, defining the initial conditions for a series of ideal simulations that are

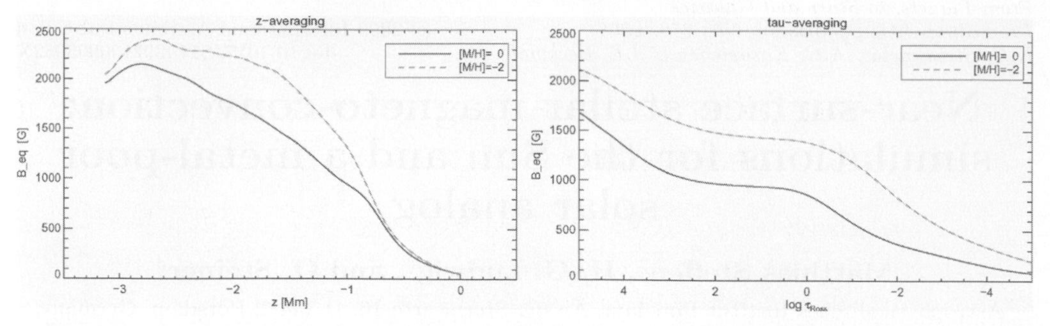

Figure 1. Equipartition field strength, $B_{\rm eq}$ (see text), on the geometrical height scale (left) and on the optical depth scale (right), computed from the two non-magnetic simulations.

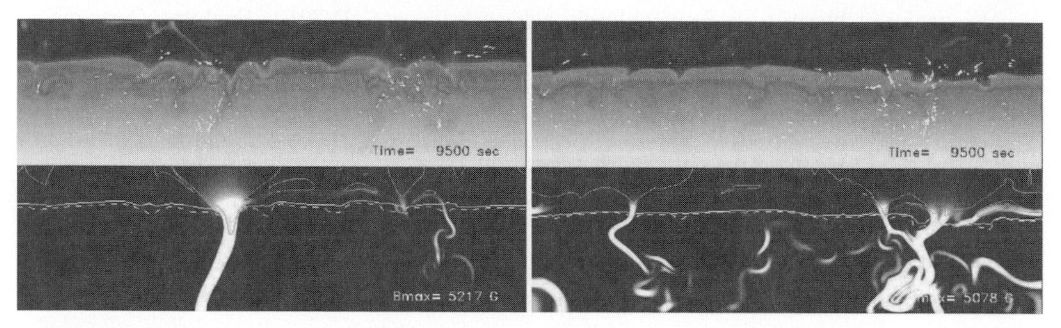

Figure 2. Snapshots from 2D MHD simulations with solar (left) and metal-poor (right) composition, taken 9 500 s after inserting the initial magnetic field ($B_{\rm z,0} = 80$ G). Upper panels: temperature structure, lower panels: distribution of magnetic field density $|B|$; the dashed (yellow) line delineates the $\tau = 1$ contour, gray lines are $\beta = 1$ contours. $|B|_{\rm max} \approx 1.5$ kG at $\tau = 1$.

used to study near-surface magneto convection at different metallicity as a function of the prescribed mean magnetic flux. At top and bottom, the magnetic field is forced to be vertical ($B_{\rm x} = B_{\rm y} = 0$); at each height, the mean vertical magnetic flux thus retains its initial value $\langle B_{\rm z} \rangle = B_{\rm z,0}$ throughout these *ideal* ($\eta = 0$) MHD simulations.

An illustration of the resulting magnetic filed configurations is given in Fig. 2. A first inspection of the simulation data suggests that, for given $B_{\rm z,0}$, the rms vertical field $B_{\rm z,rms}$ evaluated at fixed optical depth is very similar for the two metallicities, even though $B_{\rm eq}$ is significantly different (Fig. 1). However, there is a tendency for stronger horizontal fields in the photosphere of the metal-poor simulations.

3. Conclusions

These preliminary results need to be confirmed by more extensive 2D and 3D simulations with non-gray radiative transfer and alternative MHD boundary conditions. In the long run, further questions can be addressed by computing synthetic (Stokes) spectra based on the MHD model atmospheres, e.g. concerning the influence of small-scale surface magnetic fields on the micro-variability of the stellar radiation and/or on the accuracy of spectroscopic abundance determinations.

References

Carlsson, M., Stein, R. F., Nordlund, Å., & Scharmer, G. B. 2004, *ApJ* 610, L137
Schüssler, M. & Vögler, A. 2008, *A&A* 481, L5
Stein, R. F., & Nordlund, Å. 2006, *ApJ* 642, 1246
Steiner, O., Rezaei, R., Schaffenberger, W., Wedemeyer-Böhm, S. 2008, *ApJ* 680, L85
Vögler, A., Shelyag, S., Schüssler, M., Cattaneo, F., Emonet, T., & Linde, T. 2005, *A&A* 429, 335

Cosmic Magnetic Fields:
From Planets, to Stars and Galaxies
Proceedings IAU Symposium No. 259, 2008
K.G. Strassmeier, A.G. Kosovichev & J.E. Beckman, eds.

doi:10.1017/S1743921309030531

Prediction of solar magnetic cycles by a data assimilation method

Irina N. Kitiashvili[1] **and Alexander G. Kosovichev**[2]

[1]Center for Turbulence Research, Stanford University, Stanford, CA 94305, USA
email: irinasun@stanford.edu

[2]W.W. Hansen Experimental Physics Lab., Stanford University, Stanford, CA 94305, USA
email: sasha@sun.stanford.edu

Abstract. We consider solar magnetic activity in the context of sunspot number variations, as a result of a non-linear oscillatory dynamo process. The apparent chaotic behavior of the 11-year sunspot cycles and undefined errors of observations create uncertainties for predicting the strength and duration of the cycles. Uncertainties in dynamo model parameters create additional difficulties for the forecasting. Modern data assimilation methods allow us to assimilate the observational data into the models for possible efficient and accurate estimations of the physical properties, which cannot be observed directly, such as the internal magnetic fields and helicity. We apply the Ensemble Kalman Filter method to a low-order non-linear dynamo model, which takes into account variations of the turbulent magnetic helicity and reproduces basic characteristics of the solar cycles. We investigate the predictive capabilities of this approach, and present test results for prediction of the previous cycles and a forecast of the next solar cycle 24.

Keywords. Sun: activity – magnetic fields – sunspots

One of the manifestations of solar magnetic activity is the 11-year sunspot cycle, which is characterized by the fast growth and slowly decay of the sunspot number parameter. For an explanation of the magnetic field generation Parker (1955) proposed a simple $\alpha\Omega$-dynamo model, which describes the phenomenon as an action of two factors: the differential rotation and cyclonic convective vortices. For modeling the solar cycle we consider non-linear solutions of the Kleeorin-Ruzmaikin model (Kleeorin & Ruzmaikin, 1982), which in addition to the $\alpha\Omega$-dynamo process describes the evolution of the magnetic helicity based on the balance between the large-scale and turbulent magnetic helicities. To connect the dynamo model solutions with the sunspot number data we use two suggestions of Bracewell (1988): the periodical reversals of the magnetic field are represented in the sunspot number series by assigning alternating positive and negative signs to the sunspot cycles, and the sunspot number is modeled in the form of a three-halfs law: $W \sim B^{3/2}$, where W is the sunspot number, and B is the strength of the Sun's toroidal magnetic field. Our analysis of the Parker-Kleeorin-Ruzmaikin dynamo model shows the existence of nonlinear periodic and chaotic solutions for conditions of the solar convective zone (Kitiashvili & Kosovichev, 2009). For this model we obtained solutions with the profiles of W, which qualitatively reproduce the typical profile of the sunspot number variations of the solar cycles.

For predicting the solar cycle properties we make an initial attempt to use the Ensemble Kalman Filter (EnKF), a data assimilation method, which takes into account uncertainties of the dynamo model and the observed sunspot number series (Evensen, 2007). This method has been tested by calculating predictions of the past cycles using only the observational data (annual sunspot numbers) until the start of these cycles. Figures 1 a-e show examples of the EnKF method implementation for the forecasting of

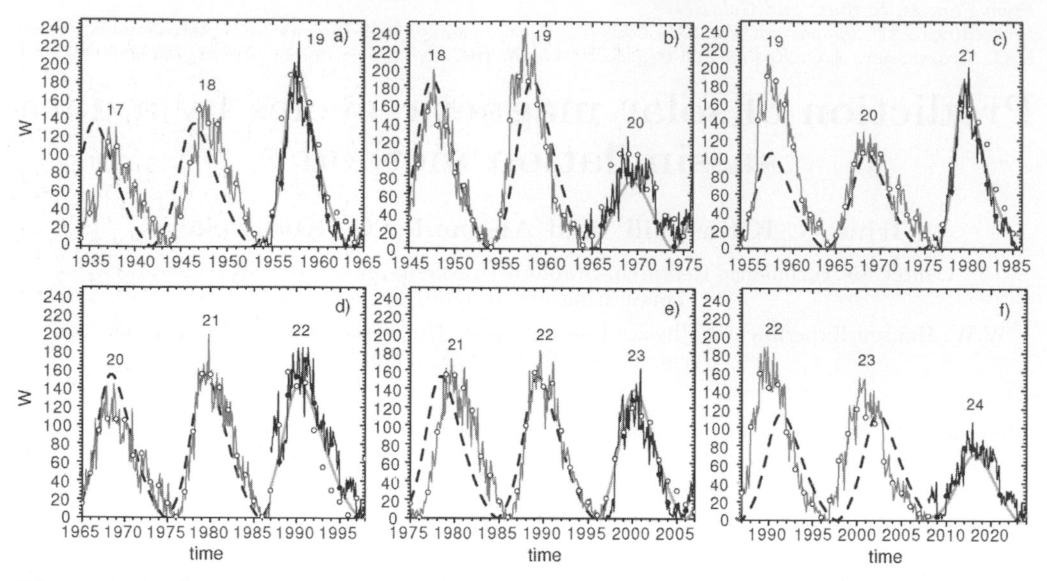

Figure 1. Predictions for solar cycles 19–24. Grey curves show variations of the sunspot number according to the known observation data (empty circles) and black curves represent the EnKF estimates for the following (predicted) cycles.

solar cycles 19–23. In this approach the exact model solution is corrected according to the observational data when these become available. This allows us to redefine the initial conditions of the model for the magnetic field components and helicity, and construct a model solution for the next time interval. To estimate the uncertainties we added noise to the model (in the form of a random forcing function) and to the simulated data, and calculated a statistical mean and error an estimate, following the EnKF procedure. We note that the errors of the predictions depend also on the number of measurements in the simulated ensemble, and also on the determining the moment of the end of the last observed cycle. The similar analysis scheme is used for predicting of the upcoming solar cycle 24 (Fig. 1f). According to our analysis, the solar cycle 24 which starts in 2008–2009 yrs. will be weaker than the current cycle by approximately 30% (Kitiashvili & Kosovichev, 2008). The estimated formal error of our prediction is $\sim 10\%$.

The application of the EnKF method for modeling and predicting solar cycles shows the power of this approach and encourages further development.

Acknowledgements

This work was supported by the Center for Turbulence Research (Stanford) and the International Space Science Institute (Bern). We thank Professors A. Brandenburg and D. Sokoloff for useful discussions.

References

Bracewell, R. N. 1988, *MNRAS* 230, 535
Evensen, G. 2007, *Data assimilation*, Springer, p. 279
Kitiashvili, I. N. & Kosovichev, A. G. 2008, *ApJ* 688, L49
Kitiashvili, I. N. & Kosovichev, A. G. 2009, *Geophys. Astrophys. Fluid Dyn.* 103, in press
Kleeorin, N. I. & Ruzmaikin, A. A. 1982, *Magnetohydrodynamics* 18, 116
Parker, E. N. 1955, *ApJ* 122, 293

Cosmic Magnetic Fields:
From Planets, to Stars and Galaxies
Proceedings IAU Symposium No. 259, 2008
K.G. Strassmeier, A.G. Kosovichev & J.E. Beckman, eds.

doi:10.1017/S1743921309030543

North-south asynchrony and long-term variations of sunspot latitudes

Nadezhda V. Zolotova† and D. I. Ponyavin

Institute of Physics, Saint-Petersburg State University, Russia
email: ned@geo.phys.spbu.ru

Abstract. The long-term records of sunspot area available separately for Northern and Southern Hemispheres have been investigated by means of cross-recurrence technique. Phase component of the north-south asymmetry was extracted. This measure demonstrates long-period systematic variations with the sign change of hemispheric leading in 1930s and 1960s. Moreover phase north-south asynchrony anticorrelates with the so called magnetic equator, which was defined as difference of the mean sunspot latitudes between two hemispheres. Relationships of the phase north-south asynchrony, magnetic equator and butterfly diagrams are presented and discussed.

Keywords. Sun: activity – sunspots – magnetic fields

1. Introduction

Variations of sunspot latitudes from 1853 to 1996 were analysed by Pulkkinen *et al.* (1999). They have introduced the so called magnetic equator as the sum of mean $\langle\lambda(N)\rangle_n + \langle\lambda(S)\rangle_n$ (the southern component is negative, n is the time epoch). Thus the magnetic equator traces a skewness in distribution of sunspot latitudes with respect to equator. A systematic variation with a period of about 90 years was revealed (Pulkkinen *et al.* 1999).

In this work we compare long-term behavior of the magnetic equator with phase north-south hemispheric asynchrony. In order to extract the last one we used the cross-recurrence plots (http://www.agnld.uni-potsdam.de/~marwan/toolbox/):

$$CR_{i,j} = \Theta(\varepsilon_i - \|\mathbf{x}_i - \mathbf{y}_j\|), \qquad i, j = 1, \ldots, N, \tag{1.1}$$

where N is the number of considered states $\mathbf{x}$, $\mathbf{y}$, $\|\cdot\|$ is a norm, and Θ is the Heaviside function (Marwan *et al.* 2007).

2. Results and conclusion

Extracted from cross-recurrence plot the Line of Synchronization (LOS) traces time delays between considered time series. LOS for the smoothed monthly northern and southern hemispheric sunspot area is presented in figure 1(a).

The change of the LOS sign indicates the change of the leading role of the hemispheres (LOS > 0 – north leading, LOS < 0 – south leading). It is seen that before the end of cycle 16 the north dominates in leading, after that till the minimum just before the cycle 20, the southern hemisphere leads, and than up to the present, the northern hemisphere leads again.

On the figure 1(b) the smoothed separate mean sunspot latitudes of northern and southern hemisphere is shown. It reveals that up to the end of the solar cycle 16 the north

† Present address: Institute of Physics, Ulyanovskaya ul., 1, Petrodvorets, 198504, Russia.

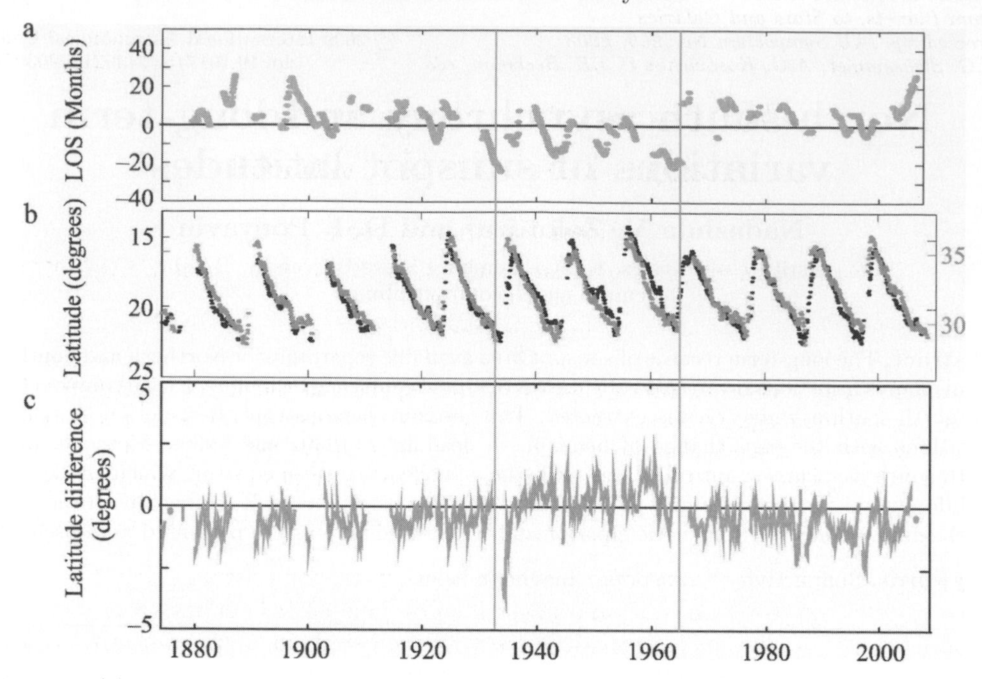

Figure 1. (a) Line of Synchronization for hemispheric sunspot area; (b) Smoothed mean sunspot latitudes for the north (black) and south (red); (c) Their difference — the magnetic equator. similar to results of Pulkkinen *et al.* (1999).

is leading in time, but with preference of sunspots to emerge at the higher latitudes in the south. After this epoch and until the beginning of cycle 20, the situation is reversed to opposite. After cycle 20 it is restored back to a situation as appeared before cycle 16.

The differences between these mean sunspot latitudes (figure 1 c) clearly reproduce variations of the magnetic equator. Moreover, the results coincide with the weighted magnetic equator derived by Pulkkinen *et al.* (1999). Analyzing time series extended to 1853 (Carrington's and Spörer's data) they found additional zero crossing just before the cycle 12. They concluded that there are regular magnetic equator variations close to the Gleissberg cycle.

In our work we observed that intervals between zero crossings are not equal to each other, probably suggesting process of stochastic or chaotic nature. Moreover it is interesting to note that asymmetry at the end of prolonged cycle 23 is unusually high. Thus we expect that some more dramatic change of the solar dynamics may occur in the nearest future.

Acknowledgements

We used the RGO USAF/NOAA data of sunspot area and their latitudes – http://solarscience.msfc.nasa.gov/greenwch.shtml.

This research has been supported by the INTAS Fellowship Grant for Young Scientists Ref. No 06-1000014-6022.

References

Marwan, N., Romano, M. C., Thiel, M., & Kurths, J. 2007, *Phys. Reports* 438, 237

Pulkkinen, P. J., Brooke, J., Pelt, J., & Tuominen, I. 1999, *A&A* 341, L43

Cosmic Magnetic Fields:
From Planets, to Stars and Galaxies
Proceedings IAU Symposium No. 259, 2008
K.G. Strassmeier, A.G. Kosovichev & J.E. Beckman, eds.

doi:10.1017/S1743921309030555

3D reconstruction of solar magnetoacoustic waves

Adrian Sabin Popescu

Astronomical Institute of the Romanian Academy, Str. Cutitul de Argint 5, RO-040557
Bucharest, Romania
email: sabinp@aira.astro.ro

Abstract. Using a new multi-wavelength technique applied on the solar corona SOHO/EIT images containing magnetoacoustic waves (EIT waves), we constrain the wave phase velocity and, as well, some β-model, height related considerations for the vertical extent of the wave, applied on the particular EIT wave triggered by the eruptive event from 15th of August, 2001.

Keywords. Sun: UV radiation – atmospheric motions – techniques: image processing – waves

---

The method employed in this paper, described in Popescu & Mierla 2008, is extremely useful in determining the solar waves phase velocity, here being applied on the eruptive event from 15th of August 2001 seen by the SOHO/EIT instrument (Delaboudinière *et al.* 1995). The EIT instrument aboard SOHO observes the Sun in four different wavelengths, namely 171 Å (Fe IX/Fe X; 1.3×10^6 K), 195 Å (Fe XII; 1.6×10^6 K), 284 Å (Fe XV; 2.0×10^6 K) and 304 Å (He II; 8.0×10^4 K), and, in consequence, from high chromosphere to mid-corona.

It has been proved by Gary (2001) that the plasma β-model plays an important role in the solar corona properties and phenomena, especially in dealing with the use of force-free magnetic fields in extrapolation over the entire coronal field range, without assuming $\beta \ll 1$ in order to overcome the full boundary conditions nonexistence in the two $\beta > 1$ regions. Comparing the Gary (2001) plasma model representation as function of height with our results, it can be observed that for the high chromosphere's 304 Å wavelength, due to a wider β allowance, we will have a greater velocity dispersion in the phase space, from where a larger number of lines in the corresponding FFT maps associated to a large number of frozen in the field plasma movements. As the height increases, with the increase of temperature in the mid-corona, from 171 Å, to 195 Å and, finally 284 Å, we have a decrease in the velocity map lines number, consistent with the narrowing of the β (where $\beta < 1$) allowed region (Gary 2001).

Without counting the central zero point line, the first and the fifth line corresponds to the velocity of the EIT wave, the second and the fourth to the velocity projection into the image plane of the CME, and the third, to the horizontal inter-shock plasma motions (see the FFT velocity map from Popescu & Mierla 2008). By comparing the evolution of the same velocity line for different wavelengths, we remark that the velocity of the phenomena giving these lines increases with height. In the case of the propagating wave, the crest is moving faster than the leading edge, this causing a steepening of the front portion of the wave and the formation of a steady shock wave in which the dissipative effects equilibrate the convective steepening effects. Into the SOHO/EIT images the dissipative effects can be identified as luminous features on the crest of the two wavefronts. This steepening is consistent with the behavior of a compression wave described by Priest (1987), and even more, with a *fast* magnetoacoustic wave, the magnetic field before the wavefronts intensifying and being seen as "dimming" regions propagating in front of the shocks.

In the FFT maps, for the lines to the left of zero, the velocity radically increases with height, while the ones to its right present just a slight increase. The reason for this behavior will be found by returning to the EIT images where we observe that one side of the wavefront is bouncing into the active region 09775, being decelerated. With the height increase, the magnetic field lines over this active region cover a wider region in the corona, rotating, as the fast shock passes, away from the shock normal (Priest 1987) and forcing the EIT and CME shock waves to adopt a larger tilt angle relative to the solar surface. On the other side, away from the active region, the EIT and CME propagation do not meet any resistance from the solar background. This difference between the two wavefronts gives rise to an excess of magnetic pressure that must relax by plasma movement from high to low pressure regions. This movement can be seen into the EIT images as arcades uniting the two EIT wavefronts, and moving away from the mentioned active region. For the 15th of August 2001 EIT wave, the velocity line close by the central zero point (the zero velocity) represents the velocity of the inter-wavefronts flows described above.

In the transition zone between the chromosphere and corona (Athay *et al.* 1980; Malherbe & Priest 1983) the horizontal flows in the proximity of a prominence are usually between 5 and 20 km/s. Scaling the velocity in all the maps with the velocity of the inter-wavefronts flow for 284 Å, considered to be 20 km/s, and translating the wavelength into the corresponding plasma temperature, we can actually follow the extent of the waves into the solar atmosphere as represented into Figure 1.

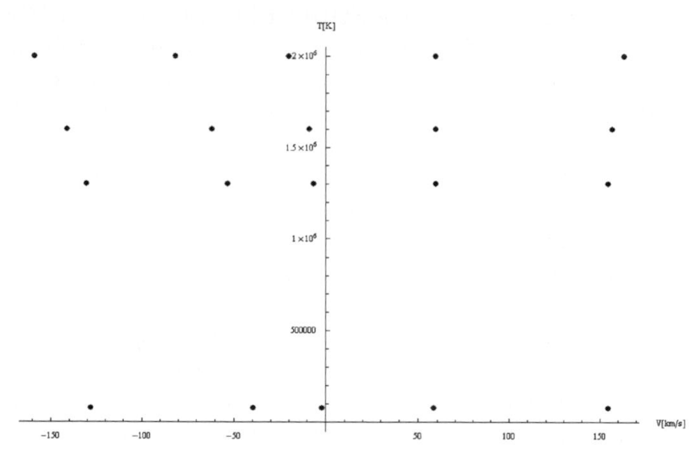

Figure 1. Velocity of the magnetoacoustic shock waves as function of temperature. The scaling is done in 284 Å considering an inter-wavefronts plasma flow velocity of 20 km/s.

We would like to thank the SOHO/EIT consortium for providing the data and the calibration software libraries. SOHO is a project of international cooperation between ESA and NASA.

References

Athay, R. G., White, O. R., Lites, B. W., & Bruner, E. C. 1980, *SP* 66, 357

Delaboudinière, J.-P., Artzner, G. E., Brunaud, J., *et al.* 1995, *SP* 162, 291

Gary, G. A. 2001, *SP* 203, 71

Priest, E. R. 1987, in: B. M. McCormac (ed.), *Solar Magneto-hydrodynamics*, Geophys. and Astrophys. Mon. 21, D. Reidel Publishing Company, Dordrecht, Holland, p. 189

Malherbe, J. H. & Priest, E. R. 1983, *A&A* 123, 80

Popescu, A. S. & Mierla, M. 2008, in: V. Mioc, C. Dumitrache, N. A. Popescu (eds.), *Exploring the Solar System and the Universe*, AIP Conference Proceedings Series, Vol. 1043, p. 309.

Cosmic Magnetic Fields:
From Planets, to Stars and Galaxies
Proceedings IAU Symposium No. 259, 2008
K.G. Strassmeier, A.G. Kosovichev & J.E. Beckman, eds.

doi:10.1017/S1743921309030567

MHD remote numerical simulations: evolution of coronal mass ejections

Liliana Hernández–Cervantes[1]†, A. Santillán[2] and A. R. González–Ponce[3]

[1]Instituto de Astronomía, UNAM, 04510, Mexico City, Mexico
email: liliana@astrosu.unam.mx

[2]Dirección General de Servicios de Cómputo Académico, UNAM, 04510, Mexico City, Mexico

[3]Instituto de Ecología, UNAM, 04510, Mexico City, Mexico

Abstract. Coronal mass ejections (CMEs) are solar eruptions into interplanetary space of as much as a few billion tons of plasma, with embedded magnetic fields from the Sun's corona. These perturbations play a very important role in solar–terrestrial relations, in particular in the spaceweather. In this work we present some preliminary results of the software development at the Universidad Nacional Autónoma de México to performe Remote MHD Numerical Simulations. This is done to study the evolution of the CMEs in the interplanetary medium through a Web–based interface and the results are store into a database. The new astrophysical computational tool is called the Mexican Virtual Solar Observatory (MVSO) and is aimed to create theoretical models that may be helpful in the interpretation of observational solar data.

Keywords. Sun: magnetic fields – Sun: corona – Sun: CMEs – Virtual Observatory

1. Introduction

The Mexican Virtual Solar Observatory (MVSO) is a set of software tools that offer global solutions for Web development. The operating system is Linux, the Web server is Apache, SQL (Structure Query Language) and the relational database management system is MySQL, everything is programmed with PHP (Hypertext Pre–Processor). The computational backbone of the MVSO is structured into three stages. The first part is the related to the graphic user interface (GUI), the second part is associated to the remote numerical simulations (RNS) and the third part is the creation of the database and associated search tools. The implementation is explained by Hernández–Cervantes *et al.* (2008).

2. Results

All calculations of the evolution of the CME in the magnetized solar wind are performed with the MHD code ZEUS-3D, which solves the 3D system of ideal MHD equations by finite differences on fixed Eulerian mesh (Stone & Norman 1992a,b). The MVSO uses a simplified model to understand the dynamics of a CME in the interplanetary space (Santillán *et al.* 2008). Initially, we produce the ambient solar wind by specifying the fluid velocity, magnetic field, density, and temperature at an inner boundary of the grid, which is located beyond the critical point (r = 18 Ro $\sim$ 0.083 AU), and then the wind is allowed to evolve and reach a stationary equilibrium. For the injection of the magnetic field we used the technique described by Stone & Norman (1992b); this consists of using

† Present address: Instituto de Astronomía, UNAM, 04510, Mexico City, Mexico

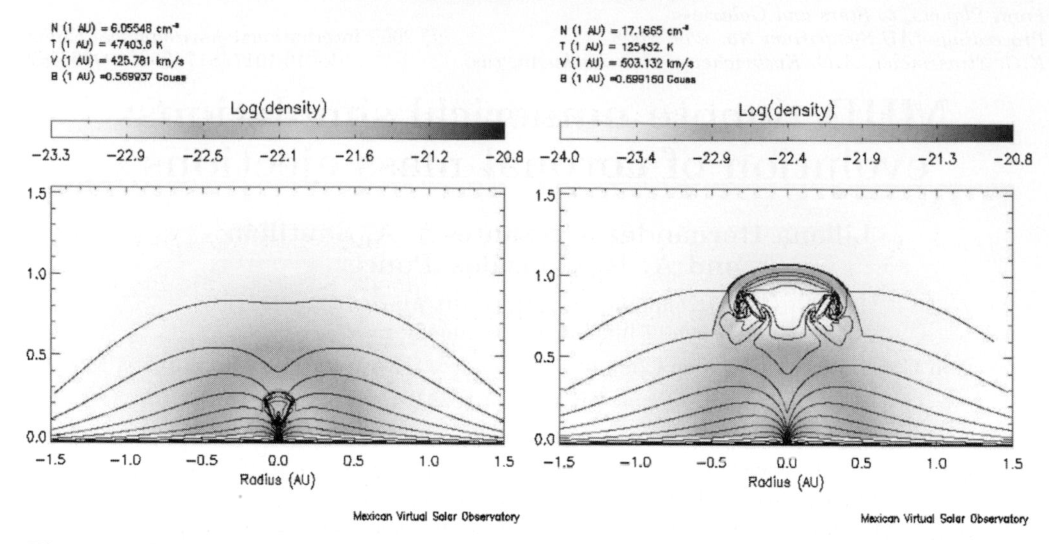

Figure 1. Evolution of the CME in the magnetized solar wind. The figure show the density (*gray logarithmic scale*) and intensity of the total magnetic field (*solid lines*) at two select times: 6 and 78 hours after the inyection of the perturbation.

time dependent analytic solutions of the Low's (1984) models. Finally we add an ejecta–like perturbation at the inner boundary to simulate the appearance of the CME into de interplanetary medium. Typical results produced by the MVSO are shown in the two snapshots displayed in the figure 1, where the density is shown in logarithmic gray-scale along with the intensity of the total magnetic field (*solid lines*). The rigidity and elasticity given to the solar wind by the magnetic field is better accentuated in 2D when the plane of motion of the CME is parallel to the field lines and the colliding gas distorts the initial field configuration. We illustrate the response of these deformed field lines in the two snapshots displayed in figure 1. The tension of the magnetic field dominates the evolution and the results are completely different from those of the purely hydrodynamic cases. The physical quantities (n, T, $\mathbf{v}$ & $\mathbf{B}$) of the medium at 1 AU change drastically, when the disturbance crosses by this point. For example, the density increase a factor $\sim$ 3 and the size of the perturbed region has grown close to 1 AU only 78 hours after the inyection of the CME. This is clearly seen in the last snapshot of the figure 1.

Acknowledgements

We are grateful to Pepe Franco for useful comments. This work has been partially supported from DGAPA-UNAM grant IN104306 and CONACyT proyect CB2006–60526.

References

Hernández–Cervantes, L., Santillán, A., & González–Ponce, A. 2008 *Geofísica Internacional* 47, 193

Low, B. C. 1984, *ApJ* 281, 392

Santillán, A., Hernández–Cervantes, L., & González–Ponce, A. 2008 *Geofísica Internacional* 47, 185

Stone, J. M. & Norman, M. L. 1992a, *ApJS* 80, 753

Stone, J. M. & Norman, M. L. 1992b, *ApJS* 80, 791

Cosmic Magnetic Fields:
From Planets, to Stars and Galaxies
Proceedings IAU Symposium No. 259, 2008
K.G. Strassmeier, A.G. Kosovichev & J.E. Beckman, eds.

doi:10.1017/S1743921309030579

Turbulent effects in flux-transport solar dynamos

G. A. Guerrero[1], E. M. de Gouveia Dal Pino[1] and M. Dikpati[2]

[1]Department of Astronomy, IAG, University of Sao Paulo, Sao Paulo, SP, Brazil
email: guerrero,dalpino@astro.iag.usp.br

[2]High Altitude Observatory, National Center for Atmospheric Research
P.O. Box 3000, Boulder, Colorado, 80307, USA
email: dikpati@ucar.edu

Abstract. The effects of turbulent pumping and η-quenching on Babcock-Leighton dynamo models are explored separately. Turbulent pumping seems to be important to solve several reported problems in these dynamo models related to the magnetic flux transport and to the parity. On the other hand, the suppression of the magnetic diffusivity, η, could help in the formation of long-lived, small and intense structures of toroidal magnetic field.

Keywords. Sun: activity – Sun: magnetic fields

1. Introduction

It is generally believed that the Solar Cycle corresponds to a hydromagnetic dynamo process operating at some place within the Solar interior. Parker (1955) was the first to build a solar dynamo model, since then there has been important improvements in the observations, theory and simulations, but a definitive model for the solar dynamo is still missing.

Two processes are necessary to close the dynamo loop: the transformation of an initial poloidal field into a toroidal field, the so called Ω effect, which is due to a large scale shear, and the transformation of the toroidal field into a new poloidal field of opposite polarity, which is a less understood process that has been the subject of intense debate and research. Two main hypotheses have been formulated in order to explain the nature of this effect (usually denominated α effect): the turbulent and the Babcock-Leighton (BL) α effect.

In the second proposed mechanism above, the large scale poloidal field is formed from the emergence and decay of bipolar magnetic regions (BMR's) which contain a net dipole moment. The new dipolar field, formed at lower latitudes, is transported by meridional circulation to the higher latitudes in order to form the observed polar field. For the meridional flow to be important in the transport of magnetic flux, the advection time must dominate upon the diffusive time, for this reason these models are often called flux-transport dynamo models.

The flux-transport dynamo model has been relatively successful in reproducing the large scale features of the solar cycle Dikpati & Charbonneau (1999), however it presents several problems that have been widely discussed in the literature Brandenburg (2005). In this work, we discuss the inclusion of two turbulent effects in a BL dynamo model – the turbulent pumping and the η-quenching – and show that, under determined conditions, they can produce results which are in better agreement with the observations.

2. Turbulent pumping and η quenching

The turbulent pumping provides the transport of magnetic flux due to the presence of density and turbulence gradients in convectively unstable layers. It has several important effects. For dynamos operating at the convection zone and at the tachocline, we find that the pumping leads to a better distribution of toroidal fields inside the convection zone when compared to a model without pumping (left panel of Figure 1). It also increases the penetration of the toroidal field formed at the convection zone in the overshoot region, allowing further field amplification by the large radial shear in that region. The strongest fields at the top of the overshoot layer are located at latitudes below 30°, as can be seen in the right panel of Fig. 1. We find that the flux transport due to the pumping can be more important than the meridional circulation at providing the correct migration of the magnetic fields, in setting the correct period of the cycle and also at providing the appropriate parity of the magnetic fields. Models including a near-surface shear layer and turbulent pumping result in butterfly diagrams that are also in agreement with the observations (Guerrero & de Gouveia Dal Pino 2008).

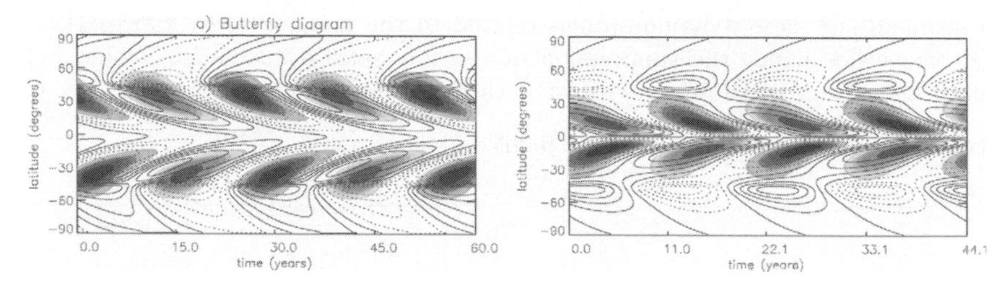

Figure 1. Butterfly diagram for models without and with turbulent pumping (left and right, respectively), the blue (red) contour scale represents positive (negative) toroidal field at the base of the convection zone, the solid (dotted) lines represent radial fields at the surface.

When the magnetic field is strong, the turbulence decreases and the turbulent diffusivity is suppressed. We have included this effect by adopting the following algebraic quenching function $\eta = \eta_T/(1+(\overline{\mathbf{B}}/B_q)^2)$ (Guerrero *et al.* 2009). We have found that, as soon as the magnetic field reaches the value B_q, the diffusivity can be strongly suppressed leading to the formation of long-lived small and intense magnetic structures. The magnitude of the maximum magnetic field can be twice as large as that of models without quenching, nevertheless these intense fields appear mainly at the center of the convection zone rather than at its base where it is usually believed to be produced. We find also that the larger the quenching efficiency the larger the period of the cycle, however for high values of η_T the models drift from the flux-transport regime to the diffusion regime.

Acknowledgements

This work has been partially supported by grants of the Brazilian agencies CNPq and FAPESP. GG thanks the hospitality of the HAO-NCAR where part of this work was developed.

References

Brandenburg, A. 2005, *ApJ* 625, 539
Dikpati, M. & Charbonneau, P. 1999, *ApJ* 518, 508
Guerrero, G., de Gouveia Dal Pino, E. M. 2008, *A&A* 485, 267
Guerrero G., Dikpati, M. & de Gouveia Dal Pino, E. M. 2009, in prep.
Parker, E. N. 1955, *ApJ* 122, 293

Carsten Denker (middle) chatting with H. Zinnecker (left) and Günther Rüdiger (right)

Elisabete de Gouveia dal Pino (right) talking to Federico Stasyszyn

Visit to the solar telescopes on Teide/Izana

The meeting place: Hotel Playa de la Arena

Session III

Planetary magnetic fields and the formation and evolution of planetary systems and planets; exoplanets

Cosmic Magnetic Fields:
From Planets, to Stars and Galaxies
Proceedings IAU Symposium No. 259, 2008
K.G. Strassmeier, A.G. Kosovichev & J.E. Beckman, eds.

doi:10.1017/S1743921309030592

The role of magnetic fields for planetary formation

Anders Johansen

Leiden Observatory, Leiden University, P.O. Box 9513, 2300 RA Leiden, The Netherlands
email: ajohan@strw.leidenuniv.nl

Abstract. The role of magnetic fields for the formation of planets is reviewed. Protoplanetary disc turbulence driven by the magnetorotational instability has a huge influence on the early stages of planet formation. Small dust grains are transported both vertically and radially in the disc by turbulent diffusion, counteracting sedimentation to the mid-plane and transporting crystalline material from the hot inner disc to the outer parts. The conclusion from recent efforts to measure the turbulent diffusion coefficient of magnetorotational turbulence is that turbulent diffusion of small particles is much stronger than naively thought. Larger particles – pebbles, rocks and boulders – get trapped in long-lived high pressure regions that arise spontaneously at large scales in the turbulent flow. These gas high pressures, in geostrophic balance with a sub-Keplerian/super-Keplerian zonal flow envelope, are excited by radial fluctuations in the Maxwell stress. The coherence time of the Maxwell stress is only a few orbits, where as the correlation time of the pressure bumps is comparable to the turbulent mixing time-scale, many tens or orbits on scales much greater than one scale height. The particle overdensities contract under the combined gravity of all the particles and condense into gravitationally bound clusters of rocks and boulders. These planetesimals have masses comparable to the dwarf planet Ceres. I conclude with thoughts on future priorities in the field of planet formation in turbulent discs.

Keywords. Diffusion – instabilities – MHD – planetary systems: protoplanetary disks – solar system: formation – turbulence

---

1. Introduction

Planets form in protoplanetary discs of gas and dust as the dust grains collide and grow to ever larger bodies (Safronov 1969). An important milestone is the formation of km-sized planetesimals. Drag force interaction between particles and gas plays a big role for the dynamics of dust particles. This way the collisional evolution of the dust grains into planetesimals is intricately connected to the physical state of the gas flow. The magnetorotational instability renders Keplerian rotation profiles linearly unstable in the presence of a magnetic field of suitable strength (Balbus & Hawley 1991). The ensuing magnetorotational turbulence is currently the best candidate for driving protoplanetary disc accretion. The relatively ease at which self-sustained magnetorotational turbulence is produced by numerical magnetohydrodynamics codes makes it an excellent test bed for analysing dust motion and formulating theories of planet formation in a turbulent environment.

An interesting constraint on the magnetic field present in the solar nebula comes from meteoritics. Most carbonaceous chondrites have a remanent magnetisation as high as a few Gauss, frozen in as the material cooled past the blocking temperature (Levy & Sonett 1978). A quote from the excellent review paper by Levy & Sonett (1978) is particularly concise on the origin of such a strong magnetic field:

"So far as we can see, there are four major candidates for the origin of the primordial magnetic field which produced the remanence in carbonaceous chondrites. They are:

1. Magnetic fields generated in very large meteorite parent bodies
2. The interstellar magnetic field compressed to high intensity by the inflowing gas
3. A strong solar magnetic field permeating the early solar system
4. A hydromagnetic dynamo field produced in the gaseous nebula itself"

Levy & Sonett (1978) continue to put forward various physical arguments to rule out possibility 1 and 2 [the undifferentiated parent bodies of carbonaceous chondrites were unlikely to harbour a magnetic field, and turbulent diffusion strongly limits the amount of field that can be dragged into the solar nebula (Lubow *et al.* 1994)]. The magnetic field of the wind emanating from the young sun can potentially be strong enough to imprint fields of several G at a few AU from the sun. But the most likely scenario remains that the magnetic field was created by the differential rotation and dynamo process in the solar nebula itself. Simulations of magnetised shear flows indeed show that a weak seed field can be amplified by the magnetorotational instability to a few percent of the thermal pressure (Brandenburg *et al.* 1995, Hawley *et al.* 1996, Sano *et al.* 2004).

In the following sections I briefly review the role of such magnetised turbulence on the motion of dust particles and on the cosmogony of planetesimal formation.

2. Diffusion of small dust grains

The magnetised turbulence in protoplanetary discs moves small dust grains around, preventing them from sedimenting to the mid-plane and transporting dusty material radially in the disc (Gail 2002, van Boekel *et al.* 2004). This section describes recent efforts to determine the turbulent diffusion coefficient $D_{\rm t}$ of magnetorotational turbulence.

If turbulent transport can be described as a diffusion process, then the evolution of the dust particle density $\rho_{\rm d}$ follows the partial differential equation

$$\frac{\partial \rho_{\rm d}}{\partial t} = \boldsymbol{\nabla} \cdot \left[D_{\rm t} \rho_{\rm g} \boldsymbol{\nabla} \left(\frac{\rho_{\rm d}}{\rho_{\rm g}} \right) \right] . \tag{2.1}$$

Here $\rho_{\rm g}$ is the gas density, its presence signifying that diffusion acts to even out differences in the solids-to-gas ratio $\epsilon_{\rm d} = \rho_{\rm d}/\rho_{\rm g}$. The vertical flux of dust particles contains contributions from the advection (sedimentation at velocity w_z) and the diffusion,

$$\mathcal{F}_z = w_z \rho_{\rm d} - D_{\rm t} \rho_{\rm g} \frac{\partial(\rho_{\rm d}/\rho_{\rm g})}{\partial z} . \tag{2.2}$$

In sedimentation-diffusion equilibrium we have $\mathcal{F}_z = 0$. Setting the velocity of the dust particles its terminal value, $w_z = -\tau_{\rm f} \Omega^2 z$ (where Ω is the Keplerian frequency and $\tau_{\rm f}$ is the friction time of the particles), gives the solution (e.g. Dubrulle *et al.* 1995)

$$\epsilon_{\rm d}(z) = \epsilon_1 \exp[-z^2/(2H_\epsilon^2)] \tag{2.3}$$

for the solids-to-gas ratio $\epsilon_{\rm d} = \rho_{\rm d}/\rho_{\rm g}$. The scale height H_ϵ follows the expression

$$H_\epsilon^2 = \frac{D_{\rm t}}{\tau_{\rm f} \Omega^2} , \tag{2.4}$$

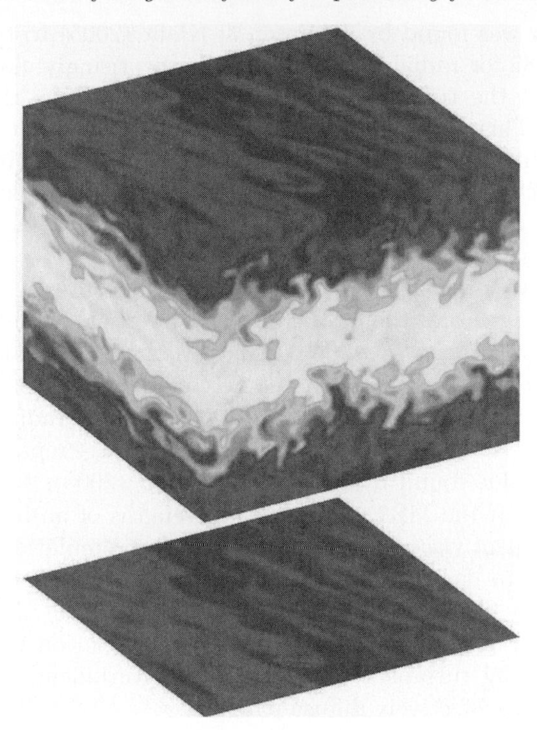

Figure 1. The dust density at the sides of a simulation box corotating with the disc at an arbitrary distance from the central star. The radial direction points right, the azimuthal direction left and up, while the vertical direction points directly up. The dust density distribution arises from an equilibrium between sedimentation and turbulent diffusion by the magnetorotational turbulence.

while the solids-to-gas ratio in the mid-plane is given by

$$\epsilon_1 = \epsilon_0 \sqrt{\left(\frac{H}{H_\epsilon}\right)^2 + 1}\,. \tag{2.5}$$

Here H is the pressure scale height of the gas. In the above derivations we have assumed (a) that the friction time is independent of height over the mid-plane and (b) that the diffusion coefficient is independent of height over the mid-plane. None of these assumptions are true in general, but if we stay within a few scale heights of the mid-plane and treat the diffusion coefficient as a suitably averaged diffusion coefficient, then the expressions are relatively good approximations.

In a real turbulent flow the observed diffusion-sedimentation equilibrium can be used to measure the turbulent diffusion coefficient of the flow. In Figure 1 we show an example of such a diffusion-sedimentation equilibrium (from Johansen & Klahr 2005) for a shearing box simulation of magnetorotational turbulence. The problem of determining the diffusion coefficient is thus reduced to measuring the scale height H_ϵ of the dust in Figure 1. Using equation 2.4 then directly yields a value of $D_{\rm t}$. Obviously the diffusion coefficient must scale with the overall strength of the turbulence. The interesting quantity to determine is thus the Schmidt number Sc, defined as the turbulent viscosity coefficient relative to the turbulent diffusion coefficient, $\mathrm{Sc} = \nu_{\rm t}/D_{\rm t}$. In a Keplerian disc the turbulent viscosity is in turn defined from the Reynolds and Maxwell stresses,

$$\nu_{\rm t} = \frac{2}{3}\frac{\langle \rho u_x u_y - \mu_0^{-1} B_x B_y \rangle}{\langle \rho \rangle} \tag{2.6}$$

The Schmidt number was found by Johansen & Klahr (2005) to be around 1.5 for vertical diffusion and 0.85 for radial diffusion. This is surprisingly close to unity and a bit mysterious given that the turbulent viscosity is dominated by the magnetic Maxwell stress $\langle -\mu_0^{-1} B_x B_y \rangle$. This stress does not directly affect the dust particles. A possible explanation is that diffusion is determined by the diagonal entries in the $u_i u_j$ correlation tensor. These are much higher than the off diagonal Reynolds stress $u_x u_y$. Thus the MRI inherently transports a passive scalar (by fluid motion) and the angular momentum (by magnetic tension) equally well.

Different groups have used various independent methods to measure the turbulent diffusion coefficient of magnetorotational turbulence. A vertical Schmidt number of around unity was measured by Turner *et al.* (2006), while Fromang & Papaloizou (2006) reported a value of approximately three. This gives some confidence that the Schmidt number is well constrained. However, Carballido *et al.* (2005) found a radial Schmidt number as high as ten in relatively strong turbulence. To address the discrepancy between this value and the much lower value found by Johansen & Klahr (2005), Johansen *et al.* (2006b) performed simulations of the MRI with various strengths of an imposed, external field, yielding a higher turbulent viscosity than in zero net flux simulations. The Schmidt number was indeed found to decrease with increasing strength of the turbulence. Stronger turbulence (such as in Carballido *et al.* 2005) is less good at diffusing dust particles relative to its stresses. The explanation is that the correlation time of the turbulence decreases with increasing turbulent energy, and that turbulent structures do not stay coherent long enough to effectively diffuse particles.

Large particles partially decouple from the turbulence and are primarily diffused by large scale eddies with relatively long correlation times. The experiments by Carballido *et al.* (2006) indeed showed that the diffusion coefficient falls rapidly for particles above a few metres in size, in good agreement with the analytical derivations of Youdin & Lithwick (2007).

3. Zonal flows

While smaller dust particles are clearly prevented from forming a very thin mid-plane layer by the magnetorotational turbulence, pebbles, rocks and boulders begin to gradually decouple from the gas. Accretion discs are radially stratified with a pressure that decreases with distance from the star. The pressure gradient acts to reduce the effect of gravity felt by the gas, and as a result the gas rotates slightly slower than Keplerian. The particles, however, do not react to gas pressure gradients and aim to orbit with the Keplerian speed. The head wind of the slower rotating gas drains the particles of angular momentum and they spiral towards the star in a few hundred orbital periods (Weidenschilling 1977).

The radial pressure profile of gas in turbulent discs need not be monotoneously falling. The presence of large scale, long-lived pressure bumps leads to concentrations of migrating dust particles into radial bands. Simulations of magnetorotational turbulence in a box gives evidence that such pressure bumps form spontaneously in the turbulent flow (Johansen *et al.* 2006, Johansen *et al.* 2009). In Figure 2 we plot the gas density and the azimuthal velocity, averaged over the azimuthal and vertical directions, as a function of radial distance from the centre of the box x and the time t. The gas density exhibits axisymmetric column density bumps with amplitude around 5% of the average density. These bumps are surrounded by a sub-Keplerian/super-Keplerian zonal flow, maintaining perfect geostrophic balance with $2\rho_0 \Omega u_y \approx \partial P/\partial r$.

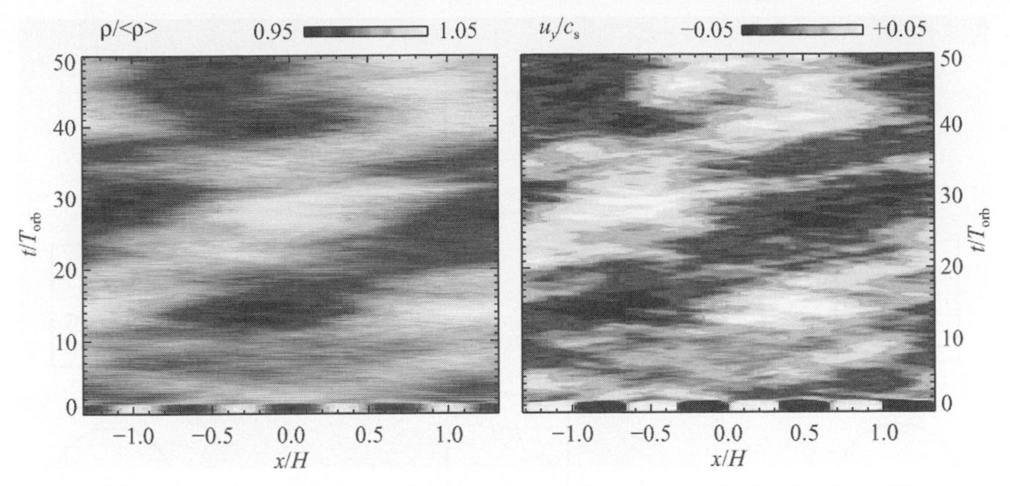

Figure 2. The gas density (left plot) and the azimuthal velocity (right plot) as a function of the radial distance from the centre of the box, H, and the time, t, measured in orbits. There is a perfect $-\pi/2$ phase difference between the pressure bump and the zonal flow, in agreement with a geostrophic balance. The zonal flow has in turned been excited by a large scale variation in the Maxwell stress.

Varying resolution, presence or non-presence of stratification, dissipation parameters and dissipation types, Johansen *et al.* (2009) find that pressure bumps and zonal flows are ubiqituous in shearing box simulations of magnetorotational turbulence, provided that the simulation box is large enough (more than one scale height in radial extent) and possibly also that the physical dissipation is high enough. What is then the launching mechanism for these zonal flows? Large scale fluctuations in the Maxwell stress lead to a differential transport of momentum. Thus the magnetic field is responsible for separating the orbital flow into regions of slightly faster and slightly slower rotation.

A model of the excitation of zonal flows and pressure bumps can be obtained from a simplified version of the dynamical equations,

$$0 = 2\Omega\hat{u}_y - \frac{c_s^2}{\rho_0}\mathrm{i}k_0\hat{\rho}\,, \tag{3.1}$$

$$\frac{\mathrm{d}\hat{u}_y}{\mathrm{d}t} = -\frac{1}{2}\Omega\hat{u}_0 + \hat{T}\,, \tag{3.2}$$

$$\frac{\mathrm{d}\hat{\rho}}{\mathrm{d}t} = -\mathrm{i}k_0\hat{u}_x - \frac{\hat{\rho}}{\tau_{\mathrm{mix}}}\,. \tag{3.3}$$

Here $\hat{u}_x$, $\hat{u}_u$ and $\hat{\rho}$ are the amplitudes of the radial and azimuthal velocity and gas density at the largest radial scale of the simulation, with wavenumber k_0. The first equation denotes geostrophic balance, while we have kept the time evolution terms in the two other equations. Non-linear terms enter through $\hat{T}$, the large scale magnetic tension, and $\hat{\rho}/\tau_{\mathrm{mix}}$, turbulent diffusion of the mass density.

We can combine the above equations into a single evolution equation for the density,

$$\frac{\mathrm{d}\hat{\rho}}{\mathrm{d}t} = \frac{1}{1+k_0^2H^2}\left(\hat{F} - \frac{\hat{\rho}(t)}{\tau_{\mathrm{mix}}}\right)\,, \tag{3.4}$$

where $\hat{F} \equiv -2\mathrm{i}k_0\rho_0\hat{T}/\Omega$ is the forcing term. The prefactor $c_k \equiv (1+k_0^2H^2)^{-1}$ is a pressure correction for small-scale modes that both decreases the amplitude of the forcing and increases the effective damping time. The coherence time-scale of the Maxwell stress (and thus of $\hat{F}$), τ_{for}, is generally much shorter than the mixing time-scale, τ_{mix}. Thus we need

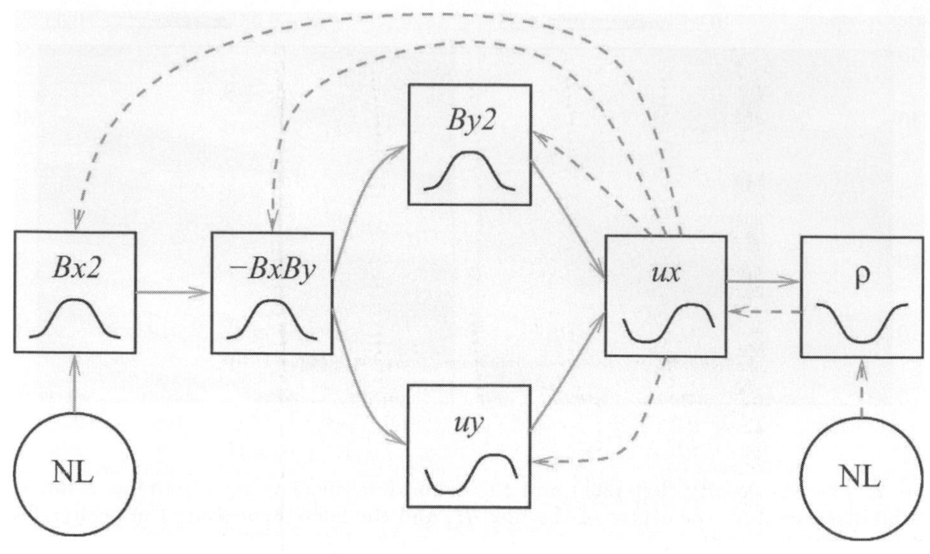

Figure 3. Diagram of how non-linear excitation of the large scale radial magnetic energy leads to the excitation of zonal flow. Green arrows label positive energy transfer, while red arrows (dashed) denote energy sinks. Non-linear interactions are responsible both for the excitation and for the balance, the latter through diffusive mixing of the gas density.

to model equation 3.4 as a stochastic differential equation (see e.g. Youdin & Lithwick 2008). The Maxwell stress gives short, uncorrelated kicks to the zonal flow. This would lead to an amplitude that grows as the square root of time. However, in presence of turbulent diffusion, the solution tends to

$$\frac{\hat{\rho}_{\rm eq}}{\rho_0} = 2\sqrt{c_k \tau_{\rm for} \tau_{\rm mix}} H k_0 \frac{\hat{T}}{c_{\rm s}} . \qquad (3.5)$$

The correlation time of the zonal flows is predicted to be equal to the mixing time-scale, in good agreement with the results. The model also predicts that $\hat{\rho}_{\rm eq} \propto k^{-2}$ for $k_0 H \gg 1$. This is in very good agreement with the very clearly sinusoidal density fluctuations seen in Figure 2.

The cause of the large scale variation in the Maxwell stress remains unknown. Johansen *et al.* (2009) argue that magnetic energy takes part in an inverse cascade from the moderate scales, excited directly by the MRI, to large scales. A diagram of the zonal flow excitation appears in Figure 3. Note that the above zonal flow excitation model assumes that the magnetic tension (i.e. $-B_x B_y$ in Figure 3) is given, whereas in fact one may go on step further back to B_x^2, which is excited directly by a non-linear term. The Maxwell stress then increases from the Keplerian stretching of the radial field. The model also predicts that the magnetic pressure should grow in anti-phase with the thermal pressure. This is indeed also observed.

4. Planetesimal formation

The zonal flows presented in the last section are very efficient at trapping particles. At the outer sub-Keplerian side the particles face a slightly stronger headwind and drift faster inwards. At the inner super-Keplerian side the particles experience a slight backwind and move out. The effect of pressure bumps on the migration of rocks and boulders goes at least back to Whipple (1972). It has later received extensive analytical treatment

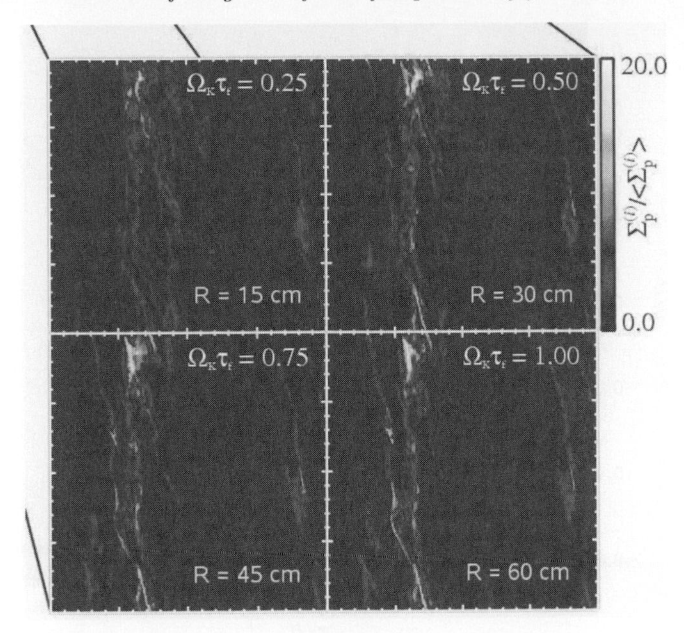

Figure 4. The column density of four different particle sizes, before self-gravity has been switched on. The particles concentrate at the same locations, but larger particles experience a higher local column density.

by Klahr & Lin (2001) and by Haghighipour & Boss (2003). The narrow box simulations of Hodgson & Brandenburg (1998) found no evidence for long-lived concentrations of relatively tighly coupled particles in magnetorotational turbulence. However, Johansen *et al.* (2006a) observed concentrations of marginally coupled dust particles (cm-m sizes), by up to two orders of magnitude higher than the average paricle density, in high pressure regions occuring in magnetorotational turbulence. In a simulation of a (part of a) global disc Fromang & Nelson (2005) reported similar concentrations in a long-lived vortex structure.

The question of how long-lived high pressure structures form and survive in magnetised turbulence is of general interest. However, their effect on planetesimal formation is no less intriguing. Johansen *et al.* (2007) expanded earlier models of boulders in turbulence by considering several particle sizes simultaneously and solving for the self-gravity of the boulders. First the turbulence is allowed to develop for 20 local rotation periods without the gravity of the particles (which is weak anyway). This way a sedimentary mid-plane layer, with a width of a few percent of the gas scale height, forms in equilibrium between sedimentation and turbulent diffusion. In Figure 4 we show the column density of the four different particle sizes – rocks and boulders with sizes 15 cm, 30 cm, 45 cm, and 60 cm. A weak zonal flow has been sufficient to create bands of very high particle overdensity. An additional instability in the coupled motion of gas and dust has further augmented the local overdensities (Goodman & Pindor 2000, Youdin & Goodman 2005, Youdin & Johansen 2007, Johansen & Youdin 2007).

As the self-gravity of the disc is activated, the overdense bands contract radially. Upon reaching the local Roche density, a full non-axisymmetric collapse occurs and a few gravitationally bound clusters of rocks and boulders condense out of the particle layer. The column density of the particles is shown in Figure 5.

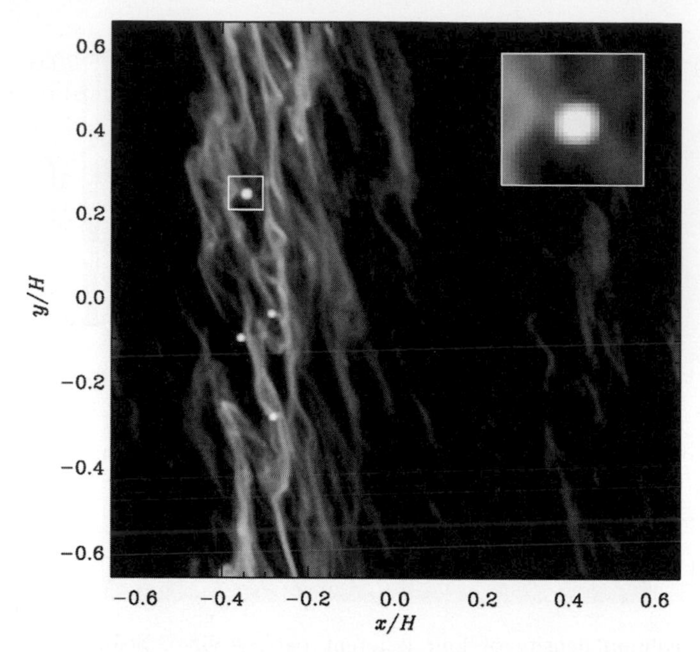

Figure 5. The column density at $\Delta t = 7T_{\rm orb}$ after self-gravity is turned on. Four gravitationally bound clusters of rocks and boulders have condensed out of the flow. The most massive cluster (see enlargement) has a mass comparable to the dwarf planet Ceres by the end of the simulation.

5. Conclusions

The presence of magnetic fields in protoplanetary discs is of vital importance for planet formation and for observational properties of protoplanetary discs. Small dust grains are transported very efficiently by the turbulence. While this counteracts sedimentation to the mid-plane, and thus prevents the razor thin mid-plane layer of Goldreich & Ward (1973) from forming, the turbulent transport underlies the presence of small dust grains many scale heights above the disc mid-plane. The presence of crystalline silicates in the cold outer regions of discs (Gail 2002, van Boekel *et al.* 2004) can likely also be attributed to turbulent diffusion (but see Dullemond *et al.* 2006 for an alternative view taking into account disc formation history).

Larger dust particles – pebbles, rocks, and boulders – slow down or reverse the radial migration as they encounter variations in the radial pressure gradient. Fluctuations in the Maxwell stress, with a coherence time of a few orbits, launch axisymmetric zonal flows. These flows in turn go into geostrophic balance with a radial pressure bump. The concentrations of solid particles in such pressure ridges can get high enough for a gravitational collapse into planetesimals to occur. However, a satisfactory mechanism for setting the scale of the pressure bumps is lacking, as the bumps grow to fill the box for all considered box sizes in Johansen *et al.* (2009). The final size may ultimately be set by global curvature effects (Lyra *et al.* 2008a).

An important problem related to the motion of dust particles in turbulence is their collision speeds. The relative speed of small particles approaches zero as the particle separation is decreased. But particles that are only marginally coupled to the turbulent eddies have a significant memory of their trajectories and can collide at non-zero speeds. Carballido *et al.* (2008) indeed found that the relative speeds of large particles is unchanged below a certain separation, giving confidence that the proper collision speed has been found. Turbulent eddies with sizes around the stopping length of the particle are most efficient at inducing relative motion. However, even for marginally coupled particles these eddies may be on the edge of the dissipative subranges of the turbulence, due to the limited resolution of numerical simulations. Johansen *et al.* (2007) found that the collision speed, measured as the velocity difference over a single grid cell, increases by 10–20% each time the resolution is doubled. Ultrahigh resolution measurements of large scale and short scale relative speeds, and comparison to analytical models (Völk *et al.* 1980, Cuzzi *et al.* 1993, Schräpler & Henning 2004, Youdin & Lithwick 2008) and to sticking experiments (Wurm *et al.* 2006, Blum & Wurm 2008), is an important future priority for our picture of how planets form in turbulent gas discs.

To our best knowledge parts of the solar nebula had so low ionisation fraction that the collisional resistivity was too high for the magnetorotational instability to develop (e.g. Gammie 1996, Sano *et al.* 2000). Kretke & Lin (2007) and Brauer *et al.* (2008) modelled the sharp increase in resistivity as the gas temperature drops below the freezing point of ice (the so-called snow line) at a few AU from the sun. The corresponding drop in turbulence activity leads to a pile up of gas and a run away growth in particle density from the influx of migrating solid particles. The lack of radial drift in such a location allows for planetesimal formation by coagulation to occur without having to compete with the radial drift time-scale. The edges of "dead zones" can be unstable to a Rossby wave instability (Inaba & Barge 2006, Varnière & Tagger 2006). The Rossby vortices efficiently trap particles, which leads to a burst of planet formation at the edge of dead zones (Lyra *et al.* 2008b). This way magnetic fields helps the planet formation process both in their presence and in their absence.

Acknowledgements

I would like to thank my collaborators Andrej Bicanski, Andrew Youdin, Axel Brandenburg, Frithjof Brauer, Hubert Klahr, Jeff Oishi, Kees Dullemond, Mordecai-Mark Mac Low, Thomas Henning, and Wladimir Lyra.

References

Balbus, S. A. & Hawley, J. F. 1991, *ApJ* 376, 21
Blum, J. & Wurm, G. 2008, *ARA&A* 46, 21
Brauer, F., Henning, T., & Dullemond, C. P. 2008, *A&A* 487, L1
Brandenburg, A., Nordlund, Å., Stein, R.F., & Torkelsson, U. 1995, *ApJ* 446, 741
Carballido, A., Stone, J. M., & Pringle, J. E. 2005, *MNRAS* 358, 1055
Carballido, A., Fromang, S., & Papaloizou, J. 2006, *MNRAS* 373, 1633
Carballido, A., Stone, J. M., & Turner, N. J. 2008, *MNRAS* 386, 145
Cuzzi, J. N., Dobrovolskis, A. R., & Champney, J. M. 1993, *Icarus* 106, 102
Dubrulle, B., Morfill, G., & Sterzik, M. 1995, *Icarus* 114, 237
Dullemond, C. P., Apai, D., & Walch, S. 2006, *ApJ* 640, L67
Fromang, S. & Nelson, R. P., 2005, *MNRAS* 364, L81
Fromang, S. & Nelson, R. P. 2006, *A&A* 457, 343
Fromang, S. & Papaloizou, J. 2006, *A&A* 452, 751
Gail, H.-P. 2002, *A&A* 390, 253
Gammie, C. F. 1996, *ApJ* 457, 355
Goldreich, P. & Ward, W. R., 1973, *ApJ* 183, 1051
Goodman, J. & Pindor, B. 2000, *Icarus* 148, 537
Haghighipour, N., & Boss, A. P., 2003, *ApJ* 598, 1301
Hawley, J. F., Gammie, C. F., & Balbus, S. A. 1996, *ApJ* 464, 690
Hodgson, L. S. & Brandenburg, A. 1998, *A&A* 330, 1169
Inaba, S. & Barge, P. 2006, *ApJ* 649, 415

Johansen, A. & Klahr, H. 2005, *ApJ* 634, 1353
Johansen, A., Klahr, H., Henning, Th. 2006, *ApJ* 636, 1121
Johansen, A., Klahr, H., & Mee, A. J. 2006c, *MNRAS* 370, L71
Johansen, A., Oishi, J. S., Low, M.-M. M., Klahr, H., Henning, T., & Youdin, A. 2007, *Nature* 448, 1022
Johansen, A. & Youdin, A. 2007, *ApJ* 662, 627
Johansen, A., Youdin, A., & Klahr, H. 2009, *ApJ*, submitted
Klahr, H. H. & Lin, D. N. C., 2001, *ApJ* 554, 1095
Kretke, K. A. & Lin, D. N. C. 2007, *ApJ* 664, L55
Levy, E. H. & Sonett, C. P. 1978, in: T. Gehrels (ed.), *IAU Colloqium 52: Protostars and Planets* (The University of Arizona Press), p. 516
Lubow, S. H., Papaloizou, J. C. B., & Pringle, J. E. 1994, *MNRAS* 267, 235
Lyra, W., Johansen, A., Klahr, H., & Piskunov, N. 2008, *A&A* 479, 883
Lyra, W., Johansen, A., Klahr, H., & Piskunov, N. 2008, *A&A* 491, L41
Safronov, V. S. 1969, *Evoliutsiia doplanetnogo oblaka* (Nakua)
Sano, T., Miyama, S. M., Umebayashi, T., & Nakano, T. 2000, *ApJ* 543, 486
Sano, T., Inutsuka, S.-i., Turner, N. J., & Stone, J. M. 2004, *ApJ* 605, 321
Turner, N. J., Willacy, K., Bryden, G., & Yorke, H. W. 2006, *ApJ* 639, 1218
van Boekel, R., *et al.* 2004, *Nature* 432, 479
Varnière, P. & Tagger, M. 2006, *A&A* 446, 13
Völk, H. J., Morfill, G. E., Roeser, S., & Jones, F. C. 1980, *A&A* 85, 316
Weidenschilling, S. J. 1977, *MNRAS* 180, 57
Whipple, F. L. 1972, in: A. Elvius (ed.), *From Plasma to Planet* (Wiley Interscience Division), p. 211
Wurm, G., Paraskov, G., & Krauss, O. 2005, *Icarus* 178, 253
Youdin, A. N. & Goodman, J. 2005, *ApJ* 620, 459
Youdin, A. N. & Johansen, A. 2007, *ApJ* 662, 613
Youdin, A. N.& Lithwick, Y. 2007, *Icarus* 192, 588

Discussion

BLACKMAN: There has been recent evidence that the Maxwell stress resulting from the MRI scales with box site for small boxes. Have you been able to see a saturation of the Maxwell stress in combining your large radial scale and multi-scale hight (in vertical direction) boxes? One would expect that boxes that allow field structures with H≈R are required.

JOHANSEN: Yes, we have seen such a saturation in the Maxwell stress. When changing the radial and azimuthal extent of the box from $L_x = L_y = 1.32H$ to $L_x = L_y = 2.64H$ the Maxwell stress increases from ≈0.005 to ≈0.01. However, another doubling at the box size yields no further change to the Maxwell stress.

Cosmic Magnetic Fields:
From Planets, to Stars and Galaxies
Proceedings IAU Symposium No. 259, 2008
K.G. Strassmeier, A.G. Kosovichev & J.E. Beckman, eds.

doi:10.1017/S1743921309030609

Planetary dynamos: from equipartition to asymptopia

Paul H. Roberts

IGPP University of California, Los Angeles, CA 90095, USA
email: roberts@math.ucla.edu

Abstract. This review focuses on three topics relevant to naturally-occurring dynamos. The first considers how a common belief, that states of equipartition of magnetic and kinetic energy are preferred in nonrotating systems, is modified when Coriolis forces are influential, as in the Earth's core. The second reviews current difficulties faced by planetary and stellar dynamo theories, particularly in representing the sub-grid scales. The third discusses recent attempts to extract scaling laws from numerical integrations of the Boussinesq dynamo equations.

Keywords. Convection – turbulence – magnetic fields – instabilities

1. Introduction

Stellar and planetary dynamo theory has exploded in the last 15 years, and the rate of expansion of this universe seems to be accelerating! It is obviously impossible to do more than touch on the two fields here, and my aim is modest: to draw attention to a few issues that seem (to me) to have relevance to both fields.

2. Equipartition

In applying magnetohydrodynamics (MHD) to cosmic contexts, order of magnitude arguments frequently appeal to equipartition: $M = K$, where $M = B^2/2\mu$ and $K = \rho V^2/2$ are the magnetic and kinetic energy densities, $\mathbf{B}$ being magnetic field, $\mathbf{V}$ fluid velocity, ρ mass density and μ magnetic permeability (here $4\pi \times 10^{-7}$H m^{-1}; SI units). This equality seems to be based on the idea that two of the principal nonlinearities in the equation of motion, the inertial force $\rho\mathbf{V}\cdot\nabla\mathbf{V}$ and the Lorentz force $\mathbf{J}\times\mathbf{B}$, should approximately balance, where $\mathbf{J}$, the electric current density, is given by Ampère's law, $\mu\mathbf{J} = \nabla\times\mathbf{B}$. If one now writes $|\rho\mathbf{V}\cdot\nabla\mathbf{V}| \approx \rho V^2/L$ and $|\mathbf{J}\times\mathbf{B}| \approx JB$ with $\mu J \approx B/L$, one at once finds that $M \approx K$.

Little thought is needed to identify weaknesses in this argument. Near a stellar surface where ρ is small, JB is a poor estimate of $|\mathbf{J}\times\mathbf{B}|$ because $\mathbf{J}$ and $\mathbf{B}$ tend to be parallel, so that $|\mathbf{J}\times\mathbf{B}| \ll JB$ and $K \ll M$. Also, deep in most stars, M is much less than K mainly because $\mathbf{V} \approx \mathbf{\Omega}_0\times\mathbf{r}$ where $\mathbf{\Omega}_0$ is the star's angular velocity and $\mathbf{r}$ is the radius vector from its center of mass. Then the largest part of $\mathbf{V}\cdot\nabla\mathbf{V}$ is $-\nabla\frac{1}{2}|\mathbf{\Omega}_0\times\mathbf{r}|^2$, which combines with the gradient of the gravitational potential Ψ in the approximate hydrostatic balance of the star, leaving only smaller terms. The kinetic energy density relative to $\mathcal{F}$ is $K^r = \rho U^2/2$, where $\mathbf{U} = \mathbf{V} - \mathbf{\Omega}_0\times\mathbf{r}$ is the velocity in frame $\mathcal{F}$. When $|\mathbf{U}| \ll |\mathbf{\Omega}_0\times\mathbf{r}|$, i.e., when the Rossby number $R_o = U/2\Omega_0 r_s$ is small where r_s is the radius of the body, the inertial force in $\mathcal{F}$ is the Coriolis force, $2\rho\mathbf{\Omega}_0\times\mathbf{U}$, which we assume is $O(2\Omega_0\rho U)$. This balances the Lorentz force of order $JB = O(B^2/\mu r_s)$ when $K^r/M = O(R_o) \ll 1$.

Irrespective of the size of R_o, the statement $K^r = R_oM$ can also be written as

$$E_\ell \approx R_m, \qquad \text{where} \qquad R_m = \frac{Ur_s}{\eta}, \qquad E_\ell = \frac{B^2}{2\Omega_0\eta\mu\rho} \tag{2.1a,b,c}$$

are the magnetic Reynolds number and Elsasser number, $\eta = 1/\mu\sigma$ being magnetic diffusivity and σ electrical conductivity. The condition for dynamo action is $R_m \geqslant R_m^c$ where R_m^c, the critical or marginal magnetic Reynolds number, is O(1). Here "O(1)" hides the fact that numerically the model–dependent R_m^c is usually of order 100. Obviously (2.1a) is inappropriate at and near the marginal state, since $R_m = \mathrm{O}(1)$ implies $E_\ell = \mathrm{O}(1)$. The statement $E_\ell = \mathrm{O}(1)$ expresses a balance between the Coriolis force $2\Omega_0\rho U$ and the Lorentz force JB when Ohm's law $\mathbf{J} = \sigma(\mathbf{E}+\mathbf{U}\times\mathbf{B})$ is used to estimate J not as $B/\mu r_s$, but as σUB, where $\mathbf{E}$ is the electric field in $\mathcal{F}$. Often E_ℓ is written as $V_A^2/2\Omega_0\eta$, where $V_A = B/\sqrt{(\mu\rho)}$ is the Alfvén velocity. In the geophysical literature E_ℓ is normally denoted by Λ, but we have a different use for Λ below.

If η and ρ are known, an estimate for U also gives R_o, R_m and K^r, so that $E_\ell = \mathrm{O}(R_m)$ and/or $K^r = \mathrm{O}(R_oM)$ provide an approximate B. Its reliability can, as for all order of magnitude arguments, be questioned. The possibility that JB overestimates $|\mathbf{J}\times\mathbf{B}|$ has already been mentioned, and $UB \approx |\mathbf{U}\times\mathbf{B}|$ and $2\Omega_0U \approx |2\mathbf{\Omega}_0\times\mathbf{U}|$ are open to the same criticism; similarly B/r_s is likely to underestimate $|\nabla\times\mathbf{B}|$. But the main objection (Christensen & Aubert 2006) is a physical one, that the magnitude of B is decided not by a force balance but by power availability. Fig. 7 of Christensen & Aubert (2006) indicates that E_ℓ/R_m is not constant but increases with R_m, roughly as $R_m^{1/2}$. See also §5 below.

3. Some geomagnitudes

The sole aim of this Section is to consider §2 in relation to the Earth. Uninterested readers are advised to skip to §4.

The observed $\mathbf{B}$ at the Earth's surface is a potential field. Because the Earth's mantle is a poor electrical conductor, spherical harmonic components of $\mathbf{B}$ of harmonic number $\ell \lesssim 13$ can be extrapolated downwards to the core surface $r = r_s = 3.480\times10^6$m. Sources of permanent magnetism in the Earth's crust prevent extrapolation of harmonics with $\ell > 13$. Extrapolation gives $B(r_s) \approx 0.39$ mT (Bloxham & Jackson 1992). As the power spectrum at $r = r_s$ is nearly flat [$P_\ell \propto \exp(-0.1\ell)$, e.g., Roberts *et al.* 2003], the missing harmonics $\ell > 13$ add significantly to the total rms B giving $B(r_s) \approx 0.46$ mT.

This is very likely to be an underestimate of B deep in the core, where $\mathbf{B}$ has a toroidal part $\mathbf{B}_T$ that does not contribute to the inferred $B(r_s)$, which is purely poloidal. Fifteen years ago, it was widely believed that the axisymmetric toroidal flow $\overline{\mathbf{U}}_T$ would dominate $\mathbf{U}$ and create from $\mathbf{B}_P$ a much larger $\mathbf{B}_T$ through the Ω–effect (usually called "the ω–effect" in the geophysical literature). The very first fully three-dimensional MHD dynamo simulations showed however that, while B_T tends to be larger than B_P, it is comparable in strength. It tends to be larger partly because $\overline{\mathbf{U}}_T$ contains a geostrophic flow, $\mathbf{U}_G = U_G(s,t)\widehat{\boldsymbol{\phi}}$; here t is time and (s,ϕ,z) are cylindrical coordinates with Oz parallel to $\mathbf{\Omega}_0$. This flow is unopposed by the Coriolis force because $2\mathbf{\Omega}_0\times\mathbf{U}_G = \nabla\psi$, where $\psi = -2\Omega_0\int U_G ds$ can be absorbed into the gravitational potential Ψ. Because it is unopposed, it tends to be larger than the ageostrophic flow $\mathbf{U}-\mathbf{U}_G$. In simulations that have a conducting inner core, there is often a large zonal shear at the tangent cylinder (the imaginary cylinder that touches the inner core on its equator). This shear also tends to enhance the toroidal field by the Ω–effect. Some notion of the overall effect of these processes can be gauged from the early simulation of Glatzmaier & Roberts (1996) for which $B(r_s)$ is about $\frac{1}{3}$ that of the Earth's, but for which the maximum B in the core is approximately 20 mT. If we take $B = 10$ mT as the mean, this exceeds $B(r_s)$ by a factor of over 60. If the same factor applies to the Earth. $B = 30$ mT would be a reasonable guestimate.

Torsional oscillations provide an indirect way of finding the rms strength of B_s. It may however underestimate B, because only $\mathbf{B}_P$ and the nonaxisymmetric part of $\mathbf{B}_T$ contribute to B_s; the zonal field $\overline{\mathbf{B}}_T$, which plausibly exceeds both, does not. This may be one reason why Zatman & Bloxham (1997) obtained their rather small mean value for B_s, approximately 0.4 mT. Another may be connected with the argument of Braginsky (1975) that the lines of force of $\overline{\mathbf{B}}_P$ in the core should tend to be parallel to the polar axis. The Alfvén velocity $\overline{V}_{As}$ for $B_s = 0.4$ mT is 3.6 mm s^{-1}, so that torsional waves cross the core in a time $2r_s/\overline{V}_{As}$ of about 60 years. There is evidence for a 60 year period in the geomagnetic secular variation and in the length of day, e.g., Roberts *et al.* (2007).

Traditionally, U is estimated from the speed of the westward drift of $\mathbf{B}(r_s)$. This is irregular and dependent on latitude, but has typically been taken as $0.2°\mathrm{yr}^{-1}$so that $\overline{U}_\phi(r_s) \approx 0.4$ mm s^{-1} at the core equator. More detailed analysis, described in detail by Holme (2007), gives a maximum $U(r_s)$, of order 1.2 mm s^{-1}. Of course, this tells nothing about U deeper in the core. It is also unclear how well Alfvén's frozen flux theorem applies and how closely the inferred motion of $\mathbf{B}$ betrays the magnitude of $\mathbf{U}$; conceivably it might be partly a wave motion (Braginsky 1964a; Hide 1966), as was seen clearly in the simulation of Glatzmaier & Roberts (1996). We take $U = 0.4$ mm s^{-1}. This gives $R_o = K^r/M = 7.9 \times 10^{-7}$, K^r= 0.8 mJ m^{-3} (taking $\rho = 10^4$ kg m^{-3}), $M = K^r/R_o = 1$ kJ m^{-3}, $R_m = E_\ell = 700$ (taking $\eta = 2$ m^{2}s^{-1}). Therefore V_A=0.45 m s^{-1} and $B = 50$ mT, which is about 50 times greater than the estimate of Christensen & Aubert (2006). Can this discrepancy be reduced? The following arguments help a little.

The Earth is cooling, currently radiating about 42TW into space. The fluid outer core (FOC) is known to be in a nearly isentropic state, implying that it is homogenized by convection that contributes $\mathcal{Q}_c$ to the outward heat flow $\mathcal{Q}(r_s)$ at the core surface. The remainder is the adiabatic heat flow of the isentropic state. The adiabatic temperature gradient $g\widetilde{\alpha}T/C_p$ is about 0.5 K km^{-1} at the core surface (from $g = 10.68$ m s^{-2} as gravitational acceleration, $\widetilde{\alpha} = 10^{-5}$ K^{-1} as thermal expansivity, $T = 4000$ K as temperature, and $C_p = 830$J kg^{-1}K^{-1} as specific heat, all at $r = r_s$). Taking the thermal conductivity as 40W m^{-1}K^{-1}, the adiabatic heat flow at the core surface is 2.8TW. Estimates of $\mathcal{Q}(r_s)$ range from 5TW to 15TW, so that 2TW $\lesssim \mathcal{Q}_c \lesssim$ 12TW.

The source of $\mathcal{Q}_c$ is partly thermal and partly gravitational. Radioactivity (^{40}K) has been estimated as providing at most 1TW. As the core cools, it becomes increasingly centrally condensed. Lighter constituents of the FOC, particularly oxygen, are preferentially rejected as the fluid freezes to form the solid inner core (SIC), These light constituents are buoyant, rising and mixing with the overlying fluid. This gravitational source of energy is much larger than the gravitational energy released in a mere contraction of the core, which is less than 1TW. Moreover it is a buoyancy source that feeds energy into the convective motions. The latent heat released in the freezing is a further (thermal) buoyancy source. Both these sources are proportional to the rate of advance, dr_i/dt, of the inner core boundary. Currently $r_i = 1.2215 \times 10^6$m and estimates of dr_i/dt are about 10^{-11}m s^{-1}, making the age of the SIC less than $\frac{1}{3}$ of the age of the Earth. Before the birth of the SIC, only primordial heat and radioactive sources (stronger then!) would provide buoyancy and one would expect the dynamo would produce a weaker field (or no field at all!). Except during brief polarity reversals, the geodipole moment has had however the same strength, within a factor of 2–3, for more than 3.4Gyr (Kono & Tanaka 1995, Fig. 6). This paradox is not faced by Christensen & Aubert (2006), since the power requirement when $B =$ 1TW is so small that dr_i/dt is less than 10^{-12}m s^{-1} and the

SIC is as old as the Earth. Presumably the FOC would be maintained on its adiabat by compositionally-driven convection, and heat would be pumped downward from the mantle (Loper 1978). This scenario might meet some geophysical opposition!

Since the Earth's core is a ferrous alloy, its magnetic Prandtl number

$$P_m = \nu/\eta \tag{3.1}$$

is very small, perhaps about 10^{-6}, where ν is the kinematic viscosity. Because $P_m \ll 1$, the total Joule loss,

$$\mathcal{Q}^\eta = \mu \int_{\text{core}} \eta J^2 dv\,, \tag{3.2}$$

is by far the larger part of the total dissipation $\mathcal{Q} = \mathcal{Q}^\eta + \mathcal{Q}^\nu$. Values of order 10^{-3} may be typical for P_m in a stellar plasma so that it is again plausible that $\mathcal{Q} \approx \mathcal{Q}^\eta$.

If we estimate the average value of J by equating $\mathcal{Q}^\eta$ and $\mathcal{Q}_c$ we obtain 0.06A m^{-2} $\lesssim J \lesssim$ 0.16A m^{-2}. An estimate of the mean strength B of the magnetic field $\mathbf{B}$ follows from $B = \mu JL$, again obtained from $\mu\mathbf{J} = \boldsymbol{\nabla}\times\mathbf{B}$. To use this, an estimate of L is required, but what should this be? Under plausible conditions, the scales mainly responsible for generating $\mathbf{B}$ are of order $R_m^{-1/2} r_s = \sqrt{(\eta r_s/U)}$, according to Christensen & Tilgner (2004) and Tobias & Cattaneo (2008). This gives 11 mT $\lesssim B \lesssim$ 28 mT.

4. Turbulent dynamos

Modeling planets and stars means coming to terms with turbulence. In the classic picture of turbulence, energy is injected on the largest scales and cascades to small scales through the inertial range, to be extracted as heat on the dissipation scale. Turbulence in the situations considered here is markedly different: energy is acquired at all scales through buoyancy. (For simplicity we consider only thermal buoyancy here.) There may be a significant sub-range of scales in which the energy cascade does not dominate, the dissipative losses being replenished directly by the buoyancy forces:

$$\rho g \widetilde{\alpha} v \theta \sim \mu \eta j^2\,. \tag{4.1}$$

Here $\mathbf{v}$ is the fluctuating part of $\mathbf{V}$, i.e., we are writing $\mathbf{V} = \overline{\mathbf{V}} + \mathbf{v}$, where $\overline{\mathbf{V}}$ is the ensemble or statistical mean; similarly $\mathbf{J} = \overline{\mathbf{J}} + \mathbf{j}$ and $T = \overline{T} + \theta$. Relation (4.1) also emphasizes another point: in MHD, the Lorentz force may be more important in the dynamical balance than the inertial force, especially when the Alfvén number, V/V_A, is small; also, ohmic dissipation may be much more significant than viscous dissipation, when $P_m \ll 1$ (see §3). This takes one even further away from the classic turbulence picture. In what follows. we shall no longer reserve $\mathbf{U}$ for velocity relative to $\mathcal{F}$, it being clear from the context whether $\mathbf{V}$ refers to the inertial or rotating frame.

The lack of an MHD turbulence theory is a void that cannot be filled by the computer, either now or in the foreseeable future. Advances in computer technology continually push back the GS/SGS frontier between the large *grid scales* (GS) of main interest and the small, numerically-unresolvable *sub-grid scales* (SGS), but the effect of the latter on the former cannot be ignored and must be represented in a physically plausible way. The first recourse is the classic Boussinesq–Reynolds ansatz (BRA) that draws an analogy between randomly moving molecules, carrying momentum and energy, and randomly moving SGS eddies or "blobs" of fluid that perform the same function for the GS, and much more effectively in strong turbulence. In this classic picture of turbulence in an incompressible fluid, the averages of the governing equations,

$$\partial_t \mathbf{V} + \mathbf{V}\cdot\boldsymbol{\nabla}\mathbf{V} = -\boldsymbol{\nabla}\Pi + \nu\nabla^2\mathbf{V}\,, \qquad \boldsymbol{\nabla}\cdot\mathbf{V} = 0\,, \tag{4.2a,b}$$

are

$$\partial_t \overline{\mathbf{V}} + \overline{\mathbf{V}}\cdot\nabla\overline{\mathbf{V}} = -\nabla\overline{\Pi} + \nu\nabla^2\overline{\mathbf{V}} + \overline{\mathbf{M}}^v\,, \qquad \nabla\cdot\overline{\mathbf{V}} = 0\,, \tag{4.2c,d}$$

where $\partial_t = \partial/\partial t$, $\Pi = P/\rho$ and P is pressure; $\overline{\mathbf{M}}^v = -\nabla\cdot\mathcal{Q}$ and $\mathcal{Q} = \overline{\mathbf{vv}}$ is the Reynolds stress tensor. The BRA represents $\overline{\mathbf{M}}^v$, as in molecular dynamics, by

$$\overline{\mathbf{M}}^v = \nabla\cdot(\nu^T\nabla\overline{\mathbf{V}})\,, \qquad \text{i.e.,} \qquad \overline{M}_i^v = \nabla_j(\nu^T\nabla_j\overline{V}_i)\,, \tag{4.2e,f}$$

where the turbulent viscosity $\nu^T \sim \frac{1}{3}\overline{u}\lambda$, in analogy with molecular dynamics; here $\overline{u} = \sqrt{\overline{v^2}}$ and λ is the correlation length. The relation (4.2e) is a godsend! The unmanageable (4.2a) is instantly converted (for constant ν^T) into the much more amenable

$$\partial_t \overline{\mathbf{V}} + \overline{\mathbf{V}}\cdot\nabla\overline{\mathbf{V}} = -\nabla\overline{\Pi} + \overline{\nu}\nabla^2\overline{\mathbf{V}}\,, \tag{4.2g}$$

where the total viscosity $\overline{\nu} = \nu^T + \nu$ is dominated by ν^T. A cynic might also say that uncertainties in $\overline{u}$ and λ and therefore in $\overline{\nu}$ afford an irresistable opportunity of choosing the $\overline{\nu}$ that makes the simulated system mimic reality best!

The BRA is clearly convenient but there are now alternative ways of incorporating the effect of the SGS on the GS. These methods are not described here but are discussed in reviews such as Meneveau & Katz (2000) and Geurts *et al.* (2008).

More relevant than (4.2) for the geodynamo are the Boussinesq equations with Coriolis, Lorentz and buoyancy forces included. To determine these, the induction and heat equations governing $\mathbf{B}$ and T must be added. Their averages determine the mean fields, $\overline{\mathbf{B}}$ and $\overline{T}$, which contain sources $\overline{\mathbf{M}}^b = \nabla\times\boldsymbol{\mathcal{E}}$ and $M^\theta = -\nabla\cdot\mathbf{I}$ where $\boldsymbol{\mathcal{E}} = \overline{\mathbf{v}\times\mathbf{b}}$ is the turbulent electromotive force and $\mathbf{I} = \overline{\theta\mathbf{v}}$ is proportional to the turbulent heat flux. The ideas behind BRA apply equally well to these and, as in (4.2e),

$$\overline{\mathbf{M}}^b = -\nabla\times(\eta^T\nabla\times\overline{\mathbf{B}})\,, \qquad M^\theta = \nabla\cdot(\kappa^T\nabla\overline{T})\,, \tag{4.2h,i}$$

where η^T and κ^T, the turbulent magnetic and thermal diffusivities, are also of order $\frac{1}{3}\overline{u}\lambda \sim \nu^T$. In stellar applications η^T and κ^T dominate $\overline{\eta} = \eta^T + \eta$ and $\overline{\kappa} = \kappa^T + \kappa$. For the Earth, there is no observational evidence that $\overline{\eta}$ differs substantially from η. This is not surprising if (see above) R_m is only about $7R_m^c$. Because $P_m \ll 1$, the corresponding dimensionless numbers quantifying ν and κ, the kinetic Reynolds number, $R_e \approx Vr_s/\nu \approx 2\times10^8$ and the Peclet number, $P_e = Vr_s/\kappa \approx 2\times10^7$, are gigantic and the SGS are important in transporting GS momentum and heat.

In stellar applications where $\overline{\rho}$ varies over many scale heights, the compressibity of the plasma must be allowed for but, if V is small compared with both $g/2\Omega_0$ and the velocity of sound c, the anelastic equations can be used and are as easy to apply as the Boussinesq equations; see Braginsky & Roberts (1995, 2007). They are now in general use in stellar dynamo simulations; e.g., Browning (2008); Browning *et al.* (2004, 2006); Brun *et al.* (2005). At the recent dynamo workshop at the Kavli Institute in Santa Barbara, an anelastic benchmark was set up so that simulators will have a ready test of their codes available. The anelastic approximation was first employed in geodynamo simulations by Glatzmaier & Roberts (1996), but the Boussinesq equations are still more commonly used.

When the Coriolis force, and of course the pressure gradient, dominate the remaining forces, the flow $\overline{\mathbf{V}}$ becomes two-dimensional with respect to the direction Oz of $\boldsymbol{\Omega}_0$, as demanded by the Proudman-Taylor theorem. [Not 'the Taylor-Proudman theorem' please! Proudman (1916) has 7 years priority over Taylor (1923)!] The early simulation of Glatzmaier (1985) used anelastic theory and in 1989 was still the most sophisticated model of solar MHD. He found that, because of the dominance of Coriolis forces, the mean angular velocity $\Omega(s,z) = \overline{V}(s,z)/s$ of the flow about Oz tended to be almost independent of z, i.e., $\Omega = \Omega(s)$, the Proudman–Taylor result. In 1989 the helioseismology bombshell burst (Libbrecht 1989): $\Omega = \Omega(\vartheta)$, where ϑ is colatitude, represents reality much better than $\Omega = \Omega(s)$. So where had the theory gone wrong?

It is now widely accepted that the BRA is too simplistic because the GS/SGS boundary is, for unavoidable numerical reasons, set at *far* too small a length scale. The SGS are significantly affected particularly by density stratification and rotation, so destroying the isotropy assumed by BRA. Lack of isotropy means that (4.2e,h,i) would better be replaced by

$$\overline{M}_i^v = \nabla_j\big(\nu^T_{ijkl}\nabla_l\overline{V}_k\big), \quad \overline{M}_i^b = -\epsilon_{ijk}\nabla_j\big(\eta^T_{klm}\nabla_m\overline{B}_l\big), \quad \overline{M}^\theta = \nabla_j\big(\kappa^T_{jk}\nabla_k\overline{T}\big), \tag{4.3a,b,c}$$

in which tensor diffusivities appear. Unfortunately even this is not enough. Mean field electrodynamics (MFE) identifies an additional term in $\boldsymbol{\mathcal{E}}$ involving the undifferentiated components of $\overline{\mathbf{B}}$, This "alpha effect" requires a supplementary $\alpha\overline{\mathbf{B}}$ to be included in $\boldsymbol{\mathcal{E}}$ for pseudo-isotropic turbulence or, in the more general non-isotropic case, $\mathcal{E}_i = \alpha_{ij}\overline{B}_j - \eta^T_{ijk}\nabla_k\overline{B}_j$. Nearly 70 years ago, before the discovery of the $\alpha-$effect [for the history, see Rüdiger (1989)], a process analogous to the $\alpha-$effect had been proposed for momentum transport. This $\Lambda-$effect includes in the Reynolds stress tensor, $\boldsymbol{\mathcal{Q}}$, a term proportional to the components of the undifferentiated $\overline{\mathbf{V}}$. Since $\mathcal{Q}_{ij} = \mathcal{Q}_{ji}$, an isotropic $\Lambda-$effect does not exist. As $\overline{\mathbf{V}}$ is usually dominated by the zonal shear, $\boldsymbol{\Omega} = \Omega(s,z)\widehat{\boldsymbol{\phi}}$, the $\Lambda-$effect is usually represented by $Q^\Lambda_{ij} = \Lambda_{ijk}\Omega_k$, where $\Lambda_{ijk} = \Lambda_{jik}$. The inclusion of the $\Lambda-$ and $\alpha-$effects, changes (4.3a,b) to

$$\overline{M}_i^v = \nabla_j\big(\Lambda_{ikl}\Omega_l - \nu^T_{ijkl}\nabla_l\overline{V}_k\big)\,, \qquad \overline{M}_i^b = \epsilon_{ijk}\nabla_j\big(\alpha_{kl}\overline{B}_l - \eta^T_{klm}\nabla_m\overline{B}_l\big)\,. \tag{4.3d,e}$$

Details of this mean field theory, together with possible forms for Λ_{ijk} and ν^T_{ijkl}, may be found in Rüdiger (1989); Rüdiger & Hollerbach (2004). The $\Lambda-$effect can have a significant impact on Ω. Combined with the influence of meridional circulation, $\overline{\mathbf{V}}_P$, the observed departure of Ω from the Proudman-Taylor $\Omega(s)$ can be successfully modeled. See for example Rüdiger & Hollerbach (2004) and Rempel (2005, 2006).

Applications of the mean field MHD apparatus just described almost invariably assume that the statistical averages $\overline{\mathbf{V}}$, $\overline{\mathbf{B}}$ and $\overline{T}$ are axisymmetric; three dimensional applications are rare. The tensor diffusivities (4.3c,d,e) contain so many "free" parameters that a cynic might again wonder whether, by their judicious choice, any desired $\overline{\mathbf{V}}$, $\overline{\mathbf{B}}$ and $\overline{T}$ follow.

This brief review of dynamo theory, as applied to stars such as the Sun in particular, but more generally to any star with a convection zone, has omitted reference to the solar tachocline and to the stellar tachoclines expected at any radiative-convective interface. These zones add another layer of complexity to an already daunting theory; see, for example, Hughes *et al.* (2007). Also omitted was a discussion of small-scale dynamos for which $\overline{\mathbf{B}} \equiv \mathbf{0}$ and MFE is inapplicable. These may be relevant to the solar convection zone and to some geodynamo models.

Do some of the lessons learned from the Sun apply to dynamos in the Earth and planets? As mentioned earlier, it seems unnecessary, from the modest value of R_m in the Earth's core, to introduce into geodynamo simulations either an $\alpha-$effect or a turbulent magnetic diffusivity, of either scalar or tensor type. Because R_e and P_e are so enormous however, it is plausible that the turbulent transport of momentum and heat (and composition) are very significant. This is sometimes used to turn the computational necessity of assuming O(1) values for the Prandtl numbers, P_r $(= \nu/\kappa)$ and P_m, into a virtue, by defining them with turbulent diffusivities. Even though a significant $\Lambda-$effect appears to be unlikely because $R_o = \overline{V}_\phi/\Omega_0 r_{\rm s} \ll 1$, it seems probable from the following discussion

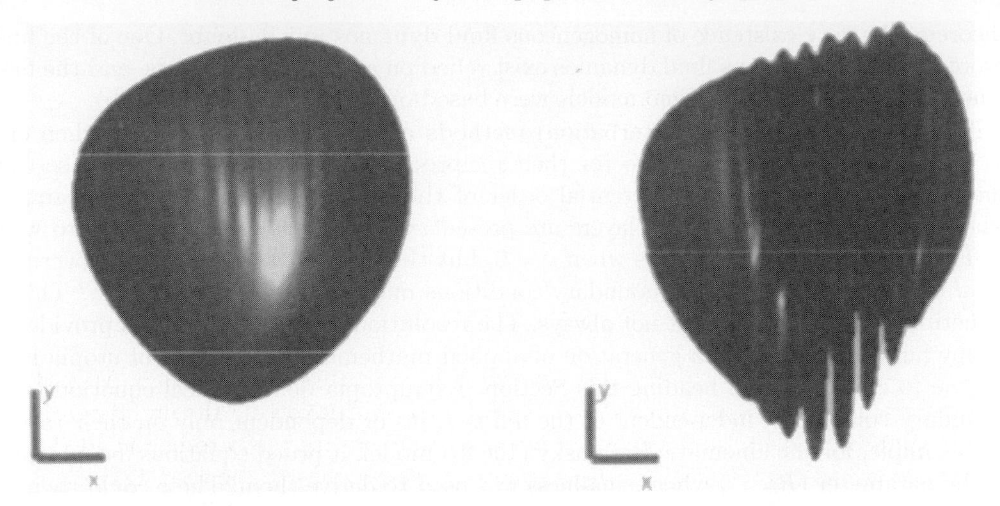

Figure 1. *View from the $z-$axis of the breakup of an initially spherical blob of buoyant fluid.* Left: at time t =1.25, plate-like structures begin to form; Right: at time t = 1.75, the blob has nearly disintegrated into individual plates. The increasing elongation of the blob in the $z-$direction is not apparent in this projection. The $z-$axis is parallel to $\mathbf{\Omega}_0$ and the $y-$axis is parallel to $\overline{\mathbf{B}}$. The dimensionless unit of time is a/v, where a = initial radius of blob, $v = g\theta_0/2\Omega_0$, θ_0 being the temperature excess of the blob. The Ekman number is 6.87×10^{-6} (St Pierre 1996). Reproduced with the permission of Taylor and Francis, publishers of *Geophysical and Astrophysical Fluid Dynamics*; http://www.informaworld.com

that isotropic diffusivities ν^T and κ^T cannot adequately describe turbulent transport of GS momentum and heat.

The nature of turbulent convection in the Earth's core has been considered in more detail by Braginsky & Roberts (1995, 2003) and Loper (2007). To explore the SGS and their anisotropies for $P_m \ll 1$, Braginsky & Meytlis (1990) devised a simple model in which $\overline{\mathbf{B}}$ is in the $y-$direction, with $\mathbf{\Omega}_0$, $\mathbf{g}$ and $\nabla\overline{T}$ in the $z-$direction, and $\beta \equiv \partial_z\overline{T} > 0$. For large E_ℓ, the instabilities of this state are highly anisotropic, being platelike, the thickness of the plates being smaller in the $x-$ or $\mathbf{\Omega}_0\times\overline{\mathbf{B}}-$direction than in the perpendicular directions by $\mathrm{O}(E_\ell^{-1})$. By an argument too lengthy to give here, Braginsky and Meytlis concluded that

$$\kappa_{xx} \sim \eta\left(\frac{g\widetilde{\alpha}\beta}{4\Omega_0^2}\right)E_\ell^2\,, \qquad \kappa_{yy} \sim \kappa_{zz} \sim \eta\left(\frac{g\widetilde{\alpha}\beta}{4\Omega_0^2}\right)E_\ell^4\,. \tag{4.4a,b}$$

These results depend strongly on $\overline{B}$. Even though κ_{xx} is much less than κ_{yy} and κ_{zz}, it generally greatly exceeds the molecular κ. The argument leading to (4.4) also gives equipartition on the smallest SGS. Matsushima *et al.* (1999) and Matsushima (2001, 2004, 2005) investigated these questions by computer models. Shimizu, reported by Loper (2007), has sugested a potentially useful scaling. That the system is highly dispersive is apparent in the simulation of St Pierre (1996) of the rise and disintegration of a buoyant blob of fluid (see Figure). Whether similar ideas have a broader, astrophysical relevance remains to be seen.

5. Asymptopia

Asymptotic methods were applied early in the history of dynamo theory at a time when powerful electronic computers were nonexistent and when, partly because of Cowling's theorem, the very existence of homogeneous fluid dynamos was in doubt. One of the first proofs that homogeneous fluid dynamos exist relied on asymptotic methods, and the first kinematic spherical geodynamo models were based on them (Braginsky 1964b).

Asymptotic (or singular perturbation) methods apply in the limiting case when one or more parameters tend to zero (or their reciprocals tend to zero), and when setting them zero would lower the differential order of the governing equations. For example, when $\nu \to 0$, viscous boundary layers are present. whose thicknesses tend to zero with ν. There are no boundary layers when $\nu = 0$, but the differential order of the governing *ideal equations* is less, so that boundary conditions must be dropped, but which? This is sometimes easily decided, but not always. The resolution of such matters has provided a happy hunting ground for a generation of applied mathematicians, a kind of utopia that led me to coin the word heading this Section. Asymptopia delivers ideal equations and boundary conditions, independent of the diffusivities or dependent only on their ratios. For example, for the kinematic Braginsky (1964b) model, it posed equations independent of the parameter $(R_m^{-1/2})$ whose smallness was used to derive them. These could then be solved by the primitive computers available at that time.

The smallness of E for naturally-occurring MHD dynamos immediately suggests asymptopia, but there are severe obstacles. Small ν means turbulence and the problems of the SGS described in §4, problems exacerbated by small κ and (except for the geodynamo) small η. Although analytic progress is impossible, one may, as an article of faith, believe that, when ν, κ and η are small, the concepts of asymptopia are valid, and that the behavior of the large GS can be characterized by parameters independent of the diffusivities (or dependent only on their ratios), even though all diffusivities are essential for the SGS and boundary layers. Such characterizations constitute *scaling theory*.

Consider, as an example of scaling theory (Jones 2007), non-rotating, non-magnetic convection in a plane layer of depth D. In the limit of infinite Rayleigh number, $R_a = g\tilde{\alpha}D^3\Delta\overline{T}/\nu\kappa$, the boundary layers on the walls are infinitely thin, and elsewhere the typical convective velocity V and departure in the temperature from $\overline{T}$ are

$$V \sim (gD)^{1/3}\left(\frac{\tilde{\alpha}F_c}{\rho C_p}\right)^{1/3}, \quad \theta \sim \frac{1}{\tilde{\alpha}(gD)^{1/3}}\left(\frac{\tilde{\alpha}F_c}{\rho C_p}\right)^{1/3}, \tag{5.1}$$

where F_c is the convective heat flux and C_p is specific heat. These results are independent of ν, κ and $\nu/\kappa = P_r$.

Scaling theory has achieved prominence recently through the influential papers of Christensen & Aubert (2006) and Christensen *et al.*, (2009). For the thermally–driven dynamo, five of their dimensionless parameters are a Rossby number R_o and

$$Ra_Q^* = \frac{r_s}{r_i}\cdot\frac{g\tilde{\alpha}F_c}{\rho C_p \Omega_0^3 D^2}, \qquad Lo = \frac{V_A}{\Omega_0 D}, \tag{5.2a,b}$$

$$Nu^* = \frac{r_s}{r_i}\cdot\frac{F_c}{\rho C_p(\Delta T)\Omega_0 D}, \quad f_{\mathrm{ohm}} = \frac{Q^\eta}{Q^\eta + Q^\nu}, \tag{5.2c,d}$$

where $D = r_s - r_i$. One of these depends on P_m but the rest are independent of the diffusivities and their ratios. In (5.2c,b), the modified Rayleigh number Ra_Q^* quantifies buoyancy, the Nusselt number Nu^* heat flux, and the Lorentz number Lo field strength.

Christensen & Aubert (2006) derive their scaling laws by analyzing 66 dynamo integrations, all in geo-geometry $r_i/r_s = 0.35$, the SIC being electrically insulating in all but 5 cases. Convection was driven between fixed, noslip boundaries by an assigned temperature difference $\Delta\overline{T}$ between them, and with $5 < R_a/R_a^c < 50$. In all except one case

($E \equiv \nu/\Omega_0 D^2 = 10^{-6}$), the Ekman number E was between 3×10^{-6} and 3×10^{-4}. The range $0.06 \leqslant P_m \leqslant 10$ was investigated.

The Christensen-Aubert study has suggested several interesting power law dependencies that pose theoretical challenges, e.g., that the dynamo fails if $P_m < 450E^{3/4}$. Unless the optimal exponents are modified, there is less hope of deducing others of their empirical laws, e.g., $R_o \propto {Ra_Q^*}^{0.43} P_m^{-0.13}$ and $Nu^* = 0.076{Ra_Q^*}^{0.53}$. The latter suggests that the exponent should be $\frac{1}{2}$, leading to the perhaps surprising conclusion that the convective heat flux is independent of κ. Their best fit for the field strength was $Lo = 0.76{Ra_Q^*}^{0.32} P_m^{0.11} f_{\rm ohm}^{1/2}$; their next best was $Lo = 0.92{Ra_Q^*}^{0.34} f_{\rm ohm}^{1/2}$. Changing 0.34 to $\frac{1}{3}$ in the latter, and setting $f_{\rm ohm} = 1$ for the Earth (see §3), leads to their proposed alternative to (2.1a):

$$V_A = 0.9(gD)^{1/3}(\widetilde{\alpha}F_c/\rho C_p)^{1/3}\,, \qquad \text{or} \qquad V_A = 0.9(gDF_b/\rho)^{1/3}\,, \tag{5.3}$$

where $F_b = \widetilde{\alpha}F_c/C_p$ is the buoyancy flux, which from their Nu^* is $6-8\times10^{-9}$kg m^{-2}s^{-1}. This makes B only about 1 mT and independent of both Ω_0 and η. It is determined mainly by the heat flux F_c (or the buoyancy flux F_b when both sources of buoyancy operate); (5.3) may be thought of as a power balance rather than a force balance such as (2.1a).

Sixty six dynamo models is a lot! And they appear to cover adequately the range of parameters that is computationally accessible. They do not, and cannot, cover the enormous parameter range over which naturally-occurring dynamos roam. Of course, the derivation of scaling laws is partly motivated by a wish to apply them outside the computationally accessible domain, so the question is one of degree: how far outside that domain can they be trusted? The Earth's P_m of 10^{-6} lies far beneath the $0.06 \leqslant P_m \leqslant 10$ range of the models, and $10^{-6} \leqslant E \leqslant 3 \times 10^{-4}$ does not include the $E \sim 10^{-15}$ of the Earth (or 10^{-9} if ν^T defines E instead of ν). There is some danger that other similarity laws may apply beyond the computationally accessible domain. An example where this may be happening is given below. Extrapolation to stars and other cosmic contexts is even more extreme and therefore even more problematic (quite apart from the unexplored effects of compressibility on the scaling laws).

The 66 models assign the core surface temperature $\overline{T}(r_{\rm s})$ but, because of the role of the mantle in transmitting heat, it is more realistic to specify the heat flux $q(r_{\rm s})$ on the core boundary [e.g., Braginsky & Roberts (2007)]. When $E \ll 1$, this may make a significant difference. In an effort to move towards geophysically more realistic parameter values, Kageyama *et al.* (2008) used the Earth Simulator to integrate a model for $E = 2.3 \times 10^{-7}$, but they assigned a uniform $\overline{T}(r_{\rm s})$. Disappointingly, the resulting field had a small scale, i.e., was less Earthlike than previous models, even though their E was more Earthlike. Sakuraba and Roberts (to appear) present results from a small E simulation ($E = 5 \times 10^{-7}$, $P_m = 0.2$) but in which $q(r_{\rm s})$ is assumed uniform rather than $\overline{T}(r_{\rm s})$. The resulting field is dipole dominated and is generally more Earthlike. We speculate that, although the characters of the constant$-q(r_{\rm s})$ and constant$-\overline{T}(r_{\rm s})$ models are similar for the values of E common in today's simulations, they will increasingly differ as E is reduced further, and possibly their scaling laws will differ too? Exciting times lie ahead!

Acknowledgments. I wish to thank Axel Brandenberg, Karl-Heinz Glassmeier, Gary Glatzmaier, Chris Jones, Ataru Sakuraba, Klaus Strassmeier and Nigel Weiss for offering helpful advice about this review. I also thank Martin St Pierre and Taylor & Francis for permission to reproduce the Figure. I am grateful to the National Science Foundation for partial support through CSEDI Grant No. 0652423.

References

Bloxham, J. & Jackson, A. 1992, *J. Geophys. Res.* 97, 19537

Braginsky, S. I. 1964a, *Geomag. & Aeron.* 4, 898

Braginsky, S. I. 1964b, *Geomag. & Aeron.* 4, 732

Braginsky, S. I. 1975, *Geomag. & Aeron.* 15, 149

Braginsky, S. I. & Meytlis, V. P. 1990, *Geophys. astrophys. Fluid Dynam.* 55

Braginsky, S. I. & Roberts, P. H. 1995, *Geophys. astrophys. Fluid Dynam.* 79, 1

Braginsky, S. I. & Roberts, P. H. 2003, in: A. Ferriz-Mas & M. Núñez (eds.), *Advances in Nonlinear Dynamos* The Fluid Mechanics of Astrophysics and Geophysics (Taylor & Francis), vol. 9, p. 60

Braginsky, S. I. & Roberts, P. H. 2007, in: D. Gubbins & E. Herreo-Brevera (eds.), *Encyclopedia of Geomagnetism and Paleomagnetism* Springer, p. 11

Browning, M. K. 2008, *ApJ* 676, 1262

Browning, M. K., Brun, A. S., & Toomre, J. 2004, *ApJ* 601, 512

Browning, M. K., Miesch, M. S., & Brun, A. S. 2006, *ApJ* 648, 157

Brun, A. S., Browning, M. K., & Toomre, J. 2005, *ApJ* 629, 461

Christensen, U. R. & Aubert, J. 2006, *Geophys. J. International* 166, 97

Christensen, U. R., Schmitt. D., & Rempel. R. 2009, *Space Sci. Rev.*, in press

Christensen, U. R. & Tilgner, A. 2004, *Nature* 429, 169

Geurts, B. J., Kuczaj, A. K., & Titi, E. S. 2008, *J. Phys. A: Math. Theor.* 41, 344008

Glatzmaier, G. A. 1985, *ApJ* 291, 300

Glatzmaier, G. A. & Roberts, P. H. 1996, *Physica D* 97, 81

Hide, R. 1966, *Phil. Trans. R. Soc. Lond. A* 259, 615

Holme, R. 2007, in: P. Olson (ed.), *Treatise on Geophysics* (Elsevier), vol. 8, p. 107

Hughes, D. W., Rosner, R., & Weiss, N. O. 2007, *The Solar Tachocline* (Cambridge UK, University Press.)

Jones, C. A. 2007, in: P. Olson (ed.), *Treatise on Geophysics* (Elsevier), vol. 8, p. 131

Kageyama, A., Miyagoshi, T., & Sato, T. 2008 *Nature* 454, 1106

Kono, M. & Tanaka, H. 1995, in: T. Yukutake (ed.), *The Earth's central Part; its Structure and Dynamics*, (Terrapub.), p. 75

Libbrecht, K. G. 1989, *ApJ* 336, 1092

Loper, D. E. 1978, *J. Geophys. Res.* 83, 5961

Loper, D. E. 2007, in: P. Olson (ed.), *Treatise on Geophysics* (Elsevier), vol. 8, p. 187

Matsushima, M. 2001, *Phys. Earth planet. Interiors* 128, 137

Matsushima, M. 2004, *Earth Planets Space* 56, 599

Matsushima, M. 2005, *Phys. Earth planet. Interiors* 159, 74

Matsushima, M. Nakajima, T., & Roberts, P. H 1999, *Earth Planets Space* 51, 277

Meneveau, C. & Katz, J. 2000, *Ann. Rev. Fluid Mech.* 32, 1

Proudman, J. 1916, *Proc. R. Soc. Lond. A* 92, 408

Rempel, M. 2005, *ApJ* 622, 1320

Rempel, M. 2006, *ApJ* 647, 662

Roberts, P. H., Jones, C. A., & Calderwood, A. R. 2003, in: C. A. Jones, A. M. Soward & K. Zhang (eds.), *Earth's Core and Lower Mantle*, The Fluid Mechanics of Astrophysics and Geophysics (Taylor & Francis), vol. 11, p. 100

Roberts, P. H. & Kono, M. 2007, *Earth Planets Space* 59, 661

Roberts, P. H., Yu, Z. J., & Russell, C. T. 2007, *Geophys. astrophys. Fluid Dynam.* 101, 11

Rüdiger, G. 1989 *Differential Rotation and Stellar Convection. Sun and Solar Type Stars*, The Fluid Mechanics of Astrophysics and Geophysics (Gordon & Breach), vol. 5

Rüdiger, G. & Hollerbach, R. 2004, The Magnetic Universe. Geophysical and Astrophysical Dynamo Theory. (Wiley VCH)

St Pierre, M. G. 1996, *Geophys. astrophys. Fluid Dynam.* 83, 293

Taylor, G. I. 1923, *Proc. R. Soc. Lond. A* 102, 180

Tobias, S. M. & Cattaneo, F. 2008, *J. Fluid Mech.* 601, 101

Zatman, S. & Bloxham, J. 1997, *Nature* 338, 760

Discussion

KOUTCHMY: Planet Mars is rotating as fast as the Earth. Why is its magnetic field much weaker?

ROBERTS: The magnetic field of Mars arises from remnant magnetization of minerals in its crust. This magnetization was probably acquired earlier in the planet's history when it operated a dynamo in its electrically conducting core. That core may have solidified; or conceivably the mantle of Mars did not allow enough heat to emerge from the core to set up sufficiently vigorous convection to permit a Martian dynamo to operate.

Paul Roberts

Anders Johansen

Yi-Jiun Su's problems are much smaller

Cosmic Magnetic Fields:
From Planets, to Stars and Galaxies
Proceedings IAU Symposium No. 259, 2008
K.G. Strassmeier, A.G. Kosovichev & J.E. Beckman, eds.

doi:10.1017/S1743921309030610

Electromagnetic interaction between Jupiter's ionosphere and the Io plasma torus

Yi-Jiun Su†

Department of Physics, the University of Texas at Arlington, Arlington, TX 76019, USA
email: yijiunsu@gmail.com

Abstract. The electromagnetic interaction between Jupiter and Io has been studied extensively since the discovery of Io-controlled decametric radio emissions (DAMs). A variety of mechanisms for electromagnetic disturbances have been considered including a unipolar inductor, the excitation of large-amplitude Alfvén waves, the generation of electrostatic electric fields parallel to the ambient magnetic field, and etc. Recently, three auroral acceleration regions categorized by terrestrial physicists have been applied to the Jupiter-Io coupling system: the Alfvénic acceleration region is associated with bright emissions at Io's magnetic footprint, whereas the quasi-static system of anti-planetward and planetward currents set up at the inner and outer edges of the torus in the downstream region of Io's wake. This review paper summarizes the current understanding of the coupling mechanisms between Jupiter's ionosphere and the Io plasma torus, as well as the electron acceleration mechanism necessary to excite Io-associated emissions.

Keywords. Acceleration of particles – waves – plasmas – planets and satellites

1. Introduction

The electrodynamic interaction of the Jupiter-Io system is unique in our solar system particularly due to the fact that (1) the strong magnetic field of Jupiter creates the largest magnetosphere, (2) Jupiter is the fastest rotating planet, and (3) Io is the most volcanically active moon (Bagenal 2007; Saur *et al.* 2004). Unlike the terrestrial electromagnetic dynamo primarily driven by the solar wind, the electrodynamics of Jupiter are dominated by the planet's fast spin with a 10-hour rotational period. The closest Galilean moon to Jupiter – Io, located at 5.9 Jovian-centric distance ($R_J \approx 71,500 km$) – is embedded deep within its vast magnetosphere, which has been known to extend to the orbit of Saturn.

The dominant neutrals generated by volcanoes on Io are sulfur and oxygen (1-3 ton s^{-1}), most likely in the form of SO_2. Approximately, 1/3-1/2 of the neutral atoms are ionized by photo-ionization, electron impact ionization, and charge exchange, thereby adding heavy ions, such as O^+, S^{++}, S^+, O^{++}, and etc., to the inner magnetosphere, where they spread out into a donut shape encircling Jupiter known as the Io plasma torus. The newly ionized neutrals are referred to as 'pickup ions' because as soon as they are ionized, they are accelerated to the speed of the bulk plasma flow, 74 kms^{-1}. Io orbits Jupiter with a velocity of 17 kms^{-1}, so the plasmas flow over the moon with a relative velocity of 57 kms^{-1} ahead of Io in its orbital motion. This flow of magnetized plasmas sweeps through the obstacle, Io, producing intense electrodynamic interaction (Kivelson *et al.* 2004, and references therein). The plasma density, momentum, and energy are modified locally through elastic and inelastic collisions with Io's atmosphere and ionization processes. This interaction also extends far away towards Jupiter's ionosphere, particularly along the magnetic

† Now at Air Force Research Laboratory/RVBXP, 29 Randolph Road, Hanscom, AFB, MA 01731, USA

field line. Auroral emissions observed at the magnetic footprint of Io and its wake or trail emissions are evidence of such interaction. An example of a Jovian auroral image taken by the Hubble Space Telescope is presented in the top panel of Figure 1 (Clarke *et al.* 2002).

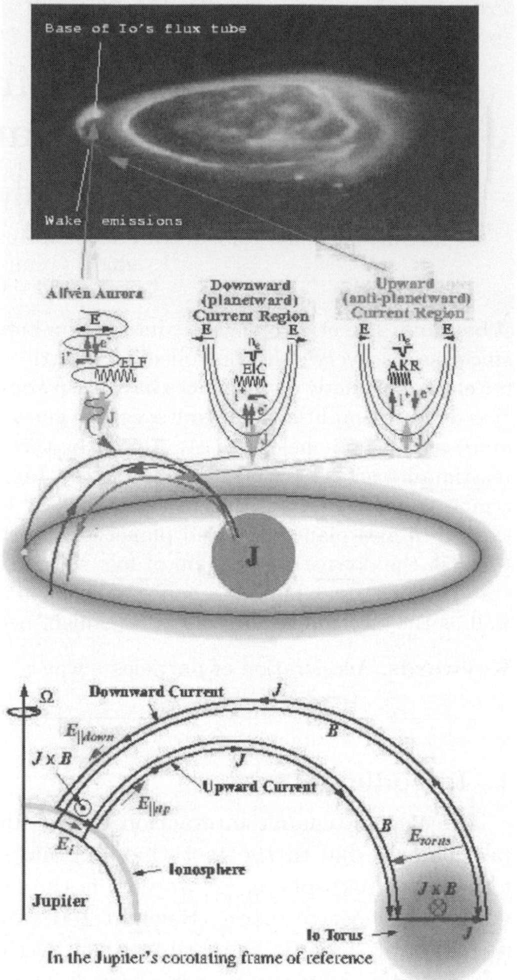

1.1. *Early theories*

The electromagnetic interaction between Jupiter and Io has been investigated extensively since the first discovery of Io-controlled DAMs by Bigg (1964). One of the early theories describing the Jupiter-Io interaction was proposed by Piddington & Drake (1968) and Goldreich & Lynden-Bell (1969) and was referred to as the unipolar inductor model: field-aligned currents link Io to Jupiter's ionosphere through magnetospheric plasmas. A schematic representation of the current loop is similar to that shown in the bottom panel of Figure 1. A $J \times B$ force is generated to accelerate plasmas downstream of Io to the corotation speed (into the page of the bottom panel). Because Jupiter's magnetic field points southward at the torus, a radially outward current at Io completes the circuit while field-aligned currents flow anti-planetward and planetward at Io's inner and outer edges, respectively. Based on the unipolar inductor theory, the induced electric potential difference between the outer and inner boundaries of Io was estimated to be 415 $kV \approx u$(57 kms^{-1}) $\times$ B(2,000 nT) $\times$ R_{Io}(3,636 km). However, the assumption of an extremely high conductivity at Io resulting in frozen-in Jovian magnetic fields with Io was not realistic.

Figure 1. The top panel shows a Jovian auroral image, where the Io-induced aurora is seen on the left with the brightest emissions at the base of the Io flux tube and an emission trail extending downstream. A depiction of the three types of auroral regions is shown in the middle panel, where the green, blue, and red lines represent the Alfvénic acceleration region, the planetward current region, and the anti-planetward current region, respectively. The quasi-static current structure downstream of Io's wake in Jupiter's corotating frame is illustrated in the bottom panel (after Su *et al.* 2003).

Electric and magnetic disturbances are generated when a flux tube of magnetospheric plasmas passes through the Io plasma torus. Due to the relative motion (u) between Io and the corotating magnetospheric plasmas, electromagnetic perturbations are propagated away from the torus with an angle $\theta_A = tan^{-1}(u/V_A)$, where V_A is the Alfvén speed. This bent magnetic field structure is known as the Alfvén wing (Drell *et al.* 1965; Goertz & Deift 1973; Goertz 1980; Neubauer 1980; Southwood *et al.* 1980). In spite of the exact nature of the perturbation in Io's vicinity, energy is carried away in the form of an Alfvén wave. Due to the density gradient of the plasma torus, a substantial amount of the wave energy may be reflected back into the torus. Exactly how much energy escapes from the Io torus remains a topic of debate.

Several major unsolved questions involving the Jupiter-Io interaction were summarized by Saur *et al.* (2004), including the coupling mechanisms between Io and Jupiter's ionosphere and the electron acceleration mechanism necessary to excite Io-associated emissions. This paper describes the author's perspective in addressing both questions.

1.2. *Recent theories*

Motivated by Jovian auroral images (Clarke *et al.* 2002; 2004), Delamere *et al.* (2003) categorized the Io-Jupiter interaction into three phases: (1) an initial mass-loading interaction; (2) the plasma acceleration in the wake of Io; and (3) the steady-state decoupling. The first phase is dominated by the combination of collision-dominated flows and mass loading processes with a time scale of ~60 sec. The second acceleration phase is due to the momentum transfer through Alfvénic disturbances with a time scale of ~8 min. Delamere *et al.* (2003) estimated that only ~20% of the input momentum is transmitted to high latitudes due to the internal reflection of the torus density gradient. Once Alfvén waves have bounced between Io and Jupiter's ionosphere a few times, quasi-static field-aligned currents are set up at Io's wake during the third phase. In this final phase, the plasma velocity is ~1 kms^{-1} deviated from the corotation speed due to the magnetic decoupling rather than the rigid corotation. From observations of trail emissions, this decoupling effect can last as long as 5 hours corresponding to 180° in longitude.

Mauk *et al.* (2002) indicated several similarities between the Earth and Io's auroras. Despite the complexity of auroral observations, terrestrial auroral physicists have simply classified the auroral acceleration process into three types based on distinct characteristics of particle and field measurements: (1) the upward current region; (2) the downward current region; and (3) the Alfvénic acceleration region (Paschmann *et al.* 2003, Chapter 4). These regions are illustrated below the auroral image in Figure 1. The prominent features of the quasi-static upward current region are up-going field-aligned ions and down-going accelerated electrons, narrowly peaked in energy but broad in pitch angle. This region consists of multiple inverted-V electron structures which are associated with converging electric fields and low-density cavities. In the downward current region, upward field-aligned electron beams are associated with diverging electric fields. The downward current region is more dynamic and complex than the upward current region. It is often combined with Alfvénic fluctuations and is rarely associated with detectable auroral emissions. At Earth, the Alfvénic acceleration region is observed near the open-closed field line boundaries which map to reconnection regions on the dayside magnetopause and on the nightside magnetotail. The most distinctive features are counter-streaming electron bursts, enhanced ion conics, and the filamentary nature of currents generated by the propagation Alfvén waves rather than quasi-static potential structures. The ratio of $\delta E/\delta B$ is approximately equal to the local Alfvén speed (Su *et al.* 2001).

Although Jovian auroral observations are currently limited to remote imaging, fundamental physics learned from Terrestrial studies should be universally applicable to any magnetized planet system. Su *et al.* (2003) proposed the three auroral acceleration processes described above for the Earth to be active magnetosphere-ionosphere coupling mechanisms on Jupiter as well (see the middle panel of Figure 1). The most intense counter-streaming electron fluxes are driven by propagating Alfvén waves associated with oscillating field-aligned electric fields (Chaston *et al.* 1999). Therefore, the bright emission at the Io magnetic footprint is caused by Alfvén dominated precipitation, whereas quasi-static currents are set up in the downstream region of Io's wake. An analogy between the Earth's midnight auroral zone during a substorm and the Jupiter-Io interaction was summarized in an illustration by Ergun *et al.* (2006). Recent theoretical studies of the upward (anti-planetward) current region are discussed in § 2 followed by a description of the physical processes of the Alfvénic acceleration region in § 3.

2. The quasi-static upward current region

The fundamental physics of the Earth's upward current region is well established, however, application of this knowledge to other magnetized planets has been limited. The aurora is generated by magnetospheric electrons slamming into the atmosphere and exciting the airglow. These precipitating electrons carry the upward field-aligned current. Due to the magnetic mirror, the current density is limited by the amount of electrons inside the loss cone. Observations have provided compelling evidence that this acceleration process is achieved by parallel electric fields, which act to accelerate electrons downward pushing more current carriers into the loss cone. Illustrations of electron distributions in three regimes of the current-voltage relation are shown in Figure 2.

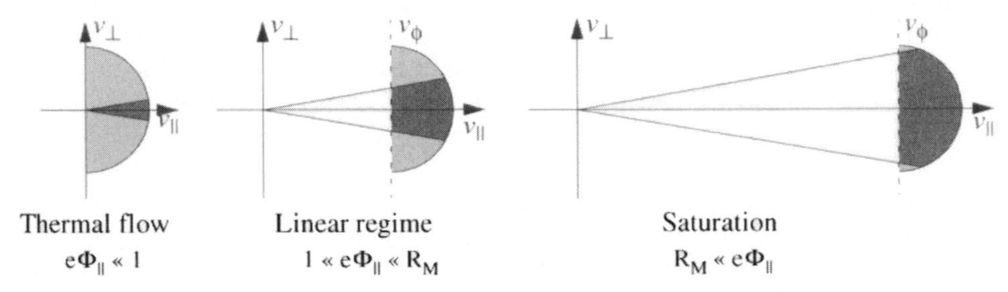

Figure 2. The thermal flow is restricted to a small region of phase space (left). Increasing $v_{||}$ due to $e\Phi_{||}$ allows more electrons to flow into the ionosphere (middle) until the current saturates (right). Dark shaded areas within the loss cone contain precipitation electrons which carry an upward current (after Paschmann *et al.* 2003).

This theoretical relationship was originally recognized by Knight (1973) and summarized by Paschmann *et al.* (2003) and Ergun *et al.* (2008).

$$J_{||Iono} = J(z)\left\{R_M(z) - \big(R_M(z)-1\big)exp\left[-\frac{e\Phi_{||}}{k_B T_e(z)\big(R_M(z)-1\big)}\right]\right\},$$

$$J(z) = en_e(z)\sqrt{\frac{k_B T_e(z)}{2\pi m_e}}, \qquad R_M(z) = \frac{B_{Iono}}{B(z)} \tag{2.1}$$

where $J_{||Iono}$ represents the parallel current density at the ionosphere, while $n_e(z)$, $T_e(z)$, $J(z)$, and $R_M(z)$ are electron density, electron temperature, electron thermal current, and the magnetic mirror ratio, respectively, at a location z between the ionosphere and magnetosphere. The electron thermal current should be at its minimum at z. $\Phi_{||}$ is the parallel electrostatic potential between ionosphere and z.

On an auroral flux tube at Earth, $J(z)$ is at its minimum in the plasmasheet. For the typical density ($0.2\ cm^{-3}$) and temperature ($1\ keV$) of plasmasheet electrons, the current-voltage relations for mirror ratios of 400 and 1000 are plotted as the solid and dashed lines, respectively, while the linear Knight relation is represented by the dotted line in

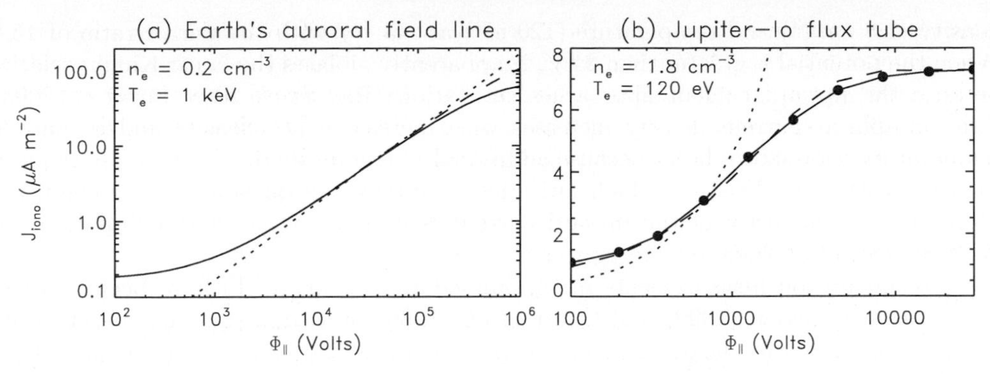

Figure 3. The current-voltage relation (a) at Earth and (b) on the Jupiter-Io flux tube.

Figure 3(a). The majority of observed data fall within the linear regime, hence, a linear Knight relation ($J_{Iono} = K\Phi_{||}$, where $K = e^2 n_e/\sqrt{2\pi m_e k_B T_e}$) is a good approximation for the upward current region of Earth's auroral field lines.

At Jupiter, $J(z)$ occurs near the minimum of the combination of gravitational and centrifugal potentials, $\sim$2.5 R_J from Jupiter's center (Su *et al.* 2003). It is very important to note that the plasma parameters at 2.5 R_J are dramatically different from those in the Io torus, because the majority of Iogenic heavy ions are confined within the torus due to Jupiter's strong centrifugal force. *In-situ* measurements of plasma parameters at high latitudes are currently unavailable due to the lack of satellite missions in Jupiter's polar region. Su *et al.* (2003) have self-consistently obtained plasma parameters along the Jupiter-Io flux tube by utilizing a quasi-static Vlasov code modified from an Earth's model (Ergun *et al.* 2000a). The basic idea of the code is to solve the Poisson's equation along the field line by specifying phase-space distributions for various species at boundaries. The lower boundary conditions for the ionospheric parameters are based on radio occultation measurements (Fjeldbo *et al.* 1975; 1976; Eshleman *et al.* 1979; Hinson *et al.* 1997), while the upper boundary conditions for the torus parameters are obtained from Voyager (Bagenal 1994) and Galileo observations (Crary *et al.* 1998).

The current-voltage relation on the Jupiter-Io flux tube is shown as the solid line with solid circles in Figure 3(b) based on the fixed boundary parameters listed in Table 1. The upper limit (32 kV) of the parallel potential in Figure 3(b) was estimated from the mean electron energy at 30° downstream of Io based on the Space Telescope Imaging Spectrograph combined with a Jovian atmosphere model (Gérard *et al.* 2002). More recent analysis suggests that downstream mean energies may be in the range of 100s eV (Gérard, private communication). The lower limit of the x-axis is, therefore, set at 100 V. In addition, the theoretical current-voltage relation (Eq. 2.1) and the linear Knight relation are plotted as the dashed and dotted lines, respectively, based on the electron

Table 1. Boundary conditions of Figure 3(b)

Species	N (cm^{-3})	T (eV)	Boundary
Ionosphere H^+	2×10^5	0.31	lower
Ionosphere e^-	2×10^5	0.31	lower
Io O^+	1750	$T_{\|\|} = 35, T_\perp = 70$	upper
Io S^+	250	50	upper
Io e^-	2000	5 eV^a	upper

^{a} a kappa distribution ($\kappa = 3$)

density (1.8 cm^{-3}) and temperature (120 eV) at $\sim$3 R_J with the mirror ratio of 13.8. When the potential is greater than 2 kV, it apparently violates the linear Knight relation because the upward current approaches saturation (Ray *et al.* 2008; Su *et al.* 2003). The ionospheric current density increases with increasing O^+ density and/or parallel temperature (not shown here) because additional magnetospheric electrons are required to balance the ion density at high latitudes to satisfy the quasi-neutrality condition. Moreover, the majority of the upward current is carried by the hotter population of electrons (Su *et al.* 2003).

Although proton measurements in the immediate vicinity of Io have been reported by Frank & Paterson (1999) and Chust *et al.* (1999), it is difficulty to separate them from the high-density heavy ions in the plasma torus. Su *et al.* (2003) explored the possibility of H^+ imposed at the upper boundary and concluded that H^+ ions became the major species at 2-3 R_J as the light ions are able to easily escape the strong centrifugal confinement of the torus. Therefore, the ionospheric upward current density carried by precipitating electrons increases with increasing H^+ density and temperature. Based on auroral images at 10°-20° downstream of Io's wake, Gérard *et al.* (2006) reported $\sim$10 kR ultraviolet emission corresponding to an energy flux of 1 mWm^{-2} resulting in an ionospheric current density of approximately 1 μAm^{-2} assuming the energy of the accelerated electrons to be 1 keV. This rough estimation suggests that the H^+ density should be much less than 1 cm^{-3} at high latitudes.

3. The Alfvénic acceleration region

The existence of Alfvén waves near Io have been established by Voyager 1 and Galileo (Saur *et al.* 2004, and references therein), however, the amount of wave energy escaping the torus is not well understood. Additionally, scientists do not know precisely where the wave energy is converted into electron energy to power the Io auroral spot. The majority of numerical studies are based on the magnetohydrodynamic (MHD) theory (e.g., Wright 1987; Dols 2001). In this paper, the propagation of dispersive Alfvén waves along the Jupiter-Io flux tube, as well as electron acceleration by these waves, are discussed in § 3.1.

Less than 10% of DAMs were observed to exhibit a distinct feature of discrete short pluses, known as S-bursts (Zarka 1992, and reference therein). S-bursts are strictly associated with an Io-dependent emission source mostly from the Io-B region in the Jupiter Central Meridian Longitude (Ergun *et al.* 2006). The generation mechanisms of S-bursts are important elements in understanding the Jupiter-Io system and are discussed in § 3.2.

3.1. *Dispersive Alfvén waves*

In the ideal MHD theory, Alfvén waves have no field-aligned electric field component and therefore provide no parallel acceleration to particles. A parallel electric field is generated by dispersive Alfvén waves when the parallel electric force is balanced by the electron inertia, electron pressure gradient, and/or the finite ion gyro-radius effect (Stasiewicz *et al.* 2000, and references therein). A dynamic model including the dispersive effects should be considered to properly address science questions associated with the Alfvénic acceleration region (Su *et al.* 2004; 2006). The general dispersion relation of dispersive Alfvén waves was summarized recently by Jones & Su (2008) as

$$v_{DAW}^2 = \frac{\omega^2}{k_\parallel^2} = v_A^2 \frac{1 + k_\perp^2 \rho^2}{1 + k_\perp^2 \lambda_e^2}; \quad \rho^2 = \rho_i^2 + \rho_s^2 \tag{3.1}$$

where $\lambda_e = \sqrt{m_e/n_e e^2 \mu_o}$, $\rho_s = \sqrt{m_i T_e}/eB$, and $\rho_i = \sqrt{m_i T_i}/eB$ are the electron skin depth, ion acoustic gyro-radius, and ion gyro-radius, respectively. v_A is the ideal Alfvén speed. When plasma $\beta > m_e/m_i$, the kinetic effect is the dominant component including the ion gyro-radius effect and the electron pressure gradient. On the other hand, the electron inertial effect becomes important when $\beta < m_e/m_i$. The parallel electric field changes polarity as dispersive Alfvén waves pass through the transition region ($\beta = m_e/m_i$). The real part of the parallel electric field reduces to zero, therefore, no particle acceleration occurs at this point. At Earth, the transition region is located at ~4-5 R_E. On the Jupiter-Io flux tube, this transition region is situated within the Io plasma torus.

Crary (1997) proposed that electrons gain energy near the edge of the torus by Fermi acceleration through parallel inertial Alfvén electric fields, hence, the wave itself carries virtually no energy to the ionosphere. Das & Ip (2000) observed kinetic Alfvén waves near Io and suggested that these waves were responsible for electron acceleration within the torus. However, Jones & Su (2008) argued that the phase speed of the dispersive Alfvén wave is too small to account for the resonant acceleration near the torus, hence, it is too weak to cause the observed bright Io spot at the magnetic footprint (Gérard *et al.* 2002). Therefore, Jones & Su (2008) suggested that electron acceleration occurs in the low density region ($\leqslant 4R_J$) outside the Io plasma torus. Moreover, the maximum $E_{||}$ is located at ~1.5 R_J, which can be two orders of magnitude higher than $E_{||}$ at 6 R_J along the Jupiter-Io flux tube.

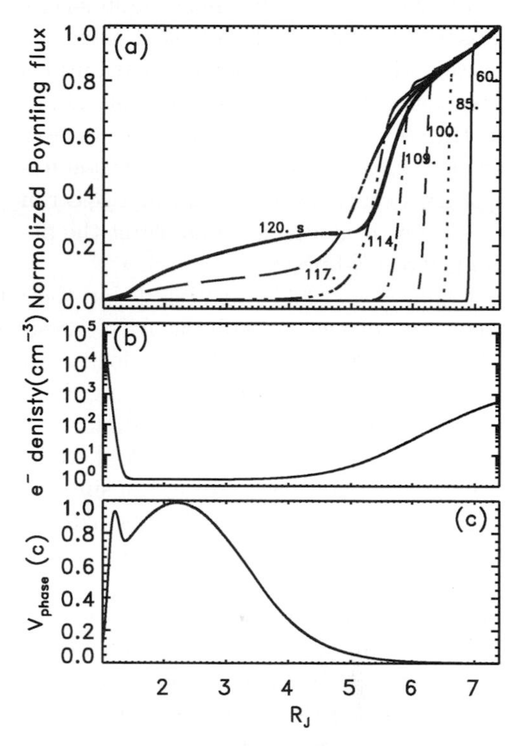

Figure 4. The simulation result based on the gyro-fluid model. (a) Normalized time-integrated Poynting flux at various times. (b) The electron density profile. (c) The phase speed of the dispersive Alfvén wave (after Su *et al.* 2006).

According to the above arguments, a certain amount of wave energy should be able to escape from the torus and exchange its energy with particles in the region of tenuous plasma. By utilizing a 1D MHD code, Wright (1987) and Dols (2001) suggested ~25% and 40%, respectively, of the Alfvén wave power leaving the torus. With a 2D fluid model, Delamere *et al.* (2003) indicated that ~20% of the input momentum is transmitted to ionosphere. Based on simulation results from a 1D linear gyro-fluid model including the dispersive effect of Alfvén waves, Su *et al.* (2006) showed that the first major reflection occurs at ~5.5 R_J from Jupiter due to the torus density gradient (Figure 4b), while ~20% of the wave energy reaches 3.5 R_J (Figure 4a). A second reflection occurs at the peak of the Aflvén speed (~2.3 R_J in Figure 4c). The majority of low-frequency, long-wavelength waves were unable to reach the Jupiter's ionosphere without wave breaking, phase mixing, or other nonlinear processes, however, a significant energy flux may be transferred via high-frequency, small-wavelength waves to the ionosphere.

3.2. *Generation mechanisms of S-bursts*

Wu & Lee (1979) were the first to explain the generation of terrestrial auroral kilometric radiation (AKR) by suggesting a loss-cone instability with a weakly relativistic treatment, which was coined the electron-cyclotron maser by Melrose & Dulk (1982). The same mechanism was applied to explain DAM from Jupiter, as well as analogous radiations from other magnetized outer planets (Zarka 1992). Radiations excited near the local electron cyclotron frequency are amplified through a gyro-resonant interaction from a perpendicular-driven instability ($\partial f/\partial v_{\perp} > 0$, where f is the electron phase-space distribution function). In the 1990s, an important modification was made to the original maser theory that strong auroral radiations are driven by a shell (or horseshoe) distribution in the auroral density cavity due to quasi-static parallel electric fields (Ergun *et al.* 2000b; 2002), because the loss-cone maser does not provide sufficient energy to power AKRs. The electron distributions responsible for the maser instability are illustrated in Figure 5. Although the maser theory explains the observed radiation frequency occurring at the local cyclotron frequency, it is unable to interpret the periodicity of S-bursts.

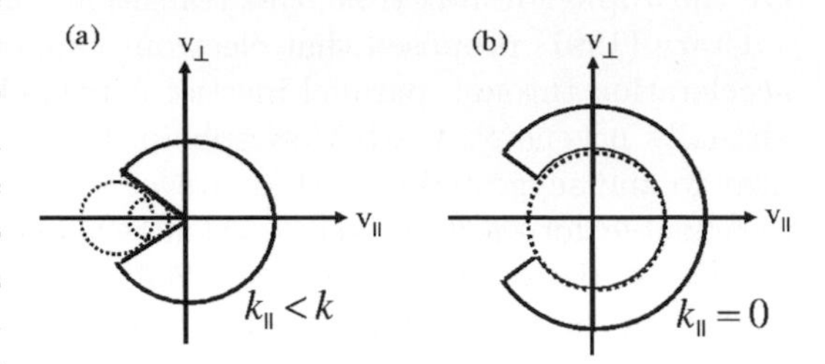

Figure 5. Schematic illustration of electron distributions for (a) the loss-cone maser and (b) shell maser, where the dotted circle represent resonance circles. (after Su *et al.* 2007)

Ergun *et al.* (2006) suggested that the S-burst periodicity is associated with the eigenfrequency of the ionospheric Alfvén resonator, $v_{AI}/2\pi H_P$, where $v_{AI} = B/\sqrt{\mu_o \rho}$ and $H_p = k_B T/mg$ are the Alfvén velocity at the ionosphere and the ionospheric scale height, respectively (Lysak 1991). By using the gyro-fluid model, Su *et al.* (2006) demonstrated that the inertial Alfvén wave originating in the torus bounces between the ionosphere and the location of the first peak of Alfvén phase velocity (at 1.2 R_J in Figure 4c). After taking a fast Fourier transform of simulated wave forms, the fundamental eigenfrequency and higher harmonics of the resonator (few-100 Hz) were found to be comparable to observed reoccurrence frequencies of S-bursts. Moreover, the eigenfrequency decreases with increasing ionospheric density and with increasing scale height. For example, with an ionospheric density of 2×10^5 cm^{-3} and a scale height of 500 km, Su *et al.* (2006) found the fundamental eigenfrequency of the resonator to be $\sim$20 Hz, comparable to the most probable reoccurrence rate of S-bursts. Ergun *et al.* and Su *et al.* (2006), therefore, suggested the ionospheric Alfvén resonator as the likely driver explaining multiple occurrences of S-bursts.

The Alfvén wave producing the ionospheric resonator is also responsible for electron acceleration (Su *et al.* 2007). Unstable electron distributions, such as shell, ring, or conics, are generated by parallel electric fields associated with this inertial Alfvén wave (Su *et al.* 2007; Hess *et al.* 2007; 2008). As stated in the maser theory, unstable electron distributions are the source of remotely observed auroral radiation with a condition that the plasma frequency is much less than the electron cyclotron frequency ($\omega_{pe}/\omega_{ce} <<$ 1), however, *in-situ* observations are currently unavailable near the ionosphere of the Jupiter-Io flux tube to confirm the Alfvén-driven maser hypothesis. Su *et al.* (2007; 2008) supported the Alfvén-driven maser theory as the generation mechanism of S-bursts on the basis of Earth-based satellite measurements.

Although rare, Su *et al.* (2008) found eight S-burst events in the Earth's Alfvénic acceleration region from three years of FAST data. An example of Earth-based S-bursts is displayed in Figure 6, while observed electron distributions associated with each of the 8 events are shown in Figure 7. Since no event was found when the satellite passed directly through the emission source region, the ratio of ω_{pe}/ω_{ce} was not able to be precisely determined. However, all events were found near apogee of the FAST orbit in the midnight local time sector during winter months suggesting a preference for low plasma density conditions. At Jupiter, the condition for the weakly relativistic cyclotron maser instability is easily satisfied above a few thousand km up to 4 R_J due to Jupiter's strong magnetic field (Ergun *et al.* 2006), which explains the higher occurrence rate of Jovian S-bursts.

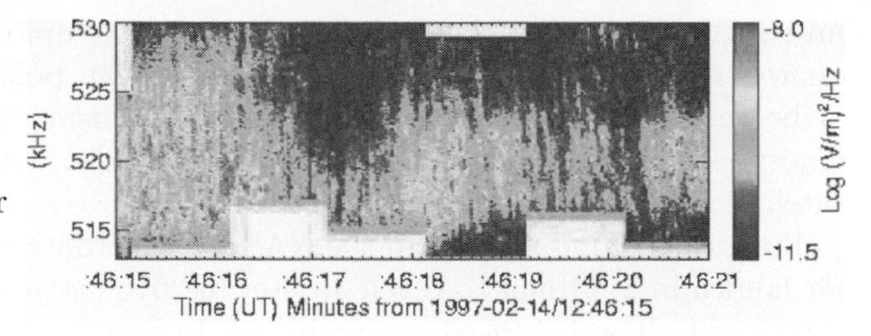

Figure 6. Earth-based S-burst (after Su *et al.* 2008).

The reoccurrence frequency of Earth-based S-bursts was found to fall between 7 and 18 Hz (Su *et al.* 2008) similar to that of Jovian S-bursts, but an order of magnitude higher than the typical frequency ($\sim$1 Hz) of Earth's ionospheric Alfvén resonator (Lysak 1991). A higher reoccurrence rate requires a smaller ionospheric scale height and/or a lower ionospheric density than typically observed. Based on simulation results from a gyro-fluid code with a test particle scheme, Su *et al.* (2008) proposed multiple Alfvénic disturbances at the upper boundary (i.e., the Io tours at Jupiter and the magnetotail at Earth) as another possible driver for the S-burst periodicity. Each discrete S-burst radiation is generated when unstable electron distributions due to each Alfvén wave pulse passes through regions with conditions matching the maser instability.

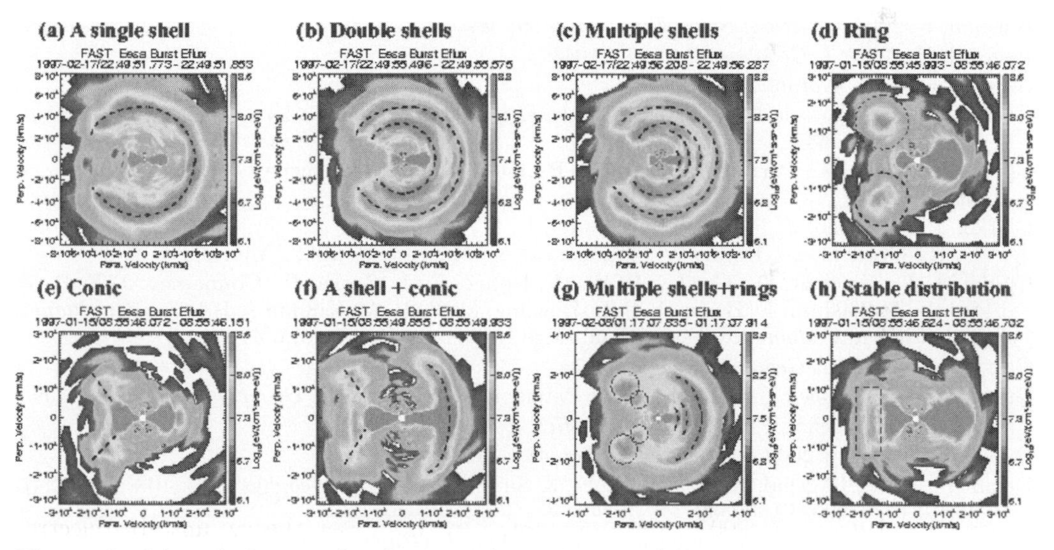

Figure 7. Selected electron distributions observed in the Alfvénic acceleration region by the FAST satellite during S-burst events. The upward direction away from the Earth's ionosphere is to the left (after Su *et al.* 2008).

4. Summary

Since the beginning of space exploration, eight spacecrafts have explored different parts of Jupiter's magnetosphere over last four decades. Seven of these have been flyby missions, including Pioneers 10 (1973) and 11 (1974), Voyagers 1 and 2 (1979), Ulysses (1992), Cassini (2000/2001), and New Horizon (2007). Galileo (1995-2005) was the only orbital mission to date to explore the inner magnetosphere of Jupiter. Many characteristics of plasmas and waves near the Io torus were provided by Io flybys during the Galileo mission, however, the plasma environment near Jupiter's polar ionosphere was not adequately sampled. In spite of the lack of a Jovian polar mission, we are able to provide a theoretical understanding of coupling mechanisms between the Io torus and Jupiter's ionosphere based on terrestrial auroral knowledge with confirmation from many Earth satellite missions and ground measurements.

Juno, the second mission in the NASA New Frontiers programs, is currently scheduled for launch in 2011 and to reach Jupiter in 2016 (http://juno.wisc.edu/). Juno's 32 polar orbits will allow for *in-situ* sampling of Jupiter's auroral acceleration regions. Many of the theoretical/numerical results summarized in this paper will be reexamined in light of new data acquired from this mission. Lessons learned from studies of the Jupiter-Io interaction, as well investigations of the Earth's auroral region, may be applicable to the Jovian main auroral oval, other less explored magnetized planets, and similar electromagnetic processes of astrophysical sources.

Acknowledgements

The author thanks Prof. Strassmeier for the invitation to talk about the Jupiter-Io interaction at IAUS259, Drs. Fran Bagenal, Bob Ergun, and Sam Jones for helpful discussions, and Ron Caton for proofreading the manuscript. She acknowledges support from NASA NNG05GM99G and NSF/CAREER ATM-0544656 to UT Arlington.

References

Bagenal, F. 2007, *J. Atmo. Solar-Terr. Phys.* 69, 387

Bagenal, F. 1994, *J. Geophys. Res.* 99, 11043

Bigg, E. K. 1964, *Nature* 203, 1008

Chaston, C. C., Carlson, C. W., Peria, W. J., Ergun, R. E., & McFadden, J. P. 1999, *Geophys. Res. Lett.* 26, 647

Chust, T., Roux, A., Perraut, S., Louarn, P., Kurth, W. S., & Gurnett, D. A. 1999, *Planet. Space Sci.* 47, 1377

Clarke, J. T., Ajello, J., Ballester, G., Jaffel, L. B., Connerney, J., Gérard, J. C., Gladstone, G. R., Grodent, D., Pryor, W., Trauger, J., & Waite, J. H. 2002, *Nature* 415, 997

Clarke, J. T., Grodent, D., Cowley, S. W. H., Bunce, E. J., Zarka, P., Connerney, J. E. P., & Satoh, T. 2004, in: F. Bagenal, T. E. Dowling, & W. B. McKinnon (eds.), *Jupiter: Planet, Satellites, and Magnetosphere*, (Cambridge University Press), p. 639

Crary, F. J. 1997, *J. Geophys. Res.* 102, 37

Crary, F. J., Bagenal, F., Frank, L. A., & Paterson, W. R. 1998, *J. Geophys. Res.* 103, 29359

Crary, F. J. & Bagenal, F. 1997, *Geophys. Res. Lett.* 24, 2135

Das, A. & Ip, W.-H 2000, *Planet. Space Sci.* 48, 127

Delamere, P. A., Bagenal, F., Ergun, R. E., & Su, Y.-J. 2003, *J. Geophys. Res.* 108(A6), 1241, doi:10.1029/2002JA009530

Drell, S. D., Foley, H. M., & Ruderman, M. A. 1965, *J. Geophys. Res.* 70, 3131

Dols, V. 2001, *M.S. Thesis*, (Univ. of Alaska, Fairbanks)

Ergun, R. E., Ray, L., Delamere, P. A., Bagenal, F., Dols, V., & Su, Y.-J. 2008, *J. Geophys. Res.*, in prep.

Ergun, R. E., Su, Y.-J., Andersson, L., Bagenal, F., Delamere, P. A., Lysak, R. L., & Strangeway, R. J. 2006, *J. Geophys. Res.* 111, A06212, doi:10.1029/2005JA011253

Ergun, R. E., Su, Y.-J., & Bagenal, F. 2002, in: H. O. Rucker, M. L. Kaiser, & Y. Leblanc (eds.), *Planetary Radio Emissions V*, (Austrian Academy of Sciences Press), p. 271

Ergun, R. E., Carlson, C. W., McFadden, J. P., Mozer, F. S., & Strangeway, R. J. 2000a, *Geophys. Res. Lett.*, 27,4053

Ergun, R. E., Carlson, C. W., McFadden, Delory, G. T., Strangeway, R. J. & Pritchett, P. L. 2000b, *ApJ* 538, 456

Eshleman, V., Tyler, L., Wood, G., Lindal, G., Anderson, J., Levy, G., & Groft, T. 1979, *Science* 204, 976

Fjeldbo, G. A., Kliore, A., Seidel, B., Sweetnam, D., & Cain, D. 1975, *A&A* 39, 91
Fjeldbo, G. A., Kliore, A., Seidel, B., Sweetnam, D., & Woiceshyn, P. 1976, in: Gehrels (ed.), *Jupiter*, (University of Arizona Press, Tucson), p. 239
Frank, L. A. & Paterson, W. R. 1999, *J. Geophys. Res.* 104, 10345
Gérard, J.-C., Saglam, A., Grodent, D., & Clarke, J. T. 2006, *J. Geophys. Res.* 111, A04202, doi:10.1029/2005JA011327
Gérard, J.-C., Gustin, J., Grodent, D., Delamere, P., & Clarke, J. T. 2002, *J. Geophys. Res.* 107, doi:10.1029/2002JA009410
Goertz, C. K. 1980, *J. Geophys. Res.* 85, 2949
Goertz, C. K. & Deift, P. A. 1973, *Planet. Space Sci.* 21, 1399
Goldreich, P. & Lynden-Bell, D. 1969, *ApJ* 156, 59
Hess, S., Mottez, F., Zarka P., & Chust, T. 2008, *J. Geophy. Res.* 113, A03209, doi:10.1029/2007JA012745
Hess, S., Mottez, F., & Zarka P. 2007, *J. Geophy. Res.* 112, A11212, doi:10.1029/2006JA012191
Hinson, D. P., Flasar, F. M., Kliore, A. J., Schinder, P. J., Twicken, J. D., & Herrera, R. G. 1997, *Geophy. Res. Lett.* 24, 2107
Jones, S. T. & Su, Y.-J. 2008, *J. Geophys. Res.*, in press
Kivelson, M. G., Bagenal, F., Kurth, W. S., Neubauer, F. M., Paranicas, C., & Saur, J. 2004, in: F. Bagenal, T. E. Dowling, & W. B. McKinnon (eds.), *Jupiter: Planet, Satellites, and Magnetosphere*, (Cambridge University Press), p. 513
Knight, S. 1973, *Planet. Space Sci.* 21, 741
Lysak, R. L. 1991, *J. Geophys. Res.* 96, 1553
Mauk, B. H., Anderson, B. J., & Thorne, R. M. 2002, in: M. Mendillo, A. Nagy, & J. H. Waite (eds.), *Atmospheres in the Solar System: Comparative Aeronomy*, (American Geophysical Union), vol. 130, p. 97
Melrose, D. B. & Dulk, G. A. 1982, *ApJ* 259, 844
Neubauer, F. M. 1980, *J. Geophys. Res.* 85, 1171
Paschmann, G., Haaland, S., & Treumann, R. (eds.) 2003, in: *Auroral Plasma Physics*, (Kluwer Academic Publishers), p. 93
Piddington, J. H. & Drake, J. F. 1968, *Nature* 217, 935
Ray, L. C., Su., Y.-J., Ergun, R. E., Delamere, P. A., & Bagenal, F. 2008, *J. Geophys. Res.*, in prep.
Saur, J., Neubauer, F. M., Connerney, J. E. P., Zarka, P., & Kivelson, M. G. 2004, in: F. Bagenal, T. E. Dowling, & W. B. McKinnon (eds.), *Jupiter: Planet, Satellites, and Magnetosphere*, (Cambridge University Press), p. 537
Southwood, D. J., Kivelson, M. G., Walker, R. J., & Slavin, J. A. 1980, *J. Geophys. Res.*, 85, 5959
Stasiewicz, K., Bellan, P., Chaston, C., Kletzing, C., Lysak, R., Maggs, J., Pokhotelov, O., Seyler, C., Shukla, P., Stenflo, L., Streltsov, A., & Wahlund, J.-E. 2000, *Space Sci. Rev.* 92, 423
Su, Y.-J., Ma, L., Ergun, R. E., Pritchett P. L., & Carlson, C. W. 2008, *J. Geophys. Res.* 113, A08214, doi:10.1029/2007JA012896
Su, Y.-J., Ergun, R. E., Jones, S. T., Ergun, R. E., Strangeway, R. J., Chaston, C. C., Parker, S. E., & Horwitz, J. L. 2007, *J. Geophys. Res.* 112, A06209, doi:10.1029/2006JA012131
Su, Y.-J., Jones, S. T., Ergun, R. E., Bagenal, F., Parker, S. E., Delamere, P. A., & Lysak, R. L 2006, *J. Geophys. Res.* 111, A06211, doi:10.1029/2005JA011252
Su, Y.-J., Jones, S. T., Ergun, R. E., & Parker, S. E. 2004, *J. Geophys. Res.* 109, A11201, doi:10.1029/2003JA010344
Su, Y.-J., Ergun, R. E., Bagenal, F., & Delamere, P. A. 2003, *J. Geophys. Res.* 108(A2), 1094, doi:10.1029/2002JA009247
Su, Y.-J., Ergun, R. E., Peterson, W. K., Onsager, T. G., Pfaff, R., Carlson, C. W., & Strangeway R. J. 2001, *J. Geophys. Res.* 106, 25595
Wright, A. N. 1987, *J. Geophys. Res.* 92, 9963
Wu, C. S. & Lee, L. C. 1979, *ApJ* 230, 621
Zarka, P. 1992, *Adv. Space Res.* 12, 99

Discussion

STRASSMEIER: Would you say that the Io-Jupiter flux tube is considered a "dynamo"?

SU: Yes. The "dynamo" is due to continuous energy exchanges between charged particles and electromagnetic waves. Io's volcanoes provide the plasma source by the ionization and mass loading processes. The electromagnetic energy is generated by the ionization and mass loading processes between Jupiter's rotating magnetic fields and the Io plasma torus. This energy propagates along the field line and accelerates electrons into Jupiter's ionosphere to create aurora emissions.

BLACKMAN: Does the energy in S-bursts, or auroral emission, scale simply with bulk properties of the magnetospheric system, e.g. surface field density in a "back of the envelope" way (e.g. for application to other systems)?

SU: Yes, the S-burst emissions are determined by the electron cyclotron frequency based on the local magnetic field. The largest frequency of S-bursts is $\approx$39 MHz, which corresponds to the surface field of $\approx$13 G. Radio emissions from other magnetized planets can serve as a good indicator for the planets' surface magnetic field.

Maxim Khodachenko

Cosmic Magnetic Fields:
From Planets, to Stars and Galaxies
Proceedings IAU Symposium No. 259, 2008
K.G. Strassmeier, A.G. Kosovichev & J.E. Beckman, eds.

doi:10.1017/S1743921309030622

The role of intrinsic magnetic fields in planetary evolution and habitability: the planetary protection aspect

Maxim L. Khodachenko[1], H. Lammer[1], H. I. M. Lichtenegger[1], J.-M. Grießmeier[2], M. Holmström[3] and A. Ekenbäck[3]

[1]Space Research Institute, Austrian Academy of Sciences, Graz, Austria
email: maxim.khodachenko@oeaw.ac.at

[2]ASTRON, Dwingeloo, The Netherlands
email: griessmeier@astron.nl

[3]Swedish Institute of Space Physics, Kiruna, Sweden
email: matsh@irf.se

Abstract. The widely used definition of a habitable zone (HZ) for planets as a circumstellar area, where the star's luminosity is sufficiently intense to maintain liquid water at the surface of a planet, is shown to be too simplified. The role of a host star's activity and the intrinsic magnetic field of a planet with respect to their influence on mass loss processes of close-in gas giants and a definition of a HZ for the terrestrial-type exoplanets are discussed. The stellar X-ray/EUV radiation and the stellar wind result in ionization, heating, chemical modification, and slow erosion of the planetary upper atmospheres throughout their lifetime. The closer the planet is to the star, the more efficient are these processes, and therefore, the more important becomes the magnetic protection of a planet as a potential habitat. Different ways for planetary magnetic dipole moment estimation, based on existing magnetic dynamo scaling laws as well as on the recent measurements of hot atomic hydrogen clouds around close-in 'Hot Jupiters' are considered, and the predictions of these estimations are compared to each other.

Keywords. Astrobiology – magnetic fields – plasmas – molecular processes – atmospheric effects – stars: planetary systems – stars: winds – stars: activity

1. Introduction

Although most of the discovered exoplanets (http://exoplanet.eu/catalog.php) have parameters more comparable to the gas giants of the Solar System rather than the Earth-like, rocky planets, nowadays, because of contemporary technical advances, observers begin to detect non-Jupiter type lower mass exoplanets. By this, the giant exoplanets family demonstrate a separation in to sub-groups: (*i*) rather short orbit Jupiter-type planets, called 'Hot Jupiters'; and (*ii*) more massive giant planets with orbits $\geqslant 1$ AU.

With the discovery of low mass exoplanets (Santos, *et al.* 2004; Rivera, *et al.* 2005), the question as to whether life could evolve on a planet outside our Solar System has taken on a new urgency. Answering this question requires a complex study of a variety of internal and external factors which may influence the conditions on a planet in order that life could evolve there. In that respect, a circumstellar area, where a planet could have the necessary conditions for development and maintaining any kind of life is called a Habitable Zone (HZ). The only criterion, used so far, for definition of boundaries of a HZ in the vicinity of a star is based on the possibility for a planet with an atmosphere to have climate and geophysical conditions which allow the existence on its surface of liquid water over geological time periods (Huang 1960; Kasting, *et al.* 1993; Kasting 1997). This approach supposes that the width and circumstellar distance of the HZ depend mainly on the stellar luminosity, which evolves during the life time of a star and influences the planetary surface temperature. It is widely accepted now, that such general definition of the HZ, based only on the "*stellar luminosity* and *on-surface liquid water*" criterion, is incomplete. It needs further specification that includes considerations of the whole complex of stellar-planetary relation factors, whereas potential habitats have to be better defined.

In this paper we discuss the potential role for a HZ definition of such factors as the activity of a host star, as well as the related to that stellar wind and plasma environment conditions around a planet. By this, the intrinsic planetary magnetic field and magnetospheric protection of a planet appear of exceptional importance in the context of possible habitability of a planet.

2. Stellar radiation and plasma environment

The stellar X-ray/EUV (XUV) radiation and the stellar wind constitute permanent forcing of upper planetary atmospheres. The effect of this forcing is ionization, heating, chemical modification, and slow erosion of the upper atmosphere throughout the lifetime of a planet. The closer a planet is to the star, the more efficient are these processes, which finally influence the whole complex of physical and climatologic conditions on a planet. At the same time, only stable and dense enough atmospheres allow water to be liquid over geological time periods and prevent the destructive action from hostile radiation on the planetary surface, increasing the chances for life to emerge there.

2.1. *Stellar activity*

The relevant physical phenomena of stellar activity on late-type stars (i.e., spectral classes G, K, M) and their observational manifestations include modulations of the stellar photospheric light due to stellar spots, intermittent and energetic flares, coronal mass ejections (CMEs), stellar cosmic rays, enhanced XUV emissions (see Scalo, *et al.* 2007 and references therein).

According to the currently accepted paradigm, the wide range of activity levels and related phenomena observed in different stars is directly connected with operation of the stellar magnetic dynamo. By this, two basic parameters: (*i*) stellar rotation rate and (*ii*) depth of the convective zone, are believed to control the stellar dynamo efficiency, which increases with increasing of any, or both of these quantities. Since the stellar convective envelope becomes thicker with decreasing stellar mass, it is straightforward to infer that, at a given rotation period (i.e. age), the low-mass M- and K- stars should be more active than a solar-type G- star. This fact has many observational confirmations. For example, a relatively old ($\sim$ 5.5 Gyr) dwarf M- star, Proxima Centauri, experiences measurable flares at a rate of about one flare per hour (Walker 1981).

Recently, Audard, *et al.* (2000) found that the energy of flares correlates with the stellar activity, characterized by L_X/L_{bol}, where L_X and L_{bol} are X-ray and bolometric luminosities of a star, respectively. The evolution of $log(L_X/L_{bol})$ with time for stars of various masses is shown in Figure 1, provided by Scalo, *et al.* (2007). According to this activity-age portrait, solar-type G- stars stay at saturated emission levels only until ages of $\sim$ 100 Myr, and then their XUV luminosities rapidly decrease with age: $\propto (t[Gyr])^{-1.72}$. On the other hand, M- stars have saturated emission periods up to 0.5–1 Gyr, and then their luminosity decreases in a way similar to the solar-type stars.

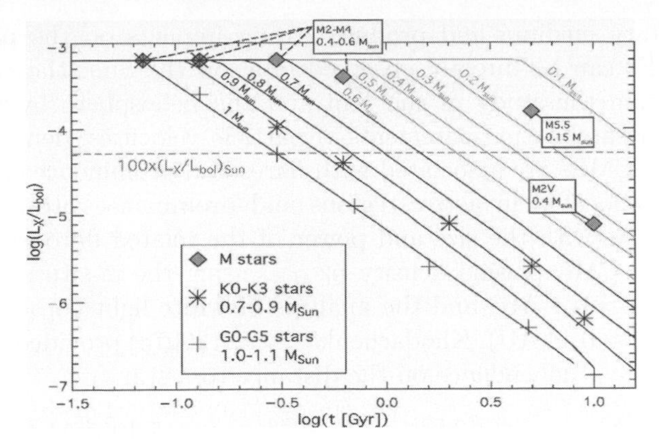

Figure 1. L_X/L_{bol} as a function of age for stars with masses $< M_{Sun}$. Symbols represent stars from the "Sun in Time" program (adopted from Scalo, *et al.* 2007).

Audard, *et al.* (2000), estimates the rate of high-energy ($E > 10^{32}$ erg) flares per day as $logN|_{E>10^{32}}$ erg $= -26.7 + 0.95logL_X$, which in the case of M- stars with a saturated activity level $L_X = 7 \times 10^{28}$ erg/s implies $\sim$ 6 strong flares per day. Altogether it has been found (Ribas, *et al.* 2005; Scalo, *et al.* 2007) that early K- stars and early M- stars may have XUV emissions level, and therefore flaring rates, of $\sim$(3–4) and $\sim$(10–100), respectively, times higher than solar-type G- stars of the same age.

2.2. *Stellar winds and CMEs*

Along with the high and long lasting stellar XUV emissions, another crucial factor of stellar activity, which may strongly affect the potential habitability of a planet, is the stellar wind plasma flow. Important parameters for characterization of the 'stellar plasma–planetary atmosphere/surface interaction' are the stellar wind density n_{sw} and velocity v_{sw}, which are highly variable with stellar age and depend also on the stellar spectral type, as well as on the orbital distance of the planet. We have currently a good knowledge of n_{sw} and v_{sw} for the Sun, whereas the amount of corresponding data related to other stars is much more limited.

While the dense stellar winds from hot stars, cool giants/supergiants and young T-Tauri stars produce detectable spectroscopic features from which the wind parameters can be derived, the tenuous winds from solar-type stars are a few orders of magnitude less massive, and cannot be detected spectroscopically. Different methods have been used to attempt direct measurements of mass loss rates in M- stars, most notably using *mm*-wavelength observations (van den Oord & Doyle 1997). Recently, there have been important developments towards indirect detections of stellar winds through their interactions with the surrounding interstellar medium. In particular, the stellar mass loss rates and related stellar winds were estimated for several nearby G- and K- stars using Hubble Space Telescope (HST) high-resolution measurements of the hydrogen Lyman-α absorption features, associated with the interaction between the stellar fully ionized coronal winds and the partially ionized local interstellar medium (Wood, *et al.* 2005). From these observations it was possible to conclude that younger solar-type G- stars have much denser and faster stellar winds as compared to the present Sun.

Furthermore, it is known from observations of our Sun that flaring activity of a star is accompanied by eruptions of coronal mass (e.g. CMEs), occurring sporadically and propagating in the stellar wind as large-scale plasma-magnetic structures. Traveling outward from the star at high speeds (up to thousands km/s), CMEs create major disturbances in the interplanetary medium and produce strong impacts on the planetary environments. Since CMEs can be directly observed only on the Sun, the current knowledge on them comes from the study of the Sun and the heliosphere by means of coronagraphs or by direct satellite measurements of particle velocities, densities and magnetic fields. On the Sun, CMEs are associated with flares and prominence eruptions and their sources are usually located in active regions and prominence sites. The likelihood of CME-events increases with the size and power of the related flare event. Based on the estimations of solar CME plasma density n_{CME}, using the in-situ spacecraft measurements (at distances > 0.4 AU) and the analysis of white-light coronagraph images (at distances $\leqslant 30R_{Sun} \approx 0.14$ AU), Khodachenko, *et al.* (2007a) provided general power-law interpolations of n_{CME} dependence on the distance to a star:

$$n_{CME}^{min}(d) = 4.88(d[\mathrm{AU}])^{-2.3}, \quad n_{CME}^{max}(d) = 7.10(d[\mathrm{AU}])^{-3.0}, \tag{2.1}$$

which identify a typical maximum-minimum range of n_{CME}. Besides of that, the average mass of solar CMEs was estimated as 10^{15} g, whereas their average duration at distances $(6\text{–}10)R_{Sun}$ is close to 8 hours.

Until now, very few indications of CME activity on other stars come mainly from absorption features in the UV range during the impulsive phase of strong flare events, or from the blue-shifted components in time series spectra (Cully, *et al.* 1994; Houdebine, *et al.* 1990). In general, Cully, *et al.* (1994) concludes that CMEs on active stars might be much stronger than the solar events. Besides of that, the existing correlation between strong flares and CMEs on the Sun may be used, assuming a solar-stellar analogy, as an argument in favour of possible high CME activity on magnetically active, late-type stars.

Because of the relatively short range of propagation of majority of CMEs, they should impact most strongly the magnetospheres and atmospheres of close orbit (< 0.1 AU) planets. Khodachenko, *et al.* (2007a) have found that for a critical CME production rate $f_{CME}^{cr} \approx 36$CMEs per day (and higher) a close orbit exoplanet appears under continuous action of the stellar CMEs. This may have crucial outcomes for the climate evolution and habitability conditions on the terrestrial exoplanets in close-in HZs of the low mass active M- and K- stars (Khodachenko, *et al.* 2007a; Lammer, *et al.* 2007). Therefore, definition of a HZ for the case of low mass M- and K- stars, besides of their longer XUV activity periods, should take also into account the effects of "short range" (in astrophysical scales) planetary impacting factors of the stellar activity such as relatively dense stellar winds, interplanetary shocks, magnetic clouds (MCs) and CMEs propagating in the stellar winds.

3. Impact of stellar radiation and plasma flows on planetary atmospheres

The action of intensive stellar radiation and stellar winds on planetary environments consists of the following effects.

1) XUV radiation of the host star affects the the planetary thermospheric heat budget, resulting in the heating and expansion of the upper atmosphere (see Figure 2), which under certain conditions could be so large that the majority of light atmospheric constituents overcome the gravitational binding and escape from the planet in the form of a hydrodynamic wind, i.e. hydrodynamic escape (Yelle 2004; Tian, *et al.* 2008; Penz, *et al.* 2008). High upper atmospheric temperatures and the resulting hydrodynamic escape have a strong impact on the atmospheric stability of terrestrial-type planets (Kulikov, *et al.* 2007; Lammer, *et al.* 2008) and the evolution of the planet's water inventory.

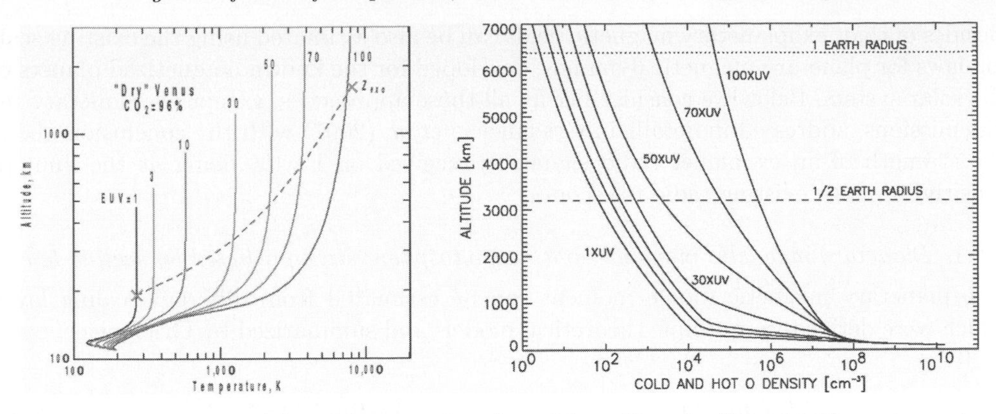

Figure 2. Temperature (a) and hot atomic oxigen (b) profiles of a CO_2-rich Venus-type atmosphere for different stellar EUV fluxes (in units of the present Sun EUV flux). The dashed line in (a) corresponds to the exobase distance at which the particle mean free path equals the scale height (adopted from Lammer, *et al.* 2007, 2008)

2) Simultaneously with the direct radiational heating of the upper atmosphere, the processes of ionization with the consequent production of energetic neutral atoms (ENAs) by various photo-chemical and charge exchange (for example $O_2^+ + e^- \rightarrow O^* + O^* + \Delta E$) reactions take place (Lammer, *et al.* 2008; Lichenegger, *et al.* 2008). Such processes result in the formation around planets of extended (in some cases) coronas, filled with hot neutral atoms.

The extended, because of the heating, upper planetary atmosphere and/or hot neutral corona may reach, and even exceed, the heights of the planetary magnetosphere. In this case they will be directly exposed to the plasmas of the stellar wind and CMEs. In this case the non-thermal atmospheric loss processes connected with ion pick-up, sputtering, and different kinds of photo-chemical energizing and escape will take place (Lichtenegger, *et al.* 2008). This makes the planetary magnetic field and connected with it, the size of the magnetosphere, as well as the parameters of the stellar wind (mainly n_{sw} and speed v_{sw}) to be important for the processes of atmospheric erosion and mass loss of a planet, affecting finally its possible habitability.

4. Exoplanet magnetic fields

Magnetic field of a planet is generated by the planetary magnetic dynamo, which is connected with the specific internal structure of a planet. Limitations of the existing observational techniques make direct measurements of the magnetic fields of exoplanets impossible. We can only judge on the strengths of extrasolar planetary magnetic fields and the sizes of exoplanetary magnetospheres, based on the sensitive to that, measurable phenomena, which may serve as their proxies. Among such indirect ways for detection of exoplanetary magnetic fields are:

1) measurement of extended ENA coronas of 'Hot Jupiters' and fit of the experimental data with the existing models, dependent on the parameters of stellar wind and planetary magnetosphere size;

2) Consideration of the detected exoplanets in the context of known planetary evolution trends, including the effects of atmospheric erosion and planetary mass loss, which are also controlled by the planetary magnetic field, and judging about possible values of the last;

3) Search for exoplanetary radioemission signatures.

Besides of that, exoplanetary magnetic fields can be also estimated using the existing scaling laws for planetary magnetic dynamos, developed for the known magnetized planets of the solar system. Below we consider briefly all these approaches, except exoplanetary radioemissions, addressed in details in Grießmeier, *et al.* (2007) with the conclusion about the strength of an exoplanet radio signal, as received on Earth, being at the limit of sensitivity of the existing radio telescopes.

4.1. *Planetary magnetic moments and magnetosphere sizes predicted by scaling laws*

The planetary magnetic dipole moment can be estimated from different scaling laws, which were derived from simple theoretical models and summarized by Grießmeier, *et al.* (2004):

$$\begin{aligned} &\mathcal{M} \propto \rho_c^{1/2}\omega r_c^4 && \text{(Busse 1976)}, \\ &\mathcal{M} \propto \rho_c^{1/2}\omega^{1/2} r_c^3 \sigma^{-1/2} && \text{(Stevenson 1983)}, \\ &\mathcal{M} \propto \rho_c^{1/2}\omega^{3/4} r_c^{7/2} \sigma^{-1/4} && \text{(Mizutani, } et\ al. \text{ 1992)}, \\ &\mathcal{M} \propto \rho_c^{1/2}\omega r_c^{7/2} && \text{(Sano 1993)}. \end{aligned} \tag{4.1}$$

Here $\mathcal{M}$ is the planetary magnetic dipole moment, r_c is the radius of the dynamo region, and ω is the angular velocity of the planet rotation around its axis. The internal properties of a planet, such as mass density and conductivity in the dynamo region are denoted by ρ_c and σ, respectively. The equations (4.1) provide a $\mathcal{M}_{min} \div \mathcal{M}_{max}$ range of reasonable planetary magnetic moment values.

It is important that all the models (4.1) yield an increase of $\mathcal{M}$ with an increasing planetary rotation rate, and vice versa. In that respect it is necessary to take into account that the close-orbit planets, such as terrestrial-type planets in the close-in HZs of the low mass stars and 'Hot Jupiters', are very likely to be tidally locked to their host stars. The angular rotation of a tidally locked planet is synchronized with its orbital motion so, that ω is equal to the orbital angular velocity, determined by Kepler's law: $\omega = \sqrt{M^{*}G/d^{3}}$, where G is gravitational constant, M^{*} is the mass of the star, and d is the semi-major axis of the planet. The time scale for tidal locking τ_{sync} depends on the planetary structure, orbital distance to the host star, and the stellar mass. By this, the planets for which $\tau_{sync} \leqslant 0.1$ Gyr, usually are assumed as tidally locked ones, since such planet's age is at least an order of magnitude longer. On the other hand, the planets with $\tau_{sync} \geqslant 10$ Gyr are almost certainly tidally unlocked. Based on (4.1), Grießmeier, *et al.* (2004, 2005) have shown that the magnetic moment $\mathcal{M}$ of a slowly rotating tidally locked planet is much smaller than it would be for a freely rotating tidally unlocked one (see Figure 3a).

A general view of an exoplanetary magnetosphere can be built on the basis of a phenomenological (non self-consistent) model which approximates the magnetosphere by a semi-infinite cylinder on the nightside and by a half-sphere on the dayside (Grießmeier, *et al.* 2004). Both have the same radius R_m. The planet is located within the half-sphere at $R_s < R_m$. Here R_s is a planetocentric standoff distance of magnetopause, which can be estimated from the pressure balance condition at the substellar point (Grießmeier, *et al.* 2004; Khodachenko, *et al.* 2007a):

$$R_s = \left[\frac{\mu_0 f_0^2 \mathcal{M}^2}{8\pi^2 \rho v^2}\right]^{1/6}, \tag{4.2}$$

where $f_0 = 1.16$ is a form-factor of the magnetosphere (Voigt 1995); $\rho = nm$ and v are the density and speed of the stellar wind, respectively. Note that for a planet with a weak magnetic moment (e.g. a tidally locked planet at a close-in orbit) exposed to a dense and/or fast stellar wind or CME plasma flow, (4.2) may yield R_s shorter then the

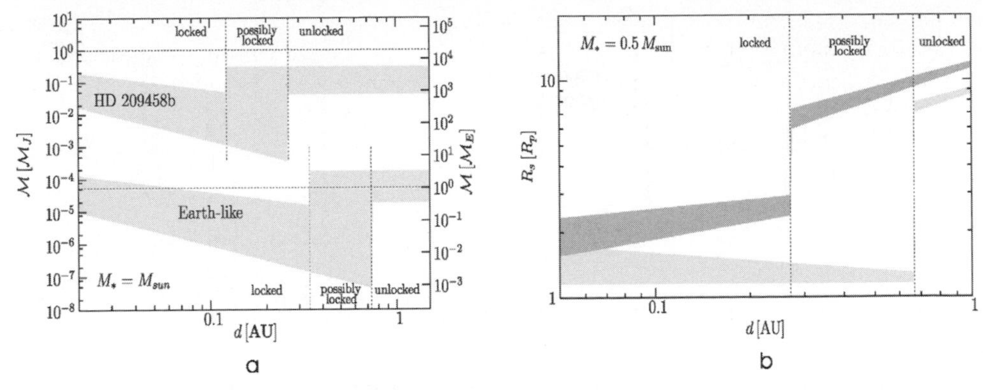

Figure 3. (a): Range of possible magnetic moments defined by (4.1) for a terrestrial-type and 'Hot Jupiter' exoplanets orbiting a solar-type star; (b): Range of the magnetopause standoff distances R_s compressed by CME plasma flow. The dark and light gray area correspond to the cases of a strongly (i.e. $\mathcal{M}_{max}$) and weakly magnetized (i.e. $\mathcal{M}_{min}$) terrestrial-type exoplanets, respectively. The size of the gray areas is determined by the difference between dense and sparse CMEs as described by (2.1). (adopted from Khodachenko, *et al.* 2007a)

planetary radius R_p. In such extreme cases R_s is usually set to $\sim R_p$, supposing that deep in the atmosphere the dense neutral gas layers deflect the plasma flow at somewhat higher altitudes above the planetary surface. Figure 3b shows the range of expected R_s, given by (4.2) with (4.1) taken into account, as a function of orbital distance for a terrestrial-type exoplanet orbiting a star with $M^* = 0.5M_{Sun}$ and affected by the flow of stellar CME plasmas. For these calculations the range of stellar CME densities, given by (2.1) as a function of orbital distance, and an average velocity of CMEs $\sim$ 500 km/s have been used. The mass of the host star also plays a certain role in these estimations because it influences the tidal locking of a planet and therefore determines its rotation rate. As can be seen from the Figure 3b, the magnetopause standoff distances of weakly magnetized terrestrial exoplanets at close orbits, which may correspond to the traditionally defined HZs, can be shrunk, under the action of CMEs, towards the planetary surface so, that the planetary atmospheres are directly exposed to the stellar CME plasma flows.

4.2. *'Hot Jupiters' magnetic moments estimations by observations of ENA coronas*

Observations of multiple transits of the 'Hot Jupiter' exoplanet HD 209458b in front of its parent star, performed with Hubble Space Telescope (HST), have shown absorption in the stellar Lyman-α line (at 1215.67$\mathring{A}$) during transit, revealing the presence of high-velocity atomic hydrogen (i.e. ENA corona) at great distances around the planet (Vidal-Madjar, *et al.* 2003). HD 209458b is a Jupiter-type gas giant with a mass of $M_p \sim 0.69M_{Jup}$ and a size of $R_p \sim 1.32R_{Jup}$ that orbits at 0.045 AU around its host star HD 209458, which is a solar-type G- star with an age of about 4 Gyr. The activity of the star is comparable to that of the present Sun during a moderately quiet phase.

There are several features of the transit Lyman-α absorption spectrum that any proposed source of the observed hydrogen atoms needs to account for: 1) the indication of presence of hydrogen atoms with velocities $\geqslant$ 130 km/s moving *away* from the star; 2) a fairly uniform absorption over the whole, outward from the star, velocity range 45–130 km/s; 3) the indication of presence of hydrogen atoms with velocities 30 - 105 km/s moving *towards* the star. The originally proposed explanation of the observed around HD 209458b ENA corona is that hydrogen atoms in the exosphere are undergoing hydrodynamic escape, and are further accelerated by the stellar radiation pressure (Vidal-Madjar, *et al.* 2003). However, there are difficulties in explaining the above mentioned three

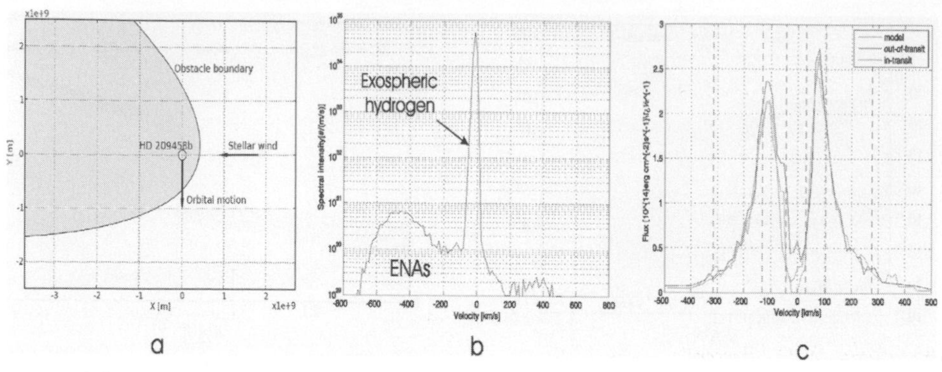

Figure 4. (a): Planetary magnetospheric obstacle in the simulation box for ENA production; (b): Hydrogen atoms velocity spectrum (high velocity population are ENAs, low-velocity narrow peak corresponds to the exospheric hydrogen); (c): Lyman-α profiles observed before and during the transit, and the modelled attenuation profile (adopted from Holmström, *et al.* 2008 and Ekenbäck, *et al.* 2008)

features of the atomic hydrogen velocity distribution. In particular, rather large radiation pressure is needed to accelerate the hydrogen atoms up to 130 km/s before they are photoionized, whereas uniform absorption in the velocity range 45 - 130 km/s, as well as the presence of atoms moving towards the star with speeds up to 105 km/s, cannot be at all reproduced within a model of an exosphere driven by radiation pressure.

In view of these difficulties, an alternative interpretation of observations was suggested by Holmström, *et al.* (2008), showing that the measured Lyman-α absorption spectrum can be explained via ENAs, produced by charge exchange between the stellar wind protons around HD 209458b and its exospheric neutrals. This mechanism is quite similar to the known formation of ENAs around Earth, Mars and Venus. In the model, proposed by Holmström, *et al.* (2008), it is assumed that charge exchange reactions take place outside a quasi-conic obstacle which represents the magnetosphere of the planet (Figure 4a), i.e. the exosphere has to be extended enough to reach the stellar wind above the planet's magnetopause. The shape of the magnetospheric obstacle is modelled phenomenologically as a surface: $X = r^2/(20R_p) + R_s$, where r is the distance to the planet – star line, aberrated by an angle of $arctan(v_p/v_{sw})$ to account for the finite stellar wind speed v_{sw} relative to the planets orbital speed v_p. Simulation of interaction between the stellar wind and planetary exosphere, extended above magnetopause, yields an exospheric cloud, along with the produced ENAs, shaped like a comet tail. In the resulting hydrogen velocity spectrum along the planet – star line (Figure 4b), the population of atoms with high velocity are the stellar wind protons that have charge-exchanged, becoming ENAs, whereas the narrower peak, centered in the low-velocity region, is due to the exospheric hydrogen atoms.

The next step is to estimate how the ENA corona affects the Lyman-α absorption spectrum of HD 209458b and to fit the estimated spectra to observations. The Lyman-α line profiles were observed outside and during the transit, and the difference between these two profiles corresponds to the attenuation by hydrogen atoms (Figure 4c). By this, the details of the neutral hydrogen distribution near HD 209458b, and the resulting attenuation spectrum, depend on the parameters of the stellar wind, geometrical characteristics of the magnetospheric obstacle and the exospheric parameters (e.g. temperature and density distributions). Taking the exospheric conditions on HD 209458b from the known models (Yelle 2004; Penz, *et al.* 2008) and assuming the stellar wind parameters similar to those of the present Sun, Ekenbäck, *et al.* (2008) determined an upper bound of the magnetospheric obstacle standoff distance $R_s^{fit} \simeq$ (4-10)$\times 10^8$ m, for which the modelled attenuation spectra provide the best fit to the observed spectra during the HD 209458b transit (Figure 4c). This value of the R_s^{fit}, according to (4.2), yields an estimation for the magnetic moment $\mathcal{M}_{HD\ 209458b} \approx 0.4\mathcal{M}_{Jup}$, where $\mathcal{M}_{Jup} = 1.56 \times 10^{27}$ Am^2 is the magnetic dipole moment of Jupiter.

Note, that the scaling laws (4.1), in view of the fact of likely tidal locking of the HD 209458b, provide for possible $\mathcal{M}_{HD\ 209458b}$ a bit smaller values: $(0.005...0.1)\mathcal{M}_{Jup}$. Thus, it may be concluded that the planetary rotation is an important factor for magnetic moment generation, however the true magnetic moment is bigger then the maximum value predicted by the scaling laws, which means that planetary rotation is not as important as previously was thought.

4.3. *Magnetospheric protection of 'Hot Jupiters'*

When considering the evolutional history of the 'Hot Jupiter' HD 209458b, it is necessary to keep in mind that close orbital location of this exoplanet to the host star may result in a strong erosion of its atmosphere by the ion pick-up mechanism caused by stellar CMEs, colliding with the planet (Khodachenko, *et al.* 2007b). Because of the sporadic character of the CME-planetary collisions, in the case of a moderately active host star of HD 209458b, these effects have not been taken into account in the considered in Section 4.2 study of ENA production on HD 209458b (Holmström, *et al.* 2008; Ekenbäck, *et al.* 2008). However, as it has been shown in Khodachenko, *et al.* (2007b), the integral action of the stellar CME impacts over the exoplanet's lifetime can produce significant effect on the planetary mass loss.

To study the interaction between the hydrodynamically driven neutral hydrogen wind in HD 209458b and the CME plasma flow, Khodachenko, *et al.* (2007b) applied the hydrodynamic numerical model described in Penz, *et al.* (2008). Distribution of the upper atmospheric neutral hydrogen, predicted by the model, was superimposed with estimations of the planetary magnetopause stand-off distance. The last was calculated for the typical maximum and minimum values of n_{CME}, given by (2.1) at the orbital distance of HD 209458b, within the assumption of the tidal locking of the planet, which according to (4.1) limits the values of possible magnetic dipole moment of the planet to the range $(0.005...0.1)\mathcal{M}_{Jup}$. It has been found that encountering CMEs plasma pushes the magnetopause stand-off distance of HD 209458b down to the heights at which the ionization and pick-up of the upper planetary neutral atmosphere by the CME plasma flow takes place. The hydrogen ion pick-up loss rates, caused by the CMEs, acting on the the extended above the magnetopause upper atmosphere of HD 209458b were calculated by Khodachenko, *et al.* (2007b) using a numerical test particle model (Lichtenegger, *et al.* 2002), which includes the effects of photo-ionization, as well as the ionization by the CME plasma flow (i.e. charge exchange and electron impact). Finally, assuming for the host star of HD 209458b the same CME occurrence rate as that on the Sun, Khodachenko, *et al.* (2007b) estimated that over its lifetime HD 209458b may have lost the mass from $\sim$0.2 (for $\mathcal{M}_{HD\ 209458b} = \mathcal{M}_{max} \equiv 0.1\mathcal{M}_{Jup}$ and dense CME flow) up to several tens (for $\mathcal{M}_{HD\ 209458b} = \mathcal{M}_{min} \equiv 0.005\mathcal{M}_{Jup}$ and sparse CME flow) of its present mass. Under certain conditions it could even be evaporated down to the core. On the other hand, the existence of rather large population of known 'Hot Jupiter' exoplanets at close orbital distances from their host stars indicates that such tremendous planetary mass losses are probably not typical. This means that the 'Hot Jupiter' exoplanets should have strong enough intrinsic magnetic fields and extended magnetospheres, which are able to protect the planetary atmospheres against of stellar winds and CME plasma flows. In view of that, the relatively small ($\sim 0.2M_p$) total mass loss of HD 209458b, caused by CMEs in the case of the predicted by the scaling laws (4.1) maximal possible value $\mathcal{M}_{HD\ 209458b} \approx 0.1\mathcal{M}_{Jup}$, may be considered as an additional indication of the consistency of even a bit higher ($0.4\mathcal{M}_{Jup}$) value for $\mathcal{M}_{HD209458b}$, estimated from the planetary hydrogen ENA corona observations.

5. Conclusions

To summarize this review we would like to emphasize that stellar XUV radiation and stellar wind plasmas exposure strongly impact the environments of close-orbit exoplanets, such as terrestrial-type exoplanets in the HZs of low mass stars and 'Hot Jupiters'. By this, complete or partial tidal locking of such exoplanets should lead to relatively weak intrinsic planetary magnetic moments. This may result in a situation when a stellar wind and encountering stellar CMEs will compress a planetary magnetosphere down to the heights at which the ionization and pick-up of the planetary neutral atmosphere by the CMEs plasma flow takes place. All this makes the stellar activity and planetary magnetospheric protection factors to play a crucial role for the whole complex of planetary evolution processes, including atmospheric erosion, mass loss and, finally, for the definition of HZ zone parameters for the terrestrial-type planets near a star. The last may significantly limit an actual HZ range, as compared to that followed from the traditional HZ definition, based on the pure climatological approach.

Acknowledgements

This work was supported by the Austrian “Fonds zur Förderung der wissenschaftlichen Forschung” (project P21197-N16) and the ÖAD Scientific and Technical Collaboration Program. The authors acknowledge the International Space Science Institute (ISSI; Bern, Switzerland) and the ISSI team “Evolution of Exoplanet Atmospheres and their Characterization”. H.L., H.I.M.L. and M.K. thank Helmholtz Association and its research alliance project "Planetary Evolution and Life".

References

Audard, M., Güdel, M., Drake, J. J., & Kashyap, V. L. 2000, *ApJ* 541, 396

Busse, F. H. 1976, *Phys. Earth Planet. Int.* 12(4), 350

Cully, S. L., Fisher, G. H., Abbott, M. J., & Siegmund, O. H. W. 1994, *ApJ* 435, 449

Ekenbäck, A., Holmström, M., Wurz, P., Grießmeier, J.-M., Lammer, H., Selsis, F., & Penz, T. 2008, *ApJ*, submitted

Grießmeier, J.-M., Stadelmann, A., Penz, T., Lammer, H., Selsis, F., Ribas, I., Guinan, E. F., Motschmann, U., Biernat, H. K., & Weiss, W. W. 2004, *A&A* 425, 753

Grießmeier, J.-M., Stadelmann, A., Motschmann, U., Belisheva, N. K., Lammer, H., & Biernat, H. 2005, *Astrobiology* 5, 587

Grießmeier, J.-M., Preusse, S., Khodachenko, M. L., Motschmann, U., Mann, G., & Rucker, H. O. 2007, *Planet. & Space Sci.* 55, 618

Holmström, M., Ekenbäck, A., Selsis, F., Penz, T., Lammer, H., & Wurz, P. 2008, *Nature* 451, 970

Houdebine, E. R., Foing, B. H., & Rodonó, M. 1990, *A&A* 238, 249

Huang, S. S. 1960, *PASP* 72, 489

Kasting, J. F., Whitmire, D. P., & Reynolds, R. T. 1993, *Icarus* 101, 108

Kasting, J. F. 1997, *Orig. Life Evol. Biosph.* 27(1/3), 291

Khodachenko, M. L., Ribas, I., Lammer, H., Grießmeier, J.-M., Leitner, M., Selsis, F., Eiroa, C., Hanslmeier, A., Biernat, H., Farrugia, C. J., & Rucker, H. 2007a, *Astrobiology* 7, 167

Khodachenko, M. L., Lammer, H., Lichtenegger, H. I. M., Langmayr, D., Erkaev, N. V., Grießmeier, J.-M., Leitner, M., Penz, T., Biernat, H. K., Motschmann, U., & Rucker, H. O. 2007b, *Planet. & Space Sci.* 55, 631

Kulikov, Yu. N., Lammer, H., Lichtenegger, H. I. M., Penz, T., Breuer, D., Spohn, T., Lundin, R., & Biernat, H. K. 2007, *Space Sci Rev.* 129, 207

Lammer, H., Lichtenegger, H., Kulikov, Yu., Grießmeier, J.-M., Terada, N., Erkaev, N., Biernat, H., Khodachenko, M. L., Ribas, I., Penz, T., & Selsis, F. 2007, *Astrobiology* 7, 185

Lammer, H., Kasting, J. F., Chassefière, E., Johnson, R. E., Kulikov, Yu. N., & Tian, F. 2008, *Space Sci Rev.* 139, 399

Lichtenegger, H. I. M., Lammer, H., & Stumptner, W. 2002, *JGR* 107 (A10), doi:10.1029/2001JA000322

Lichtenegger, H. I. M., Gröller, H., Lammer, H., Kulikov, Yu. N., & Shematovich, V. 2008, *Geophys. Res. Lett.* submitted

Mizutani, H., Yamamoto, T., & Fujimura, A. 1992, *Adv. Space Res.* 12(8), 265

Penz, T., Erkaev, N. V., Kulikov, Yu. N., Langmayr, D., Lammer, H., Micela, G., Cecchi-Pestellini, C., Biernat, H. K., Selsis, F., Barge, P., Deleuil, M., & Leger, A. 2008, *Planet. & Space Sci.* 56, 1260

Ribas, I., Guinan, E. F., Güdel, M., & Audard, M. 2005, *ApJ* 622, 680

Rivera, E. J., Lissauer, J. J., Butler, R. P., Marcy, G. W., Vogt, S. S., Fischer, D. A., Brown, T. M., Laughlin, G., & Henry, G. W. 2005, *ApJ* 634, 625

Sano, Y. 1993, *J. Geomag. Geoelectr.* 45, 65

Santos, N. C., Bouchy, F., Mayor, M., Pepe, F., Queloz, D., Udry, S., Lovis, C., Bazot, M., Benz, W., Bertaux, J.-L., Curto, G. L., Delfosse, X., Mordasini, C., Naef, D., Sivan, J.-P., & Vauclair, S. 2004, *A&A* 426, L19

Scalo. J., Kaltenegger. L., Segura, A. G., Fridlund, M., Ribas, I., Kulikov, Yu. N., Grenfell, J. L., Rauer, H., Odert, P., Leitzinger, M., Selsis, F., Khodachenko, M. L., Eiroa, C., Kasting, J., & Lammer, H. 2007, *Astrobiology* 7, 85

Stevenson, D. J. 1983, *Rep. Prog. Phys.* 46, 555

Tian, F., Kasting, J. F., Liu, H., & Roble, R. G. 2008, *JGR* 113, Issue E5, CiteID E05008 (DOI: 10.1029/2007JE002946)

van den Oord, G. H. J. & Doyle, J. G. 1997, *A & A* 319, 578

Vidal-Madjar, A., des Etangs, A. L., Désert, J.-M., Ballester, G. E., Ferlet, R., Hébrard, G., & Mayor, M. 2003, *Nature* 422, 143

Voigt, G.-H. 1995, in: H. Volland (ed.), *Handbook of atmospheric electrodynamics*, vol. II (CRC Press), p. 333

Walker, A. R. 1981, *MNRAS* 195, 1029

Wood, B. E., Müller, H.-R., Zank, G. P., Linsky, J. L., & Redfield, S. 2005, *ApJ* 628, L143

Yelle, R. V. 2004, *Icarus* 170, 167

Discussion

KOUTCHMY: Regarding 'Hot Jupiters', is not clear that they are not habitable just because they are too hot (no water), and also probably because gravity effects are too big at their surface?

KHODACHENKO: 'Hot Jupiters' cannot be considered as potential habitats by many reasons, including also those, you just have mentioned. It is unlikely that such planets may have liquid water on their surfaces. However, similarly to the known giant planets in the solar system, 'extra-solar Jupiters' may have moons (like Europa, Ganimed, Titan, Enceladus, etc.). These smaller planets, under certain circumstances, may evolve into worlds, much more suitable for life, and keep liquid water on their surfaces or in the interiors.

Serge Koutchmy (with microphone)

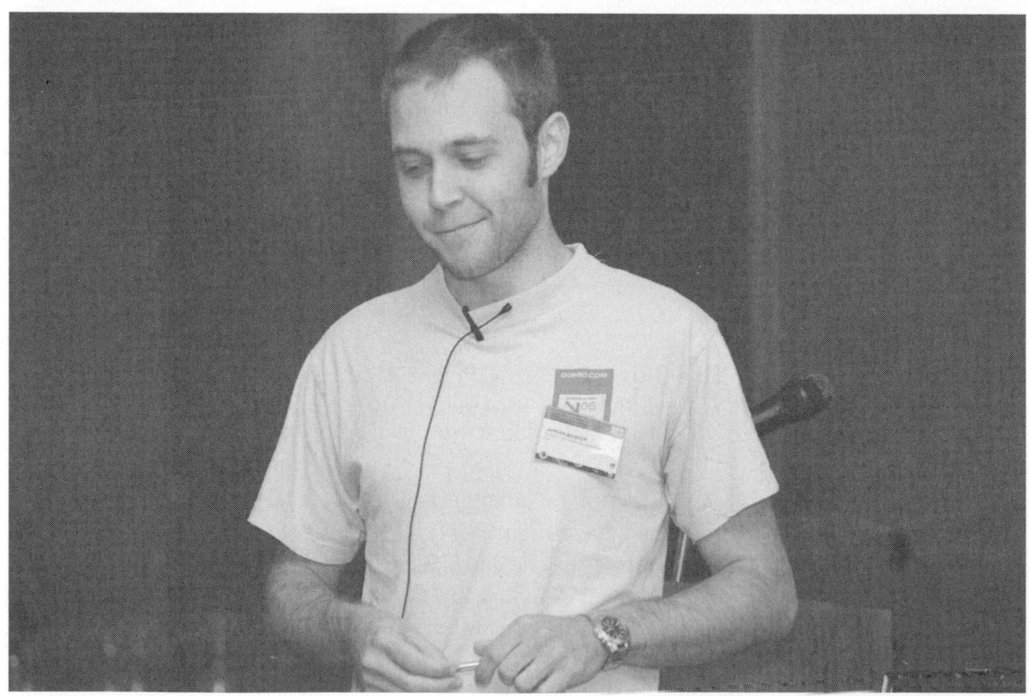

Adrian Barker

Cosmic Magnetic Fields:
From Planets, to Stars and Galaxies
Proceedings IAU Symposium No. 259, 2008
K.G. Strassmeier, A.G. Kosovichev & J.E. Beckman, eds.

doi:10.1017/S1743921309030634

Effects of magnetic braking and tidal friction on hot Jupiters

Adrian J. Barker and Gordon I. Ogilvie

Department of Applied Mathematics and Theoretical Physics, University of Cambridge,
Centre for Mathematical Sciences, Wilberforce Road, Cambridge CB3 0WA, UK
email: ajb268@cam.ac.uk

Abstract. Tidal friction is thought to be important in determining the long-term spin-orbit evolution of short-period extrasolar planetary systems. Using a simple model of the orbit-averaged effects of tidal friction (Eggleton *et al.* 1998), we analyse the effects of the inclusion of stellar magnetic braking on the evolution of such systems. A phase-plane analysis of a simplified system of equations, including only the stellar tide together with a model of the braking torque proposed by Verbunt & Zwaan (1981), is presented. The inclusion of stellar magnetic braking is found to be extremely important in determining the secular evolution of such systems, and its neglect results in a very different orbital history. We then show the results of numerical integrations of the full tidal evolution equations, using the misaligned spin and orbit of the XO-3 system as an example, to study the accuracy of simple timescale estimates of tidal evolution. We find that it is essential to consider coupled evolution of the orbit and the stellar spin in order to model the behaviour accurately. In addition, we find that for typical Hot Jupiters the stellar spin-orbit alignment timescale is of the same order as the inspiral time, which tells us that if a planet is observed to be aligned, then it probably formed coplanar. This reinforces the importance of Rossiter-McLaughlin effect observations in determining the degree of spin-orbit alignment in transiting systems.

Keywords. Planetary systems – stars: rotation – celestial mechanics – stars: XO-3

1. Introduction

Since the discovery of the first extrasolar planet around a solar-type star (Mayor & Queloz 1995), observers have now detected 322† planets around stars outside the solar system. Many of these planets have roughly Jovian masses and orbit their host stars in orbits with semi-major axes less than 0.2 AU, the so-called "Hot Jupiters" (HJs). In both of the giant planet formation scenarios, core accretion and gravitational instability, we are unable to produce HJs in situ. We require the formation of close-in planets much further out ($a \sim$ several AU) in the protoplanetary disc, before a migratory process that brings the planet in towards the star and to its present location (Lin, Bodenheimer & Richardson 1996).

The formation of systems with giant planets can be thought of as occurring in two oversimplified stages (Juric & Tremaine 2008). During stage 1 the cores of the giant planets are formed, they accrete gas and undergo migration, driven by the dynamical interaction between the planets and the gaseous protoplanetary disc (see Papaloizou *et al.* 2008 for a recent review). This stage lasts a few Myr until the gas dissipates, by which time a population of gas giants may exist. If these form sufficiently closely packed then stage 2 follows. This stage lasts from when the disc has dissipated and continues until the present, and primarily involves gravitational interactions and collisions between

† As of the day of this session, 4th Nov. 2008 – see http://exoplanet.eu/

the planets. Recent studies into stage 2 (Juric & Tremaine 2008; Chatterjee *et al.* 2008) have shown that this is a chaotic era, in which planet-planet scatterings force the ejection of all but a few ($\sim 2-3$) planets from the system in a period of large-scale dynamical instability lasting $\leqslant 10^8$yr. This mechanism can excite the eccentricities of the planets to levels required to explain observations.

Planet-planet scatterings tend also to excite the inclinations of the planets with respect to the initial symmetry plane of the system, potentially leading to observable consequences via the Rossiter-McLaughlin (RM) effect, which allows a measurement of λ, the sky-projected angle of misalignment of the stellar spin and the orbit (Winn 2008). Misaligned orbits are not predicted from stage 1 alone, so if λ is measured to be appreciably nonzero in enough systems, then it could be seen as evidence for planet-planet scattering, or alternatively Kozai migration (see Fabrycky & Tremaine 2007). This is because gas-disc migration does not seem able to excite orbital inclination (Lubow & Ogilvie 2001; Cresswell *et al.* 2007). Alternatively, if observed planets are all found with λ consistent with zero, this could rule out planet-planet scattering or Kozai migration as being of any importance. One important consideration is that at such close proximity to their parent stars, strong tidal interactions are expected to change λ (actually the true spin-orbit misalignment angle $i = \arccos(\hat{\mathbf{\Omega}} \cdot \hat{\mathbf{h}})$) over time. If tides can significantly change λ since the time of formation, we may have difficulty in distinguishing between the possible HJ formation mechanisms of planet-planet scattering, Kozai migration and gas-disc migration. Below we approach the problem of studying the effects of tidal friction on such inclined orbits.

2. Model of tidal friction & magnetic braking

The efficiency of tidal dissipation in a body is usually parametrised by a dimensionless quality factor Q, defined by:

$$Q = 2\pi E_0 \left(\oint -\dot{E} dt \right)^{-1},$$

where E_0 is the maximum energy stored in an oscillation and the integral represents the energy dissipated over one oscillation period. We find it convenient to define $Q' = \frac{3Q}{2k_2}$, where k_2 is the second-order potential Love number, since this combination appears together in the theory.

Q' is in principle a function of the tidal forcing frequency and the amplitude of the tidal disturbance, and is a result of complex dissipative processes in each body (Zahn 2008). Recent work has shown that Q' for solar-type stars depends in a highly erratic way on the tidal forcing frequency (Ogilvie & Lin 2007). In light of the uncertainties involved in calculating Q', and the difficulty of calculating the evolution when Q' depends on tidal frequency, we adopt a simplified model based on an equilibrium tide with constant lag time, using the formulation of Eggleton *et al.* 1998. This formulation is beneficial because it can treat arbitrary orbital eccentricities and stellar and planetary obliquities. We refer the reader to Barker & Ogilvie (2009) or Mardling & Lin (2002) for the full set of equations used.

Observations of solar-type stars show that the mean stellar rotational velocity decreases with main-sequence age (Skumanich 1972), following the relation $\Omega \propto t^{-1/2}$. Magnetic braking by a magnetised outflowing wind has long been recognised as an important mechanism for the removal of angular momentum from rotating stars (Weber & Davis 1967), and such a mechanism seems able to explain (on average) most of the observed stellar spin-down (Barnes 2003).

Here we include the effects of magnetic braking in the tidal evolution equations, through the inclusion of the magnetic braking torque of Verbunt & Zwaan (1981). This gives the following term into of the equation for $\dot{\mathbf{\Omega}}$ (Dobbs-Dixon *et al.* 2004):

$$\dot{\omega}_{mb} = -\alpha_{mb}\, \Omega^2\, \mathbf{\Omega}$$

where $\alpha_{mb} = 1.5 \times 10^{-14}$ yr, giving a braking timescale $\sim 2 \times 10^{11}$ yr for the Sun.

3. Analysis of the effects of magnetic braking on tidal evolution for a simplified system

We study the effects of magnetic braking on a simplified system of a circular, coplanar orbit under the influence of only the stellar tide and magnetic braking. The following set of dimensionless equations can be derived (Barker & Ogilvie 2009):

$$\frac{d\tilde{\Omega}}{d\tilde{t}} = \tilde{n}^4 \left(1 - \frac{\tilde{\Omega}}{\tilde{n}}\right) - A\,\tilde{\Omega}^3 \tag{3.1}$$

$$\frac{d\tilde{n}}{d\tilde{t}} = 3\,\tilde{n}^{\frac{16}{3}} \left(1 - \frac{\tilde{\Omega}}{\tilde{n}}\right), \tag{3.2}$$

where we have normalised the stellar rotation Ω and orbital mean motion n to the orbital frequency at the stellar surface, together with a factor C (to some power). C is the ratio of the orbital angular momentum of an orbit with semi-major axis equal to the stellar radius $R_\star$, of a mass m_p, with the spin angular momentum of an equally rapidly rotating star of radius $R_\star$, mass $m_\star$ and dimensionless radius of gyration r_g. The reduced mass is $\mu = \frac{m_\star m_p}{m_\star + m_p}$. C is important for classifying the stability of the equilibrium curve $\tilde{\Omega} = \tilde{n}$ in the absence of magnetic braking, and it can be shown that this equilibrium is stable if $\tilde{n} \leqslant 3^{-\frac{3}{4}}$ i.e. no more than a quarter of the total angular momentum can be in the form of spin angular momentum (Hut 1980). We have thus defined the following dimensionless quantities:

$$\tilde{\Omega} = \Omega \left(\frac{R_\star^3}{G(m_\star + m_p)}\right)^{\frac{1}{2}} C^{-\frac{3}{4}}, \quad \tilde{n} = n \left(\frac{R_\star^3}{G(m_\star + m_p)}\right)^{\frac{1}{2}} C^{-\frac{3}{4}}, \quad C = \frac{\mu R_\star^2}{I_\star} = \frac{\mu}{r_g^2\, m_\star},$$

$$\tilde{t} = \sqrt{\frac{G(m_\star + m_p)}{R_\star^3}} \left(\frac{9}{Q'}\right) \left(\frac{m_p}{m_\star}\right) C^{\frac{13}{4}}\, t,$$

$$A = \alpha_{mb} \sqrt{\frac{G(m_\star + m_p)}{R_\star^3}} \left(\frac{Q'}{9}\right) \left(\frac{m_\star}{m_p}\right) \left(\frac{\mu R_\star^2}{I_\star}\right)^{-\frac{7}{4}} \simeq 100 \left(\frac{Q'}{10^6}\right).$$

There is only one parameter (A) that completely characterises the solution in the $(\tilde{\Omega}, \tilde{n})$-plane, and its value may be estimated as $A \simeq 100$ for a Jupiter-mass planet orbiting a Sun-like star undergoing magnetic braking (with standard α_{mb} and with $Q' = 10^6$). The size of this term shows that in general magnetic braking dominates the stellar spin evolution. We plot some solutions on the $(\tilde{\Omega}, \tilde{n})$-plane, restricting ourselves to $0 \leqslant \tilde{\Omega} \leqslant 10, 0 \leqslant \tilde{n} \leqslant 10$. This represents the full range of orbits of the HJs, since $\tilde{n} \simeq 10$ corresponds to an orbital semi-major axis of $a \simeq 0.01$ AU, and $\tilde{n} \simeq 0.1$ corresponds to $a \simeq 0.2$ AU.

With the inclusion of magnetic braking (see Fig. 1), an initially rapidly rotating solar-type star hosting a HJ for which $\tilde{\Omega} \geqslant \tilde{n}$, will spin down as a result of the magnetic torque. During this stage of spin-down the spin frequency of the star may temporarily equal the orbital frequency of its close-in planet, but the rate of angular momentum loss

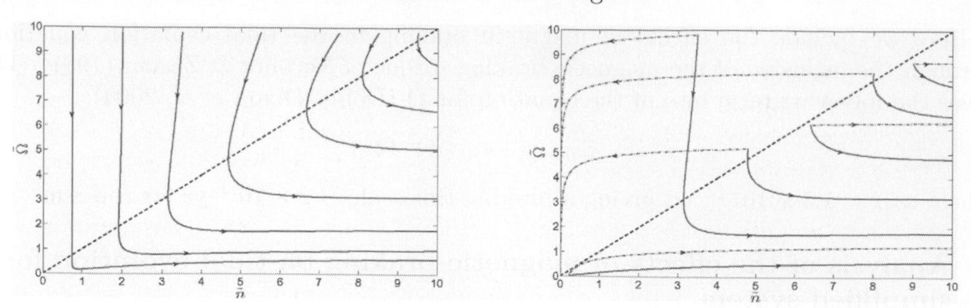

Figure 1. $(\tilde{\Omega}, \tilde{n})$-plane with $A = 100$ for a HJ orbiting a Sun-like star. The dashed line in each plot corresponds to corotation ($\tilde{\Omega} = \tilde{n}$). Left: magnetic braking spins the star down so that the planet finds itself inside corotation, where the sign of the tidal torque changes, and the planet is subject to tidally induced orbital decay. Right: solutions with and without magnetic braking for the same initial conditions. Dot-dashed lines, which are also curves of constant total angular momentum, are without braking ($A = 0$) and solid lines are with braking ($A = 100$). This shows that the inclusion of magnetic braking is extremely important in determining the secular evolution of the system, and its absence results in a very different evolutionary history unless $\tilde{\Omega} \ll \tilde{n}$ in the initial state.

through magnetic braking will exceed the tidal rate of transfer of angular momentum from orbit to spin. The stellar spin continues to drop well below synchronism until the efficiency of tidal transfer of angular momentum from orbit to spin can compensate or overcompensate for the braking. The planet continues to spiral into the star once it moves inside corotation, and $\tilde{\Omega} \simeq const$, unless the planet has sufficient angular momentum to be able to appreciably spin up the star. Note that any bound orbit will eventually decay in a finite time since the system has no stable equilibrium.

Without magnetic braking, the evolution is qualitatively different (see right plot Fig. 1), with orbits initially outside corotation ($\tilde{\Omega} > \tilde{n}$) asymptotically approaching a stable equilibrium $\tilde{\Omega} = \tilde{n}$ for $\tilde{n} \leqslant 3^{-\frac{3}{4}}$. Orbits initially inside corotation with $\frac{d\tilde{\Omega}}{dt} > \frac{d\tilde{n}}{dt} > 0$ near corotation, also approach a stable equilibrium (though no such curves are plotted in Fig. 1, since they occur only in the far bottom left of the plot, near the origin). This is when the corotation radius moves inwards faster than the orbit shrinks due to tidal friction, resulting in a final stable equilibrium state for the system if the corotation radius "catches up" with the planet. For orbits inside corotation for which this condition is not satisfied, the planet will spiral into the star under the influence of tides.

4. Tidal evolution timescales

It is common practice to interpret the effects of tidal evolution in terms of simple timescale estimates. The idea behind these is that if the rate of change of a quantity X is exponential, then $\dot{X}/X$ will be a constant, so we can define a timescale for the evolution of X, given by $\tau_X = X/\dot{X}$. If $\dot{X}/X \neq$ const, then τ_X may not accurately respresent the evolution. A tidal inspiral time can be calculated from $\tau_a \equiv -\frac{2}{13}\frac{a}{\dot{a}}$ (see Barker & Ogilvie 2009 for the full evolution equations from which these timescales are derived), since $\frac{\dot{a}}{a} \sim a^{\frac{-13}{2}}$. This assumes Ω is constant, which we have seen is unreasonable unless $\Omega \ll n$, due to magnetic braking.

If the orbital and stellar equatorial planes are misaligned by a small angle i, then dissipation of the stellar tide would align them after a time

$$\tau_i \equiv -\frac{i}{\frac{di}{dt}} \simeq 35.0\,\mathrm{Myr}\left(\frac{Q'}{10^6}\right)\left(\frac{m_\star}{M_\odot}\right)\left(\frac{M_J}{m_p}\right)^2\left(\frac{R_\odot}{R_\star}\right)^3\left(\frac{P}{1d}\right)^4\left(\frac{\Omega}{\Omega_0}\right)\left[1 - \frac{P}{2P_\star}\left(1 - \frac{1}{\alpha}\right)\right]^{-1}$$

where $\Omega_0 = 5.8\times10^{-6}\mathrm{s}^{-1}$, and $P, P_\star$ are the orbital and stellar rotational periods. α is the ratio of orbit to spin angular momentum. The validity of these timescales to accurately represent tidal evolution is an important subject of study, since they are commonly applied to observed systems. Jackson *et al.* (2008) recently found it essential to consider the coupled evolution of e and a to accurately model tidal evolution, and that both the stellar and planetary tides must be included. They showed that the actual change of e over time can be quite different from simple circularisation timescale considerations, due to the coupled evolution of a. In the following we consider the validity of τ_i to accurately model tidal evolution of i, using XO-3 as an illustrative example.

5. Application to the misaligned spin and orbit of XO-3 b

The only system currently observed with a spin-orbit misalignment is XO-3 (Hebrard *et al.* 2008), which has a sky-projected spin-orbit misalignment of $\lambda \simeq 70° \pm 15°$. This system has a very massive $m_p = 12.5M_J$ planet on a moderately eccentric $e = 0.287$, $P = 3.2$ d orbit around an F-type star of mass $m_\star = 1.3M_\odot$. Its age is estimated to be $\tau_\star \simeq (2.4-3.1)$ Gyr. Note that even if the star is rotating near breakup velocity ($P_\star \sim 1$ d), the planet is still subject to tidal inspiral, since $P_\star > P\cos i$ (where we henceforth assume $i = \lambda$, which may slightly *underestimate* i).

Hebrard *et al.* (2008) quote a spin-orbit alignment timescale of $\sim 10^{12}$ yr for this system, but we find that this is in error by $\sim 10^5$. We believe that the reason for this discrepancy is that their estimate was based on assuming that the spin-orbit alignment time for XO-3 b is the same as for HD17156 b, which is a less massive planet on a much wider orbit. We find $\tau_i \sim 15$ Myr assuming $Q' = 10^6$ to align the whole star with the orbit. Integrations for this system are given in Fig. 2 for a variety of stellar Q' values. For the system to survive and remain with its current inclination for ~ 3 Gyr we require $Q' \geqslant 10^{10}$. These integrations highlight the importance of considering coupled evolution of the orbital and rotational elements, since timescales for tidal evolution are quite different from simple estimates. Indeed, the actual spin-orbit alignment time from integrating the coupled equations is about an order of magnitude smaller than that from the simple decay estimate, due to coupled a evolution.

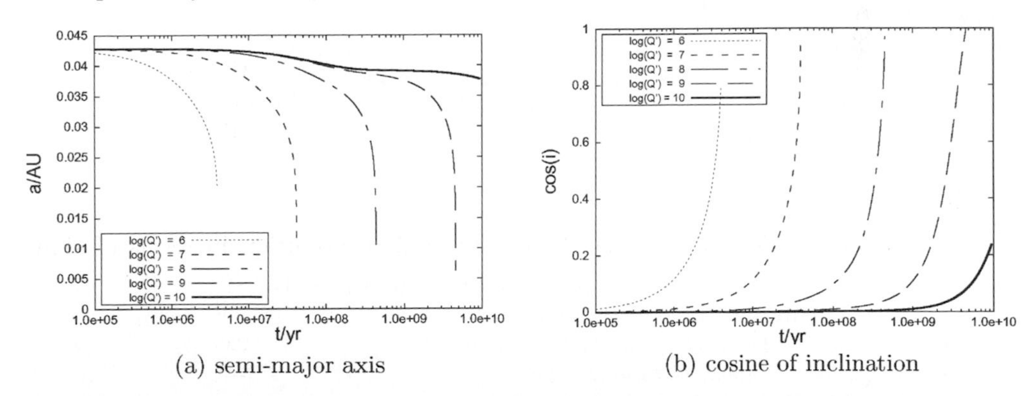

(a) semi-major axis (b) cosine of inclination

Figure 2. Tidal evolution of XO-3 b taking current values for the orbital properties of the system, except that $\cos i = 90°$ (not unreasonable since this corresponds to the upper limit on λ, which in any case gives a lower bound on i). Magnetic braking is included, and $\Omega/n = 2$ initially (results do not depend strongly on this choice). From (a) and (b) we require $Q' \geqslant 10^{10}$ for the planet to survive for several Gyr, and maintain its high inclination. Tidal dissipation in star must therefore be weak to explain the current configuration of the system.

For this system, and for typical HJs, we find that $\tau_i \sim \tau_a$. This means that if we observe a planet, then its survival implies that tides are unlikely to have aligned its orbit.

In addition, for systems on an accelerating inspiral into the star, the rate of inclination evolution will have been much lower in the past; therefore, if we observe a planet well inside corotation with a roughly coplanar orbit, we can assume that it must have started off similarly coplanar.

An explanation for the survival and remnant orbital inclination of XO-3 b could be the result of inefficient tidal dissipation in the host star. A calculation of Q' using the numerical method of Ogilvie & Lin (2007) and a stellar model appropriate for this star was performed (Barker & Ogilvie 2009). This predicts that the dissipation is weak, and $Q' \geqslant 10^{10}$ for most tidal frequencies for the host star XO-3. This can explain the survival and remnant inclination of this planet, since both inspiral and spin-orbit alignment take longer than the current age of the system.

6. Conclusions

Magnetic braking is important for calculating the long-term tidal evolution of HJs unless $\Omega \ll n$, and changes the qualitative behaviour of the evolution significantly. Tidal evolution can be much faster than simple timescale estimates predict when coupled integration of the orbital and rotational elements is considered. In addition, we find that $\tau_i \sim \tau_a$ for typical HJs, so the orbits of most close-in planets have probably not aligned, and are likely to be a relic of the migration process. This means that RM effect observations of transiting planets can potentially distinguish between planet-planet scattering, Kozai migration and gas-disc migration.

Acknowledgements

A.J.B would like to thank the STFC for a research studentship.

References

Barker, A. J. & Ogilvie, G. I. 2009, *MNRAS*, in preparation
Barnes, S. A. 2003, *ApJ* 586, 464
Chatterjee, S. *et al.* 2008, *ApJ* 686, 580
Cresswell, P. *et al.* 2007, *A&A* 473, 329
Dobbs-Dixon, I., Lin, D. N. C., & Mardling, R. A. 2004, *ApJ* 610, 464
Eggleton, P. P. and Kiseleva, L. G. and Hut, P. 1998, *ApJ* 499, 853
Fabrycky, D. and Tremaine, S., 2007, *ApJ* 669, 1298
Hebrard, G. *et al.* 2008, *A&A* 488, 763
Hut, P. 1980, *A&A* 92, 167
Hut, P. 1981, *A&A* 99, 126
Jackson, B., Greenberg, R., & Barnes, R. 2008, *ApJ* 678, 1396
Jurić, M. & Tremaine, S. 2008, *ApJ* 686, 603
Lin, D. N. C., Bodenheimer, P., & Richardson, D. C. 1996, *Nature* 380, 606
Lubow, S. H. and Ogilvie, G. I. 2001, *ApJ* 560, 997
Mardling, R. A. & Lin, D. N. C. 2002, *ApJ* 573, 829
Mayor, M. & Queloz, D. 1995, *Nature* 378, 355
Ogilvie, G. I. & Lin, D. N. C. 2007, *ApJ* 661, 1180
Ogilvie, G. I. & Lin, D. N. C. 2004, *ApJ* 610, 477
Papaloizou, J. C. B. *et al.* 2007, *Protostars and Planets V*, 655
Skumanich, A. 1972, *ApJ* 171,565
Verbunt, F. & Zwaan, C. 1981, *A&A* 100, L7
Weber, E. J. & Davis, L. J. 1967, *ApJ* 148, 217
Winn, J. N. 2008, ArXiv e-prints, 0807.4929
Zahn, J.-P. 2008, *Tidal dissipation in binary systems*, EAS Publications Series 29, 67

Discussion

JARDINE: Do your results for the banal (orbital) fate of planets depend on the assumption that there is no outer planet forcing the orbit?

BARKER: Outer planets are not included in the integrations that I have performed thus far. Their inclusion may indeed change the final results and this should be studied for any system.

KHODACHENKO: What could possibly be the outcomes of the inclusion of electrodynamic effects of interaction of planetary dipole with the conducting plasma environment in your model? This may provide an additional energy dissipation channel and introduce an additional force into the system.

BARKER: An interesting question, and one that I am planning on studying in the next few years of my PhD. It would be interesting to study whether this **B** force could be significant enough to cause any orbital changes of the planet. In addition, dissipation of time-varying stellar magnetic-field excited currents in the planet could potentially inflate the planet. This is something that needs to be studied.

Inside the lecture room

Enjoying common dynamo problems ... Paul Roberts (left) and Günther Rüdiger (right)

Irina Kitiashvili

Cosmic Magnetic Fields:
From Planets, to Stars and Galaxies
Proceedings IAU Symposium No. 259, 2008
K.G. Strassmeier, A.G. Kosovichev & J.E Beckman, eds.

doi:10.1017/S1743921309030646

Magnetic and tidal interactions in spin evolution of exoplanets

Irina N. Kitiashvili

Center for Turbulence Research, Stanford University, Stanford, CA 94305, USA
email: irinasun@stanford.edu

Abstract. The axis-rotational evolution of exoplanets on close orbits strongly depends on their magnetic and tidal interactions with the parent stars. Impulsive perturbations from a star created by periodical activity may accumulate with time and lead to significant long-term perturbations of the planet spin evolution. I consider the spin evolution for different conditions of gravitational, magnetic and tidal perturbations, orbit eccentricity and different angles between the planetary orbit plane and the reference frame of a parent star. In this report I present a summary of analytical and numerical calculations of the spin evolution, and discuss the problem of the star-planet magnetic interaction.

Keywords. Exoplanets – stars: planetary systems – stars: rotation – magnetic field

---

It is known that planetary rotation is determined by orientation of the angular momentum, L. In the case of perturbed rotation of a celestial body, the angular momentum changes its position in space relative to the evolving orbit. For planets moving at close-in orbits in addition to gravitational and tidal perturbations, magnetic field of the parent star and its activity may have significant influence (Shkolnik *et al.*, 2003).

In this report I present results of modeling the spin evolution of a dynamically symmetrical exoplanet (with equal principle moments of inertia, $A = B$) under the action of gravitational, magnetic and tidal perturbations for different values eccentricity, and propose possible scenarios for different types of exoplanet. For this study, I analyze the following system of evolutionary equations derived for the case when the angular spin velocity of a planet is significantly higher than its angular orbital velocity (Beletskii, 1981; Beletsky & Khentov 1995):

$$
\begin{aligned}
\frac{d\rho}{dt} &= \frac{1}{L\sin\rho}\left(\frac{\partial U}{\partial\psi}\cos\rho - \frac{\partial U}{\partial\Sigma}\right) + \frac{M_1}{L} - K_\Omega \sin i \cos\Sigma, \\
\frac{d\Sigma}{dt} &= \frac{1}{L\sin\rho}\frac{\partial U}{\partial\rho} + \frac{M_2}{L\sin\rho} + K_\Omega\left(\sin i \cot\rho \sin\Sigma - \cos i\right), \\
\frac{dL}{dt} &= \frac{\partial U}{\partial\psi} + M_3, \\
\frac{d\vartheta}{dt} &= \frac{1}{L\sin\vartheta}\left(\cos\vartheta\frac{\partial U}{\partial\psi} - \frac{\partial U}{\partial\phi}\right) + \frac{M_2\cos\psi - M_1\sin\psi}{L}, \\
\frac{d\psi}{dt} &= \frac{L}{A} - \frac{1}{L}\left(\frac{\partial U}{\partial\vartheta}\cot\vartheta + \frac{\partial U}{\partial\rho}\cot\rho\right) - \frac{M_1\cos\psi + M_2\sin\psi}{L}\cot\vartheta \\
&\quad - \frac{M_2}{L}\cot\rho - K_\Omega\frac{\sin i}{\sin\rho}\sin\Sigma, \\
\frac{d\phi}{dt} &= L\cos\vartheta\left(\frac{1}{C} - \frac{1}{A}\right) + \frac{1}{L\sin\vartheta}\frac{\partial U}{\partial\vartheta} + \frac{M_1\cos\psi + M_2\sin\psi}{L\sin\vartheta},
\end{aligned}
$$

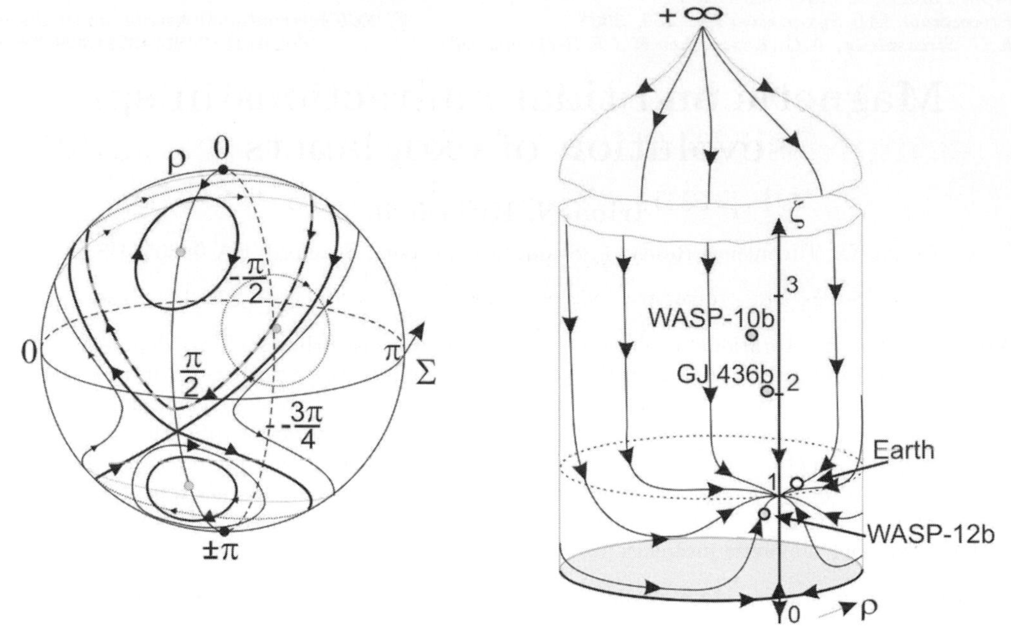

Figure 1. Left: Phase portraits in $\rho - \Sigma$ coordinates on a sphere, which show evolutionary treks of the spin angular momentum, $\vec{L}$, of several dynamically symmetrical planets under the action of gravitational and magnetic perturbations. Dashed curve shows an example of the regime when the planet spin can change sign, i.e. from direct rotation to reverse. Right: Phase portraits of $\vec{L}$ on a cylinder for ρ and $\zeta = L \cdot P/B$ coordinates (where P is the period of rotation, and B is the principle moment of inertia), including tidal perturbations and the orbital evolution. For comparison, locations of the Earth and three exoplanets with weak elliptical orbit are indicated.

where U is the potential force function of gravitational and magnetic perturbations, angles ρ, Σ describe the orientation of spin momentum $\vec{L}$ in the orbital reference frame; ψ, ϕ and ϑ are the Euler's angles, M_i are projections of the tidal perturbation forces on components of the vector $\vec{L}$; A and C are the principle moments of inertia of the planet.

The equations are investigated by using qualitative and numerical methods for dynamical systems (Kitiashvili & Gusev, 2008). Figure 1 shows different scenarios of the spin evolution under gravitational and magnetic perturbations (left panel) and a case, when tidal perturbations are essential (right panel). The results show that the joint action of gravitational and magnetic perturbations can lead to regimes, when the direct rotation of planets can change to the reverse rotation (Kitiashvili & Gusev, 2008). In addition, I have obtained estimates of the principal moment of inertia of some hot Jupiter planets, and their dynamical flattening under the action of tidal forces from the parent stars (Kitiashvili, 2008). For the close-in planets, when the tidal interaction is essential the results reproduce the flip-flop effect obtained earlier by Beletsky (1981) and determine possible states of the spin evolution of extra-solar planets.

References

Beletskii, V. V. 1981, *Celestial Mechanics* 23, 371

Beletsky, V. V. & Khentov, A. A. 1995, *Resonance rotation of celestial bodies* (Nizhnii Novgorod, Nizhegorodskii gumanitarnii centr, in Russian)

Kitiashvili, I. 2008, Proc of IAU Symposium 249, CUP, p. 197

Kitiashvili, I. N. & Gusev, A. 2008, *Cel. Mech. & Dyn. Astronomy* 100, 2, 121

Shkolnik, E., Walker, G., & Bohlender, D. A. 2003, *ApJ* 597, 1092

Cosmic Magnetic Fields:
From Planets, to Stars and Galaxies
Proceedings IAU Symposium No. 259, 2008
K.G. Strassmeier, A.G. Kosovichev & J.E. Beckman, eds.

doi:10.1017/S1743921309030658

3D structures of steady flow of ideal compressible fluid in MHD: a new model of primordial solar nebula

Vladimir V. Salmin

Siberian Federal University, Svobodny ave., 79, Krasnoyarsk, 660041, Russia
email: vsalmin@gmail.com

Abstract. Stereographic projection of Hopf field on the 3-sphere into Euclidean 3-space is used as a model of 3D steady flow of ideal compressible fluid in MHD. In such case, flow lines are Villarceau circles lying on tori corresponding to the levels of Bernoulli function. Existence of an optimal torus with minimal relative surface free energy is shown. Optimal tori are considered as precursors of planetary orbits.

Keywords. MHD – solar system: formation

---

1. Introduction

The equations of ideal incompressible magnetohydrodynamics (Landau & Lifshitz 1959) for stationary flow are clearly satisfied when the fluid moves along the magnetic field lines with velocity (see Chandrasekhar 1961). If density is constant it is easy to show that the equation of ideal magnetohydrodynamics can be rewritten in the stationary Euler equation form. In such case, $\vec{v}$ and $\nabla \times \vec{v}$ are tangential to surfaces of Bernoulli function level. Structural theorem of Arnold & Khesin (1998) determines that if α- Bernoulli function for steady flow $\vec{v}$ on orientable 3-manifold M without boundary, and $\Gamma \subset M$ is preimage of critical value $\nabla\alpha = 0$, every connected component of the set $M\backslash\Gamma$ is fibered into two-dimensional tori invariant under the flow of $\vec{v}$, the motion on each torus is quasi-periodic.

Solution of the equations of ideal magnetohydrodynamics describes a localized topological soliton with use of Hopf mapping shown by Kamchatnov (1982). Hopf field on S^3 has minimal energy among all the fields diffeomorphic to it (Arnold & Khesin 1998). Stability of Hopf field on S^3 has been proved by (Gil-Merdano & Llinares-Fuster 2001, Yampolsky 2003). Hopf fibration of S^3 with stereographic projection induces the toroidal coordinates on E^3 (Gibbons 2006).

2. Results

On a torus $\nu = \text{arcsinh}(k)/k$ stress tensor for steady flow of ideal compressible fluid in toroidal coordinates is determined by metrical tensor only

$$p^{ij}|_T = -p\delta_i^j \left(\sqrt{k^2+1} - \cos(k\alpha)\right)^2 / R^2 k^2.$$

It is easy to show that the pressure on the torus surface $P \propto r^{-2}$ where r-distance from the torus axis to the point on the torus. Let's define cartesian torus form factor $g = a/c$ where c is main torus radius, and a-tube radius. Notice that $g = \sin(\theta)$ where θ - inclination of Villarceau circles to the main plane of torus symmetry, we call this angle as "stream inclination". The form factor $g = (\sqrt{2})^{-1} \approx 0.7071...$ when the ratio of the solid torus volume to the force of pressure on it's boundary has maximum. We will call the torus meeting form factor as "optimal torus". Physically, it corresponds to minimal relative surface free energy. Beat of oscillations with wave numbers corresponding to structural radii of optimal torus leads to spatial scaling with scaling factor $K = 1 + g$. Spatial intersection of homothetic tori within one torus forms cluster with the size depending on a scaling factor.

It is easy to show that scaling factor of solar system corresponds to optimal torus precisely. Scaling factors of satellite systems have deviations depending on their axial tilt and local ratio of semi-major axes of neighbor planets. Large axial-tilt of Uranus is well explained with the suggested model. We found that the cluster size of solar, Jupiter's, Uranus' systems $C = 4$, but for Saturn's, Neptune's, Pluto's and Mars' systems $C = 3$.

At least two stages of toroidal flow development have been determined in nebula. At first stage, nebula was uniform, and tori separation brought about linear dependence of logarithm of relative planets' volumes on logarithm of relative semi-major axes. Coefficient of proportionality called dimensionality of accumulation ranged from $Z = 3.067$ for solar system to $Z = 4.413$ for Neptune's satellites. At second stage, main stream of toroidal flow located on tori has resulted in formation of paired planets with neighbor orbits and equal volumes. Number of pairs depended on a value of cluster size. If cluster size was $C = 4$, two pairs could be observed, while one pair could be seen when $C = 3$.

We found that Triton's parameters fit well the dependence of logarithm of relative volumes on logarithm of relative semi-major axes of regular Neptune's satellites. This means that formation of Triton and regular Neptune's satellites was simultaneous. We have shown that orbit inclination of Triton corresponded to its formation in opposite phase to regular satellites being similar to Pluto's system.

All regular satellites of giant planets with increasing volumes at increasing orbital semi-major axes have distribution of logarithms of relative volumes on the logarithms of relative semi-major axes approximated with linear dependence and high degree of correlation. In suggestion on absence of dissipation of particles involved in toroidal flow around planets, high correlation confirms that density of particles in all satellite systems was similar. Moon has closest parameters to concerned distribution, thereby suggesting that its formation has been processed by a common mechanism.

References

Landau, L. D. & Lifhitz, E. M. 1959, *Electrodynamics of continious media*, Pergamon Press, Oxford

Chandrasekhar, S. 1961, *Hydrodynamic and Hydromagnetic Stability*, Oxford, Clarendon Press

Kamchatnov, A. M. 1982, *Sov. JETP* 82, No1, 117

Arnold, V. I. & Khesin, B. A. 1998, *Topological Methods in Hydrodynamics*, Springer-Verlag, New York

Gil-Merdano, O. & Llinares-Fuster, E. 2001, *Math. Ann.* 320, 531

Yampolsky, A. 2003, *Acta Math. Hungar.* 101, 73

Gibbons, G. W. 2006, *Applications of Differential Geometry to Physics Part III*, Cambridge University, U.K.

First of the three poster-session chairpersons: Moira Jardine (left). First presenter: Lisa Harvey-Smith (right)

Peter Dobbie (left), and Anuj Sarma (right)

Rainer Arlt (left), and Shio Kawagoe and Nobuhiko Kusakabe (right)

Daniella Della-Giustina (left), and Natalie Dzyurkevich (right)

Uwe Motschmann (left), and Mikhail Demidov (right)

Hans Zinnecker (left), and Monika Petr-Gotzens (right)

Jeroen Stil (left), and Shih-Ping Lai (right)

Oliver Gressel (left), and Darth Vader, private (right)

Session IV

Stellar magnetic fields: cool and hot stars

Cosmic Magnetic Fields: From Planets to Stars and Galaxies
Proceedings IAU Symposium No. 259, 2008
K.G. Strassmeier, A.G. Kosovichev, J.E Beckman, eds.

doi:10.1017/S1743921309030671

The basic role of magnetic fields in stellar evolution

André Maeder, Georges Meynet, Cyril Georgy and Sylvia Ekström

Geneva Observatory, CH-1290 Sauverny, Switzerland

Abstract. Magnetic field is playing an important role at all stages of star evolution from star formation to the endpoints. The main effects are briefly reviewed. We also show that O–type stars have large convective envelopes, where convective dynamo could work. There, fields in magnetostatic balance have intensities of the order of 100 G.

A few OB stars with strong polar fields (Henrichs *et al.* 2003a) show large N–enhancements indicating a strong internal mixing. We suggest that the meridional circulation enhanced by an internal rotation law close to uniform in these magnetic stars is responsible for the observed mixing. Thus, it is not the magnetic field itself which makes the mixing, but the strong thermal instability associated to solid body rotation.

A critical question for evolution is whether a dynamo is at work in radiative zones of rotating stars. The Tayler-Spruit (TS) dynamo is the best candidate. We derive some basic relations for dynamos in radiative layers. Evolutionary models with TS dynamo show important effects: internal rotation coupling and enhanced mixing, all model outputs being affected.

Keywords. Stars – magnetic fields – stars: evolution

1. Introduction

The magnetic field of a star is, like the scent of a flower, subtle and invisible, but it plays an essential role in evolution. Magnetic field is often not accounted for in star models. However, the examples below show that from star formation to the endpoints as compact objects, magnetic fields and rotation strongly influence the course of evolution and all model outputs.

In Sect. 2, we give an overview where in evolution the fields are intervening. In Sect. 3, we emphasize some critical observations. In Sect. 4, we focus on the general dynamo equations, with examples in Sect. 5 and 6 for the Tayler–Spruit (TS) dynamo.

2. Overview on the magnetic field in star formation and evolution

Fig. 1 shows the evolutionary track of the Sun from its formation to its endpoint with indications of the various effects of the magnetic field coming into play.

–1. Collapse and ambipolar diffusion: magnetic field may contribute to cloud support. For contraction to occur it is necessary that the energy density of magnetic field $u_{\mathrm{B}} = B^2/(8\,\pi)$ is smaller than the density of gravitational energy $u_{\mathrm{G}} = \frac{3}{5}\,\frac{GM^2}{R\,(\frac{4}{3}\pi R^3)} = \frac{9}{20\,\pi}\,\frac{GM^2}{R^4}$. This defines a critical mass M_{B} above which gravitation dominates,

$$M_{\mathrm{B}} \approx \left(\frac{5}{18\,\pi^2}\right)^{\frac{1}{2}} \frac{\Phi}{\sqrt{G}} = 0.17\,\frac{\Phi}{\sqrt{G}}, \tag{2.1}$$

with the magnetic flux $\Phi = \pi\,B\,R^2$. In the original derivation (Mouschovias & Spitzer 1976), a numerical factor 0.13 was obtained instead of 0.17 in this simple derivation. If

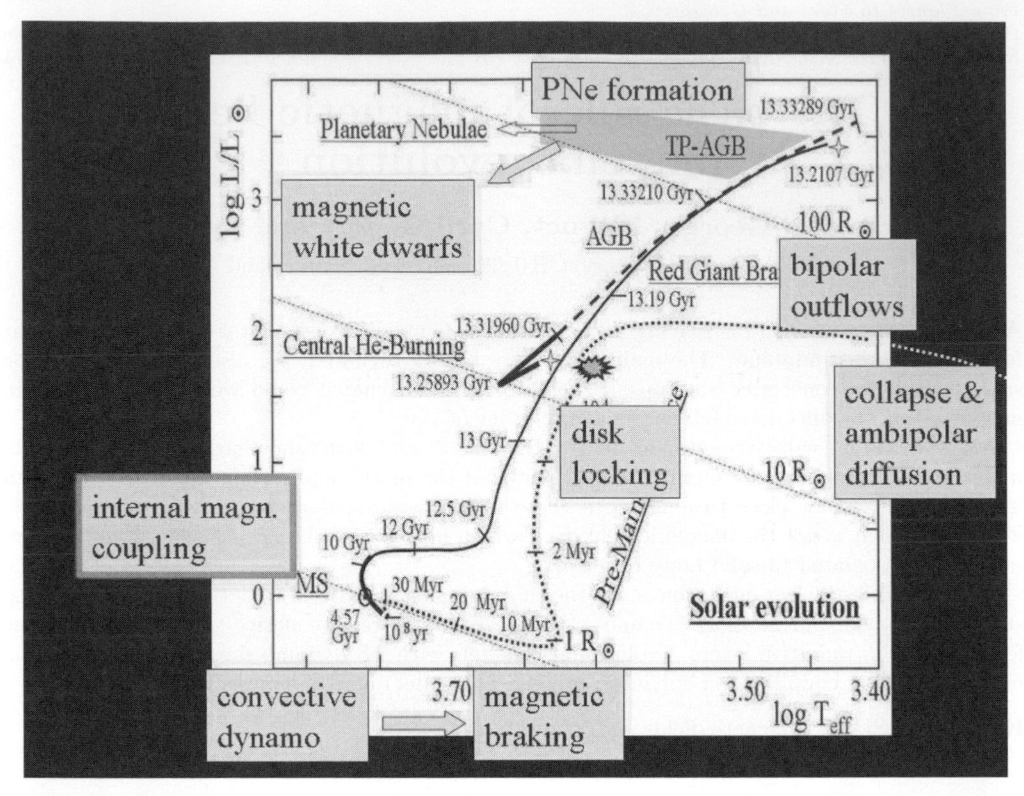

Figure 1. The evolutionary track of the Sun from the protostellar phase to the phase of planetary nebulae (courtesy from C. Charbonnel) with superposed indications of the various magnetic intervening processes

$M > M_{\rm B}$, large clusters or associations form. If $M < M_{\rm B}$, no contraction occurs until the small ionized fraction ($\sim 10^{-7}$) to which the field is attached has diffused (in about 10^7 yr) out from the essentially neutral gas forming the cloud.

–**2. Bipolar outflows:** massive molecular outflows are often detected in region of star formation. A large fraction of the infalling material is not accreted by the central object but it is ejected in the polar directions. In massive stars, the ejection cones are relatively broad. Radiative heating and magnetic field are likely driving the outflows. Remarkably the mass outflow rates correlate with the luminosities of the central objects over 6 decades in luminosity, from about 1 $L_\odot$ to 10^6 $L_\odot$, as shown by Churchwell (1998) and Henning *et al.* (2000).

–**3. Disk locking:** from the dense molecular clouds to the present Sun, the specific angular momentum decreases by $\sim 10^6$. Among the processes reducing the angular momentum in stars, disk locking is a major one (Hartmann 1998). Fields of $\sim$ 1 kG are sufficient for the coupling between the star and a large accretion disk. The contracting star is bound to its disk and it keeps the same angular velocity during contraction, thus losing a lot of angular momentum. The typical disk lifetime is a few 10^6 yr.

–**4. Convective dynamos and magnetic braking:** solar types stars have external convective zones which produce a dynamo. The resulting magnetic field creates a strong coupling between the star and the solar wind, which leads to losses of angular momentum. The relation expressing these losses as a function of the stellar parameters have been developed by Kawaler (1988). Further improvements to account for saturation effects and mass dependence have been brought (Krishnamurti *et al.* 1997).

Massive OB stars also have significant external convective zones which may represent up to 15% of the radius. Surprisingly rotation enhances these convective zones (Maeder *et al.* 2008), which may also produce magnetic braking.

–**5. Dynamo in radiative zones:** is there a dynamo working in internal radiative zones? This is the biggest question concerning magnetic field and stellar evolution, with far reaching consequences concerning mixing of the chemical elements and losses of the angular momentum. This question is also essential regarding the rotation periods of pulsars and the origin of GRBs. We devote Sect. 4 to 6 to this question.

–**6. Magnetic field in AGB stars, planetary nebulae and final stages:** red giants, AGB stars and supergiants have convective envelopes and thus dynamos. Evidences of magnetic fields up to kG in some central stars of planetary nebulae are given (see Jordan, this meeting), they contribute to shaping the nebulae (see Blackman 2009, this volume). White dwarfs have magnetic fields from about 10^4 up to 10^9 G, the highest fields likely resulting from common envelope effects in cataclysmic variables.

3. Magnetic fields and abundances: the Henrichs *et al.* results

Since OB stars also have convective envelopes, the question arises what are the possible fields created by the associated dynamos. If one considers a flux tube in magnetostatic balance in the stellar atmosphere, the equilibrium field is given by the condition $B^2/(8\,\pi) = P_{\rm ext} - P_{\rm int}$. At optical depth $\tau = 2/3$, the pressure is $P(\tau = 2/3) \approx (2/3)\,g/\kappa$. Since $P_{\rm int} > 0$, the maximum possible field $B_{\rm eq}$ for magnetic equilibrium is (Safier 1999)

$$B_{\rm eq}\,(\tau = 2/3) \approx \left(\frac{16\,\pi}{3}\,\frac{g}{\kappa}\right)^{1/2}. \tag{3.1}$$

Table 1. The equilibrium fields. The stars are on the ZAMS, except the Sun.

Spectral type	field	Spectral type	field
M0	2.8 kG	G0	1.0 kG
KO	1.5 KG	F2	0.6 kG
Sun	1.3 kG	*O9	0.2 kG

* from the author

Table 1 gives the corresponding estimates for stars of various types. The observed field intensities are often within a factor of 2 from the maximum values given in the table. Searches for magnetic fields in OB–type stars show no general evidence of fields above the level of ∼100 G (Mathys 2004). This is of the order of magnitude of the possible fields in magnetostatic equilibrium in the convective envelopes of OB stars. The strong fields of 1 kG or more are not widespread (Hubrig *et al.* 2008).

Noticeable exceptions were found and studied by Henrichs and colleagues (Henrichs *et al.* 2003a, Henrichs *et al.* 2003b). Their remarkable finding is that the few stars with high polar fields $B_{\rm p}$ also show N and He enhancements together with C and O depletions, in particular for the 4 stars listed below. The abundances are given in Table 2 as the difference in log between the observed N abundance and the solar values. These are

Table 2. Stars with intense fields and N enrichments from (Henrichs, Neiner & Geers 2003a)

β Cep	B1IVe	v sin i = 27 km s^{-1}	$B_{\rm p}$=360 G	Δ log N=1.2
V2052 Oph	B1IVe	v sin i = 63 km s^{-1}	$B_{\rm p}$=250 G	Δ log N=1.3
ζ Cas	B2IV	v sin i = 17 km s^{-1}	$B_{\rm p}$=340 G	Δ log N=2.6
ω Ori	B2IVe	v sin i = 172 km s^{-1}	$B_{\rm p}$=530 G	Δ log N=1.8

typical signatures of CNO processing, which give strong evidences of internal mixing in stars with a high magnetic field. These few results bring a lot of interesting questions.

The law of isorotation of Ferraro clearly implies that an internal polar field enforces solid body rotation, see also Sect. 6. Now, the above results suggest that even in presence of uniform rotation, there is an efficient mixing. What is the mixing process in stars with a polar magnetic field? Shear turbulence is generally the main mixing process of chemical elements. However, it is absent here, since the stars rotate uniformly.

The only process among those usually acting in massive stars is meridional circulation. Is it sufficient to produce such a mixing? In differentially rotating stars, it is usually much less efficient than shear mixing for the transport of the chemical elements (Meynet & Maeder 2000), while it is very efficient for the transport of angular momentum. However, we found that meridional circulation is strongly enhanced by solid body rotation, since uniform rotation creates a strong breakdown of radiative equilibrium.

In evolutionary models with magnetic field and meridional circulation, there is a strong interplay between meridional circulation and magnetic field (Maeder & Meynet 2005):

- Differential rotation creates the magnetic field.
- Magnetic field tends to suppress differential rotation.
- A rotation close to uniform strongly enhances meridional circulation.
- Meridional circulation increases differential rotation and produces mixing.
- Differential rotation feeds the dynamo and magnetic field (the loop is closed).

As a result, the star reaches an equilibrium rotation law close to uniform (see models in the above ref.), with always a strong thermal instability amplifying meridional circulation and thus chemical mixing. Models show that the high enrichments in magnetic stars with a dynamo (Sect. 4) are essentially due to the transport by meridional circulation. *Thus, it is not the magnetic field itself which makes the mixing, but the thermal instability associated to the solid rotation created by the field.*

The high surface magnetic field of these stars, which likely have a significant mass loss, produces a strong magnetic braking, which implies that these stars will reach a rather low rotation velocities during their evolution. The braking would tend to produce some internal differential rotation. However, the magnetic coupling is certainly strong enough to maintain a rotation law close to uniform, as illustrated by the models.

4. Dynamos in radiative layers: general properties and equations

The major question concerning magnetic fields and stellar evolution is whether a dynamo operates in radiative zones of differentially rotating stars. A magnetic field has great consequences on the evolution of the rotation velocity by exerting an efficient torque able to impose a nearly uniform rotation. This influences all the model outputs (lifetimes, chemical abundances, tracks, chemical yields, supernova types) as well as the rotation in the final stages, white dwarfs, neutron stars or black holes.

Here we first examine some general equations implied by any dynamo. The particular properties of the Tayler–Spruit (TS) dynamo have been studied by Spruit (2002) and we are using many of the equations he derived. Spruit considered the radiative zones in two cases, –1) when the μ–gradient dominates, and –2) when the μ–gradient is negligible. The more general equations of the TS dynamo have been developed by (Maeder & Meynet 2005). The TS dynamo is at present a debated subject. Some numerical simulations by Braithwaite (2006) and by Brun *et al.* (2007) confirm the existence of Tayler's instability. Braithwaite also finds the existence of a dynamo loop in agreement with Spruit's analytical developments. However, Zahn *et al.* do not find the dynamo loop proposed by Spruit and question what may close the loop.

4.1. *Energy conservation*

If a dynamo is working in a differentially rotating radiative zone, it is governed by some general relations expressing the order of magnitude of its various properties. First, the rate of magnetic energy production $W_{\rm B}$ per unit of time and volume must be equal to the rate W_ν of dissipation of rotational energy by the magnetic viscosity ν. We assume here that the whole energy dissipated is converted into magnetic energy. The differential motions are those of the shellular rotation with an angular velocity $\Omega(r)$, so that the velocity difference at radius r is $dv = r\,d\Omega$. The amount of energy corresponding to a velocity difference dv during a time dt for an element of matter dm in a volume dV is

$$W_\nu = \frac{1}{2}\,dm\,(dv)^2\,\frac{1}{dV}\,\frac{1}{dt} = \frac{1}{2}\varrho\,\nu\left(\frac{dv}{dr}\right)^2 = \frac{1}{2}\varrho\nu\Omega^2 q^2 \qquad \text{with} \quad q = r\,|\nabla\Omega|\,/\Omega, \tag{4.1}$$

because the viscous time dt over dr is given by $dt = (dr)^2/\nu$. The magnetic energy density is $u_{\rm B} = B^2/(8\pi)$, it is produced within the characteristic growth time of the magnetic field $\sigma_{\rm B}^{-1}$, thus the rate $W_{\rm B}$ of magnetic energy creation by units of volume and time is

$$W_{\rm B} = \frac{B^2}{8\pi}\sigma_{\rm B} = \frac{1}{2}\omega_{\rm A}^2 r^2 \sigma_{\rm B}\,\varrho, \tag{4.2}$$

where we have used the expression of the Alfvén frequency $\omega_{\rm A} = \frac{B}{r(4\,\pi\,\varrho)^{1/2}}$. Now, let us assume $W_\nu = W_{\rm B}$, i.e. that the excess of energy in the differential rotation (compared to an average constant rotation) is converted to magnetic energy by unit of time. This gives the following expression for the viscosity coefficient of magnetic coupling

$$\nu = \frac{\omega_{\rm A}^2\, r^2\, \sigma_{\rm B}}{\Omega^2\, q^2}. \tag{4.3}$$

This is the coefficient which intervenes in the expression for the transport of angular momentum, in the Lagrangian form as given by Eq. (5.8) below. Let us note that compared to the energy available for the solar dynamo driven by convection, the amount of energy available from differential rotation is very limited.

4.2. *The α and ω–effects: vertical instability and stretching of the field lines*

A dynamo needs both the α–effect and ω–effect. The α–effect consists in the generation of a poloidal field component from the horizontal component. In the Sun, the α–effect is created by the convective motions and by the twisting of the magnetic loop by the Coriolis force. However, other instabilities with a vertical component may produce the necessary α–effect. The ω–effect consists mainly of the stretching of a small radial field component in the East–West directions. The winding–up of the field lines generates a stronger horizontal field component, converting some kinetic energy into magnetic energy.

If due to an instability in radiative layers, some vertical displacements (necessary for the α–effect) with an amplitude $l_r/2$ occur around an average stable position, the restoring buoyancy force produces vertical oscillations with a frequency equal to the Brunt–Văisălă frequency N. The restoring oscillations will have an average density of kinetic energy $u_{\rm N} = f_N\,\rho\,\ell^2\,N^2$, where $f_N \sim 1$. In order to produce a vertical displacement, the magnetic field must overcome the buoyancy effect. In terms of energy densities, this is $u_{\rm B} > u_{\rm N}$, where $u_{\rm B}$ has been given in the previous section. Otherwise the restoring force of gravity would counteract the magnetic instability at the dynamical timescale. From this condition, one obtains $\ell^2 < \frac{1}{2 f_\mathrm{N}}\, r^2\, \frac{\omega_\mathrm{A}^2}{N^2}$. If, $f_\mathrm{N} = \frac{1}{2}$, we have the condition (Spruit 2002)

$$\ell < l_r = r\, \frac{\omega_\mathrm{A}}{N}, \tag{4.4}$$

where r is the radius at the considered level in the star.

The stretching of the field lines for the ω–effect is governed by the induction equation

$$\frac{\partial \vec{B}}{\partial t} = \vec{\nabla} \times (\vec{v} \times \vec{B}) + \eta \nabla^2 \vec{B}. \tag{4.5}$$

An unstable vertical displacement of size ℓ from the azimuthal field of lengthscale r and intensity B_φ also feeds a radial field component B_r. The relative sizes of these two field components are defined by the induction equation (4.5), which gives the following scaling over the time δt characteristic of the unstable displacement,

$$B_r \approx \delta B \approx \frac{1}{r}\, \frac{\ell}{\delta t}\, B_\varphi\, \delta t. \tag{4.6}$$

For the maximum displacement l_r given by Eq. 4.4, this gives

$$\frac{B_r}{B_\varphi} \approx \frac{l_r}{r}, \tag{4.7}$$

which provides (Spruit 2002) an estimate of the ratio of the radial to azimuthal fields.

4.3. *The magnetic and thermal diffusivities*

The magnetic diffusivity η tends to damp the instability, while the thermal diffusivity K produces heat losses from the unstable fluid elements and thus reduces the buoyancy forces opposed to the magnetic instability. Both effects have to be accounted for.

If the radial scale of the vertical instability is small, the perturbation is quickly damped by the magnetic diffusivity η (in cm^2 s^{-1}). The radial amplitude must satisfy,

$$l_r^2 > \frac{\eta}{\sigma_\mathrm{B}}, \tag{4.8}$$

where, as seen above, σ_B is the characteristic frequency for the growth of the instability. The combination of the two limits (4.7) and (4.8) gives for the case of marginal stability,

$$\eta = \frac{r^2\, \omega_\mathrm{A}^2\, \sigma_\mathrm{B}}{N^2}. \tag{4.9}$$

For given η and σ_B, this provides the minimum value of ω_A, and thus of the magnetic field B, for the instability to occur. The instability is confined within a domain, limited on the large side by the stable stratification (4.4) and on the small scales by magnetic diffusion (4.8). For the case of marginal stability, which is likely reached in evolution, this equation relates the magnetic diffusivity η and the Alfvén frequency ω_A.

The Brunt–Väisälä frequency N of a fluid element displaced in a medium with account of both the magnetic and thermal diffusivities η and K is (Maeder & Meynet 2004),

$$N^2 = \frac{\frac{\eta}{K}}{\frac{\eta}{K} + 2}\, N_{T,\,\mathrm{ad}}^2 + N_\mu^2, \tag{4.10}$$

$$\text{with} \quad N_{T,\,\mathrm{ad}}^2 = \frac{g\delta}{H_P}\,(\nabla_\mathrm{ad} - \nabla), \quad \text{and} \quad N_\mu^2 = \frac{g\,\varphi}{H_P}\nabla_\mu, \tag{4.11}$$

The ratio η/K of the magnetic to thermal diffusivities determines the heat losses. The factor of 2 is determined by the geometry of the instability, a factor of 2 applies to a thin slab, for a spherical element a factor of 6 is appropriate (Maeder & Meynet 2005).

4.4. *The magnetic coupling and the timescale* $\sigma_{\rm B}$

The momentum of force $\vec{S}$ by volume unity due to the magnetic field is obtained by writing the momentum of the Lorentz force $\vec{F}_{\rm L}$. The current density $\vec{j}$ is given by the Maxwell equation $\frac{4\pi}{c}\vec{j} = \vec{\nabla}\times\vec{B}$. Thus, one has

$$\vec{S} = \vec{r}\times\vec{F}_{\rm L} = \frac{1}{c}\vec{r}\times(\vec{j}\times\vec{B}) = \frac{1}{4\pi}\vec{r}\times\left((\vec{\nabla}\times\vec{B})\times\vec{B}\right), \tag{4.12}$$

$$\text{in modulus}\quad S \approx \frac{1}{4\pi}\,B_{\rm r}B_{\varphi} = \frac{1}{4\pi}\left(\frac{l_{\rm r}}{r}\right)B_{\varphi}^2 = \rho\, r^2\left(\frac{\omega_{\rm A}^3}{N}\right). \tag{4.13}$$

The units of S are g s^{-2} cm^{-1}, the same as for B^2 in the Gauss system. The kinematic viscosity ν (in cm^2 s^{-1}) for the vertical transport of angular momentum is

$$\nu = \frac{\eta}{\varrho} = \frac{1}{\varrho}F\,\frac{dr}{dv} = \frac{1}{\varrho}F\,\frac{dr}{rd\Omega} = \frac{1}{\varrho}F\,\frac{d\ln r}{\Omega\, d\ln\Omega}, \tag{4.14}$$

where F is a force by surface unity, which also corresponds to a momentum of force by volume unity in g s^{-2} cm^{-1}. $\vec{F}$ is applied horizontally to a slab of velocity v in a direction perpendicular to r. Considering only positive quantities, with $q = |d\ln\Omega/d\ln r|$, one has

$$\nu = \frac{S}{\rho\, q\,\Omega} = \frac{\omega_{\rm A}^3\, r^2}{N\,q\,\Omega}. \tag{4.15}$$

Now, we can compare this expression for ν to Eq. (4.3) and get

$$\sigma_{\rm B} = \frac{\omega_{\rm A}\,\Omega\, q}{N}. \tag{4.16}$$

This important expression relates the growth rate of the magnetic field to its amplitude (through $\omega_{\rm A}$). It can also be obtained by expressing the amplification time $\tau_{\rm a}$ of the field line B_r to the level of B_φ by the winding–up of the field line $B_\varphi \approx B_r\, r\left(-\frac{\partial\Omega}{\partial r}\right)\tau_a$,

$$\tau_{\rm a} = \frac{N}{\omega_{\rm A}\Omega\, q}, \tag{4.17}$$

and equaling this to $\sigma_{\rm B}^{-1}$ we also get Eq. (4.16).

4.5. *The basic equations*

Introducing the expression (4.16) of $\sigma_{\rm B}$ in Eq. (4.9), we get

$$\eta = r^2\,\Omega\, q\left(\frac{\omega_{\rm A}}{N}\right)^3. \tag{4.18}$$

Also, with the expression (4.10), we can write for $\sigma_{\rm B}^2$

$$\sigma_{\rm B}^2 = \frac{\omega_{\rm A}^2\,\Omega^2\, q^2}{\frac{\frac{\eta}{K}}{\frac{\eta}{K}+2}\,N_{T,\,\rm ad}^2 + N_\mu^2}. \tag{4.19}$$

These equations are quite general. If the growth rate σ of the instability is known, the two equations (4.18) and (4.19) form a system of 2 equations with 2 unknowns η and $\omega_{\rm A}$.

5. The case of Tayler–Spruit dynamo

In a non–rotating star the growth rate of the Tayler instabilty is the Alfvén frequency ω_{A}. In a rotating star, the instability is also present, however the characteristic growth rate σ_{B} of the instability is, if $\omega_{\mathrm{A}} \ll \Omega$,

$$\sigma_{\mathrm{B}} = \frac{\omega_{\mathrm{A}}^2}{\Omega}, \tag{5.1}$$

because the growth rate of the instability is reduced by the Coriolis force (Spruit 2002). If so, Eqs. (4.18) and (4.19) become

$$\left(\frac{\omega_{\mathrm{A}}}{\Omega}\right)^2 = \frac{\Omega^2 q^2}{N^2_{T,\,\mathrm{ad}}\frac{\eta/K}{\eta/K+2} + N^2_\mu}, \tag{5.2}$$

$$\eta = \frac{r^2\,\Omega}{q^2}\left(\frac{\omega_{\mathrm{A}}}{\Omega}\right)^6. \tag{5.3}$$

This forms a system of 2 equations for the 2 unknown quantities η and ω_{A}. With a new variable $x = (\omega_{\mathrm{A}}/\Omega)^2$, we get (Maeder & Meynet 2005) a system of degree 4,

$$\frac{r^2\Omega}{q^2 K}\left(N_{\mathrm{T}}^2 + N_\mu^2\right)x^4 - \frac{r^2\Omega^3}{K}x^3 + 2N_\mu^2\,x - 2\Omega^2 q^2 = 0. \tag{5.4}$$

The solution x provides the value of the Alfvén frequency ω_{A} and thus of the $\vec{B}$ field. By (5.3) one gets the value of η and by (4.15) the value of ν. The above equation applies to the general case where both N_μ and N_{T} are different from zero and where thermal losses may reduce the restoring buoyancy force. The solutions of this equation have been discussed (Maeder & Meynet 2005). In particular, if $N_T = 0$, one has

$$x = \left(q\,\frac{\Omega}{N_\mu}\right)^2 \quad \text{and} \quad \eta = r^2\Omega q^4\left(\frac{\Omega}{N_\mu}\right)^6, \tag{5.5}$$

which shows that the mixing of chemical elements decreases strongly for larger μ gradients and grows fast for larger q values.

The ratio $\omega_{\mathrm{A}}/\Omega$ given by the solution of (5.4) has to be equal or larger than the minimum value defined by (4.9). This leads to a condition on the minimum differential rotation for the dynamo to work (Spruit 2002),

$$q > \left(\frac{N}{\Omega}\right)^{7/4}\left(\frac{\eta}{r^2 N}\right)^{1/4}, \tag{5.6}$$

When N^2 is larger, as for example when there is a significant μ gradient, the differential rotation necessary for the dynamo to operate must also be larger. If the above condition is not fulfilled, there is no stationary solution and the dynamo does not operate. In practice, this often occurs in the outer stellar envelope.

5.1. *Equations of transport of chemical elements and angular momentum*

The equation for the transport of chemical species with mass fractions X_i is at a Lagrangian mass coordinate M_r,

$$\varrho\frac{\partial X_i}{\partial t} = \frac{1}{r^2}\frac{\partial}{\partial r}\left(\varrho r^2\left(D_{\mathrm{eff}} + \eta\right)\frac{\partial X_i}{\partial r}\right), \tag{5.7}$$

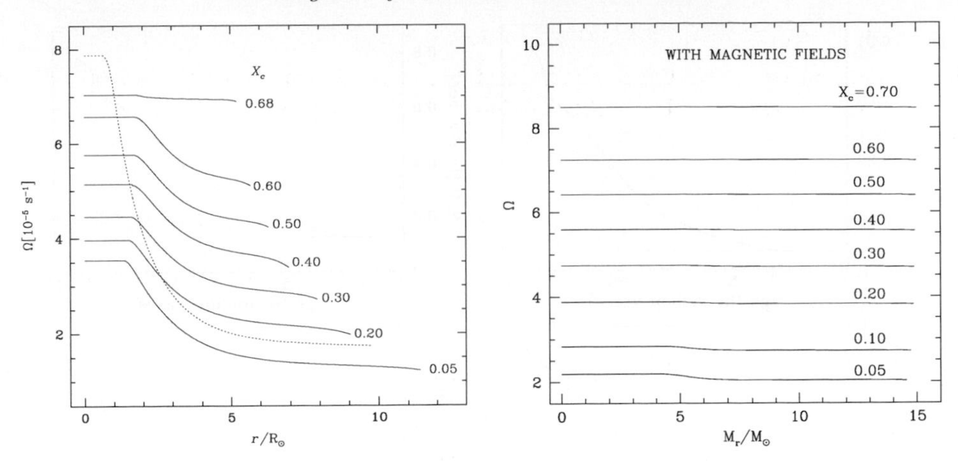

Figure 2. Left: evolution of the angular velocity Ω as a function of the distance to the center in a 20 $M_\odot$ star with $v_{\rm ini} = 300$ km s^{-1}. X_c is the hydrogen mass fraction at the center. The dotted line shows the profile when the He–core contracts at the end of the H–burning phase (Meynet & Maeder 2000). Right: rotation profiles at various stages of evolution (labeled by the central H content X_c) of a 15 $M_\odot$ model with $X = 0.705$, $Z = 0.02$, an initial velocity of 300 km s^{-1} and magnetic field from the TS dynamo (Maeder & Meynet 2005).

where $D_{\rm eff}$ is the coefficient for the transport by meridional circulation and the possible horizontal turbulence. The equation for the transport of angular momentum is

$$\varrho \frac{\partial}{\partial t}(r^2\overline{\Omega})_{M_r} = \frac{1}{5\,r^2}\frac{\partial}{\partial r}(\varrho\, r^4 \overline{\Omega}\, U_2(r)) + \frac{1}{r^2}\frac{\partial}{\partial r}\left(\varrho\,\nu\, r^4 \frac{\partial \overline{\Omega}}{\partial r}\right), \tag{5.8}$$

where $U_2(r)$ is the amplitude of the radial component of the velocity of meridional circulation and ν the value given by (4.3). This equation is currently applied in stellar models for calculating the evolution of Ω. With account of the detailed expression of $U_2(r)$, which contains terms up to the third spatial derivative of $\Omega(r,t)$, the above equation is of the fourth order and its numerical solution requires great care.

6. Numerical models

Numerical models accounting for meridional circulation and magnetic field generated by the TS dynamo have been computed (Maeder & Meynet 2005). The resulting fields are a few 10^4 G through most of the envelope, with the exception of the outer layers where differential rotation is too small to sustain the TS dynamo. The diffusion coefficient for the transport of angular momentum is large. In the Sun, it is of the order of 10^2 to 10^6 cm^2 s^{-1}, sufficient to impose solid body rotation at the age of the Sun (Eggenberger *et al.* 2005). This coefficient is much larger in more massive stars, in the range of 10^{10} to 10^{12} cm^2 s^{-1} in a 15 $M_\odot$ star. There, it imposes nearly solid body rotation during most of the MS phase, while without the field there is a high differential rotation (Fig. 2).

The nearly solid body rotation of star with magnetic field drives meridional circulation currents which are faster than the currents in differentially rotating stars. This leads to large surface enrichments in N and He together with C,O depletions in massive stars (Fig. 3). Thus, the enhanced mixing results from the thermal instability enhanced by uniform rotation. The stellar lifetimes are enlarged by the mixing and the other model outputs are also modified (Maeder & Meynet 2005). Therefore, magnetic field is also a basic ingredient of stellar evolution.

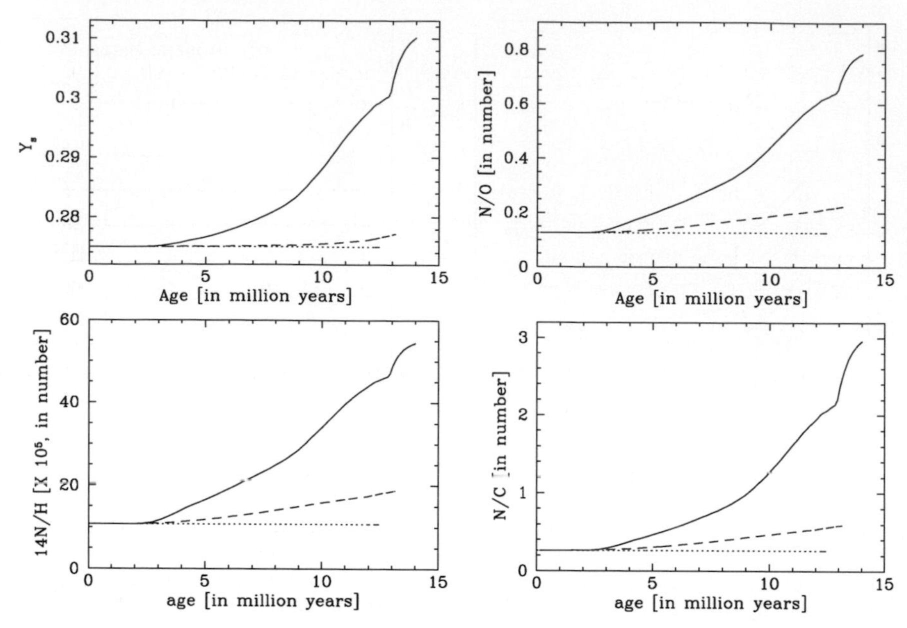

Figure 3. Time evolution of the surface helium content Y_s in mass fraction, of the N/O, N/H and N/C in mass fraction for various models: The dotted line applies to the model without rotation, the short–broken line to the model with rotation (v_{ini} = 300 km s^{-1}) but without magnetic fields, the continuous line to the model with rotation (v_{ini} = 300 km s^{-1}) and magnetic fields from the TS dynamo (Maeder & Meynet 2005).

References

Braithwaite, J. 2006, *A&A* 449, 451

Churchwell, E. 1998, *in the Origin of Stars and Planetary Systems*, Ed. C. Lada and N. Kylafis, NATO Science Ser. 540, Kluwer, p. 515

Eggenberger, P. 2005, *A&A* 440, L9

Hartmann, L. 1998, *Accretion Processes in Star Formation*, Cambridge Univ. Press.

Henning, Th. Schreyer, K, & Launhardt, R. *et al.* 2000, *A&A* 353, 211

Henrichs, H. F., Neiner, C., & Geers, V. C. 2003a, *Intl. Conf. on magnetic fields in O, B and A stars*, Eds. K. A. van der Hucht *et al.*, IAU Symp. 212, 202

Henrichs, H. F., Neiner, C., & Geers, V. C. 2003b, *A Massive Star Odysses, from Main Sequence to Supernova*, ASP Conf. Ser. 305, 301

Hubrig, S., Schöller, M., & Schnerr, R. S. *et al.* 2008, *A&A* 490, 793

Kawaler, S. D. 1988, *ApJ* 333, 236

Krishnamurti, A., Pinsonneault, M. H., & Barnes, S. *et al.* 1997, *ApJ* 480, 303

Maeder, A., Georgy, C., & Meynet, G. 2008, *A&A* 479, L37

Maeder, A. & Meynet, G. 2004, *A&A* 422, 225

Maeder, A. & Meynet, G. 2005, *A&A* 440, 1041

Mathys, G. 2004, *Stellar Rotation*, IAU Symp 215, Eds. A. Maeder, P. Eeneens, p. 270

Meynet, G. & Maeder, A. 2000, *A&A* 361, 101

Mouschovias, T. Ch. & Spitzer, L. 1976, *ApJ* 210, 326

Safier, P. N. 1999 *ApJ* 510, L127

Spruit, H. C. 2002 *A&A* 381, 923

Zahn J.-P., Brun, A. S., & Mathis, S. 2007, *A&A* 474, 145

Discussion

MORENO-INSERTIS: Is the Tayler-Spruit dynamo equally effective in the whole radiative zone of the star? How effective can the dynamo mechanism be in the convective core of the star?

MAEDER: This is an interesting point. The Tayler-Spruit dynamo needs a sufficient differential rotation to work and this prevents the field building in the outer layers. Also, a strong μ-gradient kills the instability due to the core. Thus it exists in the radiative zone, except near the edges. In the convective core, the T-S dynamo is certainly overwhelmed by the convective dynamo.

ZINNECKER: Comment: You spoke about magnetic braking of rotation in the stellar phase, but there is also a need for magnetic braking of molecular clouds in the pre-stellar phase, otherwise the angular momentum problem in star formation would be too large (Ebert, Mestel, Spitzer, Monschovias old papers); see my poster.

MAEDER: The list of authors contributing to the class of angular momentum in star formation is long. As you are saying, in addition to bipolar outflows, disk lacking, magnetic braking of T Tauri stars, binary formation, the magnetic braking may also intervene in the early stages of the collapse.

DE GOUVEIA DAL PINO: With regard to the combination of **B** plus rotation to provide efficient mass loss (e.g. in GRBs) it is an alternative solution to add an accretion disk. Do you know if there is a way to remove this degeneracy between models?

MAEDER: I would guess the analysis of the detailed spectrum might provide some indication. Indeed, we concentrated more on the severe conditions necessary for a massive star to lead to a GRB.

André Maeder

Svetlana Berdyugina

Gregg Wade

Cosmic Magnetic Fields:
From Planets, to Stars and Galaxies
Proceedings IAU Symposium No. 259, 2008
K.G. Strassmeier, A.G. Kosovichev & J. Beckman, eds.

doi:10.1017/S1743921309030683

Stellar magnetic fields across the H-R diagram: observational evidence

Svetlana V. Berdyugina[1,2]

[1]Kiepenheuer Institut für Sonnenphysik, DE-79104 Freiburg, Germany
email: sveta@kis.uni-freiburg.de

[2]Institute of Astronomy, ETH Zurich, CH-8093 Zurich, Switzerland
(EURYI Award Fellow)

Abstract. This review presents most recent measurements of magnetic fields in various types of stars and substellar objects across the H-R diagram with the emphasis on measurement methods, observational and modeling biases, and the role of magnetic fields in stellar evolution.

Keywords. Magnetic fields – molecular processes – polarization – sunspots – stars: activity – stars: imaging – stars: spots – stars: evolution – stars: fundamental parameters

1. Introduction

An extraterrestrial magnetic field was first discovered in sunspots a century ago (Hale 1908), and it took four decades since then to detect it on a distant star (Babcock 1947). Even though we evidence now a significantly more rapid progress in studying cosmic magnetic fields, it is only around 100–200 stars on which magnetic fields have been detected directly, primarily via the Zeeman effect (ZE) in spectral lines. These fields are believed to be inherited from the interstellar medium during the star formation process and are either fossilized or recycled and amplified by a magnetic dynamo in the stellar interior during subsequent evolutionary stages. Such dichotomy appears to be related to the stellar structure: fossil fields are primary suspects in hot massive stars with outer radiation zones, while dynamo generated fields are attributes of cool stars with outer convection zones (see Fig. 1). However, this simple picture should be taken with a caution as, e.g., little is known how magnetic fields are transported through the stellar interior.

Indirect evidence of magnetic fields on stars operating a magnetic dynamo comes from a multitude of activity phenomena similar to those observed on the Sun, such as spots, plages, chromospheric emission, flares, enhanced X-ray and UV radiation, coronal loops, and coronal mass ejections. Thus, a study of an extensive sample of stars of various activity levels provides key constraints for stellar and solar dynamo models. It was suggested by Skumanich (1972) and confirmed by others that rotation plays a crucial role in the generation of stellar activity and, hence, of magnetic fields, so that cool stars with more rapid rotation show a higher level of magnetic activity. Among single stars these are pre-main-sequence stars (T Tau-type) and early-age main-sequence stars both of solar type and much cooler red dwarfs. Evolved binary components which are tidally locked at fast rotation by a close companion are also strongly magnetically active (RS CVn-type, BY Dra-type, Algol-like systems as well as FK Com-type which are probably formed from coalesced binaries).

In contrast to the very dynamic appearance of cool stars, presumably fossil magnetic fields on hotter stars appear rather static and topologically uncomplicated, being often low order multipoles or even dipole-like. Most prominently such fields are observed on the main sequence Ap-Bp-type stars and white dwarfs, while the topology of magnetic fields on other hot stars remains to be discovered.

The goal of this review is to provide a brief account of the most recent measurements of stellar magnetic fields, analyze underlying assumptions in the interpretation of data, and identify requirements, both observational and theoretical, for obtaining a realistic and relatively complete overview of the magnetic H-R diagram (see Fig. 1). For earlier records see reviews by Mestel & Landstreet (2005) and Berdyugina (2005).

2. Zeeman effect in atoms and molecules

Interaction of an atom or a molecule with an external magnetic field in general leads to splitting of energy levels and, thus, to broadening or even splitting of spectral lines. For relatively weak magnetic fields (when magnetic perturbation << internal coupling) the splitting is proportional to the magnetic field strength, B, and scaled by the effective Landé factor, $g_{\rm eff}$:

$$\Delta\lambda \propto g_{\rm eff}\lambda^2 B. \tag{2.1}$$

It is also proportional to the square of the wavelength. For this reason, successful measurements of magnetic field strengths can be carried out using red or IR lines with large $g_{\rm eff}$.

Values of $g_{\rm eff}$ in the cases of pure LS-coupling for atoms and Hund's cases (a) and (b) for diatomic molecules are given by simple expressions depending only on quantum numbers of the involved energy levels (Landau & Lifshits 1991). However, there are examples of atomic transitions for which the LS-coupling approximation fails to describe the level structure, and there is almost no molecular transitions well described by the pure Hund's cases due to internal perturbations. In these cases, the effective Landé factor becomes dependent on quantum numbers of involved and perturbing energy levels as well as on coupling constants (e.g., Berdyugina & Solanki 2002). In case of stronger magnetic fields, when the magnetic perturbation is comparable to or larger than the internal coupling, i.e. in case of the Paschen-Back effect (PBE), $g_{\rm eff}$ looses its meaning as a constant in Eq. (2.1) as it becomes a function of B (e.g., Berdyugina *et al.* 2005).

Magnetic diagnostics based on atomic transitions are most often used for measuring stellar magnetic fields, largely because LS-coupling violations as well as the PBE in atoms at stellar magnetic fields (except some Ap stars and white dwarfs) are relatively rare, which significantly simplifies calculations. However, for cooler stars, whose spectra are strongly dominated by molecules, molecular magnetic diagnostics become essential. A significant progress in understanding the molecular Zeeman, Paschen-Back and Hanle effects was achieved in recent years and made it possible to utilize a novel approach for studying stellar and solar magnetism (Berdyugina & Solanki 2002, Berdyugina *et al.* 2002, Berdyugina *et al.* 2003, Berdyugina *et al.* 2005, Afram *et al.* 2007, Afram *et al.* 2008, Asensio Ramos & Trujillo Bueno 2006, Berdyugina & Fluri 2004, Shapiro *et al.* 2007). The PBE appears to be rather common in molecular lines at stellar magnetic fields. It is responsible for Stokes profile asymmetries, net polarization across line profiles, wavelength shifts and polarization sign changes depending on B as well as weakening of main branches and strengthening of satellite and forbidden lines. These pecularities make molecular lines highly sensitive diagnostics despite lower on average effective Landé factors as compared to atomic lines.

A magnetic field measured from line splitting (broadening) of a spectral line (both atomic and molecular) using the Eq. (2.1) approximation represents a mean magnetic field strength averaged over the visible stellar disk irrespective to the field polarity, $\langle|B|\rangle$, i.e. assuming that the filling factor of this field, f, is unity. In reality, the line profile can contain information on both the field distribution and f. However, because of thermal and stellar rotation broadening, this technique is limited to detection of only kG fields with large filling factors. It was employed in pioneering measurements (Robinson 1980, Saar 1988) and is still useful for surveys of large samples of stars. Reliable measurements require Zeeman splitting larger than or comparable to line widths in the absence of the field.

The most complete way to detect and study stellar magnetic fields is to use the polarimetric technique. The Stokes vector $\vec{I} = (I, Q, U, V)^T$ fully describes the polarization state of radiation and contains information on the magnetic field vector, which can be deduced from Stokes measurements by inversion techniques. Realistic estimates can only be obtained by carrying out polarized radiative transfer with detailed calculation of the Muller matrix, which describes a magnetized stellar atmosphere.

However, polarimetry of stars other than the Sun is a rather challenging task and represents a relatively small field of stellar astrophysics. This, on one hand, is due to still very limited instrumental capabilities and, on the other hand, due to disk-integrated observations of the Stokes parameters. The latter results in significant cancellation of the signal from regions of mixed polarity fields and, thus, only large-scale magnetic fields can be detected from disk-integrated polarimetric measurements, which still requires the accuracy of the order $10^{-3} - 10^{-4}$. Therefore, most of the current polarimetric measurements were made using only Stokes V parameter, which reveals only a mean longitudinal component of the magnetic field (along the line of sight and with polarity cancellations), $\langle B_z \rangle$, under the assumption that the filling factor $f = 1$ for each single-time measurement. The two quantities can be disentangled by analyzing time-series Stokes V measurements (e.g., by ZDI, see Sect. 3).

3. Zeeman-Doppler Imaging

Applying an inversion technique to time-series of the four Stokes parameters one can recover the distribution of the temperature and magnetic field vector over the stellar surface, a technique introduced by Semel 1989 and called Zeeman-Doppler Imaging (ZDI). Several numerical codes have been developed for atomic diagnostics: based on the maximum entropy method (Donati *et al.* 1989, Brown *et al.* 1991), Tikhonov regularization (Piskunov & Kochukhov 2002), and principle component analysis (Carroll *et al.* 2007). Most recently, ZDI employing both atomic and molecular lines has become available (Sennhauser *et al.* 2008, Berdyugina *et al.*, in preparation).

In practice, however, obtained Zeeman-Doppler stellar images are largely based on measurements of Stokes I and V only. To some extent, the magnetic vector contributes to the line of sight component observed in Stokes V at different rotational phases and different Doppler shifts. For instance, the radial field will dominate the Stokes V near the centre of the stellar disk, while the azimuthal field will be most noticeable in the circular polarization near the stellar limb. This allows for recovering some parts of the magnetic field components from Stokes V observations only. However, the solution is certainly not unique and strongly constrained by assumptions. For instance, in order to limit the solution, a special type of regularization based on spherical harmonic expansion was suggested (Piskunov & Kochukhov 2002). In this case the solution is forced to take the form of such an expansion which is useful for stars with clearly dominating multipole structures, like e.g. Ap stars. Thus, when interpreting ZDI results obtained only from Stokes I and V, one has to take into account that the magnetic field distribution is underdetermined for each component and that there might be a cross-talk between different components.

An additional independent constraint on the ZDI solution for cool stars is provided by molecular Stokes profiles. Since molecular lines preferrably form in cooler regions, often associated with magnetic fields, they bear the information on physical conditions in these regions and their location on the stellar surface. Simultaneous inversions of atomic and molecular lines significantly improve the quality of ZDI maps and reduce the cross-talk between magnetic field components (Sennhauser *et al.* 2008).

4. Biases in observations and interpretation

Zeeman signatures in stellar spectra are generally extremely small, with typical relative amplitudes of 0.1%. Detecting them requires measurements of polarization with noise level lower than 10^{-4}, while the current instrumentation allows for the best relative noise level of 10^{-3}. Therefore, a multi-line approach for increasing the signal-to-noise ratio of the measured polarization was proposed (Semel 1989, Semel & Li 1996) and successfully used for detecting stellar circular polarization as an indication of magnetic fields (Least Squares Deconvolution, LSD, Donati *et al.* 1997).

The underlying assumption in the LSD technique is the weak field approximation (WFA), i.e. one assumes that the magnetic splitting of spectral lines is smaller than their local Doppler broadening. In this case, the Stokes V signal is proportional to the derivative of the intensity profile $I(v)$ in the velocity domain:

$$V_i(v) \propto g_{\rm eff}(i)\lambda_i {I_i}'(v), \tag{4.1}$$

where $g_{\rm eff}(i)$ is the effective Landé factor and λ_i is the wavelength of the ith spectral line. It is assumed further that the local line profiles are self-similar and scale in depth and width with the central depth and wavelength. Finally, the LSD Stokes V profile can be obtained as a sum over many individual lines, i.e. as linear weighted average of line profiles.

A similar approach can be applied to other Stokes parameters with the gain factor in the S/N ratio as large as 30 when using more than 2000 line profiles. Drawbacks of this technique are non-linear effects in summation of Stokes profiles and effect of blends (see Sennhauser *et al.* 2008) as well as diminishing information contents on T and B due to massive averaging. This leads to the fact that LSD profiles cannot be considered anymore as observed but rather processed Zeeman signatures with strong influence of the WFA. The latter is generally not applicable for $|B| > 1\,$kG and $g_{\rm eff} \sim 1$ and in the case of the PBE. An alternative technique for increasing the S/N ratio in spectropolarimetric data is based on the principal component analysis (PCA) which does not directly rely on the WFA (Martínez González *et al.* 2008).

Usage of LSD profiles for ZDI again involves the WFA combined with the assumption on Gaussian shapes of line profiles (Donati *et al.* 1989). Furthermore, to overcome a lack of information contained in Stokes V only, a strong constraint on the magnetic field topology in form of a multipole expansion was employed (Piskunov & Kochukhov 2002, Donati *et al.* 2006).

Modeling Zeeman-broadened Stokes I profiles often involves template spectra of magnetically active and inactive stars which are weighted by a filling factor (Valenti & Johns-Krull 2001, Reiners & Basri 2006):

$$I_\lambda = (1-f)\, I_\lambda(B=0) + f\, I_\lambda(B \neq 0). \tag{4.2}$$

Such an analysis implicitly assumes that (1) the field is concentrated in patches surrounded by field free regions, (2) field is oriented radially in the photosphere, (3) magnetic regions are characterized by a single field strength, and (4) the temperature structure is the same for magnetic and non-magnetic atmosphere Valenti & Johns-Krull 2001. In this approach, the field strength being determined from the splitting is usually well constrained, but the filling factor depends on the unknown temperature structure of magnetic regions. In the case when the used spectral line becomes stronger in starspots, the filling factor can be overestimated, and vice versa. Also, only stars with strong magnetic fields and large field areas can be studied with this technique.

It is important to emphasize that the assumptions mentioned in this section introduce largely unknown biases in the determined field strengths and distributions. The most obvious improvement is to solve consistently full Stokes radiative transfer equations using stellar atmosphere models. This has been already implemented in the newest ZDI codes and should become the standard approach for interpretation of spectropolarimetric data.

Extracting weak polarization signals with multi-line techniques, such as LSD or PCA, at the level of 10^{-4} or smaller relies on the assumption that spectropolarimeters are perfect at such levels, i.e. there are no cross-talks between Stokes parameters. It is however common that a few percent cross-talks can occur between the intensity and polarization as well as between circular and linear polarization. Such cross-talks can be responsible for weak Stokes V signals in stars where linear polarization is due to other processes than magnetic fields, e.g., due to optical pumping and absorptive polarization, which can be of the order of 1% in Stokes Q and U (Kuhn *et al.* 2007, Harrington & Kuhn 2007). In such cases a cross-talk of only 1% can result in a non-magnetic 10^{-4} Stokes V signal, which may confuse a proper interpretation of a magnetic signal. The non-magnetic absorptive polarization appears to be common for stars with circumstellar material, such as Ae-Be stars with disks and winds, AGB stars with envelops, and perhaps upper-main-sequence O-B stars with winds. It is remarkable that Stokes V signals (LSD) detected in spectra of such stars are reported to be very weak are normally interpreted as indication of weak magnetic fields.

5. Pre-Main-Sequence stars

5.1. *T Tau stars*

T Tau-type stars are pre-main-sequence stars of about one solar mass at an age of a few million years, still surrounded by disks of gas and dust remaining from their formation. They are believed to be almost fully convective and maintaining vigorous magnetic dynamo. In particular, it is widely accepted that magnetic fields play an important role in the surface and flare activity of T Tau stars as well as in accretion processes and stellar wind phenomena (see reviews by Petrov 2003, Johns-Krull 2009).

Most recent magnetic field measurements in classical T Tauri stars (CTTS) were made by Johns-Krull (2007) from broadening of Zeeman-sensitive Ti II lines in the near IR. It is interesting that the line broadening could not be modeled with one single value of $\langle |B| \rangle$ and required a distribution of fields between 0 kG and 6 kG with different filling factors. When averaged over the visible stellar surface ($f = 1$), the mean field of 12 CTTS was found to be 2.5±0.3 kG (majority of the sample). It appears that this value exceeds significantly the field strength predicted by simple magnetospheric accretion theory. At the same time, these stars might be underluminous in X-rays as compared to main-sequence stars with the same mean magnetic field strengths. Relatively strong mean fields (> 1 kG) detected on all CTTS imply that they are indeed powerful magnetic generators.

A topology of the magnetic fields on CTTS remains largely unknown. Available ZDI maps of two CTTS (V2129 Oph and BP Tau) are obtained from Stokes I and V LSD profiles only and constrained by a multipole expansion (Donati *et al.* 2007, 2008a). In the two cases a combination of a relatively weak dipole (0.35 kG and 1.2 kG) and octopole (1.2 kG and 1.6 kG) was found, both slightly tilted with respect to the rotation axis. Accretion spots seem to coincide with the two main magnetic poles at high latitudes and overlap with dark photospheric spots. The average surface field strength in the case of BP Tau is however significantly underestimated compared to that obtained by Johns-Krull (2007).

5.2. *Ae-Be Herbig stars*

Herbig Ae/Be (HAeBe) stars are pre-main-sequence stars of intermediate mass (about 3 to 10 $M_\odot$) embedded in dust-gas circumstellar material similar to CTTS but with hotter photospheres and outer radiation zones. Some HAeBe stars demonstrate activity characteristics normally associated with the presence of chromospheres or coronae, which implies possible presence of magnetic fields and magnetoaccretion.

Several Stokes V surveys with different instruments (Wade *et al.* 2007, 2009, Hubrig *et al.* 2009, and references therein) reveal that about 7% of these stars may possess surface magnetic fields, $\langle B_z \rangle$, in the range of a few hundred Gauss. Note however that polarization signals in embedded stars can have non-magnetic origin too (see Sect. 4). Temporal variations of Stokes V LSD profiles can be described with a dipole field component of about 1 kG, which is quite stable over several years. Such a simple and stable topology may imply that magnetic fields are of primordial origin. There are however indications that magnetic fields on Ae stars perhaps correlate with their X-ray luminosity and become very weak or completely disappear when they arrive on the main sequence (Hubrig *et al.* 2009).

6. Main-Sequence stars

6.1. *Massive stars*

Stars in the upper part of the main-sequence with masses exceeding 10 $M_\odot$ are characterized by strong variable winds, H_α emission variations, chemical peculiarity, and non-thermal radio/X-ray emission. These phenomena can be explained by the presence of magnetic fields in such stars, which still remains a largely unexplored territory. Polarimetric surveys (e.g., Hubrig *et al.* 2008) in Stokes V found evidence of $\langle B_z \rangle \sim$200 G in only a few stars so far, which may imply that magnetism is not a very common phenomenon among such stars. It is also necessary to verify whether weak circular polarization is not contaminated by a possibly stronger non-magnetic linear polarization due cross-talks (see Sect. 4).

Less massive early B to early F stars in the upper main sequence host a subclass of chemically peculiar Bp–Ap stars with strongest known magnetic fields among nondegenerate stars, up to 20 kG. They comprise only about 5% of the population and their chemical peculiarity is related to the presence of strong magnetic fields (see a recent review by Mathys 2009). Magnetic fields of Bp–Ap stars cover their whole surface. They have a significant degree of large-scale organization and, in first approximation, their structure is resembling a simple dipole that is inclined to the stellar rotation axis. This can also be clearly seen in ZDI maps (e.g., Kochukhov *et al.* 2004). Intrinsic variations of these magnetic fields have not been definitely observed so far, which strongly suggests that they might be fossil fields.

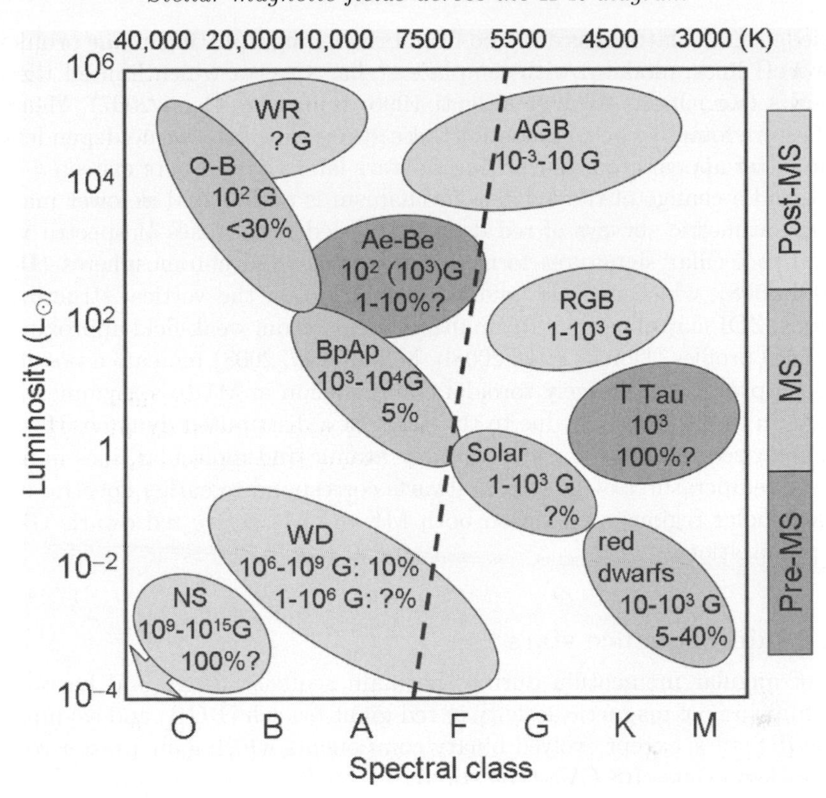

Figure 1. Occurence of magnetic fields across the H-R diagram in pre-MS, MS, and post-MS stars. Percentage indicates the fraction of stars of a given type to have such fields. The dashed line separates stars with convective (on the right) and radiative (on the left) envelops.

6.2. *Solar-type stars*

Solar-type activity on the main sequence is characteristic for stars with subsurface convection zones, roughly from F7 to K2 spectral classes. On the Sun, mean longitudinal field $\langle B_z \rangle$ does not exceed 10 G, while local magnetic fields (in plages and sunspots) are of the order of 1–3 kG. Unprecedented details of solar magnetic fields are not reviewed here, and the reader should refer to another chapter of these proceedings.

Magnetic activity in solar-type stars declines with age and is closely related to a loss of angular momentum throughout the main-sequence lifetime (see, e.g., Skumanich (1972), Güdel *et al.* 1997. Thus, young stars exhibit high average levels of activity and rapid rotation, while stars as old as the Sun and older have slower rotation rates and lower activity levels. Also, a change of the magnetic field topology with rotation rate was suggested (Petit *et al.* 2008), as deduced from ZDI maps of four stars obtained under the multipole expansion approximation from Stokes V LSD profiles. It appears that slower rotators possess mostly poloidal fields, while more rapid counterparts host a large-scale toroidal component. A rotation period of $\sim$12 days seems to be a threshold for the toroidal magnetic energy to dominate over the poloidal component for shorter periods.

6.3. *Low-mass dwarfs*

Red dwarfs constitute at least 80% of the stellar population in the Galaxy. Younger stars and binary components exhibit remarkable magnetic activity which is expressed in extremely strong optical flares, starspots, and enhanced UV, X-ray and radio emission, thus, indicating active chromospheres and coronae powered by magnetic fields. Most of the knowledge on the latter was deduced from broadening/splitting of line profiles, mainly atomic or FeH lines, modeled with template stellar spectra which limited the accuracy of the results (e.g., Johns-Krull & Valenti 1996, Reiners & Basri 2007). Mean fields of up to 4 kG were found to be common for these stars, with little or no dependence on the mass. The latter appears to be surprising as stars later than M4 are expected to be fully convective and a change of the dynamo mechanism is anticipated in lower mass stars.

Spectropolarimetric surveys of red dwarfs revealed rich Stokes V spectra with many atomic and molecular signatures formed in starspots and chromospheres (Berdyugina *et al.* 2006, 2008), which provide unique information on the vertical structure of magnetic regions. ZDI maps based on the multipole expansion, weak-field approximation and Stokes V LSD profiles (Donati *et al.* 2008b, Morin *et al.* 2008) indicate a possible change in the field topology from largely toroidal configuration in M1 to axisymmetric poloidal components in M4, supposedly due to the onset of a distributed dynamo. However, ZDI based on polarized radiative transfer in many atomic and molecular lines indicates that photosphere temperatures of spotted M dwarfs correspond to earlier spectral classes and large-scale bipolar regions are seen on both M1 and M4 active red dwarfs (Berdyugina *et al.*, in preparation).

7. Post-Main-Sequence stars

A loss of angular momentum during the main sequence stage and beyond leads to overall diminishing of magnetic activity in red giant branch (RGB) and asymptotic giant branch (AGB) stars, except evolved binary components which gain a faster rotation due to tidally locked orbits (RS CVn-type stars).

Single red giants, rotating for some reasons relatively fast, apparently possess mean longitudinal fields $\langle B_z \rangle$ (deduced from Stokes V LSD profiles) comparable to that of old solar-type stars, i.e. 1–10 G (Konstantinova-Antova *et al.* 2008, 2009). Approximately the same range of field strengths is observed near the photospheres of AGB stars, where SiO masers are formed (Herpin *et al.* 2006). It reduces with a $1/r$ law down to $10^{-3} - 10^{-1}$ G at the outskirts of the circumstellar envelope, as measured from OH masers. Such fields may play a role in shaping circumstellar envelopes in post-AGB objects.

Evolved binary components in RS CVn-type variables are distinguished by their strong chromospheric plages, coronal X-ray and microwave emissions as well as strong flares in the optical, UV, radio and X-ray. Remarkable activity and high luminosity of these stars make them favourite targets for magnetic studies. Earlier ZDI images (based on LSD and the WFA) indicated the presence of a (sub-)kG field with a dominating almost axisymmetric azimuthal component appearing as rings of opposite polarities at higher and lower latitudes (Donati *et al.* 2003). Recently, a more realistic ZDI approach has lead to detection of a bipolar region with dominant radial field component on the RS CVn-type star II Peg (Carroll *et al.* 2007). The prototype of FK Com-type stars exhibiting activity phenomena very similar to that of RS CVn stars was also confirmed to be a magnetic star with an average longitudinal field $\langle B_z \rangle$ of 250 G at the phase of maximum visibility of cool spots (Korhonen *et al.* 2009).

Degenerate stellar remnants, such as white dwarfs (WD) and neutron stars (NS), are known to be remarkable magnetic objects, especially in binary systems. Isolated magnetic white dwarfs with fields of $10^6 - 10^9$ G, are quite rare, comprising about 5–10% of all white dwarfs (Wickramasinghe & Ferrario 2000), while those with kG fields apparently are more frequent, 15-20% (Jordan *et al.* 2007). Such a distribution is reminiscent of the occurence of magnetic non-degenerated stars. Observed spectral variations of magnetic white dwarfs on a timescale of hours or days suggest a complex magnetic field distribution on their surfaces. In some cases, spot-like magnetic field enhancements superimposed on a weaker dipole magnetic field can be identified. Similar structures are most probably present on the surfaces of neutron stars with field strengths of $10^9 - 10^{15}$ G. Current theory predicts that such structures can be generated from strong subsurface toroidal fields on both white dwarfs and neutron stars.

8. Conclusions

The present review indicates that there is a tremendous progress in observations and detections of magnetic fields on various types of stars, thanks to a new generation of spectropolarimetric instruments on large telescopes. At the same time new, significantly more advanced methods for the analysis of spectropolarimetric data based on full Stokes polarized radiative transfer in atomic and molecular lines have been recently developed. They are becoming standard techniques for advancing our knowledge on stellar magnetism and should replace methods with many unrealistic approximations, so that white spots on the H-R diagram are filled with meaningful information.

Acknowledgements

This work was supported by SNF grant PE002-104552 and the EURYI Award from the ESF.

References

Afram, N., Berdyugina, S. V., Fluri, D. M., Semel, M., Bianda, M., & Ramelli, R. 2007, *A&A* 473, L1

Afram, N., Berdyugina, S. V., Fluri, D. M., Solanki, S. K., & Lagg, A. 2008, *A&A* 482, 387

Asensio Ramos, A. & Trujillo Bueno, J. 2006, *ApJ* 636, 548

Babcock, H. W. 1947, *ApJ* 105, 105

Berdyugina, S. V. 2005, *Liv. Rev. Solar Phys.* 2, 8

Berdyugina, S. V., Braun, P. A., Fluri, D. M., & Solanki, S. K. 2005, *A&A* 444, 947

Berdyugina, S. V. & Fluri, D. M. 2004, *A&A* 417, 775

Berdyugina, S. V., Fluri, D. M., & Afram, N., *et al.* 2008, in: G. van Belle (ed.), *14th Cambridge Workshop on Cool Stars, Stellar Systems, and the Sun*, ASP Conf. Ser., vol. 384, p. 175

Berdyugina, S. V., Petit, P., Fluri, D. M., Afram, N., & Arnaud, J. 2006, in: R. Casini & B. W. Lites (eds.), *Solar Polarization 4*, ASP Conf. Ser., vol. 358, p. 381

Berdyugina, S. V. & Solanki, S. K. 2002, *A&A* 385, 701

Berdyugina, S. V., Solanki, S. K., & Frutiger, C. 2003, *A&A* 412, 513

Berdyugina, S. V., Stenflo, J. O., & Gandorfer, A. 2002, *A&A* 388, 1062

Brown, S. F., Donati, J.-F., Rees, D. E., & Semel, M. 1991, *A&A* 250, 463

Carroll, T. A., Kopf, M., Ilyin, I., & Strassmeier, K. G. 2007, *AN* 328, 1043

Donati, J.-F., Semel, M., & Praderie, F. 1989, *A&A* 225, 467

Donati, J.-F., Semel, M., Carter, B. D., Rees, D. E., & Collier Cameron, A. 1997, *MNRAS* 291, 658

Donati, J.-F., Collier Cameron, A., Semel, M., *et al.* 2003, *MNRAS* 345, 1145

Donati J.-F., Howarth I., Jardine M., *et al.* 2006, *MNRAS* 370, 629

Donati J.-F., Jardine M., Gregory, S. G., *et al.* 2007, *MNRAS* 380, 1297

Donati J.-F. Jardine M., Gregory, S. G., *et al.* 2008a, *MNRAS*, 386, 1234

Donati J.-F., Morin, J., Petit, P., *et al.* 2008b, *MNRAS* 390, 545

Güdel, M., Guinan, E. F., & Skinner, S. L. 1997, *ApJ* 483, 947

Hale, G. E. 1908, *ApJ* 28, 315

Harrington, D. M. & Kuhn, J. R. 2007, *ApJ* 668, L63

Herpin, F., Baudry, A., Thum, C., Morris, D., & Wiesemeyer, H. 2006, *A&A* 450, 667

Hubrig, S., Schöller, M., Schnerr, R. S., *et al.* 2008, *A&A* 490, 793
Hubrig, S., Grady, C., Schöller, M., *et al.* 2009, this proceedings
Johns-Krull, C. M. 2007, *ApJ* 664, 975
Johns-Krull, C. M. 2009, this proceedings
Johns-Krull, C. M. & Valenti, J. A. 1996, *ApJ* 459, L95
Jordan, S., Aznar Cuadrado, R., Napiwotzki, R., Schmid, H. M. & Solanki, S. K. 2007, *A&A* 462, 1097
Kochukhov, O., Drake, N. A., Piskunov, N., & de la Reza, R. 2004, *A&A* 424, 935
Konstantinova-Antova, R., Auriére, M., Iliev, I.Kh., *et al.* 2008, *A&A* 480, 475
Konstantinova-Antova, R., Auriére, M., Schröder, K.-P., & Petit, P. 2009, this proceedings
Korhonen, H., Hubrig, S., Berdyugina, S. V., *et al.* 2009, *MNRAS*, in press
Kuhn, J. R., Berdyugina, S. V., Fluri, D. M., Harrington, D. M., & Stenflo, J. O. 2007, *ApJ* 668, L63
Landau, L. D. & Lifshits, E. M. 1991, *Quantum mechanics: Non-relativistic theory*, Oxford, Boston: Butterworth-Heinemann
Martínez González, M. J., Asensio Ramos, A., Carroll, T. A., Kopf, M., Ramírez Vélez, J. C., & Semel, M. 2008, *A&A* 486, 637
Mathys, G. 2009, in: S.V. Berdyugina, K.N. Nagendra & R.Ramelli (eds.), *Solar Polarization Workshop 5*, ASP Conf. Ser., in press
Mestel, L. & Landstreet, J. D. 2005, in: R. Wielebinski & R. Beck (eds.), *Cosmic Magnetic Fields*, Lect. Notes Phys., vol. 664, p. 183
Morin, J., Donati, J.-F., Petit, P., *et al.* 2008, *MNRAS* 390, 567
Petit, P., Dintrans, B., Solanki, S. K., *et al.* 2008, *MNRAS* 388, 80
Petrov, P. P. 2003, *Astrophysics* 46, 506
Piskunov, N. & Kochukhov, O. 2002, *A&A* 381, 736
Reiners, A. & Basri, G. 2006, *ApJ* 644, 497
Reiners, A. & Basri, G. 2007, *ApJ* 656, 1121
Robinson, R. D., J. 1980, *ApJ* 239, 961
Saar, S. H. 1988, *ApJ* 324, 441
Semel, M. 1989, *A&A* 225, 456
Semel, M. & Li, J. 1996, *Sol. Phys.* 164, 417
Sennhauser, C., Berdyugina S. V. & Fluri, D. M. 2008, in: S. V. Berdyugina, K. N. Nagendra & R.Ramelli (eds.), *Solar Polarization Workshop 5*, ASP Conf. Ser., in press
Shapiro, A. I., Berdyugina, S. V., Fluri, D. M., & Stenflo, J. O. 2007, *A&A* 475, 349
Skumanich, A. 1072, *ApJ* 171, 565
Valenti, J. A. & Johns-Krull, C. M. 2001, in: G. Mathys, S. K. Solanki, D. T. Wickramasinghe (eds.), *Magnetic Fields Across the Hertzsprung-Russell Diagram*, ASP Conf. Ser. (ASP: San Francisco), vol. 248, p. 179
Wade, G. A., Bagnulo, S., Drouin, D., Landstreet, J. D., & Monin, D. 2007, *MNRAS* 376, 1145
Wade, G. A., Alecian, E., Grunhut, J., *et al.* 2009, in: P. Bastien & N. Manset (eds.), *ASP Conf. Ser.*, in press
Wickramasinghe, D. T. & Ferrario, L. 2000, *PASP* 112, 873

Cosmic Magnetic Fields:
From Planets, to Stars and Galaxies
Proceedings IAU Symposium No. 259, 2008
K.G. Strassmeier, A.G. Kosovichev & J.E. Beckman, eds.

doi:10.1017/S1743921309030695

The MiMeS project: magnetism in massive stars

Gregg A. Wade[1], E. Alecian[1,2], D. A. Bohlender[3], J.-C. Bouret[4], J. H. Grunhut[1], H. Henrichs[5], C. Neiner[6], V. Petit[7], N. St. Louis[8], M. Aurière[9], O. Kochukhov[10], J. Silvester[1], A. ud-Doula[11] and the MiMeS Collaboration†

[1]Royal Military College of Canada, [2]LESIA, France, [3]Canadian Astronomy Data Centre, [4]LAM, France, [5]Ast. Inst. Amsterdam, Netherlands, [6]GEPI, France, [7]Université Laval, Canada, [8]Univ. de Montréal, Canada, [9]LAT, France, [10]Uppsala University, Sweden, [11]Morrisville State College, USA

Abstract. The Magnetism in Massive Stars (MiMeS) Project is a consensus collaboration among the foremost international researchers of the physics of hot, massive stars, with the basic aim of understanding the origin, evolution and impact of magnetic fields in these objects. The cornerstone of the project is the MiMeS Large Program at the Canada-France-Hawaii Telescope, which represents a dedication of 640 hours of telescope time from 2008-2012. The MiMeS Large Program will exploit the unique capabilities of the ESPaDOnS spectropolarimeter to obtain critical missing information about the poorly-studied magnetic properties of these important stars, to confront current models and to guide theory.

Keywords. Magnetic fields – massive stars – hot stars – star formation – stellar evolution – stellar winds – spectropolarimetry

1. Introduction

Massive stars are those stars with initial masses above about 8 times that of the sun, eventually leading to catastrophic explosions in the form of supernovae. These represent the most massive and luminous stellar component of the Universe, and are the crucibles in which the lion's share of the chemical elements are forged. These rapidly-evolving stars drive the chemistry, structure and evolution of galaxies, dominating the ecology of the Universe - not only as supernovae, but also during their entire lifetimes - with far-reaching consequences.

The magnetic fields of hot, higher-mass stars are qualitatively different from those of cool, low-mass stars (e.g. Wade 2003). They are detected in only a small fraction of stars, and they are structurally much simpler, and frequently much stronger, than the fields of cool stars. Most remarkably, their characteristics show no clear correlation with basic stellar properties such as age, mass or rotation (e.g. Mathys *et al.* 1997, Kochukhov & Bagnulo 2006). The weight of opinion holds that these puzzling characteristics reflect a fundamentally different field origin: that the observed fields are not generated by dynamos, but rather that they are *fossil fields* - the slowly-decaying remnants of field accumulated or generated during star formation (e.g. Mestel 1999, Moss 2001, Ferrario & Wickramasinghe 2006). This relic nature potentially provides us with a powerful and unique capability: to study how magnetic fields evolve throughout the various stages of stellar evolution, and to explore how they influence, and are influenced by, the important

† www.physics.queensu.ca/~wade/mimes

structural changes that occur during all phases of stellar evolution, from stellar birth to stellar death.

Although this fossil paradigm provides a powerful framework for interpreting the magnetic characteristics of higher-mass stars, its physical details are only just beginning to be elaborated (e.g. Braithwaite & Nordlund 2006, Auriere *et al.* 2007, Alecian *et al.* 2008a). In particular, our knowledge of the basic statistical properties of massive star magnetic fields is seriously incomplete. There is a troubling deficit in our understanding of the scope of the influence of fields on massive star evolution, and almost no empirical basis for how fields modify mass loss.

The Magnetism in Massive Stars (MiMeS) Project represents a comprehensive, multidisciplinary strategy by an international team of recognized researchers to address the big questions related to the complex and puzzling magnetism of massive stars. Recently, MiMeS was awarded "Large Program" status by both Canada and France at the Canada-France-Hawaii Telescope (CFHT), where the Project has been allocated 640 hours of dedicated time with the ESPaDOnS spectropolarimeter from late 2008 through 2012. This commitment of the observatory, its staff, its resources and expertise, allocated as a result of an extensive international expert peer review of many competing proposals, will be used to acquire an immense database of sensitive measurements of the optical spectra and magnetic fields of massive stars, which will be applied to constrain models of the origins of their magnetism, the structure, dynamics and emission properties of their magnetospheres, and the influence of magnetic fields on stellar mass loss and rotation - and ultimately the evolution of massive stars. More specifically, the scientific objectives of the MiMeS Project are:

- To identify and model the physical processes responsible for the generation of magnetic fields in massive stars;
- To observe and model the detailed interaction between magnetic fields and massive star winds;
- To investigate the role of the magnetic field in modifying the rotational properties of massive stars;
- To investigate the impact of magnetic fields on massive star evolution,and the connection between magnetic fields of non- degenerate massive stars and those of neutron stars and magnetars, with consequential constraints on stellar evolution, supernova astrophysics and gamma-ray bursts.

2. Structure of the Large Program

To address these general problems, we have devised a two-component Large Program (LP) that will allow us to obtain basic statistical information about the magnetic properties of the overall population of hot, massive stars (the Survey Component), while simultaneously providing detailed information about the magnetic fields and related physics of individual objects (the Targeted Component).

Targeted component: The MIMeS Targeted Component (TC) will provide data to map the magnetic fields and investigate the physical characteristics of a small sample of known magnetic stars of great interest, at the highest level of sophistication possible. The roughly 20 TC targets have been selected to allow us to investigate a variety of physical phenomena, and to allow us to directly and quantitatively confront models.

Each TC target will be observed many times using the ESPaDOnS spectropolarimeter, in order to obtain a high-precision and high-resolution sampling of the rotationally-modulated circular and linear polarisation line profiles. Using state-of-the-art tomographic reconstruction techniques such as Magnetic Doppler Imaging (Kochukhov &

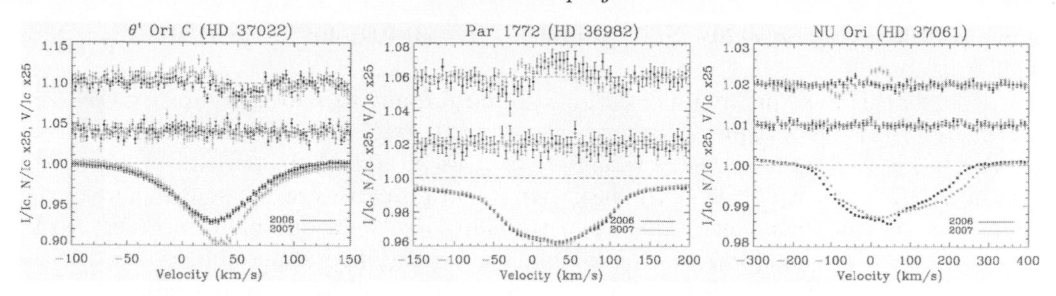

Figure 1. Least-Squares Deconvolved profiles of 3 hot stars: θ^1 Ori C (O7V, left), Par 1772 (B2V, middle) and NU Ori (B0.5V, right). The curves show the mean Stokes I profiles (bottom panel), the mean Stokes V profiles (top panel) and the N diagnostic null profiles (middle panel), black for 2006 January and red for 2007 March. Each star exhibits a clear magnetic signature in Stokes V. These results are representative of those expected from the MiMeS Survey Component. From Petit *et al.* (2008).

Piskunov 2002), detailed maps of the vector magnetic field on and above the surface of the star will be constructed.

Survey component: The MiMeS Survey Component (SC) will provide critical missing information about field incidence and statistical field properties for a much larger sample of massive stars. It will also serve to provide a broader physical context for interpretation of the results of the Targeted Component. From a much larger list of potential OB stars compiled from published catalogues, we have generated an SC sample of about 150 targets which cover the full range of spectral types from B2-O4 which are selected to be best-suited to field detection. Our target list includes pre-main sequence Herbig Be stars, field and cluster OB stars, Be stars, and Wolf-Rayet stars.

Each SC target will be observed once or twice during the Project, at very high precision in circular polarisation. From the SC data we will measure the bulk incidence of magnetic massive stars, estimate the variation of incidence versus mass, derive the statistical properties (intensity and geometry) of the magnetic fields of massive stars, estimate the dependence of incidence on age and environment, and derive the general statistical relationships between magnetic field characteristics and X-ray emission, wind properties, rotation, variability, binarity and surface chemistry diagnostics.

Of the 640 hours allocated to the MiMeS LP, 385 hours are committed to the SC and 255 hours are committed to the TC. The TC commitment includes 50 hours reserved for follow-up of targets detected in the Survey Component.

3. Precision magnetometry of massive stars

For all targets we will exploit the longitudinal Zeeman effect in metal and helium lines to detect and measure photospheric or pseudo-photospheric magnetic fields. Splitting of a spectral line due to a longitudinal magnetic field into oppositely-polarized σ components produces a variation of circular polarisation across the line (commonly referred to as a (Stokes V) Zeeman signature or magnetic signature; see Fig. 1.). The amplitude and morphology of the Zeeman signature encode information about the strength and structure of the global magnetic field. For some TC targets, we will also exploit the transverse Zeeman effect to constrain the detailed local structure of the field. Splitting of a spectral line by a transverse magnetic field into oppositely-polarized π and σ components produces a variation of linear polarisation (characterized by the Stokes Q and U parameters) across the line (e.g. Kochukhov *et al.* 2004).

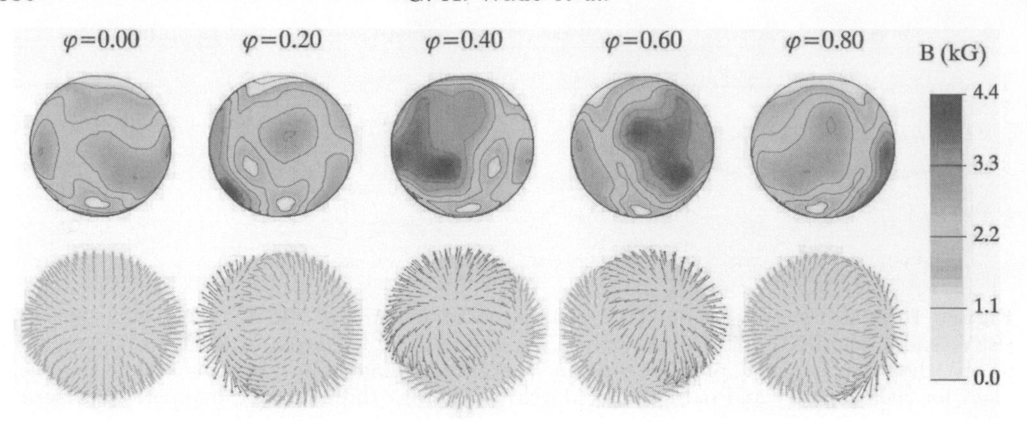

Figure 2. Magnetic Doppler Imaging (MDI) of the B9p star HD 112413 (Kochukhov *et al.*, in preparation), illustrating the reconstructed magnetic field orientation (lower images) and intensity (upper images) of this star at 5 rotation phases. The maps were obtained from a time-series of 21 Stokes *IQUV* spectral sequences. Although the field line orientation of HD 112413 is approximately dipolar, the field intensity map is far more complex. Maps similar to these will be constructed for the MiMeS Targeted Component.

3.1. *Survey Component*

For the SC targets, the detection of magnetic field is diagnosed using the Stokes V detection criterion described by Donati *et al.* (1992, 1997), and the surface field constraint characterised using the powerful Bayesian estimation technique of Petit *et al.* (2008). After reduction of the polarized spectra using the Libre-Esprit optimal extraction code, we employ the Least-Squares Deconvolution (LSD; Donati *et al.* 1997) multi-line analysis procedure to combine the Stokes V Zeeman signatures from many spectral lines into a single high-S/N mean profile (see Fig. 1), enhancing our ability to detect subtle magnetic signatures. Least-Squares Deconvolution of a spectrum requires a line mask to describe the positions, relative strengths and magnetic sensitivities of the lines predicted to occur in the stellar spectrum. The line mask characteristics are sensitive to the parameters describing the stellar atmosphere. In our analysis we employ custom line masks carefully tailored to best reproduce the observated stellar spectrum, in order to maximize the S/N gain of the LSD procedure and therefore our sensitivity to weak magnetic fields.

The exposure duration required to detect a Zeeman signature of a given strength varies as a function of stellar apparent magnitude, spectral type and projected rotational velocity. This results in a large range of detection sensitivities for our targets. The SC exposure times are based on an empirical exposure time relation derived from real ESPaDOnS observations of OB stars, and takes into account detection sensitivity gains resulting from LSD and velocity binning, and sensitivity losses from line broadening due to rapid rotation. Exposure times for our SC targets correspond to the time required to definitely detect (with a false alarm probability below 10^{-5}) the Stokes V Zeeman signature produced by a surface dipole magnetic field with a specified polar intensity. Although our calculated exposure times correspond to definite detections of a dipole magnetic field, our observations are also sensitive to the presence of substantially more complex field toplogies.

3.2. *Targeted Component*

Zeeman signatures will be detected repeatedly in all spectra of TC targets. The spectropolarimetric timeseries will be interpreted using several magnetic field modeling codes at

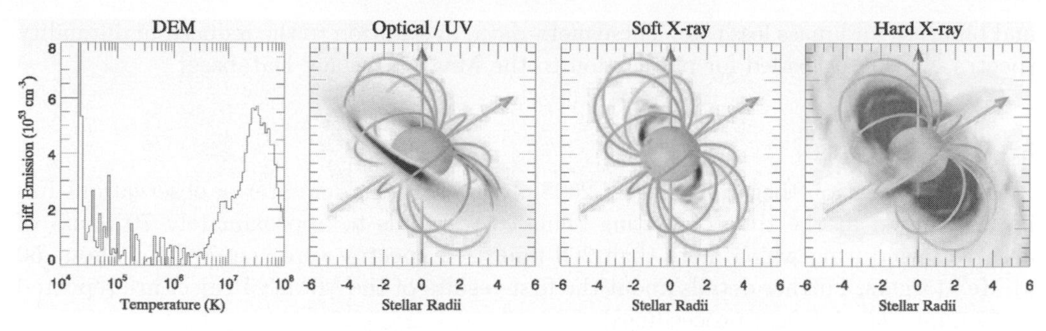

Figure 3. Example of the spectral and spatial emission properties of a rotating massive star magnetosphere modeled using Rigid Field Hydrodynamics (Townsend *et al.* 2007). The stellar rotation axis (vertical arrow) is oblique to the magnetic axis (inclined arrow), leading to a complex potential field produced by radiative acceleration, Lorentz forces and centripetal acceleration. The consequent heated plasma distribution in the stellar magnetosphere (illustrated in colour/grey scale) shows broadband emission, and is highly structured both spatially and spectrally. Magnetically-confined winds such as this are responsible for the X-ray emission and variability properties of some OB stars, and models such as this will be constructed for the MiMeS Targeted Component.

our disposal. For those stars for which Stokes V LSD profiles will be the primary model basis, the modeling codes employed by Donati *et al.* (2006) or Alecian *et al.* (2008b) will be employed. For those stars for which the signal-to-noise ratio in individual spectral lines is sufficient to model the polarisation spectrum directly, we will employ the Invers10 Magnetic Doppler Imaging code to simultaneously model the magnetic field, surface abundance structures and pulsation velocity field (Piskunov & Kochukhov 2002, Kochukhov *et al.* 2004). The resultant magnetic field models will be compared directly with the predictions of fossil and dynamo models (e.g. Braithwaite 2006, 2007, Mullan & Macdonald 2005, Arlt 2008).

Diagnostics of the wind and magnetosphere (e.g. optical emission lines and their linear polarisation, UV line profiles, X-ray photometry and spectroscopy, radio flux variations, etc.) will be modeled using both the semi-analytic Rigidly-Rotating Magnetosphere approach, the Rigid-Field Hydrodynamics (Townsend *et al.* 2007) approach and full MHD simulations using the 3D ZEUS code (e.g. Stone & Norman 1992; see Fig. 3).

4. MiMeS data pipeline

Following their acquisition in Queued Service Observing mode at the CFHT, ESPaDOnS polarised spectra are immediately reduced by CFHT staff using the Libre-Esprit reduction package and downloaded to the dedicated MiMeS Data Archive at the Canadian Astronomy Data Centre in Victoria, Canada. Reduced spectra are carefully normalized to the continuum using existing software tailored to hot stellar spectra. Each reduced ESPaDOnS spectrum is then subject to an immediate quick-look analysis to verify nominal resolving power, polarimetric performance and S/N. Preliminary LSD profiles are extracted using our database of generic hot star line masks to perform an initial magnetic field diagnosis and further quality assurance. Finally, each ESPaDOnS spectrum will be processed by the MiMeS Massive Stars Pipeline (MSP; currently in production) to determine a variety of critical physical data for each observed target, in addition to the precision magnetic field diagnosis: effective temperature, surface gravity, mass, radius, age, variability characteristics, projected rotational velocity, radial velocity and binarity, and mass loss rate. These meta-data, in addition to the reduced high-quality spectra, will be uploaded for publication to the MagIcS Legacy Database†.

5. First results

MiMeS operations began in August 2008, and nearly one semester of observations has been acquired at the time of writing. This corresponds to approximately 70 hours of observation, during which more than 200 polarised spectra were acquired for about 50 MiMeS targets. Further details about the first results of the MiMeS Project are reported by Grunhut *et al.* (these proceedings).

Acknowledgements

The MiMeS Large Program is supported by both Canada and France, and was one of 4 such programs selected in early 2008 as a result of an extensive international expert peer review of many competing proposals.

Based on observations obtained at the Canada-France-Hawaii Telescope (CFHT) which is operated by the National Research Council of Canada, the Institut National des Sciences de l'Univers of the Centre National de la Recherche Scientifique of France, and the University of Hawaii.

The MiMeS Data Access Pages are powered by software developed by the CADC, and contains data and meta-data provided by the CFHT Telescope.

References

Alecian E., Wade G. A., Catala C., Bagnulo S., *et al.*, 2008, *A&A* 481, 99

Alecian E., Catala C., Wade G. A., Donati J.-F., *et al.*, 2008, *MNRAS* 385, 391

Arlt R., 2008, Cont. Ast. Obs. Skalnaté Pleso, 38, p. 163

Auriere M., Wade G., Silvester J., Lignières F. *et al.*, 2007, *A&A* 475, 1053

Braithwaite J., 2006, *A&A* 453, 687

Braithwaite J., 2007, *A&A* 469, 275

Braithwaite J. & Nordlund A., 2006, *A&A* 450, 1077

Donati J.-F., Semel M., Carter B. D., Rees D. E., & Collier Cameron A., 1997, *MNRAS* 291, 658

Ferrario L. & Wickramasinghe D., 2006, *MNRAS* 367, 1323

Kochukhov O. & Piskunov N., 2002, *A&A* 388, 868

Kochukhov O., Bagnulo S., Wade G. A., Sangalli L., *et al.*, 2004, *A&A* 414, 613

Kochukhov O. & Bagnulo S., 2006, *A&A* 450, 763

Mathys G., Hubrig S., Landstreet J., Lanz T., & Manfroid J., 1997, *A&AS* 123, 353

Mestel L., 1999, in Magnetic Fields Across the HR Diagram, ASP Conf. Proc. vol. 248, p. 3

Moss D., 2001, Magnetic Fields Across the HR Diagram, ASP Conf. Proc. vol. 248, p. 305

Mullan D. & MacDonald J. 2005, *MNRAS* 356, 1139

Petit V., Wade G. A., Drissen L., Montmerle T., & Alecian E., 2008, *MNRAS* 387, 23

Stone J. & Norman M., 1992, ApJS 80, 753

Townsend R. H. D., Owocki S. P., & Ud-Doula A., 2007, *MNRAS* 382, 139

Wade G. A., 2003, in Magnetic Fields in O, B and A Stars, ASP Conf. Proc. vol. 305, p. 16

† The MiMeS Project is undertaken within the context of the broader MagIcS (Magnetic Investigations of various Classes of Stars) collaboration, www.ast.obs-mip.fr/users/donati/magics).

Cosmic Magnetic Fields:
From Planets, to Stars and Galaxies
Proceedings IAU Symposium No. 259, 2008
K.G. Strassmeier, A.G. Kosovichev, & J.E. Beckman, eds.

doi:10.1017/S1743921309030701

Magnetic field observations of low-mass stars

Ansgar Reiners

Universität Göttingen, Institut für Astrophysik, Friedrich-Hund-Platz 1,
D-37077 Göttingen, Germany
email: Ansgar.Reiners@phys.uni-goettingen.de

Abstract. Direct measurements of magnetic fields in low-mass stars of spectral class M have become available during the last years. This contribution summarizes the data available on direct magnetic measurements in M dwarfs from Zeeman analysis in integrated and polarized light. Strong magnetic fields at kilo-Gauss strength are found throughout the whole M spectral range, and so far all field M dwarfs of spectral type M6 and later show strong magnetic fields. Zeeman Doppler images from polarized light find weaker fields, which may carry important information on magnetic field generation in partially and fully convective stars.

Keywords. Stars: activity – stars: late-type – stars: low-mass – brown dwarfs – stars: magnetic fields

---

1. Introduction

The existence and the topology of magnetic fields on low-mass stars is a topic of great interest. On the Sun, we know that magnetic fields are generated but strong fields are concentrated in small regions. Many mechanisms that could generate these fields were suggested, but their real physical nature still is a question of lively research (see other contributions in this volume). Rotation is probably one of the fundamental parameters that plays a key role in the generation of magnetic fields (e.g., Pizzolato *et al.* (2003); Reiners *et al.* (2009)). Other paramaters that could matter are the luminosity of the stars (Christensen *et al.* (2009)), the convective velocities, the size of the convection zone, and the existence of a region of great sheer between a radiative (probably rigidly rotating) core and an outer convective envelope, the so-called tachocline (Ossendrijver (2003)).

The general paradigm of the solar dynamo assumes that at least the cyclic part of the solar magnetic field is generated close to the tachocline through a so-called $\alpha\Omega$-type dynamo (but see also Brandenburg, this volume). Any dynamo mechanism that requires a tachocline must vanish in low-mass stars around spectral type M3 because this is the boundary beyond which stars no longer maintain a radiative core but remain fully convective during their entire lifetime (very young age, more massive stars are also fully convective). It is therefore particularly interesting to investigate magnetic fields around the mass boundary between partially and fully convective stars.

2. Methods to measure magnetic fields

Magnetic fields are measured with different techniques. Here, we consider only "direct" measurements of magnetic fields, i.e., we are not considering magnetic fields that are inferred through the observation of secondary indicators. Such indicators are very useful because they trace non-thermal radiation processes, which are probably connected to the existence of magnetic fields (most prominent tracers are chromospheric Ca H&K and Hα emission, e.g., Hartmann & Noyes (1987), Mohanty & Basri (2003; coronal

X-ray emission, e.g., Pizzolato *et al.* (2003); or radio emission from high-energy electrons, Berger (2006), Hallinan *et al.* (2008)).

Direct observations of magnetic fields can be accomplished through observation of the Zeeman effect, which shows direct consequences of magnetic fields on the appearance of spectral features. Principally, two different ways must be distinguished. The first method is the measurement of the Zeeman effect in polarized light, usually detecting circular polarization in Stokes V. This method can detect relatively weak fields because it is a differential method that can be calibrated rather accurately. One caveat is that only a net polarization can be detected. Polarization signals even of strong magnetic fields can cancel out each other if they are distributed in small entities. Another problem of measurements in Stokes V is that only a fraction of the light can be used so that the usually a number of simplifications have to be applied. The second method is to determine the magnetic fields strength from the appearance of spectral lines in Stokes I, i.e., from integrated light without any polarization analysis. Here, the main problem is that the effect of Zeeman broadening is relatively small in comparison to other broadening effects in spectral lines so that calibration is very difficult and leads to large uncertainties. On the other hand, this method can detect the total mean unsigned magnetic flux including components that are distributed in small cells of opposite polarity.

3. A compilation of results

This article gives an overview of recent (direct) magnetic field measurements in low-mass stars of spectral class M. Today, the number of measurements is constantly growing and we can start to compare results from different techniques in the same stars. This opens another channel of investigation because the differences between the results from different techniques carries information about the magnetic fields and their distribution.

3.1. *Stokes I*

Measurements of stellar magnetic fields in integrated light, Stokes I, can be carried out by calculating the polarized radiative transfer in magnetically sensitive spectral lines and comparing the models to observations. Robinson (1980), has introduced the method to measure stellar magnetic fields in spectral absorption lines. This method was used by several groups, mostly in atomic spectral lines in the optical wavelength range. A summary of this effort can be found in Saar (1996), and Saar (2001). Johns-Krull & Valenti (2000), summarize results on magnetic field measurements including spectral lines in the infrared wavelength range. These authors center on developments for fully convective M dwarfs and pre-main sequence stars. In very cool atmospheres, e.g., in mid-M dwarfs at the boundary to full convection, the usefulness of atomic lines becomes limited because atomic lines are being buried under the haze of molecular absorption bands.

Valenti & Johns-Krull (2001), discuss molecular absorption bands of FeH as a Zeeman diagnostic for very cool dwarfs. FeH shows a series of narrow, strong absorption lines for example around 1 μm, a spectral region where the spectral energy distribution of M dwarfs is close to its maximum. A problem for magnetic field measurements with FeH is that magnetic splitting is difficult to calculate. Reiners & Basri (2006), introduced a semi-empirical approach in which the magnetic field of an M-star is measured by comparison of its FeH lines to FeH lines observed in stars of known magnetic field. The magnetic fields of these template stars have to be measured in atomic lines, hence only early-M stars can be used for this method. Reiners & Basri (2006), show that FeH absorption in M dwarfs follows an optical-depth scaling that allows to use early-M template spectra to

Table 1. Stokes I measurements of magnetic flux among M dwarfs.

Name	Spectral Type	$v \sin i$ [$\mathrm{km\,s^{-1}}$]	Bf [kG]	Ref
Gl 182	M0.5	9	2.5	[e]
Gl 494A	M0.5	11	3.0	[e,f]
Gl 70	M2.0	$\leqslant$3	0.0	[a]
Gl 569A	M2.5	$\leqslant$3	1.8	[e]
Gl 729	M3.5	4	2.2	[a,b]
Gl 873	M3.5	$\leqslant$3	3.9	[a,b]
Gl 388	M3.5	$\approx$3	2.9	[a,b]
GJ 3379	M3.5	$\leqslant$3	2.3	[c]
Gl 876	M4.0	$\leqslant$3	0.0	[a]
GJ 1005A	M4.0	$\leqslant$3	0.0	[a]
GJ 2069B	M4.0	6	2.7	[c]
GJ 299	M4.5	$\leqslant$3	0.5	[a]
GJ 1227	M4.5	$\leqslant$3	0.0	[a]
GJ 1224	M4.5	$\leqslant$3	2.7	[a]
Gl 285	M4.5	5	$>$ 3.9	[a,b]
Gl 493.1	M4.5	18	2.1	[c]
LHS 3376	M4.5	19	2.0	[c]
Gl 905	M5.0	$\leqslant$3	0.0	[a]
GJ 1057	M5.0	$\leqslant$3	0.0	[a]
GJ 1154A	M5.0	6	2.1	[c]
GJ 1156B	M5.0	17	2.1	[c]
LHS 1070 A	M5.5	8	2.0	[c]
GJ 1245B	M5.5	7	1.7	[a]
GJ 1286	M5.5	$\leqslant$3	0.4	[a]
GJ 1002	M5.5	$\leqslant$3	0.0	[a]
Gl 406	M5.5	3	2.1–2.4	[a,d]
GJ 1111	M6.0	13	1.7	[a]
GJ 412B	M6.0	5	$>$ 3.9	[c]
VB 8	M7.0	5	2.3	[a]
LHS 3003	M7.0	6	1.5	[a]
LHS 2645	M7.5	8	2.1	[a]
LP 412–31	M8.0	9	$>$ 3.9	[a]
VB 10	M8.0	6	1.3	[a]
LHS 1070 B	M8.5	16	4.0	[c]
LHS 1070 C	M9.0	16	2.0	[c]
LHS 2924	M9.0	10	1.6	[a]
LHS 2065	M9.0	12	$>$ 3.9	[a]

[a]Reiners & Basri (2007), [b]Johns-Krull & Valenti (2000), [c]Reiners *et al.* (2009), [d]Reiners *et al.* (2007), [e]Reiners & Basri (2009), [f]Saar (1996)

measure magnetic fields even in very-late type M dwarfs. Using this method, Reiners & Basri (2007), carried out the first survey of magnetic field measurements in M dwarfs, and Reiners *et al.* (2007), and Reiners & Basri (2009), applied the same method to another set of stars. The uncertainty of their magnetic field measurements is on the order of a kilo-Gauss so that small fields of a few hundred Gauss are difficult to detect. The method was successfull in detecting strong (several kilo-Gauss) fields in many M dwarfs, in particular all M dwarfs with spectral type M6 and later show strong magnetic fields.

The current sample of M dwarfs with measured magnetic fields is collected in Table 1. Strong magnetic fields are found throughout the whole range of spectral subclasses M0–M9. More specific, among the early-M stars with spectral type M0–M5.5, magnetic flux

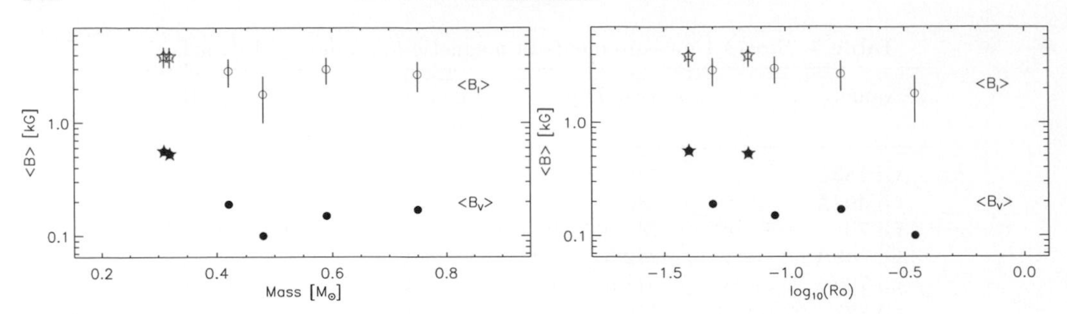

Figure 1. Mean magnetic field measurements from Stokes I (open symbols) and Stokes V (filled symbols) as a function of Mass (left panel) and Rossby number (right panel).

is found between 0 and 4 kG. Magnetic flux scales with rotation (and activity in Hα scales with magnetic flux), which is expected because of the well-known rotation-activity connection among warmer stars (e.g., Pizzolato *et al.* (2003)). The meaning of this is threefold: 1) M dwarfs (partially as well as fully convective) can generate strong magnetic fields; 2) Stellar activity is of magnetic origin; 3) The connection between rotation and magnetic field generation / magnetic activity also holds in early- to mid-M dwarfs.

Magnetic field measurements have also been carried out in young low-mass stars (see for example Johns-Krull (2007)). Measurements of magnetic fields in young objects are the subject of Johns-Krull (2009), in this volume.

3.2. *Stokes V*

The method of Zeeman Doppler Imaging in Stokes V has been very successfull during the last years providing the first information on stellar magnetic structure (e.g., Donati *et al.* (1997)). The somewhat surprising result of axisymmetric magnetic topology in a fully convective star was reported by Donati *et al.* (2006) using this technique. Recently, Donati *et al.* (2008), and Morin *et al.* (2008), measured magnetic fields in Stokes V in a sample of early- and mid-M dwarfs on both sides of the boundary to full convection. They find magnetic flux on the level of a few hundred Gauss in their sample stars. An important result of this work is that the strength of magnetic flux and the topology of magnetic fields show a marked change at spectral class M3 where stars are believed to become fully convective. This is very interesting because it was expected that magnetic fields in fully convective stars (Durney *et al.* (1993)) are less organized than in stars with a radiative core that might generate large-scale dipolar fields through an interface dynamo located at the tachocline (Charbonneau (2005); Ossendrijver (2003)).

4. A comparison of results

With the growing number of magnetic field measurements in both, integrated light and Zeeman Doppler Imaging from Stokes V, it becomes possible to compare results from both techniques. This may shed more light on the topology of M star magnetic fields because both techniques are sensitive to different aspects of the flux: While in Stokes I the mean total unsigned flux is observed, Stokes V is sensitive to the net polarized flux, i.e., to magnetic field distributions that lead to a net polarization signal in time series taken for the Doppler Images.

Currently, six M dwarfs have magnetic flux measurements from both Stokes V and Stokes I. These results are shown in Fig. 1 as a function of mass and Rossby number. In all cases, the magnetic flux measured from Stokes I is much stronger than the value from Zeeman Doppler Imaging. This is not surprising, as mentioned above, because the former is sensitive to the total flux while the latter normally is not. The ratio between magnetic flux in Stokes V and I is between 1:7 and 1:17, this means the fraction of magnetic flux seen in Stokes V is between 6 % and 15 %.

5. Summary

Magnetic field measurements are becoming available for a growing number of low-mass stars through analysis of integrated and polarized light. A compilation of results from Stokes I measurements in the FeH molecule used among M dwarfs is given in this article. Other techniques are available probing different aspecs of the magnetic field distribution and making different assumptions.

Generally, fields of kilo-Gauss strength are ubiquitously found in late-M dwars, which is consistent with their activity. Magnetic field topology is important to answer the question about the underlying dynamo mechanism, and first results on topologies are available from Zeeman Doppler imaging finding predominantly axisymmetric geometries in fully convective stars. However, the applied techniques are probably missing a substantial fraction of the total magnetic flux so that the full field distribution may not be visible.

Brown dwarfs are generally rapidly rotating and so far no useful tracer for Zeeman analysis was found (at least in old brown dwarfs), which poses a serious problems to the detection of magnetism. Whether brown dwarfs can maintain strong magnetic fields is an open question, but so far there seems to be no reason to believe they wouldn't.

I acknowledge funding through a DFG Emmy Noether Fellowship (RE1664/4-1).

References

Berger, E., 2006, *ApJ* 648, 629
Charbonneau, P., 2005, *Living Reviews in Solar Physics* 2, 2
Christensen, U., Holzwart, V., & Reiners, A., 2009, *Nature*, in press
Donati, J.-F., Semel, M., Carter, B. D., Rees, C. E., & Cameron, A. C., 1997, *MNRAS* 291, 658
Donati, J.-F., Forveille, T., Cameron, A. C., *et al.*, 2006, *Science* 311, 633
Donati, J.-F., Morin, J., Petit, P., *et al.*, 2008, *MNRAS* 390, 545
Durney, B. R., de Young, D.S., & Roxburgh, I. W., 1993, *SP* 145, 207
Hallinan, G., Antonova, A., Doyle, J. G., *et al.*, 2008, *ApJ* 684, 644
Hartmann, L. W. & Noyes, R. W., 1987, *ARAA* 25, 271
Johns-Krull, C. & Valenti, J. A., 2000, ASP Conf. Ser., 198, 371
Johns-Krull, C. M., 2007, *ApJ* 664, 975
Johns-Krull, C. M., 2009, this volume
Mohanty, S. & Basri, G., 2003, *ApJ* 583, 451
Morin, J., Donati, J.-F., Petit, P., *et al.*, 2008, *MNRAS* 390, 567
Ossendrijver, M., 2003, *A&AR* 11, 287
Pizzolato, N., Maggio, A., Micela, G., Sciortino, S., & Ventura, P., 2003, *A&A* 397, 147
Reiners, A. & Basri, G., 2006, *ApJ* 644, 497
Reiners, A. & Basri, G., 2007, *ApJ* 656, 1121
Reiners, A., Schmitt, J. H. M. M., & Liefke, C. 2007, *A&A* 466, L13
Reiners, A., Basri, G., & Browning, M., 2009, *ApJ*, in press, arXiv.org:0810.5139
Reiners, A. & Basri, G., 2009, submitted
Robinson, R. D., 1980, *ApJ* 239, 961
Saar, S. H., 1996, in IAU Symp. 176, *Stellar Surface Structure*, Strassmeier, K. G., Linsky, J. L. (eds.), Kluwer, p.237
Saar, S. H., 2001, ASP Conf. Ser. 223, 292
Valenti, J. A., & Johns-Krull, C., 2001, ASP Conf. Ser. 248, 179

Ansgar Reiners

Christopher Johns-Krull

Cosmic Magnetic Fields:
From Planets, to Stars and Galaxies
Proceedings IAU Symposium No. 259, 2008
K.G. Strassmeier, A.G. Kosovichev & J.E. Beckman, eds.

doi:10.1017/S1743921309030713

Measuring T Tauri star magnetic fields

Christopher M. Johns–Krull

Department of Physics and Astronomy, Rice University, Houston, TX 77005, USA
email: cmj@rice.edu

Abstract. Stellar magnetic fields including a strong dipole component are believed to play a critical role in the early evolution of newly formed stars and their circumstellar accretion disks. It is currently believed that the stellar magnetic field truncates the accretion disk several stellar radii above the star. This action forces accreting material to flow along the field lines and accrete onto the star preferentially at high stellar latitudes. It is also thought that the stellar rotation rate becomes locked to the Keplerian velocity near the radius where the disk is truncated. This paper reviews recent efforts to measure the magnetic field properties of low mass pre-main sequence stars, focussing on how the observations compare with the theoretical expectations. A picture is emerging indicating that quite strong fields do indeed cover the majority of the surface on these stars; however, the dipole component of the field appears to be alarmingly small. The current measurements also suggest that given their strong magnetic fields, T Tauri stars are somewhat faint in X-rays relative to what is expected from simple main sequence star scaling laws.

Keywords. Accretion – accretion disks – stars: formation – stars: magnetic fields – stars: pre-main-sequence

1. Introduction

It is now generally accepted that accretion of circumstellar disk material onto the surface of a classical T Tauri star (CTTS) is controlled by a strong stellar magnetic field (e.g. see review by Bouvier *et al.* 2007). The first detailed magnetospheric accretion model for CTTSs was developed by Uchida & Shibata (1984). This model includes both accretion of disk material onto the star as well as the formation of a shock driven bipolar outflow; however, rotation is ignored. Camenzind (1990) first considered the rotational equilibrium of a CTTS with a kilogauss strength dipolar magnetic field accreting matter from a circumstellar disk. Electric currents in the stellar and disk magnetospheres are found to offset the angular momentum accreted with the disk material, producing an equilibrium rotation rate with the disk truncated close to the corotation radius. A wind is then driven off the disk outside the corotation radius. Variations of this magnetospheric accretion model have been studied analytically or semi-analytically, sometimes without an attendant outflow (Königl 1991; Collier Cameron & Campbell 1993) and sometimes with (Shu *et al.* 1994). In all cases, the field truncates the inner disk at or close to corotation and an equilibrium rotation rate (P_{rot}) is established which depends on the (assumed) dipolar field strength, the stellar mass (M_*), radius (R_*), and the mass accretion rate ($\dot{M}$) in the disk. The relationships published in these papers can be used to predict the stellar field strength on CTTSs for which measurements for the other parameters exist. The predicted field variations from star to star correlate extremely well from study to study, even though the magnitude of the predicted fields can vary substantially from one study to another due to different assumptions regarding the efficiency of the field and disk coupling, ionization state in the disk, and so on (Johns–Krull *et al.* 1999b; Johns–Krull 2007).

Observationally, support for magnetospheric accretion in CTTSs is significant and is reviewed elsewhere in this volume. Despite these successes, open issues remain. Most current theoretical models assume the stellar field is a magnetic dipole with the magnetic axis aligned with the rotation axis. As discussed below, spectropolarimetric measurements are often at odds with this assumption. On the other hand, it is expected that even for complex magnetic geometries, the dipole component of the field should dominate at distance from the star where the interaction with the disk is taking place, so the dipole assumption may not seriously contradict current theory. In the case of the Sun, the dipole component appears to become dominant at $2.5R_\odot$ or closer (e.g. Luhmann *et al.* 1998). For expected disk truncation radii of $3 - 10\ R_*$ in CTTSs, this suggests the dipole component will govern the stellar interaction with the disk. Additionally, Gregory *et al.* (2006) show that accretion can occur from a truncated disk even when the stellar field geometry is quite complex; however, no study has considered the torque balance between a star and its disk in the case of a complex stellar field geometry. Another concern is the work of Stassun *et al.* (1999) who find no correlation between rotation period and the presence of an infrared (IR) excess indicative of a circumstellar disk in a sample of 254 stars in Orion. However, IR excess alone is not a good measure of the accretion rate. Muzerolle, Calvet & Hartmann (2001) note that current theory predicts a correlation between rotation period and mass accretion rate which they do not observe. Muzerolle *et al.* (2001) suggest that variations in the stellar magnetic field strength from star to star may account for the lack of correlation. Indeed, there are several stellar and accretion parameters that enter into the equilibrium relationship, and the stellar magnetic field remains the quantity measured for the fewest number of CTTSs. Here, we review magnetic field measurements on TTSs, paying particular attention to how the magnetic field data agrees or not with the predictions of magnetospheric accretion models for young stars. We refer the reader to the contribution by Wade in this volume for a review of magnetic field measurements on higher mass stars.

2. Techniques

Virtually all measurements of stellar magnetic fields make use of the Zeeman effect. Typically, one of two general aspects of the Zeeman effect is utilized: (1) Zeeman broadening of magnetically sensitive lines observed in intensity spectra, or (2) circular polarization of magnetically sensitive lines. When an atom is in a magnetic field, different projections of the total electron angular momentum are no longer degenerate, shifting the energy levels taking part in the transition. In the simple Zeeman effect, a spectral line splits into 3 components: 2 σ components split to either side of the nominal line center and 1 unshifted π component. The wavelength shift of a given σ component is

$$\Delta\lambda = \frac{e}{4\pi m_e c^2}\lambda^2 g B \tag{2.1}$$

where g is the Landé g-factor of the specific transition, B is the strength of the magnetic field, and λ is the wavelength of the transition. Evaluating the constants, the wavelength shift is

$$\Delta\lambda = 4.67 \times 10^{-7}\lambda^2 g B \text{mÅ} \tag{2.2}$$

where λ is in Å and B is in kG. One thing to note from this equation is the λ^2 dependence of the Zeeman effect. Compared with the λ^1 dependence of Doppler line broadening mechanisms such as rotation and turbulence, this means that observations in the IR are generally more sensitive to the presence of magnetic fields than optical observations.

The simplest model of the spectrum from a magnetic star assumes that the observed line profile can be expressed as $F(\lambda) = F_B(\lambda) \times f + F_Q(\lambda) \times (1 - f)$; where F_B is the spectrum formed in magnetic regions, F_Q is the spectrum formed in non-magnetic (quiet) regions, and f is the flux weighted surface filling factor of magnetic regions. The magnetic spectrum, F_B, differs from the spectrum in the quiet region not only due to Zeeman broadening of the line, but also because magnetic fields can in principle affect atmospheric structure, causing changes in both line strength and continuum intensity at the surface. Most studies assume that the magnetic atmosphere is in fact the same as the quiet atmosphere because there is no theory to predict the structure of the magnetic atmosphere. If the stellar magnetic field is very strong, the splitting of the σ components is a substantial fraction of the line width, and it is easy to see the σ components sticking out on either side of a magnetically sensitive line. Such is the case for many active M dwarfs (e.g. Johns–Krull & Valenti 1996). In this case, it is relatively straightforward to measure the magnetic field strength, B. Differences in the atmospheres of the magnetic and quiet regions primarily affect the value of f. If the splitting is a small fraction of the intrinsic line width, then the resulting observed profile is only subtly different from the profile produced by a star with no magnetic field, and more complicated modelling is required to be sure all possible non-magnetic sources (e.g. rotation, pressure broadening, turbulence) have been properly constrained.

In cases where the Zeeman broadening is too subtle to detect directly in line profile analysis, it is still possible to diagnose the presence of magnetic fields through their effect on the equivalent width of magnetically sensitive lines. For strong lines, the Zeeman effect moves the σ components out of the partially saturated core into the line wings where they can effectively add opacity to the line and increase the equivalent width. The exact amount of equivalent width increase is a complicated function of the line strength and the true Zeeman splitting pattern (Basri *et al.* 1992). This method is primarily sensitive to the product of B multiplied by the filling factor f. Since this method relies on relatively small changes in the line equivalent width, it is very important to be sure other atmospheric parameters which affect equivalent width (particularly temperature) are accurately measured.

Measuring circular polarization in magnetically sensitive lines is perhaps the most direct means of detecting magnetic fields on stellar surfaces, but it is also subject to several limitations. When viewed along the axis of a magnetic field, the Zeeman σ components are circularly polarized, but with opposite helicity; and the π component(s) is(are) absent. The helicity of the σ components reverses as the polarity of the field reverses. Thus, on a star like the Sun that typically displays equal amounts of + and - polarity fields on its surface, the net polarization is very small. If one magnetic polarity does dominate the visible surface of the star, net circular polarization is present in Zeeman sensitive lines, resulting in a wavelength shift between the line observed through right- and left-circular polarizers. The magnitude of the shift represents the surface averaged line of sight component of the magnetic field (which on the Sun is typically less than 4 G even though individual magnetic elements on the solar surface range from $\sim$ 1.5 kG in plage to $\sim$3.0 kG in spots). Several polarimetric studies of cool stars have generally failed to detect circular polarization, placing limits on the net magnetic field strength present of 10 - 100 G (e.g. Vogt 1980; Brown & Landstreet 1981; Borra, Edwards & Mayor 1984). The interpretation resulting from these studies is that the late-type stars studied (primarily main sequence and RS CVn types) likely have complicated surface magnetic field topologies which display approximately equal amounts of opposite polarity field which

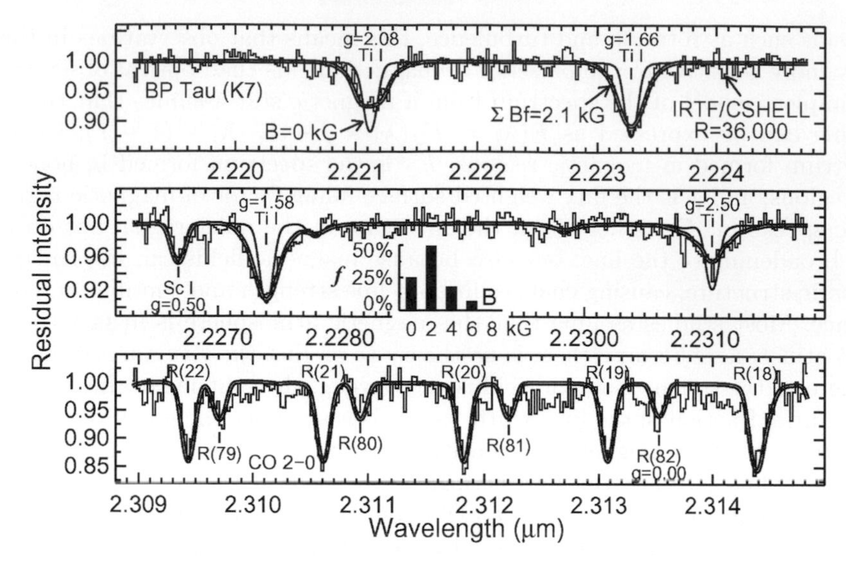

Figure 1. An IRTF/CSHELL spectrum of the K7 CTTS BP Tau (histogram) is compared with synthetic spectra based on magnetic (doubled curve) and nonmagnetic (single curve) models. Zeeman insensitive CO lines are well fit by both models. The Zeeman sensitive Ti I lines are much broader than predicted by the nonmagnetic model. The magnetic model reproduces the observed spectrum, using a distribution of magnetic field strengths (inset histogram) with a mean of 2.1 kG (Johns–Krull 2007) over the entire stellar surface. The effective Landé-g factor is given for each atomic line.

results in no detectable net magnetic field. On the other hand, stars with strong dipole components, such as the magnetic Ap stars, show quite strong circular polarization in their photospheric absorption lines (e.g. Mathys 2004 and references therein). If CTTSs do have strong dipole components, circular polarization should be detectable in photospheric absorption lines.

3. Zeeman broadening measurements

3.1. *The equivalent width method*

TTSs typically have $v\sin i$ values of 10 km s^{-1}, which means that observations in the optical typically cannot detect the actual Zeeman broadening of magnetically sensitive lines because the rotational broadening is too strong. Nevertheless, optical observations can be used with the equivalent width technique to detect stellar fields. Basri *et al.* (1992) were the first to detect a magnetic field on the surface of a TTS. They studied two TTSs showing no evidence for accretion, the so-called weak line or naked TTSs (WTTSs or NTTSs). Basri *et al.* find a value of $Bf = 1.0$ kG on the NTTS Tap 35. In addition to Tap 35, Basri *et al.* also observed the NTTS Tap 10, finding only an upper limit of $Bf < 0.7$ kG. Guenther *et al.* (1999) apply the same technique to spectra of 5 TTSs, with apparent significant field detections on two stars; however, these authors analyze their data using models different by several hundred K from the expected effective temperature of their target stars, which can introduce artifacts in equivalent width analyses. In principle, the equivalent width technique can separately measure B and f; however, in practice this is quite difficult and the technique primarily gives a measure of the product Bf (see Basri *et al.* 1992; Guenther *et al.* 1999). While measurements of actual Zeeman broadening as described below can give more detailed information about the magnetic fields on

TTSs, that method is biased towards stars with intrinsically narrow line profiles, and hence is generally less usefull when studying rapidly rotating stars. Line blending makes equivalent width measurements more difficult in rapidly rotating stars as well; however, the equivalent width method used on IR lines (where the density of lines is lower in many regions) is likely to be the only way to get robust mean field measurements on high $v\sin i$ TTSs.

3.2. *Zeeman broadening of infrared lines*

As described above, observations in the IR help solve the difficulty in detecting direct Zeeman broadening. There are two principle IR diagnostics that have been utilized for magnetic field measurements in late-type stars. The first is a series of Zeeman sensitive Fe I lines at 1.56 μm, including one with a Landé-g value of 3.00 at 1.5649 μm (e.g. Valenti *et al.* 1995, Rüedi *et al.* 1995). These Fe I lines have a relatively high excitation potential, and as a result are best used to study G and early K type stars. To date, no TTS magnetic field measurements have been made using these lines; however, Guenther & Emerson (1996) demonstrate the suitability of these lines for TTS magnetic field work and present observations of these lines in Tap 35 which give an upper limit of $Bf < 2000$ G, consistent with the result of Basri *et al.* (1992) described above. For later spectral types such as the majority of TTSs with field measurements, lower excitation potential lines are best. There are several Ti I lines near 2.2 μm which are suitable for magnetic field work on late K and M stars. Saar & Linsky (1985) first made use of these lines to study the magnetic field on the dMe flare star AD Leo. By far, observations of these K band Ti I lines have yielded the most information on the magnetic fields of TTSs, starting with the measurement of the magnetic field on BP Tau given by Johns–Krull *et al.* (1999b). These authors found that the broadening of the Ti I lines in BP Tau could not be well fit assuming a single magnetic field component with some value of B and f. Instead, they find that a distribution of magnetic field strengths is required. For example, one fit includes atmospheric components with field strengths of 0, 2, 4, and 6 kG magnetic fields, with individually determined filling factors which sum to 1.0. This distribution of magnetic field strengths can be characterized by the mean field $\bar{B} = \Sigma B_i f_i = 2.6 \pm 0.3$ kG for BP Tau.

Robust Zeeman broadening measurements require Zeeman insensitive lines to constrain nonmagnetic broadening mechanisms. Fortunately, numerous CO lines at 2.31 μm have negligible Landé-g factors, making them an ideal null reference. These CO lines are well fit by synthetic stellar models with only rotational and turbulent broadening. In contrast, the 2.2 μm Ti I line spectra are best fit by a model with a distribution of field strengths as described above (and see Figure 1). A total of about two dozen TTSs now have magnetic field measurements based on observations of the K band Ti I lines (Johns–Krull *et al.* 1999b; Johns–Krull, Valenti & Saar 2004; Yang, Johns–Krull & Valenti 2005, 2008; Johns–Krull 2007). These studies show that strong magnetic fields appear to be ubiquitous on TTSs. The mean magnetic field strength, $\bar{B}$, of most TTSs is ~ 2.5 kG. Thus, on these low surface gravity stars, the magnetic pressure dominates the photospheric gas pressure (see Johns–Krull *et al.* 2004; Johns–Krull 2007).

4. Spectropolarimetry and magnetic field geometry

Zeeman broadening measurements are sensitive to the distribution of magnetic field strengths, but they have limited sensitivity to magnetic geometry. In contrast, circular polarization measurements for individual spectral lines are sensitive to magnetic

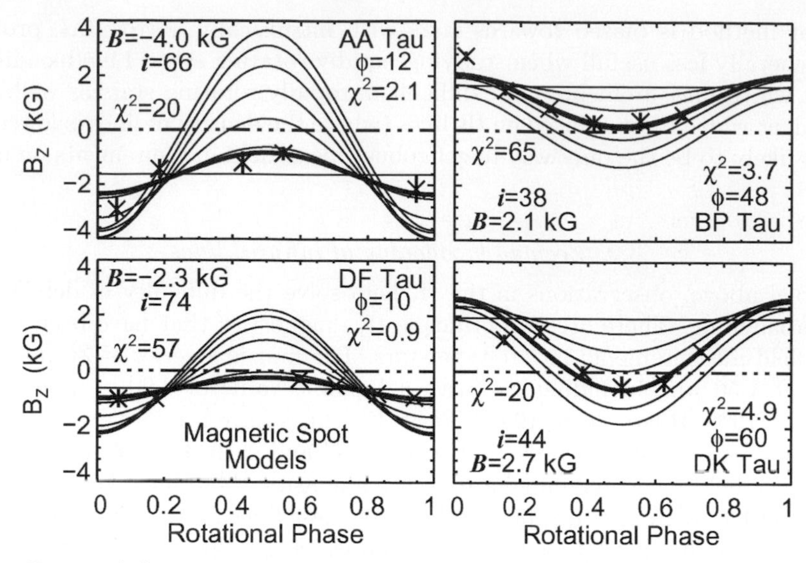

Figure 2. Crosses (×) with vertical error bars indicate the net longitudinal magnetic field (B_Z), measured on six consecutive nights using the He I 5876 Å accretion diagnostic. The family of curves show predicted B_Z values for a simple model with radial magnetic field lines concentrated in a spot a latitude ϕ. Magnetic field strengths are constrained by independent Zeeman broadening measurements. Reduced χ^2 is 1-5 for the best fitting model, shown as a long dashed curve for each star. Reduced χ^2 is 20-60 for a model that assumes $B_Z = 0$.

geometry, but they provide limited information about field strength. The two techniques complement each other well, as we demonstrate below.

4.1. *The photospheric fields of T Tauri stars*

As mentioned above, existing magnetospheric accretion models assume that intrinsic TTS magnetic fields are dipolar; however, this would be unprecedented for cool stars. Nevertheless, the typical mean field (2.5 kG) measured is similar in magnitude to the dipole field strength required to truncate the accretion disk and enforce disk locking. Higher order multi-polar components of the magnetic field should fall off more rapidly with distance, so if the surface field on the star is dominated by higher order components, even stronger surface fields would be required on the star to produce the required field strength at the inner edge of the disk a few stellar radii from the surface of the star. If we assume for the moment that the mean fields described above are in fact dipolar, we can then ask what net longitudinal magnetic field, B_Z, should be measured using spectropolarimetry? The answer depends on the angle the dipole field axis makes with the line of sight. If the field axis is 90° from the line of sight, $B_Z = 0$. If the dipole axis is aligned with the line of sight, $|B_Z| \sim 0.64 B_e$ where B_e is the equatorial value of the dipole field strength (B_e is the predicted field value tabulated in Johns–Krull *et al.* 1999b and Johns–Krull 2007). The exact value of the coefficient depends a little on the value of the limb darkening coefficient used. Assuming a dipolar field geometry observed at an angle of 45° bewteen the field axis and the line of sight, $|B_Z| \sim 800$ G if the mean field strength on the stellar surface is 2.5 kG.

Overall, there are relatively few measurements of B_Z for TTSs. Until recently, T Tau was the only TTS observed polarimetrically, with a 3σ upper limit of $|B_Z| < 816$ G set by Brown & Landstreet (1981). T Tau has been the focus of more recent study: Smirnov *et al.* (2003) report a detection of a net field of 160 ± 40 G on T Tau which was not confirmed by Smirnov *et al.* (2004) or Daou *et al.* (2006). Johnstone & Penston (1986, 1987) set 3σ upper limits on $|B_Z|$ on 3 TTSs: 494 G (RU Lup), 1110 G (GW Ori), and 2022 G (CoD-34 7151). Donati *et al.* (1997) used the rapid rotation of the diskless NTTS V410 Tau to effectively isolate strips on the stellar surface and detect net circular polarization from the star; however, no field strength was ascribed to these results. In addition, Donati *et al.* do not detect polarization on two other rapidly rotating TTSs. Yang *et al.* (2007) detect a net field of $B_Z = 149 \pm 33$ G on TW Hya on one night of their 6 night monitoring campaign on this star, finding only (3σ) upper limits of ~ 100 G on the other nights. Most additional studies also only find upper limits (3σ) of 100 – 200 G on 4 additional CTTSs (Johns–Krull *et al.* 1999a; Valenti & Johns–Krull 2004). Donati *et al.* (2007) do detect a weak global field on the TTS V2129 Oph, and more recently, Donati *et al.* (2008) surprisingly detect a strong global field on the CTTSs BP Tau. In light of the strong magnetic fields measured using Zeeman broadening techniques, the general absence of polarimetric detections strongly suggest the magnetic fields on TTSs are not dipolar, at least at the stellar surface. Again, as higher order terms will fall off more rapidly with distance, it is expected that the dipole component of the field will indeed dominate at distances of several stellar radii. However, measuring the fields at these distances is quite difficult. The only direct field measurement above the surface of a TTSs is the recent detection of circular polarization in the line profiles of FU Ori (Donati *et al.* 2005). Here, the fields detected are likely in the accretion disk, and the measured fields may not be anchored in the star at all.

4.2. *Magnetic fields in accretion shocks on CTTSs*

Johns–Krull *et al.* (1999a) made the surprising discovery of circular polarization in emission line diagnostics that form predominantly in the accretion shock at the surface of CTTSs. This circular polarization signal is strongest in the narrow component of the He I 5876 Å emission line, but it is also present in the Ca II infrared triplet lines (e.g. Yang *et al.* 2005). Valenti & Johns–Krull (2004) detect He I polarization in four CTTSs: AA Tau, BP Tau, DF Tau, and DK Tau. Symington *et al.* (2005) also detect He i polarization at greater than the 3σ level in three stars (BP Tau, DF Tau, and DN Tau) in their survey of seven CTTSs, and Yang *et al.* (2005) detect polarization in this line in the CTTS TW Hya. All these stars are characterized by He I emission lines which have strong narrow components (NCs) to their line profiles (see Edwards *et al.* 1994 or Alencar & Basri 2000 for a discussion of NC and broad component, BC, emission in CTTSs). Smirnov *et al.* (2004) reported detections of circular polarization in the He I 5876 Å emission line of T Tau on all 3 nights they observed the star, though with significant variability from one night to the next (field measurements range from +350 G to +1100 G with no uncertainty estimates). T Tau's He I line is dominated by BC emission. Daou *et al.* (2006) observed T Tau on 2 nights, finding field values in the He I line of -29 ± 116 G on one night and -43 ± 300 G on the second. TW Hya's He I line has a significant broad component, but Yang *et al.* (2005) do not report any polarization in this part of the line.

The NC of the He I emission is commonly associated with the accretion shock itself at the stellar surface, whereas the BC may have contributions from the magnetospheric accretion flow and/or a hot wind component (e.g. Beristain, Edwards & Kwan 2001). Since the BC of the He I emission line forms over a large, extended volume, its magnetic field strength should be weaker than at the stellar surface. In addition field line curvature may enhance polarization cancellation in the BC. As a result, circular polarization in the BC of the He I 5876 Å emission is predicted to be less than in the NC. Therefore, the result of Smirnov *et al.* (2004) for T Tau is quite surprising. Additional observations of T Tau and other CTTSs dominated by BC emission are needed to confirm the polarization

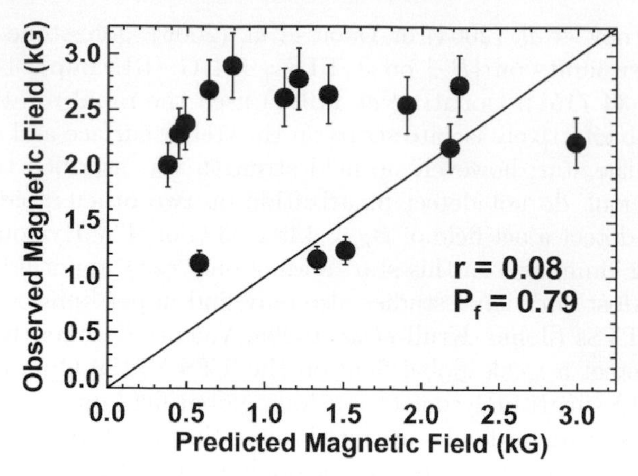

Figure 3. Measured mean magnetic fields as diagnosed by K band Ti I line profiles versus predicted fields using magnetospheric accretion models which assume disk-locking in CTTSs (taken from Johns–Krull 2007).

detections. Such observations can strongly constrain the formation region of the BC emission. For example, it is difficult to see how formation of this component over an extend region such as in a hot wind can produce significant polarization characteristic of magnetic field strengths in excess of 1000 G.

The polarization measured in the He I lines is observed to be variable. Figure 2, taken from Valenti & Johns–Krull (2004), shows measurements of B_Z determined from the He I line on 6 consecutive nights for 4 CTTSs. The field values in the He I line vary smoothly on rotational timescales, suggesting that uniformly oriented magnetic field lines in accretion regions sweep out a cone in the sky as the star rotates. Rotational modulation implies a lack of symmetry about the rotation axis in the accretion or the magnetic field or both. For example, the inner edge of the disk could have a concentration of gas that corotates with the star, preferentially illuminating one sector of a symmetric magnetosphere. Alternatively, a single large scale magnetic loop could draw material from just one sector of a symmetric disk. Many variants are possible. Figure 2 shows one interpretation of the He I polarization data. Predicted values of B_Z are shown for a simple model consisting of a single magnetic spot at latitude ϕ that rotates with the star. The magnetic field is assumed to be radial with a strength equal to the measured values of B. Inclination of the rotation axis is constrained by measured $v\sin i$ and rotation period, except that inclination (i) is allowed to float when it exceeds 60° because $v\sin i$ measurements cannot distinguish between these possibilities. Predicted variations in B_Z are plotted for spot latitudes ranging from 0° to 90° in 15° increments. The best fitting model is the heavy curve. The corresponding spot latitude and reduced χ^2 are given on the right side of each panel. Large values of χ^2 on the left side of each panel rule out the hypothesis that no polarization signal is present. In all four cases, this simple magnetic spot model reproduces the observed He I time series. Similar results are found by Symington *et al.* (2005).

5. Comparing the measurements with the theory

At first glance, it might appear that magnetic field measurements on TTSs are generally in good agreement with theoretical expectations. IR Zeeman broadening measurements indicate mean fields on several TTS of ~ 2 kG, similar in value to those predicted by magnetospheric accretion models (Johns–Krull *et al.* 1999b; Johns–Krull 2007).

However, in detail the field observations do not agree with some aspects of the theory. This is shown in Figure 3 where the measured mean magnetic field strengths are plotted versus predicted field strengths. Clearly, the measured field strengths show no correlation with the predicted ones. The field topology measurements give some indication to why there may be a lack of correlation: the magnetic fields on TTSs are not dipolar. On the other hand, the smoothly varying polarization detected in the He i accretion shock lines suggests that the region where the disk interacts with the stellar field is dominated by a simple magnetic field geometry. Since the dipole component of the field falls off the least rapidly with distance, it may well be that the stellar field at the disk truncation radius is dominated by the dipole component. The disk material then loads onto these field lines and accretes onto the star, landing at the surface in those regions which contribute the the large scale dipolar field. Perhaps then, the correct correlation to look for is between the predicted fields and the dipole component, or the predicted fields and the field in the He I region? Currently, there are not enough reliable measurements of either of these field diagnostics to look for such a correlation.

In addition to comparisons with magnetospheric accretion predictions, we can also use the magnetic field measurements of TTSs to study magnetic heating on these young stars. Using solar active regions, X-ray bright points, and other solar features in combination with observations of main sequence G and K stars, Pevtsov *et al.* (2003) show that there is an excellent correlation between total X-ray luminosity and total magnetic flux. The strong mean magnetic fields found on TTSs combined with the large radius of these pre-main sequence stars implies quite large total magnetic flux, and using the Pevtsov *et al.* relationship, correspondingly large X-ray luminosities for TTSs. Indeed, TTSs are generally X-ray bright; however, their observed X-ray luminosities fall systematically below the level predicted by the Pevtsov *et al.* (2003) relationship, given the large magnetic flux on these stars (Johns–Krull 2007, Yang et a. 2008). This indicates that magnetic heating on TTSs is somewhat less efficient on these stars compared to the heating on main sequence stars. This may actually be a back reaction of the very strong fields on these stars hindering the rate at which convective gas motions on TTSs can push around magnetic fields on their surfaces, building up magnetic stresses and producing heating. This could explain why TTSs are observed to saturate at X-ray emission levels that are lower (though with a great deal of scatter) than is typically found for low mass main sequence stars (Feigelson *et al.* 2003).

6. Conclusion

The current magnetic field measurements show that the strong majority of TTSs are covered by kilogauss magnetic fields. The observations also suggest these fields manifest themselves in a complicated surface topology and that the dipole component of the field is likely small on TTSs. Despite this surface complication, fields measured in the accretion shock on CTTSs suggest that the disk interacts with a primarily dipolar field geometry several stellar radii above the star. However, it is likely the strength of this dipole component is substantially weaker than current models require.

Acknowledgements

I am pleased to acknowledge numerous, stimulating discussions with J. Valenti on all aspects of the work reported here. I also wish to acknowledge partial support from the NASA Origins of Solar Systems program through grant numbers NAG5-13103 and NNG06GD85G made to Rice University.

References

Alencar, S. & Basri, G. 2000, *AJ* 119, 1881
Bouvier, J., Alencar, S. H. P., Harries, T. J., Johns–Krull, C. M., & Romanova, M. M. 2007, Protostars and Planets V, 479
Brown, D. N. & Landstreet, J. D. 1981, *ApJ* 246, 899
Camenzind, M. 1990, *Rev. Mod. Ast.* 3, 234
Collier Cameron, A. & Campbell, C. G. 1993, *A&A* 274, 309
Daou, A. G., Johns–Krull, C. M., & Valenti, J. A. 2006, *AJ* 131, 520
Donati, J.-F., Paletou, F., Bouvier, J., & Ferreira, J. 2005, *Nature* 438, 466
Donati, J.-F., Semel, M., Carter, B. D., Rees, D. E., & Collier Cameron, A. 1997, *MNRAS* 291, 658
Donati, J.-F., *et al.* 2007, *MNRAS* 380, 1297
Donati, J.-F., *et al.* 2008, *MNRAS* 386, 1234
Edwards, S., Hartigan, P., Ghandour, L., & Andrulis, C. 1994, *AJ* 108, 1056
Feigelson, E. D., Gaffney, J. A., Garmire, G., Hillenbrand, L. A., & Townsley, L. 2003, *ApJ* 584, 911
Gregory, S. G., Jardine, M., Simpson, I., & Donati, J.-F. 2006, *MNRAS* 371, 999
Guenther, E. W. & Emerson, J. P. 1996, *A&A* 309, 777
Guenther, E. W., Lehmann, H., Emerson, J. P., & Staude, J. 1999, *A&A* 341, 768
Johns-Krull, C. M. 2007, *ApJ* 664, 975
Johns-Krull, C. M. & Valenti, J. A. 1996, *ApJ* 459, L95
Johns-Krull, C. M., Valenti, J. A., & Koresko, C. 1999b, *ApJ* 516, 900
Johns-Krull, C. M., Valenti, J. A., Hatzes, A. P., & Kanaan, A. 1999a, *ApJL* 510, L41
Johns-Krull, C. M., Valenti, J. A., & Saar, S. H. 2004, *ApJ* 617, 1204
Johnstone, R. M. & Penston, M. V. 1986, *MNRAS* 219, 927
Johnstone, R. M. & Penston, M. V. 1987, *MNRAS* 227, 797
Königl, A. 1991, *ApJL* 370, L39
Luhmann, J. G., Gosling, J. T., Hoeksema, J. T., & Zhao, X. 1998, *JGR* 103, 6585
Mathys, G. 2004, The A-Star Puzzle: Proc. IAUS 224, 225
Muzerolle, J., Calvet, N., & Hartmann, L. 2001, *ApJ* 550, 944
Pevtsov, A. A., *et al.* 2003, *ApJ* 598, 1387
Rüedi, I., Solanki, S. K., & Livingston, W. 1995, *A&A* 302, 543
Saar, S. H. & Linsky, J. L. 1985, *ApJL* 299, L47
Shu, F. H., Najita, J., Ostriker, E., Wilkin, F., Ruden, S., & Lizano, S. 1994, *ApJ* 429, 781
Smirnov, D. A., Fabrika, S. N., Lamzin, S. A., & Valyavin, G. G. 2003, *A&A* 401, 1057
Smirnov, D. A., Lamzin, S. A., Fabrika, S. N., & Chuntonov, G. A. 2004, *Astronomy Letters* 30, 456
Stassun, K. G., Mathieu, R. D., Mazeh, T., & Vrba, F. J. 1999, *AJ* 117, 2941
Symington, N. H., Harries, T. J., Kurosawa, R., & Naylor, T. 2005, *MNRAS* 358, 977
Valenti, J. A., Basri, G., & Johns, C. M. 1993, *AJ* 106, 2024
Valenti, J. A. & Johns-Krull, C. M. 2004, *Ap&SS* 292, 619
Valenti, J. A., Marcy, G. W., & Basri, G. 1995, *ApJ* 439, 939
Vogt, S. S. 1980, *ApJ* 240, 567
Uchida, Y. & Shibata, K. 1984, *PASJ* 36, 105
Yang, H., Johns-Krull, C. M., & Valenti, J. A. 2005, *ApJ* 635, 466
Yang, H., Johns-Krull, C. M., & Valenti, J. A. 2007, *AJ* 133, 73
Yang, H., Johns-Krull, C. M., & Valenti, J. A. 2008, *AJ* 136, 2286

Discussion

De Gouveia dal Pino: The polarimetric observations seem to suggest that the stellar **B** has a more complex structure than a dipole one. However, is it possible that this could be due to the presence of the magnetized-disk contribution in the inner disk region?

JOHNS-KRULL: I don't think so. The observations to date do not show a significant difference between the stellar fields of T Tauri stars with disks compared to flows of T Tauri stars without disks.

ZINNECKER: You mentioned high resolution infrared observations at Gemini-South with the Phoenix spectrograph. Can you summarize for us which are the most important infrared lines for magnetic field measurements in T Tauri stars in the era of 8m class telescopes?

JOHNS-KRULL: For late K and early M stars (to about M3), there are approximately 6 TiO lines near 2.2μm that are very good. For early K and G stars, these lines get weak, but there are some very Zeeman sensitive Fe I lines near 1.56μm that are strong in warmer stars.

HUSSAIN: When you measure the magnetic fields of stars you fit a distribution of magnetic fields. How unique is this distribution?

JOHNS-KRULL: Limiting the solution to field regions separated by 2 kG, so 0, 2, 4, 6 kG, the solutions are quite unique. For finer magnetic field spacing, the solutions become degenerate. Note though, we can't really tell the difference between a 0, 2, 4, 6 kG solution and a 1, 3, 5, 7-kG solution, but both will give very similar mean fields, different by less than 10%.

BLACKMAN: You showed the tight correlation between X-ray luminosity and magnetic flux. What does the $L_{\rm bol}$ vs. L_x plot look like ($L_x/L_{\rm bol}$ for the Sun is 10^{-6} whereas for T-Tauri stars I thought it to be as high as 10^{-3})?

JOHNS-KRULL: $L_{\rm bol}$ and L_x are well correlated. For ZAMS stars (Pleiades, Hyades), $L_x/L_{\rm bol}$ saturates at 10^{-3}. For the COUP data in Orion from Feigelson *et al.*, there is a large spread, but $L_x/L_{\rm bol}$ appears to saturate at a mean of $\approx 10^{-4}$ or so, so TTS do appear a little low in their peak X-ray emission compared to other cool stars.

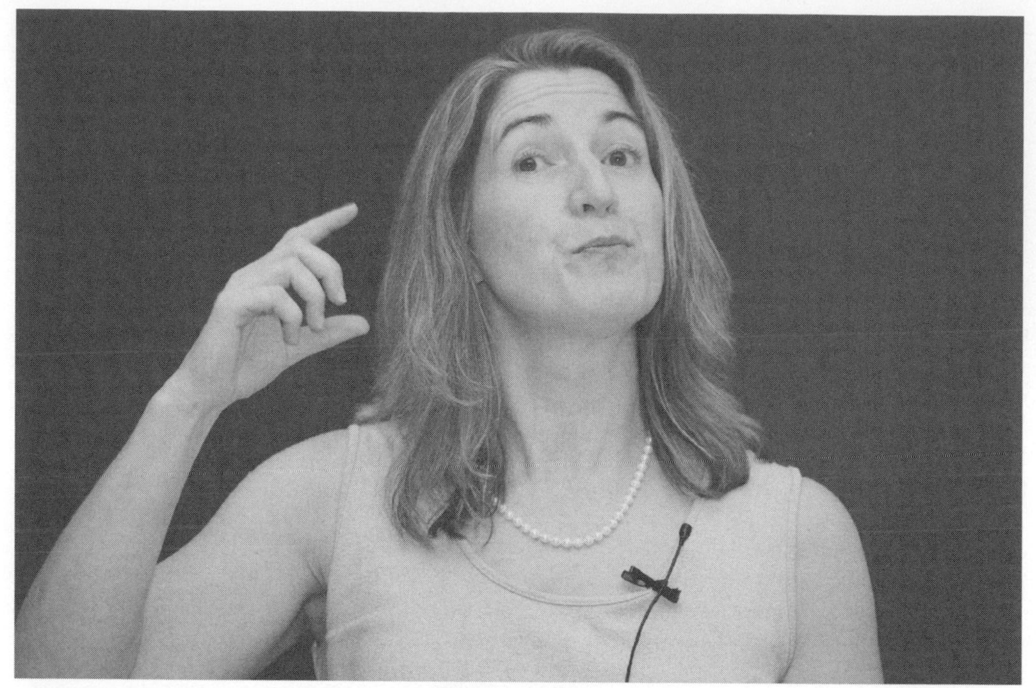

Moira Jardine

Stefan Jordan (left), and Matthias Steffen (right)

Cosmic Magnetic Fields:
From Planets, to Stars and Galaxies
Proceedings IAU Symposium No. 259, 2008
K.G. Strassmeier, A.G. Kosovichev & J.E. Beckman, eds.

doi:10.1017/S1743921309030725

Magnetic coronae of active main-sequence stars

Moira Jardine[1] **and Jean-Francois Donati**[2]

[1]SUPA, School of Physics and Astronomy, University of St Andrews, North Haugh, St Andrews, KY16 9SS, UK
email: mmj@st-andrews.ac.uk

[2]LATT, CNRS–UMR 5572, Obs. Midi-Pyrénées, 14 Av. E. Belin, F-31400 Toulouse, France
email: donati@ast.obs-mip.fr

Abstract. The coronal structure of main sequence stars continues to puzzle us. While the solar corona is relatively well understood, it has become clear that even stars of the same mass as the Sun can display very non-solar coronal behaviour, particularly if they are rapid rotators or in a binary system. At masses greater than and also less than that of the Sun, the non-solar internal structure appears to affect both the geometry and dynamics of the stellar corona and the nature of the X-ray and radio emission. In this talk I will describe some recent advances in our understanding of the structure of the coronae of some of the most active (and interesting) main sequence stars.

Keywords. Stars: magnetic fields – stars: coronae – stars: imaging – stars: spots

---

1. Introduction

During the course of this Symposium we have already heard about the range of surface magnetic fields that can be found on different types of stars across the HR diagram. In this talk I want to address the slightly different question of the nature of the coronae of active stars on the main sequence. In the last 5 to 10 years it has become apparent that the change in internal structure with mass along the main sequence can have profound effects on the types of magnetic fields that can be generated. At the high-mass end of the main sequence, the presence of significant magnetic fields on some stars with radiative interiors challenges current dynamo theories that rely on convective processes. One possibility is that they generate fields in their convective cores, although this raises the question of how to transport the flux to the surface (Charbonneau & MacGregor 2001; Brun *et al.* 2005). They can also generate magnetic fields in the radiative zone, but a very non-solar dynamo process (Spruit 2002; Tout & Pringle 1995; MacDonald & Mullan 2004; Mullan & MacDonald 2005; Maeder & Meynet 2005), Alternatively, the fields may be fossils, left over from the early stages of the formation of the star (Moss 2001; Braithwaite & Spruit 2004; Braithwaite & Nordlund 2006).

The fossil field explanation is perhaps the one most likely to explain the complex field detected on Tau Sco (Donati *et al.* 2006b). At 15 $M_{\odot}$ it has a radiative interior, and yet as shown in Fig. 1 it displays a complex, strong field. If this is a fossil field, it might be expected to be a simple dipole, but the very youth of this star, at only a million years, may be the reason why the higher-order field components have not yet decayed away. Interestingly, Tau Sco shows H_{α} absorption features that are very similar to prominence signatures in lower mass stars. These prominences are clouds of cool, mainly neutral hydrogen, confined within the million-degree gas of the stellar corona. They are observed as transient absorption features crossing the H-α profile when the prominence crosses in

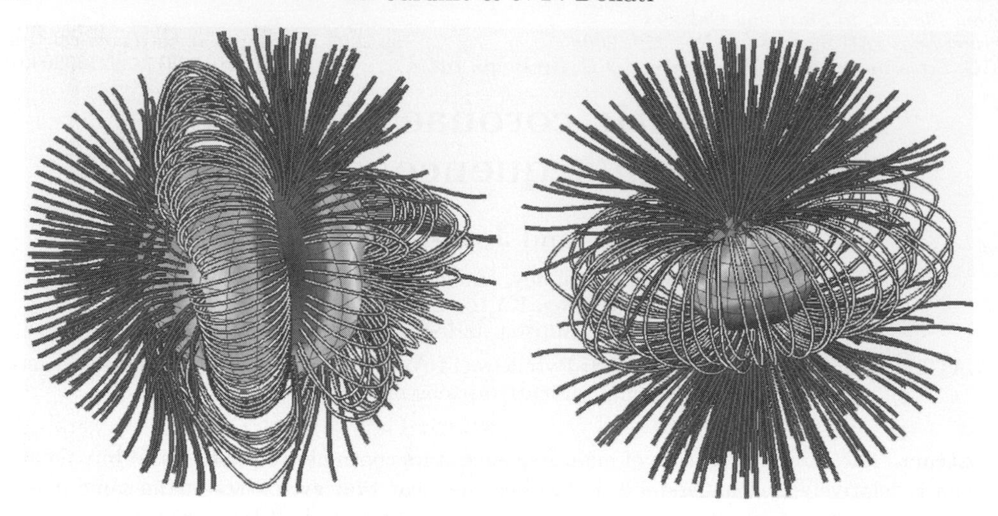

Figure 1. Closed field lines (white) and open field lines (blue) extrapolated from a Zeeman-Doppler image of Tau Sco (left) and V374 Peg (right).

front of the stellar disk as seen by an observer (Collier Cameron & Robinson 1989b,a; Collier Cameron & Woods 1992; Jeffries 1993; Byrne *et al.* 1996; Eibe 1998; Barnes *et al.* 2000; Donati *et al.* 2000). High mass stars like Tau Sco also show similar H_α absorption features, but in this case they are attributed to a "wind-compressed disk" that forms when sections of the very massive wind emanating from different parts of the stellar disk collide and cool (Townsend & Owoki 2005).

Very low mass stars also have an internal structure that is very different from that of the Sun in that convection may extend throughout their interiors. In the absence of a tachocline, these stars cannot support a solar-like interface dynamo, yet they, like the high mass stars, exhibit observable magnetic fields. The mechanism by which they generate these magnetic fields has received a great deal of attention recently. While a decade or so ago, it was believed that these stars could only generate small-scale magnetic fields (Durney *et al.* 1993; Cattaneo 1999), more recent studies have suggested that large scale fields may be generated. These models differ, however, in their predictions for the form of this field and the associated latitudinal differential rotation. They predict that the fields should be either axisymmetric with pronounced differential rotation (Dobler *et al.* 2006), non-axisymmtric with minimal differential rotation (Küker & Rüdiger 1997, 1999; Chabrier & Küker 2006).

In contrast, as shown in Fig. 1 the very low mass fully-convective star V374 Peg has a very simple, dipolar field (Donati *et al.* 2006a). The highly-symmetric nature of the field and the absence of a measureable differential rotation are consistent with the recent models of Browning (2008). It is unfortunately not possible at present to detect any prominences that might be present on these very low mass stars because they stars are intrinsically too faint. Their detection would, however, be a very clear test of the magnetic structure, since in a simple dipole any prominences should, by symmetry, form in the equatorial plane.

2. How do we model stellar coronae?

In this short talk I will not describe any further the magnetic fields of stars at the extreme ends of the main sequence. I will concentrate instead on the solar mass stars for

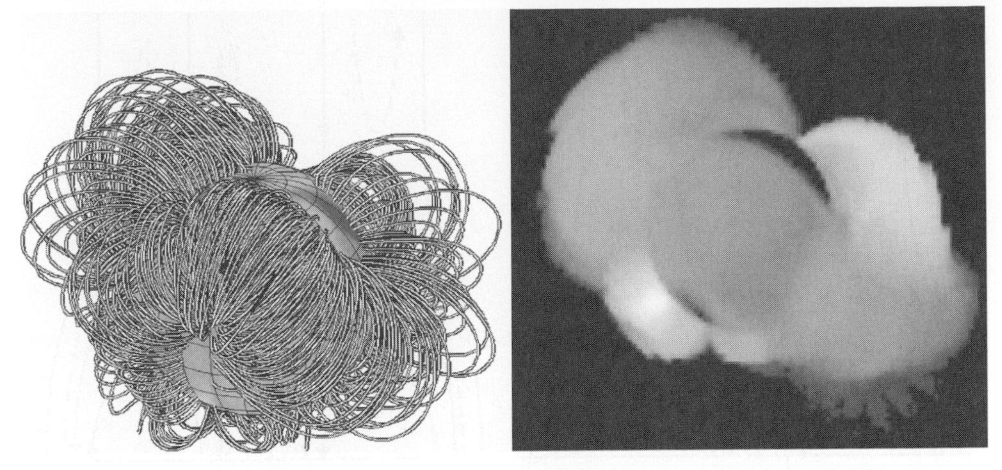

Figure 2. Closed field lines (left) and corresponding X-ray image (right) for the rapidly-rotating star LQ Hya. A coronal temperature of 10^6K is assumed.

which many more observations exist. So how do we construct models of stellar coronae? The first step is to obtain a map of the surface magnetic field. This is most commonly done using the technique of Zeeman-Doppler imaging (Donati & Collier Cameron 1997; Donati *et al.* 1999). These maps typically show a complex distribution of surface spots that is often very different from that of the Sun, with spots and mixed polarity flux elements extending over all latitudes up to the pole (Strassmeier 1996).

From this magnetogram we can extrapolate the coronal magnetic field using a *Potential Field Source Surface* method (Altschuler & Newkirk, Jr. 1969; Jardine *et al.* 1999, 2001, 2002a; McIvor *et al.* 2003), or using non-potential fields (Donati 2001; Hussain *et al.* 2002). By assuming that the gas trapped on these field lines is in isothermal, hydrostatic equilibrium, we can determine the coronal gas pressure, subject to an assumption for the gas pressure at the base of the corona. We assume that it is proportional to the magnetic pressure, i.e. $p_0 \propto B_0^2$, where the constant of proportionality is determined by comparison with X-ray emission measures (Jardine *et al.* 2002b, 2006; Gregory *et al.* 2006). For an optically thin coronal plasma, this then allows us to produce images of the X-ray emission, as shown in Fig.2.

This immediately highlights one of the greatest puzzles about active solar mass stars, which is the extent of their coronae. X-ray spectra reveal coronal densities so high that they can only be confined in a compact corona (Dupree *et al.* 1993; Schrijver *et al.* 1995; Brickhouse & Dupree 1998; Maggio *et al.* 2000; Güdel *et al.* 2001; Sanz-Forcada *et al.* 2003). Despite this, as discussed in the introduction, these stars often harbour so-called "slingshot prominences". In many instances these prominences re-appear on subsequent stellar rotations, often with some change in the time taken for the absorption feature to travel through the line profile. As many as six may be present in the observable hemisphere. What is most surprising about them is their location, which is inferred from the time taken for the absorption features to travel through the line profile. Values of several stellar radii from the stellar rotation axis are typically found, suggesting that the confinement of these clouds is enforced out to very large distances. Indeed the preferred location of these prominences appears to be at or beyond the equatorial stellar co-rotation radius, where the inward pull of gravity is exactly balanced by the outward pull of centrifugal forces. Beyond this point, the effective gravity (including the centrifugal acceleration) points outwards and the presence of a restraining force, such as the

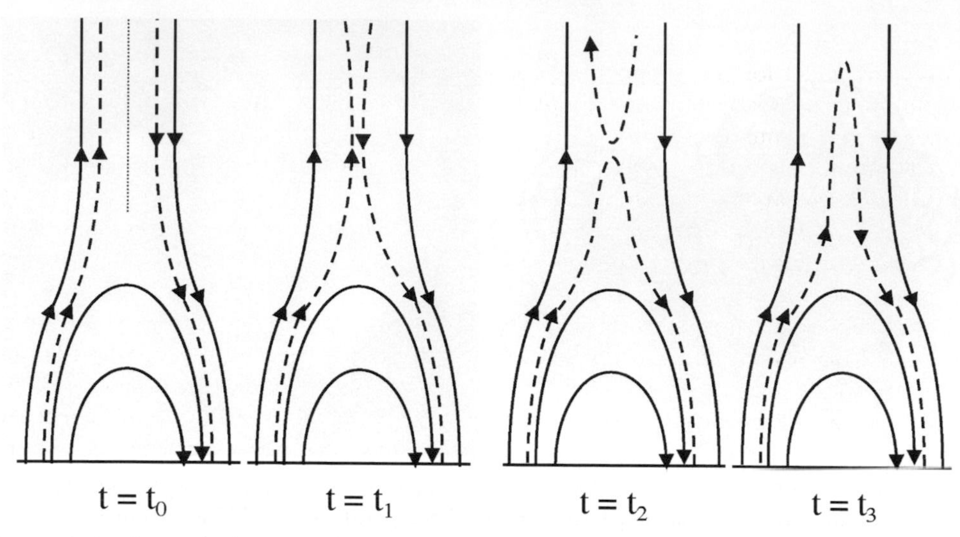

Figure 3. A schematic diagram of the formation of prominence-bearing loops. Initially, at $t = t_0$ a current sheet is present above the cusp of a helmet streamer. Reconnection at in the current sheet at $t = t_1$ produces a closed loop at $t = t_2$. The stellar wind continues to flow until pressure balance is restored, thus increasing the density in the top of this new loop. Increased radiative losses cause the loop to cool and the change in internal pressure forces it to a new equilibrium at $t = t_3$.

tension in a closed magnetic loop, is required to hold the prominence in place against centrifugal ejection. The presence of these prominences therefore immediately requires that the star have many closed loop systems that extend out for many stellar radii. It is difficult to reconcile this need for an extended corona with the high densities inferred from X-ray studies.

One way out of this problem is to confine the prominences in the wind region *beyond* the closed corona. Jardine & van Ballegooijen (2005) have produced a model for this that predicts a maximum height y_m for the prominence as a function of the co-rotation radius, y_K where

$$\frac{y_m}{R_\star} = \frac{1}{2}\left(-3 + \sqrt{1 + \frac{8GM_\star}{R_\star^3\omega^2}}\right) \tag{2.1}$$

$$= \frac{1}{2}\left(-3 + \sqrt{1 + 8\left[\frac{y_K}{R_\star} + 1\right]^3}\right). \tag{2.2}$$

Fig. 3 shows the sequence of events that might lead to the formation of one of these "slingshot" prominences. The stellar wind flows along the open field lines that bound a closed field region, forming a helmet streamer. If the current sheet that forms between these oppositely-directed field lines reconnects, then a loop of magnetic field will be formed. The stellar wind will continue to flow for a short time, until pressure balance is re-established with a new field configuration. Jardine & van Ballegooijen (2006) showed that a new, cool equilibrium was possible which could reach out well beyond the co-rotation radius. The distribution of prominence heights shown in Dunstone *et al.* (2008a,b) for the ultra-fast rotator Speedy Mic shows prominences forming up to (but not significantly beyond) this maximum height.

3. Conclusions

This new model for the support of prominences in rapid rotators may indeed resolve the conflicting pieces of evidence about the extent of the stellar corona, allowing for a compact X-ray corona *and* the support of many prominences at large distance from the stellar rotation axis. More observations are clearly needed, however, to test this theory. The nature of the large-scale field is difficult to determine because it may well be relatively dark in X-ray emission. This is the case on the Sun, where the X-ray corona is compact, but the white-light corona (as seen in eclipses) is much more extended. On very young stars that are still accreting, we have the opportunity to observe the large-scale field since it is "illuminated" by the accretion process. On such young stars, it seems that the large-scale field is very simple, even when the X-ray corona is highly structured (Donati *et al.* 2007, 2008; Jardine *et al.* 2008; Gregory *et al.* 2008). The degree of complexity of stellar coronae and the nature of their large-scale field may therefore be a function of their evolutionary state, as well as with their mass when on the main sequence. The change in internal structure that takes place when young stars develop a radiative core may be reflected in the different geometries of the magnetic fields that they generate and hence the structures of their coronae. Low mass stars, of course, remain fully convective and so may retain the simple field structures that they had in their youth even through their main-sequence lifetimes. This may have consequences not only for their X-ray emission, but also for their rotational histories, since their magnetic fields are crucial to their ability to spin down from the pre-main sequence phase. Over the next few years, as information on distributions of stellar rotation rates becomes available through CoRoT, this will undoubtedly prove to be a fruitful line of research.

References

Altschuler, M. D. & Newkirk, Jr., G. 1969, *SP* 9, 131
Barnes, J., Collier Cameron, A., James, D. J., & Donati, J.-F. 2000, *MNRAS* 314, 162
Braithwaite, J. & Nordlund, A. 2006, *A&A* 450, 1077
Braithwaite, J. & Spruit, H. C. 2004, *Nature* 431, 819
Brickhouse, N. & Dupree, A. 1998, *ApJ* 502, 918
Browning, M. 2008, *ApJ*, in press
Brun, A. S., Browning, M. K., & Toomre, J. 2005, *ApJ* 629, 461
Byrne, P., Eibe, M., & Rolleston, W. 1996, *A&A* 311, 651
Cattaneo, F. 1999, *ApJ* 515, L39
Chabrier, G. & Küker, M. 2006, *A&A* 446, 1027
Charbonneau, P. & MacGregor, K. B. 2001, *ApJ* 559, 1094
Collier Cameron, A. & Robinson, R. D. 1989a, *MNRAS* 238, 657
Collier Cameron, A. & Robinson, R. D. 1989b, *MNRAS* 236, 57
Collier Cameron, A. & Woods, J. A. 1992, *MNRAS* 258, 360
Dobler, W., Stix, M., & Brandenburg, A. 2006, *ApJ* 638, 336
Donati, J.-F. 2001, LNP Vol. 573: Astrotomography, Indirect Imaging Methods in Observational Astronomy, 573, 207
Donati, J.-F. & Collier Cameron, A. 1997, *MNRAS* 291, 1
Donati, J.-F., Collier Cameron, A., Hussain, G., & Semel, M. 1999, *MNRAS* 302, 437
Donati, J.-F., Forveille, T., Cameron, A. C., *et al.* 2006a, *Science* 311, 633
Donati, J.-F., Howarth, I. D., Jardine, M. M., *et al.* 2006b, *MNRAS* 370, 629
Donati, J.-F., Jardine, M. M., Gregory, S. G., *et al.* 2007, *MNRAS* 380, 1297
Donati, J.-F., Jardine, M. M., Gregory, S. G., *et al.* 2008, *MNRAS* 386, 1234
Donati, J.-F., Mengel, M., Carter, B., Cameron, A., & Wichmann, R. 2000, *MNRAS* 316, 699
Dunstone, N. J., Hussain, G. A. J., Cameron, A. C., *et al.* 2008a, *MNRAS* 387, 1525
Dunstone, N. J., Hussain, G. A. J., Collier Cameron, A., *et al.* 2008b, *MNRAS* 387, 481

Dupree, A., Brickhouse, N., Doschek, G., Green, J., & Raymond, J. 1993, *ApJ* 418, L41
Durney, B. R., De Young, D. S., & Roxburgh, I. W. 1993, *SP* 145, 207
Eibe, M. T. 1998, *A&A* 337, 757
Gregory, S. G., Jardine, M., Cameron, A. C., & Donati, J.-F. 2006, *MNRAS* 373, 827
Gregory, S. G., Matt, S. P., Donati, J.-F., & Jardine, M. 2008, *MNRAS*, in press
Güdel, M., Audard, M., Briggs, K., *et al.* 2001, *A&A* 365, L336
Hussain, G. A. J., van Ballegooijen, A. A., Jardine, M., & Collier Cameron, A. 2002, *ApJ* 575, 1078
Jardine, M., Barnes, J., Donati, J.-F., & Collier Cameron, A. 1999, *MNRAS* 305, L35
Jardine, M., Collier Cameron, A., & Donati, J.-F. 2002a, *MNRAS* 333, 339
Jardine, M., Collier Cameron, A., Donati, J.-F., Gregory, S. G., & Wood, K. 2006, *MNRAS* 367, 917
Jardine, M., Collier Cameron, A., Donati, J.-F., & Pointer, G. 2001, *MNRAS* 324, 201
Jardine, M., Gregory, S. G., & Donati, J.-F. 2008, *MNRAS* in press
Jardine, M. & van Ballegooijen, A. A. 2005, *MNRAS* 361, 1173
Jardine, M., Wood, K., Collier Cameron, A., Donati, J.-F., & Mackay, D. H. 2002b, *MNRAS* 336, 1364
Jeffries, R. 1993, *MNRAS* 262, 369
Küker, M. & Rüdiger, G. 1997, *A&A* 328, 253
Küker, M. & Rüdiger, G. 1999, in ASP Conf. Ser. 178: Workshop on stellar dynamos, Vol. 178, 87
MacDonald, J. & Mullan, D. J. 2004, *MNRAS* 348, 702
Maeder, A. & Meynet, G. 2005, *A&A* 440, 1041
Maggio, A., Pallavicini, R., Reale, F., & Tagliaferri, G. 2000, *A&A* 356, 627
McIvor, T., Jardine, M., Cameron, A. C., Wood, K., & Donati, J.-F. 2003, *MNRAS* 345, 601
Moss, D. 2001, in ASP Conference Series, Vol. 248, Magnetic fields across the Hertzsprung-Russell diagram, ed. S. G. Mathys & D. Wickramasinghe (San Francisco), 305
Mullan, D. J. & MacDonald, J. 2005, *MNRAS* 356, 1139
Sanz-Forcada, J., Maggio, A., & Micela, G. 2003, *A&A* 408, 1087
Schrijver, C., Mewe, R., van den Oord, G., & Kaastra, J. 1995, *A&A* 302, 438
Spruit, H. 2002, *A&A* 381, 923
Strassmeier, K. 1996, in IAU Symposium 176: Stellar Surface Structure, ed. Strassmeier, K.G. & Linsky, J.L., Kluwer, Dordrecht, p.289
Tout, C. A. & Pringle, J. E. 1995, *MNRAS* 272, 528
Townsend, R. H. D. & Owoki, S. P. 2005, *MNRAS* 357, 251

Cosmic Magnetic Fields:
From Planets, to Stars and Galaxies
Proceedings IAU Symposium No. 259, 2008
K.G. Strassmeier, A.G. Kosovichev & J.E. Beckman, eds.

doi:10.1017/S1743921309030737

Starspots: signatures of stellar magnetic activity

Klaus G. Strassmeier

Astrophysikalisches Institut Potsdam, An der Sternwarte 16, D-14482 Potsdam, Germany
email: kstrassmeier@aip.de

Abstract. Starspots, just as Sunspots, are among the most obvious tracers and signatures of stellar surface magnetic activity. Emphasized already several decades ago as the origin for the rotationally modulated brightness of cool late-type stars, it is just now that we start to trace individual surface features in great enough detail to understand their magnetic behavior and interaction. Starspots also became the most important "noise" for detecting extra-solar planets and could possibly be decisive when it comes to detect another Earth. Since this is not a review, and because indirect imaging techniques are covered in other papers in this volume, I focus in this paper on some specific detections of starspots and introduce four new facilities particularly suited for starspot research in the near future.

Keywords. Magnetic fields – stars: spots – stars: magnetic fields – stars: rotation – methods: photometric – methods: spectroscopic – instrumentation: spectrographs – instrumentation: polarimeters

1. Introduction

Magnetic processes just like those seen in the space environment of the Earth and, of course, on the Sun have now slowly moved to center stage in other astrophysical areas. The application to other stars opened up a new field of research that became widely known as the solar-stellar connection. We emphasize that all previous attempts to disentangle the impact of magnetic fields onto stellar evolution had focused on higher-mass stars (see Maeder, this proceeding) or favored specific target classes in few subregions of the H-R diagram, e.g. Ap-stars like α^2 CVn or 53 Cam, or T-Tauri stars (see Johns-Krull, this proceeding). Magnetic fields can play a crucial role in the accretion process as well as in the acceleration and collimation of jet-like flows in young stellar objects in general. Only just recently were, e.g., O-star magnetic-field measurements published (Wade *et al.* 2006, see also this proceeding). Finally, solar-type stars with convective envelopes and their rich but subtle variety of magnetic phenomena were added to the zoo just most recently (Catala *et al.* 2007, Donati *et al.* 2007, Petit, this proceeding).

Another area is the physics of accreting black holes, where magnetic activity is now believed to be responsible for most of the behavior of these objects, including their X-ray spectrum, their notoriously dramatic variability, and the powerful relativistic jets they produce. Another is the physics of the central engines of cosmical gamma-ray bursts, the most powerful explosions in the current universe. The main ingredient for explaining these events invokes powering and focusing of the bursts by magnetic fields. It is fair to say that virtually all the physics of magnetic fields exploited in these areas of astrophysics is based directly on our understanding of the Sun's magnetic field (see, e.g., Stenflo 2002, Solanki 2004, Rüdiger & Hollerbach 2004, Thomas & Weiss 2008).

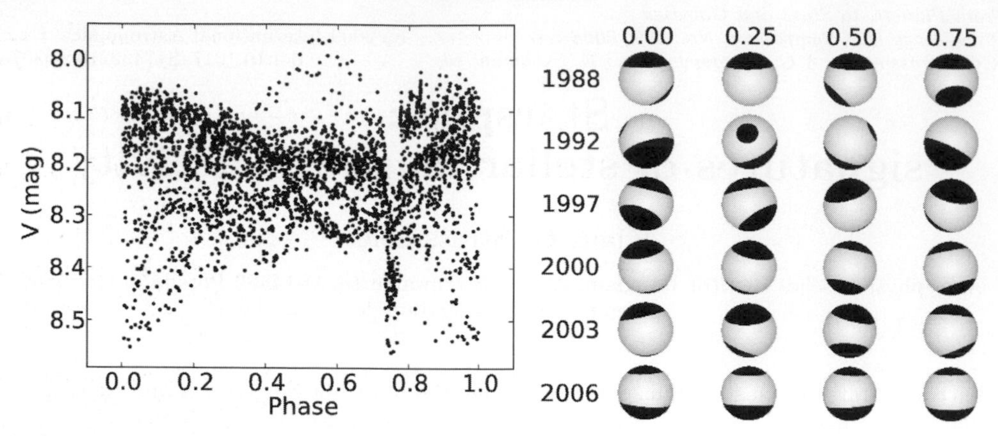

Figure 1. Almost all of the photometric variability of a magnetically active star is related to starspots. The light curve (left panel) of the spotted binary star HD 6286 shows 20 years of V-band data from the 0.75m Vienna-T7 and the 0.40m Tennessee-T3 Automatic Photoelectric Telescope, both located in southern Arizona. All data were phased with the rotation period of the primary component and demonstrate the continuous and systematic variability of the star as shown in the spot-model solutions to the right. Note that HD 6286 is an eclipsing binary with primary minimum at phase 0.75.

2. Starspot detections and some obvious limitations

2.1. *Broad-band photometry*

Indirect detections and techniques, like the tracing and timing of sunspots moving across the solar disk due to rotation, led not only to the discovery of solar rotation but also to the discovery of the famous solar butterfly diagram. For stellar analogs, Vogt & Penrod (1983) were the first to obtain an indirect image of a cool and heavily spotted star (V711 Tau = HR 1099) from data in late 1981 and discovered a prominent polar spot. Because the Sun does not show spots in excess of $\pm 40°$ (but see Tsuneta 2008 for the recent discovery of high-latitude kG magnetic fields), this discovery spurred significant interest in these types of stars, and in starspots in particular. Today, dark photospheric spots and bright chromospheric plages are still the main tracers for detecting stellar rotation periods using photometry and, if long time series are available, also for detecting differential rotation and magnetic-cycle periods (Donahue *et al.* 1996, Rodonó *et al.* 2001, Strassmeier 2005).

Figure 1 is a recent data example and shows almost 20 consecutive years of automated V-band light curves of the spotted star HD 6286 (Strassmeier *et al.* 2008a). Plotting all data versus its 35-day rotation period is not a typical representation but is intended to emphasize the long-term variability due to changes of the star's spottedness. From this signal, systematic changes of the spot's surface positions, sizes, and possibly temperatures are derived and interpreted accordingly.

Figure 2 shows several simulated light curves and the stellar images obtained from them. Various spot configurations are adopted, as shown in the left columns, while the basic input parameters are kept fixed (see Savanov & Strassmeier 2008 for details). Because photometric light curves represent only one-dimensional time series, opposite to spectral line profiles for Doppler imaging, the reconstructed stellar image contains only poor information in the direction of stellar latitude. Usually, the inversions easily resolve low-latitude details but practically fail to recover polar and high-latitude features. The code always tries to concentrate spots at the "sub-observer" latitude, where the photometric amplitudes become strongest.

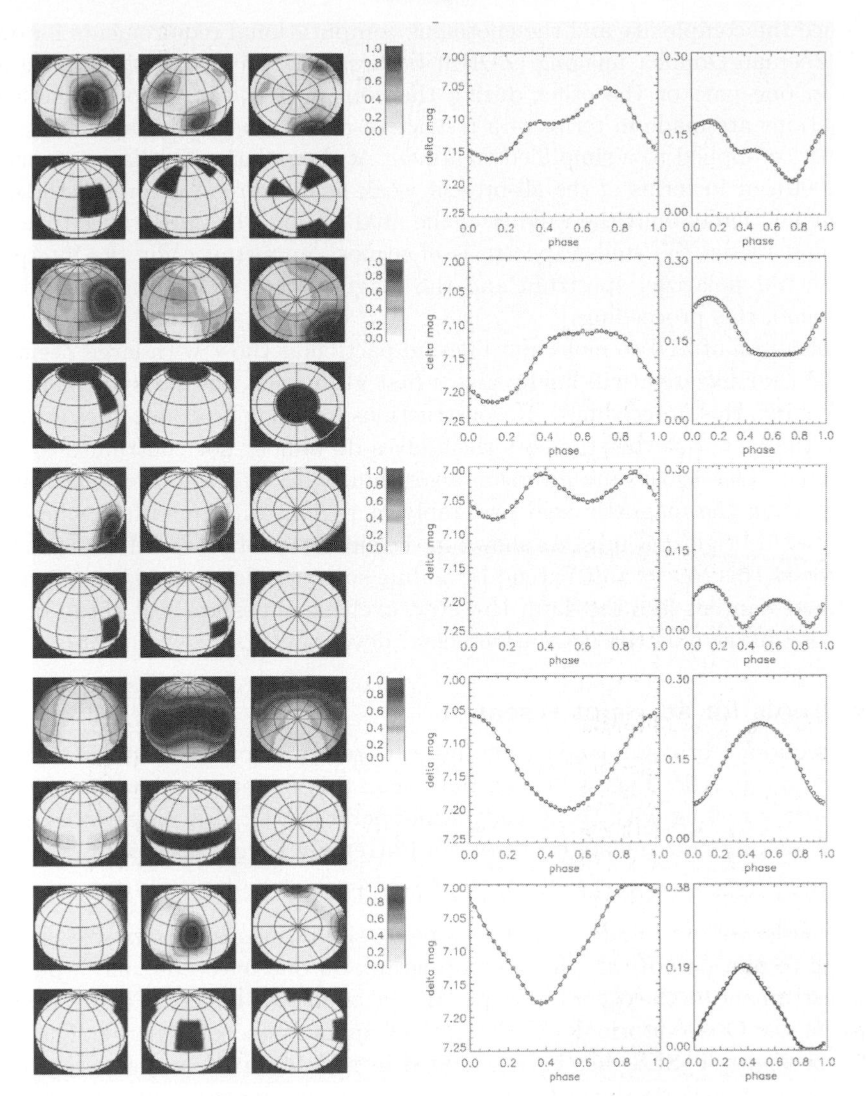

Figure 2. Light-curve inversions. The left columns show the image of the input test star and, above, its respective reconstruction. In the right columns are the simulated and reconstructed light curves and the change of the visible spottedness plotted versus rotational phase.

The tests in Fig. 2 illustrate another problem with the use of light curves. Two equal spots separated by 90° in longitude produce a single symmetric light minimum, as would a single spot of slightly different shape. If such a light curve is used to search for "active longitudes" just by measuring the light-curve minimum, even the spot longitudes would be wrongly estimated while an inversion technique, as shown in Fig. 2, reproduces the correct longitudes.

2.2. *High-resolution spectroscopy and Stokes polarimetry*

A big deal was heard during this conference on spectropolarimetry. It appears clear now that full four-Stokes vector observations at high spectral resolution are the next future steps (replace high resolution with wavelength coverage for the class of degenerate stars)(see Carroll, this proceedings and Jordan, this proceedings).

To reduce the complexity and the enormous computational requirements for even just Stokes-V Zeeman-Doppler imaging (ZDI) it became practice to use simplifying approximations in one part or the other during the course of the inversion. Whether these approximations are made in terms of a restrictive underlying stellar model (e.g. Donati *et al.* 1994), or applied to a simplified treatment of the polarized radiative transfer (e.g. Hussain 2000), or in terms of the all-present weak-field approximation, all these simplifications may severely limit the validity of the final result. The importance to accurately calculate and model the stellar spectra is of utmost importance for the interpretation of the observed polarized spectrum and the diagnostic of stellar magnetic fields (see Trujillo-Bueno, this proceedings).

The application of ZDI to molecular lines, in particular those with large Zeeman splittings in the near infrared, will finally give a first glimpse at the starspot magnetic field (see Berdyugina, this proceedings). Reconstructions from solely atomic lines are currently limited by the fact that the starspots themselves do almost not contribute to the overall ZDI signal. This leaves any inversion algorithm, with or without regularization, the choice of placing the magnetic field preferably at regions on the surface where there is also light, i.e. in bright features. As shown in the simulation by Carroll, this proceedings, one can bypass this to a certain extend by adding some artificial inversion weighting but, of course, one can not fool the data. However, exciting times are still ahead of us and I focus in this talk now on the new instrumental developments.

3. New tools for starspot research

In this section, I briefly introduce the projects relevant for starspot research that we are working on at AIP. This is biasing but, since this is not a review, it maybe useful for the reader. Starting with the already semi-operable STELLA observatory I will end with the, hopefully not so distant, European Extremely Large Telescope.

3.1. *STELLA*

STELLA consists of two robotic 1.2m telescopes to simultaneously monitor stellar activity with a high resolution optical echelle spectrograph on one telescope, and a photometric imaging instrument on the other telescope (Weber *et al.* 2008). The STELLA observatory is located at the Observatorio del Teide on the Canary island of Tenerife. The STELLA Echelle spectrograph (SES) has been operated in robotic mode for two years now, and produced over 10,000 spectra of the entire optical range between 390 and 900 nm at a spectral resolution of 55,000 with a peak shutter-open time of 95%. Although we do not use an iodine cell nor an actively stabilized chamber, its average radial velocity precision over the past two years was 60 to 150 m/s rms, depending on target. The Wide-Field STELLA Imaging Photometer (WIFSIP) is currently being tested and will enter operation in mid 2009.

3.2. *PEPSI for the LBT*

PEPSI is the fiber-feed high-resolution echelle spectrograph for the 11.8m Large Binocular Telescope (LBT) in Arizona (Strassmeier *et al.* 2008c). It is designed to utilize the two 8.4m apertures of the LBT in a unique spectropolarimetric mode. With two identical but independent polarimeters in each of the direct f/15 Gregorian foci, it will allow the simultaneous observation of circularly and linearly polarized light with high spectral and temporal resolution. Non-polarized "integral" light is fed to the spectrograph via two permanently mounted focal stations, thereby providing a standby spectrograph for quick-reaction science. Its thumbnail specifications are (see also Ilyin *et al.*, this proceedings):

• Fiber-fed optical integral-light spectroscopy in the wavelength range 380-1050nm with spectral resolving powers, R, of 40,000 (2.2″ aperture), 130,000 (1.5″ aperture) and 310,000 (0.75″ aperture). The R=40,000 mode with a sky aperture of 2.2″ allows efficient spectroscopy even in bad seeing.

• Full wavelength coverage in three exposures. Pixel binning in dispersion direction can be used for R=20,000 work. S/N of 10:1 of a V=19mag star is reached at R=130,000 and seeing 0.7″ in 1 hour integration time.

• Full four-Stokes polarimetry in the wavelength range 450–1050nm at a fixed spectral resolution of R=130,000.

• Radial-velocity stability of order m/s over an observing season in integral light.

• A polarimetric precision, $\delta P/P$ (P the degree of polarization), of 10^{-5} for targets brighter than V=4mag in 1 hour integration time, of $10^{-3...4}$ for targets brighter than V=10th mag, 10^{-2} for V <14mag, and 10^{-1} for V <18mag.

• Fast reaction time for "Targets of Opportunity" due to the stand-by modus.

• The R=310,000 mode will be fed with Sunlight during daytime by a "Sun-as-a-star" telescope.

3.3. *GREGOR@night*

Gregor@night is the fiber-fed double echelle spectrograph (Strassmeier *et al.* 2007) for the night-time use of the new 1.5m German solar telescope GREGOR. The telescope is foreseen to start routine day-time operation in 2010. The spectrograph's design specifications are driven by a science case based on the solar-stellar connection, in particular from the search and characterization of solar analogs but also ranging from the characterization of late-type stars with exoplanets to asteroseismology with high time resolution and cadence, and the long-term monitoring of stellar activity cycle analogs and related topics. The spectrographs are based on a white-pupil design with a 110mm beam, two refractive cameras and collimators, both optimized for the wavelength ranges 360-490nm and 510-870nm for the two arms, respectively. The design achieves a two-pixel resolution of 100,000 for an entrance aperture of 3″ at 20% total throughput. The instrument would be fully automated and no on-site night observer required.

3.4. *ICE-T*

ICE-T (the International Concordia Explorer Telescope) is a twin telescope for Dome C in Antarctica and consists of two 60cm f/1.1 ultra-wide-field Schmidt telescopes optimized for Sloan g (402-552nm) and Sloan i (691-818nm), respectively (Strassmeier *et al.* 2008b). The technical goal is to perform time-series photometry of a million stars between 9–18.5 mag. A photometric precision of up to 100 micro-mag for the brightest stars is envisioned. Such ultra-high precision for stars brighter than 10th mag is possible because the site has 3.6-times reduced scintillation which makes a 60cm telescope in the Antarctic equivalent to a global network of 1.5m telescopes at temperate sites. Nominal time resolution would be between 10sec and 600sec, depending upon the choice of co-adding and the on-site disk space. The two parallel-mounted 60/80cm telescopes each feed a monolithic 10.3k×10.3k thinned, back-illuminated 9-micron CCD, operated in frame-transfer mode. Its total field of view (FOV) would be 12° diameter, of which 65 square degrees would be seen by each CCD (8.1°×8.1°). The image scale is 2.75″ per pixel with a nominal defocus to 6 pixels (18″) for optimal point-spread-function sampling. Opposite to many other highly automated wide-field surveys, e.g. SDSS (Ivezić *et al.* 2004) or Pan-STARRS (Hodapp *et al.* 2004), ICE-T would operate in a truly robotic mode, i.e. with no human attendance at all.

3.5. *SFPP: a Smart-Focal-Plane Polarimeter for the E-ELT*

We are currently conducting a design study of a spectropolarimetric light feed for the European Extremely Large Telescope (E-ELT) (Strassmeier & Ilyin 2009). Such a polarimeter may feed an UV, an optical, and a near-IR spectrograph simultaneously, thus covering the 320–2500nm range. Our design concept is tentatively called SFPP, Smart Focal Plane Polarimeter. Its aim is to detect a differential polarimetric precision of 10^{-6} in selected spectral lines. The polarimeter itself shall be part of a smart focal plane for the symmetric beam of the telescope in its intermediate f/4.15 focus. Among the future scientific applications is a search for the linearly-polarized signal (or the lack thereof) from light reflected off a close-by planet in the combined Stokes spectra. We stress the importance of measuring stellar and planetary magnetic fields because these are prerequisites for planet habitability with life as we know it.

Acknowledgements

I would like to thank my colleagues Rainer Arlt, Thorsten Carroll and Ilya Ilyin for many discussions and Manfred Woche and Emil Popow for sharing their overall technical wisdom with me. I acknowledge grant STR645-1 from the Deutsche Forschungsgemeinschaft (DFG) and the support from the Bundeministerium für Bildung & Forschung (BMBF) through the Verbundforschung grant 05AL2BA1/3 and 05A09IPA.

References

Catala, C., Donati, J.-F., Shkolnik, E., Bohlender, D., & Alecian, E. 2007, *MNRAS* 374, L42
Donahue, R. A., Saar, S. H., & Baliunas, S. L. 1996, *ApJ* 466, 384
Donati, J.-F., Achilleos, N., Matthews, J. M., & Wesemael, F. 1994, *A&A* 285, 285
Donati, J.-F., Moutou, C., Farés, R., *et al.* 2008, *MNRAS* 385, 1179
Hodapp, K. W., Kaiser, N., Aussel, H., *et al.* 2004, *AN* 325, 636
Hussain, G. A. J., Donati, J.-F., Collier Cameron, A., & Barnes, J. R. 2000, *MNRAS* 318, 961
Ivezić, Z., Lupton, R. H., Schlegel, D., *et al.* 2004, *AN* 325, 583
Rodonó, M., Cutispoto, G., Lanza, A. F., & Messina, S. 2001, *AN* 322, 333
Rüdiger, G. & Hollerbach, R. 2004, *The Magnetic Universe*, Wiley-VCH, Berlin
Savanov, I. S. & Strassmeier, K. G. 2008, *AN* 329, 364
Solanki, S. K. 2004, in G. Poletto & S. T. Suess (eds.), *The Sun and the Heliosphere as an Integrated System*, A&SS 317, p.373
Stenflo, J. O. 2002, in F. Moreno-Insertis, and F. Sánchez (eds.), *Polarized radiation diagnostics of stellar magnetic fields*, Astrophysical spectropolarimetry. XII Canary Islands Winter School, Cambridge University Press, p.55
Strassmeier, K. G. 2005, *AN* 326, 269
Strassmeier, K. G., Bartus, J., Fekel, F. C., & Henry, G. W. 2008a, *A&A* 485, 233
Strassmeier, K. G., Granzer, T., & Denker, C., *et al.* 2008b, *EAS Publications Series* 33, 199
Strassmeier, K. G. & Ilyin, I. V. 2009, in Moorwood, A. (ed.), *Science with the VLT in the ELT era*, Ap&SS, Springer, p.255
Strassmeier, K. G., Woche, M., Granzer, T., Andersen, M. I., Schmidt, W., & Koubsky, P. 2007, in Kneer, F., Puschmann, K. G., Wittmann, A. (eds.), *Modern solar facilities - advanced solar science*, Univ.-Verlag Göttingen, p.51
Strassmeier, K. G., Woche, M., & Ilyin, I., *et al.* 2008c, *SPIE* 7014, 21
Thomas, J. H. & Weiss, N. O. 2008, *Sunspots and Starspots*, Cambridge University Press
Tsuneta, S. 2008, in S. A. Matthews, J. M. Davis & L. K. Harra (eds.), *First Results From Hinode*, ASP Conference Series, Vol. 397, p.19
Vogt, S. S. & Penrod, G. D. 1983, *PASP* 95, 565
Wade, G. A., Fullerton, A. W., Donati, J.-F., Landstreet, J. D., Petit, P., & Strasser, S. 2006, *A&A* 451, 195
Weber, M., Granzer, T., Strassmeier, K. G., & Woche, M. 2008, *SPIE* 7022

Cosmic Magnetic Fields:
From Planets, to Stars and Galaxies
Proceedings IAU Symposium No. 259, 2008
K.G. Strassmeier, A.G. Kosovichev & J. Beckman, eds.

doi:10.1017/S1743921309030749

Magnetic fields in White Dwarfs and their direct progenitors

Stefan Jordan

Astronomisches Rechen-Institut, Zentrum für Astronomie, Mönchhofstr. 12-14,
D-69120 Heidelberg, Germany
email: jordan@ari.uni-heidelberg.de

Abstract. The paper provides an overview on the results of the analyses of spectro-polarimetric observations of white dwarfs, subdwarfs, and central stars of planetary nebulae. It will also discuss the question of the origin of the magnetic fields in white dwarfs.

Keywords. White Dwarfs – magnetic fields – subdwarfs – Planetary Nebulae

---

1. Introduction and history

Blackett (1947) predicted that very strong magnetic fields ($\approx 1\,$MG) could exist in white dwarfs if the magnetic moment of a star is proportional to its angular momentum, which he assumed to be conserved during the stellar evolution and the collapse. This is, however, probably not the case since most isolated white dwarfs seem to be relatively slow rotators ($v\lesssim 40\,$km/sec, see Karl *et al.* 2005 and references thererein), although a few exceptions from this rule exist (e.g. RE J 0317-853 with a rotational period of 725 sec, Barstow *et al.* 1995); if angular momentum were completely conserved during the evolution we would expect the white dwarf remnant to have $v_{rot} \approx 10,000\,$km/sec.

Another possibility was proposed by Ginzburg (1964) and Woltjer (1964). They argued that if the magnetic flux, which is proportional to BR^2, is conserved during evolution and collapse, very strong magnetic fields can be reached in degenerate stars. A main sequence star with a radius $R \approx 10^{11}$ cm and a surface magnetic field of 1-10 kG can therefore become a white dwarf ($R \approx 10^9$ cm) with a magnetic field strength of 10^7-10^8 G.

The search for magnetic white dwarfs began in 1970 when Preston looked for quadratic Zeeman shifts in the spectra of DA white dwarfs. Due to the extremely strongly Stark broadened Balmer lines and the limited spectral resolution he was only able to place upper limits of about 0.5 MG for the magnetic fields in several white dwarfs. A rather sensitive method to detect magnetic fields in white dwarfs is the measurement of circular polarization. Kemp (1970) detected circular polarization of several percent in Grw +70°8247, an object that was known for its rather shallow and unidentified "Minkowski bands" (Minkowski 1938). Although his derived value for the magnetic field strength was wrong this proved the existence of a magnetic field in this white dwarf. Nevertheless, all attempts to identify the Minkowski bands with various atoms or molecules in magnetic fields of a few MG failed.

Even for the simplest atoms, hydrogen and helium, accurate calculations for the line components did not exist at that time for field strengths above 20-100 MG (Kemic 1974a,b); only for extremely intense fields (10^9-10^{10} G) data were available again Garstang (1977), but none of the predicted line positions were in agreement with the wavelengths of the Grw +70°8247 features. For this reason Angel (1979) proposed that the star must possess a field strength above 100 MG (but below the intense-field regime).

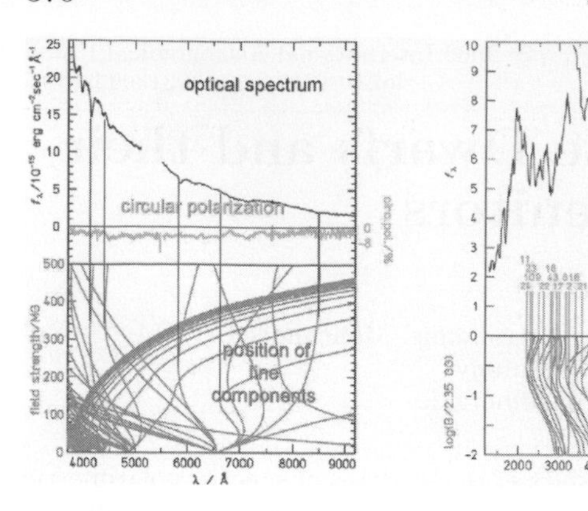

Figure 1. Spectrum and circular polarisation of the famous magnetic white dwarf Grw +70°8247 (left) can the identification of the spectral features with stationary line components; the dipole field strength is ≈ 320 MG (Jordan 1992). On the right hand side, the spectrum of GD 229 is compared to stationary line components of neutral helium (Jordan *et al.* 1998).

For hydrogen the intermediate-field gap has been closed more than twenty years ago with calculations of energy level shifts and transition probabilities for bound-bound transitions by groups in Tübingen and Baton Rouge (Forster *et al.* 1984; Rösner *et al.* 1984; Henry & O'Connell 1984).

2. Stationary line components

Since the magnetic field on the surface of a white dwarf normally is not homogeneous but e.g. better described by a magnetic dipole or more complicated field geometries, the variation of the field strengths from the pole to the equator (a factor of two for a pure dipole field) smears out most of the absorption lines; this explains why the spectral features on Grw +70°8247 are so shallow. However, a few of the line components become stationary, i.e. their wavelengths go through maxima or minima as functions of the magnetic field strength. These stationary components are visible in the spectra of magnetic white dwarfs despite a considerable variation of the field strengths.

It was a great confirmation for the correctness of the theoretical calculations that indeed the unidentified features in the optical and UV spectrum of Grw +70°8247 could be attributed to stationary components of hydrogen (see Fig. 1) in fields between about 150 and 500 MG (Greenstein 1984; Greenstein *et al.* 1985).

For helium reliable atomic data for arbitrary magnetic field strengths became available only in the late nineties (see Jordan *et al.* 1998, 2001) when the existence of helium could be proven (Jordan *et al.* 1998).

Table 1: List of all known MWDs, extending the one by Kawka *et al.* (2007), containing the name of the object, the effective temperature, the chemical composition, the magnetic field strength (usually the dipole field strength), the rotational period. References are provided only if not provided by Kawka *et al.* (2007). For uncertainties of the values please refer to the original papers.

Object	$T_{\rm eff}$/kK	Comp.	B/MG	$P_{\rm rot}$	References
SDSS J000555.91−100213.4	29000	He/C	?	···	
LHS 1038	6540	H	$\lesssim$0.2	> 2 h	
LHS 1044	6010	H	16.7	···	
SDSS J001742.44+004137.4	15000	He	8.30	···	
SDSS J002129.00+150223.7	7000	H	530.69	···	Külebi *et al.* (08)
SDSS J004248.19+001955.3	11000	H	2.00	···	Külebi *et al.* (08)
Feige 7	20000	H/He	35.00	12 m	

Table 1: continued.

Object	$T_{\rm eff}$/kK	Comp.	B/MG	$P_{\rm rot}$	References
SDSS J014245.37+131546.4	15000	He	4.00	···	
SDSS J015748.15+003315.1	$\sim$ 6000	He (DZ)	3.70	···	
MWD 0159−032	26000	H	6.00	···	
SDSS J021116.34+003128.5	9000	H	341.31	···	Külebi *et al.* (08)
SDSS J021148.22+211548.2	12000	H	168.20	···	Külebi *et al.* (08)
SDSS J023609.40−080823.9	10000	H (DQA)	5.00	···	
HE 0236−2656	6500	He	?	···	
HE 0241−155	12000	H	200	···	Reimers *et al.* (04)
KPD 0253+5052	15000	H	13.5	···	
LHS 5064	6680	H	$\sim$ 0.1	···	
SDSS J030407.40−002541.7	15000	H	10.95	···	Külebi *et al.* (08)
MWD 0307−428	25000	H	10.00	···	
SDSS J031824.19+422651.0	10500	H	10.12	···	Külebi *et al.* (08)
REJ J0317−853	33000	H	$>$ 180	725 s	
KUV 03292+0035	26500	H	12.10	···	
HE 0330−0002	6500	He	?	···	
SDSS J033145.69+004517.0	15500	H	13.13	···	Külebi *et al.* (08)
SDSS J034308.18−064127.3	13000	H	19.78	···	Külebi *et al.* (08)
SDSS J034511.11+003444.3	8000	H	1.96	···	Külebi *et al.* (08)
40 Eri B	16490	H	0.0023	···	
BPM 3523	23450	H	0.00428	···	
LHS 1734	5300	H	7.3	···	
G 99−37	6070	C_2/CH	7.5	4.1 h	Berd. *et al.* (07)
G 99−47	5790	H	20	0.97 h	
EUVE J0616−649	50000	H	14.80	···	
GD 77	14870	H	1.2	···	
G 234−4	4500	H	0.0396	···	
SDSS J074850.48+301944.8	22000	H	6.75	···	Külebi *et al.* (08)
SDSS J075234.96+172525.0	9000	H	10.30	···	Külebi *et al.* (08)
SDSS J075819.57+354443.7	22000	H	26.40	···	Külebi *et al.* (08)
G 111−49	8500	H	220	···	
SDSS J080440.35+182731.0	11000	H	48.47	···	Külebi *et al.* (08)
SDSS J080502.29+215320.5	28000	H	6.11	···	Külebi *et al.* (08)
SDSS J080743.33+393829.2	13000	H	65.75	···	Külebi *et al.* (08)
SDSS J080938.10+373053.8	14000	H	39.74	···	Külebi *et al.* (08)
SDSS J081648.71+041223.5	11500	H	7.35	···	Külebi *et al.* (08)
GD 90	14000	H	9.00	···	
EUVE J0823−254	43200	H	$2.8-3.5$	···	
SDSS J082835.82+293448.7	19500	H	33.40	···	Külebi *et al.* (08)
EG 61	17100	H	$\sim$ 3	···	
SDSS J084008.50+271242.7	12250	H	3.38	···	Külebi *et al.* (08)
SDSS J084155.74+022350.6	7000	H	5.00	···	Külebi *et al.* (08)
SDSS J083945.56+200015.7	15000	H	3.38	···	
SDSS J084716.21+484220.4	19000	H	$\sim$ 3	···	

Table 1: continued.

Object	$T_{\rm eff}$/kK	Comp.	B/MG	$P_{\rm rot}$	References
SDSS J085106.12+120157.8	11000	H	2.03	···	Külebi *et al.* (08)
LB 8915	24000	H/He	0.75 − 1.0	···	
SDSS J085523.87+164059.0	15500	H	12.23	···	Külebi *et al.* (08)
SDSS J085830.85+412635.1	7000	H	3.38	···	Külebi *et al.* (08)
SDSS J090632.66+080716.0	17000	H	10.00	···	
SDSS J090746.84+353821.5	16500	H	22.40	···	Külebi *et al.* (08)
SDSS J091124.68+420255.9	10250	H	35.20	···	Külebi *et al.* (08)
SDSS J091437.40+054453.3	17000	H	9.16	···	Külebi *et al.* (08)
G 195−19	7160	He	$\sim$ 100	1.3 d	
SDSS J091833.32+205536.9	14000	H	2.04	···	Külebi *et al.* (08)
SDSS J092527.47+011328.7	10000	H	2.04	···	Külebi *et al.* (08)
SDSS J093313.14+005135.4	···	He (C_2H)?	?	···	
SDSS J093356.40+102215.7	8500	H	2.11	···	Külebi *et al.* (08)
SDSS J093447.90+503312.2	8900	H	9.50	···	Külebi *et al.* (08)
SDSS J094458.92+453901.2	15500	H	15.91	···	
LB 11146	16000	H	670.00	···	
SDSS J095442.91+091354.4	···	DQ	?	···	
SDSS J100005.67+015859.2	9000	H	19.74	···	Külebi *et al.* (08)
SDSS J100356.32+053825.6	23000	H	10.13	···	Külebi *et al.* (08)
SDSS J100657.51+303338.1	10000	H	1.00	···	Külebi *et al.* (08)
SDSS J100715.55+123709.5	18000	H	5.41	···	Külebi *et al.* (08)
LHS 2229	4600	He (C_2H)	$\sim$ 100	···	
SDSS J101529.62+090703.8	7200	H	4.09	···	Külebi *et al.* (08)
SDSS J101618.37+040920.6	10000	H	7.94	···	Külebi *et al.* (08)
PG 1015+014	14000	H	70.00	99 m	Euchner *et al.* (06)
GD 116	16000	H	65	···	
SDSS J102239.06+194904.3	9000	H	2.94	···	Külebi *et al.* (08)
LHS 2273	7160	H	18.00	···	
PG 1031+234	$\sim$ 15000	H	$\lesssim$1000	3.4 h	
SDSS J103655.38+652252.0	···	DQ	0.17	···	
LP 790−29	7800	He	50.00	26 y?	
HE 1043−0502	$\sim$ 15000	He	$\sim$ 820	···	
HE 1045−0908	10000	H	16.00	2.7 h	
SDSS J105404.38+593333.3	9500	H	17.63	···	Külebi *et al.* (08)
SDSS J105628.49+652313.5	16500	H	29.27	···	Külebi *et al.* (08)
LTT 4099	15280	H	0.0039	···	
SDSS J111010.50+600141.4	30000	H	6.37	···	Külebi *et al.* (08)
SDSS J111341.33+014641.7	···	He ?	?	···	
SDSS J111812.67+095241.4	10500	H	3.38	···	Külebi *et al.* (08)
SDSS J112257.10+322327.8	12500	H	11.38	···	Külebi *et al.* (08)
SDSS J112852.88−010540.8	11000	H	2.00	···	Külebi *et al.* (08)
SDSS J112924.74+493931.9	10000	H	5.31	···	
SDSS J113357.66+515204.8	22000	H	8.64	···	Külebi *et al.* (08)
SDSS J113756.50+574022.4	7800	H	5.00	···	Külebi *et al.* (08)

Table 1: continued.

Object	$T_{\rm eff}$/kK	Comp.	B/MG	$P_{\rm rot}$	References
LBQS 1136−0132	10500	H	22.71	···	
SDSS J114006.37+611008.2	13500	H	50.19	···	Külebi *et al.* (08)
SDSS J114829.00+482731.2	27500	H	32.47	···	Külebi *et al.* (08)
SDSS J115418.14+011711.4	1125	H	33.47	···	Külebi *et al.* (08)
SDSS J115917.39+613914.3	23000	H	20.10	···	Külebi *et al.* (08)
SDSS J120150.10+614257.0	10500	H	11.35	···	Külebi *et al.* (08)
SDSS J120609.80+081323.7	13000	H	760.63	···	Külebi *et al.* (08)
SDSS J120728.96+440731.6	16750	H	2.03	···	Külebi *et al.* (08)
SDSS J121209.31+013627.7	10000	H	10.12	···	Külebi *et al.* (08)
HE 1211−1707	∼ 12000	He	50.00	∼ 2 h	
LHS 2534	6000	He (DZ)	1.92	···	
SDSS J121635.37−002656.2	20000	H	59.70	···	Külebi *et al.* (08)
SDSS J122209.44+001534.0	20000	H	14.70	···	Külebi *et al.* (08)
PG 1220+234	26540	H	3.00	···	
SDSS J122249.14+481133.1	9000	H	8.05	···	Külebi *et al.* (08)
SDSS J122401.48+415551.9	9500	H	22.36	···	Külebi *et al.* (08)
SDSS J123414.11+124829.6	8200	H	4.32	···	Külebi *et al.* (08)
SDSS J124806.38+410427.2	7000	H	7.03	···	Külebi *et al.* (08)
SDSS J124851.31−022924.7	13500	H	7.36	···	Külebi *et al.* (08)
SDSS J125044.42+154957.4	10000	H	20.71	···	Külebi *et al.* (08)
SDSS J125416.01+561204.7	13250	H	38.86	···	Külebi *et al.* (08)
HS 1254+3440	15000	H	9.5	···	
SDSS J125434.65+371000.1	10000	H	4.10	···	Külebi *et al.* (08)
SDSS J125715.54+341439.3	8500	H	11.45	···	Külebi *et al.* (08)
G 256−7	∼ 56000	H	4.9	···	
PG 1312+098	∼ 20000	H	10.00	5.4 h	
SDSS J132002.48+131901.6	14750	H	2.02	···	Külebi *et al.* (08)
SDSS J132858.20+590851.0	25000	H (DQA)	18.00	···	
G165-7	6440	He (DZ)	0.65	···	
G 62−46	6040	H	7.36	···	
SDSS J133359.86+001654.8	···	He (C_2H)?	?	···	
SDSS J133340.34+640627.4	13500	H	10.71	···	Külebi *et al.* (08)
SDSS J134043.10+654349.2	15000	H	4.32	···	Külebi *et al.* (08)
SDSS J134820.79+381017.2	35000	H	13.65	···	Külebi *et al.* (08)
SBS 1349+5434	11000	H	760	···	
LP 907−037	9520	H	$\lesssim$0.3	···	
SDSS J140716.66+495613.7	20000	H	12.49	···	Külebi *et al.* (08)
SDSS J141906.19+254356.5	9000	H	2.03	···	Külebi *et al.* (08)
SDSS J142625.71+575218.3	19830	He/C_2 (DQ)	∼ 1.2	···	Dufour *et al.* (08)
SDSS J142703.40+372110.5	19000	H	27.04	···	Külebi *et al.* (08)
SDSS J143019.05+281100.8	9000	H	9.34	···	Külebi *et al.* (08)

Table 1: continued.

Object	$T_{\rm eff}$/kK	Comp.	B/MG	$P_{\rm rot}$	References
SDSS J143218.26+430126.7	24000	H	2.04	···	Külebi *et al.* (08)
SDSS J143235.46+454852.5	16750	H	12.29	···	Külebi *et al.* (08)
EUVE J1439+750	$>$ 20000	H	$14-16$	···	
SDSS J144614.00+590216.7	12500	H	4.42	···	Külebi *et al.* (08)
SDSS J145415.01+432149.5	11500	H	2.35	···	Külebi *et al.* (08)
GD 175	6990	H	2.30	···	
SDSS J150813.20+394504.9	17000	H	13.23	···	Külebi *et al.* (08)
SDSS J151130.20+422023.0	9750	H	22.40	···	Külebi *et al.* (08)
SDSS J151745.19+610543.6	9500	H	13.98	···	Külebi *et al.* (08)
GD 185	18620	H	0.035	···	
PG 1533−057	20000	H	31	···	
SDSS J153532.25+421305.6	18500	H	5.27	···	Külebi *et al.* (08)
SDSS J153829.29+530604.6	13500	H	13.99	···	Külebi *et al.* (08)
SDSS J154213.48+034800.4	8500	H	8.35	···	Külebi *et al.* (08)
SDSS J160437.36+490809.2	9000	H	59.51	···	Külebi *et al.* (08)
GD 356	7510	He	13.00	0.1 d	
SDSS J164357.02+240201.3	16500	H	2.00	···	Külebi *et al.* (08)
SDSS J164703.24+370910.3	16250	H	2.10	···	Külebi *et al.* (08)
SDSS J165029.91+341125.5	9750	H	3.38	···	Külebi *et al.* (08)
SDSS J165203.68+352815.8	11500	H	7.37	···	Külebi *et al.* (08)
PG 1658+440	30510	H	2.3	···	
SDSS J170400.01+321328.7	23000	H	50.11	···	Külebi *et al.* (08)
NLTT 44447	6260	H	1.30	···	
SDSS J171556.29+600643.9	13500	H	2.03	···	Külebi *et al.* (08)
SDSS J172045.37+561214.9	22500	H	19.79	···	Külebi *et al.* (08)
SDSS J172329.14+540755.8	16500	H	32.85	···	Külebi *et al.* (08)
SDSS J172932.48+563204.1	10500	H	27.26	···	Külebi *et al.* (08)
BPM 25114	$\sim$ 20000	H	36.00	2.8 d	
G 240−72	5590	He	$\gtrsim$100	$\gtrsim$20 y	
G 183−35	6500	H	$\sim$ 14	$\sim$ 50 m?	
G 141−2	6340	H	$\sim$ 3	···	
G 227−35	6280	H	175	···	
Grw +70°8247	16000	H	320	···	
G 92−40	7920	H	$\gtrsim$0.1	1.4 d	
GD 229	18000	He	700	···	Jordan *et al.* (98)
LTT 7987	15360	H	0.001	···	Jordan *et al.* (07)
SDSS J202501.10+131025.6	17000	H	10.10	···	Külebi *et al.* (08)
SDSS J204626.15−071037.0	8000	H	2.03	···	Külebi *et al.* (08)
SDSS J205233.52−001610.7	19000	H	13.42	···	Külebi *et al.* (08)
L 24−52	10200	H	0.043 ?	···	
SDSS J214900.87+004842.8	11000	H	10.09	···	Külebi *et al.* (08)
SDSS J214930.74−072812.0	22000	H	44.71	···	Külebi *et al.* (08)
SDSS J215135.00+003140.5	9000	H	$\sim$ 300	···	Külebi *et al.* (08)
SDSS J215148.31+125525.5	14000	H	20.76	···	Külebi *et al.* (08)
SDSS J220435.05+001242.9	22000	H	1.02	···	Külebi *et al.* (08)
SDSS J221828.59−000012.2	15500	H	225.00	···	

Table 1: continued.

Object	$T_{\rm eff}$/kK	Comp.	B/MG	$P_{\rm rot}$	References
SDSS J224741.46+145638.8	18000	H	42.11	···	Külebi *et al.* (08)
SDSS J225726.05+075541.7	40000	H	16.17	···	Külebi *et al.* (08)
KUV 813−14	11000	H	45	18 d	
SDSS J231951.73+010909.3	8300	H	9.35	···	Külebi *et al.* (08)
SDSS J232248.22+003900.9	39000	H	21.40	···	Külebi *et al.* (08)
SDSS J232337.55−004628.2	15000	He	4.80	···	
PG 2329+267	9400	H	2.31	···	
SDSS J234623.69−102357.0	8500	H	9.17	···	Külebi *et al.* (08)
SDSS J234605.44+385337.7	26000	H	1000.00	···	
LTT 9857	8570	H	0.0031	···	

3. Modelling spectra and polarisation of magnetic white dwarfs

These identifications with stationary line components allowed an estimation of the approximate range of field strengths covering the stellar surface. However, the detailed field structure could not be inferred. This was only possible by simulating the radiative transfer through a magnetized stellar atmospheres, as e.g. described by Jordan (1992). In order to calculate theoretical flux and polarisation spectra for a given geometry of the magnetic field, theoretical Stokes parameters have to be evaluated on a few hundred surface elements taking into account the local magnetic field strength and the orientation of the magnetic field with respect to the observer and the normal of the stellar surface. Afterward, the results from the respective surface elements must be summed up.

Such a forward calculation for a given magnetic field geometry is relatively simple. The goal, however, is to determine the magnetic field geometry from the observed spectra and polarisation data. Since it turned out that most magnetic white dwarfs could not be described by simple dipole fields (see eg. Achilleos & Wickramasinghe 1989), the number of free parameters to describe the field geometry can be very large. Therefore, finding a magnetic field geometry that adequately reproduces the observed spectro-polarimetric data is extremely tedious and practically impossible when time resolved observations of rotating magnetic white dwarfs are used that have to reproduce the observation for all different rotational phases.

For the first time Burleigh *et al.* (1999) used an automatic least-squares fit procedure to determine the magnetic field geometry of RE J 0317-853 from phase resolved HST data (see Fig. 3) They used a downhill simplex method in multidimensions to account for the rotational geometry and an assumed offset-dipole configuration of the magnetic field. However, the approach has the disadvantage of easily running into local minima rather than the global χ^2 minimum of the high-dimensional parameter space. It also turned out that the optical spectro-polarimetric data of RE J 0317-853 are not fully consistent with the geometry derived from the UV data so that more sophisticated fit procedures need to be used.

A rather successful approach is the application of genetic (Jordan) and evolutionary (Euchner) algorithms to find a magnetic field geometry that is consistent with the spectro-polarimetric data of for all rotational phases (Euchner *et al.* 2002; Gänsicke *et al.* 2002; Euchner *et al.* 2005, 2006; Beuermann *et al.* 2007).

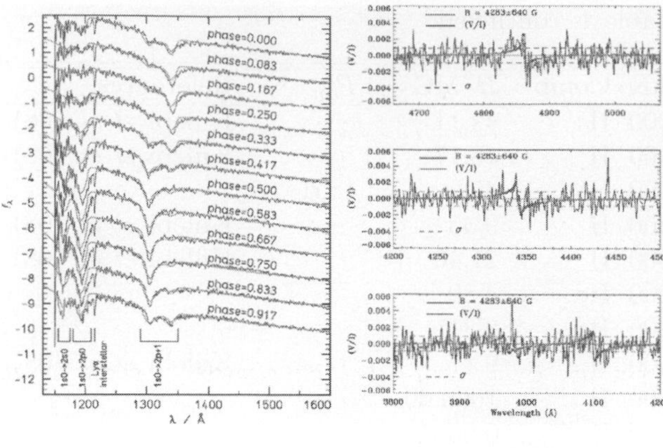

Figure 2. Left side: Observed and theoretical spectra UV spectra for 12 different phase across the 725 s rotation period of RE J 0317-853 for a magnetic dipole with a polar field strength of 363 MG, shifted by 0.19 stellar radii along the dipole axis (Burleigh *et al.* 1999). Right: Measurement of a 4.3 kG magnetic field in the white dwarf WD 446−789.

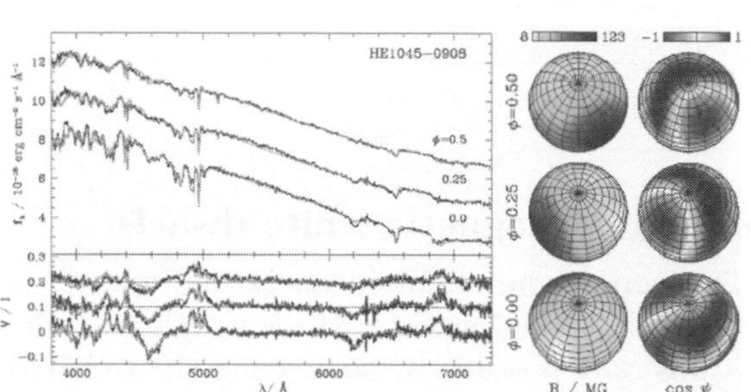

Figure 3. Zeeman tomography analysis of the magnetic field configuration of HE 1045−0908 using a dipole/quadrupole combination. B is the magnetic field strength and ψ is the angle between the local magnetic field vector and the observer.

4. Chemical species other than hydrogen

All really detailed analyses of magnetic white dwarfs were limited to those with hydrogen-rich atmospheres. The reason is that atomic data for helium have not consistently been included into our radiative transfer models yet. First tentative analyses using the new helium data were, however, performed by Wickramasinghe *et al.* (2002).

In some cool ($T_{\rm eff} < 9000$ K) helium rich magnetic white dwarfs, carbon molecules (C_2, CH) are responsible for the absorption. Bues & Pragal (1989) have interpreted such spectra with very simple approximations of the molecular physics of the Swan bands of C_2 in magnetic fields between 10 and 150 MG. Berdyugina *et al.* (2007) has used a Paschen-Back approximation for the C_2 absorption but has not yet analyse such objects with state-of-the-art radiation transfer models.

One of the most exciting recent discoveries is the detection of non-radial pulsations in a magnetic white dwarf with a hot carbon-rich atmosphere (Dufour *et al.* 2008).

5. Magnetic white dwarfs with high magnetic fields

About 10% of all isolated white dwarfs posses surface magnetic fields of more that 10^6 G (Liebert J. *et al.* 2005). Generally, magnetic white dwarfs with MG fields tend to be more massive than non-magnetic ones (Liebert J. 1988), but the number of objects with reliable mass determinations is rather small.

The Sloan Digital Sky Survey (SDSS) has almost tripled the number of known white dwarfs with MG fields (Schmidt *et al.* 2003; Vanlandingham *et al.* 2005). Külebi *et al.* (2008) have analysed all hydrogen-rich magnetic white dwarfs and found that almost all objects show indications field geometries which are more complicated than centered magnetic dipoles (see also Külebi *et al.* in these proceedings). The list of all currently know magnetic white dwarfs is provided in Table. 1.

6. Kilogauss magnetic fields in white dwarfs and their progenitors

Until recently, magnetic fields below 30 kG could not be detected, with the exception of the very bright white dwarf 40 Eri B ($V = 8.5$), in which Fabrika *et al.* (2003) found a magnetic field of 4 kG. However, by using the ESO VLT, we could push the detection limit down to about 1 kG in two investigations of 22 DA white dwarfs with $11 < V < 14$ (Aznar Cuadrado *et al.* 2004; Jordan *et al.* 2007) by means of polarimetry. The degree of polarization in the vicinity of the spectral lines caused by 1 kG is only about 0.1% (see Fig. 3 as an example).

Magnetic fields of kilogauss strength have also been detected in the direct progeny of white dwarfs – central stars of planetary nebulae Jordan *et al.* (2005), probably the channel leading to about 99% of all white dwarfs, and hot subdwarfs O'Toole *et al.* (2005), accounting for about 1%. The magnetic fields found were between 1 and 3 kG. If the magnetic flux is fully conserved during the transition phase to white dwarfs – these objects will become magnetic white dwarfs with fields up to 2 MG.

However, great care is necessary, since these detections in central stars and subdwarfs are at the very limit of what is observationally possible. Together with new observations of central stars of planetary nebular we are also checking the data analysis again in order to be convinced that the detections are real.

7. Origin of magnetic white dwarfs

The strong magnetic fields can be fossil if the magnetic flux is conserved during stellar evolution. The very high electric conductivity of degenerate electron gas in the interior of white dwarfs leads to very long decay times ($\gtrsim 10^{10}$ yr). For a long time it was assumed that the higher modes decay more rapidly than the fundamental (Chanmugam *et al.* 1972; Fontaine *et al.* 1973). However, Muslimov *et al.* (1995) have shown that a weak quadrupole (or octupole, etc.) component on the surface magnetic field of a white dwarf may survive the dipole component and specific initial conditions: Particularly the evolution of the quadrupole mode is very sensitive (via Hall effect) to the presence of internal toroidal field. Under certain conditions, the higher-order components of the magnetic field my survive as long as the dipole component.

This shows that, in principle, higher-order multipoles may give us some information about internal magnetization of white dwarfs and the initial conditions from the pre-white dwarf evolution. Therefore, further investigations of the complex magnetic fields of white dwarfs remain important.

The magnetic fluxes of the main-sequence stars with the highest magnetic fields are very similar to the highest magnetic fields found in magnetic white dwarfs (10^9 G). However, it is not easy to identify the main-sequence progenitors if we assume the fossil-field hypothesis, because magnetic fields below a few hundred Gauss are currently not detectable; moreover, magnetic fields in later type do not contain large-scale magnetic fields so that they also escape detection. The situation is also complicated by observational biases in the selection of the magnetic white dwarfs themselves. By applying initial-final mass relations and studying the number statistics Wickramasinghe & Ferrario (2005) concluded that magnetic white dwarfs cannot come from A_p and B_p stars alone.

Acknowledgements

I thank Baybars Külebi for his help preparing Table 1. The work on magnetic white dwarfs was supported by DLR project 50 OR 0802.

References

Achilleos, N. & Wickramasinghe, D. T. 1989, *ApJ* 346, 444
Angel J. R. P. 1979, in White Dwarfs and Variable Deg. Stars, Univ. of Rochester Press, p.313
Aznar Cuadrado, R., Jordan, S., Napiwotzki, R., Schmid, H. M., Solanki, S. K., & Mathys, G. 2004, *A&A* 423, 1081
Barstow, M. A., Jordan, S., O'Donoghue, D., Burleigh, M. R., Napiwotzki, R., & Harrop-Allin, M. K. 1995, *MNRAS* 277, 971
Berdyugina, S. V., Berdyugin, A. V., & Piirola, V. 2007, Phys. Rev. Lett. 99, 1101
Berdyugin, A. V. & Piirola, V. 1999, *A&A* 352, 619
Beuermann K., Euchner F., Reinsch K., Jordan S., & Gänsicke B. T. 2007, *A&A* 463, 647
Blackett P. M. S. 1947, *Nature* 159, 658
Bues, I. & Pragal, M. 1989, in White Dwarfs, Lecture Notes in Physics 328, ed. G. Wegner, p. 329
Burleigh, M. R., Jordan, S., & Schweizer, W. 1999, *ApJ* 510, L37
Chanmugam G. & Gabriel M. 1972, *A&A* 16, 149
Dufour, P., Fontain, G., Liebert, J., Williams, K., & Lai, D. K. 2008, *ApJ* 683, 167
Euchner F., Jordan S., Beuermann K., Gänsicke B., & Hessman F. V. 2002, *A&A* 390, 633
Euchner, F., Reinsch, K., Jordan, S., Beuermann, K., & Gänsicke, B. T. 2005, *A&A* 442, 651
Euchner, F., Jordan, S., Beuermann, K., Reinsch, K., & Gänsicke, B. T. 2006 *A&A* 451, 671
Fabrika S. N., Valyavin G. G., & Burlakova T. E. 2003, *Astronomy Letters* 29, 737
Fontaine G., Thomas J. H., & Van Horn H. M. 1973, *ApJ* 184, 911
Forster H., Strupat W., Rösner W., Wunner G., Ruder H., & Herold H. 1984, *J.Phys.* V 17, 1301
Garstang R. H. 1977, *Rep.Prog.Phys.* 40, 105
Gänsicke, B. T., Euchner, F., & Jordan, S. 2002, *A&A* 394, 957
Ginzburg V. L. 1964, *Sov. Phys. Dokl.* 9, 329
Greenstein J. L. 1984, *ApJ* 281, L47
Greenstein J. L., Henry R. J. W., & O'Connell R. F. 1985, *ApJ* 289, L25
Henry R. J. W. & O'Connell R. F. 1984, *ApJ* 282, L97
Jordan, S. 1992, *A&A*, 265, 570
Jordan, S., Schmelcher, P., & Becken W. 1998, *A&A*, 336, 33
Jordan, S., Schmelcher, P., Becken W., & Schweizer, W. 2001, *A&A* 376, 614
Jordan, S., Werner, K., & O'Toole, S. J. 2005, *A&A* 432, 273
Jordan, S., Aznar Cuadrado, R., Napiwotzki, R., Schmid, H. M., & Solanki, S. K. 2007, *A&A* 462, 1101
Karl, C. A., Napiwotzki, R., Heber, U., Dreizler, S., Koester, D., & Reid, I. N. 2005, *A&A* 434, 637
Kawka, A, Vennes, S, Schmidt G. D., Wickramasinghe D. T., & Koch R. 2007, *ApJ* 654, 499
Kemp J. C. 1970, *ApJ* 162, 169
Kemic S. B. 1974a, *ApJ* 193, 213
Kemic S. B. 1974b, JILA Rep. 133
Külebi, B., Jordan, S., Euchner F., Hirsch, H., & Löffler, W. 2008, Proceedings of the 16th European Workshop on White Dwarfs, Barcelona, in press
Liebert J. 1988, *PASP* 100, 1302
Liebert J. *et al.* 2005, *AJ* 129, 2376
Muslimov A. G., Van Horn H. M., & Wood M. A. 1995, *ApJ* 442, 758
O'Toole, S. J., Jordan, S., Friedrich, S., & Heber, U. 2005, *A&A* 437, 227
Preston G. W. 1970, *ApJ* 160, L143
Reimers, D., Jordan, S., & Christlieb, N. 2004, *A&A* 414, 1105
Rösner W., Wunner G., Herold H., & Ruder H. 1984, *J.Phys.* V 17, 29
Schmidt, G. D. *et al.* 2003, *ApJ* 595, 1101
Vanlandingham, K. M. *et al.* 2005, *AJ* 130, 734
Wickramasinghe, D. T., Schmidt, G. D., Ferrario, L., & Vennes, S. 2002, *MNRAS* 332, 29
Wickramasinghe, D. T. & Ferrario, L. 2005, *MNRAS* 356, 1576
Woltjer L. 1964, *ApJ* 140, 1309

Cosmic Magnetic Fields:
From Planets, to Stars and Galaxies
Proceedings IAU Symposium No. 259, 2008
K.G. Strassmeier, A.G. Kosovichev & J.E Beckman, eds.

doi:10.1017/S1743921309030750

Analysis of the hydrogen-rich magnetic White Dwarfs in the SDSS

Baybars Külebi[1], Stefan Jordan[1], Fabian Euchner[2], and Heiko Hirsch[3]

[1]Astronomisches Rechen-Institut, Mönchhofstrasse 12-14, D-69120 Heidelberg, Germany
[2]Swiss Seismological Service, ETH Zürich, Sonneggstrasse 5, CH-8092 Zürich, Switzerland
[3]Dr.-Remeis-Sternwarte, Bamberg, Sternwartestrasse 7, D-96049 Bamberg

Abstract. We have calculated optical spectra of hydrogen-rich (DA) white dwarfs with magnetic field strengths between 1 MG and 1000 MG for temperatures between 7000 K and 50000 K. Through a least-squares minimization scheme, we have analyzed the spectra of 114 magnetic DAs from the Sloan Digital Survey (SDSS; 95 previously published plus 14 newly discovered within SDSS).

Keywords. Stars: white dwarfs – magnetic fields

1. Introduction

White dwarfs with magnetic fields between 10^4 and 10^9 G are thought to represent more than 10% of the total population of white dwarfs (Liebert *et al.* 2003). In this work we present the re-analysis of the 95 DA Magnetic White Dwarfs (MWDs) discovered by Schmidt *et al.* (2003) and Vanlandingham *et al.* (2005), plus the analysis of 19 additional objects from SDSS up to Data Release 6 (9583 deg^2; http://www.sdss.org/dr6/).

Schmidt *et al.* (2003) and Vanlandingham *et al.* (2005) determined the field strengths and the inclinations of magnetic dipoles by comparing the observed spectra visually with model spectra calculated using a simplified radiation transfer code (Latter *et al.* 1987). Their and our results are compared on Fig. 1.

2. Analysis

The magnetic field geometry of the MWDs was determined with a modified version of the code developed by Euchner *et al.* (2002). This code calculates the total flux (and circular polarization) spectra for an arbitrary magnetic field topology by adding up appropriately weighted model spectra for a large number of surface elements. Magnetic field geometries are accounted for by a centered or an offset dipoles in this work. The model parameters are the magnetic dipole field strength B_d, the effective temperature T_{eff}, the inclination of the dipole axis i, and an offset along the magnetic dipole axis z_{off}, if offset dipoles are used. The observed spectra are fitted using an evolutionary algorithm (Rechenberg 1994) with a least-squares quality function and all of our fits resulted in reduced χ^2 values between 0.8 and 3.0 except for some high-field objects.

3. Discussion

In many cases offset dipole models resulted in significantly better fits than the models with centered dipoles. In particular some MWDs with high field strengths ($> 50\,\text{MG}$), where the spectra become very sensitive to the details of the magnetic field geometry, are not accounted for by our simple models.

SDSS has nearly tripled the number of MWDs; therefore the completeness of the total known MWD population is strongly affected by the selection biases of SDSS (priority selection of the spectroscopic targeting; Stoughton *et al.* 2002).

High-field MWDs are thought to be remnants of magnetic Ap and Bp stars. If flux conservation is assumed, population synthesis models predict the majority of MWDs' polar field strengths to be in the interval 50–500 MG (Wickramasinghe & Ferrario 2005). On the pther hand in our sample objects with magnetic field strengths lower than 50 MG are more numerous than the objects with higher magnetic field strengths (see Fig. 2), partly due to the selection biases. Nevertheless our result is consistent with previous ones and supports the hypothesis that magnetic fossil fields from Ap/Bp stars alone are not sufficient to produce the observed number of MWDs (Wickramasinghe & Ferrario 2005). Possible progenitor populations are A and B stars with magnetic field strengths below 100 G or magnetic F stars (Schmidt *et al.* 2003), which both are currently unobserved.

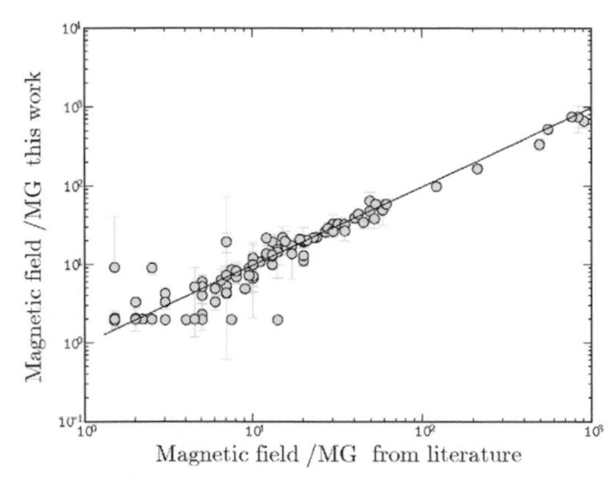

Figure 1. Comparison of dipole magnetic field fit values in this work versus Schmidt *et al.* (2003), Vanlandingham *et al.* (2005).

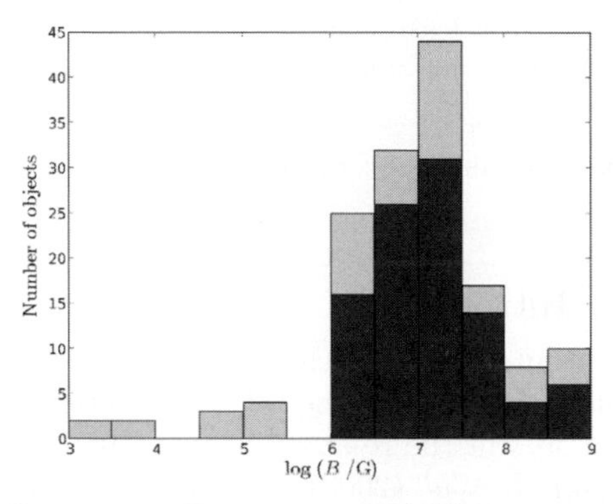

Figure 2. Histogram of magnetic white dwarfs in equal intervals of $\log B$. Gray columns represent the number of all DA MWDs and black shades represent the the contribution of SDSS to DA MWDs.

References

Euchner, F., Jordan, S., Beuermann, K., Gänsicke, B. T., & Hessman, F. V. 2002, *A&A* 390, 633

Latter, W. B., Schmidt, G. D., & Green, R. F. 1987, *ApJ* 320, 308

Liebert, J., Bergeron, P., & Holberg, J. B. 2003, *AJ* 125, 348

Rechenberg, I. 1994, Werkstatt Bionik und Evolutionstechnik No. 1 (Stuttgart: frommann-holzboog)

Schmidt, G. D., Harris, H. C., Liebert, J., *et al.* 2003, *ApJ* 595, 1101

Stoughton, C., Lupton, R. H., Bernardi, M., *et al.* 2002, *AJ* 123, 485

Vanlandingham, K. M., Schmidt, G. D., Eisenstein, D. J., *et al.* 2005, *AJ* 130, 734

Wickramasinghe, D. T. & Ferrario, L. 2005, *MNRAS* 356, 1576

Cosmic Magnetic Fields:
From Planets, to Stars and Galaxies
Proceedings IAU Symposium No. 259, 2008
K.G. Strassmeier, A.G. Kosovichev & J.E. Beckman, eds.

doi:10.1017/S1743921309030762

Magnetic fields in O-type stars measured with FORS 1 at the VLT

Swetlana Hubrig[1], **M. Schöller**[2], **R. S. Schnerr**[3], **I. Ilyin**[4], **H. F. Henrichs**[5], **R. Ignace**[6] **and J. F. González**[7]

[1]ESO, Santiago, Chile; email: shubrig@eso.org; [2]ESO, Garching, Germany; [3]Inst. for Solar Physics, Royal Swedish Academy of Sciences, Stockholm, Sweden: [4]AIP, Potsdam, Germany; [5]University of Amsterdam, The Netherlands [6]East Tennessee State University, Johnson City, USA; [7]Complejo Astronomico El Leoncito, San Juan, Argentina

Abstract. The presence of magnetic fields in O-type stars has been suspected for a long time. The discovery of such fields would explain a wide range of well documented enigmatic phenomena in massive stars, in particular cyclical wind variability, Hα emission variations, chemical peculiarity, narrow X-ray emission lines and non-thermal radio/X-ray emission. Here we present the results of our studies of magnetic fields in O-type stars, carried out over the last years.

Keywords. Stars: early-type – stars: magnetic fields – techniques: polarimetric – stars: HD 36879, HD 148937, HD 152408, HD 164794, HD 191612

1. Introduction

Direct measurements of the magnetic field strength in massive stars using spectropolarimetry to determine the Zeeman splitting of the spectral lines are difficult, since only a few spectral lines are available for these measurements, which are usually strongly broadened by rapid rotation. Before our study (Hubrig *et al.* 2008), a magnetic field had only been found in the three O-type stars, θ^1 Ori C, HD 155806, and HD 191612 (Donati *et al.* 2002; Hubrig *et al.* 2007; Donati *et al.* 2006).

2. Observations and analysis

More than 50 spectropolarimetric observations of 15 O-type stars were obtained from 2005 to 2008 using the multi-mode instrument FORS 1 at the 8.2-m Kueyen telescope. GRISMs 600B and 600R were used with the 0.4″ slit to obtain $R \approx 2000$ and $R \approx 3000$, respectively. Longitudinal magnetic fields were measured in two ways: using only the absorption hydrogen Balmer lines or using the entire spectrum including all available absorption lines of hydrogen, He I, He II, C III, C IV, N II, N III, and O III. Most of the targets were observed on three or four different nights to take into account the strong dependence of the longitudinal magnetic field on rotational aspect. Interestingly, a large part of the observed stars exhibited a change of polarity over certain nights. Four stars of our sample, HD 36879, HD 148937, HD 152408, and HD 164794, showed evidence for the presence of a weak magnetic field in the measurements using all spectral absorption lines. The uncertainties in the mean longitudinal field determination are obtained from the formal uncertainty in the linear regression of V/I versus the quantity $-\frac{g_{\rm eff} e}{4\pi m_e c^2} \lambda^2 \frac{1}{I} \frac{{\rm d}I}{{\rm d}\lambda} \langle B_z \rangle + V_0/I_0$.

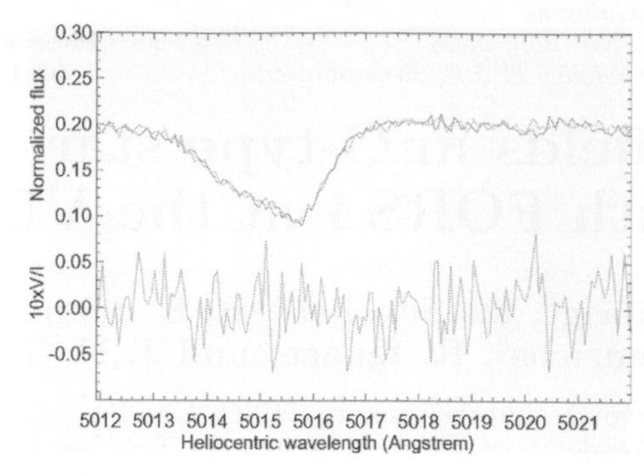

Figure 1. Low signal-to-noise Stokes I and V spectra of HD 191612 obtained with the echelle spectrograph SOFIN. Due to the rather fast rotation of the star the contribution of blends is not easily recognizable in the Stokes I spectrum. On the other hand, the blends become detectable in the Stokes V spectrum due to the Zeeman signatures produced in magnetically sensitive lines.

3. Discussion

This is the first time that magnetic field strengths were determined for such a large sample of stars, with an accuracy comparable to the errors obtained for the three previously known magnetic O-type stars, θ^1 Ori C, HD 155806, and HD 191612. For the magnetic Of?p star HD 191612, Donati *et al.* (2006) measured a magnetic field of $\langle B_z \rangle = -220 \pm 38$ G, by averaging a total of 52 exposures obtained over 4 different nights. This is similar to our typical errors of a few tens of G. The new high-resolution observations ($R \approx 30\,000$) of the O7 V(n) star HD 36879 and the O8fpe star HD 191612 with the SOFIN echelle spectrograph mounted at the 2.56 m Nordic Optical Telescope indicate the presence of weak magnetic fields of positive polarity in both stars. In spite of a rather low signal-to-noise ratio achieved in these observations (S/N≈200–270), caused by bad weather conditions, it was still possible to detect Zeeman features at the positions of He II, C IV, O II, and N III lines. An example of our observation of HD 191612 is presented in Fig. 1.

The four new magnetic O-type stars have different spectral types, luminosity classes, and behavior in various observational domains. The study of the evolutionary state of one of the Galactic Of?p stars, HD 191612, indicates that it is significantly evolved with an ∼O8 giant-like classification (Howarth *et al.* 2007). The youth of the most carefully studied magnetic O-type star θ^1 Ori C and the older age of the Of?p star HD 191612 suggest that the presence of magnetic fields in O-type stars is unrelated to their evolutionary state. We note that it is unclear yet whether more complex, smaller scale fields play a role in the atmospheres of hot stars. In the case of a more complex magnetic field topology, the longitudinal magnetic field integrated over the visible stellar surface will be smaller (or will even cancel) and will not be easily detected with the low-resolution FORS 1 measurements. However, high resolution spectropolarimeters (like ESPaDOnS, Narval, or SOFIN) should be able to detect such complex fields.

References

Donati, J.-F., Babel, J., Harries, T. J., *et al.* 2002, *MNRAS* 333, 55
Donati, J.-F., Howarth, I. D., Bouret, J.-C., *et al.* 2006, *MNRAS* 365, L6
Howarth, I. D., Walborn, N. R., Lennon, D. J., *et al.* 2007, *MNRAS* 381, 433
Hubrig, S., Yudin, R. V., Pogodin, M., *et al.* 2007, *AN* 328, 1133
Hubrig, S., Schöller, M., Schnerr, R. S., *et al.* 2008, *A&A* 490, 793

Cosmic Magnetic Fields:
From Planets, to Stars and Galaxies
Proceedings IAU Symposium No. 259, 2008
K.G. Strassmeier, A.G. Kosovichev & J.E. Beckman, eds.

doi:10.1017/S1743921309030774

Search for the magnetic field of the O7.5 III star ξ Persei

Huib F. Henrichs[1], R.S. Schnerr[2], J.A. de Jong[3], L. Kaper[1], J.-F. Donati[4] and C. Catala[5]

[1]Astronomical Institute, University of Amsterdam, Amsterdam, Netherlands
[2]Inst. for Solar Physics, Royal Swedish Academy of Sciences, Stockholm, Sweden
[3]Max Planck Institute for Extraterrestrial Physics, Garching, Germany
[4]Observatoire Midi-Pyrénées, Toulouse, France
[5]LESIA, Observatoire de Paris, CNRS, Université Paris Diderot, Meudon, France

Abstract. Cyclical wind variability is an ubiquitous but as yet unexplained feature among OB stars. The O7.5 III(n)((f)) star ξ Persei is the brightest representative of this class on the Northern hemisphere. As its prominent cyclical wind properties vary on a rotational time scale (2 or 4 days) the star has been already for a long time a serious magnetic candidate. As the cause of this enigmatic behavior non-radial pulsations and/or a surface magnetic field are suggested. We present a preliminary report on our attempts to detect a magnetic field in this star with high-resolution measurements obtained with the spectropolarimeter Narval at TBL, France during 2 observing runs of 5 nights in 2006 and 5 nights in 2007. Only upper limits could be obtained, even with the longest possible exposure times. If the star hosts a magnetic field, its surface strength should be less than about 300 G. This would still be enough to disturb the stellar wind significantly. From our new data it seems that the amplitude of the known non-radial pulsations has changed within less than a year, which needs further investigation.

Keywords. Stars: magnetic fields – techniques: polarimetric – stars: atmospheres – stars: ξ Per – stars: early-type – stars: winds – outflows – stars: rotation – stars: pulsations

1. Introduction

Like many O and B stars, the O7.5III(n)((f)) star ξ Per shows very prominent cyclical wind variability in the UV resonance lines (Fig. 1a), manifested by discrete absorption components (DACs) which migrate from red to blue, and narrow when they approach (but not reach) the terminal velocity as measured from saturated wind profiles (*e.g.* Kaper *et al.* 1999). In ξ Per the DAC period is 2.09 d. Multiwavelength observations of a number of OB stars, including ξ Per, have shown that the cyclic behavior is present down to the surface of the star (de Jong *et al.* 1997), and that the typical timescale varies with the (estimated) rotational timescale. This strongly argues in favor of a surface phenomenon which perturbs the base of the flow. Two scenarios are proposed: in the so-called Corotating Interaction Region (CIR) model a perturbation at the surface of the star causes a local increase (or decrease) of the radiative force driving the stellar wind, or surface magnetic fields may disturb the outflow, both resulting in a rotationally modulated stellar wind (see Cranmer and Owocki 1996). A number of coordinated UV and optical observations have confirmed the CIR model, including for the case of ξ Per, but the origin of the perturbations is not known, which is one of the most challenging problems in stellar wind research of the last decades. We present here our most recent efforts to measure the magnetic field and the pulsation properties.

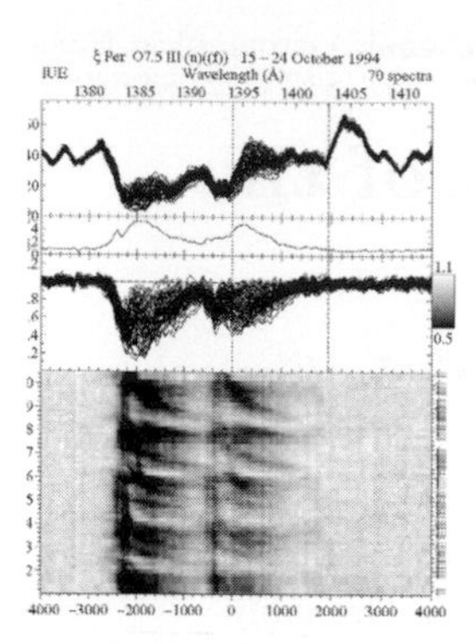

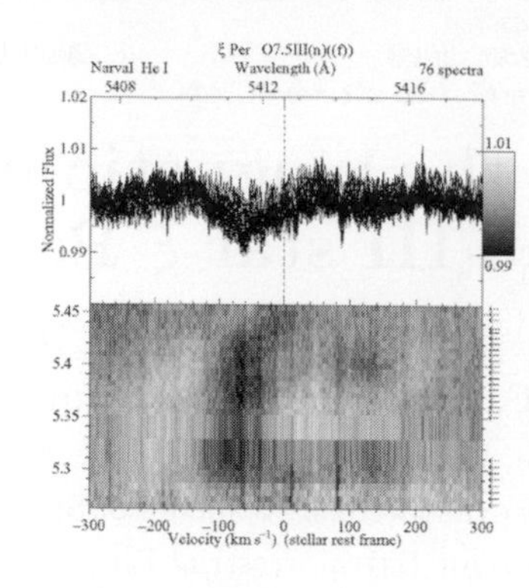

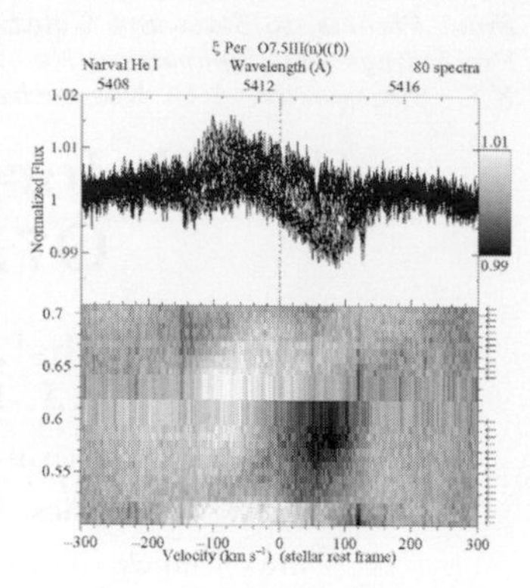

Figure 1. (a) Progressing DACs in the Si IV wind lines in 1994, (b) Non radial pulsation (period 3.5 h) in the He I 5411 line, December 2006. (c) September 2007; the amplitude of the moving features seems to have changed after 9 months.

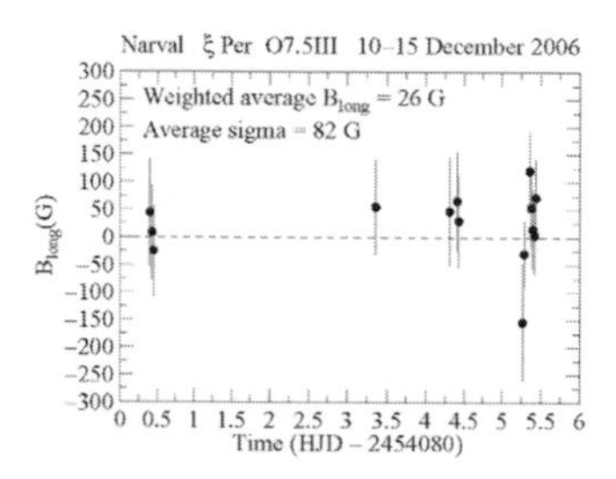

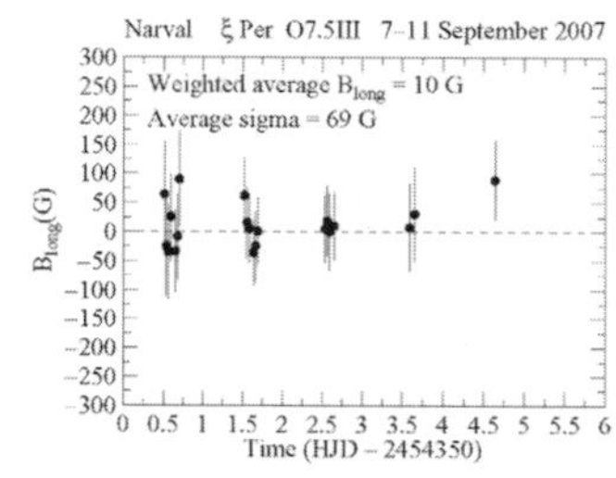

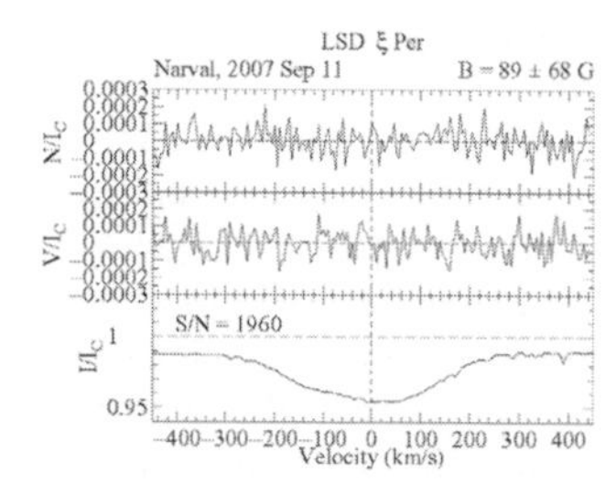

Figure 2. Preliminary magnetic results of Narval spectropolarimetry as a function of time. (a) December 2006. (b) September 2007. (c) LSD Stokes V profile of the last point in September 2007. No significant Zeeman signature was found.

2. Data analysis and discussion

Previous Musicos magnetic measurements of ξ Per were presented by de Jong *et al.* (2001), with no detection. For Narval data the magnetic analysis is essentially the same, applying the least-squares deconvolution method (Donati *et al.* 1997) to the spectral lines sensitive to magnetic effects. This yields the longitudinal component of the field, averaged over the facing hemisphere of the star. In Fig. 2a and 2b the 45 results are plotted as a function of time. No Zeeman signature was found. As an example, Fig. 2c shows the LSD profile of the best exposed spectrum of September 2007 with a S/N = 1960. The magnetic values are preliminary, as the used spectral linelist can still be optimized.

We also analysed a number of spectral lines for non-radial pulsations with known period of 3.5 h found in a previous Musicos campaign (de Jong *et al.* 1999). The amplitude may have changed during the 9 months between our Narval runs, see Fig. 1b and 1c. This obviously needs further investigation, as the impact on the stellar wind may have changed.

Observations with a higher efficiency should be able to detect a possible field, unless the geometry is such that the magnetic properties of the two hemispheres cancel out.

References

de Jong, J. A., Henrichs, H. F., Schrijvers, C., Gies, D. R. *et al.* 1999, *A&A* 345, 172
de Jong, J. A., Henrichs, H. F., Kaper, L., Nichols, J. S. *et al.* 2001, *A&A* 368, 601
Cranmer, S. R. & Owocki, S. P. 1996, *ApJ* 462, 469
Donati, J.-F., Semel, M., Carter, B. D. *et al.* 1997, *MNRAS* 291, 658
Kaper, L., Henrichs, H. F., Nichols, J. S., & Telting, J. H. 1999, *A&A* 344, 231

Cosmic Magnetic Fields:
From Planets, to Stars and Galaxies
Proceedings IAU Symposium No. 259, 2008
K.G. Strassmeier, A.G. Kosovichev & J.E. Beckman, eds.

doi:10.1017/S1743921309030786

Is the wind of the Oe star HD 155806 magnetically confined?

Véronique Petit[1], A. W. Fullerton[2], S. Bagnulo[3], G. A. Wade[4] and the MiMeS Collaboration

[1]Département de Physique, Université Laval, Québec, Canada
[2]Space Telescope Science Institute, Baltimore, USA
[3]Armagh Observatory, Armagh, Northern Ireland
[4]Department Physics, Royal Military College of Canada, Kingston, Canada

Abstract. Oe stars are a subset of the O-type stars that exhibit emission lines from a circumstellar disk. The recent detection of magnetic fields in some O-type stars suggests a possible explanation for the stability of disk-like structures around Oe stars. According to this hypothesis, the wind of the star is channeled by a dipolar magnetic field producing a disc in the magnetic equatorial plane. As a test of this model, we have obtained spectropolarimetric observations of the hottest Galactic Oe star HD 155806. Here we discuss its results and implications.

Keywords. Stars: magnetic fields – stars: mass loss – stars: early-type – stars: emission-line

---

Oe stars are a subset of the O-type stars that exhibit emission in spectra of their Balmer lines, indicating the presence of a circumstellar disk. It is generally believed that the low incidence of the Oe stars (only 4 are known in the Milky Way) is caused by the increasingly powerful radiation-driven winds of the O-type stars, which inhibit the formation of long-lived circumstellar structures. However, all the *bona fide* Oe stars exhibit typical wind profiles in their ultraviolet resonance lines, showing that these stars have winds similar to those of other O-type stars. The mystery is therefore not "why are there so few Oe stars?", but "why are there any at all?".

The recent detection of magnetic fields in O-type stars suggests a possible explanation for the stability of disk-like structures around Oe stars. From MHD simulations, ud-Doula *et al.* (2008) have shown that under high magnetic confinement and high rotation, a "magnetically confined wind" can trap material in its magnetosphere. The circumstellar structure is composed of different material at different times, but its longevity is mediated by the magnetic field of the star. Consequently, this model neatly avoids problems associated with the disruption of the disks of Oe stars, and suggests that Oe stars might represent extreme cases of a more widespread phenomenon that depends on the balance between the strength of the stellar wind and the strength of the magnetic field.

HD 155806 is the Galactic Oe star with the most powerful wind. We estimated that a surface dipolar magnetic field of at least 235 G is required to confine its wind. It exhibits a large, single-peaked Hα emission profile that implies that the disk is not being viewed "edge-on". Encouragingly, the detection of a weak magnetic field (115 ± 37 G) in HD 155806 with VLT/FORS1 has recently been reported by Hubrig *et al.* (2007), although this detection was not reproduced in further measurements (Hubrig *et al.* 2008). In addition, a re-analysis of the FORS1 observations (available in ESO archives) as a component of this present study led to less conclusive results. New high-resolution spectropolarimetric observations of HD 155806 were obtained with ESPaDOnS at CFHT in June and July 2008. No magnetic Stokes V signature is detected in our observations

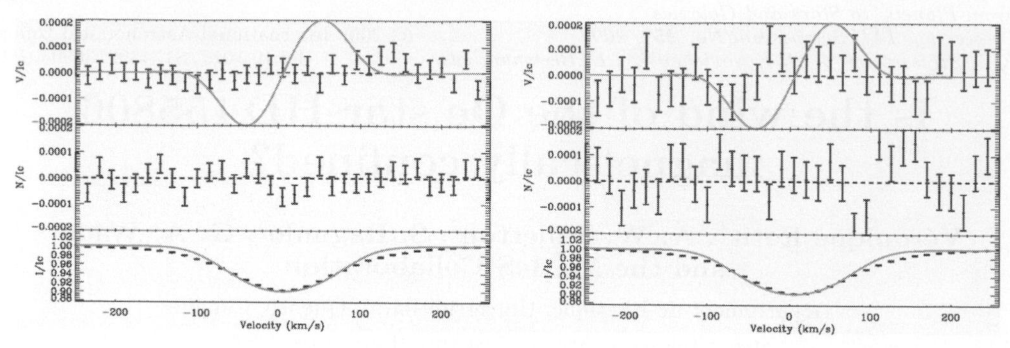

Figure 1. Least Squares Deconvolved profiles in June (left) and July (right). The curves are the mean Stokes I profiles (bottom), the mean Stokes V profiles (top) and the N diagnostic null profiles (middle). The bold line represents the smallest-amplitude Stokes V profile corresponding to the longitudinal field of 115 G reported by Hubrig *et al.* (2007), which is a pure dipole with a polar strength of 380 G.

(Figure 1). In order to extract the surface field characteristics constrained by the observed Stokes V profiles, we are in the process of comparing them with theoretical profiles derived from a large grid of dipolar magnetic field configurations that were calculated with the polarised LTE radiative transfer code Zeeman2 (Landstreet 1988). With the Bayesian method described by Petit *et al.* (2008), we will obtain an upper limit on the strength of the dipolar magnetic field at the stellar surface that is independent of the magnetic configuration. We shall also use the FORS1 measurements as additional constraints.

Although detailed modeling still needs to be performed, we estimate that the upper limit on the surface magnetic dipole will be about 200 G, according to our preliminary analysis. A stronger magnetic field would need to be in a configuration that produces a weak or null Stokes V signature. For dipoles, these rare configurations have well-defined geometries and correspond to the observer looking directly at the magnetic equator. According to MHD simulations of an aligned dipole rotator (ud-Doula *et al.* 2008), an accumulation at the equator of the magnetosphere will occur when both the confinement parameter and the rotation rate are large. Rigid-field hydrodynamics modelling of a tilted dipole (Townsend *et al.* 2007) predicts that the disk will have an average inclination that lies somewhere between the rotation and magnetic equatorial planes, with plasma concentrated at the intersection between them. If the disk of HD 155806 is produced by a large magnetic field hidden in a Stokes V-free configuration, we would expect to see a more edge-on disk signature in the Hα emission line than that which is observed at the phase when the Stokes V signature is null.

Since a field with a surface dipolar strength lower than 200 G would not confine the wind of HD 155806 sufficiently to produce a disk; and since a stronger (but undetected) field would be in a highly improbable configuration that would in any case be unable to produce the emission structure seen in Hα, we conclude that it is unlikely that the disk of HD 155806 is caused by a large-scale magnetic field confining the stellar wind.

References

Hubrig, S., Yudin, R. V., Pogodin, M., Schller, M., & Peters, G. J. 2007, *AN* 328, 1133
Hubrig, S., Schöller, M., Schnerr, R. S., González, J. F. *et al.* 2008, *A&A* 490, 793
Landstreet, J. D. 1988, *ApJ* 326, 967
Petit, V., Wade, G. A., Drissen, L., Montmerle, T., & Alecian, E. 2008, *MNRAS* 387, L23
Townsend, R. H. D., Owocki, S. P., & ud-Doula, A. 2007, *MNRAS* 382, 139
ud-Doula, A., Owocki, S. P., & Townsend, R. H. D. 2008, *MNRAS* 385, 97

Cosmic Magnetic Fields:
From Planets, to Stars and Galaxies
Proceedings IAU Symposium No. 259, 2008
K.G. Strassmeier, A.G. Kosovichev & J.E. Beckman, eds.

doi:10.1017/S1743921309030798

The MiMeS project: first results

Jason H. Grunhut[1], E. Alecian[1,2], D.A. Bohlender[3], J.-C. Bouret[4], H. Henrichs[5], C. Neiner[6], V. Petit[7], N. St. Louis[8], M. Aurière[9], O. Kochukhov[10], J. Silvester[1], G.A. Wade[1], A. ud-Doula[11] and the MiMeS Collaboration†

[1]Royal Military College of Canada, [2]LESIA, France, [3]Canadian Astronomy Data Centre,[4]LAM, France,[5]Ast. Inst. Amsterdam, Netherlands, [6]GEPI, France, [7]Université Laval, Canada, [8]Univ. de Montréal, Canada, [9]LAT, France, [10]Uppsala University, Sweden, [11]Morrisville State College, USA

Abstract. Massive stars are those stars with initial masses above about 8 times that of the sun, eventually leading to catastrophic explosions in the form of supernovae. These represent the most massive and luminous stellar component of the Universe, and are the crucibles in which the lion's share of the chemical elements are forged. These rapidly-evolving stars drive the chemistry, structure and evolution of galaxies, dominating the ecology of the Universe - not only as supernovae, but also during their entire lifetimes - with far-reaching consequences. Although the existence of magnetic fields in massive stars is no longer in question, our knowledge of the basic statistical properties of massive star magnetic fields is seriously incomplete. The Magnetism in Massive Stars (MiMeS) Project represents a comprehensive, multidisciplinary strategy by an international team of recognized researchers to address the "big questions" related to the complex and puzzling magnetism of massive stars. This paper present the first results of the MiMeS Large Program at the Canada-France-Hawaii Telescope.

Keywords. Magnetic fields – massive stars – hot stars – star formation – stellar evolution – stellar winds – spectropolarimetry

---

The Magnetism in Massive Stars (MiMeS) Project represents a comprehensive, multidisciplinary strategy by an international team of recognized researchers to address the big questions related to the complex and puzzling magnetism of massive stars. Recently, MiMeS was awarded "Large Program" status by both Canada and France at the Canada-France-Hawaii Telescope (CFHT), where the Project has been allocated 640 hours of dedicated time with the ESPaDOnS spectropolarimeter from late 2008 through 2012.

The structure of the MiMeS Large Program includes 255 hours for a 20-target "Targeted Component" (TC) which will obtained time-resolved high-resolution spectropolarimetery of known magnetic massive stars in Stokes I and V and for some targets also Stokes Q and U, in addition to 385 hours dedicated to the $\sim$ 150-target "Survey Component" (SC), with the goal to provide critical missing information about field incidence and statistical field properties of a much larger sample of massive stars. For more information about the goals and structure of the MiMeS Large Program, see Wade *et al.* (these proceedings).

MiMeS observations first began in August 2008, since which time over 200 polarised spectra of approximately 50 targets have been acquired. The signal-to-noise ratio of the spectra have been in good agreement with the MiMeS exposure model (see Wade *et al.*, these proceedings), with 80% of spectra achieving more than 0.8 times the predicted S/N. The precision of the magnetic diagnosis has also generally exceeded expectations.

† www.physics.queensu.ca/~wade/mimes

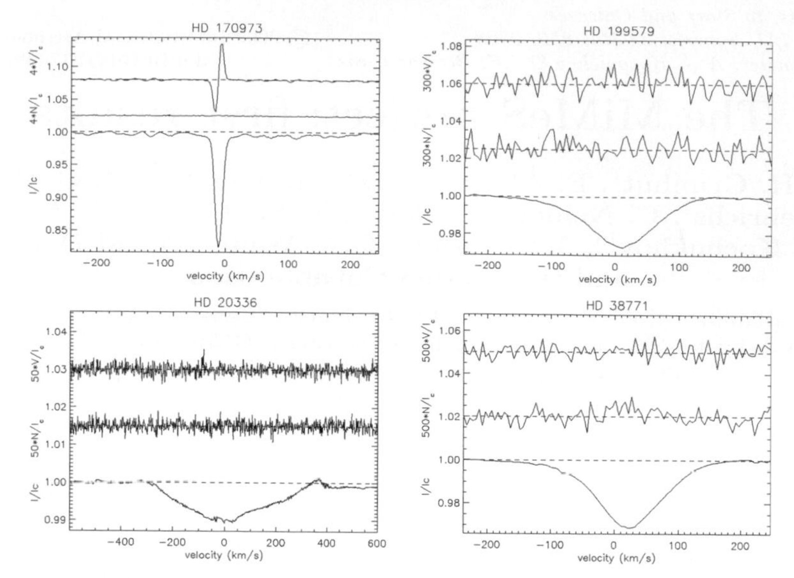

Figure 1. Least-Squares Deconvolved (LSD) profiles of four MiMeS Survey Component targets observed during the first semester of observations. *Top left:* HD 170973 (A0p), $v \sin i \simeq 20$ km s^{-1}, $B_\ell = -538 \pm 11$ G (Stokes V magnetic signature detected, dipole strength $B_{\rm d} > 1.6$ kG); *Top right:* HD 199579 (06Ve), $v \sin i \simeq 75$ km s^{-1}, $B_\ell = 1 \pm 29$ G (no detection, $B_{\rm d} \lesssim 100$ G); *Bottom left:* HD 20336 (B2.5Vne), $v \sin i \simeq 175$ km s^{-1}, $B_\ell = 32 \pm 63$ G (no detection, $B_{\rm d} \lesssim 200$ G; *Bottom right:* HD 38771 (B01ab), $v \sin i \simeq 65$ km s^{-1}, $B_\ell = -1 \pm 8$ G (no detection, $B_{\rm d} \lesssim 30$ G). Each frame shows Stokes I (bottom), scaled Stokes V (top) and N diagnostic null (middle). Note the large range of different scaling factors.

Seven of the observed targets are TC stars, while the remainder are SC objects (primarily field O, B and Be stars). The first MiMeS observation, of the A0p star HD 170973, was also its first detection. Preliminary Least-Squares Deconvolved (LSD; Donati *et al.* 1997) Stokes I and V LSD profiles, as well as the N diagnostic null profile, of HD 170973, along with those of other SC targets, are shown in Fig 1. The SC targets presented in Fig. 1 span much of the range of spectroscopic types, rotational velocities and emission properties of stars to be observed within the context of MiMeS, and provide a reasonable overview of the typical precision expected from the Survey Component. In addition to some ~ 40 observed SC targets in which no magnetic field is detected, one new magnetic massive star appears to have been identified. The physical characteristics of this new detection make it a fascinating and unique object for further detailed study. However, the announcement of the details of this discovery must await a confirming observation to be acquired very soon (in December 2008 or January 2009).

The low rate of field detections was fully expected, based on the small number of magnetic massive stars reported in the literature. This suggests that the incidence of magnetic stars begins to decrease with increasing mass somewhere past $\sim 4-5\ M_\odot$ (see Power *et al.* 2008).

Further results of the MiMeS SC are reported by Petit *et al.* (these proceedings).

References

Donati J.-F., Semel M., Carter B. D., & Rees D. E., Collier Cameron A. 1997, *MNRAS* 291, 658

Power J., Wade G. A., Aurière M., Silvester J., & Hanes D. 2007, *Physics of Magnetic Stars*, Special Astrophysical Observatory, p. 89

Cosmic Magnetic Fields:
From Planets, to Stars and Galaxies
Proceedings IAU Symposium No. 259, 2008
K.G. Strassmeier, A.G. Kosovichev & J.E. Beckman, eds.

doi:10.1017/S1743921309030804

New magnetic field measurements of β Cephei stars and slowly pulsating B stars

Swetlana Hubrig[1], M. Briquet[2], P. De Cat[3], M. Schöller[4], T. Morel[5] and I. Ilyin[6]

[1]ESO, Santiago, Chile; email: shubrig@eso.org;
[2]Katholieke Universiteit Leuven, Belgium
[3]Koninklijke Sterrenwacht van België, Brussel, Belgium
[4]ESO, Garching, Germany
[5]Université de Liège, Belgium
[6]Astrophysikalisches Institut Potsdam, Potsdam, Germany

Abstract. We present the results of the continuation of our magnetic survey with FORS 1 at the VLT of a sample of B-type stars consisting of confirmed or candidate β Cephei stars and Slowly Pulsating B stars. Roughly one third of the studied β Cephei stars have detected magnetic fields. The fraction of magnetic Slowly Pulsating B and candidate Slowly Pulsating B stars is found to be higher, up to 50%. We find that the domains of magnetic and non-magnetic pulsating stars in the H-R diagram largely overlap, and no clear picture emerges as to the possible evolution of the magnetic field across the main sequence.

Keywords. Stars: variables: β Cephei – stars: variables: SPBs – stars: oscillations – stars: Hertzsprung-Russell diagram, stars: magnetic fields, techniques: polarimetric – stars: ξ^1 CMa

1. Introduction

In our first publication on the magnetic survey of pulsating B-type stars (Hubrig *et al.* 2006), we announced detections of weak mean longitudinal magnetic fields of the order of a few hundred Gauss in 13 Slowly Pulsating B (hereafter SPB) stars and in the β Cephei star ξ^1 CMa. ξ^1 CMa showed a mean longitudinal magnetic field of the order of 300 G. Since the role of magnetic fields in modeling oscillations of B-type stars remained to be studied, we collected additional 98 magnetic field measurements in a sample of 60 pulsating and candidate pulsating stars.

2. Observations and analysis

The spectropolarimetric observations have been carried out in the years 2006 to 2008 at the European Southern Observatory with FORS 1 mounted on the 8-m Kueyen telescope of the VLT. For the major part of the observations we used the GRISM 600B in the wavelength range 3480–5890 Å to cover all hydrogen Balmer lines from Hβ to the Balmer jump. One additional high resolution observation of ξ^1 CMa has been obtained with the SOFIN echelle spectrograph installed at the 2.56 m Nordic Optical Telescope. The Stokes I and V spectra are presented in the left panel of Fig. 1. In the right panel of Fig. 1 we present magnetic field measurements of the same star using FORS 1 over the last 4.4 years. Among the SPBs and β Cephei stars, the detected magnetic fields are mainly of the order of 100–200 G.

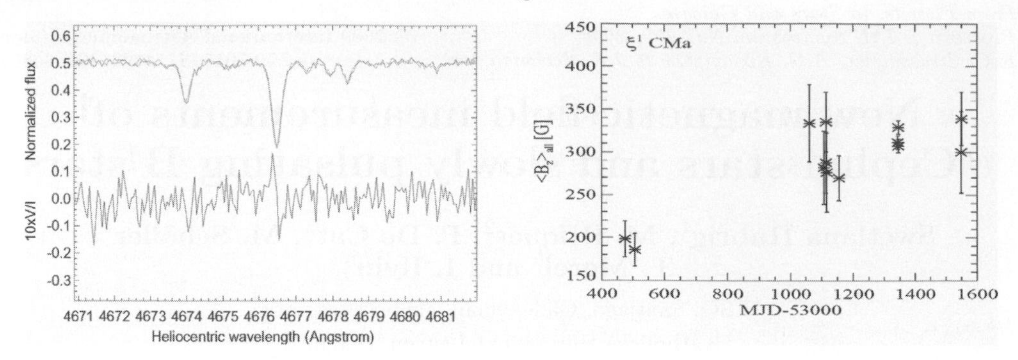

Figure 1. Left panel: High-resolution (R≈30 000) SOFIN polarimetric spectrum of ξ^1 CMa. A clear Zeeman feature is detected at the position of the unblended line O II λ4676.2. Right panel: Magnetic field measurements with FORS 1 of ξ^1 CMa over the last 4.4 years

3. Discussion

Roughly one third of the β Cephei stars have detected magnetic fields: Out of 13 β Cephei stars studied to date with FORS 1, four stars possess weak magnetic fields, and out of the sample of six suspected β Cephei stars two show a weak magnetic field. The fraction of magnetic SPBs and candidate SPBs is found to be higher: roughly half of the 34 SPB stars have been found to be magnetic and among the 16 candidate SPBs, eight stars possess magnetic fields.

About a dozen pulsating stars discussed in our previous study Hubrig *et al.* (2006) have been subsequently observed by Silvester et al. (this proceedings) with ESPaDOnS at CFHT and NARVAL at TBL. The interesting result is that they could confirm our previous non-detections in eight β Cephei and SPB stars but failed to detect weak magnetic fields in the other five magnetic SPB stars selected from our study. On the other hand, as has been shown at the present meeting by Henrichs *et al.* (this proceedings) who confirmed our detection of a magnetic field in the B3V star 16 Peg, the integrated Zeeman features obtained from the application of a cross-correlation method, the Least-Squares Deconvolution, are extremely weak for ∼100 G fields. This means, that even using high-resolution spectrographs like ESPaDOnS or Narval, very high signal-to-noise observations are mandatory to be able to detect magnetic fields in hot stars where only a limited number of metal lines are suitable for the measurements.

In an attempt to understand why only a fraction of the pulsating stars exhibit magnetic fields, we studied the position of magnetic and non-magnetic pulsating stars in the H-R diagram. We find that their domains in the H-R diagram largely overlap, and no clear picture emerges as to the possible evolution of the magnetic field across the main sequence. It is possible that stronger fields tend to be found in stars with lower pulsating frequencies and smaller pulsating amplitudes. A somewhat similar trend is found if we consider a correlation between the field strength and the $v \sin i$-values, i.e. stronger magnetic fields tend to be found in more slowly rotating stars.

References

Hubrig, S., Briquet, M., Schöller, M., *et al.* 2006, *MNRAS* 369, L61

Cosmic Magnetic Fields:
From Planets, to Stars and Galaxies
Proceedings IAU Symposium No. 259, 2008
K.G. Strassmeier, A.G. Kosovichev & J.E. Beckman, eds.

doi:10.1017/S1743921309030816

Magnetic observations of pulsating B and Be stars with ESPaDOnS and Narval

James Silvester[1,2], C. Neiner[3], H. F. Henrichs[4], G. A. Wade[2], E. Alecian[2] and V. Petit[5]

[1]Department of Physics, Queen's University, Kingston, Ontario, Canada, K7L 3N6
[2]Department of Physics, Royal Military College of Canada, Kingston, ON, Canada, K7K 7B4
[3]GEPI, Observatoire de Paris, CNRS, 5 place Jules Janssen, 92190 Meudon, France
[4]Astronomical Institute "Anton Pannekoek", University of Amsterdam, The Netherlands
[5]Départment de physique, CRAQ, Université Laval, Québec, Canada, G1K 7P4

Abstract. Discoveries of magnetic fields in pulsating B and Be stars have been claimed from low-resolution spectropolarimetric observations with FORS1 at VLT. We used the new generation of high-resolution spectropolarimeters, ESPaDOnS at CFHT and NARVAL at TBL, to check for the existence of these fields. We find that most of the claimed magnetic stars do not host a magnetic field. This work shows the importance of a critical analysis of FORS1 data when searching for weak magnetic fields in early-type stars and the advantage of using ESPaDOnS and NARVAL to study such type of stars.

Keywords. Stars: magnetic fields – techniques: polarimetric – stars: early-type – stars: emission-line, Be

1. Introduction

Slowly Pulsating B (SPB), β Cephei and Be stars are pulsating variables classically considered to be non-magnetic stars. Up to now only two β Cephei stars, β Cep itself (Henrichs *et al.* 2000) and V2052 Oph (Neiner *et al.* 2003a), and one SPB star, ζ Cas (Neiner *et al.* 2003b), are known to host a magnetic field, discovered using the MuSiCoS spectropolarimeter at TBL. Amongst the Be stars only ω Ori (Neiner et al 2003c) showed evidence suggestive of the presence of a magnetic field. The longitudinal fields in the detected stars are relatively weak, of the order of a hundred gauss.

However, recent observations by Hubrig *et al.* (2006) have suggested that the occurrence of observable fields in SPB stars is significant ($\sim$50%) and they should be considered a class of magnetic pulsators. In addition Hubrig *et al.* (2007) report that 3 out of a sample of 15 Be stars (χ Oph, υ Sgr and EW CMa) are magnetic with weak fields detected at the 3σ level. To verify this potentially important result, an observing program was undertaken using the ESPaDOnS and NARVAL spectropolarimeters installed at CFHT (Hawaii) and TBL (Pic du Midi, France) respectively.

2. Results

The spectra have been reduced using the libre-ESPRIT data reduction package (Donati et al 1997). The peak signal-to-noise ranges from 300 to 1100 and in all cases the spectra were normalised using an automated code with a polynomial fit of 5 or 6 degrees. To obtain a longitudinal field measurement, Least-Squares Deconvolution (LSD; Donati *et al.* 1997) was implemented. In each case a suitable atmosphere line mask based on solar abundances was used. Input atomic line data for the spectrum synthesis were extracted

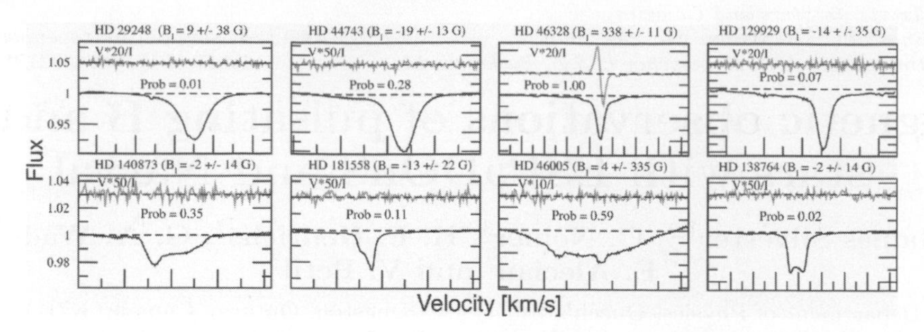

Figure 1. LSD profiles for a sample of stars included in this study. Each square represents a different star, with the bottom line showing the LSD mean intensity (Stokes I) profile and the top line showing the Stokes V profile.

from the VALD database (Kupka *et al.* 1999). For the Be star line masks, only lines not showing emission were used in order to minimize the impact of the circumstellar environment.

The LSD results are shown in Fig. 1. With the exception of the marginal detection in 16 Peg, we detect no field in any of the other SPB targets, with the highest detection probability (excluding 16 Peg) being 0.59 (HD 46005). In agreement with Hubrig *et al.* (2006), of the β Cephei stars we only detect a field in ξ^1 CMa, greater than the 3σ level. Of the Be stars we detect no field in either objects υ Sgr and χ Oph over 4 and 3 observations respectively.

3. Conclusions

The incidence of detected magnetic field in SPBs, β Cephei and Be stars appears to reflect what is seen in other stellar groups (a few percents), such as magnetic Ap/Bp stars compared to normal A and B stars (Power *et al.* 2008), or non-magnetic to magnetic Herbig AeBe stars (Alecian *et al.* 2008). It may also be that the objects detected as magnetic may represent the extreme magnetic cases of these stellar groups. Regardless of the reasoning, our observations certainly indicate that, to date, magnetic SPB, β Cep and Be stars remain rather rare. Our study also emphasises the necessity of using high-resolution data for a robust magnetic field diagnosis of these types of stars.

References

Alecian, E., Wade, G. A., Catala, C., *et al.* 2008, *CoSka* 38, 235

Donati, J.-F., Semel, M., Carter, B. D., Rees, D. E., & Collier, A. C. 1997, *MNRAS* 313, 658

Hubrig, S., Briquet, M., Schöller, M., De Cat, P., Mathys, G., & Aerts, C. 2006, *MNRAS* 369, L61

Hubrig, S., Yudin, R.V., Pogodin, M., Sch́oller, M., & Peters, G. J. 2007, *AN* 328, 1133

Henrichs, H. F., de Jong, J. A., Donati, J.-F., Catala, C., Wade, G. A., Shorlin, S. L. S., Veen, P. M., Nichols, J. S., & Kaper, L. 2000, The Be Phenomenon in Early-Type Stars, IAU Colloquium 175, ASP Conference Proceedings, Vol. 214, p.324

Kupka, F., Piskunov, N. E., Ryabchikova, T. A., Stempels, H. C., & Wiess, W. W. 1999, *A&AS* 138, 119

Neiner, C., Geers, V. C., Henrichs, H. F., Floquet, M., Frémat, Y., Hubert, A.-M., Preuss, O., & Wiersema, K. 2003b, *A&A* 406, 1019

Neiner, C., Henrichs, H. F., Floquet, M., Frémat, Y., Preuss, O., Hubert, A.-M., Geers, V. C., Tijani, A. H., Nichols, J. S., & Jankov, S. 2003a, *A&A* 411, 565

Neiner, C., Hubert, A.-M., Frémat, Y. *et al.* 2003c, *A&A* 409, 275

Power, J., Wade, G. A., Aurière, M., Silvester, J., & Hanes, D. 2008, *CoSka* 38, 443

Cosmic Magnetic Fields:
From Planets, to Stars and Galaxies
Proceedings IAU Symposium No. 259, 2008
K.G. Strassmeier, A.G. Kosovichev & J.E. Beckman, eds.

doi:10.1017/S1743921309030828

The magnetic field of the B3V star 16 Pegasi

Huib F. Henrichs[1], C. Neiner[2], R. S. Schnerr[3], E. Verdugo[4], A. Alecian[5], C. Catala[6], F. Cochard[7], J. Gutiérrez[2], A.-L. Huat[2], J. Silvester[5], and O. Thizy[7]

[1]Astronomical Institute, University of Amsterdam, Amsterdam, Netherlands
[2]GEPI, Observatoire de Paris, CNRS, Université Paris Diderot, 92190 Meudon, France
[3]Inst. for Solar Physics, Royal Swedish Academy of Sciences, Stockholm, Sweden
[4]European Space Astronomy Centre (ESAC), Madrid, Spain
[5]Dept. of Physics, Royal Military College of Canada, Kingston, Canada
[6]LESIA, Observatoire de Paris, CNRS, Université Paris Diderot, 92190 Meudon, France
[7]Shelyak Instruments, Revel, France

Abstract. The slowly pulsating B3V star 16 Pegasi was discovered by Hubrig *et al.* (2006) to be magnetic, based on low-resolution spectropolarimetric observations with FORS1 at the VLT. We have confirmed the presence of a magnetic field with new measurements with the spectropolarimeters Narval at TBL, France and Espadons at CFHT, Hawaii during 2007. The most likely period is about 1.44 d for the modulation of the field, but this could not be firmly established with the available data set. No variability has been found in the UV stellar wind lines. Although the star was reported once to show Hα in emission, there exists at present no confirmation that the star is a Be star.

Keywords. Stars: magnetic fields – techniques: polarimetric – stars: atmospheres – stars: 16 Peg – stars: early-type – stars: winds – stars: rotation – stars: pulsations

1. Introduction

The B3V star 16 Peg received our particular attention because more than 50 years ago it was reported at a single occasion to show emission in Hα, giving the star its Be status. However, this emission was never observed again, also not in our data set, and hence this star is probably not a Be star.

We describe the magnetic analysis leading to the confirmation of the presence of a magnetic field. We also summarize the period search in these data, and conclude that the most likely rotation period is near 1.44 d, but the data set is actually too poor to firmly constrain the period.

2. Magnetic Analysis

We applied the least-squares deconvolution method (LSD, Donati *et al.* 1997) to the spectral lines sensitive to magnetic effects in the 38 obtained Stokes V spectra (see *e.g.* Fig. 1 left), yielding the longitudinal component of the field, averaged over the star, B_ℓ. In Fig. 1 (right) the magnetic data are plotted as a function of time, including the two FORS1 data points from Hubrig *et al.* (2006). We found a varying magnetic field, with absolute maximum values of about 170 $\pm$ 45 G. The smallest error bars are about 30 G. The results are rather robust against integration limits in the Zeeman signature, but we note that these are preliminary in the sense that the used linelist can still be optimized.

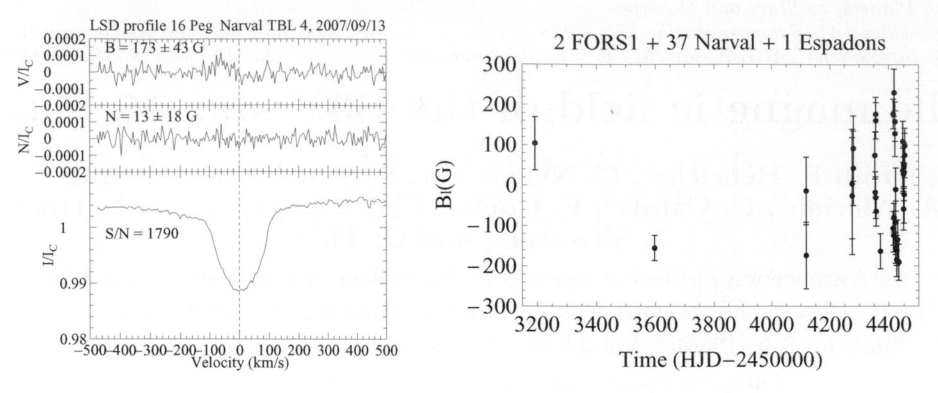

Figure 1. *Left:* Example of a Stokes V Zeeman signature. Top to bottom: LSD circularly polarised Stokes V, (null) N, and unpolarised I profiles of a Narval spectrum. *Right:* Magnetic data history of 16 Peg, including the first 2 discovery datapoints with FORS1 by Hubrig *et al.*

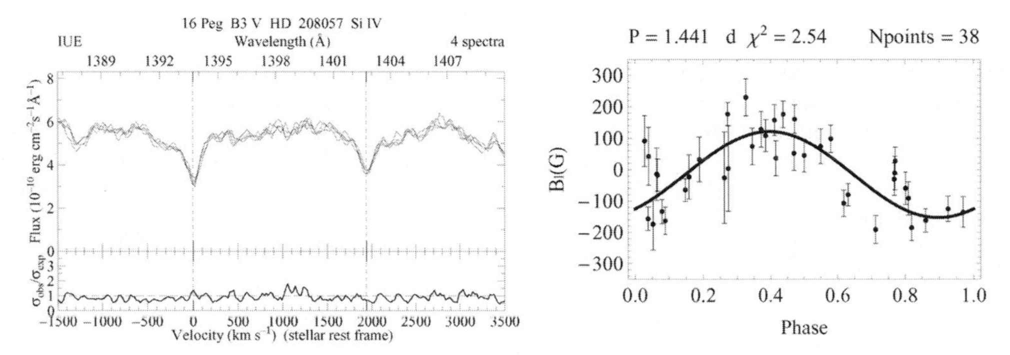

Figure 2. *Left:* Overplot of ultraviolet Si IV resonance lines of 16 Peg. The significance of the variability, expressed as the square root of the ratio of the measured to the expected variances in the lower panel shows no significant variability. *Right:* Phase plot of Narval + Espadons magnetic data overplotted with the best weighted fit period of 1.441 day.

3. Period search and conclusions

We analysed the 5 available high-resolution UV spectra, taken over 9 years and searched for variability in stellar wind lines. Although such a variability is expected for magnetic B stars, no variability at the S/N level of about 15 could be found (see Fig. 2 left).

A weighted sinusoidal fit to the data with starting values taken at maximum power of a (unweighted) CLEAN analysis resulted for the 2007 data in $P = 1.441$ d with a reduced $\chi^2 = 2.5$. Including the two earlier FORS1 points changed the fit significantly, showing the poor coverage and varying quality of the data set. This period is consistent with $P_{\rm rot}/\sin i = 1.6 \pm 0.6$ d from the model stellar parameters given by Hubrig *et al.* (2006): $v\sin i = 104 \pm 6$ km/s and $R/R_\odot$ 3.2 ± 1.0, which constrains the inclination angle.

De Cat *et al.* (2007) found three photometric periods in this SPB star: 0.8905 d, 1.24668 d and 0.4039 d. Our proposed magnetic period is not consistent with a rotational splitting.

New spectropolarimetric data with a proper coverage of the suggested rotation period is obviously needed.

References

De Cat, P., Briquet, M., Aerts, C., Goossens, K., Saesen, S., *et al.* 2007, *A&A* 463, 243

Donati, J.-F., Semel, M., Carter, *et al.* 1997, *MNRAS* 291, 658

Hubrig, S., Briquet, M., Schöller, M., De Cat, P., Mathys, G., & Aerts, C. 2006, *MNRAS* 369, L61

Cosmic Magnetic Fields:
From Planets, to Stars and Galaxies
Proceedings IAU Symposium No. 259, 2008
K.G. Strassmeier, A.G. Kosovichev & J.E Beckman, eds.

doi:10.1017/S174392130903083X

Searching for a link between the magnetic nature and other observed properties of Herbig Ae/Be stars

Swetlana Hubrig[1], C. Grady[2], M. Schöller[3], O. Schütz[1], B. Stelzer[4], M. Pogodin[5], M. Cure[6] and R. Yudin[5]

[1]ESO, Santiago, Chile; email: shubrig@eso.org
[2]Eureka Scientific, Oakland, USA [3]ESO, Garching, Germany
[4]INAF-Osservatorio Astronomico di Palermo, Italy
[5]Pulkovo Observatory, St.-Petersburg, Russia
[6]Univ. de Valparaiso, Chile

Abstract. We present the results of a new magnetic field survey of Herbig Ae/Be and A debris disk stars. They are used to determine whether magnetic field properties in these stars are correlated with the mass-accretion rate, disk inclinations, companion(s), Silicates, PAHs, or show a more general correlation with age and X-ray emission as expected for the decay of a remnant dynamo.

Keywords. Stars: pre-main-sequence – stars: magnetic fields – X-rays: stars – techniques: polarimetric – stars: HD 101412, HD 139614, HD 144668, HD 152404, HD 190073

1. Introduction

Protoplanetary disks are where planets form, and where the pre-biotic materials, which produce life-bearing worlds, are assembled or produced. We need to understand them, how they interact with their central stars, and their evolution; both to reconstruct the Solar System's history, and to account for the observed diversity of exo-planetary systems. Our most detailed view of protoplanetary disks is for those surrounding intermediate-mass stars, the Herbig Ae/Be stars (e.g. Herbig 1960) where the disks are revealed by their thermal emission, and in scattered light in the optical and near-IR.

2. Observations and measurements

The observations were carried out in May 2008 at the European Southern Observatory with FORS 1 mounted on the 8 m Kueyen telescope of the VLT. New magnetic field detections were achieved in eight stars. For three Herbig Ae/Be stars, we confirm the previous magnetic field detections. The star HD 101412, with the largest magnetic field strength measured in our sample stars using hydrogen lines, $\langle B_z \rangle = -454 \pm 42$ G , shows a change of the field strength by $\sim$100 G during two consecutive nights. In Fig. 1 (left panel) we present distinct Zeeman features detected at the positions of the hydrogen Balmer lines and the Ca II H&K lines. The Hβ line in the Stokes I spectrum is contaminated by the presence of a variable emission in the line core and was not included in our measurements. Strong distinct Zeeman features at the positions of the Ca II H&K lines are detected in four Herbig Ae/Be stars, HD 139614, HD 144668, HD 152404, and HD 190073. In Fig. 1 (right panel) we present the Stokes V spectra for these stars in the region around the Ca II doublet, together with our previous observation of HD 190073. As we already reported

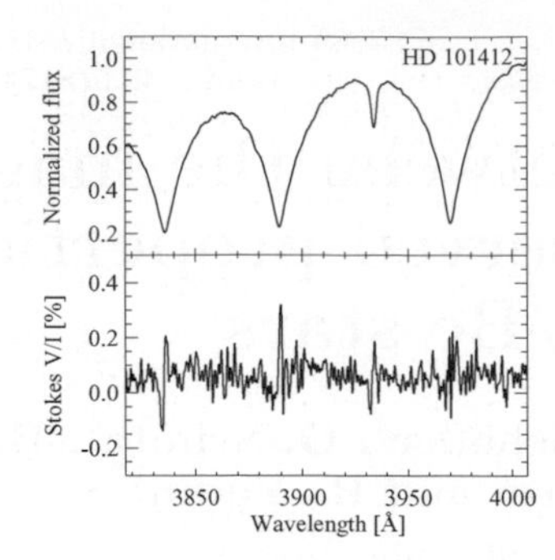

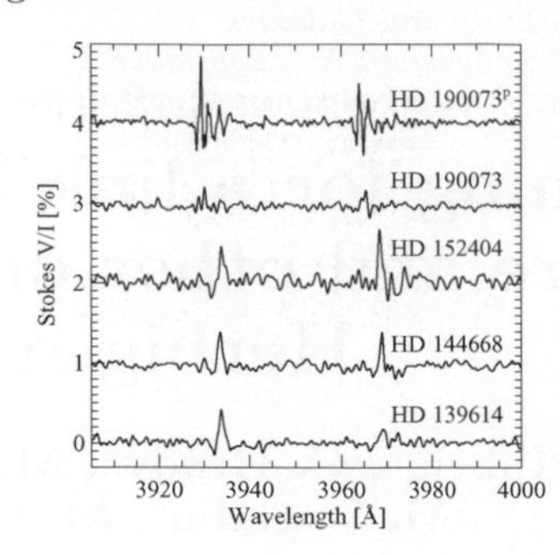

Figure 1. Left panel: Stokes I and V spectra in HD 101412; Right panels: Stokes V spectra in the vicinity of the Ca II H&K lines of the Herbig Ae/Be stars HD 139614, HD 144668, HD 152404, and HD 190073. At the top we present the previous observation of HD 190073, obtained in May 2005. The amplitude of the Zeeman features in the Ca II H&K lines observed in our recent measurement has decreased by $\sim$0.5% compared to the previous observations.

in our earlier studies (Hubrig *et al.* 2004; Hubrig *et al.* 2006; Hubrig *et al.* 2007a) these lines are very likely formed at the base of the stellar wind, as well as in the accretion gaseous flow, and frequently display multi-component complex structures in both the Stokes V and Stokes I spectra.

3. Looking for the links

Most of our sample targets are Herbig Ae stars with masses of $3\,M_\odot$ and less. Since the observations of the disk properties of intermediate mass Herbig Ae stars suggest a close parallel to T Tauri stars, revealing the same size range of the disks, similar optical surface brightness and similar structure consisting of inner dark disk and a bright ring, it is quite possible that especially the magnetic fields play a crucial role in controlling accretion onto, and winds from, Herbig Ae stars, similar to the magnetospheric accretion observed in T Tauri stars. We do not find any trend between the presence of a magnetic field, disk inclination angles and mass-accretion rates. Also the membership in binary or multiple systems does not seem to have any impact on the presence of a magnetic field, whereas we find a hint that the appearance of magnetic fields is more frequent in Herbig stars with flared disks and hot, inner gas. The stronger magnetic fields tend to be found in younger Herbig stars. The magnetic fields become very weak or completely disappear in stars when they arrive on the ZAMS, clearly confirming the conclusions of Hubrig *et al.* (2000) and Hubrig *et al.* (2007b) that magnetic Ap stars with masses less than $3\,M_\odot$ are only rarely found close to the ZAMS. We also find a hint for an increase of the magnetic field strength with the level of the X-ray emission, suggesting a dynamo mechanism to be responsible for the coronal activity in Herbig Ae stars.

References

Herbig, G. H. 1960, *ApJS* 4, 337
Hubrig, S., North, P., & Mathys, G. 2000, *ApJ* 539,352
Hubrig, S., Schöller, M., & Yudin, R. V. 2004, *A&A* 428, L1
Hubrig, S., Yudin, R. V., Schöller, M., & Pogodin, M. A. 2006, *A&A* 446, 1089
Hubrig, S., North, P., & Schöller, M. 2007b, *AN* 328, 475
Hubrig, S., Pogodin, M. A., Yudin, R. V., *et al.* 2007a, *A&A* 463, 1039

Cosmic Magnetic Fields:
From Planets, to Stars and Galaxies
Proceedings IAU Symposium No. 259, 2008
K.G. Strassmeier, A.G. Kosovichev & J.E. Beckman, eds.

doi:10.1017/S1743921309030841

Magnetic fields in classical Be stars

Ruslan Yudin[1], S. Hubrig[2], M. Pogodin[1], I. Savanov[3], M. Schöller[4], G. Peters[5] and M. Cure[6]

[1]Pulkovo Observatory, St.-Petersburg, Russia;
email: ruslan61@hotmail.com; [2]ESO, Santiago, Chile; [3]Institute of Astronomy, Russian Academy of Science, Moscow, Russia [4]ESO, Garching, Germany; [5]Univ. Southern California, USA; [6]Univ. de Valparaiso, Chile

Abstract. We report the results of our study of magnetic fields in a sample of Be stars using spectropolarimetric data obtained at the European Southern Observatory with the multi-mode instrument FORS 1 installed at the 8 m Kueyen telescope. The detected magnetic fields are rather weak, not stronger than ∼150 G. A few classical Be stars display cyclic variability of the magnetic field with periods of tens of minutes.

Keywords. Stars: Be – polarization – magnetic fields

1. Introduction

Over an extended period of years many attempts to obtain reliable direct measurements of magnetic fields of Be stars have been rather unsuccessful. Recently, weak field detections in a few Be stars were reported by Hubrig *et al.* (2007). Below, we report some preliminary results of our most recent observing runs with the multi-mode instrument FORS1 installed at the 8m Kueyen telescope at the VLT. We also present first results of our magnetic field time series in the Be star λ Eri.

2. Observations

As it was already reported by Hubrig *et al.* (2007), weak photospheric magnetic fields have been detected in four stars, HD 56014, HD 148184, HD 155806, and HD 181615. We also pointed out the likely presence of distinct circular polarisation features in the circumstellar components of Ca II H&K in the three stars, HD 58011, HD 117357, and HD 181615, indicating a probable presence of magnetic fields in the circumstellar mass loss disks. The FORS 1 observations of HD 181615 are presented in the left panel of Fig. 1. New observations of classical Be stars have been been carried out at the VLT from 2006 to 2008. The Be stars have been observed in 2006 in service mode with the GRISM 600B in the wavelength range 348-589 nm and with the GRISM 1200B in the spectral range 373-497 nm. The last observations we report here have been carried out in visitor mode in 2008 in the framework of the study of 15 B-type members of the open cluster NGC 3766.

3. Results

Weak magnetic fields have been detected in eight classical Be stars. The magnetic field strengths of the studied Be stars are found to be in the range 40 G to 150 G. A few Be stars with detected magnetic fields show non-detections at some other observing dates, indicating possible temporal variability of their magnetic fields. To confirm this suggestion we performed time-resolved magnetic field measurements over one hour for a

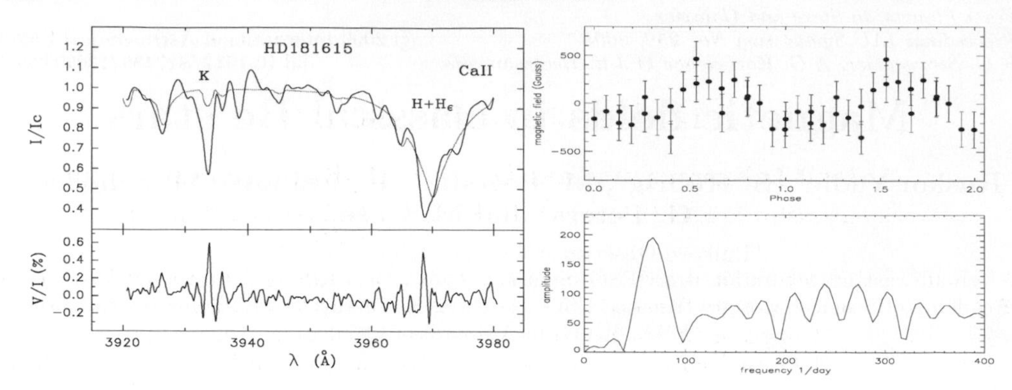

Figure 1. Left panel: Stokes I and V spectra in HD 181615; Right panels: Cyclic variability of the magnetic field in λ Eri with a period of 21.1 min and Fourier spectrum for the magnetic field data derived from hydrogen lines.

Table1. Measurements of NGC 3766 cluster stars with detected magnetic fields.

NGC3766	MJD	$\langle B_z \rangle_{\rm all}$	$\langle B_z \rangle_{\rm hydr}$
45	54550.066	−78±41	−194±62
47	54549.020	−146±43	−129±58
94	54550.327	294±53	310±65
111	54549.020	**112±34**	89±38
170	54550.186	1522±34	1559±38
176	54550.016	**121±36**	**141±41**
200	54550.375	**128±40**	**115±38**

few Be stars to get information about the behaviour of the magnetic field over at least a part of the stellar surface. The service mode observations revealed periodic changes of magnetic fields on the time scale of tens of minutes in a few classical Be stars. Among these stars, the star λ Eri displays a cyclic variability of the magnetic field with a period of 21.1 min (see the right panel of Fig. 1). The results of the study of fifteen early B-type members of the open cluster NGC 3766 have already been reported by McSwain (2008) who noted two definite and two possible detections in the studied sample. We detected three additional stars with weak magnetic fields which are members of this cluster (highlighted by bold face in Table 1).

4. Conclusions

Our search for magnetic fields in Be stars revealed that their magnetic fields are rather weak, but fields of less than ∼150 G are not rare. The magnetic fields are clearly variable and a non-detection of magnetic fields in some stars may be explained by temporal variability. A few classical Be stars display cyclic variability of the magnetic field with periods of tens of minutes. The cluster NGC 3766 seems to be extremely interesting, where we find clear evidence for the presence of a magnetic field in seven early B-type cluster members out of fifteen members.

References

Hubrig, S., Yudin, R., Pogodin, M., Schöller, M., & Peters, G. 2007, *AN* 238, 1133
McSwain, M. V., 2008, *ApJ* 686, 1269

Cosmic Magnetic Fields:
From Planets, to Stars and Galaxies
Proceedings IAU Symposium No. 259, 2008
K.G. Strassmeier, A.G. Kosovichev & J.E. Beckman, eds.

doi:10.1017/S1743921309030853

First magnetic stars

Iosif I. Romanyuk and Dimitry O. Kudryavtsev

Special Astrophysical Observatory RAS, Nizhny Arkhyz, 369167, Russia

Abstract. This contribution dedicated to the analysis of the magnetism of chemically peculiar (CP) stars of the upper Main Sequence. We use our own measurements and published data to compile a catalog of magnetic CP stars containing a total of 326 objects with confidently detected magnetic fields and 29 stars which are very likely to possess magnetic field. Our analysis shows that the number of magnetic CP stars decreases with increasing field strength in accordance with exponential law, hotter and faster rotating stars have stronger fields. Intensity of depressions in the continua correlates with the magnetic field strength.

Keywords. Stars: magnetic field – stars: chemically peculiar

1. Introduction

First (after the Sun) astronomical objects with measured magnetic fields were CP stars (Babcock 1947). The technique proposed by Babcock allowed only large-scale regular magnetic fields with simple structure to be studied. The first catalog of magnetic stars was published by Babcock (Babcock 1958). It contains the results of his own studies. He found a total of 89 magnetic stars, majority of which were Ap/Bp stars.

2. The catalog

In the last 10 years more than 200 new magnetic stars were found. Now there are more then 350 stars known to be magnetic CP ones (see our catalog (Romanyuk and Kudryavtsev 2008)), it is the most numerous group of stars with measured magnetic fields. The data of magnetism of 300 stars are based on measurements of longitudinal field component B_e, and the surface field is measured from split Zeeman components only for 50 of these stars. Thus present-day concepts of magnetism of CP stars are based on results of measurements of longitudinal field component.

It was shown that the greatest number of the magnetic observations were performed by H. Babcock, G. Preston and their co-authors, J. Landstreet and his coauthors (E. Borra, I. Thompson, D. Bohlender, G. Wade and others), G. Mathys and his coauthors, S. Hubrig *et al.*, new teams (S. Bagnulo and his co-authors, M. Auriere and his co-authors), and also by a team of researchers from the Special Astrophysical Observatory working on the 6-m telescope of the Russian Academy of Sciences (Yu.V. Glagolevskij, I.I. Romanyuk, V.G. Elkin, D.O. Kudryavtsev, G.A. Chountonov, E.A. Semenko *et al.*).

Absolute majority of CP stars have magnetic field of simple dipole configuration and only 3% of these objects have more complex field.

The longitudinal component B_e averaged over the entire surface of the star varies with the phase of star's rotation. Its extreme value $B_e(extr)$ (in the case of the simplest dipole configuration of the field) is, on the average, equal to 1/3 of the surface field B_s and thus quantitatively is a quite adequate measure of the actual field strength at the surface of the CP stars. However, $B_e(extr)$ can be correctly inferred only for the objects for which the curve of variations of B_e phased with rotation period is known.

If observational data are scare and/or the period of star's rotation is unknown then its magnetic field must be estimated using an average parameter – the so-called root-mean square field (Bohlender *et al.* 1993).

3. Discussion

The number of magnetic CP stars decrease with increasing field strength with exponential law (in the $B_e(extr)$ interval from 0.7 to 5 kG):

$$N = \exp(5.2 - 0.0008 B_e(extr))$$

with the correlation coefficient 0.988.

It is easy to understand why exponential dependence breaks is not valid the interval mentioned above: 1) not all magnetic CP stars with $B_e(extr)$ smaller 600-700 G have been discovered (and hence not all such stars were included into our list) and therefore fall of the histogram in the domain of weak fields can be explained by observational selection and 2) there too few objects in the domain of strong (greater than 5 kG) fields and hence statistical approach cannot be applied in this case.

Our data show that the extreme field strength B_e exceeds 4 kG only in 6% of all magnetic CP stars. We can thus estimate that at least 90% of all magnetic CP stars in our list (Romanyuk and Kudryavtsev 2008)) have average surface field $B_s$weaker than or equal to 10 kG. We find only a few objects with stronger fields (with B_s greater than 20 kG) in our list. It appears that at the level of several tens of kG the threshold lies beyond which stronger large-scale fields cannot form in the atmospheres of Main Sequence stars.

There's a weak dependence: hotter and faster rotating stars have stronger fields. Nevertheless, strong fields (about 1 kG strength) are observed in very slow rotators, with rotation period of tens of years. The observational data give an evidence on the relic nature of the magnetic field of these objects.

Depressions in the continua of chemically peculiar stars are observed, intensity of depressions correlates with the magnetic field strength. Chemical composition is anomalous, the distribution of elements over the surface is non-uniform. Recently, first reliable observational data on the connection between location of spots of chemical elements and magnetic field configuration were obtained; that confirmed efficiency of magnetic field separation mechanism in the atmospheres of CP stars.

There is large number of nearly and bright stars among these objects so high quality observations with high S/N ratio is possible to obtain for them. Detail study of Zeeman effect using such spectra gives us new possibilities for searching the inaccessible before fine features in physical processes taking place in the atmospheres of stars with strong magnetic fields.

References

Babcock H. W. 1947, *ApJ* 105, 105
Babcock H. W. 1958, *ApJS* 20, 141
Bohlender D. A., Landstreet J. D., & Thompson I. B. 1993, *A&A* 269, 355
Romanyuk I. I. & Kudryavtsev D. O. 2008, *Astrophys. Bull.* 63, n.2, 139

Cosmic Magnetic Fields:
From Planets, to Stars and Galaxies
Proceedings IAU Symposium No. 259, 2008
K.G. Strassmeier, A.G. Kosovichev & J.E. Beckman, eds.

doi:10.1017/S1743921309030865

Searching for a link between the presence of chemical spots on the surface of HgMn stars and their weak magnetic fields

Igor S. Savanov[1], S. Hubrig[2], J. F. González[3] and M. Schöller[2]

[1]Institute of Astronomy, Russian Academy of Science, Moscow, Russia;
email: isavanov@rambler.ru

[2]ESO, Santiago, Chile & Garching, Germany

[3]Complejo Astronomico El Leoncito, San Juan, Argentina

Abstract. We present the results of mapping the HgMn star AR Aur using the Doppler Imaging technique for several elements and discuss the obtained distributions in the framework of a magnetic field topology.

Keywords. Stars: atmospheres – stars: chemically peculiar – stars: magnetic fields – stars: spots

1. Introduction

Late B-type stars with HgMn peculiarity are characterized by low rotational velocities and weak or non-detectable magnetic fields. The most distinctive features of their atmospheres are the extreme overabundance of Hg and Mn. More than 2/3 of HgMn stars belong to spectroscopic binaries. The presence of an inhomogeneous distribution of some elements over the surface of HgMn stars was discussed for the first time by Hubrig & Mathys (1995). The first definitively identified spectrum variability has been reported for the binary star α And by Wahlgren *et al.* (2001) and Adelman *et al.* (2002).

2. Doppler imaging

We present the first results of our preliminary Doppler Imaging (DI) modeling of abundance distributions for four elements on the surface of the HgMn star AR Aur (Fig. 1). The analysis of the spectra is made with the DI inversion code iAbu, which reconstructs the stellar surface abundance inhomogeneities from the series of spectral line profiles using the Tikhonov regularization algorithm.

Previously, Hubrig *et al.* (2006) concluded that the employment of a partially fractured equatorial ring presents a realistic distribution of Y on the stellar surface. Our new map based on the inversion of the Y II λ4900 Å line is in good agreement with this conclusion. The Y is overabundant and strongly concentrated in the equatorial ring, where the average Y abundance reaches $\log\varepsilon(\mathrm{Y}) = 6.0$ (in the scale $\log\varepsilon(\mathrm{H}) = 12.0$). One large fraction of the ring is missing exactly on the surface area which is permanently facing the secondary. An additional polar detail can be seen at longitudes 180°–270°. The surface distribution of strontium was studied with the ion line Sr II λ4215 Å. The Sr surface distribution in AR Aur is not completely similar to the map reconstructed for Y II, although it also presents significant equatorial details of Sr overabundance in the range of longitudes from 70° to 250°.

In the case of the Hg II λ3984 Å line we normalized our spectra relative to the pseudo-continuum of the Hε wing, but accounted for blending by the hydrogen line in the

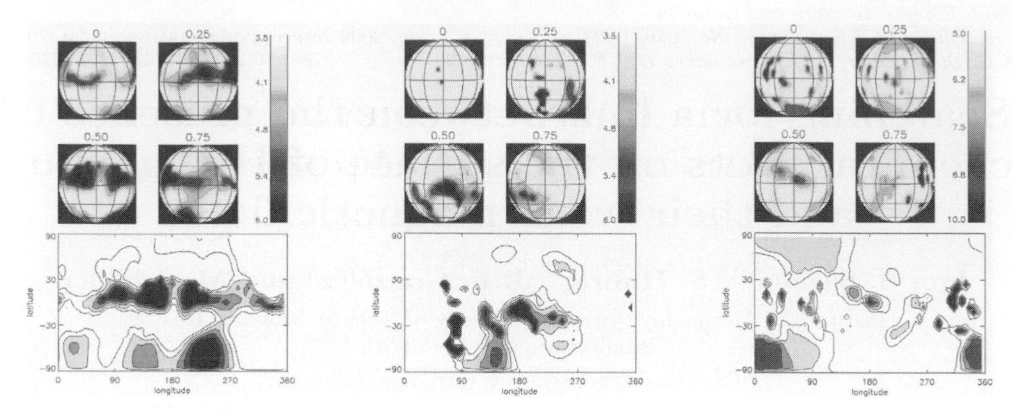

Figure 1. Results of the DI reconstruction of the Y, Sr and Hg surface distributions (from left to right). The rectangular plots (lower panels) show a pseudo-Mercator projection of the surface maps. Spherical DI maps are presented at four equidistant rotational phases (upper panels).

calculations of the synthetic spectra by adding corresponding opacity sources. The line list used for the DI consists of isotopic and hyperfine structure components of the Hg II $\lambda3984$ Å line and a Y II $\lambda3982.59$ Å blending line. We parameterized the mercury isotope composition using the isotopic-fractionation model of White *et al.* (1976). Our observational data are insufficient for a simultaneous determination of both the Hg abundance and the surface distribution of the q-parameter, hence we accept q=0 for the terrestrial mixture. The mercury overabundance is strongly concentrated in several equatorial details on the surface of the star. Further we find polar appendages at longitudes 0°–90°

For the first time we reconstructed the manganese surface distribution in a HgMn star. The Mn-ion map is based on modeling of the Mn II $\lambda4292$ Å line. This map is not shown in Fig. 1 due to lack of space. We demonstrated that the Mn surface distribution shows similarities with those of Y and Sr, namely both equatorial and polar features.

3. Conclusions

What is the explanation for the discovered inhomogeneities? Taking into account that more than 2/3 of the HgMn stars are known to belong to spectroscopic binaries, a scenario how a magnetic field can be built up in binary systems has been presented some time ago by Hubrig (1998) who suggested that a tidal torque varying with depth and latitude in a star induces differential rotation. Differential rotation in a radiative star can, however, be prone to the magneto-rotational instability (e.g., Arlt *et al.* 2003). Magneto-hydrodynamical simulations revealed a distinct magnetic field topology similar to the latitudinal fractured rings observed on the surface of α And and AR Aur.

References

Adelman, S. J., Gulliver, A. F., Kochukhov, O. P., & Ryabchikova, T. A. 2002, *ApJ* 575, 449
Arlt, R., Hollerbach, R., & Rüdiger G. 2003, *A&A* 401, 1087
Hubrig, S. & Mathys, G. 1995, *Comments Astrophys.* 18, 167
Hubrig, S. 1998, *CoSka* 27, 296
Hubrig, S., Gonzales, J. F., Savanov, I., Schöller, M., Ageorges, N., Cowley, C. R., & Wolff, B. 2006, *MNRAS* 371, 1953
Wahlgren, G. M., Ilyin, I., & Kochukhov, O. 2001, *A&A* 33, 1506
White, R. E., Vaughan, A. H.,Jr., Preston, G. W., & Swings, J. P. 1976, *ApJ* 204, 131

Cosmic Magnetic Fields:
From Planets, to Stars and Galaxies
Proceedings IAU Symposium No. 259, 2008
K.G. Strassmeier, A.G. Kosovichev & J.E. Beckman, eds.

doi:10.1017/S1743921309030877

Cartography of the magnetic fields and chemical spots of Ap stars

James Silvester[1,2], O. Kochukhov[3], G. A. Wade[1], N. Piskunov[3], J. D. Landstreet[4] and S. Bagnulo[5]

[1]Department of Physics, Royal Military College of Canada, Kingston, ON, Canada K7K 4B4
[2] Department of Physics, Queen's University, Kingston, Ontario, Canada
[3] Department of Physics and Astronomy, Uppsala University, Box 515, 751 20 Uppsala, Sweden
[4] Dept. of Physics & Astronomy, Univ. of Western Ontario, London, ON, Canada, N6A 3K7
[5] Armagh Observatory, College Hill, Armagh BT61 9DG, Northern Ireland

Abstract. We will introduce a project using Magnetic Doppler Imaging (MDI) to create assumption-free vector magnetic field maps and chemical surface structure maps of chemically peculiar A and B type (or Ap) stars. We are exploiting the latest generation of spectropolarimeters (NARVAL at the Pic du Midi observatory, and ESPaDOnS at the Canada-France-Hawaii telescope), to obtain high-resolution time series of Stokes IQUV spectra of a selection of Ap stars. The spectra have superior signal-to-noise ratio, resolution and wavelength coverage to those used previously. This combined with the ground-breaking inversion techniques introduced by Kochukhov *et al.* (2002) results in maps which represent the state-of-the-art in the field of stellar cartography. These maps will allow us to better understand the links between the magnetic field and the physical processes leading to the formation of chemical structures in the photosphere and allow us to address questions surrounding the detailed magnetic field geometry of Ap stars.

Keywords. Stars: magnetic fields – techniques: polarimetric – stars: early-type – stars: chemically peculiar

1. Introduction

Magnetic fields play a fundamental role in the physics of the atmospheres of a significant fraction of stars on the H-R diagram. In early-type stars, the large-scale surface magnetic field is static on timescales of at least many decades, and appears to be "frozen" into a rigidly rotating atmosphere. The magnetic field is globally organised, permeating the entire stellar surface, with a relatively high field strength. The presence of the magnetic field strongly influences energy and mass transport (e.g., diffusion, convection and weak stellar winds) within the atmosphere of a star, and results in the presence of strong chemical abundance nonuniformities in photospheric layers, although to what level the magnetic field plays a role in these processes is still unclear.

The very first measurements of rotationally-modulated Zeeman circular and linear polarisation resolved within stellar line profiles were reported by Wade *et al.* (2000). This data set was later used with the new Magnetic Doppler Imaging technique (MDI), described by Piskunov & Kochukhov (2002) and Kochukhov & Piskunov (2002), to construct the very first assumption-free, high resolution maps of the surface vector magnetic field of the Ap stars 53 Cam (Kochukhov *et al.* 2004) and α^2 CVn (Kochukhov *et al.* in prep). The unique maps of 53 Cam and α^2 CVn reveal that the detailed magnetic topology departs significantly from the commonly-assumed low-order multipolar geometry, and that the abundances were not distributed uniformly with respect to the magnetic

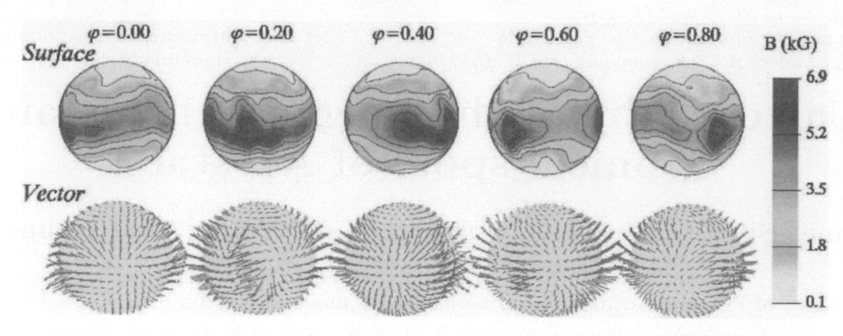

Figure 1. Preliminary magnetic field orientation and intensity maps of 49 Cam, as recovered using the MDI technique and the new IQUV spectropolarimetric data.

field, but localised in complex structures. The data used in these early maps represented the best data set obtained from several years of MuSiCoS observations. However, with this data set the Stokes Q and U signatures were only really detectable in 3 strong lines, with a S/N of 5 or less in a sample of 2 or 3 stars. Due to the low signal-to-noise ratio and resolving power achievable with the MuSiCoS instrument, only an extremely limited range of stellar properties (rotation, mass, temperature, magnetic field, etc.) which may influence the phenomena of interest could be studied using the MuSiCoS data.

2. The Project

With the vastly improved data now available with ESPaDOnS and NARVAL we are creating the new generation of MDI maps using the INVERS10 inversion code, with these maps we will be able to characterize the magnetic field geometry of Ap stars, further refine the current model of the magnetic field structure and characterize the abundance surface structure. To test the quality of this new data and to confirm consistency with previous studies, the longitudinal field and net linear polarisation observed in 49 Cam were compared with measurements obtained by Wade *et al.* (2000). The data proved to be consistent with the previous MuSiCoS measurements, with uncertainties on average 3 to 4 times smaller than those obtained with MuSiCoS. A preliminary surface and vector field map for 49 Cam constructed using ESPaDOnS and NARVAL observations and using INVERS10 is shown in Fig. 1.

Once complete phase coverage is obtained for 49 Cam, HD 32633 and α^2 CVn, the next generation of MDI maps can be produced. This will be followed by a multipolar analysis on the resulting magnetic maps to determine the true magnetic field structure of each star and an investigation into any relationships between the magnetic field and abundance structures by performing a pixel by pixel analysis.

References

Kochukhov, O., Bagnulo, S., Wade, G. A., Sangalli, L., Piskunov, N., Landstreet, J. D., Petit, P., & Sigut, T. A. A. 2004, *A&A* 414, 613

Kochukhov O., Piskunov N., Ilyin I., Ilyina S., & Tuominen I. 2002, *A&A* 389, 420

Kochukhov O. & Piskunov N. 2002, *A&A* 288, 868

Kochukhov, O. & Bagnulo, S. 2006, *A&A* 450, 763

Piskunov N. & Kochukhov O. 2002, *A&A* 381, 736

Wade G. A., Donati J.-F., Landstreet J. D., & Shorlin S. L. S. 2000b, *MNRAS* 313, 823

Cosmic Magnetic Fields:
From Planets, to Stars and Galaxies
Proceedings IAU Symposium No. 259, 2008
K.G. Strassmeier, A.G. Kosovichev & J.E. Beckman, eds.

doi:10.1017/S1743921309030889

Analysis of magnetic pressure effects in atmospheres of CP stars

Denis Shulyak[1], G. Valyavin[2], O. Kochukhov[3], and T. Burlakova[4]

[1]Institute for Astronomy, Vienna University, Türkenschanzstrasse, 17, 1180 Vienna, Austria
email: denis@jan.astro.univie.ac.at

[2]Korea Astronomy and Space Science Institute, 61-1, Whaam-Dong, Youseong-Gu, Taejeon, Korea 305-348
email: gendoz@boao.re.kr

[3]Department of Astronomy and Space Physics, Uppsala University, Box 515, 751 20, Uppsala, Sweden
email: Oleg.Kochukhov@fysast.uu.se

[4]Special Astrophysical Observatory, Russian Academy of Sciences, Nizhnii Arkhyz, Karachai Cherkess Republic, 369167, Russia

Abstract. Several dynamical processes may induce considerable Lorentz forces in atmospheres of mCP stars, thus modifying the hydrostatic structure of their atmospheres. This modification can be seen as characteristic rotational variability of certain spectral features such as hydrogen Balmer lines. In this work we present the first results of modeling the magnetic pressure effects in atmospheres of mCP stars in the framework of model atmosphere analysis with accurate treatment of microscopic properties of atmospheric plasma. We show that at least part of the rotational variability of hydrogen lines seen in high-resolution spectra of mCP stars could be attributed to the non-zero electrical currents flowing along stellar surfaces.

Keywords. Stars: chemically peculiar – stars: magnetic fields – stars: atmospheres

1. Introduction

Magnetic chemically peculiar (mCP) stars display the presence of stable large-scale magnetic fields. The slow evolution of these magnetic fields (as well as other dynamical processes) may induce considerable electric currents, thus modifying the pressure balance of stellar atmospheres via the induced Lorentz forces. These forces can be detected observationally analysing the pressure sensitive spectral features like hydrogen lines (Kroll 1989). Here we present the study of hydrogen lines variability in two mCP stars θ Aur and 56 Ari within the framework of model atmospheres with magnetic pressure included.

2. Model predictions

To model the variation of the Balmer line profiles, we follow the approach outlined in Shulyak *et al.* (2007); Valyavin*et al.* (2004). The hydrostatic equation of stellar atmosphere in the presence of the magnetic pressure can be written as:

$$\frac{\partial P_{\rm total}}{\partial r} = -\rho g \pm \frac{1}{c}\lambda_{\perp} \sum_n c_n P^1_n(\mu) \sum_n B^{(n)}_{\theta}, \tag{2.1}$$

where c_n–the effective *e.m.f.* generated by the n-th magnetic field component at the stellar magnetic equator and B_θ is the horizontal field component. $\lambda_\perp$ is the plasma conductivity across magnetic field lines and $P^1_n(\mu)$ are the Legendre polynomials.

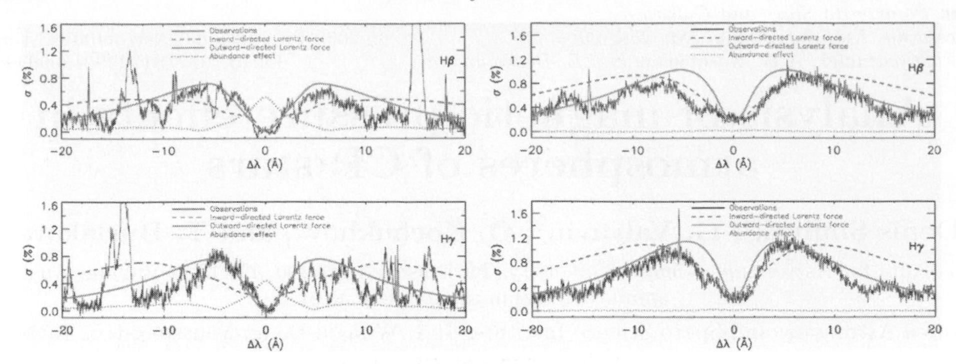

Figure 1. Standard deviation around Hβ and Hγ lines of θ Aur (left panel) and 56 Ari (right panel) obtained under different assumptions about the direction of the Lorentz force.

Table 1. Main stellar and magnetic field parameters of θ Aur and 56 Ari. Here i–inclination angle, β–magnetic obliquity, $B_{\rm d}$–polar strength of the dipole component, $c_{\rm q}/c_{\rm d}$–relative strength of induced $e.m.f.$

	$T_{\rm eff}$	$\log g$	i°	$v\sin i$	β°	$B_{\rm d}$ kG	$B_{\rm q}/B_{\rm d}$	$c_{\rm q}/c_{\rm d}$
θ Aur	10 400	3.6	51	55	78	1.4	−2	2.5
56 Ari	12 800	4.0	70	160	82	1.3	0	0

We applied the LLMODELS stellar model atmosphere code (Shulyak *et al.* 2004) to compute $\lambda_\perp$ and magnetic pressure for each rotational phase of the two stars according to Eq. 2.1. The methodology of analysis are summarized as follows:

- Finding stellar parameters from photometric and spectroscopic data available.
- Determination of magnetic field geometry via analysis of available magnetic measurements.
- Verifying the effect of inhomogeneous surface distribution of chemical elements.
- Model atmosphere calculations with Lorentz force included for each rotational phase adjusting values of induced $e.m.f.$ to match observed amplitude of standard deviation.

Table 1 lists the best fitted parameters used for modeling the Lorentz force effects in atmospheres of θ Aur and 56 Ari. Substantial errors in observations of the magnetic field variation found for 56 Ari did not allow at present to model magnetic field geometry other than dipole.

The best fit to the amplitude of the observed variations was obtained with $c_1 = 1\times10^{-10}$ CGS (inward-directed force) and $c_1 = 1\times10^{-11}$ CGS (outward-directed force) for θ Aur and $c_1 = 5\times10^{-10}$ CGS (inward-directed force) and $c_1 = 7.5\times10^{-11}$ CGS (outward-directed force) in the case of 56 Ari.

Acknowledgements

This work was supported by FWF Lisa Meitner grant Nr. M998-N16 to DS.

References

Kroll, R. 1989, *Rev. Mexicana AyA* 2, 194
Shulyak, D., Valyavin, G., Kochukhov, O., *et al.* 2007, *A&A* 464, 1089
Shulyak, D., Tsymbal, V., Ryabchikova, T., Stütz, Ch., & Weiss, W. W. 2004, *A&A* 428, 993
Valyavin, G., Kochukhov, O., & Piskunov, N. 2004, *A&A* 420, 993

Cosmic Magnetic Fields:
From Planets, to Stars and Galaxies
Proceedings IAU Symposium No. 259, 2008
K.G. Strassmeier, A.G. Kosovichev & J.E. Beckman, eds.

doi:10.1017/S1743921309030890

Advanced model atmospheres with magnetic field effects included

Denis Shulyak[1], **S. Khan**[2], **and O. Kochukhov**[3]

[1]Institute for Astronomy, Vienna University, Türkenschanzstrasse, 17, 1180 Vienna, Austria
email: denis@jan.astro.univie.ac.at

[2]Institute for Computational Astrophysics, Saint Mary's University, 923 Robie Street, Halifax, B3H 3C3, Nova Scotia, Canada
email: serg@starsp.org

[3]Department of Astronomy and Space Physics, Uppsala University, Box 515, 751 20, Uppsala, Sweden
email: Oleg.Kochukhov@fysast.uu.se

Abstract. The atmospheres of magnetic chemically peculiar (mCP) stars display the presence of magnetic fields of different geometry and strength, ranging from a few hundred G up to tens of kG. Except several very approximate attempts there were no detailed studies of magnetic field effects on model atmospheres structure, possibly leading to errors in the stellar parameter determination and abundance analysis routines. We present the magnetic model atmospheres based on LLmodels code which accounts for the detailed treatment of anomalous Zeeman splitting and polarized radiative transfer.

Keywords. Stars: atmospheres – stars: magnetic fields – stars: chemically peculiar

---

1. Introduction

The atmospheres of magnetic chemically peculiar (mCP) stars exhibit the presence of large-scale magnetic fields that can have the strength about tens of kG. However, the magnetic field effects are ignored in standard-model atmosphere calculations. To circumvent this longstanding problem, we have developed a new model atmosphere code LLMODELS (Shulyak *et al.* 2004) which account for detailed treatment of anomalous Zeeman splitting (Kochukhov *et al.* 2005), polarized radiative transfer (Khan & Shulyak 2006a) and inclination of the magnetic field vector (Khan & Shulyak 2006b).

2. Model predictions

We have computed a grid of model atmospheres and investigated the effect of magnetic field on model structure, energy distribution, hydrogen line profiles, and photometric colors. The main results can be summarized as follows:

- Magnetic opacity leads to the heating of certain atmospheric layers.
- Flux redistribution from UV to visual region with the increase of the magnetic field strength.
- Flux depression at 5200Å appears for lower temperatures and vanishes for the higher ones for a given magnetic field modulus.
- Relation between photometric indexes and the magnetic field strength is not linear due to the saturation effect (except Δa index for cool models) and depends strongly upon the metallicity of the models assumed.

Table 1. Observed and calculated photometric parameters of HD 137509. Suffixes "m", "n" and "s" refer to magnetic, non-magnetic (both with individual abundances), and non-magnetic scaled-solar abundance model from Kochukhov (2006). The improved models are presented in last two rows.

	$b-y$	c_1	Δa	X	Y	Z
observations	**−0.095**	**0.411**	**0.066**	**0.762**	**0.076**	**−0.067**
t12750g3.8m	−0.056	0.573	0.054	0.936	0.107	−0.043
t12750g3.8n	−0.043	0.588	0.032	0.957	0.066	−0.028
t12750g3.8s	−0.026	0.581	0.014	0.942	0.032	−0.016
t13750g4.2m	−0.081	0.453	0.062	0.799	0.081	−0.051
t13750g4.2n	−0.064	0.470	0.037	0.818	0.040	−0.036

• The Δa system is not able to clearly distinguish CP stars with metallicity other than solar for low $T_{\rm eff}$, thought it remains the most sensitive one to the field strength among the others for hotter stars.

• Magnetic opacity does not introduce significant errors in the photometric estimates of CP-stars parameters.

• The effect of magnetic opacity on model structure do not influence much the hydrogen lines profiles.

• The widely used models computed with pseudo-microturbulence parameter to simulate the magnetic field effects can only be used as a very rough guess.

• The inclination of the magnetic field vector has small impact on model atmosphere structure and energy distribution.

3. Strong magnetic field star HD 137509

HD137509 is a B–type star with the second-largest magnetic field ever found among CP stars with $B_{\rm z} = 29\,{\rm kG}$, which is the excellent test-ground for newly developed magnetic models. The fundamental parameters $T_{\rm eff} = 12\,750\,{\rm K}$, $\log g = 3.8$ were previously derived using standard non-magnetic model computed with $[M/H] = +1$ (Kochukhov 2006). To improve the fundamental parameters of the star in a consistent way we redetermined the abundances using model atmospheres computed with LLmodels with both magnetic field and individual abundances taken into account. Some photometric colors of models computed with initial abundances from Kochukhov (2006) and models with improved fundamental parameters and abundances are presented in Table 1. We find that the simultaneous fit to the H-line profiles and photometric indexes required as much as 1000 K correction to the $T_{\rm eff}$ and 0.4 dex correction to the $\log g$ compared to the results obtained using simple scaled-solar models (for more details, see Shulyak *et al.* 2008).

Acknowledgements

This work was supported by FWF Lisa Meitner grant Nr. M998-N16 to DS.

References

Khan, S. & Shulyak, D. 2006b, *A&A* 454, 933
Khan, S. & Shulyak, D. 2006a, *A&A* 448, 1153
Kochukhov, O. 2006, *A&A* 454, 321
Kochukhov, O., Khan, S., & Shulyak, D. 2005, *A&A* 433, 671
Shulyak, D., Kochukhov, O., & Khan, S. 2008, *A&A* 487, 689
Shulyak, D., Tsymbal, V., Ryabchikova, T., Stütz, Ch., & Weiss, W. W. 2004, *A&A* 428, 993

Cosmic Magnetic Fields:
From Planets, to Stars and Galaxies
Proceedings IAU Symposium No. 259, 2008
K.G. Strassmeier, A.G. Kosovichev & J.E. Beckman, eds.

doi:10.1017/S1743921309030907

Simulations of magneto-hydrodynamic waves in atmospheres of roAp stars

Elena Khomenko[1,2] **and Oleg Kochukhov**[3]

[1]Instituto de Astrofísica de Canarias, 38205, C/ Vía Láctea, s/n, Tenerife, Spain
[2]Main Astronomical Observatory, NAS, 03680, Kyiv, Ukraine
[3]Department of Physics and Astronomy, Uppsala University, Box 515, SE-751 20, Sweden
email: khomenko@iac.es

Abstract. We report 2D time-dependent non-linear magneto-hydrodynamical simulations of waves in the atmospheres of roAp stars. We explore a grid of simulations in a wide parameter space. The aim of our study is to understand the influence of the atmosphere and the magnetic field on the propagation and reflection properties of magneto-acoustic waves, formation of shocks and node layers.

Keywords. MHD – stars: magnetic fields – stars: oscillations

RoAp stars are late-A, chemically peculiar stars with effective temperatures between 6500–8100 K and global dipolar-like magnetic fields with strengths between 1–25 kG. They show low-order angular degree p-mode pulsations with periods between 4 and 22 minutes. These non-radial pulsations are aligned with dipolar field axis. Magneto-acoustic oscillations in peculiar A stars are of particular interest due to unique opportunities to study the interaction of pulsations, chemical inhomogeneities, and strong magnetic fields. Recent reviews on the properties of these stars and their pulsations can be found in Kurtz (2008); Cunha (2007); Kochukhov (2007, 2008).

We solve the basic 2D non-linear adiabatic equations of the ideal MHD by means of the numerical code described by Khomenko & Collados (2006); Khomenko et al. (2008). We assume that: (1) the magnetic field varies on spatial scales much larger than the typical wavelength, allowing the problem to be solved locally for a plane-parallel atmosphere with a homogeneous inclined magnetic field; (2) waves in the atmosphere are excited by low-degree pulsation modes with radial velocities exceeding horizontal velocities. The unperturbed atmospheric model has an effective temperature of $T_{\rm eff} = 7750$ K and gravitational acceleration at the surface $\log_{10} g = 4.0$. The simulation grid covers the magnetic field strength $B = 1 - 7$ kG; magnetic field inclination to the local vertical $\gamma = 0 - 60$ degrees and driving periods $T = 6 - 13$ minutes. An example of the wave pattern developed in the simulations is given in Fig. 1. We use the longitudinal and transversal field projections of the velocity in order to separate clearly the fast and slow MHD waves. Our first results pick up some observed properties of the roAp stars pulsations, such as: rapid growth of the wave amplitude with height, presence of the node surfaces, large variations in pulsation properties depending on the parameters of the model. Velocity signal observed in the upper atmospheric layers of roAp stars is mostly due to running slow mode (acoustic) waves propagating along the inclined magnetic field lines. The node structures and the rapid phase variations at the lower atmospheric layers are due to multiple reflections and interference of the slow and fast mode waves. The disc-integrated velocity signal produced by the atmospheric pulsations of such a star will depend in a complex way on the inclination of the magnetic axis with respect to the observational line of sight and will be a subject of our further study.

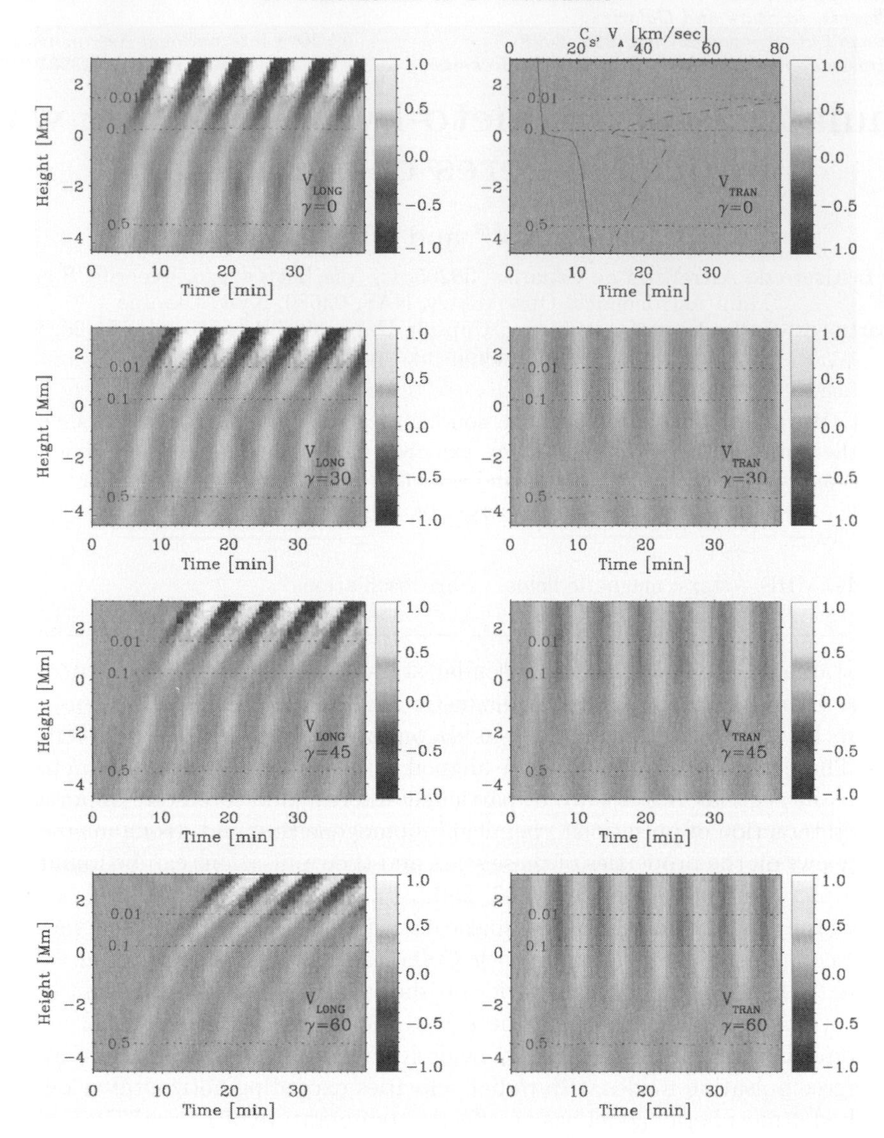

Figure 1. Height-time variations of the longitudinal (parallel to the field, left) and transversal (perpendicular to the field, right) velocities for B=1 kG, T=360 sec at latitudes where the inclination equals 0, 30, 45 and 60 degrees. The color bars give velocity scale in km/sec. Zero height corresponds to the photospheric base. Dotted lines marked with numbers are contours of constant c_S^2/v_A^2 . Height dependences of the sound speed c_S (solid line) and Alfvén speed v_A (dashed line) are plotted over the top panel, the scale is given by the upper axis.

References

Cunha, M. S. 2007, *Comm. in Astroseismology* 150, 48

Khomenko, E. & Collados, M. 2006, *ApJ* 653, 739

Khomenko, E., Collados, M., & Feliipe, T. 2008, *SP* 251, 589

Kochukhov, O. 2007, *Comm. in Astroseismology* 150, 39

Kochukhov, O. 2008, *Comm. in Astroseismology* arXiv:0810.1508, in press

Kurtz, D. W. 2008, *SP* 251, 21

Cosmic Magnetic Fields:
From Planets, to Stars and Galaxies
Proceedings IAU Symposium No. 259, 2008
K.G. Strassmeier, A.G. Kosovichev & J.E. Beckman, eds.

doi:10.1017/S1743921309030919

Gradient of the stellar magnetic field in measurements of hydrogen line cores

Dimitry O. Kudryavtsev and Iosif I. Romanyuk

Special Astrophysical Observatory RAS, Nizhny Arkhyz, 369167, Russia

Abstract. We report the observed systematic differences in longitudinal magnetic field values, obtained from measurements of metal lines and the core of the H_β line for a number of Ap stars, having strong global magnetic fields. In overwhelming majority of cases the magnetic field values, obtained from measurements of hydrogen lines cores, is smaller then the ones obtained from metal lines. We discuss some possible explanations of this effect, the most probable of which is the existence of the gradient of the magnetic field in stellar atmospheres.

Keywords. Stars: magnetic field – stars: chemically peculiar – stars: 53 Cam

1. Observations and measurements

The Zeeman spectra have been obtained in three different observational runs in 2006, 2008. We used the Main Stellar Spectrograph of the 6m telescope equipped with the image slicer and the circular polarization analyzer (Chountonov (2004)). The spectral resolution $R = 15000$. For our observations we used the spectral range 4760–5000Å, which includes 50–100 metal lines and the line H_β. Longitudinal magnetic field values were measured by the Zeeman shift between lines positions in right–hand and left–hand circularly polarized spectra, using standard Babcock's formula. Positions of spectral lines were determined by the fitting of profiles with a Gaussian.

2. The effect

In Fig. 1 we show a region of a spectrum of the star 53 Cam in right–hand and left–hand circular polarizations, and the parameter Stokes V for this region. One can see that the Stokes V values in metal lines is greater than the Stokes V in H_β. Particularly this effect is caused by the fact that metall lines have generally higher Lande factors ($z = 1.23$ on average), and therefore are more sensitive to the magnetic field, while H_β has $z = 1.0$. But in Zeeman measurements we take this effect into account, and nevertheless the measured longitudinal magnetic field B_e shows sufficiently different values for metal lines and for the core of H_β.

We have about 50 measurements for 30 stars, which unfortunately cannot be presented here due to the page limits. The core of the H_β line shows systematically lower values of B_e for almost all the cases when the longitudinal magnetic field differs from the zero.

3. Possible explanations

Partial Zeeman splitting. Metal lines generally have more complex picture of Zeeman splitting than the H_β. For stars with strong fields this may cause the asymmetry or the partial splitting of spectral lines. In those cases our procedure of the line center determination by fitting a Gaussian may cause a fault. But with the moderate spectral

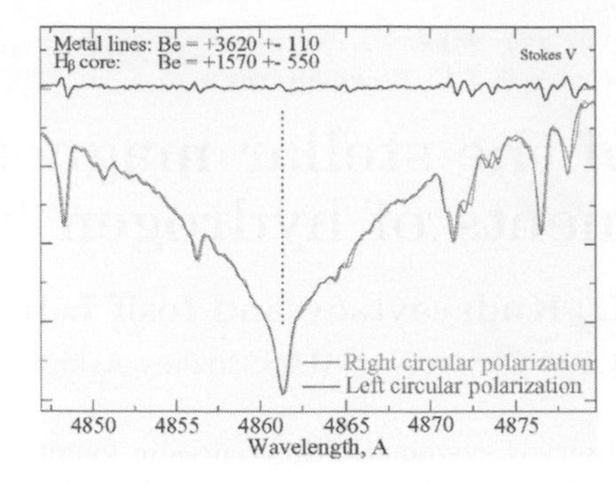

Figure 1. A spectrum of right, left circular polarizations and Stokes V for the star 53 Cam.

resolution we used, this could be true only for stars with a very strong surface magnetic fields. But for most our stars the metal lines are symmetrical for the resolution $R = 15000$ and no partial Zeeman splitting is observed.

Spots. Ap stars are well known as the objects with the inhomogeneous distribution of chemical elements over the surface. This could influence the magnetic field measurements. If the measured lines belong to an element concentrated near magnetic poles than the measured B_e will be higher than the B_e, obtained from hydrogen lines. Metal lines, we used for measurements, are mostly the lines of Fe and Cr. Though these elements could have inhomogeneous distribution, they are present enough over all the surface of a star and actually for most cases the spectral variability is not so great during rotational period. As an evidence of low influence of spots to our measurements we refer to the measurements of 53 Cam and γ Equ, where our measurements of metal lines corresponds well with the measurements of other authors, obtained by measuring of polarization in wings of hydrogen lines.

Gradient of the magnetic field. Another way to explain why the measured longitudinal magnetic fields differ, is to suggest that we observe the gradient of the magnetic field in stellar atmosphere. The forming of metall lines and the cores of hydrogen lines takes place in different layers of the stellar atmosphere. So if there is a difference in magnetic field strengths between these layers we must observe different values of the longitudinal magnetic field. Accepting this explanation, we should note that the observed gradient of the magnetic field is not represent usual magnetic field decreasing with the moving out of the center of a dipole as the thickness of the stellar atmosphere is too small for this effect to be observed. Thus, probably we observe a strong deviation from the standard dipole model of the magnetic fields of Ap stars within the rather small radial scale.

4. Resume

Although the partial Zeeman splitting and inhomogeneous chemical abundances could explain the difference between measured magnetic field values in some special cases, there is good reason to believe that they are not sufficient to describe all our measurements. We suppose that there is the influence of the radial gradient of the magnetic field in the stellar atmosphere having place in our measurements of cores of hydrogen lines, at least partially.

References

Chountonov, G. A. 2004, in: Yu. V. Glagolevskij, D. O. Kudryavtsev, & I. I. Romanyuk (eds.), *Magnetic stars*, Proc. International Conf. (Nizhny Arkhyz: SAO Press), p. 289

Cosmic Magnetic Fields:
From Planets, to Stars and Galaxies
Proceedings IAU Symposium No. 259, 2008
K.G. Strassmeier, A.G. Kosovichev & J.E. Beckman, eds.

doi:10.1017/S1743921309030920

Probing the size of a magnetosphere of a young solar-like star

Catrina M. Hamilton[1], **C. M. Johns-Krull**[2], **R. Mundt**[3], **W. Herbst**[4] **and J.N. Winn**[5]

[1]Department of Physics and Astronomy, Dickinson College, Carlisle, PA 17013, USA
email: hamiltoc@dickinson.edu

[2]Department of Physics and Astronomy, Rice University, Houston, TX 77005, USA
email: cmj@rice.edu

[3]Max-Planck-Institut für Astronomie, Königstuhl 17, D-69117 Heidelberg, Germany
email: mundt@mpia.de

[4]Astronomy Department, Wesleyan University, Middletown, CT 06459, USA
email: wherbst@wesleyan.edu

[5]Department of Physics, Massachusetts Institute of Technology, Cambridge, MA 02139, USA
email: jwinn@space.mit.edu

Abstract. We have obtained high resolution spectra of the pre-main sequence binary system KH 15D (V582 Mon) while the star is fully visible, fully occulted, and during several ingress and egress events over the course of five contiguous observing seasons. The Hα line profile is a standard probe of the magnetospheric accretion flows on young stars such as KH 15D. We use these time series data to map out the size of the magnetosphere and find that it changes size from one observing season to the next.

Keywords. Line: profiles – stars: individual (KH 15D, V582 Mon) – stars: pre-main-sequence

1. Introduction

KH 15D (V582 Mon) is an eclipsing pre–main-sequence binary system with a period of 48.37 days and an eccentric orbit (Hamilton *et al.* 2005). The eclipses occur whenever the motion of a star carries it behind a circumbinary ring (Chiang & Murray-Clay 2004; Winn *et al.* 2004). As the edge of the occulting matter cuts across the visible star and its magnetosphere, it acts like a "natural coronagraph", progressively covering or revealing structure within a few stellar radii of the photosphere (Hamilton *et al.* 2003).

2. The Hα Emission Line Profiles

Out of eclipse (large +R), one sees a redshifted absorption component in the Hα line profile that extends below the stellar continuum. During eclipse, the line profile is narrower and double-peaked, with an underlying broad component, believed to arise in the magnetosphere, which ranges from about -350 to +350 km/s. The profiles that have "shoulders" are considered evidence for magnetospheric accretion (see profiles in Figures 1 and 2). From these profiles and the distance the star is from the screen at each epoch, we infer that the magnetosphere is changing in size.

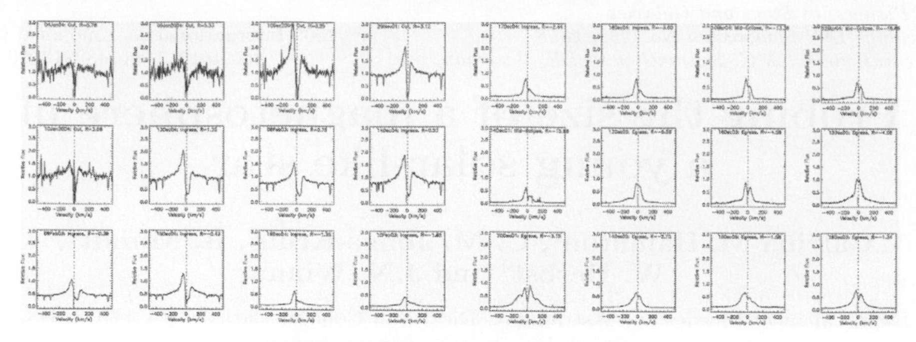

Figure 1. Hα emission line profiles.

3. Magnetospheric Toy Model for Hα

A simple dipole model is computed (Alencar *et al.* (2001)) where material freely falls from a disk to the star along field lines. We consider a model that employs a linear flux distribution and includes rotation. We vary the inclination and match the output to an out-of-eclipse profile. We then vary the distance that the star is from the screen and scale the model profile by the photometry appropriate for that night. Our results are shown in Figure 3. In all cases, the inclination was set to 15°.

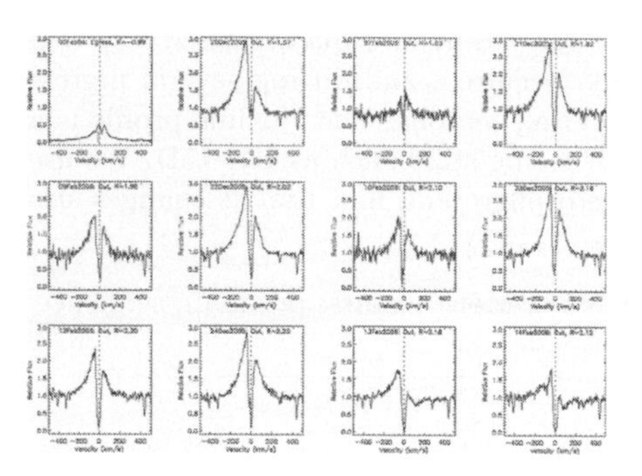

Figure 2. Hα emission line profiles.

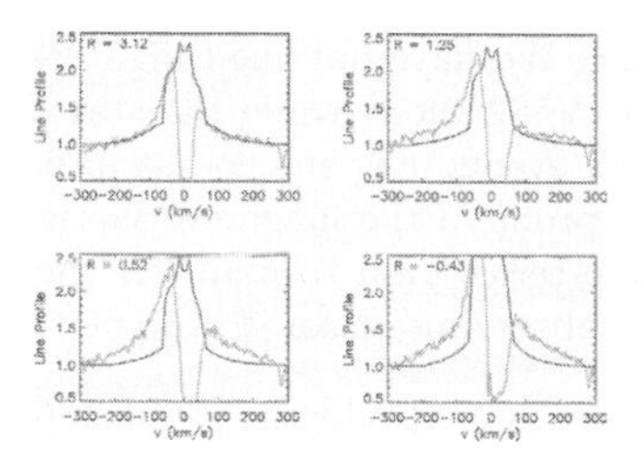

Figure 3. Hα line profiles as calculated by a magnetospheric toy model employed by Alencar *et al.* (2001) to model the broad wings observed in the KH 15D Hα line profiles, which are believed to arise in the magnetosphere.

4. Conclusions

Our spectroscopic studies have revealed intrinsic variability and amazing structure in the Hα line profiles caused by high velocity gas flows in the vicinity of the stars. As the model profile is eclipsed, the shape is generally consistent with the observed profile, however, the flux level does not match the observations. We intend to further investigate the influence the many geometrical parameters and different flux distributions has on the model profiles.

References

Alencar, S. H. P, Johns-Krull, C. M., & Basri, G. 2001, *AJ* 122, 3335
Chiang, E. I. & Murray-Clay, R. A. 2004, *ApJ* 607, 913
Hamilton, C. M., *et al.* 2003, *ApJ* 591, L45
Hamilton, C. M., *et al.* 2005, *AJ* 130, 1896
Winn, J. N., *et al.* 2004, *ApJ* 603, L45

Cosmic Magnetic Fields:
From Planets, to Stars and Galaxies
Proceedings IAU Symposium No. 259, 2008
K.G. Strassmeier, A.G. Kosovichev & J.E. Beckman, eds.

doi:10.1017/S1743921309030932

Numerical simulations of magnetized winds of solar-like stars

Aline A. Vidotto[1,2], M. Opher[2], V. Jatenco-Pereira[1] and T. I. Gombosi[3]

[1]University of São Paulo, Rua do Matão 1226, São Paulo, SP, Brazil, 05508-090
email: aline@astro.iag.usp.br

[2]George Mason University, 4400 University Drive, Fairfax, VA, USA, 22030-4444

[3]University of Michigan, 1517 Space Research Building, Ann Arbor, MI, USA, 48109-2143

Abstract. We investigate magnetized solar-like stellar winds by means of self-consistent three-dimensional (3D) magnetohydrodynamics (MHD) numerical simulations. We analyze winds with different magnetic field intensities and densities as to explore the dependence on the plasma-β parameter. By solving the fully ideal 3D MHD equations, we show that the plasma-β parameter is the crucial parameter in the configuration of the steady-state wind. Therefore, there is a group of magnetized flows that would present the same terminal velocity despite of its thermal and magnetic energy densities, as long as the plasma-β parameter is the same.

Keywords. Stars: coronae – stars: late-type – stars: magnetic fields – stars: mass loss

1. Introduction

Observations have shown that the solar corona (SC) is a highly complex system, consisting of both long-lived magnetic features (e.g., streamers, coronal holes) and short-lived ones (coronal mass ejections, solar flares, sunspots). Although indirect detections of stellar coronal winds have been performed (Wood *et al.* 2005), direct measurements of tenuous coronal winds for other stars rather than the Sun are very difficult to do. However, as it occurs in the SC, it is quite certain that the magnetic field is playing an important role in coronal winds of solar-like stars.

Several studies have been made toward the understanding of an expanding magnetized corona (*e.g.*, Pneuman & Kopp 1971, Low & Tsinganos 1986, Washimi & Shibata 1993). However, despite all the notable evolution of both analytical and numerical studies performed in the last decades, we are far from a satisfactory 3D MHD description of a magnetized wind. Several approximations are usually made in order to make the system analytically and numerically tractable.

The present work investigates the influence of the magnetic field in solar-like stellar winds with different β's (the ratio between thermal and magnetic energy densities). We solve the fully 3D MHD equations with the temporal evolution of the energy equation. Therefore, the topology of the field is not restricted and the steady-state arises from the dynamical interplay of the outflow and the field. We neglect the stellar rotation.

2. The Numerical Model

To perform the simulations, we make use of BATS-R-US, a 3D ideal MHD numerical code developed at University of Michigan (Powell *et al.* 1999). We adopted the same, non-uniform grid resolution for all the simulations: the smallest cell size is 0.018 r_0 (near the central body), the maximum cell size is 4.68 r_0. The center of the star is placed at the origin of the grid. The axes x, y, and z extend from $-75\ r_0$ to $75\ r_0$.

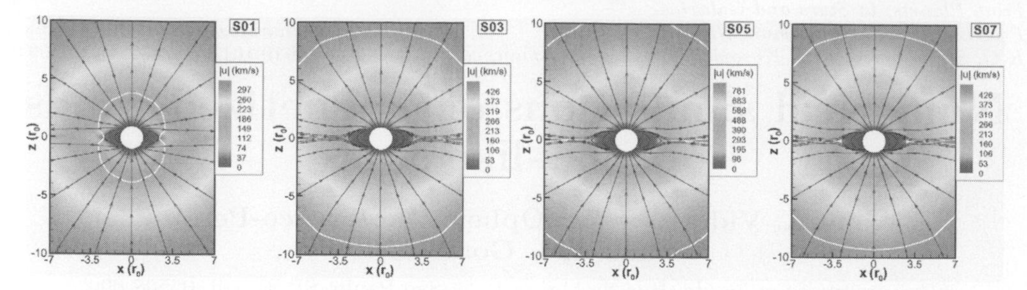

Figure 1. Meridional cuts of the steady-state configurations for cases S01 - S07.

We consider a star with $1M_\odot$, $r_0 = 1R_\odot$. We initialized the simulations with an isothermal HD wind permeating the grid, and a dipolar magnetic field anchored on the stellar surface. The system was then evolved in time until steady-state was achieved. The inner boundary of the system is considered to be the base of the wind at r_0, where fixed boundary conditions were adopted. The outer boundary has outflow conditions.

3. Results

We present here four simulations. We consider a corona with temperature and density at the base of the wind of 1.56 MK and 1.544×10^{-16} g cm^{-3}, respectively. The magnitude of the magnetic field was varied in each case. For case S01, $B_0 = 1$ G; 10 G in S03; and 20 G in S05. In case S07, $B_0 = 20$ G, and density was increased 20 times the previous value. This choice of parameters leads to a plasma-β of: 1 (S01), 0.01 (S03, S07), and 0.0025 (S05). Figure 1 presents meridional cuts of the steady-state configurations: the contours represent the velocity of the flow, black streamlines are the final configuration of the magnetic field, and the white line is the Alfven surface.

From Figure 1, we note that the crucial parameter in determining the wind velocity profile is the β-parameter (compare S03 and S07). Therefore, there is a group of magnetized flows that would present the same terminal velocity despite of its thermal and magnetic energy densities, as long as the plasma-β parameter is the same. We also note that the steady-state magnetic field topology for all cases is similar, presenting a configuration of helmet streamer-type, with zones of closed field lines and open field lines coexisting. The wind is not spherical, it is latitude-dependent: higher velocities are achieved at high latitudes. Cases with lower β show more accelerated winds (Fig. 1).

The resultant bi-modality of the wind is due to the nature of the magnetic force. A purely HD (non-rotating) wind is spherically symmetric but in the MHD case, this symmetry is lost because the magnetic force has a meridional component.

The authors thank FAPESP (04-13846-6, 07/58793-5), CAPES (BEX4686/06-3), CNPq (305905/2007-4), NSF CAREER (ATM-0747654), and the staff at NASA Ames Research Center for the use of the Columbia supercomputer.

References

Low, B. C., & Tsinganos, K. 1986, *ApJ* 302, 163
Pneuman, G. W. & Kopp, R. A. 1971, *SP* 18, 258, PK71
Powell, K. G. *et al.* 1999, *J. of Comp. Phys.* 154, 284
Washimi, H. & Shibata, S. 1993, *MNRAS* 262, 936
Wood, B. E. *et al.* 2005, *ApJ* 628, L143

Cosmic Magnetic Fields: From Planets, to Stars and Galaxies
Proceedings IAU Symposium No. 259, 2008
K.G. Strassmeier, A.G. Kosovichev & J.E. Beckman, eds.

doi:10.1017/S1743921309030944

Stellar nonlinear dynamos: observations and modelling

Ilkka Tuominen[1], Maarit J. Korpi[1], Petri J. Käpylä[1], Marjaana Lindborg[1], and Ilya Ilyin[2]

[1]Observatory, University of Helsinki, PO BOX 14, FI-00014 University of Helsinki, Finland
[2]Astrophysical Institute Potsdam, An der Sternwarte 16, D-14482 Potsdam, Germany

Abstract. Recent numerical modelling of mean-field stellar dynamos with induction and Reynolds' equations show that, with increasing rotation, field symmetry changes from an axisymmetric solar type to a nonaxymmetric one, where the so-called active longitudes in the same stellar hemisphere are predicted to be of opposite polarities. It was originally named Active star Hale rule in Tuominen *et al.* (2002), being an analogue to the famous bipolar sunspot polarity rule but different in scale and being a global phenomenon. In addition to long timeseries of temperature mapping and photometry, during the last few years we have been able to measure accurately polarization spectra of an active late-type star II Peg with the upgraded spectropolarimetric option of the high-resolution spectrograph SOFIN at the Nordic Optical Telescope, La Palma. The magnetic inversions (Carroll *et al.*, Kochukhov *et al.*) can be compared to the dynamo models and the preliminary results show some resemblance to the dynamo solutions.

Keywords. Stars:late-type – stars: spots – MHD – methods: numerical

---

1. Introduction

Recently Käpylä *et al.* (2006) performed local convection calculations to determine the α-tensor resulting from turbulent convection. They utilized the so-called imposed field method, where weak uniform magnetic fields into certain direction are imposed on the flow, and the corresponding α-tensor components solved from the series expansion of the turbulent electromotive force. Calculations at varying latitude and angular velocity were performed. At the highest angular velocity, corresponding to the Coriolis number of ≈ 10, the $\alpha_{\phi\phi}$ component is observed no longer to peak at the pole as conventionally assumed, but to be concentrated near the equatorial region, following approximately a $\sin^2\theta\cos\theta$-dependence. Here we test corresponding profiles in a mean-field dynamo calculation.

2. Nonaxisymmetric mean-field model

We solve for the mean-field induction and Reynolds'equations in $2\frac{1}{2}$ dimensions, meaning that the azimuthal direction is treated in Fourier space (see e.g. Moss *et al.* 1995). The equations are solved in an incompressible fluid shell $R_0 < r < 1, \pi/2 < \theta < -\pi/2$, for the lowest four Fourier modes $m = 0, \ldots, 3$. Here we take $R_0 = 0.5$. We adopt a simple model for the Lambda-effect, namely $\Lambda_V = \nu_T V^0$ while other components vanish; here we use $V^0 = 1$. Different types of latitudinal α-profiles are used, motivated by the local convection simulations. Similar study has been conducted for axisymmetric α^2-dynamo model by Rüdiger *et al.* (2003); our model differs in that we also consider nonaxisymmetric modes, include some (but only very little) differential rotation, and consider only isotropic α-effect.

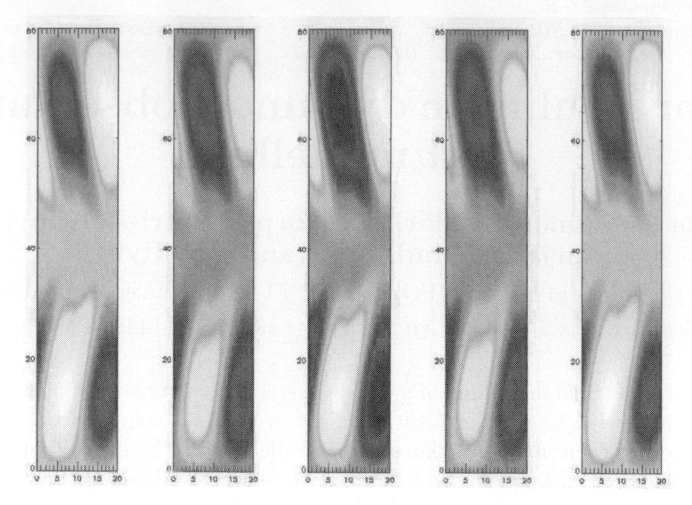

Figure 1. Radial field at the surface for the run $C_{\alpha 1}$=100, Ta=10 000 at five different timepoints covering the whole oscillation cycle. The small and rapid oscillation is related to the energy levels of the spots, while the polarity of the spots remains unchanged. The overall appearance of the instantaneous dynamo solution shows some resemblance to the observed magnetic field configuration (Carroll *et al.* 2009, Kochukhov *et al.* 2009).

3. Results

As reported in our earlier papers (Tuominen *et al.* 1999, 2002), the simplest α-profile leads to $A1$-type solutions at rapid rotation i.e. high Taylor numbers (Ta). Due to $V^0 \neq 0$, some radial differential rotation is generated; the absolute differential rotation, however, is small, $|\Delta\Omega| < 100$ for the largest values of Ta investigated. The dominating Fourier mode $m = 1$ is always drifting in azimuth with respect to the rotational frame so that for $Ta < 1000$ the mode is lagging and for $Ta > 1000$ going ahead the rotational frame. The higher Ta, the closer the poles the spots appear. There is a small oscillation at high Ta related to the relative energy levels of the spots; the spots with negative radial field get simultaneously weaker while the positive ones gets stronger and vice versa; this behaviour is depicted in Fig.1. The timescale of this variation is very short compared to the diffusion time (typically < 0.1 diffusion times).

Local convection calculations suggest that the α-effect would be concentrated to the equatorial regions with rapid rotation. Implementing profiles of this type to the nonaxisymmetric mean-field dynamo model including dynamics and small amounts of differential rotation showed that axisymmetric modes can coexist with the nonaxisymmetric modes even in the rapid rotation regime, mixed modes and parities are typical, and solutions can be oscillatory.

References

Carroll, T., Kopf, M., Strassmeier, K., Ilyin, I., & Tuominen, I., 2009, *this proceedings*

Kochukhov, O., Piskunov, N., Ilyin, I., & Tuominen, I., 2009, *this proceedings*

Käpylä, P.J., Korpi, M.J., Ossendrijver, M., & Stix, M. 2006, *A&A* 455, 401

Moss, D., Barker, D. M., Brandenburg, A., & Tuominen I. 1995, *A&A* 294, 155

Rüdiger, G., Elstner, D., & Ossendrijver, M. 2003, *A&A* 406, 15

Tuominen, I., Bergyugina, S. V., Korpi, M. J., & Rönty, T. 1999, in *Stellar dynamos: Nonlinearity and Chaotic Flows*, Nunez & Ferriz-Mas (eds.), ASP Conference Series 178

Tuominen, I., Berdyugina, S. V., & Korpi, M. J. 2002, *AN* 323, 367

Cosmic Magnetic Fields:
From Planets, to Stars and Galaxies
Proceedings IAU Symposium No. 259, 2008
K.G. Strassmeier, A.G. Kosovichev & J.E. Beckman, eds.

doi:10.1017/S1743921309030956

Low-dimensional models of stellar and galactic dynamos

Dimitry Sokoloff and S. Nefyodov

Department of Physics, Moscow State University, Moscow, 119991, Russia
email: sokoloff@dds.srcc.msu.su; email: rainbow84@rambler.ru

Abstract. A regular method how to simplify dynamo model for a particular celestial body up to a dynamical system is suggested. Dynamical system obtained occurs specific for a thin galactic disc, a fully convective star and a thin convective shell.

Keywords. Magnetic fields – convection – stars: low-mass

---

1. Introduction

Traditional dynamo models for magnetic field generation in various celestial bodies including Sun, stars, planets and galaxies presume a substantial knowledge concerning spatial distribution of magnetic field generators, i.e. differential rotation and so-called α-effect connected with mirror asymmetry of turbulence. Solar differential rotation is known quite in details due to impressive progress of helioseismolody, rotation curves of many galaxies come directly from observations however we still know some integral quantities for differential rotation of many stars and planets. Even less complete are our knowledge concerning α-effect which come from order of magnitude estimates and are supported by current helicity observations in solar active regions Kleeorin, Kuzanyan, Moss *et al.*, 1983). Because of that, it looks reasonable to simplify dynamo model for a celestial bodies with poorly known hydrodynamics up to dynamical system with coefficients depending on integral quantities representing hydrodynamical flow driving dynamo.

Dynamo models in form of a dynamical system was considered in several papers, e.g. Ruzmaikin (1981) suggested to reduce solar dynamo model to the Lorentz attractor. Kitiashvili & Kosovichev (2008) used a dynamo model in form of a dynamical system for a predictions of properties for the further solar activity cycle. A choice of a particular dynamical system in such papers is motivated by rather general similarity of terms in the full set of equations and that ones participating in the dynamical system so a more regular method to obtain low-dimensional dynamo models in form of dynamical systems looks desirable. Here we present such a method.

2. Constructing dynamical system

We follow here an old idea of Rädler & Wiedemann (1989) who suggested to present the desired solution as a combination of several first free decay modes. The number of the modes exploited is chosen as a minimal one which gives an adequate dynamo action (steady growth for a galactic disc and oscillatory one for spherical models) in a kinematic (linear) case. The desired solution of the dynamo system is a set of Fourier coefficient as functions of time. Coefficients of the system are calculated as various integrals which involves these free decay modes and rotation curve or spatial distribution of α. Rädler & Wiedemann (1989) stressed that the number of the free decay modes required for the dynamo self-excitation occurs to be unexpectedly large and the Fourier coefficients obtained are rather unstable. We suggest that this happens because quite a lot of first free decay modes do not contribute much into the dynamo action. We identify such useless modes taking the free decay modes one after the other from the set used to construct the solution. If the excitation occurs to be independent on the mode selected we exclude it from the set. The procedure of selection is described in details by Sokoloff & Nefyodov (2007).

Spectrum of free decay modes depends of geometrical shape of dynamo region. Correspondingly, dynamical system obtained depends on the geometry of dynamo region. In particular, dynamical system for a dynamo generation in thin galactic disc nearby a given radius (local disc dynamo problem) requires two modes only what determines a remarkable robustness of galactic dynamos known from experiences with more detailed galactic dynamo models. The low-dimensional model of galactic dynamo is presented by Moss, Shukurov & Sokoloff (1999). Dynamical system for a fully convective star (Sokoloff, Nefedov, Ermash, *et al.*, 2008) requires 5 free decay modes. We present below the dynamical system obtained for a star with a thin convective shell from classical Parker migratory dynamo model:

$$\frac{da_1}{dt} = \frac{R_\alpha b_1}{2} - a_1 - \xi^2 \frac{3R_\alpha b_1}{8}(b_1^2 + 2b_2^2)\,,$$

$$\frac{da_2}{dt} = \frac{R_\alpha}{2}(b_1 + b_2) - 9a_2 - \xi^2 \frac{3R_\alpha (b_1 + b_2)}{8}(b_1^2 + b_1 b_2 + b_2^2)$$

$$\frac{db_1}{dt} = \frac{R_\omega}{2}(a_1 - 3a_2) - 4b_1\,, \qquad \frac{db_2}{dt} = \frac{3R_\omega a_2}{2} - 16b_2\,.$$

Here b_1 and b_2 are Fourier amplitudes of first toriodal free decay modes, a_1 and a_2 are first poloidal modes. A simple algebraic α-quenching in form $\alpha = \alpha_0(1 - B^2/B_0^2)$ is supposed (B_0 is the equipartition magnetic field strength and toroidal magnetic field B is measured in units of B_0). Dimensionless numbers R_α and R_ω represent normalized intensities of α-effect and differential rotation correspondingly.

The dynamical system obtained is quite remote from naive expectation based on usual cartoon for Parker migratory dynamos which includes toroidal and poloidal magnetic fields represented by first free decay modes. Nevertheless, the model demonstrates a dynamo self-excitation in form of traveling wave of toroidal magnetic field and realistic nonlinear behaviour in form of nonlinear activity waves.

Acknowledgements

Financial support from RFBR under grant 07-02-00127 and from IAU in form of a travel grant is acknowledged.

References

Rädler, K.-H. & Wiedemann, E. 1989, *Geophys. Astrophys. Fluid Mech.* 49, 71
Ruzmaikin, A. A. 1981, *Comm. Astrophys.* 9, 86
Kitiashvili, I. & Kosovichev, A. G. 2008, *ApJ* 688, L49
Sokoloff, D. & Nefyodov, S. 2007, *Numeric. Mod. Advance Progr.* 8, 195
Sokoloff, D. D., Nefedov, S. N., Ermash, A. A., & Lamzin S. A. 2008, *Astron. Lett.* 34, 761
Moss D., Shukurov A., & Sokoloff, D. 1977, *A&A* 343, 120
Kleeorin, N., Kuzanyan, K., Moss, D., Rogachevskii, I., Sokoloff, D., & Zhang, H. 2003, *A&A* 409, 1097

Cosmic Magnetic Fields:
From Planets, to Stars and Galaxies
Proceedings IAU Symposium No. 259, 2008
K.G. Strassmeier, A.G. Kosovichev & J.E. Beckman, eds.

doi:10.1017/S1743921309030968

On MHD rotational transport, instabilities and dynamo action in stellar radiation zones

Stéphane Mathis[1,2], **A.-S. Brun**[1,2], **J.-P. Zahn**[2]

[1] Laboratoire AIM, CEA/DSM-CNRS-Université Paris Diderot, IRFU/SAp, F-91191 Gif-sur-Yvette Cedex, France
email: stephane.mathis@cea.fr, allan-sacha.brun@cea.fr

[2] LUTH, Observatoire de Paris-CNRS-Université Paris Diderot, 5 Place Jules Janssen, F-92195 Meudon Cedex, France
email: jean-paul.zahn@obspm.fr

Abstract. Magnetic field and their related dynamical effects are thought to be important in stellar radiation zones. For instance, it has been suggested that a dynamo, sustained by a $m = 1$ MHD instability of toroidal magnetic fields (discovered by Tayler in 1973), could lead to a strong transport of angular momentum and of chemicals in such stable regions. We wish here to recall the different magnetic transport processes present in radiative zone and show how the dynamo can operate by recalling the conditions required to close the dynamo loop ($B_{\rm Pol} \rightarrow B_{\rm Tor} \rightarrow B_{\rm Pol}$). Helped by high-resolution 3D MHD simulations using the ASH code in the solar case, we confirm the existence of the $m = 1$ instability, study its non-linear saturation, but we do not detect, up to a magnetic Reylnods number of 10^5, any dynamo action.

Keywords. MHD – Sun: magnetic fields – Sun: interior – stars: magnetic fields – stars: interiors

1. MHD instabilities and possible dynamo in stellar radiation zones

Purely axisymmetric poloidal and toroidal fields are unstable (see Pitts & Tayler 1985 and references therein). Moreover, Spruit (2002) suggests that the instability of such toroidal field could sustain a dynamo in stellar radiation zones. This idea is quite interesting, but we argue that this dynamo cannot operate as he describes it. According to him, the non-axisymmetric instability-generated small-scale field, which has zero average, is wound up by the differential rotation "into a new contribution to the azimuthal field. This again is unstable, thus closing the dynamo loop." But this shear induced azimuthal field has the same azimuthal wavenumber as the instability-generated field, i.e. $m \neq 0$ and predominantly $m = 1$: it has no mean azimuthal component, and thus it cannot regenerate the mean toroidal field that is required to sustain the instability. For the same reason, the instability-generated field cannot regenerate the mean poloidal field, as was suggested by Braithwaite (2006). Therefore, the Pitts & Tayler instability cannot be the cause of a dynamo, as it was described by Spruit and Braithwaite. In fact, the dynamo loop can only be achieved through the azimuthal average of the fluctuation-fluctuation term of the induction equation $\langle \vec{\nabla} \times (\vec{v}^{'} \times \vec{B}^{'}) \rangle_{\varphi}$ (cf. Zahn, Brun & Mathis 2007).

2. Numerical simulations

We perform 3D-numerical simulations of the problem using the global ASH code (Clune *et al.* 1999, Brun *et al.* 2004) to solve the relevant anelastic MHD equations in a spherical shell representing the upper part of the solar radiation zone ($0.35 \leqslant r/R_{\odot} \leqslant 0.70$) using a resolution of $N_r \times N_\theta \times N_\varphi = 193 \times 128 \times 256$. A detailed discussion of the set-up is given in Brun & Zahn (2006) and in Zahn, Brun & Mathis (2007). We study the case A

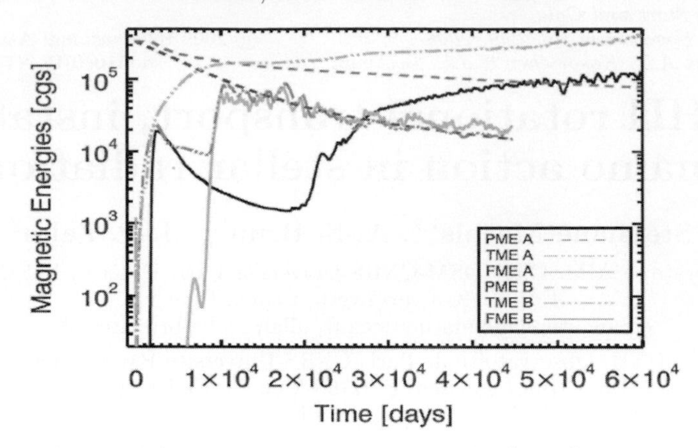

Figure 1. Time evolution of the energies of the mean poloidal (PME), mean toroidal (TME) and non-axisymmetric (FME) components of the magnetic field. Cases A and B refer respectively to higher and lower magnetic diffusivity. Note the steady decline of the poloidal field, which is not affected by the irruption of the $m = 1$ Pitts & Tayler instability (at t$\approx$ 8,000 days in case A and $\approx$ 20,000 days in case B). (Zahn, Brun & Mathis 2007, courtesy A&A)

discussed in Brun & Zahn (2006) and we performed an additional series of simulations with a lower Ohmic diffusivity (by a factor of 10, case B), in order to reach a higher magnetic Reynolds number in favor of a dynamo. In our simulations (cf. Fig. 1), the α-effect plays a negligible role since no regeneration of the mean poloidal field is found, at least up to the magnetic Reynolds number $Rm = R^2\Delta\Omega/\eta \sim 10^5$ for Prandtl number $P_m = \nu/\eta = 1$. On the other hand, the β-effect, i.e. the turbulence-enhanced diffusivity, is absent here. Hence, one should not expect much mixing of the stellar material and the magnetic transport of angular momentum is mainly due to the Lorentz torque that leads to Ferraro's law ($\vec{B}\cdot\vec{\nabla}\Omega = 0$) (cf. Brun & Zahn 2006). In fact, the smallest resolved scales do not act on the mean poloidal field as a turbulent diffusivity: they seem to behave rather as gravito-Alfvén waves. Finally, there is no sign either of a small-scale fluctuation dynamo. To check this point, we suppressed the mean poloidal field at the latest stage of our simulation. Then, the mean toroidal field decreases rapidly, because it is no longer produced by the Ω-effect, and the instability-generated field accompanies its decline. Thus the fluctuating field does not maintain itself. Therefore, we conclude that in our simulations the Pitts & Tayler instability is unable to sustain a large-scale mean field dynamo, in the parameter domain that we have explored (see also Gellert, Rüdiger & Elstner 2008).

References

Braithwaite, J. 2006, *A&A* 449, 451
Brun, A.-S. & Zahn, J.-P. 2006, *A&A* 457, 665
Brun, A.-S., Miesch, M. S. & Toomre, J. 2004, *ApJ* 614, 1073
Clune, T. L. *et al.* 1999, *Parallel Comput.* 25, 361
Gellert, M., Rüdiger, G., & Elstner, D. 2008, *A&A* 479, L33
Pitts, E. & Tayler, R. J. 1985, *MNRAS* 216, 139
Spruit, H. C. 2002, *A&A* 381, 923
Zahn, J.-P., Brun, A.-S., & Mathis, S. 2007, *A&A* 474, 145

Cosmic Magnetic Fields:
From Planets, to Stars and Galaxies
Proceedings IAU Symposium No. 259, 2008
K.G. Strassmeier, A.G. Kosovichev & J.E. Beckman, eds.

doi:10.1017/S174392130903097X

Angular momentum loss and stellar spin-down in magnetic massive stars

Asif ud-Doula[1], Stanley P. Owocki[2] and Richard H.D. Townsend[3]

[1]Department of Physics, SUNY-Morrisville State College, Morrisville, NY 13408, USA
e-mail: uddoula@morrisville.edu

[2]Department of Physics and Astronomy, University of Delaware,
Newark, DE 19716, USA

[3]Department of Physics, University of Wisconsin-Madison,
Madison, WI 53706, USA

Abstract. We examine the angular momentum loss and associated rotational spin-down for magnetic hot stars with a line-driven stellar wind and a rotation-aligned dipole magnetic field. Our analysis here is based on our previous 2-D numerical MHD simulation study that examines the interplay among wind, field, and rotation as a function of two dimensionless parameters, W(=Vrot/Vorb) and 'wind magnetic confinement', η_* defined below. We compare and contrast the 2-D, time variable angular momentum loss of this dipole model of a hot-star wind with the classical 1-D steady-state analysis by Weber and Davis (WD), who used an idealized monopole field to model the angular momentum loss in the solar wind. Despite the differences, we find that the total angular momentum loss averaged over both solid angle and time follows closely the general WD scaling $\dot{J} \sim \dot{M}\Omega R_A^2$. The key distinction is that for a dipole field Alfvèn radius R_A is significantly smaller than for the monopole field WD used in their analyses. This leads to a slower stellar spin-down for the dipole field with typical spin-down times of order 1 Myr for several known magnetic massive stars.

Keywords. MHD – Stars: winds – Stars: magnetic fields – Stars: early-type – Stars: rotation – Stars: mass loss

---

1. Introduction

An outflowing wind carries away angular momentum and thus spins down the stellar rotation. Winds with magnetic fields exert a braking torque that is significantly larger than for non-magnetic cases, due to the larger lever arm of magnetic field lines that extend outward from the stellar surface. A seminal analysis of this process was carried out by Weber & Davis (WD, 1967), who modelled the angular momentum loss of the solar wind for the idealized case of a simple monopole magnetic field from the solar surface. In terms of the surface angular velocity Ω and wind mass loss rate $\dot{M}$, Weber & Davis concluded that the total angular momentum loss rate scales as:

$$\dot{J} = \frac{2}{3}\dot{M}\Omega R_A^2 , \tag{1.1}$$

with R_A the Alfvén radius,defined by where the radial components of the field and flow have equal energy density. However, WD did not provide a prescription for computing such an Alfvén radius even for the idealized monopole field geometry.

For any radius r, the energy density ratio between radial field and flow is given by

$$\eta(r) \equiv \frac{B_r^2/8\pi}{\rho v_r^2/2} . \tag{1.2}$$

The Alfvén radius is then defined implicitly by $\eta(R_A) \equiv 1$. We can derive approximate *explicit* expressions in terms of fixed values for the equatorial field strength B_{eq} at the surface radius R_*, and for the wind mass loss rate $\dot{M}$ and terminal flow speed v_∞. Specifically, following ud-Doula & Owocki (paper 1, 2002) and ud-Doula *et al.* (paper 2, 2008a), if we define here a wind *magnetic confinement parameter*,

$$\eta_* \equiv \frac{B_{eq}^2 R_*^2}{\dot{M} v_\infty}, \tag{1.3}$$

then we can write the energy density ratio in the form

$$\eta(r) = \eta_* \left[\frac{r}{R_*}\right]^{2-2q} \frac{v_\infty}{v_r(r)} \tag{1.4}$$

where q is the power-law exponent for radial decline of the assumed magnetic field and terminal speed v_∞. It turns out that in the strong magnetic confinement limit $\eta_* \gg 1$, for a monopole field $R_A \sim \sqrt{\eta_*}$ whereas for a dipole field the scaling is significantly weaker $R_A \sim {\eta_*}^{1/4}$.

2. Key Results

In our recent paper 3 (ud-Doula, Owocki & Townsend (2008b)), which this poster summarizes, we examine the wind magnetic spin-down of massive stars with a rotation-aligned dipole field based on previous magnetohydrodynamic (MHD) simulation parameter study presented in paper 2. Despite key differences, we find that the total angular momentum loss from these massive stars follow the general WD scaling 1.1. However, because for dipole fields the Alfvén radius has a stronger field scaling than for the idealized monopole case, the net stellar spin-down time is also significantly longer. Our numerical simulations show that this spin-down time can be expressed as:

$$\frac{\tau_{spin}}{\tau_{mass}} \approx \frac{\frac{3}{2}k}{\left[0.29 + (\eta_* + 0.25)^{1/4}\right]^2}. \tag{2.1}$$

where τ_{mass} is the characteristic mass loss rate ($= M_*/\dot{M}$), k is the moment of inertia constant with a typical value $k \approx 0.1$. This leads to typical spin-down time of ~1 Myr for several know massive stars. Details of our full calculations can be found in ud-Doula, Owocki & Townsend (2008b).

Acknowledgements

This work was supported by NASA Grants Chandra/TM7-8002X and NNX08AY04G.

References

ud-Doula A. & Owocki S. P., 2002, *ApJ* 576, 413
ud-Doula A., Owocki S. P., & Townsend R. H. D., 2008, *MNRAS* 385, 97
ud-Doula A., Owocki S. P., & Townsend R. H. D., 2008, *MNRAS*, in press
Weber E. J. & Davis L. J., 1967, *ApJ* 148, 217

Cosmic Magnetic Fields:
From Planets, to Stars and Galaxies
Proceedings IAU Symposium No. 259, 2008
K.G. Strassmeier, A.G. Kosovichev & J.E. Beckman, eds.

doi:10.1017/S1743921309030981

Radial differential rotation vs. surface differential rotation: investigation based on dynamo models

Heidi Korhonen[1] and Detlef Elstner[2]

[1] ESO, Karl-Schwarzschild-Str. 2, D-85748 Garching bei München, Germany

[2] Astrophysical Institute Potsdam, An der Sternwarte 16, D-14482 Potsdam, Germany

Abstract. Differential rotation plays a crucial role in the alpha-omega dynamo, and thus also in creation of magnetic fields in stars with convective outer envelopes. Still, measuring the radial differential rotation on stars is impossible with the current techniques, and even the measurement of surface differential rotation is difficult. In this work we investigate the surface differential rotation obtained from dynamo models using similar techniques as are used on observations, and compare the results with the known radial differential rotation used when creating the dynamo model.

Keywords. Stars: rotation – stars: magnetic fields – stars: activity

1. Model and analysis methods

The model used in this work was originally used to investigate the flip-flop phenomenon (Elstner & Korhonen 2005, Korhonen & Elstner 2005). In the current work is is used to study the correlations between surface and radial differential rotation. The dynamo is modelled with a turbulent fluid in a spherical shell. The internal rotation is similar to the solar one with equatorial regions rotating faster than the polar regions, but here a smaller difference between core and surface rotation is used. The inner boundary of the convective zone is at the radius $0.4R_\star$. The mean electromotive force contains an anisotropic alpha-effect and a turbulent diffusivity. The nonlinear feedback of the magnetic field acts on the turbulence only. The boundary conditions describe a perfect conducting fluid at the bottom of the convection zone and at the stellar surface the magnetic field matches the vacuum field.

We use standard cross-correlation methods to study the changes in the magnetic pressure maps obtained from the dynamo calculations. As these maps can be treated the same way as the temperature maps obtained from Doppler imaging, we can use the same techniques as in the case of the real observations to analyse the snapshots from the dynamo calculations. We have taken magnetic pressure maps from 36 time points over the activity cycle. The chosen time pointsare separated by about 50 days (Fig. 1).

2. Results

The results from cross-correlating the 36 maps are shown in Fig. 2. The plots give the shift in °/day for each latitude on the 'northern' hemisphere. The field migration in the model is removed by normalising to the shift at the lowest latitude used in the investigation (2°). The last plot in Fig. 2 shows the average of the measurements from the cross-correlations, with standard deviation of the measurements as the error. The

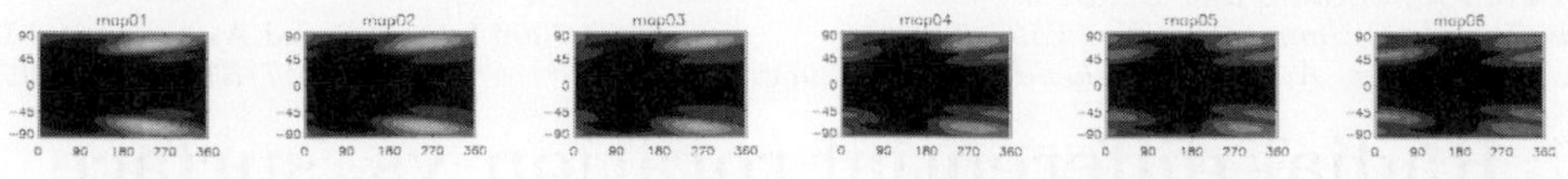

Figure 1. Six examples of the 36 snapshots of the Dynamo model.

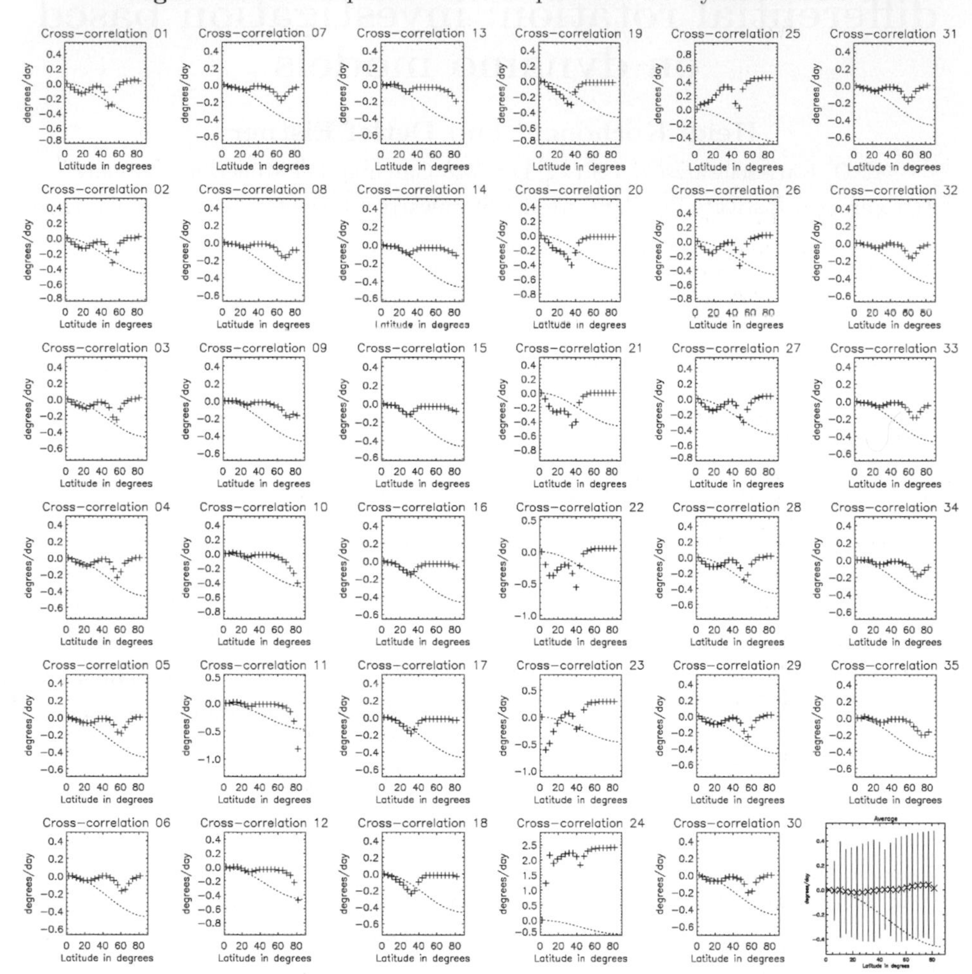

Figure 2. The results from cross-correlating the 36 snapshots.

dashed line is the input rotation at the stellar surface. The surface differential rotation pattern clearly changes during the activity cycle. It is also evident that the measured surface differential rotation at low latitudes is what one would expect from the model, but at higher latitudes, where most of the magnetic flux is, the correlation is very poor.

Acknowledgements

HK acknowledges the grant from IAU to help participating in the symposium.

References

Elstner, D. & Korhonen, H. 2005, *AN* 326, 278
Korhonen, H. & Elstner, D. 2005, *A&A* 440, 1161

Cosmic Magnetic Fields: From Planets, to Stars and Galaxies
Proceedings IAU Symposium No. 259, 2008
K.G. Strassmeier, A.G. Kosovichev & J.E. Beckman, eds.

doi:10.1017/S1743921309030993

Time series Doppler imaging using STELLA

Michael Weber, Klaus G. Strassmeier, and Thomas Granzer

Astrophysikalisches Institut Potsdam, An der Sternwarte 16, D–14482 Potsdam, Germany
email: mweber@aip.de

Abstract. One of the core programs of the STELLA robotic observatory is to monitor the stellar activity on a sample of stars using Doppler imaging. We present first preliminary results of the rapidly rotating, single giant star HD 31993 from the first two years of operation. We confirm the presence and orientation of differential rotation on the stellar surface.

Keywords. Instrumentation: spectrographs – techniques: spectroscopic – stars: imaging – stars: rotation – stars: spots

1. Observations

All observations were carried out at the STELLA robotic observatory using the STELLA echelle spectrograph (SES) which is fed with an optical fiber from one of the two STELLA 1.2 m telescopes (Weber *et al.* 2008). We accumulated approximately 170 spectra of HD 31993 during the last 2 years. Since it is a winter target, it was first observed in winter 2007, again during the winter 2007/2008 and is now being observed for the third round. The average S/N ratio is 250, but has been significantly lower during a period of bad fiber alignment (spring to summer 2007). We derived heliocentric radial velocities by cross-correlating 20 orders of each spectrum with a template spectrum of similar spectral type. Since the line-profiles are considerably deformed due to the surface activity of the star, it is difficult to reliably derive radial velocities without manual intervention. Therefore we spun-up the template spectrum before doing the cross-correlations, which resulted in very reliable measurements in unattended mode.

2. Doppler imaging and differential rotation

The data consists of three seasons so far. The first season (beginning of 2007) covers a bit more than 2 months, the second season half a year, and the third season has just begun. We concentrate on the second half of the second season, since the beginning of that season had many bad nights and also suffered from lower efficiency due to instrument alignment issues.

The data set used for the following analysis consists of 57 spectra taken between the third of December 2007 and the 19th of March 2008. This is a bit more than three stellar rotations at a rotation period of 25.3 days. The phase coverage is generally very good, depending only on the weather on the site, which was exceptionally good this last winter.

As in Strassmeier *et al.* (2003) we used `TempMap` (Rice & Strassmeier 2000) to calculate the Doppler images. The primary goal was to compare the results with the previous ones, which were based on two spectral lines only. SES delivers the full optical spectrum, but in the red half of the spectrum inter-order gaps exist. Due to this, we did this comparison only with one spectral line (Ca I 6439) for now, since the other line (Fe I 6430) is right on the edge of two orders.

In Strassmeier *et al.* (2003), which was based on data from late 1996, the surface was dominated by spots at latitudes from 0° to 40°. There was no feature at or near the pole

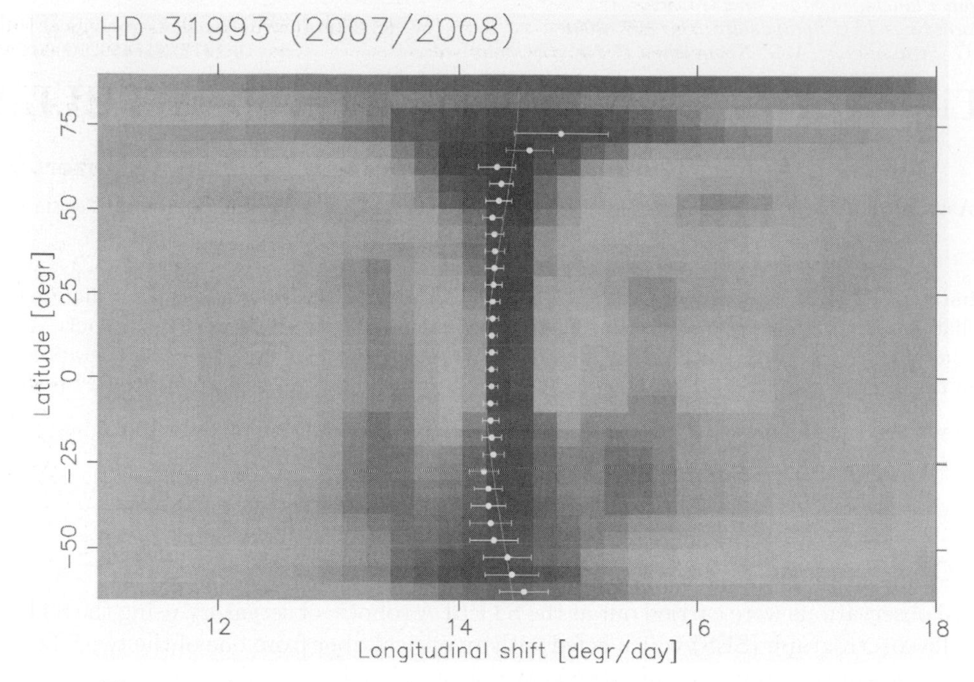

Figure 1. Average cross correlation of HD 31993. The red line is the fit $\Omega(b) = 14.22 + 0.28 \sin^2(b)°/\text{day}$.

then. Our new Doppler image is dominated by spots at lower latitudes, but also shows dark features at the pole and a bright ring at 65°. Since the latter corresponds exactly to the sub-observer latitude, it is likely that these features are caused by non-optimal values for the projected rotational velocity and/or the instrumental profile.

To derive a possible differential rotation, we cross-correlated all consecutive Doppler images of our data set. Using just two individual Doppler images does not give conclusive results, but is dominated by the noise in the spectra and/or short-term variations of spots on the stellar surface. We thus divided the data set, which consists of 57 spectra, into 16 individual chunks, and calculated a Doppler image for each of those. We then cross-correlated the individual consecutive images with each other (e. g. the first with the sixth, the second with the seventh, etc.), which gave eleven cross-correlation images. The average of these eleven images, each adjusted for small variations in the time difference between the data chunks, is shown in Fig. 1.

The image reveals almost rigid rotation, the relative differential rotation is only 2%, lap time is 360./0.28 = 1285 days. This is much weaker than the 12% found by Strassmeier *et al.* (2003), but the values derived here are based on only one spectral line and are thus preliminary. Using many of the lines available in the SES spectra will give a more precise measurement of this star's differential rotation.

References

Rice, J. B. & Strassmeier, K. G. 2000, *A&AS* 147, 151

Strassmeier, K. G., Kratzwald, L., & Weber, M. 2003, *A&A* 408, 1103

Weber, M., Granzer, T., Strassmeier, K. G., & Woche, M. 2008, in Advanced Software and Control for Astronomy II, ed. A. Bridger & N. M. Radziwill, *SPIE* 7019, 70190L

Cosmic Magnetic Fields:
From Planets, to Stars and Galaxies
Proceedings IAU Symposium No. 259, 2008
K.G. Strassmeier, A.G. Kosovichev & J.E. Beckman, eds.

doi:10.1017/S1743921309031007

Polarized hydrogen emission lines in Mira stars: a mystery behind the shock

Nicolas Fabas[1], Agnès Lèbre[1] and Denis Gillet[2]

[1]GRAAL, UMR 5024,Université Montpellier II/CNRS, France
[2]Observatoire de Haute-Provence, USR 2207, OAMP/CNRS, France

Abstract. We present a full spectropolarimetric study (in the Stokes parameters I, Q, U and V) on omicron Ceti (the prototype of Mira stars), and focus on the strong polarization detected in Balmer emission lines. This study is made with observations from NARVAL instrument at TBL (Telescope Bernard Lyot) in Pic du Midi Observatory, France.

Keywords. Shock waves – AGB and post-AGB – line: formation

1. Introduction

Mira stars are cool, evolved and variable stars, the prototype being omicron Ceti (as known as Mira). They are radially pulsating stars with long period of luminosity (e.g., the period of o Ceti is 332 days). From maximum to minimum light, the spectral type of o Ceti varies from M5 to M9. A cool and very extended stellar atmosphere is present, which is typical from this kind of late-type stars. Those atmospheres are surrounded with a circumstellar envelope. Among the peculiar features of Mira stars' spectra, emission lines are detected. The hydrogen Balmer lines are observed in emission during about 80% of the luminosity period and are supposed to be formed in the radiative wake of a hypersonic shock wave propagating periodically throughout the stellar atmosphere (Fadeyev & Gillet, 2004).

2. Spectropolarimetric analysis

Spectro-polarimetric signatures are detected in the 4 Stokes parameters (IQUV) associated to Balmer hydrogen lines (from $H\alpha$ to $H\delta$). Those signatures appear to be time variable, being much more visible and structured at the maximum luminosity, when the shock is emerging from the photosphere and is propagating with a high intensity. Below (Fig. 1) are presented the signatures for $H\beta$ and $H\delta$ at the three observed phases (ϕ=0.58, 1.00 and 1.06). From Stokes parameters Q & U, linear polarization in the Balmer lines can be estimated (Fig. 2).

3. Results

We confirm the earlier and single detection in the $H\beta$ line (McLean & Coyne, 1978) on o Ceti when observed at its maximum luminosity: Balmer hydrogen emission lines are strongly polarized, much more than the local continuum. Besides, the linear polarization rate, in all the Balmer lines, appears to be time variable, and null at the minimum light. We find that this polarization in Balmer emission lines is likely to be linked with the emission line formation mechanism occurring within the de-excitation post shock wake of the strong shock wave propagating throughout the stellar atmosphere at each pulsation

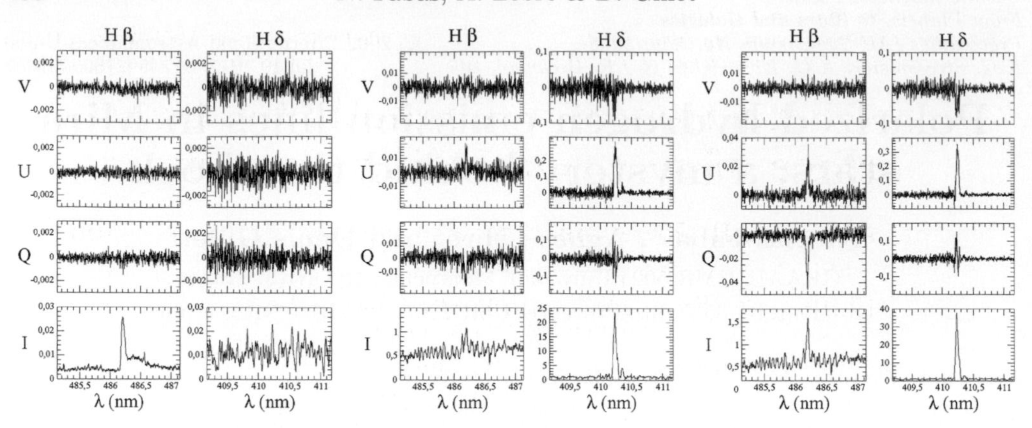

Figure 1. Signatures for $H\beta$ and $H\delta$ at the three observed phases (ϕ=0.58, 1.00 and 1.06)

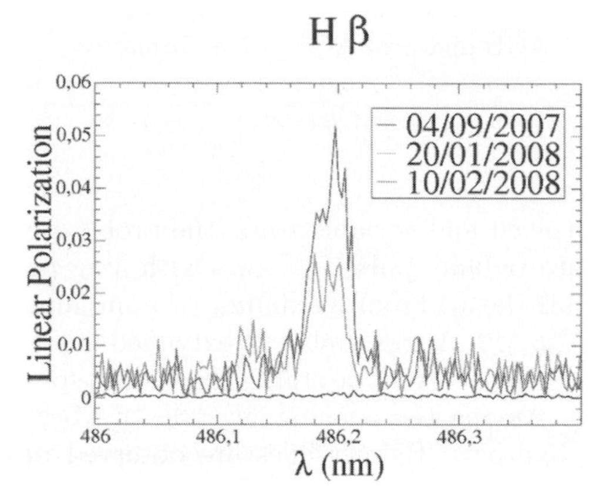

Figure 2. Linear polarization for $H\beta$ at the three observed phases (ϕ=0.58, 1.00 and 1.06)

cycle. Moreover, according to shock wave theory, an electromagnetic field should appear just behind the shock front (where emission lines are formed), inducing polarization. To date, this field has never been conclusively detected on Mira stars (neither on any other type of radially pulsating star). Our spectro-polarimetric observations constitute the first attempt to fully characterize and understand the associated physical mechanism. Eventually, this study would be also useful for the characterization of the shock wave with respect to the central star.

Acknowledgements

N. Fabas would like to thank the IAU for the grant he was given to take part to IAU Symposium 259.

References

Fadeyev, Y. A. & Gillet, D. 2004, *A&A* 420, 423
McLean, I. S. & Coyne G. V. 1978, *ApJ* 226, 145

Cosmic Magnetic Fields:
From Planets, to Stars and Galaxies
Proceedings IAU Symposium No. 259, 2008
K.G. Strassmeier, A.G. Kosovichev & J.E. Beckman, eds.

doi:10.1017/S1743921309031019

Direct detection of a magnetic field on the surface of slowly rotating giant stars

Michel Aurière[1], R. Konstantinova-Antova[2], P. Petit[1], G. Wade[3] and T. Roudier[1]

[1]LATT, CNRS, University of Toulouse, 57 Avenue d'Azereix, 65008 Tarbes, France,
email:name@ast.obs-mip.fr

[2]Institute of Astronomy, BAS, 72 Tsarigradsko shose, 1784 Sofia,Bulgaria,
email:renada@astro.bas.bg

[3]Department of Physics, RMC, PO Box 17000, Station "Forces", Kingston, K7K4B4 Ontario, Canada
email:gregg.wade@rmc.ca

Abstract. We present first results of the magnetic survey of a sample of slow rotating giant stars for which an X-ray emission or variations of CaII H& K lines have been already detected.

Keywords. Stars: late type – stars: magnetic fields

1. Observations

Using the twin spectropolarimeters NARVAL (Telescope Bernard Lyot, Pic du Midi, France) and ESPaDOnS (Canada France Hawaii Telescope) and the LSD technic (Donati *et al.*, 1997), we undergo a sensitive program of detection and measurement of magnetic fields at the surface of slowly rotating single giants for which an X-ray emission (Huensch *et al.* 1998, Schroeder *et al.* 1998, Tarasova *et al.* 2002) or variations of CaII H & K lines (Choi *et al.*, 1995) have been already detected. The selected giant stars have $vsini < 5km/s$ or rotational periods greater than 60 days. They are intermediate mass (1.7 - 3.5 $M_{\odot}$ giants or subgiants, situated near the base of the RGB, in the left part of the X-ray dividing line (Gondoin, 1999).

Table 1 gives information for the part of our sample stars for which magnetic measurements were performed.

2. First results

Up to now Stokes V signal is significantly detected on 12 slow-rotating single giants of our sample and the inferred longitudinal magnetic field B_l is measured to be of the order of a few G.

EK Eri appears to host a surface magnetic field with an extraordinary strength. This supports the suggestion of Stepień (1993) and Strassmeier *et al.* (1999) that EK Eri could be the descendant of a strongly magnetic Ap star. Modeling the magnetic field of EK Eri using the ZDI inversion model of Donati *et al.* (2006), we obtained a mean surface magnetic field of 270 G (Aurière *et al.*, 2008). Figure 1 shows Stokes V at its maximum as well as the fit of its variations with our model (photometric period of 306.9 d).

Definitive detection of the surface magnetic field of Pollux was obtained using both ESPaDOnS and NARVAL's observations. The Stokes V polarization signal is as weak as $2x10^{-5}$ of the continuum. Figure 1 shows the LSD Stokes V profile obtained when averaging all the obtained spectra during one year.

Table 1. Slow rotating giant stars detected with ROSAT and with magnetic measurements performed. L_x values are by Gondoin (1999), or calculated using the ROSAT fluxes (Huensch *et al.* 1998).

Name	HD	Sp	vsini (km/s)	P d	L_x (10^{27}erg/s)	$\lvert B_l\rvert^{max}+3\sigma$ $\leqslant$ G
iot Cap	203387	G8III	7.0	68	4482	4.9
77 Tau	28307	K0IIIb	1.5	140	1996	2.6
del Crb	141714	G3.5III	5.6	59	1456	3.7
bet Cet	4127	K0III	3.5		1138	8.0
EK Eri	27536	G8III-IV	1.5	307	1000	101.6
24 UMa	82210	G4III-IV	5.5		901	10.5
14 Cet	3229	F5IV	5.0		336	35.0
bet Boo	133208	G8IIIa	2.5		153	ND
eta Her	150997	G7.5IIIb	1.7		63	8.4
Pollux	62509	K0III	1.7		5	0.9

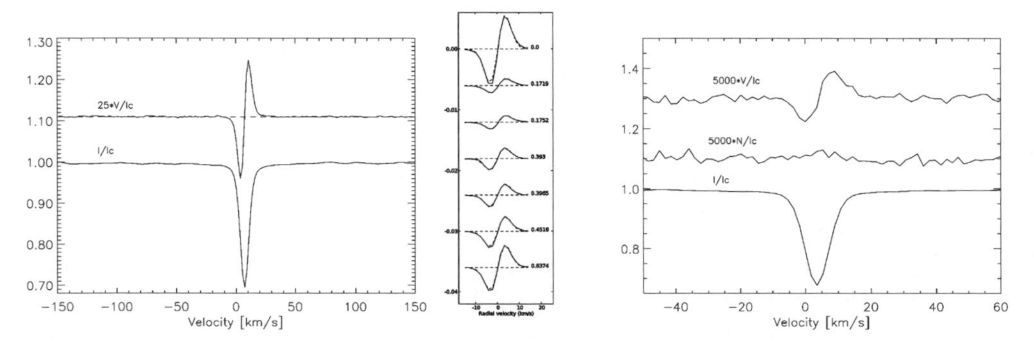

Figure 1. Left: LSD profiles of EK Eri as observed on 20 Sept. 2007 with NARVAL. From bottom to top, Stokes I and Stokes V are presented. For display purposes, the Stokes V profile is enlarged by a factor of 25. Center: Variation of Stokes V profile for EK Eri with rotational phase, as observed with NARVAL during the 2007/2008 season. Right: Mean LSD profiles of Pollux from 51 spectra taken with ESPaDOnS or NARVAL in the September 2007- September 2008 period. From bottom to top are the Stokes I, Null polarization and Stokes V profiles. For display purposes, the Stokes V and Null polarization profiles are enlarged by a factor of 5000.

Acknowledgements

A significant part of the observations were supported by OPTICON.

References

Aurière, M., Konstantinova-Antova, R., Petit, P., *et al.* 2008, *A&A* 491, 499
Choi, H.-J., Soon, W, Donahue, R. A., *et al.* 1995, *PASP* 107, 744
Donati, J.-F., Semel, M., Carter, B. D., *et al.* 1997, *MNRAS* 291, 658
Donati, J.-F., Howarth, I. D., Jardine, M. M., *et al.* 2006, *MNRAS* 370, 629
Gondoin P. 1999, *A&A* 352, 217
Huensch M., Schmitt, J. H. M. M., & Voges,W. 1998, *A&AS* 127, 251
Schroeder, K.-P., Huensch, M., & Schmidtt, J. H. M. M., 1998, *A&A* 335, 591
Stepień, K., 1993, *ApJ* 416, 368
Strassmeier, K.G., Stepień, K., & Henry, G. W., 1999, *A&A* 343, 175
Tarasova, T. N., 2002, *Astron. Rep.* 46, 474

Cosmic Magnetic Fields:
From Planets, to Stars and Galaxies
Proceedings IAU Symposium No. 259, 2008
K.G. Strassmeier, A.G. Kosovichev & J.E. Beckman, eds.

doi:10.1017/S1743921309031020

Dynamo-generated magnetic fields in fast rotating single giants

Renada Konstantinova-Antova[1], Michel Aurière[2], Klaus-Peter Schröder[3] and Pascal Petit[2]

[1]Inst. of Astronomy, Bulgarian Acdy of Sciences, Sofia, 72 Tsarigradsko shosse, Bulgaria
email: renada@astro.bas.bg

[2]Laboratoire d'Astrophysique de Toulouse-Tarbes, Université de Toulouse, CNRS, Observatoire Midi Pyrénés, 57 Avenue d'Azereix, 65008 Tarbes, France
email: michel.auriere@ast.obs-mip.fr

[3]Departmento de Astronomia, Universidad de Guanajuato, GTO, Academic Street, Camford, CF3 5QL, Mexico
email: kps@astro.ugto.mx

Abstract. Red giants offer a good opportunity to study the interplay of magnetic fields and stellar evolution. Using the spectro-polarimeter NARVAL of the Telescope Bernard Lyot (TBL), Pic du Midi, France and the LSD technique we began a survey of magnetic fields in single G-K-M giants. Early results include 6 MF-detections with fast rotating giants, and for the first time a magnetic field was detected directly in an evolved M-giant: EK Boo. Our results could be explained in the terms of α–ω dynamo operating in these giants.

Keywords. Magnetic fields – stars: evolution – stars: late-type – stars: activity

1. Introduction

Magnetic fields (MF) in single evolved stars are still poorly studied. Most of the G–K–M giants presently been known as active are fast rotators. (Fekel&Balachandran, 1993; Huensch *et al.*, 2004). Angular momentum dredge-up has been suggested to provide the fast rotation, driven by the convective zone reaching near the stellar core (Simon&Drake, 1989). In this way, MFs could be generated by a new dynamo.

2. Observations and data processing

The new generation spectro-polarimeters like NARVAL at TBL, Pic du Midi, France (Aurière, 2003) and the LSD technique (Donati et al., 1997) are very suitable for precision detections of MFs in giants (Konstantinova-Antova *et al.*, 2008; Aurière *et al.*, 2008). We observed 7 fast rotating giants with NARVAL in the period of November 2006 to April 2008. A precise computation of the longitudinal MF B_l using the First Moment Method (Donati *et al.*, 1997; Rees & Semel, 1979) was carried out, and the time-behavior of the activity indicator CaII K (S index) was studied.

3. First results

MF structures, indicative of a dynamo, were detected for 6 of the 7 giants (exception: HD233517), see Table 1 listing the S index and B_l. We determined the evolutionary status and masses of these 6 giants, using Hipparcos parallaxes, $T_{\rm eff}$ from the Wright catalogue (2003), and matching evolutionary tracks from Schröder *et al.* (1997) (Fig. 1).

All 6 giants with MF-detection have masses $\geqslant$1.5 $M_\odot$ and their convective zones are currently deepening while the stars are evolving. Their evolutionary stages reach from the

Table 1. Data for the studied giants.

Star	Sp class	$M/M_{\odot}$	Vsini km/s	L_x 10^{30} erg/s	CaII K S index	B_l Gauss	error B_l Gauss
V390 Aur	G8III	1.85	29	5.04	0.64	-5 – -15	4.4
FI Cnc	G8III	2.25	17	26	1.04 – 1.35	-16.48 – +3.45	1.87
37 Com	G9III-II	4.2	11	5.20	0.34	+5.62	0.63
7 Boo	G5III	3.6	14.5	3.72	0.24	+1.83	0.84
κHerA	G8III	2.8	9.4	2.98	0.29	-3.94	0.67
HD233517	K2III	1.5?	15			no detect.	
EK Boo	M6III	1.9	11	14.12	0.21–0.26	-3.19 – -6.76	0.62

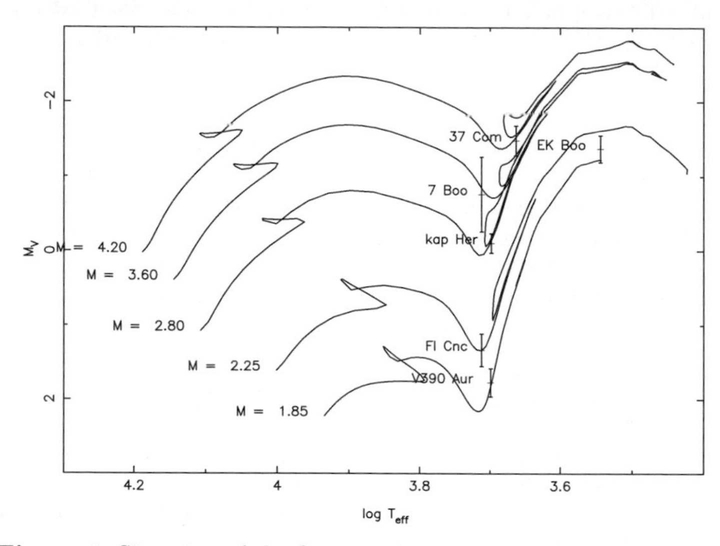

Figure 1. Situation of the fast rotating giants on the HR diagram.

Hertzsprung gap to the AGB. While the particular reasons for the fast rotation could be different, depending on mass and evolutionary history, an α–ω dynamo presents a likely reason for the detected magnetic activity.

Acknowledgements

We thank the TBL team for providing the observations, a substantial part was supported by the OPTICON programme. The Hipparcos database was used.

References

Aurière M. 2003, *EAS Publ. Series* 9, 105

Aurière, M., Konstantinova-Antova, R. Petit, P., *et al.* 2008, *A&A* 491, 499

Donati, J.-F., Semel, M., Carter, B. D. *et al.* 1997, *MNRAS* 291, 658

Fekel F. C. & Balachandran S. 1993, *ApJ* 403, 708

Hünsch M., Konstantinova-Antova R., Schmitt J. H. M. *et al.* 2004, in Proc. IAU Symp. 219, A. Dupree and A. Benz (eds.), p.223

Konstantinova-Antova R., Aurière, M., Iliev I. *et al.* 2008, *A&A* 480, 475

Rees D. E. & Semel M. 1979, *A&A* 74, 1

Schröder K.-P., Pols O. P., & Eggleton P. P. 1997, *MNRAS* 285, 696

Simon T. & Drake S. A. 1989, *ApJ* 346, 303

Wright C. O., Egan M. P., Kraemer, K. E. *et al.* 2003, *AJ* 125, 359

Cosmic Magnetic Fields:
From Planets, to Stars and Galaxies
Proceedings IAU Symposium No. 259, 2008
K.G. Strassmeier, A.G. Kosovichev & J.E. Beckman, eds.

doi:10.1017/S1743921309031032

HD 232862: a magnetic and Lithium-rich giant field star

Agnès Lèbre[1], A. Palacios[1], J. D. do Nascimento Jr[2], J. R. De Medeiros[2], R. Konstantinova-Antova[3], D. Kolev[3] and M. Aurière[4]

[1]GRAAL, Université de Montpellier,CNRS, Montpellier, France
email: lebre@graal.univ-montp2.fr

[2]DFTE, Federal University of Rio Grande do Norte, Natal, Brazil

[3]Institute of Astronomy, Bulgarian Academy of Sciences, Sofia, Bulgaria

[4]LATT, Université de Toulouse, CNRS, Tarbes, France

Abstract. From spectropolarimetric data and spectral synthesis analysis, we report the serendipitous discovery of an unusually high lithium content field giant. HD 232862, classified as a G8II star, appears to be the first Lithium-rich field giant star hosting a surface magnetic field.

Keywords. Stars: abundances – stars: evolution – magnetic fields

---

1. HD 232862, a Li–rich giant star

This X-ray source in the ROSAT data basis is a G8II (field) giant, namely a bright giant star (mass range from ~2.5 to 9 $M_\odot$), with a high rotational velocity ($v \sin i =$ 20.6 $\mathrm{km\,s^{-1}}$). Specific features in its IUE spectra point to a coronal and chromospheric activity. It has also been resolved as a visually tight binary (Couteau, 1988).

We have collected spectropolarimetric data of HD 232862, with ESPaDOnS (Donati *et al.*, 2006) at the CFHT (Mauna Kea), using the circular polarimetric mode to collect informations in Stokes V and I parameters. The two components of HD 232862 were easily separated on the guiding camera during our subarcsec seeing observations ensuring to access to the first high resolution and high S/N spectra of the giant star alone.

With synthetic spectra computed from MARCS models of stellar atmosphere (Gustafsson *et al.* 2008), we have derived stellar parameters and Lithium abundance of HD 232 862. Adopting $T_{\rm eff}$ = 5000 K, $\log g = 3.0$ and $[Fe/H] = -0.3$, and adopting the solar abundances from Grevesse and Sauval (1998), we measured a high Li content: $\mathrm{A_{Li}} = 2.45 \pm 0.1$ dex. Comparison of these data to predictions from evolutionary models computed with the STAREVOL code (Siess, 2006) helped to precise the evolutionary status of HD 232862. It appears to be a low mass star ($1.5\ M_\odot < \mathrm{M} < 2.5\ M_\odot$) at the bottom of the Red Giant Branch. Its Li abundance is hence far in excess from the values predicted by standard theory. Indeed, the deepening of the convective envelope on the lower RGB dilutes the preserved surface Li with material from Li free regions. The Li abundance derived in HD 232862 is unusually high, as expected values for its mass range and evolutionary stage are less than 1.5 dex.

This high Li content is also quite peculiar for a bright giant star, as to date there is no clear register of Li-rich objects among the luminosity class II stars (Lèbre *et al.* 2006).

2. Magnetic field detection in HD 232 862

From the ESPaDOnS data, we have also detected the presence of a magnetic field at the surface of HD 232862. Complex and time variable Stokes V signatures are detected

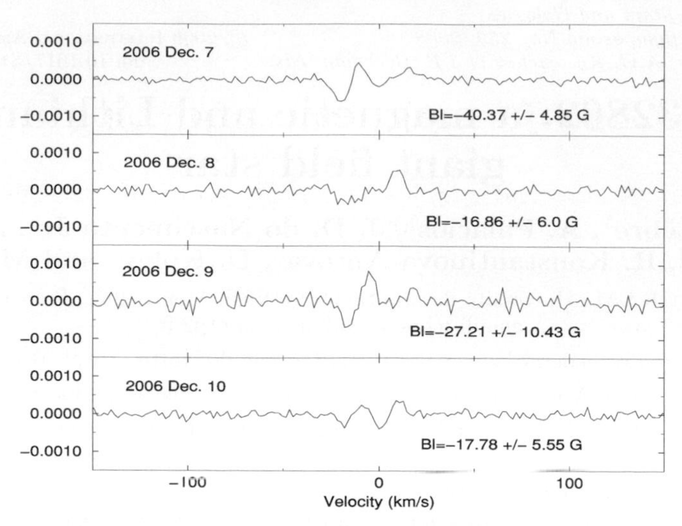

Figure 1. Variation of Stokes V profiles (extracted with LSD method) over a 4-days timescale. A clear signal, hosting a complex structure strongly variable from one night to another, is obtained. Date is indicated in each panel, as well as the longitudinal magnetic field component B_l (in G).

(Fig. 1), pointing to a dynamo origin for the field. We also suspect a complex structure for the parent field, with likely topology variations over a period that has to be specified. Longitudinal B_l magnetic field has been derived from LSD analysis (Donati *et al.*, 1997), using a specific mask computed with $T_{\rm eff}$=5000 K and $\log g$=3.0 (involving ∼12,000 lines). The B_l component shows strong variations over 4-days timescales and points – under dipolar field assumption – to an intense magnetic field (∼100 G) at the surface of HD 232862.

This is the very first detection of such an intense magnetic field in a Lithium rich giant. Further observations are now needed to infer the mean magnetic field of HD 232862, to clarify its origin, and to assess the star's progenitor (that may be a magnetic A star). Only then, will it be possible to disentangle whether the high Li content of HD 232862 is due to its binary status, to the presence of a surface magnetic field, or to another process.

Acknowledgements

We thank the french PNPS/INSU-CNRS for financial support.

References

Couteau, P. 1988, *A&AS* 75, 163

Donati, J. F. *et al.* 2006, in: R. Casini & B. W. Lites (eds.), *Solar Polarization 4*, ASP Conf. Series, Vol. 358, p. 362

Donati, J.-F., Semel, M., Carter, B. D., Rees, D. E., & Cameron, A. C. 1997, *MNRAS* 291, 658

Grevesse, N., & Sauval, A. J. 1998, *SSRv* 85, 161

Gustafsson, B., Edvardsson, B., Eriksson, K., *et al.* 2008, *A&A* 486, 951

Lèbre, A., de Laverny, P., do Nascimento, J. D., & de Medeiros, J. R. 2006, *A&A* 450, 1173

Siess, L. 2006, *A&A* 448, 717

Cosmic Magnetic Fields:
From Planets, to Stars and Galaxies
Proceedings IAU Symposium No. 259, 2008
K.G. Strassmeier, A.G. Kosovichev & J.E. Beckman, eds.

doi:10.1017/S1743921309031044

Zeeman-Doppler imaging of II Peg

Thorsten A. Carroll[1], Markus Kopf[1], Klaus G. Strassmeier[1], Ilya Ilyin[1] and Ilkka Tuominen[2]

[1]Astrophysikalisches Institut Potsdam, An der Sternwarte 16, D-14882 Potsdam, Germany
email: tcarroll@aip.de

[2]Observatory, University of Helsinki, PO Box 14, FI-00014 Helsinki, Finland

Abstract. We present Zeeman-Doppler images of the active K2 star II Peg for the years 2004 and 2007. The surface magnetic field was reconstructed with our new ZDI code *iMap* which provides a full polarized radiative transfer driven inversion to simultaneously reconstruct the surface temperature and magnetic vector field distribution. II Peg shows a remarkable large scale magnetic field structure for both years. The magnetic field is predominantly located at high latitudes and is arranged in active longitudes. A dramatic evolution in the magnetic field structure is visible for the two years, where a dominant and largely unipolar field in 2004 has changed into two distinct and large scale bipolar structures in 2007.

Keywords. Stars: activity – stars: magnetic fields – stars: spots – radiative transfer – methods: data analysis

1. Introduction

Zeeman-Doppler imaging (ZDI) also known as Magnetic-Doppler imaging (MDI) is a powerful inversion method to reconstruct stellar magnetic surface fields and has significantly contributed to our current understanding of stellar magnetic fields (e.g. Donati *et al.* 1997; Donati 1999; Kochukhov *et al.* 2004; Donati *et al.* 2006, 2007). As ZDI is a non-linear inverse problem it critically depends on the underlying forward modeling and particular attention must be given to a correct modeling of radiative transfer effects (see Carroll *et al.* this proceeding). In an effort to pursue a more rigorous modeling we have developed our ZDI code *iMap* which incorporates a full polarized radiative transfer based inversion (Carroll *et al.* 2007, 2008). The inverse module is based on either a conjugate gradient or Levenberg-Marquardt method. The regularization that we apply is based on a new local maximum entropy (Carroll *et al.* 2007). One of the great obstacles for a radiative transfer based interpretation of observed Stokes profiles is the noise level. Although there exist a powerful multi-line reconstruction technique, i.e. the least-square deconvolution method of Donati *et al.* (1997), the interpretation of the reconstructed *mean* line profiles is not straight forward. Another method which provides a way of reconstructing individual Stokes line profiles and thus allows a proper radiative transfer modeling was introduced by Carroll *et al.* (2007); Martínez González *et al.* (2008) and is based on a principal component analysis reconstruction technique. (see also Carroll *et al.* this proceeding). Our code was applied to Stokes I and V observations of the K2 subgiant II Pegsi. obtained with the SOFIN spectrograph at the Nordic Optical Telescope (La Palma). The data were recorded in Summer 2004 and 2007 and cover 8 and 10 rotational phases respectively. The original S/N of the data were 270 and after the application of our multi-line PCA reconstruction (with 880 spectral lines) we could enhance the S/N of our target lines to 2000.

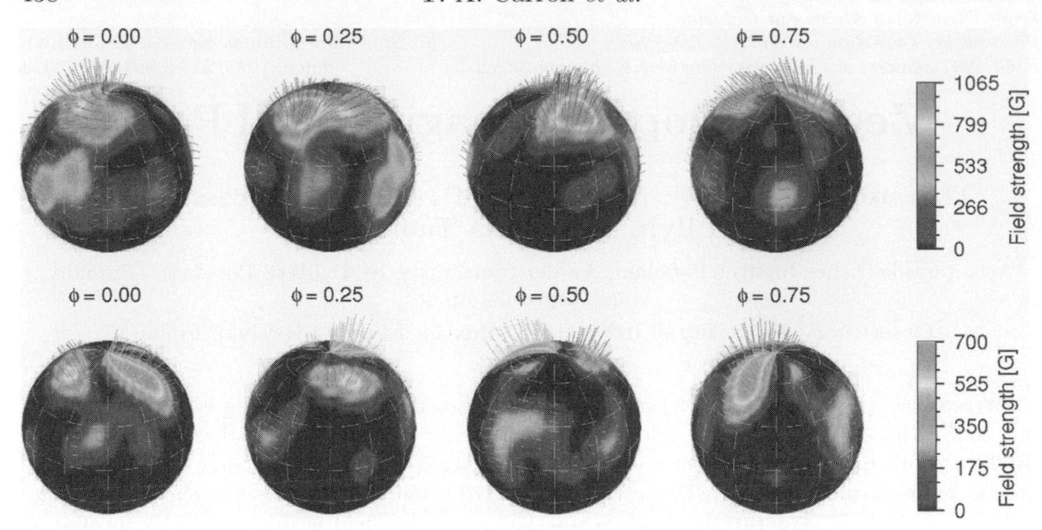

Figure 1. The reconstructed surface magnetic field of II Peg, 2004 (top) and 2007 (bottom). Bright and dark field lines indicate the polarities.

2. The surface magnetic field of II Peg in 2004 and 2007

From the reconstruction (figure 1) we can readily identify that the magnetic field of II Peg consist of a small number of large scale magnetic structures which predominantly located at high latitudes and arranged in active longitudes. It is conspicuous that the field is mainly radially oriented an effect which we attribute to our improved S/N ratio that allows our inversion to better constrain the field orientation (see also Kochukhov *et al.* this proceeding). A very interesting evolution can be seen from figure 1, the field structure has changed from a more or less coherent unipolar region in 2004 to two separate field structures of different polarity. This characteristic change of the surface magnetic field topology resembles that of a flip-flop event as found from temperature inversions (Berdyugina 2007) and also found in theoretical modeling of an $\alpha - \Omega$ dynamo (Elstner & Korhonen 2005). Although, it is to early to confirm the presence of a real flip-flop event from our two year data, this work provides the first real indication of a magnetic reorganization on II Peg.

References

Berdyugina, S. V. 2007, *Mem. Soc. Astr. It.* 78, 242

Carroll, T. A., Kopf, M., & Strassmeier, K. G. 2008, *A&A* 488, 781

Carroll, T. A., Kopf, M., Ilyin, I., & Strassmeier, K. G. 2007, *AN* 328, 1043

Donati, J.-F., Jardine, M. M., Gregory, S. G., Petit, P., Bouvier, J., Dougados, C., Ménard, F., Cameron, A. C., Harries, T. J., Jeffers, S. V., & Paletou, F. 2007, *MNRAS* 380, 1297

Donati, J.-F., Forveille, T., Cameron, A. C., Barnes, J. R., Delfosse, X., Jardine, M. M., & Valenti, J. A. 2006, *Science* 311, 633

Donati, J.-F. 1999, *MNRAS* 302, 457

Donati, J.-F., Semel, M., Carter, B. D., Rees, D. E., & Collier Cameron, A. 1997, *MNRAS* 291, 658

Elstner, D. & Korhonen, H. 2005, *AN* 326, 278

Kochukhov, O., Bagnulo, S., Wade, G. A., Sangalli, L., Piskunov, N., Landstreet, J. D., Petit, P., & Sigut, T. A. A. 2004, *A&A* 414, 613

Martínez González, M. J., Asensio Ramos, A., Carroll, T. A., Kopf, M., Ramírez Vélez, J. C., & Semel, M. 2008, *A&A* 486, 637

Cosmic Magnetic Fields:
From Planets, to Stars and Galaxies
Proceedings IAU Symposium No. 259, 2008
K.G. Strassmeier, A.G. Kosovichev & J.E. Beckman, eds.

doi:10.1017/S1743921309031056

Magnetic Doppler imaging of II Peg

Oleg Kochukhov[1], N. Piskunov[1], I. Ilyin[2] and I. Tuominen[3]

[1]Department of Physics and Astronomy, Uppsala University, 751 20 Uppsala, Sweden
[2]Astrophysikalisches Institut Potsdam, An der Sternwarte 16, D-14482 Potsdam, Germany
[3]University of Helsinki, PO Box 14, 00014 Helsinki, Finland

Abstract. Rotational modulation of the intensity and polarization spectra of magnetic stars offers a unique possibility to reconstruct the structure of surface magnetic fields and to investigate their relation to cool starspots. We have developed a new magnetic Doppler imaging code which aims at self-consistent temperature and magnetic mapping of cool active stars. Here we present magnetic Doppler imaging analysis of high-resolution circular polarization observations of the active star II Peg. We demonstrate that a self-consistent approach to magnetic inversion unveils stronger magnetic fields than found previously through disjoint analyses of polarization and intensity observations of active stars.

Keywords. Stars: activity – stars: atmospheres – stars: magnetic fields – stars: II Peg

1. Introduction

Rotational modulation of the intensity and polarization spectra of active stars offers a possibility to reconstruct 2-D maps of the photospheric magnetic fields and temperature spots, thus providing a key constraint for the theoretical modelling of the stellar activity and dynamo. Until now *Zeeman Doppler imaging* of the active late-type stars (Brown, Donati, Rees, *et al.* 1991; Donati 1999; Hussain, Donati, Cameron, *et al.* 2000) was based upon a non-simultaneous and simplified interpretation of the average (least-squares deconvolved) Stokes I and Stokes V spectra. In contrast, detailed and physically consistent methods of the stellar magnetic field mapping have been developed (Piskunov & Kochukhov 2002; Kochukhov & Piskunov 2002) and successfully applied (Kochukhov, Piskunov, Ilyin, *et al.* 2002; Kochukhov, Bagnulo, Wade, *et al.* 2004) to the high-resolution circular and linear polarization observations of the early-type magnetic stars.

Improvements in the observational techniques and instrumentation have recently made it possible to detect polarization signatures inside individual atomic (Petit 2006) and molecular (Berdyugina, Petit, Fluri, *et al.* 2006) lines of the late-type stars. With these observational data in mind, we have developed a new *Magnetic Doppler imaging* (MDI) code `Invers13`. Our magnetic inversion code reconstructs iteratively two-dimensional distributions of temperature and three components of the magnetic field vector using high-resolution spectropolarimetric time series recorded in the Stokes I and V or all four Stokes parameters. Inversions employ the Levenberg-Marquardt minimization procedure and are constrained with the Tikhonov regularization (Piskunov & Kochukhov 2002).

The modelling of the magnetic field and cool spots is self-consistent, i.e. the local intensity and polarization profiles are computed taking into account the Zeeman effect and the presence of temperature inhomogeneities at the same time. This allows us to deduce correct field strength inside cool spots.

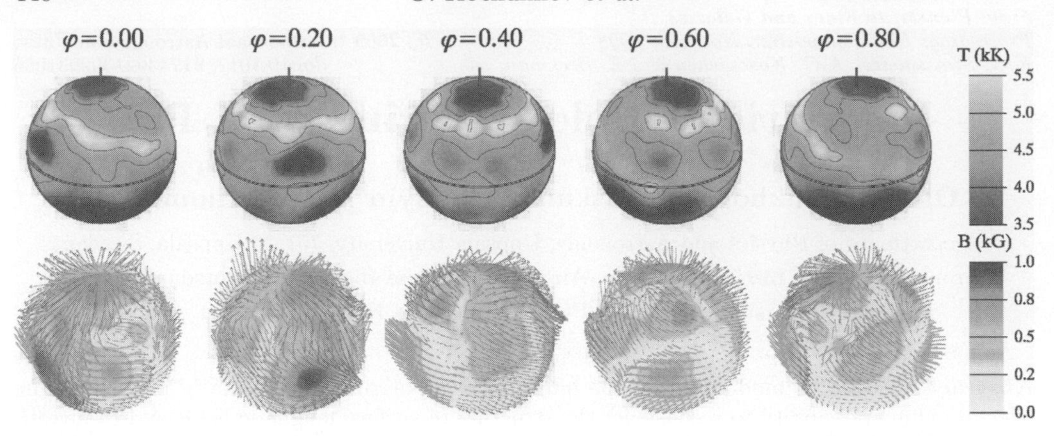

Figure 1. Temperature spot distribution (*upper panel*) and magnetic field geometry (*lower panel*) inferred from the magnetic inversion of the Stokes I and V spectra of II Peg.

2. Magnetic field topology of II Peg

We have applied the new MDI code to the circular polarization observations of II Peg obtained with the SOFIN spectrograph at the Nordic Optical Telescope (La Palma). High-resolution spectra in the Stokes I and V parameters were obtained at 8 rotational phases in June-August 2004. These observations are characterized by the resolving power $R = 70\,000$ and have typical $S/N > 400$. The spectra were reduced with the 4A software package (Ilyin 2000).

Adopting the rotation period and atmospheric parameters of II Peg determined in previous model atmosphere and Doppler imaging studies of this star (Ottmann, Pfeiffer, Gehren 1998; Berdyugina, Jankov, Ilyin, *et al.* 1998), we have carried out magnetic inversion simultaneously with the reconstruction of temperature map using profiles of the three magnetically sensitive Fe I lines at λλ 5495–5507 Å and two strong Ca I lines 6102.7 and 6122.2 Å. Resulting magnetic and temperature distributions are presented in Fig. 1. We find that at the surface of II Peg the field strength reaches up to one kG. Some strong-field regions coincide with low-temperature regions. The magnetic field inferred in our analysis is stronger than typically found in previous ZDI mapping of cool active stars based on the LSD technique and weak field approximation (e.g., Donati 1999).

References

Berdyugina, S. V., Jankov, S., Ilyin, I., Tuominen, I., & Fekel, F. C. 1998, *A&A* 334, 863

Berdyugina, S. V., Petit, P., Fluri, D. M., Afram, N., & Arnaud, J. 2006, in *Astronomical Society of the Pacific Conference Series*, eds. R. Casini, & B. W. Lites, *ASP Conf. Ser.* 358, 381

Brown, S. F., Donati, J.-F., Rees, D. E., & Semel, M. 1991, *A&A* 250, 463

Donati, J.-F. 1999, *MNRAS* 302, 457

Hussain, G. A. J., Donati, J.-F., Cameron, A. C., & Barnes, J. R. 2000, *MNRAS* 318, 961

Ilyin, I. V. 2000, Ph.D. thesis, University of Oulu Finland

Kochukhov, O. & Piskunov, N. 2002, *A&A* 388, 868

Kochukhov, O., Piskunov, N., Ilyin, I., Ilyina, S., & Tuominen, I. 2002, *A&A* 389, 420

Kochukhov, O., Bagnulo, S., Wade, G. A., Sangalli, L., Piskunov, N., Landstreet, J. D., Petit, P., Sigut, T. A. A. 2004, *A&A* 414, 613

Ottmann, R., Pfeiffer, M. J., & Gehren, T. 1998, *A&A* 338, 661

Petit, P. 2006, in *Astronomical Society of the Pacific Conference Series*, eds. R. Casini, and B. W. Lites, *ASP Conf. Ser.* 358, 335

Piskunov, N. & Kochukhov, O. 2002, *A&A* 381, 736

Cosmic Magnetic Fields:
From Planets, to Stars and Galaxies
Proceedings IAU Symposium No. 259, 2008
K.G. Strassmeier, A.G. Kosovichev & J.E. Beckman, eds.

doi:10.1017/S1743921309031068

Magnetic geometries of Sun-like stars: exploring the mass-rotation plane

Pascal Petit[1], B. Dintrans[1], M. Aurière[1], C. Catala[2], J.-F. Donati[1], R. Fares[1], T. Gastine[1], F. Lignières[1], A. Morgenthaler[1], J. Morin[1], F. Paletou[1], J. Ramirez[2], S.K. Solanki[3], and S. Théado[1]

[1]Laboratoire d'Astrophysique de Toulouse-Tarbes, Universit de Toulouse, CNRS, France
email: petit@ast.obs-mip.fr, dintrans@ast.obs-mip.fr, auriere@ast.obs-mip.fr, donati@ast.obs-mip.fr, rim.fares@ast.obs-mip.fr, thomas.gastine@ast.obs-mip.fr, francois.lignieres@ast.obs-mip.fr, amorgent@ast.obs-mip.fr, julien.morin@ast.obs-mip.fr, fpaletou@ast.obs-mip.fr, stheado@ast.obs-mip.fr

[2]LESIA, Observatoire de Paris-Meudon, 92195 Meudon, France
email: claude.catala@obspm.fr, julio.ramirez@obspm.fr

[3]Max-Planck-Institut fr Sonnensystemforschung, Max-Planck-Str. 2, 37191 Katlenburg-Lindau, Germany, email: solanki@mps.mpg.de

Abstract. Sun-like stars are able to continuously generate a large-scale magnetic field through the action of a dynamo. Various physical parameters of the star are able to affect the dynamo output, in particular the rotation and mass. Using the NARVAL spectropolarimeter (Observatoire du Pic du Midi, France), it is now possible to measure the large-scale magnetic field of solar analogues (i.e. stars very close to the Sun in the mass-rotation plane, including strict solar twins). From spectropolarimetric time-series, tomographic inversion enables one to reconstruct the field geometry and its progressive distortion under the effect of surface differential rotation. We show the first results obtained on a sample of main-sequence dwarfs, probing masses between 0.7 and 1.4 solar mass and rotation rates between 1 and 3 solar rotation rate.

Keywords. Stars: activity – stars: atmospheres – stars: imaging – stars: late-type – stars: magnetic fields – stars: rotation

---

Spectropolarimetric observations of a sample of late-type main-sequence dwarfs have been obtained with the NARVAL spectropolarimeter (Petit *et al.* 2008). The full stellar sample at our disposal is now constituted of about 20 objects probing a range of stellar masses and rotation rates.

A single, average photospheric line profile was extracted from each spectrum using the LSD technique (Donati *et al.* 1997), according to line-lists matching various stellar photospheric models calculated for the spectral types of our target list. Using this cross-correlation method, the noise level of the mean Stokes V profiles is reduced by a factor of about 40 with respect to the initial spectrum. The resulting noise levels are in the range $2 \times 10^{-5} - 10^{-4}$ I_c (where I_c denotes the intensity of continuum), depending on the stellar magnitude, spectral type, exposure time and observing conditions.

Assuming that the observed temporal variability of Stokes V profiles is controlled by the stellar rotation (including a possible latitudinal shear, Petit *et al.* 2002), we reconstruct the magnetic geometry of our targets by means of Zeeman-Doppler Imaging. We employ here the modelling approach of Donati & Brown (1997), including also the spherical harmonics expansion of the surface magnetic field implemented by Donati *et al.* (2006) in order to easily distinguish between the poloidal and toroidal components of the reconstructed magnetic field distribution.

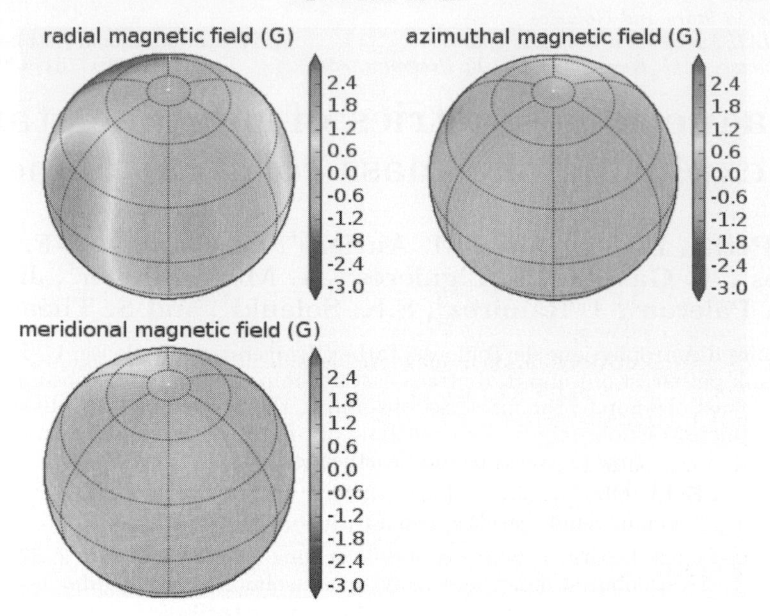

Figure 1. Magnetic map of the solar twin 18 Sco. Each chart illustrates the field projection onto one axis of the spherical coordinate frame. The magnetic field strength is expressed in Gauss.

From the set of reconstructed magnetic maps, we find that stars with high Rossby numbers (defined as $P_{\rm rot}/\tau_c$, where $P_{\rm rot}$ is the rotation period and τ_c the convective turnover time) feature weak, mostly poloidal surface fields. Stars with smaller Rossby numbers are able to generate stronger magnetic fields, with a mainly toroidal field geometry at photospheric level.

We plan to monitor the selected targets over 5 to 10 years to estimate the long-term variability of their magnetic topologies and the dependence of magnetic cycles on various stellar parameters. The temporal evolution of the total magnetic energy, the poloidal/toroidal distribution of the surface field or the distribution of the magnetic energy between the axisymmetric and non-axisymmetric components will then provide us with a new set of surface observables that will help to constrain numerical models of stellar dynamos.

References

Donati, J.-F., *et al.* 1997, *MNRAS* 291, 658
Donati J.-F. & Brown S. F., 1997, *A&A* 326, 1135
Donati, J.-F., *et al.* 2006, *MNRAS* 370, 629
Petit, P., *et al.* 2002, *MNRAS* 334, 374
Petit, P., *et al.* 2008, *MNRAS* 388, 80
Sanderson, T. R., *et al.* 2003, *JGR* 108, SSH 7-1
Wright, J. T., *et al.* 2004, *ApJS* 152, 261

Cosmic Magnetic Fields:
From Planets, to Stars and Galaxies
Proceedings IAU Symposium No. 259, 2008
K.G. Strassmeier, A.G. Kosovichev & J.E. Beckman, eds.

doi:10.1017/S174392130903107X

Fossil fields in early stellar evolution

Rainer Arlt

Astrophysikalisches Institut Potsdam, An der Sternwarte 16, D-14482 Potsdam, Germany
email: rarlt@aip.de

Abstract. Favored explanations for the presence of magnetic fields on CP stars and the presence of the solar tachocline below the convection zone both imply fossil magnetic fields in the radiative zones. The initial, convective evolution of magnetic fields in a proto-star is studied by numerical, global simulations. The computations are to be extended by a change of the convection zone depth on an evolutionary time-scale.

Keywords. Sun: magnetic fields – stars: magnetic fields – turbulence – MHD

1. Introduction

While magnetic fields can often be attributed to a dynamo process, we are concerned with the possible long-term existence of magnetic fields in stellar radiation zones which are unlikely to host a dynamo. There are two indications for the presence of fossil magnetic fields in radiative zones. One of them is the presence of strong magnetic fields in chemically peculiar A and B stars (CP stars, *cf.* Arlt 2008 for a review). Since the convectively stable radiation zones covering the outer $\sim$ 80% of the stellar radius do not provide a turbulent dynamo, the fields are often explained by fossil fields – left-overs from the contraction of the ambient magnetic field together with the proto-star. The second indication stems from the presence of the tachocline in the solar interior. The differential rotation of the convection zone (CZ), covering the outer 30% of the solar radius, turns into a nearly uniform rotation underneath the CZ on a length-scale of a few percent of the solar radius. A favoured explanation postulates the presence of a fossil magnetic field which maintains the uniform rotation in the solar interior, thus reducing the width of the tachocline to its small observed value (Rüdiger & Kitchatinov 1997; Gough & McIntyre 1998). The question arises to which extent magnetic fields can survive the early convective phase of star formation, *i.e.* which magnetic field strengths and geometries can be expected in the radiative zones of the Sun or a CP star after their proto-stellar phases.

2. Numerical simulations

Global, compressible simulations of stars are beyond the capabilities of computational facilities. When simplifying the problem to the Boussinesq approximation in which sound waves and the strong density contrast are eliminated, we may obtain results for reasonably long physical times. The computations thus employ the spectral spherical code by Hollerbach (2000) and integrate the velocity $\boldsymbol{u}$, the magnetic field $\boldsymbol{B}$, and the temperature fluctuations Θ:

$$\frac{\partial \boldsymbol{u}}{\partial t} = \nu\nabla^2\boldsymbol{u} - (\boldsymbol{u}\cdot\nabla)\,\boldsymbol{u} - \nabla p - \alpha\Theta\boldsymbol{g} + \frac{1}{\mu_0\rho_0}(\nabla\times\boldsymbol{B})\times\boldsymbol{B}, \tag{2.1}$$

$$\frac{\partial \boldsymbol{B}}{\partial t} = \eta\nabla^2\boldsymbol{B} + \nabla\times(\boldsymbol{u}\times\boldsymbol{B}), \tag{2.2}$$

$$\frac{\partial \Theta}{\partial t} = \kappa\nabla^2\Theta - \boldsymbol{u}\cdot\nabla\Theta - \boldsymbol{u}\cdot\nabla T, \tag{2.3}$$

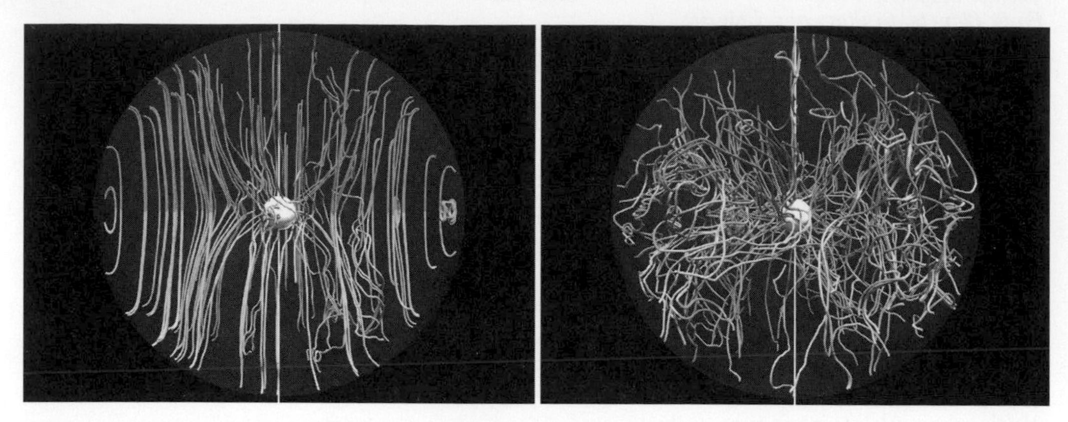

Figure 1. Magnetic field lines inside a convective sphere after roughly 1000 yr (left) and after about 2000 yr (right). The lines are rendered as tubes to enhance clarity in 3D; they do not represent flux-tubes. Dark parts of the lines denote $B_r > 0$, light parts denote $B_r < 0$.

where p is the pressure. The the kinematic viscosity ν, the magnetic and thermal diffusivities η and κ, the thermal expansion coefficient α, the magnetic permeability μ_0, and the density ρ_0 are constants in this setup. T is the non-convective background temperature profile. The runs are initialized by a uniform angular velocity Ω and a uniform magnetic field of strength B_0 parallel to the rotation axis, $\boldsymbol{B} = (B_0 \cos\theta, -B_0 \sin\theta, 0)$. When writing the equations in dimensionless form, the magnetic Reynolds number Rm, the magnetic Prandtl number Pm, the Lundquist number S, and a modified Rayleigh number $\widetilde{\mathrm{Ra}}$ govern our system. The boundary conditions for the magnetic field are those of a perfect conductor at $r_{\mathrm{i}} = 0.1$ and those of a vacuum exterior at $r_{\mathrm{o}} = 1$.

Note that simulations of non-ideal MHD always bear the problem of not reaching the extremely large microscopic magnetic Reynolds numbers in stellar plasma. Although we are able to compute flows with $\mathrm{Rm} = 10^4$, the time-scales of rotation and microscopic diffusion are not as widely separated as in real stars.

Two snap-shots of the magnetic-field evolution inside a convective sphere are shown in figure 1. The left panel is an early distribution of random field lines, while the right panel shows a turbulent state in which the magnetic field is highly complex. The initial antisymmetry with respect to the equator is barely visible; the magnetic quadrupole and non-axisymmetric modes quickly gain more than 50% of the total magnetic energy.

3. Conclusions

We show that global, direct numerical simulations of the magnetic evolution of a protostar are doable in a simplified setup. The ongoing simulations of magneto-convection will include the opening up of a radiative core and the turbulent pumping of magnetic field into the core. The magnetic field strength and geometry will be determined after the proto-stellar phase of either a solar-mass star or an intermediate-mass star of say $2M_\odot$, and the incidence angle of the initial magnetic field can be varied.

References

Arlt, R. 2008, *Contr. Astron. Obs. Skalnaté Pleso* 38, 163
Gough, D. & McIntyre, M. E. 1998, *Nature* 394, 755
Hollerbach, R. 2000, *Int. J. Numer. Meth. Fluids* 32, 773
Rüdiger, G. & Kitchatinov, L. L. 1997, *AN* 318, 273

Cosmic Magnetic Fields:
From Planets, to Stars and Galaxies
Proceedings IAU Symposium No. 259, 2008
K.G. Strassmeier, A.G. Kosovichev & J.E. Beckman, eds.

doi:10.1017/S1743921309031081

Discovery of fossil magnetic fields in the intermediate-mass pre-main sequence stars

Evelyne Alecian[1,2], Gregg A. Wade[1] and Claude Catala[2]

[1]Royal Military College of Canada, PO Box 17000, Stn Forces, Kingston K7K 7B4, Canada

[2]Observatoire de Paris, LESIA, place Jules Janssen, 92190 Meudon, France

Abstract. It is now well-known that the surface magnetic fields observed in cool, lower-mass stars on the main sequence (MS) are generated by dynamos operating in their convective envelopes. However, higher-mass stars (above 1.5 $M_\odot$) pass their MS lives with a small convective core and a largely radiative envelope. Remarkably, notwithstanding the absence of energetically-important envelope convection, we observe very strong (from 300 G to 30 kG) and organised (mainly dipolar) magnetic fields in a few percent of the A and B-type stars on the MS, the origin of which is not well understood. In this poster we propose that these magnetic fields could be of fossil origin, and we present very strong observational results in favour of this proposal.

Keywords. Stars: magnetic field – Stars: pre-main-sequence – Instrumentation: spectropolarimetry

---

1. Introduction

The fossil field theory assumes that the magnetic fields observed in the main sequence (MS) A/B stars are remnants from the galactic fields observed among the molecular clouds. Magnetic fields are observed in the molecular clouds and in (MS) intermediate mass stars. Until recently no magnetic fields has been observed in intermediate stages of stellar formation, especially during the pre-main sequence (PMS) phase (except in HD 104237, Donati *et al.* 1997). We therefore tried to find observational evidence that some PMS intermediate mass stars, the so-called Herbig Ae/Be stars (HAeBe), possess magnetic fields.

2. Observations and data reduction

We used the high-resolution spectropolarimeters ESPaDOnS installed on the Canada-France-Hawaii Telescop(CFHT, Hawaii) and Narval installes on the Bernard Lyot Telescope(TBL, Pic du Midi, France). We obtained Stokes I and V spectra of 65000 spectral resolution for a sample of 130 HAeBe stars in the field and in three very young clusters: NGC 2264 (age$\sim$2.6 Myr), NGC 2244 (age$\sim$2.3 Myr), and NGC 6611 (age$<$1 Myr).

We used the Libre ESpRIT reduction package, and we applied the Least Square Deconvolution Method (LSD, Donati *et al.* 1997), using ATLAS9 masks of appropriate effective temperature and surface gravity, excluding hydrogen Balmer lines and lines contaminated by emission. Figure 1 show resulting LSD I and V profiles for one of these stars, HD 200775, in which we detect a strong dipolar magnetic field.

3. Results

We have detected magnetic fields in 7 HAeBe stars (Wade et al 2005, Catala *et al.* 2007, Alecian *et al.* 2008ab). For 4 of them, we have characterised their magnetic fields

Table 1. Results of the fitting procedures for 4 magnetic HAeBe stars. References: 1: Alecian *et al.* 2008a, 2: Folsom *et al.* 2008, 3: Alecian *et al.*, in prep., 4: Catala *et al.* 2007

Star	Sp.T.	P (d)	i (°)	β (°)	$B_{\rm d}$ (kG)	$d_{\rm dip}$ (R_*)	$B_{\rm dZ}$ (kG)	Ref.
HD 200775	B3	4.3281 ± 0.0010	60 ± 11	125 ± 8	1.00 ± 0.15	0.05 ± 0.04	3.6	1
HD 72106	A0	0.63995 ± 0.00009	24 ± 10	57 ± 5	1.25 ± 0.08	0	1.25	2
V380 Ori	A2	4.31 ± 0.08	154 ± 30	110 ± 30	2.6 ± 1.5	0	4.5	3
HD 190073	A2				> 0.3		> 1.2	4

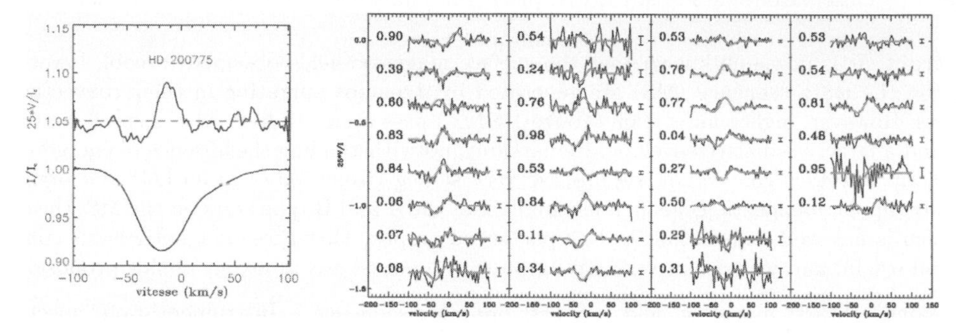

Figure 1. *Left:* Stokes I (bottom) and V (up) LSD profils of HD 200775. *Right:* Observed (noisy black) and modelled (red smooth) Stokes V profiles of HD 200775.

as follows. We monitored these stars in Stokes V to densely sample their rotation cycles. Then we used the oblique rotator model, as described in Stift (1975), to reproduce the variations of their Stokes V profiles (see also Alecian *et al.* 2008a). The free parameters are: the rotation period (P), the reference time (T_0), the rotation inclination (i), the magnetic obliquity (β), the dipole strength ($B_{\rm d}$), and the dipole shift ($d_{\rm dip}$) from the center of the star expressed in stellar radii (R_*). We performed a χ^2 minimisation in order to reproduce our observations. The Fig. 1 shows the result of this fitting procedure for one star, while Table 1 gives the values of the fitted parameters.

4. Conclusion

We find that around 5% of HAeBe stars are magnetic, as predicted from the fossil field theory. These stars have strong and large scale magnetic fields stable over more than 3 years, as observed in the MS A/B stars. Finally, assuming the conservation of magnetic flux during the PMS phase, we find that when the magnetic HAeBe stars will reach the ZAMS their magnetic strengths will be between 1.2 to 3.6 kG, similar to the magnetic MS A/B stars. Therefore the magnetic fields of the A/B stars is very likely fossil.

References

Alecian, E., Catala, C., Wade, G. A., *et al.* , 2008a, *MNRAS* 385, 391
Alecian, E., Wade, G. A., Catala, C., *et al.* , 2008b, *A&A* 481, L99
Catala, C., Alecian, E., Donati, J.-F., *et al.* , 2007, *A&A* 462, 293
Donati, J.-F., Semel, M., Carter, B. D., *et al.* , 1997, *MNRAS* 291, 658
Folsom, C.P., Wade, G.A., Kochukhov, O., *et al.* , 2008, *MNRAS* 391, 901
Stift, M. J., 1975, *MNRAS* 172, 133
Wade, G. A., Drouin, D., Bagnulo, S., *et al.* , 2005, *A&A* 442, L31

Cosmic Magnetic Fields:
From Planets, to Stars and Galaxies
Proceedings IAU Symposium No. 259, 2008
K.G. Strassmeier, A.G. Kosovichev & J.E. Beckman, eds.

doi:10.1017/S1743921309031093

The surface magnetic fields of T Tauri stars

Gaitee A.J. Hussain[1], Andrew Collier Cameron[2], Moira Jardine[2] and Jean-Francois Donati[3]

[1]ESO, Karl-Schwarzschild-Strasse 2, 85748 Garching bei München, Germany.
email: ghussain@eso.org

[2]School of Physics and Astronomy, University of St Andrews, North Haugh, Fife KY16 9SS, Scotland, UK
email:acc4@st-andrews.ac.uk, mmj@st-andrews.ac.uk

[3]LATT, CNRS-UMR 5572, Obs. Midi-Pyrénées, 14. Av. E. Belin, F-31400 Toulouse, France
email: donati@ast.obs-mip.fr

Abstract. We present surface magnetic field maps of the two accreting T Tauri stars, CV Cha and CR Cha. Our magnetic field maps show evidence for strong, complex multi-polar fields similar to those obtained on young rapidly rotating main sequence stars. Both CV Cha and CR Cha show magnetic field patterns that are more complex than those recovered for the lower mass, fully convective T Tauri star, BP Tau.

By comparing our maps with previously published maps of classical T Tauri stars, we infer that magnetic field patterns on T Tauris, and their underlying magnetic field generation mechanisms evolve quickly as they develop radiative cores. This may have implications for the efficiency with which T Tauri stars can effectively lock onto their surrounding disks under the magnetospheric accretion model scenario. This, in turn, has implications for the angular momentum evolution of T Tauri stars as they evolve towards the main sequence.

Keywords. Stars: magnetic fields – stars: imaging – stars: accretion – stars: formation – techniques: spectro-polarimetry

1. Background

The main properties of classical T Tauri stars (CTTS) can be explained using the "magnetospheric accretion scenario" (e.g., Camenzind 1990). In this scenario, the magnetosphere of the central star is responsible for truncating the inner edge of its circumstellar disk; material is channelled along magnetic field lines onto the stellar surface, causing accretion shocks at the site of impact on the stellar surface.

We present the first surface magnetic field maps of the accreting pre-main sequence (PMS) stars, CV Cha and CR Cha, using the technique of Zeeman Doppler imaging. Only two other similar systems have been imaged before using this technique: V2129 Oph and BP Tau (Donati *et al.* 2007, 2008a). Our results suggest that PMS stars with radiative cores have more complex field distributions than fully convective stars. These results have implications for magnetospheric accretion and for the evolution of magnetic fields in intermediate mass stars as they evolve onto and along the main sequence.

2. Summary

• CV Cha and CR Cha are the first intermediate-mass T Tauri stars to be imaged using the technique of Zeeman Doppler imaging.

• Our maps show that T Tauri stars with radiative cores have more complex magnetic field distributions than the fully convective T Tauri, BP Tau (Donati *et al.* 2008a).

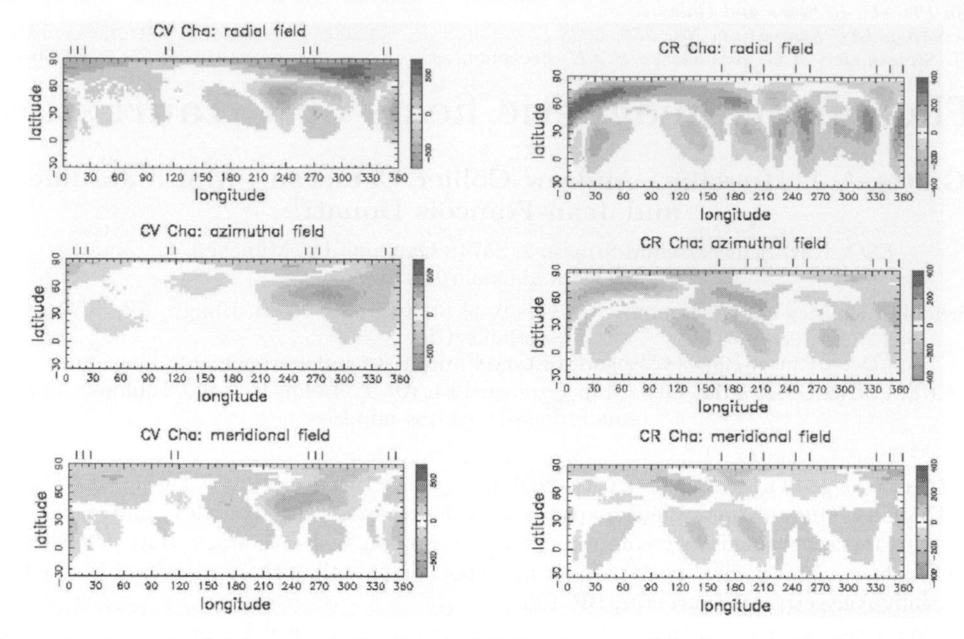

Figure 1. Magnetic field maps of the classical T Tauri stars, CV Cha and CR Cha. Our maps suggest magnetic fields in higher mass stars are more complex than those in the lower mass star, BP Tau (Donati *et al.* 2008a). These maps were obtained using the technique of Zeeman Doppler imaging and circularly polarised data that were acquired using the UCLES/SemelPol instrument configuration on the Anglo-Australian Telescope (AAT). The phases of observations are represented by the tick marks above each plot.

- This may be analogous to the behaviour observed across the fully convective boundary in main sequence M-type stars. Morin *et al.* (2008) find that M4 stars, i.e. stars that are nearly fully convective, have simpler axisymmetric large-scale poloidal fields. In contrast, higher mass cool stars, show significant toroidal field (Donati *et al.* 2008b).
- CV Cha and CR Cha are progenitors of A-type stars. They will take under 9 Myr to become fully radiative. Only 10% of A-type stars are found to be magnetic (Ap stars). Over the next 5-6 Myr, this magnetic field will either dissipate or be frozen in. Further details are discussed in Hussain *et al.* (2009).

Acknowledgements

The assistance of the staff at the Anglo-Australian Telescope was instrumental in obtaining this data and their help is gratefully acknowledged.

References

Camenzind, M., 1990, *Rev. Modern Astronomy* 3, 234
Donati, J.-F. *et al.* 2007, *MNRAS* 380, 1297
Donati, J.-F. *et al.* 2008a, *MNRAS* 386, 1234
Donati, J.-F. *et al.* 2008b, *MNRAS* 390, 545
Hussain, G. A. J. *et al.* 2009, *MNRAS*, submitted
Morin, J. *et al.* 2008, *MNRAS* 390, 567

Cosmic Magnetic Fields:
From Planets, to Stars and Galaxies
Proceedings IAU Symposium No. 259, 2008
K.G. Strassmeier, A.G. Kosovichev & J.E. Beckman, eds.

doi:10.1017/S174392130903110X

Magnetic fields, winds and X-rays of massive stars in the Orion Nebula Cluster

Véronique Petit[1], G. A. Wade[2], L. Drissen[1], T. Montmerle[3] and E. Alecian[2]

[1]Département de Physique, Centre de recherche en Astrophysique du Québec, Université Laval, Québec, Canada

[2]Department Physics, Royal Military College of Canada, Kingston, Canada

[3]Laboratoire d'Astrophysique de Grenoble, Université Joseph Fourier, CNRS, Grenoble, France

Abstract. In massive stars, magnetic fields are thought to confine the outflowing radiatively-driven wind, resulting in X-ray emission that is harder, more variable and more efficient than that produced by instability-generated shocks in non-magnetic winds. Although magnetic confinement of stellar winds has been shown to strongly modify the mass-loss and X-ray characteristics of massive OB stars, we lack a detailed understanding of the complex processes responsible. The aim of this study is to examine the relationship between magnetism, stellar winds and X-ray emission of OB stars. In conjunction with a Chandra survey of the Orion Nebula Cluster, we carried out spectropolarimatric ESPaDOnS observations to determine the magnetic properties of massive OB stars of this cluster.

Keywords. Stars: magnetic fields – stars: mass loss – stars: early-type – X-rays: stars – techniques: polarimetric

---

Magnetic fields are well known to produce X-rays in late-type convective stars like the Sun. On the other hand, X-ray emission from massive stars is thought to have a different origin. Their powerful winds are driven by by radiation pressure through spectral lines, which is an unstable process. These instabilities result in collisions between wind streams of different velocity, resulting in small shocks that generate X-ray emission (Owocki & Cohen 1999, Lucy & White 1980).

The Chandra Orion Ultradeep Project (COUP) was dedicated to observe the Orion Nebula Cluster (ONC) in X-rays. The sample of 20 O, B and A-type stars was studied with the goal of disentangling the respective roles of winds and magnetic fields in producing X-rays (Stelzer *et al.* 2005). The production of X-rays by radiative shocks should be the dominant mechanism for the subsample of 9 O to early-B stars. However, aside from 2 of those stars, all targets showed X-ray intensity and/or variability which were inconsistent with the small shock model predictions (Figure 1).

To examine the effect of magnetic fields on the winds of the massive stars in the ONC, we conducted spectropolarimetric observations in order to directly detect and characterise potential magnetic fields in these stars, through the Stokes V circular polarisation induced by the Zeeman effect. We used the ESPaDOnS spectropolarimeter at CFHT in 2006 and 2007 to observe 8 of the 9 ONC massive stars. Additional measurements of θ^1 Ori C and HD 36982 were taken with ESPaDOnS in December 2007 and with ESPaDOnS's twin Narval, installed at Télescope Bernard Lyot, in November 2007.

From our observations, we found clear Stokes V magnetic signatures for three stars: the previously-detected θ^1 Ori C, as well as HD 36982 and HD 37061 (Petit *et al.* 2008). The magnetic stars are encircled in Figure 1. The Stokes V signatures were directly modelled with the polarised LTE radiative transfer code ZEEMAN2 (Landstreet 1988), leading to

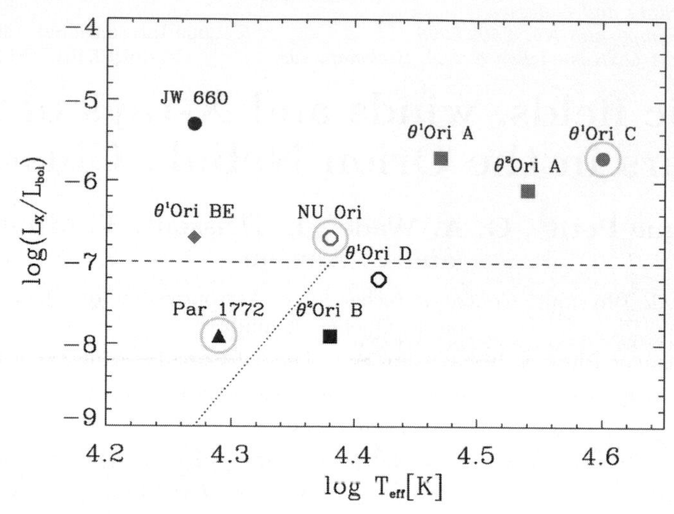

Figure 1. X-ray efficiency of the ONC massive stars as a function of effective temperature (Stelzer *et al.* 2005). The 3 detected stars are circled. Filled symbols are for stars with indirect indications of the presence of a magnetic field, and grey symbols are for confirmed or suspected binaries. Plotting symbols indicate the following properties: circles are for stars showing possible X-ray rotational modulation, squares are for T Tauri type X-ray emission, triangles are chemically peculiar (CP) stars, and the diamond star was not observed. The dotted lines indicates the typical efficiency for massive O-type stars, and a qualitative illustration of the sharp decrease in the B-type stars range.

an inferred surface dipolar field strength of 1150^{+320}_{-200} G and 620^{+220}_{-170} G for HD 39682 and HD 37061 respectively.

This study of the Orion stellar cluster represents a complete magnetic survey of a co-evolved and co-environmental population of massive stars. The 3 magnetic stars of the ONC have fields that should be strong enough to dynamically influence their stellar winds at a significant level. However, this field-wind interaction is not reflected in any systematic way in the X-ray properties of these stars. Furthermore, no fields are found (with typical upper limits on the surface dipole strength of the order of 100 G) in other ONC massive stars that Stelzer *et al.* (2005) considered to be "prime candidates" for magnetism. From this we conclude that: (i) X-ray variability, intensity and hardness enhancement are not systematically correlated with the presence of a magnetic field. More detailed studies of the field geometries of these magnetic stars will serve as inputs to new models (Townsend, Owocki & ud-Doula 2007) and 3D MHD simulations of magnetic wind confinement (ud-Doula, Owocki & Townsend 2008), to better understand the mechanisms that lead to this variety of X-ray properties. (ii) The classical Lucy & White (1980) mechanism for soft X-ray production in non-magnetic winds requires significant revision to explain the characteristics of some stars in the ONC.

References

Landstreet, J. D. 1988, *ApJ* 326, 967
Lucy, L. B. & White, R. L. 1980, *ApJ* 241, 300
Owocki, S. P. & Cohen, D. H. 1999, *ApJ* 520, 833
Petit, V., Wade, G. A., Drissen, L., Montmerle, T., & Alecian, E. 2008, *MNRAS* 387, L23
Stelzer, B., Flaccomio, E., Montmerle, T., Micela, G., *et al.* 2005, *ApJS* 160, 557
Townsend, R. H. D., Owocki, S. P., & ud-Doula, A. 2007, *MNRAS* 382, 139
ud-Doula, A., Owocki, S. P., & Townsend, R. H. D. 2008, *MNRAS* 385, 97

Session chair Swetlana Hubrig (left), and Michael Weber (right)

Iosif Romanyuk (left), and Ilya Ilyin & Mrs. and Mr. Stenflo (right)

Ilkka Tuominen & Katrin Götz (from right to left), and a nameless dolphin (picture by R. Arlt)

Jason Grunhut & Shih-Ping Lai (left), and Klaus Puschmann & Valeria Buenrostro (right)

A happy LOC at the banquet

During the banquet late-night slide show

Session V

From stars to galaxies and the intergalactic space

Cosmic Magnetic Fields:
From Planets, to Stars and Galaxies
Proceedings IAU Symposium No. 259, 2008
K.G. Strassmeier, A.G. Kosovichev & J.E. Beckman, eds.

doi:10.1017/S1743921309031123

The magnetic structure of our Galaxy: a review of observations

JinLin Han

National Astronomical Observatories, Chinese Academy of Sciences
Jia-20 DaTun Road, Chaoyang District, Beijing 100012, China
email: hjl@bao.ac.cn

Abstract. The magnetic structure in the Galactic disk, the Galactic center and the Galactic halo can be delineated more clearly than ever before. In the Galactic disk, the magnetic structure has been revealed by starlight polarization within 2 or 3 kpc of the Solar vicinity, by the distribution of the Zeeman splitting of OH masers in two or three nearby spiral arms, and by pulsar dispersion measures and rotation measures in nearly half of the disk. The polarized thermal dust emission of clouds at infrared, mm and submm wavelengths and the diffuse synchrotron emission are also related to the large-scale magnetic field in the disk. The rotation measures of extragalactic radio sources at low Galactic latitudes can be modeled by electron distributions and large-scale magnetic fields. The statistical properties of the magnetized interstellar medium at various scales have been studied using rotation measure data and polarization data. In the Galactic center, the non-thermal filaments indicate poloidal fields. There is no consensus on the field strength, maybe mG, maybe tens of μG. The polarized dust emission and much enhanced rotation measures of background radio sources are probably related to toroidal fields. In the Galactic halo, the antisymmetric RM sky reveals large-scale toroidal fields with reversed directions above and below the Galactic plane. Magnetic fields from all parts of our Galaxy are connected to form a global field structure. More observations are needed to explore the untouched regions and delineate how fields in different parts are connected.

Keywords. ISM: magnetic fields – Galaxy: structure – radio continuum: ISM

1. Introduction

Radio observations measure the total intensity, polarized intensity and rotation measure (RM). Sky maps of these quantities are all related to the magnetic fields in our Galaxy. Because the Milky Way is the largest edge-on galaxy in the sky, the synchrotron emission from relativistic electrons gyrating in the Galactic magnetic field is dominant in the radio sky. The stronger radio emission is observed nearer the Galactic plane, and strongest towards the Galactic central regions. The Galactic radio emission is polarized (Reich 2007), if it is not depolarized, best shown at tens of GHz (Page *et al.* 2007). The polarized emission undergoes Faraday rotation due to the Galactic magnetic fields and ionized electrons. The Faraday sky, i.e. the RM sky, is strikingly antisymmetric to the Galactic coordinate (Han *et al.* 1997, see Fig. 5).

The magnetic field of our Galaxy is more important in astrophysics and astroparticle physics than the fields in other galaxies. Magnetic fields are certainly one of key ingredients of the interstellar medium. Large-scale magnetic fields contribute to the hydrostatic balance and stability of the interstellar medium (Boulares & Cox 1990), and even disk dynamics (Battaner & Florido 2008). Magnetic fields in molecular clouds, which are closely related to the Galactic fields (Han & Zhang 2007), play an important role in the star formation process (see a review by Heiles & Crutcher 2005). More important is that the magnetic fields of our Galaxy are the main agent for transport of charged cosmic-rays (e.g. Tinyakov & Tkachev 2002, Prouza & Smída 2003). It is impossible to understand the origin and propagation of cosmic rays without adequate knowledge of Galactic magnetic fields. The Galactic radio emission and its polarization, which result from the Galactic magnetic fields, is found to heavily (up to 95% in polarization) "pollute" the measurements of the cosmological microwave background (CMB, e.g. Page *et al.* 2007). Galactic magnetic fields have suddenly become very important in the CMB studies of cosmology!

To understand magnetic fields, we have first to measure them and learn their properties. The galactic scale is in the middle between the stellar scale and the cosmological scale. It is on this scale that the magnetic fields can be well *measured* at present, at least from our Galaxy. Observations of galactic-scale magnetic fields provide the most important hints for and constraints on the origin and maintenance of the magnetic fields of Galactic objects and the fields in the universe. In the following, I will review the observed magnetic structures in the Galactic disk, the Galactic center and the Galactic halo.

2. Magnetic fields in the Galactic disk

Magnetic fields pervade the diffuse interstellar medium, molecular clouds, and very dense cloud cores or HII regions. When interstellar gas contracts to form a cloud or a cloud core the field is enhanced. Therefore, the observed field strength in clouds increases with gas density (Crutcher 1999). Here, I review the observational results of large-scale magnetic fields in the Galactic disk which are related to the diffuse medium and spiral structure and have a scale greater than 1 kpc. Magnetic fields on smaller scales will be mentioned only if they are related to the large-scale fields.

There are five observational tracers of the Galactic magnetic fields: polarization of starlight, polarized thermal dust emission from clouds, Zeeman splitting of lines, diffuse synchrotron radio emission, and Faraday rotation of polarized sources.

Polarization of starlight

Starlight becomes polarized when it passes through the interstellar medium and is absorbed or scattered by interstellar dust grains preferentially aligned by magnetic fields. The observed polarization is the integrated effect of scattering between the star and the Sun, and the "polarization vectors" show the averaged field orientation (weighted by the unknown local dust content). Starlight polarization data have been obtained for about 10,000 stars, mostly within 2 or 3 kpc of the Sun (see Mathewson & Ford 1970, Heiles 2000). Analysis of these data show that the local field is parallel to the Galactic plane and follows the local spiral arms (Heiles 1996). Starlight polarization data are difficult to use for detection of Galactic magnetic fields in a much larger region than 2 or 3 kpc. However, recent developments in instruments help to get a lot of new starlight polarization data for revealing magnetic fields in given objects (e.g. Feinstein *et al.* 2008).

Polarized thermal dust emission from clouds

In recent years, with development of instruments and backend technology (e.g. Hildebrand *et al.* 2000), observations of polarized thermal dust emission at mm, sub-mm and infrared bands have been used to detect the transverse orientation of magnetic fields in molecular clouds (see review by Heiles & Crutcher 2005) on scales from 1 pc to tens of pc. The observed magnetic fields always have an hourglass shape, which indicates that the fields were enhanced when the clouds were formed by compressing the diffuse interstellar medium. Recently, Li *et al.* (2006) found that magnetic fields in molecular clouds seem to be preferentially parallel to the Galactic plane, indicating that the magnetic fields in the clouds preserve the fields frozen into the diffuse medium.

Zeeman splitting of lines

Zeeman splitting of spectral lines can measure *in situ* field strength of the line-of-sight component in molecular clouds or maser spots with a scale size < 1 AU. To date, Zeeman splitting of emission or absorption lines has been detected from about 20 clouds (see a list in Crutcher 1999, Han & Zhang 2007) and in OH masers associated with 140 HII or star-formation regions using single dishes (e.g. Caswell 2003) or interferometers (e.g. Fish *et al.* 2003). From collected Zeeman splitting data of OH masers of HII regions and OH or HI absorption or emission lines of molecular clouds themselves, Han & Zhang (2007) found large-scale reversals in the sign of the line-of-sight component of the median field, indicating field reversals in a pattern similar to reversals revealed from pulsar RM data (see below). Evidently, magnetic fields on such a small scale are related to the large-scale Galactic magnetic fields to a very surprising extent (Reid & Silverstein1990, Fish *et al.* 2003). Interstellar magnetic fields are apparently preserved as fossil fields through the cloud formation and star formation process, and even in stars (see G. Wade's talk in this volume). How can such coherent magnetic field directions be preserved from the low density medium ($\sim 1\,\mathrm{cm}^{-3}$) to higher density clouds ($\sim 10^3\,\mathrm{cm}^{-3}$), even to the highest density in maser regions ($\sim 10^7\,\mathrm{cm}^{-3}$), with compression of 3 or even 10 orders of magnitude? It is a puzzle. The turbulent and violent processes in molecular clouds apparently cannot significantly alter the mean magnetic field.

Diffuse synchrotron radio emission

The total and polarized intensities of synchrotron emission are usually used to estimate the total and *ordered* field strength. For nearby galaxies, the strength of the *regular* large-scale field is probably much overestimated from the polarization percentage, because the so-called *ordered* fields consist of the regular large-scale fields and anisotropic random fields (Han *et al.* 1999a) which both produce polarized emission.

We cannot have a face-on view of the global magnetic field structure in our Galaxy through polarized synchrotron emission, as is possible for nearby spiral galaxies (see R. Beck's talk in this volume). Polarization surveys of the Galactic plane have been comprehensively reviewed by Reich (2007). The observed polarized emission from the Galactic plane is the sum of all contributions with different polarization properties (i.e. polarization angles and polarization percentages) coming from various regions at different distances along a line of sight. Emission from more distant regions suffers from more Faraday effect produced by the foreground interstellar medium. Polarized emissions from different regions should "depolarize" each other when they are summed, which is more obvious at lower frequencies. Observations at higher frequencies (e.g. at 6 cm by Sun *et al.* 2007) show polarized structures at larger distances because depolarization is less severe. The polarized structures are closely related to the magnetic field structure where the emission is generated. Some large-angular-scale polarized features are seen emerging from the Galactic disk, for example, the North Polar Spur (e.g. Junkes *et al.* 1987).

Faraday rotation of pulsars and radio sources

Faraday rotation of linearly polarized radiation from pulsars and extragalactic radio sources (EGRs) is the most powerful probe of the diffuse magnetic field in the Galaxy (e.g. Han *et al.* 2006, Brown *et al.* 2007). Magnetic fields in a large part of the Galactic disk have been revealed by RM data of pulsars, which gives a measure of the line-of-sight component of the field. EGRs have the advantage of large numbers but pulsars have the advantage of being spread through the Galaxy at approximately known distances, allowing direct three-dimensional mapping of the magnetic fields. For a pulsar at distance D (in pc), the RM (in radians m^{-2}) is given by $\mathrm{RM} = 0.810 \int_0^D n_e \mathbf{B} \cdot d\mathbf{l}$, where n_e is the electron density in cm^{-3}, $\mathbf{B}$ is the vector magnetic field in μG and $d\mathbf{l}$ is an elemental vector along the line of sight toward us (positive RMs correspond to fields directed toward us) in pc. With the pulsar dispersion measure, $\mathrm{DM} = \int_0^D n_e dl$, we obtain a direct estimate

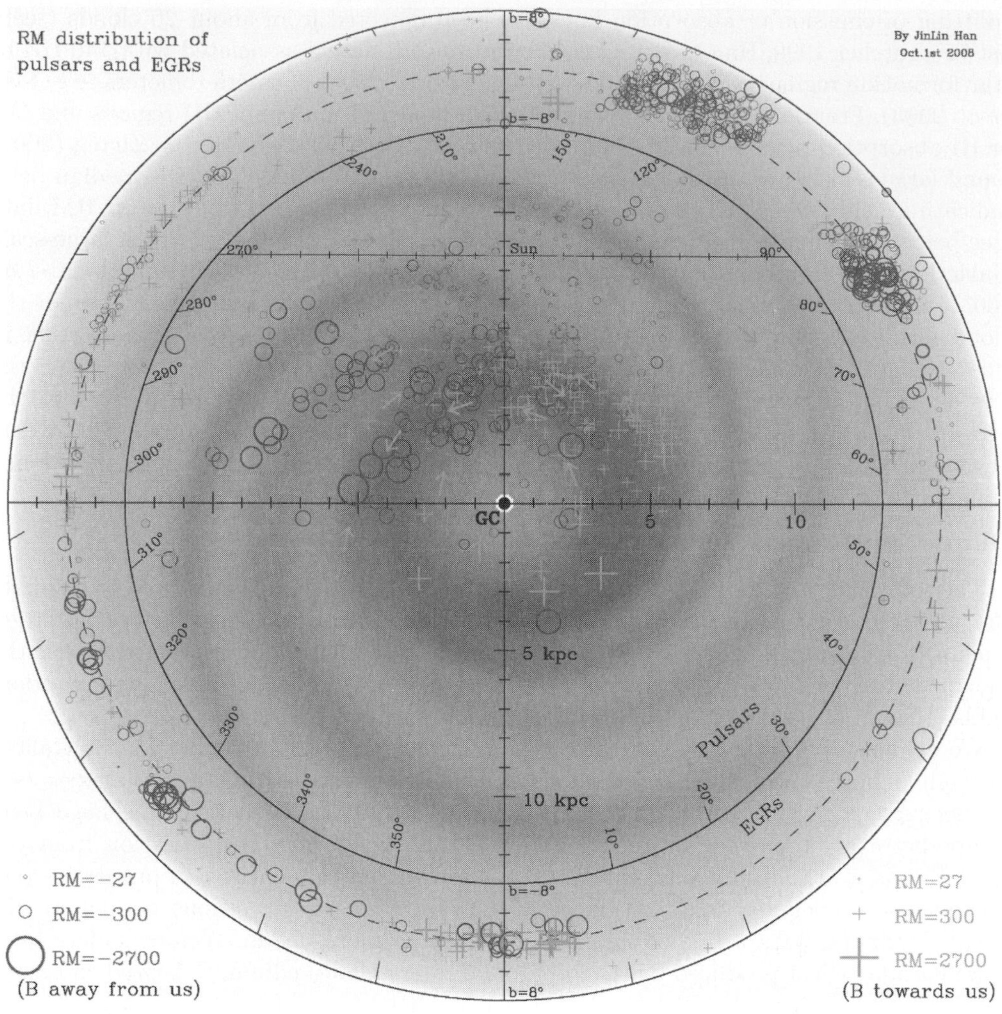

Figure 1. The RM distribution of 736 pulsars with $|b| < 8°$ projected onto the Galactic plane, including new data of Han *et al.* (2009, in preparation). The linear sizes of the symbols are proportional to the square root of the RM values with limits of ± 27 and ± 2700 rad m^{-2}. Positive RMs are shown by plus signs and negative RMs by open circles. The background shows the approximate locations of spiral arms used in the NE2001 electron density model (Cordes & Lazio 2002). RMs of 1285 EGRs of $|b| < 8°$ (data mainly from Clegg *et al.* 1992, Gaensler *et al.* 2001, Brown *et al.* 2003, Roy *et al.* 2005, Brown *et al.* 2007 and other RM catalogs) are displayed in the outskirt ring according to their l and b, with the same convention of RM symbols and limits. The large-scale structure of magnetic fields indicated by arrows was derived from pulsar RMs and comparison of them with RMs of background EGRs (details in Han *et al.* 2009). The averaged RM fluctuations with Galactic longitudes of EGRs are consistent with magnetic field directions derived from pulsar data, for example, in the 4th Galactic quadrant.

of the field strength weighted by the local free electron density

$$\langle B_{||} \rangle = \frac{\int_0^D n_e \mathbf{B} \cdot d\mathbf{l}}{\int_0^D n_e dl} = 1.232 \,\frac{\text{RM}}{\text{DM}}. \tag{2.1}$$

where RM and DM are in their usual units of rad m^{-2} and cm^{-3} pc and $B_{||}$ is in μG.

Previous analyses of pulsar RM data have often used the model-fitting method (Manchester 1974, Thomson & Nelson 1980, Rand & Kulkarni 1989, Han & Qiao 1994, Indrani

& Deshpande 1999, Noutsos *et al.* 2008), i.e., to model magnetic field structures in all of the paths from the pulsars to us (observer) and fit them, together with the electron density model, to the observed RM data. *Significant improvement* can be obtained when both RM and DM data are available for many pulsars in a given region with similar lines of sight. Measuring the gradient of RM with distance or DM is the most powerful method of determining both the direction and magnitude of the large-scale field local to that particular region of the Galaxy (Lyne & Smith 1989, Han *et al.* 1999b, Han *et al.* 2002, Weisberg *et al.* 2004, Han *et al.* 2006). Field strengths in the region can be *directly derived* from the slope of trends in plots of RM versus DM. Based on Eq. (2.1), we get

$$\langle B_{||} \rangle_{d1-d0} = 1.232 \frac{\Delta \mathrm{RM}}{\Delta \mathrm{DM}} \tag{2.2}$$

where $\langle B_{||} \rangle_{d1-d0}$ is the mean line-of-sight field component in μG for the region between distances $d0$ and $d1$, $\Delta\mathrm{RM} = \mathrm{RM}_{d1} - \mathrm{RM}_{d0}$ and $\Delta\mathrm{DM} = \mathrm{DM}_{d1} - \mathrm{DM}_{d0}$.

i) Field structure

So far, RMs of 1021 pulsars have been observed (Hamilton & Lyne 1987, Rand & Lyne 1994, Qiao *et al.* 1995, Han *et al.* 1999b, Weisberg *et al.* 2004, Han *et al.* 2006, Noutsos *et al.* 2008), if new RMs of 477 pulsars by Han, van Straten, Manchester & Demorest (2009, in preparation) are included (see Fig. 1). This enables us to investigate the structure of the Galactic magnetic field over a larger region than that previously possible. We have detected counterclockwise magnetic fields in the innermost arm, the Norma arm (Han *et al.* 2002). A more complete analysis for the fields of our Galaxy (Han *et al.* 2006 and Han *et al.* 2009, in preparation) from both RMs of pulsars and EGRs gives a picture for the coherent large-scale fields aligned with the spiral-arm structure, as shown in Fig.1: magnetic fields in all inner spiral arms are counterclockwise when viewed from the North Galactic pole. On the other hand, at least in the local region and in the inner Galaxy in the fourth quadrant, there is good evidence that the fields in interarm regions are similarly coherent, but reversed to be clockwise. There are at least four or five reversals in the fourth quadrant, probably occurring near the boundary of the spiral arms. In the Galactic central region interior to the Norma arm, new RM data of pulsars indicate that the fields are clockwise, reversed again from the counterclockwise field in the Norma arm. In the first Galactic quadrant, because the separations between spiral arms are so small, the RM data are dominated by counterclockwise fields in the arm regions though some (not many) negative pulsar RMs indicate clockwise fields in the interarm regions. The magnetic field in the Perseus arm cannot be determined well.

The averaged RM of EGRs in a given sky region reflect the common foreground Galactic RM contribution, which is the integration of $n_e B$ from the Sun to the outskirts of the Galactic disk. Comparison of the mean of RMs of background EGRs with RMs of foreground pulsars can reveal the magnetic fields behind the pulsars (see Fig. 1). However, the dominant contribution of RMs of EGRs behind the Galactic disk comes from the interstellar medium mostly in tangential regions. The fluctuations in the RM distribution of extragalactic radio sources (Clegg *et al.* 1992, Gaensler *et al.* 2001, Brown *et al.* 2003, Roy *et al.* 2005, Brown *et al.* 2007) with Galactic longitude, especially these of the fourth Galactic quadrant, are consistent with magnetic field directions derived from pulsar data (see Fig. 1). The negative RMs of EGRs in the 2nd quadrant suggest that the interarm fields both between the Sagittarius and Perseus arms and beyond the Perseus arm are predominantly clockwise.

Fitting various models to the new RM dataset shows different results (e.g. Noutsos *et al.* 2008). The most important is to have a model consistent with the detected field reversals. Looking at Table 2 of Men et al. (2008), one can see that any single-mode model is not enough to fit RM data though the bisymmetrical spiral model with field reversals is the best and gives the smallest χ^2.

ii) Field strength

With much more pulsar RM data, for the first time, Han *et al.* (2006) were able to *measure* the regular field strength near the tangential regions in the 1st and 4th Galactic quadrants, and then plot the dependence of regular field strength as a function of galactocentric radius (Fig. 2). Although the "uncertainties", which in fact reflect the random fields, are large, the tendency is clear that fields get stronger at smaller Galactocentric radius and weaker in interarm regions. To parameterize the radial variation, an exponential function was used. This was chosen to give the smallest χ^2 value and to avoid a singularity at $R = 0$ (for $1/R$) and unphysical values at large R (for the linear gradient). The function is, $B_{\rm reg}(R) = B_0 \ \exp\left[\frac{-(R-R_\odot)}{R_{\rm B}}\right]$, with the strength of the large-scale field at the Sun, $B_0 = 2.1 \pm 0.3$ μG, and the scale radius $R_{\rm B} = 8.5 \pm 4.7$ kpc.

iii) Field statistics on small scales and spatial B-energy spectrum

Magnetic fields in our Galaxy exist on all scales. For the large-scale field, we can determine the field structure and field strength. To study small-scale magnetic fields, the only approach is to make statistics and have a description of their statistical properties.

Pulsar RMs have been used to study the small-scale random magnetic fields in the Galaxy. Some pairs of pulsars close in sky position have similar DMs but very different RMs, indicating an irregular field structure on scales of about 100 pc (Lyne & Smith 1989). Some of these irregularities may result from HII regions in the line of sight to a pulsar (Mitra *et al.* 2003). It has been found from pulsar RMs that the random field has a strength of $B_r \sim 4 - 6\mu$G independent of cell-size in the scale range of 10 – 100 pc (Rand & Kulkarni 1989, Ohno & Shibata 1993). From pulsar RMs in a very large region of the Galactic disk, Han *et al.* (2004) obtained a power law distribution for magnetic field fluctuations of $E_B(k) = C \ (k/{\rm kpc}^{-1})^{-0.37\pm0.10}$ at scales from $1/k = 0.5$ kpc to 15 kpc, with $C = (6.8 \pm 0.8) \ 10^{-13}$erg cm^{-3} kpc, corresponding to an rms field of $\sim 6\mu$G in the scale range.

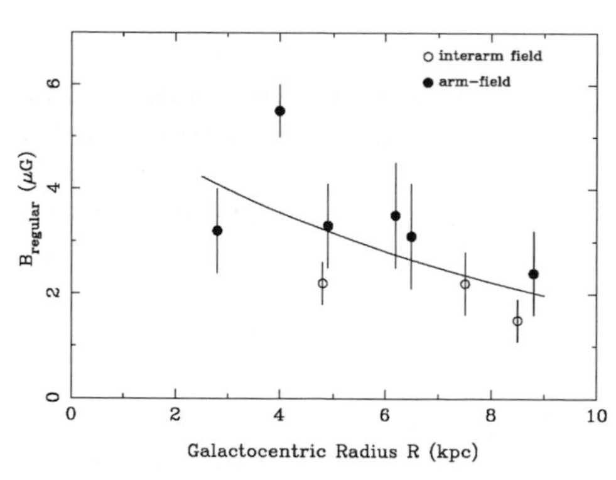

Figure 2. Variation of the large-scale regular field strength with Galactocentric radius derived from pulsar RM and DM data near the tangential regions (Han *et al.* 2006). Note that the "error bars" are not caused by the uncertainty of the pulsar RM or DM data, but reflect the random magnetic fields in the regions.

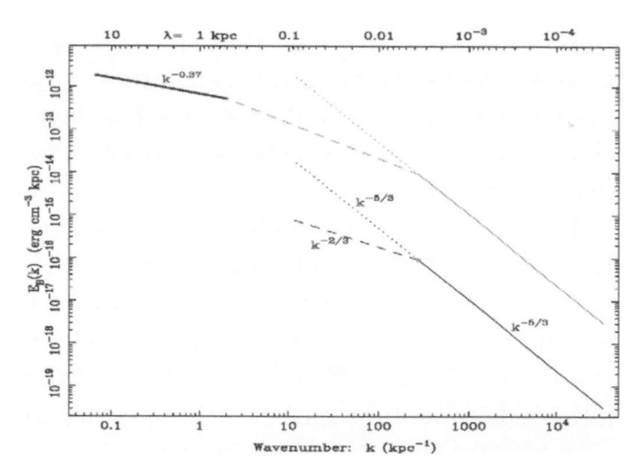

Figure 3. Composite spatial magnetic-energy spectrum in our Galaxy. See Han *et al.* (2004) for details. The thin solid and dashed/dotted lines at smaller scales are the Kolmogorov and 2D-turbulence spectra derived from Minter & Spangler (1996), and the upper ones from new measurements of Minter (2004, private communication).

Some results of magnetic field statistics were also obtained from RMs of EGRs. Armstrong *et al.* (1995) showed that the spatial power spectrum of electron density fluctuations from small scales up to a few pc could be approximated by a single power law with a 3D spectral index −3.7, very close to the Kolmogorov spectrum. Magnetic fields on such small scales should follow the same power-law, since magnetic fields are frozen into the ionized interstellar medium. Minter & Spangler (1996) found that structure functions of rotation measure and emission measure were consistent with a 3D-turbulence Kolmogorov spectrum of magnetic fields up to 4 pc, but with a 2D turbulence between 4 pc and 80 pc. Haverkorn *et al.* (2006) found that the structure functions of RMs in the directions tangential to the arms have much larger slopes than those in the inter-arm directions, indicating that the arm regions are more turbulent. Sun & Han (2004) found that the structure function for RMs at the two Galactic poles is flat, but at lower latitudes it becomes inclined with different slopes at different Galactic longitudes.

Combining the above information, one can get the spatial energy spectrum of Galactic magnetic fields (see Han *et al.* 2004) which should constrain the theoretical simulations (e.g. Balsara & Kim 2005) of the generation and maintenance of Galactic magnetic fields. Evidently, on small scales, the distribution follows the Kolmogorov power-law spectrum, but at larger scales, it becomes flat and probably has two breaks at a scale of few pc and several tens of pc (Fig. 3). The interstellar magnetic fields probably become strongest at the scales of energy injection by supernova explosions and stellar winds (e.g. 1 to 10 pc).

3. Magnetic fields in the Galactic center

Within a few hundred pc of the Galactic center, both poloidal and toroidal magnetic fields have been observed. The non-thermal radio filaments (Fig. 4) discovered within 1° from the Galactic center (Yusef-Zadeh *et al.* 1984, Lang *et al.* 1999, Yusef-Zadeh *et al.* 2004, Nord *et al.* 2004, LaRosa *et al.* 2004) indicate poloidal magnetic fields in the region within a few hundred parsecs of the center of the Galaxy. These filaments are highly polarized (Lang *et al.* 1999) and almost perpendicular to the Galactic plane, although some newly found examples are not so perpendicular to the Galactic plane (LaRosa *et al.* 2004). RM studies show that the magnetic field is aligned along the filaments (Lang *et al.* 1999). The filaments are probably illuminated flux tubes, with a field strength of about 1 mG (Morris & Serabyn 1996). LaRosa *et al.* (2005) detected diffuse radio emission of extent 400 pc and on this basis argued for a weak pervasive field of tens of μG in the central region. However, this is the volume-averaged field strength in such a large region. The poloidal fields are possibly limited to a smaller central region. The newly discovered "double helix" nebula (Morris *et al.* 2006), with an estimated field strength of order 100 μG, reinforces the presence of strong poloidal magnetic fields in tube format merging from the rotating circumnuclear gas disk near the Galactic Center.

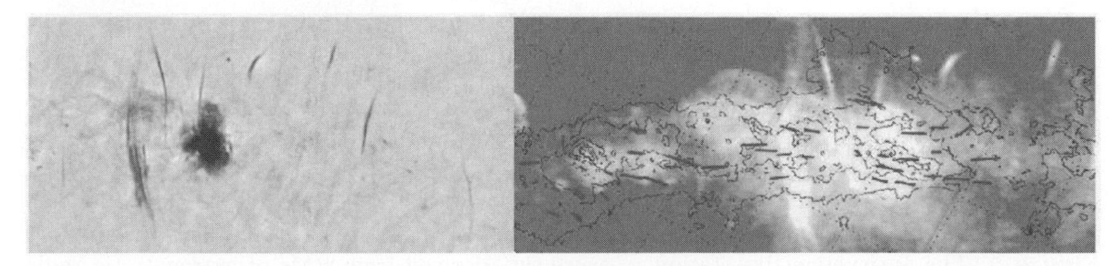

Figure 4. Non-thermal radio filaments discovered in the Galactic center (λ90 cm data from LaRosa *et al.* 2004) and the polarized thermal dust emission detected in the molecular cloud zone (bars in the right panel, see details in Novak *et al.* 2003).

Polarized thermal dust emission has been detected in the molecular cloud zone at sub-mm wavelength (see Fig. 4, Novak *et al.* 2003, Chuss *et al.* 2003), which is probably related to the toroidal fields parallel to the Galactic plane and complements the poloidal fields shown by the vertical filaments. The observed molecular cloud zone has a size of a few hundred pc, and is possibly a ring-like cloud outside the central region with poloidal field (see Fig.1 of Chandran 2001). The sub-mm polarization observations of the cloud zone offer information only about the field orientations. Zeeman splitting measurements of HI absorption against Sgr A (e.g. Plante *et al.* 1995) or of the OH maser in the Sgr A region (Yusef-Zadeh *et al.* 1999) give a line-of-sight field strength of a few mG in the clouds. It is possible that toroidal fields in the clouds are sheared from the poloidal fields, so that the RM distribution of radio sources in this very central region could be antisymmetric (Novak *et al.* 2003).

Outside the central region of a few hundred pc to a few kpc, the structure in the stellar and gas distributions and the magnetic structure are all mysterious. There probably is a bar. The large-scale magnetic fields should be closely related to the material structure but have not been revealed yet. The large positive RMs of background radio sources within $|l| < 6°$ of the Galactic Center (Roy *et al.* 2005) are probably related to magnetic fields following the bar (Roy *et al.* 2008). Comparison of the RMs of these background radio sources with RMs of foreground pulsars (see Fig. 1) should be helpful in delineating the field structure.

4. Magnetic fields in the Galactic halo

Magnetic field structure in the Galactic halo can be revealed from RMs of EGRs if allowance can be made for sources with outstanding intrinsic RMs. The foreground Galactic RM is the common contribution to the observed RMs of all EGRs within a small patch of sky. After "anomalous" RMs are eliminated, the pattern for the Galactic RM can be obtained. Han *et al.* (1997) discarded any source if its RM deviated from the average RM of neighbouring sources by more than 3σ, and obtained a "cleaned" RM sky.

A striking antisymmetry in the inner Galaxy with respect to Galactic coordinates (see Fig. 5) has been identified from the cleaned RM sky (Han *et al.* 1997, Han *et al.* 1999b). The antisymmetry must result from azimuthal magnetic fields in the Galactic halo with reversed field directions below and above the Galactic plane. Such a field can be produced by an 'A0' dynamo mode. The observed filaments near the Galactic Center should result from the dipole field in this dynamo scenario. The local vertical field component of

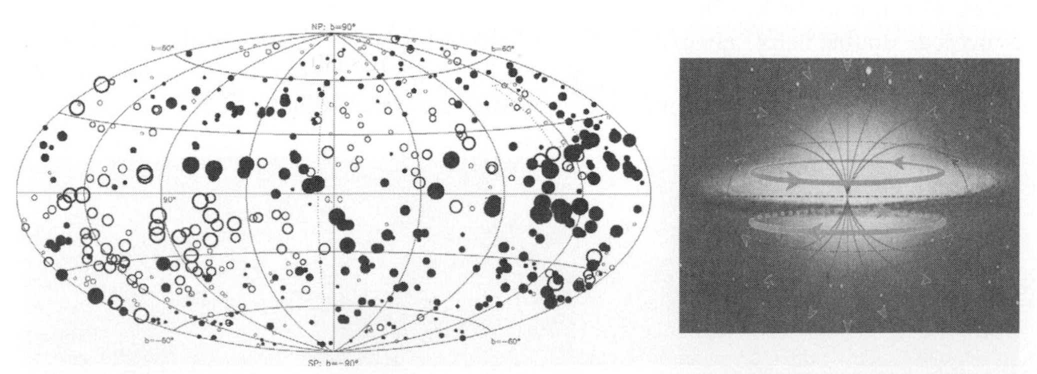

Figure 5. The antisymmetric rotation measure sky, derived from RMs of extragalactic radio sources after filtering out the outliers with anomalous RM values. The distribution corresponds to magnetic structure in the Galactic halo as illustrated on the right. See Han *et al.* (1997, 1999).

0.2 μG (Han & Qiao 1994, Han *et al.* 1999b) may be a part of this dipole field in the solar vicinity.

Han (2004) has shown that the RM amplitudes of extragalactic radio sources in the mid-latitudes of the inner Galaxy are systematically larger than those of pulsars, indicating that the antisymmetric magnetic fields are not local but are extended towards the Galactic center, far beyond the pulsars. Model-fitting by Sun et al. (2008) has confirmed this conclusion. The antisymmetry has been shown more clearly after adding newly observed RMs of more mid- or high latitude EGRs.

5. Concluding remarks

Only after the magnetic field, $\vec{B}(r, \theta, z)$, is known at all positions in our Galaxy will we have a complete picture of the magnetic structure of the Milky Way. With only a small number of measurements we must rely on fitting models to the data. "Partial" measurements in a greater number of regions, including quite large regions, are now available and we are able to "connect" the available measurements (pulsar RMs, maps of the RM sky) and outline some basic features of the Galactic magnetic field.

However, as seen from the above observational review, many regions of our Galaxy have not been observed very well for magnetic fields. For example, the farther half of the Galactic disk and the RM sky are not well known. Data in these regions are still scarce, either because of lack of field probes or limited capability of current instruments. In addition, when we try to get the large-scale field structure, the small-scale "random" fields, which are equally strong or even stronger, "interfere" with our measurements to a large extent. In the first Galactic quadrant, we obtained a large sample of pulsar RMs recently for the Galactic magnetic fields, but the spiral structure and pulsar distances are very uncertain. Statistical properties of fields on the energy-injection scales are not yet available, but are crucial for theoretical studies of magnetic field origin and maintenance. Magnetic fields in the disk-halo interface regions are certainly complex but few observations are available. To fully observe and understand the Galactic magnetic fields, there is a long way to go.

Acknowledgements

Thanks to Drs. Tom Landecker and A. Nelson for reading the manuscript. I am very grateful to colleagues who have collaborated with me for a long time on work on Galactic magnetic fields: Dr. R.N. Manchester from the Australia Telescope National Facility, CSIRO, Prof. G.J. Qiao from Peking University (China), Dr. Willem van Straten from Swinburne University (AU). The author is supported by the National Natural Science Foundation (NNSF) of China (10521001, 10773016 and 10833003) and the National Key Basic Research Science Foundation of China (2007CB815403).

References

Armstrong, J. W., Rickett, B. J., & Spangler, S. R. 1995, *ApJ* 443, 209
Balsara, D. & Kim, J. 2005, *ApJ* 634, 390
Battaner, E. & Florido, E., 2007, *AN* 328, 92
Boulares, A. & Cox, D. 1990, *ApJ* 365, 544
Brown, J. C., Taylor, A. R., & Jackel, B. J. 2003, *ApJS* 145, 213
Brown, J. C., Haverkorn, M., Gaensler, B. M., *et al.* 2007, *ApJ* 663, 258
Caswell, J. L. 2003, *MNRAS* 341, 551
Chandran, B. D. G. 2001, *ApJ* 562, 737
Chuss, D. T., Davidson, J. A., Dotson, J. L., & *et al.* 2003, *ApJ* 599, 1116
Clegg, A. W., Cordes, J. M., Simonetti, J. M., & Kulkarni, S. R. 1992, *ApJ* 386, 143

Cordes, J. M. & Lazio, T. J. W. 2002, preprint (arXiv:astro-ph/0207156)
Crutcher, R. M. 1999, *ApJ* 520, 706
Feinstein, C., Vergne, M. M., Martńez, R., & Orsatti, A. M., 2008, *MNRAS* 391, 447
Fish, V. L., Reid, M. J., Agron, A. L., & Menten, K. M., 2003, *ApJ* 596, 328
Gaensler, B. M., Dickey, J. M., McClure-Griffiths, N. M., & *et al.* 2001, *ApJ* 549, 959
Hamilton, P. A. & Lyne, A. G. 1987, *MNRAS* 224, 1073
Han, J. L., 2004, In: The Magnetized Interstellar Medium, eds: Uyaniker, B., *et al.*, GmbH, p. 3,
Han, J. L. & Qiao, G. J. 1994, *A&A* 288, 759
Han, J. L., Beck, R., Ehle, M., Haynes, R. F., & Wielebinski, R., 1999a, *A&A* 348, 405
Han, J. L., Ferriere, K., & Manchester, R. N. 2004, *ApJ* 610, 820
Han, J. L., Manchester, R. N., Berkhuijsen, E. M., & Beck, R. 1997, *A&A* 322, 98
Han, J. L., Manchester, R. N., & Qiao, G. J. 1999b, *MNRAS* 306, 371
Han, J. L., Manchester, R. N., Lyne, A. G., & Qiao, G. J. 2002, *ApJ* 570, L17
Han, J. L., Manchester, R. N., Lyne, A. G., Qiao, G. J., & van Straten, W. 2006, *ApJ* 642, 868.
Han, J. L., Zhang, J. S. 2007, *A&A* 464, 609
Haverkorn, M., Gaensler, B. M., Brown, J. C., *et al.* 2006, *ApJ* 637, L33
Heiles, C. 1996, *ApJ* 462, 316
Heiles, C., 2000, *AJ* 119, 923
Heiles, C, & Crutcher, R. 2005, In: Cosmic Magnetic Fields, LNP 664, 137
Hildebrand, R. H., Davidson, J. A., Dotson, J. L., *et al.*, 2000, *PASP* 112, 1215
Indrani, C. & Deshpande, A. A. 1999, *New Astronomy* 4, 33
Junkes, N., Fürst, E., & Reich, W. 1987, *A&AS* 69, 451
Lang, C. C., Morris, M., & Echevarria, L. 1999, *ApJ* 526, 727
LaRosa, T. N., Brogan C. L., Shore S. N., *et al.* 2005, *ApJ* 626, L23
LaRosa, T. N., Nord, M. E., Lazio, T. J. W., & Kassim, N. E. 2004, *ApJ* 607, 302
Li, H., Griffin, G. S., Krejny, M., & *et al.* 2006, *ApJ* 648, 340
Lyne, A. G. & Smith, F. G. 1989, *MNRAS* 237, 533
Manchester, R. N. 1974, *ApJ* 188, 637
Mathewson, D. S. & Ford, V. L. 1970a, *MemRAS* 74, 139.
Men, H., Ferrière, K., & Han, J. L. 2008, *A&A* 486, 819
Minter, A. H. & Spangler, S. R. 1996, *ApJ* 458, 194
Mitra, D., Wielebinski, R., Kramer, M., & Jessner, A. 2003, *A&A* 398, 993
Morris, M., Serabyn, 1996, *ARA&A* 34, 645
Morris, M., Uchida, K.,& Do, T. 2006, *Nature* 440, 308
Nord, M. E., Lazio, T. J. W., Kassim, N. E., *et al.* 2004, *AJ* 128, 1646
Novak, G., Chuss, D. T., Renbarger, T., & *et al.* 2003, *ApJ* 583, L83
Noutsos, A., Johnston, S., Kramer, M., & Karastergiou, A. 2008, *MNRAS* 386, 1881
Ohno, H., Shibata, S., 1993, *MNRAS* 262, 953
Page, L., *et al.*, 2007, *ApJS* 170, 335
Plante, R.. L., Lo, K. Y., & Crutcher, R.. M., 1995, *ApJ* 445, L113
Prouza, M. & Smída, R. 2003, *A&A* 410, 1
Qiao, G. J., Manchester, R. N., Lyne, A. G., & Gould, D. M. 1995, *MNRAS* 274, 572
Rand, R. J. & Kulkarni, S. R. 1989, *ApJ* 343, 760
Rand, R. J. & Lyne, A. G. 1994, *MNRAS* 268, 497
Reich, W. 2007, in: Cosmic Polarization, (astro-ph/0603465), p.91
Reid, M. J. & Silverstein, E. M., 1990, *ApJ* 361, 483
Roy, S., Rao, A. P., & Subrahmanyan, R. 2005, *MNRAS* 360, 1305
Roy, S., Rao, A. P., & Subrahmanyan, R. 2008, *A&A* 478, 435
Sun, X. H. & Han, J. L. 2004, in: Magnetized Interstellar Medium, eds: Uyaniker, B., *et al.*, GmbH, p.25
Sun, X.H., Han, J.L., Reich, W., & *et al.* 2007, *A&A* 463, 993
Sun, X. H., Reich, W., Waelkens, A., & Enßlin, T. A. 2008, *A&A* 477, 573
Thomson, R. C. & Nelson, A. H. 1980, *MNRAS* 191, 863
Tinyakov, P. G. & Tkachev, I. I. 2002, *Astroparticle Physics* 18, 165
Weisberg, J. M., Cordes, J. M., Kuan, *et al.* 2004, *ApJS* 150, 317
Yusef-Zadeh, F., Hewitt, J. W., & Cotton, W. A. 2004, *ApJS* 155, 421
Yusef-Zadeh, F., Morris, M., & Chance, D. 1984, *Nature* 310, 557
Yusef-Zadeh, F., Roberts, D. A., Goss, W. M., *et al.* 1999, *ApJ* 512, 230

Discussion

NELSON: Can we obtain any information on the field strength in the outer disk of our Galaxy from the extragalactic rotation measures, say from 20-30kpc radius?

HAN: Hard. From pulsar RMs, we can determine the field strength up to $R \sim 10$ kpc. For the outer Galaxy, the RMs of extragalactic radio sources have to be fitted with a model of B and a model of n_e. Given the many uncertainties in these models for the outer disk, it is really hard to determine the field strength. Nevertheless, from the determined variation of the field strength with the Galactocentric radius, you can extrapolate the field strength to the outer Galaxy.

DE GOUVEIA DAL PINO: Do you see any indication of an X-shaped magnetic structure similar to the one we see in other edge-on star-forming galaxies?

HAN: For this purpose, one needs to map the polarization in a large field-of-view around the Galactic center. At present, the available good maps with a field-of-view of 1 or 2 degrees were made with the VLA, and show the non-thermal filaments indicating the poloidal fields. It is impossible to restore the absolute level of U and Q in these maps. Therefore we can not map it now even if there may be such an X-shaped field structure. It is possible that the double helix nebula seen by Morris *et al.* (2006) and the filaments observed in the Galactic center region are somehow the foot or leg part of such a structure.

LANDECKER: I am concerned that the interpretation of RMs of pulsar and extragalactic radio sources is heavily dependent on the models of electron density distribution of Taylor & Cordes (1993) or Cordes & Lazio (2002). These models incorporate implicitly a spiral pattern for the Galaxy, going back to the work of Georgelin & Georgelin (1976) whose basic assumption is that there is a neat spiral pattern in the structure of the Milky Way. Please comment.

HAN: You are right on the electron density model. For interpretation for pulsar RMs, we work on the RM variations against not only the pulsar distances but also the observed DM. Considering that the relative positions of pulsars are more closely related to the DM rather than the electron density model, we believe that the magnetic structure derived from pulsar RMs and DMs are less model-dependent. For interpretation for RMs of extragalactic radio sources, yes, it is very heavily dependent on the electron density model and non "DM" to use.
From all observations available, evidence for spiral arms is so strong, therefore it is a correct idea to incorporate the spiral structure into any electron density model. Now, the spiral arms in the fourth Galactic quadrant have been better determined, while in the first quadrant, it is much less clear. Also, the electron density within a few kpc from the Galactic center is not clear at all. The electron density model and hence the pulsar distances therefore have a large uncertainty in these regions. Future electron density models will be probably improved there.

JinLin Han (left), and Tom Landecker

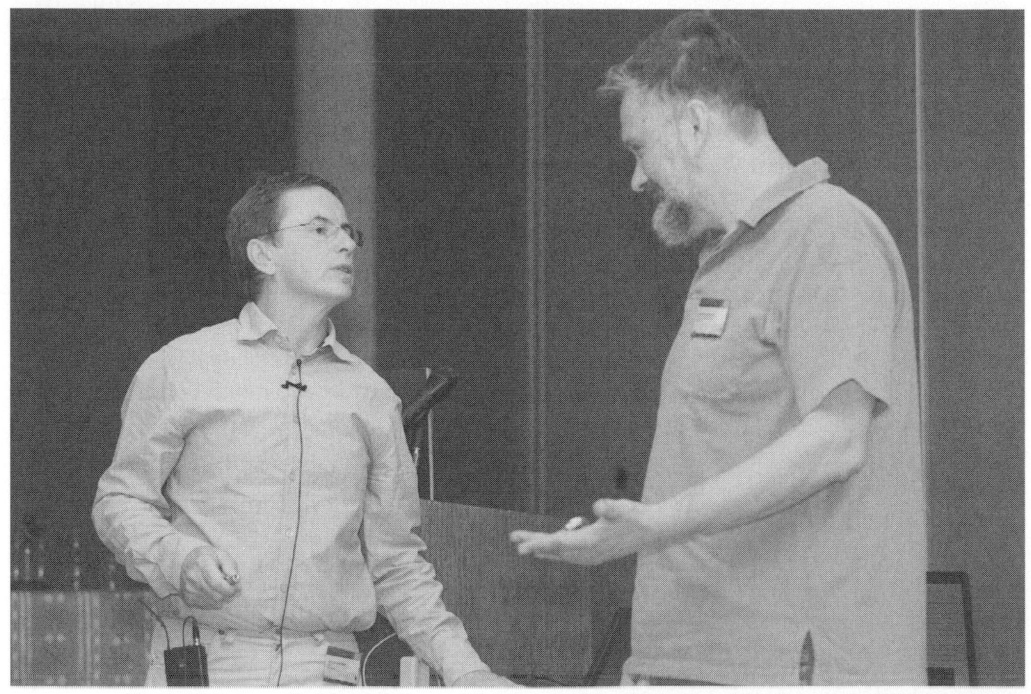

Detlef Elstner (left), arguing with the session chairman Dimitry Sokoloff

Cosmic Magnetic Fields:
From Planets, to Stars and Galaxies
Proceedings IAU Symposium No. 259, 2008
K.G. Strassmeier, A.G. Kosovichev & J.E. Beckman, eds.

doi:10.1017/S1743921309031135

Galactic dynamo simulations

Detlef Elstner, Oliver Gressel and Günther Rüdiger

Astrophysikalisches Institut Potsdam, An der Sternwarte 16, D-14482 Potsdam, Germany
email: elstner@aip.de

Abstract. Recent simulations of supernova-driven turbulence within the ISM support the existence of a large-scale dynamo. With a growth time of about two hundred million years, the dynamo is quite fast – in contradiction to many assertions in the literature. We here present details on the scaling of the dynamo effect within the simulations and discuss global mean-field models based on the adopted turbulence coefficients. The results are compared to global simulations of the magneto-rotational instability.

Keywords. ISM: magnetic fields – galaxies: magnetic fields

1. Introduction

Large-scale coherent magnetic fields are observed in many spiral galaxies (Beck *et al.* 1996; Beck 2009). It is now widely accepted that a turbulent dynamo process is responsible for the sustained amplification of the mean magnetic field within galaxies. The turbulence driven by supernova explosions together with the differential rotation should in fact lead to the action of the classical $\alpha\Omega$ dynamo (Parker 1971). First steps to globally model flat objects in an ellipsoidal geometry were done by Stix (1975); White (1978). Starting from the early nineties, first multidimensional numerical mean-field models assuming a disk geometry appeared (Donner & Brandenburg 1990; Elstner *et al.* 1992). These models applied simple estimates for the turbulent transport coefficients in the mean-field induction equation.

Rüdiger & Kitchatinov (1993), in the framework of second order correlation approximation (SOCA), derived the full α tensor and turbulent diffusivity for stratified turbulence. Fröhlich & Schultz (1996), under the assumption of hydrostatic equilibrium, computed the turbulent velocity distribution $u'(z)$ based on the observed density stratification and an observationally motivated gravitational potential from stars and dark matter. Applying SOCA theory, one could now specify the full α tensor, resulting in a model with the correlation time $\tau_{\rm c}$ of the turbulence as the only free parameter.

With the increased computer power of the nineties, Kaisig *et al.* (1993) and Ziegler (1996), in an alternative approach, computed the turbulence coefficients by averaging the results from simulations of a single supernova remnant. It, however, turned out that the values from isolated explosions were about one magnitude smaller then the estimated numbers from the previous considerations. Similar results, based on semi-analytical models, were obtained by Ferrière (1998). Extending the investigation to super-bubbles and including the stratification of the ISM, she finally found turbulence coefficients of expected magnitude. However, with a huge escape velocity, which could switch off the dynamo process. Ferrière and Schmitt (2000) used these results as input for an axisymmetric dynamo model of our Galaxy. They reached growth times of about 450 Myr but with a ten times larger toroidal field than the radial component, i.e., with a rather low pitch angle.

A lot of models, but mostly axisymmetric, were published during that period. Most of the models were investigating different aspects of the turbulence, looking also for parameters where non-axisymmetric solutions could appear. Another application was the effect of radial flows on the dynamo properties driven by the Maxwell stress of the dynamo generated magnetic field. Models including a wind were investigated, especially to explain vertical fields above the disk – notably Brandenburg *et al.* (1993); Elstner *et al.* (1995). Both models used different assumptions about the halo diffusivity and the wind structure. In the model of Brandenburg *et al.* (1993), diamagnetism was included, and because of the positive turbulence gradient, the pumping was not an escape velocity but an inward transport term. In Elstner *et al.* (1995), a passive halo with low conductivity was assumed. These models already showed a fast growth time below 0.1 Gyr. All these models could in principle explain the main features of the observed magnetic fields in nearby spiral galaxies. But the connection between optical and magnetic spirals was still unclear.

With the advent of the new century, full-blown 3D models became feasible, and one could now attack the spiral structure. Simple artificial velocity models adopted from density-wave theory were now taken into account. One spectacular observational result was the anti-correlation of optical and magnetic arms, found by Beck & Hoernes (1996). A turbulent dynamo could reproduce such a behaviour due to the different properties of the turbulence in the spiral arms and in the inter-arm regions – see Rohde & Elstner (1998), Shukurov (1998) and Rohde *et al.* (1999).

More elaborated models used the velocity from numerical galaxy models computed with SPH or sticky-particle methods (Elstner *et al.* 2000). Models for spiral galaxies produced no inter-arm fields by the large-scale spiral motion alone. Only with a variation of the turbulence, one could have inter-arm fields. The situation was different for bar galaxies (Otmianowska-Mazur *et al.* 2002). In these simulations, one could observe inter-arm magnetic fields in the outer spiral arm, and the field lagged behind the optical arm (Moss *et al.* 1998, cf.). More recently, Beck *et al.* (2005) and Moss *et al.* (2007) conducted models for barred galaxies with stationary flows.

Another problem was raised in the context of the so-called catastrophic quenching: Because of helicity conservation, in the ideal MHD case, the α term will be quenched (with the square-root of the magnetic Reynolds number) if there is no additional transport of magnetic helicity in the box. Under the assumption of a galactic wind or fountain flow, these problems may disappear. But a too strong wind would also result in a suppression of the dynamo (Sur *et al.* 2007). Recent investigations have shown that already the galactic shear flow can suppress the catastrophic quenching of the dynamo (Brandenburg 2009; Käpylä *et al.* 2008).

The recent box model of the turbulent ISM driven by multiple clustered supernova explosions (Gressel *et al.* 2008), for the first time demonstrated the presence of a dynamo process by direct numerical simulations. It turned out that the turbulence coefficients derived from the SOCA models are in good agreement with the new results from the direct simulations. In a similar way, the box simulations of Hanasz *et al.* (2004) have shown an exponential growth of the magnetic field. Their simulations, however, do not resolve the supernova remnant in itself. Instead, the explosion site only serves as a source for the released cosmic ray energy. Due to the fast diffusion of cosmic rays, the pressure redistribution is fast enough to allow for similarly short growth times of several 100 Myr. The role of a magnetic instability of the Parker type could be essential for this process, too.

In general, the effect of magnetic instabilities in releasing kinetic energy should be considered as a possible mechanism for magnetic field amplification in galaxies – especially in regions with low star formation activity. The primary candidate of this type is the magneto-rotational instability (MRI), which has a growth time on the order of the galactic rotation frequency and thus can become efficient on very short timescales. With todays computational resources, global models of the MRI are already within reach (Dziourkevitch *et al.* 2004; Nishikori *et al.* 2006). However, these models still have to assume quite strong initial fields to properly resolve the fastest growing modes of the instability.

2. The pitch angle problem

When it comes to the explanation of the observed magnetic fields in spiral galaxies, there is one main reason for favouring the dynamo hypothesis over the primordial field hypothesis, namely the existence of fields – of equipartition strength – with a large pitch angle. If the field amplification were mainly due to the shearing motion of the differential rotation, the field would have to be wound up several times to explain its observed strength. This, however, would lead to an amplification of the toroidal field only, resulting in a configuration with an intrinsically small pitch angle.

For the dynamo scenario, which furthermore implies exponential growth, we additionally have an amplification of the poloidal field. The pitch angle p of the growing field in the kinematic regime of the $\alpha\Omega$ dynamo can be estimated from the dimensionless dynamo numbers C_α and C_Ω (for a definition see below) via

$$p = \arctan\left(\sqrt{\frac{C_\alpha}{C_\Omega}}\right). \tag{2.1}$$

This leads to values between 5° and 30°, for reasonable values of $C_\alpha = 5$–10 and $C_\Omega = 25$–800. For a given rotational velocity, the angular frequency is defined by the turnover radius, i.e., the point where the rotation curve becomes flat. Small turnover radii lead to rather large C_Ω and thus to low pitch angles. In the saturated regime, the situation becomes worse – at least for the conventional $\alpha\Omega$ dynamo. This is because α will then be reduced by non-linear effects and the pitch angle fades.

Direct simulations of the ISM turbulence show a possible way out of this dilemma, namely by another saturation process, which we will demonstrate in the following section. Moreover, the action of the magneto-rotational instability may be a reasonable candidate for the explanation of large pitch angles as they are observed in maps of polarised synchrotron emission for instance in the ringed galaxy NGC 4736 (Chyży & Buta 2008).

3. Global models with turbulence coefficients from direct simulations

Global simulations of interstellar turbulence remain out of reach, at least if one is interested in resolving the relevant scales. Therefore, we apply a hybrid approach, where we consider the mean induction equation

$$\frac{\partial \vec{B}}{\partial t} = \mathrm{curl}(\vec{u} \times \vec{B} + \alpha \circ \vec{B} - \eta_\mathrm{T} \circ \mathrm{curl}\vec{B}) \tag{3.1}$$

for the mean magnetic field $\vec{B}$ and apply closure parameters obtained from direct simulations. The background rotation serves as the main part of the mean flow

$$u_\varphi = r\Omega_0 \left(1 + \left(\frac{r}{r_\Omega}\right)^2\right)^{-1/2}, \tag{3.2}$$

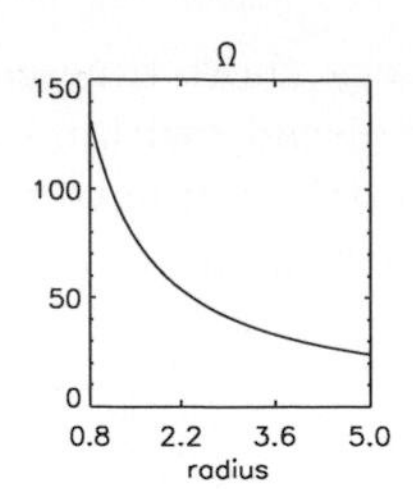

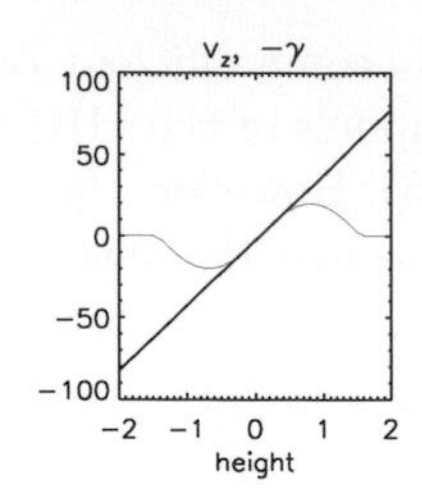

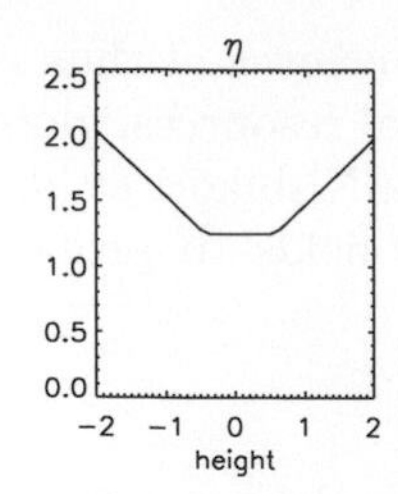

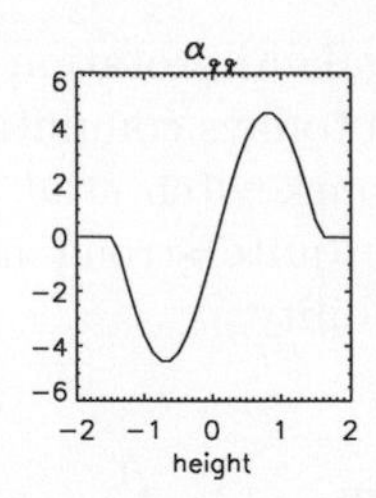

Figure 1. Radial dependence of Ω and vertical profiles of the fountain flow and turbulent transport coefficients.

and is supplemented by a vertical wind $u_z = u_0 z$ which is a result of the density-stratified turbulence driven by SNe (cf. Fig. 2 in Gressel *et al.* 2009). The test-field method, furthermore, yields an α tensor of the form

$$\alpha = \alpha_0 \begin{pmatrix} \alpha_{rr} & -\gamma_z & 0 \\ \gamma_z & \alpha_{\varphi\varphi} & 0 \\ 0 & 0 & \alpha_{zz} \end{pmatrix}, \tag{3.3}$$

where the dependence of these quantities on the height z is modelled according to the profiles in Gressel *et al.* (2008).

For a realistic global galaxy model one further needs the radial dependence of the relevant effects. The radial profiles of the rotational frequency, SN rate, gravitational potential and density are in principle known. By means of a parameter study, we try to obtain the α effect from our box simulations, accordingly. From these, we find a dependence of the diagonal terms of the α tensor on rotation, namely $\alpha_{\varphi\varphi} \propto \Omega^{0.5}$. This is weaker than expected from SOCA theory, which predicts a scaling linear with Ω and only a minor rotational quenching. Consistent with SOCA, we find the turbulent diamagnetism γ_z to be nearly independent of Ω. For the turbulent diffusivity $\eta_{\rm t}$, we apply only horizontal components which also vary according to $\Omega^{0.5}$. The dependence on the supernova rate is given in Gressel *et al.* (2009).

After proper normalisation with a length of $l = 1\,{\rm kpc}$ and a diffusion time $t_{\rm D} = l^2/\eta_0$ with $\eta_0 = 1\,{\rm kpc}^2\,{\rm Gyr}^{-1}$, the dynamo is characterised by the Reynolds numbers

$$C_\alpha = \alpha_0 l/\eta_0\,, \quad C_\Omega = \Omega_0 l^2/\eta_0 \quad \text{and} \quad C_w = u_0 l/\eta_0\,. \tag{3.4}$$

We assume a disk of radius $R = 5\,{\rm kpc}$ and with a height of $H = 2\,{\rm kpc}$, according to our box dimension in the direct simulations. We approximate the profiles of the turbulence coefficients by smooth functions (see Fig. 1). The vertical dependence of the α tensor can simply be reproduced by a sine function

$$\alpha(z) \propto \sin(z\pi/{\rm h}_\alpha) \tag{3.5}$$

with a scale height $h_\alpha = 1.5\,{\rm kpc}$. The mean vertical velocity is linearly growing with height. Only the diagonal terms of the α tensor have a radial dependence proportional to Ω, whereas the other terms are constant. We, furthermore, use a simple α quenching

$$\alpha = \alpha_0/(1 + B^2/B_{\rm eq}^2) \tag{3.6}$$

for the full tensor, i.e., including the diamagnetic term γ_z.

In our first run, we neglect turbulent diamagnetism and mean vertical velocity, as it is done in the classical $\alpha\Omega$ dynamo. In this model, we find a growth time of 700 Myr, a field strength for the toroidal component $B_\varphi = 20 B_{\rm eq}$ and the radial component $B_r = 0.6 B_{\rm eq}$. If we neglect only the fountain flow, the field decays. For the *real* model, we find a

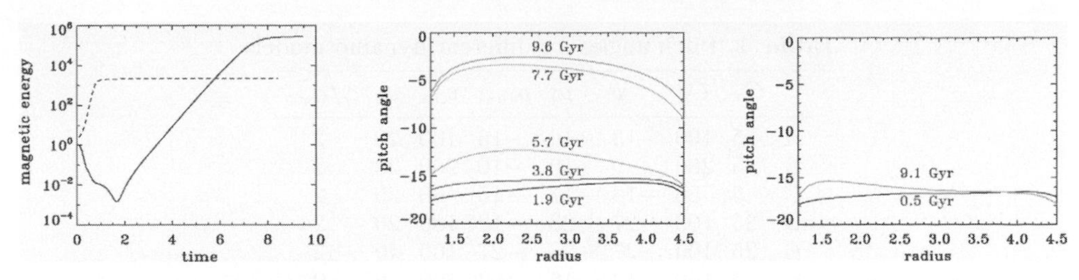

Figure 2. Left: Time evolution of magnetic energy for model A, with wind (dashed) and model F, without wind and pumping (solid). Middle: Time evolution from 1.9 Gyr until 9.6 Gyr of the radial pitch angle distribution for model F. Right: Pitch angle for model A at 0.5 Gyr and 9.1 Gyr.

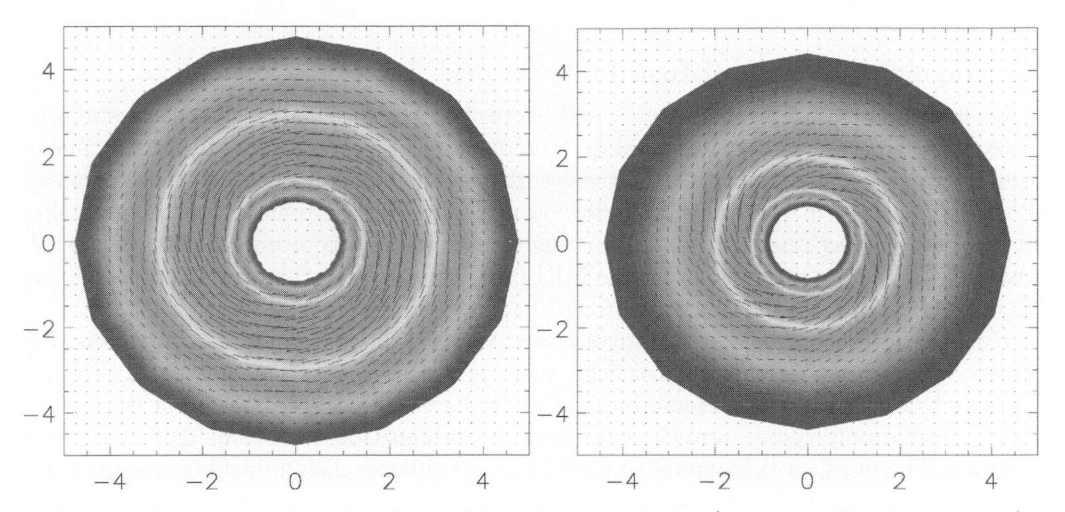

Figure 3. Magnetic vectors as observed in polarised emission (neglecting Faraday rotation) for the final fields in model F (left) and model A (right)

growth rate of 250 Myr – similar to the box simulation. The field now saturates earlier with $B_\varphi = 1.1 B_{\rm eq}$ and $B_r = 0.3 B_{\rm eq}$.

The final state is not determined by the usual α quenching in this case. Instead, a combination of fountain flow and quenching of the diamagnetic term leads to the saturation of the dynamo. This means, the dynamo stops growing because of field losses by the fountain flow. As a nice consequence of this non-linear feedback, the pitch angle remains large in the saturated regime (see Fig. 2).

Moreover, in Fig. 3 we present directions of the magnetic field resulting from polarised emission of synchrotron radiation, where we assumed a scale height of 500 pc for the relativistic electron distribution.

In Table 1, the pitch angle is given for a choice of different models. Comparing $p = \arctan(\sqrt{C_\alpha / C_\Omega})$ with the initial pitch angle p_0 during the kinematic growth phase, we find reasonable agreement. The final value $p_{\rm final}$ in the saturated state for the first three models is similar to the value during the kinematic growth phase.

In order to have a higher pitch angle for our rotation law with $r_\Omega = 1$ kpc, we need larger values of C_α and, consequently, a larger vertical outflow. Otherwise the α quenching would reduce the final pitch angle – as is the case in model E. This increase of the mean outflow velocity with increasing SN rate is indeed observed in the box simulations

Table 1. Pitch angles for different dynamo models

	C_α	C_Ω	p	p_0	p_{final}	v_{rot}	u_0	$\gamma/\alpha_{\varphi\varphi}$
A	5	100	−13	−16	−16	100	20	2
B	5	200	−9	−10	−10	200	20	2
C	5	50	−18	−17	−15	50	20	2
D	25	100	−27	−28	−18	100	20	2
E	25	100	−27	−28	−27	100	40	2
F	5	100	−13	−16	−3	100	0	0

(cf. Gressel *et al.* 2009). For model F, one can see a dramatic reduction of the pitch angle due to the missing mean vertical velocity and diamagnetic pumping.

4. Magneto-Rotational Instability

Galaxies are known to be unstable with respect to the magneto-rotational instability. This instability, discovered by Velikhov (1959) and Chandrasekhar (1960), has meanwhile been studied in great detail in the astrophysical context (see e.g. Balbus & Hawley 1991; Kitchatinov & Mazur 1997). Sellwood & Balbus (1999) pointed to the possibility of turbulence generation by the MRI in regions of low star formation activity. Piontek & Ostriker (2007) recently simulated the MRI in a vertically stratified two-phase ISM box model.

Kitchatinov & Rüdiger (2004) investigated the stability of global galactic disks. For a disk threaded by a weak vertical field, they estimated a field strength of 10^{-25} G for the onset of the instability. Accordingly, the MRI should already be present during the galaxy formation process – and therefore a good candidate for the seed field of a turbulent dynamo. The most unstable mode in their idealised galactic disk model is of quadrupolar type. A very interesting result is the large pitch angle of 45° for the fastest growing mode. For weak fields, the unstable modes have a small spatial scale but with the property of a uniform large pitch angle. This leads to a smooth map of magnetic vectors for the polarised radio emission, but the Faraday rotation measure will be small for these types of magnetic fields.

The instability, moreover, has a very short growth time proportional to the rotation time. A proof for a large scale dynamo driven by the MRI in the sense, that a small initial field can be amplified over several magnitudes has not been given yet. This is because it is numerical difficult to resolve the small spatial scales for the unstable modes connected to the weak initial field. Nevertheless, there are hints that the process is possible. Especially the turbulence excited by MRI in a stratified disk gives a non-vanishing α effect, but with negative sign in the northern hemisphere (Arlt & Rüdiger 2001; Brandenburg & Sokoloff 2002).

Since we are interested in configurations with large pitch angles, we here consider global models of a density-stratified galactic disk with a uniform vertical field. Tests with toroidal fields did not produce a comparable radial field component. For simplicity, we assumed an isothermal disk where the rotation and vertical density stratification is supported by the gravitational potential. Without magnetic fields the configuration was tested to be in a stationary equilibrium over 10 Gyr with only small velocity fluctuations below 0.01 km s^{-1}, which are slowly decaying with time.

The applied density profile is

$$\rho = \rho_0 \exp\left(\frac{-4z^2}{z^2 + 3h_\rho^2}\right) \tag{4.1}$$

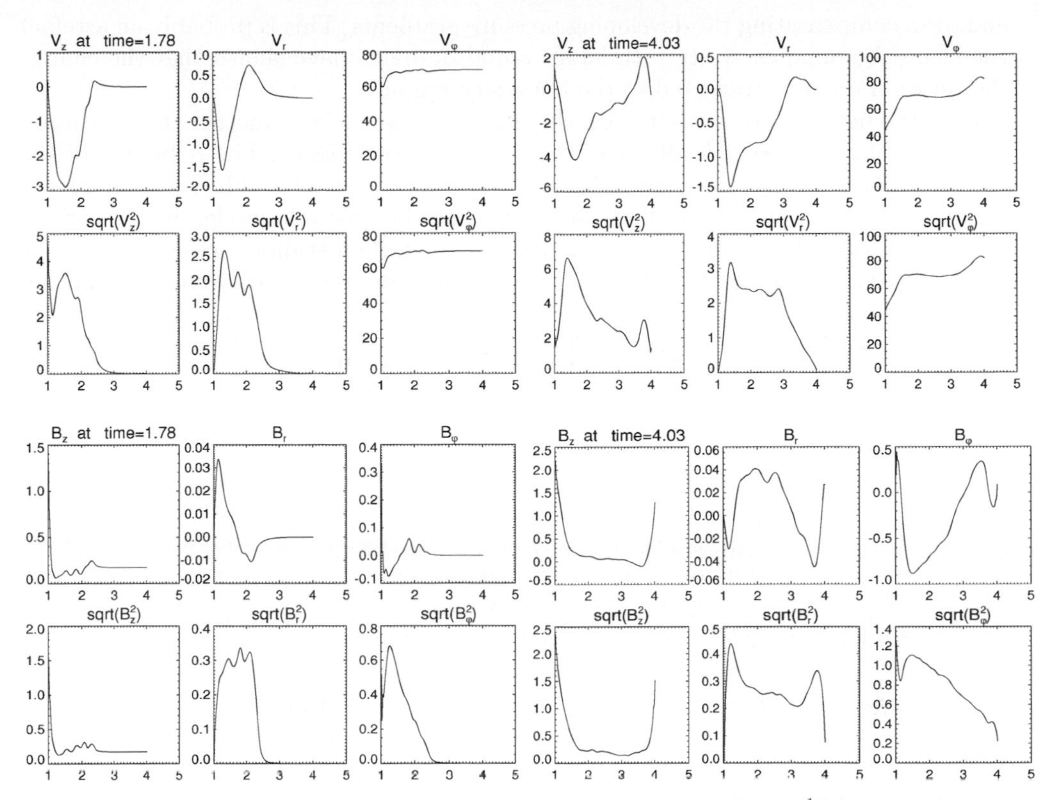

Figure 4. Radial profiles of φ, z averaged velocity components in $[\,\mathrm{km\,s^{-1}}\,]$ (first row), rms-velocity components in $[\,\mathrm{km\,s^{-1}}\,]$ (second row), φ, z-averaged magnetic field components in $[\,\mu\mathrm{G}\,]$ (third row) and rms magnetic field components in $[\,\mu\mathrm{G}\,]$ (fourth row) at 1.8 Gyr.

with the scale height $h_\rho = 0.3\,\mathrm{kpc}$ and $\rho_0 = 10^{-24}\mathrm{gcm^{-3}}$. Note that the density minimum is about $0.02\ \rho_0$. The rotation curve is, furthermore, defined by (3.2) with the turnover radius $r_\Omega = 0.35\,\mathrm{kpc}$ and $\Omega_0 = 200\,\mathrm{Gyr^{-1}}$. The inner radius of our disk model is $r_{\mathrm{in}} = 1\,\mathrm{kpc}$ and the outer radius $r_{\mathrm{out}} = 4\,\mathrm{kpc}$. The vertical extent is $\pm 5\,\mathrm{kpc}$. Within $\pm 1\,\mathrm{kpc}$, we use a constant grid spacing with 256 grid points and a non-uniform grid for the outer regions with 128 points at each side. This means, the inner disk has a resolution of about 7 pc and for the outer disk we yield about 80 pc. This is consistent with a density variation of a factor 100, in order to get a comparable growth of the unstable modes for a uniform magnetic field strength within our restricted resolution.

The radial boundary is reflecting and the vertical boundary is reflecting for the velocity only, while the tangential component of the magnetic field and the gradient of the vertical component are set to zero. Under these conditions, the vertical flux is conserved but a development of horizontal flux is allowed. Applying a weak and uniform vertical magnetic field of $0.1\,\mu\mathrm{G}$, we find an exponential growth of the unstable mode after a transient relaxation phase of about 200 Myr. The field starts to grow at the inner radial boundary, where Ω has a maximum. As the field grows further, it spreads out radially. At the same time, the Maxwell stress causes an inflow in the inner part of the magnetised region. At the outer magnetic region, we observe an outflow of mass, which is a consequence of angular momentum conservation.

Because of our closed boundaries, the density increases strongly at the inner boundary and moderately at the outer boundary. We see a weak change of the rotation curve at the boundaries, compensating the developing pressure gradients. This is probably an artefact of the closed boundaries. Tests with larger radial domains have shown that the change of the rotation curve is restricted to the boundary region.

Finally, the field growth saturates. At the moment it is unclear whether the saturation is caused by the non-linear feedback of the developed turbulence. The reduction of the pitch angle could in fact be a hint that the system reaches the diffusive limit of the instability, where we know from linear analysis that the unstable mode, in this regime, is mainly toroidal. On the other hand, we see a large-scale meridional circulation in the disk. Moreover, our model disk has a limited mass reservoir, which could be another reason for the early saturation with rather low magnetic field strength. As long as the toroidal and radial field strength are below the field strength of the initial vertical field, the field parallel to the disk grows with a time of 0.05 Gyr; later we observe growth times of 0.8 Gyr until the saturation sets in. It should be noted that the growth-time of a possible MRI-driven dynamo has not to be as fast as the growth time of the instability.

In Fig. 4, we plot radial profiles which have been averaged along the vertical and azimuthal direction for $-5\,\mathrm{kpc} < \mathrm{z} < 0\,\mathrm{kpc}$ and $0 < \varphi < 2\pi$. The values shown are the components of the velocity, magnetic field and their corresponding rms values. A large-scale meridional circulation develops with an amplitude of roughly 0.5 of the rms value ($\simeq 6\,\mathrm{km\,s^{-1}}$ for the vertical component and $\simeq 3\,\mathrm{km\,s^{-1}}$ for the radial component).

The mean magnetic field along the z and φ direction are of the same order as the rms values of the fields. Only the mean radial field is by an order of magnitude smaller than the rms field. We find $B_r^{\mathrm{rms}}/B_\varphi^{\mathrm{rms}} = 0.3$ and for the mean-field $B_r/B_\varphi = 0.05 - 0.1$, corresponding to pitch angles of 17° and 3° − 6° , respectively. In contrast, the pitch angle in the polarisation map varies from 10° to 20° if the boundaries are excluded ($1.3\,\mathrm{kpc} < \mathrm{r} < 3.6\,\mathrm{kpc}$).

Initially, at time 1.8 Gyr, the pitch angle varies from 20° to 40° for $1.3\,\mathrm{kpc} < \mathrm{r} < 2\,\mathrm{kpc}$. In the outer region the field is still too weak, and we do not observe any polarised emission. Fig. 5 shows a slice through the magnetic field at $z = -0.8\,\mathrm{kpc}$, the vertically averaged field along the lower disk half ($-5\,\mathrm{kpc} < \mathrm{z} < 0\,\mathrm{kpc}$) and the magnetic direction of polarised emission (neglecting Faraday rotation). Varying the scale height height of the relativistic electron distribution from 0.5 kpc to 2.0 kpc gave no significant differences. Note the smooth polarisation map for an incoherent magnetic field (i.e., with many reversals along the vertical direction). The pitch angle is here much larger than the pitch angle for the averaged field.

5. Summary

New direct simulations of supernova-driven turbulence in a Cartesian box have shown the existence of the turbulent dynamo in a galactic environment. We find the resulting balance of the turbulent diamagnetism with the mean vertical outflow to be crucial for the operation of the dynamo. Using the turbulence of the direct simulations for the determination of the transport coefficients, one can show that the fast dynamo growth time of a few hundred million years is due to the fast transport by the mean vertical fountain flow and the downward pumping of the turbulent diamagnetism. This is in contrast to the slow conventional $\alpha\Omega$ dynamo with growth times of the order of 1 Gyr, where only diffusion contributes to a vertical field transport.

We have shown the possibility of an alternative saturation process other than the simple quenching of the diagonal α term. In the case of quenching via the transport mechanisms, the pitch angle of the final saturated field will not deplete. But still, there

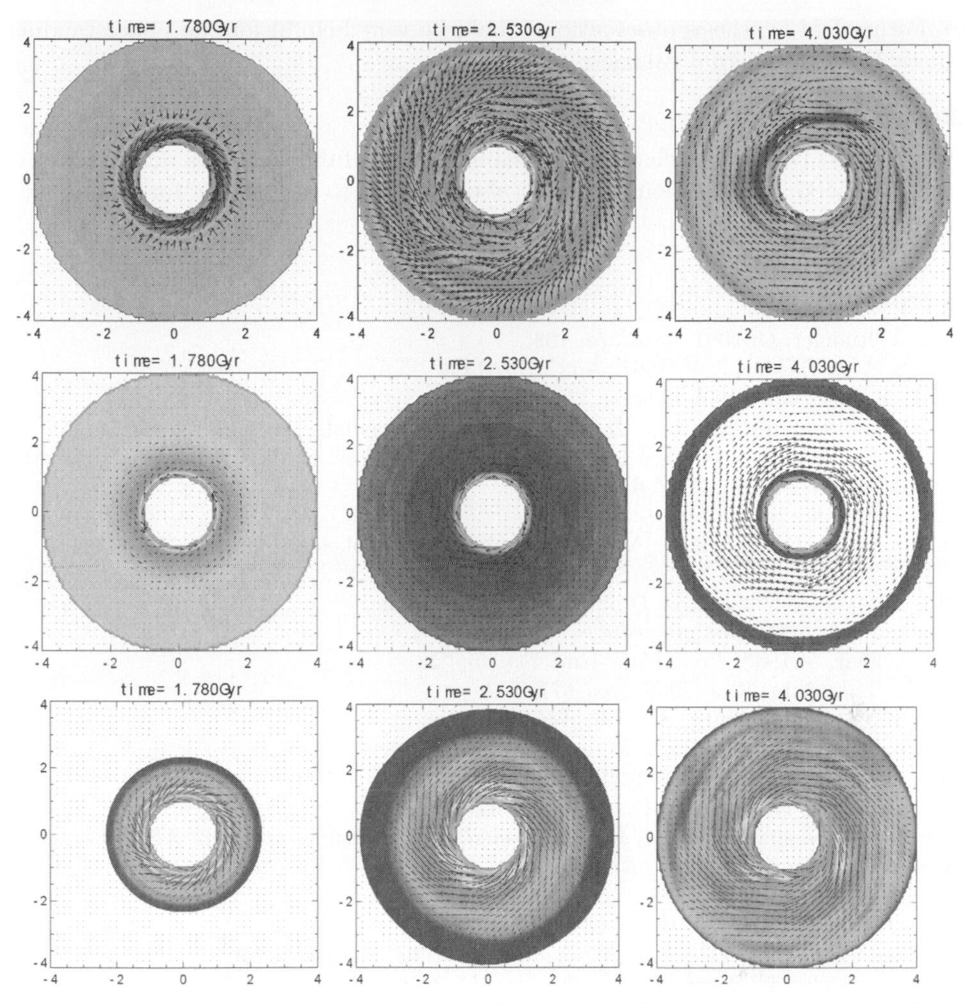

Figure 5. Magnetic field slices at $z = -0.8\,\mathrm{kpc}$ (first row) and averaged field over one disk half (second row) with colour coded z component. Magnetic field direction from polarised emission with colour coded total intensity (third row).

are cases of very strong differential rotation where it may be difficult to get observed pitch angles of up to 40° with actual SN-driven dynamo models.

More detailed investigations of the ISM model must be performed. For instance, cosmic rays could further improve the situation. Also the effects of strong magnetic fields have to be considered, as magnetic instabilities could contribute to further amplification processes. The magneto-rotational instability would be a good candidate – not only as a generator for the seed field, but also for the generation of magnetic fields with large pitch angles. It is not clear, however, whether an MRI driven dynamo can generate enough vertical flux through the disk in order to get solutions with large enough pitch angles, which are known from the linear unstable modes. But at least under a given vertical flux, the MRI leads to incoherent magnetic fields, which appear in the polarised synchrotron emission (neglecting Faraday rotation) as smooth spiral fields. The mean-field generated in such models has a smaller pitch angle than the polarised emission vectors, so a significant contribution from a small scale field enhances the pitch angle in the polarisation map. Future RM synthesis observations would be very helpful for the understanding of magnetic fields in external galaxies.

Acknowledgements

We thank N. Dziourkevitch and A. Bonanno for helpful discussions and support concerning the MRI computations. We are also grateful to U. Ziegler for supplying the NIRVANA code.

References

Arlt, R. & Rüdiger, G. 2001 *A&A* 374, 1035
Balbus, S. A. & Hawley, J. F. 1991, *ApJ* 376, 214
Beck, R. & Hoernes, P. 1996, *Nature* 379, 47
Beck, R., Brandenburg, A., Moss, D., Shukurov, A., & Sokoloff, D. 1996, *ARAA* 34, 155
Beck, R., Fletcher, A., Shukurov, A., Snodin, A., Sokoloff, D. D., Ehle, M., Moss, D., & Shoutenkov, V. 2005, *A&A* 444, 739
Beck, R. 2009, *this proceedings*
Brandenburg, A., Donner, K. J., Moss, D., Shukurov, A., Sokoloff, D. D., & Tuominen, I. 1993, *A&A* 271, 36.
Brandenburg, A. & Sokoloff, D. D. 2002, *GApFD* 96, 319
Brandenburg, A. 2009, *this proceedings*
Chandrasekhar, S. 1960, *Proc. Nat. Acad. Sci.* 46, 253
Chyży, K. T. & Buta, R. J. 2008, *ApJ* 677, L17
Donner, K.J.& Brandenburg, A. 1990, *A&A* 240, 289
Dziourkevitch, N., Elstner, D., & Rüdiger, G. 2004, *A&A* 423, L29
Elstner, D., Meinel, R., & Beck, R. 1992, *A&AS* 94, 587
Elstner, D., Golla, G., Rüdiger, G., & Wielebinski, R. 1995, *A&A* 297, 77
Elstner, D., Otmianowska-Mazur, K., von Linden, S., & Urbanik, M. 2000, *A&A* 357, 129
Ferrière, K. & Schmitt, D. 2000, *A&A* 358, 125
Ferrière, K. 1998, *A&A* 335, 488
Fröhlich, H.E. & Schultz, M. 1996, *A&A* 311, 451
Gressel, O., Elstner, D., Ziegler, U., & Rüdiger, G. 2008, *A&A* 486, L35
Gressel, O., Ziegler, U., Elstner, D., & Rüdiger, G. 2009, *this proceedings*
Hanasz, M., Kowal, G., Otmianowska-Mazur, & K. Lesch, H. 2004, *ApJ* 605, L33
Kaisig, M., Rüdiger, G., & Yorke, H. W. 1993, *A&A* 274, 757
Käpylä, P. J., Korpi, M. J., & Brandenburg, A. 2008, *A&A* 491, 353
Kitchatinov, L.L. & Mazur, M.V. 1997, *A&A* 324, 821
Kitchatinov, L.L. & Rüdiger, G. 2004, *A&A* 424, 565
Moss, D., Korpi, M., Rautiainen, P., & Salo H. 1998, *A&A* 329, 895
Moss, D., Snodin, A., Englmaier, P., Shukurov, A., Beck, R., & Sokoloff, D. 2007, *A&A* 465, 157
Nishikori, H., Machida, M., & Matsumoto, R. 2006, *ApJ* 641, 862
Otmianowska-Mazur, K., Elstner, D., Soida, M., & Urbanik, M. 2002, *A&A* 384, 48
Parker, E. N. 1971, *ApJ* 164, 491
Piontek, R A. & Ostriker, E. C. 2007, *ApJ* 663, 183
Rohde, R. & Elstner, D. 1998, *A&A* 333, 27
Rohde, R., Beck, R., & Elstner, D. 1999, *A&A* 350, 423
Rüdiger, G. & Kitchatinov, L.L. 1993, *A&A* 269, 581
Sellwood, J.A. & Balbus, S.A. 1999, *ApJ* 511, 660
Shukurov, A. 1998, *MNRAS* 299, L21
Stix, M. 1975, *A&A* 42, 85-89.
Sur, S., Shukurov, A., & Subramanian, K. 2007, *MNRAS* 377, 874.
Velikhov, E.P. 1959, *Sov. Phys. JETP* 9, 995
White, M. 1978, *AN* 299, 209
Ziegler, U. 1996, *A&A* 313, 448

Discussion

DE GOUVEIA DAL PINO: We have recently published results of global 3D simulations of galactic fountain evolution and we have found that for the typical SN rates of our galaxy, the fountain flows reach a maximum rate of $\approx$4 kpc and they attain a "steady" state (with the amount of matter going up becoming equal to the amount going down and returning to the disk) in about 150 Myr (Melioli *et al.* 2008, MNRAS). So my question is, how do the dynamo parameters depend on these properties, i.e. the SN rate and height of the fountain flow?

ELSTNER: We investigated the dependence on the SN-rate for smaller value than the galactic value and found a square-root dependence. The relaxation time for the system to develop a stationary state is between 100-200 Myr.

FLETCHER: Have you compared your mean-field models to the recently published models developed by Shukurov, Subramanian, Sur and Brandenburg involving the outward transport of small-scale magnetic helicity by a galactic wind?

ELSTNER: Yes we compared that. But due to the turbulent diamagnetism, we have a broader dynamo regime.

REICH: Do you think you can simulate the galactic magnetic field in terms of your simulations, or are there features left not fitting well?

ELSTNER: In order to get the galactic magnetic field, you probably have to include some evolution information like detailed star formation mass. Also initial condition and the environment field may be necessary. The main question remains, why our galaxy with it's reversals should be exceptional.

OTMIANOWSKA-MAZUR: Do you get the α and η coefficients from EMF? Do you think that your process is connected with the Parker instability?

ELSTNER: The turbulence coefficients are determined by the test-field method. The process is probably not connected with the Parker instability because the turbulence parameters are founded already in the hydro regime

SOKOLOFF: It is very pity that because of shortage of time the speaker omitted to remark that the galactic dynamo models suggested in 1990th by Shukurov *et al.*, Brandenburg *et al.*, Moss et al., Beck *et al.* were much more successful in reporting observations than the model by Ferrière and they resulted in something very similar to what was presented in this impressive detailed simulations.

ELSTNER: Yes I did not mention all models based on SOCA estimates for the mean field simulations in the 90th. They were already successful, because SOCA is a good prediction, which we now learn from the direct simulations.

Michal Hanasz

Michael Kramer

Cosmic Magnetic Fields:
From Planets, to Stars and Galaxies
Proceedings IAU Symposium No. 259, 2008
K.G. Strassmeier, A.G. Kosovichev & J.E. Beckman, eds.

doi:10.1017/S1743921309031147

Cosmic-ray driven dynamo in galactic disks

Michał Hanasz[1], K. Otmianowska-Mazur[2], H. Lesch[3], G. Kowal[2,4], M. Soida[2], D. Wóltański[1] K. Kowalik[1], R. K. Pawłaszek[1] and B. Kulesza-Żydzik[2]

[1]Centre for Astronomy, Nicholas Copernicus University, PL-87148 Piwnice/Toruń, Poland,
mhanasz@astri.uni.torun.pl

[2]Astronomical Observatory, Jagiellonian University, ul. Orla 171, 30-244 Kraków

[3]Astronomical Observatory, Munich University, Scheinerstr. 1, D-81679, Germany

[4]Department of Physics and Astronomy, McMaster University, 1280 Main St. W., Hamilton, ON L8S 4M1, Canada.

Abstract. We present new developments on the Cosmic–Ray driven, galactic dynamo, modeled by means of direct, resistive CR–MHD simulations, performed with ZEUS and PIERNIK codes. The dynamo action, leading to the amplification of large–scale galactic magnetic fields on galactic rotation timescales, appears as a result of galactic differential rotation, buoyancy of the cosmic ray component and resistive dissipation of small–scale turbulent magnetic fields. Our new results include demonstration of the global–galactic dynamo action driven by Cosmic Rays supplied in supernova remnants. An essential outcome of the new series of global galactic dynamo models is the equipartition of the gas turbulent energy with magnetic field energy and cosmic ray energy, in saturated states of the dynamo on large galactic scales.

Keywords. Galaxies: ISM – magnetic fields – ISM: cosmic rays – magnetic fields – MHD

---

1. Introduction

Numerous processes have been proposed to explain initial magnetic fields in early galaxies. However all these processes, like phase transitions in the early universe or Biermann battery in protogalactic objects provide only very week magnetic fields of the order of $B \sim 10^{-20}$ G at the beginning of galactic evolution. On the other hand, Rees (1987) proposed that initial magnetic fields in galaxies might have been generated in first stars and then scattered in the interstellar medium (ISM) by SN explosions, and subsequently amplified in plerionic (Crab–type) supernova remnants (SNRs). The mean magnetic field on the galactic scale has been estimated to be of the order of 10^{-9} G. Radio observations (see the review by R. Beck, this volume) indicate that typical contemporary magnetic fields in spiral galaxies are of the order of few up to few tens of μG, which means that magnetic fields have undergone amplification by at least four orders of magnitudes within the galactic lifetime. Therefore, a model for efficient magnetic field amplification during galactic evolution is necessary. The standard model of magnetic field amplification in disk galaxies is based on the theory of turbulent mean field dynamo (see Widrow 2002 for a recent review).

The mean–field dynamo theory has been successful in explaining magnetic fields in various astrophysical objects, however, the classical, kinematic dynamo seems to be rather slow in galaxies. Magnetic field amplification timescale $t_{\rm dynamo} \sim (0.5 \div 1) \times 10^9$ yrs is too long to explain $\sim 1\,\mu$G magnetic fields in galaxies which are only few 10^9 yrs old

($z \sim 1 \div 3$) (Wolfe, Lanzetta & Oren 1992, Kronberg *et al.* 2008). These circumstances seem to indicate a need of an alternative approach. We suggest that direct numerical MHD modeling of ISM dynamics, through considerations of detailed physical processes, involving major ISM components: gas, magnetic fields and cosmic rays, is necessary to follow magnetic history of galaxies.

2. Cosmic Rays in the interstellar medium

The dynamical role of CRs was recognized for the first time by Parker (1966), who noticed that a vertically stratified ISM, consisting of thermal gas magnetic field and CRs is unstable due to buoyancy of the weightless components: magnetic fields and CRs. The CR component appears to be an important ingredient of this process. According to the diffusive shock acceleration models CRs are continuously supplied to ISM by SN remnants. Therefore, strong buoyancy effects due to CRs are unavoidable.

Theories of diffusive shock acceleration predict that about 10 % of the $\sim 10^{51}$ erg of the SN II explosion energy is converted to the CR energy (see e.g. Jones et al 1998 and references therein). Observational data indicate that gas, magnetic fields and CRs appear in approximate energetic equipartition, which means that all the three components are dynamically coupled. Moreover, numerical experiments by Giaccalone & Jokipii (1999) indicate that CRs diffuse anisotropically in the ISM, along the mean magnetic field direction.

3. Cosmic Ray transport

To incorporate CR propagation in MHD considerations we use the diffusion - advection equation (e.g. Schlickeiser & Lerche 1985)

$$\frac{\partial e_{\rm cr}}{\partial t} + \boldsymbol{\nabla}(e_{\rm cr}\boldsymbol{v}) = -p_{\rm cr}\boldsymbol{\nabla}\cdot\boldsymbol{v} + \boldsymbol{\nabla}(\hat{K}\boldsymbol{\nabla}e_{\rm cr}) + \text{CR sources (SN remnants)} \tag{3.1}$$

where the diffusion term in written in the tensorial form to account for anisotropic diffusivity of CRs. The sources of CRs on the rhs. of Eqn. 3.1 correspond to the CR production in Supernova remnants.

The dynamics of ISM including the energetics of CRs can be now described by the CR transport Eqn. (3.1) and the system of MHD equations. We note that in the presence of cosmic rays additional source term: $-\nabla P_{\rm CR}$ should be included in the gas equation of motion (see e.g. Berezinski *et al.* 1990), in order to incorporate the effects of CRs on gas dynamics. The cosmic ray diffusion–advection Eqn. (3.1) has been supplemented to the MHD algorithm of Zeus–3D MHD code by Hanasz & Lesch (2003a) with the aim of studying ISM physical processes, e.g. Parker instability, in which CRs play a dynamically important role.

4. CR-driven dynamo

Referring back to the issue of galactic magnetic fields, we are primarily interested in processes leading to the amplification of large scale magnetic fields. The original idea of CR–driven dynamo has been raised by Parker (1992). Our first local (shearing–box) CR–driven dynamo numerical experiments (Hanasz *et al.* 2004, 2006; Otmianowska-Mazur *et al.* 2007) rely on the following ingredients: (1.) the cosmic ray component described by the diffusion–advection transport equation (3.1), including localized sources of cosmic

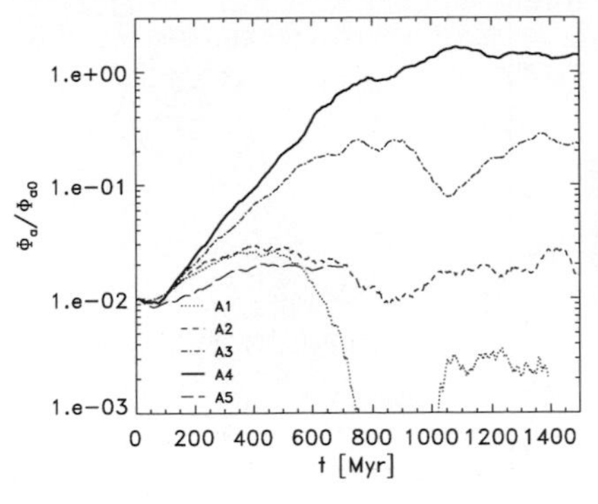

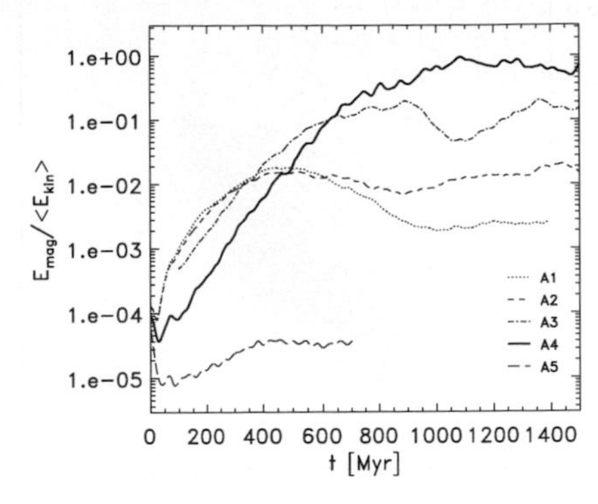

Figure 1. Time evolution of azimuthal magnetic flux and total magnetic energy for different values of magnetic diffusivity in simulation series A. The curves represent respectively cases of $\eta = 0$ (A1), $\eta = 1$ (A2), $\eta = 10$ (A3), $\eta = 100$ (A4) and $\eta = 1000$ (A5) in units $\mathrm{pc}^2\ \mathrm{Myr}^{-1}$.

rays — supernovae remnants, exploding randomly in the disk volume, while SN shocks and thermal effects are neglected (see Gressel *et al.* 2008a, 2008b) for a recent dynamo model relying on the thermal energy output from SNe); (2.) resistivity of the ISM (see Hanasz, Otmianowska–Mazur and Lesch 2002), leading to topological evolution of magnetic fields; (3) shearing boundary conditions (Hawley, Gammie and Balbus 1995) together with Coriolis and tidal forces, aimed at modeling of differentially rotating disks in the local approximation; (4) realistic vertical disk gravity and rotation following the model of ISM in the Milky Way by Ferrière (1998).

In the most recent paper (Hanasz *et al.* 2009) we present a parameter study of the CR–driven dynamo model by examining the dependence of magnetic field amplification on magnetic diffusivity, supernovae rate determining the CR injection rate, temporal modulation of SN activity, grid resolution, and CR diffusion coefficients. We find the dominating influence of magnetic diffusivity (treated as a free parameter), among other parameters, (Fig. 1) on the efficiency of magnetic field amplification.

The fastest magnetic field amplification, observed in the experiments, coincides with the magnetic diffusivity comparable to $\eta \simeq \frac{1}{3}\eta_{turb}(obs)$, where $\eta_{turb}(obs) \simeq \frac{1}{3}100\ \mathrm{pc} \times 10\ \mathrm{km\ s^{-1}} \simeq 10^{26}\ \mathrm{cm\ s^{-1}}$. The question of which physical process can be responsible for that large magnetic diffusivity coefficients points out towards the current studies on magnetic reconnection (Lazarian *et al.* 2004, Cassak *et al.* 2006).

To investigate observational properties of current dynamo models we perform a series of numerical simulations, placing the local shearing–boxes at different galactocentric radii. We replicate the contents of local cubes into rings, and then combine rings to build up a disk (Otmianowska–Mazur *et al.* 2009). A synthetic polarization radio–map of synchrotron emission constructed, on the base of simulated magnetic fields and CR distribution is shown in Fig. 4.

A polarized emission map together with superposed polarization vectors of synchrotron emission of an edge–on galaxy NGC 5775 is shown for comparison. As it is apparent the X–shaped structures of magnetic field, present in the real galaxy, form also in the synthetic radio image. The presented results demonstrate that the CR–MHD simulations provide a way of verification of CR–dynamo models against the observational data.

To summarize the overall outcome of the local simulations of CR–driven dynamo, we note that the model provides efficient amplification of large–scale magnetic fields, which

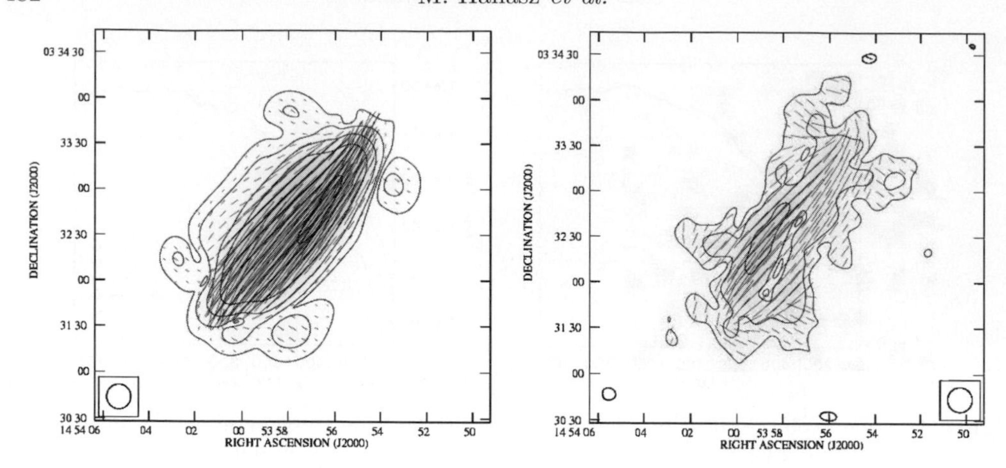

Figure 2. Synthetic radio–map of polarized radio emission for our model at the inclination $i = 80^\circ$ and a map of real galaxy NGC 5775.

attain, in the saturated state, equipartition with turbulent gas motions. However,we note also a drawback of the local models, since the CR energy density remains more than an order of magnitude larger than gas and magnetic energy densities. We recognize that improvements of the current models are necessary, since the possible reason of the excess of cosmic rays may result from the effect of trapping of cosmic rays by predominantly horizontal magnetic field in the computational box which is periodic in horizontal directions. Therefore, global galactic disk simulations should be performed to enable escape of CRs along the predominantly horizontal magnetic field in the galactic plane, in order to avoid the CR excess.

5. Global disk simulations

Recently we started a new series of global simulations of CR–driven dynamo with the aid of PIERNIK MHD code (see Hanasz *et al.* 2009a, 2009b, 2009c, 2009d), which is a grid–based MPI parallelized, resistive MHD code based on the Relaxing TVD (RTVD) scheme by Jin & Xin (1995) and Pen *et al.* (2003). The original scheme is extended to deal with the diffusive CR component (see Hanasz & Lesch 2003a).

In Fig. 3 we show results of one of the first semi–global simulations of CR–driven dynamo. The simulations have been performed for a quarter of galactic disk resembling Milky Way, with parallel CR diffusion coefficient $K_\parallel = 3 \times 10^{28}$ cm^2 s^{-1}, in a computational domain 20 kpc $\times$ 20 kpc $\times$ 20 kpc, and resolution of $400 \times 400 \times 160$ grid cells, in 16 MPI blocks.

The semi–global simulations confirm the fast amplification rate of the mean magnetic field: growth time of the mean magnetic flux is approximately 210 Myr in the whole galaxy, which is close to the galactic rotation period near the orbit of Sun. As expected, the global simulations introduce a significant improvement with respect to the former shearing–box simulations. It is apparent in Fig. 3 that the ratio of magnetic to cosmic ray energy is close to one in the synchrotron emitting volume (compare panels 4.and 5.), while over–equipartition of CR energy with respect to magnetic energy still holds outside the disk.

The latest development of our CR–dynamo model is a fully global galactic disk simulation (see the complementary paper by Hanasz, *et al.* this volume), where we demonstrate that dipolar magnetic fields supplied on small SN–remnant scales, can be amplified

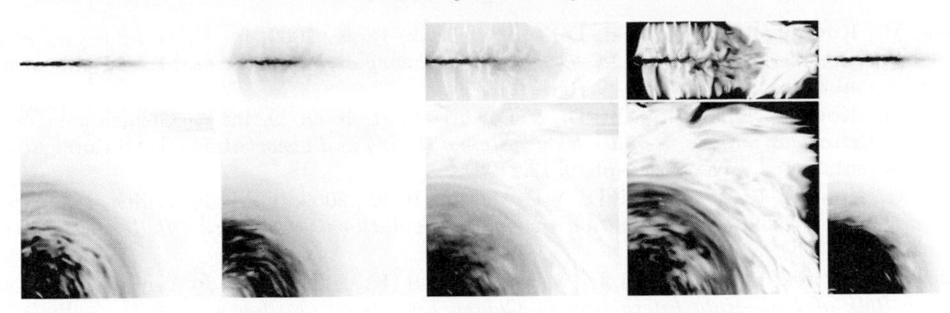

Figure 3. Horizontal and vertical slices through the computational box for a simulation of CR–driven dynamo in a semi–global galactic disk. The color maps reflect: 1: gas density, 2: CR energy density, 3: magnetic field strength, 4. the ratio of magnetic to CR energy densities and 5. synchrotron emissivity for a saturated state of the dynamo. The gray–scales in all panels, except 4. are chosen arbitrarily. The darker shades correspond to larger values of displayed quantities. The greyscale of panel 4. is such that black areas represent values of $e_{\rm mag}/e_{\rm cr} \simeq 1$.

exponentially by the CR–driven dynamo to the present equipartition values, and transformed simultaneously to large galactic–scales.

6. Conclusions

We have shown that the CR contribution to the dynamics of ISM, studied by means of CR–MHD simulations, in both local and global scales, leads to a very efficient magnetic field amplification in galactic disks.

The cosmic ray driven dynamo amplifies efficiently galactic magnetic fields on a timescale of galactic rotation. The resulting growth of the large–scale magnetic field by 4 orders of magnitude within 2 Gyr (Hanasz *et al.* 2004), is fast enough to expect $\sim 1\mu$G magnetic field in galaxies at $z \sim 1 \div 3$.

We point out the advantage of global disk models, which on the contrary to shearing–box models, provide CR energy equipartition with magnetic field in the synchrotron emitting part of the disks. This work was supported by Polish Ministry of Science and Higher Education through the grants 92/N–ASTROSIM/2008/0 and PB 0656/P03D/2004/26 and by Nicolaus Copernicus University through Rector's grant No. 516–A.

References

Beck R. 2008, *Ap&SS*, in press (arXiv:astro-ph/0711.4700)

Berezinskii, V. S., Bulanov, S. V., Dogiel, V. A., Ginzburg, V. L., & Ptuskin, V. S. *Astrophysics of cosmic rays*, Amsterdam: North-Holland, 1990.

Cassak, P. A., Drake, J. F., & Shay, M. A. 2006, *ApJ* 644, L145

Giacalone, J. & Jokipii, R. J. 1999, *ApJ* 520, 204

Ferrière, K. 1998, *ApJ* 497, 759

Gressel, O., Ziegler, U., Elstner, D., & Rüdiger, G. 2008, *AN* 329, 61

Gressel, O., Elstner, D., Ziegler, U., & Rüdiger, G. 2008, *A&A* 486, L35

Hanasz, M., Otmianowska–Mazur, K., & Lesch, H. 2002, *A&A* 386, 347

Hanasz, M. & Lesch, H. 2003, *A&A* 412, 331

Hanasz, M., Kowal, G., Otmianowska–Mazur, K., & Lesch, H. 2004, *ApJ* 605, L33

Hanasz, M., Kowal, G., Otmianowska–Mazur, K., & Lesch, H. 2006, *AN* 327, 469

Hanasz, M., Otmianowska–Mazur, K., Kowal, G., & Lesch, H. 2009, *A&A*, in press (arXiv:astro-ph/0812.3906)

Hanasz, M., Kowalik, K., Wóltański, D., & Pawłaszek, R. K. 2009, in: K. Goździewski *et al.* (eds.) *Extrasolar planets in multi–body systems: theory and observations*, EAS Publications Series (submitted, arXiv:astro-ph/0812.2161)

Hanasz, M., Kowalik, K., Wóltański, D., & Pawłaszek, R. K. 2009, in: K. Goździewski *et al.* (eds.) *Extrasolar planets in multi–body systems: theory and observations*, EAS Publications Series (submitted, arXiv:astro-ph/0812.2799)

Hanasz, M., Kowalik, K., Wóltański, D., & Pawłaszek, R. K., 2009, in: M. de Avillez *et al.* (eds.), *The Role of Disk–Halo Interaction in Galaxy Evolution: Outflow vs Infall?*, (submitted, arXiv:astro-ph)

Hanasz, M., Kowalik, K., Wóltański, D., & Pawłaszek, R. K., 2009, in: M. de Avillez *et al.* (eds.), *The Role of Disk–Halo Interaction in Galaxy Evolution: Outflow vs Infall?*, (submitted, arXiv:astro-ph)

Hawley, J. F., Gammie, C. F., & Balbus, S. A. 1995, *ApJ* 440, 442

Jones, T. W., Rudnick, L., Jun, B.–I., Borkowski, K. J., Dubner, G., Frail, D. A., Kang, H., Kassim, N. E., & McCray, R. 1998, *PASP* 110, 125

Jin, S. & Xin, Z. 1995, *Comm. Pure Appl. Math.* 48, 235

Kronberg, P. P., Bernet, M. L., Miniati, F., Lilly, S. J., Short, M. B., & Higdon, D. M. 2008, *ApJ* 676, 70

Lazarian, A., Vishniac, E. T., & Cho, J. 2004, *ApJ* 603, 180L

Otmianowska-Mazur, K., Kowal, G., & Hanasz, M., 2007, *ApJ* 668, 1100

Otmianowska–Mazur, K., Soida, M., Kulesza–Żydzik, B., Hanasz, M., & Kowal, G. 2009, *ApJ*, in press (arXiv:astro-ph/0812.2150)

Parker, E. N. 1966, *ApJ* 145, 811

Parker, E. N. 1992, *ApJ* 401, 137

Pen, U.–L., Arras, P., & Wong, S. 2003, *ApJS* 149, 447

Rees, M. J. 1987, *QJRAS* 28, 197

Ryu, D., Kim, J., Hong, S. S., & Jones, T. W. 2003, *ApJ* 589, 338

Schlickeiser, R. & Lerche, I. 1985, *A&A* 151, 151

Widrow, L. M. 2002, *Rev. Mod. Phys.* 74, 775

Wolfe, A. M., Lanzetta, K. M., & Oren, A. L. 1992, *ApJ* 388, 17

Discussion

Gressel: 1) In your simulations, what is the ratio of the turbulent and regular magnetic field strength? 2) With a strong turbulent component of the field, wouldn't you expect the cosmic-ray diffusion to be suppressed/diminished due to the highly tangled field geometry?

Hanasz: 1) The ratio of the turbulent to regular magnetic field strength is of the order of 1, i.e. unity. 2) The buoyancy of cosmic rays continuously supplied to the system, and trapped by the tangled magnetic field, becomes strong enough to stretch magnetic field lines in the vertical direction. Then, the magnetic field should become less tangled, allowing cosmic rays to leave the disk along the stretched-out vertical field lines.

Cosmic Magnetic Fields:
From Planets, to Stars and Galaxies
Proceedings IAU Symposium No. 259, 2008
K.G. Strassmeier, A.G. Kosovichev & J.E. Beckman, eds.

doi:10.1017/S1743921309031159

Pulsars & Magnetars

Michael Kramer

University of Manchester, Jodrell Bank Centre for Astrophysics,
Alan-Turin Building, Oxford Road, Manchester M13 9PL, UK
email: Michael.Kramer@manchester.ac.uk

Abstract. The largest magnetic field encountered in the observable Universe can be found in neutron stars, in particular in radio pulsars and magnetars. While recent discoveries have slowly started to blur the distinction between these two classes of highly magnetized neutron stars, it is possible that both types of sources are linked via an evolutionary sequence. Indications for this to be the case are obtained from observations of the spin-evolution of pulsars. It is found that most young pulsars are heading across the top of the main distribution of radio pulsars in the $P - \dot{P}$-diagram, suggesting that at least a sub-class of young pulsars may evolve into objects with magnetar-like magnetic field strengths. Part of this evolutionary sequence could be represented by RRATs which appear to share at least in parts properties with both pulsars and magnetars.

Keywords. Pulsars: general – stars: neutron – stars: magnetic fields

---

1. Introduction

The highest magnetized objects in the known Universe are neutron stars (NSs) with magnetic fields estimated to reach 10^{15} Gauss or above. The creation and maintenance of such large field strengths remains an important and unsolved astrophysical problem. It is therefore useful to review the evidence for such estimates and to critically study the made assumptions when deriving such numbers. The importance of the answers anticipated and sought for in this process go beyond the mere understanding of neutron stars, but they have relevance for the evolution of cosmic magnetic fields in general, stellar evolution, and studies of the Milky Way and cosmic ray production.

2. Manifestations of magnetized neutron stars

Neutron stars are observed in many flavours, including those as dim isolated NSs, as normal and radio millisecond pulsars, as magnetars, as Rotating Radio Transients (RRATs) or as part of X-ray binary systems. We will concentrate here only on those manifestations that are believed† to have very large magnetic fields. In all discussed cases, the identification of the observed object class with NSs is safe, and general properties can be assumed (see e.g. Lyne & Smith 2005 for a summary): With possible variations due to the unknown Equation-Of-State, the radius of a NS lies between 10 and 15 km. Masses have been measured very accurately for an increasing number of binary NSs and are found in a range between $1.2M_{\odot}$ and $\sim 2M_{\odot}$. NSs therefore represent the most extreme matter in the observable Universe with densities in their interior that vastly exceeds that of any known matter on Earth. The result is a peculiar high-temperature super-fluid super-conductor. We can expect that the super-conductivity plays an important role in the maintenance of the enormous magnetic field strengths which range from

† or "observed" – the reader can make up her/his mind after reading this review!

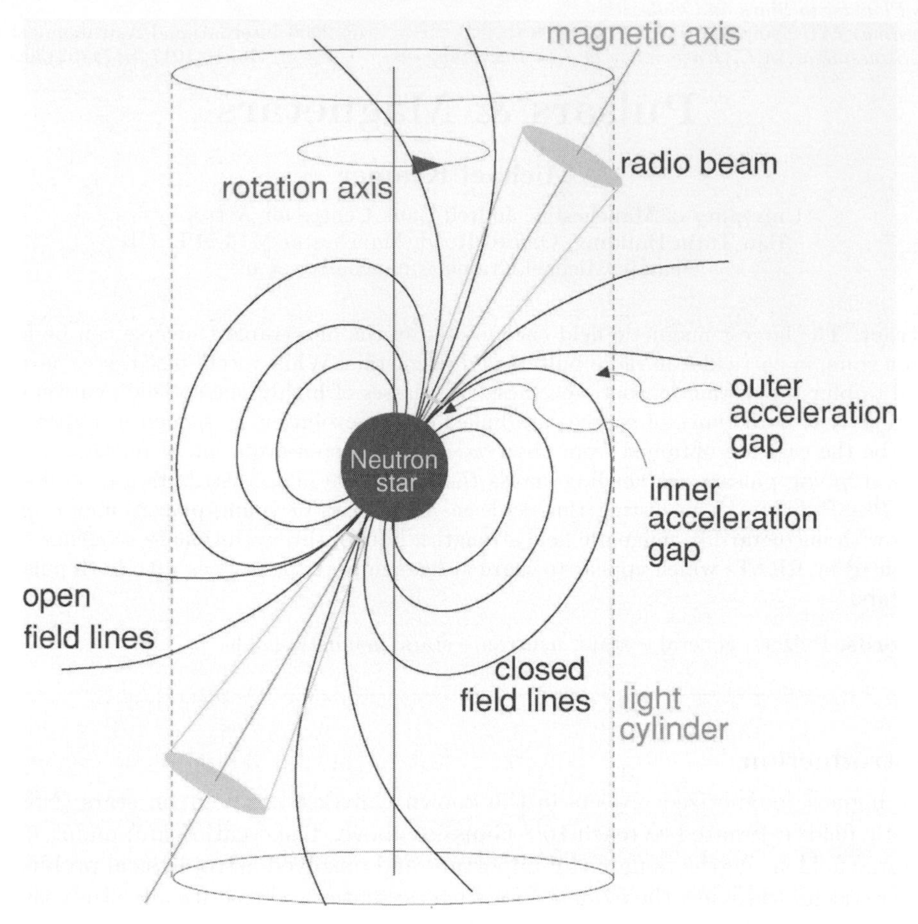

Figure 1. Toy model for the pulsar-like otating neutron star and its magnetosphere (not drawn to scale!).

10^8 Gauss in millisecond pulsars, over typically 10^{12} Gauss in normal pulsars to 10^{15} Gauss in magnetars.

2.1. *Pulsars*

Most of the NSs known are observed as radio pulsars. The nearly 2,000 objects currently known are classified in two groups, i.e. the old recycled or millisecond pulsars and the younger, normal pulsars. Periods range from 1.4 ms to 8.5 s † and, in all cases, are found to increase slowly with time. Spin-down rates are typically observed from $\dot{P} \sim 10^{-12}$ for the very youngest pulsars to $\dot{P} \sim 10^{-21}$ for the oldest pulsars. This increase in rotation period reflects the loss of rotational energy that is converted in electromagnetic radiation.

Even though radio emission accounts for only a tiny fraction (say, 10^{-6} to 10^{-4}) of the total "spin-down luminosity" of these *rotation-powered NSs*, most pulsars are only visible at radio frequencies. While pulsars are normally rather weak radio sources, the emission is coherent and highly polarized, and the result of an emission process that is only poorly understood at best. In our basic understanding (see Lorimer & Kramer 2005 for more details, Fig. 1), the magnetized rotating neutron star induces an electric quadrupole field which is strong enough to pull out charges from the stellar surface (the

† see for instance `http://www.atnf.csiro.au/research/pulsar/psrcat/`

electrical force exceeds the gravitational force by a factor of $\sim 10^{12}$!). The magnetic field forces the resulting dense plasma to co-rotate with the pulsar. This *magnetosphere* can only extend up to a distance where the co-rotation velocity reaches the speed of light‡. This distance defines the so-called light cylinder which separates the magnetic field lines into two distinct groups, i.e. *open and closed field lines.* The plasma on the closed field lines is trapped and co-rotates with the pulsar. In contrast, plasma on the open field lines can reach highly relativistic velocities and can leave the magnetosphere, creating the observed radio beam at a distance of a few tens to hundreds of km above the pulsar surface.

The location of the pulsar beam within the open dipolar field line region allows us to relate the beam radius, ρ, to the emission height, $r_{\rm em}$ and the rotation period, i.e.

$$\rho = \sqrt{\frac{9\pi r_{\rm em}}{2cP}}. \tag{2.1}$$

The scaling of $\rho \propto P^{-1/2}$ is indeed observed (see e.g. Lorimer & Kramer 2005) and can be viewed as evidence that the field structure is dipolar in the emission region.

2.2. *Magnetars*

Magnetars are slowly rotating neutron stars for which magnetic fields are inferred that are even stronger than those estimated for pulsars (see Section 3). Currently, 13 magnetars are confirmed† and are observed as variable X-ray/γ-ray sources which show common, short energetic bursts (i.e. in case of the so-called *Soft-Gamma-Ray Repeaters* (SGRs) of which 4 are known) or long flares and rare and soft short bursts (i.e. in case of the so-called *Anomalous X-ray Pulsars* (AXPs) of which 9 are known). Observations during bursts allow the detection of rotation periods which range from 2 to 12 s. Spin-down is also detected with values from $\dot{P} \sim 10^{-13}$ to 10^{-10}. The burst luminosity commonly exceeds the spin-down luminosity which is given by

$$\dot{E} = -I\Omega\dot{\Omega} = 4\pi^2 I\dot{P}P^{-3} = 3.95 \times 10^{31} \text{ erg s}^{-1} \left(\frac{\dot{P}}{10^{-15}}\right) \left(\frac{P}{\rm s}\right)^{-3}, \tag{2.2}$$

where $\Omega = 2\pi/P$ is the rotational angular frequency, I the moment of inertia, and the last equality holds for a canonical moment of inertia of $I = 10^{45}$ g cm^2. Consequently, the objects cannot be rotation powered but a different source of energy must exist. In the most commonly applied model of Duncan & Thompson (1992), the source of energy results from the energy stored in huge magnetic fields. Indications for this are also obtained from the giant flares observed in SGRs: the bright, spectrally hard initial spike is followed by a tail of thermal X-ray emission which is strongly modulated at the spin frequency of the neutron star. This feature is interpreted as a "trapped fireball" after a significant shift in the NS crust occurred. Estimating the magnetic field necessary to contain the plasma one indeed obtains fields in excess of 10^{14} G. Furthermore, features in X-ray spectra interpreted as cyclotron lines support such estimates. Finally, magnetars are considered to be young sources as associations of several magnetars with supernova remnants suggests. (For a recent more detailed review of magnetar properties, see Kaspi

‡ Strictly speaking, the Alfvén velocity will determine the co-rotational properties of the magnetosphere.

† see `www.physics.mcgill.ca/ pulsars/magnetar/main.html` for updated information

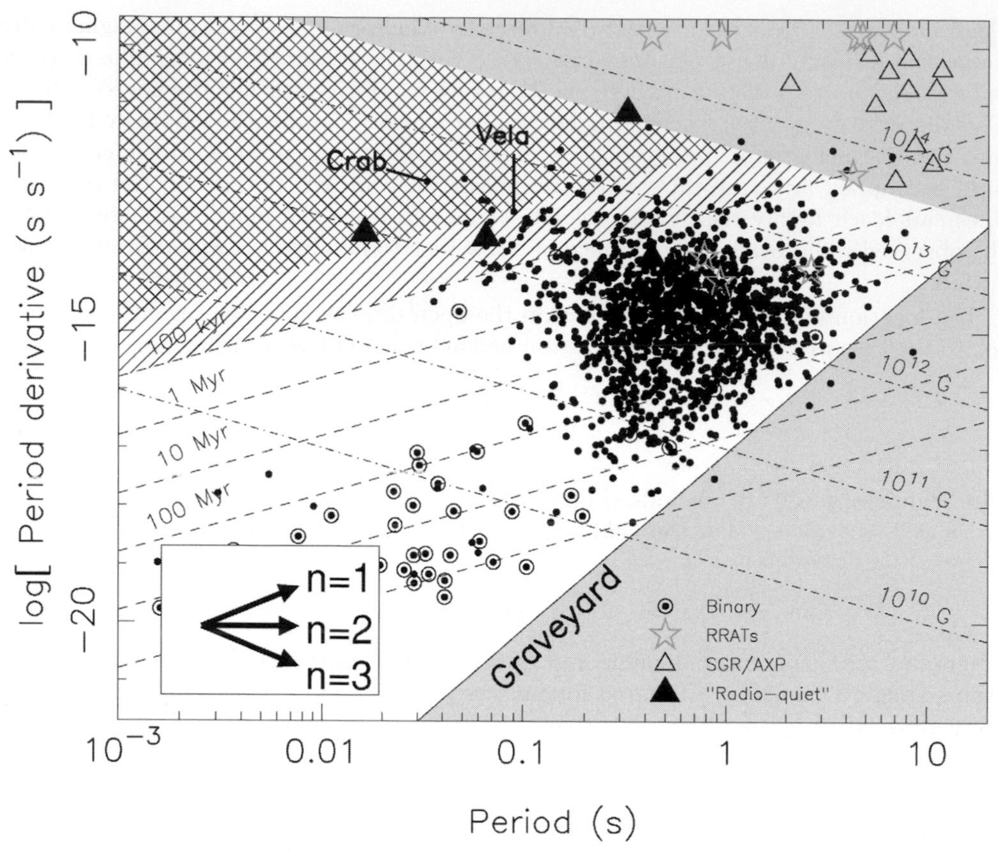

Figure 2. The P–$\dot{P}$ diagram shown for a sample consisting of radio pulsars, 'radio-quiet' pulsars and magnetars, i.e. soft-gamma repeaters (SGRs) and anomalous X-ray pulsars (AXPs). Lines of constant characteristic age τ_c and magnetic field B are also shown. The single hashed region shows 'Vela-like' pulsars with ages in the range 10–100 kyr, while the double-hashed region shows 'Crab-like' pulsars with ages below 10 kyr. The grey regions are areas where radio pulsars are not predicted to exist by theoretical models. The inset at the bottom-left indicates the expected direction of movement for pulsars with a braking index of $n = 1$, 2 and 3, respectively.

2007.) Small ages are also obtained from *characteristic age* estimates, given by

$$\tau_{\rm c} \equiv \frac{P}{2\dot{P}} \simeq 15.8\ {\rm Myr} \left(\frac{P}{\rm s}\right) \left(\frac{\dot{P}}{10^{-15}}\right)^{-1} \tag{2.3}$$

which is also applied to normal pulsars. This estimate assumes that the spin period at birth is much shorter that the present value and that the spin-down is dominated by magnetic dipole radiation. We discuss these assumptions in the following.

3. Magnetic field estimates

In addition to cyclotron lines in isolated NS and magnetars, the most common estimate for the magnetic field is obtained from the observed spin-down of the NSs, using

$$B = 3.2 \times 10^{19}{\rm G}\ \sqrt{P\dot{P}} \simeq 10^{12}\ {\rm G} \left(\frac{\dot{P}}{10^{-15}}\right)^{1/2} \left(\frac{P}{\rm s}\right)^{1/2}. \tag{3.1}$$

which assumes a canonical NS with a moment of inertia $I = 10^{45}$ g cm^2 and radius $R = 10$ km. More importantly, it assumes that the spin-down is dominated by the emission magnetic dipole radiation and that the magnetic axis is perpendicular to the rotation axis.

3.1. *Spin-evolution & braking indices*

If the spin-down is dominated by "dipole braking", the magnetized NS behaves like a rotating bar magnet, and the change in angular spin-frequency can be written as

$$\dot{\Omega} = -\left(\frac{2|\mathbf{m}|^2 \sin^2 \alpha}{3Ic^3}\right)\Omega^3 \propto -\Omega^n \tag{3.2}$$

where $\mathbf{m}$ is the magnetic dipole moment and α the angle between magnetic and rotation axis. The usage of the *braking index* n is a general way of describing the spin-evolution of NSs. Here, it takes a value of $n = 3$. Ideally, the assumption of magnetic dipole braking can be verified by determining the braking index from observations which is in principle possible by measuring $\ddot{\Omega}$ since

$$n = \frac{\Omega\ddot{\Omega}}{\dot{\Omega}^2}. \tag{3.3}$$

However, this is an extremely difficult task, as significant $\ddot{\Omega}$-values can only be expected for young pulsars where the spin-down is largely influenced by timing noise and the recovery from glitches rather than by a regular long-term spin-down. Eleven braking indices have been measured reliably at the time of writing, ranging from $n = 1.1$ to $n = 2.9$, i.e. in all cases $n < 3$ (Espinoza *et al.* in prep.).

3.2. *Divas and increasing fields*

The spin characteristics of pulsars can be displayed in a *$P - \dot{P}$-diagram* where we plot logarithms of P and $\dot{P}$ and lines of constant magnetic field and age using Eqns. 3.1 & 2.3 (see Fig. 2). Young pulsars are located in the top-left corner while the majority of NSs can be found in the central region near $P \sim 0.6$s and $\dot{P} \sim 10^{-15}$. Magnetars are in the top-right corner overlapping in parts with high-B-field pulsars which we discuss in Section 4.1. If the braking index $n = 3$, NSs evolve along constant magnetic field lines, while for $n = 2$ they move horizontally and for $n < 2$ they move up in the diagram. Note that for $n < 3$ the magnetic field appears to increase with time. Figure 3 demonstrates that this increase is particularly rapid for $n < 2$ which is the case for four measured pulsars. But also the inferred characteristic age is affected. Figure 3 also shows that NSs lie about their age in terms of displayed characteristic age which becomes smaller than the true age for $n < 3$. In fact, pulsars do not seem to age at all for $n = 1$.

In summary, a large number of young pulsars are heading across the top of the main distribution of radio pulsars in the $P - \dot{P}$-diagram. All young pulsars with measured braking index have $n < 3$, so that the effective magnetic fields increase with time. At the same time, the characteristic ages increase only slowly or even decrease with time. We therefore conclude that the evolution of some young pulsars is consistent with travel to region occupied by magnetars (see Lyne 2004)!

3.3. *Assumptions revisited*

In order to verify the startling conclusions that some young pulsars may evolve to possibly become magnetars, we ought to review some of the made assumptions in the above analysis. Firstly, all braking indices measured so far suggest that the spin-down is not solely determined by magnetic dipole radiation. Indeed, using the discovery of intermittent pulsars, Kramer *et al.* (2006) provided first observational evidence that a significant

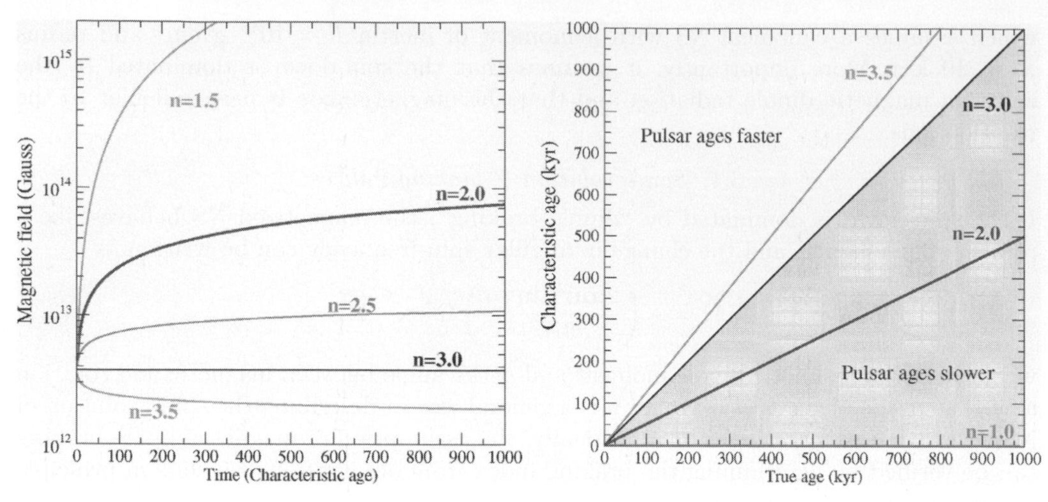

Figure 3. Evolution of the inferred magnetic field and the characteristic age for different values of the braking index. The left panel shows that the evolution of the magnetic field appears to increase for $n < 3$. The right panel demonstrates the deviation of characteristic age from true age.

fraction of spin-down torque is also provided by a particle wind and its magnetospheric current. One can only assume that the estimated field is therefore overestimated. In case of intermittent pulsars, the error may be as small as $\sim$ 20%. Secondly, the mentioned result that the beam radii of pulsars scale with $\propto P^{-1/2}$ suggests that the assumption of dipolar fields is correct. This is also supported by modelling the eclipses of pulsar A in the Double Pulsar where the observations are consistent with a simple dipolar magnetosphere of pulsar B (Breton *et al.* 2006). However, the movement of the pulsars in the $P - \dot{P}$-diagram can also be interpreted as a change in magnetic inclination angle α. Such changes should, however, become visible in changes of the pulse profile and its polarisation characteristics.

4. Population overlap?

Another way of learning more about the possible evolution from pulsars to magnetars is to study the properties of objects which lie at the "border" of the particular population by sharing some of each other properties. Here we encounter pulsars with magnetar-like spin properties, magnetars that radiate radio emission and RRATs which may be considered as a possible missing link.

4.1. *High-B field pulsars*

A number of young radio pulsars is now known (e.g. McLaughlin *et al.* 2003) that occupy a region in the $P - \dot{P}$ diagram which was traditionally associated with magnetars (see Fig. 2). In particular, the magnetic field estimate lies above the quantum critical field where standard pulsar radio emission is not necessarily expected (e.g. Lorimer & Kramer 2005). However, none of the pulsar properties, e.g. the emission characteristics at X-ray or radio frequencies, does show really similarities with those of magnetars (e.g. Gonzale *et al.* 2007).

4.2. *Radio-loud magnetars*

A more interesting development of recent years is the discovery of radio emission from two magnetars (Camilo *et al.* 2006, 2007). In both cases, for AXP J1810−197 and for 1E 1547.0−5408, the radio emission is detected with the magnetar's rotation period of $P = 5.5$ s and $P = 2.1$ s, respectively. Both sources had a previous X-ray burst that may have triggered the radio emission, as pre-burst radio observations did not reveal a source at the magnetars' position. Therefore, magnetic field rearrangements may be responsible for the radio switch-on but a failed failed radio detection of another AXP after burst (SGR 1627-41, Camilo & Sarkissian 2008), shows that such a rearrangement may be necessary but not sufficient (cf. Halpern *et al.* 2008). A comparison of magnetar radio and pulsar radio emission is intriguing, i.e. some properties are similar but many features are very different: unlike pulsars, the magnetar radio spectrum is flat and changing with time; unlike pulsars, there is a rough alignment between radio and X-ray arrival times; unlike pulsars, pulse profiles are not stable but change dramatically with time and frequency; the emission is $\sim 100\%$ polarized but unlike pulsars the polarisation position angle swing is changing with time. Some emission properties may be explainable by non-dipolar fields (e.g. Kramer *et al.* 2007) but overall it is unlikely that magnetars are simply variations of normal radio pulsars. However, it is indeed not inconceivable that they are linked in terms of evolution. This would solve another problem, namely the notion that current birth rate estimates suggest that the derived number of neutron stars is too large to be sustained by the Galactic core collapse supernova rate (Keane & Kramer 2008).

4.3. *RRATs*

The missing link in an evolutionary scenario could be provided by the recently discovered Rotating Radio Transients (RRATs, McLaughlin *et al.* 2006): RRATs appear as transient radio sources (more than 11 are known at the moment) which show bursts with durations between 2 and 30 ms. The average interval between to events ranges from 4 min to 3 hr. While the signals are hence repeating, the total emission time is less than 1 s per day, suggesting that many more RRATs remain undiscovered in the Galaxy. Even though the signals arrive Poisson-distributed, underlying periods have been discovered, ranging from $P = 0.4$ s to $P = 6.9$ s (mean $= 3.5 \pm 0.8$ s). These periods and the detection that they increase with time lead to the identification of RRATs with rotating NSs. This was confirmed later when RRAT J1819−1458 was also discovered as an X-ray source which was modulated with the radio period of $P = 4.3$ s and which showed features of a young, cooling NS (McLaughlin *et al.* 2007). Moreover an also detected spectral feature may be interpreted as a cyclotron line, leading to $B \sim 10^{14}$ G which is consistent with the value derived from the source's spin-down properties, $B = 0.5 \times 10^{14}$ G. However, in order to proof the speculation that RRATs are the evolutionary link between radio pulsars and magnetars, magnetar-like high-energy bursts need to be detected. So far, this has not been possible, in particular since the position of most RRATs is yet too uncertain for an appropriate analysis.

5. Conclusions

Most young pulsars are heading across the top of the main distribution of radio pulsars in the $P - \dot{P}$-diagram. This begs the question how the crowded area in this diagram is populated. Either these young pulsars have no evolutionary link to the population of old pulsars (i.e. separate populations) or magnetic fields start to decay before $\tau \sim 100$ kyr. All young pulsars have $n < 3$, hence the effective magnetic fields increase with time while characteristic ages increase only slowly or even decrease with time As a result, the evolution of some young pulsars is consistent with travel to the $P-\dot{P}$-region occupied by magnetars (see also Lyne 2004). For $n = 1$, the travel time from the current position to those of magnetars is about 10 kyr for the Crab pulsars. If that is the case, magnetars are much older than their characteristic ages indicate. This would also explain the relative paucity of SGR/AXP SNR associations and the large offset of SGRs from SNR centres, without invoking massive velocities. However, questions remain such as to whether there is a continuum of pulsars along the evolutionary tracks, or as to whether RRATs are an intermediate phase? It is clear that LOFAR and later the SKA surveys will solve this problem!

Acknowledgements

I would like to thank the organizers for a successful and enjoyable meeting.

References

Breton, R. P. *et al.* 2008, *Science* 321, 104
Camilo, F. & Sarkissian, J. 2008, *The Astronomer's Telegram*, 1558
Camilo, F., Ransom, S. M., Halpern, J. P., Reynolds, J., Helfand, D. J., Zimmerman, N., & Sarkissian, J. 2006, *Nature* 442, 892
Camilo, F., Ransom, S. M., Halpern, J. P., & Reynolds, J. 2007, *ApJ* 666, L93
Duncan, R. C. & Thompson, C. 1992, *ApJ* 392, L9
Gonzalez, M. E., Kaspi, V. M., Camilo, F., Gaensler, B. M., & Pivovaroff, M. J. 2007, *Ap&SS* 308, 89
Halpern, J., Gotthelf, E., Reynolds, J., Ransom, S., & Camilo F. 2008, *ApJ* 676, 1178
Kaspi, V. 2007, *Ap&SS* 308, 1
Keane, E. & Kramer, M. 2008, *MNRAS* 391, 2009
Kramer, M., Lyne, A. G., O'Brien, J. T., Jordan, C. A., & Lorimer, D. R. 2006, *Science* 312, 549
Kramer, M., Stappers, B. W., Jessner, A., Lyne, A. G., & Jordan, C. A. 2007, MNRAS 377, 107
Lorimer, D. R. & Kramer, M. 2005, *Handbook of Pulsar Astronomy*, Cambridge University Press
Lyne, A. G. & Smith, F. G. 2005, *Pulsar Astronomy*, 3rd ed. Cambridge University Press, Cambridge
Lyne, A. G., 2004, in Camilo, F., Gaensler, B. M., eds, *Young Neutron Stars and Their Environments*, IAU Symposium 218, Astronomical Society of the Pacific, San Francisco, p. 257
McLaughlin, M. A. *et al.* 2003, *ApJ* 591, L135
McLaughlin, M. A. *et al.* 2006, *Nature* 439, 817
McLaughlin, M. A. *et al.* 2007, *ApJ* 670, 1307

Discussion

REICH: If the apparent increase in **B**-field can also be interpreted as a change in α, do you see changes in the profile of the Crab pulsar?

KRAMER: No. We monitor the Crab now for 40 years and have looked carefully at possible profile changes. Apart from changes potentially caused by changing instrumentation, we have found no evidence for profile changes.

Cosmic Magnetic Fields:
From Planets, to Stars and Galaxies
Proceedings IAU Symposium No. 259, 2008
K.G. Strassmeier, A.G. Kosovichev & J.E. Beckman, eds.

doi:10.1017/S1743921309031160

The magnetic field in luminous star-forming galaxies

Timothy Robishaw[1] and Carl Heiles[2]

[1]Sydney Institute for Astronomy, School of Physics, The University of Sydney
email: robishaw@physics.usyd.edu.au

[2]Astronomy Department, University of California, Berkeley
email: heiles@astro.berkeley.edu

Abstract. An ongoing search for Zeeman splitting in the 1667 MHz OH megamaser emission from luminous star-forming galaxies has yielded numerous detections. These results, in addition to being the first extragalactic measurement of the Zeeman effect in an emission line, suggest that OH megamasers are excellent extragalactic magnetometers. We review the progress of our survey and discuss future observations.

Keywords. Masers – galaxies: magnetic fields – galaxies: starburst – ISM: molecules

1. What we knew then

The study of magnetic fields in luminous star-forming galaxies has been limited to optical, infrared, and submillimeter polarimetric imaging of the nearby starburst galaxy M82. This process is sensitive to the plane-of-sky component of the magnetic field and does not permit the estimation of field strengths. Using this method, Jones (2000) was able to infer that the field in the nucleus of M82 is configured in a polar geometry aligned with the bipolar superwind seen to be emanating from the nuclear region, while the field in the disk of the galaxy is arranged in a toroidal configuration.

Wielebinski (2006) used the 100 m Effelsberg radio telescope to map the plane-of-sky magnetic field in M82 using polarized 32 GHz emission. These observations confirmed the azimuthal field structure in the disk and vertical field structure along the poles but provide no estimate of the magnetic field strength in M82.

To the knowledge of the authors, no direct field strength measurements had been made towards nearby luminous star-forming galaxies until the results of Robishaw, Quataert, & Heiles (2008), who needed a new tool to help them.

2. ULIRGs and OH Megamasers

Ultraluminous infrared galaxies (ULIRGs) are a population of galaxies that emit far-infrared (FIR) radiation with energies comparable to those of the most luminous quasars ($\log(L_{\rm FIR}/L_\odot) > 12$). Nearly every ULIRG appears to have undergone a merger or interaction and contains massive star formation and/or an active galactic nucleus (AGN) induced by gravitational interactions. Diamond *et al.* (1999) conducted very long baseline interferometry (VLBI) observations of the 1667 MHz hydroxyl (OH) transition in the nuclear regions in ULIRGs and found multiple masing regions with $1 < \log(L_{\rm OH}/L_\odot) < 4$; these regions are known as OH megamasers (OHMs). Each OHM has a spectral linewidth of between 50 and 150 $\rm km\,s^{-1}$; when viewed by a single dish, these spectral components are superimposed. The 1667 MHz OHM flux is always a few to many times the flux of the 1665 MHz transition and in many cases the 1665 MHz line is absent (Darling & Giovanelli

2002); this is an interesting contrast to the case of OH masers in the Galaxy in which the 1665 MHz is usually the dominant transition (Reid & Moran 1988). The starbursts and AGNs in ULIRGs create strong FIR dust emission as well as a strong radio continuum; the OHMs are generally believed to be pumped by the FIR radiation field (e.g., Randell *et al.* 1995) although collisional excitation may be important as well (e.g., Lonsdale *et al.* 1998). Given the conditions that exist in ULIRGs and considering that every OH maser in our Galaxy is associated with massive star-forming regions, it is not surprising that the entire OHM sample finds homes in LIRGs, strongly favoring the most FIR-luminous, the ULIRGs (Darling & Giovanelli 2002).

ULIRGs are natural locations to expect very strong magnetic fields given their high gas and energy densities. Much of the radio emission in ULIRGs is resolved on scales ~ 100 pc with VLA observations (Condon *et al.* 1991). In Arp 220, high-resolution observations (Rovilos *et al.* 2003) show that the OHMs arise in this region as well. With this size scale and the observed radio fluxes, minimum energy arguments suggest *volume averaged* field strengths of ≈ 1 mG (e.g., Condon *et al.* 1991,Thompson *et al.* 2006), significantly larger than the $\sim 10\mu$G fields measured in normal spirals. The field strengths in ULIRGs cannot be much below a mG or else inverse Compton cooling would dominate over synchrotron cooling, making it energetically difficult to explain the radio fluxes from ULIRGs and the fact that ULIRGs lie on the FIR-radio correlation. The field strengths could, however, in principle be larger than the minimimum energy estimate if, as in our Galaxy, the magnetic energy density is in approximate equipartition with the total pressure (Thompson *et al.* 2006). This can be estimated from the observed surface density; CO observations of Arp 220 and several other systems reveal $\sim 10^9 M_\odot$ of molecular gas in the central ~ 100 pc (e.g., Downes & Solomon 1998) implying gas surface densities $\Sigma \sim 1{-}10$ g cm^{-2}, 10^3–10^4 times larger than in the Milky Way. The equipartition field scales as $B \propto \Sigma$ implying that the mean field in ULIRGs could approach ~ 10 mG.

3. Observations

Motivated by the above considerations, Robishaw, Quataert, & Heiles (2008) used both the 300 m Arecibo† telescope and the 100 m Robert C. Byrd Green Bank Telescope‡ in full-Stokes mode in an attempt to detect Zeeman splitting in megamaser emission of the 1667 MHz OH transition. Eight of the brightest OHMs were observed by spending equal time at on-source and off-source positions. Every VLBA observation of OHM emission in ULIRGs has shown that the spatial extent of the masing spots is of order at most a few arc seconds, much smaller than the beam of either Arecibo or the GBT; there are no instrumental polarization effects due to beam structure or sidelobes. The observations were made in full-Stokes mode and the Mueller matrix was calibrated using the standard technique of observing the linearly-polarized continuum sources 3C 138 and 3C 286 (Heiles *et al.* 2001, Heiles & Troland 2004).

4. What we know now

Five of the eight sources that we observed exhibit significant Stokes V emission that can be interpreted as Zeeman splitting of the 1667 MHz line. We present the results for the

† The Arecibo Observatory is part of the National Astronomy and Ionosphere Center, which is operated by Cornell University under a cooperative agreement with the National Science Foundation.

‡ The National Radio Astronomy Observatory is a facility of the National Science Foundation operated under cooperative agreement by Associated Universities, Inc.

canonical ULIRG Arp 220 in the left panels of figure 1; the right panels of figure 1 show the results for III Zw 35. The top panel for each source shows the Stokes I spectrum¶ as a function of frequency and optical heliocentric velocity. It is clear that the OHM emission profile for each source is a composite of many narrower components.

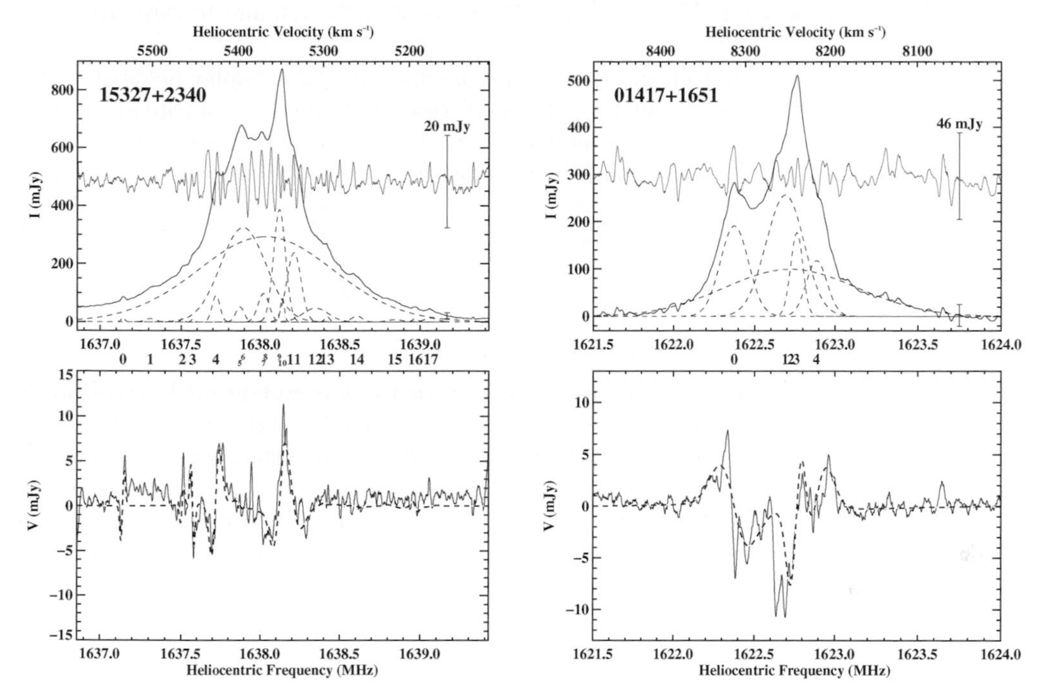

Figure 1. Stokes I spectrum (*top left*) for Arp 220. The profile of each Gaussian component is plotted as a dashed line. The residuals of the composite Gaussian profile from the data are plotted on an enhanced scale through the center of the Stokes I plot. The Stokes V (*bottom left*) spectrum of Arp 220; the composite fit for Zeeman splitting of each corresponding Gaussian component is plotted as a dashed line. (*Top right*) Stokes I profile for III Zw 35. (*Bottom right*) Stokes V spectrum for III Zw 35.

4.1. *Arp 220*

The Stokes I profile was decomposed into 18 Gaussian components; we attempted to use the fewest number of components to obtain reasonable residuals (which are plotted with an enhanced scale through the middle height of the Stokes I profile) while allowing enough components to reproduce multiple splittings in the Stokes V spectrum. The complexity of the VLBA spectra of Rovilos *et al.* (2003) prevents a clear association of single-dish components with the individual maser spots. The bottom panel shows the Stokes V spectrum: it is clear that the detectable circular polarization corresponds well with many of the individual Stokes I peaks. Six components show Zeeman signatures in the Stokes V spectrum and the absolute values of the fitted magnetic field strengths range from 0.7 to 4.7 mG. The magnetic field reverses within Arp 220 since four fields point towards us and two away.

¶ We use the classical definition of Stokes I, which is the sum (not the average) of two orthogonal polarizations. Thus, fluxes in Stokes I are twice the usual flux density given in catalogs.

4.2. *III Zw 35*

The Stokes I profile for III Zw 35 required only 5 Gaussian components to obtain reasonable residuals. Pihlström *et al.* (2001) and Parra *et al.* (2005) model high-resolution observations of the OHM emission as a torus of multiple maser clouds inclined at an angle of 60° and rotating at 57 km s^{-1} away from us at the north and towards us at the south.

From the circular polarization spectrum, components 0, 3, and 4 display fields of $+2.94$, -2.73, and -3.59 mG, respectively. From the high-resolution mapping of Pihlström *et al.* (2001), component 0 appears to be associated with the northern maser spots, while components 3 and 4 seem to originate in the south. In terms of the torus model, these results suggest that the magnetic field might be constrained to the OHM torus with the magnetic field following the velocity field, pointing away from us in the north and towards us in the south.

4.3. *Summary of results*

Robishaw, Quataert, & Heiles (2008) report a total of 14 independent field detections in multiple components within five extragalactic sources including Arp 220 and III Zw 35. The median absolute field strength is ~ 3 mG and the direction of the field reverses within 3 of the galaxies. The distribution is strikingly similar to that in the OH masers that probe the star-forming regions in the Milky Way (Fish *et al.* 2003), suggesting that the conditions in regions of massive star formation in ULIRGs are similar to the conditions in our own Galaxy's star-forming regions. The measured magnetic field strengths are also consistent with those inferred from sychrotron emission in ULIRGs by Thompson *et al.* (2006).

5. OH Megamasers: the new extragalactic magnetometers

The Zeeman detection rate for our sample of ULIRGs suggests that OHMs are excellent extragalactic magnetometers. Since it is not possible to unambiguously associate VLBI spectra of the OHM spots within any ULIRG with the Gaussian components in our single-dish spectra, it will be necessary to follow up all detections with full-Stokes VLBI observations in order to detect Zeeman splitting directly in the OHM spots. This would allow us to make high-resolution maps of the magnetic field strength and line-of-sight field direction in star-forming galaxies. Such high-resolution field maps would allow us to infer whether the field structure in the nuclei of these starburt galaxies is similar to that of M82.

A full-Stokes, flux-limited Arecibo survey of OHMs (including the OH satellite lines) is currently underway at Arecibo. Additional Zeeman detections have been revealed in the preliminary results. Finally, we have proposed to carry our hunt for Zeeman splitting to the southern OHM population by incorporating the 64 m Parkes telescope.

Acknowledgements

This research was supported in part by NSF grant AST-0406987. Support for this work was also provided to TR by the NSF through award GSSP 06-0003 from the NRAO. TR would like to thank: the SOC for an invitation to present these results, the IAU for providing financial support to attend this conference, and Guayota for not becoming angry.

References

Condon, J. J., Huang, Z.-P., Yin, Q. F., & Thuan, T. X. 1991, *ApJ* 378, 65
Darling, J. & Giovanelli, R. 2002, *AJ* 124, 100
Diamond, P. J., Lonsdale, C. J., Lonsdale, C. J., & Smith, H. E. 1999, *ApJ* 511, 178
Downes, D. & Solomon, P. M. 1998, *ApJ* 507, 615
Fish, V. L., Reid, M. J., Argon, A. L., & Menten, K. M. 2003, *ApJ* 596, 328
Heiles, C., *et al.* 2001, *PASP* 113, 1274
Heiles, C. & Troland, T. H. 2004, *ApJS* 151, 271
Jones, T. J. 2000, *AJ* 120, 2920
Lonsdale, C. J., Lonsdale, C. J., Diamond, P. J., & Smith, H. E. 1998, *ApJL* 493, L13
Parra, R., Conway, J. E., Elitzur, M., & Pihlström, Y. M. 2005, *A&A* 443, 383
Pihlström, Y. M., Conway, J. E., Booth, R. S., Diamond, P. J., & Polatidis, A. G. 2001, *A&A* 377, 413
Randell, J., Field, D., Jones, K. N., Yates, J. A., & Gray, M. D. 1995, *A&A* 300, 659
Reid, M. J. & Moran, J. M. 1988, in Galactic and Extragalactic Radio Astronomy, ed. G. .L. Verschuur & K. I. Kellermann (2nd ed.; Berlin: Springer), 255
Robishaw, T., Quataert, E., & Heiles, C. 2008, *ApJ* 680, 981
Rovilos, E., Diamond, P. J., Lonsdale, C. J., Lonsdale, C. J., & Smith, H. E. 2003, *MNRAS* 342, 373
Thompson, T. A., Quataert, E., Waxman, E., Murray, N., & Martin, C. L. 2006, *ApJ* 645, 186
Wielebinski, R. 2006, *AN* 327, 510

Discussion

SARMA: From the I profiles, one can see that several masers are present in the Arecibo beam, and unresolved. Is it the expectation that VLBA observations will spatially resolve these into single maser "spots"?

ROBISHAW: Yes! Only a handful of the brightest OHMs have been imaged with VLBI. In most cases, the VLB maps recover all of the single-dish flux. For the measured milligauss field strengths in these OHMs, we should be able to directly map the field in these ULIRGs in a reasonable integration time by directly detecting the Zeeman splitting of the megamaser spots using the high-sensitivity VLBA.

Shea Brown (left), and Tim Robishaw (middle)

Timothy Robishaw

Philipp Kronberg

Cosmic Magnetic Fields:
From Planets, to Stars and Galaxies
Proceedings IAU Symposium No. 259, 2008
K.G. Strassmeier, A.G. Kosovichev & J.E. Beckman, eds.

doi:10.1017/S1743921309031172

Magnetic field transport from AGN cores to jets, lobes, and the IGM

Philipp P. Kronberg[1] †

Department of Physics, University of Toronto, Toronto, ON M5S 1A7, Canada
email: kronberg@physics.utoronto.ca

Abstract. I describe various stages of energy flow along an extragalactic jet, which subsequently evolves into an extended lobe which is visible in radio and X-rays. The sizes of the lobes vary from kpc scales to several megaparsec, so that the largest lobes are clearly injecting *back hole* energy into the IGM on scales comparable with a galaxy-galaxy separation. This is sometimes loosely referred to as Black hole-IGM "feedback". My talk begins with a well-formed jet, and avoids the complex and unclarified physics at less than a few Schwarzschild radii that cause the initial launching the jet.

This presentation focuses on recent thinking and supercomputer simulations that appear to clarify the fundamental nature of these remarkable jets and lobes. The energy transport process appears to be electrodynamic, rather than particle beam−driven. A new observational verification of a 10^{18} Ampère current in an actual jet is concordant with the predictions and simulations of poynting flux-dominated electromagnetic jets. In this model the current is tightly related to the BH mass and angular energy.

The magneto-plasma properties of the lobes must obviously match to the jets which feed them. The "energy sink" phase is when BH energy is ultimately deposited on supra-galactic scales. The process from the BH to the lobe production happens with remarkable efficiency. The presence or absence of a galaxy cluster environment creates laboratory conditions that help to calibrate the energy flow paths, and the magnetic rigidity of these jet-lobe systems.

I conclude by describing recent, sensitive radio observations on supra-cluster scales that test for final magnetic energy deposition - the "sink" phase - into the intergalactic medium.

Keywords. Galaxies: jets – galaxies: active – galaxies: magnetic fields – intergalactic medium

1. The nearby, black hole-driven jet in M87

Our ability to diagnose extragalactic radio jets is currently limited by instrumental sensitivity, and especially by limited resolution transverse to the jet. Currently, jets can be probed on scales of a few parsecs with very long baseline interferometry (VLBI), and on scales above ~ 100 parsecs with connected interferometers such as the NRAO VLA. The best cm λ VLBI images have been made at $\sim 10^{-3}$ arcsec resolution. The VLA's resolution is typically ~ 0.1 arcsec or larger. Figure 1 shows state-of-the-art images of the M87 jet in each resolution range.

At only 15Mpc distance, M87 is the nearest "laboratory" jet, emanating from a nuclear 10^9 $M_\odot$ black hole. By common intergalactic standards M87, despite its high BH mass, is a relatively small and low luminosity radio galaxy.

It is evident that in both régimes of angular resolution in Figure 1, the best available resolution at cm wavelengths fails to show the brightest jet features with high resolution *across* the jet. To study the jet in the desired detail, the transverse resolution should ideally be ~ 10 times higher, and at several wavelengths so that the Faraday rotation

† Present address: Los Alamos National Laboratory P.O Box 1663, MS T006, Los Alamos NM 87545, USA

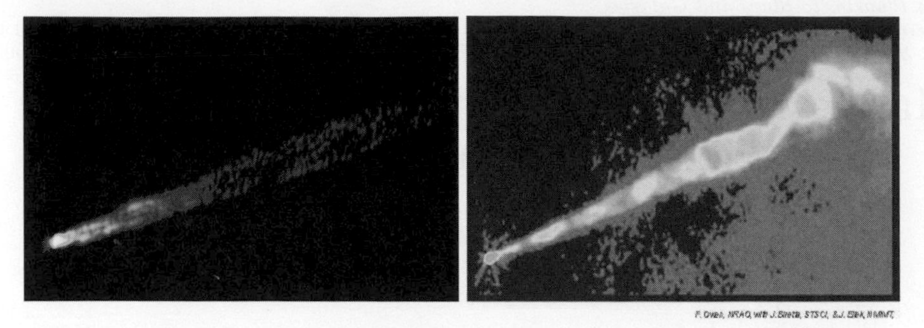

Figure 1. *Left*: VLBI image of the inner 0.01 arcsec of the M87 jet at a resolution of 0.8 × 1.3 milliarcsec (Kovalev *et al.*, 2007). *Right*: A VLA 15 GHz image of a 15 arcsec jet segment at a resolution of 0.1 arcsec Owen *et al.* 1989).

and polarization structure can be well studied. Yet M87 is one of the most easily "accessible" jets because of its unique proximity. This illustrates why extragalactic jets have not yielded easy answers about their detailed physics over the past four decades. Unfortunately there is little immediate prospect of exceeding the VLA's resolution at the required cm - λ wavelength range. However increasing interferometer capability down to $\lambda \sim 3$ mm ($\sim$ 90GHz) will partially improve this resolution deficiency. Partially, because higher resolution is also required at the longer cm wavelengths where Faraday rotation is needed to probe the plasma parameters. For kpc scale jets, the plasma parameters are such that sufficient Faraday depths occur mainly at the longer wavelengths, as will be illustrated in the discussion of 3C303 below.

2. 3C303 a kpc-scale jet at z = 0.141

The moderately distant 3C303 offers is a more typical example of a radio galaxy in terms of its radio luminosity and dimensions. Figure 2 shows the the central ~1/3 of the radio image, where we see a repeating "knot" structure similar to M87, and seen in several other kpc-scale jets. However the knot lengths are about 20 times those of M87's knots, and their radiating volumes are $\sim 10^4$ times those in M87. This illustrates, as does Figure 1, that the magneto-plasma structures in these jets are a *highly scalable* phenomenon; The linear extents of the largest, "giant" radio galaxies (GRG) are a further factor of ~20 bigger than 3C303.

3C303 originally attracted interest because of its somewhat unusual morphology (Kronberg 1976a,b, 1986, Leahy & Perley 1991), and its unusual optical field (Kronberg *et al.* 1977). It had been imaged in I, Q, and U in 2 bands with the NRAO Green Bank Interferometer in 1970/71 (Kronberg, 1976b) and in 3 bands (1.4, 6 and 15 GHz, each in Stokes I, Q, and U) in 1981 with the newly completed VLA (Kronberg, 1986; see also Leahy & Perley, 1991). In an X-ray study of this system, Kataoka *et al.* (2003) found that the Chandra X-ray image of 3C303 has remarkable morphological similarities to the VLA 1.4 GHz image, including the jet "knots", and at the similar resolution of ∼ 2".

It has become apparent meanwhile that 3C303 shares common properties with many AGN-jet radio galaxies and quasars (see Bridle *et al.*, 1994). Its magnetic and plasma properties discussed below should be regarded as probably typifying a broad class of extragalactic jet systems.

2.1. *Jet plasma parameters*

The synchrotron luminosity of 3C303's jet, its associated equipartition field strength, the overall source luminosity, the regular magnetic field structure in the knots, lack of

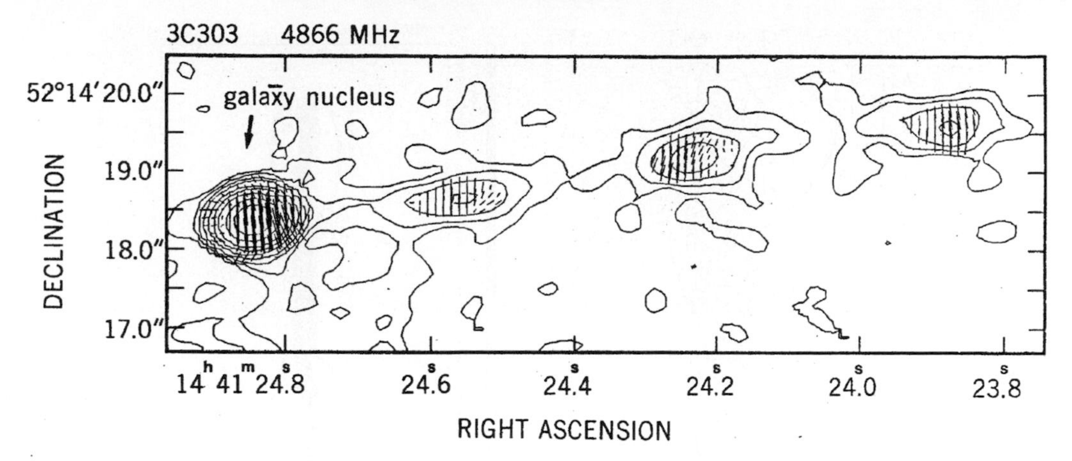

Figure 2. A VLA image of 3C303 at λ 6 cm in Stokes' parameters I, Q, and U. degree of polarization in the knots is high, of order 20%, and the polarization structure tends to repeat from knot to knot (Lapenta & Kronberg, 2005).

depolarization in the knots, and low depolarization between $\lambda\lambda$ 6 and 20 cm in the knots, are informative for constructing and testing physical models of the jet.

Details of the analysis methods for 3C303 are given in Lapenta and Kronberg (2005), and are summarized in the following results: The total source luminosity from ~100 MHz to ~10 keV is $\sim 1.7 \times 10^{42}$ erg s^{-1}. The equipartition-assumed jet magnetic field strength at $\simeq$ 400 pc from the jet axis is $|B| \sim 3mG$. This, combined with the Faraday rotation and depolarization limits gives $n_{th} \leqslant 1.4\times\ 10^{-5}$ cm^{-3} within the knot zones around the jet. The combination of |B| and n_{th} gives an approximate lower limit to the Alfvén speed within the knot volumes, which is a significant fraction of c.

$$V_A^{knot} \propto \frac{B^{knot}}{\sqrt{n_{th}}} \approx c \tag{2.1}$$

This means that the plasma $\beta = n_{th} k \mathrm{T}_8/(\mathrm{B}^2/8\pi)$ is very low ($\leqslant 10^{-5}$), T_8 being the plasma temperature in units of 10^8K. The correspondingly low n_{th} also suggests that the 1- 5 keV X-ray emission from the knots cannot be thermal bremsstrahlung, nor IC emission because of the mG level magnetic field, and must, solely on these arguments, be X-ray synchrotron radiation. The highly polarized radio synchrotron radiation indicates a well-organized magnetic field. It is consistent with, thought does not prove, a low pitch angle helical structure about the jet axis.

2.2. *First assumption-free measurement of an extragalactic jet current*

A particular magnitude and direction of a Faraday rotation measure gradient can, in the right circumstances, indicate an electric current. The resolved knot "C", third from the AGN in Figure 2 shows a RM gradient (∇RM), a 2−D vector that is exactly perpendicular to the jet axis. This would correspond to a net current, within a particular radius from the jet axis, if, and only if, RM = 0 exactly on the jet axis. To test this requires an absolute calibration of RM in the vicinity of 3C303's jet. This has recently been possible with a new large independent sample of extragalactic radio source RM's close to the (l,b) location of 3C303 where, in fact, the RM sky is relatively "quiet". Indeed RM $\approx$ 0 within measurement errors after subtracting the smoothed RM of nearby background

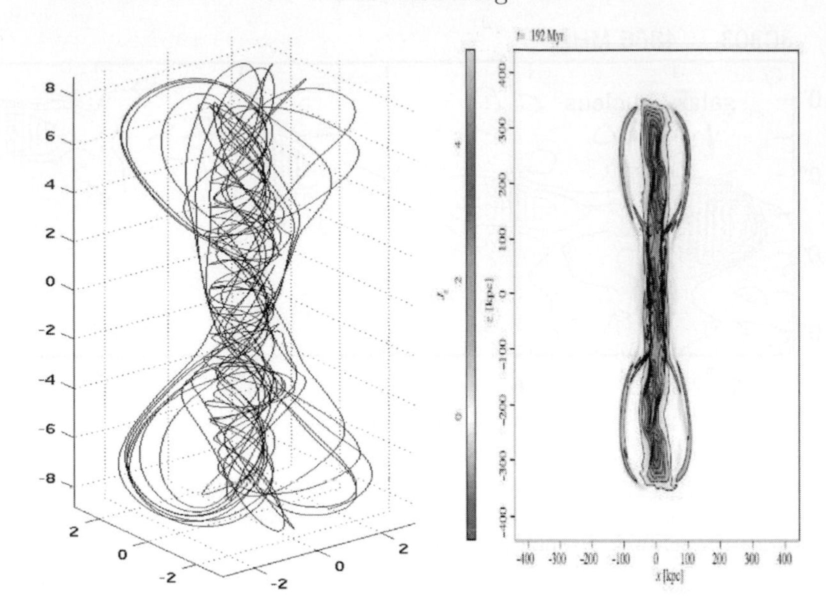

Figure 3. *Left*: The helical structure of a simulated "magnetic tower" poynting flux jet, after Li *et al.* (2006). *Right*: Simulated current structure of a combined magnetic tower poynting flux jet + lobe system (Nakamura *et al.*, 2008).

sources. Adding this information to the above indicates that the 3C303 jet has an average current within $\simeq$ 400 pc of its axis, which is 7×10^{17} ampères, and directed away from the central BH (AGN) (Ji *et al.*, 2008). It is the largest electrical current ever measured. This result is of particular interest in the context of poynting flux energy flow models for extragalactic jets, as we shall discuss in the next section.

3. Poynting flux jets

The idea that the jet power from a rotating black hole is delivered not via a particle beam, but electromagnetically, was proposed by Lovelace (1976), and developed in several succeeding papers: Blandford & Znajek (1977), Benford (1978), Lesch, Appl & Camenzind (1989) and others.

The extraction of this energy from the central BH involves complex magneto-plasma processes, and possibly the coupling into the BH rotational energy at the ergosphere in a Kerr metric-distorted space-time. Partly due to lack of access to observation, the physics of the jet launching at this level has not yet been completely clarified. However the already-formed jet better lends itself to modeling, analysis and observation as illustrated above. A considerable literature exists on jets and their confinement, and in this limited space I concentrate solely on the class of electromagnetic, poynting flux jets. Why? – because impressive recent progress has been made in understanding the relevant magneto-plasma processes, and in their simulation. Figure 3 shows a supercomputer simulation by Li *et al.* (2006) of an electromagnetically dominated jet and of its current structure (Nakamura *et al.* 2008).

In the electromagnetic model the jet current is coupled to the BH. It is integral to both the stability, and instabilities in the jet and lobe. The jet and lobe are part of the same magneto-plasma system. The magnitude of $\sim 10^{18}$ ampères for the electric current associated with a $10^8 M_\odot$ BH, was originally predicted by Lovelace (1976). It is integral to the simulations of Li *et al.* (2006) in Figure 3, and now tentatively confirmed in an actual measurement in (by Kronberg in Ji *et al.*, 2008) as discussed above.

Another success of this class of model is that the current-driven instabilities are more able than hydrodynamic, *e.g.* Kelvin-Helmholz, instabilities to recover at locations of instability and still retain the energy in the jet-lobe "system", rather than to scatter and dissipate the power. This robustness to complete disruption appears *both* observationally and theoretically confirmed in the multi-bubble (lobe) system in the Hydra A cluster(Wise *et al.*, 2007, Diehl *et al.*, 2008). Here the jet energy flow appears to be "pinched" at the "end" of one lobe, only to re-appear and energize a successive coherent lobe! Such behaviour appears only to be possible if the jet and lobe are magnetically dominated, and maintained by a (current driven) reverse field pinch process (*e.g.* Benford & Protheroe 2008).

The electromagnetic jet is likely to push shocked ambient ahead of it, causing shock-induced particle acceleration and a visible outer lobe beyond the magnetic structure in Fig 3a. This outer shock structure is broadly similar to what is predicted in particle beam/hydrodynamic models, and this has tended to mask the observational evidence for poynting flux dominated power flow. The jet's visibility in synchrotron radiation is expected from particle acceleration around the periphery of the electromagnetic jet.

4. Calibration of BH magnetic energy output into the IGM

4.1. *Quantification of BH energy release into the IGM*

The following graph, reproduced from Kronberg *et al.* (2001) shows measured total energy contents of lobe images for a large sample of giant radio galaxies (GRG), and compares these with the (gravitational) formation energy of the host galaxy's central black hole, taking $10^8 M_\odot$ as a fiducial BH mass. The upper envelope of GRG energy contents is, remarkably, within about an order of magnitude of the putative gravitational energy reservoir. This verifies the typical high efficiency of energy conversion from gravitational to magnetic η_B (and particle, η_P) energy (Kronberg *et al.* 2001). Radio lobes approximately within a cluster core radius (open squares in Fig. 4) have markedly lower internal energies, and the energy deficit, compared with the GRG's, is approximately the PdV work done against the intracluster gas, whose pressure can be independently estimated from the intracluster X-ray emission!

4.2. *Basic calculations of expectation*

The average density of supermassive black holes is now fairly well known, and shown to correlate with the bulge mass of the host galaxy. Given this average intergalactic BH density, ρ_{BH}, we expect that, sometime within a Hubble time, every central BH releases an enormous amount of magnetic energy $\sim 10^{61-62}$ergs into the IGM via a jet-lobe system. Thus, globally over a few $\times$ 10^9 yr this magnetic energy diffuses to an intergalactic volume.

$$
\begin{aligned}
&< \rho_{BH} > \approx 2 \times 10^5 M_\odot / Mpc^3 \\
&M_{BH} c^2 = 1.8 \times 10^{62} \tfrac{M_{BH}}{10^8 M_\odot} ergs
\end{aligned}
\tag{4.1}
$$

where $M_{BH}c^2$ is the gravitational energy per BH associated with infall down to the BH's Schwarzschild radius. Let us now assume that a fraction, $f^{VOL}_{\text{FILAMENTS}}$ of the entire IGM volume is *available* to be filled by magnetic flux from aggregate magnetic energy outflow. The available volume constitutes the filaments and walls of galaxy overdensity in large

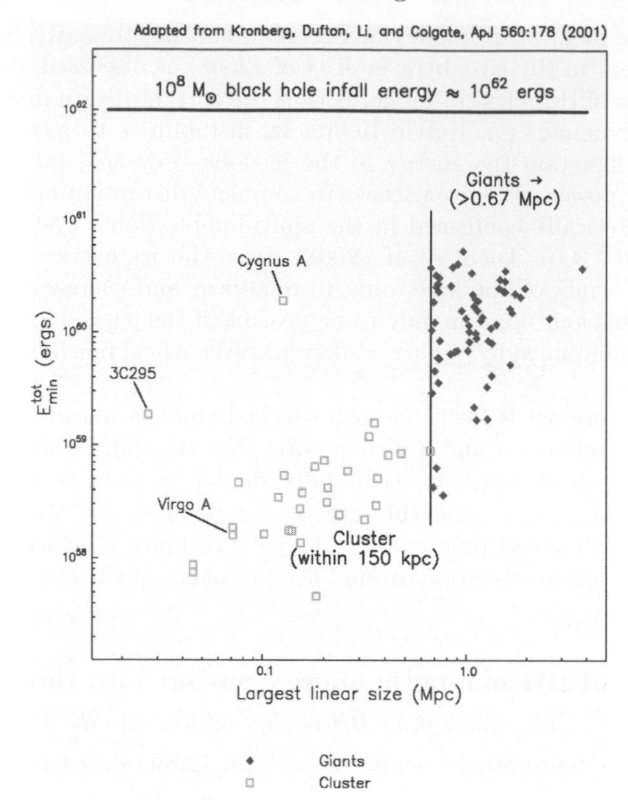

Figure 4. The total radio lobe energy content for (a) a 70-source sample of giant radio galaxies, and (b) for radio lobes within 150 kpc of the centre of a galaxy cluster. For sample (a) a lower size cutoff of 700 kpc was applied. Reproduced from Kronberg *et al.* (2001).

scale structure (LSS).

$$\varepsilon_{\rm B} = 1.36 \times 10^{-15} \left(\frac{\eta_{\rm B}}{0.1}\right) \times \left(\frac{f_{\rm RG}}{0.1}\right) \times \left(\frac{f^{VOL}_{\rm FILAMENTS}}{0.1}\right)^{-1} \times \left(\frac{M_{\rm BH}}{10^8 M_\odot}\right) {\rm erg\,cm^{-3}} \quad (4.2)$$

The fraction of $M_{BH}c^2$ released as magnetic energy is η_B, and fraction of all elliptical galaxies that become radio sources at some point in their lifetime is defined as $f_{\rm RG}$. $\varepsilon_{\rm B}$ is the average injected intergalactic magnetic energy density. The corresponding magnetic field strength is

$$B^{BH}_{IG} = \sqrt{8\pi\varepsilon_{\rm B}} = 1.8 \times 10^{-7}{\rm G} \quad (4.3)$$

for the normalization chosen in (4.2). This estimate is high enough to have some chance of being observable, by exploiting the capabilities of current instruments. I discuss this in the next section.

5. New attempts to verify BH magnetic energy "sinks" in the IGM

This section describes two attempts to detect intergalactic magnetic fields that might have been infused by central BH's. In the first, B^{BH}_{IG} would be revealed in faint diffuse synchrotron radiation. In the second, we searched for a Faraday rotation signal due to a hypothesized B^{BH}_{IG} in a nearby, large supergalactic filament – the Perseus-Pisces supercluster.

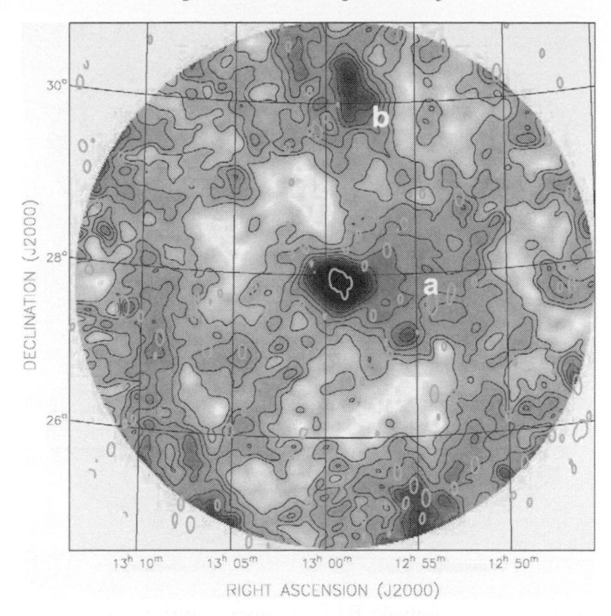

Figure 5. Diffuse radiation at 0.4GHz from the first combination of the Arecibo Telescope with the DRAO Interferometer. The resolution of this image is 10' - that of the Arecibo 305m telescope. The 2.7K CMB, and discrete sources have been removed, as has a linear-plane component of the Milky Way foreground radiation. Ellipses (proportional to source flux density)show the locations of the strongest discrete sources that were removed. T_b contours are shown at 1.4, 1.9, 2.4, 2.9, 3.4, 3.9, 4.4, 10, and 40K. The rms noise is 250mK, and is limited by confusion. Thermal noise is ~4 mK. Adapted from the colour original in Kronberg, Kothes, Salter, & Perillat (2007).

5.1. *Diffuse synchrotron radiation detected with an unprecedented combination of T_b sensitivity and resolution*

The detection of a zone of extragalactic magnetic field by synchrotron radiation requires a very high instrumental surface brightness sensitivity, and a resolution that is sufficiently high to ensure that the emission is diffuse, and not a blend of faint discrete radio sources. Finally, since faint diffuse radiation patches can also be expected from Milky Way foregrounds, some way is needed to plausibly ascertain if the diffuse emission in question truly intergalactic.

A combined Arecibo-DRAO Interferometer image of a 9° field within the Coma supercluster at $\simeq$ 100 Mpc distance is shown in Figure 5. Here, the output of deep scans with the 10' beam of the Arecibo telescope were Fourier transformed into u-v space, and combined with double, full synthesis observations of the same wide field with the DRAO Interferometer (Kronberg *et al.* 2007). The full resolution of the resulting combined image $\simeq$ 3.5', that of the DRAO interferometer.

The area "a" contains widespread, diffuse glow that appears "connected" to the strong synchrotron halo of the Coma cluster itself. Within its projected size of 2Mpc, this feature contains about 7 low luminosity radio galaxies and all have redshifts comparable with the Coma cluster. This large region has comparable dimensions to a giant radio galaxy, but it is not associated with any single galaxy. Its synchrotron luminosity can be interpreted as due to the collective energization of several AGN-powered radio sources. The total energy content of region "a" is $\sim 10^{60}$ ergs, and B_{IG} is 10^{-7}G or greater, both of which are consistent with this interpretation. The unusually high concentration of radio galaxies in the Coma region of the "Great Wall" supercluster enables a kind of test

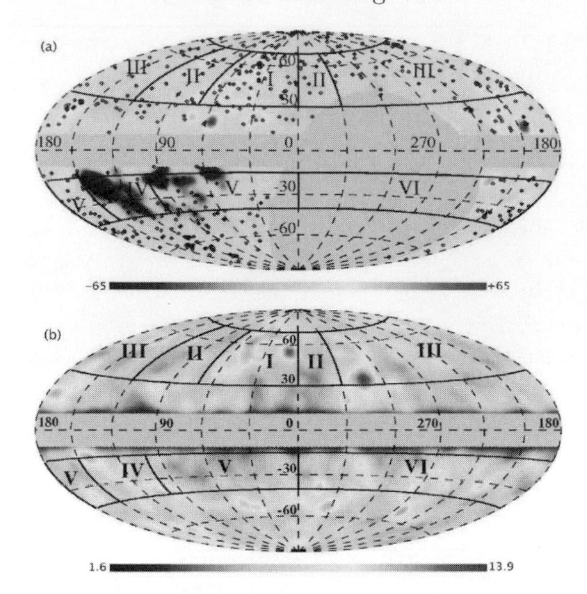

Figure 6. Smoothed extragalactic source RM's (*upper*), and smoothed 2MASS galaxy column densities (*lower*) compared at the same resolution of 7°. The p-p supercluster "chain" is approx. between l 120° and 150°, and from b -18° to -50°. Further details can be found in Xu *et al.* (2006).

laboratory for equation (4.2). It is likely the first direct confirmation of the magnetization of the IGM due to *a collectivity* of radio galaxies. For further details, and discussion of other new radio features in Figure 5 see Kronberg *et al.* (2007).

5.2. *Search for a detectable* B_{IG}^{BH} *via Faraday rotation in a nearby filament of galaxy overdensity*

As with the synchrotron glow test for intergalactic fields at the 10^{-7}G level, A Faraday RM test also presses the limits of current technology. Using a suitable subset of extragalactic source RM's with a high measurement accuracy, Xu *et al.* (2006) compared the RM's with the (l,b)- dependent galaxy densities in the 2MASS galaxy survey, and separately for the CfA2 survey which had spectroscopic redshifts. The more accurate z's of the CfA galaxies permitted construction of 3-D intergalactic zones of galaxy overdensity, and the galaxy density - weighted pathlengths through these supercluster volumes. For the less accurate z's of the 2MASS survey, corresponding galaxy column densities were calculated in an independent exercise.

Xu *et al.* 2006 investigated 3 superclusters at locations suitably far from the higher RM zones of the Galactic plane. These were the Virgo, Hercules and Perseus-Pisces superclusters, and the test was to ascertain if any excess Faraday rotation could be seen through the supercluster volumes. The Virgo supercluster is too dilute in the sky, and the Hercules supercluster was subject to confusion with higher latitude Galactic RM features. The galaxy distribution in the Perseus-Pisces supercluster is the most clearly defined, and probably, though not with absolute certainty, less confused by a Galactic foreground RM. Assuming the latter not to be the case, Xu *et al.* found evidence for B_{IG} of few $\times 10^{-7}$G for assumed magnetic field reversal scales of 200-800 kpc. The fields must be scaled, of course, to n_{IG} the intergalactic electron density. Values of n_{IG} on these scales are still very uncertain, but were assumed to be a few $\times 10^{-5}$cm^{-3} on the basis of X-ray absorption and EUV -based estimates. More, and comparably accurate RM's, and more detailed galaxy counts and spectra are needed to carry this type of probe a stage further in the future. However these B_{IG} values are independently consistent with equation (4.2).

6. Conclusions

In these few pages I have selected some recent computational and observational results that appear to outline a self-consistent physical picture of the physics of magnetized extragalactic jets and radio/X-ray lobes. I have argued that current dominated jets and lobes, as simulated in several recent papers are coupled to each other, and that they are poynting flux dominated. I also reviewed their global energetics, with the conclusion that central BH's alone contribute much of their enormous formation energy to intergalactic space. They contribute significantly to a magnetized intergalactic medium.

Acknowledgements

I thank Stirling Colgate, Hui Li, and Masanora Nakamura for beneficial discussions, and Raul Cunha, Quentin Dufton, and Phyllis Orbaugh for their assistance. I also thank the Natural Sciences and Engineering Research Council of Canada (NSERC) and the U.S. Department of Energy for support.

References

Benford, G. & Protheroe, R. J. 2008, *MNRAS* 383, 663
Blandford, R. D. & Znajek, R. L. 1977, *MNRAS* 197, 433
Bridle, A. H., Hough, D. H., Lonsdale, C. J., Burns, J. O., & Laing, R. A. 1994, *AJ* 108, 766
Li, H., Colgate, S. A., Wendroff, B., & Liska, X. Z. 2001, *ApJ* 581, 874
Diehl, S., Li, H., Fryer, C. L., & Rafferty, D. 2008, *ApJ*, in press
Kataoka, J., Edwards, P., Georganopoulos, M., Takahara, F., & Wagner, S. 2003, *A&A* 399, 91
Kovalev, V. V., Lister, M. L., Homan, D. C., & Kellermann, K. I. 2007, *ApJ* 668, L27
Kronberg, P. P. 1976a, *ApJ* 203, L47
Kronberg, P. P. 1976b, *I.A.U. Symp.* 74
Kronberg, P. P., Burbidge, E. M., Smith, H. E., & Strom, R. G. 1977, *ApJ* 218, 8
Kronberg, P. P. 1986, *Can.J. Phys.* 64, 449
Kronberg, P. P., Dufton, Q. W., Li, H., & Colgate, S. A 2001, *ApJ* 560, 178
Kronberg, P. P., Kothes, R., Salter, C. J., & Perillat, P. 2007, *ApJ* 659, 267
Lapenta G. & Kronberg, P. P. 2005, *ApJ* 625, 37
Leahy, P. & Perley, R. A. 1991, *AJ* 102, 537
Lesch, H., Appl, S., & Camenzind, M. 1989, *A&A* 225, 341
Li, H., Lapenta, G., Finn, J. M., Li, S., & Colgate, S. A. 2006, *ApJ* 643, 92
Lovelace, R. V. E. 1976, *Nature* 262, 649
Nakamura, M., Tregillis, I. L., Li, H., & Li, S. 2008, *ApJ* 686, 843
Owen, F. N., Biretta, J., & Eilek, J. 1989, *ApJ* 698, 707
Robinson, K., Dursi, L. J., Ricker, P. M., Rosner, R., Calder, A. C., Zingale, J. W., Truran, J. W., Linde, T., Cacares, A., Fryxell, B., Olson, K., Riley, K., Siegel, A., & Vladimirova, N. 2004, *ApJ* 601, 623
Wise, M. W., McNamara, B. R., Nulsen, Houck, P. E. J., & David, L. P. 2007, *ApJ* 659, 1153
Xu, Y., Kronberg, P. P., Habib, S., & Dufton, Q. W. 2007, *ApJ* 637, 19

Discussion

DURRER: Are the magnetic fields in PP super clusters coherent? On which scales?

KRONBERG: If confirmed, they must have a large coherent component, up to ≈800 kpc.

DE GEOUVEIA DAL PINO: There is a poster upstairs where we have explored the role of SN-turbulent feeding on the ICM central region of the Perseus cluster and we have found that this can be a competing and even a dominant effect when compared to the AGN feedback (Falceta-Gonçalves, de Gouveia dal Pino, Gallagher, Lazarian (Poster 1.10, this conf.). Could you comment on the relative importance of these two mechanism for the central ICM structure and evolution?

KRONBERG: That is entirely possible, I agree, especially in the ICM. I avoided star/SN-driven outflow in my talk *only* to restrict my focus to massive-BH driven energy flows within the limited time.

SCHLICKEISER: 1) Unless there is a return current, your quoted outflow current of $\approx 10^{18}$ violates the Alfven current limit by 14 orders of magnitude. 2) The presence of a pair outflow would naturally explain the small Faraday RM in 3C303. Why did you exclude this interpretation? 3) If the Alfven speed is of order $v_{\rm A} \simeq c$ you can not form super-Alfvenic shock waves to explain your lobes.

KRONBERG: 1) Yes, there is a return current, but for 3C303 it is likely more diffuse, and outside $r \approx 250$ pc from the jet axis. Certain variations of the CDJ models *could*, however, have the return current closer to the axis than in the 3C303 system. 2) Possibly, but not the ∇RM measurement. 3) The electron acceleration in the highly magnetized lobes might have other causes. However, I only discussed $v_{\rm A}$ in the jet "cocoon" zone.

Volker Heesen

Cosmic Magnetic Fields:
From Planets, to Stars and Galaxies
Proceedings IAU Symposium No. 259, 2008
K.G. Strassmeier, A.G. Kosovichev & J.E. Beckman, eds.

doi:10.1017/S1743921309031184

The magnetic field structure in NGC 253 in presence of a galactic wind

Volker Heesen[1], M. Krause[2], R. Beck[2], and R.-J. Dettmar[1]

[1]Astronomisches Institut der Ruhr-Universität Bochum, 44780 Bochum, Germany
email: heesen@astro.rub.de, dettmar@astro.rub.de

[2]Max-Planck-Institut für Radioastronomie, 53121 Bonn, Germany
email: rbeck@mpifr-bonn.mpg.de, mkrause@mpifr-bonn.mpg.de

Abstract. We present radio continuum polarimetry observations of the nearby edge-on galaxy NGC 253 which possesses a very bright radio halo. Using the vertical synchrotron emission profiles and the lifetimes of cosmic-ray electrons, we determined the cosmic-ray bulk speed as $300 \pm 30\,\mathrm{km\,s^{-1}}$, indicating the presence of a galactic wind in this galaxy. The large-scale magnetic field was decomposed into a toroidal axisymmetric component in the disk and a poloidal component in the halo. The poloidal component shows a prominent X-shaped magnetic field structure centered on the nucleus, similar to the magnetic field observed in other edge-on galaxies. Faraday rotation measures indicate that the poloidal field has an odd parity (antisymmetric). NGC 253 offers the possibility to compare the magnetic field structure with models of galactic dynamos and/or galactic wind flows.

Keywords. Galaxies: halos – galaxies: individual (NGC253) – galaxies: magnetic fields – galaxies: spiral – radio continuum: galaxies

---

1. Introduction

Gaseous halos are a common property of star forming galaxies. The abundance of hot coronal gas in the halo requires heating from the star forming disk and a transport of energy from the disk into the halo. This energy transport has been discussed as galactic fountains or chimneys (Field *et al.* 1969, Norman & Ikeuchi 1989). The relativistic cosmic-ray gas has an adiabatic index of 4/3 compared to 5/3 of the (single atomic) hot gas; this results in a larger pressure scaleheight of the cosmic-ray gas. The magnetic field has an even larger pressure scaleheight (Beck 2009). The pressure contributions of the cosmic rays and the magnetic field are thus dominating in the halo and can accelerate the outflow. This led to the model of a cosmic-ray driven wind that causes acceleration of the flow accelerates further away from the disk (Breitschwerdt *et al.* 1991).

The existence of radio halos around many nearby galaxies shows the presence of cosmic rays and magnetic fields. Cosmic rays spiral around magnetic field lines and follow them from the disk into the halo. Hence, a vertical component of the large-scale magnetic field can enhance the vertical cosmic-ray transport. How to generate vertical fields is still an open question. Parker (1992) suggested that field lines start to overlap and reconnect to form "open" field lines that connect the disk with the halo. Another possibility is the mean field $\alpha\Omega$-dynamo, which creates dipolar or quadrupolar field configurations. Furthermore, the galactic wind may shape the magnetic field and align with it. This may particularly hold for *superwinds*, which originate in a region of very active star formation (starburst).

As shown by X-ray observations, NGC 253 has a superwind and is an ideal object to study the interaction between the superwind and the magnetic field. There is a nuclear plume of Hα and X-ray emitting gas (Strickland *et al.* 2000, Bauer *et al.* 2007) where the outflow velocity was directly measured by spectroscopy as 390 km s^{-1} (Schulz & Wegner 1992). The connection between the heated gas in the halo, which extends far from the disk, and the the superwind is likely as suggested by numerical simulations of a centrally driven wind (Suchkov *et al.* 1994). A detailed study of this galaxy is possible because it is nearby ($D = 3.94$ Mpc, Karachentsev *et al.* 2003, $30'' = 600$ pc). With an inclination angle of 78° the galaxy's disk is only mildly edge-on. But it contains a very bright nucleus which complicates the data reduction of radio continuum observations.

2. Observations

Our new results are based on a deep mosaic with the VLA at λ6.2 cm. This wavelength is optimal to study magnetic fields in galactic halos, because the polarization is not too much affected by Faraday rotation and depolarization. On the other hand, the contribution from the thermal emission is still small ($< 10\%$) and the vertical extent is large. The largest angular scale that can be well imaged with the VLA at λ6.2 cm is $5'$. We used a λ6.3 cm map observed with the 100-m Effelsberg telescope to fill in the missing zero-spacing flux.

Additionally we used VLA maps at λ20 cm (Carilli *et al.* 1992) and at λ90 cm (Carilli 1996). In order to correct the magnetic field for Faraday rotation we observed the galaxy with the 100-m Effelsberg telescope at λ3.6 cm. A detailed description of the observations and the data reduction can be found in Heesen (2008).

3. Morphology

In Fig. 1 we present the λ6.2 cm total power radio continuum emission from the combined VLA and Effelsberg observations. The vectors show the Faraday corrected orientation of the large-scale magnetic field. The emission shows a prominent dumbbell shaped radio halo. The vertical extent is smallest near the center and increases further out in the disk. A detailed analysis of the vertical emission profiles shows that the scaleheights are smallest near the center.

The distribution of the polarized emission shown in Fig. 2 is more concentrated to the inner disk than the total power emission. It is extending into the halo particularly in the inner disk. The orientation of the magnetic field is disk-parallel in the inner part of the disk but turns away from the disk in the outer part. The vertical component even dominates at the "radio spur" (R.A. = $00^{\mathrm{h}}47^{\mathrm{m}}50^{\mathrm{s}}$, Dec. = $-25°17^{\mathrm{m}}00^{\mathrm{s}}$), where the magnetic field orientation is almost perpendicular to the disk.

4. Cosmic-ray bulk speed

The dumbbell shaped radio halo shows that the vertical extension is smallest in the inner part of the disk where the magnetic field strength is highest. Synchrotron losses are the dominant energy loss of the cosmic-ray electrons (Heesen *et al.* 2009a). The synchrotron lifetime

$$t_{\mathrm{syn}} = \frac{8.352 \times 10^9\ \mathrm{yr}}{(\nu/16.1\,\mathrm{MHz})^{1/2}(\mathrm{B}_\perp/\mu\mathrm{G})^{3/2}} \tag{4.1}$$

depends on the total magnetic field strength $B_\perp$ perpendicular to the line-of-sight and on the observation frequency ν. Thus, it is smallest in the central part of the disk. To

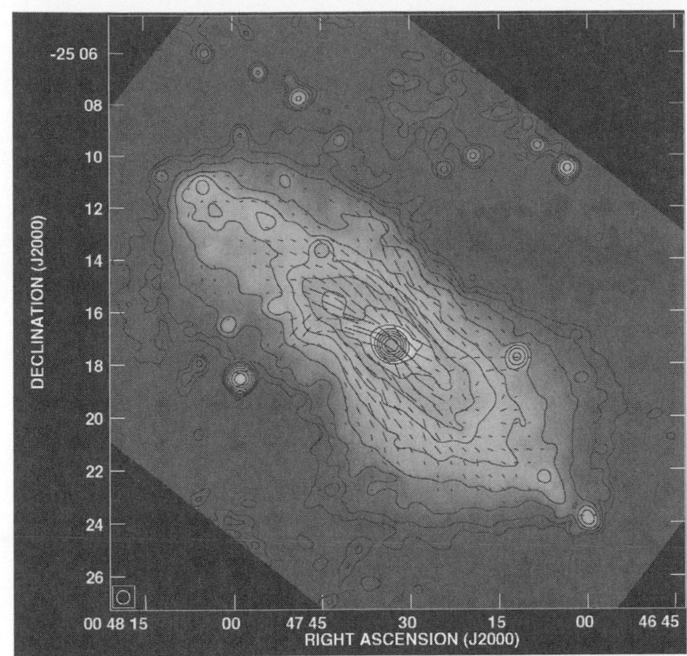

Figure 1. Total power radio continuum at $\lambda 6.2$ cm obtained from the combined VLA + Effelsberg observations with $30''$ resolution. Contours are at 3, 6, 12, 24, 48, 96, 192, 384, 768, 1536, 3077, 6144, 12288, and 24576 $\times$ the rms noise of $30\,\mu$Jy/beam. The overlaid vectors indicate the orientation of the Faraday corrected regular magnetic field. A vector length of $1''$ is equivalent to $12.5\,\mu$Jy/beam polarized intensity.

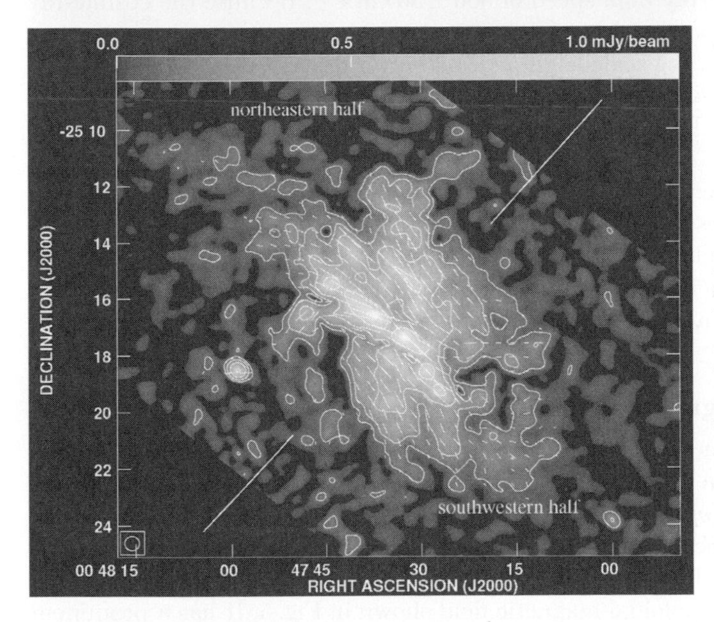

Figure 2. Polarized intensity obtained from the combined VLA + Effelsberg observations at $\lambda 6.2$ cm with $30''$ resolution. Contours are at 3, 6, 12, and 24 $\times$ the rms noise level of $30\,\mu$Jy/beam. The overlaid vectors indicate the orientation of the Faraday corrected regular magnetic field. A vector length of $1''$ is equivalent to $12.5\,\mu$Jy/beam polarized intensity.

calculate the electron lifetime we have to add the timescale $t_{\rm a}$ of the adiabatic losses:

$$\frac{1}{t_{\rm e}} = \frac{1}{t_{\rm a}} + \frac{1}{t_{\rm syn}}. \tag{4.2}$$

Figure 3 shows the scaleheight of the synchrotron emission as a function of the electron lifetime. The linear dependence, particularly in the northeastern halo of the galaxy (c.f. Fig. 2), suggests a definition of an average cosmic-ray bulk speed

$$\mathrm{v} = \frac{h_{\rm e}}{t_{\rm e}}, \tag{4.3}$$

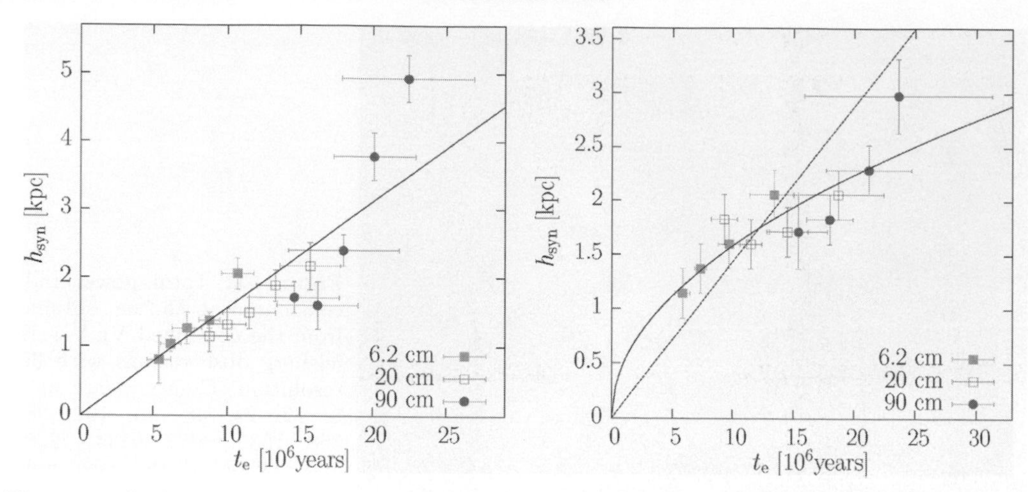

Figure 3. Scaleheight h_{syn} of the thick radio disk as a function of the electron lifetime t_e in the northeastern halo (left) and in the southwestern halo (right). The linear fit is the theoretical expectation for a convective cosmic-ray transport with a constant bulk speed. The fit $h_e \propto \sqrt{t_e}$ is the theoretical expectation for a diffusive cosmic-ray transport.

where h_e is the electron scaleheight (twice the synchrotron scaleheight). In the northeastern halo, we find a cosmic-ray bulk speed of $300 \pm 30\,\mathrm{km\,s^{-1}}$. Because the cosmic-ray bulk speed does not depend on the electron energy, the cosmic-ray transport is convective. In the southwestern halo, the scaleheight can be better described by $h_e \propto \sqrt{t_e}$. The cosmic-ray bulk speed increases with the electron energy, which indicates a larger role of diffusion in this halo part. In case of pure diffusion we have a diffusion coefficient of $\kappa = 2.0 \pm 0.2 \times 10^{29}\,\mathrm{cm^2\,s^{-1}}$.

Because the cosmic-ray electrons lose their energy via synchrotron radiation, their energy spectral index increases during their lifetime. This is visible as a steepening of the radio spectral index in the halo. We detected an almost linear dependence of the radio spectral index on the distance from the disk. From this we derived $170\,\mathrm{km\,s^{-1}}$ as a lower limit for the average cosmic-ray bulk speed.

5. Structure of the magnetic field

Our model of an axisymmetric spiral, disk-parallel *toroidal* magnetic field (Fig. 4) represents the observations well. We chose the inclination angle of the optical disk and a best-fit spiral pitch angle of 25°, which is similar to the pitch angle of 20° of the optical spiral arms. At the locations where the toroidal magnetic field is weak we also observe a vertical *poloidal* field (Fig. 2). We subtracted the toroidal field model from the observations and obtained the poloidal magnetic field shown in Fig. 5. It has a prominent X-shaped structure centered on the nucleus.

The structure of the poloidal magnetic field seems to be connected to the distribution of the heated gas in the halo. This gas forms a horn-like structure with lobes that are thought to be the walls of a superbubble filled with dilute hot gas. This can be seen in Fig. 5 where the sensitive XMM observations show X-ray emitting gas also inside the bubble. The similarity of the structure of the poloidal magnetic field to that of the superbubble and their alignment can be explained by an interaction between the superwind, driven by the nuclear starburst, and the surrounding medium transported by the disk wind, possibly enhanced by shock waves of the expanding superbubble.

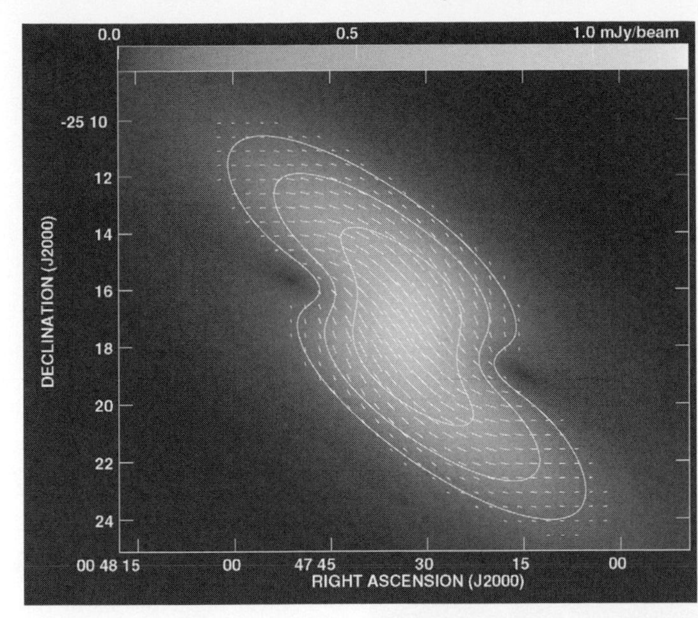

Figure 4. Polarized emission and magnetic field orientation expected for an axisymmetric spiral model for the toroidal magnetic field at 180″ resolution. Contours and vectors as in Fig. 2.

The difference between the polarization angles at λλ 6.2 cm and 3.6 cm is due to Faraday rotation. This provides information about the line-of-sight component of the large-scale magnetic field. Krause *et al.* (1989) showed that the azimuthal variation of rotation measure (RM) can be used to determine the contributions from individual dynamo modes. In NGC 253, the RM is the superposition of the toroidal magnetic field in the disk and the poloidal magnetic field in the halo. Subtracting the model for the disk magnetic field leaves the RM distribution of the poloidal magnetic field. It has the maximum amplitude along the minor axis. We propose a conical configuration for the poloidal magnetic field with an opening angle of 66° of odd symmetry: the field direction points towards the disk in the southern hemisphere (c.f. 5) and away from the disk in the northern hemisphere (Heesen *et al.* 2009). The field lines are aligned with the lobes of hot gas in the halo.

6. Conclusions

NGC 253 possesses a bright radio halo. The scaleheight of the synchrotron emission depends on the lifetime of the cosmic-ray electrons. This requires a vertical cosmic-ray transport from the disk into the halo. The disk wind has an average velocity of $300 \pm 30\,\mathrm{km\,s^{-1}}$ and is surprisingly constant over the extent of the disk. The disk wind transports material from the disk into the halo. If this material is the origin of the luminous, heated gas in the halo, it can explain the asymmetry between the northeastern and southwestern halo, as the cosmic-ray transport is much more efficient in the convective northeastern halo than in the diffusive southwestern halo.

In the disk the magnetic field is parallel to the disk. A model of an axisymmetric spiral disk field can explain the observed asymmetries in polarization by a geometrical effect. After subtracting the disk field model from the observations we obtain the poloidal magnetic field in the halo. This field has a prominent X-shaped structure that is also seen in several other edge-on galaxies. For the first time the parity of the large-scale field in any galaxy could be determined, which is odd (antisymmetric) in NGC 253. The X-shape may be explained by interaction of the halo gas transported by the disk wind with the

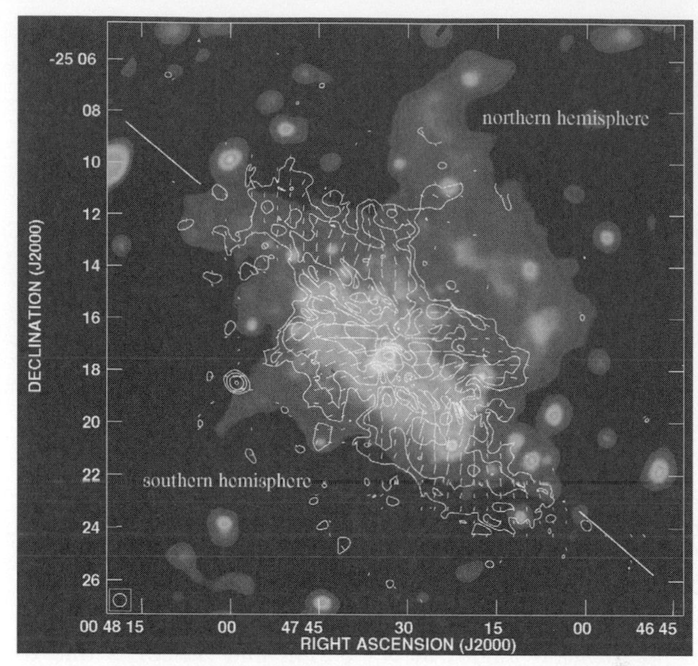

Figure 5. Polarized emission and magnetic field orientation of the poloidal magnetic field overlaid onto diffuse X-ray emission. Contours and vectors as in Fig. 2. The X-ray map is from XMM observations in the energy band 0.5–1.0 keV (Bauer *et al.* 2008).

superwind from the starburst center. However, several galaxies without a superwind also show X-shaped halo fields (Krause 2008), which needs to be further investigated.

References

Bauer, M., Pietsch, W., Trinchieri, G., Breitschwerdt, D., Ehle, M., & Read, A. 2007, *A&A* 467, 979

Bauer, M., Pietsch, W., Trinchieri, G., Breitschwerdt, D., Ehle, M., Freyberg, M. J., & Read, A. M. 2008, *A&A* 489, 1029

Beck, R. 2009, *these proceedings*

Breitschwerdt, D., McKenzie, J. F., & Voelk, H. J. 1991, *A&A* 245, 79

Carilli, C. L., Holdaway, M. A., Ho, P. T. P., & de Pree, C.G. 1992, *ApJ* 399, L59

Carilli, C. L. 1996, *A&A* 305, 402

Field, G. B., Goldsmith, D. W., & Habing, H. J. 1969, *ApJ* 155, L149

Heesen, V. 2008, *PhD thesis*, Ruhr-Universität Bochum, Germany

Heesen, V., Beck, R., Krause, M., & Dettmar R.-J., 2009a, *A&A*, in press; arXiv:0812.0346

Heesen, V., Krause, M., Beck, R., & Dettmar R.-J. 2009b, *to be submitted*

Karachentsev, I. D., Grebel, E. K., Sharina, M. E., Dolphin, A. E., Geisler, D., Guhathakurta, P., Hodge, P. W., Karachentseva, V. E., Sarajedini, A., & Seitzer, P. *A&A* 404, 93

Krause, M., Hummel, E., & Beck, R. 1989, *A&A* 217, 4

Krause, M. 2008, *Magnetic Fields in the Universe II*, eds. A. Esquivel *et al.*, arXiv:0806.2060

Norman, C. A. & Ikeuchi, S. 1989, *ApJ* 345, 372

Parker, E. N. 1992, *ApJ* 401, 137

Schulz, H. & Wegner, G. 1992, *A&A* 266, 167

Strickland, D. K., Heckman, T. M., Weaver, K. A., & Dahlem, M. 2000, *AJ* 120, 2965

Suchkov, A. A., Balsara, D. S., Heckman, T. M., & Leitherner, C. 1994, *ApJ* 430, 511

Cosmic Magnetic Fields:
From Planets, to Stars and Galaxies
Proceedings IAU Symposium No. 259, 2008
K.G. Strassmeier, A.G. Kosovichev & J.E. Beckman, eds.

doi:10.1017/S1743921309031196

What do we really know about the magnetic fields of the Milky Way?

Richard Wielebinski

Max-Planck-Institut für Radioastronomie, Auf dem Hügel 69, 53121 Bonn, Germany
email: rwielebinski@mpifr-bonn.mpg.de

Abstract. We have several methods of measuring magnetic fields in the Milky Way. We can study optical polarization, radio polarization, rotation measures of pulsars and extragalactic radio sources as well as include Zeeman results. Each of the above mentioned methods was at times used to make a model of the magnetic fields of the Milky Way. However one or two of the data sets by themselves cannot tell us the whole story. Any model of the magnetic fields must be able to fit all the observational results. At the present time a lot of progress has been made. We have increased our data bases in most of the observational areas. However a robust model of the magnetic field of the Milky Way has not yet emerged. We must possibly wait to the era of SKA.

Keywords. Galaxies: Milky Way – galaxies: magnetic fields – galaxies: structure – ISM: magnetic field

1. Introduction

We have been studying the magnetic fields of the Milky Way ever since the first optical polarization observations were published. Radio observations confirmed the optical results and gave a new impetus to the study of cosmic magnetic fields. Extragalactic sources became potential probes of the magnetic fields in the Milky Way. Finally pulsars became the probes par excellence for the determination of Milky Way's magnetic field. The Zeeman effect, the logical method of observing magnetic fields, has succeeded only in molecular clouds so far.

Models of the magnetic fields in the Milky Way followed the observational results. From the very beginning it was clear that the magnetic field was directed along the spiral arms. Optical polarization data suggested that this pertained only to the nearby magnetic fields, i.e. $l < 1$ kpc. The 'B' vectors were generally aligned along the Galactic plane with some suggestion of extensions along the North Polar Spur. The general discussion followed the original description of the spiral structure of the Galaxy proposed by Georgelin & Georgelin (1976). The discussion then evolved around the question whether the magnetic fields of the Milky Way were bi-symmetrical or axi-symmetrical. The bi-symmetrical idea bases on the early suggestion by Piddington (1964) that compression of a primordial field by a rotating galaxy would create such a magnetic field pattern. The alternative interpretation was offered by the dynamo theory (e.g. Parker 1955; Steenbeck & Krause 1969) especially after the multimode extension of the dynamo process was published (Baryshnikova, Shukurov, Ruzmaikin, *et al.* 1987). In this theory a multiplicity of modes is possible, including the axi-symmetric mode. In the present contribution I will discuss the recent observational results and conclude that we still have no robust model of the magnetic fields of the Milky Way. I will also point out that any model must satisfy the data available from the different observational methods.

2. Optical polarization

After the early observations, dating back to 1949, various observers collected data on the polarization of stars. A synthesis of all the available data was made by Mathewson & Ford (1970) that showed the general structure of the nearby magnetic field. The local magnetic field was aligned along the Galactic Plane with some additional loops that indicated local objects (large Supernova Remnants). More recent catalogues collected by Heiles (2000) increased the number of objects but did not go much beyond the original limitation of observing the nearby magnetic field.

3. Radio continuum polarization

The polarization of the diffuse radio emission that was finally discovered in 1962 had a long period of quiescence after the surveys of Wielebinski & Shakeshaft (1964), Mathewson & Milne (1965) and Brouw & Spoelstra (1976) have been published. The follow-up investigation by Spoelstra (1984) concluded that the Rotation Measure (RM) in the ISM towards $l = 140°$ was low, $RM < 18$ rad m^{-2}. The realization that polarization changes dramatically with higher angular resolution due to high RM (e.g. Wieringa, de Bruyn, Jansen, *et al.* 1993; Uyanıker, Fürst, Reich, *et al.* 1999) led to new work that has added a lot of significant observations. Radio polarization observations at higher frequencies allowed investigations deeper in the Galaxy, showing that even at the Galactic centre continuum polarization can be observed (e.g. Seiradakis, Lasenby, Yusef-Zadeh, *et al.* 1985). The northern sky was mapped in polarization by Wolleben, Landecker, Reich, *et al.* (2006) and the southern sky by Testori, Reich & Reich (2008) both with good angular resolution. These surveys showed that the Faraday effects are large and not only confined to the narrow Galactic plane (see the contribution of W. Reich & P. Reich, these Proceedings). This was also confirmed by Gaensler, Madsen, Chatterjee, *et al.* (2008) who analysed pulsar dispersion measure and the diffuse Hα emission. The three-year WMAP data (Page, Hinshaw, Komatsu, *et al.* 2007) gave us new insight of the radio polarization distribution at high frequencies. The new thermal electron model of Cordes & Lazio (2002) was also available. As a result of these surveys a simulation was made by Sun, Reich, Waelkens, *et al.* (2008) which is an advance on previous such investigations. The work of Sun, Reich, Waelkens, *et al.* (2008) suggests an axi-symmetric magnetic field of the Galaxy. To confirm this result we need a dense grid of RMs (an RM all-sky survey) with all the anomalous values due to individual HII regions sorted out.

4. Extragalactic radio sources

The collection of data on the Rotation Measure of extra-galactic radio sources (EGRS) has a long tradition. The survey of Simard-Normandin & Kronberg (1980) was used to model the magnetic fields in the Milky Way. In this work the first identification of a magnetic field reversal was made. The discussion about reversals has been taken up by a number of authors, e.g. Vallée (1983, 2005). However in the data base of Simard-Normandin & Kronberg (1980) very few EGRS were observed through the plane of the Galaxy, where the magnetic fields are concentrated. This investigation of RM of the EGRS in the Galactic plane was taken up by Brown & Taylor (2001); Brown *et al.* (2003, 2007). The northern sky data shows a rather regular distribution of RMs towards the Perseus arm with some rather high values of RM, e.g. RM $\sim$ 700 rad m^{-2}. The southern sky survey of EGRS in the Galactic plane showed that values of up to RM $\sim$ 1000 rad m^{-2} are observed. Also the observed reversals of the sign of the RM led Brown, Haverkorn, Gaensler, *et al.* (2007) to conclude that there is possibly a magnetic field reversal in the inner Galaxy. The study of the RM of EGRS towards Galactic centre by Roy, Pramesh

Rao & Subrahmanyan (2008) led to the suggestion of a bi-symmetrical magnetic field. To confirm this interpretation we need a much denser sampling of sources.

5. Pulsars

Pulsars give an excellent method of studying magnetic fields since the combination of the Dispersion Measure DM and the RM gives us a value of the mean magnetic field along the line of sight. From the DM and a model of electrons (e.g. Cordes & Lazio 2002) a distance of the pulsar can be determined. The discussion about the magnetic fields of the Galaxy based on pulsar observations dates back to a number of authors (e.g. Rand & Kulkarni 1989; Han & Qiao 1994). The addition of new pulsar observations let to even more detailed models (e.g. Han, Manchester, Lyne, *et al.* 2002, 2006) suggesting a number of large-scale reversals as could be expected in a bi-symmetrical system. Recently however a number of new investigations questioned this model. The field reversal beyond the Perseus arm was shown (Mitra, Wielebinski, Kramer, *et al.* 2003) to be due to two pulsars observed behind an HII region, where the magnetic field is reversed. Also studies of further pulsars by Noutsos, Johnston, Kramer, *et al.* (2008) questioned the accepted interpretation. Men, Ferrière & Han (2008) investigated the available pulsar data and ruled out all the simple models that have been published to date.

6. The Zeeman effect

The Zeeman effect is the method par excellence to measure magnetic fields. However we are limited here to dense molecular clouds in the Milky Way (e.g. Heiles & Crutcher 2005). A recent investigation of the Zeeman results by Han & Zhang (2007) suggests that molecular clouds "remember" the direction of the local magnetic field. Since we do not know exactly the real directions of the local magnetic field we must wait until more data becomes available.

7. Discussion with some conclusions

There are many local reversals seen in the pulsar and EGRS data sets. Some of the reversals seem to be associated with HII regions. A large, but possibly local reversal is seen towards $l = 75°$. Most of the pulsars that show this reversal are at a distance of $d \sim 1$ kpc. This would be a local rather than large scale reversal. There are directions where the agreement between pulsar and EGRS data is good. On the other hand serious disagreement is seen, for instance in the direction $l = 10°$, $b = 25°$. The values of RMs in the anti-centre region are not very different to the RMs seen towards the Galactic centre. Also the pulsars and EGRS in the Galactic plane seem to be seen through similar Faraday screens - are these effects all local? This would also put the standard method of estimation of pulsar distances in question. It would also suggest that Faraday Rotation takes place in localised regions of high magnetic fields and that we cannot expect to map the magnetic fields of the Milky Way by these methods.

I have collected all the recently published literature pertaining to modelling of the magnetic fields of the Milky Way and have pointed out the limitations of the conclusions reached by many authors. An important fact that must be remembered is that all the different data sets must agree with a proposed robust model of the magnetic field of the Milky Way. It is not realistic to build up a model using only pulsars but disregarding radio continuum polarization data. We also must have a good sampling of the sources across the sky before we can make robust models. In particular we have insufficient data on the RM of southern EGRS. Pulsars are concentrated close to the Galactic plane and the inner Galaxy. The observations of RM of EGRS along the Galactic plane is still incomplete. There is a great need of all-sky surveys (with polarization) in the $\lambda \sim 6$ cm wavelength range. We must have information (all-sky surveys) of the RM of the Galactic radio continuum with good angular resolution. Some of these 'needs' will be addressed by new projects like LOFAR or SKA. Some projects must be executed by dedicated special instruments. One of the 'small' problems that must be addressed is the absolute calibration of polarization surveys.

References

Baryshnikova, I., Shukurov, A., & Ruzmaikin, A., *et al.* 1987, *A&A* 177, 27
Brouw, W. N. & Spoelstra, T. A. Th. 1976, *A&AS* 26, 129
Brown, J. C. & Taylor, A. R. 2001, *ApJ* 563, L31
Brown, J. C., Taylor, A. R., Wielebinski, R., & Müller, P. 2003, *ApJ* 592, L29
Brown, J. C., Haverkorn, M., & Gaensler, B. M., *et al.* 2007, *ApJ* 663, 258
Cordes, J. M. & Lazio, T. J. W. 2002, astro-ph/0207156v2
Gaensler, B. M., Madsen, G. J., & Chatterjee, S., *et al.* 2008, astro-ph:0808.2550
Georgelin, Y. M. & Georgelin, Y. P. 1976, *A&A* 49, 57
Han, J. L. & Qiao, G.J. 1994, *A&A* 288, 759
Han, J. L. & Zhang, J. S. 2007, *A&A* 464, 609
Han, J. L., Manchester, R. N., Lyne, A, G., & Qiao, G. J. 2002, *ApJ* 570, L17
Han, J. L., Manchester, R. N., & Lyne, A. G., *et al.* 2006, *ApJ* 642, 868
Heiles, C. 2000, *AJ* 119, 923
Heiles, C. & Crutcher, R. 2005, in: R. Wielebinski & R. Beck (eds.), *Cosmic Magnetic Fields*, LNP 664 (Heidelberg: Springer), p. 137
Mathewson, D. S. & Ford, V. L. 1970, *MemRAS* 74, 139
Mathewson, D. S. & Milne, D. K. 1965, *Australian J. Phys.* 18, 635
Men, H., Ferrière & Han, J. L. 2008, *A&A* 486, 819
Mitra, D., Wielebinski, R., Kramer, M., & Jessner, A. 2003, *A&A* 398, 993
Noutsos, A., Johnston, S., & Kramer, M., *et al.* 2008, *MNRAS* 386, 1881
Page, L., Hinshaw, G., & Komatsu, E., *et al.* 2007, *ApJS* 170, 335
Parker, E.N. 1955, *ApJ* 122, 293
Piddington, J. H. 1964, *MNRAS* 128, 345
Rand, R. J. & Kulkarni, S. R. 1989, *ApJ* 343, 760
Roy, S., Pramesh Rao, A. & Subrahmanyan, R. 2008, *A&A* 478, 435
Seiradakis, J. H., Lasenby, A. N., & Yusef-Zadeh, F., *et al.* 1985, *Nature* 317, 697
Simard-Normandin, M. & Kronberg, P. P. 1980, *ApJ* 242, 74
Spoelstra, T. A. Th. 1984, *A&A* 135, 238
Steenbeck, M. & Krause, F. 1969, *AN* 291, 49
Sun, X. H., Reich, W., Waelkens, A., & Enßlin, T. A. 2008, *A&A* 477, 573
Testori, J. C., Reich, P., & Reich, W. 2008, *A&A* 484, 733
Uyanıker, B., Fürst, E., & Reich, W., *et al.* 1999, *A&AS* 138, 31
Vallée, J. P. 1983, *A&A* 124, 147
Vallée, J. P. 1996, *A&A* 308, 433
Vallée, J. P. 2005, *ApJ* 619, 297
Wielebinski, R. & Shakeshaft, J. R. 1964, *MNRAS* 128, 19
Wieringa, M. H., de Bruyn, A, G., & Jansen, D., *et al.* 1993, *A&A* 268, 215
Wolleben, M., Landecker, T. L., Reich, W., & Wielebinski, R. 2006, *A&A* 448, 411

Discussion

STIL: Comment: The high rotation-measure (RM) pulsar near the region $\ell = 5$ $b = +24$ coincides with an HII region around the runaway O star ζ Oph. The HII region causes bandwidth depolarization of NVSS sources, and therefore has a high rotation measure.

WIELEBINSKI: Yes, you must look in detail in such areas. The RM varies rapidly even on scales of 1 arcmin – as seen in the CGPS – aee poster P5.10.

Cosmic Magnetic Fields:
From Planets, to Stars and Galaxies
Proceedings IAU Symposium No. 259, 2008
K.G. Strassmeier, A.G. Kosovichev & J.E. Beckman, eds.

doi:10.1017/S1743921309031202

Magnetic fields and cosmic rays in galaxy clusters and large scale structures

Klaus Dolag, F. Stasyszyn, J. Donnert and R. Pakmor

MPI for Astrophysics, Karl-Schwarzschild-Str. 1, D-85741 Garching, Germany
E-mail: kdolag@mpa-garching.mpg.de

Abstract. In galaxy clusters, non-thermal components such as magnetic field and high energy particles keep a record of the processes acting since early times till now. These components play key roles by controlling transport processes inside the cluster atmosphere and beyond and therefore have to be understood in detail by means of numerical simulations. The complexity of the intra cluster medium revealed by multi-frequency observations demonstrates that a variety of physical processes are in action and must be included properly to produce accurate and realistic models. Confronting the predictions of numerical simulations with observations allows us to validate different scenarios about origin and evolution of large scale magnetic fields and to investigate their role in transport and acceleration processes of cosmic rays.

Keywords. Magnetic fields – cosmology – cosmic microwave background – large-scale structure of universe

---

1. Introduction

Magnetic fields have been detected in galaxy clusters by radio observations, via the Faraday rotation signal of the magnetized cluster atmosphere towards polarized radio sources in or behind clusters (see Carilli & Taylor 2002 for a recent review) and from diffuse synchrotron emission of the cluster atmosphere (see Govoni & Feretti 2004; Ferrari *et al.* 2008, for recent reviews). However, our understanding of their origin is still very limited. Furthermore most questions about their evolution and structures are still unanswered. Recent developments in interpretation of rotation measures help to understand the properties of magnetic fields in galaxy clusters but simulations are needed to overcome degeneracies in the model parameters needed to interpret the observations. Furthermore, the origin and the evolution of the population of cosmic rays within galaxy clusters are tightly connected to the dynamics of the system and to the evolution of the magnetic field. Therefore, cosmological MHD simulations are a valuable tool to investigate and distinguish different scenarios. See Dolag *et al.* (2008) for a recent review.

2. Observations

For a small sample of galaxy clusters, Faraday rotation can be observed towards several radio galaxies located at different radial distances with respect to the cluster center or along very elongated sources located at the center of galaxy clusters. Such examples can be used to infer the magnetic field structure over a range of length scales. Figure 1 is showing various examples of observations of rotation measure towards elongated radio sources within galaxy clusters, covering scales ranging from kpc to Mpc. Motivated by numerical simulations (Dolag *et al.* 2001), the observed magnetic field is often modeled with a radially-declining field strength and a power law spectral structure. From such interpretation of the observations, one can constrain the power law spectral index (Murgia

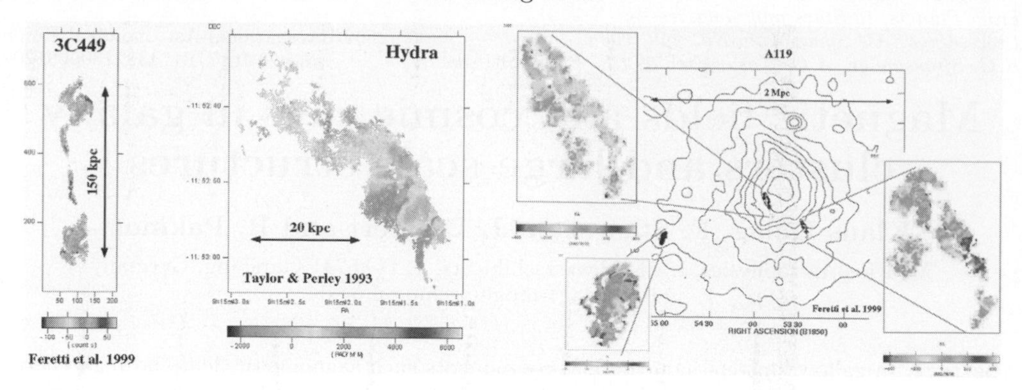

Figure 1. Various rotation measure maps. Left panel shows the central source in 3C449 (Feretti *et al.* 1995). Middle panel shows the central source in the strong cooling core of the Hydra cluster (Taylor & Perley 1993). Right panel shows 3 elongated radio sources within the galaxy cluster A119 (Feretti *et al.* 1999).

et al. 2004; Govoni *et al.* 2006) or directly reconstruct the power spectrum of the magnetic field (Vogt & Ensslin 2003, 2005). Given the sparse observational data available at the moment, a degeneracy exists between various parameters describing the assumed magnetic field model. Specially between the central value of the magnetic field and its rate of radial decline (see for example Guidetti *et al.* 2008, Bonafede *et al.* 2008). Therefore detailed predictions from simulations can be very useful to break these degeneracies. Simulations must therefore examine different possible magnetic field origins in galaxy clusters in order to test the robustness of such inferred magnetic field properties.

3. Simulations

In previous work, non radiative simulations of galaxy clusters within a cosmological environment following the evolution of a primordial magnetic seed field were performed using Smooth-Particle-Hydrodynamics (SPH) codes (Dolag *et al.* 1999, 2002, 2005) as well as Adaptive Mesh Refinement (AMR) codes (Brueggen *et al.* 2005; Dubois & Teyssier 2008; Li *et al.* 2008). Although these simulations are based on different numerical techniques they show good agreement in the predicted properties of the magnetic fields in galaxy clusters, when the evolution of an initial magnetic seed field is followed. This work has also demonstrated, that the properties of the final magnetic field in galaxy clusters do not depend on the detailed structure of the assumed initial magnetic field. The spatial distribution and the structure of the predicted magnetic field in galaxy clusters is primarily determined by the dynamics of the velocity field imprinted by cluster formation (Dolag *et al.* 1999, 2002) and compares well with measurements of Faraday rotation.

Figure 2 shows a zoom-in from the full cosmological box down to the cluster. The structures in the outer parts get less pronounced due to the decrease in resolution, which is designed to capture only the very largest scales of the simulation volume. Each panel shows (in clockwise order) a zoom-in by a factor of ten. Finally the elongated box in the lower left panel marks the size of the observational frame shown on the left. For comparison we produced a synthetic Faraday Rotation map from the simulation and clipped it to the shape of the actual observations to give an indication of the structures resolved by such simulations. The simulation follows the evolution of a primordial magnetic seed field and the dynamical range spans over more than five orders of magnitude in spatial dimension. The gravitational force resolution of this MHD-SPH simulation is $\approx$ 3kpc and

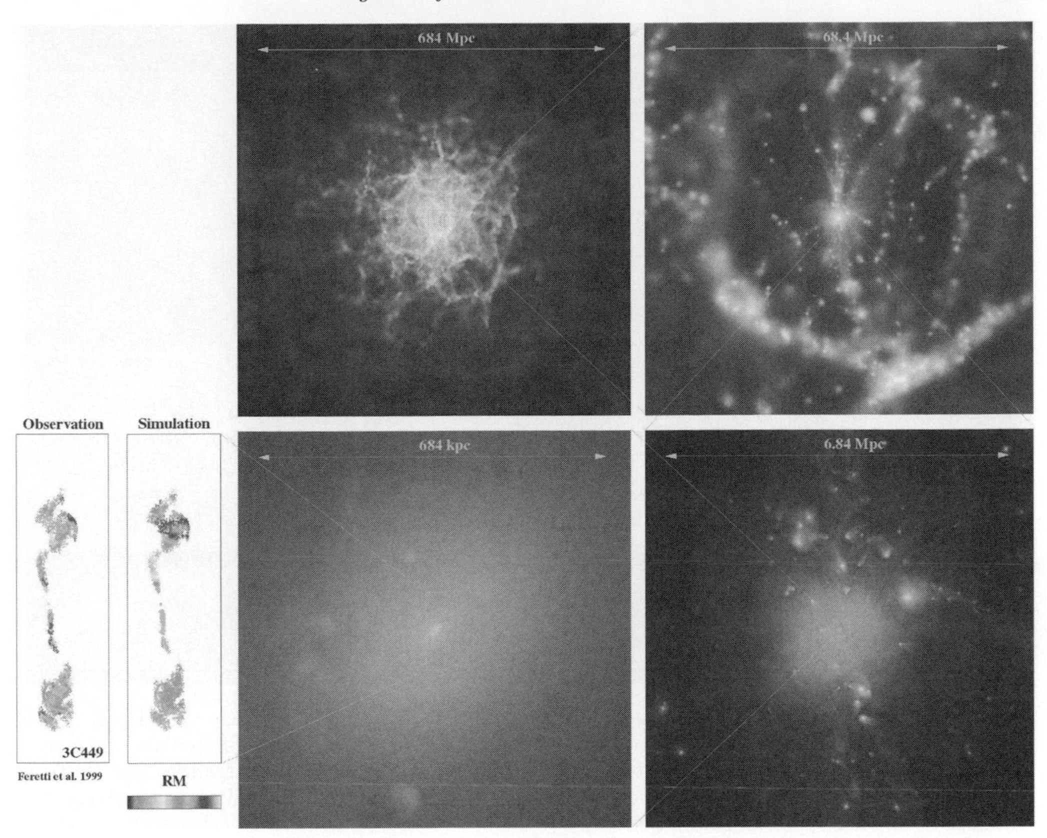

Figure 2. Zoom into the cluster simulated within the cosmological box. Clockwise, each panel displays a factor 10 increase in imaging magnification, starting from the full box (684 Mpc) down to the cluster center (680 kpc). On the very large scale, the density of the dark matter particles are shown, whereas in the high resolution region the temperature of the gas is rendered to emphasize the presence and dynamics of the substructure. The last zoom extracts a region of the same size of an observed radio jet 3C449 (Feretti *et al.* 1999) with infered rotation measure. Taken from Dolag & Stasyszyn (2008).

the galaxy cluster at redshift zero is resolved by several millions of particles within the virial radius.

Recently Donnert *et al.* 2008 performed cosmological, magneto-hydrodynamical simulations to follow the evolution of magnetic seed fields originating from galactic outflows during the star-burst phase, further processed by structure formation. Several simulations where performed, exploring the effect of various parameters of the adapted, semi-analytic model, relevant for the strength of the magnetic seed field from the galactic outflows. Also two control runs where performed, exploring the effect of the detailed magnetic field configuration assumed within the galactic outflows as well as on details of the seeding and galaxy identification strategy. It was found that the strength and structure of magnetic fields observed in galaxy clusters are well reproduced for a wide range of model parameters for the magnetized, galactic winds and do only weakly depend on the exact magnetic structure within the assumed galactic outflows. Figure 3 shows the final magnetic field for various models of the galactic outflows and a reference simulation following the evolution of a primordial magnetic field. Although the evolution of a galactic wind originating magnetic seed fields within the galaxy clusters shows no significant differences to that obtained by previous studies, it is clear to see that the magnetic field pollution

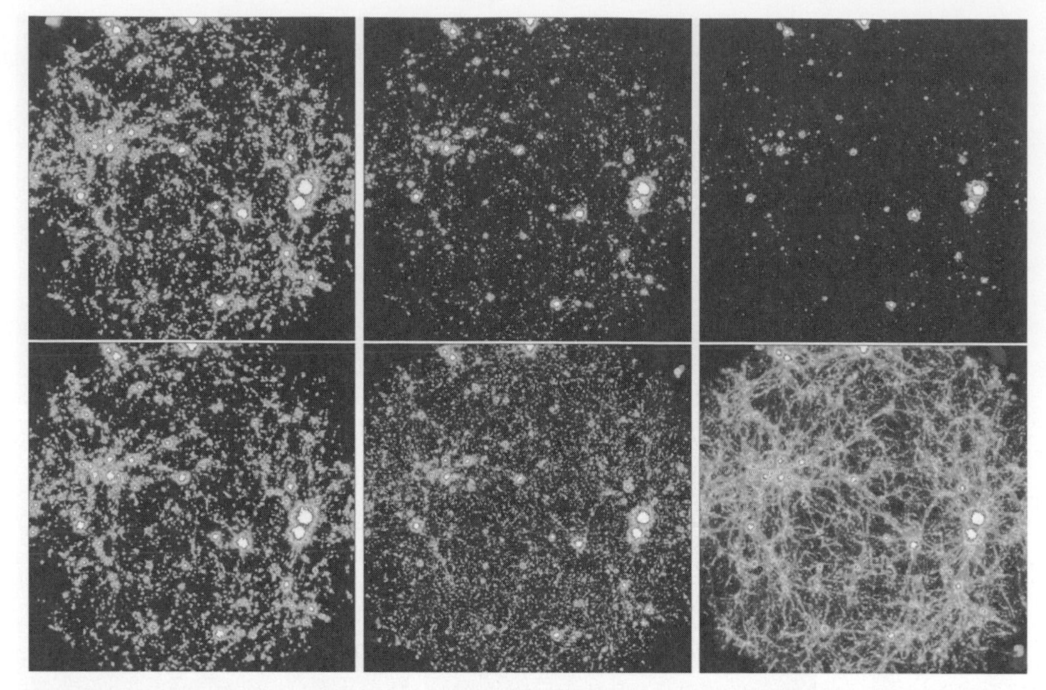

Figure 3. Visualization of the magnetic field strength in the simulation box at redshift $z = 0$. Every image shows a region of 204 Mpc, using the same arbitrary color bar. Shown are the results of the *Dipole* (top left), *0.1 Dipole* (top middle), *0.01 Dipole* (top right), *Quadrupole* (bottom left), *Multi Seed* (bottom middle), and the *Control* simulation (bottom right), respectively. See Donnert *et al.* (2008) for more details.

in the diffuse medium within filaments varies strongly between the models and in general is below the level predicted by scenarios with pure primordial magnetic seed field. Figure 4 shows a comparison of the predicted rotation measure signal of galaxy clusters for two different simulations, following magnetic seed fields from galactic outflows and from primordial origin, respectively.

4. Magnetic Field Structure

The complexity of the atmosphere of galaxy clusters reflects their hierarchical buildup within the large scale structure. The infall of thousands of objects with various sizes and their subsequent disruption within the cluster potential is being the source of shocks and turbulence, steering up the intra cluster medium. All these processes directly act on the magnetic field causing its re-distribution and amplification. Therefore all cosmological MHD simulations predict that the final structure of the magnetic field in galaxy clusters reflects these process of structure formation, and no measurable memory on the initial magnetic field configuration survives within galaxy clusters. In general, such models predict a magnetic field profile similar to the density profile. Thereby the predicted rotation measure profile agrees well with the observed one (see Fig. 4). Early findings of the shape of the magnetic field profiles based on MHD-SPH simulations (Dolag *et al.* 2001,2002,2005) are in good agreement with more recent simulations using different numerical methods – as Adaptive Mesh Refinement (AMR) codes – (Brueggen *et al.* 2005; Dubois & Teyssier 2008; Li *et al.* 2008).

Such numerical experiments predict a slope of the magnetic field power spectra similar to what would be expected for a Kolmogorov spectra (Dolag *et al.* 2002, Brueggen *et al.*

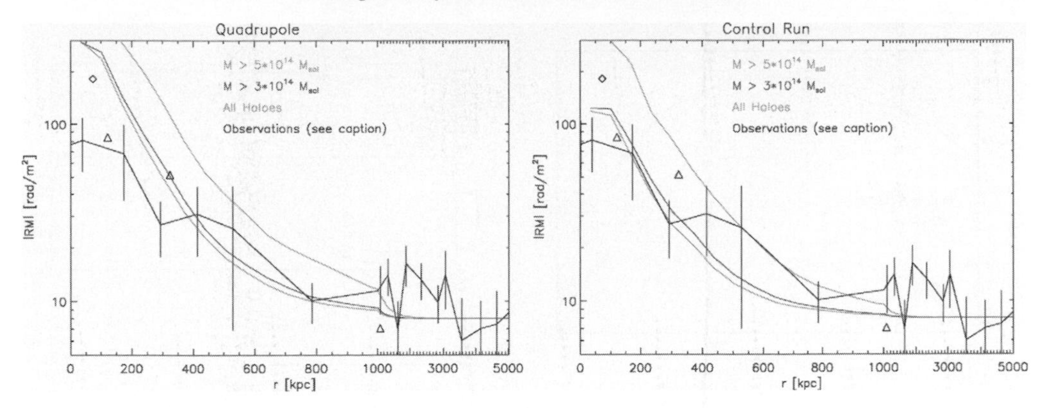

Figure 4. Rotation Measure as function of distance to the cluster center averaged over a sample of simulated clusters compared with observations. Solid line is drawn from combining three observational samples based on Abell clusters (Kim *et al.* 1991; Clarke *et al.* 2001; Johnston-Hollitt & Ekers 2004). We also included in the plot the values inferred from three elongated sources (triangles) observed in the single galaxy cluster A119 (Feretti *et al.* 1999) and one elongated source within the Coma cluster (diamond) (Feretti *et al.* 1995). Left panel is for a run which follows galactic outflows, right panel for one which follows a primordial magnetic seed field. See Donnert *et al.* (2008) for more details.

2005). In general this is in line with observations (Vogt & Ensslin 2003, 2005), however there are indications that the magnetic field spectra in observations might be more complex and on average has a slightly different slope in different galaxy clusters. There are even indications that the power spectra within individual galaxy clusters is either not a strict power law or the power law index varies with position (Murgia *et al.* 2004; Govoni *et al.* 2006).

The left part of Fig. 5 shows the magnetic field power spectrum obtained from a high resolution galaxy cluster simulation. As these simulations – due to the adaptive nature of the method used – resolve much smaller scales in the central region than in the outer parts, the power spectra is constructed from the measurement within a consecutive sequence of boxes with increasing resolution and decreasing size, centered on the center of the galaxy cluster. For guidance, the dotted straight line marks the expectations from a Kolmogorov like power spectra. In general the slope of the power spectra in the region of interest (e.g. tens of kpc) is predicted to be close to Kolmogorov like slope but with some indications of a curved shape, reflecting the complex dynamics acting during structure formation.

The right part of Fig. 5 shows the evolution of various quantities for a galaxy cluster, starting from early times on. For example, the middle panel shows the evolution of the virial mass and the mean, mass weighted temperature. The sudden increase of mass at a age of the universe of 5 respectively 11 G Years mark the two mayor merging events the cluster undergoes in its evolution. This is also reflected in the increase in temperature which happens delayed by $\approx 0.5 - 1$ G Year, which correspond to the time delay between start of the merger event and the core passage. The lower panel shows the fraction of the cluster material which undergoes shocks as measured by the build in shock detection scheme (Pfrommer *et al.* 2006). Interestingly, in general the cluster is just more and more relaxing, as indicated by the overall decrease of the amount of shocked gas in the cluster, independent of the major merging events (black line). However, if one looks at the fraction of high Mach number shocks inside the cluster, a clear excess, driven by the merger events is visible which are mainly initiated short after core passage (green line).

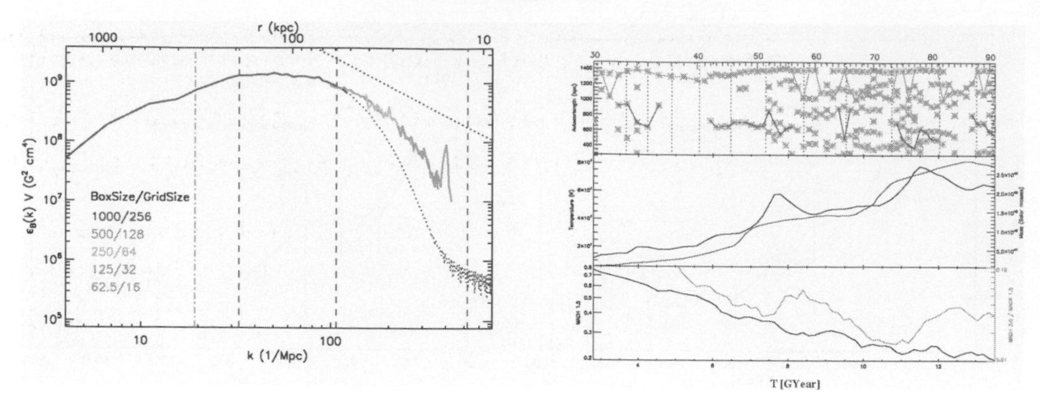

Figure 5. Left panel shows the magnetic field power spectrum from a simulated galaxy cluster. To scope with the adaptive resolution of the simulation the spectrum is constructed from calculating the power spectra within a series of boxes centered on the cluster with decreasing size and increasing resolution. The straight, dotted line marks the slope expected from Kolmogorov spectrum. The right panel shows the evolution of a galaxy cluster undergoing two major merging events. Here, every point in the upper panel marks the corresponding length scale of a extrema detected within the computed autocorrelation function at each time-step. The middle panel shows the evolution of the cluster mass and the averaged (mass weighted) cluster temperature. The lower panel shows the evolution of shocks with different Mach numbers within the cluster atmosphere. For more details see Pakmor & Dolag 2009.

It is interesting to know that it takes a relatively long time (e.g. more than 2 GYears) until this excess is completely gone. This has strong implications on the magnetic field structure, as can be seen in the upper panel. Here every dot correspond to the associated length scale of an extrema in the auto-correlation function of the magnetic field. So each point marks a length scale of individual structures present within the magnetic field. Clear to see that the first major merging event initiates a whole bunch of structures in the magnetic field, which – given by the low numerical dissipation of the SPH MHD code – does not vanish until the next major merger happens. By investigating a whole set of simulated galaxy clusters we find that only within a small number of clusters the magnetic field can relax to a more ordered configuration between merger events, which otherwise always initiate new structures within the magnetic field. For more details see Pakmor & Dolag 2009.

5. Radio Emission

The diffuse radio emission within galaxy clusters is produced by synchrotron radiation of relativistic electrons with the cluster magnetic fields. Such diffuse emission – often refered to as giant radio haloes – is detected over regions spanning Mpc in size. For recent reviews see Govoni & Feretti 2004 and Ferrari *et al.* 2008. One basic problem in explaining this phenomena is that the cooling time of such relativistic electrons is much shorter than their diffusion time over the region of interest. Therefore they basically have to be produced locally within the whole radio emitting region. One, often discussed mechanism to produce such relativistic electrons is the so called secondary model, where the relativistic electrons are a product by scattering of cosmic ray protons with thermal protons. Cosmic ray protons can for example be produced within accretion shocks and then advected into the cluster, or directly produced within merger shocks. Due to their larger mass compared to the cosmic ray electrons they can diffuse throughout the radio emitting region within the galaxy cluster without undergoing significant energy losses.

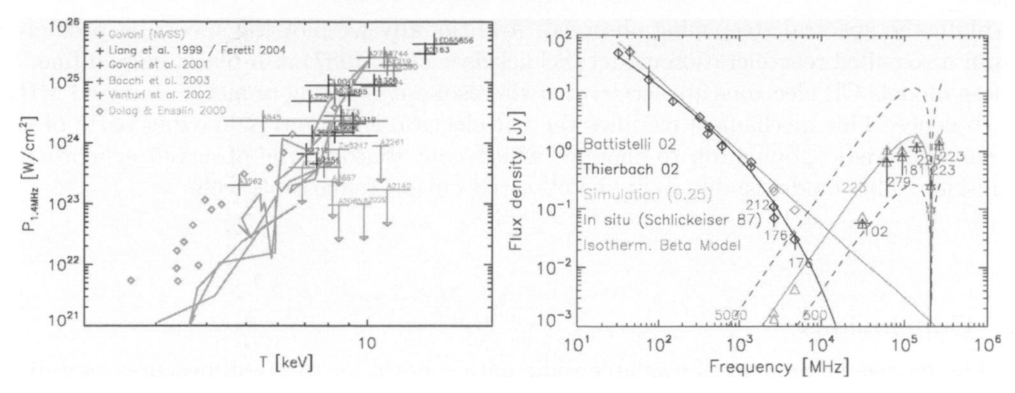

Figure 6. The left panels shows the total power of radio halos observed at 1.4 MHz vs. cluster temperature. We plot the data and upper limits from literature as indicated. For the simulations we plotted the predictions based on a secondary model (Dolag & Ensslin 2001) for individual clusters at redshift zero (diamonds) as well as the evolutionary track of an individual clusters undergoing two major merging events (solid line). The right panel shows the spectrum of a simulated Coma cluster compared to observations. The declining solid line marks the radio flux expected from a secondary model. The rising line marks the predicted (negative) flux from the SZ effect of the cluster atmosphere. The diamonds and triangles are the expected total flux from the clusters, evaluated over the size of the individual observations at the different frequencies. See Pakmor & Dolag 2009 and Donnert *et al.* 2009 for more details.

The left panel of figure 6 shows a comparison of observed radio luminosity of clusters as function of the mean cluster temperature (data points with error bars) with the predicted relations from a set of simulations (diamonds) assuming a simple, secondary models where the local energy density of cosmic ray protons is a fixed fraction of the thermal energy (see Dolag & Ensslin 2001). Although the scaling between radio luminosity and temperature of the simulated clusters agree well with the observed ones, there are no indications that simulations would be able to produce the class of galaxy clusters, for which no radio emission is observed. In fact, the scatter in the predicted scaling relation is very small, as also found in previous studies (Dolag & Ensslin 2001, Miniatti *et al.* 2001, Pfrommer *et al.* 2007). Additional we show a evolutionary track of the galaxy cluster which is undergoing two major merger events (see also right panel of Fig. 5). Clear to see that the merger events lead to very elongated loops along the scaling relation which can not bridge the gap between the clusters with and without observed radio emission. It is not unexpected that the amplification of the magnetic field during the merging event can not be accounted for the presence/non presence of radio emission as also the temperature gets boosted during the merger event and therefore the cluster evolves nearly along the observed correlation.

The right panel of Fig. 6 shows the expected radio spectrum for a simulated Coma cluster from a secondary model. The observed spectrum (diamonds with error bars) show a power law like behavior with a steepening above $\approx$ 1GHz. At higher frequencies we also plotted the observed (negative) flux caused by the Sunyaev Zeldovich (SZ) effect (triangles with error bars), which, if strong enough, could lead to such a spectral signature (dotted line). The solid red lines show the prediction of the radio emission from the secondary model and the expected (negative) SZ flux at higher frequencies for a simulated Coma cluster. The diamonds and the triangles are the expected, total flux convolved with the observed area for each frequency. Clear to see that the predicted (negative) SZ flux gives a perfect match to the observational data points for the SZ measurements, but the influence on the spectra of the radio observations is negligible and therefore can not explain the spectral steepening observed. Additionally we plot the spectrum predicted from a so-called reacceleration model (Schlickeiser *et al.* 1987) as a black, curved line. In these models CR electrons are accelerated via resonant coupling to merger induced MHD turbulence. This mechanism modifies the CR electron spectrum by moving parts of the trans-relativistic population to energies which contribute to the observed synchrotron emission. This model is able to fit the observed cut-off remarkable well.

6. Conclusions

The increasing amount of available radio data – both, for rotation measures as well as for diffuse radio emission – are driving our understanding of magnetic fields and cosmic rays in galaxy clusters. The improvements in the interpretations of these data over the last years are revealing a quite complex structure of the magnetic fields within galaxy clusters. Also the improvements in the numerical methods are producing more robust predictions for the magnetic field in galaxy clusters, which are helping to interpret the observations. Therefore, in the last years, a consistent picture of the magnetic fields in clusters of galaxies has been emerged from both, numerical work and observations.

Simulations of individual processes like shear flows, shock/bubble interactions or turbulence/merging events predict consistently a super-adiabatic amplification of magnetic fields within such processes. This now has been largely confirmed through direct cluster simulations within a cosmological context. It is worth mentioning that this common result is obtained by using a variety of different codes, which are based on different numerical schemes. Within this context, various observational aspects are reproduced. Moreover, the overall amount of amplification of the magnetic field driven by the structure formation process lead to a final magnetic field strength at a level, sufficient to link models that predict magnetic field seed by various different processes with the magnetic fields observed in galaxy clusters. In fact, the imprint of structure formation onto the magnetic field within galaxy clusters is such strong, that no measurable properties of the initial magnetic seed fields remain inside galaxy clusters. Therefore the only place we can hope to still find signs of the original process of magnetization are mildly non linear regimes of structure formation like filaments (Dolag *et al.* 2001, 2005; Donnert *et al.* 2008). The detailed structures of the magnetic field within individual clusters are driven by the actual merger history, where major merger events initiate significant structures within the magnetic field which are – in absence of any dissipation – only slowly relaxed. The induced magnetic field power spectra appears to be not a strict power law, however, when approximated locally as a power law, the slope is close to the expectations from a Kolmogorov like spectra (see Dolag *et al.* 2001, Brueggen *et al.* 2005, Pakmor & Dolag 2009).

Such models of magnetic field in galaxy clusters allow also to constrain the origin of cosmic rays within galaxy clusters when confronted with observations of the diffuse radio emission. Although so called secondary models are able to produce sufficient radio emission, a detailed comparison shows that they fail to produce some key observational aspects. Most striking they overproduce the number of galaxy clusters which are expected to show radio emission as well as the fail to produce the observed spectral shape for the diffuse radio emission (see Dolag & Ensslin 2001, Donnert *et al.* 2009). All this demonstrate the power of such cosmological MHD simulations to learn more about non thermal components like magnetic fields and cosmic rays within galaxy clusters and the large scale structure.

Acknowledgments

KD acknowledge supported by the DFG cluster of excellence "Origin and Structure of the Universe".

References

Bonafede, A., Govoni, F., Murgia, M., Feretti, L., Dallacasa, D., Giovannini, G., Dolag, K., & Taylor, G. 2008, Proceedings "Magnetic Fields in the Universe II", *Rev. Mex. Astron. Astrof.*, in press*Rev. Mex. Astron. Astrof.*,

Brueggen, M., Ruszkowski, M., Simunescu, A., Hoeft, M., & Dalla Vecchia, C. 2005, *ApJ* 631, L21

Carilli, C. T., Taylor, G. B. 2002, *ARA&A* 40, 319

Clarke, T. E., Kronberg, P. P., & Boehringer, H. 2001, *ApJ* 547, L111

Dolag, K., Bykov, A. M., & Diaferio, A. 2008, *Space Science Reviews* 134, 311

Dolag, K., Bartelmann, M., & Lesch, H. 1999, *A&A* 348, 351

Dolag, K., Bartelmann, M., & Lesch, H. 2002, *A&A* 387, 383

Dolag, K. & Ensslin, T. 2000 *A&A* 362, 151

Dolag, K., Grasso, D., Springel, V., & Tkachev, I. 2005, *Journal of Cosmology and Astro-Particle Physics* 1, 9

Dolag, K., Schindler, S., Govoni, F., & Feretti, L. 2001, *A&A* 378, 777

Dolag, K. & Stasyszyn, F. 2008, *arXiv* arXiv:0807.3553

Donnert, R., Dolag K., Lesch, H., & Mueller, E. 2008, *MNRAS*, in press

Donnert, R., Dolag K., Cassano, R., & Brunetti, G., *in prep*

Dubois, Y. & Teyssier, R. 2008, *A&A* 482, L13

Feretti, L., Dellacasa, D., Giovannini, G., & Tagliani, A. 1995, *A&A* 302, 680

Feretti, L., Dellacasa, D., & Govino, F. *et al.* 1999, *A&A* 344, 472

Ferrari, C., Govoni, F., Schindler, S., Bykov, A. M., & Rephaeli, Y. 2008, *Space Science Reviews* 134, 93

Govoni, F. & Feretti, L. 2004, *International Journal of Modern Physics D* 13, 1549

Govoni, F., Murgia, M., & Feretti, L. *et al.* 2006, *A&A* 460, 425

Guidetti, D., Murgia, M., Govoni, F., Parma, P., Gregorini, L., de Ruiter, H. R., Cameron, R. A., & Fanti, R. 2008, *A&A* 483, 699

Johnsoton-Hollitt, M. & Ekers, R. D. 2004, *arXiv* arXiv:0401.045

Kim, K. T., Kronberg, P. P., & Tribble, P. C. 1991, *ApJ* 379, 80

Li, S., Li, H. & Cen, R. 2008, *ApJS* 174, 1

Miniatti, F., Johns, T. W., Kang, H., & Ryu, D. 2001, *ApJ* 512, 233

Murgia, M., Govoni, F., & Feretti, L. *et al.* 2004, *A&A* 460, 425

Pakmor, R. & Dolag, K. 2009, *MNRAS*, in prep.

Pfrommer, C., Ensslin, T. A., Jubelgas, M., Springel, & Dolag, K. 2007 *MNTAS* 378, 385

Pfrommer, C., Springel, V., Ensslin, T. A., & Jubelgas, M. 2006 *MNRAS* 367, 113

Schlickeiser, R., Sievers, A., & Thiemann, H 1987, *A&A* 182, 21

Stasyszyn, F. & Dolag, K. 2009, *MNRAS*, in prep.

Taylor, G. B. & Perley, R. A. 1993, *ApJ* 416, 554

Vogt, C. & Ensslin, T. A. 2003, *A&A* 412, 373

Vogt, C. & Ensslin, T. A. 2005, *A&A* 434, 67

Discussion

GAENSLER: What can your models predict about diffuse radio emission *between* clusters of galaxies?

DOLAG: In these models such diffuse emission is extremely small. Such emission is thought to be produced by shocks and directly involve phenomena which are (yet) not included in the presented simulations.

DE GOUVEIA DAL PINO: I missed a point regarding the cosmic-ray (CR) flux injection hypothesis. Could you clarify this point? Also, it seems you have a nice tool to explore cosmic-ray propagation and more important ultra-high energy CR propagation.

DOLAG: We assumed a simple model in which we inject CR protons as fixed fraction of thermal energy. Using results from direct simulations (which, at the moment, come with different predictions by different groups working on this) do not significantly change the main conclusions.

KRONBERG: In your models can you identify a "cluster-pause" – i.e., the cluster equivalent of the solar/ISM magnetopause, and the galaxy's interface "galactopause" to the ISM?

DOLAG: No.

Klaus Dolag

On the boat trip

Cosmic Magnetic Fields:
From Planets, to Stars and Galaxies
Proceedings IAU Symposium No. 259, 2008
K.G. Strassmeier, A.G. Kosovichev & J.E. Beckman, eds.

doi:10.1017/S1743921309031214

Magnetic fields in the early Universe

Eduardo Battaner[1,2] and Estrella Florido[1,2]

[1]Departamento de Física Teórica y del Cosmos, Universidad de Granada
Avenida Fuentenueva s/n, 18071 Granada

[2]Institute Carlos I for Theoretical and Computational Physics. Universidad de Granada
email: battaner@ugr.es,estrella@ugr.es

Abstract. There is increasing evidence that intense magnetic fields exist at large redshifts. They could arise after galaxy formation or in very early processes, such as inflation or cosmological phase transitions, or both. Early co-moving magnetic strengths in the range 1-10 nG could be present at recombination. The possibilities to detect them in future CMB experiments are discussed, mainly considering their impact in the anisotropy spectra as a result of Faraday rotation and Alfven waves. Magnetic fields this magnitude could also have a non-negligible influence in determining the filamentary large scale structure of the Universe.

Keywords. Magnetic fields – cosmology – cosmic microwave background – large-scale structure of universe

---

1. Introduction

There are previous excellent reviews on this topic. For elaborating this one the authors have found particularly instructive those by Giovannini 2002, Giovannini 2004, Giovannini 2006 and Grasso & Rubinstein (2001).

We could advance that we have no measurement at all of early universe magnetic fields, therefore, the topic is highly speculative. Nevertheless, there is a large amount of theoretical work, even at an embryonic state due to the lack of observational parallel work. There is also the reasonable hope that measurements will become available in the next decade. If early magnetic fields really exist our conception of the Universe should be modified and the number of parameters defining it should be enriched.

In fact, the higher redshift at which we have a reliable rotation measure is $z \approx 2$. This is the case of 3C191 with an absorption line spectrum due to a wind-driven shell of gas. Rotation measures higher than 200 rad/m^2 corresponding to field strengths in the range 0.4-4 μG (Kronberg 2005, Kronberg, Bernet, Miniati *et al.* 2008, this book) were detected. Radioastronomical observations of point sources limit our possibilities to Lyman-α forests or absorbers in front of quasars. See also Bernet, Miniati, Lilly *et al.* (2008). Beck (2005; this book) have comprehensive reviewed the measurements of galactic and extragalactic fields.

The present matter density is very inhomogeneous and magnetic fields should be as well inhomogeneously distributed. In particular, no measurements of magnetic fields in large scale voids have been obtained (and this would be out of present observational capabilities).

In order to observe and respect the Cosmological Principle of isotropy, we could disregard a homogeneous magnetic field within the present Hubble radius. To be specific about, we would assume $<\vec{B}>=0$ even if $<B^2>\neq 0$, i.e. the mean magnetic energy density is not vanishing. The magnetogenesis mechanism, whatever it was, could produce fields this scale, but this is unlike. Upper limits exist to disregard an ordered field on the scale of the observable universe (e.g. Widrow 2002).

Because of the solenoidal character of the field, their distribution in space would consist in closed loops of different sizes, being their topological and array properties uncertain.

It is not surprising the fact that so many theoretical work has been developed about magnetic fields in the early universe, despite the absolute lack of measurements. There are basically three reasons: the first one is more or less sentimental: if we find magnetic fields in all astrophysical subsystems (planets, stars and interplanetary, interstellar and intergalactic media) why should a cosmological field be absent? Second, several forthcoming experiments could provide an observational detection, particularly PLANCK, but also the SKA (Beck 2008) QUIJOTE (Rubiño-Martin *et al.* 2008) and others. A Planck Project *Constraints on Primordial Magnetic Fields* is now running within the Planck Working Group on Non-Gaussianity (e.g. Battaner & Rubiño-Martin 2008). There is a Planck Project devoted to this study. The third is that the introduction of magnetic fields in the interpretation of CMB and the large scale structure could modify the actual values of the parameters defining our universe, what is of higher and exciting cosmological interest.

An interesting argument was raised by Fermi. Taking into account that in a perfectly conducting fluid, magnetic fields were long lived, and considering the isotropy of the cosmic rays, he was able to obtain an order of magnitude of the galactic magnetic field. He then considered the possibility of a primordial origin of this field. Cosmic rays are confined into the galaxy if their Larmor radius is lower than the size of the galaxy, i.e. when

$$\frac{c}{\nu_{giro}} = \frac{mc^2}{eB} = \frac{E}{eB} = L \tag{1.1}$$

(e is the charge of the cosmic proton, E its energy, and L the "size" of the galaxy, for this purpose about twice the width of the disc, say 1 kpc). If B is taken to be as $5 \times 10^{-6}\mu$G, energies below $E = eBL \approx 5 \times 10^9$ Gev will correspond, in general, to bounded chaotic orbits, isotropically detected at the Earth.

We could use a similar argument to obtain a very rough estimate of the extragalactic magnetic field. Now, we find that energies above 10^{10} Gev correspond to extragalactic cosmic rays. Their arrival is far from isotropic, but let us ignore this important fact, take $L \approx 10$ Mpc and obtain 1 nG for the extragalactic field.

Ly-α forests or other pre-galactic structures, already had μG fields. The dynamo is then necessary to order magnetic lines and achieve a coherent distribution at the galactic scale, and it is necessary as well to avoid dissipation of the field as a result of the turbulent magnetic diffusion, but not for amplifying the strength. See Brandenburg & Subramanian 2005 as an extensive analysis of dynamos. The pregalactic field should be only three orders of magnitude lower, i.e. 5×10^{-9}G to account for the galactic relative collapse. This is approximately the field that magnetogenesis must obtain and the strength we are looking for in the early universe. In order to compare the field strength B at any cosmic scale factor, a, with the present values of the strength, it is customary to define the "equivalent-to-present" or "comoving" strength, B_0 as

$$B_0 = Ba^2 \tag{1.2}$$

in order to take into account the expansion effect; therefore, B_0 is not the field strength today; it would be the present strength if no variation of the field other than that induced by expansion would take place since z.

2. Conductivity

In all epochs of the Universe, the conductivity can be assumed to be infinite and the field can be neither created nor eliminated, just amplified and ordered, therefore the diffusion time becomes very large, larger than the Universe lifetime. This is only a simplified picture. There is an additional diffusion enhanced by turbulence, and a battery, of the type of the Bierman battery, can create small fields without a seed. But the question is if the pre-galactic universe had a high conductivity, so that we are able to envisage fields at very early times.

Consider for instance the post-Recombination epoch, the most resistive one, because the number density of electrons was evidently very low. There are however residual electrons that survived Recombination. Kolb & Turner (1990) estimated the density of free electrons as about $10^{-10}(1+z)^3\mathrm{cm}^{-3}$, assuming that these free electrons had no opportunity to recombine, once the recombination rate became smaller than the expansion rate. The conductivity is

$$\sigma = \frac{n_e^2 e^2 \tau_T}{m_e} \tag{2.1}$$

where τ_T is the time between collisions of the electric chargers which are assumed to be electrons. Collisions are mainly due to CMB photons, through Thomson scattering, therefore

$$\tau_T = \frac{1}{n_\gamma \sigma_T} \tag{2.2}$$

and therefore

$$\sigma = \frac{n_e^2 e^2}{m_e n_\gamma \sigma_T} \tag{2.3}$$

Here n_γ/n_e should be proportional to the specific entropy of photons per baryons, which is constant if the universe fluid is ideal. Finally we find $\sigma \approx 10^{11} s^{-1}$. Due to the constancy of n_γ/n_e the conductivity is a constant after Decoupling. With this value of the conductivity we can find the maximum size of a magnetic coherence cell that can survive until today. The diffusion time is calculated with

$$\tau_{diff} = \frac{4\pi\sigma L_B^2}{c^2} \tag{2.4}$$

L_B is the characteristic length of the coherence cell. Substituting τ_{diff} by the life-time of the Universe t_0, we obtain that the characteristic diffusion length is only about 1 AU. Therefore, even in the most resistive post-Recombination era, the most adverse for the maintenance of magnetic fields, the assumption of infinite conductivity is very reasonable.

3. Comments on magnetogenesis

We must then find a mechanism able to create magnetic fields; i.e. without a seed. In general, these mechanisms should produce a positive-negative charge separation, producing a small electric field with non vanishing curl. The magnetic fields created at any epoch should have the strength and the scale large enough to reach the pre-galactic time satisfying the astrophysical requirements.

The many different magnetogenesis theories can be classified into four groups (e.g. Battaner & Lesch (2000), Battaner & Florido (2000) and references therein):

a) After Recombination, b) During the Radiation Dominated era, c) in Cosmological Phase Transitions, d) in Inflation. Let us briefly comment on them:

- After Recombination.

Rees (2005) has proposed that the field could be generated rather recently in material ejected by a first generation of supernovae or by population III stars. Intergalactic magnetic fields can be the result outflows from starburst galaxies, like M82 (Kronberg 2005). A generation of M82-type galaxies could inject important quantities of magnetic fields

into the cluster medium, as it is directly appreciated (Reuter, Klein, Lesch *et al.* 1994). Massive black holes in the center of galaxies have also been considered by Kronberg as seeding intergalactic fields too.
Therefore, the possibility of a very recent origin of the intergalactic fields from galactic and stellar ejections is very appealing. It does not exclude, however, other ultra-early magnetogenetic processes, as these recent mechanisms do not forcedly rule out others.

• During the Radiation Dominated era.
Harrison (1973) considered a relation between vorticity and magnetic field. The Jeans mass after annihilation is very small, rendering turbulence at the beginning of this era rather peculiar.

• In cosmological phase transitions. Hogan (1983) early proposed cosmological phase transitions as a source of primordial magnetic fields. The electroweak phase transition (at 10^{16} K $= 100$ GeV at $a = 10^{-16}$, approximately at $10^{-12}s$, being the Hubble radius about 3 cm, have considered by many authors (see the review by Enqvist 1998). The QCD phase transition (at 3×10^{12} K $= 100$ McV at $a = 10^{-12}$ with a Hubble radius of 4×10^4 m) has also been extensively studied (e.g. Cheng & Olinto 1994). Even GUT phase transition has been proposed (e.g. Brandenberger, Davis, Matheson *et al.* 1992). In general, phase transitions produce large enough strengths but very small lengths of the coherence cells. A very general argument can illustrate this fact.
If the mechanism was completely efficient at the *ith* phase transition to provide the maximum energy density, then:

$$\frac{B_i}{8\pi} = a_r T_i^4 \tag{3.1}$$

where a_r is the radiation constant. The real magnetic strength should be less than the value obtained with this equipartition. Then:

$$B_{0i} = B_i a_i^2 = (8\pi a_r)^{1/2} T_i^2 a_i^2 = (8\pi a_i)^{1/2} T_0^2 \approx 4 \times 10^{-6} G \tag{3.2}$$

therefore and rather interestingly, independent of i, that is, independent of the phase transition considered. There is a compensation: the earlier the phase transition, the higher the temperature, but the higher the dilution of the field by expansion. The value of the equivalent to present field strength B_{0i} is higher than requested, but we cannot expect that the mechanism was so efficient.

Concerning the coherence cell size, however, the results is worst. As mechanisms based on phase transitions are causal, the maximum length would correspond to the horizon at that epoch. Before Recombination the Universe expanded as $t^{1/2}$, therefore, we have for this maximum size

$$\lambda_i = ct_i = c(a_i/a_R)^2 t_R = c(T_R/T_i)^2 t_R = (cT_R^2 t_R) T_i^{-2} \tag{3.3}$$

where the subindex R means Recombination. After expansion:

$$\lambda_{0i} = \lambda_i/a_i = \lambda_i T_i/T_0 = (cT_R^2 t_R/T_0) T_i^{-1} \approx 10^{30} T_i^{-1} \tag{3.4}$$

For the electroweak phase transition with $T_i = 10^{16}$ K we obtain $\lambda_{0i} = 10^{16} cm = 3 \times 10^{-3} pc$ which is very small for cosmic lengths of interest. In general, the correlation length will be much less than this. For the QCD transition, with a temperature three orders of magnitude lower, the scale is higher but, in any case, insufficient.

This problem could be, in part, alleviated if an inverse cascade transfers the magnetic energy from lower turbulence scales to larger. This inverse cascade has been applied and numerically reproduced in other astrophysical systems (see, for instance, Brandenburg & Subramanian 2005). The inverse cascade effect is more efficient when the magnetic field has helicity. Helicity is a topological quantity which is conserved along the history of the

Universe. Some magnetogenesis process have been proposed producing helical magnetic fields (see Kahniashvili (2005), for references). Giovannini 2006 estimate that even with inverse cascade in helical fields the present coherence cells are too small (<100 pc).

• In Inflation. This possibility is extremely attractive. It can give fields at any scale in the same way that super-horizon energy density structures are created and observed. Turner & Widrow (1988) first proposed an Inflation scenario for the creation of primordial magnetic fields. The idea is exactly the same to explain CMB anisotropies on angular scales larger than the horizon. A coherence cell with present size λ has had at any epoch a size of $a\lambda$. This must be compared with the horizon that is a varying function of a, During the first phase of Inflation the horizon is independent of a, varies as $a^{3/2}$ during reheating, as a^2 along the Radiation era, as $a^{3/2}$ along the Matter dominated era. The cell could be sub-horizon when it was created, become super-horizon as a result of the inflationary exponential expansion, and sub-horizon again at photon decoupling. As an example, quantum-mechanical fluctuations could produce electromagnetic waves. The oscillating electric and magnetic fields will appear as static fields when the fast expansion catch up with the wavelength. In later eras electric fields will be canceled by the high conductivity. Gasperini, Giovannini & Veneziano 1995 first considered the potential relation between inflation fields and CMB.

Very promising is the supersymmetric theory and pre-Big-Bang type models (Gasperini, Giovannini & Veneziano 1995). The reader is addressed to excellent recent reviews (Giovannini 2002, Giovannini 2006). Supersymmetric theories are able to provide field strengths as high as 10^{-8} G and scales any length, including Sachs-Wolfe scales, only limited in the very small ones by diffusion and Silk damping. This scenario can even provide the initial spectrum of primordial fields (Giovannini 2006).
Though considering the exoticism of theories in the ultra-early Universe, the inflationary magnetogenesis may satisfy the astrophysical requirements. However, Kahniashvili, Maravin & Kosowsky (2008) have deduced an upper limit of 0.7 μG for scales of 100 Mpc.

4. The large scale structure

The paradigm ΛCDM gives a very good overall description of the evolution of large scale structures until the formation of galaxies, clusters and superclusters, but some failures are to be considered, which suggest that the physics should be enriched, or even substantially modified. Probably, magnetic fields are not ignorable and may solve some of these failures. The new data about very large magnetic fields (84 μG in a pregalactic structure at $z \approx 0.7$, Wolfe, Jorgenson, Rodishwa *et al.* (2008) and those of Kronberg, Bernet, Miniati *et al.* 2008 prevent us about a premature ignorance of magnetic effects).

Some failures or unexpected results are well known, in particular concerning galaxy formation. The rotation curve is unsatisfactorily explained (Navarro & Steinmetz 2000), being the simulated Tully-Fisher relation far from the observational one by more than an order of magnitude, the simulated distribution of galactic DM do not well fit the observed rotation curves (de Block, Bosma & McGaugh 2003). Probably, rotation curves of spiral galaxies, with their important cosmological implications, cannot be understood without the inclusion of magnetism in the dynamics (Battaner and Florido 1992, 1995, 2000, 2007).

The influence of magnetism may be non negligible since very early times, thus not only affecting the birth and structure of galaxies and clusters but the large structure itself (e.g. Wasserman 1978; Giovannini 2004). The variations in the barotropic index (w in the so called equation of state) has an important influence on the metric perturbations (Giovannini 2007). Battaner, Florido & Garcia-Ruiz (1997), Battaner,

Florido & Jiménez-Vicente, J. (1997), Battaner (1998), Florido & Battaner (1997), and Battaner & Florido (2000) have considered the effects of magnetic fields in the radiation dominated era strengths in the range $10^{-9}-10^{-8}$ G. The magnetic field has its own barotropic index equal to 1/3. Fields larger than these excessively accelerate galaxy formation, and lower than these, render their effects unimportant. Along this epoch a linear perturbation analysis in the RG Maxwell, fluid and Einstein Field equations is perfectly justified.

Primordial magnetic fields originated at Inflation could acquire a filamentary structure at a large scale. These filaments could connect to form loops that could be the siege of present large scale filaments and voids. Magnetic fields affect the motion of the tightly coupled photon-electron-baryon fluid not only through the Lorentz force, but mainly because they must be included in the energy-momentum tensor, thus introducing an anisotropic gravitating effect. This gravitating magnetic field would produce radiation filaments along this era, DM will fall in these potential wells and after decoupling magnetic fields-DM-baryon filaments would remain. Well after decoupling, non-linear effects would tangle the fields, increasing the strength by about three orders of magnitude, rendering the early distribution nearly impossible to detect today. Superclusters, clusters and galaxies could then be formed out of relic magnetic flux tubes, being distributed along fossil interconnected filaments, which are today observed. It is to be emphasized that gravity alone is able to produce filaments too. Pure gravitational driven filaments cannot possess scales larger than about 50 Mpc.

WMAP found an unexpected high Re-Ionization redshift (Kogut *et al.* 2003). It is possible that magnetic fields can have accelerated the ionizing stellar formation responsible of the premature re-ionization. Subramanian (2006) estimated that at z>15 objects with dwarf galaxy masses and smaller could be the results of cosmic 0.1nG fields.

Small scale effects can also be produced by PMF. At $z = 1$, a typical redshift is the Hubble Deep Field, the intergalactic magnetic energy density was larger by a factor $(1 + z)^4 \approx 16$, thus been able to affect galaxy structure, for instance contributing for producing warps (Reshetnikov, Battaner, Combes *et al.* 2002).

5. Magnetic fields and CMB

If the generation of magnetic fields took place before photon decoupling there is the possibility and the hope that they can be observed in CMB. At present no clear signs of magnetic fields in CMB experiments have been found, but there is a interesting expectation in forthcoming data.

Before collapsing the field was $10^{-10}-10^{-8}$ G. Hence Ω_B is about $10^{-5}\Omega_\gamma$. As both energies vary as a^{-4}, this relation should hold at Decoupling. As 10^{-5} is the level of observed CMB anisotropies, there is the reasonable hope that the cosmic magnetic field has left observable traces in the CMB.

The physics is more exotic when compared with terrestrial standards, but not so much. The magnetic field strength is, as we have seen, of the order of $10^{-8}(z + 1)^2$ about 10^{-2} G which is close to a typical value in the terrestrial high atmosphere. The density at Decoupling is of the order of the critical density times $(1 + z)^3$ about 10^{-20}gr cm^{-3}, also very similar to the atmospheric density at, say, 1000 km high and the temperature at $z \sim 1000$ was also similar to the upper thermosphere. It is to be noticed that the first detections of the CMB were obtained (though not with this purpose) in the high atmosphere. To emphasize even more the familiarity of the CMB medium, note that the anisotropy power spectrum, with the main Doppler peak and other smaller peaks reminds more music than noise, and given the fact that the anisotropies are found with μK over temperatures of $\sim 2.7K$, we could measure the amplitude of this music corresponding to few decibels, i.e. a barely perceptible music.

These coincidences being noticeable, the physics in the last scattering surface and in the terrestrial upper thermosphere also present important differences, such as the Hubble expansion, the initial conditions corresponding to the radiation dominated era and earlier, the influence of dark matter and energy, the collapse of inhomogeneities, and other facts that render our knowledge of the terrestrial atmosphere of little help for studying CMB.

The most direct measure of magnetic fields present in the last scattering surface would be achieved by direct detection of Faraday Rotation (FR), which is very difficult, but the hope is not completely lost. In order to estimate the order of magnitude of the rotation angle can be obtained by an approximated argument (Kosowsky & Loeb 1997):

In our case the source and the FR medium is the same. The polarization mechanism is Thomson scattering (under certain quadrupole distribution of the electron density) and the optical depth is:

$$\tau_T = \int \sigma_T n_e dl \tag{5.1}$$

What are the limits of the integral, i.e. how deep is the source and the transmission medium?. It cannot be Δz, the width of the decoupling transition, because only the closest layers within Δz drive out polarized light. For an oder of magnitude we could set $\tau_T = 1$, hence:

$$\int n_e dl = \frac{1}{\sigma_T} \tag{5.2}$$

For the rotation angle we would then get:

$$\varphi = \lambda^2 K \int B_{\|} n_e dl = \lambda^2 K B_{\|} \frac{1}{\sigma_T} \tag{5.3}$$

But $B_{\|}\lambda^2$ is invariant ($B \propto a^{-2}$, $\lambda \propto a$) hence, $B_{\|}\lambda^2 = [B_{\|}\lambda^2]_0$ at present. For 30 GHz, $\lambda = 1cm$ if $B \approx= 10^{-8}$ G we obtain $\varphi = 0.36rad = 20^o$. Kosowsky & Loeb 1997, Kosowsky & Loeb (2005) carried out a detailed calculation and obtained $\varphi = 1$ for $B = 10^{-9}$ G. An angle of 20 degrees is perfectly measurable by Planck. However the main difficulty is that the signal is too noisy, as the polarization intensity is very low. Another important difficulty is that we need to observe the primordial Faraday Rotation across the Milky Way. A very precise model of the galactic magnetic field distribution at the galactic large scale is required in order to find clean windows for observing primordial FR, and to decontaminate the galactic contribution.

However, FR could affect the polarization spectra, identifiable as due to magnetic field driven perturbations because of its λ^2-dependence. In particular, FR can convert E polarization modes into B modes (E and B are polarization modes, similar to the Stokes parameters but rotational invariants) and could give appreciable traces into the polarization power spectra. The polarization in the E and B modes has been calculated by Scoccola, Harari & Mollerach (2004) and Lewis (2004). Kosowsky & Loeb (2005) and Kosowsky & Loeb (2005) have detaily calculated this effect by considering stochastic magnetic fields. FR should produce low power polarization spectra, unless frequencies lower than 30 GHz and arcmin angular resolutions are used. These values renders the detection of BMF FR in E and B polarization spectra rather challenging. Helical fields produce no FR at all. Faraday Rotation can also have a net depolarization effect (Harari, Hayward & Zaldarriaga 1997).

To reproduce the effect of magnetic fields, mainly on the angular power spectrum a rather large list of possible choices complicate the models. Initial conditions may be of isocurvature or of adiabatic type. The initial spectrum may be assumed to be a power low or not. Some authors adopt a mean field homogeneous, other completely inhomogeneous, i.e. stochastic magnetic fields, or any other intermediate arrangement. The field may be helical or not. The perturbations induced in the metric may be scalar, vector or tensor modes, being magnetic fields able to produce the three types. The gauge may be synchronous, longitudinal or others, etc. To account for so many choices would render this review too extended.

Some works by Caprini & Durrer (2001), and Caprini & Durrer (2006) have shown that the continuous production of gravity waves, render magnetic strengths very low at Recombination. If primordial magnetic fields are produced by a causal process, as the electroweak phase transitions is, they estimate that the strength should be less than 10^{-27} G and even lower if they are produced at Inflation. Only large scales would be free of this constraint for invariant scale produced fields close to -3, being n+3 the exponent of the assumed power low spectrum of primordial fields. A controversy (Kosowsky & Loeb (2005); Caprini & Durrer 2005) have followed this work.

Durrer, Kahniashvili & Yates (1998) have shown that in an homogeneous magnetic field, Alfvèn waves should produce a testable correlation between different coefficients of the spherical harmonic analysis, in particular

$$D_{l(m)} = < a_{l-1,m} a^{*}_{l+1,m} > \neq 0 \tag{5.4}$$

which should vanish under perfect Gaussian conditions. This would constitute a testable prediction with no other interpretation than magnetic fields, indicating a preferred direction in the Universe. This would hold for any multipol scale lower than Silk damping scale at about l=500.

There are some recent codes that are an important tool to analyze the forthcoming Planck data considering the effects of magnetic fields. In a series of papers Giovannini and Kunze have contributed with a valuable work to understand Planck results. Giovannini 2007 first developed a semianalitical model for the temperature spectrum, for a nearly scale invariant fully inhomogeneous primordial field, covering the Sachs-Wolfe zone (l<30) and the Doppler zone (l<100). This is an efficient strategy to analytically try to obtain results before dealing with numerical computations. This semianlytical model was followed by a modification of the popular CMBFAST to include magnetic fields (Giovannini and Kunze, 2008 a,b,c,d,e) in which they estimated the anisotropy temperature and polarization spectra simulated maps and the Faraday Rotation. The first and third Doppler peaks increase, the second is distorted and the field increases the number of both hot and cold peaks, among other reliable results.

The non magnetic parameters were kept the same as in the so called concordance model. Relatively large magnetic strengths are needed to modify the standard interpretation, but measurable in most cases. The authors found possible remarkable and warning degeneracies. For example, a lower DM could be compensated with magnetic fields.

Clearly, a blind simultaneous estimation of the cosmological parameters –the standard ones plus two magnetic– would be the final aim of this research about what these authors designate as the mΛCDM, where m states for magnetic.

Finelli, Paci & Paoletti (2008) and Paoletti, Finelli & Paci (2008) have developed another code to study scalar, vector and tensor perturbations from stochastic primordial magnetic fields and their impact on the temperature and polarization spectra, which will become a useful tool to understand the forthcoming Planck maps. The scalar modes are more important for low multipols and the vector modes dominate at large multipols. Magnetic fields are another potential source of non-gaussianity (e.g. Brown & Crittenden (2005), Brown (2008), Naselsky, Chiang, Obsen *et al.* 2004).

6. Concluding thoughts

The possibility of a magnetized Universe is not new. We have quoted the early works by Fermi (1949). Even so, this is not the first idea. This merit corresponds to Lemaitre in 1933, as quoted by Peebles (1993). Lemaitre suggested that magnetic flux was conserved in the bounces of the oscillating universe, so that the last generation of galaxies before a bounce were able to became the nucleation of the next generation after the bounce, thus establishing a naive connection between the unending turns of the cycloidic Universe.

Many years before, Faraday, looking for an unification of light, electricity, magnetism and gravity, put a powerful electro-magnet in the path of a plane polarized light ray and observed the today called Faraday Rotation, establishing the basis of the most available tools to measure B at large $z's$.

It has been demonstrated that magnetic fields are dynamically important in a large variety of astrophysical systems. May be, in months, we can complete the list by measuring magnetic fields at z = 1000, even witness of much earlier fields. The difficulties being recognized, the mΛCDM is in progress. As stated by Peebles (1993): *If magnetic fields existed before galaxies, it will be a fascinating hint to what happened in the early universe and a considerable challenge to conventional ideas.*

Acknowledgements

Projects: FQM108, FQM792, AYA2004-08251-C02-02, AYA2007-67625-C02-02.

References

Battaner, E. 1998, *A&A* 334, 770

Battaner, E. & Florido, E. 1995, *MNRAS* 277, 1129

Battaner, E. & Florido, E. 2000, *Fund. Cosmic Phys.* 21, 1

Battaner, E. & Florido, E. 2007, *AN* 328, 92

Battaner, E., Florido, E., & Garia-Ruiz, J. M. 1997, *A&A* 327, 8

Battaner, E., Florido, R., & Jiménez-Vicente, J. 1997, *A&A* 326, 13

Battaner, E., Garrido, J. L., Membrado, M., & Florido, E. 1992, *Nature* 360, 652

Battaner, E. & Lesch, H. 2000, *Anal. de Fisica* 95, 2000 (astro-ph/0003370)

Battaner, E. & Rubiño-Martin, J. A. 2008, Constraints on primordial magnetic fields. *Project Satus, Planck Working Group 4.*

Beck, R. 2005, in: R. Wielebinski & R. Beck (eds.) *in Cosmic Magnetic Fields.* Lecture Notes in Physics (Heidelberg: Springer), vol. 664

Beck, R. 2002, in: A. Esquivel *et al.* (eds.) *Magnetic Fields in the Universe II* Rev. Mex. Astron. Astrof. (arXiv 0804.4594)

Bernet, M. L., Miniati, F., Lilly, S. J., Kronberg, P. P., & Dessauges-Zavadski, M. 2008, *Nature* 454, 302

de Block, W. J. G., Bosma, A., & Mc Gaugh, S. 2003, *MNRAS* 340, 657

Bosma, A. 2004, in: S. D. Ryder, D. J. Pisano, M. A. Walker & K. C. Freeman (eds.), *Astron. Soc. Pacific,* IAU Symp. 220

Brandenberger, R. H., Davis, A. C., & Matheson, A. M. *et al.* 1992, *Phys. Lett. B* 293, 287

Brandenburg, A. 2005, in: R. Wielebinski & R. Beck (eds.) *in Cosmic Magnetic Fields.* Lecture Notes in Physics (Heidelberg: Springer), vol. 664

Brandenburg, A. & Subramanian, K. 2005, *Phys. Rep.* 417, 1

Brown, I. A. 2008, *Ph.D.* University of Portsmouth

Brown, I. & Crittenden, R. 2005, *Phys. Rev. D* 72, 063002

Caprini, C. & Durrer, R. 2001, *Phys. Rev. D* 65, 023517

Caprini, C.& Durrer, R. 2005, *Phys. Rev. D* 72, 088301

Caprini, C. & Durrer, R. 2006, *Phys. Rev. D* 74, 321

Cheng, B. & Olinto, A. 1994, *Phys. Rev. D* 50, 2421

Durrer, R., Kahniashvili, T., & Yates, A. 1998, *Phys. Rev. D* 58, 123004

Enqvist, K. 1998, *Int. J. Mod. Phys. D* 7, 331
Finelli, F., Paci, F., & Paoletti, D. 2008, *Phys. Rev. D* 78, 023510
Florido, E. & Battaner, E. 1997, *A&A* 327, 1
Gasperini, M., Giovannini, M., & Veneziano, G. 1995, *Phys. Rev. D* 52, 6651
Giovannini, M. 2002, *hep-ph/0208152*
Giovannini, M. 2004, *Int. J. Modern Phys. D* 13, 391
Giovannini, M. 2006, *astro-ph/12378*
Giovannini, M. 2007, *Phys. Rev. D* 76, 103508
Giovannini, M. & Kunze, E. 2008a, *Phys. Rev. D* 77, 063003
Giovannini, M. & Kunze, E. 2008b, *Phys. Rev. D* 77, 061301
Giovannini, M. & Kunze, E. 2008c, *arXiv 0804.2238*
Giovannini, M. & Kunze, E. 2008d, *Phys. Rev. D* 77, 123001
Giovannini, M. & Kunze, E. 2008e, *Phys. Rev. D* 78, 023010
Grasso, D. & Rubinstein, H. R. 2001, *Phys. Rep.* 348, 163
Harari, D. D., Hayward, J. D., & Zaldarriaga, M. 1997, *Phys. Rev. D* 55, 1841
Harrison, E. H. 1973, *MNRAS* 165, 185
Hogan, C. J. 1983, *Phys. Rev. Lett.* 51, 1488
Kahniashvili, T. 2005, *New Astron.* 49, 79
Kahniashvili, T., Maravin, Y., & Kosowsky, A. 2008, *arXiv 0806.1876*
Kogut, A. *et al.* 2003, *ApJ* 665, 355
Kolb, E. W. & Turner, M. S. 1990, *The early Universe.* Adison-Wesley
Kosowsky, A., Kahniashvili, T., Lavrelashvili, G., & Ratra, B. 2005, *Phys. Rev. D* 71, 043006
Kosowsky, A. & Loeb, A. 1997, *ApJ* 469, 1
Kosowsky, A. & Loeb, A. 2005, *Phys. Rev. D* 70, 043011
Kronberg, P. P. 2005, in: R. Wielebinski & R. Beck (eds.) *in Cosmic Magnetic Fields.* Lecture Notes in Physics (Heidelberg: Springer), vol. 664, p. 9
Kronberg, P. P., Bernet, M. L., Miniati, F., Lilly, S., Short, M. B., & Higdon, D. M. 2008, *ApJ* 676, 70
Lewis, A. 2004, *Phys. Rev. D* 70, 043518
Naselsky, P., Chiang, L. Y., Obsen, P., & Verkhodanov, O. V. 2004, *ApJ* 615, 45
Navarro, J. & Steinmetz, M. 2000, *ApJ* 528, 607
Paoletti, D., Finelli, F., & Paci, F. 2008, *astrp-ph/0811.0230*
Peebles, P. J. E. 1993, *Principles of Physical Cosmology*, Princeton Series in Physics
Rees, M. 2005, in: R. Wielebinski & R. Beck (eds.) *in Cosmic Magnetic Fields.* Lecture Notes in Physics (Heidelberg: Springer), vol. 664
Reshetnikov, V., Battaner, E., Combes, F., & Jiménez-Vicente, J. 2002, *A&A* 382, 513
Reuter, H. P., Klein, U., Lesch, H., Wielebinski, R., & Kronberg, P. P. 1994, *A&A* 282, 724
Rubiño-Martín, J. A. *et al.* 2008, *arXiv 0810.3141*
Scoccola, C., Harari, D., & Mollerach, S. 2004, *Phys. Rev. D* 70, 063003
Subramanian, K. 2006, *AN* 327, 403
Turner, M. S. & Widrow, L. M. 1988, *Phys. Rev. D* 37, 2743
Wasserman, I. 1978, *ApJ* 224, 337
Widrow, L. M. 2002, *Rev. Mod. Phys.* 74, 775
Wolfe, A. M., Jorgenson, R. A., Robishaw, T., Heiles, C., & Prochaska, J. X. 2008, *Nature* 455, 638

Discussion

DURRER: Might it not be that just the magnetic field spectrum is too blue to observe them on the large scales probed by CMB experiments like *Planck*?

BATTANER: I agree, considering the limits you have commented on in your talk. But theory is still poorly supported by observations. I think that *Planck* remains an interesting possibility to observe PMF (or to establish precise observational bounds). Let's wait for *Planck*.

Cosmic Magnetic Fields: From Planets, to Stars and Galaxies
Proceedings IAU Symposium No. 259, 2008
K.G. Strassmeier, A.G. Kosovichev & J.E. Beckman, eds.

doi:10.1017/S1743921309031226

The polarized emission from the galactic plane at arcminute angular resolution

Tom L. Landecker[1], W. Reich[2], R. I. Reid[1]†, P. Reich[2], M. Wolleben[1,2], R. Kothes[1,3], D. Del Rizzo[1], B. Uyanıker[1]‡, A. D. Gray[1], E. Fürst[2], A. R. Taylor[3], and R. Wielebinski[2]

[1]National Research Council of Canada, Herzberg Institute of Astrophysics, Dominion Radio Astrophysical Observatory, P.O. Box 248, Penticton, British Columbia, V2A 6J9, Canada

[2]Max-Planck-Institut für Radioastronomie, Auf dem Hügel 69, 53121 Bonn, Germany

[3]Department of Physics and Astronomy, University of Calgary, 2500 University Drive N.W., Calgary, AB, Canada

Abstract. As part of the Canadian Galactic Plane Survey (CGPS) we have imaged the polarized emission from the plane of the Milky Way at 1420 MHz, covering 1200 square degrees with arcminute resolution. Structure on all scales is represented by combining aperture-synthesis data with single-antenna data. The survey depicts the Magneto-Ionic Medium at a resolution that matches images of other components of the Interstellar Medium within the CGPS database (http://www4.cadc.hia.nrc.gc.ca).

Keywords. Polarization – Techniques: polarimetric – Surveys – ISM: magnetic fields

1. A new survey

We describe a survey of the polarized emission at 1420 MHz, the first extensive survey to combine aperture-synthesis data (from the DRAO Synthesis Telescope - Landecker *et al.* 2000) with single-antenna data (from the Effelsberg 100-m Telescope – Reich *et al.* 2004 – and from the DRAO 26-m Telescope – Wolleben *et al.* 2006). The survey images accurately portray all polarized emission features from the broadest scales to the limit imposed by the angular resolution, $\sim 1'$. Data on structures of size $S > 40'$ comes mainly from the 26-m Telescope, $40' > S > 15'$ from the 100-m Telescope, and $15' > S > 1'$ from the Synthesis Telescope. The survey covers $\ell = 65°$ to $\ell = 175°$ over a range $-3.5° < b < 5.5°$ along the northern Galactic plane, with a high-latitude extension from $\ell = 105°$ to $\ell = 120°$ up to $b = 20°$. This survey (Landecker *et al.* in prep.) is a component of the Canadian Galactic Plane Survey (CGPS - Taylor *et al.* 2003).

2. The data

With 1.7×10^7 independent data points, this is the largest polarization survey made to date, but we can present only a small region here (see Fig. 1). Polarized intensity (PI) is high in the top right corner of Fig. 1, and structure is smooth: this is part of the Fan region seen prominently in single-antenna surveys (*e.g.* Wolleben *et al.* 2006). Superposition of other features lowers PI across the rest of the image. We focus our attention on three

† Present address: National Radio Astronomy Observatory, 520 Edgemont Rd., Charlottesville, VA, 22903, USA

‡ Present address: 35/3737 Gellatly Road, Westbank, British Columbia, V2T 2W8, Canada

Figure 1. Polarized intensity along the Galactic plane from $\ell = 156°$ to $\ell = 175°$. The intensity scale is linear from zero (white) to 500 mK (black).

distinct objects seen in Fig. 1. They are different in nature, and illustrate the diversity of phenomena that can be studied with polarization data of this quality.

The supernova remnant (SNR) HB9 (G160.9+2.6) is seen prominently as a polarized feature, but as a *decrease* in PI relative to its surroundings. The superposition of polarized SNR emission on a polarized background (or foreground) leads to this reduction.

The planetary nebula Sharpless 2-216 (G158.4+0.2) generates a polarization feature through Faraday rotation in its ionized shell (Ransom *et al.* 2008). The ionized shell, of density $n_e \approx 8$ cm^{-3}, generates a barely detectable feature in total intensity, but a strong Faraday rotation signature, with $\Delta\psi \approx 110°$, produced in a field whose line-of-sight component is $\sim 5\mu$G. This bears witness to the sensitivity of polarization measurements.

The third object would not have been discovered by any means other than its Faraday rotation signature. It is seen in Fig. 1 as a large shell, at least 8° in extent, centered at $\ell = 166°, b = -1°$. Association with an atomic hydrogen shell at $v_{lsr} = -20\,\mathrm{km\,s^{-1}}$ places the object in the Perseus Arm at a distance of ~2 kpc. Kothes *et al.* (in preparation) interpret this object as a stellar-wind bubble whose extent is some 350 pc. Once identified through its polarization signature, the object can be recognized in other wavebands.

We conclude from this remarkable result that (a) Faraday rotation is a powerful tool for the detection of ionized gas, (b) there are Galactic objects many degrees in extent which can only be detected in data with arcminute resolution, and (c) the data for this object would be impossible to interpret without the incorporation of single-antenna data into the polarization images. Further work on the survey will include correlations of polarization features with other ISM tracers, studies of depolarization by foreground HII regions, and statistical (power-spectrum and structure-function) studies.

References

Landecker, T. L., Dewdney, P., Burgess, T. A., *et al.* 2000, *A&AS* 145, 509

Ransom, R., Uyanıker, B., Kothes, R., & Landecker, T.L. 2008, *ApJ* 684, 1009

Reich, W., Fürst, E., Reich, P., *et al.* 2004, in The Magnetized Interstellar Medium, ed. B. Uyanıker, W. Reich, & R. Wielebinski, 45

Taylor, A. R., Gibson, S. J., Peracaula, M., *et al.* 2003, *AJ* 125, 3145

Wolleben, M., Landecker, T. L., Reich, W., & Wielebinski, R. 2006, *A&A* 448, 411

Cosmic Magnetic Fields:
From Planets, to Stars and Galaxies
Proceedings IAU Symposium No. 259, 2008
K.G. Strassmeier, A.G. Kosovichev & J.E. Beckman, eds.

doi:10.1017/S1743921309031238

Why we need to measure turbulent magnetic fields in HII regions

John E. Beckman[1,2] **and Mónica Relaño**[1,3,4]

[1]Instituto de Astrofísica de Canarias, Calle Vía Láctea s/n,
E-38200 La Laguna, Tenerife, Spain
email:jeb@iac.es

[2]Consejo Superior de Investigaciones Científicas, Madrid, Spain

[3]Departamento de Física Teóretica y del Cosmos, Universidad de Granada, Spain

[4]Institute of Astronomy, University of Cambridge, UK
email: mrelano@ast.cam.ac.uk

Abstract. We have measured the emission line profiles in Hα from populations of HII regions in nearby spiral galaxies, and extracted their non-thermal line widths. These are supersonic for the more luminous regions. We infer from plots of line width, σ, against Hα luminosity, a set of regions in virial equilibrium, and derive their masses summing all known components. The virial masses are considerably larger. If this discrepancy, and the supersonic line widths, are due to the presence of magnetic fields, we can estimate their strengths at a few microgauss. Observational confirmation is clearly required.

Keywords. HII regions – magnetic fields – turbulence

1. Deriving the non-thermal line widths

Using Fabry-Perot spectrometers (see Hernandez *et al.* 2008) we have measured emission line profiles in Hα from large populations of HII regions in a number of nearby galaxies (Relaño & Beckman 2005; Relaño *et al.* 2005; Fathi *et al.* 2008). These have characteristic shapes: twin weak peaks in the wings (due to expanding shells, not the topic here), and a strong central peak. The widths of the latter include contributions from, (a) the instrumental width, (b) the natural line width, (c) the thermal width, and (d) the turbulent width. To derive (d) we subtract in quadrature components (a), (b), and (c) from the total width. For many HII regions these turbulent widths are considerably larger than the sound speed, (Rozas, Beckman & Knapen 1996) of order 13 km s^{-1}

2. Turbulent line width v. Luminosity

When we plot the turbulent line widths, σ, against the H luminosities, L, of the HII regions, (see Relaño & Beckman 2005) we find: (a) A large scatter in for a given value of L, but (b) a clear lower envelope in , which we assign to HII regions in virial equilibrium, while the others, the majority, have additional impulsive velocity components. Even for these virialized regions the virial masses are considerably higher than the masses estimated by summing over all known components (dark matter can be neglected on these scales). Finally we find that (c), the turbulent line widths on the virial envelope are in general supersonic.

It is important to note that we have gone far to include all mass components in our sum. Most relevant is that, in the more luminous HII regions the greater part of the gas is not ionized (Giammanco *et al.* 2004). Estimates of the HI contribution are of order 80%

of the total gas mass, and over 70% of the total HII region mass (Chu & Kennicutt 1994; Relaño *et al.* 2005). Even so the sum falls short of the virial mass by factors typically of 2 to 3.

3. The requirement for magnetic support

An obvious way to account both for the apparently supersonic line widths and the excess mass derived using them is to postulate the presence of turbulent magnetic fields with the HII regions. It is at least conceptually plausible that turbulent plasmas give rise to turbulent fields. If we assume equipartition between kinetic and magnetic energies we obtain fields of between a few, and a few tens of microgauss. There is easily enough energy input from the ionizing stars to maintain fields of these strengths (see Myers & Goodman 1988). It would clearly be of great interest to measure these, and some recent techniques are promising in that respect.

References

Beckman, J. E. & Relaño, M. 2004, *ApJS* 292, 111
Chu, Y.-H. & Kennicutt, R. C., Jr. 1994, *ApJ* 425, 720
Fathi, K., *et al.* 2008, *ApJ* 675, L17
Giammanco, C., Beckman, J. E., Zurita, A., & Relaño, M. 2004, *A&A* 424, 877
Hernandez, O., *et al.* 2008, *PASP* 120, 665
Myers, P. C. & Goodman, A. A. 1988, *ApJ* 329, 392
Relaño, M., Beckman, J. E., Zurita, A., Rozas, M., & Giammanco, C. 2005, *A&A* 431, 235
Relaño, M. & Beckman, J. E. 2005, *A&A* 430, 911
Rozas, M., Beckman, J. E., & Knapen, J. H. 1996, *A&A* 307, 735

Cosmic Magnetic Fields:
From Planets, to Stars and Galaxies
Proceedings IAU Symposium No. 259, 2008
K.G. Strassmeier, A.G. Kosovichev & J.E. Beckman, eds.

doi:10.1017/S174392130903124X

The integrated polarization of spiral galaxies

Jeroen M. Stil[1], Marita Krause[2], Lydia Mitchell[1], Rainer Beck[2] and A. Russell Taylor[1]

[1] Centre for Radio Astronomy, The University of Calgary, Calgary AB T2N 1N4, Canada
[2] Max Planck Institut für Radioastronomie, Auf dem Hügel 69, Bonn, Germany

Abstract. We present deep observations of polarized radio emission of distant spiral galaxies with the Effelsberg 100-m telescope† at 4.8 GHz. These cross scans with sensitivity of 50 μJy or better open the possibility of a statistical study of magnetic field properties and internal Faraday rotation as a function of inclination for a large sample of unresolved galaxies. The Square Kilometre Array (SKA) will be able to detect polarization of spiral galaxies to high redshift, probing the evolution of magnetic fields in disk galaxies over cosmic time.

Keywords. Galaxies: magnetic fields – galaxies: spiral – radio continuum: galaxies – polarization

1. Integrated polarization of spiral galaxies

The regular azimuthal magnetic field in spiral galaxy disks leads to polarized synchrotron emission, with locally observed values up to 40% polarized at 4.8 GHz. When an unresolved galaxy is observed at an angle with respect to the plane of the disk, the regular magnetic field projects into a net component in the plane of the sky, resulting in net polarized emission for the galaxy as a whole. Stil *et al.* (2008) presented observations and models of the polarization of the integrated radio emission of spiral galaxies at 4.8 GHz, with fractional polarization up to 17%. Figure 1 shows the sample of spiral galaxies without a bar (a) and with a bar (b) from Stil *et al.* (2008).

The curves in Figure 1 (a) and (b) represent axially symmetric models for the integrated polarization of spiral galaxies by Stil *et al.* (2008) that include depolarization by wavelength-independent depolarization, Faraday dispersion, and differential Faraday rotation (Sokoloff *et al.* 1998). Parameters in the model are the regular magnetic field B, the ratio of random to regular magnetic field strength $f_{\rm B}$, electron density n_e with filling factor f_i, thickness of the disk $2h$, and a turbulent scale length for rotation measure fluctuations $l_{\rm turb}$. A mean thermal fraction of 23% is assumed at 4.8 GHz (Condon & Yin 1990). From inclination 0° (face-on), to higher inclination, the degree of polarization initially increases as the regular magnetic field projects into a net component of the magnetic field in the plane of the sky. At inclinations higher than $\sim 60°$, Faraday depolarization dominates and the degree of polarization decreases with inclination.

The axially symmetric models successfully reproduce the observations in Figure 1a, but not in Figure 1b where large-scale magnetic fields along the bar add the orientation of the bar as an additional parameter that is currently not included in the models. An interesting feature of the models is the convergence of curves with the same value of $f_{\rm B}$ for inclinations less than 50°. The uniformity of the magnetic field defines the polarized intensity almost everywhere in the disk, and thus the degree of polarization of

† The Effelsberg 100-m telescope is operated by the MPIfR on behalf of the Max-Planck-Gesellschaft

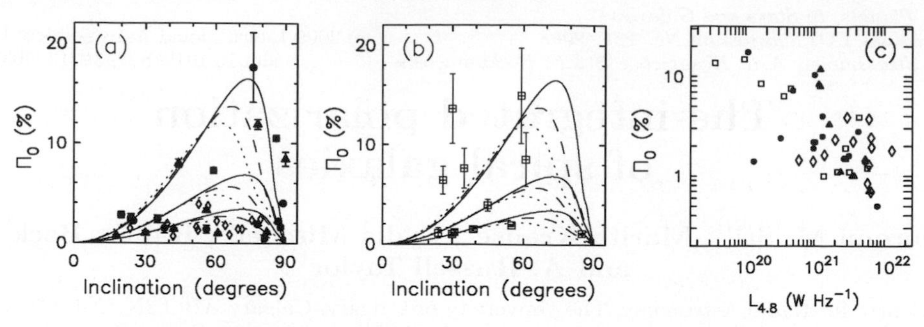

Figure 1. (*a*) Integrated fractional polarization at 4.8 GHz of nearby spiral galaxies (dots), Virgo cluster spirals (triangles), and new Effelsberg observations of distant UGC spiral galaxies (diamonds). The curves represent axially symmetric models of the integrated polarization that include wavelength-independent depolarization Faraday dispersion and differential Faraday rotation. Upper solid curve: $B = 5$ μG, $f_{\rm B} = 1.0$, $n_e = 0.03$ cm^{-3}, $f_i = 0.5$, $l_{\rm turb} = 50$ pc, $\langle 2h \rangle = 1000$ pc, thermal fraction is 0.23. Dashed curve: double the magnetic field B. Dotted curve: triple the electron density n_e. (*b*) The same as (a) for barred galaxies (squares). The model curves in (b) are shown for comparison with (a) only. (*c*) Anti-correlation between fractional polarization and 4.8 GHz luminosity.

the integrated emission. This is a feature of models at frequencies more than a few GHz, where Faraday dispersion and differential Faraday rotation are relatively unimportant for low inclinations.

Figure 1c shows an anti-correlation of fractional polarization of the integrated emission with radio luminosity. This anti-correlation suggest that the magnetic field of luminous galaxies is less uniform, possibly because of a higher star formation rate.

Figure 1 also shows 14 galaxies observed in deep cross scans at 4.8 GHz with the Effelsberg telescope shown as diamonds. These galaxies are Milky Way size spiral galaxies at distances of several tens of Mpc selected from the Uppsala General Catalog (UGC) and detected in the NVSS with a 1.4 GHz flux density of 20 mJy or higher. The angular size of these galaxies is smaller than the 2.5′ beam of the Effelsberg telescope at 4.8 GHz. A sensitivity of 50 μJy beam^{-1} can be obtained with 1 hour total observing time per galaxy. This makes it possible to observe a large sample of galaxies for a statistical analysis of magnetic field uniformity and Faraday depolarization.

2. Evolution of Magnetic fields in disk galaxies

The Square Kilometre Array (SKA) will be able to detect polarization of radio sources fainter than 100 μJy, most of which are distant star forming galaxies. Although high-redshift galaxies will be mostly unresolved, polarization of their integrated emission provides information on the uniformity of their magnetic field and the amount of Faraday rotation. Our Effelsberg observations, when completed, will be part of a local comparison sample for galaxies at high redshift. Such observations will test large-scale dynamo models such as the turbulent dynamo at high redshift (Arshakian *et al.* 2008).

References

Arshakian, T., Beck, R., Krause, M., & Sokoloff, D., 2008, *A&A*, in press, `arXiv 0810.3114`
Condon, J. J. & Yin, Q. F. 1990, *ApJ* 357, 97
Condon, J. J., Cotton, W. D., Greisen, E. W., *et al.* 1998, *AJ* 115, 1693
Sokoloff, D. D., Bykov, A. A., Shukurov, A., *et al.* 1998, *MNRAS* 299, 189
Stil, J. M., Krause, M., Beck, R., & Taylor, A. R. 2008, *ApJ*, in press `arXiv 0810.2303`

Cosmic Magnetic Fields:
From Planets, to Stars and Galaxies
Proceedings IAU Symposium No. 259, 2008
K.G. Strassmeier, A.G. Kosovichev & J.E. Beckman, eds.

doi:10.1017/S1743921309031251

Magnetism in the nearby galaxy M 33

F. S. Tabatabaei[1], M. Krause[1], R. Beck[1], and A. Fletcher[2]

[1]Max-Planck-Institut für Radioastronomie, Auf dem Hügel 69, 53121 Bonn, Germany
email: tabataba@mpifr-bonn.mpg.de, mkrause@mpifr-bonn.mpg.de, rbeck@mpifr-bonn.mpg.de

[2]School of Mathematics and Statistics, Newcastle University, Newcastle upon Tyne, NE1 7RU, U.K., email: andrew.fletcher@ncl.ac.uk

Abstract. Using high-resolution data of the linearly polarized intensity and polarization angle at 3.6, 6.2, and 20 cm together with a 3-D model of the regular magnetic field, we study variations of the structure, strength, and energy density of the magnetic field in the Scd galaxy M 33. The regular magnetic field consists of a horizontal component (represented by an axisymmetric mode from 1 to 3 kpc radius and a superposition of axisymmetric and bisymmetric modes from 3 to 5 kpc radius) and a vertical component. However, the inferred 'vertical field' may be partly due to a galactic warp. We estimate the average total and regular magnetic field strengths as $\simeq 6.4$ and $2.5\,\mu$G, respectively. Generation of interstellar magnetic fields by turbulent gas motion in M 33 is indicated as the turbulent and magnetic energy densities are about equal.

Keywords. Galaxies: individual: M33 – radio continuum: galaxies – galaxies: magnetic field – galaxies: ISM

---

1. Introduction

M 33, the nearest Scd galaxy at a distance of 840 kpc, with its large angular size and medium inclination, allows determination of the magnetic field components both parallel and perpendicular to the line of sight equally well. The RM studies of M 33 by Beck (1979) and Buczilowski *et al.* (1991) suggested a bisymmetric regular magnetic field structure in the disk of M 33. However, due to the low-resolution (1.8 kpc) and low-sensitivity of their observations, these results were affected by high uncertainty particularly in the southern half of M 33. Our recent observations of this galaxy provide high-resolution (0.7 kpc) maps of total power and linearly polarized intensity at 3.6 cm, 6.2 cm, and 20 cm presented by Tabatabaei *et al.* (2007a). These data are ideal to study the rotation measure (RM), the structure and strength of the magnetic field, and depolarization effects in detail. By combining an analysis of multi-wavelength polarization angles with modeling of the wavelength-dependent depolarization, Fletcher *et al.* (2004) and Berkhuijsen *et al.* (1997) derived the 3-D regular magnetic field structures in M 31 and M 51, respectively. The high sensitivity of our new observations allows a similar study for M 33.

2. Nonthermal degree of polarization

Using the polarized intensity maps of Tabatabaei *et al.* (2007a) and the nonthermal maps obtained by Tabatabaei *et al.* (2007b), we derived maps of the nonthermal degree of polarization at different wavelengths. Integrating the polarized and nonthermal intensity maps in the galactic plane out to a galactocentric radius of R $\leqslant$ 7.5 kpc, we obtained the flux densities of the nonthermal and linearly polarized emission and the average nonthermal degrees of polarization of $10.3\% \pm 2.0\%$, $11.3\% \pm 1.9\%$, and $6.6\% \pm 0.6\%$ at 3.6 cm, 6.2 cm, and 20 cm, respectively, indicating considerable wavelength-dependent depolarization by Faraday effects at 20 cm.

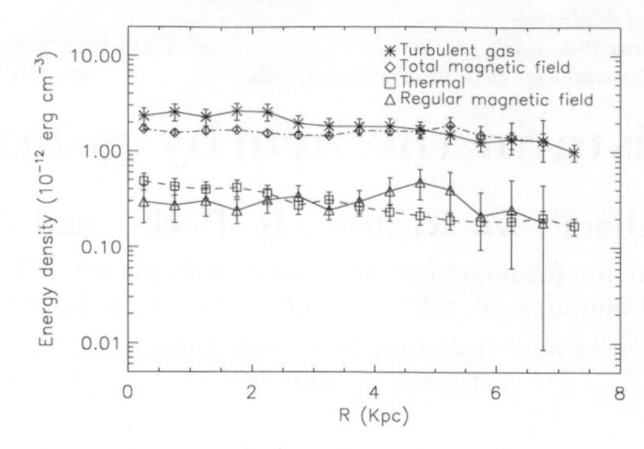

Figure 1. Energy densities and their variations with galactocentric radius in M 33.

3. The regular magnetic field structure

In order to identify the 3-D structure of the *regular* magnetic field $B_{\rm reg}$ we fit a parameterized model of $B_{\rm reg}$ to the observed polarization angles at different wavelengths. We find that the Fourier modes $m = 0 + z0 + z1$ ($z0$ and $z1$ are the first and second Fourier modes of the vertical field) in the 1–3 kpc ring and $m = 0 + 1 + z1$ in the 3–5 kpc ring can best reproduce the observed pattern of polarized intensity at 6.2 cm (see Tabatabaei *et al.* (2008) for details). The horizontal magnetic field component follows an arm-like pattern with pitch angles smaller than those of the optical arm segments, indicating that large-scale gas-dynamical effects such as compression and shear are not solely responsible for the spiral magnetic lines. The dominant axisymmetric mode ($m = 0$) in the disk in both rings indicates that galactic dynamo action is present in M 33. We also find that the fitted 'vertical field', in the outer ring, could be mainly due to the severe warp of M 33 and hence apparent. However, a real vertical field of a broadly comparable strength to the disk field can exist in the inner ring.

4. Magnetic field strengths and energy densities

The strengths of the total magnetic field $B_{\rm tot}$ and its regular component $B_{\rm reg}$ can be found from the total synchrotron intensity and its degree of linear polarization. Assuming equipartition between the energy densities of the magnetic field and cosmic rays leads to $B_{\rm tot} = 6.4 \pm 0.5\,\mu$G and $B_{\rm reg} = 2.5 \pm 1.0\,\mu$G for the disk of M 33 ($R < 7.5$ kpc).

The energy densities of the equipartition magnetic fields in the disk ($B^2_{\rm tot}/8\pi$ and $B^2_{\rm reg}/8\pi$ for the total and regular magnetic fields, respectively) are shown in Fig. 1. The energy densities of the magnetic field and turbulence are about the same, confirming the theory of generation of interstellar magnetic fields from turbulent gas motion. Furthermore, it seems that the ISM in M 33 can be characterized by a low-β plasma and is dominated by supersonic turbulence, as the energy densities of the magnetic field and turbulence are both much higher than the thermal energy density.

References

Beck, R. 1979, *Ph.D. Thesis*, Rheinische Friedrich-Wilhelms-Universitaet, Bonn
Berkhuijsen, E. M., Horellou, C., Krause, M., Neininger, N., *et al.* 1997, *A&A* 318, 700
Buczilowski, U. R. & Beck, R. 1991, *A&A* 241, 47
Fletcher, A., Berkhuijsen, E. M., Beck, R., & Shukurov, A. 2004, *A&A* 414, 53
Tabatabaei, F. S., Krause, M., & Beck, R. 2007a, *A&A* 472, 785
Tabatabaei, F. S., Beck, R., Krügel, E., Krause, M., Berkhuijsen, *et al.* 2007b, *A&A* 475, 133
Tabatabaei, F. S., Krause, M., Fletcher, A., & Beck, R. 2008, *A&A* 490, 1005

Cosmic Magnetic Fields:
From Planets, to Stars and Galaxies
Proceedings IAU Symposium No. 259, 2008
K.G. Strassmeier, A.G. Kosovichev & J.E. Beckman, eds.

doi:10.1017/S1743921309031263

Interplay of CR-driven galactic wind, magnetic field, and galactic dynamo in spiral galaxies

Marita Krause

Max-Planck-Institut für Radioastronomie, Auf dem Hügel 69, 53121 Bonn, Germany
email: mkrause@mpifr-bonn.mpg.de

Abstract. From our radio observations of the magnetic field strength and large-scale pattern of spiral galaxies of different Hubble types and star formation rates (SFR) we conclude that – though a high SFR in the disk increases the total magnetic field strength in the disk and the halo – the SFR does not change the global field configuration nor influence the global scale heights of the radio emission. The similar scale heights indicate that the total magnetic field regulates the galactic wind velocities. The galactic wind itself may be essential for an effective dynamo action.

Keywords. Galaxies: spiral – magnetic fields – halos – radio continuum: galaxies

1. Magnetic field strength and star formation

Observations of a sample of three late-type galaxies with low surface-brightness and the radio-weak edge-on galaxy NGC 5907 (all with a low SFR) revealed that they all have an unusually high thermal fraction and weak total and regular magnetic fields (Chyży *et al.* 2007, Dumke *et al.* 2000). However, these objects still follow the total radio-FIR correlation, extending it to the lowest values measured so far. Hence, these galaxies have a lower fraction of synchrotron emission than galaxies with higher SFR. It is known that the thermal intensity is proportional to the SFR. Our findings fit to the equipartition model for the radio-FIR correlation (Niklas & Beck 1997), according to which the nonthermal emission increases $\propto SFR^{1.3\pm0.2}$ and the *total* magnetic field strength B_t increases $\propto SFR^{0.34\pm0.14}$.

No similar simple relation exists for the *regular* magnetic field strength. We integrated the polarization properties in 41 nearby spiral galaxies and found that (independently of inclination effects) the degree of polarization is lower ($< 4\%$) for more luminous galaxies, in particular those for $L_{4.8} > 2 \times 10^{21}$ WHz^{-1} (Stil *et al.* 2008). The radio-brightest galaxies are those with the highest SFR. Though a dynamo action needs star formation and supernova remnants as the driving force for velocities in vertical direction, we conclude from our observations that stronger star formation seems to reduce the magnetic field regularity. On kpc-scales, Chyży (2008) analyzed the correlation between magnetic field regularity and SFR locally within one galaxy, NGC 4254. While he found that the total and random field strength increase locally with SFR, the regular field strength is locally uncorrelated with SFR.

2. Vertical scale heights and CR-driven galactic wind

We determined the exponential scale heights of the total power emission at $\lambda 6$ cm for four edge-on galaxies (NGC 253, NGC 891, NGC 3628, NGC 4565) for which we have combined interferometer and single-dish data (VLA and the 100-m Effelsberg). In spite of their different intensities and extents of the radio emission, the vertical *scale heights* of the thin disk and the thick disk/halo are similar in this sample (300 pc and 1.8 kpc) (Dumke & Krause 1998, Heesen *et al.* 2009). We stress that our sample includes the brightest halo observed so far, NGC 253, with a very high SFR, as well as one of the weakest halos, NGC 4565, with a small SFR.

For NGC 253 Heesen *et al.* (this volume) argued that the synchrotron lifetime (which is $\propto B_t^{-2}$) mainly determines the vertical scale height of the synchrotron emission and estimated the cosmic ray bulk velocity to 300 ± 30 km/s. As this is similar to the escape velocity, it shows the presence of a galactic wind in this galaxy. The fact that we observe similar averaged scaleheights at $\lambda 6$ cm for the four galaxies mentioned above imply that the galactic wind velocity is proportional to B_t^2, and hence proportional to $SFR^{0.7 \pm 0.3}$.

3. Magnetic field structure, dynamo action, and galactic wind

In a larger sample of 11 edge-on galaxies we found in all of them (except the inner part of NGC 4631, see Krause 2009) mainly a disk-parallel magnetic field along the galactic midplane together with an X-shaped poloidal field in the halo. Our sample includes spiral galaxies of different Hubble types and SFR, ranging from $0.5\ M_\odot yr^{-1} \leqslant SFR \leqslant 27\ M_\odot yr^{-1}$. The disk-parallel magnetic field is the expected edge-on projection of the spiral magnetic field within the disk as observed in face-on galaxies. It is generally thought to be generated by a mean-field $\alpha\Omega$-dynamo for which the most easily excited field pattern is the axismmetric spiral (ASS) field (e.g. Beck *et al.* 1996). The poloidal part of the ASS dynamo field alone, however, cannot explain the observed X-shaped structures in edge-on galaxies as the field strength there seems to be comparable to that of the large-scale disk field. Model calculations of the mean-field $\alpha\Omega$-dynamo for a disk surrounded by a spherical halo including a *galactic wind* (Brandenburg *et al.* 1993) simulated similar field configurations as the observed ones. New MHD simulations are in progress (see e.g. Gressel *et al.* this volume, Hanasz *et al.* this volume) which include a galactic wind implicitely. A galactic wind can also solve the helicity problem of dynamo action (e.g. Sur *et al.* 2007). Hence, a galactic wind may be essential for an effective dynamo action, and to explain the observed similar vertical scale heights and X-shaped magnetic field structure in edge-on galaxies.

References

Beck, R., Brandenburg, A., Moss, D., Shukurov, A., & Sokoloff, D. D. 1996, *ARAA* 34, 155

Brandenburg, A., Donner, K. J., Moss, D., Shukurov, A., Sokoloff, & D. D., Tuominen, I. 1993, *A&A* 271, 36

Chyży, K. T., Bomans, D. J., Krause, M., Beck, R., Soida, M., & Urbanik, M. 2007, *A&A* 462, 933

Chyży, K. T. 2008, *A&A* 482, 755

Dumke, M. & Krause, M. 1998, in: D. Breitschwerdt, M. Freyberg & J. Trümper (eds.), *The Local Bubble and Beyond*, Lecture Notes in Physics (Heidelberg: Springer), vol. 166, p. 555

Dumke, M., Krause, M., & Wielebinski, R. 2000, *A&A* 355, 512

Heesen, V., Beck, R., Krause, M., & Dettmar, R.-J. 2009, *A&A*, in press, arXiv:astro-ph/0812.0346

Krause, M. 2009, *Rev. Mexicana AyA*, in press, arXiv:astro-ph/0806.2060

Niklas, S. & Beck, R. 1997, *A&A* 320, 54

Stil, J. M., Krause, M., Beck, R., & Taylor, A. R. 2008, *A&A*, in press, arXiv:astro-ph/0810.2303

Sur, S., Shukurov, A., & Subramanian, K. 2007, *MNRAS* 377, 874

Cosmic Magnetic Fields:
From Planets, to Stars and Galaxies
Proceedings IAU Symposium No. 259, 2008
K.G. Strassmeier, A.G. Kosovichev & J.E. Beckman, eds.

doi:10.1017/S1743921309031275

Global simulations of a galactic dynamo driven by cosmic-rays and exploding magnetized stars

Michał Hanasz, Dominik Wóltański, Kacper Kowalik, and Rafał Pawłaszek

Centre for Astronomy, Nicholas Copernicus University, PL-87148 Piwnice/Toruń, Poland,
mhanasz@astri.uni.torun.pl

Abstract. We conduct global galactic–scale magnetohydrodynamical (MHD) simulations of the cosmic–ray driven dynamo. We assume that exploding stars deposit small–scale, randomly oriented, dipolar magnetic fields into the differentially rotating ISM, together with a portion of cosmic rays, accelerated in supernova shocks. Our simulations are performed with the aid of a new parallel MHD code PIERNIK. We demonstrate that dipolar magnetic fields supplied on small SN–remnant scales, can be amplified exponentially by the CR–driven dynamo to the present equipartition values, and transformed simultaneously to large galactic–scales by an inverse cascade promoted by resistive processes.

Keywords. Galaxies: ISM – magnetic fields – ISM: cosmic rays – magnetic fields – MHD

It has been suggested by Rees (1987) that galactic seed fields were created and amplified in stars during early stages of galactic evolution, and then spread into the interstellar medium (ISM) by stellar explosions, and subsequently amplified in plerionic (Crab–type) supernova remnants (SNRs). Rees (1987) estimates that a contribution of 10^6 randomly oriented plerionic SNRs may lead to 10^{-9}G mean magnetic fields on galactic scales. The initial setup of our galactic disk is based on the model by Ferriere (1998), with the gravitational potential by Allen & Santillán (1991). The global CR–driven dynamo model involves basic elements of local dynamo models presented by Hanasz *et al.* (2004): (1.) Cosmic rays supplied in randomly distributed SNRs, which are described as relativistic gas diffusing anisotropically along magnetic field lines, according to the diffusion–advection transport equation, supplemented to the standard set of resistive MHD equations. (2.) A finite resistivity of the ISM, responsible for dissipation of small–scale magnetic fields. Moreover, we assume that no magnetic field is present in the initial configuration, and that each SN supplies a weak, randomly oriented, dipolar magnetic field within the supernova remnant, together with the portion of CRs, while the thermal energy output from supernovae is neglected. Simulations have been performed with the aid of PIERNIK MHD code (see Hanasz *et al.* 2009a,b and references therein), which is a grid–based MPI parallelized, resistive MHD code based on the Relaxing TVD (RTVD) scheme by Jin & Xin (1995) and Pen *et al.* (2003). The original scheme is extended to deal with the diffusive CR component (see Hanasz & Lesch 2003). The simulation has been performed with the spatial resolution of $1000 \times 1000 \times 160$ grid cells, in the domain spanning 25 kpc$\times$25 kpc$\times$8 kpc in x, y and z directions, respectively. We show that the CR driven dynamo, seeded by small–scale magnetic dipoles and cosmic rays supplied in supernova remnants, amplifies magnetic fields exponentially by several orders of magnitude (Fig. 1), up to the saturation level, and develops large scale magnetic fields in the disk and the surrounding galactic halo (Fig. 2). The horizontal slice demonstrates the spiral structure of the amplified field. Formation of large lobes of unipolar magnetic

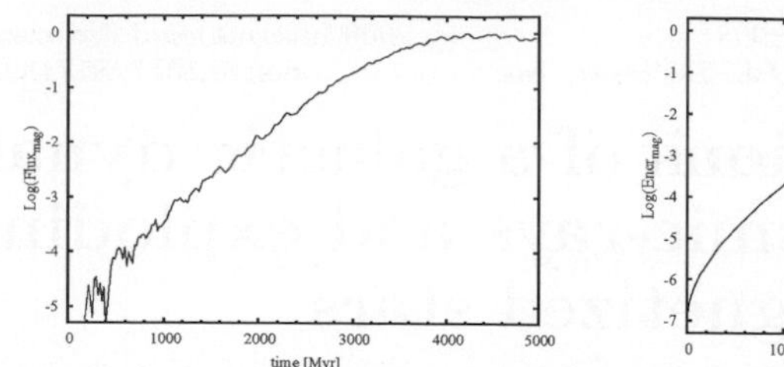

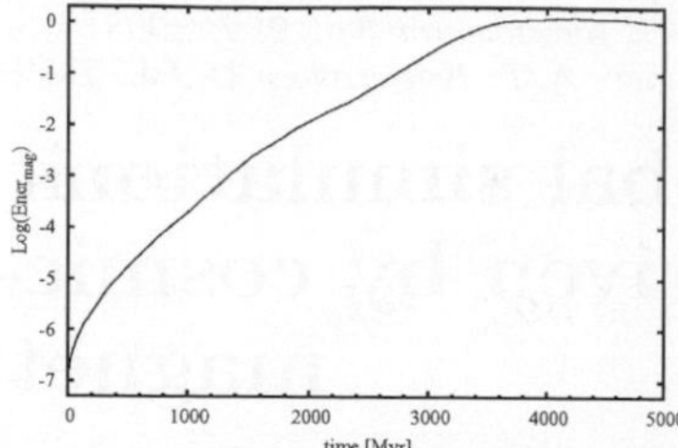

Figure 1. Temporal evolution of toroidal magnetic flux (left) and total magnetic energy (right). The final saturation level corresponds to the equipartition magnetic fields

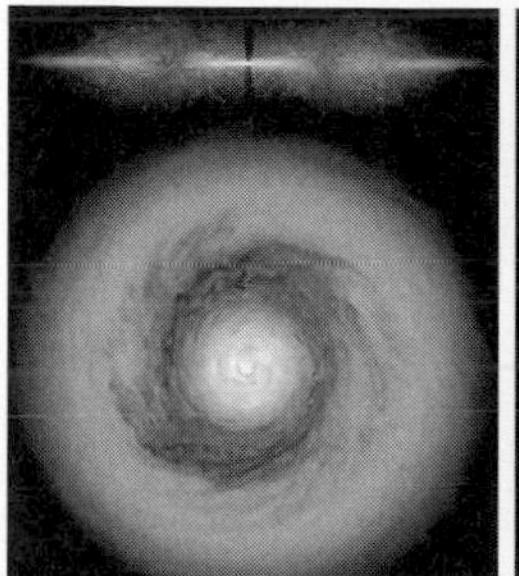
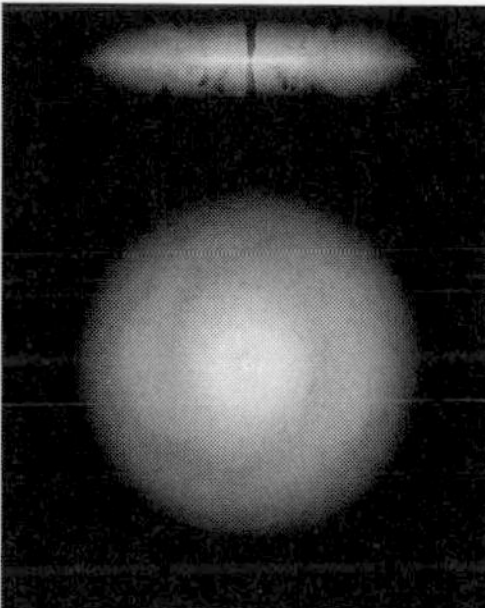
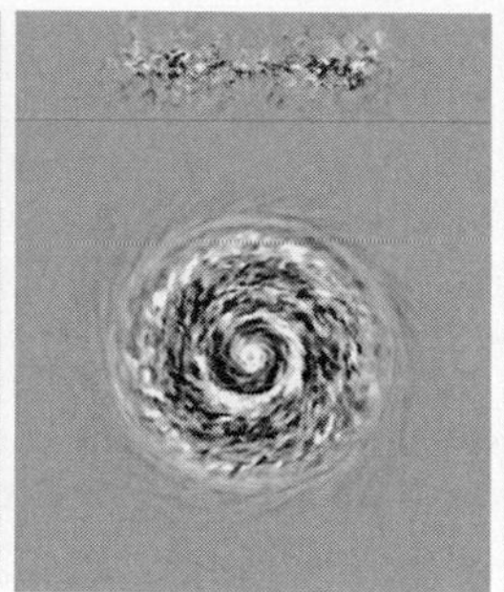
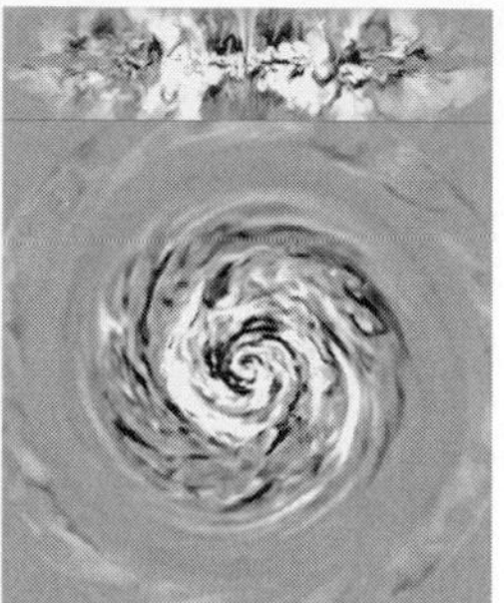

Figure 2. Logarithm of gas density (1st panel) and cosmic ray energy density (2nd panel) at $t = 4.2$Gyr . Distribution of toroidal magnetic field at $t = 0.5$ Gyr (3rd panel) and $t = 5$ Gyr (4th panel). Unmagnetized regions of the volume are marked with the gray colour, positive and negative toroidal magnetic fields are marked with white and black colours, respectively. Initially the toroidal magnetic field strength is a few 10^{-8}G and at the saturated state it is a few μG.

fields is apparent in vertical slices through the disk volume. The magnetic field large–scale structure forms the X–shaped configuration. The experiment supports strongly the idea by Rees (1987) that galactic dynamos may have been initiated by small–scale magnetic fields of stellar origin.

Acknowledgements

The computations were performed on the GALERA supercomputer in TASK Academic Computer Centre in Gdańsk. This work was supported by Polish Ministry of Science and Higher Education through the grant 92/N–ASTROSIM/2008/0.

References

Allen, C. & Santillán, A. 1991, *Rev. Mexicana Astron. Astrofis.* 22, 255

Ferriere, K. 1998, *ApJ* 497, 759

Hanasz, M. & Lesch, H. 2003, *A&A* 412, 331

Hanasz, M., Kowal, G., Otmianowska–Mazur, K., & Lesch, H. 2004, *ApJ* 605, L33

Hanasz, M., Kowalik, K., Wóltański, D., & Pawłaszek, R. K., 2009a, in *The Role of Disk–Halo Interaction in Galaxy Evolution: Outflow vs Infall?* , M. de Avillez *et al.* (eds.), submitted

Hanasz, M., Kowalik, K., Wóltański, D., & Pawłaszek, R. K., 2009b, in *The Role of Disk–Halo Interaction in Galaxy Evolution: Outflow vs Infall?* , M. de Avillez *et al.* (eds.), submitted

Jin, S. & Xin, Z. 1995, *Comm. Pure Appl. Math.* 48, 235

Pen, U.-L., Arras, P., & Wong, S. 2003, *ApJS* 149, 447

Rees, M. J. 1987, *QJRAS* 28, 197

Cosmic Magnetic Fields:
From Planets, to Stars and Galaxies
Proceedings IAU Symposium No. 259, 2008
K.G. Strassmeier, A.G. Kosovichev & J.E. Beckman, eds.

doi:10.1017/S1743921309031287

Structure of magnetic fields in spiral galaxies

Hanna Kotarba[1,3]†, H. Lesch[1], K. Dolag[2], T. Naab[1], P. H. Johansson[1] and F. A. Stasyszyn[2]

[1]University Observatory Munich, Scheinerstr. 1, D-81679 Munich, Germany
email: kotarba@usm.lmu.de

[2]MPI for Astrophysics, Karl-Schwarzschild-Str. 1, D-85741 Garching, Germany

[3]MPI for Extraterrestrial Physics, Giessenbachstrasse, D-85748 Garching, Germany

Abstract. We present a set of global, self-consistent N-body/SPH simulations of the dynamic evolution of galactic discs with gas and including magnetic fields. We have implemented a description to follow the ideal induction equation in the SPH part of the code VINE. Results from a direct implementation of the field equations are compared to a representation by Euler potentials, which pose a $\nabla \cdot \boldsymbol{B}$-free description, a constraint not fulfilled for the direct implementation. All simulations are compared to an implementation of magnetic fields in the code GADGET. Starting with a homogeneous field we find a tight connection of the magnetic field structure to the density pattern of the galaxy in our simulations, with the magnetic field lines being aligned with the developing spiral pattern of the gas. Our simulations clearly show the importance of non-axisymmetry of the dynamic pattern for the evolution of the magnetic field.

Keywords. Methods: n-body simulations – magnetic fields – galaxies: evolution – galaxies: spiral

1. Numerical Methods and Setup

Hydrodynamics including the ideal induction equation are treated by the SPH method implemented within the OpenMP parallel N-body code VINE (Wetzstein *et al.* 2008, Nelson *et al.* 2008). We have also implemented a description of magnetic fields using the Euler potentials, which are free of $\nabla \cdot \boldsymbol{B}$ per construction ($\boldsymbol{B} = \nabla\alpha_E \times \nabla\beta_E$, where α_E and β_E are scalar potentials).

A similar treatment of magnetic fields is implemented in the MPI parallel N-body/SPH code GADGET (Springel 2005, Dolag & Stasyszyn 2008) which includes also cleaning methods for $\nabla \cdot \boldsymbol{B}$. The initial conditions for our Milky Way like galaxy are realized using the method described by Springel *et al.* (2005) which is based on Hernquist (1993). The galaxy consists of an exponential stellar disc and a flat extended gas disc, a stellar bulge and a dark matter halo of collisionless particles. We adopt an isothermal equation of state with a fixed sound speed of ≈ 15 km/s.

2. Results

The magnetic field strength pattern is tightly connected to the density pattern of the gas in our simulated galaxy for all implementations (figure 1). Using the direct implementation of the induction equation, the largest magnetic field strengths are reached in the center (not shown), whereas in the Euler implementation the largest values are reached in the spiral arms. The morphology shows great similarity for both implementations. Using Euler potentials, the magnetic field gets amplified by one order of magnitude during the simulation, which is still one to two orders of magnitude below the amplification for the

† Present address: University Observatory, Scheinerstr. 1, D-81679 Munich, Germany

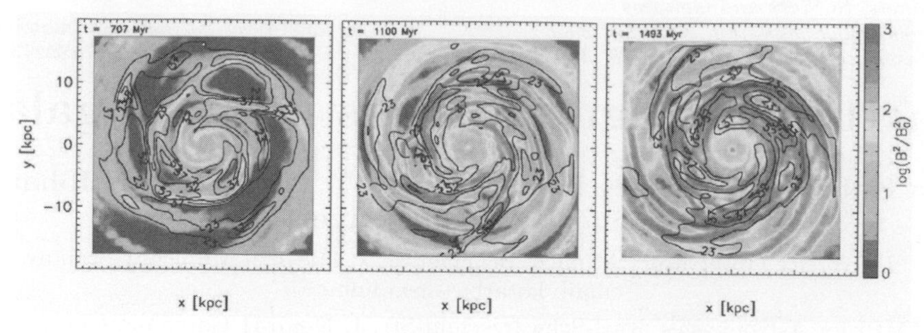

Figure 1. Face-on magnetic field energy and gas density as a function of time for a simulation performed with VINE using the Euler potentials. The colors correspond to the magnetic field energy $B^2/8\pi$ on a logarithmic scale, normalized to the initial value of $\frac{1}{8\pi} \cdot 10^{-18}$ erg cm^{-3}. The contour lines indicate physical densities of 23, 37 and 52 $M_{\odot}$ pc^{-3}, respectively.

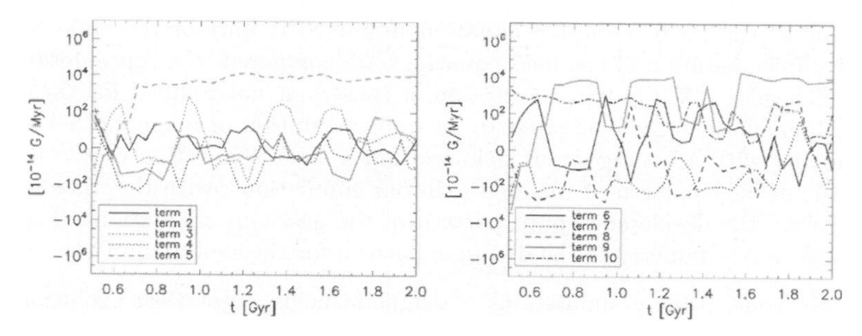

Figure 2. Values of terms responsible for the evolution of the radial (left graphic) and toroidal magnetic field component (right graphic) as a function of time for a GADGET simulation using Euler potentials. Positive values imply amplification, and negative attenuation of the corresponding $\boldsymbol{B}$-component. Non-axisymmetric terms are shown in red.

direct implementation. In all cases, the growth of $\nabla \cdot \boldsymbol{B}$ correlates with the growth of the total magnetic field, i.e. the higher the divergence, the stronger the growth of the magnetic field. Applying smoothing of $\boldsymbol{B}$ lowers the divergence (Dolag & Stasyszyn 2008) and lowers also the field amplification, leading to an amplification of the total field similar to that using the Euler potentials (not shown). Thus, for an ideal treatment, Euler potentials are a good method to follow the evolution of magnetic fields in spiral galaxies.

Dropping all dependencies on z, the equations for the temporal evolution of the radial and toroidal magnetic field depend on five terms each. Figure 2 displays the values of the terms responsible for the evolution of the radial (left graphic) and the toroidal magnetic field component (right graphic) as a function of time for a GADGET simulation using Euler potentials. Terms depending on azimuthal derivatives are shown in red. Although azimuthal derivatives are comparatively small, these terms have the largest values, thus showing that even small deviations from axial symmetry are of crucial importance for the evolution of the magnetic field.

References

Dolag, K. & Stasyszyn, F. A. 2008, *ArXiv e-prints* arXiv:0807.3553
Wetzstein, M., Nelson, A. F., Naab, T., & Burkert A. 2008, *ArXiv e-prints* arXiv:0802.4245
Nelson, A. F., Wetzstein, M., & Naab, T. 2008, *ArXiv e-prints* arXiv:0802.4253
Springel, V., Di Matteo, T., & Hernquist, L. 2005, *MNRAS* 361, 776
Hernquist, L. 1993, *ApJ* 86, 389
Springel, V. 2005, *MNRAS* 364, 1105

Cosmic Magnetic Fields:
From Planets, to Stars and Galaxies
Proceedings IAU Symposium No. 259, 2008
K.G. Strassmeier, A.G. Kosovichev & J.E. Beckman, eds.

doi:10.1017/S1743921309031299

3D MHD simulations of magnetic fields and radio polarization of barred galaxies

B. Kulesza-Żydzik[1], K. Kulpa-Dybeł[1], K. Otmianowska-Mazur[1], G. Kowal[1,2] and M. Soida[1]

[1]Astronomical Observatory, Jagiellonian University, ul Orla 171, 30-244 Kraków, Poland

[2]Department of Astronomy, University of Wisconsin, 475 North Charter Street, Madison, WI 53706, USA

Abstract. We present results of three-dimensional, fully nonlinear MHD simulations of a large-scale magnetic field evolution in a barred galaxy. The model does not take into consideration the dynamo process. We find that the obtained magnetic field configurations are highly similar to the observed maps of the polarized intensity of barred galaxies, because the modeled vectors form coherent structures along the bar and spiral arms. Due to the dynamical influence of the bar the gas forms spiral waves which go radially outward. Each spiral arm forms the magnetic arm which stays much longer in the disk, than the gaseous spiral structure. Additionally the modeled total energy of magnetic field grows due to strong compression and shear of non-axisymmetrical bar flows and differential rotation, respectively.

Keywords. MHD – numerical simulations – barred galaxies – magnetic fields

1. Model description

We investigate the evolution of barred galaxy solving the resistive set of MHD equations. We apply an isothermal equation of state. Our galaxy is composed of four components: the large and massive halo, the central bulge, rotating disc of stars and finally the bar. The rotation curve of the stellar disc we derived from the isochrone potential. The bar component is described by the second order Ferrers ellipsoid (Ferrers 1877) with semi-axes $a = 4$ kpc, $b = 2$ kpc, $c = 2$ kpc. It is initiated into the galaxy gradually in time until it reaches its final mass $M_{bar} = 10^{10}\mathrm{M}_{\odot}$. In order to conserve the total mass of the galaxy we reduce the bulge mass, so we have $M_{bar}(t) + M_b(t) =$ const. during the calculations. The bar angular velocity Ω_{bar} is set to be 25 km s^{-1} kpc^{-1}. We assume that the initial magnetic field is azimuthal ($B_z = 0$, $B_r = 0$, $B_\varphi(z, r)$) and its distribution strictly depends on the gas distribution via the following condition: $\alpha = p_{mag}/p_{gas}$. The computational domain extends from -10 kpc to 10 kpc in the x and y direction, and from -2.5 kpc to 2.5 kpc in the z direction. In all models we use the same value of the resistivity coefficient $\eta = 3 \cdot 10^{25}$ cm^{2}s^{-1} and the resolution $n_x = n_y = 256$, $n_z = 65$. We perform simulations of magnetic field evolution with a constant isothermal sound of speed $c_s = 5$ km/s and $\alpha = 0.001$ ($B_{\varphi\ 0} = 0.1\mu$G).

2. Results

Below we discuss the time evolution of the distributions of polarization angle and polarized intensity superimposed onto the column density. In order to show the magnetic field behaviour in our model we present three crucial time steps (Fig. 1):

• At time $t_1 = 0.42$ Gyr (Fig. 1, left), as a consequence of the non-axisymmetric gravitational potential of a bar, the gaseous and magnetic arms are formed. In the inner

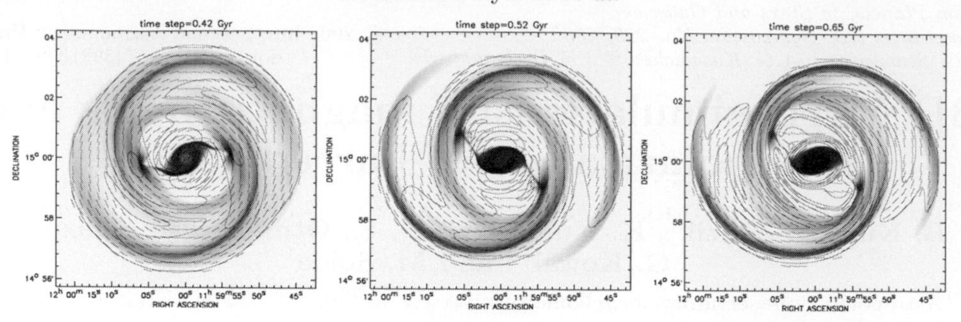

Figure 1. Face-on polarization maps at $\lambda = 6.2$ cm at selected times steps superimposed onto gaseous map. All maps have been smoothed to the resolution $25''$

part of the disk, where the bar is presented, we obtain the highest density region and the strongest magnetic field. Moreover the magnetic field maxima are aligned along the gaseous ones.

• As the simulation proceed, magnetic arms start to detach from gaseous spirals into the interarm regions (see $t_2 = 0.52$ Gyr Fig. 1, in the middle).

• At $t_3 = 0.65$ Gyr (Fig. 1, right) our magnetic arms are also visible in the interarm region. This is because the magnetic arms do not co-rotate with gaseous spirals but have a slightly lower angular velocity. The process of drift of magnetic structures into the interarm area have also been obtained by Otmianowska-Mazur *et al.* (2002).

We started our simulations with mean magnetic field equal 0.1 μG. During the whole simulation time we observe the growth of the total magnetic energy. This amplification is caused by a local compression accompanied with leading sides of the bar and inner edges of spiral arms. As we do not apply the dynamo effect the mean value of the B_ϕ flux in the galactic midplane drops.

3. Conclusions

1. We obtained the magnetic field vectors distributed along the bar, spiral arms and also in the interarm region.

2. Magnetic arms are developing in the gaseous ones, but are detached from the density waves. In the consequence the magnetic arms are shifted to the interarm regions, what is in agreement with observations (e.g. NGC 1356 (Beck *et al.* 2005))

3. The magnetic field energy in barred galaxies can be amplified without any dynamo action but only due to non-axisymmetrical velocity.

Acknowledgements

This work was supported by Polish Ministry of Science and Higher Education through grants: 92/N-ASTROSIM/2008/0, 2693/H03/2006/31 and 3033/B/H03/2008/35.

References

Beck, R., Fletcher, A., Shukurov, A., Snodin, A., Sokoloff, D. D., Ehle, M., Moss, D., & Shoutenkov, V. 2005, *A&A* 444, 739

Ferrers, N. M. 1877, *Quart. J. Pure Appl. Math.* 14, 1

Otmianowska-Mazur, K., Elstner, D., Soida, M., & Urbanik, M. 2002, *A&A* 384, 48

Cosmic Magnetic Fields:
From Planets, to Stars and Galaxies
Proceedings IAU Symposium No. 259, 2008
K.G. Strassmeier, A.G. Kosovichev, & J.E Beckman, eds.

doi:10.1017/S1743921309031305

Magnetic fields in irregular galaxies

Amanda Kepley[1]†, Eric Wilcots[1], Ellen Zweibel[1], Stefanie Mühle[2], John Everett[1], Timothy Robishaw[3], Carl Heiles[4], and Uli Klein[5]

[1]Department of Astronomy, University of Wisconsin–Madison, 475 N. Charter St., Madison, WI 53704, USA
email: kepley@virginia.edu, ewilcots@astro.wisc.edu, zweibel@astro.wisc.edu, everett@physics.wisc.edu

[2]Stichting JIVE, Postbus 2, 7990 AA Dwingeloo, The Netherlands
email: muehle@jive.nl

[3]School of Physics, The University of Sydney, NSW 2006, Australia
email: robishaw@physics.usyd.edu.au

[4]Astronomy Department, University of California, Berkeley, California 94720-3411, USA
email: heiles@astro.berkeley.edu

[5]Argelander-Institut für Astronomie, Universität Bonn, Auf dem Hügel 71, D-53121 Bonn, Germany
email: uklein@astro.uni-bonn.de

Abstract. The low masses of irregular galaxies change the behavior of their interstellar medium (ISM) compared to that of normal spirals, so the role of magnetic fields in the ISM in irregulars may be very different than in spirals. We present high-resolution and high-sensitivity observations of the magnetic fields of two irregular galaxies: NGC 4214 and NGC 1569.

Keywords. Magnetic fields – polarization – ISM: magnetic fields – galaxies: dwarf – galaxies: ISM – galaxies: irregular – galaxies: magnetic fields – galaxies: starburst – radio continuum: galaxies

---

The chaotic interstellar medium (ISM) of irregular galaxies is very different from ISM of spirals. Irregulars rotate as solid bodies and their ISM is prone to disruption because of their low masses. What role does the magnetic field play in the ISM of these galaxies? To date only five irregulars have detailed observations of their magnetic fields (NGC 4449: Chyży *et al.* 2000; IC 10 and NGC 6822: Chyży *et al.* 2003; LMC: Klein *et al.* 1993, Gaensler *et al.* 2005; SMC: Mao *et al.* 2008). Our goal is to increase the number of irregulars with observed magnetic fields to address the following questions: 1) what generates and sustains large-scale magnetic fields in irregulars and 2) what causes the range of magnetic field structure seen in irregulars? Here we present observations of two well studied, actively star-forming irregular galaxies: NGC 1569 and NGC 4214.

For NGC 4214, we estimate from the 6 cm synchrotron flux that the field strength is 50 μG in the center and 18 μG at the edges. The magnetic pressure is comparable to the pressures of the other ISM components and thus does not dominate the ISM. We do not detect much polarization, implying that the field is random. To see if an α-ω dynamo could generate large-scale field in NGC 4214, we calculated the dynamo number using an HI rotation curve (Allsopp 1979). It is above the critical value only for radii of 2 to 4 kpc, so NGC 4214 is not able to produce a large-scale field over its entire disk with an α-ω dynamo.

† Present address: Department of Astronomy, University of Virginia, P.O. Box 400325, Charlottesville, VA 22904-4325

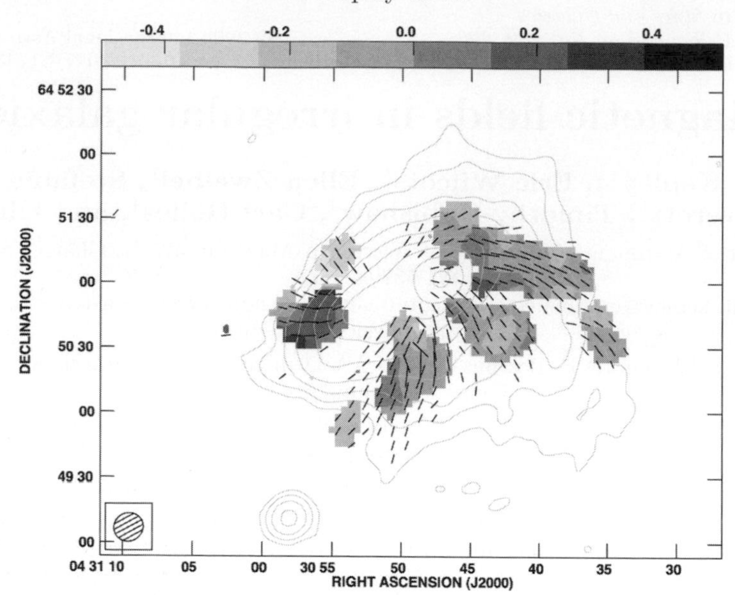

Figure 1. The magnetic field structure of NGC 1569 (Kepley *et al.* 2009, in preparation). The contours show the 3 cm radio continuum emission, the pseudo-vectors show the magnetic field orientation, and the greyscale shows the rotation measures. The contours are 3, 6, 12, 24, 48, 96, 192, 384, and 768 times 29 μ Jy beam^{-1}, a pseudo-vector with a length of 1′ has a polarized intensity of 12.8 μ Jy beam^{-1}, and the rotation measures are in units of 1000 rad m^{-2}.

NGC 1569 is one of the most extreme starbursts in the local universe. This highly inclined galaxy has a strong outflow (Martin 1998, Westmoquette *et al.* 2008). Its magnetic field strength, estimated from the 3 cm synchrotron flux, ranges from 35 μG in the center to 10 μG at the edges. The magnetic pressure is comparable to the pressure of the X-ray gas, so the field does not dominate the ISM. The field in the disk is mostly random, which reflects the turbulent ISM there. The plane of sky halo field is roughly perpendicular to the disk. The polarized regions are correlated with Hα bubbles. The observed field has a bubble-like morphology, e.g. the southwestern most region of polarization points away from the observer near the center of the galaxy, but is mostly in the plane of the sky in the halo of the galaxy. These structures are analogous to the α-ω dynamo model of K. Ferrière, e.g., Ferrière & Schmitt (2000).

References

Allsopp, N. J. 1979, *MNRAS* 188, 765

Chyży, K. T., Beck, R., Kohle, S., Klein, U., & Urbanik, M. 2000, *A&A* 355, 128

Chyży, K. T., Knapik, J., Bomans, D. J., Klein, U., Beck, R., Soida, M., & Urbanik, M. 2003, *A&A* 405, 513

Ferrière, K. & Schmitt, D. 2000, *A&A* 358, 125

Gaensler, B. M., Haverkorn, M., Staveley-Smith, L., Dickey, J. M., McClure-Griffiths, N. M., Dickel, J. R., & Wolleben, M. 2005, *Science* 307, 1610

Klein, U., Haynes, R. F., Wielebinski, R., & Meinert, D. 1993, *A&A* 271, 402

Mao, S. A., Gaensler, B. M., Stanimirovic, S., Haverkorn, M., McClure-Griffiths, N. M., Staveley-Smith, L., & Dickey, J. M. 2008, arXiv:0807.1532

Martin, C. L. 1998, *ApJ* 506, 222

Westmoquette, M. S., Smith, L. J., & Gallagher, J. S. 2008, *MNRAS* 383, 864

Cosmic Magnetic Fields:
From Planets, to Stars and Galaxies
Proceedings IAU Symposium No. 259, 2008
K.G. Strassmeier, A.G. Kosovichev & J.E. Beckman, eds.

doi:10.1017/S1743921309031317

Cosmic-ray driven dynamo in the medium of irregular galaxy

Hubert Siejkowski[1], **Marian Soida**[1], **Katarzyna Otmianowska-Mazur**[1], **Michał Hanasz**[2] **and Dominik Bomans**[3]

[1]Astronomical Observatory, Jagiellonian University, ul. Orla 171, 30-244 Kraków, Poland
email: h.siejkowski@oa.uj.edu.pl

[2]Toruń Centre for Astronomy, Nicolaus Copernicus University, 87-148 Toruń/Piwnice, Poland

[3]Astronomical Institute of Ruhr-University Bochum, Univeristätsstr. 150/NA7, D-44780 Bochum, Germany

Abstract. We investigate the cosmic ray driven dynamo in the interstellar medium of irregular galaxy. The observations (Chyży *et al.* 2000, 2003) show that the magnetic field in irregular galaxies is present and its value reaches the same level as in spiral galaxies. However the conditions in the medium of irregular galaxy are very unfavorable for amplification the magnetic field due to slow rotation and low shearing rate.

In this work we present numerical model of the interstellar medium in irregular galaxies. The model includes magnetohydrodynamical dynamo driven by cosmic rays in the interstellar medium provided by random supernova explosions. We describe models characterized by different shear and rotation. We find that even slow galactic rotation with low shearing rate gives amplification of the magnetic field. Simulations have shown that high amount of the magnetic energy flow out off the simulation region becoming an efficient source of intergalactic magnetic fields.

Keywords. MHD – methods: numerical – galaxies: magnetic fields – galaxies: irregular

---

1. Model and input parameters

The CR-driven dynamo model consists of the following elements (based on Hanasz *et al.* 2004, 2006): (1) the cosmic ray component is an relativistic gas described by diffusion-advection transport equation; (2) anisotropic diffusion of CR along magnetic field lines; (3) localized sources of CR: random explosions of supernova remnants in the disk volume, which supply the CR energy density; (4) uniform resistivity of ISM; (5) shearing boundary conditions and tidal forces are implemented to reproduce the differentially rotating disk in the local approximation (Hawley *et al.* 1995); (6) realistic vertical disk gravity following the model of Milky Way (Ferriére 1998) with scaled contribution of disk and halo to irregular galaxies. We used resistive MHD equations in 3D Cartesian domain of $25 \times 50 \times 400$ grid points. The initial conditions of system assumes the magnetohydrostatic equilibrium with horizontal, purely azimuthal magnetic filed corresponding to $p_{mag}/p_{gas} = 10^{-4}$.

2. Results

We perform a two series of numerical experiments for different values of angular velocity Ω and shearing parameter q (see caption of Fig. 1). In experiments A1–A5 we found that the efficiency of magnetic field amplification depends strongly on the angular velocity (Fig. 1 left), but for $\Omega \geqslant 0.03$ Myr^{-1} the amplification rate stabilizes and becomes

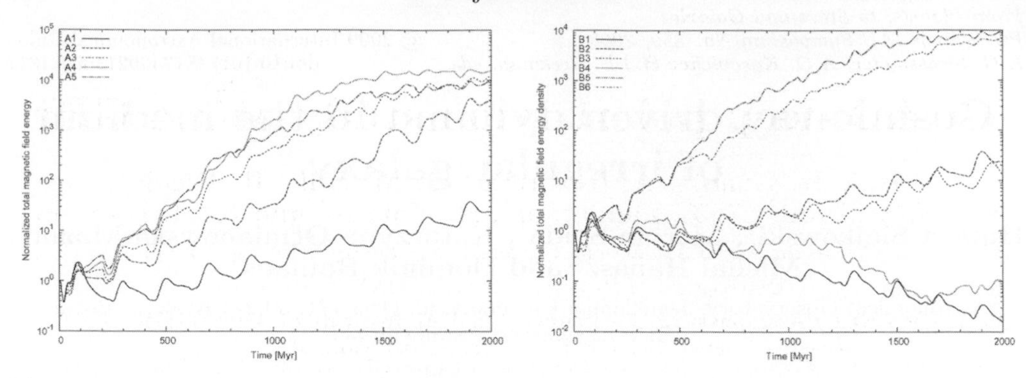

Figure 1. Plots of total magnetic field energy evolution during 2 Myr. The left panel shows the evolution for models (A1–A5) with different Ω (respectively 0.01–0.05 Myr^{-1}) and the right with different q: B1–B3 for $q = 0, 1, 1.5$ with $\Omega = 0.01$ Myr^{-1} and B4–B6 with $\Omega = 0.05$ Myr^{-1}.

roughly independent on Ω. In the second set of experiments we change the shearing rate (Fig. 1 right), for a constant angular velocity. We find no amplification for experiments with $q = 0$ (B1,B4). The other simulations show that the dynamo process does not depend on shearing rate, as long as q is finite. We find a similar evolution of the total magnetic field energy for $q = 1$ and 1.5 (B2,B3 and B5,B6, respectively).

To estimate the total production rate of magnetic field energy during simulation time, we calculate the outflowing magnetic energy density through the xy top and bottom domain boundaries. Our results show that large amounts of magnetic energy are lost through the open boundaries of the computational box, suggesting that irregular galaxies due to their weaker gravitation can be efficient sources of intergalactic magnetic fields.

3. Conclusions

In our work we found: a) the amplification of the total magnetic field energy in irregular galaxies is possible even with slow rotation and a weak shear; b) for dynamo action the shearing is needed, but the amplification rate on the shear is weak; c) the larger angular velocity the higher efficiency of the dynamo process; d) due to the weaker gravitation in irregular galaxies the outflow of magnetic field off the simulation domain is large, suggesting that they may effectively magnetize the intergalactic medium (Kronberg *et al.* 1999).

Acknowledgements

This work was supported by Polish Ministry of Science and Higher Education through grants: 92/N-ASTROSIM/2008/0, 2693/H03/2006/31 and 3033/B/H03/2008/35.

References

Chyży, K. T., Beck, R., Kohle, S., Klein, U., & Urbanik, M. 2000, *A&A* 355, 128
Chyży, K. T., Knapik, J., Bomans, D. J., Klein, U., Beck, R., Soida, M., & Urbanik, M. 2003, *A&A* 405, 513
Ferrière, K. 1998, *ApJ* 497, 759
Hanasz, M., Kowal, G., Otmianowska-Mazur, K., & Lesch, H. 2004, *ApJ* 605, L33
Hanasz, M., Kowal, G., Otmianowska-Mazur, K., & Lesch, H. 2006, *AN* 327, 469
Hawley, J. F., Gammie, C. F., & Balbus, S. A. 1995, *ApJ* 440, 742
Kronberg, P. P., Lesch, H., & Hopp, U. 1999, *ApJ* 551, 56

Cosmic Magnetic Fields: From Planets, to Stars and Galaxies
Proceedings IAU Symposium No. 259, 2008
K.G. Strassmeier, A.G. Kosovichev & J.E. Beckman, eds.

doi:10.1017/S1743921309031329

DRAO deep polarization study at 1.4 GHz

Julie K. Grant[1], A. R. Taylor[1], J. M. Stil[1], R. Ricci[1], S. P. O'Sullivan[2], T. L. Landecker[3], R. Kothes[3] and R. R. Ransom[3]

[1]Centre for Radio Astronomy, University of Calgary, Alberta, Canada

[2]Department of Physics, University of Cork, Ireland

[3]Dominion Radio Astrophysical Observatory, Herzberg Institute of Astrophysics, National Research Council Canada, Penticton, British Columbia, Canada

Abstract. The Dominion Radio Astrophysical Observatory synthesis telescope (DRAO-ST) was used to produce a deep polarization mosaic at 1.4 GHz to a noise level of 45 microJy beam^{-1} for both Stokes Q and U at $1'$ resolution. The DRAO deep field covers 8.6 sq. degrees in polarization centered on the ELAIS N1 field. We identified over 1700 total intensity (Stokes I) radio sources of which 197 are linearly polarized down to a flux density level of 203 microJy. The fractional polarization of faint polarized sources are flat down to a polarized flux density of about 4 mJy, at which point the numbers increase, until the counts drop for polarized flux densities below 1 mJy. These faint polarized radio sources are mostly AGNs with luminosities below the traditional FRI/FRII boundary. Follow-up observations with the VLA show that the origin of the polarization of the radio sources down to a polarized flux of 1 mJy comes from both the lobes and central region of these objects.

Keywords. Galaxies: magnetic fields – galaxies: ELAIS N1 – techniques: polarimetric

1. Introduction

Very little is known about the polarization properties of the faint radio source population. Previous studies of polarized radio sources were done by Mesa *et al.* (2002) and Tucci *et al.* (2002) who used NVSS (Condon *et al.* 1998) polarized sources with $S_{1.4} > 100$ mJy. These results indicate that for steep-spectrum radio sources, the mean fractional polarization increases as flux density decreases. This was confirmed by Taylor *et al.* (2007) for sources with $P > 500$ microJy.

The European Large Area *ISO* Survey North 1 (ELAIS N1) is the area for the DRAO ELAIS N1 deep field (Taylor *et al.* 2007). This area was selected for deep polarization imaging as a window on the extragalactic sky previously observed by *ISO* (Oliver *et al.* 2000) and the *Spitzer* Wide Area Extragalactic Survey (SWIRE, Lonsdale *et al.* 2003). This region is ideal for studies of the faint extragalactic polarized radio source population because of the large amount of data available at other frequencies that cover the ELAIS N1 region.

The deepest observations of the polarized extragalactic sky was achieved with the Dominion Radio Astrophysical Observatory synthesis telescope (DRAO ST) which reached a noise level of 45 microJy beam^{-1} for both Stokes Q and U at $1'$ resolution. Observations begun in August 2004 and completed in July 2008, creating the final mosaic of 40 fields at 1.4 GHz in both total intensity and linearly polarized intensity.

2. The Polarized Source Population

The euclidean-normalized polarized source counts (left plot in Figure 1) for the DRAO deep field are shown down to 800 microJy in polarized intensity. In order to understand

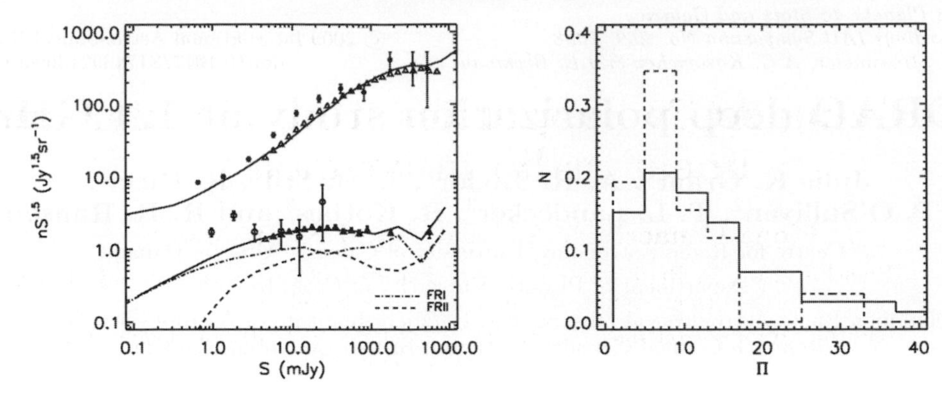

Figure 1. Left: Euclidean-normalized radio source counts in both total intensity (•) and polarized intensity (∘) from the DRAO deep field while the △ are the NVSS source counts. The upper solid black line is the Stokes I model from Wilman *et al.* (2008) and the lower solid black line is the polarized intensity model from Stil *et al.* (2008). Right: Histogram of DRAO deep field polarized sources in polarized flux density ranges of $0.7 < P < 2.0$ mJy (—) and $4.0 < P < 100$ mJy (- - -).

the nature of this faint polarized source population, the observed ELAIS N1 source counts and the NVSS polarized source counts (Ricci *et al.* 2008) were modelled using a semi-empirical simulation of the extragalactic Stokes I continuum developed by Wilman *et al.* (2008) and theoretical percent polarization distributions for FRI and FRII radio sources from Stil *et al.* (2008). The models fit well down to $P \sim 3$ mJy, but are unable to fit the polarized source counts below 3 mJy.

The right plot in Figure 1 shows the histogram of polarized sources in a polarized flux density range of $0.7 < P < 2.0$ mJy (—) and $4.0 < P < 100$ mJy (- - -). The distribution of these two histograms indicates an increase in percentage polarization ($\Pi = P/S$) for sources with a lower flux density. Polarized sources in a flux density range of $0.7 < P < 2.0$ mJy have a median Π of 12.3% while sources in a flux density range of $4.0 < P < 100$ mJy have a median Π of 8.8%.

Observations were made of a sample of these polarized sources at 1.4 GHz in the A configuration, in full polarization mode. The total polarized emission is seen to be dominated by emission from the central region in some cases and from lobes in others.

Acknowledgments: Observations and research on the DRAO Planck Deep Fields are supported by the Natural Science and Engineering Council of Canada (NSERC).

References

Condon, J. J., Cotton, W. D. *et al.* 1998, *AJ* 115, 341
Oliver, S., Rowan-Robinson, M., Alexander, D. M. *et al.* 2000, *MNRAS* 316, 749
Lonsdale, C., Smith, H. E., Rowan-Robinson, M., Surace, J. *et al.* 2003, *PASP* 115, 897
Mesa, D., Baccigalupi, C. *et al.* 2002, *A&A* 396, 463
Ricci, R., Stil, J. M., Taylor, A. R. *et al.* 2008, in prep.
Stil, J. M., O'Sullivan, S. P., & Taylor, A. R. 2008, in prep.
Taylor, A. R., Stil, J. M., Grant, J. K. *et al.* 2007, *ApJ* 666, 201
Tucci, M., Martinez-Gonzalez, E. *et al.* 2004, *MNRAS* 349, 1267
Wilman, R. J., Miller, L., Jarvis, M. J. *et al.* 2008, *MNRAS* 388, 1335

Cosmic Magnetic Fields:
From Planets, to Stars and Galaxies
Proceedings IAU Symposium No. 259, 2008
K.G. Strassmeier, A.G. Kosovichev & J.E. Beckman, eds.

doi:10.1017/S1743921309031330

Deriving AGN properties from circular and linear radio polarimetry

Elena Cenacchi[1]†, A. Kraus[1] and K.-H. Mack[2]

[1]Max-Planck-Institut für Radioastronomie, Auf dem Hügel 69, D-53121 Bonn, Germany
email: cenacchi@mpifr.de, akraus@mpifr.de

[2]Istituto di Radioastronomia, INAF, Via P. Gobetti 101, I-40129 Bologna, Italy
email: mack@ira.inaf.it

Abstract. We report multi-frequency circular polarization measurements for the radio source 0056-00 taken at the Effelsberg 100-m radio telescope. The data reduction is based on a new calibration procedure that allows the contemporary measurement of the four Stokes parameters with single-dish radio telescopes.

Keywords. AGN – magnetic fields – polarization

1. Introduction

Multi-frequency full Stokes polarimetry is a powerful tool to study the radiation emission and transfer processes in the observed sources and determine the dominant mechanism for circular polarization (CP) production. CP, linear polarization (LP) and spectral information can be used to constrain the low energy end of the relativistic particle distribution (Beckert 2003), derive magnetic field strength and geometry (Gabuzda 2006) and make assumptions about the composition of the relativistic plasma within jets (Wardle *et al.* 1998).

We have developed a new procedure to calibrate full-Stokes polarimetric data obtained with the Effelsberg 100-m radio telescope (Cenacchi *et al.* 2008) and have applied it at 2.7, 5, 8.5 and 10-GHz.

2. Results

Here we present, as an example, the results of our observations on 0056-00. This source, point-like to the Effelsberg beams, has been observed cross-scans, resulting in the simultaneous measurement of all four Stokes parameters. The data were reduced using suited calibrators to correct the 4×4 Müller matrix. 0056-00 exhibits a change in sign of CP between 5 and 8.5 GHz, consistent with the minimum in LP and with the change in polarization angle. We have compared our results with those obtained with a model developed by Beckert & Falcke (2002) that provides the radiative transfer coefficients for polarized synchrotron radiation applied to the standard model for relativistic radio jets. The model assumes an extended unpolarized synchrotron source, which dominates the flux density below nearly 5 GHz which is modeled with energy equipartition between B-field and particles, $B = 4$ mG, $n_e = 10^{-3}$ cm^{-3}, and a typical power-law for electrons above $\gamma = 100$, with $p = 2.45$ (power-law index). The size of the emitting region is $L = 8 \cdot 10^{21}$ cm. The polarized emission would be produced by a compact jet component of $L = 1.5 \cdot 10^{19}$ cm,

† E.C. is a member of the International Max-Planck Research School for Radio and Infrared Astronomy

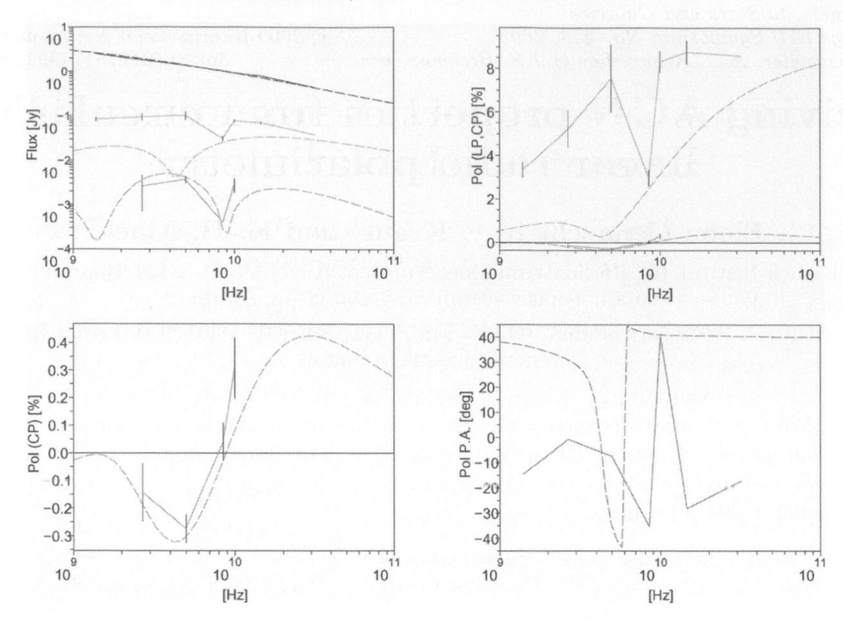

Figure 1. 0056-00, each plot contains a comparison of the observed radio continuum spectra (continuum line) and the simulated one (dashed line). Upper left, top to bottom: flux density I (the predicted values are almost coincident with the observed one), LP, CP (absolute value). Upper right, top to bottom: LP, CP. Lower left: CP. Lower right: PA.

$B = 90$ mG, $n_e = 0.5$ cm^{-3} with a well-ordered magnetic field (a tightly wound spiral) seen at an angle of 85°. This component becomes self-absorbed below 6 GHz and the emission is relativistically boosted with $\Gamma = 6$. This combination reproduces the observed level of circular polarization, the sign flip at nearly 8 GHz and the observed flux density (Fig. 1). With respect to the observed behaviour, the minimum in LP is at too low a frequency. Also the overall level of the simulated LP is too low, which indicates that there might be an additional component (even more compact) that dominates LP and produces the turn in polarization angle at higher frequencies. The observed levels of CP are: $(-0.14 \pm 0.11\%)$ at 2.8 GHz, $(-0.28 \pm 0.05\%)$ at 5 GHz, $(0.04 \pm 0.07\%)$ at 8.5 GHz, $(0.31 \pm 0.11\%)$ at 10.GHz.

To summarize, the comparison between the observed polarimetric parameters and the modeled simulation shows that the values observed in 0056-00 are consistent with those predicted by the model, that is currently under refinement. Further full Stokes measurements are planned at 14.5 and 32 GHz.

Acknowledgements

This research was supported by the EU Framework 6 Marie Curie Early Stage Training programme under contract number MEST-CT-2005-1966 "ESTRELA".

References

Beckert, T. 2003, *Ap&SS* 288, 123

Beckert, T. & Falcke, H. 2002, *A&A* 388, 1106

Cenacchi, E., Kraus, A., Orfei, A., & Mack, K.-H. 2008, *A&A*, subm.

Gabuzda, D. 2006, in Proceedings of the 8th European VLBI Network Symposium, Marecki, A. *et al.* (eds.)

Wardle, J. F. C., Homan, D. C., Ojha, R., & Roberts, D. H. 1998, *Nature* 395, 457

Wardle, J. F. C. & Homan, D. C. 2003, *Ap&SS* 288, 143

Cosmic Magnetic Fields:
From Planets, to Stars and Galaxies
Proceedings IAU Symposium No. 259, 2008
K.G. Strassmeier, A.G. Kosovichev & J.E. Beckman, eds.

doi:10.1017/S1743921309031342

Evolution of magnetic fields in the IGM: kinetic MHD turbulence

Reinaldo S. de Lima[1], E. M. de Gouveia Dal Pino[1], A. Lazarian[2] and D. Falceta-Gonçalves[3]

[1]Instituto de Astronomia, Geofísica e C. Atmosféricas, Universidade de São Paulo, Brazil
email: rlima@astro.iag.usp.br, dalpino@astro.iag.usp.br

[2]Núcleo de Astrofísica Teórica, Universidade Cruzeiro do Sul, Brazil

[3]Astronomy Department, University of Wisconsin, USA

Abstract. In this work, we present 3D MHD simulations of non-helical, forced turbulence, with an anisotropic thermal pressure with respect to the orientation of the local magnetic field. Such anisotropy arises when the plasma is weakly collisional, i.e., when the Larmor frequency is much greater than the ion-ion collision frequency. In this Kinetic MHD regime (KMHD), there are instabilities that give rise to fast growing magnetic fluctuations in the smallest scales. The plasma that fills the intergalactic and intracluster media has small density ($n \sim 10^{-3}$ cm^{-3}), hence the effects of these instabilities could be important in the turbulent amplification of the magnetic fields there. In order to study the KMHD turbulence, we have performed 3D numerical simulations employing a godunov-MHD code (e.g., Kowal, Lazarian & Beresnyak 2007; Falceta-Gonçalves, Lazarian & Kowal 2008). The power spectrum of the velocity and magnetic fields were calculated for two cases: when there is a pre-existing mean magnetic field, and when there is only an initial weak magnetic field.

Keywords. Magnetic fields – plasmas – intergalactic medium

Kinetic MHD description

Due the low density, an MHD description of the IGM and ICM is not adequate: the ions Larmor radius r_L is much smaller than its mean free path λ. For instance, in the Hydra A cluster, $r_L \sim 10^5$ km and $\lambda \sim 10^{15}$ km (Ensslin & Vogt 2006). A collisionless plasma can be described by an MHD set of equations under the kinematic approximation (the KMHD equations), where the scalar pressure in the ideal MHD equations is replaced by the following tensor:

$$P_{ij} = p_\perp \delta_{ij} + (p_\parallel - p_\perp) B_i B_j / \mathbf{B}^2 \tag{0.1}$$

where $p_\parallel$ and $p_\perp$ are the components of the pressure in the directions parallel and perpendicular to $\mathbf{B}$. The standard approach for each pressure component is the double-adiabatic law:

$$\frac{d}{dt}\left(\frac{p_\perp}{\rho B}\right) = 0 \qquad \frac{d}{dt}\left(\frac{p_\parallel B^2}{\rho^3}\right) = 0 \tag{0.2}$$

However, in the preliminary numerical simulations shown below, we have adopted an isothermal description for the pressure components in each direction, with sound velocities c_{perp} and c_{par}.

In all simulations, the turbulence is forced in a 3D periodic box (with 128^3 resolution) by a random, non-helical, solenoidal force acting around the forcing wave number k_f. The velocity of the forcing scale is $u_f \sim 1$.

Turbulence in the presence of an external magnetic field: In these simulations, the external magnetic field $\mathbf{B}_{ext}$ is uniform with Alfénic Mach number $M_A = 1$ (see Fig. 1).

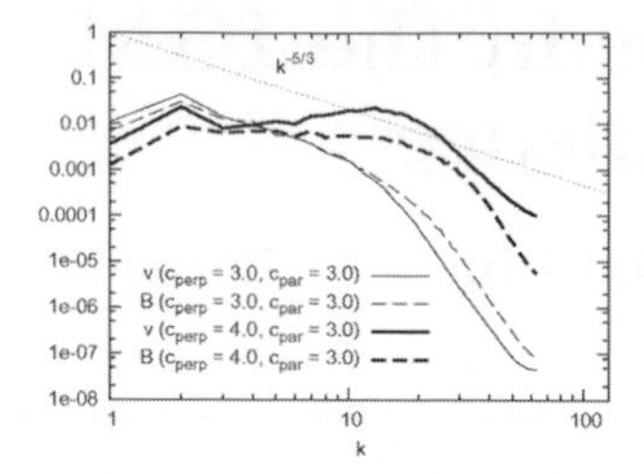

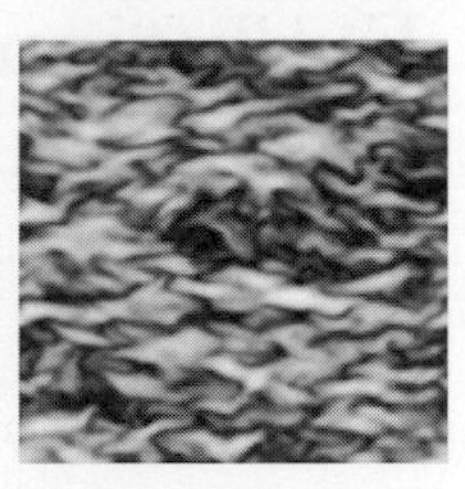

Figure 1. Left: power spectrum of the velocity and magnetic fields for a collisional MHD simulation with $c_{perp} = c_{par} = 3.0$ (MHD) and for a collisionless kinetic MHD simulation $c_{perp} = 4.0, c_{par} = 3.0$ (KMHD). The images in the right show $|\mathbf{B}|$ in the central slices of the cubes, with horizontal $\mathbf{B}_{ext}$ (middle: MHD, right: KMHD). In these simulations $k_f = 2.5$.

Growth of a weak initial magnetic field: For different degrees of pressure anisotropy and the same initial weak magnetic field (see Fig. 2).

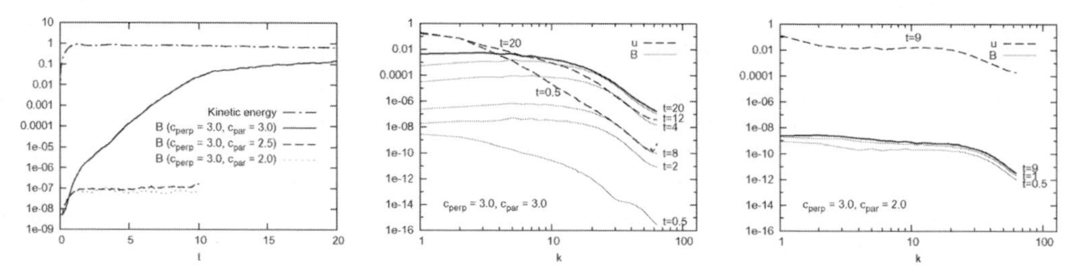

Figure 2. Left: Magnetic energy evolution. Middle and right: magnetic power spectrum evolution for the MHD case and KMHD case, respectively. In these simulations $k_f = 1.5$.

The instabilities that develop due to the anisotropic pressure accumulate energy at the smallest scales (limited by the numerical dissipation), changing the inclination of the spectrum in the inertial range. The small-scale fluctuations of the fields due to these instabilities give rise to a more "wrinkled" field distribution in the KMHD case.

In the simulations with anisotropic pressure, the dynamo action seems to be inhibited. This issue however, still requires deeper investigation with simulations spanning a larger parameter space.

Further simulations including a double-adiabatic law must be performed in order to study the turbulence development with spontaneously raised anisotropic pressures.

Acknowledgements

This work is partially suported by the Brazilian agency Fundação de Amparo à Pesquisa do Estado de São Paulo (FAPESP).

References

Ensslin, T. A. & C. Vogt, C. 2006, *A&A* 453, 447

Falceta-Gonçalves, D., Lazarian A., & Kowal, G. 2008, *ApJ* 679, 537

Kowal, G., Lazarian, A., & Beresnyak, A. 2007, *ApJ* 658, 423

Cosmic Magnetic Fields:
From Planets, to Stars and Galaxies
Proceedings IAU Symposium No. 259, 2008
K.G. Strassmeier, A.G. Kosovichev & J.E. Beckman, eds.

doi:10.1017/S1743921309031354

MHD simulations of cool core clusters

Frederico Stasyszyn and Klaus Dolag
MPI for Astrophysics, Karl-Schwarzchild-Str. 1, D-85741 Garching, Germany
e-mail: fstasys@mpa-garching.mpg.de

Abstract. Using Smoothed Particle Magneto Hydrodynamics (Dolag & Stasyszyn), we study the effects of magnetic fields in galaxy clusters with the aim to infer their dynamical role within the cool core region. Therefore we we investigate the role of regularization as well as divergence cleaning schemes (Stasyszyn & Dolag 2009). We run cosmological simulations of a reference cluster in order to evaluate our various implementation in a realistic scenario. The preliminary results indicate that the final magnetic field profile in the simulations depends only on the amount of artificial dissipation, but not in the amount of numerical div(B) present in the different implementation. We also present first results from simulations which are including radiative cooling and star formation. Even at the this low resolution we find a strong additional amplification of the magnetic field within the cool core region indicating that magnetic pressure could become comparable to the thermal pressure in theses regimes.

Keywords. MHD – methods: n-body simulations – galaxies: clusters – galaxies: magnetic fields

1. Introduction

The SPH MHD implementation (Dolag & Stasyszyn) in Gadget (Springel) allows to use several regularization schemes to stabilize the numerical MHD implementation and to suppress the grow of irregularities as well as numerical div(B) in the simulations. We extended these regularization schemes by a div(B) cleaning method originally proposed by Dedner. In this method a scalar field is evolved, which propagates and damps the numerical divergence. To obtain good stability of the Dedner method, some modifications have been made. The most important are the suppression of the correction within shocks fronts and a self consistent formulation which includes the corresponding terms in the MHD equations. We performed several tests as the Fast Rotor (see Price & Monaghan, Lodrillo & Del Zanna) and Orzang-Tang Vortex (see Price & Monaghan, Orzang & Tang, Balsara) to demonstrate the performance and robustness of our implementation and all MHD SPH simulations compare well with the results obtained with Athena (Stone). In our tests, the regularization schemes which involve some numerical dissipation are very effective in suppressing irregularities evolving in the magnetic field but also

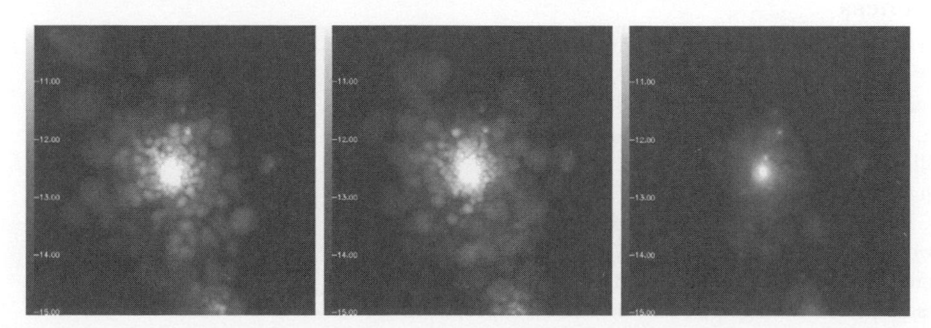

Figure 1. B^2 of a slice through our test cluster for non radiative simulations. The normal implementation at left, Dedner in the center and Euler potentials in the right.

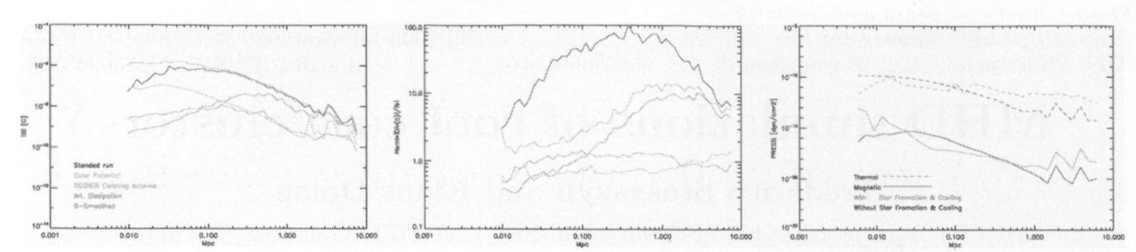

Figure 2. The profiles for the $|B|$ (left), $h * |\mathrm{div}(B)|/|B|$ (center), thermal and magnetic pressure for the Euler Potential runs with and without cooling (right).

tend to significantly smear sharp features. In contrary the new div(B) cleaning is able to significantly suppress the presence of numerical div(B) without any visible smearing of sharp features. For some test we also used so called Euler Potential (see Rosswog & Price), where B is calculated from the evolution of two potentials.

2. The reference cluster

In the figure 1 we show a slice of $|B^2|$ from or reference cluster, simulated with the different schemes. The mass of the cluster is $\approx 10^{14} M_{\odot}/h$ and the particles masses are $1.1 \times 10^9 M_{\odot}/h$ and $1.7 \times 10^8 M_{\odot}/h$ for the dark matter and the gas particles respectively. In Fig. 2 we show the profiles of the magnetic field, a measure of div(B) and the pressure profiles of several runs for this cluster. The magnetic field profiles show how the regularization schemes which involve different numerical dissipation are (over-)suppressing the magnetic field in different regions. In contrary, the Dedner scheme results in a quite similar profile to the run without regularization, but with significant less numerical div(B). First test runs using the Euler Potentials with and without cooling and star formation indicate that magnetic pressure could become comparable to the thermal one in the cluster center.

3. Conclusions

We implemented a SPH-MHD method with different regularization and divergence cleaning schemes in the cosmological code Gadget3. We explore the these methods in a cosmological frame using a low mass cluster. We demonstrate that the resulting magnetic field profiles are sensitive to the amount of numerical dissipation involved in the different regularization schemes. Using a div(B) cleaning scheme further demonstrate that the resulting magnetic field profile is not driven by div(B) errors. First test simulations including cooling and star-formation indicate an additional growth of magnetic field in cool core regions, where the magnetic pressure can eventually become significant.

References

Balsara, D. S. *et al.* 1998, *JCP* 149, 270
Borbe, S. *et al.* 2001, *ApJ* 561, 82
Dedner, J. *et al.* 2002, *JCP* 175, 645
Dolag, K. & Stasyszyn, F. 2008, *arXiv* arXiv:0807.3553
Lodrillo, P. & Del Zanna, L. 2000, *ApJ* 530, 508
Orzang, S. & Tang, C. 1978, *JFM* 90, 128
Price, D. & Monaghan, J. 2005, *MNRAS* 264, 384
Rosswog, S. & Price, D. 2007, *MNRAS* 379, 915
Springel, V. 2005, *MNRAS* 364, 1105
Stasyszyn, F. & Dolag K., 2009, *MNRAS*, in prep.
Stone, J. *et al.* 2008, *ApJS* 178, 137

Cosmic Magnetic Fields:
From Planets, to Stars and Galaxies
Proceedings IAU Symposium No. 259, 2008
K.G. Strassmeier, A.G. Kosovichev & J.E. Beckman, eds.

doi:10.1017/S1743921309031366

MHD numerical simulations of Perseus A: formation of filaments and magnetic loops

Diego Falceta-Gonçalves[1], **E. M. de Gouveia dal Pino**[2], **J. Gallagher III**[3] and **A. Lazarian**[3]

[1]Núcleo de Astrofísica Teórica, Universidade Cruzeiro do Sul, Brazil
email: diego.goncalves@cruzeirodosul.edu.br

[2]Instituto de Astronomia, Geofísica e Ciências Atmosféricas, Universidade de São Paulo, Brazil

[3]Astronomy Department, University of Wisconsin, USA

Abstract. The Perseus Cluster (A426) is the brightest cluster of galaxies observed in X-rays in the sky and its giant central galaxy (NGC1275) hosts the extended double radio source 3C84. There is a spectacular H-alpha nebulosity surrounding NGC1275 with loops and filaments that are probably magnetized and extend over 100 kpc. The continuous blowing of bubbles leading to the propagation of shock fronts is also evident and more recently, outflow and infall velocities of several 1000 km/s have been detected associated to the surrounding filaments. We here present preliminary results of 2.5D MHD simulations of the Perseus cluster central region assuming that the production of the outflow structures and loops that arise from the surface of NGC1275 are due to turbulent injection triggered by recent star formation and SNe activity in the galaxy. This is in turn, probably induced by a continuous gas infall from the satellite galaxies around NGC1275. Our simulations which include both, the turbulent gas outflow and gas infall from the surroundings, have revealed a continuous formation of the observed features, like the filaments, the gigantic magnetic loops and weak shock fronts that propagate into the ICM medium with the observed velocities of 1000–5000 km/s. After 10 Myr, a nearly steady state is established between the outflow material emerging from the central galaxy and the inflow gas from the surrounds. The outflow activity seems to retard the cannibalism action of the central galaxy over the surrounding galaxies. This result may have important implications over the evolution of the whole cluster as it seems to indicate that the SF and SNe production that are induced by the cannibalism may help to decelerate the later due to turbulence and outflow production. These results also offer important clues to the hot halo formation in the center of the cluster and in the suppression of cooling flows.

Keywords. Galaxies: clusters: general – cooling flows – methods: numerical

1. Motivation and Modelling

One of the most interesting paradigms on the physics of galaxy clusters is related to the cooling flows. It was expected that the intracluster medium (ICM) would indefinitely cool down and be pulled towards the core of the cluster by gravity. As density increases inwards cooling would be more efficient. However, cool cores are quite rare in galaxy clusters. Abell 426 (Perseus Cluster) is the brightest cluster in X-rays. At its core the giant galaxy NGC 1275 has been extensively studied in the past few years, revealing giant filamentary structures surrounding the central source (Conselice *et al.* 2001, Salome *et al.* 2006). The suppression of cooling flows and the rise of filaments are usually associated to giant bubbles of hot gas, possibly excited by the AGN (Fabian *et al.* 2003). However, numerical simulations have showed that the AGN's energy feedback is insufficient to reduce the cooling flow effects to the observed values (Gardini 2007), as well as inefficient in creating

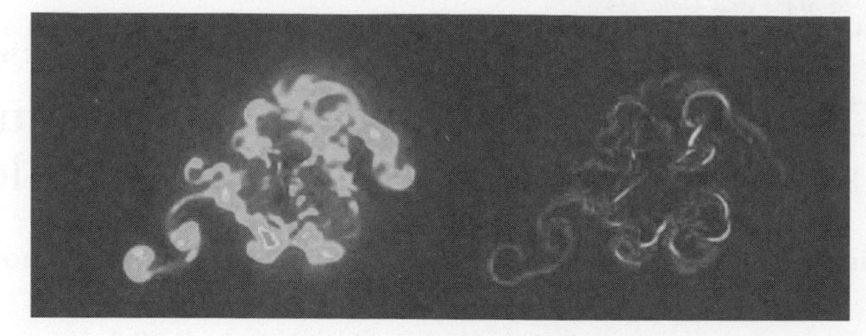

Figure 1. Density (left) and magnetic energy (right) maps for the simulated cluster core with active star formation.

isotropic distribution of filaments, as seen in NGC 1275, and additional mechanisms are needed.

In this work, we assume that SNe and turbulence excited by mergings may be suplementary energy feedback sources in galaxy clusters. To test this hypothesis, we perfomed a number of 2.5D magneto-hydrodynamical (MHD) simulations including turbulent motions at the central region. We implemented the problem in the well-tested Godunov-MHD code studied in Falceta-Gonçalves, Kowal & Lazarian (2008). The simulations were performed solving the set of ideal MHD equations, in conservative form, including turbulence. Gravity is implemented as a Navarro-Frenk-White potential for the distribution of dark matter. The turbulent energy was injected at a rate of 10^{56}ergs/Myr, equivalent to $\sim 10^{-1}$ SN/yr.

2. Results and Conclusions

Fig. 1 shows the density and magnetic energy maps. As a consequence of the SNe shocks, acoustic waves propagate outwards. Interactions of shocked material form filamentary structures that are confined by magnetic fields. The plasma flowing outwards also drags the magnetic energy as the loops and archs expands.

Initially, a strong cooling flow is observed in the simulation, due to the cooling and gravitaty. From the simulations, we obtained an initial infall rate of about $600 M_\odot$/yr. The effect of turbulence in the core counter balanced the cooling flow. The turbulence is responsible for both an increase in thermal and wave/kinetic pressures that prevent the cooling flow at later stages of the run. The infall ceases after $\sim$10 Myr. This effect may be responsible for the absence of cooling flows in several observed galaxy clusters.

In this work we presented preliminary results of the role of MHD turbulence in the formation of large scale structures in central galaxies of clusters. Here we assumed that SNe explosions in the starburst regions of clusters of galaxies trigger shocks that propagate outwards, dragging material and heating the medium. The numerical simulations have shown that most of the peculiar structures observed in Perseus A galaxy may be explained in this scenario. Also, the energy input of SNe occurs at a rate that is compatible with a starburst region.

References

Conselice, C. J., Gallagher, J. S., & Wyse, R. F. G. 2001, *AJ* 122, 2281
Fabian, A. C., Sanders, J. S., Allen, S. W., *et al.* 2003, *MNRAS* 344, 43
Gardini, A. 2007, *A&A* 464, 143
Salomé, P., Combes, F., Edge, A. C., *et al.* 2006, *A&A* 454, 437

Cosmic Magnetic Fields:
From Planets, to Stars and Galaxies
Proceedings IAU Symposium No. 259, 2008
K.G. Strassmeier, A.G. Kosovichev & J.E. Beckman, eds.

doi:10.1017/S1743921309031378

Cosmic magnetic fields in galaxies, groups and clusters

Joern Geisbuesch and Paul Alexander

Astrophysics Group, University of Cambridge, Cambridge, CB3 0HE, UK
email: joern@mrao.cam.ac.uk

Abstract. The distribution of the Faraday depth induced by galaxies, groups and clusters on a patch of sky is investigated. For instance, we utilise a halo model approach to obtain synthetic Faraday skies. Moreover, our modelling includes cluster physical as well as cosmological aspects. A SKA sky survey will provide a large sample of rotation measures of polarised sources. Hence, we examine to what extent statistics of rotation measures of these sources can yield information about cosmic magnetic fields and cosmology.

Keywords. Galaxies: clusters: general – galaxies: magnetic fields – magnetic fields – radio continuum: galaxies – polarization – methods: data analysis

1. Faraday sky simulations

The origin and evolution of magnetic fields in the Universe is still little understood. Generally, it is assumed that seed magnetic fields are amplified by dynamo mechanisms. These seed fields, which can later be fostered by dynamo actions, might be generated by Weibel instabilities or the Biermann battery mechanism. A powerful tool for studying magnetic fields in a variety of environments is to use Faraday rotation against background and embedded polarised sources. New and up-coming instruments, such as the SKA and its pathfinders, will allow for the first time to perform detailed studies of the magnetic Universe via Faraday rotation measure (RM) techniques. In the presented work, we investigate which insights can be gained on magnetic field evolution from observations by these instruments on the basis of cosmological simulations of magnetic fields and electron gas distributions in large scale structures (LSS), such as clusters, groups and galaxies. In particular, our Faraday sky simulations build on a halo model approach. We employ N-body simulations as well as Lagrangian perturbation theory in order to derive the spatial distribution of halos within chosen mass and redshift ranges. Further halo properties, which can not be directly derived from these simulations of gravitational growth of LSS, are obtained via scaling relations. For example, we assume different scalings between central halo mean magnetic field amplitudes and halo masses and also alter their redshift evolutions. These scalings and their evolutions are physically and/or observationally motivated. The structure of the magnetic field inside clusters is realised by a power-law power spectrum of the magnetic field vector components and our modelling ensures that the vector field is divergence free. Furthermore, we scale the cluster magnetic field strength by a profile so that it decreases radially outwards. Note that our applied magnetic field modelling is backed by observations (see e.g. Murgia *et al.* 2004). The electron gas density profile is derived from the distribution of dark matter in the halo, which is modelled by a NFW profile, by assuming hydrostatic equilibrium. Figure 1 shows Faraday sky patches for different lower halo mass limits and different slopes of the cluster magnetic field power spectra.

Figure 1. Faraday sky realisations: The left and mid panel show the projected Faraday depth obtained from massive cluster sized halos on a patch of sky assuming different power-law power spectrum slopes of the magnetic field structure inside clusters (colour bar units: rad/m^2). The Faraday sky patch on the right includes galaxy and group sized halos.

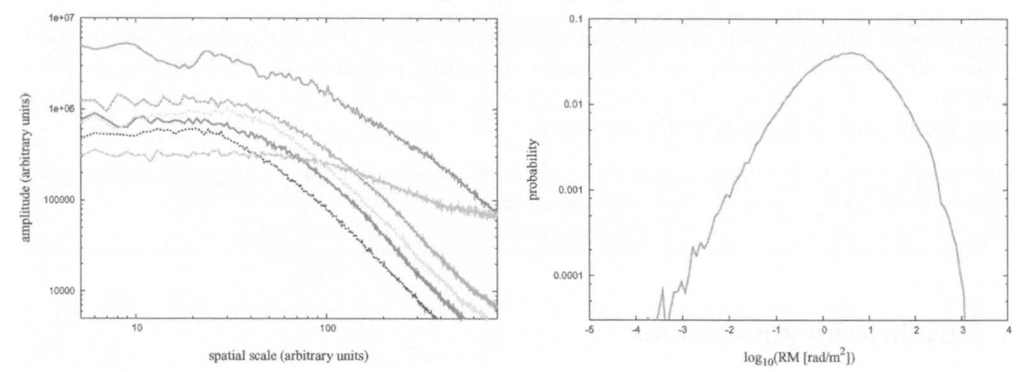

Figure 2. (a) Faraday sky power spectra for different halo mass cuts, magnetic field evolutions, magnetic field amplitude to halo mass scalings and cosmologies (σ_8). (b) Probability distribution of LSS background source RMs for a 'high' σ_8 simulation including galaxy and group halos.

2. Studying the magnetic Universe

Future radio telescopes (especially the SKA) will yield large samples of RMs of polarised extragalactic sources (for a discussion of RM grids and RM synthesis techniques see Geisbuesch & Alexander (2009) (this volume)). RM distributions are sensitive to the evolution of the magnetic field and baryon densities in the LSS and also dependent on cosmology. Especially, the normalisation of the matter power spectrum, σ_8, has an impact on RM statistics. Note further that also the total number of detectable polarised sources, their redshift distribution and spatial correlations depend besides on the integration time on cosmology and LSS physics. Moreover, the variation of the RM distribution for sources at different redshifts is as well dependent on the cosmological evolution of structures and their embedded magnetic fields (see e.g. Kronberg *et al.* (2008) for a recent study of the 'RM-redshift' variation). In Figure 2a we show the power spectra of the line-of-sight projected Faraday depth of generated sample patches. The realisations assume different scaling relations and magnetic field and cosmological evolutions. The Faraday depth distribution can be probed by the polarised background source population, which is spatially uncorrelated with the Faraday foreground. The probability distribution of RMs for lines of sight to high redshift LSS background sources is shown in Figure 2b.

References

Geisbuesch, J. & Alexander, P. 2009, *IAU S259 - Cosmic Magnetic Fields: From Planets, to Stars and Galaxies*, this volume

Kronberg, P. P., Bernet, M. L., Miniati, F. *et al.* 2008, *ApJ* 676, 70

Murgia, M., Govoni, F., Feretti, L. *et al.* 2004, *A&A* 424, 429

Cosmic Magnetic Fields:
From Planets, to Stars and Galaxies
Proceedings IAU Symposium No. 259, 2008
K.G. Strassmeier, A.G. Kosovichev & J.E. Beckman, eds.

doi:10.1017/S174392130903138X

Cosmic magnetic fields in clusters of galaxies and their analysis

Joern Geisbuesch and Paul Alexander

Astrophysics Group, University of Cambridge, Cambridge, CB3 0HE, UK
email: joern@mrao.cam.ac.uk

Abstract. We discuss how measurements of linear polarisation of cluster background and embedded sources can be used to study cluster magnetic fields via Faraday rotation. We make forecasts for up-coming radio instruments on the basis of synthetic radio sky observations. Our mock polarised sky is modelled to agree with the sparse available data. By applying Bayesian statistical analysis methods, such as Markov Chain Monte Carlo and nested sampling techniques, we investigate which constraints can be placed on cluster magnetic field properties.

Keywords. Galaxies: clusters: general – galaxies: magnetic fields – magnetic fields – radio continuum: galaxies – polarization – methods: data analysis

---

1. The polarised source sky and clusters as Faraday screens

A powerful tool for studying magnetic fields in a variety of environments is to use Faraday rotation against background and embedded polarised sources. These source populations consist of normal and starburst galaxies as well as different types of radio AGN. To obtain realistic models of radio source populations, we use the publicly available SKADS synthetic radio source catalogue (Wilman *et al.* 2008). The catalogue is derived from semi-empirical simulations of different populations of radio continuum sources and extends down to faint flux limits in order to allow observation simulations of high-sensitive future radio facilities, such as the SKA and LOFAR. To determine the degree of polarisation of individual sources, we base our modelling on statistics of the NVSS catalogue (Condon *et al.* 1998) as derived by Tucci *et al.* (2003) and semi-analytic source models. Apart from semi-analytic predictions of probability distributions of fractional source polarisations, we fit the observed 1.4 GHz distribution at the high fractional polarisation tail by a lognormal function. Moreover, we adopt the depolarisation correction recipe of Tucci *et al.* (2003) to obtain the intrinsic source polarisation. A map of Stokes U of a simulated polarised sky realisation is shown in Figure 1a.

Observed magnetic field strengths in clusters are typically $0.1 - 1\mu$G. Cooling flow clusters have up to an order of magnitude stronger fields (see e.g. Taylor & Carilli 2002). However, observational data on the structure of magnetic fields in clusters is presently rather sparse. To simulate cluster Faraday profiles we utilise a power law spectrum to model correlations of magnetic field amplitudes on different scales. The spectrum is cut on a scale below which the magnetic field strength shows no spatial fluctuations. Moreover, the cluster magnetic vector field is modelled in a divergence-free manner. For the electron density distribution we adopt a King profile and also scale the magnetic field strength in a similar fashion (i.e. decreasing with cluster radius). Figure 1b shows a Faraday depth image of a cluster for a magnetic field power spectrum of Kolmogorov type. To model the cluster number and distribution within an observed patch of sky, we employ N-body simulations.

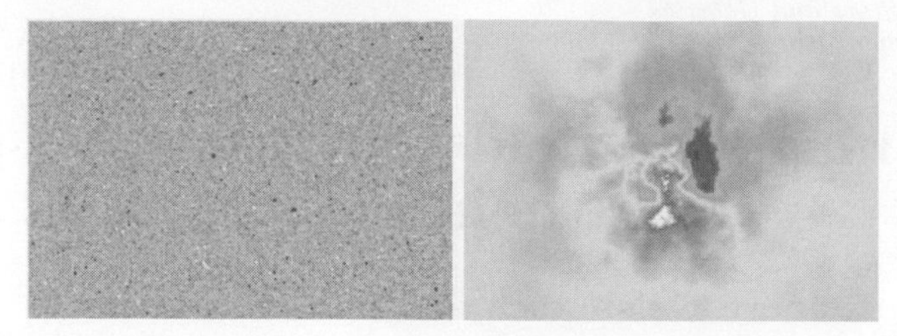

Figure 1. (a) Sky simulation of linearly polarised sources – Stokes U is shown. (b) Faraday depth profile of a galaxy cluster.

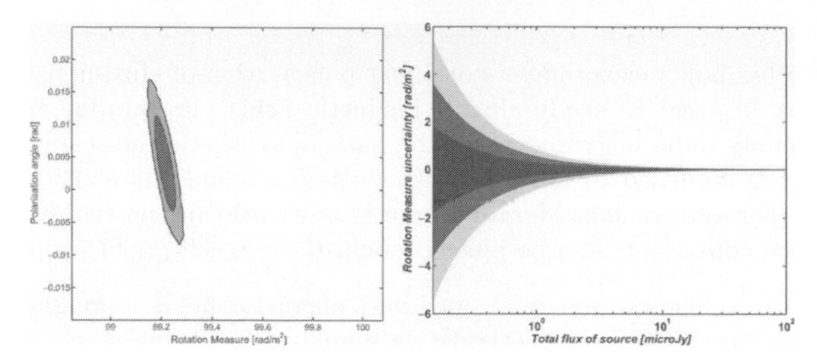

Figure 2. (a) Confidence contours for the foreground RM value and the intrinsic source polarisation angle. The dark blue shaded area gives the 68% and the light blue shaded one the 95% confidence region. (b) Uncertainty on the reconstructed RM of the foreground screen.

2. Rotation Measure grids and Bayesian analysis

Cluster X-ray or SZ data in combination with rotation measure (RM) observations can yield an estimate of cluster magnetic fields (see e.g. Geisbuesch *et al.* 2008). However, current RM grids are sparse and the sample of clusters which have a larger number of polarisation measurements of embedded or background sources is small. The SKA with its unique collection area reaches unmatched sensitivities for reasonable integration times. Thus, the number of source detections within a cluster field will significantly increase.

To investigate the potential of future SKA data for determining source intrinsic and foreground screen magneto-ionic (Faraday) properties, we model SKA mid-band observations based on the latest SKA reference designs. We apply then Bayesian techniques based on Markov Chain Monte Carlo and nested sampling methods to the simulated data to explore model posterior distributions and to evaluate model evidences (see e.g. Skilling 2006 for a detailed description of the techniques). We find that for our simulated SKA linear polarisation data the model selection ratio gives reliable information on the projected cluster line-of-sight Faraday depth even for less bright background sources with high fractional polarisation. In Figure 2a and 2b confidence contours and reconstruction uncertainties are shown.

References

Condon, J., Cotton, W., Greisen, E., Yin, Q., Perley, R., Taylor, G., & Broderick J. 1998, *AJ* 115, 1693

Geisbuesch, J., Alexander, P., Krause, M., & Bolton R. 2008, *Il Nuovo Cimento*, 122 B

Skilling, J. 2006, *Bayesian Analysis*, vol. 4, p. 833

Taylor, G. B. & Carilli, C. L. 2002, *ARA&A* 40, 319

Tucci, M., Martinez-Gonzalez, E., Toffolatti, L., Gonzalez-Nuevo, J., & de Zotti, G. 2003, *NewAR* 47, 1135

Wilman, R., Miller, L., Jarvis, M., Mauch, T., Levrier, F., Abdalla, F., Rawlings, S., Kloeckner, H.-R., Obreschkow, D., Olteanu, D., & Young, S. 2008, *MNRAS* 388, 1335

Cosmic Magnetic Fields:
From Planets, to Stars and Galaxies
Proceedings IAU Symposium No. 259, 2008
K.G. Strassmeier, A.G. Kosovichev & J.E. Beckman, eds.

doi:10.1017/S1743921309031391

A study of the regular structure of the galactic magnetic field using WMAP5 polarization data at 22 GHz

Beatriz Ruiz-Granados[1], J. A. Rubiño-Martín[2] and E. Battaner[1]

[1]Dpto. Física Teórica y del Cosmos. Edif. Mecenas. Campus Fuentenueva, E-18071. Universidad de Granada & Instituto de Física Teórica y Computacional Carlos I, Granada (Spain)
email: bearg@ugr.es , battaner@ugr.es

[2]Instituto de Astrofísica de Canarias (IAC), C/Vía Láctea, s/n, E-38200, La Laguna (Spain)
email: jalberto@iac.es

Abstract. We study the spatial structure of the 3-dimensional large-scale pattern of the Galactic Magnetic Field using the polarization maps obtained by the WMAP satellite at 22 GHz. By using five different models of the large-scale magnetic field of the Milky Way and a model for the cosmic rays distribution, we predict the expected polarized synchrotron emission. Those maps are compared to the observed 22 GHz polarization data using a Maximum Likelihood method. For each model, we obtain the parameter values which better reproduce the data and obtain their marginal probability distribution functions. We find that the model that best reproduces the observed polarization maps is an "axisymmetric" model.

Keywords. Magnetic fields – galaxies: structure – polarization

1. Model of the galactic magnetic field

Since the discovery of the polarized synchrotron emission from our Galaxy (Wielebinski & Shakeshaft 1962), several works have attempted to constrain the pattern of the regular component of the Galactic Magnetic Field (GMF). Recent studies of the GMF are given in Sun *et al.* (2008), Noutsos *et al.* (2008) and Men *et al.* (2008). At 22 GHz, the physical mechanism which produces the dominant contribution to the large-scale pattern of the radio-emission is the synchrotron radiation. Two ingredients are needed to predict the polarized synchrotron emission of the Galaxy. The first one is a model describing the cosmic rays electron distribution; in this work, we use the model by Drimmel & Spergel (2001). The second ingredient is a model for the regular component of the GMF. Here we have explored five different models, namely: (a) Axisymmetric Spiral (ASS; with and without radial dependence of the strength); (b) Logarithmic Spiral Arms (LSA; as proposed by Page *et al.* 2007), (c) Bi-symmetric Spiral (BSS) with and without radial dependence of the strength (Han & Qiao (1994)); (d) Concentric Circular Ring (CCR; Rand & Kulkarni 1989); and (e) Bi - Toroidal (BT; based on Han 1997).

2. Method

For every considered model, we simulate the Stokes's U and Q parameters and the polarization angle (PA) following the procedure described in Ruiz-Granados *et al.* (in prep). The basic assumption is that the dominant emission process is synchrotron with an average spectral index of ~ -2.7. For each model, we perform a blind exploration

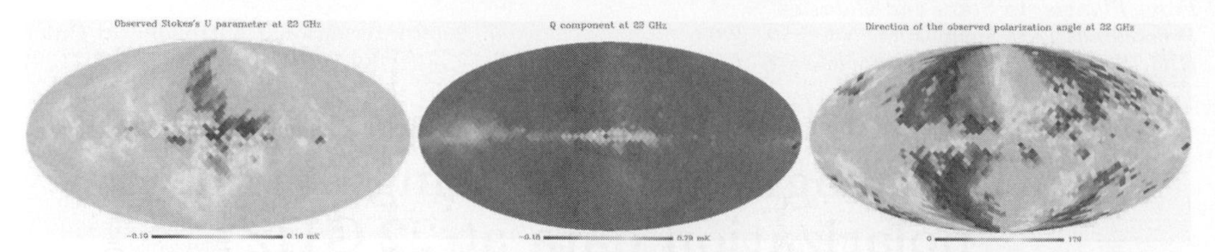

Figure 1. Observed U, Q and PA maps at 22 GHz (WMAP 5-year maps).

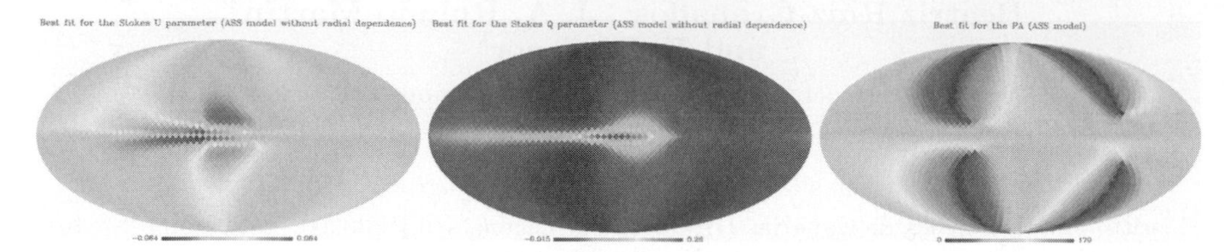

Figure 2. Predicted 22 GHz U, Q and PA maps from our best-fit model for the GMF.

of the parameter space by obtaining the χ^2 statistics both from the (Q,U) and PA maps. Using a Bayesian approach, we build the likelihood function as $\ln L = -1/2\chi^2$, and compute the marginalized posteriors for each parameter. The comparison between different models is done using the reduced χ^2 as the "goodness-of-fit" statistic.

3. Conclusions

The model that best reproduces the large-scale pattern of the observed WMAP polarization data at 22 GHz is ASS without any radial dependence of the strength of the field. The parameter values for the best-fit are: $B_0 = (2.5\pm^{1.7}_{0.8})\mu$G, pitch angle $p = (24\pm^{0.8}_{0.8})^\circ$ and tilt angle $\chi_0 = (24\pm^{1.2}_{1.3})^\circ$. Fig. 1 shows the observed U, Q and PA maps, while Fig. 2 presents the corresponding simulated maps for the best-fit model. We note that the best-fit model is obtained in all cases when masking the loops (the NPS and other spurs) and the Galactic center region, due to the disturbance of the emission introduced by these regions with respect to the regular pattern.

In general, all axisymmetric-like models considered here provide reasonable fits. In particular, the best-fit LSA model practically coincides with the ASS one, since the data do not seem to require a radial variation of the pitch angle. Both the BSS, CCR and BT models do not provide good fits to the data. However, the BT could reproduce the halo component.

References

Drimmel, R. & Spergel, D. N. 2001, *ApJ*, 556, 181
Han, J. L. & Qiao, G. J. 1994 *A&A* 288, 759
Han, J.-L. 1997 *Chinese Astronomy and Astrophysics* 21, 130
Men, H., Ferriere, K., & Han, J. L. 2008 *A&A* 486, 819
Noutsos, A., Johnston, S., Kramer, M., & Karastergiou, A. 2008 *MNRAS* 386, 1881
Page, L., *et al.* 2007 *ApJ. Suppl.* 170, 335
Rand, R. J. & Kulkarni, S. R. 1989 *ApJ.* 343, 760
Sun, X. H., Reich, W., Waelkens, A., & Enßlin, T. A. 2008 *A&A* 477, 573
Wielebinski, R. and Shakeshaft, J. R. 1962 *Nature* 195, 982

Session chair Dimitry Sokoloff arguing with Philipp Kronberg (left), and Tim Gledhill (right)

Richard Wielebinski (left), and Jeroen Stil & Julie Grant (right)

Ryo Kandori, Felipe Alves & Gabriel Franco (left to right), and the Teide (picture by R. Arlt)

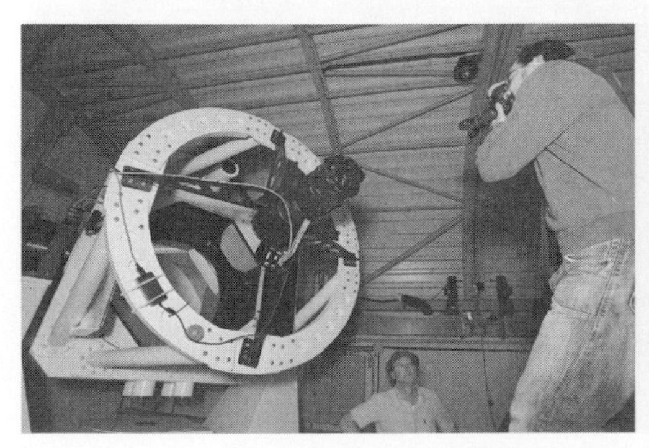

Visit at STELLA (left), and Moira Jardine & Heidi Korhonen (right)

Ready to leave the harbor with the "pirates" ship

The crew

Session VI

Advances in methods and instrumentation for measuring magnetic fields across all wavelengths and targets

Cosmic Magnetic Fields:
From Planets, to Stars and Galaxies
Proceedings IAU Symposium No. 259, 2008
K.G. Strassmeier, A.G. Kosovichev & J.E. Beckman, eds.

doi:10.1017/S174392130903141X

Zeeman splitting in the diffuse interstellar medium—The Milky Way and beyond

Carl Heiles[1] and Timothy Robishaw[2]

[1]Astronomy Department, UC Berkeley
email: heiles@astro.berkeley.edu

[2]School of Physics, The University of Sydney
email: robishaw@physics.usyd.edu.au

Abstract. We begin with a brief review of Zeeman-splitting fundamentals and the importance of circular polarization, i.e. Stokes V. We then turn to modern results in several areas, emphasizing the diffuse interstellar medium in the Galaxy. The median field in the Cold Neutral Medium is determined from HI absorption lines and is about 6 μG; the magnetic and turbulent pressures are comparable. Using HI emission lines the field has been mapped in several areas: the field reverses across the Orion Molecular Cloud; the 3-d field structure has been determined in the ρ Oph region; and in regions having shock-like morphology the field is generally stronger, strong enough to limit further compression. We briefly present new field measurements for: photo-dissociation regions at the edges of HII regions, determined from carbon recombination lines; Ultra Luminous Infrared Galaxies, from OH megamasers; and the 3C 286 damped Lyman-α absorption system, determined from the 21-cm line in absorption. We show the sidelobe response of the Green Bank Telescope, which is surprisingly severe and makes the telescope less than optimum for Zeeman-splitting measurements of HI emission lines. Finally, we compare two techniques for determining field strengths, i.e. Zeeman splitting and the Chandrasekhar-Fermi method, and show why the latter usually gives higher field strengths – and sometimes unrealistically high fields.

Keywords. ISM: magnetic fields – galaxies: magnetic fields – polarization – techniques: polarimetric

---

Zeeman splitting probes magnetic fields in many different environments. In this meeting we have heard about its use in probing magnetic fields of stars (Berdyugina, Herpin, Johns-Krull, Reiners, Wade), the X-ray binary Cyg X-1 (Karitskaya), and OH megamasers (Robishaw). Ferland told us about magnetic fields revealed by Zeeman splitting of the 21-cm line in the star-forming regions Orion and M17. The literature on the role of magnetic fields in other star-forming regions is . . . not quite vast, but it's quite impressive; for a review, one that includes the physical interpretations of equipartition among turbulent, gravitational, and magnetic energies, see Heiles & Crutcher (2006).

What *hasn't* been covered so far in this meeting, nor much in the literature, is the impressive knowledge Zeeman splitting has provided about diffuse regions in the Galaxy. These observations probe the HI line in absorption, which is produced by the Cold Neutral Medium (CNM), and in emission, which is produced by HI gas of all temperatures. We will also mention very briefly some other new, unpublished 3σ results in dense regions (photodissociation regions—PDRs) and also extragalactic magnetic fields seen in Zeeman splitting of OH megamasers and damped Lyman-α (DLA) absorption systems. We begin, however, by briefly reviewing some fundamentals; these are intimately familiar to those working in the Zeeman business, but not so to many who study magnetic fields using other techniques.

1. Zeeman splitting fundamentals and Stokes parameters

When a simple atom like hydrogen (HI) sits in a magnetic field, the lines suffer Zeeman splitting. Here we shall focus on the 21-cm line, which radio astronomers use to study the diffuse interstellar medium. In a magnetic field of 1 μG, the line splits into two components centered at different frequencies; the separation is 2.8 Hz, i.e. ± 1.4 Hz from line center. These so-called σ components are polarized. If the field lies along the line of sight, then they are 100% circularly polarized, with opposite senses for the two frequencies. Classically, this corresponds to the electrons gyrating around the field in the clockwise and counterclockwise directions. Now turn the field so it is perpendicular to the line of sight; then the gyrating electrons appear to oscillate in straight lines across the field lines, producing linear polarization oriented perpendicular to the field. So the polarization of the σ components changes from pure circular to pure linear as the field changes from line-of-sight to plane-of-sky. Finally, electrons can oscillate in straight lines oriented *along* the field lines, and then the field has no effect on the frequency. The so-called π component has no frequency shift and is linearly polarized parallel to the plane-of-sky field; the intensity goes to zero when the field becomes line-of-sight, because then the classical electrons are oscillating to and from along the line of sight.

First consider the Stokes parameter V, which is the difference between the two circular polarizations, for the case when the ratio of Zeeman splitting to line width—call this ratio R—is small. For a line-of-sight field, Stokes V sees the intensity difference between two σ components that are centered at different frequencies. So the Stokes-V *shape* is the frequency derivative of the line, and the *amplitude* is proportional to $R \cos\theta$, where θ is the angle of the field to the line of sight. The $\cos\theta$ arises because the circular polarization intensity $\propto \cos\theta$. When R is small, Stokes V is small; Zeeman observations are usually sensitivity-limited and we need lots of observing time.

Now consider the linear-polarization Stokes parameters Q and U, for which the sensitivity situation is much worse: the difference between linearly polarized intensities for position angles parallel and perpendicular to the plane-of-the-sky field is a difference in line *width*, which means that the *shape* looks like the *second* derivative of the line with an amplitude $\propto R^2 \sin^2\theta$. With $R \ll 1$, this makes the intensity of linear polarization impossibly weak. The watchword for Zeeman-splitting specialists is that, for weak lines having $R \ll 1$, circular polarization is the only way to go.

Radio astronomers use cross-correlation techniques to measure all four Stokes parameters simultaneously (Heiles 2002, Heiles *et al.* 2001). With native linear polarization from the feeds (call them X and Y), Stokes U is the XY product and Stokes V is the XY product with a 90° phase shift. Cross-Correlation is an exquisitely sensitive and robust technique, so this situation is highly desirable compared to using feeds with native circular polarization (call them R and L). In this latter case, Stokes V is the difference $R^2 - L^2$; this difference between two large numbers is subject to small gain variations in the electronics. For measuring Zeeman splitting with the $R \ll 1$ case, we have the seemingly paradoxical situation that native linear feeds are better.

2. Zeeman splitting in HI absorption: history and current status

Zeeman splitting of the 21-cm HI line was discovered by Verschuur (1969) by looking at the line in absorption against two strong radio sources, Cas A and Tau A. Amazingly enough, no further work was done in HI *absorption* for 32 years—not for lack of suitable equipment (Arecibo would have been superb, even before its final upgrade in the late 1990s) but more because (1) it's not easy to measure weak polarization (it takes lots of time and careful attention to detail) and (2) the few observers willing to spend the effort concentrated on mapping the field in HI *emission*. We cover emission in the next section.

Finally, just before the new millennium, Troland and Heiles—over some glasses of Jack Daniel's whiskey—conceived the Millennium Survey of Zeeman splitting in HI absorption of all feasible sources in the Arecibo sky. They observed 79 sources (different directions in the sky) and examined 69 Gaussian components for which the derived fields and errors were well-defined. This is the largest statistical body of magnetic field information on the diffuse ISM. A big bonus is that the Gaussian components give line widths, kinetic temperature, and column density. Heiles & Troland (2005) discuss the scientific implications of the statistics. The principal conclusions are that the median CNM magnetic field is about 6 μG, the typical temperature 50 K, the median HI column density for $|b| > 10^\circ$ is $\sim 0.5 \times 10^{20}$ cm^{-2} (far smaller than the generally accepted "standard Spitzer cloud," which is about 3×10^{20} cm^{-2}), and—as seems to be usual in astrophysical situations—the turbulent and magnetic energy densities are comparable, both much larger than the gas thermal energy density (i.e., it's a low-β plasma).

3. Zeeman splitting in HI emission: history and current status

The field of ISM Zeeman observing wasn't dead during thosee 32 years between absorption line studies. Heiles and Troland used the Hat Creek 85-ft, and Verschuur used the NRAO 140-footer, to look at HI in emission in the hope of making magnetic field maps. Others used Arecibo, Westerbork, the 140-foot, and the VLA to measure Zeeman splitting, mostly in molecular clouds and OH masers. Verschuur's work was limited by the time one can hope to get from a national-observatory multipurpose telescope. Heiles and Troland had the Hat Creek 85-ft telescope almost to themselves and were able to make real progress in dealing with sensitivity and instrumental concerns (Troland & Heiles 1982; Heiles 1989). And technically, the 85-foot's electronics were fully state-of-the-art, particularly with the advent of a 1024-channel digital correlator after Troland earned his Ph.D. This led to a series of 5 papers (Heiles 1988, Heiles 1989, Goodman & Heiles 1994, Myers *et al.* 1995, Heiles 1997) in which a total of 429 positions had sufficiently reliable field strengths to publish. Some of these yielded maps of large regions: in Orion, where a large-scale reversal was seen across the Orion Molecular Cloud (Heiles 1997); in Ophiuchus, where the stellar polarization and HI Zeeman splitting offered a 3-dimensional characterization of the field morphology (Goodman & Heiles 1994); and the North Celestial Pole (Myers *et al.* 1995), where there is a 25-degree diameter expanding shell with enhanced field (~ 10 μG) in the compressed shell.

Figure 1 projects the totality of 429 positions onto an image of the integrated HI line. The coverage is sparse in most regions, and for weak fields the results are somewhat uncertain because of instrumental polarized sidelobes (mainly beam squint; see below), but they offer a robust conclusion: the observed fields are high in regions where the HI morphology suggests we see an edge-on shock. This, then, is consistent with our expectation of field amplification in dense gas because of flux freezing. Moreover, the field is strong and limits the gas compression.

These measurements are regarded with considerable suspicion by many astronomers, especially European astronomers, because of *Verschuur's Indictment.* In a pair of papers (Verschuur 1995a, 1995b), amplified by a series of colloquia given mainly to European astronomers in the mid 1990s, Verschuur claimed that every single Zeeman-splitting detection in HI emission was bogus because of polarized sidelobes interacting with the HI velocity field. Most serious is "beam squint," in which the two circular polarizations point in slightly different directions on the sky. For most telescopes [the Hat Creek

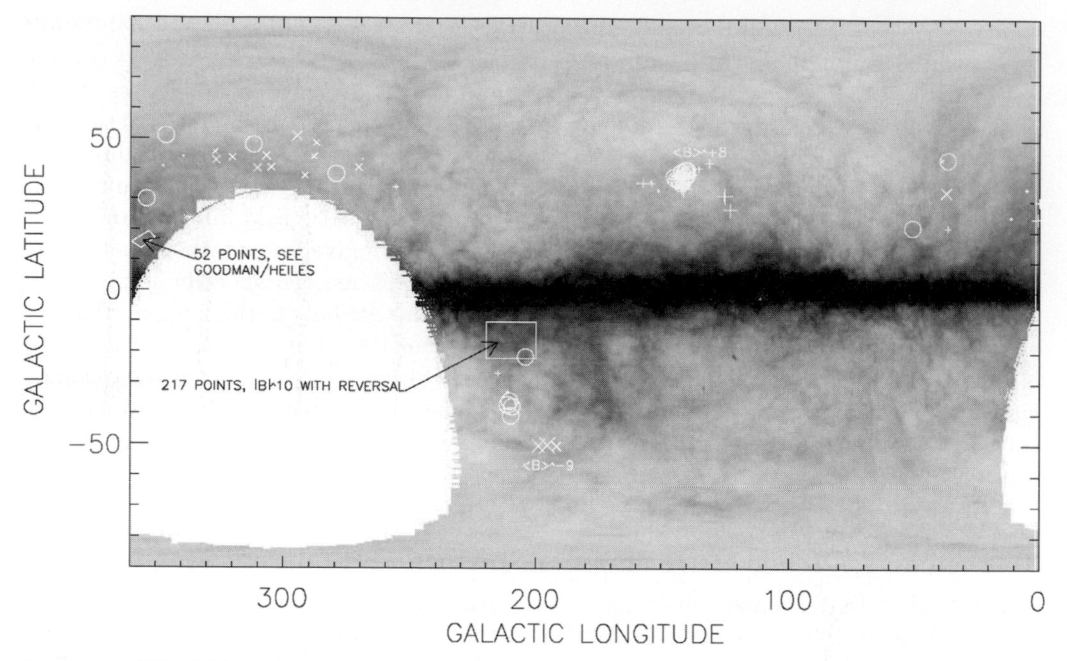

Figure 1. The 429 positions at which the 5 Hat Creek papers reported magnetic fields derived from HI in emission.

85 foot (Heiles 1996), the Arecibo telescope (Heiles 1999), the Green Bank Telescope (Robishaw & Heiles 2009)], this position difference is ~ 1 arcsec. For the 140 foot, the squint changes with time, probably as a result of feed-mounting details, and is usually bigger: ~ 7 arcsec (Verschuur 1969); unmeasurably small (Verschuur 1989); and ~ 2 arcsec (Verschuur 1995a). A 1-arcsec beam squint doesn't seem like much, but it can be serious: a velocity gradient of 1 km s^{-1} deg^{-1} that is aligned with the direction of beam squint produces a frequency difference of 5 Hz between the two circular polarizations, which is equivalent to 1.6 μG. Verschuur claimed that all results were simply fake, caused by beam squint. This, however, is simply not the case. Most telling, many GBT results match the Hat Creek ones—in particular, at the well-studied North Celestial Pole.

Heiles (1996) thoroughly studied the response of the Hat Creek telescope to Zeeman splitting at the NCP and provides a detailed measurement of the uncertainties in the derived field for that position, for which the HI velocity gradient is somewhat larger than usual. Finally, Heiles (1998), in an unfamiliar and rarely-quoted journal, thoroughly discusses Verschuur's criticisms and explains why they are exaggerated. The bottom line: most of the emission results are reasonably accurate.

4. The Green Bank Telescope as a case study in polarized sidelobes

Beam squint isn't all there is, as Robishaw and Heiles (2009) discovered for the GBT—much to our dismay. We pointed the GBT at the NCP, staring at it for a full 24 hours. There were significant variations in both Stokes I and V with time. For a telescope with small beam squint, and also a clear aperture so that there should be no other sidelobes, this was not only a disappointment but also raised a puzzling question: *Why?* How did these time variations arise?

They arise from spillover of the feed's illumination of the secondary mirror. This produces three different sidelobe components: (1) a $\sim 40°$-diameter doughnut-like pattern

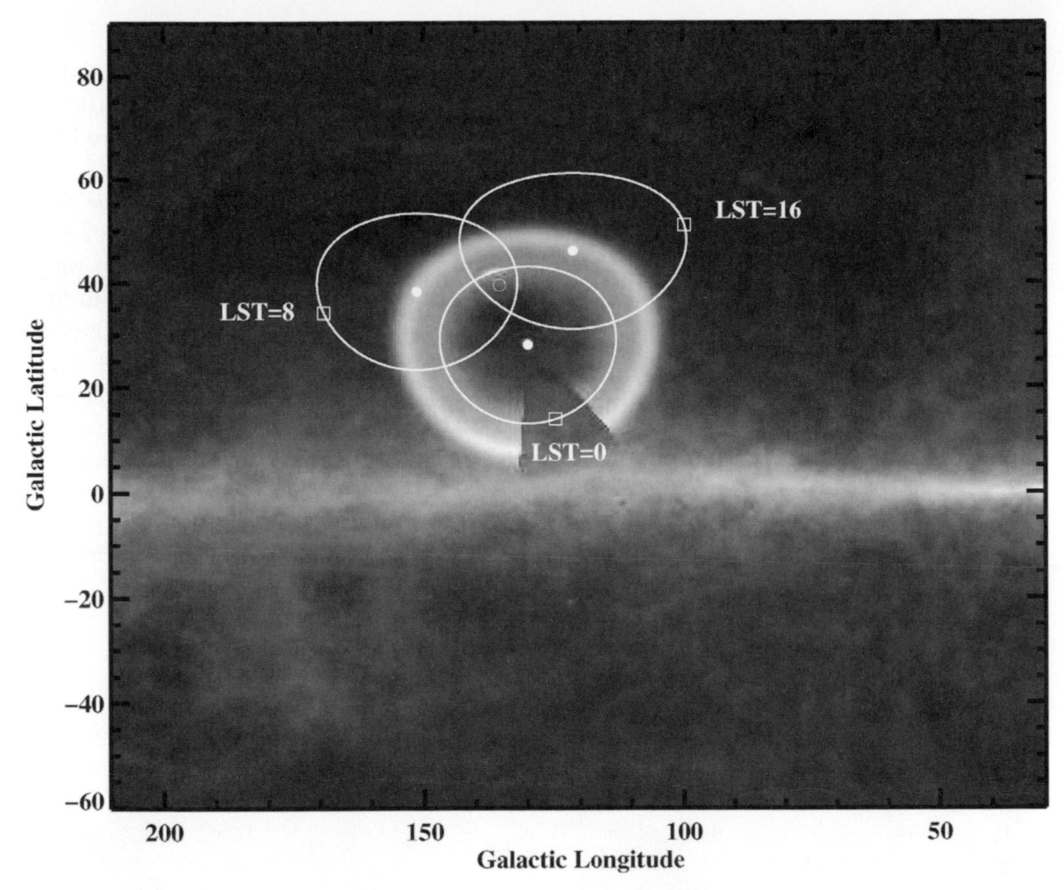

Figure 2. Images of the secondary-reflector sidelobes at three LSTs for the position G135.5+39.5, superposed on the integrated 21-cm line intensity from the LAB survey. The empty circle is the source position. The filled circles are the Arago spot. The solid 30°-diameter circles approximately show the subreflector edge as seen by the feed. Each of the three squares lies on its subreflector circle just inside the feed arm. The circular greyscale image has whiteness proportional to the spillover pattern; this is shown only for LST = 0. The blanked-out portion of the spillover pattern is the angular slice of the screen.

centered on the line joining the centers of the feed and the secondary mirror; (2) the "Arago spot," which is the diffraction spike at the center of an occulting disk; and (3) a "screen" component associated with the spillover radiation interacting with the feed leg, which has a protective screen intended to control this aspect of the spillover radiation. Components (1) and (3), especially (1), tend to be the most serious for Stokes I; component (3) is most serious for Stokes V. Component (1) is well-enough known so that its Stokes-I contribution is quite accurately predictable. Unfortunately, the same is not true for component (3), especially in Stokes V.

Figure 2 depicts these feed/secondary sidelobe components on the sky, superposed on an image of the total HI line intensity. As the sky turns, component (1) sweeps around the source, covering a total sky area that is bounded by a small circle of radius $\sim 40°$. The other components move around the observed position as the parallactic angle changes. This polarized-sidelobe contribution to Stokes V cannot be accurately predicted, which makes the GBT a less desirable telescope than we would have hoped for measuring Zeeman-splitting of HI in emission.

5. Three new results in HI emission

One: the NCP shell. Above we mentioned the high field strengths in the NCP shell. This shell has fascinating structure on the ∼3 arcmin scale, as revealed by the 100 μm IRAS emission; moreover, the measured field strengths from Zeeman splitting are high, easily reaching 10 μG—but this is with 36-arcmin resolution, which is much worse than the typical structural scale. A few positions measured with the Effelsberg telescope show higher fields (Myers *et al.* 1995). We have used the GBT to map a very large region containing the NCP shell. These mapping data also include Stokes V; while we won't have the sensitivity to map the field with the full GBT resolution, by averaging in area we will obtain a large-scale map of the field strength, albeit limited in angular resolution. Currently we have mapped only the Stokes I, which is itself intriguing. Figure 3 shows this map at one particular velocity, 2.6 km s^{-1}, together with Zeeman-derived magnetic field measurements for a few individual positions. Unfortunately, we don't have space to include other velocities. The structure changes markedly with velocity—a circumstance that indicates that the structure is *not* filamentary, but rather sheetlike: what seems like filaments is simply the cusps where the sheet lies tangential to the line of sight.

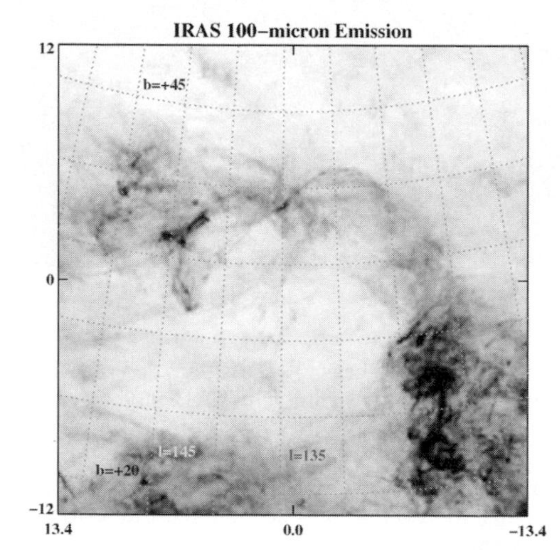

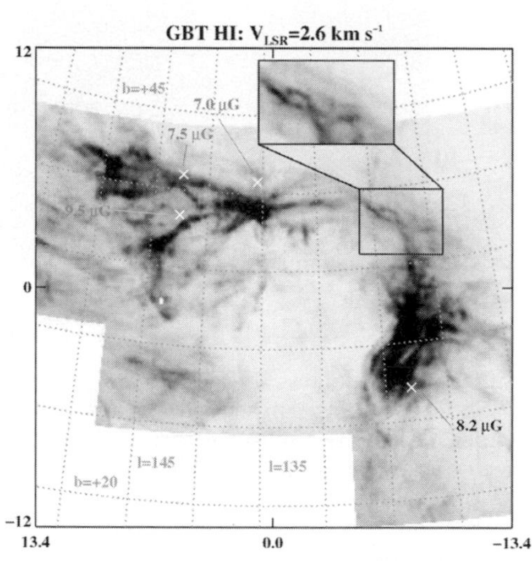

Figure 3. Left, the 100 μm IRAS image of the North Celestial Pole (NCP) loop; right, the HI image for $V_{\rm LSR} = 2.6$ km s^{-1}. The numbers on the right panel are $B_{||}$, which is strong. All images are negatives, in which black means higher intensity. The graticule spacing is 5°.

Two: The Orion Molecular Cloud. Above we mentioned the Hat Creek result for the Orion Molecular Cloud, which may indicate that a helical field envelopes the cloud. Our GBT results amplify this possibility, because they show that the reversal is very sharp: it's limited by the angular resolution of the GBT.

Three: Enhanced Faraday Rotation Spots near the Taurus Molecular Cloud. Wolleben & Reich (2004) made a beautiful map of Faraday rotation measure (RM) for an area around the Taurus Molecular Cloud (TMC) and found nine well-defined spots having small RM. They used Hα data to place an upper limit on the electron density, which provided a lower limit on the line-of-sight field $B_{||} \gtrsim 20$ μG. They suggested that these spots are associated with the TMC, possibly associated with recently-formed stars. We obtained 21-cm line Zeeman-splitting data for 3 of these positions and do not find such strong fields for velocities near to the TMC velocity. However, we do find fields of ∼50 μG for a −50 km s^{-1} component. We emphasize that these derived fields have not been corrected for the GBT sidelobes, so these detections *may not be real.* Nevertheless, it is interesting to take these measurements at face value—i.e., as being real—and to superpose the 9 RM-minimum positions on a grey scale map of the -50 km s^{-1} component. Figure 4 shows these spots projected against both the TMC and this component. Morphologically, the case for their association with the -50 km s^{-1} component is at least as good as with the TMC. If so, the shock could have produced the enhanced field and, also, an enhanced electron density.

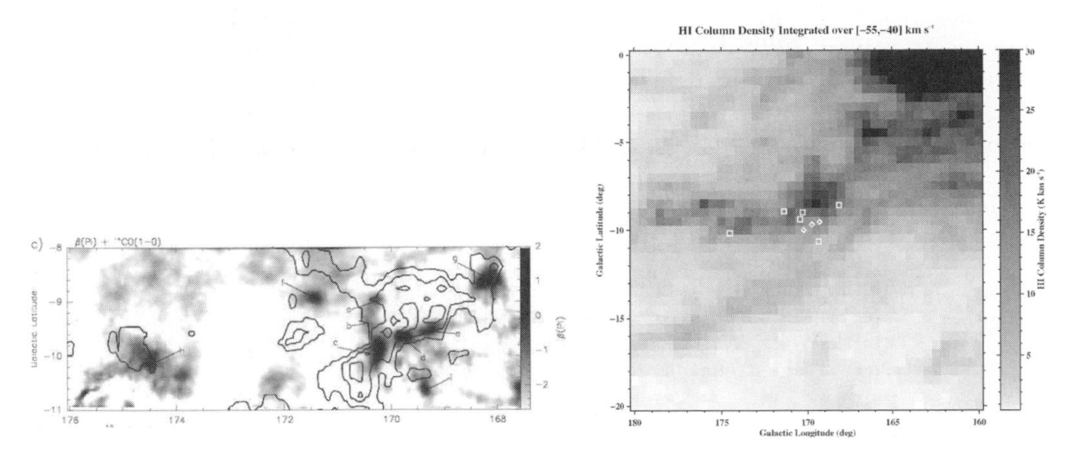

Figure 4. Left, in *greyscale* is the Wolleben & Reich (2004) RM map and the *contours* show the boundary of the Taurus Molecular Cloud. The 9 RM minima (dark spots) are labelled. Right, the -50 km s^{-1} HI emission intensity (area of the Gaussian) versus position, with the same 9 minimum-RM spots.

6. New results for small sources

Zeeman splitting measurements of unresolved sources suffer no contamination from polarized sidelobes, so one is limited only by signal-to-noise. Here we mention three recent results.

One: Magnetic Fields in PDRs. The interface between an HII region and its neutral surroundings is bounded on the outside by a shock and on the inside by an ionization front. The pressure is comparable to that in the HII region. Within the PDR, the ambient gas has been compressed by the shock and the density has increased by roughly the ratio of the temperatures in the HII region and the ambient medium—a factor $\sim$100. In a one-dimensional shock, the field strength should increase by the same factor, so the field strength in PDRs should be in the mG range. We used the GBT to observe 8 CRRLs simultaneously to measure Zeeman splitting. We examined 5 sources and obtained 3σ detections for two: DR21 (-916 ± 305 μG) and M17 (2530 ± 864 μG). In the Zeeman-splitting game one is not comfortable with just 3σ detections, but these are very suggestive. In particular, these sources are small and should have small angular structure in the density and magnetic field structure. Rather than pursue higher sensitivity with a single dish like the GBT, we feel the current results justify a significant time commitment with the VLA.

Two: OH Megamasers. It is surprisingly easy to detect Zeeman splitting in OH megamasers. This offers a unique probe into the magnetic properties of the host galaxy, which is always an Ultra Luminous IR Galaxy (ULIRG), usually at moderate redshift. We are pursuing an active program to observe these galaxies; Robishaw (this meeting) discusses Zeeman splitting in OHMs.

Three: DLAs. We have looked for Zeeman splitting in four DLA systems against 3 sources, with the results given in Table 1. The first entry has an impressively small upper limit and the next two have less impressive upper limits. What's *really* impressive is the fourth entry for 3C 286, an 84 μG field for a $z = 0.692$ DLA. Figure 5 shows this result.

Table 1. Zeeman-Splitting Observations of DLAs

System	$\frac{N(HI)}{10^{20}}$	$B_{\|\|}$, μG	Ref
0738+313, z=0.0912	15	1.8 ± 2.4	1, 2
0738+313, z=0.2212	8	12.5 ± 9.0	1, 2
0235+164, z=0.524	30	$\lesssim 10$	3
3C 286, z=0.692	30	84 ± 9.0	4

Ref 1: Turnshek *et al.* (2001) (DLA properties)
Ref 2: Lane & Heiles (2008) (Zeeman splitting)
Ref 3: Wolfe *et al.* (1982) (DLA properties)
Ref 4: Wolfe *et al.* (2008) (Zeeman splitting)
Note: Field strengths are unpublished except for Ref 4.

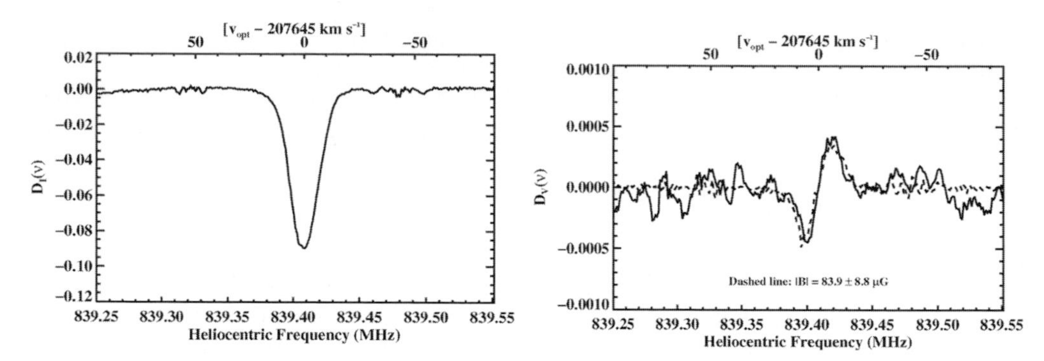

Figure 5. The 3C 286 DLA in the 21-cm line. Left, Stokes I; right, Stokes V. On the right, the solid line is the data and the dashed line is the best fit.

This field is exceptionally high: the magnetic pressure $\sim 10^8$ cm^{-3}K, more than 10^4 higher than the anticipated CNM thermal pressure. But there are additional amazing oddities about this DLA: (1) its column density $N(HI) \sim 3 \times 10^{21}$ cm^{-2}, about 100 times bigger than Galactic CNM clouds; (2) VLBI studies (Wolfe *et al.* 1976) show that the DLA cloud is at least 200 pc in size, about 100 times bigger than Galactic CNM clouds; (3) its velocity dispersion ~ 3.8 km s^{-1}, so the velocity gradient across the cloud $\lesssim 0.02$ km s^{-1} pc; this is tiny, *especially* in view of the huge magnetic pressure!

This cloud is very weird, even without the strong field. The strong field makes it even weirder, so one might wish to discount the observation as being bogus. This is possible, and the result needs confirmation, but it does seem real for the following reasons: (1) the result repeats from day to day for our 8-day series of observations; (2) the result is statistically robust, with the average and median giving comparable results; and (3) most telling, Stokes U has no signal. This is important: Stokes V and Stokes U are both derived from cross-correlations of native linear feeds, the only difference being the 90° phase shift for Stokes V.

7. Zeeman splitting versus Chandrasekhar-Fermi

A popular method for obtaining magnetic field strengths in atomic or molecular emission lines is the Chandrasekhar-Fermi (CF) method. This relies on linear polarization and is sensitive to the plane-of-sky field $B_\perp$. The idea is this: the field is frozen into the gas; turbulence in the gas produces a dispersion in the field orientation, and hence the orientation of linear polarization; stronger fields make the system stiffer, so the dispersion is smaller. The bottom line: the square of the plane-of-sky field $(B_\perp)$ is proportional to the variance of the measured position angles.

The CF method almost always gives stronger field strengths than Zeeman splitting. There are at least three reasons for this:

(*a*) Statistically, $B_\perp$ is stronger than $B_{||}$ because it is closer to the total field strength $B_{\rm tot}$. In essence, it is more probable for the field to lie in the plane of the sky than along the line of sight. Figure 6 shows this quantitatively by plotting the probability density functions (PDFs) and cumulative distribution functions (CDFs) for various ratios of field components. The CDFs are the most instructive: for half of all measurements, $B_{||}/B_\perp < 0.58$. Moreover, half the time $B_\perp/B_{\rm tot} > 0.87$, which makes $B_\perp$ a reasonably good tracer of $B_{\rm tot}$.

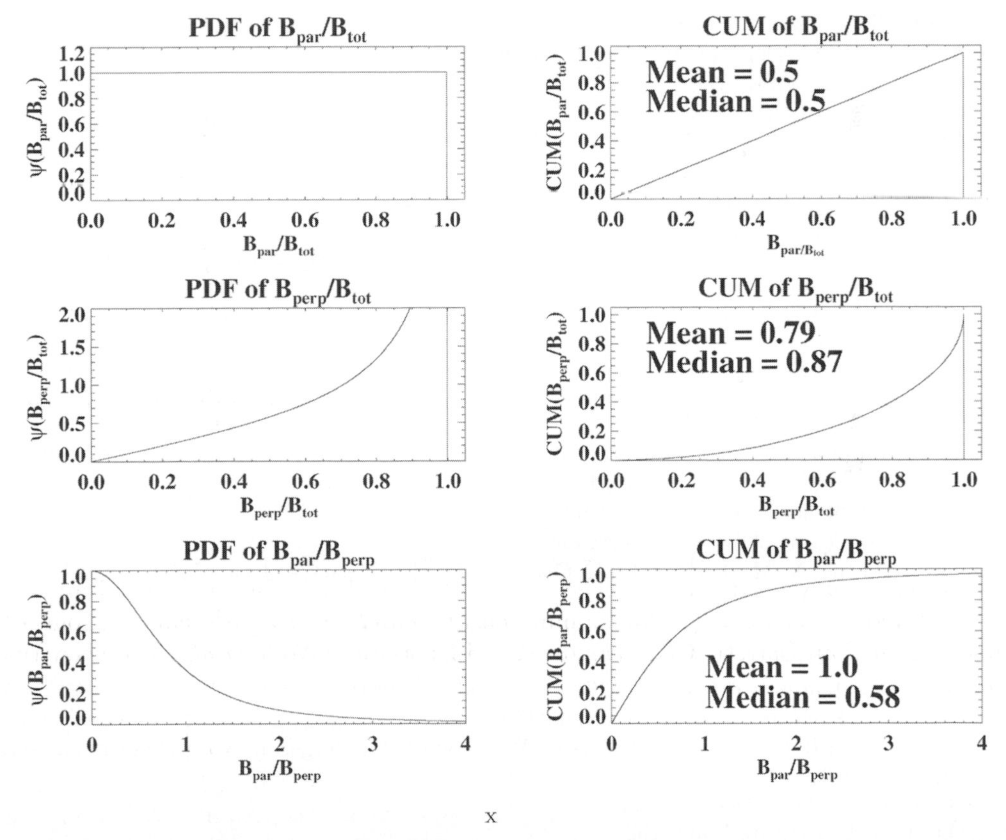

Figure 6. The left column of panels exhibits plots of the probability density functions (PDFs) of $B_{||}/B_{\rm tot}$, $B_\perp/B_{\rm tot}$, and $B_{||}/B_\perp$. The right column exhibits the cumulative distribution functions of the same quantities.

(*b*) The CF method assumes that there is only a single polarizing cloud along the line of sight. This doesn't seem likely. With N clouds and random field orientations among clouds, the derived field $\propto N^{1/2}$. Theorists have considered the question of how many elements should typically exist (e.g. Wiebe & Watson 2004) and conclude that the effect is not too serious, but these predictions rely on turbulence models and theory. How much can we trust them?

(c) The CF method assumes that velocity fields are turbulent and random. However, interstellar space is subject to strong shocks from sources such as supernovae. These shocks will sweep up the field, remove the kinks, and make it look very smooth. An incredible example is the McClure-Griffiths *et al.* (2007) high-resolution HI map of the Riegel-Crutcher cloud, which shows a beautiful set of aligned angel-hair-like filaments; their Figure 6 shows that these filaments are aligned with stellar polarization. The CF method gives $B_\perp \approx 60\ \mu$G. This is huge and could easily be detected with Zeeman splitting. If so, it would be the strongest field detected in HI emission. It's worth a try!

Acknowledgements

This work was supported in part by NSF grant AST-0406987 and, also, by the awards GSSP 02-0011, 05-0001, 05-0003, 05-0004, and 06-0003 from the NRAO. It is a pleasure to acknowledge support from the IAU for this invited review.

References

Goodman, A. A. & Heiles, C. 1994, *ApJ* 424, 208
Heiles, C. 1988, *ApJ* 324, 321
Heiles, C. 1989, *ApJ* 336, 808
Heiles, C. 1996, *ApJ* 466, 224
Heiles, C. 1997, *ApJS* 111, 245
Heiles, C. 1998, *Astrophysics Letters and Communications*, 37, 85. The title of this paper, "Zeeman Splitting Opportunities and Techniques at Arecibo", gives no hint that it discusses Veerschuur's Indictment. Moreover, this article is difficult to find on the internet. Copies are available at the authors' websites, currently `http:astro.berkeley.edu/˜heiles` and `http://www.physics.usyd.edu.au/˜robishaw)`
Heiles, C. 1999, Arecibo Tech Memo 99-02 (`http://www.naic.edu/science/techmemos_set.htm`)
Heiles, C. 2002, in *Single-Dish Radio Astronomy: Techniques and Applications*, ASP Conf. Proceedings, Vol. 278. Eds S. Stanimirovic, D. Altschuler, P. Goldsmith, and C. Salter, 131
Heiles, C. & Crutcher, R. 2005, *LNP Vol. 664: Cosmic Magnetic Fields* 664, 137
Heiles, C., Perillat, P., Nolan, M., Lorimer, D., Bhat, R., Ghosh, T., Lewis, M., O'Neil, K., Salter, C., & Stanimirovic, S. 2001, *PASP* 113, 1274
Heiles, C. & Troland, T. H. 2005, *ApJ* 624, 773
Lane, W. & Heiles, C. 2008, in preparation
McClure-Griffiths, N. M., Dickey, J. M., Gaensler, B. M., Green, A. J., & Haverkorn, M. 2006, *ApJ* 652, 1339
Myers, P. C., Goodman, A. A., Güsten, R., & Heiles, C. 1995, *ApJ* 442, 177
Robishaw, T. & Heiles, C. 2009, *PASP*, in press
Robishaw, T., Quataert, E., & Heiles, C. 2008, *ApJ* 680, 981
Troland, T. H. & Heiles, C. 1982, *ApJ* 252, 179
Turnshek, D. A., Rao, S., Nestor, D., Lane, W., Monier, E., Bergeron, J., & Smette, A. 2001, *ApJ* 553, 288
Verschuur, G. L. 1969, *ApJ* 156, 861
Verschuur, G. L. 1989 *ApJ* 339, 163
Verschuur, G. L. 1995a, *ApJ* 451, 624
Verschuur, G. L. 1995b, *ApJ* 451, 645
Wiebe, D. S. & Watson, W. D. 2004, *ApJ* 615, 300
Wolfe, A., Broderick, J., Condon, J., & Johnston, K. 1976, *ApJ* 208, L47
Wolfe, A., Jorgenson, R., Robishaw, T., Heiles, C., & Prochaska, J. 2008, *Nature* 455, 638
Wolleben, M. & Reich, W. 2004, *A&A* 427, 537

Discussion

ZINNECKER: I was intrigued by your result that the measured magnetic fields are stronger in shock-compressed regions. Does this imply that the mass-to-flux ratio also changes upon shock-compression with consequences for star formation?

HEILES: Usually not; because during the shock compression which is one-dimensional, $\mathbf{B} \propto n$.

BECK: The detection of a $\approx$80μG field in a $z \approx$0.6 galaxy is amazing, especially in view of the absence of any Faraday rotation. What is your physical picture of this object?

HEILES: It is very hard to understand. VLBI shows it is at least 200pc big, yet it has a narrow line with. This, by itself, is a puzzle, adding the magnetic pressure – 1000 times higher than thermal – makes it even harder. For a physical model we might think of two extremes: 1) The HI cloud is gravitationally confined. This needs a large surface density of stars (which would imply lots of star formation; not observed though) or Dark Matter. 2) An unconfined slab that was produced by a collision between two clouds, each having a moderate field strength of a few μG, the shock waves compress the gas and the field; what is the correct magnitude? We have to find out with more data.

LAI: Can you talk about the first two problems of the CF method?

HEILES: CF is sensitive to $\mathbf{B}_\perp$, Zeeman is sensitive to $\mathbf{B}_\parallel$. 1) The PDF (probability density function) is such that a) half the time, $\mathbf{B}_\perp/\mathbf{B}_{\rm total} > 0.87$, so $\mathbf{B}_\perp$ is a good tracer of $\mathbf{B}_{\rm total}$, b) half the time, $\mathbf{B}_\parallel/\mathbf{B}_{\rm total} > 0.5$, these two together give, c), half the time $\mathbf{B}_\parallel/\mathbf{B}_\perp < 0.58$. 2. Stellar polarization averages elements over the line of sight. With N randomly magnetic fields over many lines of sight the variance in position angles is reduced by a factor N. This increases the derived field by a factor N.

Carl Heiles, with the session chair Tomm Landecker in the background

George Heald

Cosmic Magnetic Fields:
From Planets, to Stars and Galaxies
Proceedings IAU Symposium No. 259, 2008
K.G. Strassmeier, A.G. Kosovichev & J.E. Beckman, eds.

doi:10.1017/S1743921309031421

The Faraday rotation measure synthesis technique

George Heald
ASTRON, P.O. Box 2, 7990 AA Dwingeloo, The Netherlands
email: heald@astron.nl

Abstract. We discuss practical aspects of the novel Faraday Rotation Measure Synthesis technique, first described by Burn (1966), and recently extended and implemented by Brentjens & de Bruyn (2005). The method takes advantage of the excellent spectral coverage provided by modern radio telescopes to reconstruct the intrinsic polarization properties along a line of sight, using a Fourier relationship between the observed polarization products and a function describing the intrinsic polarization (the Faraday dispersion function). An important consequence of the Fourier relationship and discrete frequency sampling is the need, in some cases, to deconvolve the sampling response from the reconstructed Faraday dispersion function. Practical aspects of the deconvolution procedure are discussed. We illustrate the use of the technique by summarizing a recent investigation carried out with the WSRT. We conclude by briefly describing the applicability to future programs which will be carried out with the next generation of radio telescopes such as LOFAR.

Keywords. Magnetic fields – polarization – methods: data analysis – techniques: polarimetric

1. Introduction

Polarimetric radio observations enable the study of synchrotron emission radiated by relativistic electrons as they are accelerated by magnetic fields. The plane of polarization of synchrotron radiation is perpendicular to the component of the magnetic field in the plane of the sky ($B_\perp$). The polarization vector can be modified by the Faraday rotation effect. Faraday rotation is induced by thermal electrons, coincident with magnetic fields which are at least partially oriented along the line of sight (LOS) between the source and the observer. The amount of Faraday rotation is characterized by the Faraday rotation measure (RM):

$$\mathrm{RM} = 0.81 \int_{\mathrm{source}}^{\mathrm{observer}} n_e \vec{B} \cdot \mathrm{d}\vec{l}, \tag{1.1}$$

where the electron density n_e is expressed in units cm^{-3}, the magnetic field B is in μG, and the pathlength l is in parsecs. The projection of the field along the line of sight ($\vec{B} \cdot \mathrm{d}\vec{l}$) is referred to as $B_\parallel$. Faraday rotation modifies the polarization angle via

$$\chi(\lambda) = \chi_0 + \mathrm{RM} \cdot \lambda^2, \tag{1.2}$$

where χ_0 is the intrinsic polarization angle, and $\chi(\lambda)$ is the polarization angle observed at wavelength λ.

Traditionally, RM is determined by plotting the observed polarization angle as a function of the square of the observing wavelength, and performing a least-squares fit to the data. There are three potential problems with this approach,

• The observed polarization angle is only known modulo π radians; thus, with measurements in only a few wavelength bands, the RM fit is often arbitrary. This is commonly referred to as the $n\pi$ ambiguity. For an illustration of the problem (and how closely spaced adjacent $\chi(\lambda)$ measurements can help to resolve the ambiguity), see Figure 1 of Rand & Lyne (1994).

• Polarized emission with different RM values can be present in a single line of sight. The signal from these different regions mixes, and makes a linear fit inappropriate.

• Faint sources with high rotation measure will be undetectable in individual channels due to low signal-to-noise, and will remain undetectable even after integrating all channels due to bandwidth depolarization. Thus, no $\chi(\lambda)$ data points would be available for a traditional linear fit.

One way to deal with the $n\pi$ ambiguity is to rely on resolving smooth spatial gradients in the polarization angle at each frequency. With this assumption, the appropriate value of n can be determined in each pixel, yielding the correct polarization angle at each frequency, and thus the true value of RM. This is the basis of the PACERMAN routine developed by Dolag *et al.* (2005).

However, routines such as PACERMAN cannot deal with the second and third problems mentioned above, because it is ultimately based on calculating a single RM in each LOS. In this contribution, we describe a novel technique introduced by Burn (1966), and recently extended and implemented by Brentjens & de Bruyn (2005), who coined the term RM synthesis to describe it, and discuss its application in various astrophysical situations. The RM synthesis technique is summarized in § 2. A related deconvolution issue is described in § 3. The use of RM-synthesis is illustrated in § 4 with results of a recent observational program carried out with the Westerbork Synthesis Radio Telescope (WSRT). We also briefly discuss future applications of this technique with the Low Frequency Array (LOFAR) in § 5.

2. RM synthesis: theory and practice

2.1. *In theory ...*

As shown by Burn (1966), one can start by writing the observed complex polarization vector as $P(\lambda^2) = pIe^{2i\chi}$, where p is the fractional polarization. We now substitute Eqn. 1.2 for χ, replacing RM with a generalized quantity ϕ, the Faraday depth. Since the observed polarization vector originates from emission at *all* possible values of ϕ,

$$P(\lambda^2) = \int_{-\infty}^{+\infty} pIe^{2i[\chi_0 + \phi\lambda^2]}\mathrm{d}\phi \tag{2.1}$$

which can be rewritten as

$$P(\lambda^2) = \int_{-\infty}^{+\infty} F(\phi)e^{2i\phi\lambda^2}\mathrm{d}\phi, \tag{2.2}$$

where $F(\phi)$, the Faraday dispersion function, describes the intrinsic polarized flux, as a function of the Faraday depth. Thus, we have a simple expression relating the intrinsic polarized radiation along the LOS, $F(\phi)$, to the observed quantities, $P(\lambda^2)$. The relation takes the form of a Fourier transform. The equation can be inverted to express the intrinsic polarization in terms of observable quantities:

$$F(\phi) = \int_{-\infty}^{+\infty} P(\lambda^2)e^{-2i\phi\lambda^2}\mathrm{d}\lambda^2. \tag{2.3}$$

However, one is confronted with a problem: namely, that we do not observe at wavelengths where $\lambda^2 < 0$. Nor do we observe at all values of $\lambda^2 > 0$. These issues are addressed by Brentjens & de Bruyn (2005).

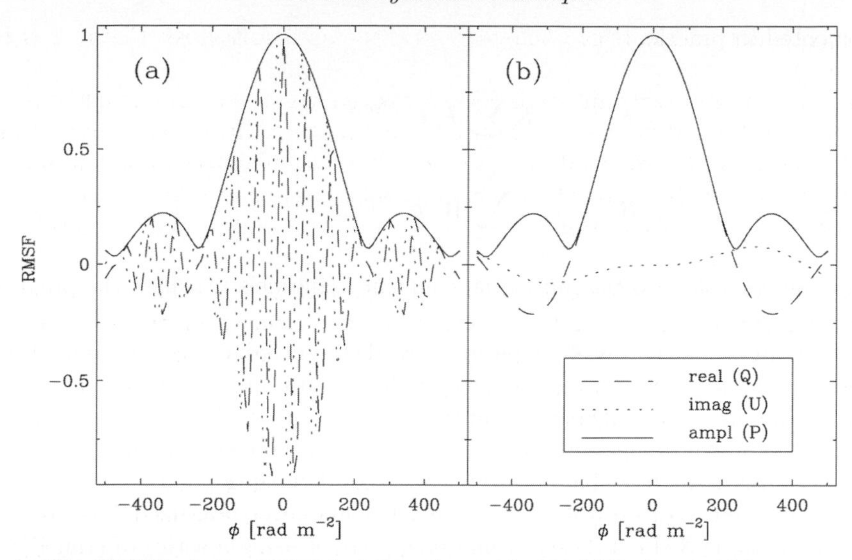

Figure 1. The RMSF of an artificial observation with 200 MHz bandwidth, centered at 1.4 GHz, using 512 frequency channels. The RMSF is shown (a) with the factor $\lambda_0^2 = 0$, and (b) with λ_0^2 set to the weighted mean λ^2 value as advocated by Brentjens & de Bruyn (2005). The real part of the RMSF (i.e., Stokes Q) is shown with dashed lines, the imaginary part (Stokes U) with dotted lines, and the amplitude ($P = \sqrt{Q^2 + U^2}$) with solid lines.

2.2. *In practice ...*

Brentjens & de Bruyn (2005) introduce a window function, $W(\lambda^2)$, which is nonzero only at values of λ^2 which are sampled by the telescope. They proceed to show that Eqns. 2.2 and 2.3 can be rewritten to express the *observed* polarized emission as

$$\tilde{P}(\lambda^2) = W(\lambda^2)P(\lambda^2) = W(\lambda^2)\int_{-\infty}^{+\infty} F(\phi)e^{2i\phi(\lambda^2-\lambda_0^2)}\mathrm{d}\phi \tag{2.4}$$

and, after some intermediate steps, the reconstructed Faraday dispersion function as

$$\tilde{F}(\phi) = K\int_{-\infty}^{+\infty} \tilde{P}(\lambda^2)e^{-2i\phi(\lambda^2-\lambda_0^2)}\mathrm{d}\lambda^2 = F(\phi) \star R(\phi), \tag{2.5}$$

where K is the inverse of the integral over $W(\lambda^2)$, the $\star$ denotes convolution, and $R(\phi)$ is the RM spread function (RMSF)†. The RMSF is a crucially important quantity, and is defined by

$$R(\phi) \equiv K\int_{-\infty}^{+\infty} W(\lambda^2)e^{-2i\phi(\lambda^2-\lambda_0^2)}\mathrm{d}\lambda^2. \tag{2.6}$$

The factor λ_0^2 has been introduced in Eqns. 2.4−2.6 in order to improve the behavior of the RMSF (see Figure 1). Brentjens & de Bruyn (2005) show that the optimal choice of λ_0^2 is the mean of the sampled λ^2 values, weighted by $W(\lambda^2)$.

Brentjens & de Bruyn (2005) next move on to show that Eqns. 2.5 and 2.6 can be written as sums; these are the equations which define the RM synthesis technique as it

† Brentjens & de Bruyn (2005) actually called this function the RM transfer function (RMTF). However, in analogy to telescope optics, the quantity $R(\phi)$ is more similar to the point spread function (PSF) than to the optical transfer function (OTF). It has therefore since been renamed the RM spread function (RMSF).

is implemented in practice.

$$\tilde{F}(\phi) \approx K \sum_{c=1}^{N} \tilde{P}_c \, e^{-2i\phi(\lambda_c^2 - \lambda_0^2)}, \tag{2.7}$$

$$R(\phi) \approx K \sum_{c=1}^{N} W_c \, e^{-2i\phi(\lambda_c^2 - \lambda_0^2)}, \tag{2.8}$$

where the index c refers to the individual frequency channels in which the polarized flux is observed at the radio telescope.

One of the main motivations for performing RM synthesis is to minimize the effects of the $n\pi$ ambiguity. This is best done by splitting up the observing bandwidth into many individual narrow frequency channels. By observing in this way, only the brightest polarized emission will be detected above the noise level in each narrow channel. Adding up the individual frequency channels may cause bandwidth depolarization. But the RM synthesis operation is often able to recover such low-level polarized flux. In fact, one interpretation of the RM synthesis technique is that of using a series of trial RM values, and finding the one which maximizes the signal level resulting from the coaddition of the polarized flux from all channels. In this interpretation, the flux as a function of ϕ – the reconstructed Faraday dispersion function – will peak at the value of ϕ corresponding to the actual RM of the source. At other values of ϕ, the polarization vector rotates at the wrong rate through λ^2 space, the polarization vector will not constructively interfere throughout the band, and the total flux will thus be lower.

The other main reason for performing RM synthesis is to recover emission at multiple Faraday depths along a particular LOS. This circumstance can occur, for example, when a background polarized source shines through a foreground medium which both generates Faraday rotation, and also produces its own synchrotron radiation in the same volume – an excellent example is the Milky Way. In such a case, polarized flux appears at different values of ϕ, as illustrated in Figure 2.

The idealized physical situations shown in Figure 2 illustrate some of the key effects which determine the form of the Faraday dispersion function. In the top panel, a polarized background source is shown. The source has an intrinsic rotation measure of $+100\,\mathrm{rad\,m^{-2}}$. In this case, no magnetoionized medium is present along the LOS between the source and the observer, so the Faraday dispersion function is simply a delta function, with amplitude equal to the polarized flux density of the background source, and centered at $\phi = +100\,\mathrm{rad\,m^{-2}}$. In the middle panel, the situation is the same, except that a region of ionized gas (A) is introduced along the line of sight. The region contains an ordered magnetic field inclined with respect to the LOS (but does not contain cosmic-ray electrons) in such a way that it provides an additional rotation measure of $-50\,\mathrm{rad\,m^{-2}}$. Thus, the emission from the background source is shifted to $\phi = +50\,\mathrm{rad\,m^{-2}}$. In the bottom panel, cosmic-ray electrons are added to the intervening region (B), so that in addition to the extra Faraday rotation, the region also emits its own synchrotron radiation. That which is emitted at the back of the region (furthest from the observer) is produced without a rotation measure, but accumulates a net rotation measure of $-50\,\mathrm{rad\,m^{-2}}$ by the time it propagates to the front of the region. The emission from the front of the region never accumulates any Faraday rotation on its way to the observer. The radiation emitted in the middle of the region picks up a rotation measure between those two values. In the simplest possible case, where the field distribution and particle densities are uniform, the resulting Faraday dispersion function is a tophat function, or Burn slab, as illustrated. It is present in addition to the radiation from the background source, for

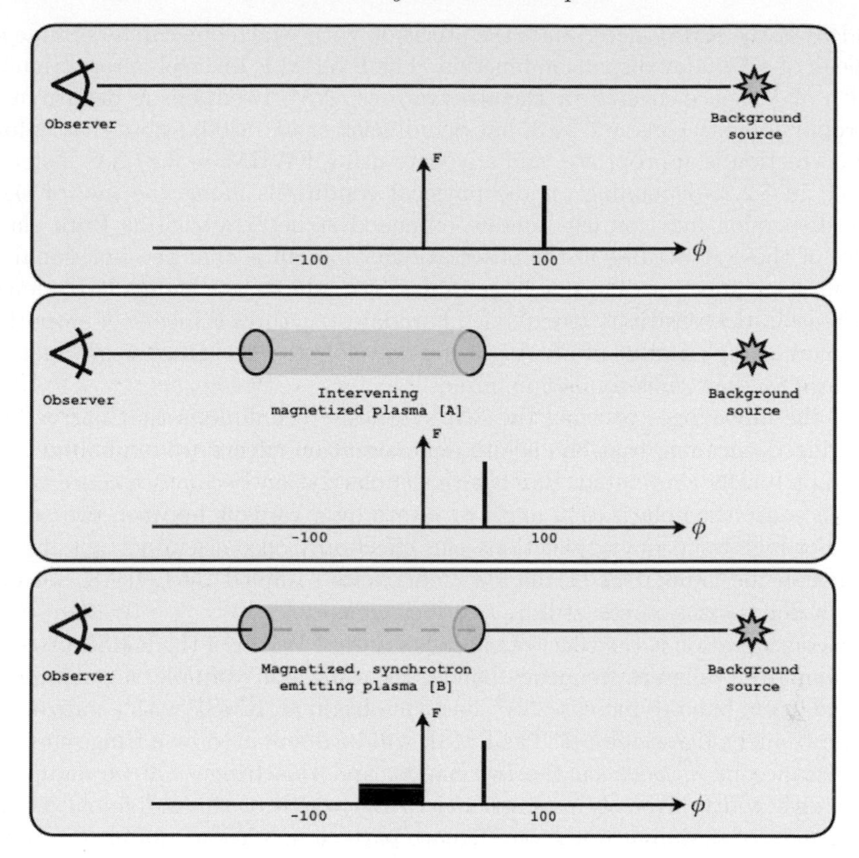

Figure 2. Three idealized physical situations along an imaginary LOS, and the Faraday dispersion functions which correspond to those situations. Refer to the text for a discussion of the individual images.

which nothing has changed relative to the picture in the middle panel (the intervening medium still changes its RM by $-50\,\mathrm{rad\,m^{-2}}$ as before).

Real objects which could make up such a picture include unresolved radio galaxies or pulsars as the background source, and the diffuse Milky Way, galaxy clusters, or galaxies as the foreground medium. Distinguishing the contributions from different objects along the LOS can be helped by including additional information about the spatial scales of, and distances to, the different objects. This is referred to as Faraday tomography.

The RM synthesis technique was originally conceived for analyzing data along a single LOS. However, many polarized sources are resolved, and it is now common to perform the inversion in Eqn. 2.7 for every spatial pixel in full Stokes Q and U data cubes. The output of such an operation is referred to as an "RM cube", with ϕ as the third axis. Frequency-dependent instrumental effects must be calibrated before generating an RM cube. For example, the synthesized beam (for an interferometer) must be the same at all frequencies. Also, the single dish primary beam attenuation must be corrected for at all frequencies. Otherwise, the frequency-dependent instrumental signal will be picked up by the RM synthesis operation, and appear in the RM cube.

2.3. *Practical issues*

There are a number of practical issues which were noted by Brentjens & de Bruyn (2005) and which must be taken into account when doing an RM synthesis experiment. First, the FWHM of the RMSF determines the precision with which one can determine the RM at the peak of a Faraday dispersion function. The FWHM is inversely proportional to the full width of λ^2 space covered by the observations, $\Delta\lambda^2$. Brentjens & de Bruyn (2005) use a proportionality constant $2\sqrt{3}$, but Schnitzeler *et al.* (2008) note that empirically, a small correction is appropriate, and advocate using FWHM $= 3.8/\Delta\lambda^2$ instead.

As seen in § 2.2, depending on the physical conditions along the line of sight, the Faraday dispersion function can contain extended structures such as Burn slabs. The signature of these structures in the observational domain is that the functional form of $||P(\lambda^2)||$ is a sinc function. Thus, at large values of λ^2, the degree of depolarization is high. For this reason, the sensitivity to extended Faraday structures is inversely proportional to the minimum sampled value of λ^2. At shorter wavelengths, the amount of depolarization due to Faraday thick emission is minimized.

One of the advantages of using the RM synthesis technique is that narrow channels can be utilized, meaning that bandwidth depolarization effects are minimized. However, they are not totally eliminated. Bandwidth depolarization becomes a concern at values of ϕ which cause the polarization angle to rotate by π radians between two adjacent λ^2 samples. For high frequency applications, this effect only becomes important at extremely large rotation measures. But at the low frequencies sampled by LOFAR, for example, bandwidth depolarization can still be a serious concern.

A final consideration is the effect of minimizing the FWHM of the RMSF by combining observations from different frequency bands. By using, for example, a combination of a λ20cm and λ6cm band to increase $\Delta\lambda^2$, one will obtain an RMSF with a narrow FWHM, but also extremely high sidelobes. The RMSF will be dominated by a fringe characterized by the distance in λ^2 between the two bands, and this fringe will be damped by an envelope with width inversely proportional to the width of the individual bands. This can be thought of as similar to the interference pattern in a Young double slit experiment – the fringe spacing is set by the distance between the two slits (observing bands), and the heights of the individual fringes are determined by the width of the slits. In an observation where there is a large gap in λ^2 coverage, the effect is qualitatively the same.

2.4. *To weight or not to weight?*

The window function $W(\lambda^2)$ introduced earlier can be used to weight the individual values of $P(\lambda^2)$ that go into the calculation of the Faraday dispersion function. By weighting each value by the square of the signal-to-noise ratio (S/N), one can hope to maximize the precision with which we determine the RM at the peak of the profile (at the expense of broadening the RMSF).

We have performed a Monte Carlo simulation to assess the effect of using such a weighting scheme. We generated mock observations of a polarized point source with a Faraday dispersion function described by ($Q = 1/\sqrt{2}$, $U = 1/\sqrt{2}$) at $\phi = 30\,\mathrm{rad\,m^{-2}}$, and zero elsewhere. We used the same frequency coverage as was used to produce the RMSF in Figure 1. Then we added noise to the artificial observation, such that the noise level varies quadratically from channel to channel. The resulting S/N ranged from 1 in the central channel, to about 1/12 in the first and last channels. Then we reconstructed the Faraday dispersion function using RM synthesis, with and without S/N weighting, and determined the RM at the peak of the Faraday dispersion function, $\hat{\phi}$, in each case. We ran 2500 realizations of the simulation.

The simulation results show that the weighted version of the RM synthesis calculation provides a significantly better determination of the value of $\hat{\phi}$. In Figure 3, we plot the distribution of the $\hat{\phi}$ values determined in each run of the simulation, with and without the signal-to-noise weighting. Without the weighting, the mean $\hat{\phi}$ determination was

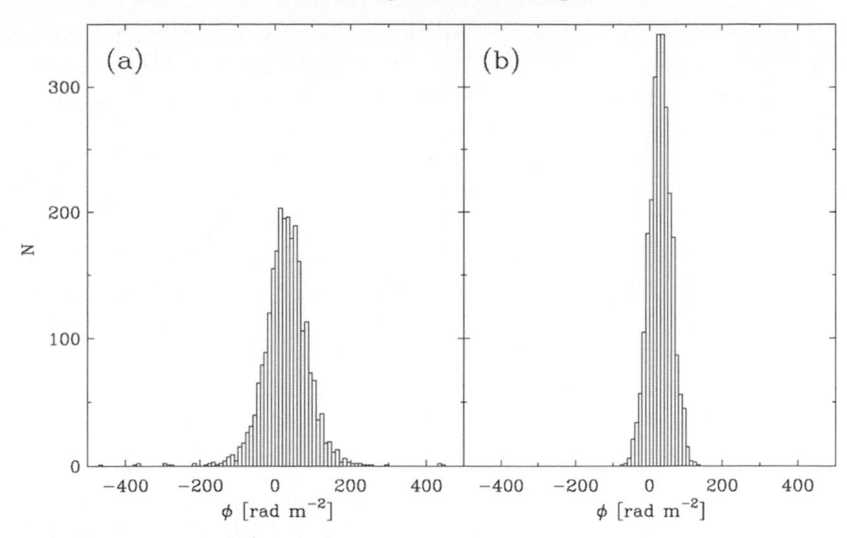

Figure 3. The distribution of $\hat{\phi}$ values determined during the Monte Carlo simulation described in the text. Results are shown from the simulated Faraday dispersion functions which were constructed (a) without weighting, and (b) with $4 \times (S/N)^2$ weighting. The weighting scheme serves to narrow the distribution of values. Both are centered at the actual value, $30\,\mathrm{rad\,m^{-2}}$, but the unweighted distribution is ~ 2 times broader.

$29.4 \pm 61.2\,\mathrm{rad\,m^{-2}}$, whereas with weighting, the determination was 29.6 ± 30.0, yielding a factor of 2 improvement in the precision. Note that with this frequency coverage, the expected FWHM of the RMSF is $\approx 284\,\mathrm{rad\,m^{-2}}$.

By weighting the $P(\lambda^2)$ values in this way, the RMSF is significantly broadened. Thus, despite the increase in precision gained, the ability to distinguish two separate features in the Faraday dispersion function will be markedly decreased. One should use different weighting schemes to optimize for the detection of different types of polarized emission, just as the weighting of visibilities measured by an aperture synthesis telescope should be optimized for the characteristics of the emission being studied.

3. Faraday dispersion function deconvolution

As pointed out in § 2.2, after performing RM synthesis, the reconstructed Faraday dispersion function, $\tilde{F}(\phi)$, is the convolution of the actual Faraday dispersion function, $F(\phi)$, with the RMSF, $R(\phi)$. When it appears that there may be multiple features in a reconstructed Faraday dispersion function, confusion with RMSF sidelobes can make the interpretation difficult. The situation can often be improved by performing a deconvolution operation.

The algorithm is rather similar to the deconvolution routine `CLEAN` developed for use with aperture synthesis radio telescope images and described by Högbom (1974). The differences are (1) the deconvolution takes place in one dimension (Faraday depth) rather than two (spatial) dimensions; and (2) the functions involved are complex quantities.

The implementation of RM synthesis deconvolution proceeds as follows. First, the location of the peak of the reconstructed Faraday dispersion function, ϕ_p, is searched for (either by locating the peak of $||\tilde{F}(\phi)||$, or alternatively the peak of cross-correlation of that function with the RMSF). Once ϕ_p is determined, the values of the real and imaginary parts of $||\tilde{F}(\phi_p)||$ are scaled by a loop gain parameter g, typically taken to be 0.1. This is stored as a "clean component". Next, a version of the RMSF, shifted and

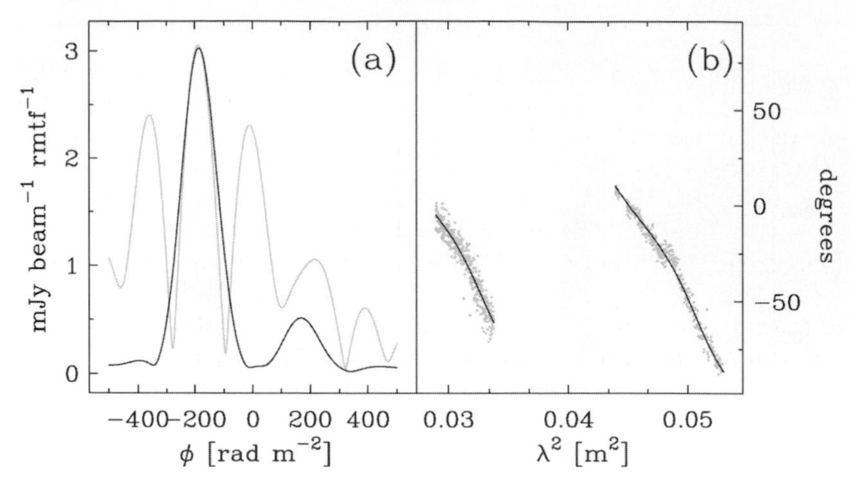

Figure 4. An example of `RMCLEAN` results. In panel (a) we display the Faraday dispersion function of a relatively bright polarized point source in the field of NGC 7331, observed during the WSRT-SINGS survey. The gray line is the original reconstructed Faraday dispersion function, and the black line is the deconvolved Faraday dispersion function. The resulting profile clearly shows two distinct components. In panel (b), the observed polarization angle at each frequency is plotted with gray points. The polarization angle predicted by the clean components extracted by `RMCLEAN` are displayed with the black line. Clearly, the model matches the data quite well. A traditional linear fit to the polarization angle vs. λ^2 would not have been possible in this case.

scaled to be equal to $g||\tilde{F}(\phi_p)||$ at $\phi = \phi_p$, is subtracted from the Faraday dispersion function. The residuals are searched for a new peak, and the loop (scale, store, subtract, and search) is repeated until the residuals are all below a user-specified threshold, or a maximum number of iterations have been performed. Finally, the clean components are convolved with a restoring function – we use a real-valued gaussian with FWHM approximately the same as the main lobe of the RMSF – and added to the residuals. The result is the deconvolved Faraday dispersion function.

This algorithm has been implemented within the `MIRIAD` software package, as a task called `RMCLEAN`, during analysis of the WSRT-SINGS survey data described in § 4.1. The routine is available upon request.

An example of this operation is shown in Figure 4. In the left panel, the original reconstructed Faraday dispersion function, of a bright polarized point source in the WSRT-SINGS survey, is compared to the deconvolved version. In the right panel, the observed $\chi(\lambda^2)$ values are compared to those predicted by the clean components extracted during the deconvolution operation, as a check that the behavior predicted by the model matches the data. It can also be seen that a linear fit to the $\chi(\lambda^2)$ data would be insufficient to characterize the situation. RM synthesis is particularly powerful in similar cases where the S/N level is much lower.

4. Applications

In recent years, several research projects have made use of the RM synthesis technique. Many of them have been performed using data from the WSRT, which is equipped with a powerful and flexible correlator, and which can provide many channels over a large simultaneous bandwidth in all four Stokes parameters. Those projects have mostly made use of data at low frequencies, near 350 MHz; see for example the work by Brentjens (2008), Schnitzeler *et al.* (2008), and Bernardi *et al.* (2008).

4.1. *The WSRT-SINGS survey*

One of the data sets to which RM synthesis has recently been applied is the WSRT-SINGS survey. The survey has been described by Braun *et al.* (2007), who also give an atlas of the observed galaxies in Stokes I emission. The targets were selected mainly from the Spitzer SINGS survey (Kennicutt *et al.* 2003), if they have optical diameters $D_{25} > 5'$ and declination $\delta > 12.5°$. Each galaxy in the survey was observed for a total of 12 hours, split between two wide 160 MHz frequency bands centered at 22 cm and 18 cm. All four polarization products were recorded. The polarization data has been analyzed using RM synthesis (Heald *et al.*, in prep.), which is particularly useful for this survey in order to recover faint diffuse polarized emission and its associated RM. The final noise level in Stokes Q and U after RM synthesis was typically $25 - 35\mu\text{Jy beam}^{-1}$ depending on the particular target. Because of the large gap between the 18 cm and 22 cm bands, the first sidelobes of the RMSF reach $\sim 78\%$ of the main lobe. Thus, deconvolution of the Faraday dispersion functions was performed (see Figure 4 for an example which also illustrates the high sidelobes caused by the gap between the 18 cm and 22 cm bands).

RM cubes were constructed for 28 of the galaxies in the WSRT-SINGS survey. Linear polarization detections were made in 21 of 24 observed spiral galaxies, but no detections were made in the Magellanic and elliptical type galaxies. One of the fundamental products of this project is a rotation measure map for each galaxy detected in diffuse polarized emission. The RM maps were constructed by determining the value of ϕ at the peak of each deconvolved Faraday dispersion function. An example for NGC 6946 is shown in Figure 5. The distribution of rotation measures shows a clear gradient across the disk. A clear sinusoidal pattern is revealed by determining the average RM in wedges, and plotting as a function of the azimuthal angle in the galaxy. This sinusoidal pattern may be indicative of an axisymmetric dynamo mode, as shown in Figure 1 of Krause (1990). Using the rotation measure maps which we have determined, the observed polarization angles were corrected to their intrinsic values. This reveals the magnetic field structure in the galaxy, as shown in the last panel of Figure 5.

5. Conclusions and future applications

Use of the RM synthesis technique has recently become practical, thanks to the flexible correlators of modern radio telescopes such as the WSRT. The technique is very powerful in eliminating the $n\pi$ ambiguity which plagues the traditional least-squares fitting technique, and moreover provides a simple way to recover multiple polarized structures along a single LOS. In many cases, deconvolution is necessary, especially in observational programs in which large gaps in frequency coverage lead to high RMSF sidelobe levels.

Forthcoming radio telescopes, such as LOFAR (Falcke *et al.* 2007), which is currently being built in the Netherlands, will benefit greatly from the techniques reviewed here. At the low frequencies which will be observed (30 – 240 MHz), the corresponding $\Delta\lambda^2$ coverage is very broad, and this leads to a very narrow RMSF response. We expect to achieve an RM precision which is more than two orders of magnitude better than in the combination of the 18 cm and 22 cm bands used in the WSRT-SINGS survey, for example. One of the LOFAR Key Science Projects (KSPs), the Magnetism KSP, will make use of the RM synthesis technique to study weak magnetic fields in the Milky Way, as well as in nearby galaxies.

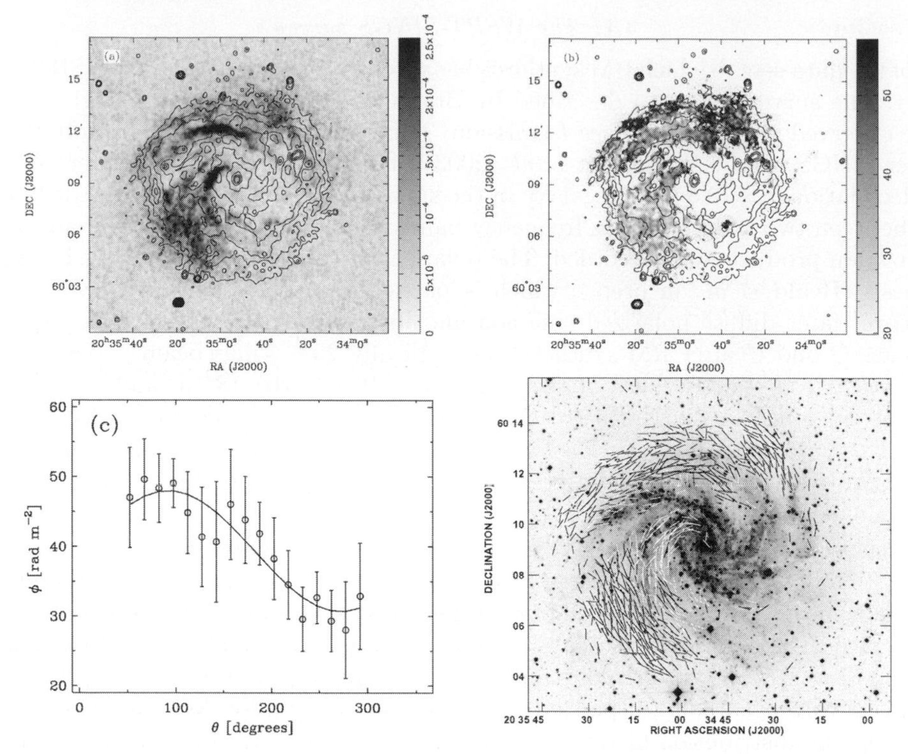

Figure 5. Linearly polarized emission in the spiral galaxy NGC 6946, observed in the WSRT-SINGS survey. (a) The distribution of the linearly polarized emission. (b) The distribution of ϕ values, determined as described in the text. (c) Azimuthal variation in the mean rotation measure value, together with a sinusoidal fit. (d) Magnetic field map, generated from the Faraday rotation-corrected polarization angles and overlaid on the DSS image.

Acknowledgements

The Westerbork Synthesis Radio Telescope is operated by ASTRON (Netherlands Institute for Radio Astronomy) with support from the Netherlands Foundation for Scientific Research (NWO). I would like to acknowledge my collaborators on the WSRT-SINGS work shown here, R. Braun and R. Edmonds. Many thanks to M. Brentjens for numerous enlightening conversations on these topics, and to G. de Bruyn for helpful discussions and a critical reading of this paper.

References

Bernardi, G. *et al.* 2008, *A&A*, submitted
Braun, R., Oosterloo, T. A., Morganti, R., Klein, U., & Beck, R. 2007, *A&A* 461, 455
Brentjens, M. A. 2008, *A&A* 489, 69
Burn, B. J. 1966, *MNRAS* 133, 67
Brentjens, M. A. & de Bruyn, A. G. 2005, *A&A* 441, 1217
Dolag, K., Vogt, C., & Enßlin, T. A. 2005, *MNRAS* 358, 726
Falcke, H. D. *et al.* 2007, *HiA* 14, 386
Högbom, J. A. 1974, *A&AS* 15, 417
Kennicutt, R. C. *et al.* 2003, *PASP* 115, 928
Krause, M. 1990, *IAUS* 140, 187
Rand, R. J. & Lyne, A. G. 1994, *MNRAS* 268, 497
Schnitzeler, D. H. F. M., Katgert, P., & de Bruyn, A. G. 2008, *A&A*, in press (arXiv:0810.4211)

Discussion

KEPLEY: How does RFI affect the RM-synthesis technique?

HEALD: One should take care to remove strongly affected channels from the analysis. But as for weak RFI, aside from increasing the noise level, it should not degrade the results (as long as the RFI is not structured in the frequency domain).

FLETCHER: George clearly shows (in M51) a possible $m = 1$ azimuthal mode in the 18/20 cm RMs. A note of caution on the interpretation should be added (in my opinion): depolarization by unresolved magnetic fluctuations in the synchrotron emitting regions can lead to complete depolarization along part of the observed line of sight. Rotation measures at higher frequencies show a clear sign of a $m = 0 + 2$ azimuthal pattern near the disc of M51 (Fletcher *et al.* in prep.), with a $m = 1$ pattern (as shown by George) in a layer nearer to the observer (a thick disc or halo). Dynamo modellers should take note!

HEALD: We see a hint of a contribution from an $m = 0$ mode, which may correspond to the disk component of your model, but your point about depolarization effects at these frequencies is well taken.

HAN: 1) Comment: when we do pulsar observations with 512/1024 channels inside a 256 MHz band, `PSRCHIVE` software (developed by Willem van Straten) also uses RM synthesis technique for RM. Afterwards, we still check the ΔPA from the upper and lower half bands for the uncertainties of the RM measurements. 2) In your observations of SINGS galaxies, your RM image is very impressive. Do you see different RM layers during the process? I think maybe there are different layers in each part inside a galaxy? You did average on that? How did you compose the final RM map?

HEALD: 1) When signal-to-noise permits, one should always check for consistency in the measurement domain. However, the real power of the RM-synthesis technique derives from the capability to extract Faraday dispersion functions from data with signal levels below the noise in each channel. 2) As Andrew Fletcher pointed out, depolarization at these frequencies likely prevents us from detecting polarization from deeply within the galaxies. But by using this technique at higher frequencies, where Faraday effects are less severe, one could probe the different layers of disk and halo.

Wolfgang Reich

Tim Gledhill

Cosmic Magnetic Fields:
From Planets, to Stars and Galaxies
Proceedings IAU Symposium No. 259, 2008
K.G. Strassmeier, A.G. Kosovichev & J.E. Beckman, eds.

doi:10.1017/S1743921309031433

Measuring and calibrating galactic synchrotron emission

Wolfgang Reich and Patricia Reich

Max-Planck-Institut für Radioastronomie, Auf dem Hügel 69, D-52121 Bonn, Germany
email: wreich, preich@mpifr-bonn.mpg.de

Abstract. Our position inside the Galaxy requires all-sky surveys to reveal its large-scale properties. The zero-level calibration of all-sky surveys differs from standard 'relative' measurements, where a source is measured in respect to its surroundings. All-sky surveys aim to include emission structures of all angular scales exceeding their angular resolution including isotropic emission components. Synchrotron radiation is the dominating emission process in the Galaxy up to frequencies of a few GHz, where numerous ground based surveys of the total intensity up to 1.4 GHz exist. Its polarization properties were just recently mapped for the entire sky at 1.4 GHz. All-sky total intensity and linear polarization maps from WMAP for frequencies of 23 GHz and higher became available and complement existing sky maps. Galactic plane surveys have higher angular resolution using large single-dish or synthesis telescopes. Polarized diffuse emission shows structures with no relation to total intensity emission resulting from Faraday rotation effects in the interstellar medium. The interpretation of these polarization structures critically depends on a correct setting of the absolute zero-level in Stokes U and Q.

Keywords. Techniques: polarimetric – surveys – ISM: radio continuum

1. Introduction

All-sky radio continuum surveys provide basic information on our local environment and the large-scale properties of the Galaxy. They are required to model the Galactic emission components in 3-D (Sun *et al.* 2008) and guide more sensitive higher-angular resolution observations of the Galactic plane or other regions or objects of interest. All-sky surveys are quite time consuming projects, which require special observing methods to accurately measure large-scale sky emission. They need similar telescopes in the northern and southern hemisphere and have to adapt calibration data from additional instruments to provide the absolute level of sky emission. Thus all-sky surveys are rare, in particular in linear polarization, where the signals are much weaker than in total intensities.

In the following we describe the basic methods to calibrate and adjust total intensity surveys as well as the more complex requirements being applied for polarization surveys. We do not discuss the instrumental corrections to be taken into account during the data reduction process, which can be found in the original publications.

2. Radio continuum measurements

When pointing a radio telescope to a certain sky direction we record - beside the signal of interest - a variety of other much stronger signals. The observed signal can be expressed as a temperature T_{obs}:

$$T_{\mathrm{obs}} = T_{\mathrm{sys}} + T_{\mathrm{atm}} + T_{\mathrm{ground}} + T_{\mathrm{CMB}} + T_{\mathrm{conf}} + T_{\mathrm{gal}} + T_{\mathrm{sou}} \tag{2.1}$$

where T_{sys} is the contribution from all components of the receiving system, typically 20 K to 30 K for a cooled receiver for cm-wavelength observations. T_{atm} is the contribution from the atmosphere, typically a few K, depending on elevation and the actual weather conditions. $T_{\rm ground}$ is the emission picked up by the sidelobes of a telescope from the ground, thus depending on azimuth and elevation. $T_{\rm CMB}$ is the wavelength independent isotropic radiation from the cosmic microwave background (CMB) of 2.73 K (Mather *et al.* 1994). $T_{\rm conf}$ also is an isotropic component resulting from unresolved weak extragalactic sources. $T_{\rm conf}$ depends on wavelength and beam size. $T_{\rm gal}$ is the diffuse Galactic background emission and, finally, $T_{\rm sou}$ the signal of interest from a specific Galactic or extragalactic source. $T_{\rm sou}$ and - in the case of Galactic surveys - $T_{\rm gal}$ are to be separated from the other components in an appropriate way. The problem is the weakness of the sky emission compared to the other contributions to $T_{\rm obs}$.

2.1. *Absolute versus relative measurements*

"Relative measurements" are the standard mode for continuum and polarization observations, when the signal from a source or a certain region of the sky is of interest. In that case the region surrounding the source or the source complex is referred to as the zero-level and the difference signal is taken as $T_{\rm sou}$. This is justified in case the measurements of the source signal and its adjacent zero-level are not suffering from time or telescope position dependent changes of the unwanted contributions in an unpredictable way. Galactic diffuse emission shows intrinsic variations, which might become a problem when they mix up with emission from extended faint sources, e.g. a large diameter supernova remnant (SNR), and can not be separated. Thus large errors for integrated flux densities reported for extended SNRs are very common.

An "absolute measurement" means that the entire signal from the sky is recovered and the three terms $T_{\rm sys} + T_{\rm atm} + T_{\rm ground}$ are subtracted. This is required for Galactic total intensity and polarization all-sky surveys, but not easy to obtain with high precision at decimeter and shorter wavelengths. The two isotropic sky components $T_{\rm CMB}$ and $T_{\rm conf}$ don't matter in this context. Special observing and reduction procedures are applied for survey observations. For instance, very long scans across the sky are observed along the same azimuth or elevation range to ensure that $T_{\rm atm}$ and/or $T_{\rm ground}$ are as constant as possible, while the sky emission varies as a function of time. From an appropriate analysis of a large number of scans individual corrections for each scan against the mean of all scans can be calculated. The final all-sky survey map is constructed from these corrected scans. Nevertheless, the maps often show 'striping effects' along the scan direction at the noise level or exceeding it, which indicate residuals from the correction procedures. In general distortions of the measurements on short time scales, e.g. by weather changes or varying low-level RFI, are more difficult to correct than smooth changes of the baseline level by temperature effects over a night.

2.2. *Sky horn measurements*

Continuum all-sky surveys receive their zero-level calibration by using low-resolution sky horn data. Their antenna diagrams are well known to enable a proper far-sidelobe correction. Reference loads at precisely known temperatures are required to find $T_{\rm sys}$. The all-sky surveys need to be convolved to the beam size of the sky horns, which are typically several degrees, to find the temperature offset. This way the 1.42 GHz survey (Fig. 1) was calibrated to an accuracy of 0.5 K $T_{\rm B}$. The 408 MHz survey by Haslam *et al.* (1982) has a zero-level accuracy of about 3 K $T_{\rm B}$. This accuracy is lower than the sensitivity (3×r.m.s-noise) of both surveys of 0.05 K $T_{\rm B}$ and 2K $T_{\rm B}$, respectively.

The probably most famous sky horn observations are those by Penzias & Wilson (1965) leading to the discovery of the 3 K cosmic microwave background radiation. This led to the award of the Nobel Prize in 1978. Their absolute measurements revealed an isotropic

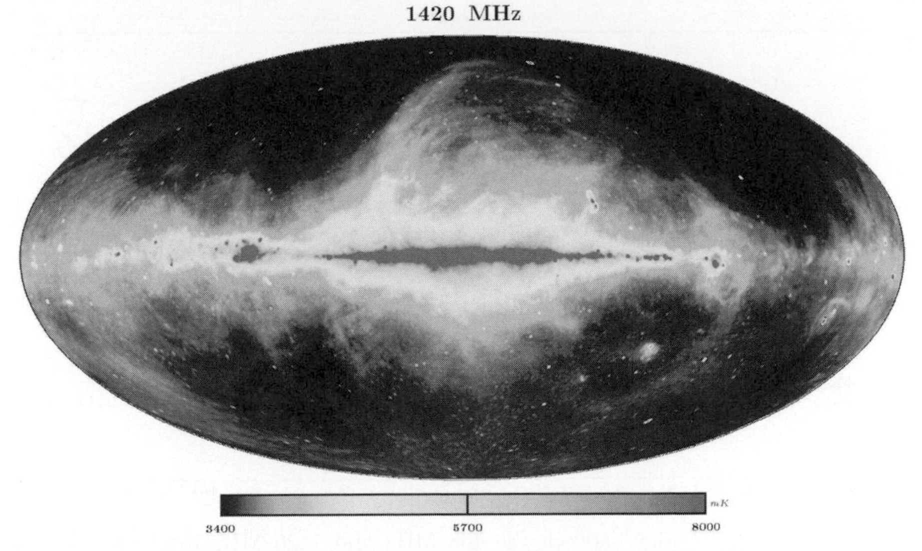

Figure 1. All-sky survey at 1.42 GHz combined from northern sky data observed with the Stockert 25-m telescope near Bonn/Germany (Reich 1982; Reich & Reich 1986) and southern sky observations observed with a 30-m dish at Villa Elisa/Argentina (Reich *et al.* 2001).

sky component of 3.5 K, interpreted by Dicke *et al.* (1965) to originate from the CMB. This 4.08 GHz measurement of 3.5 K happened at a time, when 'relative' measurements in the mK range were regularly made. This clearly reflects the technical challenges for these kind of measurements.

2.3. *Adjusting surveys by TT-plots*

Ground based surveys carried out at different frequencies with their zero-level adapted from sky horn measurements may be further adjusted relative to each other by using the so-called TT-plot method according to Turtle *et al.* (1962). This method was discussed in some detail by Reich *et al.* (2004). The surveys to be adjusted were convolved to a common angular resolution of 15° and the TT-plots were performed for a strip in right ascension for declinations between 30° and 45°, where no contamination from local structures like the giant loops is evident. As an example we show the TT-plot between the 408 MHz survey by Haslam *et al.* (1982) and the 1420 MHz survey by Reich (1982) in Fig. 2 (left panel) after a zero-level correction of -2.7 K for the 408 MHz survey. This correction is within its quoted zero-level error of 3 K. The mean of the fitted lines passes 0 K at both wavelengths. This result assumes the 1420 MHz survey to be correct, although the sky horn data are uncertain by 0.5 K. Using several pairs of surveys improves and constrains the corrections (see Reich *et al.* 2004 for details).

The WMAP total intensity all-sky surveys provide valuable high-frequency maps of Galactic emission and are of similar low-angular resolution compared to the ground based surveys up to 1.4 GHz. Unfortunately the WMAP total intensity maps have not yet set on an absolute zero-level. The three year's release (Hinshaw *et al.* 2007) at 22.8 GHz shows numerous small patches at high latitudes with temperatures below -100 μK, which is incorrect. We used the TT-plot technique to adjust the WMAP 22.8 GHz map in respect to 1420 MHz (see Fig. 2, right panel) and find an offset of 250$\pm$70 μK. For the recent WMAP release of five years observations (Hinshaw *et al.*, preprint), an offset of

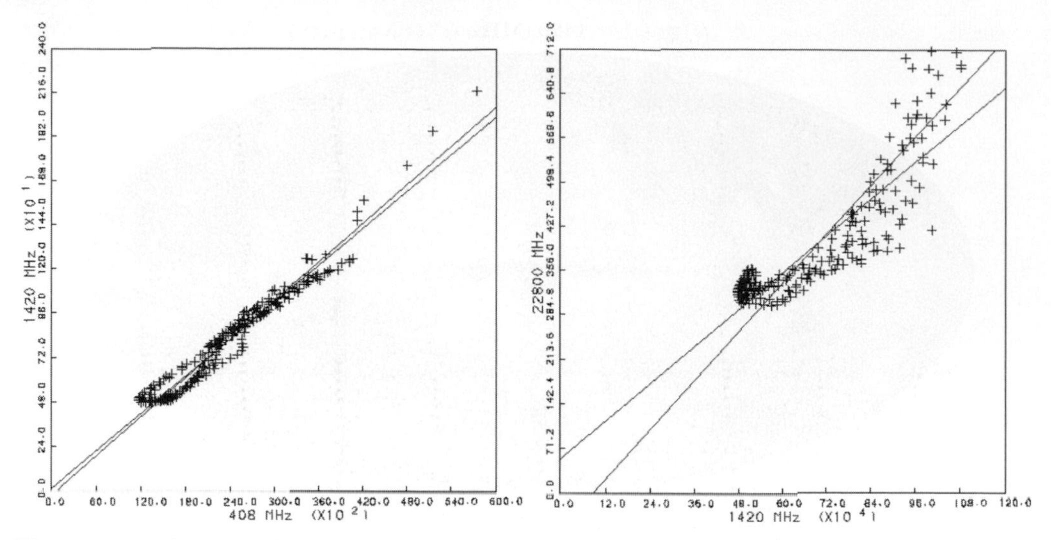

Figure 2. Left panel: *TT*-plot between the 408 MHz and 1420 MHz total intensity surveys leading to a zero-level correction of -2.7 K for the 408 MHz survey. Right panel: *TT*-plot between 1420 MHz and 22.8 GHz leading to an offset correction of the WMAP 3-yr survey by +250 μK.

260 ± 70 μK using the same method. This is quite a significant correction and has a clear effect on the spectral index between 1420 MHz and 22.8 GHz. The most frequent high-latitude temperature spectral indices increase from about –3.1 (without correction) to about –2.8, which is just slightly steeper (by about 0.1) compared to the spectra between 408 MHz and 1420 MHz.

Absolute sky measurements with high precision are needed to calibrate the WMAP total intensity maps. This is expected from ARCADE (Kogut *et al.* 2006), a balloon-borne instrument designed to measure the absolute sky temperature between 3.3 GHz and 90 GHz in a number of channels. This experiment will provide high-precision low-resolution data of Galactic emission.

2.4. *Galactic plane surveys*

Current Galactic plane surveys were carried out with large single-dish telescopes at arcmin angular resolution to resolve diffuse Galactic emission structures from individual sources like SNRs or HII-regions. Synthesis telescope surveys achieve even higher angular resolutions. These surveys use all-sky surveys to add the missing large-scale information and adjust their zero-level. This has been done using the total intensity northern sky 1.4 GHz survey (Fig. 1) to adjust 1.4 GHz Effelsberg total intensity surveys (Kallas & Reich 1980; Reich *et al.* 1990, 1997; Uyanıker *et al.* 1999). These combined data at about 9' angular resolution were used to calibrate the Canadian Galactic Plane Survey (CGPS) carried out with the synthesis telescope at DRAO/Penticton as described by Taylor *et al.* (2003), thus providing a survey including all structural information larger than 1'. Similarly the CGPS survey at 408 MHz uses the all-sky survey carried out by Haslam *et al.* (1982) with the Effelsberg, Parkes and Jodrell Bank telescopes at that frequency.

3. Calibration of polarization data

The scheme for "relative" and "absolute" zero-level calibration of polarization data, which means the calibration of the observed Stokes U and Q maps, in principle follows that of total intensity data: The absolute zero-level is required for all-sky maps and a

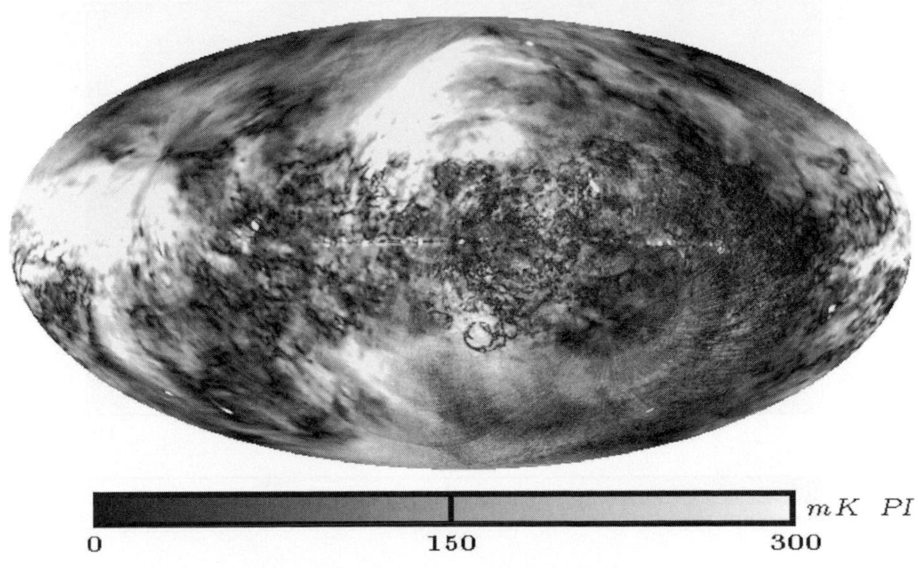

Figure 3. 1.4 GHz polarization all-sky survey combined from northern sky observations by Wolleben *et al.* (2006) and southern sky observations by Testori *et al.* (2008) at an absolute zero-level. The angular resolution of the survey is about 36' and its sensitivity about 45 mK (3×r.m.s-noise).

relative zero-level setting is sufficient for polarized sources. The only ground-based all-sky polarization survey was made at 1.4 GHz by Wolleben *et al.* (2006) for the northern hemisphere using the DRAO 26-m telescope and by Testori *et al.* (2008) using one of the Villa Elisa 30-m telescopes for the southern sky, where the different calibration steps are discussed. The 'absolute zero-level' of this polarization survey is adapted to 1.4 GHz measurements with the Dwingeloo 25-m telescope published by Brouw & Spoelstra (1976), where rotating dipoles were used allowing to separate sky emission from ground radiation contamination. The polarization intensity scale of these surveys, however, was adjusted by using smoothed polarization data from the Effelsberg or the Parkes telescopes, respectively. We show the combined all-sky 1.4 GHz polarization survey (Reich *et al.*, in prep.) in Fig. 3.

For the interpretation of Galactic polarization emerging from Faraday rotation in the interstellar medium, also the step of absolute zero-level calibration is required. Otherwise any interpretation becomes problematic, as it was discussed in some detail by Reich (2006). Polarization features emerging from Faraday rotation are in general not or very weakly related to total intensity sources or emission structures. The effect of zero-level calibration is different for polarized emission, because vectors are added, while for total intensities it is a scalar.

Polarized emission PI and polarization angle ϕ are calculated from the observed Stokes parameters U and Q, assumed to be corrected for all kinds of instrumental effects, by:

$$\mathrm{PI}^2 = \mathrm{U}^2 + \mathrm{Q}^2 \quad \text{and} \quad \phi = 0.5\,\mathrm{atan}(\mathrm{U}/\mathrm{Q}) \tag{3.1}$$

Adding missing components in $\mathrm{U_{zero}}$ and $\mathrm{Q_{zero}}$ gives:

$$\mathrm{PI}^2_{\mathrm{abs}} = (\mathrm{U}+\mathrm{U_{zero}})^2 + (\mathrm{Q}+\mathrm{Q_{zero}})^2 \quad \phi = 0.5\,\mathrm{atan}((\mathrm{U}+\mathrm{U_{zero}})/(\mathrm{Q}+\mathrm{Q_{zero}})) \tag{3.2}$$

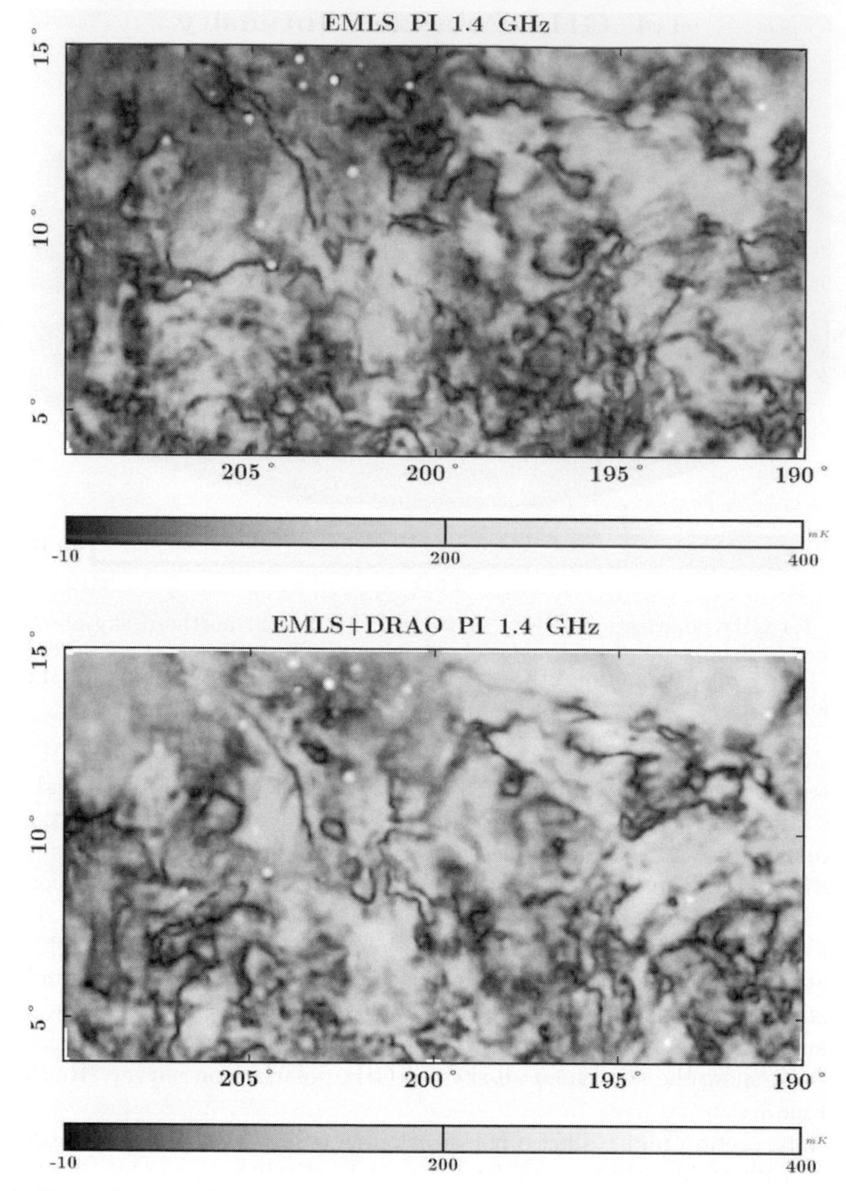

Figure 4. Example map showing polarized intensities from the 1.4 GHz "Effelsberg Medium Latitude Survey" (Uyanıker *et al.* (1999)) with and without absolute zero-level adjustment. The upper panel shows the original Effelsberg map of polarized intensities. The lower panel shows the Effelsberg map with added large-scale components in U and Q exceeding the map size.

PI and ϕ depend on U(Q) and $U_{zero}(Q_{zero})$ in a non-linear way. U(Q) and $U_{zero}(Q_{zero})$ may have rather different levels and signs, that the inclusion of $U_{zero}(Q_{zero})$ does not only change the level of PI, but also largely influences the morphology of the observed structures. This is demonstrated in Fig. 4, where 1.4 GHz polarized emission observed with the Effelsberg 100-m telescope is shown. The numerous small-scale polarization structures ('canals' and 'rings') have almost no counterpart in the very smooth total intensity map showing a smooth positive temperature gradient towards the Galactic plane and many compact extragalactic sources (see Uyanıker *et al.* 1999). Fig. 4 shows the polarized emission as it was observed and after a correction for missing large-scale components provided by the northern sky polarization survey by Wolleben *et al.* (2006). The polarization level increases including large-scale components, but most remarkable is the change of morphology of the small-scale structures, which may change from absorption into emission structures and vice versa. Without the large-scale correction any physical interpretation or radiation transfer modelling of the observed features is strongly limited.

While the WMAP total intensity surveys need a zero-level correction as described above, this does not hold for the polarization data, where the maps are at a correct zero-level. WMAP observes difference signals from two feeds with about 140° viewing angle difference (Bennett *et al.* 2003). Other than for Stokes I, the Stokes parameter U and Q have no isotropic component.

3.1. *Polarization surveys of the Galactic plane*

With the availability of the 1.4 GHz all-sky polarization survey, Galactic plane surveys at 1.4 GHz can be properly calibrated as demonstrated in Fig. 4 for the a section of the Effelsberg 1.4 GHz survey. The Effelsberg maps combined with the DRAO 26-m survey were then used to correct the 1.4 GHz polarized emission component of the CGPS, which results in a Galactic plane survey with 1' angular resolution showing unprecedented details of the magnetized interstellar medium (Landecker *et al.*, in prep.). This survey covers the Galactic plane in the longitude range $65° \leqslant l \leqslant 175°$ and latitudes between $-3.5° \leqslant b \leqslant 5.5°$. The filtering applied to the three surveys before merging the U and Q maps is shown in Fig. 5.

The WMAP polarization survey at 22.8 GHz (Page *et al.* 2007) proves to be very valuable for the correction of high-resolution polarization data at a few GHz. Faraday rotation of the polarization angles at high latitudes and towards the anti-center direction are small, that an extrapolation of the WMAP U and Q maps from 22.8 GHz towards lower frequencies is possible. This requires the correct spectral index for polarized emission, which should be close to that observed for total intensity synchrotron emission. For example a rotation measures (RM) of 50 rad m^{-2} causes a polarization angle rotation of about 10° at λ6 cm. This technique was applied to the running Sino-German λ6 cm (4.8 GHz) polarization survey of the Galactic plane, which is carried out with the Urumqi 25-m telescope of NAOC/China. As described by Sun *et al.* (2007) this survey aims to cover the Galactic plane for latitudes between $\pm 5°$ with an angular resolution of 9.5' and a rms-sensitivity of 1.4(0.7) mK $T_{\rm B}$ for total (polarized) intensities. Measurements of an absolute polarization level of a few milli-Kelvin is a very ambitious task for a multi-purpose instrument like the Urumqi 25-m telescope and should be done with dedicated small instruments. The C-BASS project (Pearson & C-BASS collaboration 2007) might be able to provide this required information.

Sun *et al.* (2007) published the first survey map from the Urumqi λ6 cm survey and demonstrated the application of the WMAP zero-level correction for an area centered at Galactic longitude 125.5°. The zero-level offset extrapolated from the smoothed U and Q 22.8 GHz maps was found to be –0.3 mK $T_{\rm B}$ and 8.3 mK $T_{\rm B}$ for U and Q, respectively. The large value for Q indicates that the magnetic field runs almost parallel to the Galactic plane, as expected. The observed maps show a lot of diffuse polarized structures partly associated with SNRs, but also in the direction of some HII-regions, which act as Faraday screens and prove the presence of well-ordered magnetic fields of a few μG along the line-of-sight. Of particular interest is a large polarized structure in the field, G125.6–1.8, with a diameter of about 70' showing no signature in total intensity nor in Hα-emission. This structure changes from a polarized emission structure as seen in the original Urumqi map

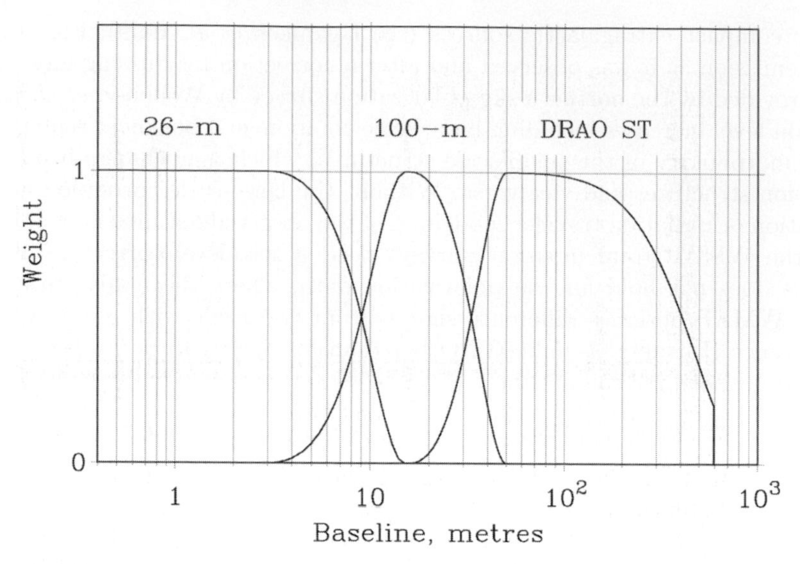

Figure 5. Weighting scheme of spatial structures applied to the three surveys as indicated (Landecker *et al.* in prep.). The combined maps in U and Q result in a Galactic plane survey, which is complete for all polarized structures larger than 1' (see text).

into an absorption structure after WMAP calibration. The properties of this feature were modeled by Sun *et al.* (2007). For a size of 58 pc and a RM of 200 rad m^{-2} the calculated magnetic field strength exceeds 6.9 μG along the line-of-sight for an upper limit of the thermal electron density of 0.84 cm^{-3} and assumed spherical symmetry. Such regular magnetic fields exceed the strength of the regular Galactic magnetic field, which is about 2μG in this area according to Han *et al.* (2006). The origin of these magnetic bubbles is not known, and it is also yet unclear how numerous they are and to what extent they influence the properties of the magnetized interstellar medium. Structures with such a high RM can not be studied at low frequencies, because small RM fluctuations cause depolarization across the observing beam. G125.6–1.8 is outstanding at λ6 cm, but remains undetected in the λ21 cm Galactic plane polarization survey by Landecker et al. (in prep.).

3.2. *Faraday screen observations*

Numerous Galactic polarized features were already discussed in the literature, but a number of them clearly suffer from insufficient calibration. As an example for a misinterpretation of that kind we discuss the case of the dust complex G159.5–18.5 in Perseus (Reich & Gao, in prep.). G159.5–18.5 is an extended source of about 1° in size and very well studied at optical and infrared wavelengths, star light polarization and molecular line emission (Ridge *et al.* 2006 and references therein).

Recently G159.5–18.5 received particular attention, because it is one of the very few well established cases, where emission from spinning dust particles was clearly detected. Watson *et al.* (2005) presented observations made with the COSMOSOMAS telescope in the frequency range between 11 GHz and 17 GHz, which were combined with WMAP data at higher frequencies, showing a spectrum as expected for spinning dust emission, which is clearly in excess over the weak thermal emission. Battistelli *et al.* (2006) presented 11 GHz polarization observations with the COSMOSOMAS telescope of a large area in Perseus and detected faint polarized emission of 3.4% +1.5%/−1.9% from G159.5–18.5, which they interpreted as polarized emission emerging from spinning dust grains, which

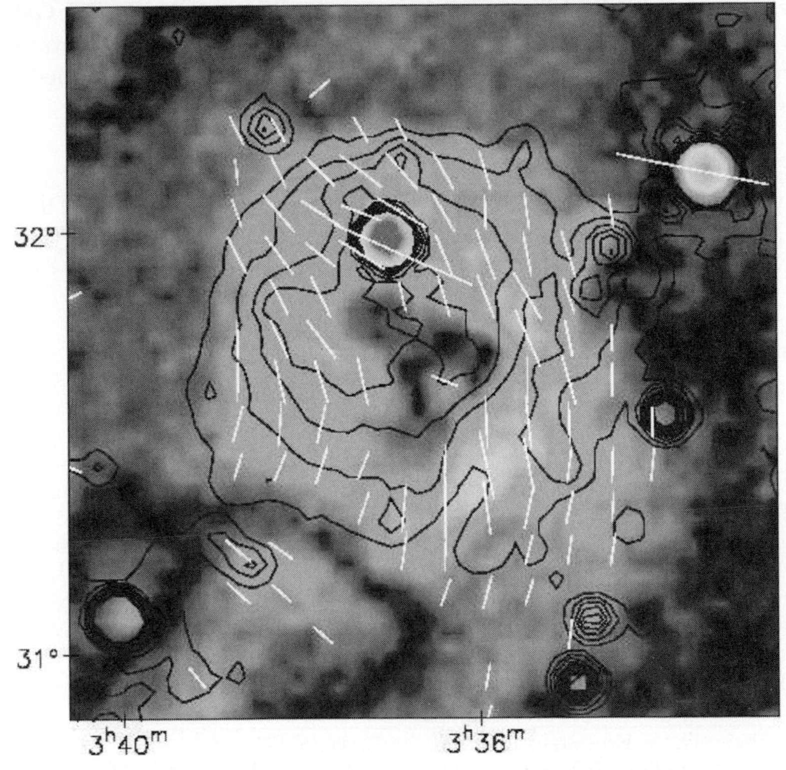

Figure 6. λ11 cm Effelsberg observations of the Perseus dust cloud G159.5–18.5 in total intensity (contours) and polarised intensity (greyscale and vectors). The linear polarization data indicate strong emission apparently originating from the thermal gas of G159.5–18.5. This is a clear indication that G159.5–18.5 acts as a Faraday screen rotating the polarization angle of background emission (see text).

would be the first detection of that kind. However, the polarization data are on a relative zero-level in U and Q.

Watson *et al.* (2005) extracted information of the thermal emission component of G159.5–18.5 from existing ground based all-sky surveys. Recent pointed observations with the Effelsberg 100-m telescope at λ11 cm and the Urumqi 25-m telescope at λ6 cm with higher quality clearly confirm the existence of optically thin thermal gas in G159.5–18.5 with a flux density of about 3.8 Jy, more than four times below the flux density of spinning dust seen at COSMOSOMAS wavelengths. Surprisingly strong polarized emission was observed at λ11 cm (Fig. 6) and λ6 cm, which is not expected to originate from thermal gas. Spinning dust plays no role at λ11 cm wavelength, that the only valid interpretation is that G159.5–18.5 acts as a Faraday screen hosting a strong regular magnetic field along the line-of-sight. The Faraday screen rotates the background polarized emission of G159.5–18.5, which then adds to foreground polarization in a different way than in the area outside of G159.5–18.5. In fact, correcting the Battistelli *et al.* (2006) polarization data of G159.5–18.5 for missing large-scale emission as observed by WMAP reduces the polarized emission signal below that of its surroundings.

The Faraday screen model as discussed by Reich & Gao (in prep.) fits all available polarization observations with a RM of the order of 190 rad m^{-2} implying a line-of-sight magnetic field of about 10 μG.

4. Conclusions

We have discussed calibration methods being applied to all-sky and Galactic plane continuum surveys in total intensity and linear polarization. The interpretation of polarized emission emerging from Faraday rotation effects in the interstellar medium requires special attention, in particular an absolute zero-level setting is essential.

Acknowledgements

We like to thank all our colleagues at MPIfR/Bonn, IAR/Argentina, DRAO/Canada and NAOC/China for their invaluable contributions to the various surveys projects discussed in this paper.

References

Battistelli, C. S., Rebolo, R., & Rubino-Martin, J. A. 2006, *ApJ* (Letters) 645, L141
Bennett, C. L., Bay, M., & Halpern, M. *et al.* 2003, *ApJ* 583, 1
Brouw, W. & Spoelstra, T. 1976, *A&AS* 26, 129
Dicke, R. H., Peebles, P. J. E., Roll, P. G., & Wilkinson, D. T. 1965, *ApJ* 142, 414
Han, J. L., Manchester, R. N., Lyne, A. G., Quiao, G. J., & van Straten, W. 2006, *ApJ* 642, 868
Haslam, C. G. T., Salter, C. J., Stoffel, H., & Wilson, W. E. 1982, *A&AS* 47, 1
Hinshaw, G., Nolta, M. R., & Bennett, C. L., *et al.* 2007, *ApJS* 170, 288
Hinshaw, G., Weiland, J. L., & Hill, R. S., *et al.* astro-ph 0803.0732
Kallas, E. & Reich, W. 1980, *A&AS* 42, 227
Kogut, A., Fixsen, D., & Fixen, S., *et al.* 2006, *New Astron. Revs.* 50, 925
Mather, J. C., Cheng, E. S., & Cottingham, D. A., *et al.* 1994, *ApJ* 420, 439
Page, L., Hinshaw, G., & Komatsu, E., *et al.* 2007, *ApJS* 170, 335
Pearson, T. J. & C-BASS collaboration 2007, *AAS-Meeting* 211.9003
Penzias, A. A. & Wilson, R. W. 1965, *ApJ* 142, 419
Reich, P. & Reich, W. 1986, *A&AS* 63, 205
Reich, P., Reich, W., & Fürst, E. 1997, *A&AS* 126, 413
Reich, P., Testori, J. C., & Reich, W. 2001, *A&A* 376, 861
Reich, P., Reich, W., & Testori, J. C. 2004 in B. Uyanıker, W. Reich & R. Wielebinski (eds.), *The Magnetized Interstellar Medium*, Katlenburg-Lindau: Copernicus GmbH, p. 63
Reich, W. 1982, *A&AS* 48, 219
Reich, W. 2006 in R. Fabbri (ed.), *Cosmic Polarization*, Kerala/India: Research Signpost, p. 91 (astro-ph 0603465)
Reich, W., Reich, P., & Fürst, E. 1990, *A&AS* 83, 539
Ridge, N. A., Schnee, S. L., Goodman, A. A., & Foster, J. B. 2006, *ApJ* 643, 932
Sun, X. H., Han, J. L., & Reich, W., *et al.* 2007, *A&A* 463, 993; erratum: 469, 1003
Sun, X. H., Reich, W., Waelkens, A., & Enßlin, T. A. 2008, *A&A* 477, 573
Taylor, A. R., Gibson, S. J., & Peracaula, M., *et al.* 2003, *AJ* 125, 3145
Testori, J. C., Reich, P., & Reich, W. 2008, *A&A* 484, 783
Turtle, A. Y., Pugh, G. F., Kenderdine, S., & Pauliny-Toth, I. I. K. 1962, *MNRAS* 124, 297
Uyanıker, B., Fürst, E., Reich, W., Reich, P., & Wielebinski, R. 1999, *A&AS* 138, 31
Watson, R. A., Rebolo, R., & Rubino-Martin, J. A., *et al.* 2005, *ApJ* (Letters) 624, L89
Wolleben, M., Landecker, T. L., Reich, W., & Wielebinski, R. 2006, *A&A* 448, 441

Cosmic Magnetic Fields: From Planets, to Stars and Galaxies
Proceedings IAU Symposium No. 259, 2008
K.G. Strassmeier, A.G. Kosovichev & J.E. Beckman, eds.

doi:10.1017/S1743921309031445

Imaging polarimetry as a diagnostic tool

Tim M. Gledhill

School of Physics, Astronomy & Mathematics, University of Hertfordshire, College Lane, Hatfield, AL10 9AB, UK
email: t.gledhill@herts.ac.uk

Abstract. Some of the earliest polarimetric measurements made in astronomy were concerned with the polarization of the interstellar medium resulting from dust grains aligned in the Galactic magnetic field. More than 50 years later, polarimetry continues to be an important diagnostic of field structure on size scales ranging from planetary to galactic. The use of both linear and circular polarimetry at optical and infrared wavelengths can provide additional insights into the nature of dust particles, their alignment in magnetic fields and the field topology. Given the science benefits that polarimetry offers it is perhaps surprising that the continued existence of polarimetric facilities on current and next generation large telescopes needs to be ensured.

Keywords. Magnetic fields – polarization – scattering – instrumentation: polarimeters

1. Introduction

Polarization is a natural tracer of asymmetry, and the interaction between light and matter in many astrophysical environments is inherently asymmetric. Often, the presence of a magnetic field imparts the necessary asymmetry, an example being electrons spiralling around field lines and emitting polarized synchrotron radiation. Even where the emission is unpolarized at source, subsequent interactions often produce polarization before the light reaches Earth; scattering of starlight by dust grains in circumstellar environments can produce high degrees of linear (and sometimes circular) polarization at optical and infrared wavelengths; extinction by elongated and aligned dust grains in the ISM can induce polarization up to a few per cent at optical wavelengths; thermal emission from dust in star-forming regions results in polarization from the mid-infrared through to the sub-millimetre; radio emission from extragalactic sources is polarized due to Faraday rotation in our Galaxy. Even the CMB radiation pervading the entire Universe is polarized. Indeed, it is hard to think of an astronomical source that is not polarized to some degree at some wavelength. It may seem surprising then that this polarimetric information is often discarded and that many instrument and telescope systems which are insensitive to polarization continue to be built.

This review will concentrate on polarization produced by dust, principally at optical and infrared wavelengths; polarization due to the Zeeman effect, Faraday rotation and synchrotron emission are reviewed elsewhere in these proceedings. The technique of imaging polarimetry in particular will be described, along with its history and applications. For typical astrophysical dust grain sizes ($\sim 10^{-1}$ μm), at wavelengths less than 5 μm, polarization due to scattering usually dominates, especially in optically thin environments. However, if the optical depth is significant, then dichroic polarization can be important if the grains are aligned. At wavelengths greater than 5 μm, where the scattering cross-section for sub-micron grains becomes negligible, any polarization must come from dichroic absorption and emission of radiation from aligned grains. Emissive polarization is particularly important in the mid- and far-IR.

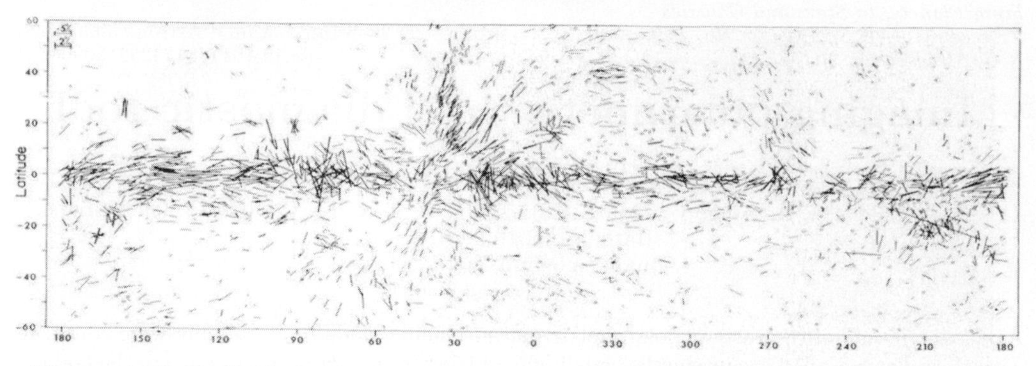

Figure 1. Optical polarization vectors for $\sim$ 7000 stars, indicating the projected Galactic magnetic field orientation. From Mathewson & Ford (1970).

Polarimetry provides an important tool for investigating the structure of magnetic fields. Aligned dust grains seem to be present in many astrophysical environments. Although the details of the alignment process may not be known exactly, grains with rotational frequencies of 10^5-10^6 Hz develop a magnetic moment due to the Barnett effect (Aitken *et al.* 2004; Hough & Aitken 2003) so that their spin axes precess around the B-field giving a net polarization. Where grains do have a preferential alignment over the resolution element of the observation and where that alignment is reasonably coherent along a line-of-sight, then a net linear polarization (LP) can be observed due to dichroic extinction or emission, depending on the wavelength/grain-size and optical depth. If the alignment orientation changes along a line-of-sight, due to structure in the magnetic field for example, then LP can be converted to circular polarization (CP), and vice versa due to the circular birefringence of the medium. Scattering from aligned grains can produce degrees of CP in the region of tens of per cent, much larger than those produced by birefringence. In addition to being an excellent diagnostic of grain alignment and magnetic fields, polarimetry can also be used to place limits on grain properties, such as the size distribution, shapes, compositions and structures. This is especially the case where observations are obtained at several wavelengths and information on the chemical composition of the dust is available, for example from spectroscopy of its emission features or from a general knowledge of the environmental chemistry.

2. Early polarimetric work

The link between polarimetry and magnetic fields is illustrated by the fact that some of the earliest astronomical polarimetric measurements of stars were used to infer the large-scale structure of the Galactic magnetic field. These observations used photoelectric photometers with polaroid sheets as analysers. Typically the instrument as a whole was rotated through various angles to recover the polarimetric information (Hiltner 1949; Hall & Mikesell 1949). Hiltner (1949) observed stars in the Perseus cluster, noted the tendency for the polarization E-vector to lie parallel to the Galactic plane and speculated that this could be related to a large scale Galactic magnetic field, which would require the dust grains to be both elongated and aligned. Hiltner (1951) measured the polarization of 841 stars and arrived at the same conclusions. This was all put onto a firmer theoretical footing by the publication in 1951 of Davis & Greenstein's famous paper explaining how spinning dust grains could align in the presence of a magnetic field due to the torque induced by paramagnetic relaxation (Davis & Greenstein 1951).

The first optical polarization survey was published by Mathewson & Ford (1970) and combined a southern hemisphere survey of 1800 stars observed at Siding Springs, with

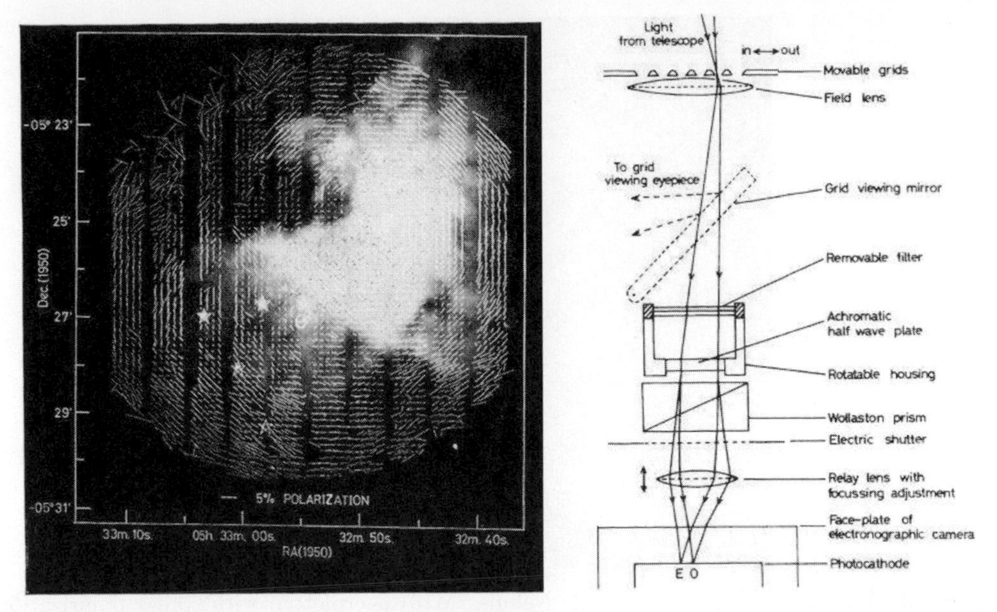

Figure 2. a) Imaging polarimetry of the Orion Nebula region by Pallister *et al.* (1977) showing an 8 arcmin square region. The scattering pattern indicates that much of the region is illuminated by sources in the central Trapezium Cluster, principally Θ_1 and Θ_2 Orionis. b) A schematic illustration of a two-channel polarimeter design, from Scarrott *et al.* (1983)

the northern hemisphere observations of Hiltner, Hall and others. This resulted in maps of the projected Galactic B-field (Fig. 1) at various distance intervals, showing agreement with radio observations and providing evidence for kpc-scale fields.

3. Imaging polarimetry

Before the mid-1970s, polarimeters were effectively single-pixel devices, summing intensities over an aperture. Imaging of a sort could be carried out by repeatedly stepping the aperture on the sky (e.g. Fig. 1) but this is time consuming and limited by the accuracy with which apertures can be reliably positioned. Bingham *et al.* (1976) published one of the first imaging polarimetric observations, of the galaxy M82, shortly followed by observations of M104 (Scarrott *et al.* 1977) and M42 (Pallister *et al.* 1977), shown in Fig. 2a. These optical wavelength results used a two-channel imaging polarimeter incorporating a rotating half-waveplate to rotate the plane of polarization of the incident light, and a Wollaston prism to separate orthogonal polarisation states (Fig. 2b). This system, which has now become a standard configuration, had already been used by Öhman (1939) to measure the polarization of the Moon and comets, and originally dates back to Pickering (1886) who used his 'polarigraph' to measure atmospheric polarization.

Apart from efficiency considerations, an important advantage of imaging polarimetry is that it allows polarization to be measured on a spatial scale determined by the seeing-limited resolution of the telescope. This is necessary in order to recover complex structure, for example in scattering patterns around protostars or in the detailed structure of a magnetic field. As polarization is a vector quantity, poor spatial resolution may result in a decreased polarimetric signal; an extreme example would be the summation of a centro-symmetric polarization pattern to a net polarization of zero in an aperture centred on the source.

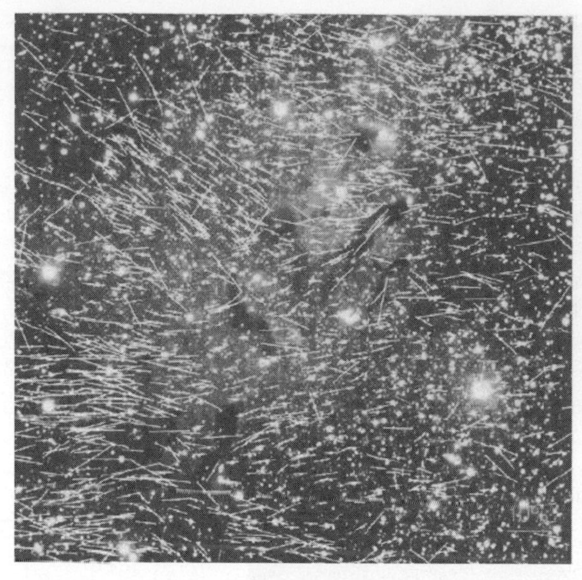

Figure 3. A H-band image of the Eagle Nebula (M16) overplotted with stellar polarizations indicating the presence of dust grains aligned in an ordered magnetic field. From Sugitani *et al.* (2008).

CCD detectors began to be used regularly for optical polarimetry from the early 1980s (e.g. Scarrott *et al.* 1983) and advances in infrared detector technology also allowed imaging polarimetry up to 5 um using InSb arrays. The first common-user infrared camera was IRCAM, commissioned on the UK Infrared Telescope in 1986 (McClean *et al.* 1986), which was later fitted with a Wollaston prism and waveplate system in 1995 to allow imaging and spectropolarimetry between 1 and 5 μm. Although perhaps of less interest to European astronomers, a near-infrared polarimetry module is also planned for the 6.5-m MMT telescope in Arizona (Packham & Jones 2008). MMT-POL will use a Wollaston prism and take advantage of the MMT's adaptive secondary, providing AO correction with no off-axis reflections, to perform high-precision imaging polarimetry between 1 and 5 μm. Despite the relative maturity of the technique, imaging polarimetry continues to be under-used, which must be at least in part due to the scarcity of common-user facilities, especially on the 8-m class telescopes.

4. Exploring magnetic fields with imaging linear polarimetry

There are many applications in which polarimetry provides a novel approach to a particular observational problem. Examples are its use as a differential imaging technique for the detection of faint objects around bright stars (particularly circumstellar discs, debris discs and, ultimately, exoplanets), and its use to probe hidden sources, such as embedded protostars or the nuclei of AGN. In this section, we briefly highlight some examples of the use of linear polarimetry in the detection and investigation of magnetic fields.

4.1. *Polarimetric surveys*

For a number of years, large format CCDs have been used at optical and near-infrared wavelengths for major imaging surveys. Ongoing surveys include the UKIRT Infrared Deep Sky Survey (UKIDSS; www.ukidss.org) in the J, H and K bands, and in 2009 the dedicated survey telescopes VISTA (www.vista.ac.uk) and VST (www.eso.org/sci/) are due to begin operation. However these facilities do not include polarimetry. A linear and circular polarimetry survey of the southern sky is being undertaken using the Japanese Infrared Survey Facility (IRSF)†, a 1.4-m telescope, located at the SAAO in South Africa. The sky can look very different in polarized light, revealing a wealth of information not visible in direct images, such as diffuse reflection nebulae, circumstellar discs and, by their lack of polarization, regions dominated by intrinsic emission lines. In addition, the large number of point sources available in a wide-field image can be investigated using simulated aperture polarimetry to look for unresolved circumstellar discs or to study the foreground polarization due to dichroic absorption by dust grains aligned in a magnetic field. The SIRIUS camera attached to the IRSF telescope has three HgCdTe Hawaii arrays allowing simultaneous J-, H- and K-band imaging using dichroics, and is equipped with a single channel (wire-grid) polarimeter (SIRPOL; Kandori *et al.* 2006). The image shown in Fig. 3 is from Sugitani *et al.* (2008) and shows a H-band polarization map of point sources in an 8×8 arcmin region of the Eagle Nebula (M16). M16 is a star-forming region, with numerous very young objects, and is famous for the gas and dust pillars seen with HST. The polarization vectors appear to be well aligned, suggesting an ordered large-scale magnetic field, with the dominant field direction at position angle 80-90 deg.

4.2. *Mid-infrared imaging polarimetry*

The arrival of mid-infrared arrays made imaging polarimetry possible from the ground in the 10 and 20 μm (N and Q) bands, allowing polarized absorption and thermal emission from magnetically aligned dust grains to be detected. A dedicated N-band mid-infrared polarimeter, NIMPOL, is described by Smith *et al.* (1997) and is based on a single-channel design with a rotating CdS half-waveplate and a fixed cold wire-grid analyser. The instrument has been used to map magnetic field structure at arcsecond resolution in the Galactic Centre (Aitken *et al.* 1998), in the Orion star-forming region (Aitken *et al.* 1997), and in individual objects such as the massive evolved star η Carinae (Aitken *et al.* 1995). See Hough & Aitken (2003) for a review of infrared polarimetry.

Using NIMPOL on the UK Infrared Telescope, Aitken *et al.* (1998) mapped the magnetic field structure in the SgrA* region of the Galactic centre at 12.5 μm. They noted that the field direction followed the northern arm and appeared unperturbed even in the region of embedded OB stars and clusters, suggesting a field strength of at least 2 mG in the vicinity of IRS1. The polarimetry was used to construct a 3-D representation of the field topology.

Common-user facilities capable of imaging polarimetry in the mid-infrared are rare, one example being Michelle, on Gemini-North. There is currently no equivalent polarimetric facility on VLT although CanariCam should be operational on the 10.4-m GranTeCan telescope on La Palma in 2009. CanariCam will be unique in being the first dual-channel mid-infrared polarimeter, incorporating a large CdSe Wollaston prism, offering much-improved efficiency over previous single-channel designs (Packham, Hough & Telesco 2005). At wavelengths longer than 20 μm, the use of polarizing prisms becomes problematic. A possible novel solution to this issue is the use of a polarization grating as a two-channel analyser, a concept that is being explored for a proposed mid-infrared polarimeter for SOFIA (Packham *et al.* 2008). This instrument would function between 5 and 40 μm. The ability of SOFIA to extend the wavelength coverage beyond that available from the ground is particularly important when we realise than none of the current

† http://www.z.phys.nagoya-u.ac.jp/~irsf/index_e.html

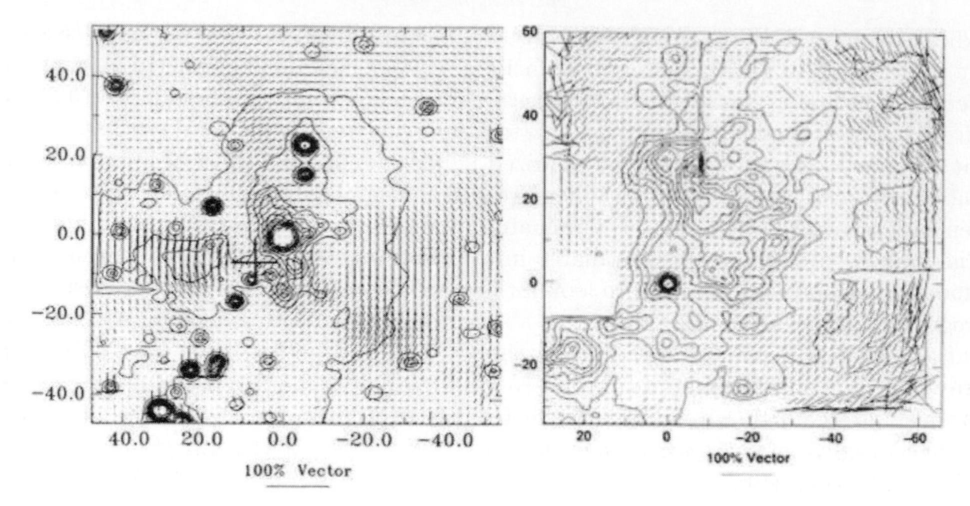

Figure 4. *Left:* A K-band image and polarization map from Minchin *et al.* (1991) showing continuum emission from IRc2, scattered in the surrounding dust cloud. *Right:* Imaging polarimetry in the narrow-band light of H_2 shows the signature of the magnetic field threading the dust sheet in front of the IRc2 outflow. From Chrysostomou *et al.* (1994).

and planned infrared space missions (e.g. Spitzer, JWST, Herschel) will have polarimetric options.

4.3. H_2 *imaging polarimetry*

A standard method for investigating the polarizing properties of a dust cloud is to image the polarization of background light that has passed through the cloud. Where a diffuse unpolarized background source is available, then imaging polarimetry allows the magnetic field in the cloud to be mapped. The main body of the OMC-1 molecular cloud in Orion is illuminated by IRc2, a higly embedded protostar, which appears to be powering a bipolar outflow. CO mapping shows the blue and red-shifted components of the high-velocity outflow, which is interacting with the surrounding molecular material creating shocks. The shocked H_2 1-0 S(1) emission is unpolarized and can be used to probe the structure of the overlying magnetic field (Chrysostomou *et al.* 1994). Broad-band imaging polarimetry in the near-IR continuum (e.g. K-band, see Fig. 4 *left*) shows a reflection nebula (resulting from scattered light) illuminated principally by IRc2 (Minchin *et al.* 1991). However, narrow band Fabry-Perot imaging polarimetry of the H_2 line shows polarization consistent with dichroic extinction in a foreground medium of magnetically aligned grains and so traces out the morphology of the field (Fig. 4 *right*). The field is roughly aligned with the outflow axis, although there is evidence for a twist in the region of IRc2 (also see Aitken, Hough & Chrysostomou 2006).

5. Imaging circular polarimetry

Polarization, especially determined as a function of wavelength, can be used to derive properties of the dust particles such as their composition, size, shape and degree of alignment. This is especially the case if the CP is measured as well as the LP. However, circular polarimetry is a less commonly used technique than linear polarimetry, and again this is to some extent dictated by the available facilities. In the optical, CP can be measured using ISIS on WHT and FORS on VLT. In the infrared, the UIST imager/spectrometer on UKIRT can be used.

In general, in the presence of magnetic fields, and hence aligned grains, we need to consider a full Stokes radiation transport solution including differential extinction (dichroism) and birefringence as well as scattering. Although multiple scattering from non-aligned grains (i.e. spherical or randomly oriented) can produce CP, it is inefficient and results usually in less than 1 per cent polarization. These low polarizations have been observed in a number of pre-main sequence objects in the near-infrared, such as the Chamaeleon Infrared Nebula (Gledhill, Chrysostomou & Hough 1996) and GSS30 (Chrysostomou *et al.* 1997). These objects are often linearly polarized at several tens of per cent, so care must be taken to ensure that CP is not erroneously measured, due to 'cross-talk' (conversion of LP to CP) in the polarimeter. This can occur if the retardance of the quarter-waveplate is not exactly quarter-wave for the input wavelength, or if the fast axis of the quarter-waveplate is not aligned ± 45 degrees with the analyser axis (e.g. Hough & Aitken 2002). In the ISIS and UKIRT circular polarimeters, this is achieved by continuously rotating a half-waveplate up-stream of the polarimeter, to average the incident LP to close to zero.

Large degrees of both LP and CP can be produced in the presence of aligned grains. Consider the study of Whitney & Wolff (2002) where they take a simple spherical nebula with a uniform density of oblate spheroids with axis ratios of 2:1. The spheroidal grains are aligned with their symmetry axes along the vertical axis of the nebula, corresponding to a magnetic field along this same axis. They find that these aligned grains can produce $\pm 25 - 40$ per cent CP, whereas non-aligned grains would produce a maximum CP of less that 1 per cent. This suggests that, in environments where light propagates through regions of aligned dust grains, especially along optically thick paths, then CP produced by scattering and dichroism can be an important diagnostic of the field.

5.1. *Circular polarization from aligned grains*

Degrees of CP of up to 17 per cent in the K-band were reported in the OMC-1 region in Orion by Chrysostomou *et al.* (2000), and to date this is the highest degree of CP seen in diffuse nebulosity. Also see Buschermohle *et al.* (2005) for a wide-field CP survey of the OMC-1 region. Chrysostomou *et al.* (2000) found that high CP occurred in regions where high LP was also seen (although the reverse was not the case) and speculated that the CP resulted from the scattering of infrared light, originating from IRc2, by dust grains aligned in a structured magnetic field within the OMC-1 cloud. A simple model was proposed to show that the required degrees of CP could indeed be produced with the proposed field geometry. More detailed modelling by Lucas *et al.* (2005) was able to reproduce both the CP and LP in the region, as well as placing constraints on the field configuration and grain axis ratio, concluding that the CP results primarily from dichroic extinction.

5.2. *Circular polarization and helical field structure*

Magnetic fields have been thought for many years to play a crucial role in regulating accretion onto protostars, both in powering and shaping outflows and removing angular momentum from disc material, to allow the protostar to gain mass. Getting evidence for the morphology of these fields has been tricky though – and this is an area in which polarimetry can help. In particular, CP can provide evidence for changing grain/field alignment directions along the line-of-sight and hence the presence of twisting fields.

The HH135/136 outflow is associated with a young stellar object (YSO), thought to be an intermediate mass Herbig Ae/Be star, in the Carina nebula at a distance of approx. 2.7 kpc. Fig. 4 shows total intensity (I) and fractional circular polarization (V/I) in the H- and K-bands (Chrysostomou, Lucas & Hough 2007). The peak polarization in the top

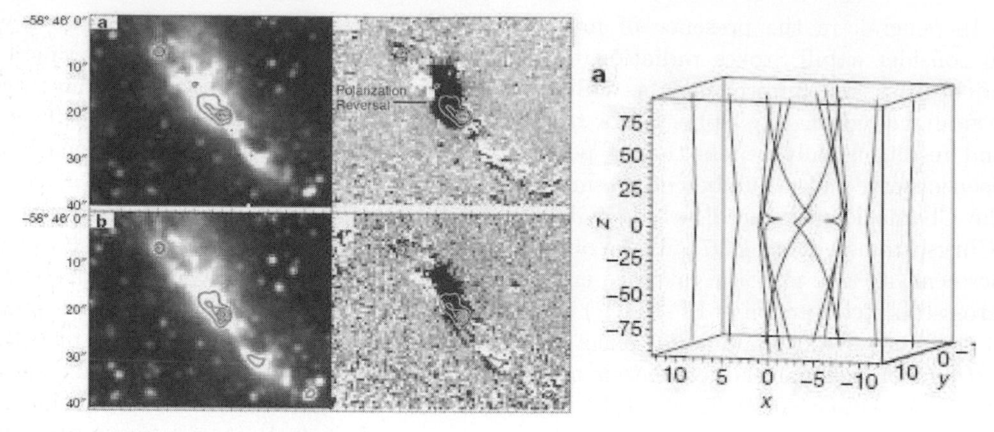

Figure 5. *Left:* K-band (upper) and H-band (lower) images of the HH135/136 outflow, with intensity (I) on the left and fractional CP (V/I) on the right. The CP peaks at −8 per cent (black) in the K-band and −3 per cent (black) in the H-band. *Right:* Helical field models used to reproduce the observed CP data. From Chrysostomou, Lucas & Hough (2007).

K-band image is −8 percent (black) and −3 per cent in the lower H-band image. This is intermediate between the low (less than 1 percent) CP seen in a number of low mass YSO outflows (e.g. ChaIRN, see above), which can be produced by multiple scattering from non-aligned particles, and the higher polarizations of 17 per cent seen in the high-mass star-forming region of Orion, which have been attributed to a combination of scattering and dichroic extinction by aligned grains. This makes sense if we assume that stronger magnetic fields (and hence more efficient grain alignment) tend to be associated with higher-mass star formation.

In HH135/136, the north-east lobe is mostly negatively polarized and the south-west lobe mostly positively polarized, and the sign of CP flips between quadrants. However, the sign flipping occurs on the limb, rather than on the axis. The latter is expected for an axial aligning field (as shown in the models of Whitney & Wolff 2002). The degree and pattern of CP and also LP, which is closely centrosymmetric, have been modelled using a pinched and twisted field morphology as shown on the right in Fig. 4 (Chrysostomou, Lucas & Hough 2007; also see Lucas *et al.* (2005)). The majority of CP produced in this model comes from dichroic extinction.

6. Summary

Astronomical polarimetry has a long history and has contributed enormously to our understanding of many astrophysical processes, particularly where a magnetic field introduces an asymmetry, such as the alignment of elongated dust particles. Once a niche area undertaken only by aficionados with their own private instruments, polarimetry is now a common-user activity. This has resulted not just from an appreciation of the power of the technique, but also from the availability of reliable instrumentation and pipelining software that removes the need for detailed specialist understanding of the technical details. Imaging polarimetry in particular has come of age and should benefit enormously from the increased spatial resolution and light-gathering power of the next generation of large telescopes. However, the continued availability of polarimetric facilities on future large telescopes is in some doubt. The standard location for a polarimeter has always been the Cassegrain focus, where instrumental polarization can be minimised by avoiding off-axis reflections (see Hough 2007). Cassegrain focci are now less common on 8-m and larger telescopes and future instruments are likely to be so large that they must necessarily be located at a Nasmyth focal platform or at a gravitationally invariant focus. This immediately means that a significant, and often time-variable, systemic polarization is introduced by the off-axis reflection at the M3 mirror. At the VLT, for example, the infrared instrument NACO provides AO-corrected imaging polarimetry in the H- and K-bands, but the modulator and analyser are located within the instrument, after the M3 reflection. This results in 'instrumental' polarization of several per cent, which is at the level of the intrinsic polarization of many astronomical sources, so that accurate calibration is required if a meaningful detection can be realised. Detection of polarization much below 1 per cent is difficult with such a system.

Some progress can be made by the use of compensating mirrors and optics, which act to cancel the telescope polarization (e.g. see Tinbergen 2007). The next generation of larger telescopes are likely to be even less polarimeter-friendly. The current concept for the E-ELT, for example, involves 5 mirrors, three of which are off-axis. In addition, the adaptive nature of the mirrors, required to achieve the best image quality possible with a 42-m primary, means that the induced polarization of the telescope system is likely to be very difficult to characterise and compensate for.

In the case of space-based observatories, there is an obvious imperative to keep mechanical systems as simple and trouble-free as possible; repairing a malfunction due to a stuck modulator may be impossible and may jeopardise the whole instrument. However, both ISO and HST implemented polarimetric facilities. In particular HST has seen impressive results with many programs benefiting from polarimetry in the optical and near-infrared. Amongst the recent and near-future infrared space missions though, there is no polarimetry. With a properly designed polarimeter, then polarimetry can be achieved effectively from the ground, as long as the desired wavelength is in a region of atmospheric transmission. This makes it all the more important to include polarimetric facilities on future ground-based telescopes.

Acknowledgements

Jim Hough is thanked for comments on this paper.

References

Aitken, D. K., Smith, C. H., Moore, T. J. T, & Roche, P. F., 1995 *MNRAS* 273, 359

Aitken, D. K., Smith, C. H., Moore, T. J. T., Roche, P. F., Fujiyoshi, T., & Wright C. M. 1997, *MNRAS* 286, 85

Aitken, D. K., Smith, C. H., Moore, T. J. T, & Roche, P. F., 1998, *MNRAS* 299, 743

Aitken, D. K., Hough, J. H., Roche, P. F., Smith, C. H., & Wright, C. M. 2004, *MNRAS* 348, 279

Aitken, D. K., Hough, J. H., & Chrysostomou, A. 2006, *MNRAS* 366, 491

Bingham, R. G., McMullan, D., Pallister, W. S., White, C., Axon, D. J., & Scarrott, S. M. 1976, *Nature* 259, 463

Buschermohle, M., Whittet, D. C. B., Chrysostomou, A., Hough, J. H., Lucas, P. W., Adamson, A. J., Whitney, B. A., & Wolff, M. J. 2005, *ApJ* 624, 821

Chrysostomou, A., Hough, J. H., Burton, M. G., & Tamura, M. 1994, *MNRAS* 268, 325

Chrysostomou A., Menard, F., Gledhill, T. M., Clark, S., Hough, J. H., McCall, A., & Tamura, M. 1997, *MNRAS* 285, 750

Chrysostomou, A., Gledhill, T. M., Menard, F., Hough, J. H., Tamura M., & Bailey, J. A. 2000, *MNRAS* 312, 103

Chrysostomou, A., Lucas, P. W., & Hough, J. H. 2007, *Nature* 450, 71

Davis, L. Jr. & Greenstein, J. L. 1951, *ApJ* 114, 206

Gledhill, T. M., Chrysostomou, A., & Hough, J. H. 1996, *MNRAS* 282, 1418

Hall, J. S. & Mikesell, A. H. 1949, *AJ* 54, 187
Hiltner, W. A. 1949, *ApJ* 109, 471
Hiltner, W. A. 1951, *ApJ* 114, 241
Hough, J. H. & Aitken, D. K. 2002, *Proc. SPIE* 4843, 200
Hough, J. H. & Aitken, D. K. 2003, *Journal of Quantitative Spectroscopy & Radiative Transfer* 79-80, 733
Hough, J. H. 2007, *Journal of Quantitative Spectroscopy & Radiative Transfer* 106, 122
Kandori, R. *et al.* 2006, *Proc. SPIE* 6269, 159
Lucas, P. W., Hough, J. H., Bailey, J. A., Chrysostomou, A., Gledhill, T. M., & McCall, A. 2005, *Origins of Life and Evolution of Biospheres*, 35, 29
Mathewson, D. S. & Ford, V. L. 1970, *MemRAS* 74, 139
McClean, I. S., Chuter, T. C., McCaughrean, M. J., & Rayner, J. T. 1986, *Proc. SPIE* 637, 430
Minchin, N. R. *et al.* 1991, *MNRAS* 248, 715
Öhman, Y. 1939, *MNRAS* 99, 624
Packham, C., Hough, J. H., & Telesco, C. M. 2005, in: A. Adamson, C. Aspin, C. J. Davis & T. Fujiyoshi (eds.), *Astronomical Polarimetry: Current Status and Future Directions*, ASP Conference Series, Vol. 343, p. 38
Packham C. & Jones, T. J. 2008, *Proc. SPIE* 7014, 193
Packham, C. *et al.* 2008, *Proc. SPIE* 7014, 86
Pallister, W. S., Perkins, H. G., Scarrott, S. M., Bingham, R. G., & Pilkington J. D. H. 1977, *MNRAS* 178, 93p
Pickering, E. C. 1889, American Academy of Arts and Sciences, Proceedings, New Series 13, p.294
Scarrott, S. M., White, C., Pallister, W. S., & Solinger, A. B. 1977, *Nature* 265, 32
Scarrott, S. M., Warren-Smith, R. F., Pallister, W. S., Axon, D. J., & Bingham, R. G. 1983, *MNRAS* 204, 1163
Smith, C. H., Moore, T. J. T., Aitken, D. K., & Fujiyoshi, T. 1997, *PASA* 14, 179
Sugitani, K. *et al.* 2008, *PASJ* 59, 507
Tinbergen, J. 2007, *PASP* 119, 1371

Discussion

KOUTCHMY: Could you comment on the observations of neutral points (singularities on the polarization map where the polarization reaches a zero value like it is in the case of the Crab Nebula) with patterns like 'focus', 'saddle' etc

GLEDHILL: Neutral points and other effects often result due to an averaging effect when a lack of spatial resolution, for example due to seeing, causes the Stokes intensities to sum close to zero. Alternatively a neutral point could result if the aligning field is tangled on the scale of the measurement.

ZINNECKER: This was a wonderful summary of polarimetric techniques and results. I hope the forthcoming extremely large telescopes (ELTs) will be equipped with polarimetric capabilities, given the strong community interest in magnetic field science. I think ESO is considering polarimetric instrumentation for the E-ELT, but what about the Americans for their TMT?

GLEDHILL: Indeed, I hope polarimetry goes ahead on E-ELT, but it will be challenging given the number of off-axis reflections before the first usable focus in the planned design. I haven't heard of any polarimetry plans for TMT

Cosmic Magnetic Fields: From Planets, to Stars and Galaxies
Proceedings IAU Symposium No. 259, 2008
K.G. Strassmeier, A.G. Kosovichev & J. Beckman, eds.

doi:10.1017/S1743921309031457

Diagnostic methods based on scattering polarization and the joint action of the Hanle and Zeeman effects

Javier Trujillo Bueno[1,2]

[1]Instituto de Astrofísica de Canarias, 38205 La Laguna, Tenerife, Spain
[2]Consejo Superior de Investigaciones Científicas (Spain)
email: jtb@iac.es

Abstract. Polarized light provides the most reliable source of information at our disposal for diagnosing the physical properties of astrophysical plasmas, including the magnetic fields of the solar atmosphere. The interaction between radiation and hydrogen plus free electrons through Rayleigh and Thomson scattering gives rise to the polarization of the stellar continuous spectrum, which is very sensitive to the medium's thermal and density structure. Anisotropic radiative pumping processes induce population imbalances and quantum coherences among the sublevels of degenerate energy levels (that is, atomic level polarization), which produce polarization in spectral lines without the need of a magnetic field. The Hanle effect caused by the presence of relatively weak magnetic fields modifies the atomic polarization of the upper and lower levels of the spectral lines under consideration, allowing us to detect magnetic fields to which the Zeeman effect is blind. After discussing the physical origin of the polarized radiation in stellar atmospheres, this paper highlights some recent developments in polarized radiation diagnostic methods and a few examples of their application in solar physics.

Keywords. Polarization – scattering – Sun: magnetic fields – stars: magnetic fields

1. Polarization of the stellar continuum radiation

In the atmospheres of the stars, the dominant contribution to the linear polarization of the visible continuous spectrum comes from scattering at neutral hydrogen in its ground state (Lyman scattering) and Thomson scattering at free electrons (Chandrasekhar 1960). The contribution of these processes to the total absorption coefficient (χ_c) is quantified by $\sigma_c = \sigma_T N_e + \sigma_R n_1(H)$, where $\sigma_T = 6.653\times10^{-25}$ cm^2 is the Thomson scattering cross section, N_e the electron number density, $n_1(H)$ the population of the ground level of hydrogen, and σ_R the wavelength-dependent Rayleigh cross section. The total absorption coefficient is given by $\chi_c = \kappa_c + \sigma_c$, where κ_c contains all the relevant non-scattering contributions to the continuum absorption coefficient (e.g., the bound-free transitions in the H$^-$ ion). Since at visible wavelengths the continuum absorption coefficient does not depend on the polarization of the incident radiation, the transfer equation for the Stokes parameter X (with $X = I, Q, U$) at a given frequency ν and direction of propagation $\vec{\Omega}$ is given by $\frac{d}{d\tau}X = X - S_X$, where τ (with $d\tau = -\chi_c\, ds$) is the monochromatic optical distance along the ray and S_X (with $X = I, Q, U$) are the source function components (see Trujillo Bueno & Shchukina 2009). For example,

$$S_Q = \frac{\sigma_c}{\kappa_c+\sigma_c}\Big\{\frac{3}{2\sqrt{2}}(\mu^2-1)J^2_0 - \sqrt{3}\mu\sqrt{1-\mu^2}(\cos\chi\tilde{J}^2_1 + \sin\chi\hat{J}^2_1) \\ -\frac{\sqrt{3}}{2}(1+\mu^2)(\cos 2\chi\,\tilde{J}^2_2 + \sin 2\chi\,\hat{J}^2_2)\Big\}, \tag{1.1}$$

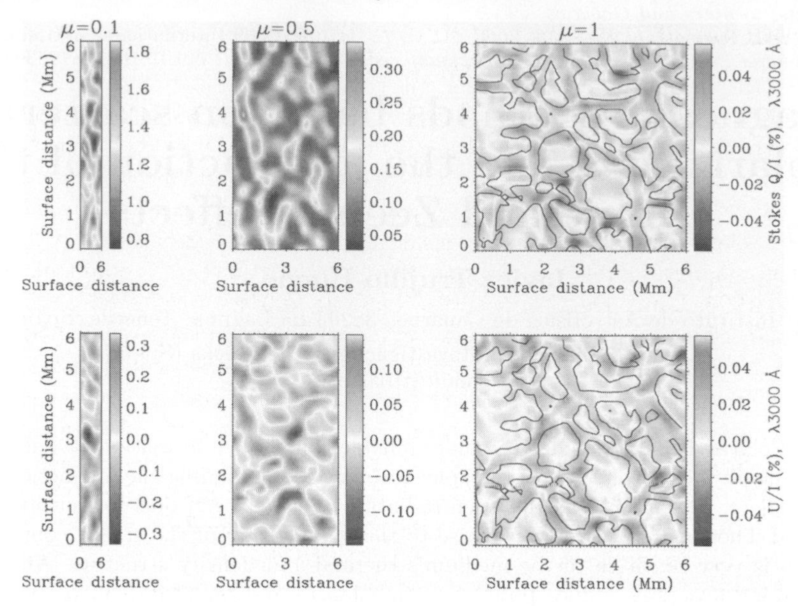

Figure 1. The emergent Q/I (top panels) and U/I (bottom panels) at 3000 Å calculated for three line of sights in a realistic 3D hydrodynamical model of the solar photosphere and accounting for the diffraction limit effect of a 1-m telescope. The reference direction for Stokes Q lies along the vertical direction of the corresponding panel. Note that at 3000 Å the Q/I and U/I images show sizable values (e.g., at $\mu = 0.1$ Q/I varies between 0.8% and 1.8%, while U/I fluctuates between ±0.3%). For more information see Trujillo Bueno & Shchukina (2009).

and

$$S_U = \frac{\sigma_c}{\kappa_c + \sigma_c}\sqrt{3}\left\{\sqrt{1-\mu^2}(\sin\chi \tilde{J}^2_1 - \cos\chi \hat{J}^2_1) + \mu(\sin 2\chi\, \tilde{J}^2_2 - \cos 2\chi\, \hat{J}^2_2)\right\}, \qquad (1.2)$$

where the orientation of the ray is specified by $\mu = \cos\theta$ (with θ the polar angle) and by the azimuthal angle χ. In these source function expressions the J^K_P quantities (with $K = 0, 2$ and $P = 0, 1, 2$) are the spherical components of the radiation field tensor (see § 5.11 in Landi Degl'Innocenti & Landolfi 2004), which quantify the symmetry properties of the radiation field at the spatial point under consideration. Thus, J^0_0 is the familiar mean intensity, J^2_0 quantifies its anisotropy, while the real and imaginary parts of J^2_P (with $P = 1, 2$) (i.e., $\tilde{J}^2_P$ and $\hat{J}^2_P$, respectively) measure the breaking of the axial symmetry. Obviously, J^2_1 and J^2_2 are zero in a plane-parallel or spherically symmetric model atmosphere, but they can have significant positive and negative values in a 3D model of stellar surface convection (see figure 3 of Trujillo Bueno & Shchukina 2009). Therefore, Eqs. (1.1) and (1.2) tell us that $U = 0$ in 1D models and that the key observational signatures of the symmetry breaking effects in a 3D model (caused by its horizontal atmospheric inhomogeneities) are non-zero U signals at any on-disk position and non-zero Stokes Q and U signals for the line of sight (LOS) with $\mu = 1$, which corresponds to forward-scattering geometry.

Figure 1 shows the Q/I and U/I images that we would see at 3000 Å if we could observe the solar continuum polarization at very high spatial and temporal resolution. Obviously, without spatial and/or temporal resolution $U/I = 0$ and the only observable quantity would be Q/I, whose wavelength variation at a solar disk position close to the limb has been determined semi-empirically by Stenflo (2005).

2. Polarization of the spectral line radiation

2.1. *The Zeeman effect*

The spectral line polarization produced by the Zeeman effect is caused by the *wavelength shifts* between the π ($\Delta M = M_u - M_l = 0$) and $\sigma_{b,r}$ ($\Delta M = \pm 1$) transitions, whose wavelength positions and strengths should (in general) be calculated within the framework of the Paschen-Back effect theory. Such wavelength shifts are of course due to the presence of a magnetic field, which causes the atomic and molecular energy levels to split into different magnetic sublevels characterized by their magnetic quantum number M (e.g., Stenflo 1994; Landi Degl'Innocenti & Landolfi 2004).

The Zeeman effect is most sensitive in circular polarization (quantified by the Stokes V parameter), with a magnitude that for not too strong fields scales with the ratio, $\mathcal{R}$, between the Zeeman splitting and the Doppler broadened line width (which is usually very much larger than the natural width of the atomic levels!), and in such a way that the emergent Stokes $V(\lambda)$ profile changes its sign for opposite orientations of the magnetic field vector. This so-called *longitudinal* Zeeman effect responds to the line-of-sight component of the magnetic field. In contrast, the *transverse* Zeeman effect responds to the component of the magnetic field perpendicular to the line of sight, so that the linear polarization Stokes Q and U profiles change sign when the direction of the transverse component changes by $\pm 90°$. Given that for not too strong fields the Stokes Q and U signals produced by the transverse Zeeman effect scale as $\mathcal{R}^2$, their amplitudes are normally below the noise level of present observational possibilities for intrinsically weak fields (typically B$\lesssim$100 gauss in solar spectropolarimetry). A good new is that the mere detection of the Zeeman effect polarization signature(s) implies the presence of a magnetic field. One disadvantage of the polarization of the Zeeman effect as a diagnostic tool is that it is blind to magnetic fields that are tangled on scales too small to be resolved.

2.2. *Anisotropic radiation pumping and atomic level polarization*

The illumination of the atoms in a stellar atmosphere is anisotropic. This is easy to understand if we consider the case of a plasma structure embedded in the optically thin outer layers of a stellar atmosphere (e.g., a solar coronal filament), because the incident radiation comes mainly from the underlying quiet photosphere and is contained within a cone of half aperture $\alpha \leqslant 90°$, with the vertex centered on the point under consideration. The larger the height above the visible stellar "surface" the smaller α and the larger the anisotropy factor $w = \sqrt{2}\mathcal{A} = \sqrt{2}J^2_0/J^0_0$, where J^0_0 is the familiar mean intensity and $J^2_0 \approx \oint \frac{d\vec{\Omega}}{4\pi}\frac{1}{2\sqrt{2}}(3\mu^2-1)I_{\nu,\vec{\Omega}}$. Neglecting the $\vec{\Omega}$ dependence of the incident intensity $I_{\nu,\vec{\Omega}}$, and assuming that it is unpolarized, it is easy to find that $w = [1+\cos\alpha]\cos\alpha/2$, which shows that in this case where the $\vec{\Omega}$-dependence of $I_{\nu,\vec{\Omega}}$ is neglected $0 \leqslant w \leqslant 1$, with $w = 1$ for the limiting case of a unidirectional unpolarized light beam that propagates along the vertical direction. The radiation field is also anisotropic within a stellar atmosphere itself (i.e., at heights where the overlying atmospheric plasma is not optically thin), but in this case w can be positive or negative. As shown in the right panel of Fig. 2, at such heights the outgoing radiation shows limb darkening (i.e., it is predominantly *vertical*) while the incoming radiation shows limb brightening (i.e., it is predominantly *horizontal*). Therefore, there is competition, because "vertical" rays (i.e., with $|\mu| > 1/\sqrt{3}$) make positive contributions to w, while "horizontal" rays (i.e., with $|\mu| < 1/\sqrt{3}$) make negative contributions to w. It wins the subset of intensities (outgoing or incoming) having the largest variation with μ. Figure 4 in the review paper by Trujillo Bueno (2001) shows how is $\mathcal{A} = J^2_0/J^0_0$ within a Milne-Eddington model atmosphere for increasing values of its source function gradient.

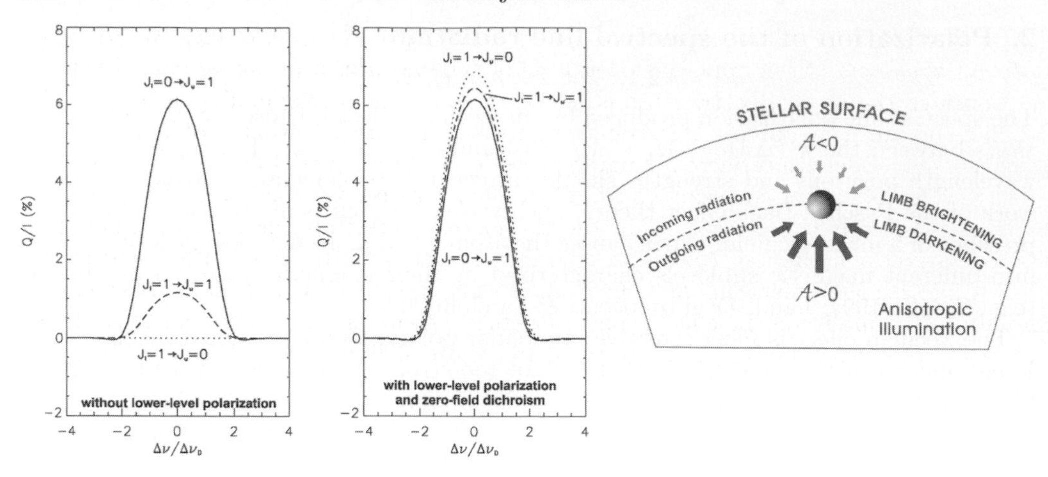

Figure 2. The emergent Q/I profiles (for a LOS with $\mu = 0.1$) of three line transitions calculated in a model atmosphere with $T = 6000$ K and $B = 0$ G. All these Q/I signals are solely due to the atomic level polarization that results from the anisotropic illumination illustrated in the right panel. Left panel: assuming that the lower level is unpolarized. Middle panel: taking into account the full impact of lower-level polarization. Like in figures 3 and 4 below, the reference direction for Stokes Q is the parallel to the closest stellar limb. From Trujillo Bueno (1999).

Why do we worry about the anisotropy of the radiation field? The reason lies in that anisotropic radiation pumping processes in a stellar atmosphere tend to induce population imbalances among the magnetic sublevels of the atomic levels (i.e., atomic level polarization), in such a way that the populations of substates with different values of $|M|$ are different. There are two key mechanisms capable of producing *directly* atomic level polarization through the absorption of anisotropic radiation (Happer 1972; Trujillo Bueno 2001): *upper-level* selective population pumping (which occurs when some substates of the upper level have more chances of being populated than others) and *lower-level* selective depopulation pumping (which occurs when some substates of the lower level have more chances of being depopulated than others).

Why do we worry about the atomic level polarization? The reason lies in that spectral line polarization can be produced by the mere presence of *atomic level polarization*, i.e., by the existence of population imbalances among the sublevels pertaining to the upper and/or lower atomic levels of the line transition under consideration. Upper-level polarization produces *selective emission* of polarization components (i.e., the emitted light is polarized, even in the absence of a magnetic field), while lower-level polarization causes *selective absorption* of polarization components or "zero-field" dichroism (i.e., the transmitted beam is polarized, even in the absence of a magnetic field).

A useful expression to estimate the amplitude of the emergent fractional linear polarization is the following generalization of the Eddington-Barbier formula (Trujillo Bueno 2001), which establishes that the emergent Q/I at the line center of a sufficiently strong spectral line when observing along a line of sight specified by $\mu = \cos\theta$ is approximately given by

$$Q/I \approx \frac{3}{2\sqrt{2}}(1-\mu^2)[\mathcal{W}\,\sigma_0^2(J_u) \; - \; \mathcal{Z}\,\sigma_0^2(J_l)], \tag{2.1}$$

where $\mathcal{W}$ and $\mathcal{Z}$ are numerical factors that depend on the angular momentum values (J) of the lower (l) and upper (u) levels of the transition (e.g., $\mathcal{W} = 1$ and $\mathcal{Z} = 0$ for a line with $J_l = 0$ and $J_u = 1$, $\mathcal{W} = 0$ and $\mathcal{Z} = 1$ for a transition with $J_l = 1$ and $J_u = 0$, and

$\mathcal{W} = \mathcal{Z} = -1/2$ for a line with $J_l = J_u = 1$), while $\sigma_0^2 = \rho_0^2/\rho_0^0$ quantifies the *fractional atomic alignment* of the upper or lower level of the line transition under consideration†

Consider the three line transitions of Fig. 2, and the corresponding emergent Q/I profiles obtained by solving numerically the scattering polarization problem in an unmagnetized model atmosphere assuming a two-level atomic model for each line independently. The left panel corresponds to calculations carried out assuming that the lower level is completely unpolarized, while the middle panel takes into account the full impact of lower-level polarization. Note that when the atomic polarization of the lower level is taken into account then the "null line" (i.e., that with $J_l = 1$ and $J_u = 0$) shows the largest Q/I amplitude. In conclusion, "zero-field" dichroism is a very efficient mechanism for producing linear polarization in the spectral lines that originate in a stellar atmosphere.

2.3. *The Hanle effect*

In order to understand what the Hanle effect is it is first necessary to clarify that in the general case where polarization phenomena are taken into account, the full description of an atomic system requires to specify, for each J-level, a density matrix with $(2J + 1)^2$ elements. The diagonal ones, $\rho_J(M, M)$, quantify the populations of the individual sublevels and the non-diagonal ones, $\rho_J(M, M')$, the quantum coherences between each pair of them. We say that the quantum coherence $\rho_J(M, M')$ is non-zero when the wave function presents a well defined phase relationship between the pure quantum states $|JM\rangle$ and $|JM'\rangle$. The law of transformation of the density-matrix under a rotation of the reference system chosen for the specification of its elements indicates that it is actually very common to find non-zero coherences when describing the excitation state of an atomic system under the influence of anisotropic radiative pumping (e.g., Landi Degl'Innocenti & Landolfi 2004).

Consider a reference system whose z-axis (the quantization direction of total angular momentum) is chosen along the direction of the applied magnetic field and J-levels whose sublevels are not affected by possible crossings and/or repulsions with the sublevels pertaining to other levels. In this simplest case, the Hanle effect tends to reduce and dephase the quantum coherences with respect to the non-magnetic case, without modifying the population imbalances. For the Hanle effect to operate the magnetic field must be inclined with respect to the symmetry axis of the pumping radiation field. What happens is that as the sublevels are split by the magnetic field, the degeneracy of the J-level under consideration is lifted and the coherences are modified. This gives rise to a characteristic magnetic-field dependence of the linear polarization of the emergent spectral line radiation that provides an attractive diagnostic tool of magnetic fields in astrophysics.

Approximately, the amplitude of the emergent spectral line polarization is sensitive to magnetic strengths between $0.1\,B_H$ and $10\,B_H$, where the critical Hanle field intensity (B_H, in gauss) is that for which the Zeeman splitting of the J-level under consideration is similar to its natural width: $B_{\rm H} = (1.137 \times 10^{-7})/(t_{\rm life}\, g_J)$ (with g_J the level's Landé factor and $t_{\rm life}$ its radiative lifetime in seconds). If the line's lower level is the ground level or a metastable level, its $t_{\rm life}(J_l) \approx 1/B_{lu} J_0^0$ (with J_0^0 the mean intensity of the spectral line radiation and B_{lu} the Einstien coefficient for the absorption process), which for relatively strong solar spectral lines is typically between a factor 10^2 and 10^3 larger than the upper-level lifetime ($t_{\rm life}(J_u) \approx 1/A_{ul}$, where A_{ul} is Einstein's coefficient for the spontaneous emission process). For this reason, in solar-like atmospheres *the lower-level*

† For example, $\rho_0^0(J = 1) = (N_1 + N_0 + N_{-1})/\sqrt{3}$ and $\rho_0^2(J = 1) = (N_1 - 2N_0 + N_{-1})/\sqrt{6}$, where N_1, N_0 and N_{-1} are the populations of the magnetic sublevels.

Hanle effect is normally sensitive to magnetic fields in the milligauss range, while *the upper-level Hanle effect* is sensitive to fields in the gauss range.

Typically, in 90° scattering geometry (e.g., when observing off the solar limb) the largest polarization occurs for the unmagnetized case, with the direction of the linear polarization perpendicular to the scattering plane. In the presence of a magnetic field pointing either towards the observer (case *a*) or away from him/her (case *b*) the polarization amplitude is significantly reduced with respect to the previously discussed unmagnetized case. Moreover, the direction of the linear polarization is rotated with respect to the zero field case. Normally, this rotation is counterclockwise for case (a) and clockwise for case (b). Therefore, when opposite magnetic polarities coexist within the spatio-temporal resolution element of the observation the direction of the linear polarization is like in the unmagnetized reference case, simply because the rotation effect cancels out. However, the polarization amplitude is indeed reduced with respect to the zero field reference case, which provides an "observable" that can be used to obtain information on hidden, tangled magnetic fields at subresolution scales in the solar atmosphere (Stenflo 1994; Trujillo Bueno *et al.* 2004).

On the other hand, in forward scattering geometry (e.g., when the line of sight points to the solar disk center) we have zero polarization in the absence of magnetic fields, while *non-zero* linear polarization in the presence of an inclined field. The ensuing Q/I amplitude reaches its maximum possible value for a magnetic strength such that the ensuing Zeeman splitting is much larger than the level's natural width (i.e., for $B > 10B_H$, approximately). In forward scattering geometry the linear polarization is created by the Hanle effect, a physical phenomenon that has been demonstrated via spectropolarimetry of solar coronal filaments in the He I 10830 Å multiplet (Trujillo Bueno *et al.* 2002).

3. Diagnostic methods based on the continuum polarization

As §1 suggests, without spatial resolution the fractional linear polarization of the stellar continuum radiation is largely determined by the effective polarizability, $\sigma_c/(\kappa_c + \sigma_c)$, and by the radiation field's anisotropy, $\mathcal{A} = J^2_0/J^0_0$, which in turn depend on the density and thermal structure of the stellar atmosphere under consideration. Collisional and/or magnetic depolarization do not play any role on the polarization of the continuum radiation of the Sun's visible spectrum. Therefore, the more realistic is the thermal and density structure of a given solar atmospheric model, the closer to the empirical data will be the calculated linear polarization of the solar continuum radiation. Interestingly, Fig. 3 demonstrates that 3D radiative transfer modeling of the polarization of the Sun's continuous spectrum in a well-known 3D hydrodynamical model of the solar photosphere shows a notable agreement with Stenflo's (2005) semi-empirical data, significantly better than that obtained via the use of 1D atmospheric models.

4. Diagnostic methods based on the spectral line polarization

4.1. *The scattering polarization of the Ca* II *IR triplet*

A suitable diagnostic window for investigating the thermal and magnetic structure of the "quiet" solar chromosphere is the scattering polarization in the Ca II IR triplet (see Manso Sainz & Trujillo Bueno 2003a, 2009). The radiative transfer modeling results of Fig. 4 indicate that while the scattering polarization in the 8498 Å line shows a strong sensitivity to inclined magnetic fields with strengths between 1 mG and 10 G, the emergent linear polarization in the 8542 Å and 8662 Å lines is sensitive to magnetic fields in the milligauss

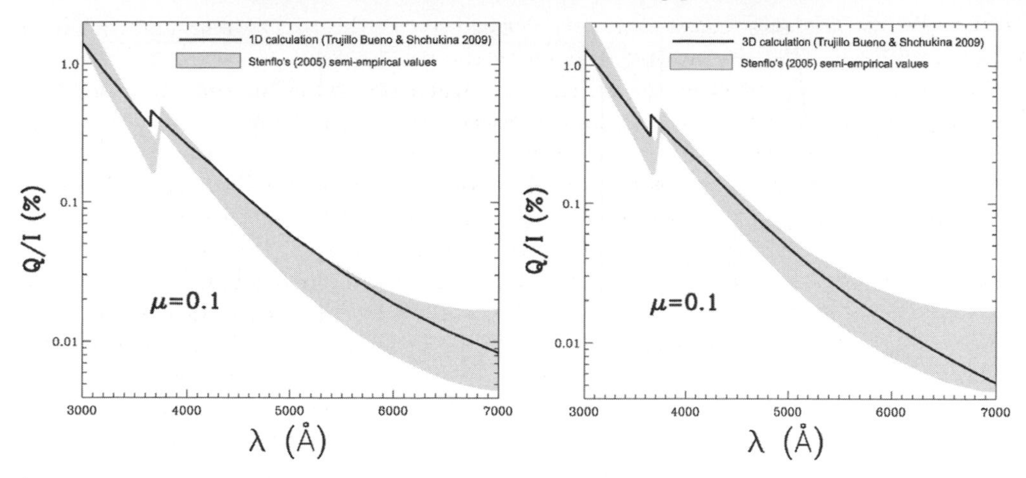

Figure 3. The wavelength variation of the polarization of the Sun's continuous spectrum. The gray shaded areas corresponds to Stenflo's (2005) semi-empirical data. The solid lines show the results of radiative transfer calculations in a well-known 1D semi-empirical model (left panel) and in a well-known 3D hydrodynamical model (right panel).

range. The reason for this very interesting behavior is that the scattering polarization in the 8498 Å line gets a significant contribution from the selective emission processes that result from the atomic polarization of the short-lived upper level, while that in the 8542 Å and 8662 Å lines is dominated by the selective absorption processes that result from the atomic polarization of the metastable (long-lived) lower levels. Therefore, in "quiet" regions of a stellar atmosphere the magnetic sensitivity of the linear polarization of the 8542 Å and 8662 Å lines is controled by the lower-level Hanle effect, which implies that in regions with $1 \lesssim B \lesssim 50$ G the Stokes Q and U profiles are only sensitive to the orientation of the magnetic field vector. In such regions the 8498 Å line is however sensitive to both the orientation and the strength of the magnetic field through the upper-level Hanle effect. In summary, "zero-field" dichroism is the key physical origin of the enigmatic scattering polarization observed by Stenflo *et al.* (2000) in the Ca II IR triplet (see Manso Sainz & Trujillo Bueno 2003a). Interestingly, this linear dichroism caused by the mere presence of population imbalances in the lower levels of the line transitions may also be operating in other astrophysical objects (e.g., supernovae) and should be taken into account when interpreting spectropolarimetric observations in other spectral lines besides the Ca II IR triplet itself.

4.2. *The Hanle and Zeeman effects in the He* I *10830 Å multiplet*

A suitable diagnostic tool for determining the magnetic field of plasma structures embedded in the solar chromosphere and corona (e.g., spicules, prominences and filaments) can be achieved by observing and interpreting the polarization signals produced by the joint action of atomic level polarization and the Hanle and Zeeman effects in the He I 10830 Å triplet. This multiplet originates between the metastable state, $2^3\mathrm{S}_1$, and the first excited term, $2^3\mathrm{P}_{2,1,0}$, of the triplet system of He I. Therefore, its blue component has $J_l = 1$ and $J_u = 0$ (i.e., it is a "null" line), while its red blended components have $J_u = 2$ and $J_u = 1$.

For field strengths $B \lesssim 100$ G the linear polarization of the He I 10830 Å triplet is fully dominated by the atomic level polarization that is produced by anisotropic radiation pumping processes. Since the blue component has $J_u = 0$ its linear polarization can only

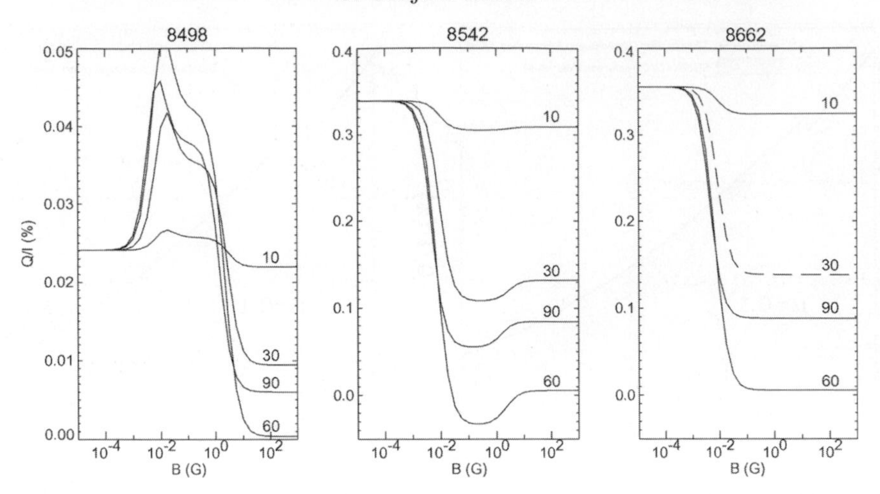

Figure 4. The emergent fractional linear polarization of the Ca II IR triplet calculated for a LOS with $\mu = 0.1$ in a semi-empirical model of the solar atmosphere. Each curve corresponds to the indicated inclination, with respect to the solar local vertical direction, of the assumed random-azimuth magnetic field. From Manso Sainz & Trujillo Bueno (2009).

be due to selective absorption of polarization components (i.e., "zero-field" dichroism), so that its detection requires to observe a light beam after it has passed through an optically pumped plasma (e.g., when observing a solar coronal filament against the bright background of the solar disk). The left panels of Fig. 5 show an example of the Stokes I and Q profiles of a solar coronal filament observed in forward scattering geometry at the solar disk center. The detection of linear polarization in the blue line implies that the metastable state, $2^3\mathrm{S}_1$, was significantly polarized. Even more interesting is the conclusion that in the forward scattering geometry of this observation the mere detection of linear polarization in any of the components of the He I 10830 Å triplet implies the presence of a magnetic field *inclined* with respect to the solar radius vector through the observed point (Trujillo Bueno *et al.* 2002).

Another example of a "weakly" magnetized plasma (i.e., with $B \lesssim 100$ G) is that of the dynamic jets that we call spicules, needle-shaped plasma structures emanating from the solar network regions and reaching heights as large as 10000 km in the chromosphere. The determination of the magnetic field that channels the spicular motions can be achieved by modeling observations of the polarization of the He I 10830 Å and/or 5876 Å (or D_3) multiplets in 90° scattering geometry at various heights above the visible solar limb (see Centeno *et al.* 2009; and more references therein). The right panels of Fig. 5 show an example of one of our most recent spectropolarimetric observations in the He I 10830 Å triplet. In this 90° scattering geometry observation the linear polarization of the scattered light provides *direct* information only about the polarization of the excited states. For this reason, we do not see now any linear polarization signal in the blue line ($J_l = 1$ and $J_u = 0$), in spite of the fact that its lower level is probably polarized. However, we do see linear polarization in the red blended component, which implies that its upper levels are polarized. Interestingly, the observed non-zero Stokes U signal is produced by the Hanle rotation of the direction of linear polarization, which indicates the presence of a magnetic field inclined with respect to the local vertical direction. The presence of a magnetic field is also indicated by the observed Stokes V profile, which in the He I 10830 Å triplet is caused by the longitudinal Zeeman effect. Its detection is indeed crucial for the determination of the full magnetic field vector via theoretical modeling based on the

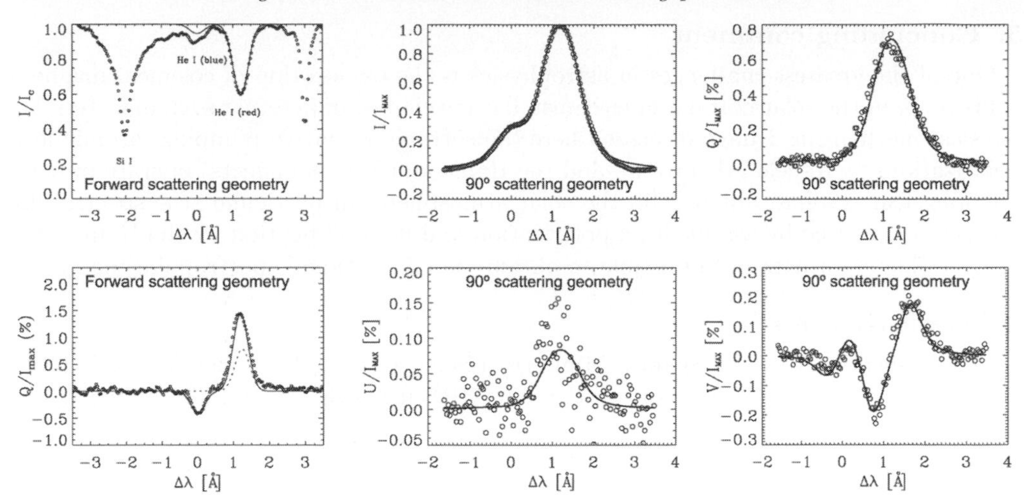

Figure 5. Forward scattering panels (from Trujillo Bueno *et al.* 2002): The open circles show the I and Q profiles of the He I 10830 Å triplet observed in a coronal filament at the solar disk center. The solid lines show the theoretical profiles corresponding to a magnetic field with $B = 20$ G and inclination $\theta_B = 105°$. The dotted line shows the theoretical Stokes Q profile when the lower level is assumed to be unpolarized. The positive direction of Stokes Q is parallel to the projection of the magnetic field vector on the solar surface. 90° scattering panels (from Centeno *et al.* 2009): The open circles show the Stokes profiles observed in spicules of the very quiet Sun. The solid lines show the theoretical profiles corresponding to $B \approx 36$ G, $\theta_B \approx 38°$ and azimuth $\chi_B \approx -2°$. The Stokes Q reference direction is the parallel to the observed solar limb.

quantum theory of polarization (see the solid lines fit), because for magnetic strengths larger than only a few gauss the He I 10830 Å multiplet is in the saturation regime of the upper-level Hanle effect ($B > 10B_H \approx 8$ G), where the linear polarization signals are only sensitive to the inclination and azimuth of the magnetic field.

For field strengths $100 < B \lesssim 2000$ G the linear polarization of the He I 10830 Å triplet is caused by the joint action of atomic level polarization and the transverse Zeeman effect (see Fig. 2 of Trujillo Bueno & Asensio Ramos 2007). As shown by these authors, the impact of atomic level polarization on the linear polarization of the emergent radiation in the He I 10830 Å multiplet can indeed be very significant, even for magnetic field strengths as large as 1000 G. Finally, for field strengths $B > 2000$ G the linear polarization of the He I 10830 Å triplet turns out to be dominated by the transverse Zeeman effect.

4.3. *Some computer programs for modeling the Hanle and Zeeman effects*

In order to carry out the above-mentioned type of investigations in an efficient way, it is useful and important to have at our disposal reliable codes for the synthesis and/or inversion of Stokes profiles caused by atomic level polarization and the Hanle and Zeeman effects. To this end, we have developed MULTIPOL (Manso Sainz & Trujillo Bueno 2003b) and HAZEL (Asensio Ramos *et al.* 2008), which are based on the quantum theory of spectral line polarization described in the monograph by Landi Degl'Innocenti & Landolfi (2004). The multilevel radiative transfer calculations of Fig. 4 were carried out with MULTIPOL, while the four right panels of Fig. 5 show an example of the application of the inversion option of HAZEL to spectropolarimetric observations of solar spicules in the He I 10830 Å triplet. HAZEL (from HAnle and ZEman Light) is a public, user-friendly computer program for the synthesis and inversion of Stokes profiles caused by the joint action of atomic level polarization and the Hanle and Zeeman effects in plasma structures embedded in a stellar atmosphere.

5. Concluding comment

One of the greatest challenges in astrophysics is the exploration of cosmical magnetic fields (e.g., in the solar corona, in circumstellar envelopes, in acreting systems, etc.) The physical mechanisms I have discussed here (anisotropic radiative pumping, atomic level polarization, "zero-field" dichroism, and the Hanle and Zeeman effects) operate in many astrophysical systems, not only in the solar atmosphere. In particular, the spectral line polarization caused by atomic level polarization and its modification by the Hanle effect provides key information, impossible to obtain via conventional spectropolarimetry.

Acknowledgements

Finantial support by the Spanish Ministry of Science through project AYA2007-63881 and by the European Commission via the SOLAIRE network is gratefully acknowledged.

References

Asensio Ramos, A., Trujillo Bueno, J., & Landi Degl'Innocenti, E. 2008, *ApJ* 683, 542
Centeno, R., Trujillo Bueno, J., & Asensio Ramos, A. 2009, *ApJ*, in preparation
Chandrasekhar, S. 1960, *Radiative Transfer*, Dover Publications, Inc.
Happer, W. 1972, *Rev. Mod. Phys.* 44, 169
Landi Degl'Innocenti, E. & Landolfi, M. 2004, *Polarization in Spectral Lines*, Kluwer
Manso Sainz, R. & Trujillo Bueno, J. 2003a, *Phys. Rev. Lett.* 91, 111102
Manso Sainz, R. & Trujillo Bueno, J. 2003b, in ASP Conf. Ser. Vol. 307, Solar Polarization, ed. J. Trujillo Bueno & J. Sánchez Almeida, 251
Manso Sainz, R. & Trujillo Bueno, J. 2009, *ApJ*, in preparation
Stenflo, J. O. 1994, Solar Magnetic Fields: Polarized Radiation Diagnostics (Dordrecht: Kluwer)
Stenflo, J. O. 2005, *A&A* 429, 713
Stenflo, J. O., Keller, C. U., & Gandorfer, A. 2000, *A&A* 355, 789
Trujillo Bueno, J. 1999, in ASSL 243, Solar Polarization, ed. K. N. Nagendra & J. O. Stenflo, Dordrecht, Kluwer, p.73
Trujillo Bueno, J. 2001, in ASP Conf. Ser. Vol. 236, Advanced Solar Polarimetry: Theory, Observation and Instrumentation, ed. M. Sigwarth (San Francisco: ASP), 161
Trujillo Bueno, J. & Asensio Ramos, A. 2007, *ApJ* 655, 642
Trujillo Bueno, J. & Shchukina, N. 2009, *ApJ*, in press
Trujillo Bueno, J., Landi Degl'Innocenti, E., Collados, M., Merenda, L., & Manso Sainz, R. 2002, *Nature* 415, 403
Trujillo Bueno, J., Shchukina, N., & Asensio Ramos, A. 2004, *Nature* 430, 326

Discussion

BECKMAN: Could you explain how you use the Hanle effect to distinguish between a tangled field and an aligned field?

TRUJILLO BUENO: For the case of an aligned field see Trujillo Bueno *et al.* (2002). For the case of a tangled field see Trujillo Bueno *et al.* (2004). See also §2.3.

HEILES: Radio astronomers are familiar with the Goldreich-Kylafis effect. Is this the same as the Hanle effect?

TRUJILLO BUENO: Since they were considering a hypothetical molecular line with $J_l = 0$ and $J_u = 1$ (that is, without the possibility of the lower-level polarization I have discussed in §2.2 and §2.3) the so-called Goldreich-Kylafis effect is nothing but the upper-level Hanle effect in its saturation limit (because of their additional assumption that the Zeeman splitting is much larger than the natural width of level J_u.).

Cosmic Magnetic Fields:
From Planets, to Stars and Galaxies
Proceedings IAU Symposium No. 259, 2008
K.G. Strassmeier, A.G. Kosovichev & J.E. Beckman, eds.

doi:10.1017/S1743921309031469

Zeeman-Doppler imaging: old problems and new methods

Thorsten A. Carroll, Markus Kopf, Klaus G. Strassmeier and Ilya Ilyin

Astrophysikalisches Institut Potsdam, An der Sternwarte 16, D-14482 Potsdam, Germany
email: tcarroll@aip.de

Abstract. Zeeman-Doppler Imaging (ZDI) is a powerful inversion method to reconstruct stellar magnetic surface fields. The reconstruction process is usually solved by translating the inverse problem into a regularized least-square or optimization problem. In this contribution we will emphasize that ZDI is an inherent non-linear problem and the corresponding regularized optimization is, like many non-linear problems, potentially prone to local minima. We show how this problem will be exacerbated by using an inadequate forward model. To facilitate a more consistent full radiative transfer driven approach to ZDI we describe a two-stage strategy that consist of a principal component analysis (PCA) based line profile reconstruction and a fast approximate polarized radiative transfer method to synthesize local Stokes profiles. Moreover, we introduce a novel statistical inversion method based on artificial neural networks (ANN) which provide a fast calculation of a first guess model and allows to incorporate better physical constraints into the inversion process.

Keywords. Stars: activity – stars: magnetic fields – stars: spots – radiative transfer – methods: data analysis

1. Introduction

Zeeman-Doppler Imaging (ZDI) was pioneered by Semel (1989), although first described as a mere detection method for surface magnetic fields on rapid rotating stars, the method quickly evolved to a real mapping (inversion) technique (Donati *et al.* 1990; Brown *et al.* 1991). Today ZDI or Magnetic-Doppler Imaging (MDI) is a synonym for the inversion of time and phase resolved spectropolarimetric observations to reconstruct stellar magnetic surface fields. Since that time ZDI was a success story of its own and has enormously contributed to our current understanding of stellar magnetism (e.g. Donati 1999; Hussain *et al.* 2002; Kochukhov *et al.* 2004; Donati *et al.* 2006, 2007; Petit *et al.* 2008). However, there are a number of fundamental problems which has to be considered when dealing with ZDI as an inverse problem. One of these problems comes from an inadequate formulation and approximation of the forward problem while another comes from the inherent non-linearity of the inverse problem.

2. The forward problem

Inverse problems arise whenever one searches for causes of observed data or desired effects. The most general way of solving the inverse problem (directly or indirectly) is to utilize a model of the underlying forward problem. The forward problem constitutes our physical theory and/or knowledge about the system to make predictions about the outcome and result of measurements. The direct or forward problem in ZDI is typically the polarized radiative transfer based on a model of the stellar atmosphere. Here we may encounter the first fundamental problem, owing to our limited theoretical knowledge (e.g.

lack of atomic and/or atmospheric parameters) as well as numerical and computational constraints we are restricted to an approximate description of the *real* physical system. Critical in this context is to provide an appropriate and adequate parameterization of the model , i.e the minimal number of free variable that determine the measurable outcome. The determination of a particular parameterization is not always obvious and requires a careful evaluation. Formally the forward problem of ZDI can be written in a compact way, by using the formal solution of the Stokes vector as described by Landi Deglinnocenti & Landi Deglinnocenti (1985). This allows us to write the disk-integrated Stokes vector $\boldsymbol{I}^*$ at a wavelength λ and for a particular rotational phase ϕ as follows :

$$\boldsymbol{I}^*(\lambda,\phi) = \int_{\hat{M}} \int_0^{\infty} \boldsymbol{O}\left(0, s(M',\theta), \boldsymbol{X}(M'), M', \phi\right) \, \boldsymbol{j}\left(s(M',\theta), \boldsymbol{X}(M'), M', \phi\right) ds \cos\theta dM. \tag{2.1}$$

The vector $\boldsymbol{X}$ provides the parameterization of the model and comprises parameters like the magnetic field vector, temperature, abundance, etc., dM denotes the infinitesimal surface element at the position M' (given in spherical coordinates), θ denotes the angle between the surface normal at M' and the direction to the observer, s is the geometrical path length towards the observer, $\boldsymbol{j}$ is the emission vector (see Stenflo 1994), and $\boldsymbol{O}$ is the evolution operator of the formal solution which incorporates the absorption matrix. Following Landi Deglinnocenti & Landi Deglinnocenti (1985), we can write the evolution operator, in the case of a piecewise constant absorption matrix $\boldsymbol{K}$, as

$$\boldsymbol{O}(s, s', M, \boldsymbol{X}(M)) = e^{-\boldsymbol{K}(M,\boldsymbol{X}(M))\,|s'-s|}\,. \tag{2.2}$$

Equation 2.1 can also be written symbolically in a more compact form as

$$\boldsymbol{I}^*(\lambda,\phi) \;-\; \boldsymbol{\Omega}(\boldsymbol{X})\,, \tag{2.3}$$

where $\boldsymbol{\Omega}$ represents the formal non-linear integral operator of the forward problem. Several things must be kept in mind here, both integrals, the outer surface integration over the model atmosphere as well as the inner transport integral through the atmosphere, must be discretized in order to cope with the problem numerically. We should be aware that this already means that we transform our originally continuous problem into a discrete one with finite dimension. As mentioned above, the finite parameterization of our model $\boldsymbol{X}(M)$ deserves particular attention because the limitation to those variables that we consider as necessary and sufficient, will directly determine the dimension of the model space, we will see later how the neglect of the temperature may have drastic consequences for the inversion. The radiative transfer model we use is still formulated in local thermodynamic equilibrium (LTE) which is quite sufficient for the majority of photospheric spectral lines we are using but certainly has its limitations. Moreover, it is customary to describe the atmosphere as height independent, i.e. neglecting all height gradients in most of the model parameter, an approximation which becomes questionable in particular for stars with extended atmospheres like giants. All these approximations of the *real* physical situation in stellar atmospheres will therefore have direct consequences on the inversion process. Finally, it should be also clear from equation (2.1) and (2.2) that our forward problem is in general non-linear in $\boldsymbol{X}$, an aspect that will be discussed further in the following section.

3. The inverse problem

The retrieval of model parameters from measured data constitutes the general definition of a inverse problem. In the particular case of ZDI the measured data are given in the form of phase resolved Stokes profile observations. In general there is a distinction between continuous and discrete inverse problems which means that we want to retrieve either a continuous functions or a parameterized model. In that sense it is already the formulation of the forward problem which defines the kind of inverse problem. The same holds for the definition of linear or non-linear inverse problems, which is determined by the linearity or non-linearity of the forward problem. Inverse problems, like ZDI, are often said to be ill-posed by which, in general, we mean that i.) the inverse problem might be non-unique and ii.) the inverse problem might not depend continuously on the data, i.e. the solution is unstable against small perturbation of the observed data. The non-uniqueness may have different reasons which can be a real intrinsic non-uniqueness of the problem, or uncertainties in the observed data, or an ill-defined forward model, or a lack of data that sufficiently constrain the problem. The problem of ill-posedness can be effectively addressed by a regularization approach, i.e. adding additional a priori assumptions (i.e. constraints) to the inverse operator. A powerful method in this context is the Tikhonov regularization but many other forms of regularizations are also possible. For linear problems it can indeed be shown that the regularization leads to a unique and stable solution (see e.g. Engl *et al.* 1996). However, the forward problem we deal with in ZDI is non-linear, and at first, it is not obvious how this will affect the uniqueness and stability of our solution. The general way to proceed is to formulate and minimize a regularized least-square problem. This approach, relies on the linearization of the forward operator, which can be obtained in our case from a first order expansion of (2.3),

$$\boldsymbol{\Omega}(\boldsymbol{X}) = \boldsymbol{\Omega}(\boldsymbol{X}_0) + \boldsymbol{J}(\boldsymbol{X} - \boldsymbol{X_0}) , \tag{3.1}$$

where $\boldsymbol{J}$ is the Jacobian matrix of $\boldsymbol{\Omega}$. Using the sum of square error and a regularization functional $G(\boldsymbol{X})$ the objective functional $F(\boldsymbol{X})$ can be written as

$$F(\boldsymbol{X}) = \|\boldsymbol{I}_{obs} - \boldsymbol{I}^*\|^2 + \alpha G(\boldsymbol{X}) \tag{3.2}$$

where α is the regularization parameter which determines the influence of the regularization. The regularization functional G can be expressed in the from $\|L(\boldsymbol{X}_{ref} - \boldsymbol{X}_0)\|^2$ where $\boldsymbol{X}_{ref}$ is a parameter vector which represents our a priori knowledge (assumption) about the solution, and L is a suitable matrix approximation of a differential operator. The essence of the non-linear solution is then to repeat the process of solving the linearized solution in terms of a gradient descent method. Using a Gauss-Newton type form of linearization (Engl *et al.* 1996) we obtain from (3.2) for an increment $\Delta\boldsymbol{X}_{n+1}$ at the iteration $n+1$ the following solution,

$$\Delta\boldsymbol{X}_{n+1} = (\boldsymbol{J}^*\boldsymbol{J} + \alpha\boldsymbol{L}^*\boldsymbol{L})^{-1}(\boldsymbol{J}^*(\boldsymbol{I}_{obs} - \boldsymbol{I}^*)) + \alpha\boldsymbol{L}^*\boldsymbol{L}(\boldsymbol{X}_{ref} - \boldsymbol{X}_n) , \tag{3.3}$$

where $\boldsymbol{J}^*$ is the conjugate transpose of $\boldsymbol{J}$. The stabilizing effect of the regularized solution is then ensured by the positive definiteness of the inverse $(\boldsymbol{J}^*\boldsymbol{J} + \alpha\boldsymbol{L}^*\boldsymbol{L})^{-1}$. As the regularization parameter is held constant throughout the iteration, the regularization part in (3.2) will have an increasing influence as we approach the solution and thus provides the additional benefit of preventing the inversion from fitting arbitrary details (i.e. noise) in the spectra. However, although we gain a stabilizing effect at each iteration step, it should be clear from (3.2) that even when G provides a convex constraint this needs not necessarily be true for $F(\boldsymbol{X})$ which, in general, is still a non-convex function due to the non-linearity of (2.1). The inverse problem may therefore have many local minima in the parameter space in which a gradient descent method can get stuck and a regularization alone can not guarantee to find the unique global minimum. Furthermore, the determination of an appropriate regularization parameter α is by no means an easy task and can require high computational efforts (Engl *et al.* 1996).

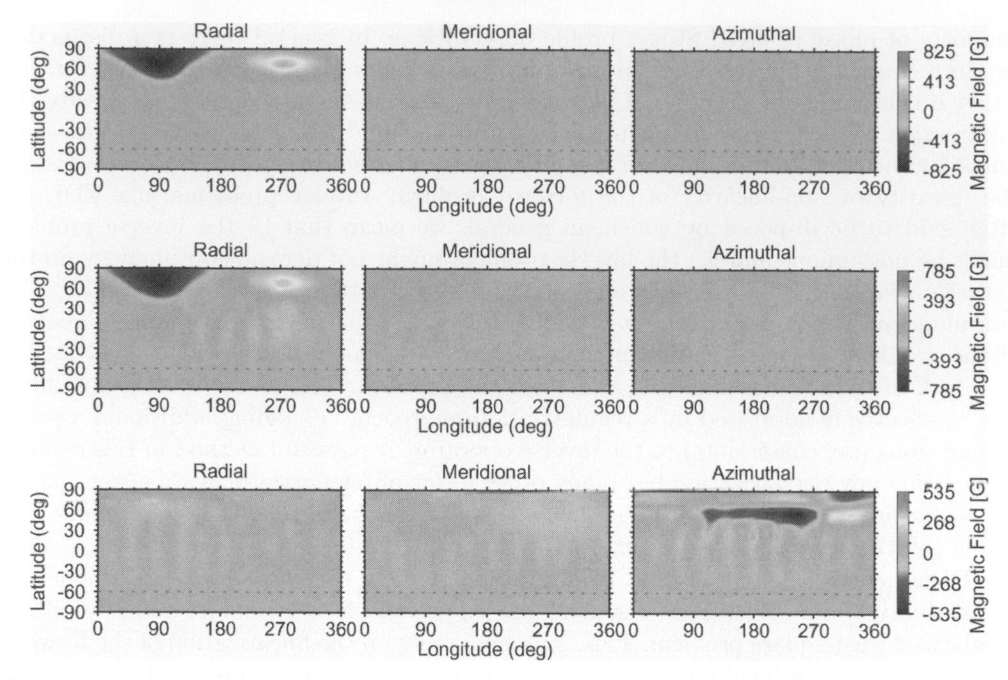

Figure 1. Mercator plots of a simulated magnetic field distribution. Top: the synthetic reference model. Middle: the reconstructed magnetic field from a simultaneous temperature and magnetic field inversion. Bottom: the result from the restricted inversion (fixed temperature). See text.

4. More than one solution – an inadequate forward model

To show how much the solution of the ZDI problem depends on the forward model and the chosen parameterization we made a simple synthetic experiment. The stellar model surface consists of a large bipolar magnetic field region. The stellar atmosphere is that of a typical fast rotating K subgiant star (i.e. $\log(g) = 3.5$, $T_{eff} = 4600$ K, [M/H] $= -0.5$, vsini $= 40$ km/s). One of the magnetic regions is located in a cool spot like structure while the other polarity is embedded in a hot and more dispersed like structure to mimic a faculae or plage region, see top of Figure 2 The magnetic field is exclusively radial oriented in both magnetic regions. Two inversion were performed, the first one worked on the basis of the correct parameterization with the magnetic field vector and temperature as free parameter, while the second inversion runs with a fixed temperature. This allows the second inversion only to seek in a subspace for a possible solution. For the inversion we have our ZDI code *iMap* (Carroll *et al.* 2007) with a surface segmentation of 1800 elements, and the Zeeman-sensitive iron line FeI λ 6173. Although we have all four Stokes components available we restricted the number of available Stokes components to Stokes I and Stokes V of the Zeeman-sensitive iron line FeI λ 6173. We have added noise with a signal-to-noise level of 1000 to the original model spectra. Both inversion run with the same value for the regularization parameter. The first inversion converged to a solution that is close to the reference field structure, see Figure 1 middle. The surface structure is well reconstructed from the simultaneous temperature and magnetic field inversion and also the Stokes V spectra (Figure 2, left) are well reproduced. In the second inversion where we have neglected the temperature as a free parameter the fit to the reference spectra is again very good but the resulting magnetic surface field is drastically different (Figure 1, bottom). The surface field shows a characteristic high

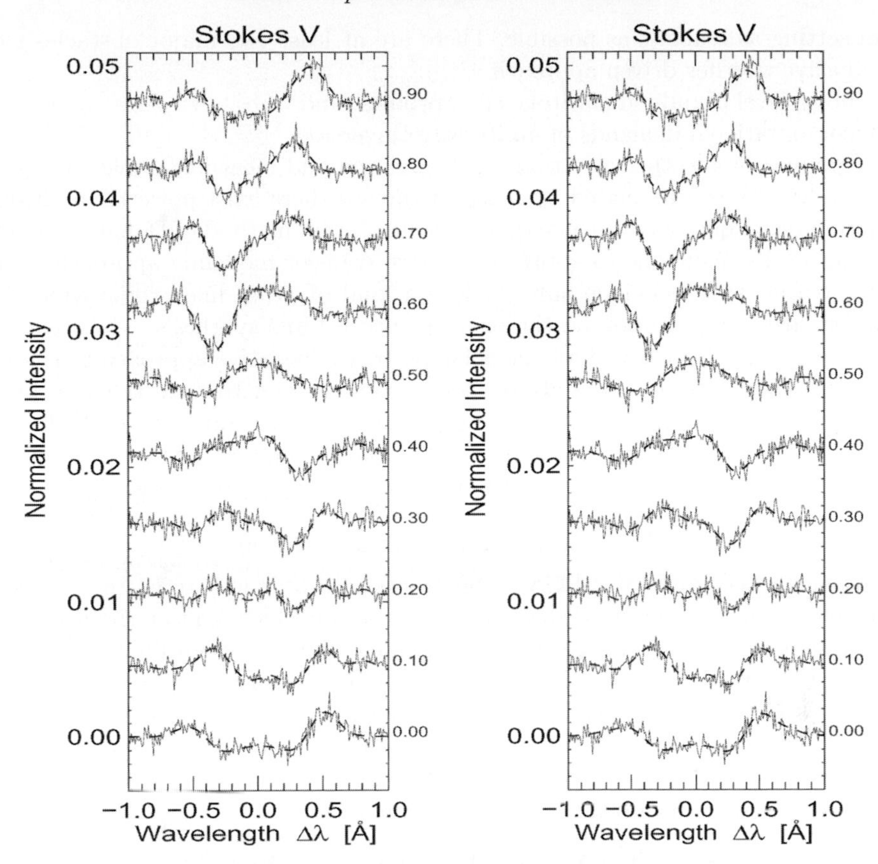

Figure 2. The fits (dashed lines) to the noisy reference Stokes V profiles from the two test inversions : on the left the fits of the simultaneous temperature and magnetic field inversion, $\chi^2_{red} = 1.01$ and on the right the fits of the restricted inversion, $\chi^2_{red} = 1.02$.

latitude band of a unipolar azimuthal field with an longitudinal extension of about 180^0 which is in sharp contrast to the original reference field. The inversion in the restricted parameter regime has apparently converged to a false solution, although the fits are very good (Figure 2, right). The quality (goodness) of the fit (Figure 2) itself do not provide enough information in this case to distinguish between the competing solutions. This issue could readily be resolved if the corresponding Stokes Q and U profiles are taken into account (not shown here). Due to the critical role of the temperature and the associated radiative transfer effects it will be of particular importance to provide an accurate and appropriate forward model for the inverse problem. We will investigate the impact of radiative transfer effects and inadequate modeling for ZDI inversions in greater detail in a forthcoming paper (Carroll *et al.* 2009).

5. Towards a full polarized radiative transfer approach

We have seen from the foregoing synthetic example how the underlying model can affect the outcome of the inversion. Moreover we have seen the critical role of the temperature which is of course tightly correlated to the radiative transfer effects. It is therefore worthwhile to include a full polarized radiative transfer in the inversion process to make the problem setting as realistic as possible. There are at least two major obstacles towards a full radiative transfer driven approach :

- the noise level of individual Stokes line profiles, and
- the computational demands of an iterative inversion.

The first point refers to the cases that most observed and measured Stokes V, Q and U spectra are heavily contaminated by noise. Although there exist powerful multi-line reconstruction techniques as the least-square deconvolution method of Donati *et al.* (1997), they do not easily allow one to apply a radiative transfer modeling approach. The crucial point is that this technique only retrieve a kind of *mean* line profile which has no straight forward interpretation on the basis of spectral line synthesis. The second point highlights the computational effort one is facing by the iterative approach to the inverse problem. To address both of the above listed problems and to pave the way for a full radiative transfer modeling approach we propose a two-step strategy, a multi-line reconstruction method which enables the reconstruction of individual Stokes profiles and a fast approximate radiative transfer calculation by artificial neural networks.

5.1. *Multi-Line PCA reconstruction*

The basic idea of the multi-line PCA reconstruction technique is to extract the systematic features within the observed data set, i.e. Stokes profiles, and at the same time to reduce the noise. This is done by identifying the redundancy within the dataset and describing the dataset with a minimal number of characteristic parameter (i.e. dimensions) as possible. The redundancy among the observed data can be described by the correlation between the individual spectra such that we can express the correlation of entire observational dataset by the covariance matrix $\boldsymbol{C}_x$ which can be written, after transforming the spectra into the velocity domain, as

$$\boldsymbol{C}_x = \sum_n \left(\boldsymbol{X}_n(v) - \bar{\boldsymbol{X}}(v)\right)\left(\boldsymbol{X}_n(v) - \bar{\boldsymbol{X}}(v)\right)^T \ , \tag{5.1}$$

where $v = c\Delta\lambda/\lambda$ and n is the number of individual spectral line profiles used for the analysis and $\bar{\boldsymbol{X}}$ the mean Stokes profiles of all spectral lines. We then seek a set of orthonormal direction in the data space (given by the spectral resolution) along which the variance in $\boldsymbol{X}$ is maximized in descending order. These new set of coordinate axes which accounts for the maximum variance in the observed data (Stokes spectra) can be determined by calculating the eigenvectors of the covariance matrix Eq. (5.1). The so called Principal Component Analysis (PCA) or Karhunen-Loeve transformation (Bishop 1995) will provide us with the calculation of the eigenvectors of the covariance matrix of the observation whereby the eigenvectors are ordered according to their associated eigenvalues (variances). We use the PCA method to decompose the entire set of observed Stokes spectra into a new coordinate system. This procedure will project the most coherent and systematic features in the observed Stokes profiles into the first few eigenvectors with the largest eigenvalues while the incoherent features (i.e. noise or blends) will be mapped to the less significant eigenvectors (with low eigenvalues). After having calculated the set of orthogonal eigenvectors we can use this new basis to decompose all observed Stokes spectra into the new basis of eigenvectors $\boldsymbol{u}_l$, as

$$\boldsymbol{x}_k(v) = \sum_l \alpha_{k,l} \boldsymbol{u}_l(v) \ , \tag{5.2}$$

where $\alpha_{k,l} = \boldsymbol{x}_k(v)\boldsymbol{u}_l(v)$ is the scalar product (i.e. the projection or the cross-correlation) between the observed $k-th$ Stokes profile $\boldsymbol{x}_k(v)$ and the $l-th$ eigenvector $\boldsymbol{u}_l(v)$. If we use a limited number of the first few l eigenvectors for this decomposition the reconstruction will be made only with those components who carry the most significant information about the systematic features and inherent line characteristics. We therefore reconstruct individual Stokes profiles to an extent that the majority of the individual line characteristics are preserved and only a small contribution of uncorrelated effects are present. To determine the maximum number of meaningful eigenvectors we calculate the covariance matrix of pure noise profiles with a S/N ratio that corresponds to the noise level of the observations. This allows us to estimate the magnitude of the eigenvalues from which the corresponding eigenvectors will mainly contain noise. Since the PCA multi-line method now allows the reconstruction of *individual* line profiles, all the known line parameters of a particular spectral line can be used in a subsequent DI and ZDI inversion. See Carroll *et al.* (2007); Martínez González *et al.* (2008) for a more detailed description of the method.

5.2. *A fast Stokes profile synthesis*

The basic idea of our proposed approach is to emulate the process of polarized line formation by using an adaptive regression model, that is fast to evaluate, and which provides the required accuracy. The adaptive model we seek must provide a sufficient complexity to describe the non-linear mapping of Eq. (2.1), between the most prominent atmospheric parameters and the resulting Stokes spectra. For this purpose, we used a supervised machine learning algorithm, e.g., an artificial neural network (ANN) model. A popular type of of ANN, also used in this work, is the so called multilayer-perceptron (MLP) which is known as a universal function approximator. The MLP can be regarded as a class of nonlinear function, which performs a continuous and multivariate mapping between an input vector $\boldsymbol{x}$ and an output vector $\boldsymbol{y}$. The network function represents a function composition of elementary non-linear functions $g(a)$. These elementary functions are arranged in layers whereby each of these functions in one layer is connected via an adaptive weight vector $\boldsymbol{w}$ to all elementary functions in the neighboring layers. The $l-th$ output (component of the output vector $\boldsymbol{y}$) of a two layer (of weights) MLP for example can be written as

$$y_l(\boldsymbol{x};\boldsymbol{w}) = g_k\left(\sum_{j=0}^{J} w_{kj}^{(2)}\, g_j\left(\sum_{i=0}^{I} w_{ji}^{(1)}\, x_i\right)\right) , \tag{5.3}$$

where x_i represents the i-th component of the input vector $\boldsymbol{x}$ and $w_{ji}^{(1)}$ the connecting weight from the i-th input component to the j-th elementary function g_j in the first unit layer. The weights $w_{kj}^{(2)}$ then connecting all the functions g_j with the functions g_k in the second unit layer. The capital letters (I,J) giving the numbers of elementary functions (units) in the respective unit layer. The elementary functions $g(a)$, which are also called activation functions, are given by the following type of sigmoid function,

$$g(a) = \frac{1}{1 + exp(-a)} \ . \tag{5.4}$$

The network function $\boldsymbol{y}(\boldsymbol{x})$ will thus process a given input vector $\boldsymbol{x}$ by propagating this vector (via multiplication with the individual weight values and subsequent evaluation of the different activation functions) through each layer of the network. The particular function that will be implemented by the MLP is determined by the overall structure of the network and the individual adaptive weight values. The process of determining these weight values for the MLP is called (supervised) training and is formulated as a non-linear optimization process. This training is performed on the basis of a representative dataset, which includes the input to target (i.e. training output) relations of the underlying problem. This process is similar to a non-linear regression for a given data set, but as the underlying model (i.e., MLP) is much more general, the regression function is not restricted to a specific predetermined (or anticipated) model. In fact, it can be shown that MLPs provide a general framework for approximating arbitrary non-linear functions (Bishop 1995). In our case the input-to-target relation is dictated by our synthesis problem, for the input vectors we have chosen the following atmospheric parameters : the temperature and pressure structure of model atmospheres (described by the effective temperature), logarithmic gravitation, iron abundance, the local bulk velocity of the plasma, microturbulence, macroturbulence, magnetic field strength, magnetic field inclination, magnetic field azimuth and the LOS angle between the observer and the local normal. For the output parameter we have used the full Stokes vector profiles of the Zeeman-sensitive iron lines FeI λ 6173 Å and FeI λ 5497 Å which have an effective Landé factor of $g_{\rm eff} = 2.5$ and $g_{\rm eff} = 2.25$ respectively. Once the network is successfully trained and has converged in terms of minimizing the error between the calculated output vectors and the target vectors of the training database, the network weights are frozen and the MLP, which now represents the desired approximation of the underlying problem, can be applied to new and unknown data coming from the same parameter domain as the training data. To test the accuracy of the trained MLPs we created a large number of input vectors with randomly chosen combinations for the atmospheric parameters. The statistical evaluation shows that the MLPs are able to calculate the corresponding Stokes profiles with a high degree of accuracy. The comparison of the MLP calculation with the results from the conventional numerical integration of the polarized radiative transfer, performed with the DELO method of Rees *et al.* (1989), demonstrates the impressive results of the MLP synthesis. For the Stokes I profile calculation we obtain a rms error as low as 0.11 %, for Stokes V 0.17 %; and for Stokes Q and U, slightly above 1 % relative to the DELO solution. The MLP has in fact *learned* to disentangle the different and sometimes competitive effects of the various atmospheric input parameters to calculate accurate local Stokes profiles for both iron lines. But it is not only the accuracy of the method which makes this synthesis method attractive for ZDI inversion, it is first and foremost the speed of the calculation. A benchmark test with several workstations confirmed that the MLP synthesis is more than a factor 1000 faster then the conventional numerical approach, for more details see Carroll *et al.* (2008).

6. An artificial neural network approach to ZDI

In this section we want to introduce our new statistical inverse approach to ZDI which is based on artificial neural networks. Instead of using the ANNs in a forward modeling approach as in the preceding section, we now try to use ANNs to provide a direct approximation of the inverse mapping between the observed Stokes spectra and the corresponding magnetic surface distribution. Again we rely on the non-linear approximation capabilities of the multilayer-perceptron as given in Eq. 5.3, to find a continuous approximation of the underlying inverse problem. ANNs have been already successfully used in solar Stokes profile inversion problems (Carroll & Staude 2001; Carroll & Kopf 2008) which have shown their great potentials in approximating the inverse mappings between observed Stokes profiles and the underlying atmospheric parameters. One of the key issues in the following ZDI inversion is the reduction of the input (i.e. Stokes spectra) as well as the output dimension (i.e. surface images) by a PCA decomposition. The learning task of the neural network is then to approximate the inverse mapping in the reduced eigenspace between the eigenprofiles and the corresponding eigenimages of the surface. The reduction to the subspace of eigenspectra and images has two main reasons, first, we reduce the topology of the MLP structure which allows a more stable and faster

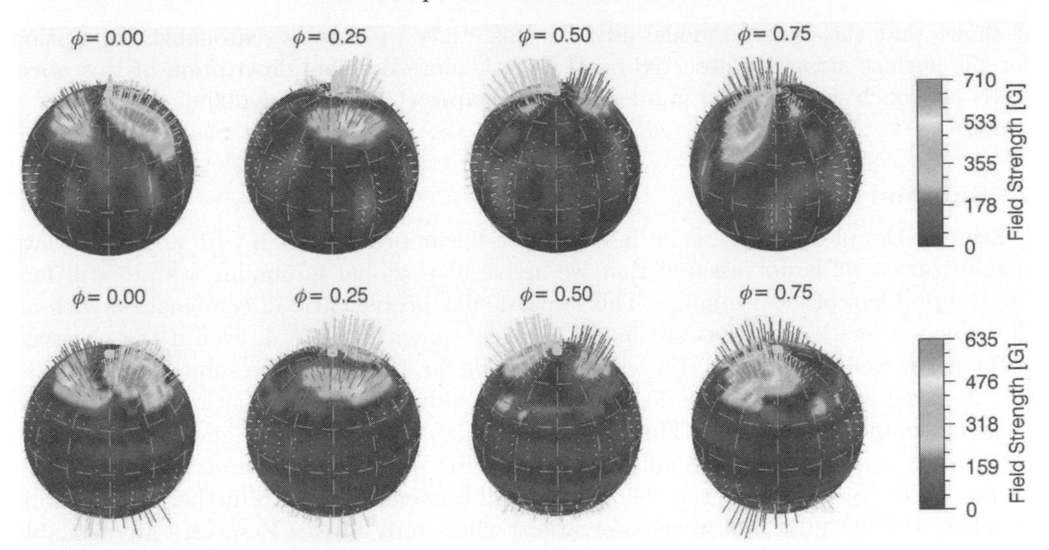

Figure 3. The reconstructed radial magnetic field of II Peg for the year 2007. In the top the reconstruction based on the conventional inversion is shown, whereas, the ANN inversion based on the $\alpha^2 - \Omega$ dynamo model is shown in the bottom.

convergence in the training process and second, the restriction to a subspace of the problem, has a regularizing effect on the solution due to the reduced redundancy in the input spectra and the noise reducing effects by describing the solution as a superposition of eigenimages. Because the transformation and training of the network is based upon a training sample, this will have a decisive impact on the resulting eigenspectra and eigenimages. A finite training sample will only provide a good statistical modeling in a limited region of the parameter domain and an approximation of the inverse function will therefore only provide meaningful results in that restricted domain. To apply the ANN inversion on real observed Stokes V spectra (Carroll *et al.* 2007) we trained our MLP on synthesized Stokes profiles which are based on a theoretical $\alpha^2 - \Omega$ dynamo (Elstner & Korhonen 2005). The $\alpha^2 - \Omega$ dynamo originally developed to study and explain the appearance of flip-flop events on active stars provides not only a good model representation since our target star II Pegasi is known to be a candidate for a flip-flop cycle (Berdyugina 2007), it also allows us to incorporate (implicitly) the physics provided by this model into our inversion approach. From this simulation we have synthesized approximately 1000 Stokes V profiles at different phases and evolution stages of the dynamo. Assuming typical stellar and atmospheric parameters for the subgiant K2 star II Peg (Berdyugina *et al.* 1998) we have synthesized approximately 1000 disk-integrated Stokes V profiles for different phases and different evolution stages of the dynamo simulation. The training process follows again a large-scale optimization of the network parameters of the MLP to find the best possible approximation for the inverse mapping between the decomposed Stokes V profiles and the corresponding decomposed surface images After the training process the observed II Peg Stokes V spectra applied to the trained MLP to estimate the magnetic surface structure. In Figure 3 we see a comparison of the surface magnetic field structure of II Peg retrieved with the conventional ZDI inversion made with our ZDI code *iMap* and the new ANN inversion. The surface structure inferred from this first ANN approach is in remarkable good agreement with the conventional inversion. The good convergence of the training process and the good results of the inversion indicate that the MLP has in fact found a well-behaved inverse mapping to that problem and moreover it shows that the dynamo model used for this study provides a reasonable assumption for the surface structure observed on II Peg. A more detailed description of this novel ANN approach will be given in a forthcoming paper (Carroll *et al.* 2009).

7. Summary

Zeeman-Doppler imaging is an inherent non-linear problem even with an appropriate regularization, it is not assured that we arrive at a global minimum, we are still facing the problem of local minima. This is particular problematic in combination with an ill-defined or inadequate forward model. As was shown in Sect. 4, even if the observed spectra are well reproduced, the solution can be far from the true solution. The fit to the observed Stokes V profiles does often not provide enough information to distinguish between competing solutions. The additional availability of Stokes Q and U profiles with the next generation of spectropolarimeter (PEPSI) at the 8.4 m Large Binocular Telescope (LBT) (Strassmeier *et al.* 2007, 2008) will immensely help to further mitigate this problem. But for all the current observations where only Stokes V spectra are available we have to rely on the best possible forward modeling. In an effort to provide a more systematic radiative transfer modeling we have introduced a new PCA based multi-line reconstruction technique. This method allows us to recover Stokes profiles of individual spectra lines and thus facilitates their modeling by means of radiative transfer calculations. Moreover, we have introduced a fast Stokes profile synthesis which is not only accurate but accelerates the process of calculating the Stokes profiles by more than three orders of magnitude. In a preliminary study we have investigated the capabilities of a new artificial neural network approach to ZDI. ANNs have a number of favorable features, they provide a direct inversion in terms of approximating the inverse mapping between the observed Stokes profiles and the magnetic surface distribution. And moreover, ANNs as a statistical learning tool, allows one to incorporate theory based knowledge in the training process which is then utilized in the subsequent inversion. Although further investigations are needed, the ANN approach already provides a new and promising way of a direct inversion or for a hybrid approach with a conventional ZDI inversion.

References

Berdyugina, S. V. 2007, *Mem. Soc. Astr. It.* 78, 242
Berdyugina, S. V., Jankov, S., Ilyin, I., Tuominen, I., & Fekel, F. C. 1998, *A&A* 334, 863
Bishop, C. M., 1995, *Neural Networks for Pattern Recognition*, Oxford University Press
Brown, S. F., Donati, J.-F., Rees, D. E., & Semel, M. 1991, *A&A* 250, 463
Carroll, T. A., Kopf, M., Ilyin, I., & Strassmeier, K. G. 2009, *A&A*, in prep.
Carroll, T. A. & Kopf, M. 2008, *A&A* 481, L37
Carroll, T. A., Kopf, M., & Strassmeier, K. G. 2008, *A&A* 488, 781
Carroll, T. A., Kopf, M., Ilyin, I., & Strassmeier, K. G. 2007, *AN* 328, 1043
Carroll, T. A. & Staude, J. 2001, *A&A* 378, 316
Donati, J.-F., Jardine, M. M., Gregory, S. G., Petit, P., Bouvier, J., Dougados, C., Mnard, F., Cameron, A. C., Harries, T. J., Jeffers, S. V., & Paletou, F. 2007, *MNRAS* 380, 1297
Donati, J.-F., Forveille, T., Cameron, A. C., Barnes, J. R., Delfosse, X., Jardine, M. M., & Valenti, J. A. 2006, *Science* 311, 633
Donati, J.-F. 1999, *MNRAS* 302, 457
Donati, J.-F., Semel, M., Carter, B. D., Rees, D. E., & Collier Cameron, A. 1997, *MNRAS* 291, 658
Donati, J.-F., Semel, M., Rees, D. E., Taylor, K., & Robinson, R. D. 1990, *A&A* 232, L1
Elstner, D. & Korhonen, H. 2005, *AN* 326, 278

Engl, H. W., Hanke, M., & Neubauer, A. 1996, *Regularization of Inverse Problems*, Kluwer Academic Publishers Group, Dordrecht, The Netherlands

Hussain, G. A. J., van Ballegooijen, A. A., Jardine, M., & Collier Cameron, A. 2002, *ApJ* 575, 1078

Kochukhov, O., Bagnulo, S., Wade, G. A., Sangalli, L., Piskunov, N., Landstreet, J. D., Petit, P., & Sigut, T. A. A. 2004, *A&A* 414, 613

Landi Deglinnocenti, E. & Landi Deglinnocenti, M. 1985, *SP* 97, 239

Martínez González, M. J., Asensio Ramos, A., Carroll, T. A., Kopf, M., Ramírez Vélez, J. C., & Semel, M. 2008, *A&A* 486, 637

Petit, P., Dintrans, B., Solanki, S. K., Donati, J.-F., Aurire, M., Lignires, F., Morin, J., Paletou, F., Ramirez Velez, J., Catala, C., & Fares, R. *et al.* 2008, *MNRAS*, 388, 80

Piskunov, N. & Kochukhov, O. 2002, *A&A* 381, 736

Press, W. H., Teukolsky, S. A., Vetterling, W. T., & Flannery, B. P. 1992, *Numerical Recipes*, Cambridge: University Press, 2nd ed.

Rees, D. E., Durrant, C. J., & Murphy, G. A. 1989, *ApJ* 339, 1093

Semel, M. 1989, *A&A* 225, 456

Strassmeier, K. G., Woche, M., Andersen, M., & Ilyin, I. 2007, *AN* 328, 627

Strassmeier, K. G., Woche, M., Ilyin, I., Popow, E., Bauer, S.-M., Dionies, F., Fechner, T., Weber, M., Hofmann, A., Storm, J., Materne, R., Bittner, W., Bartus, J., Granzer, T., Denker, C., Carroll, T., Kopf, M., DiVarano, I., Beckert, E., Lesser, M. *et al.* 2008, *Ground-based and Airborne Instrumentation for Astronomy*, Proc. of the SPIE, Vol. 7014, p. 70140N

Stenflo, J. O. 1994, *Solar Magnetic Fields – Polarized Radiation Diagnostics*, Astrophysics and Space Science Library, 189, Dordrecht, Boston: Kluwer Academic Publishers

Thorsten Carroll

Ilya Ilyin gets special treatment on the pirates ship

Javier Trujillo Bueno

Cosmic Magnetic Fields:
From Planets, to Stars and Galaxies
Proceedings IAU Symposium No. 259, 2008
K.G. Strassmeier, A.G. Kosovichev & J.E. Beckman, eds.

doi:10.1017/S1743921309031470

Cosmic magnetism with the Square Kilometre Array and its pathfinders

Bryan M. Gaensler

Sydney Institute for Astronomy (SIfA), School of Physics,
The University of Sydney, NSW 2006, Australia
email: bgaensler@usyd.edu.au

Abstract. One of the five key science projects for the Square Kilometre Array (SKA) is "The Origin and Evolution of Cosmic Magnetism", in which radio polarimetry will be used to reveal what cosmic magnets look like and what role they have played in the evolving Universe. Many of the SKA prototypes now being built are also targeting magnetic fields and polarimetry as key science areas. Here I review the prospects for innovative new polarimetry and Faraday rotation experiments with forthcoming facilities such as ASKAP, LOFAR, the ATA, the EVLA, and ultimately the SKA. Sensitive wide-field polarisation surveys with these telescopes will provide a dramatic new view of magnetic fields in the Milky Way, in nearby galaxies and clusters, and in the high-redshift Universe.

Keywords. Galaxies: magnetic fields – instrumentation: interferometers – polarimeters – intergalactic medium – ISM: magnetic fields – magnetic fields – radio continuum: general – techniques: polarimetric

---

1. Introduction

The Square Kilometre Array (SKA)† is a concept for a next-generation radio telescope, with a total approximate collecting area of 1 km^2 (Schilizzi *et al.* 2008); an artist's impression of the central core of the SKA is shown in Figure 1. From the outset, the SKA project has been a global effort, and is currently governed by an international consortium of 18 member countries. Two potential sites are being considered for the SKA: near Boolardy station in Western Australia, and in the Karoo region of South Africa.

The SKA will be a highly flexible instrument, covering a broad range of frequencies (0.07–25 GHz), with many different operating modes. At an observing frequency of 1.4 GHz, the SKA will have a maximum angular resolution of $0\farcs02$ and a field of view of 20 square degrees. This latter specification provides incredible survey capability, far exceeding that of any other sensitive radio telescope. A unique capability of radio arrays is that they can begin taking science data science long before the full facility is complete. Operations for the SKA are thus projected to begin in around 2016, with an approximate construction cost of €1.5 billion.

Five key science projects have been designated for the SKA, as discussed by Gaensler (2004) and Carilli & Rawlings (2004). Below I summarise recent progress and developments on one of these five key projects, "The Origin and Evolution of Cosmic Magnetism" (see Gaensler *et al.* 2004).

† http://www.skatelescope.org

Figure 1. An artist's impression of the SKA, showing the central core of steerable dishes and passive aperture tiles. Image created by XILOSTUDIOS for the SKA Project Development Office.

2. SKA Polarisation Pathfinders

Although the full SKA is still some years away, a large number of pathfinder facilities are either under construction or are already beginning to take data. Many of these will be carrying out exciting new experiments on polarimetry, Faraday rotation and magnetic fields. New telescopes with such capabilities include:

- The Galactic Arecibo L-Band Feed Array Continuum Transit Survey (GALFACTS),† a 1.4-GHz survey that began in late-2008, and which will map the entire polarised sky visible to Arecibo;
- The Low Frequency Array (LOFAR),‡ currently under construction in the Netherlands and Germany, which will study polarisation over the whole northern sky at very low frequencies ($\nu = 30 - 80, 110 - 240$ MHz) (Beck 2009);
- The Allen Telescope Array (ATA)¶ in northern California, a newly operational facility that has a wide field of view (5 deg^2 at 1.4 GHz) and can carry out very large continuum surveys;
- The Square Kilometre Array Molonglo Prototype (SKAMP),‖ a refurbishment of the Molonglo Observatory Synthesis Telescope in south-eastern Australia, which will provide 18 000 m^2 of collecting area for studying diffuse polarisation at an observing frequency of ~ 1 GHz over wide fields;
- The Murchison Widefield Array (MWA),†† an interferometer being built in Western Australia, which will study polarised emission over wide fields in the frequency range 80–300 MHz;
- The Expanded Very Large Array (EVLA),‡‡, a substantial upgrade to the VLA in

† http://www.ucalgary.ca/ras/GALFACTS
‡ http://www.lofar.org
¶ http://ral.berkeley.edu/ata
‖ http://www.physics.usyd.edu.au/sifa/Main/SKAMP
†† http://www.haystack.mit.edu/ast/arrays/MWA
‡‡ http://www.aoc.nrao.edu/evla

New Mexico, providing greatly improved continuum sensitivity, frequency coverage and correlator capability;

- The Karoo Array Telescope (MeerKAT),† an array of 80 12-metre dishes, each equipped with a wideband feed covering the frequency range 0.7–10 GHz;
- The Australian SKA Pathfinder (ASKAP)‡, an array of 36 12-metre antennas to be built on the Western Australian SKA site. ASKAP will be a very wide-field survey instrument (30 $\deg^2$ at 1.4 GHz), and will be able to study polarisation at a range of spatial scales in the frequency range 700–1800 MHz (Johnston *et al.* 2007).

3. Cosmic Magnetism with the SKA

The SKA key science project on cosmic magnetism focuses on three themes: structure, evolution and origin of magnetic fields (Gaensler *et al.* 2004). The questions we hope to address for each theme can be summarised as follows:

(*a*) **Structure:** What is the strength and structure of magnetic fields in the Milky Way, in other galaxies, and in galaxy clusters?

(*b*) **Evolution:** How have magnetic fields evolved in galaxies and clusters over cosmic time?

(*c*) **Origin:** When and how was the Universe magnetised?

In the following discussion, we outline experiments with the SKA and its pathfinders that can address each of these topics.

3.1. *Structure: The Rotation Measure Grid*

Recent surveys of polarised extragalactic sources, carried out with the Australia Telescope Compact Array and the DRAO Synthesis Telescope, have yielded background Faraday rotation data at a sky density of ~ 1 RM per $\deg^2$. These studies have started to reveal the large-scale magnetic field geometry of the Milky Way and Magellanic Clouds, allowing us to infer parameters such as the pitch angle of the magnetic field, the presence and location of field reversals, and the overall dynamo mode (e.g., Gaensler *et al.* 2005; Brown *et al.* 2007; Mao *et al.* 2008).

Observations with the SKA can spectacularly improve the sky density of background RMs. An hour of integration with the full SKA will yield approximately 2000 RMs per $\deg^2$. Over the full 20 $\deg^2$ field of view, we will obtain an order of magnitude more background RMs in the first hour of observations than what has been accumulated over the last 40 years!

This will allow many exciting new applications of the "rotation measure grid". For our own Milky Way, we will be able to combine these data with pulsar RMs to derive a full three-dimensional map of the Galactic magnetic field, on scales ranging from the overall geometry of the field in the disk and halo, down to the properties of magnetised turbulence on sub-parsec scales. Deep observations of nearby galaxies will yield $> 10^5$ RMs per target, as shown in Figure 2 — these data will allow a full reconstruction of the large-scale magnetic field in these systems. An all-sky survey of polarised continuum emission with the SKA will provide ~ 10 background RMs for each of the ~ 5000 nearest galaxies, for each of which we will be able to fit for the overall geometry and structure of the magnetic field (see Stepanov *et al.* 2008, for a full discussion). These studies will deliver a definitive census on the magnetic field properties of typical galaxies, and will allow us to explore how field properties (e.g., field strength, pitch angle, number

† http://www.ska.ac.za/meerkat
‡ http://www.atnf.csiro.au/projects/askap

Figure 2. A simulation of the rotation measure grid behind the nearby galaxy M31, as seen with the SKA in a 30-hour observation. The colours are overlays from three observations: red indicates optical emission, blue corresponds to total intensity radio emission at 5 GHz, and green indicates polarised radio emission at 5 GHz. The white dots show the locations of 50 000 simulated polarised background sources for which the SKA could extract RMs. Optical image: Digitized Sky Survey; radio images: courtesy of Rainer Beck.

of reversals) depend on parameters such as galaxy type, presence/absence of a bar, or degree of interaction.

In the lead-up to the SKA, pathfinder instruments also have an important role to play. In particular, the new wide-field sensitive surveys that will be carried out by ASKAP and the EVLA will allow us to derive catalogues of polarised extragalactic source counts down to fluxes of a microjansky or lower. The polarisation properties of these sources will allow us to separate starburst populations from low-luminosity AGNs (Taylor *et al.* 2007), and to characterise the statistical properties of magnetic field geometries in unresolved spiral galaxies (Arshakian *et al.* 2009; Stil *et al.* 2009).

3.2. *Evolution: Rotation Measure vs. Redshift*

For Faraday rotation at high redshifts, RMs in the observer frame will be reduced in magnitude by a factor $(1+z)^2$ compared to the RM in the region where rotation occurs. Thus if extragalactic RMs are dominated by Faraday rotation intrinsic to the emitting source, and if all sources have similar intrinsic RMs, then we expect that an ensemble of sources for which we have measured both RM and z will demonstrate a dependence $|\mathrm{RM}| \propto (1+z)^{-2}$. However, a sample of 268 high-latitude quasars in the redshift range $0 < z < 3.7$ show a slight *increase* in RM as a function of z (Kronberg *et al.* 2008). This result, along with other new measurements of Faraday rotation (Bernet *et al.* 2008) and Zeeman splitting (Wolfe *et al.* 2008) against sources at $z \sim 0.5-2$, provides strong evidence that the observed RMs have substantial contributions from intervening absorbing galaxies, and that throughout the last 10 Gyr of cosmic time, typical field strengths in these absorbers have been at the level of a few microgauss or more. This is at odds with the expectation from mean-field dynamo theory that the galactic magnetic fields observed today have steadily grown over time, and favours models in which the field undergoes relatively rapid amplification.

Current instrumental capabilities introduce three limitations on these intriguing new results. First, since the observed RM from an extragalactic source is usually dominated by the contribution from the Milky Way's interstellar medium (ISM), any correlation between RM and z must use not the total observed RM, but the *residual* rotation measure (RRM), i.e., the RM after a contribution from the Galactic foreground has been estimated and subtracted. Currently, the relatively spare sampling of RMs over the sky limits the accuracy with which we can calculate RRMs (Dineen & Coles 2005; Short *et al.* 2007). The greatly improved RM grid data that will come from pathfinder surveys with ASKAP and the ATA will provide a dramatic improvement in our ability to model and remove the Galactic foreground. The corresponding RRMs will have much smaller uncertainties.

Second, we simply lack the sample sizes needed to build up sufficient statistics of RM vs. z. An all-sky polarisation survey with the SKA down to sub-μJy sensitivities will yield $\approx 10^8$ background RMs. Provided that a few percent of these sources also have spectroscopic or photometric redshifts, we will be able to invert the distribution of (RM, Dec., RM, z) to derive the magnetic power spectrum of the IGM as a function of time, out to $z \sim 3$ (Kolatt 1998; Blasi *et al.* 1999).

Finally, if indeed the evolution of magnetic field strengths in galaxies is relatively slow over the last 10 Gyr, then we need RM data on very distant sources. Such measurements can allow us to push back to early enough times to reveal how magnetic fields in galaxies were created and amplified. Deep SKA observations of quasars or gamma-ray burst afterglows at $z > 6$ can provide information on Faraday rotation at these early epochs.

3.3. *Origin: The Cosmic Web*

Deep observations of galaxy clusters at low frequencies have begun to reveal filaments and diffuse synchrotron emission seen on the periphery of, or even between, clusters (Kronberg *et al.* 2007; Pizzo *et al.* 2008). While the detections so far have been at relatively low signal-to-noise, such studies have enormous potential, because they serve as direct probes of relativistic particles and magnetic fields in the pristine intergalactic medium (IGM) (Hoeft & Brüggen 2007; Ryu *et al.* 2008). Furthermore, Xu *et al.* (2006) have suggested that the magnetic field of the IGM might be measurable through the Faraday rotation of the background RM grid, as discussed in §3.1 above.

Observations of total intensity, polarisation and Faraday rotation with the MWA, LOFAR and SKA can extend such studies far beyond their current preliminary levels, and will provide superb insights into the magnetised large-scale structure of the Universe (Keshet *et al.* 2004a,b; Battaglia *et al.* 2009). In particular, the structure of the magnetised IGM may serve as a potential discriminant between primordial models vs. outflows as the origin of the IGM magnetic field (Donnert *et al.* 2009).

3.4. *Multi-wavelength Synergies*

All the above experiments will have important overlap with new experiments being carried out in other wavebands and using other techniques. At much higher radio frequencies than will be observed by the SKA, the *Planck* satellite will make a superb Faraday-free map of Galactic synchrotron emission and its polarisation (Enßlin *et al.* 2006). At optical and infrared wavelengths, forthcoming wide-field survey facilities such as SkyMapper, PanSTARRS and the LSST will provide complementary information on large-scale structure and on photometric redshifts. And finally, deflection of the trajectories of ultra-high-energy cosmic rays measured by Auger will help map the strength and structure of magnetic fields in the local IGM.

4. Conclusions

There have been many new discoveries and ideas relating to cosmic magnetism in the last few years, Using these new results as a starting point, the SKA and its pathfinders promise a suite of unique experiments aimed at revealing the structure, evolution and origin of magnetic fields in the Universe. Specifically, the dense RM grid that the SKA will provide will probe the structure of magnetic fields in the Milky Way, nearby galaxies and clusters; studies of RM vs. redshift will allow us to measure the evolution of magnetic fields in galaxies and in the IGM over cosmic time; and radio imaging of the relativistic cosmic web provides a direct view of intergalactic magnetic fields, providing information on the origin of magnetism in the Universe.

The richness of the polarised sky is already beginning to be revealed by pathfinder experiments such as GALFACTS and the ATA, with even more powerful facilities such as the EVLA and ASKAP now under construction. These activities will culminate in the next decade with the arrival of the SKA, and a consequent exploration of the full magnetic Universe. This endeavour will address major unanswered issues in fundamental physics and astrophysics, and will almost certainly yield new and unanticipated results.

Acknowledgements

I thank the organizers for financial support, and my many colleagues with whom I have collaborated on polarimetry and magnetic fields. I acknowledge the support of a Federation Fellowship from the Australian Research Council through grant FF0561298.

References

Arshakian, T. G., Beck, R., Krause, M., & Sokoloff, D. 2009, *A&A*, in press (arXiv:0810.3114)

Battaglia, N., Pfrommer, C., Sievers, J. L., Bond, J. R., & Ensslin, T. A. 2009, *MNRAS*, in press (arXiv:0806.3272)

Beck, R. 2009, in Magnetic Fields in the Universe II, in press (arXiv:0804.4594)

Bernet, M. L., Miniati, F., Lilly, S. J., Kronberg, P. P., & Dessauges-Zavadsky, M. 2008, *Nature* 454, 302

Blasi, P., Burles, S., & Olinto, A. V. 1999, *ApJ* 514, L79

Brown, J. C., Haverkorn, M., Gaensler, B. M., Taylor, A. R., Bizunok, N. S., McClure-Griffiths, N. M., Dickey, J. M., & Green, A. J. 2007, *ApJ* 663, 258

Carilli, C. & Rawlings, S. 2004, *New Astron. Rev.* 48, 979

Dineen, P. & Coles, P. 2005, *MNRAS*, 362, 403

Donnert, J., Dolag, K., Lesch, H., & Müller, E. 2009, *MNRAS*, in press (arXiv:0808.0919)

Enßlin, T. A., Waelkens, A., Vogt, C., & Schekochihin, A. A. 2006, *AN* 327, 626

Gaensler, B. M. 2004, Key Science Projects for the SKA (SKA Memo Series, No. 44)

Gaensler, B. M., Beck, R., & Feretti, L. 2004, *New Astron. Rev.* 48, 1003

Gaensler, B. M., Haverkorn, M., Staveley-Smith, L., Dickey, J. M., McClure-Griffiths, N. M., Dickel, J. R., & Wolleben, M. 2005, *Science* 307, 1610

Hoeft, M. & Brüggen, M. 2007, *MNRAS* 375, 77

Johnston, S., Bailes, M., Bartel, N., Baugh, C., Bietenholz, M., Blake, C., Braun, R., Brown, J., Chatterjee, S., Darling, J., Deller, A., Dodson, R., Edwards, P. G., Ekers, R., Ellingsen, S., Feain, I., Gaensler, B. M., Haverkorn, M., Hobbs, G., Hopkins, A., Jackson, C., James, C., Joncas, G., Kaspi, V., Kilborn, V., Koribalski, B., Kothes, R., Landecker, T. L., Lenc, E., Lovell, J., Macquart, J.-P., Manchester, R., Matthews, D., McClure-Griffiths, N. M., Norris, R., Pen, U.-L., Phillips, C., Power, C., Protheroe, R., Sadler, E., Schmidt, B., Stairs, I., Staveley-Smith, L., Stil, J., Taylor, R., Tingay, S., Tzioumis, A., Walker, M., Wall, J., & Wolleben, M. 2007, *PASA* 24, 174

Keshet, U., Waxman, E., & Loeb, A. 2004a, *ApJ* 617, 281

Keshet, U., Waxman, E., & Loeb, A. 2004b, *New Astr.* 48, 1119

Kolatt, T. 1998, *ApJ* 495, 564

Kronberg, P. P., Bernet, M. L., Miniati, F., Lilly, S. J., Short, M. B., & Higdon, D. M. 2008, *ApJ* 676, 70
Kronberg, P. P., Kothes, R., Salter, C. J., & Perillat, P. 2007, *ApJ* 659, 267
Mao, S. A., Gaensler, B. M., Stanimirović, S., Haverkorn, M., McClure-Griffiths, N. M., Staveley-Smith, L., & Dickey, J. M. 2008, *ApJ* 688, 1029
Pizzo, R. F., de Bruyn. A. G., Feretti, L., & Govoni, F. 2008, *A&A* 481, L91
Ryu, D., Kang, H., Cho, J., & Das, S. 2008, *Science* 320, 909
Schilizzi, R., Dwedney, P. E. F., & Lazio, T. J. W. 2008, Proc. SPIE 7012, 52
Short, M. B., Higdon, D. M., & Kronberg, P. P. 2007, Bayesian Analysis, 2, 665
Stepanov, R., Arshakian, T. G., Beck, R., Frick, P., & Krause, M. 2008, *A&A* 480, 45
Stil, J. M., Krause, M., Beck, R., & Taylor, A. R. 2009, *ApJ*, in press (arXiv:0810.2303)
Taylor, A. R., Stil, J. M., Grant, J. K., Landecker, T. L., Kothes, R., Reid, R., Gray, A. D., Scott, D., Martin, P. G., Boothroyd, A., Joncas, G., Lockman, F. J., English, J., Sajina, A., & Bond, J. R. 2007, *ApJ* 666, 201
Wolfe, A. M., Jorgenson, R. A., Robishaw, T., Heiles, C., & Prochaska, J. X. 2008, *Nature* 455, 638
Xu, Y., Kronberg, P. P., Habib, S., & Dufton, Q. W. 2006, *ApJ* 637, 19

Discussion

KRONBERG: The "Extended Very Large Array (EVLA) II", whose funding request was rejected, would serve as an obvious nucleus of a future "northern SKA" operating at the higher frequencies.

DE GOUVEIA DAL PINO: After this excellent talk on SKA and the coming projects for cosmic magnetism search, I'd like to make a comment that there is presently a project for building a millimeter VLBI apparatus in the southern hemisphere. It is an Argentinian-Brazilian collaboration for installing two antennas in desert areas in Argentina at distances of $\sim$ 100 km from ALMA, allowing VLBI with ALMA at mm frequencies.

Bryan Gaensler

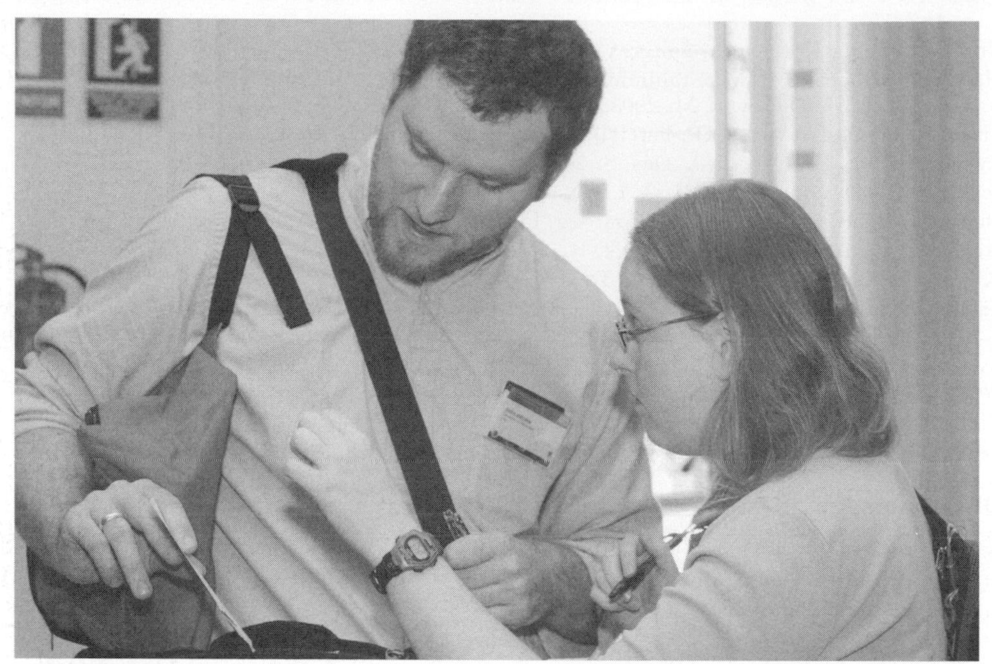

Shea Brown (left), and Amanda Kepley

Oleg Kochukhov (facing the photographer)

Cosmic Magnetic Fields:
From Planets, to Stars and Galaxies
Proceedings IAU Symposium No. 259, 2008
K.G. Strassmeier, A.G. Kosovichev & J.E. Beckman, eds.

doi:10.1017/S1743921309031482

Measuring cosmic magnetic fields with very large telescopes

Oleg Kochukhov and Nicolai Piskunov

Department of Physics and Astronomy, Uppsala University, 751 20 Uppsala, Sweden

Abstract. We review general properties and capabilities of the instrumentation employed to diagnose cosmic magnetic fields using medium-size and large optical telescopes. During the last decade these spectropolarimeters and high-resolution spectrographs have been successfully used to detect and characterize magnetic fields in stars across the H-R diagram. A new generation of high-resolution spectropolarimeters will benefit from the large collecting area of the future E-ELT and currently operating 8-m class telescopes. We review plans to develop spectropolarimeters for these very large telescopes and outline a number of science cases where new spectropolarimetric instrumentation is expected to play a key role.

Keywords. Instrumentation: polarimeters – instrumentation: spectrographs – telescopes – stars: activity – stars: magnetic fields

1. Introduction

For a very long time astronomical polarization measurements were some of the most promising and the most frustrating. Promising – because polarization often carries geometrical information about the observed object and because it allows tracing magnetic fields, invisible otherwise. Frustrating – because polarization signal in most cases is rather small and is easily corrupted by the instrumental effects (such as instrumental polarization, chromatism of the polarimeter, sensitivity and noise properties of the detector).

Recent progress in the quality of detectors but most importantly a major progress in understanding the instrumental polarization effects made it possible to develop the new strategies for observations, calibrations and data reduction. The net results are reproducible polarization measurements of various astronomical objects carried out to unprecedented precision.

The corresponding astronomical observations are commonly divided into measurements of circular polarization, induced by the line-of-sight component of the magnetic fields, and linear polarization also produced by scattering. Scattering creates broadband continuum polarization which requires different type of instrumentation compared to high-resolution spectropolarimeters aimed at studying polarization across spectral line profiles.

In the following we review the instruments for polarimetric measurements. We start with the most advanced existing instruments followed by the projects under development.

2. Current instrumentation

2.1. *High-resolution spectropolarimetry at medium-size telescopes*

The ESPaDOnS spectropolarimeter at the 3.6-m CFHT and its twin instrument NARVAL at the 2-m TBL telescope of the Pic du Midi observatory represent the current state of the art in high-resolution, optical spectropolarimetry. ESPaDOnS is a bench-mounted echelle spectrograph designed to record a complete optical spectrum (370–1050 nm) at the resolution of $R = 65000$ in polarimetric mode (Manset & Donati 2003). The instrument is

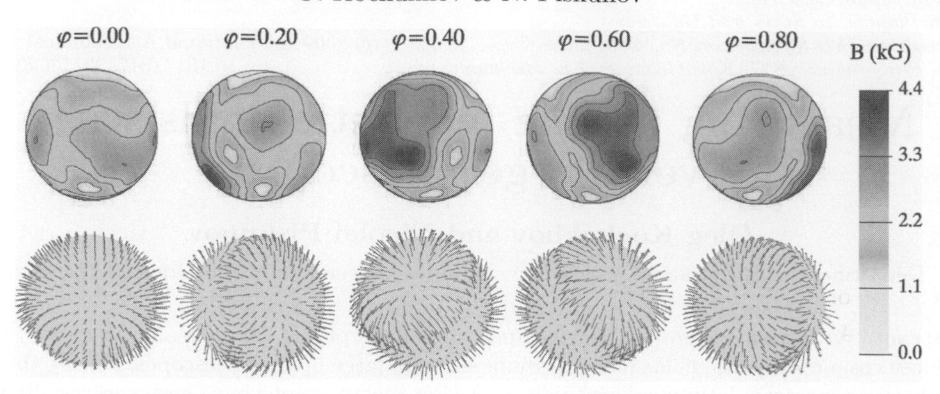

Figure 1. Magnetic field structure of the Ap star α^2 CVn reconstructed with magnetic Doppler imaging modelling of the phase-resolved observations in all four Stokes parameters (Kochukhov *et al.*, in preparation). The upper panel shows the maps of the field strength. The lower panel illustrates distribution of the field orientation.

installed in a thermally-stabilized room and is fibre-linked to a four Stokes parameter polarimeter mounted at the Cassegrain focus. An achromatic polarization analysis is made possible by the Fresnel rombs, acting as quarter-wave and half-wave plates, coupled to a Wollaston prism. The measured cross-talk from circular to linear polarization does not exceed 1% for ESPaDOnS. Analysis of circular polarization in the line profiles of weakly magnetic stars demonstrates the absence of polarization artifacts down to the level of 10^{-5}. ESPaDOnS provides $\approx 13\%$ peak total throughput, corresponding to $S/N = 100$ in 1 h exposure of a $V = 13^{\mathrm{m}}$ star.

Observations carried out with ESPaDOnS and NARVAL have contributed significantly to the progress in understanding of stellar magnetism. For magnetic chemically peculiar A- and B-type stars, these instruments provided a unique opportunity to obtain high-quality spectra in all four Stokes parameters. Detection of linear polarization signatures in the profiles of individual lines was previously possible only for 2–3 strongest metal lines in a few very bright Ap stars using now decommissioned MuSiCoS spectropolarimeter (Wade, Donati, Landstreet, *et al.* 2000). Thanks to the improved sensitivity of ESPaDOnS and NARVAL, the $S/N = 500$–1000, required to study linear polarization in meaningful number of spectral lines, can be achieved for many more magnetic Ap stars. Ongoing phase-resolved observations of a sample of magnetic chemically peculiar stars (Silvester, Wade, Kochukhov, *et al.* 2008 and in this conference) will provide the data for detailed modelling of the stellar field geometries using magnetic Doppler Imaging technique (Piskunov & Kochukhov 2002). Generally, magnetic inversions require observations in all four Stokes parameters to achieve a unique reconstruction of the surface magnetic field topology (Kochukhov & Piskunov 2002). Availability of such data for several Ap stars has allowed us to produce the first assumption-free magnetic maps of early-type stars (Kochukhov, Bagnulo, Wade, *et al.* 2004). This novel, full Stokes vector approach to the problem of imaging stellar magnetic fields has demonstrated that, in addition to well-known dipolar-like, global field component seen in modulation of the circular polarization spectra and traditionally modelled with a low-order multipolar approximation, Ap stars possess unexpected, small-scale magnetic structures (Fig. 1). The latter contribute significantly to the morphology of the Stokes QU profiles but cannot be detected otherwise.

High-resolution spectropolarimetry has provided new insights into magnetism of massive stars. Due to large geometrical sizes of early-B and O stars, even fairly weak fields in these objects yield astrophysically important magnetic fluxes. Magnetic fields in massive stars control the mass loss (Babel & Montmerle 1997), creating complex and dynamic magnetospheric structures (e.g., Townsend, Owocki & Ud-Doula 2007). Magnetism is an important ingredient of the evolutionary models of massive stars (Maeder & Meynet 2000) and is possibly linked to the presence of magnetic fields in supernovae and neutron stars. Until now only a handful of magnetic O-type stars were identified and thoroughly studied (Donati, Babel, Harries, *et al.* 2002; Donati, Howarth, Bouret, *et al.* 2006). Their numbers are steadily increasing with the discovery of weak fields in other hot stars (Petit, Wade, Drissen, *et al.* 2008 and in this conference). The MiMeS Large Program (Grunhut *et al.*, this conference), recently started at CFHT, will perform a comprehensive magnetic survey using ESPaDOnS for all O-type stars in the solar neighbourhood.

For late-type stars, ESPaDOnS and NARVAL are employed for the magnetic field detection and Doppler imaging analysis of the strong fields in faint active low-mass and young stars and of very weak fields in bright solar-type stars. This work is based upon circular polarization measurements and is focused on the interpretation of the mean line polarization signature derived with the help of multi-line Least-Squares Deconvolution technique (Donati, Semel, Carter, *et al.* 1997). This powerful line co-addition method provides a factor of 10–40 gain in S/N by combining information from thousands spectral lines at the price of adopting a greatly simplified description of the intensity and polarization spectra, valid in the limit of low field strength and weak lines.

A combination of the multi-line approach with the increased sensitivity and wavelength coverage of the new high-resolution spectropolarimeters has allowed to detect global fields weaker than 10 G in solar-like stars (Petit, Ditrans, Solanki, *et al.* 2008 and in this conference). Reconstruction of the large-scale magnetic geometries shows that both the total magnetic flux and the relative contribution of toroidal field component in the surface field strongly depends on the stellar rotation rate and mass. The global fields were also reconstructed for the exoplanet-host stars HD 189733 (Moutou, Donati, Savalle, *et al.* 2007) and τ Boo (Catala, Donati, Shkolnik, *et al.* 2007). The field polarity reversals seen in the latter star may be a signature of activity cycle driven by the star-planet magnetic interaction (Donati, Moutou, Fares, *et al.* 2008).

For T Tauri stars, the ESPaDOnS spectropolarimeter provided observations that made possible reconstruction of the large-scale magnetic field topology (Donati, Jardine, Gregory, *et al.* 2008) and subsequent investigation of the role of magnetic fields in accretion process and star-disk coupling using more realistic field topologies than the commonly assumed dipolar fields (Gregory, Matt, Donati, *et al.* 2008). These models of stellar magnetic field could provide a realistic boundary condition for detailed MHD simulations of the magnetospheric accretion on T Tauri stars (Long, Romanova & Lovelace 2008).

Significant progress was also achieved for low-mass active stars. Despite the well-known evidence for the presence of strong magnetic fields in M dwarfs (e.g., Johns-Krull & Valenti 1996), the global geometry of fields in these stars remained unknown until recently. The study by Morin, Donati, Petit, *et al.* (2008) suggested that active mid-M dwarfs host relatively strong, mainly poloidal, axisymmetric, global fields which are stable on the time scale of many rotation periods. At the same time, these stars show no differential rotation. This underlines significant difference of the field generation processes operating in the fully-convective envelopes of M dwarfs compared to the solar tachocline dynamo. The large strength and simple organization of the large-scale fields in active M stars suggests that it should be possible to detect and interpret linear polarization in their spectral lines.

The application of the multi-line LSD method to the problem of magnetic field detection and modelling in late-type stars succeeds in extracting useful information from the entire stellar spectrum. However, it also introduces certain complications and poorly understood systematic biases when LSD profiles are interpreted using Zeeman Doppler Imaging codes. More robust and detailed information about magnetic fields and their relation to the temperature inhomogeneities can be obtained by modelling individual atomic and molecular lines and exploiting their differential sensitivity to temperature and magnetic field at different atmospheric layers (Kochukhov *et al.*; Carroll *et al.*; Berdyugina *et al.*, this conference). The current instrumentation at the 3–4-m class telescopes is suitable to accomplish the task of reaching a high S/N necessary for the direct detection of circular polarization signatures in the spectra of most active late-type stars (Petit 2006; Berdyugina, Petit, Fluri, *et al.* 2006). Thus, some effort should be directed towards acquiring and interpreting such data in order to complement, verify and, possibly, re-calibrate the LSD-based results.

2.2. *Low-resolution spectropolarimetry at VLT*

Among the instruments at the 8–10-m class telescopes equipped with polarization analysing optics only FORS1 (Focal reducer Optical Range Spectrograph) at the ESO VLT is widely used to test the presence of magnetic fields in different types of stars. FORS1 is a multi-mode spectrograph, capable of performing low-resolution ($R \leqslant 2000$) optical spectropolarimetry in the wavelength range from the Balmer jump to Hα. Bagnulo, Szeifert, Wade, *et al.* (2002) developed a technique to measure the mean longitudinal magnetic fields with FORS1. At the low spectral resolution provided by this instrument only the broad hydrogen Balmer line wings can be resolved. Consequently, Bagnulo *et al.* (2002) proposed to employ a magnetic diagnostic technique similar to the photopolarimetric method of Borra & Landstreet (1980). Using a weak-field approximation, which is an adequate assumption for the hydrogen line formation in all types of non-degenerate magnetic stars, one can obtain longitudinal magnetic field from a correlation between the Stokes V signal and the derivative of Stokes I. Extending this regression analysis to the spectral regions containing unresolved metal line blends yields formally a more precise measure of the longitudinal field, but may introduce some systematic errors (Bagnulo *et al.* 2002).

Magnetic field measurements with FORS1 rely on obtaining $S/N \geqslant 1000$ in the hydrogen line wings. In practice, one has to collect photons for ≈ 1 h with the 8-m VLT to achieve a 50 G field measurement accuracy for a $V = 10^{\rm m}$ star. Due to the large intrinsic width of the hydrogen lines, the method is not particularly sensitive to stellar rotation and, thus, can be applied to essentially all classes of early-type stars.

The FORS1 instrument was used to search for links between magnetic fields and stellar evolution in open cluster A- and B-type stars (Bagnulo, Landstreet, Mason, *et al.* 2006). Several young Ap stars with remarkably strong longitudinal fields were discovered in this survey (Bagnulo, Landstreet, Lo Curto, *et al.* 2003; Bagnulo, Hensberge, Landstreet, *et al.* 2004). Based on FORS1 observations, global, kG-strength fields were also detected in DA white dwarfs (Aznar Cuadrado, Jordan, Napiwotzki, *et al.* 2004), hot subdwarfs (O'Toole, Jordan, Friedrich, *et al.* 2005), and in central stars of planetary nebulae (Jordan, Werner & O'Toole 2005). Systematic FORS1 observations of large samples of A and B chemically peculiar stars allowed to increase substantially the number of stars with known magnetic field properties and investigate possible variation of the field intensity with stellar age (Hubrig, Szeifert, Schöller, *et al.* 2004a; Kochukhov & Bagnulo 2006; Hubrig, North, Schöller, *et al.* 2006a).

FORS1 at VLT was also used to measure weak magnetic fields in O-type stars (Hubrig, Schöller, Schnerr, *et al.* 2008 and in this conference), hot pulsating stars (Hubrig, Briquet, Schöller, *et al.* 2006b), and Herbig Ae/Be stars (Hubrig, Schöller & Yudin 2004b). It should be noted, however, that more precise high-resolution spectropolarimetric analyses by Wade, Drouin, Bagnulo, *et al.* (2005), Wade, Abecassis, Auriere, *et al.* (2006) and Silvester, *et al.* (this meeting) cast doubt on some of these claimed FORS1 detections of fields weaker than 1–2 hundred gauss.

In addition to measurement of the mean line of sight magnetic field component in early-type main sequence stars, the FORS1 polarimeter provided data for detailed Stokes line profile analysis of several magnetic white dwarfs with fields exceeding 100 kG (Euchner, Reinsch, Jordan, *et al.* 2005; Valyavin, Wade, Bagnulo, *et al.* 2008; Jordan, this meeting).

2.3. *Diagnosing magnetic fields with high-resolution infrared spectra*

Analysis of high-resolution circular polarization observations across a wide wavelength range has been successfully used for detection of weak stellar magnetic fields. Spectropolarimetric time series, interpreted using magnetic inversion codes, provided constraints on the global field topologies for a variety of magnetic stars, from massive magnetic O stars to late-M dwarfs. However, circular polarization measurements suffer from two important limitations. On the one hand, the Stokes V spectrum provides information only on the line of sight projection of the stellar magnetic field. The mean field strength and magnetic flux cannot be inferred in a straightforward and model-independent manner for the majority of magnetic stars which host weak or moderate-strength fields. Secondly, the amplitude of the observed disk-integrated line of sight field component is very sensitive to the geometry of the magnetic field. The presence of complex field structures, such as bipolar groups frequently seen on the sun, leads to cancellation of the polarization signal and no detectable net longitudinal field.

Magnetic broadening and splitting of atomic and molecular lines in the intensity spectra can supply missing information about small-scale, complex magnetic fields. The relative strength of the π and σ components of the Zeeman split spectral lines is generally weakly sensitive to the field orientation, while the separation of the Zeeman components depends only on the field strength. In this way, interpretation of the Stokes I spectra of magnetic stars can complement and enhance polarization measurements.

The fundamental difficulty of the Stokes I analysis is the necessity of disentangling small magnetic perturbations from the intrinsic, often poorly known, line shape. Therefore, analysis of magnetic broadening in active late-type stars is often considered as not particularly robust and strongly model-dependent (Robinson 1980; Basri & Marcy 1994; Guenther, Lehmann, Emerson, *et al.* 1999). But as the Zeeman splitting increases and eventually exceeds the other line broadening mechanisms (thermal and rotational Doppler broadening), one can securely measure the mean field strength in a simple and direct way. The optical spectra of slowly rotating strongly magnetic Ap stars provide such a possibility (Mathys, Hubrig, Landstreet, *et al.* 1997; Freyhammer, Elkin, Kurtz, *et al.* 2008). For low-mass stars, especially the mid- and late-M active dwarfs, the optical and near-infrared spectra yield evidence of kG-strength fields (Johns-Krull & Valenti 1996; Reiners & Basri 2007; Reiners, this conference). However, even for M stars with the fields in the range of 3–4 kG none of the molecular or atomic diagnostic lines accessible to common optical spectrographs exhibits resolved Zeeman split pattern.

Saar & Linsky (1985) showed that observations in the infrared K-band are of enormous potential for the magnetic diagnosis of active M dwarfs and other cool stars. Due to the λ^2 wavelength dependence of the Zeeman effect, the magnetic splitting of lines at 2.2 μm is 5–10 times larger than for most Zeeman-sensitive lines below 1 μ. Moreover, in the infrared wavelengths the intensity contrast between the photosphere and a cool starspot decreases, leading to substantial increase of the contribution of cool, strongly magnetic regions to the disk-integrated stellar spectrum.

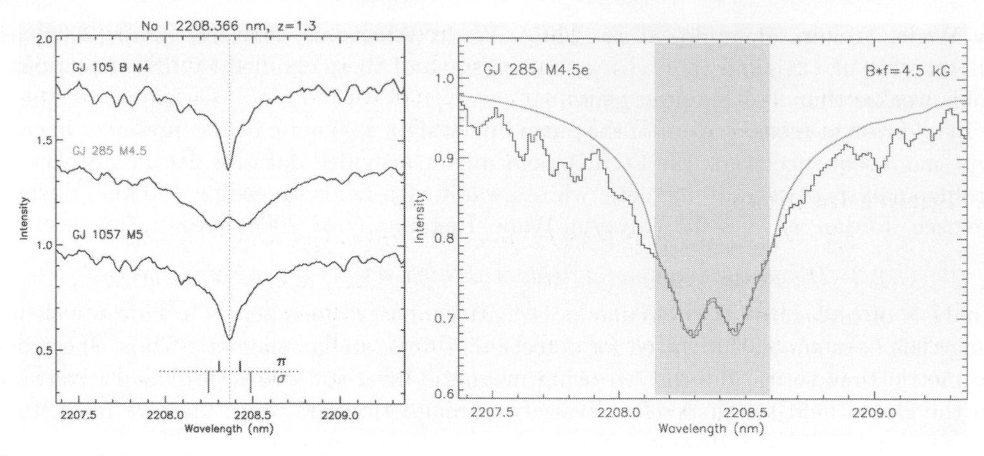

Figure 2. Direct detection and measurement of the strong small-scale magnetic field in M4.5 flare star YZ CMi (GJ 285) using infrared observations with the CRIRES instrument at the ESO 8-m VLT (Kochukhov *et al.* 2008). *Left*: comparison of the Na I 2208 nm line profile in the spectra of inactive M stars (top and bottom) and in YZ CMi (middle). Zeeman splitting is clearly detected in the spectrum of the active star. *Right*: magnetic spectrum synthesis fit to the observations of YZ CMi yields magnetic flux, $\sum B \cdot f$, of 4.5 kG.

The promising diagnostic content of the infrared magnetic observations was utilized to detect the presence of 1–2.5 kG magnetic fields in T Tauri stars (Johns-Krull, Valenti & Koresko 1999; Johns-Krull, Valenti & Saar 2004). These observations, summarized by Johns-Krull (2007 and in this conference), were performed with the help of the CSHELL spectrometer (Greene, Tokunaga, Toomey, *et al.* 1993) at the NASA Infrared Telescope Facility. The $S/N \approx 100$ spectra of T Tauri stars were recorded for several 6 nm wavelength windows, at the resolving powers of 21 000–37 000.

Magnetic analysis of the infrared spectra can be significantly improved and extended to fainter stars with the availability of the CRIRES (CRyogenic high-resolution InfraRed Echelle Spectrograph) instrument (Kaeufl, Ballester, Biereichel, *et al.* 2004) at the ESO VLT. CRIRES works in the 0.9–5.4 μm range and has a wavelength coverage similar to CSHELL, but allows to reach $R = 10^5$ using an adaptive optics system. These high-quality infrared spectra of cool stars open access to a much more refined analysis of magnetic topologies than possible with the optical spectra.

Kochukhov, Heiter, Piskunov, *et al.* (2008) have used CRIRES to carry out a survey of a large sample of M dwarfs with the aim to increase the number of stars with known surface field strengths and to determine small-scale field characteristics for several stars with strong fields. Such observations are essential for understanding non-solar dynamo mechanism operating in these fully convective stars (Dobler, Stix & Brandenburg 2006; Browning 2008). We found that in many active M dwarfs Zeeman splitting is readily detectable in the strong Na I line at λ 2208.4 nm (Fig. 2). The mean field modulus can be directly inferred from the resolved Zeeman split line profile. A more elaborate spectrum synthesis fitting, illustrated in Fig. 2 for the M4.5e star YZ CMi, can be used to determine a distribution of field strengths over the stellar surface.

Magnetic fluxes measured with the infrared lines are significantly stronger than $\leqslant 0.5$ kG fields determined by the ZDI analysis of the circular polarization time series obtained for the same active M dwarfs (Morin *et al.* 2008). This demonstrates that for this interesting type of low-mass magnetic stars the field topologies are dominated by the complex, small-scale structures, unaccessible to circular polarimetry in the optical. On the other hand, the latter constrains the large-scale field component and allows to determine stellar rotation periods. Thus, high-resolution spectropolarimetry in the K-band could be extremely useful for combining the two types of magnetic observations. This may become possible with the implementation of the circular polarization modulator currently considered for the CRIRES instrument.

3. Future facilities

3.1. *The upgrade of HARPS to a full-Stokes spectropolarimeter*

HARPS (High Accuracy Radial velocity Planet Searcher) is one of the most successful ESO instruments dedicated to search of exoplanets and studies of stellar oscillations. The instrument is installed in environmentally and vibrationally isolated tank while the light is transmitted by optical fibers from the Cassegrain adaptor of the La Silla 3.6m telescope. This results in extremely stable PSF in time and across the focal plane. As part of the HARPS upgrade ESO plans to replace the iodine cell (which was barely used) with a 4-Stokes polarimeter (Snik, Jeffers, Keller, *et al.* 2008) installed in the Cassegrain adaptor. The possibility of polarization splitting before accumulating any significant instrumental polarization, the availability of two fibers and the extreme stability of the HARPS make it an ideal instrument for measuring full Stokes vector with a very high precision ($\sim 10^{-4}$ and better). The polarimeter is currently in the manufacturing phase with the commissioning scheduled for the middle of 2009. Polarization measurements will be provided in the ESO standard as a sequence of raw frames and pipeline-reduced data.

The HARPS polarimeter is using a novel beam-splitter – a Foster prism – which unlike the conventional Wollaston prism allows the rays to cross all air-glass boundaries at nearly normal angle. This makes the polarimeter truly achromatic. The design includes two identical beam splitters (one for linear and one for circular polarization) with rotatable super-achromatic retarder plates installed in front of them. The full Stokes vector is obtained in 6 sub-exposures where the two beams are switched between the fibers and the two beam splitters sequentially positioned on the optical axis of the telescope. This allows to exclude CCD pixel sensitivity variations from the relative polarization measurements.

The HARPS polarimeter is primarily aimed at detection and modelling of stellar magnetic fields measuring polarization in spectral lines, similar to ESPaDOnS.

3.2. *High-resolution spectropolarimetry with PEPSI at LBT*

The PEPSI (Potsdam Echelle Polarimetric and Spectroscopic Instrument, Strassmeier, Woche, Ilyin, *et al.* 2008; Ilyin *et al.*, this conference) instrument for the 2×8.4-m Large Binocular Telescope will represent the next major step in the high-resolution astronomical spectropolarimetry. The instrument, scheduled for first light in 2011, is a fibre-fed, dual-arm echelle spectrograph capable of reaching an unprecedented resolution of $R = 310\,000$. In polarimetric mode PEPSI will be able to perform full Stokes vector observations in the wavelength range 450–1050 nm at a spectral resolution of $R = 130\,000$. A unique feature of the polarization observations with PEPSI is the use of two independent polarimeters in each of the two Gregorian foci of LBT. The spectrograph will be fibre-linked to both polarimeters and will have spacing of the echelle orders sufficient for recording of four spectra for each order. This will allow simultaneous observation of circular and linear polarization with spectral and temporal resolution surpassing that of any existing or planned instruments.

The design of PEPSI polarimetric modules (Strassmeier, Hofmann, Woche, *et al.* 2003) is based on a combination of a Wollaston prism and a new type of super-achromatic quarter-waveplate retarder. Measurements of circular polarization will be performed by rotating the quarter-waveplate in 45° steps. No half-wave retarders are foreseen for observations in linear polarization. Instead the entire polarimeter package will be rotated in steps of 45° with respect to the optical axis. This will minimize the polarization cross-talks and lead to a significant improvement in the throughput compared to the Fresnel-rhomb design adopted for ESPaDOnS and NARVAL.

PEPSI at LBT is expected to achieve $S/N \approx 100$ for a $V = 14^{\rm m}$ star in 1 hour integration time. The relative polarimetric precision of $\delta P/P \sim 10^{-2}$ will be obtained for objects in the $V = 10$–$14^{\rm m}$ brightness range. This remarkable performance will allow making a significant progress in detecting magnetic fields in different types of active stars and reconstructing stellar surface field topologies. Furthermore, the range of topics which could be addressed with PEPSI extends far beyond stellar magnetism. The instrument has a potential to make an important contribution to the research on accreting binary systems, AGNs, supernovae, interstellar and intergalactic magnetic fields, etc. (see Strassmeier, Pallavicini, Rice *et al.* 2004 for an overview of PEPSI science cases).

3.3. *Spectropolarimetry at E-ELT*

Several instruments for the future E-ELT are currently going through phase A studies and some of them include spectropolarimetry as a main or optional operation mode.

CODEX – high-resolution single-object fibre-fed spectrometer with a spectral format and stability similar to that of HARPS and the throughput comparable or higher than that of the UVES (18 %). The polarimetric capability is not included in the baseline design but provision is made for such option by including two fibers feeding the light from the intermediate focus of the ELT and allowing sufficient spacing between orders in the detector. The availability of polarimetric option depends on the budget to be allocated for phase B.

EPICS – imager and spectropolarimeter for exoplanets. This instrument will have two arms – IR coronograph with extreme adaptive optics for imaging young exoplanets and a low-resolution spectropolarimeter to detect reflected light from exoplanets on short orbits. Linear polarization analysis will be used to discriminate between the stellar and the reflected light. This is a very specialized instrument with the emphasis on detection of the broad-band linear polarization.

METIS – mid-infrared imager and medium/high resolution spectrometer. This instrument is aimed at working in 3–17 μm region. The polarimetry will be supported for the imaging and the medium-resolution spectroscopy modes in the N band. The main goal of the polarimetry is to analyze extended objects where polarization is created by scattering and not magnetic fields. The accuracy of linear polarization measurements should be around 10^{-3}.

SIMPLE – is a near-infrared cross-dispersed echelle spectrometer with $R \sim 10^5$. This spectrometer will work in the 0.8–2.5 μm range and will provide the stability and the wavelength coverage similar to the optical high-resolution echelle spectrometers. The current baseline includes fibre feed from the intermediate focus of the ELT which offers a great opportunity for implementing a full Stokes polarimeter. The consortium is currently considering a totally novel type of polarimeter where the light passes a highly-chromatic retarder plate and the reduction procedures must recover spectral and polarization information.

4. Conclusions

In Fig. 3 we summarize performance of a typical high-resolution, optical spectropolarimeter for different telescope mirror diameters. The throughput of ESPaDOnS at CFHT is taken as a reference. The noise levels necessary for meaningful four Stokes parameter observations of a typical Ap star and circular polarization measurements of a

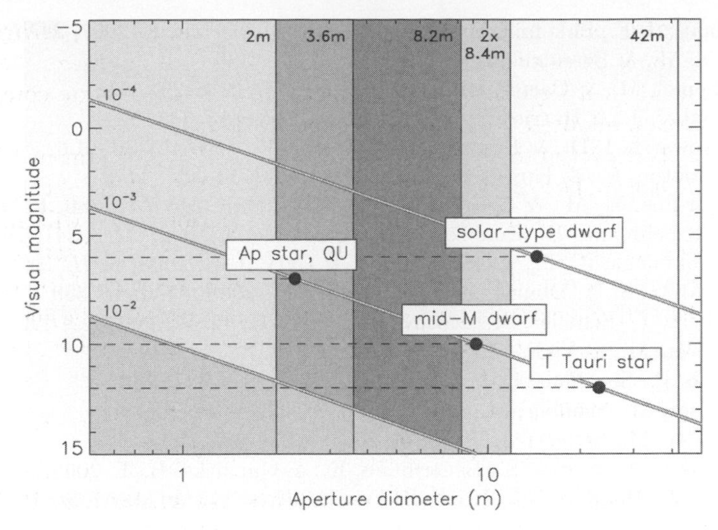

Figure 3. Performance of a generic high-resolution spectropolarimeter for different telescope diameters. Thick double lines correspond to a noise level that could be achieved in 1 h exposure time for a star of a given magnitude, assuming the total instrument throughput similar to that of ESPaDOnS at CFHT. Noise levels required for magnetic field detection and modelling are indicated for the full Stokes vector analysis of a typical Ap star and for the circular polarization observations of a typical bright solar-type dwarf, mid-M dwarf, and T Tauri star.

bright solar-type dwarf, mid-M dwarf and T Tauri star are also indicated. It is clear that only relatively bright Ap stars can be studied using direct analysis of the polarization signatures in individual spectral lines. For late-type stars the LSD multi-line methods have to be applied to decrease the noise level by a factor of 10–40 and allow magnetic field studies at 2–4-m class telescopes.

A major performance improvement that can be provided by a larger collecting area of 8–10-m class telescopes and, eventually, by the E-ELT will allow expanding high-resolution magnetic field studies along the two main directions. On the one hand, it will become possible to apply the multi-line polarization analysis to much fainter objects, such as very low-mass M and L dwarfs, late-type stars in well-studied clusters of different age, various accreting systems with magnetized components, etc. On the other hand, the bright active stars, currently studied with approximate, average-line treatment of the Stokes V spectra, can be analysed with a better accuracy and a physical completeness matching that of the solar magnetic field studies. This can be accomplished by dealing with individual atomic and molecular absorption lines and extending polarization analysis to all four Stokes parameters.

References

Aznar Cuadrado, R., Jordan, S., Napiwotzki, R., *et al.* 2004, *A&A* 423, 1081

Babel, J. & Montmerle, T. 1997, *A&A* 323, 121

Bagnulo, S., Szeifert, T., Wade, G. A., Landstreet, J. D., & Mathys, G. 2002, *A&A* 389, 191

Bagnulo, S., Landstreet, J. D., Lo Curto, G., Szeifert, T., & Wade, G. A. 2003, *A&A* 403, 645

Bagnulo, S., Hensberge, H., Landstreet, J. D., Szeifert, T., & Wade, G. A. 2004, *A&A* 416, 1149

Bagnulo, S., Landstreet, J. D., Mason, E., *et al.* 2006, *A&A* 450, 777

Basri, G. & Marcy, G. W. 1994, *ApJ* 431, 844

Berdyugina, S. V., Petit, P., Fluri, D. M., Afram, N., & Arnaud, J. 2006, in *Astronomical Society of the Pacific Conference Series*, eds. R. Casini, and B. W. Lites, *ASP Conf. Ser.* 358, 381

Borra, E. F. & Landstreet, J. D. 1980, *ApJS* 42, 421

Browning, M. K. 2008, *ApJ* 676, 1262

Catala, C., Donati, J.-F., Shkolnik, E., Bohlender, D., & Alecian, E. 2007, *MNRAS* 374, L42
Dobler, W., Stix, M., & Brandenburg, A. 2006, *ApJ* 638, 336
Donati, J.-F., Semel, M., & Carter, B. D., *et al.* 1997, *MNRAS* 291, 658
Donati, J.-F., Babel, J., & Harries, T. J., *et al.* 2002, *MNRAS* 333, 55
Donati, J.-F., Howarth, I. D., & Bouret, J.-C., *et al.* 2006, *MNRAS* 365, L6
Donati, J.-F., Moutou, C., & Fares, R., *et al.* 2008a, *MNRAS* 385, 1179
Donati, J.-F., Jardine, M. M., & Gregory, S. G., *et al.* 2008b, *MNRAS* 386, 1234
Guenther, E. W., Lehmann, H., Emerson, J. P., & Staude, J. 1999, *A&A* 341, 768
Greene, T. P., Tokunaga, A. T., Toomey, D. W., & Carr, J. B. 1993, *SPIE* 1993, 1946
Gregory, S. G., Matt, S. P., Donati, J.-F., & Jardine, M. 2008, *MNRAS* 389, 1839
Hubrig, S., Szeifert, T., Schöller, M., Mathys, G., & Kurtz, D. W. 2004a, *A&A* 415, 685
Hubrig, S., Schöller, M., & Yudin, R. V. 2004, *A&A* 428, L1
Hubrig, S., North, P., Schöller, M., & Mathys, G. 2006a, *AN* 327, 289
Hubrig, S., Briquet, M., Schöller, M., *et al.* 2006b, *MNRAS* 369, L61
Hubrig, S., Schöller, M., Schnerr, R. S., *et al.* 2008, *A&A* 490, 793
Euchner, F. Reinsch, K. Jordan, S., Beuermann, K., & Gänsicke, B. T. 2005, *A&A* 442, 651
Freyhammer, L. M., Elkin, V. G., Kurtz, D. W., Mathys, G., & Martinez, P. 2008, *MNRAS* 389, 441
Johns-Krull, C. M. & Valenti, J. A. 1996, *ApJ* 459, L95
Johns-Krull, C. M. Valenti, J. A., & Koresko, C. 1999, *ApJ* 516, 900
Johns-Krull, C. M. Valenti, J. A., & Saar, S. H. 2004, *ApJ* 617, 1204
Johns-Krull, C. M. 2007, *ApJ* 664, 975
Jordan, S., Werner, K., & O'Toole, S. J. 2005, *A&A* 432, 273
Kaeufl, H.-U., Ballester, P., Biereichel, P., *et al.* 2004, *SPIE* 5492, 1218
Kochukhov, O. & Piskunov, N. 2002, *A&A* 388, 868
Kochukhov, O., Bagnulo, S., Wade, G. A., *et al.* 2004, *A&A* 414, 613
Kochukhov, O. & Bagnulo, S. 2006, *A&A* 450, 763
Kochukhov, O., Heiter, U., Piskunov, N., *et al.* 2008, in *Cool Stars 15*, ed. H.C. Stempels, in press
Long, M., Romanova, M. M., & Lovelace, R. V. E. 2008, *MNRAS* 386, 1274
Maeder, A. & Meynet, G. 2000, *ARA&A* 38, 143
Manset, N. & Donati, J.-F. 2003, *SPIE*, 4843, 425
Mathys, G., Hubrig, S., Landstreet, J. D., Lanz, T., & Manfroid, J. 1997, *A&AS* 123, 353
Morin, J., Donati, J.-F., Petit, P., *et al.* 2008, *MNRAS* 390, 567
Moutou, C., Donati, J.-F., Savalle, R., *et al.* 2007, *A&A* 473, 651
O'Toole, S. J., Jordan, S., Friedrich, S., & Heber, U. 2005, *A&A* 437, 227
Petit, P. 2006, in *Astronomical Society of the Pacific Conference Series*, eds. R. Casini, and B. W. Lites, *ASP Conf. Ser.* 358, 335
Petit, V., Wade, G. A., Drissen, L., Montmerle, T., & Alecian, E. 2008, *MNRAS* 387, L23
Piskunov, N. & Kochukhov, O. 2002, *A&A* 381, 736
Reiners, A. & Basri, G. 2007, *ApJ* 656, 1121
Robinson, Jr., R. D. 1980, *ApJ*, 239, 961
Saar, S. H. & Linsky, J. L. 1985, *ApJ* 299, L47
Silvester, J., Wade, G. A., Kochukhov, O., Landstreet, J. D., & Bagnulo, S. 2008, *CoSka* 38, 341
Snik, F., Jeffers, S., Keller, Ch., *et al.* 2008, *SPIE* 7014, 22
Strassmeier, K. G., Hofmann, A., Woche, M. F., *et al.* 2003, *SPIE* 4843, 180
Strassmeier, K. G., Pallavicini, R., Rice, J. B., & Andersen, M. I. 2004, *AN* 325, 278
Strassmeier, K. G., Woche, M.., Ilyin, I., *et al.* 2008, *SPIE* 7014
Townsend, R. H. D., Owocki, S. P., & Ud-Doula, A. 2007, *MNRAS* 382, 139
Valyavin, G., Wade, G. A., Bagnulo, S., *et al.* 2008, *ApJ* 683, 466
Wade, G. A., Donati, J.-F., Landstreet, J. D., & Shorlin, S. L. S. 2000, *MNRAS* 313, 823
Wade, G. A., Drouin, D., Bagnulo, S., *et al.* 2005, *A&A* 442, L31
Wade, G. A., Auriere, M., Bagnulo, S., *et al.* 2006, *A&A* 451, 293

Cosmic Magnetic Fields:
From Planets, to Stars and Galaxies
Proceedings IAU Symposium No. 259, 2008
K.G. Strassmeier, A.G. Kosovichev & J.E. Beckman, eds.

doi:10.1017/S1743921309031494

Spectropolarimetry with PEPSI at the LBT: accuracy vs. precision in magnetic field measurements

Ilya Ilyin, Klaus G. Strassmeier, Manfred Woche and Axel Hofmann

Astrophysikalisches Institut Potsdam, An der Sternwarte 16, D-14482 Potsdam, Germany
email: ilyin@aip.de

Abstract. We present the design of the new PEPSI spectropolarimeter to be installed at the Large Binocular Telescope (LBT) in Arizona to measure the full set of Stokes parameters in spectral lines and outline its precision and the accuracy limiting factors.

Keywords. Instrumentation: spectrographs – instrumentation: polarimeters – techniques: polarimetric – polarization

---

1. Spectropolarimeter

The PEPSI *IQUV* spectropolarimeter being designed for the 2×8.4 m LBT (Strassmeier *et al.* 2008) on Mt. Graham in Arizona has the following principal components:

• The PMMA (acrylic glass) super-achromatic quarter-wave retarder (Samoylov *et al.* 2004) in front of the Wollaston prism is to record circular polarization $I \pm V$ in one exposure. A turn of the retarder by 90° swaps the polarity of the two beams.

• The MgF_2 Wollaston prism (with the retarder being retracted from the optical beam) is to record linear polarization $I \pm Q$ in one exposure and $I \pm U$ once it turned by 45°.

• The two polarized beams emerged from the Wollaston prism are corrected for the atmospheric dispersion and focused on two 200 μm entrance fibers of the spectrograph ($R = 130,000$). A fixed fiber viewing CCD camera for accurate centering and guiding on the target star sees the residual light rendered from the rotating beam splitter unit with a flexible single mode fiber bundle.

• Two such identical polarimeters mounted at the two Gregorian focuses of LBT are forming four polarized spectra on a 10.3×10.3k CCD per arm displaced in cross-dispersion alone by the successive échelle orders. Four subsequent exposures with different polarization configurations at three fixed spectral settings result in the Stokes *IQUV* spectra in the range of 450–1050 nm.

• Calibration unit in front of the science retarder: a Glan-Thompson prism, two super-achromatic retarders optimized for the blue and red, and the wedge de-polarizer. The calibration light of the known polarization state allows to model and restore necessary parameters of the polarimeter versus wavelength: phase delay and axis orientation of the retarder, orientation and relative transmission of the Wollaston prism, and transmission modulation function of the fibers.

2. Precision and accuracy

With the 2×8.4 m LBT telescope at $R = 130,000$ and 10% total efficiency in 1 hour integration time, the polarimeter would attain a precision in Stokes *IQUV* measurements of 10^{-4} for a star of 4th magnitude, and 10^{-3} for 9th magnitude. The accuracy of such measurements will be limited by a number of factors:

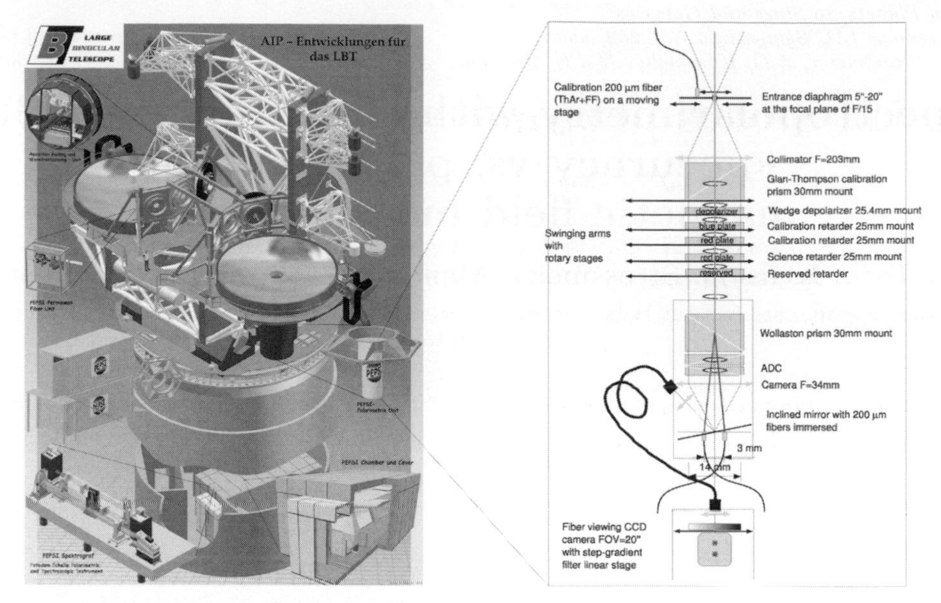

Figure 1. Location of the PEPSI polarimeter on the LBT telescope and its schematic layout.

• The cross-talk between Stokes Q and U induced by a slight misalignment of the Wollaston prism: the relative accuracy in QU of 10^{-3} requires 1 arcmin calibration accuracy of the prism angle.

• The cross-talk between V and QU can be canceled with the turn of the retarder but the residual quadratic terms define the relative accuracy in V up to 10^{-5} if the retardation and axis orientation angles of the retarder are calibrated with the accuracy of $0.1°$.

• The entrance collimator lens birefringence can be caused by the mechanical and thermal stresses, anti-reflection coating, and the intrinsic structure of the glass which may lead to a severe cross-talk between Stokes U and V. The use of polarization grade glass (Sun & Adlou 2006) with the path difference less than 1 nm/cm would confine the cross-talk down to 10^{-4}.

• The thermal stability of the polarization elements over a period of time within 1°C is necessary to keep the calibration parameters constant, though, we allow for a slow seasonal variations with the new set of parameters derived.

• Spurious polarization may arise due to CCD fringes and cannot be canceled out: a high quality flat fielding is essential, although, no optical fringes were detected from the PMMA retarders at the level of 10^{-4}.

• The Earth's atmosphere changes the polarization state of the refracted light at large angles of incidence; the effect is of order of only 10^{-5} at zenith distance of 87°.

• The spectrograph stability ($\pm 0.01°$ C in temperature and ± 0.1 mbar in pressure) is essential for the subsequent combination of the polarized spectra obtained in exposures with different configurations of the polarization optics: a drift between two spectra of 10 m/s is equivalent to the Zeeman splitting of a 5G longitudinal magnetic field at 700 nm.

References

Samoylov, A. V., Samoylov, V. S, Vidmachenko, A. P., & Perekhod, A. V. 2004, *JQSRT* 88, 319

Strassmeier, K. G., Woche, M., Ilyin, I., *et al.* 2008, *SPIE* 7014

Sun, L. & Adlou, S. 2006, *SPIE* 6288, 62890

Cosmic Magnetic Fields:
From Planets, to Stars and Galaxies
Proceedings IAU Symposium No. 259, 2008
K.G. Strassmeier, A.G. Kosovichev & J.E. Beckman, eds.

doi:10.1017/S1743921309031500

A full-Stokes polarimeter for the GREGOR Fabry-Perot interferometer

Horst Balthasar[1], N. Bello González[2]†, M. Collados[3], C. Denker[1], A. Hofmann[1], F. Kneer[2] and K. G. Puschmann[3]

[1]Astrophysikalisches Institut Potsdam, An der Sternwarte 16, D-14482 Potsdam, Germany
email: hbalthasar;cdenker;ahofmann@aip.de

[2]Institut für Astrophysik, Friedrich-Hund-Platz 1, D-37077 Göttingen, Germany
email: nazaret;kneer@astro.physik.uni-goettingen.de

[3] Instituto de Astrofísica de Canarias, Vía Láctea s/n, E-38205 La Laguna, Tenerife, Spain
email: mcv;kgp@iac.es

Abstract. One of the first post-focus instruments of the new solar telescope GREGOR will be a Fabry-Perot spectrometer, which is an upgrade of the Göttingen Fabry-Perot interferometer at the Vacuum Tower Telescope (VTT) on Tenerife. This spectrometer is equipped with a full-Stokes polarimeter. The modulation is performed with two ferroelectric liquid crystals, one acting nominally as quarter-wave plate, and the other as half-wave plate. A modified Savart plate serves as polarimetric beam splitter. With the present liquid crystals, the optimum wavelength range of this polarimeter is between 580 and 660 nm. The spectro-polarimeter will benefit from the capabilities of the new telescope GREGOR which will provide a spatial resolution of about $0\overset{''}{.}1$ (75 km on the solar surface). Thus we will be able to investigate small magnetic features, and we will study their development with high cadence.

Keywords. Instrumentation: polarimeters – instrumentation: spectrographs – Sun: magnetic fields

---

1. Introduction

The new 1.5 m solar telescope GREGOR (see Volkmer, von der Lühe, Kneer, *et al.*, 2007), which will become operational in the near future, will be able to resolve solar features with a size of 75 km. At this time, an upgrade of the Göttingen Fabry-Perot Interferometer (GFPI) described by Puschmann, Kneer, Seelemann & Wittmann (2006) will be moved to GREGOR. Spectrometers based on Fabry-Perot interferometers scan through spectral line profiles, thus allowing to record a two-dimensional image at each wavelength position. They have a high transmission enabling faster cadences compared to grating spectrographs, which allow us to study rapid processes on the Sun. Small solar magnetic features such as filigrees, G-band bright points, penumbral filaments and umbral dots undergo changes within a few minutes. To understand the physical processes behind these changes, it is important to observe the development of the features with short time intervals. The GFPI originally described by Bendlin, Volkmer & Kneer (1992) is presently installed at the Vacuum Tower Telescope (VTT) on Tenerife where a long experience with FPIs exists. Recently, the GFPI has been equipped with a full-Stokes polarimeter, and first results have been presented by Bello González & Kneer (2008).

† Present address: Kiepenheuer-Institut für Sonnenphysik, Schöneckstr. 6, D-79104 Freiburg, Germany

2. The Polarimeter

The modulation of the polarimeter is performed with two ferroelectric liquid crystal retarders (FLCR). FLCRs can be switched between two fixed orientations of their fast axis. Nominally, the first FLCR in the beam has a retardation of $\lambda/2$ and the second one has a retardation of $\lambda/4$. FLCRs can be switched with kHz-rates, which is an advantage over nematic liquid crystals. High modulation rates are required in applications with post-facto image reconstructions, *e.g.* speckle reconstruction. The presently used pair of FLCRs can be used in the spectral range from 580 to 660 nm with reasonable efficiency. We intend to purchase another pair of FLCRs for a different wavelength range.

Behind the FLCRs, positive and negative polarization components are separated by a modified Savart plate which consists of two calcites rotated by 180° with a half-wave plate inbetween, as suggested by Keller (2006). The half-wave plate exchanges ordinary and extraordinary beams, so that they have both the same amount and orientation of astigmatism caused by the calcites. Thus, we achieve a wide separation and the same optical length for both beams. For each of the four modulation states of the FLCRs an exposure is taken. We obtain a linear combination of the four Stokes parameters which must be demodulated by software.

The position of the polarimeter is just in front of the CCD detector in the converging beam where the requirements on optical quality are relaxed. Instrumental polarization must be corrected. For this purpose, a calibration unit, which is described by Hofmann, Rendtel & Arlt (2008), is mounted near GREGOR's secondary focus. The optical beam before this point is axisymmetric and introduces almost no instrumental polarization.

The spectral resolution power of GFPI is 250,000 at 600 nm. A photospheric line profile is typically covered within about 40 scanning steps, and eight exposures per step and Stokes parameter enable image reconstruction (see below). The CCD detector has a sensitive area of 1376×1040 pixels. The typical readout speed is 15 images per second, and the single exposure time is 10 ms. Thus, scanning the line profile and recording the full Stokes vector takes about 85 s. Recording only circular polarization is possible. The field-of-view of the polarimeter will be about $25'' \times 39''$. Binning and readout of a selected part of the image can be used to increase the frame rate.

Simultaneously to the narrow-band spectral data, images in a broad-band channel are taken. These images allow a speckle reconstruction, and the deconvolution function obtained this way can be applied to the polarized images (see Keller & von der Lühe, 1992). An application to VTT observations was presented by Bello González & Kneer (2008). A userfriendly software package for all basic data reductions is under development.

In combination with the high spatial resolution of the 1.5 m GREGOR telescope, the new GREGOR spectro-polarimeter will be a powerful instrument to study the dynamics of small-scale magnetic features on the Sun.

References

Bello González, N. & Kneer, F. 2008, *A&A* 480, 265

Bendlin, C., Volkmer, R., & Kneer, F. 1992, *A&A* 257, 817

Keller, C. U. 2006, *private communication*

Keller, C. U. & von der Lühe, O. 1992, *A&A* 261, 321

Hofmann, A., Rendtel, J., & Arlt, K. 2008, *Cent. Eur. Astrophys. Bull.*, submitted

Puschmann, K. G., Kneer, F., Seelemann, T., & Wittmann, A. D. 2006, *A&A* 451, 1151

Volkmer, R., von der Lühe, O., Kneer, F., and 17 coauthors 2007, in F. Kneer, K. G. Puschmann, A. D. Wittmann (eds.) *Modern Solar Facilities - Advanced Solar Science*, Universitätsverlag Göttingen, p. 39.

Cosmic Magnetic Fields:
From Planets, to Stars and Galaxies
Proceedings IAU Symposium No. 259, 2008
K.G. Strassmeier, A.G. Kosovichev & J.E. Beckman, eds.

doi:10.1017/S1743921309031512

Testing the cosmological evolution of magnetic fields in galaxies with the SKA

Tigran G. Arshakian[1], Rainer Beck[1], Marita Krause[1] and Dimitry Sokoloff[2]

[1]Max-Planck-Institut für Radioastronomie, Bonn, Germany
email: [tarshakian;rbeck;mkrause]@mpifr-bonn.mpg.de

[2]Department of Physics, Moscow State University, Russia
email: sokoloff@dds.srcc.msu.su

Abstract. We investigate the cosmological evolution of large- and small-scale magnetic fields in galaxies at high redshifts. Results from simulations of hierarchical structure formation cosmology provide a tool to develop an evolutionary model of regular magnetic fields coupled to galaxy formation and evolution. Turbulence in protogalactic halos generated by thermal virialization can drive an efficient turbulent dynamo. The mean-field dynamo theory is used to derive the timescales of amplification and ordering of regular magnetic fields in disk and dwarf galaxies. For future observations with the SKA, we predict an anticorrelation at fixed redshift between galaxy size and the ratio between ordering scale and galaxy size. Undisturbed dwarf galaxies should host fully coherent fields at $z < 1$, spiral galaxies at $z < 0.5$.

Keywords. Techniques: polarimetric – galaxies: evolution – galaxies: magnetic fields – radio continuum: galaxies

1. Introduction: importance of magnetic evolution in galaxies

The observed polarized synchrotron emission and Faraday rotation showed the presence of regular large-scale magnetic fields with spiral patterns in the disks of nearby spiral galaxies (Beck 2005), which were successfully reproduced by mean-field dynamo theory (Beck *et al.* 1996, Shukurov 2005). It is therefore natural to apply dynamo theory also in predicting the generation of magnetic fields in young galaxies at high redshifts.

We now have sufficient evidence that strong magnetic fields were present in the early Universe ($z < 3$; Bernet *et al.* 2008, Seymour *et al.* 2008) and that synchrotron emission from distant galaxies should be detected with future radio telescopes such as the Square Kilometre Array (SKA). The SKA will allow us to observe an enormous number of distant galaxies at similar resolution to that achievable for nearby galaxies today (van der Hulst *et al.* 2004). The formation and evolution of regular large-scale magnetic fields is intimately related to the formation and evolution of disks in galaxies in terms of geometrical and physical parameters. A more robust understanding of the history of magnetism in young galaxies may help to solve fundamental cosmological questions about the formation and evolution of galaxies (Gaensler *et al.* 2004).

2. Three-phase model for the evolution of magnetic fields in galaxies

We have used the dynamo theory to derive the timescales of amplification and ordering of magnetic fields in disk and quasi-spherical galaxies (Arshakian *et al.* 2008). This has provided a useful tool in developing a simple evolutionary model of regular magnetic fields, coupled with models describing the formation and evolution of galaxies. In the hierarchical structure formation scenario, we identified three main phases of magnetic-field evolution in galaxies. In the epoch of *dark matter halo formation*, seed magnetic fields of $\approx 10^{-18}$ G strength could have been generated in protogalaxies by the Biermann battery or Weibel instability (first phase). Turbulence in the protogalactic halo generated by *thermal virialization* could have driven the turbulent (small-scale) dynamo and amplify the seed field to the equipartition level of $\approx 20\ \mu$G within a few 10^8 yr (second phase). In the epoch of *disk formation*, the turbulent field served as a seed for the mean-field (large-scale) dynamo in the disk (third phase).

We defined three characteristic timescales for the evolution of galactic magnetic fields: one for the amplification of the seed field, a second for the amplification of the large-scale regular field, and a third for the field ordering on the galactic scale (Arshakian *et al.* 2008). Galaxies similar to the *Milky Way* formed their disk at $z \approx 10$. Regular fields of equipartition (several μG) strength and a few kpc coherence length were generated within 2 Gyr (until $z \approx 3$), but field ordering up to the coherence scale of the galaxy size took another 6 Gyr (until $z \approx 0.5$). *Giant galaxies* had already formed their disk at $z \approx 10$, allowing more efficient dynamo generation of equipartition regular fields (with a coherence length of about 1 kpc) until $z \approx 4$. However, the age of the Universe is too young for fully coherent fields to have already developed in giant galaxies larger than about 15 kpc. *Dwarf galaxies* formed even earlier and should have hosted fully coherent fields at $z \approx 1$. *Major mergers* excited starbursts with enhanced turbulence, which in turn amplified the turbulent field, whereas the regular field was disrupted and required several Gyr to recover. Measurement of regular fields can serve as a clock for measuring the time since the last starburst event. Starbursts due to major mergers enhance the turbulent field strength by a factor of a few and drive a fast wind outflow, which magnetizes the intergalactic medium. Observations of the radio emission from distant starburst galaxies can provide an estimate of the total magnetic-field strength in the IGM.

This evolutionary scenario can be tested by measurements of polarized synchrotron emission and Faraday rotation with the SKA. We *predict*: (i) an anticorrelation at fixed redshift between galaxy size and the ratio between ordering scale and galaxy size, (ii) undisturbed dwarf galaxies should host fully coherent large-scale fields at $z < 1$, spiral galaxies at $z < 0.5$, (iii) weak regular fields (small Faraday rotation) in spiral galaxies at $z < 3$, but possibly associated with strong anisotropic fields (strong polarized emission), would be signatures of major mergers.

Acknowledgements

This work is supported by the EC Framework Program 6, Square Kilometre Array Design Study (SKADS) and the DFG-RFBR project under grant 08-02-92881.

References

Arshakian, T. G., Beck, R., Krause, M., & Sokoloff, D. 2008, *A&A*, in press, arXiv:0810.3114

Beck, R. 2005, in: R. Wielebinski & R. Beck (eds.), *Cosmic Magnetic Fields*, Lecture Notes in Physics (Berlin: Springer), vol. 664, p. 41

Beck, R., Brandenburg, A., Moss, D., Shukurov, A., & Sokoloff, D. 1996, *Ann. Rev. Astron. Astrophys.* 34, 155

Bernet, M. L., Miniati, F., Lilly, S. J., Kronberg, P. P., & Dessauges-Zavadsky, M. 2008, *Nature* 454, 302

Gaensler, B. M., Beck, R., & Feretti, L. 2004, *New Astr. Revs* 48, 1003

Seymour, N., Dwelly, T., Moss, D., *et al.* 2008, *MNRAS* 386, 1695

Shukurov, A. 2005, in: R. Wielebinski & R. Beck (eds.), *Cosmic Magnetic Fields*, Lecture Notes in Physics (Berlin: Springer), vol. 664, p. 113

van der Hulst, J. M., Sadler, E. M., Jackson, C. A., *et al.* 2004, *New Astr. Revs* 48, 1221

Cosmic Magnetic Fields:
From Planets, to Stars and Galaxies
Proceedings IAU Symposium No. 259, 2008
K.G. Strassmeier, A.G. Kosovichev & J.E. Beckman, eds.

doi:10.1017/S1743921309031524

Detecting magnetic fields in large-scale structure with radio polarization

Shea Brown, Lawrence Rudnick and Damon Farnsworth
Department of Astronomy, University of Minnesota, Minneapolis, MN 55455, USA

Abstract. We present our attempts to detect magnetic fields in filamentary large-scale structure (LSS) by observing polarized synchrotron emission emitted by structure formation shocks. Little is known about the strength and order of magnetic fields beyond the largest clusters of galaxies, and synchrotron emission holds enormous promise as a means of probing magnetic fields in these low density regions. We report on observations taken at the Green Bank Telescope which reveal a possible Mpc extension to the Coma cluster relic. We also highlight the major obstacle that diffuse galactic foreground emission poses for any search for large-scale, low surface-brightness extragalactic emission. Finally we explore cross-correlation of diffuse radio emission with optical tracers of LSS as a means to statistically detecting magnetic fields in the presence of this confounding foreground emission.

Keywords. Magnetic fields – polarization – radiation mechanisms: nonthermal

1. Introduction

One of the goals of the next generation(s) of radio telescopes (LOFAR, MWA, EVLA, SKA) will be to determine the origin of cosmic magnetism. As noted by Donnert *et al.* (2008), the large amplification of fields to several μG in rich galaxy cluster cores largely erases the signatures of their origins, whereas the 0.1μG fields expected in filaments can retain indicators of their origins. Thus, it is critical to study the strength and structure of filament fields. Radio synchrotron emission is a powerful means of detecting magnetic fields in these regions, and shocks from infall into and along the filamentary structures between clusters are now widely expected to generate relativistic plasmas (e.g., Ryu *et al.* 2008, Skillman *et al.* 2008). We are searching for the synchrotron signatures of these shocks.

2. Results

The Coma cluster's radio relic is evidence of shock activity due to infall from a smaller cluster. We performed 1.4 GHz observations of this cluster with the Green Bank Telescope (GBT). Fig. 1 shows a Stokes I image of the Coma region after subtracting point sources and applying a $20'$ median weight filter. Diffuse emission, both polarized (not shown) and unpolarized, extends a full 2 Mpc aligned with the relic. It is unclear if this is all related to the shock/filament, though the fact that it is polarized is consistent with shock compression. We are currently analyzing 350 MHz RM-Synthesis (Brentjens & de Bruyn 2005) data from the WSRT which could resolve the ambiguity.

We have also identified high galactic-latitude "Faraday screens" as a significant source of confusion for low frequency polarization studies (Brown, Rudnick, & Farnsworth in prep). These Faraday rotating clouds modulate the smooth galactic background to produce polarized power in an interferometer that has no total-intensity counterpart (Fig. 1). These features will likely be ubiquitous at frequencies < 200 MHz, and will present a foreground contamination to extragalactic studies. Intrinsic angle changes in galactic

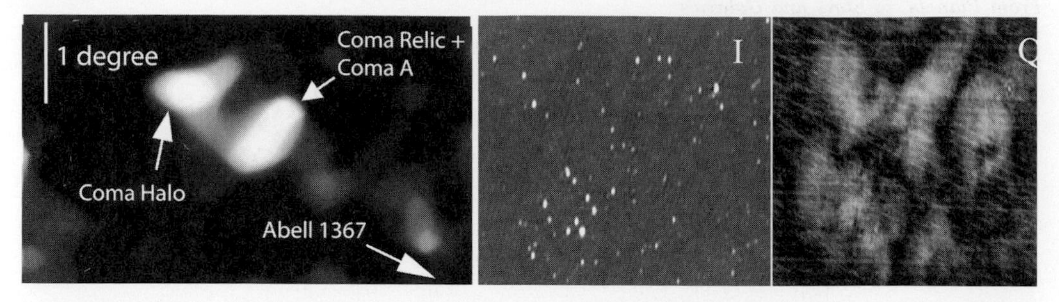

Figure 1. Left: GBT Stokes I image of Coma cluster; Center/right: Westerbork Synthesis Radio Telescope (WSRT) Stokes I/Q images of field north of Coma cluster where Kronberg *et al.* (2007) report diffuse source "B". We do not confirm this source in I, although Faraday modulated Q and U patches are seen in this region.

emission, independent of Faraday rotation, will also add polarized power into an interferometer while the total intensity remains invisible because it is smooth on large scales.

Motivated by these foreground issues, we are exploring cross-correlation methods for detecting synchrotron emission in filamentary LSS. Similar to detections of the Integrated Sachs-Wolfe effect (ISW), we cross-correlate large FOV radio maps with optical/IR tracers of LSS. As a first step, we performed a zero-shift cross correlation of a 34×34 degree area of the 1.4 GHz Bonn survey with the corresponding distribution of 2MASS galaxies ($0.03 < z < 0.04$). To assess the significance of a positive correlation, we also correlated the 2MASS galaxies with 24 random fields of the same size from the rest of the Bonn survey. Though we obtained a null result (as expected), we found that adding a signal weighted by the 2MASS image with a mean surface brightness of $\sim$1 mK was sufficient to produce a 3σ positive correlation. This injected signal is below the rms noise of the Bonn survey and demonstrates the power of this technique.

3. Conclusions

Measuring the weak magnetic fields within filaments of galaxies is critical to determining the origins of cosmic magnetism, and synchrotron radio emission can be a powerful tracer of magnetic fields in these regions. Diffuse Galactic synchrotron emission and Faraday rotating plasma will present a significant foreground problem for upcoming low-frequency surveys, particularly in polarization. We demonstrate that cross-correlation of synchrotron maps and tracers of large-scale structure provides a powerful method for detecting faint emission in filaments.

Acknowledgements

The GBT is a facility of the National Science Foundation, operated by NRAO under contract with AUI, Inc. The WSRT is operated by the ASTRON (Netherlands Foundation for Research in Astronomy) with support from the Netherlands Foundation for Scientific Research (NWO). Partial support for this work at the University of Minnesota comes from the U.S. National Science Foundation grant AST 0607674.

References

Brentjens, M. A., & de Bruyn, A. G. 2005, *A&A* 441, 1217
Donnert, J., Dolag, K., Lesch, H., & Müller, E. 2008, arXiv:0808.0919
Kronberg, P. P., Kothes, R., Salter, C. J., & Perillat, P. 2007, *ApJ* 659, 267
Ryu, D., Kang, H., Cho, J., & Das, S. 2008, *Science* 320, 909
Skillman, S. W., O'Shea, B. W., Hallman, E. J., Burns, J. O., & Norman, M. L. 2008, *ApJ* 689, 1063

Cosmic Magnetic Fields:
From Planets, to Stars and Galaxies
Proceedings IAU Symposium No. 259, 2008
K.G. Strassmeier, A.G. Kosovichev & J.E. Beckman, eds.

doi:10.1017/S1743921309031536

Effects of turbulence on magnetic reconnection: 3D numerical simulations

Katarzyna Otmianowska-Mazur[1], **G. Kowal**[1,2], **A. Lazarian**[2] **and E. Vishniac**[3]

[1]Astronomical Observatory, Jagiellonian University, Kraków, Poland
[2]Department of Astronomy, University of Wisconsin, Madison, USA
[3] Department of Physics and Astronomy, McMaster University, Hamilton, ON, Canada

Abstract. Turbulent reconnection is studied by means of three dimensional (3D) compressible magnetohydrodynamical numerical calculations. The process of homogeneous turbulence is set up by adding three-dimensional solenoidal random forcing implemented in the spectral space at small wave numbers with no correlation between velocity and forcing. We apply the initial Harris current sheet configuration together with a density profile calculated from the numerical equilibrium of magnetic and gas pressures. We assume that there is no external driving of the reconnection. The reconnection develops as a result of the initial vector potential perturbation. We use open boundary conditions. Our main goal is to find the dependencies of reconnection rate on different properties of turbulence. The results of our simulations show that turbulence significantly affects the topology of magnetic field near the diffusion region. We present that the reconnection speed does not depend on the Reynolds numbers as well the magnetic diffusion. In addition, a fragmentation of current sheet decreases the disparity in inflow/outflow ratios. When we apply the large scale and more powerful turbulence the reconnection is faster.

Keywords. MHD – numerical simulations – magnetic fields – turbulence

1. Introduction and numerical model of turbulent reconnection

Fast reconnection should have speeds close to the Alfvén speed V_A. Sweet-Parker (1958, 1957) found that their model of magnetic reconnection applied to the astrophysical bodies is very slow. The reconnection rate, according to this model, depends on the magnetic Reynolds number R_M (as $V_A R_M^{-1/2}$). Petschek (1964) introduced a fast mechanism of magnetic reconnection, which is proportional to $(\log R_M)^{-1}$. Later on, it was shown that the X-point region required in Petschek mechanism collapsed to Sweet-Parker geometry for large R_M (Biskamp, 1996). Lazarian & Vishniac (1999, 2000) suggested that the presence of a stochastic magnetic field component enhances the reconnection rate enabling fast reconnection.

Numerical Model of Turbulent Reconnection is calculated in a box with open boundary conditions. We use Harris current sheet setup as a initial configuration and we set V=0 initially. Reconnection develops as a result of initial vector potential perturbation. We do not drive reconnection! Input parameters: B_x=1.0 above the Y=0 plane and B_x=-1 below it, B_z varies from 0.0 to 1.0. We numerically solve 3D non-ideal normalized isothermal MHD equations varying the power of turbulence, its injection scales and magnetic resistivity. For details refer to Kowal *et al.* (2008).

2. Results

In the left panel of Figure 1 we demonstrate XY-cut across our computational box at time 12. The map shows current density (visible as color plot) with magnetic field vectors superimposed onto the figure. The stable in the beginning Sweet-Parker configuration

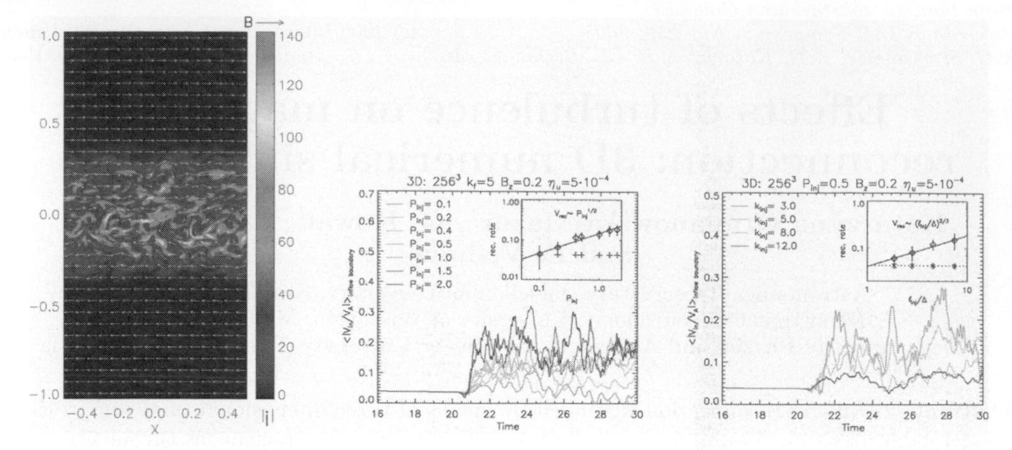

Figure 1. *Left*: XY-cut through the box showing the colors of absolute value of current density and vectors of magnetic field at time t=12 for model with P=1.0 and k=8. Resistivity is set to 10^{-3}. *Middle*: Dependence of the reconnection rate $\langle V_{in}/V_A \rangle$ on the power of turbulence P_{inj}. *Right*: Dependence of the reconnection rate $\langle V_{in}/V_A \rangle$ on the injection scale l_{inj}.

after introduction of turbulence quickly changes its character. We observe fragmentation of current sheets. The middle and right panels of Figure 1 show the time dependence of the reconnection rate defined as $< V_{in}/V_A >$ on the power of turbulent motions (middle) and the injection scale l_{inj} (right). We observe the growth of the reconnection rate with the increasing power strength P_{inj} and injection scale l_{inj}. We also report no dependence of the reconnection rate on the resistivity (not presented here, see Kowal, 2008).

3. Conclusions

(*a*) Numerical studies of stochastic reconnection are finally possible, even though the reconnection in numerical simulations is always fast.

(*b*) Turbulence significantly affects the topology of B near the diffusion region.

(*c*) Fragmentation of current sheet decreases the disparity in inflow/outflow ratios.

(*d*) For large scale turbulence, the reconnection rate V_{rec} scales as V_T^2 and $(l_{inj}/\Delta)^{2/3}$ with the strength and injection scale

(*e*) Reconnection rate is independent on resistivity

Acknowledgements

This work was supported by Polish Ministry of Science and Higher Education through grants: 92/N-ASTROSIM/2008/0, 2693/H03/2006/31 and 3033/B/H03/2008/35.

References

Biskamp, D., 1996, *Ap&SS* 242, 165

Kowal, G, Lazarian, A., Vishniac, E. T., & Otmianowska-Mazur, K., 2008, *Rev. Mex. de Astron. y Astrof.*, in press

Lazarian, A. & Vishnniac, E. 1999, *ApJ* 517, 700

Lazarian, A. & Vishnniac, E. 2000, *Rev. Mex. de Astron. y Astrof.* 9, 55

Parker, E. N. , 1957, *J. Geophys. Res.* 62, 509

Petschek, H. E., 1964, The Physics of Solar Flares, AAS-NASA Symp., NASA SP-50, ed. W. H. Hess, Greenbelt, Maryland, p. 425

Sweet, P. A., 1958, in IAU Symp. 6, Electromagnetic Phenomena in Cosmical Plasma, ed. B. Lehnert, p. 123

Puerto Santiago

Meeting room

Hotel lobby

Farewell Tenerife

Author Index

Object Index

Subject Index